11th International Conference on Advanced Manufacturing Technology and Materials Engineering 2024

Journal of Physics: Conference Series Volume 2842

Guangzhou, China
22-23 May 2024

ISBN: 979-8-3313-0396-9
ISSN: 1742-6588

Printed from e-media with permission by:

Curran Associates, Inc.
57 Morehouse Lane
Red Hook, NY 12571

Some format issues inherent in the e-media version may also appear in this print version.

Institute of Physics
Dirac House, Temple Back
Bristol BS1 6BE UK

Phone: 44 1 17 929 7481
Fax: 44 1 17 920 0979

techtracking@iop.org

Additional copies of this publication are available from:

Curran Associates, Inc.
57 Morehouse Lane
Red Hook, NY 12571 USA
Phone: 845-758-0400
Fax: 845-758-2633
Email: curran@proceedings.com
Web: www.proceedings.com

TABLE OF CONTENTS

Journal of Physics: Conference Series **2842** (2024) 011001 doi:10.1088/1742-6596/2842/1/011001

Preface

The 2024 11th International Conference on Advanced Manufacturing Technology and Materials Engineering (AMTME 2024) was grandly held in Guangzhou, China from May 22nd to 23rd, 2024. Scholars, experts and enterprise representatives from all over the world gathered together to discuss the latest research results, technological advances and innovative applications in the fields of advanced manufacturing technology and materials engineering.

This conference was organized in a variety of formats, including opening ceremony, keynote speech, oral presentation, poster presentation, Q&A session, academic investigation, etc., aiming to provide a platform for participants to present their research achievements, have in-depth exchanges, and discuss collaborations. As the conference kicked off, the 13th "Jinbo Award" Global High-level Talent Science and Technology Innovation Competition, initiated by Guangdong Doctoral Innovation and Development Promotion Association and hosted by Doctoral Science and Technology, was simultaneously launched. It's not only committed to mining and displaying scientific and technological innovation achievements, but also achieving win-win results through diversified cooperation modes. About 200 representatives attended the conference and its two sub-forums, including experts with outstanding achievements in the fields of advanced manufacturing technology and materials engineering as well as young researchers and graduate students interested in the domains. They are actively committed to advancing the fields and promoting academic exchanges and collaborations through this international event. Attendees listened to the presentations of outstanding researchers from all over the world and engaged in in-depth discussions and exchanges with them.

The Proceedings of AMTME 2024 contains excellent academic papers collected from the conference and submitted from all over the globe, which covers a wide range of topics and cutting-edge researches in the fields of advanced manufacturing technology and materials engineering and their relevant disciplines, including but not limited to: Material Processing and Forming Technology, Metamaterial Structure Design, Smart Construction Technology, Intelligent Manufacturing Technology, Robots and Intelligent Systems, Machine Vision and Sensor Technology, etc. The editorial board, a group of experts with rich experience and outstanding achievements in related fields, were responsible for reviewing and selecting the best papers to ensure the high quality and academic impact of this Proceedings.

By sharing the latest research results and technology applications, we are confident that AMTME 2024's Proceedings will be an important literature resource in the fields of advanced manufacturing technology and materials engineering, providing scholars and researchers with an extensive knowledge base that will help them keep abreast of the latest research development and technology trends.

Finally, we would like to thank all the people and organizations that participated in and supported the conference. Sincere thanks to all the authors and editorial board members who contributed to this Proceedings. Through the publication and dissemination of the Proceedings, we can facilitate the cooperation between academia and industry, promote the transformation and application of scientific research results, and contribute to the solution of practical problems in relevant areas and industries.

The Committee of AMTME 2024

Committee Member

General Conference Chair
Prof. Liejun Li, South China University of Technology, China
Prof. Ali Reza Kamali, University of Cambridge, UK

Publication Chair
Prof. Jixiang Gao, Guangdong Polytechnic Normal University, China

Technical Program Committee Chair
Prof. Tongguang Zhai, Shandong Jianzhu University, China

Organizing Committee
Prof. Jixiang Gao, Guangdong Polytechnic Normal University, China
Prof. Xuegong Yu, Zhejiang University, China
Prof. Zude Zhou, Wuhan University of Technology, China
Prof. Xin Chen, Xi'an Jiaotong University, China
Prof. Cong Wang, Beihang University, China
A. Prof. Kai Wu, Tongji University, China
Prof. Weihua Dan, Sichuan University, China
Prof. Sirong Yu, China University of Petroleum, China
Prof. Shilu Dai, South China University of Technology, China
Prof. Sheng-Rui Jian, I-Shou University, Taiwan, China
Prof. Mingyang Chen, Jiangsu University, China
Prof. Mingxin Huang, The University of Hong Kong, China
Prof. Feng Wen, Hainan University, China
Dr. Hao Peng, Yangtze Normal University, China
A. Prof. Kai Wu, Tongji University, China
Prof. Chuang Feng, Nanjing University of Technology, China
Prof. Weibing Zhu, Xihua University, China
Prof. Weihua Dan, Sichuan University, China

Technical Program Committee
Prof. Dr. Yasuto Hijikata, Saitama University, Japan
Prof. Eckart H. Meiburg, University of California, USA
Prof. DAVID BASSIR, French University of Technology, France
Prof. Volkan Kahya, Karadeniz Technical University, Turkey
Prof. David E. Esezobor, University of Lagos, Nigeria
A. Prof. Nour F. Attia, National Institute of Standards (NIS), Egypt
A. Prof. Yipin Wang, Chang'an University, China
A. Prof. Mohamed Ahmed Ramadan, Helwan University, Egypt
Dr. Prashant J. Patil, Abhinav Education Society's College of Engineering and Technology Wadwadi, India
Dr. Quah Hock Jin, University Sains Malaysia, Malaysia
Dr. Alexander Lukin, Western-Caucasus Research Center, Russian Federation
Dr. R. RANJITH KUMAR, Viyen Biotech LLP, India
Dr. Xiao Qi, Northeastern University, China
Dr. Yang Luo, Swiss Federal Laboratories, China
Dr. Hicham Bakkali, University of Cadiz, Spain
Dr. P K Dash, NMIT, India

Peer Review Statement

All papers published in this volume have been reviewed through processes administered by the Editors. Reviews were conducted by expert referees to the professional and scientific standards expected of a proceedings journal published by IOP Publishing.

- **Type of peer review:** Single Anonymous
- **Conference submission management system:** Morressier
- **Number of submissions received:** 239
- **Number of submissions sent for review:** 194
- **Number of submissions accepted:** 139
- **Acceptance Rate (Submissions Accepted / Submissions Received × 100):** 58.2
- **Average number of reviews per paper:** 3
- **Total number of reviewers involved:** 30
- **Contact person for queries:**
  Name: Xuexia
  Email: xx.ye@keoaeic.org
  Affiliation: AEIC Academic Exchange Information Centre

Journal of Physics: Conference Series **2842** (2024) 012001 doi:10.1088/1742-6596/2842/1/012001

Experimental verification of characteristic parameter analysis method for pipeline leak before break

Liming Zhang [1,2,3], Chen Wang [1*], Tian Tan [1], Ling Chen [1] and Yanlong Xu [1]

[1] School of Nuclear Science and Technology, Naval University of Engineering, Wuhan 430033, China
[2] Chongqing Pump Factory Co., Ltd. Chongqing 400033
[3] Chongqing Machine Tool Group Chongqing 401336

[*] Corresponding author's email: 15072080609@163.com

Abstract: The failure mode of the nuclear reactor main pipeline is often the first break and then leakage, i.e., Leak-Before-Break (LBB). If the LBB analysis technique of the pipeline is used in the design, the double-end shear fracture accident of the pipeline can be avoided. In this paper, it is assumed that an annular penetration crack already exists in the pipeline, and the tip-opening displacement of the crack and the leakage rate of the pipeline under high pressure are calculated by using LBB theoretical analysis. Then, according to the assumed conditions, the LBB test design of the pipeline under high pressure is carried out, and the crack tip opening displacement δ and the leakage rate Q of the pipeline under high pressure are measured through the experiments.

1. Introduction

The Leak-Before-Break (LBB) criterion is an important assessment criterion for preventing catastrophic fractures of pressure pipes in nuclear power plants [1]. Since the early 1980s, the research findings of LBB have been widely applied in nuclear power plants, bringing significant economic benefits to both safety assessments of existing nuclear power plants and the design of new ones, while also promoting further research on LBB [2]. Double-ended shear fracture refers to the sudden complete fracture of a pipeline (such as a primary loop main pipeline) under high pressure, resulting in a severe Loss-of-Coolant Accident (LOCA). In numerous coolant leakage accidents, the high-pressure liquid and gas generate powerful jet forces, causing harm to surrounding structures and equipment [3].

However, theoretical studies and experiments since the 1980s have shown that the probability of double-ended shear fracture accident occurring in nuclear reactor main pipelines is extremely low. Using this as the design basis accident could lead to some disproportionate issues. By adopting Leak-Before-Break (LBB) analysis techniques in design, the occurrence of double-ended shear fracture accident in pipelines can be avoided, ensuring the safety of nuclear reactors while also saving unnecessary facilities, reducing the complexity of nuclear reactor structures, and lowering construction costs of nuclear power plants [4].

Furthermore, design bases must specify the essential capabilities of nuclear power plants to adapt to operational states and design basis accidents determined within specified radiation protection requirements [5]. Among these, the design basis accident for nuclear reactor pipelines is the double-ended shear fracture accident, which is the most conservative assumption. For example, in determining safety shell capacity, core emergency cooling system capability, radiation source item doses, etc., pressurized water reactor nuclear power plants use the occurrence of double-ended shear fracture accidents in the primary loop main pipeline as the basis, which later extends to the protection against the dynamic effects of this assumed accident [6].

Reactors made of high-toughness materials exhibit high resistance to unstable crack propagation, making the probability of double-ended shear fracture accidents extremely low even under severe accident conditions. Before the sudden and catastrophic fracture of a pipeline, there will always be subcritical stable crack propagation, and some cracks will stabilize after reaching a certain length, forming subcritical stable cracks; however, more cracks will undergo unstable propagation after reaching the critical unstable fracture length, resulting in the complete fracture of the pipeline. There is sufficient time between the discovery of leakage and the catastrophic fracture of the pipeline to take safety measures, thus avoiding serious accidents. Therefore, it is believed that such pipelines meet the LBB condition.

As long as defects are discovered during in-service inspections before they become through-wall cracks, or before the size of through-wall cracks reaches the critical length for unstable fracture, the leakage caused by through-wall cracks can be detected by relevant leakage monitoring systems. In that case, timely detection of leakage-by-leakage monitoring systems, safe shutdown by nuclear power plant operators, and maintenance by repair personnel all become possible. Hence, the occurrence of double-ended shear fracture accident in nuclear reactor pipelines can be avoided. This paper will experimentally verify the characteristics and parameters of the leak-before-break analysis method for pipelines.

In this paper, based on the analysis of the forward leakage (LBB) of the main pipeline of the pressurized water reactor (PWR) nuclear power plant, combined with the related fracture mechanics theory, the forward leakage (LBB) phenomenon of the pipeline under high pressure is investigated by combining the theoretical and experimental methods. The main contributions are as follows:

1. An experimental rig for front leakage analysis of the main pipeline of the pressurized water reactor (PWR) nuclear power plant has been established, and front leakage experiments of the pipeline under high pressure have been carried out.

2. The results of crack tip opening displacement δ derived from LBB analysis are compared with real front leakage experiments, and a series of conclusions are drawn.

3. The leakage rate of the experimental pipeline calculated by the LBB theory is much higher than the experimental results, mainly due to the faster pressure release during the leakage rate measurement process, resulting in the actual pressure being lower than the theoretical calculation conditions. The theoretical calculations have been corrected.

2. Types of CRACKS and crack propagation

2.1 Types of cracks

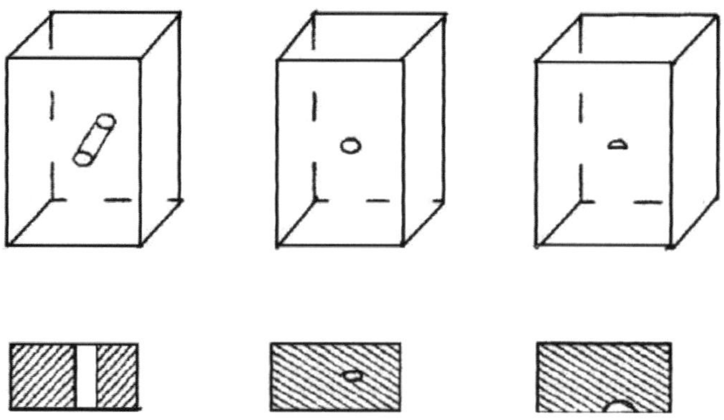

Figure 1. Types of cracks: (a) Penetrating cracks; (b) Buried cracks; (c) Surface cracks.

A large number of tests have proved that it is the presence of cracks that significantly reduces the actual strength of structural materials. In fracture mechanics, cracks are categorized into three types: penetrating cracks, buried cracks, and surface cracks, depending on the location of the crack in the member.

Penetrating cracks are cracks that run through the entire thickness of the member, as shown in Fig. 1(a); buried cracks are cracks that are within the material of the member and do not come into contact with the surface of the material, as shown in Fig. 1(b); surface cracks are cracks that come into contact with one of the surfaces of the member, as shown in Fig. 1(c). Defects in pipelines can generally be simplified to be treated as one of these three types of cracks [7].

2.2 Types of crack propagation

The strength of cracked materials depends on the ability of the material to resist crack propagation. The form of crack propagation is complex, but it can be summarized into three basic forms, namely, open type, slip type, and tear type, as shown in Fig. 2. Any crack propagation is one of these three basic forms or a combination of them.

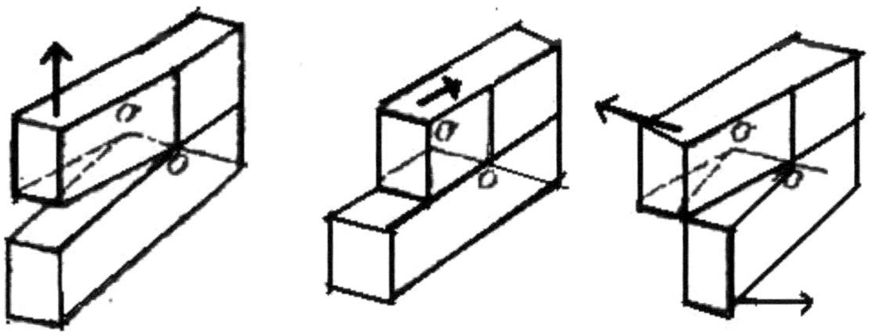

Figure 2. Expansion types of cracks: (a) Tensioned; (b) Slip-open; (c) Tear-open.

The open-type propagation is also called type I propagation. It is the crack tip opening under the action of tensile stress perpendicular to the crack surface, and the propagation form of the propagation is shown in Fig. 2(a). The propagation direction is perpendicular to the direction of the tensile stress. The open-type crack is also known as type I crack. In high-pressure pipelines, axial cracks along the axial direction and circumferential cracks perpendicular to the axial direction are typical type I cracks.

The sliding mode propagation is also called type II propagation, which is the propagation form of the crack sliding and expanding under the shear stress parallel to the crack surface, as shown in Fig. 2(b). The expansion direction is parallel to the shear stress direction.

The tearing propagation, also known as type III propagation, is a form of propagation in which the crack surfaces are torn apart and expanded under the action of shear stress parallel to the crack surface, as shown in Fig. 2(c). The extension direction is perpendicular to the shear stress direction.

3. Theoretical calculations

3.1 Leakage rate through cracks

The calculation of the leakage rate of cracks in pipelines is based on thermodynamic analysis and fluid mechanics [8]. The amount of fluid leakage from a penetrating crack under pressure depends mainly on the pressure inside the pipeline, the temperature, and the resistance during the flow. If the pressure inside the pipe exceeds a certain level relative to the ambient pressure, the critical flow will occur, the fluid velocity will reach the speed of sound and the flow rate will be maximized. A sudden decrease in pressure at the opening causes a phase change and the formation of water vapor.

Leakage calculations under different conditions can be done in different ways. Depending on the thermodynamic conditions, methods such as Bernoulli's equation equations can be chosen to analyze and calculate. For the problem of water loss from a small breach in the reactor's first circuit, the leakage rate is often calculated using the two-phase flow theory, and the models often used are the homogeneous flow model, slip model, etc. [9].

For the leakage of high-pressure pipelines under different thermal conditions, the leakage calculation can take different methods. In this paper, the two-phase critical flow model is used to calculate the leakage rate of the paper. This model is simpler and has better calculation accuracy, and it is suitable for supercooled water, superheated steam, and two-phase flow. The leakage rate calculation can be obtained using the following equation:

$$m = GA = A \times 0.61\sqrt{2P\rho\left(1-\frac{p(T)}{p_0}\right)} \tag{1}$$

where ρ is the mass density of the fluid, T is the temperature of the medium in the pipe, and p(T) is the saturation pressure at temperature T.

Tada and Paris are solved for the dimensionless crack tensile area α_p of the pipe when subjected to an internal pressure P, using one approximate solution for the stress intensity factor as well as Eq. (2).

$$\alpha_p = \begin{cases} 1+0.16\lambda_1^2 & 0 \le \lambda_1 \le 5 \\ \frac{1}{50\lambda_1^2}+0.81+0.3\lambda_1^2+0.03\lambda_1 & 1 \le \lambda_1 \le 5 \end{cases} \tag{2}$$

where λ_1 is calculated as:

$$\lambda_1 = \frac{\alpha}{\sqrt{Rt}} \tag{3}$$

The formula is strict in the case of R/t=10, and it is approximately valid in the range of 5≤R/t≤15. The accuracy of the calculation is good.

3.2 COD theory

The COD method is the Crack Opening Displacement (COD) method. Proposed by Wells in 1963, it is an analytical method based on an empirical foundation; however, it has been widely used in the engineering world [10].

For a cracked body made of a very ductile metallic material, when loaded, the high-stress concentration at the crack tip results in a plastic slip of the material in that region, which leads to blunting (i.e., rounding) of the crack tip and the opening up of crack surface opens. The so-called COD is the crack surface tensioning displacement of the crack tip expressed as δ. The crack tip tensioning displacement δ is derived from the D-M model.

$$\delta = \pi a \times \frac{\sigma_s}{E} \times \left(\frac{\sigma}{\sigma_s}\right)^2 \tag{4}$$

where E is the modulus of elasticity of the material, 2a is the crack length, σ_S is the yield stress of the material, and σ is the stress to which the material is subjected. At the experimental pressure of 16 MPa, δ = 9.525um is calculated from the above equation.

4. LBB test of pipes under high pressure

4.1 LBB test bench design

The high-pressure pipeline LBB test bench mainly consists of four parts: the experimental pipeline, the test bench frame, the pressurization system, and the LBB experimental measurement system. As shown in Fig. 3.

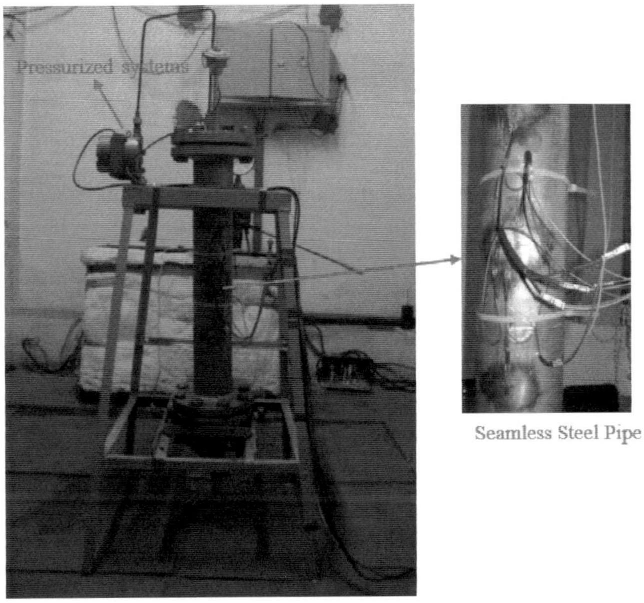

Figure 3. The high-pressure pipeline LBB test bench.

The LBB experimental pipeline is made of No. 10 seamless steel pipe with an inner diameter of 100.60mm, a wall thickness of 4.22mm, and a length of 700mm. The test bench frame is welded from angle steel and serves to support and fix the experimental pipeline. The pressurization system consists of two sets of connecting flanges, three stop valves, stainless steel connecting pipes, a pressurization pump, and pressure gauges. The experimental measurement system includes a computer, 35951B and 35951C IMP models, pressure sensors, strain gauges, thermocouples, wires, and cables. It measures the pressure of the pipeline LBB test, the opening displacement of the crack tip, temperature, and records and processes experimental data. The schematic diagram of the experiment is shown in Fig. 4.

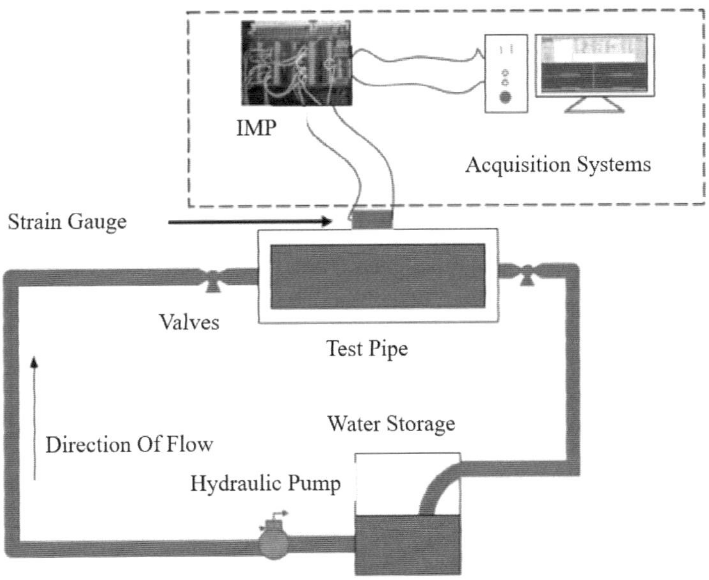

Figure 4. Experimental schematic diagram of high-pressure pipe LBB.

The design parameters of the test bench are as follows:
(1) Test pressure range: 0-18 MPa;
(2) Test pressure measurement range: 0-18 MPa; measurement accuracy 0.1 MPa;
(3) Measurement range of crack tip opening displacement: 0-30 μm; measurement accuracy 0.1 μm;
(4) Temperature measurement range: 0-350°C;
(5) Data acquisition: 10 times per second.

4.2 Design of the test measurement system
In the LBB pipeline test, three physical quantities need to be measured: test pressure (MPa), crack tip opening displacement of the LBB test pipeline (μm), and the temperature of the water during the test (°C). The specific measurement plan is as follows:

(1) **Test Pressure Measurement:** To ensure accurate measurement of test pressure, a capacitive pressure sensor with a range of 18 MPa will be utilized. Capacitive pressure sensors are known for their high accuracy and reliability in pressure measurement applications. However, before use, it's crucial to calibrate the pressure sensor to ensure precise readings throughout the experiment.

(2) **Crack Tip Opening Displacement Measurement:** Monitoring the crack tip opening displacement is vital for assessing the integrity of the pipeline during the LBB test. This measurement will be conducted indirectly using strain gauges. The strain gauges will be strategically placed along the pipeline to detect the strain generated as the crack opens. Utilizing a 1/4 bridge circuit configuration enhances the sensitivity and accuracy of the strain measurement, allowing for precise determination of the crack opening displacement.

(3) **Water Temperature Measurement:** Since water temperature can influence material properties and the behavior of the pipeline, it's essential to monitor it throughout the experiment. Standardized Type E thermocouples will be employed for temperature measurement. Type E thermocouples are suitable for a wide temperature range and offer high accuracy, making them ideal for precise temperature monitoring during the test.

The experiments require the experimental measurement system to be able to carry out real-time measurements, i.e., measurements are carried out in the process of pipeline crack expansion, and the experimental data can be automatically recorded and stored. Therefore, 35951B and 35951C precision measurement modules (IMP) are used in the measurement system for real-time monitoring, and a set of data is measured every 0.1 seconds and processed to automatically store the data in the computer.

4.3 Experimental data processing
During the experiment, the pressure, temperature, and strain produced by the strain gauges due to the crack opening of the test pipe were measured and collected by the 35951B IMP and the 35951C IMP every 0.1 second and stored in the computer.

The leakage rate measurement data of the experimental pipeline, which needs to be recorded by the experimenter, was set to 180s; the weight of the leaked water was measured to be 1.7kg, and the leakage rate was 33.84%.

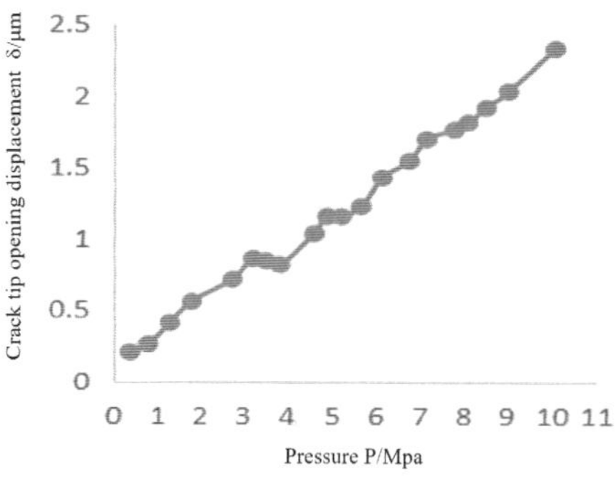

Figure 5. Relation between crack tip opening displacement and pressure.

Taking the pressure inside the experimental pipeline from 0MPa to 10MPa as an example, the relationship between the crack tip opening displacement δ and the pressure P of the experimental pipeline is shown in Fig. 5. From the analysis of the experimental data, it can be seen that when the pressure in the experimental pipeline increases, the crack tip opening displacement of the pipeline also increases gradually. It shows that the pressure in the experimental pipe has a non-negligible effect on the cracks of the pipe.

Three surface cracks with crack depths of t/4, t/2, and 3t/4 (t was the wall thickness of the experimental pipe) were measured on the experimental pipe, and the tensile displacements of the cracks with different depths were measured at the same pressure.

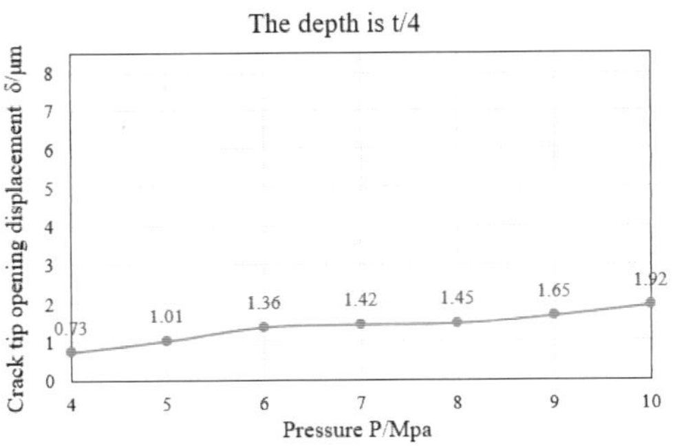

Figure 6. Variation with pressure for crack depth of t/4.

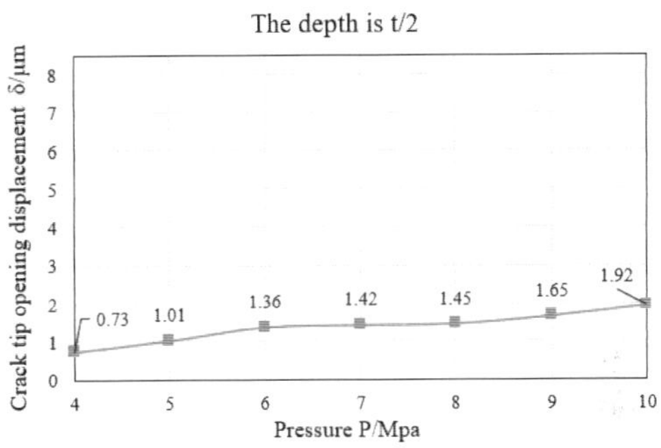

Figure 7. Variation with pressure for crack depth of t/2.

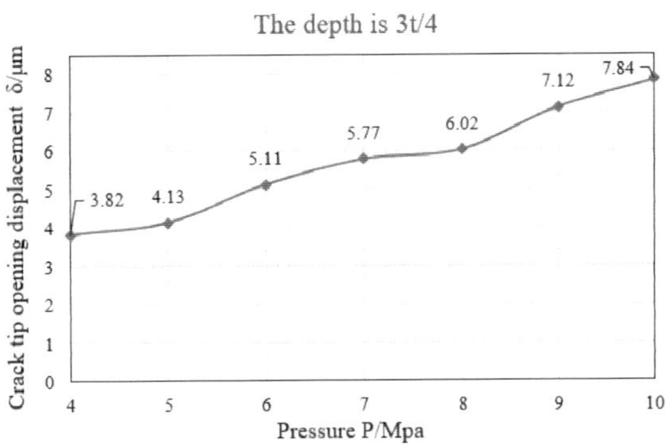

Figure 8. Variation with pressure for crack depth of 3t/4.

As can be seen from Figs. 6 and 7, the tensile displacement of the cracks increases slowly with increasing pressure when the crack depths are t/4 and t/2, respectively. From Fig. 8, it can be seen that the tensile displacement of the crack increases rapidly with increasing pressure when the crack depth is 3t/4. This indicates that the effect of pressure is greater when the crack is deeper. In addition, comparing the results of the three sets of experiments, it is found that the tensile displacement of the crack increases with the increase of crack depth at the same pressure.

5. Conclusions

Based on the Leak-Before-Break (LBB) analysis of the main pipeline in pressurized water reactor (PWR) nuclear power plants and combined with relevant fracture mechanics theories, the phenomenon of pre-leakage (LBB) of pipelines under high pressure was investigated through a combined method of theory and experiment.

1. The experimental pipelines did not experience unstable crack propagation, and there was no brittle double-ended shear fracture in the experimental pipelines. Instead, the cracks slowly extended

with increasing pressure, consistent with the theoretical analysis results of LBB for the experimental pipelines.

2. The results of the crack tip opening displacement δ obtained from the LBB theoretical analysis were in good agreement with the experimental results, indicating that the LBB theory is effective in analyzing the crack propagation of pipelines.

3. The leakage rate of the experimental pipelines calculated using the LBB analysis theory was much higher than the experimental results, mainly due to the rapid pressure release during the leakage rate measurement process, resulting in actual pressures lower than the conditions calculated theoretically.

4. In the experiments, it was found that the radial expansion of pipeline cracks followed certain patterns:

(1) The crack opening displacement of the experimental pipelines increased with increasing pressure.

(2) Under the same pressure conditions, the crack opening displacement of the pipelines increased with increasing crack depth.

References

[1] Zhang J, Chen R H, Wang M J, et al. A code development for leak before break (LBB) leakage from supercritical to subcritical conditions [J]. Progress in Nuclear Energy, 2018, 103: 217-228.

[2] Da Silva I G F, De Andrade A H P, and Monteiro W A. Leak-before-break methodology applied to different piping materials: a performance evaluation [J]. Frattura ed Integrità Strutturale, 2019, 13(50): 46-53.

[3] Hu L, Zhang Y, Li L, et al. Investigation of severe accident scenario of PWR response to LOCA along with SBO [J]. Progress in Nuclear Energy, 2015, 83: 159-166.

[4] Feng H, Chen H, Yang J, et al. Double-Ended Shear Fracture Accident of Short Casing Inner Tube in Small Modular Reactor [C]//Proceedings of the 23rd Pacific Basin Nuclear Conference, Volume 2: PBNC 2022, 1-4 November, Beijing & Chengdu, China. Singapore: Springer Nature Singapore, 2023: 144-154.

[5] Chun J H, Lim S W, Chung B D, et al. Safety evaluation of accident-tolerant FCM fueled core with SiC-coated zircalloy cladding for design-basis-accidents and beyond DBAs [J]. Nuclear Engineering and Design, 2015, 289: 287-295.

[6] Hung Z Y, Ferng Y M, Hsu W S, et al. Analysis of AP1000 containment passive cooling system during a loss-of-coolant accident [J]. Annals of Nuclear Energy, 2015, 85: 717-724.

[7] Chen K, Huangfu Y, Ma H, et al. Calculation of mesh stiffness of spur gears considering complex foundation types and crack propagation paths [J]. Mechanical Systems and Signal Processing, 2019, 130: 273-292.

[8] Park J H, Cho Y K, Kim S H, et al. Estimation of leak rate through circumferential cracks in pipes in nuclear power plants [J]. Nuclear Engineering and Technology, 2015, 47(3): 332-339.

[9] Kalani A, Kandlikar S G. Effect of taper on pressure recovery during flow boiling in open microchannels with manifold using homogeneous flow model [J]. International Journal of Heat and Mass Transfer, 2015, 83: 109-117.

[10] Fett T, Njiwa A B K, and Rödel J. Crack opening displacements of Vickers indentation cracks [J]. Engineering Fracture Mechanics, 2005, 72(5): 647-659.

Journal of Physics: Conference Series **2842** (2024) 012002 doi:10.1088/1742-6596/2842/1/012002

Effect of Cr_3C_2 on wear behavior of Ni45 laser cladding coating

Limeng Liang[1], Dawei Feng[1], Shuli Ma[1], Jie Zhang[1] and Guohua Ding[2,*]

[1]School of Mechanical and Vehicular Engineering, Bengbu University, Bengbu 233030, China

[2]Anhui Engineering Research Center of Additive Manufacturing, Bengbu 233030, China

[*]Corresponding author's email: dgh@bbc.edu.cn

Abstract. To improve the performance of engineering components subjected to severe wear, temperature fluctuations, and corrosive conditions in industries such as chemical, aerospace, and power plants, Ni45-based coatings with varying Cr_3C_2 contents (0%, 10%, 20%, and 30%) were applied to a 45-steel surface. The microstructure of these coatings was examined by scanning electron microscopy (SEM), and their wear resistance was compared. As the Cr_3C_2 content in the cladding layer increased, there was a notable improvement in its wear resistance. Specifically, when the Cr_3C_2 content reached 30% by weight, the wear resistance of the cladding layer exceeded that of the GCr15 steel. Furthermore, the adhesive wear between the cladding layer and the GCr15 steel diminished, while the oxidation wear of the GCr15 steel intensified. This study aims to elucidate the primary wear mechanisms of the Cr_3C_2/Ni45 cladding layer and offer valuable insight into improving the durability of engineering components in harsh environments.

1. Introduction

Engineering components may be deemed invalid if their performance or functionality falls short of required standards, with four primary failure types being wear, deformation, fracture, and corrosion. Statistical data from actual engineering instances indicate that wear contributes around 55% of all failures, and abrasive wear alone accounts for up to 20% [1]. Within the agricultural sector, abrasive wear is predominantly responsible for the mechanical failure of agricultural tools, which influences tool quality. The effectiveness of these tools is determined by the working edge properties, such as surface roughness and hardness [1]. Consequently, the material utilized to build the device must be allowed to sustain severe force and corrosive environmental conditions [2]. These results highlight the need for new abrasive wear solutions.

There are two main approaches for improving the operational surface of any factor: external support and matrix strengthening. The use of external addition may cause chemical incompatibilities between the substrate and the reinforcer, resulting in unsatisfactory interface bonding. Furthermore, differences in thermal behaviour may enhance the danger of cracking at the weakly bonded junction. Conversely, under suitable in-situ thermodynamic conditions, it is possible to foster the growth of relatively complex phases within the matrix, effectively eliminating the interface issue [3]. Laser cladding technology has emerged as a pivotal method in in-situ matrix strengthening, enhancing the hardness, heat resistance, wear resistance, and corrosion resistance of workpiece surfaces. This technique not only facilitates the repair and remanufacturing of workpiece surfaces but also conserves costly materials, aligning with sustainable development strategies.

Reinforcing materials like WC-based rigid alloy, Co-base alloy, Ni-based alloy, and chromium carbide are commonly used in wear applications for their high hardness, resistance, and thermal stability. Owing to the exceptional oxidation resistance, corrosion resistance, and high-temperature resilience of chromium carbide, as well as its good wettability with Ni alloy, it can be used to prepare a laser cladding coating with high hardness, wear resistance, and corrosion resistance.

In this study, we prepared the Cr_3C_2/Ni45 cladding layer using optimal technological parameters. We then systematically investigated the influence of Cr_3C_2 content on the mechanical properties of the Cr_3C_2/Ni45 cladding layer and established its variation rule.

2. Experimental materials and methods

The experimental substrate selected was a 45-tempered structural steel, measuring 50mm × 50mm × 10mm. Its composition is detailed in Table 1. Before the laser cladding process, the 45 steel underwent a quenching procedure at 830 °C for 60 minutes, followed by a tempering process at 560 °C for an additional 60 minutes. These steps were essential to eliminate internal stress and ensure uniform microstructure.

Table 1. The chemical composition of the 45 steel used as the substrate is presented in weight percent (wt.%).

Element	C	Si	Mn	S	P	Cr	Ni	Fe
Amount	0.42~0.5	0.17~0.3	0.50~0.8	⩽0.035	⩽0.035	⩽0.25	⩽0.25	Bal.

A pulsed laser, AXL-600AW, was used to fabricate a cladding coating on the 45 steel matrix surface, with a preset layer thickness of 0.8mm. Before cladding, the 45 steel matrix was preheated to 200 °C using an external heat source. After cladding, the sample was subjected to annealing for 60 minutes to mitigate the propensity for cracking within the cladding layer. Table 2 presents the optimized process parameters.

Table 2. Laser cladding process parameters.

Electric current (A)	Scanning speed (mm/min)	Frequency (Hz)	Lap ratio	Argon flow rate (L/min)
150	200	30	60%	10

The Ni45 powder exhibits self-melting properties, with a particle size ranging from 45 to 109μm. It has a melting point that falls within the range of 980-1050°C and a hardness measurement of 45HRC. The chemical composition of this material is detailed in Table 3.

Table 3. The chemical composition of Ni45 powder expressed in weight percent (wt.%).

Fe	Cr	C	B	Si	Ni
10	13	0.45	2.5	3	Bal.

Four types of Cr_3C_2/Ni45 cladding layers, containing 0, 10wt.%, 20wt.%, and 30wt.% Cr_3C_2 respectively, were fabricated on a 45 steel substrate. Before the cladding process, the composite powder was mixed uniformly using a mixer, followed by moisture removal in a drying oven. After cladding, the samples were polished, ultrasonically cleaned, and etched with aqua regia. The microstructure of the cladding layer was analyzed using a Phenom ProX desktop scanning electron microscope.

For the wear test, a wire-cutting machine was employed to segment multiple cladding layers into cylindrical specimens with a diameter of 30mm. The wear behavior of the cladding layer was evaluated using the MPX-2000 disc and pin type friction and wear tester. The pin, made of GCr15 steel grinding material with a hardness of 62HRC, and the disc, which is the polished cladding layer sample from the grinder, were used in this test. The test speed was set at 370 r/min, with a load of 500 N. The weight loss of both the cladding layer and the GCr15 steel post-wear was measured to assess the wear resistance of the cladding layer. Weight loss from wear was recorded on an electronic scale with an accuracy of 0.1mg every 15 minutes, totaling a measurement time of 120 minutes for both materials. Before measurement, the samples were cleaned with alcohol and dried to minimize errors. The worn morphology was examined using a desktop scanning electron microscope equipped with an integrated energy spectrometer (Phenom proX), followed by energy spectrum analysis.

3. Results and discussion

Coatings with varying mass fractions of Cr_3C_2 are abraded against the hardened high-carbon chromium-bearing steel GCr15, which has a microhardness of 760 HV. The temporal relationship between weight loss for the different Cr_3C_2/Ni45 cladding layers and GCr15 is depicted in Figure 1.

As illustrated in Figure 1, the Ni45 coating exhibits a lower hardness compared to the GCr15 steel, resulting in a higher weight loss for the Ni45 layer. Furthermore, at a fixed Cr_3C_2 content, an increase in the grinding time leads to a marginal increase in weight loss for GCr15, while it significantly increases for the Cr_3C_2/Ni45 cladding layers. Interestingly, as the Cr_3C_2 content increases, weight loss across different Cr_3C_2/Ni45 cladding layers remains consistent. Similarly, the weight loss of GCr15 increases markedly, exceeding that of the Cr_3C_2/Ni45 cladding layers when the Cr_3C_2 content exceeds 25% by weight. In essence, while the weight loss of the Cr_3C_2/Ni45 cladding layers is influenced by grinding time alone, that of GCr15 is governed by both grinding time and Cr_3C_2 content, as depicted in Figures 2(a) and 2(b), respectively.

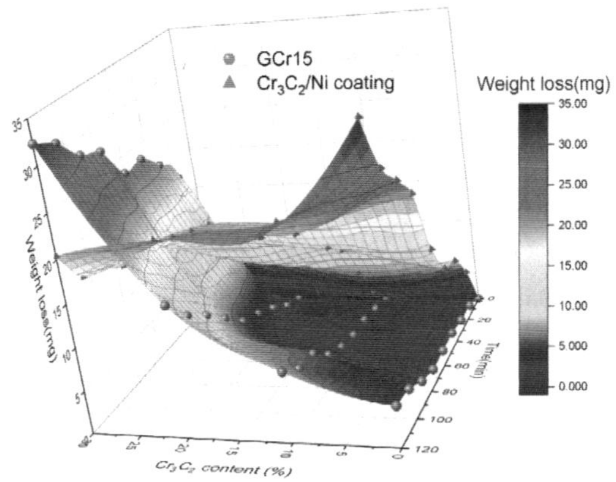

Figure 1. Weight loss of Cr_3C_2/Ni45 coating and GCr15 steel during the grinding process over time.

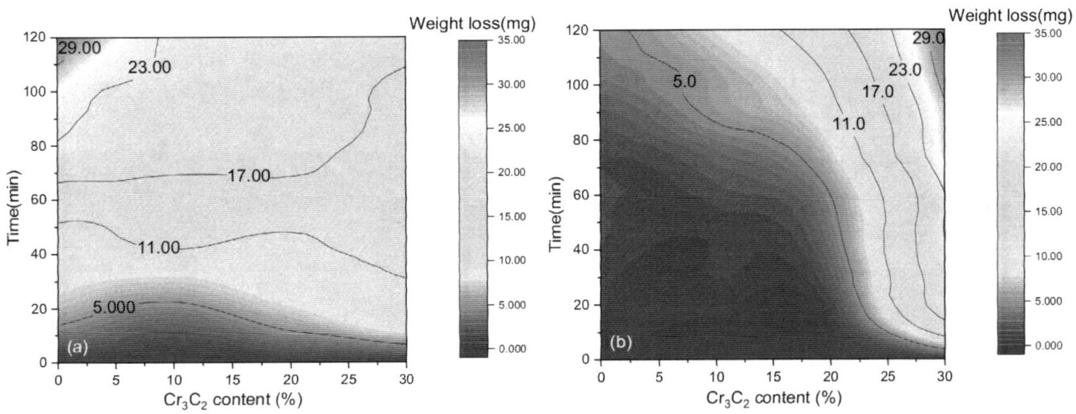

Figure 2. The distribution of weight loss for both the Cr_3C_2/Ni45 coating (a) and GCr15 steel (b) during grinding is examined with time and Cr_3C_2 content.

This study quantitatively examines the wear resistance of the Cr_3C_2/Ni45 coating with changes in Cr_3C_2 content. The relative weight loss ratio of the Cr_3C_2/Ni45 coating to GCr15 steel was calculated after a two-hour grinding period, as depicted in Figure 3. The wear process revealed that the wear resistance of the Cr_3C_2/Ni45 coating improved significantly and exceeded that of the GCr15 steel as the Cr_3C_2 content increased. In particular, when the Cr_3C_2 content increased from 20wt.% to 30wt.%, the wear resistance of the coating increased by 57%, while its hardness only increased by 21%. This can be attributed to several factors [4]: (1) During the cladding process, Cr_3C_2 melts into C and Cr elements, forming a solid solution in the dendrite. This enhances the solid solution strengthening effect and the second phase strengthening and fine-grain strengthening, which precipitate new carbides during solidification. (2) Following laser cladding, the Cr element content in the coating increases with the addition of Cr_3C_2 particles, triggering an oxidation reaction and the formation of an oxide Cr_3C_2 film at high wear temperatures. The compact structure and high melting point of the Cr_3C_2 oxide film protect the cladding layer from wear, thereby reducing the oxidation rate of the coating. (3) Given that Cr_3C_2 particles possess high hardness and wear resistance, the cutting wear effect on GCr15 steel is effectively reduced as the amount of Cr_3C_2 particles increases.

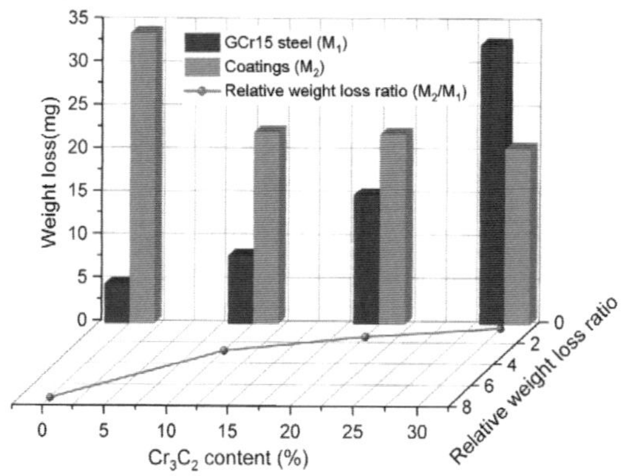

Figure 3. Weight loss ratio of Cr_3C_2/Ni45 coating with different Cr_3C_2 content to GCr15 steel.

Figure 4 presents the results of the wear morphology and energy spectrum analysis of Cr_3C_2/Ni45 coatings with varying Cr_3C_2 contents after a two-hour grinding process. Energy spectrum analysis of white area 1 and black area 2 in Figure 4 (a) reveals that the elemental contents of Ni and Cr are higher in white area 1, while the elements of O and Fe predominate in black area 2, with lower contents of Ni and Cr. Consequently, it can be deduced that the white space represents the cladding layer and the black site signifies the oxidized adhesive produced by the interaction between the cladding layer and hardened GCr15 steel during the grinding process [5]. The wear mechanism observed between the Cr_3C_2/Ni45 laser cladding coating and the GCr15 steel is adhesion wear, which is accompanied by oxidation wear.

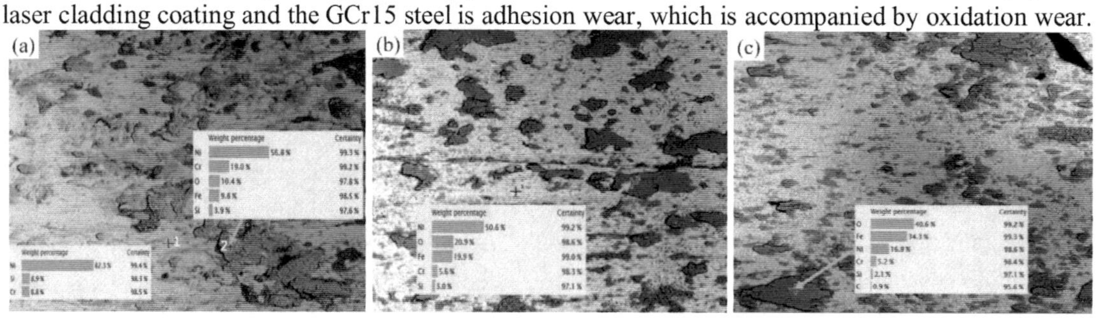

Figure 4. Wear morphology and energy spectrum analysis of the Cr_3C_2/Ni45 coating: (a) Ni45; (b) Ni45+20wt.%Cr_3C_2; (c) Ni45+30wt.%Cr_3C_2.

As illustrated in Figures 4(b) and 4(c), the wear morphology of Cr_3C_2/Ni45 cladding layers with varying Cr_3C_2 content exhibits similarities. However, an increase in the Cr_3C_2 content within the coating results in a reduction in the black area in the wear morphology. The black spot present in the wear morphology of three types of layers was analyzed using the energy spectrum. The findings indicate that as the Cr_3C_2 content increases, so does the Fe and O content in the black area. Consequently, this leads to a decrease in the degree of adhesion wear between the cladding layer and the GCr15 steel, while increasing the degree of oxidation wear [6]. In particular, the content of elements Fe and O in the black area of the Ni45+30wt.% Cr_3C_2 coating is significantly high, which contributes to an increase in the wear of GCr15 steel. This once again demonstrates that the wear resistance of the coating improves markedly when Cr_3C_2 increases from 20wt.% to 30wt.%.

4. Conclusions

In the present study, we examine the impact of varying Cr_3C_2 content on the mechanical properties of a Cr_3C_2/Ni45 cladding layer applied to the 45-steel. The following conclusions can be drawn:

(1) Within a specified Cr_3C_2 content, the weight loss of the GCr15 cladding layers showed a slight increase with the grinding time. On the contrary, the weight loss of the Cr_3C_2/Ni45 cladding layers showed a significant increase. However, as the Cr_3C_2 content increased, the weight loss in the different Cr_3C_2/Ni45 cladding layers remained nearly identical. It should be noted that the weight loss of GCr15 increased significantly, even surpassing that of the Cr_3C_2/Ni45 cladding layers when the Cr_3C_2 content exceeded 25wt.%.

(2) During the wear process, the wear resistance of the Cr_3C_2/Ni45 coating improves significantly, surpassing that of the GCr15 steel as the Cr_3C_2 content increases. Specifically, when the Cr_3C_2 content escalates from 20wt.% to 30wt.%, there is a marked increase in the wear resistance of the coating by 57%.

(3) The wear mechanism between the Cr_3C_2/Ni45 composite laser cladding coating and the GCr15 steel involves a combination of adhesive and oxidative wear.

Acknowledgments

This study was funded by the Anhui Province Natural Science Foundation for Higher Education Institutions (Grant No. KJ2021A1118), the Bengbu University High-quality Scientific Research Cultivation Project (Grant No. 2021pyxm03), the Bengbu University Scientific Research Foundation for Talented Scholars (Grant No. BBXY2018KYQD01), and the Innovative Entrepreneurial Training Plan Program of College Students (Grant Nos. S202211305042 and S202211305033).

References

[1] R. Bendikiene, A. Ciuplys, R. Sertvytis, A. Surzhenkov, D. Tkachivskyi, M. Viljus, R. Traksmaa, M. Antonov, P. Kulu, Wear behaviour of Cr3C2–Ni cermet reinforced hardfacings, Journal of Materials Research and Technology 9(4) (2020) 7068-7078.

[2] A. Singh Kang, G. Singh, G. Singh Cheema, Improving wear resistance via hardfacing of cultivator shovel, Materials Today: Proceedings 4(8) (2017) 7991-7999.

[3] D. Sharma, P. K. Ghosh, S. Kumar, S. Das, R. Anant, N. Kumar, Surface hardening by in-situ grown composite layer on microalloyed steel employing TIG arcing process, Surf. Coat. Technol. 352 (2018) 144-158.

[4] Q. Wu, W. Long, L. Zhang, H. Zhao, A review on ceramic coatings prepared by laser cladding technology, Optics & Laser Technology 176 (2024) 110993.

[5] Y. P. Liu, Y. L. Liang, H. G. Fu, Improvement properties of laser cladding Ni45-Cr3C2 coatings by adding B4C and V, Mater. Sci. Technol. 39(4) (2023) 443-453.

[6] A.M. Hebbale, M. Kumar, M.E.M. Soudagar, T. Ahamad, M.A. Kalam, N.M. Mubarak, A. Alfantazi, M. Khalid, A comparative study on characteristics of composite (Cr3C2-NiCr) clad developed through diode laser and microwave energy, Scientific Reports 13(1) (2023) 10778.

Preparation and adsorptive performance study of amino modified-crosslinked composite aerogel for VOCs in wooden furniture

Qian Li, Panpan Feng[*] and Mingyuan Qian

Shanghai Institute of Quality Inspection and Technical Research, Shanghai, 201114, China

[*]Corresponding author's e-mail: fengpp@sqi.org.cn

Abstract: There is a lack of standards for harmful VOC substances in furniture products in China, and furniture's peculiar smell is still a hot concern for society. Amino modified-crosslinked composite aerogel was prepared using a modified sol-gel method by incorporating chitosan and inorganic silica as the inorganic phase, and it was added to polyurethane coating (PU paint) and wooden substrate. Results indicated that the prepared amino modified-crosslinked composite aerogel had a pore size of 12 μm to 16 μm, retaining the lightweight structure and mesoporous characteristics of the gel while containing the active organic groups of chitosan. It exhibited a certain adsorption capacity for VOCs in both PU paint and wooden substrate, with the highest adsorption efficiency observed in fir, reaching 72.86%. These results demonstrate the feasibility of this method and suggest a new approach to pollution prevention and control in wooden furniture. It holds promising potential for scalable applications.

1. Introduction

In recent years, new types of furniture manufacturing materials and decorative materials have emerged in the market. However, along with these trends, there have been an increasing number of complaints about pollution, such as odors. The production process, materials, and scope of use of wooden furniture have become a concern and a hot topic for the general public [1]. These pollutants mainly come from raw and auxiliary materials, including solid wood, artificial boards, adhesives, and paints [2]. During the production and processing procedures, these materials may introduce volatile organic compounds (VOCs) [3-4], but China has yet to publish limits for VOCs in harmful substances in furniture products. Due to their use of water as a solvent and reduced pollution, water-based paints are increasingly utilized abroad. Apart from Beijing and Shenzhen, policies regarding water-based paints have not been fully implemented in all regions. Additionally, adopting water-based paints requires modifications to the entire production process, making it challenging for many companies to switch fully. Moreover, water-based paints for furniture often fall short in abrasion resistance and decorative performance compared to solvent-based coatings [5]. Therefore, this project plans to develop a nano-additive and incorporate it into PU paints to reduce VOCs in wooden furniture, presenting a new method for pollution prevention and control in the field of wooden furniture.

In this study, an additive was investigated for its adsorption effect on volatile organic compounds (VOCs) in PU paints used in wooden furniture. This additive offers a new pollution prevention and control method in wooden furniture and has the potential for large-scale industrial application. The additive is prepared using a cost-effective and easily accessible high molecular material, silica [6-7] as the inorganic phase and chitosan [8-10] as the organic phase through a co-sol-gel process. The resulting amino modified-crosslinked chitosan-silica composite aerogel (AMCCA) is low-cost, highly operable, and has great potential for application.

2. Experiment

2.1 Reagents, materials, and instruments

The materials are as follows: Chitosan (Sinopharm Chemical Reagent Co., Ltd., BR, 80%~95% deacetylation degree); Hydrofluoric acid (Shanghai Shenbo Chemical Co., Ltd., AR, 40%); HMDS (Sinopharm Chemical Reagent Co., Ltd., AR≥98.0%); Tetraethyl orthosilicate (Sinopharm Chemical Reagent Co., Ltd., AR≥28.4%); Epichlorohydrin (Sinopharm Chemical Reagent Co., Ltd., AR≥99.0%); Hexane (Sinopharm Chemical Reagent Co., Ltd., AR≥97.0%); Hydrochloric acid (Sinopharm Chemical Reagent Co., Ltd., analysis of AR, 36.0%~38.0%); and PU semi-glossy topcoat (Guangdong Huarun Coatings Co., Ltd., YAM214).

The instruments are as follows: SEM515 scanning electron microscope (Philip); TriStar 3000 surface area and pore size analyzer (Micro Instrument Corporation, USA); 7890A/5975C gas chromatograph-mass spectrometer (Agilent).

2.2 Preparation of AMCCA

We dissolved 9 g of chitosan in 500 mL of 0.15 mol/L diluted hydrochloric acid and filtered it. We slowly added the chitosan solution into a mixture of tetraethyl orthosilicate and anhydrous ethanol in a volume ratio of 1: 1 (250 mL). Then, we added 10 mL of glutaraldehyde, sealed the container, and stirred for 12 hours at room temperature. We added 9 mL of 5% HF solution while stirring continuously. After standing for 12 hours, a semi-transparent chitosan-silica composite aerogel (g1) was obtained. We washed the obtained sample g1 thrice with deionized water and anhydrous ethanol, each for 6 hours. We added the washed sample g1 to a mixture of 45 mL of epichlorohydrin, 2 mL of perchloric acid (70%), and 400 mL of anhydrous ethanol. We let it stand for 12 hours at room temperature and then washed the sample with anhydrous ethanol to remove residual epichlorohydrin and perchloric acid. We added the washed sample to 50 mL of ethylenediamine $C_2H_8N_2$ and 400 mL of anhydrous ethanol. We allowed it to stand at 50°C in a water bath for 12 hours, and then exchanged it three times with anhydrous ethanol and n-hexane C_6H_{14}, each time for 6 hours. Finally, the sample was added to 65 mL of HMDS C6H19NSi2 solution and 300 mL of n-hexane. After soaking for 12 hours, we washed the gel with n-hexane to remove residual silane and reaction products. It was dried at room temperature for 12 hours and then in a 60°C oven for 6 hours to obtain the final product, amino modified-crosslinked chitosan-silica composite aerogel (AMCCA).

3. Results and discussion

3.1 Analysis of AMCCA mechanism and surface morphology characterization

3.1.1 Mechanism of AMCCA. Under the action of HF, tetraethyl orthosilicate slowly hydrolyzes to form -Si-O-Si- units in the form of a sol matrix. Chitosan chains contain abundant -OH and $-NH_2$ groups, and the -OH groups form covalent bonds with exposed silica, while the -H in $-NH_2$ forms hydrogen bonds with the O in Si-O. Under the influence of the van der Waals force (VDW), the silica nanoparticles stably bind to the chitosan chains, forming a stable hybrid structure of a three-dimensional network.

Glutaraldehyde crosslinks chitosan through reaction with the amino groups on chitosan. In the presence of perchloric acid, the -OH groups on chitosan undergo ring-opening reactions with epichlorohydrin, and the propyl chloride group is attached to the oxygen of the hydroxyl group. Under alkaline conditions, the modifying agent ethylenediamine replaces the -Cl groups to obtain more $-NH_2$ groups. The reaction process is illustrated in Figure 1.

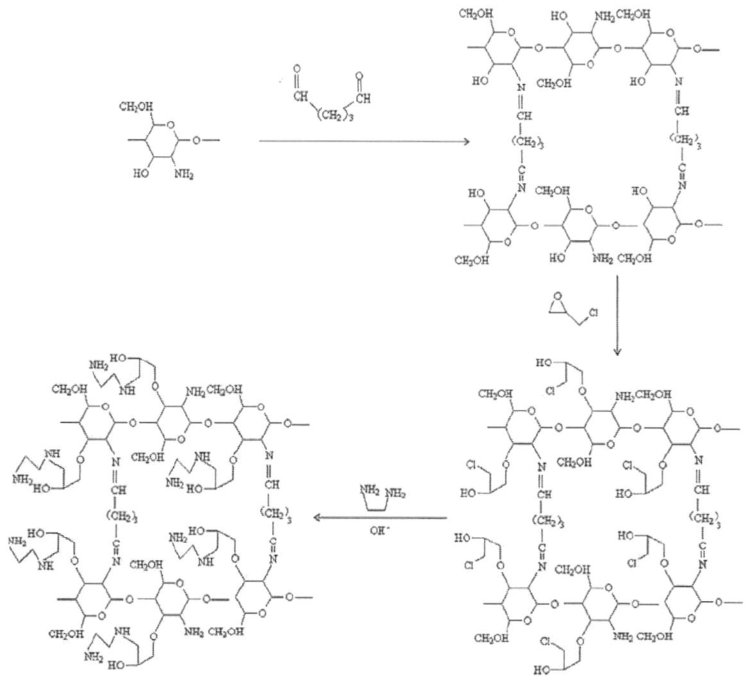

Figure 1. Amino modified-crosslinked reaction on the composite aerogel.

3.1.2 Surface morphology characterization of AMCCA. Figure 2 shows that AMCCA products are generally spherical with slight accumulation, and their diameters range from 12 μm to 16 μm. The modified samples exhibit numerous surface irregularities, wrinkles, and loose structures with increased porosity. The nano-level pore structure plays a significant role in the adsorption of volatile organic compounds.

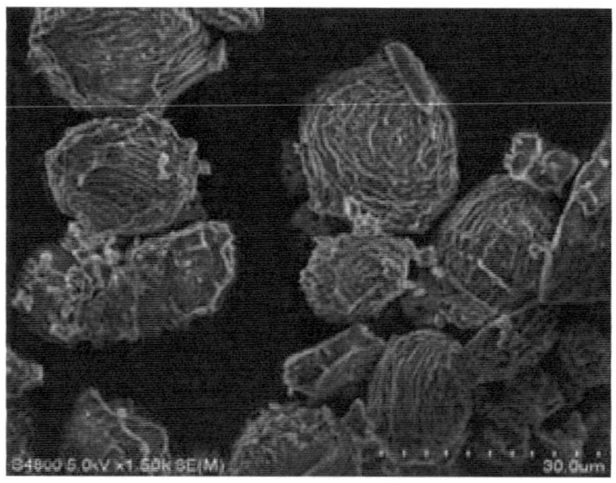

Figure 2. Typical SEM images of AMCCA.

3.2 BET analysis of AMCCA surface area and pore size distribution

Figure 3 shows that the nitrogen adsorption-desorption isotherm of AMCCA exhibits a typical type IV isotherm, indicating that it is a typical mesoporous material. Additionally, it shows significant nitrogen adsorption at low pressure, suggesting the presence of a large number of micropores in AMCCA. The hysteresis loop of AMCCA is of the H_2 type, indicating that the mesopores may be inkbottle-shaped pores formed by accumulating relatively uniform particles. The mesopores mainly originate from the aggregation of silica nanoparticles, while the assembly of amino acid-modified macromolecular structures on the particle surface forms the micropores.

Analysis of the desorption branch (Type D) in the adsorption-desorption isotherm reveals the possibility of typical cone-shaped adsorption, forming folded cone-shaped mesoporous structures based on the silica mesopores and the amino acid-modified material as the scaffold.

According to the BET analysis, AMCCA exhibits a high specific surface area of up to 588 m^2/g, a pore volume of 0.53 cm^3/g, and an average pore size of 3.62 nm. This indicates that AMCCA retains the surface characteristics of SiO_2 gel, consistent with the scanning electron microscope results. The pore size distribution curve obtained from the desorption branch using the BJH simulation shows that the pore size of AMCCA is mainly concentrated around 2.4 nm in the mesoporous region, which is consistent with the analysis of the adsorption-desorption isotherm.

In practical applications, the mesopores of AMCCA can effectively accelerate the diffusion of adsorbates into the internal micropores of the material, leading to rapid attainment of adsorption equilibrium. The abundant micropores can efficiently adsorb and immobilize the adsorbates, achieving high-efficiency and large-scale adsorption.

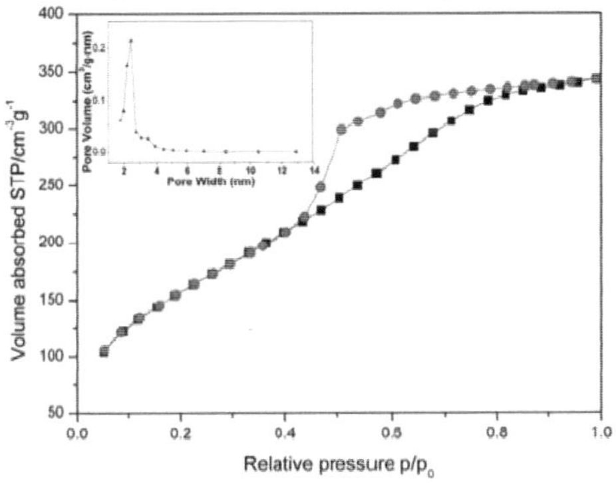

Figure 3. Nitrogen adsorption-desorption isotherms and pore size distribution of AMCCA.

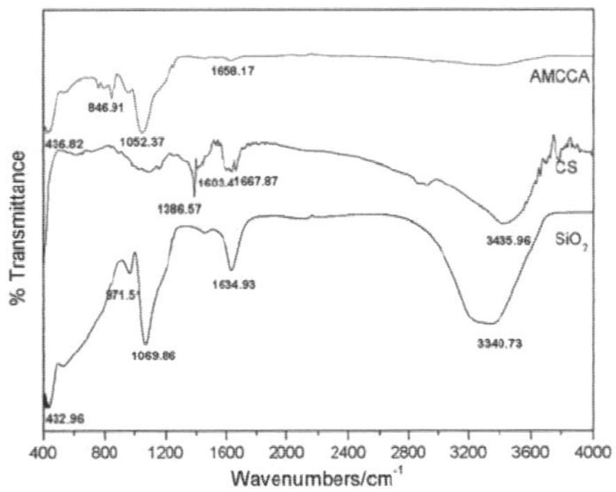

Figure 4. The FT-IR spectra of AMCCA, CS, and SiO_2.

3.3 Infrared spectroscopic analysis of AMCCA

The infrared absorption peaks of AMCCA are observed at 1658.17 cm^{-1} (stretching vibration of -C=N), 436.82 cm^{-1} (symmetric stretching vibration of Si-O bond), and 1052.37 cm^{-1} (anti-symmetric stretching vibration of Si-O-Si bond). The first peak is mainly attributed to the formation of imine bonds (-C=N-) by the condensation reaction of amino groups ($-NH_2$) on chitosan, resulting in the formation of a gel with high intensity. The latter two peaks mainly correspond to the silica component. The infrared spectrum confirms the formation of amino modified-crosslinked chitosan-silica composite aerogel (Figure 4).

3.4 Reduction of harmful substances in wooden materials by AMCCA

To evaluate the reduction of volatile organic compounds (VOCs) in wooden materials by AMCCA, two 1.5 m^3 test chambers (A and B) were used. Chamber A was set as the blank chamber, while Chamber B was the experimental chamber. Two beakers (A_1 and B_1) were filled with PU paints, each containing 200 mL of the solution. Untreated base paper or wood boards (fir wood, rubber wood, and pine wood) were suspended in the blank Chamber A, while three pieces of 1 m^2 base paper or wood boards (fir wood, rubber wood, and pine wood) coated with AMCCA diluted in acrylic acid were evenly sprayed and suspended in the experimental Chamber B. The chamber doors were tightly sealed. At room temperature, the fans in Chambers A and B were turned on and stirred for 1 minute to ensure air mixing, and then the fans were turned off simultaneously. After 24 hours, air samples were collected from both chambers using Tenax tubes. The sampling volume was set to 10L (V_1) at a sampling rate of 0.2 L/min for 50 minutes to capture as many VOCs as possible. The actual temperature (T_1) and pressure (P_1) were recorded to calculate the standard volume (V_0) in Equation (1). The concentration was analyzed using a gas chromatograph-mass spectrometer instrument, and the concentration in the blank chamber A was denoted as C_A. In contrast, the concentration in the sample chamber B was denoted as C_B. The reduction rate (y) was calculated using Equation (2):

$$\text{Standard Volume}(V_0)=\frac{T_0\times P_1\times V_1}{T_1\times P_0} \quad (1)$$

where P_0 is the standard atmospheric pressure (101.3 kPa) and T_0 is 273 K.

$$\text{And y (reduction rate \%)}=\frac{C_A-C_B}{C_A}\times 100\% \quad (2)$$

(1) For the PU paint, the reduction rates were as follows: 44.07% for VOCs, 82% for benzene, and 71.7% for methylbenzene, xylene, and ethylbenzene.

(2) For fir wood, rubber wood, and pine wood, the reduction rates by AMCCA were obtained, as shown in Table 1 and Figures 5 and 7. Fir wood and rubber wood mainly contain terpenoids, and they are formed by the reaction of multiple terpene groups and several functional groups through ethylene linkage and carboxyl groups, with cyclopentane and hemiacetal structure. The AMCCA's amino group did a maillard reaction with the carbonyl group, thereby playing an adsorption role. So, the effect of AMCCA was more pronounced on fir wood and rubber wood, and the maximum reduction rate for VOCs in fir wood reached 72.86%. However, pine wood consisted of alkanes and olefins, with relatively lower VOC concentrations, and the effect of AMCCA was not good.

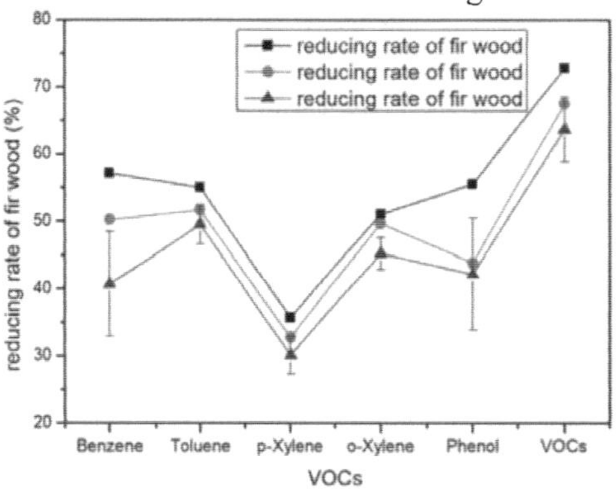

Figure 5. The reduction rate of fir wood's VOCs by AMCCA.

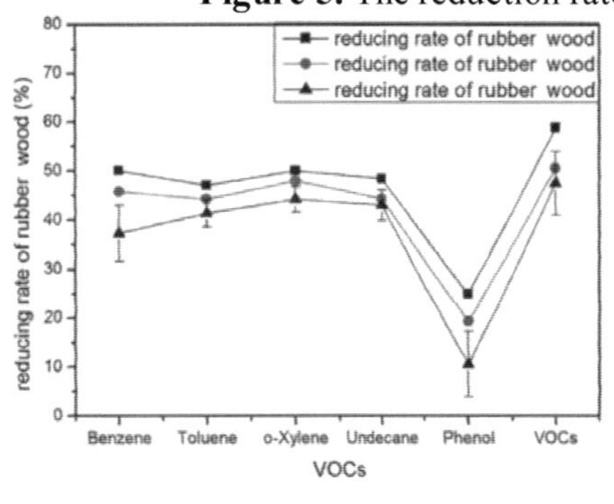

Figure 6. The reduction rate of rubber wood's VOCs by AMCCA.

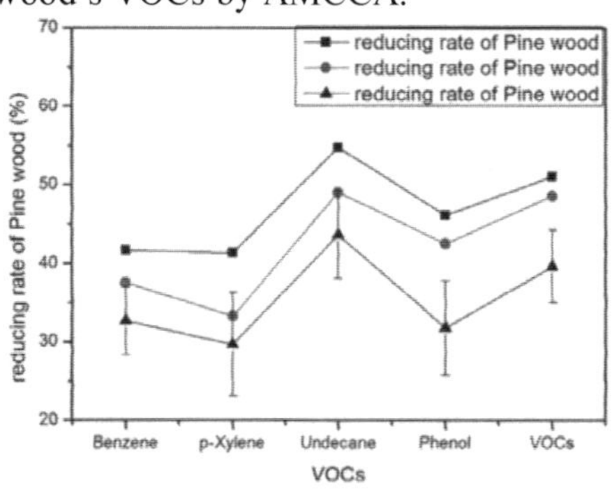

Figure 7. The reduction rate of pine wood's VOCs by AMCCA.

Table 1. The maximum reduction rate of fir, rubber, and pine wood by AMCCA.

No.	Organics	Reduction Rate (%)		
		Fir wood	Rubber wood	Pine wood
1	Benzene	57.14	50	41.67
2	Methylbenzene	55	47.06	-

3	P-xylene	35.71	-	41.38
4	O-xylene	51.06	50	-
5	Hendecane	-	48.39	54.76
6	Phenol	55.56	25	46.15
7	VOCs	72.86	58.79	51.10

4. Conclusions

The composite of AMCCA was prepared by sol-gel method with the inorganic phase of silica particles and further modified and cross-linked. SEM, infrared, and BET analysis showed that AMCCA was mainly spherical with slight aggregation and colloidal phenomenon. The microsphere diameter ranged from 12 μm to 16 μm, with many surface irregularities and wrinkles, loose surfaces, increased gaps, and nano-sized pore structures. The reduction rate of VOCs in PU paint was 44.07%, the benzene reduction rate was 82%, and the methylbenzene, xylene, and ethylbenzene reduction rates were 71.7%. AMCCA showed a more significant effect on fir wood and rubber wood materials. The adsorption rate of VOCs in fir wood reached 72.86%, demonstrating that the prepared composite gel had good adsorption capacity, laying the foundation for adding to PU paint and becoming wood furniture.

This study provides a new and relatively inexpensive method for the pollution prevention and control of wood furniture and has the potential for large-scale industrial application.

References

[1] Yan Shi. Study on volatile organic compounds emitted from wooden furniture and optimization of furniture production technology. Beijing Forestry University. 2021 DOI: 10.26949/d.cnki.gblyu.2020.000833.

[2] Yang Weiwei, Zhou Xueli, Zeng Dongdong, et al. Review on Research Progress of Heavy Metal Harmful Substances in Solid Wood Furniture in Gannan. China Quality and Standards Review [J]. 2022, 2: 65-67, 71. https: //kns.cnki.net/kcms2/article/abstract?v=UQzSFoOd3Sf ZyoiLbEWhREKDVauaTI9M3on_VDEGL6TsyaiggiWDE28LTUA0Kh07CHPZoY9A5GA A1Q5gG57cVzbLsn85rfUmktQGkzqiVaLL456oIN_YO9mzPpdkHQo3z3eKCwrBkUcNvlP -skR1Tw==&uniplatform=NZKPT&language=CHS.

[3] Xue Caifeng. Emission characteristics, environmental impacts and health risk assessment of VOCs from typical wooden furniture manufacturing industry. Taiyuan University of Science and Technology. DOI: 10.27721/d.cnki.gyzjc.2022.000306.

[4] Liu Zhongyi. Application of Inductively Coupled Plasma Mass Spectrometry in Heavy Metal Detection in Furniture. Chemical Engineering & Equipment [J]. 2022, 09: 275-276. DOI: 10.19566/j.cnki.cn35-1285/tq.2022.09.062.

[5] Yu Changjiang. Environmentally Friendly Application Solutions for Waterborne Solid Wood Furniture Coatings. CHINA COATINGS [J]. 2022, 37(8): 59-63. DOI: 10.13531/j.cnki.china. coatings.2022.08.012.

[6] Rao A.V., Kulkarni M. Hydrophobic Properties of TMOS/TMES-Based Silica Aerogels. Materials Research Bulletin [J]. 2002, 37: 1667-1677.

[7] Dorcheh A S, Abbasi M H. Silica aerogel: synthesis, properties and characterization. Journal of Materials Processing Technology [J]. 2008, 199: 10-26. DOI: 10.1016/j.jmatprotec.2007. 10.060.

[8] Ochoa M, Duraes L, Beja A M, et al. Study of the suitability of silica based xerogels synthesized using ethyltrimethoxysilane and/or methyltrimethoxysilane precursors for aerospace applications. Journal of Sol-Gel Science and Technology[J]. 2012, 61(1): 151-160.

[9] Lynn A, Capadona, et al. Flexible, low-density polymer crosslinked silica aerogels. Polymer[J]. 2006, 47(16): 5754-5761.DOI: 10.1016/jopolymer.2006.07.019.

[10] Kanamori K. Low-density, transparent aerogels and xerogels. Journal of Sol-Gel Science and Technology[J]. 2017, 81: 42-51. DOI: 10.1007/s10971-016-4077-1.

Nanometer-thick Ga/Al substituted yttrium iron garnet ultra-thin films by liquid phase epitaxial method

Yuanjing Zhang and Qinghui Yang *

State Key Laboratory of Electronic Thin Films and Integrated Devices, University of Electronic Science and Technology of China, Chengdu, Sichuan, 611731, China

* Corresponding author's e-mail: yangqinghui@uestc.edu.cn

Abstract. In microwave and spintronic applications, the Gilbert damping coefficient and ferromagnetic resonance (FMR) line width play a decisive role in the component losses. Yttrium iron garnet (YIG) film is a good candidate due to the extremely small damping factor. 4.8 and 51.3 nm ultra-thin $Y_3(GaAlFe)_5O_{12}$ (GaAl-YIG) films are grown on gadolinium gallium garnet (GGG) substrate using liquid phase epitaxial (LPE) technology. The films show a (111) orientation and tensile strain, and the roughness of the film surface is small. The films show a low saturation magnetization because of Ga and Al substituted, and the comparison of magnetization properties with film thickness is analyzed. The FMR line width is tested, and the calculated damping constant of the 51.3 nm film is on the order of 10^{-4}. The structural and magnetization properties analysis demonstrates that LPE technology has the potential to provide nanometer-thick garnet films for low-loss spintronic devices.

1. Introduction

For microwave and spintronic applications, the Gilbert damping parameter, and ferromagnetic resonance (FMR) line width determine their loss. Yttrium iron garnet (YIG) films showing almost intrinsic line-width and smallest damping parameter α=3×10^{-5} [1] have played an important role in the microwave [2], spintronic [3, 4], and magnetic-optic devices [5, 6], such as resonators, oscillators, and isolators. Nanometer-thick and low-damping YIG films are needed for these applications and spintronic fundamental studies. Liquid phase epitaxial (LPE) [7] technique is the best way to prepare micron scale YIG films on gadolinium gallium garnet (GGG) substrates, pulsed laser deposition (PLD) [8], molecular beam epitaxy (MBE) [9], and RF magnetron sputtering [10] methods are easier for the preparation of submicron and nanometer scale YIG films. However, the films fabricated by these techniques are quasi-single crystals and show slightly worse performance in spintronic studies.

The YIG film and GGG substrate interface have been studied. Vittoria and Schelleng [11] analyzed the effects of the diffusion at the LPE and chemical vapor deposited (CVD) YIG/GGG interface on spin wave excitations. The CVD films show a more prominent interdiffusion of the Fe and Ga elements. Suturin et al. [9] clarified a magnetically dead layer in nanometer YIG films prepared by MBE. The Ga-rich area at the interface is investigated by composition depth, and the resonance disappears when the thickness of the film is reduced to the nanometer level. Carsten et al. [7] studied the interface diffusion between 100 nm YIG and GGG. The interface is sharp, the FMR line width is narrow, and they want to conduct further experiments to optimize the FMR property for thinner YIG film.

In this work, 4.8 and 51.3 nm $Y_3(Fe_{5-x-y}Ga_xAl_y)O_{12}$ (GaAl-YIG) films are manufactured by the LPE technique. The microstructural and magnetic characterizations of the two GaAl-YIG films are investigated and compared in detail.

2. Experiment

The GaAl-YIG ultra-thin films are fabricated on (111) oriented GGG substrates by a horizontal impregnation LPE process based on a high-temperature melt of PbO-B_2O_3. Detailed procedures are as follows: firstly, high-purity Y_2O_3, Fe_2O_3, Ga_2O_3, Al_2O_3, B_2O_3, and PbO (99.999%) are mixed and heated in a platinum (Pt) crucible. Secondly, homogeny is melted at about 1100°C and then stirred with a Pt paddle for 12 h. Thirdly, the well-mixed melt is gradually cooled to proper growth temperatures according to previous experiments, and 870-880°C is selected for this study. Fourthly, the GGG substrates rotating at 60 rpm are immersed into the melt, then, the GaAl-YIG films could be grown on the substrates. Then, it is necessary to soak the holder in a hot diluted acidic acid solution to remove the melt remnants on the film surface. The thicknesses of these films are detected by X-ray reflectivity (XRR) and the lattice configuration is obtained from high-resolution X-ray diffraction (HR-XRD). The roughness of the film surface is tested from atomic force microscopy (AFM) over a 5 × 5 μm^2 range. The magnetostatic property is measured by a vibration sample magnetometer (VSM), and the ferromagnetic resonance (FMR) property is analyzed using vector network analyzer ferromagnetic resonance (VNA-FMR) at ambient temperature.

3. Results

3.1 Structural characteristics

Figure 1 is the film's thickness profile determined by the XRR technique. From the XRR patterns, the film thickness can be calculated from two adjacent interference fringes oscillating peaks by Equation 1:

$$t = \frac{[(n+1)-n]\lambda}{2(\sin\theta_{n+1}-\sin\theta_n)} \approx \frac{\lambda}{2\cos\theta\Delta\theta} \approx \frac{\lambda}{2\Delta\theta} \quad (1)$$

Among these two GaAl-YIG ultra-thin films, the thinner one is 4.8 nm, another is 51.3 nm. The films' parameters are listed in Table 1. From Figure 1, the faster the decrease in reflective intensity, the greater the surface roughness of the film. The curves show a similar decline, which means these films' surface roughness is close, and this can be confirmed by the AFM measurement. The surface morphology is shown in Figure 2, the surface roughness R_a is 1.5 nm for the 4.8 nm film, and 0.4 nm for the 51.3 nm film. The rate of weakening of the interference fringe vibration amplitude is proportional to the garnet/substrate interface roughness. The reflectivity shows less noise when 2θ is smaller than 4°, and the 4-7° noise can be viewed as equipment noise.

Table 1. The parameters of the GaAl-YIG films.

No.	Thickness (nm)	T_g (°C)	R_a (nm)	$4\pi M_S$ (G)	α (×10^{-4})	ΔH_0 (Oe)
1	4.8	877	1.506	207.76	43.2	15.9
2	51.3	876	0.429	75.74	4.35	13.9

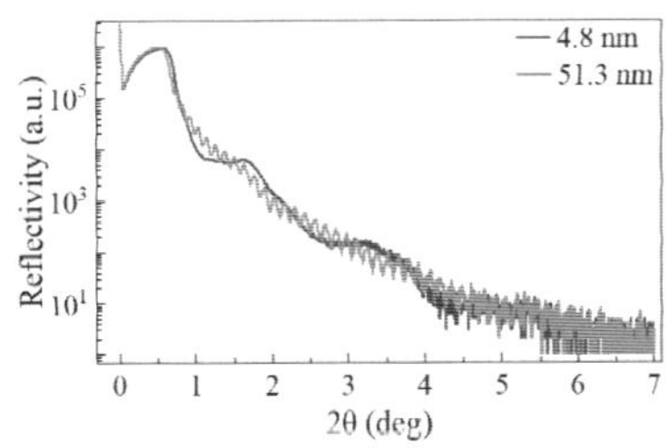

Figure 1. XRR plot for the GaAl-YIG films.

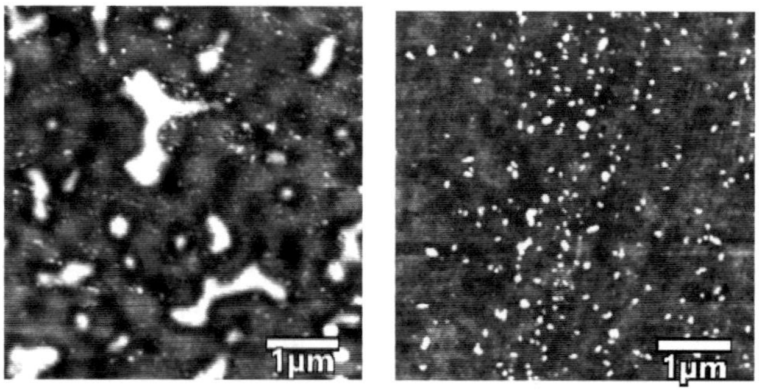

Figure 2. AFM images of the LPE GaAl-YIG monocrystalline films.

Figure 3 shows the $2\theta/\omega$ scanning curve of the prepared GaAl-YIG films. There are two diffraction peaks, corresponding to the (444) peaks for the GGG substrates and GaAl-YIG films. It is known that the ionic radii Pb^{2+}(119 Å) > Y^{3+}(90 Å) > Fe^{3+}(64.5 Å) > Ga^{3+}(62 Å) > Al^{3+}(53.5 Å), so the substitution of Ga/Al can decrease the lattice constant. As a result, these films are under tensile strain. In contrast to our previous study in [12], the Pb concentration in this study is smaller due to higher growth temperature, so these GaAl-YIG films are under more tension. The diffraction peaks of GaAl-YIG films (the gray shaded area) are broad because of the presence of the interface where the element content is unstable. As the film thickness increases, the peak of the film becomes higher.

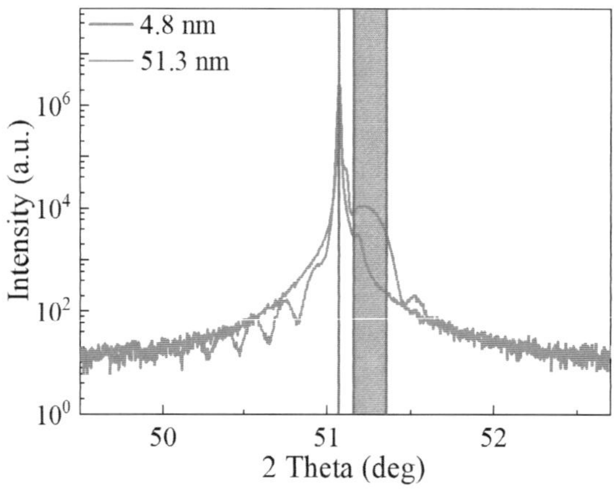

Figure 3. The XRD curves of GaAl-YIG films are around 51°.

3.2 Magnetic characterization

The magnetostatic characterization for these GaAl-YIG films is tested by VSM by portraying hysteresis loops shown in Figure 4, and the applied magnetic fields are both perpendicular and parallel to the (111) orientation for the films. Ga^{3+} and Al^{3+} enter both tetrahedral 24(d) and octahedral 16(a) positions, and these non-magnetic ions replaced magnetic ions Fe^{3+}, resulting in a decrease of saturation magnetization $4\pi M_S$. At the same time, the Fe^{3+} diffusion into the substrate leads to magnetic ion dilution, further contributes to the decrease in $4\pi M_S$. The 4.8 nm film show a higher saturation magnetization probably owing to the thickness calculation deviation and $4\pi M_S$ test error. With the GaAl-YIG film thickness increase, the shape of hysteresis loops changes from curve to straight lines. This is related to the effect of the interface, where the concentrations of elements are inconsistent and magnetization is difficult, and thinner films are more susceptible. The magnetic field required to magnetize to saturation in-plane $H_{S\parallel}$ and out-of-plane $H_{S\perp}$ are different for the films, and the 51.3 nm film is easier to be magnetized in-plane.

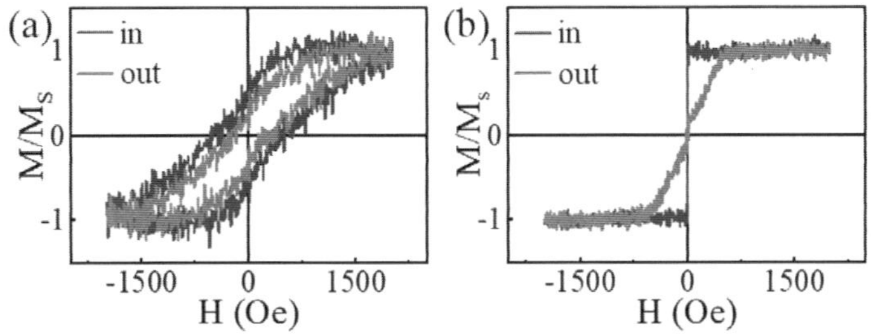

Figure 4. The normalized magnetic hysteresis curves for the GaAl-YIG films.

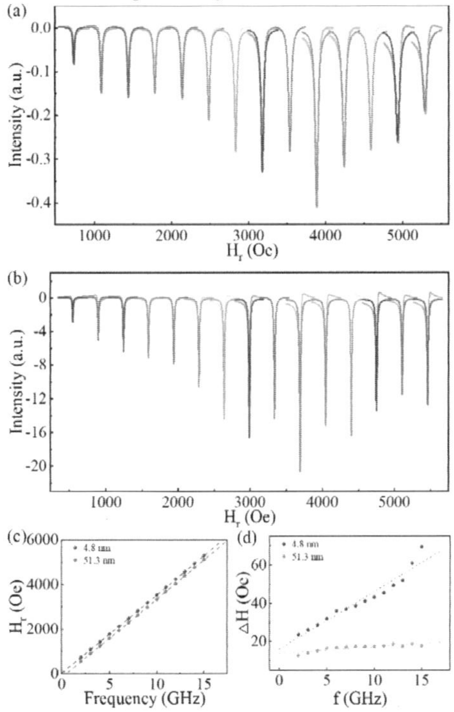

Figure 5. The FMR curves for these GaAl-YIG films.

Dynamic magnetization properties for the GaAl-YIG films are characterized by the VNA-FMR spectrum, which is tested on a coplanar waveguide by VNA-FMR. The measured frequency is 2 to 15 GHz, and the swept magnetic field is parallel to the films. The FMR curves are shown in Figure 5(a) and (b), the test data are hollow points, and the solid lines are the corresponding Lorentz fit data from 2 to 15 GHz. The line widths of the 51.3 nm film are much narrower than the 4.8 nm film. The relationship between the frequency and resonant fields is shown in Figure 5(c), and the frequency dependence of line width is shown in Figure 5(d). The value of damping factor α and non-uniform line width ΔH_0 are calculated by Kittel equations and are listed in Table 1.

$$f = \frac{\gamma \mu_0}{2\pi} \sqrt{H_r(H_r + 4\pi M_S)} \tag{2}$$

$$\Delta H = \Delta H_0 + \frac{4\pi\alpha}{\gamma} f \tag{3}$$

4. Conclusion

The 4.8 and 51.3 nm ultra-thin Ga/Al substituted YIG garnet films can be successfully fabricated by LPE. According to HRXRD and AFM measurements, the great single-crystal and little surface roughness indicate that the films are prepared with high quality. The doping of Ga^{3+} and Al^{3+}

decreases both the lattice constant and the saturation magnetization. The normalized magnetic hysteresis loops show the magnetization process of the films with different thicknesses, and the GaAl-YIG/GGG interface influences the shape of the loops a lot. The FMR line width of these ultra-thin films is on a competitive level compared with the films prepared by MBE or PLD. All the structural and magnetization characteristics corroborate the GaAl-YIG films prepared by LPE have potential promise for microwave and spin wave applications. A series of thicknesses for the films are needed for further preparation and analysis.

Acknowledgment
This work is underpinned by the National Natural Science Foundation of China (U2341286, 52072062, 61831012, 62235004, 51472046), Sichuan Science and Technology Support Program (2021JDTD0026, 2023YFG0072), and National Key Research and Development Program (SQ2018YFE0205600).

References
[1] Sparks, M (1964) Ferromagnetic-relaxation theory. McGraw-Hill, New York.
[2] Krupka, J., et al., Electrodynamic study of YIG filters and resonators. Scientific Reports, 2016. 6(1): p. 34739.
[3] Kajiwara, Y., et al., Transmission of electrical signals by spin-wave interconversion in a magnetic insulator. Nature, 2010. 464(7286): p. 262-266.
[4] Evelt, M., et al., Spin Hall-induced auto-oscillations in ultrathin YIG grown on Pt. Scientific Reports, 2018. 8(1): p. 1269.
[5] Yang, Y., et al., Recent advances in the development of magnetic garnet thin films for applications in spintronics and photonics. Journal of Alloys and Compounds, 2021. 860: p. 158235.
[6] Yu, J., et al., Study on laser-induced damage of TbBiIG crystal at 1064 nm. Optics Express, 2022. 30(17): p. 29991-30000.
[7] Dubs, C., et al., Sub-micrometer yttrium iron garnet LPE films with low ferromagnetic resonance losses. Journal of Physics D: Applied Physics, 2017. 50(20): p. 204005.
[8] Onbasli, M.C., et al., Pulsed laser deposition of epitaxial yttrium iron garnet films with low Gilbert damping and bulk-like magnetization. APL Materials, 2014. 2(10).
[9] Suturin, S.M., et al., Role of gallium diffusion in the formation of a magnetically dead layer at the Y3Fe5O12/Gd3Ga5O12 epitaxial interface. Physical Review Materials, 2018. 2(10): p. 104404.
[10] Ding, J., et al., Sputtering Growth of Low-Damping Yttrium-Iron-Garnet Thin Films. IEEE Magnetics Letters, 2020. 11: p. 1-5.
[11] Vittoria, C. and J.H. Schelleng, Effects of diffusion on magnetic excitations in films of yttrium iron garnet. Physical Review B, 1977. 16(9): p. 4020-4031.
[12] Zhang, Y., et al., Structural and magnetic properties of Y3(GaAlFe)5O12 liquid-phase epitaxy films with low ferromagnetic resonance losses. Acta Crystallographica Section B, Structural science, crystal engineering and materials, 2023. 79(Pt 2): p. 157-163.

Journal of Physics: Conference Series **2842** (2024) 012005 doi:10.1088/1742-6596/2842/1/012005

Adsorption of tryptophan by covalent organic frameworks

Chen Li and Mengchu Feng*

School of Chemistry and Chemical Engineering, Beijing Institute of Technology, Beijing, China

*Corresponding author's email: 1106535670@qq.com

Abstract: Hemoperfusion based on adsorption has been widely used to remove the blood toxins of patients with hepatic encephalopathy. It is important to develop adsorbents with high adsorption capacity, and specific adsorption performance for toxins is highly desired for hemoperfusion. Herein, we evaluate the tryptophan adsorption performance of four kinds of covalent organic frameworks (COFs) by powder X-ray diffraction (PXRD), high-performance liquid chromatography (HPLC), and so on. Benefitting from the abundant pore, suitable pore size, and attractive effect between tryptophan and hydroxy groups in COFs, TAPB-DHTA-COF and TAPT-DHTA-COF with abundant hydroxy groups show remarkable adsorption performance for tryptophan, which is superior to commercial activated carbon. Furthermore, the selective adsorption of COF does not affect other components in the blood, thus avoiding the occurrence of other diseases. Therefore, COFs offer a promising platform for the adsorption of toxins in hemoperfusion.

1. Introduction

Hepatic failure caused by severe hepatitis can lead to the accumulation of aromatic amino acids in the patient's body, further leading to death from hepatic encephalopathy[1]. Tryptophan is one of the main aromatic amino acids and strongly binds to albumin, which can cause hepatic encephalopathy. Hemoperfusion, a blood purification technology based on adsorption and separation, is an effective clinical treatment for the removal of excess tryptophan in the blood by using the extracorporeal blood circulation device[2]. Crucially, the key to hemoperfusion lies in the choosing and design of adsorption materials. Top-performing adsorption material should favor the highly efficient toxins capture and possess hemocompatibility[3]. Various materials, such as activated carbon[4], polysaccharides [5], and resin[6], have been utilized to adsorb toxins in hemoperfusion. However, these materials may exhibit drawbacks such as poor blood compatibility, limited adsorption specificity, or low adsorption capacity.

COFs are emerging porous crystalline materials, which are especially desirable as an alternative adsorbent due to their high porosity, vastly tunable structure, and so on[7]. Exceptional porosity, along with finely tailorable and uniform channels, endow COFs with sufficient adsorption space and selective adsorption. The easy functionalization of COFs favors further modification, thus enhancing interaction with toxins. In addition, COFs are stable in an aqueous environment, making them suitable for being used in blood.

Herein, we applied several COFs with different functional groups and pore structures to adsorb tryptophan in a simulated blood environment. For comparison, commercial activated carbon was also tested for adsorbing tryptophan in identical conditions. Particularly, COFs, containing hydroxyl functional groups, exhibited satisfactory adsorption performance because of their selective permeability and high tryptophan affinity and could adsorb tryptophan efficiently in some biological solutions, including bovine serum albumin (BSA).

2. Experiment

2.1 Preparation of TAPB-DHTA-COF

A mixture of n-butyl alcohol/o-dichlorobenzene (1.2 mL/0.8 mL), acetic acid solution (6 M, 0.2 mL), 1,3,5-Tris(4aminophenyl) benzene (TAPB, 28 mg, 0.08mmol), and 2,5-dihydroxyterephthalaldehyde (DHTA, 20 mg, 0.12 mmol) was degassed in a 10 mL Pyrex tube for three-time by freeze-pump-thaw cycles. Sealed the tube and heated it at 120 °C for 72 hours. Then, the precipitates were washed with DMF and ethanol. Finally, they were activated via supercritical CO_2 drying for further use. The synthesis method and process of TAPB-DHTA-COF can be seen in Figure 1.

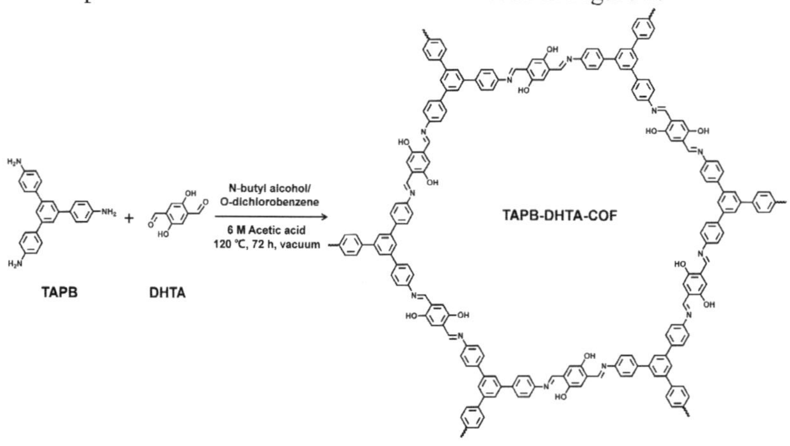

Figure 1. Synthesis reaction of TAPB-DHTA-COF.

2.2 Preparation of TAPT-DHTA-COF

20 mL glass vials were preloaded with 4,4',4''-(1,3,5-Triazine-2,4,6-triyl) trianiline (TAPT, 53 mg) and DHTA (37.5 mg). Then, a mixture of n-butyl alcohol/o-dichlorobenzene (4.5 mL/2 mL) was added to each vial and sonicated in order to get a homogenous dispersion. After that, each vial added 0.6 mL of acetic acid (3 M), which served as a catalyst. Then, gently shake the mixture by hand for 10 s and heat it at 100 °C for 72 hours. The red precipitates were washed with THF, DMF, and ethanol. Finally, they were activated via supercritical CO_2 drying for further use. The synthesis method and process of TAPT-DHTA-COF are shown in Figure 2.

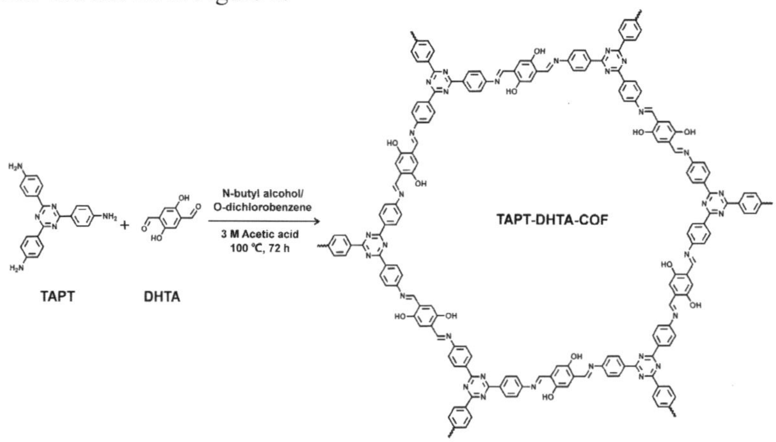

Figure 2. Synthesis reaction of TAPT-DHTA-COF.

2.3 Preparation of TAPB-DMA-COF

Typically, TAPB (14.1 mg, 0.04 mmol) and 2,5-dimethoxyterephthalaldehyde (DMA, 16.7 mg, 0.06 mmol) were added to a glass vial (20 mL). Next, 5 mL acetonitrile was added to each vial and sonicated until the organic monomers were completely dissolved. After that, each vial added 0.5 mL of acetic acid (12 M), which served as a catalyst. Then, gently shake the mixture by hand for 10 s and leave at room temperature for 72 hours. The yellow precipitates were washed with THF. Finally, they were activated via supercritical CO_2 drying for further use. The synthesis method and process of TAPB-DMA-COF are shown in Figure 3.

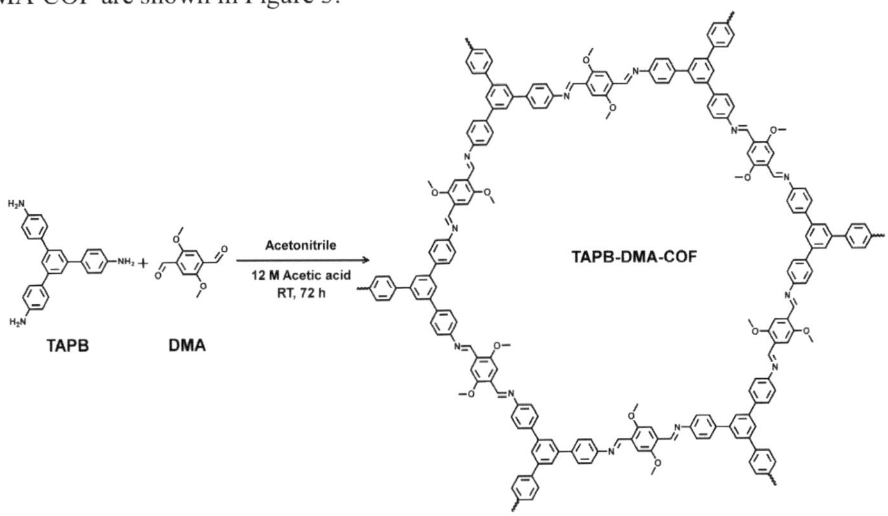

Figure 3. Synthesis reaction of TAPB-DMA-COF.

2.4 Preparation of Tp-Pa-COF

A mixture of mesitylene/dioxane (1 mL/2 mL), acetic acid solution (3 M, 0.5 mL), *p*-phenylenediamine (Pa, 48.6 mg, 0.45 mmol), and 1,3,5-triformylphloroglucinol (Tp, 63 mg, 0.3 mmol) was degassed in a 10 mL Pyrex tube for three times by freeze-pump-thaw cycles. Sealed the tube and heated it at 120 °C for 72 hours. The precipitates were washed with anhydrous acetone and DMF. Finally, they were dried under a vacuum at 85 °C for 12 hours to obtain Tp-Pa-COF. The synthesis method and process of Tp-Pa-COF are shown in Figure 4.

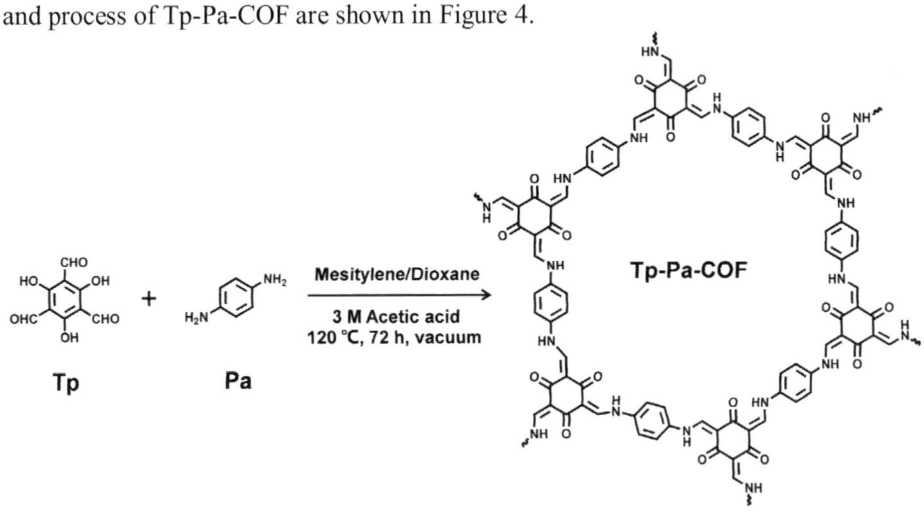

Figure 4. Synthesis reaction of Tp-Pa-COF.

2.5 Static adsorption experiment
First, 30 mg of COF was dispersed in 6 mL PBS (pH=7.4). Next, tryptophan (3.4 mg) and BSA (1.5 g) were dissolved in 30 mL PBS and stirred for 1 hour. Subsequently, the above solution (24 mL) was taken and added to the COF dispersion, where the tryptophan concentration was 90 mg/L, and the albumin concentration was 40 g/L to mimic the real blood environment. The adsorption procedure was carried out on a rotating shaker (175 rpm) at 37 °C. To determine the tryptophan concentration left in this system, aliquots (100 μL) were taken from the solution at different reaction times and removed BSA by 5% perchloric acid, finally analyzed by HPLC. The adsorption procedures for pure BSA or tryptophan are the same as above, except for the use of pure BSA or tryptophan as adsorbate.

3. Results and discussion

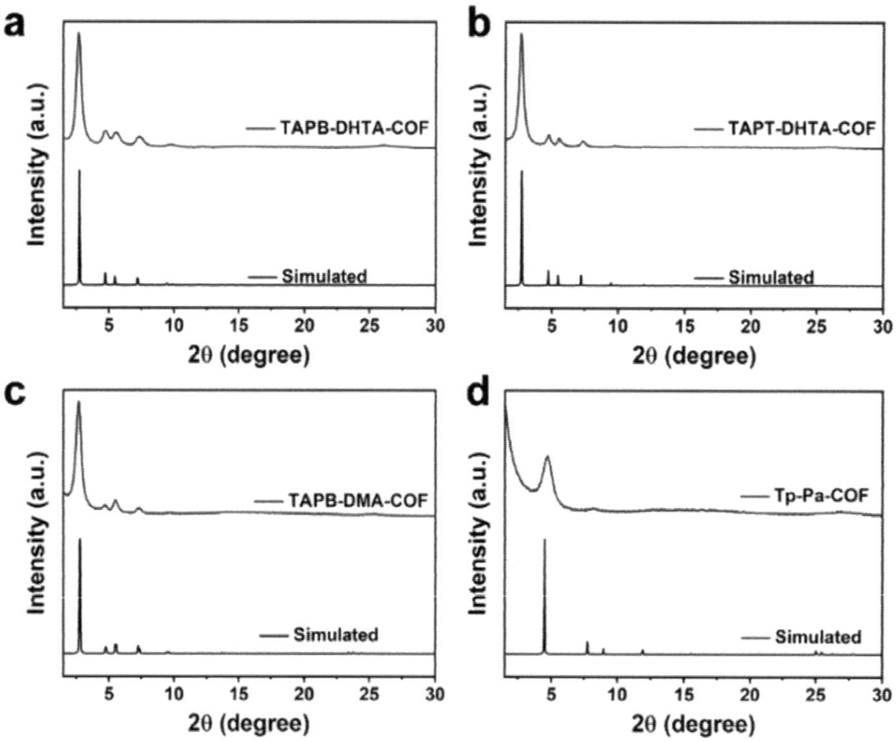

Figure 5. PXRD patterns of (a) TAPB-DHTA-COF, (b) TAPT-DHTA-COF, (c) TAPB-DMA-COF, (d) Tp-Pa-COF.

To determine the crystallinity of these four COFs, PXRD measurements were performed. Each COF's obvious intense diffraction peaks indicated that these four COFs were highly crystalline and agreed well with the theoretical simulations, proving that the four COFs' crystal structures were well maintained (Figure 5). The porosity and architectural stability of the four COFs were evaluated by N_2 sorption isotherms (77 K), revealing the surface area of 2197 m^2/g, 2095 m^2/g, 2101 m^2/g, and 315 m^2/g, corresponding to TAPB-DHTA-COF, TAPT-DHTA-COF, TAPB-DMA-COF, and Tp-Pa-COF, respectively. The results showed that these four COFs had high surfaces (Figure 6). The pore size distributions of these COFs were centered at 3.4 nm, 3.4 nm, 3.3 nm, and 1.6 nm, respectively (Figure 7), indicating these COFs all possess large one-dimensional straight porous channels. These COFs can serve as decent adsorbents for hemoperfusion in light of high surface and porosity, and regular and open 1D channels, which facilitate selective access and adsorption of toxins.

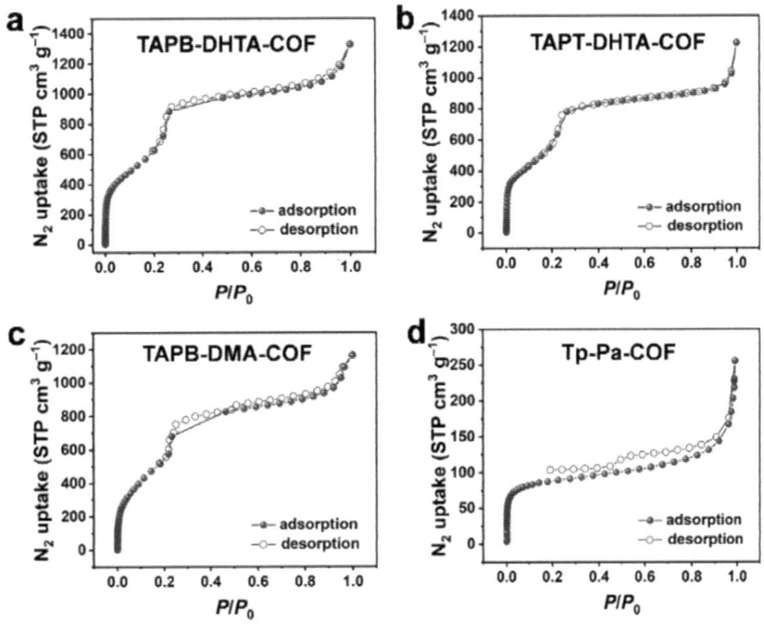

Figure 6. Nitrogen sorption isotherms of (a) TAPB-DHTA-COF, (b) TAPT-DHTA-COF, (c) TAPB-DMA-COF, (d) Tp-Pa-COF.

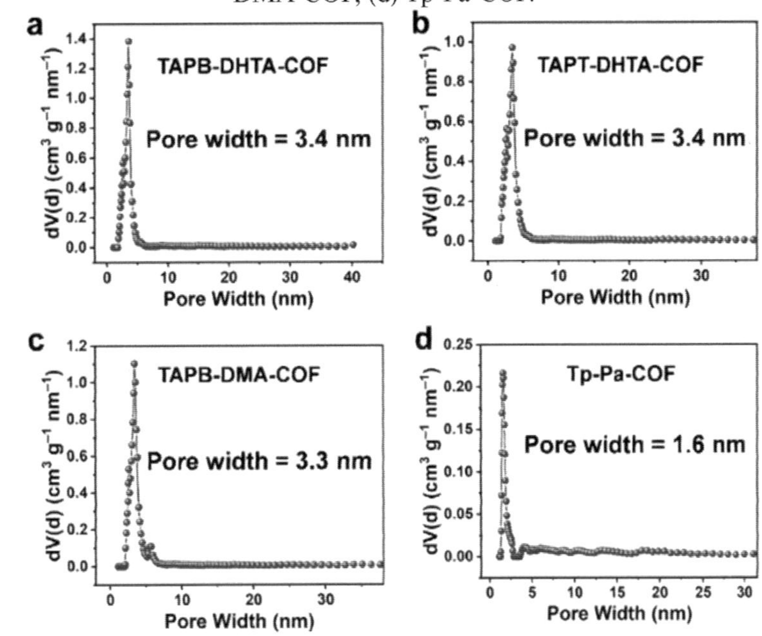

Figure 7. Pore size distribution curves of (a) TAPB-DHTA-COF, (b) TAPT-DHTA-COF, (c) TAPB-DMA-COF, (d) Tp-Pa-COF.

Tryptophan adsorption of COFs, as well as commercial activated carbon, was measured, and the results are illustrated (Figure 8a). In the first minute of this reaction, all materials could quickly adsorb tryptophan, and when the reaction lasted for 30 minutes, the adsorption reached equilibrium. Notably, TAPB-DHTA-COF and TAPT-DHTA-COF showed higher tryptophan adsorption ability (i.e., adsorption capacity and adsorption rate) than commercial activated carbon, whereas the adsorption ability of TAPB-DMA-COF with methoxy functional group and Tp-Pa-COF with smaller pore size is inferior to commercial activated carbon. Specifically, the equilibrium adsorption capacities of tryptophan for TAPB-DHTA-COF (19.2 mg/g) and TAPT-DHTA-COF (10.8 mg/g) are higher than that of commercial activated carbon (9.6 mg/g), while the equilibrium adsorption capacities of tryptophan for TAPB-DMA-COF and Tp-Pa-COF is only 7.0 mg/g and 6.0 mg/g, respectively, making TAPB-DHTA-COF and TAPT-DHTA-COF attractive candidates in hemoperfusion adsorbent (Figure 8b). It was then deduced that TAPB-DHTA-COF and TAPT-DHTA-COF can better adsorb tryptophan because these two COFs have hydroxyl functional groups, which can form hydrogen bonds and intermolecular forces with amino or carboxyl groups in tryptophan.

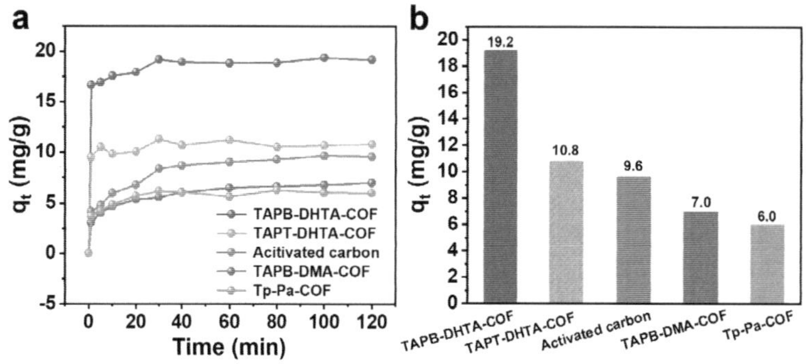

Figure 8. (a) Time-dependent adsorption quantity for tryptophan by COFs and activated carbon in two hours. (b) Saturation adsorption quantity for tryptophan by COFs and activated carbon in two hours.

Tryptophan, a protein-binding toxin, is carried by albumin in the blood. However, removing albumin from the blood can lead to severe albuminemia. Selective adsorption is significant so that tryptophan can be removed, whereas albumin is left in the blood. BSA (7.4 × 6.3 × 8.3 nm) was used to replace human serum albumin (HSA) in this work[8]. Based on the above experiments, we chose TAPB-DHTA-COF, having the highest adsorption capacity of tryptophan, as the experimental object to carry out the follow-up experiments. Initially, we investigated the BSA adsorption of TAPB-DHTA-COF at the identical experiment condition except for the exitance of tryptophan. The result showed that the concentration of BSA is almost unchanged because the pore size (3.4 nm) of TAPB-DHTA-COF is much smaller than the molecular size (7.4 × 6.3 × 8.3 nm) of BSA, BSA cannot enter the COF pore, that is, COF does not adsorb BSA (Figure 9a). Subsequently, we assessed the effect of BSA on tryptophan adsorption. In the existence of BSA, TAPB-DHTA-COF still had effective adsorption toward tryptophan (Figure 9b), which proves that TAPB-DHTA-COF has selective adsorption, that is, it can efficiently remove tryptophan without affecting HSA in blood.

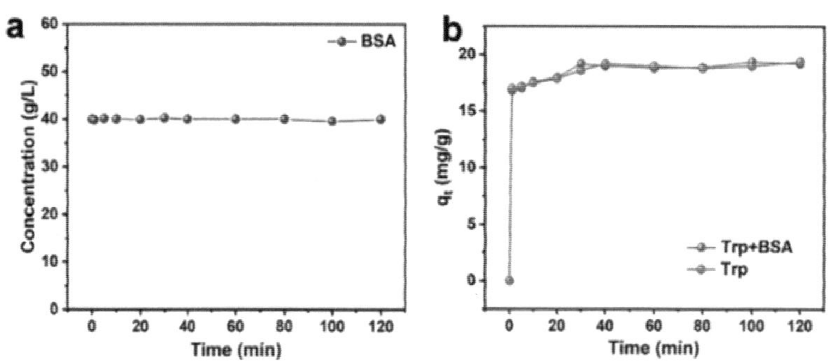

Figure 9. (a) Time-dependent adsorption quantity for tryptophan by TAPB-DHTA-COF in two hours. (b) Time-dependent adsorption quantity for tryptophan (Trp) by COFs in biological solutions (blue curve) and PBS (pink curve) in two hours.

4. Conclusion

In summary, we reported the application of COFs for hemoperfusion to adsorb tryptophan. Firstly, we synthesized four kinds of COFs, having excellent crystallinity, high surface area, and large pore size. Secondly, TAPB-DHTA-COF and TAPT-DHTA-COF with abundant hydroxyl functional groups showed excellent tryptophan adsorption capacity in PBS solution, which is compared with commercial activated carbon for hemoperfusion under the same conditions, proving that it has great application value in hemoperfusion. Thirdly, TAPB-DHTA-COF, having the highest adsorption capacity for tryptophan among the four COFs and commercial activated carbon, showed relatively high tryptophan adsorption selectivity at normal albumin concentration. Therefore, the above three conclusions are promising for the applications of COFs in blood purification.

Acknowledgments

The author thanks the Analysis & Testing Center, Beijing Institute of Technology for HPLC measurement.

References

[1] Pares, A., Deulofeu, R., Cisneros, L., Escorsell, A., Salmerón, J. M., Caballería, J., Mas, A. (2009) Albumin dialysis improves hepatic encephalopathy and decreases circulating phenolic aromatic amino acids in patients with alcoholic hepatitis and severe liver failure, Critical Care, 13: R8.

[2] Ronco, C., Bellomo, R. (2022) Hemoperfusion: technical aspects and state of the art, Critical Care, 26: 135.

[3] Li, Q., Guo, H., Yang, J., Zhao, W., Zhu, Y., Sui, X., Xu, T., Zhang, J., Zhang, L. (2020) MOF-Based Antibiofouling Hemoadsorbent for Highly Efficient Removal of Protein-Bound Bilirubin, Langmuir, 36: 8753–8763.

[4] Cai, N., Li, Q., Zhang, J., Xu, T., Zhao, W., Yang, J., Zhang, L. (2017) Antifouling zwitterionic hydrogel coating improves hemocompatibility of activated carbon hemoadsorbent, Journal of Colloid and Interface Science, 503: 168–177.

[5] Wei, H., Han, L., Tang, Y., Ren, J., Zhao, Z., Jia, L. (2015) Highly flexible heparin-modified chitosan/graphene oxide hybrid hydrogel as a super bilirubin adsorbent with excellent hemocompatibility, J. Mater. Chem. B., 3: 1646–1654.

[6] Weber, V., Linsberger, I., Hauner, M., Leistner, A., Leistner, A., Falkenhagen, D. (2008) Neutral Styrene Divinylbenzene Copolymers for Adsorption of Toxins in Liver Failure, Biomacromolecules, 9: 1322–1328.

[7] Shedrack, G. A., Imteaz, A., Pillaiyar, P., Kwangsun, Y., Wha-Seung, A. (2020) Microporous Organic Polymers for Efficient Removal of Sulfamethoxazole from Aqueous Solutions, Microporous and Mesoporous Materials, 296: 109979.

[8] Guo, L., Zhang, L., Zhang, J., Zhou, J., He, Q., Zeng, S., Cui, X., Shi, J. (2009) Hollow mesoporous carbon spheres-an excellent bilirubin adsorbent, Chem. Commun., pp: 6071–6073.

Study of the band alignment at the CZTS/ZrS_2 interface based on first-principles calculation

Haoning Li*

School of Physical Science and Technology, Kunming University, Kunming, Yunnan, 650214, China

* Corresponding author's e-mail: l15566513289@163.com

Abstract. To explore whether ZrS_2 monolayer has the potential to become a buffer layer material for Cu_2ZnSnS_4 (CZTS) solar cells, the band arrangement of CZTS/ZrS_2 heterojunction was predicted using density functional theory(DFT) first principles. In this paper, the hybrid functional HSE06 is used in all processes. The calculated results reveal a Type III band alignment at the CZTS/ZrS_2 heterojunction interface, with the maximum valence band (VBM) of CZTS positioned 1.67 eV higher than the minimum conduction value (CBM) of ZrS_2. The results indicate that the CZTS/ZrS_2 interface may not be well-suited for utilization in solar cells, pointing towards potential areas for further investigation.

1. Introduction

To address the issues associated with non-renewable energy, optimizing the energy mix and vigorously developing green and renewable energy sources have become common goals among all countries in the world. Compared with traditional fossil fuels, solar energy can be said to be an inexhaustible energy source, and has the greatest development potential among many renewable energy sources, attracting the attention of many researchers. Solar cells play a critical role in the renewable energy sector as they harness sunlight to generate electricity. Due to its potential to offer sustainable and clean energy sources, this technology has attracted considerable interest in the scientific community. Due to the consistent advancement in photovoltaic cell efficiency and the reduction of preparation costs, photovoltaic power generation is on track to significantly contribute to the global energy supply in the coming years.

CZTS is a compound semiconductor that presents a viable option for materials used in solar cell manufacturing, specifically as an alternative to Cu (In, Ga) Se_2 (CIGS). By using elements like Zn and Sn instead of In and Ga, CZTS reduces the reliance on expensive materials, making it a more cost-effective option [1]. CZTS has an optimal band gap of approximately 1.5 eV and exhibits a strong light absorption capability of $10^4\ cm^{-1}$ within the visible spectrum [2], and a theoretically predicted photoelectric conversion efficiency (PCE) of more than 30% [3], which is an environmentally friendly clean energy source for solar cells. Chen's team used first-principles calculations to prove that the energy of the kesterite (KS) structure is lower and more stable [2]. In this paper, the KS structure cell is constructed, and this structure is used for subsequent calculation. The primary barrier hindering the industrialization of CZTS solar cells is their low PCE. The best-documented efficiency for CZTS solar cells is around 11% [4]. According to previous studies, the type of energy band alignment between the absorber layer and the buffer layer significantly impacts the PCE of photovoltaic cells. Two types of band alignment exist in thin film solar cells: spike barrier (spike) and cliff barrier (cliff). The spike band arrangement is optimal when it is less than 0.4 eV. The negative conduction band offset (CBO) of -0.28 eV between the CZTS and CdS has a significant impact on the recombination of photons and electrons at the buffer-absorber interface. This offset contributes to a depression in photocurrent in the photovoltaic cell device. The mismatch in energies between the conduction band levels of the two layers hinders the efficient collection and transport of charge carriers, affecting the overall performance of the device. This highlights the importance of optimizing the band alignment between different layers in solar cell structures to enhance their photovoltaic efficiency. Further research and development efforts are needed to address these band alignment issues and improve the efficiency of CZTS solar cells [5]. For this reason, it is crucial to explore other potential materials for the buffer layer instead of CdS.

In recent years, the research of transition metal dichalcogenides (TMDs) has become mature. As a typical TMD, the ZrS_2 monolayer has been successfully fabricated in experiments, demonstrating clear n-type transport properties and high mobility. Ahmad's research team suggests that the ZrS_2 nanostructure shows promise as a potential option for large-area solar cell applications due to its ability to generate high short-circuit current [6]. Under the application of uniaxial strain within the elastic range, the ZrS_2 monolayer's band gap can be systematically adjusted from 0 to 2.47 eV. Furthermore, a shift from an indirect band gap to a direct band gap is noted upon the application of uniaxial strain in the zigzag orientation [7]. The potential of becoming a buffer layer material for CZTS solar cells is greatly enhanced by these benefits. In a study by Moustafa et al., a numerical assessment of ZrS_2 as a buffer layer material was conducted using SCAPS-1D. The results showed a promising PCE of 17.6%. This indicates that ZrS_2 could serve as a viable alternative buffer layer material for the development of environmentally friendly CZTS solar cells [8]. However, due to their simulation analysis of band arrangement characteristics and other properties through assumptions, the attempt to use ZrS_2 as a buffer layer material for CZTS solar cells needs to be further calculated and analyzed.

2. Calculation methods

Based on the density functional theory method, CASTEP in Materials Studio is used to calculate the band gap, state density of CZTS, and the ZrS_2 monolayer, as well as the energy band alignment types of CZTS/ZrS_2 heterojunctions, to calculate and analyze the properties at the interface. Firstly, we utilized the hybrid functional HSE06 to conduct structure optimization and electronic structure calculations on both CZTS and ZrS_2 monolayers. In this process, we employed the norm-conserving pseudopotential method. For CZTS, the cutoff energy was specified as 400 eV, with a k-point setting of $4 \times 4 \times 2$. On the other hand, for the ZrS_2 monolayer, we set the cutoff energy at 390 eV and the k point at $7 \times 7 \times 4$. To ensure minimal interlayer interaction, a vacuum layer thickness of 18 Å was established for the ZrS_2 monolayer. The relaxation criteria were met when the energy convergence reached 10^{-6} eV/atom, the maximum force converges to 0.03 eV/Å, and the maximum stress converges to 0.05 Gpa.

3. Results and discussion

3.1. Construction of CZTS/ZrS_2 heterojunction

The lattice constant for CZTS was determined to be $a = b = 5.397$ Å and $c = 10.802$ Å, with a band gap of $E_g = 1.6$ eV. For the ZrS_2 monolayer, the lattice constant was determined to be $a = b = 3.666$ Å, the Zr-S bond length was 2.567 Å, and the $E_g = 1.9$ eV. The calculated structure and band state density of both materials closely matched previous research findings [2,4,6,7]. To address the lattice mismatch between CZTS and ZrS_2, a lattice rearrangement was performed on both materials. The CZTS surface A was transformed to (3 1 0) and B to (2 2 0), while the ZrS_2 monolayer had A as (5 1 0) and B as (5 3 0), with CZTS serving as the reference surface. Using the lattice constant of the CZTS surface, a parallel heterojunction interface was created to form the CZTS/ZrS_2 heterojunction, with a vertical interface vacuum layer of 18 Å, as illustrated in Figure 1. The spacing of the heterojunction layer was 3.4 Å, with lattice constants of a = 17 Å, b = 15 Å, and c = 34 Å. The lattice angles were $\alpha = \beta = 90°$, and $\gamma = 26°$. To calculate the band alignment of heterojunctions, it is necessary to consider deep-level atomic orbitals due to the large number of atoms involved, leading to increased computational costs. Consequently, the band structure and state density of the CZTS/ZrS_2 heterojunction, CZTS, and ZrS_2 were determined by handling the Perdew-Burke-Ernzerhof (PBE) functional of the generalized gradient approximation (GGA) and OTFG ultrasoft pseudopotential. The valence electron configurations for the elements involved were Cu-$3d^{10}4s^1$, Zn-$3d^{10}4s^2$, Sn-$4d^{10}5s^25p^2$, S-$3s^23p^4$, and Zr-$4s^24p^64d^25s$. Additionally, the van der Waals Force was contemplated by applying the Grimme method for GGA-PBE calculations.

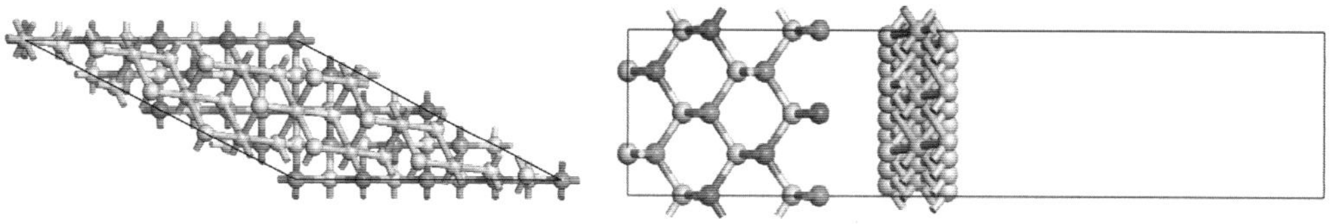

Figure 1. The CZTS/ZrS_2 heterojunction exhibits top (left) and side (right) perspectives.

3.2. Calculation of CZTS surface energy

The thickness and exposed atomic surface of the CZTS (0 0 1) layer are determined by calculating the surface energy (γ), which is obtained by summing the cutting energy (γ_u) and relaxation energy (γ_r).

$$\gamma = \gamma_u + \gamma_r \tag{1}$$

The cutting energy can be obtained by the following equation

$$\gamma_u = (E_{slab}^u(CuSn) + E_{slab}^u(S_2) - NE_{bulk})/4A \tag{2}$$

where $E_{slab}^u(CuSn)$ and $E_{slab}^u(S_2)$ are the unrelaxed $CuSn$ and S_2 terminated slab energies. E_{bulk} are the total energy of the bulk cells, N is the formula unit of the surface model, and A is the surface area. Factor 4 in the denominator comes from the four cutting surfaces of the two ends. Next, we can calculate the relaxation energy of $CuSn$, as shown in the following equation

$$\gamma_r = (E_{slab}^r(CuSn) - E_{slab}^u(CuSn))/2A \tag{3}$$

where $E_{slab}^r(CuSn)$ is the energy of $CuSn$ after surface relaxation. The calculation of layer thickness is shown in Table 1. The difference in surface energy of different layer thicknesses is less than 0.01 eV. Considering the calculation cost, the CZTS surface is selected as one layer in order to save the calculation amount. It can be observed that the influence of γ_r is negligible, so only γ_u is considered when calculating the surface energy of the severed position, and it is concluded that the surface energy is lowest when the bond is broken between the $CuSn$ atomic plane and the S_2 atomic plane. Here, the $CuSn$ atomic plane is chosen as the surface of CZTS in the heterojunction.

Table 1. Enhanced convergence tests of CZTS (0 0 1) surface with varying layer thicknesses.

Thickness	γ_u (eV/Å^2)	γ_r (eV/Å^2)	γ (eV/Å^2)
1	0.06375	0.00470	0.06845
1.5	0.06986	-0.00745	0.06241
2	0.06553	0.00386	0.06939

3.3. Calculation of heterojunction band migration

For the calculation of heterojunction band offset, the orbital level that forms a separation band with the continuous valence band can be used as the core level, and the valence band offset (VBO) of CZTS/ZrS_2 is given by the following equation

$$\Delta E_v = \Delta E_{VBM-1}^{CZTS} - \Delta E_{VBM-2}^{ZrS_2} - \Delta E_{core}^{CZTS/ZrS_2} \tag{4}$$

ΔE_{VBM-1}^{CZTS} serve as the discrepancy in energy levels at intervals between the VBM of CZTS and its core level. $\Delta E_{VBM-2}^{ZrS_2}$ represents the difference in energy levels between the VBM of the ZrS_2 monolayer and its core level. $\Delta E_{core}^{CZTS/ZrS_2}$ refers to the energy discrepancy between the core levels of CZTS and ZrS_2 in their heterojunction. The conduction band offset is

$$\Delta E_c = |\Delta E_g - \Delta E_v| \quad (5)$$

The difference in band gap (ΔE_g) is illustrated between bulk CZTS and ZrS_2 monolayer. As depicted in Figure 2, the states close to the VBM primarily arise from the 3d orbitals of Cu and the 3p orbitals of S, along with the 5s and 5p orbitals of Sn at the bottom of the CBM. Generally, the band gap separation alongside the continuous valence band serves as a fundamental reference level. Within the lower energy range, a portion of the separation band at the 4d orbital of Sn is combined with the continuous valence band. The Partial density of states (PDOS) for the ZrS_2 monolayer protocell is depicted as illustrated in Figure 3, with VBM being influenced by the 3p orbitals of S and the 4d orbitals of Zr, while CBM is mainly influenced by the 4d orbitals of Zr. The Zr-4s orbital located at a deeper level is denoted as the core level.

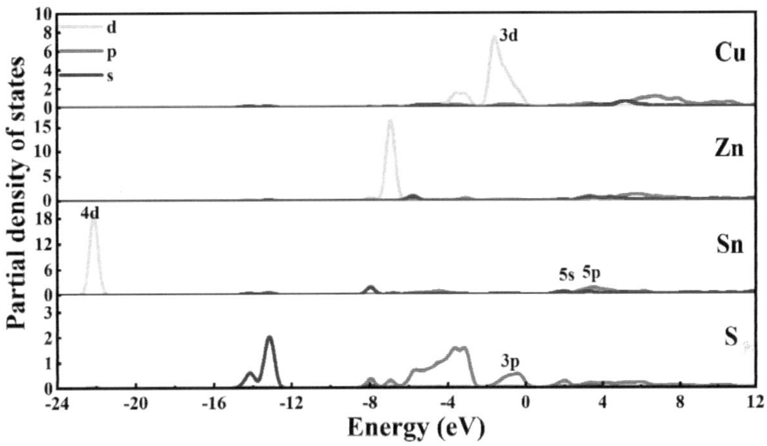

Figure 2. The PDOS of CZTS.

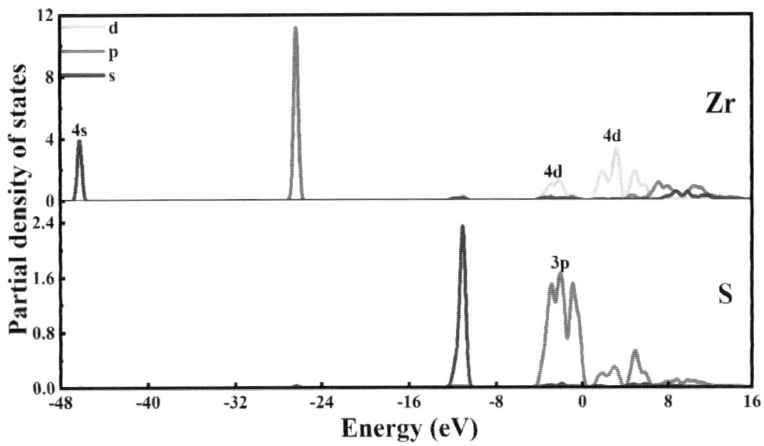

Figure 3. The PDOS of ZrS_2.

3.4. Band arrangement of CZTS/ZrS_2 heterojunction

The core level energy of each layer in the CZTS/ZrS_2 heterojunction is shown in Figure 4 (a). The "Ev" energy marked by the star shows the VBM position of the ZrS_2 monolayer and CZTS, which ΔE_v is 3.57 eV calculated from the average VBM energy. The CBM of the ZrS_2 layer is diminished than the VBM of the CZTS layer, that is, the CZTS/ZrS_2 heterojunction is classified as a Type III interface. The band bias of the CZTS/ZrS_2 heterogeneous interface is shown in Figure 4 (b), revealing a significant energy difference of 1.67 eV between the VBM of CZTS and the CBM of ZrS_2. This substantial offset demonstrates a typical Type III band alignment that is unsuitable for photovoltaic applications.

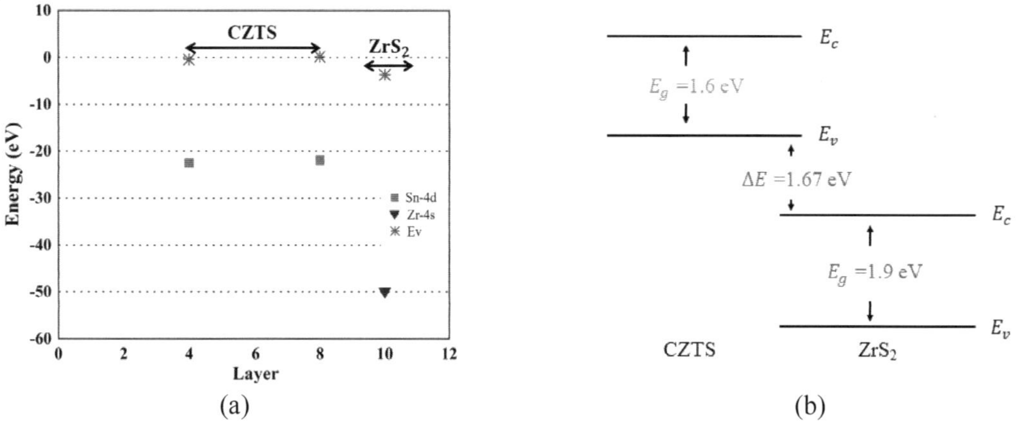

Figure 4. (a) Computed energy levels and estimated VBM positions, (b) Predicted band alignment. (E_c is CBM, E_v is VBM.).

3.5. Electron density difference of CZTS/ZrS_2 heterojunction

To observe the charge transfer at the interface, the electron density difference of the CZTS/ZrS_2 heterojunction was calculated, by averaging the density difference along the z-axis and dividing it by the volume. Comparing it with the crystal structure, it can be found that the density changes are mainly concentrated near the interface as shown in Figure 5. The interface near the ZrS_2 side has a peak of charge accumulation, while the interface loses charge near the CZTS side at about 10 Å, on the order of 10^{-3}.

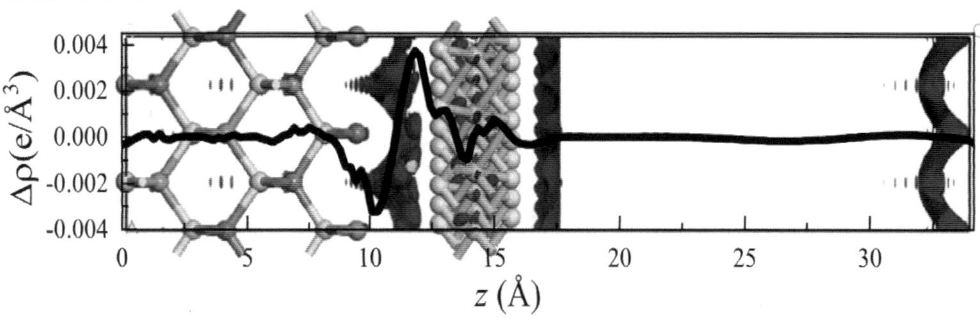

Figure 5. Electron density difference of CZTS/ZrS_2 heterojunction.

4. Conclusion

The band alignment and differential charge density of CZTS/ZrS_2 heterojunction are calculated based on first principles. The results show that CZTS/ZrS_2 heterojunction is a Type III band arrangement with little diffusion and drift of electrons at the interface, which is not suitable for photovoltaic devices, ZrS_2 monolayer needs to be regulated to obtain the best band arrangement. Herein, this study offers insights into enhancing the efficiency of CZTS solar cells.

Acknowledgments
This job was financially backed by the General Projects from the Yunnan Education Department (No. 2023Y0893).

References

[1] Zhao Y, Yuan S, Kou D, Zhou Z, Wang X, Xiao H, Deng Y, Cui C, Chang Q, and Wu S (2020) High-Efficiency CIGS Solar Cells by Bulk Defect Passivation through Ag Substituting Strategy. ACS Applied Materials & Interfaces, 12: 12717-12726.
[2] Ravindran M and Praveenkumar C (2018) Status review and the prospects of CZTS based solar cell — A novel approach on the device structure and material modeling for CZTS based photovoltaic device. Renewable and Sustainable Energy Reviews, 94: 317-329.
[3] Shockley W and Queisser H J (1961) Detailed Balance Limit of Efficiency of p-n Junction Solar Cells. Journal of Applied Physics, 32: 510-519.
[4] Sun K, Yan C, Liu F, Huang J, Zhou F, Stride J A, Green M, and Hao X (2016) Over 9% Efficient Kesterite Cu2ZnSnS4 Solar Cell Fabricated by Using Zn1–xCdxS Buffer Layer. Advanced Energy Materials, 6.
[5] Kaur K, Kumar N, and Kumar M (2017) Strategic review of interface carrier recombination in earth-abundant Cu-Zn-Sn-S-Se solar cells: current challenges and prospects. Journal of Materials Chemistry A, 5: 3069-3090.
[6] Ahmad S, D'Souza R, and Mukherjee S (2018) Band gap modulation of ZrX2 (X = S, Se, Te) mono-layers under biaxial strain and transverse electric field and its lattice dynamic properties: a first-principles study. Materials Research Express, 6.
[7] Hasan S, Ahmed M T, Roman A A, Islam S, Ahmed F, and Gad S A (2023) Investigation of Structural, Electronic, and Optical Properties of Chalcogen-Doped ZrS2: A DFT Analysis. Advances in Materials Science and Engineering, 2023: 1-10.
[8] Moustafa M, Al Zoubi T, and Yasin S (2022) Exploration of CZTS-based solar using the ZrS2 as a novel buffer layer by SCAPS simulation. Optical Materials, 124.

Journal of Physics: Conference Series **2842** (2024) 012007 doi:10.1088/1742-6596/2842/1/012007

Effect of surface state on wear performance of low alloy high strength marine steel

Chaoyi Wang[1,2,*], Ling Yan[1], Xueting Chang[3] and Wenrui An[2]

[1]State Key Laboratory of Metal Materials for Marine Equipment and Application, Anshan, Liaoning, 114009, China
[2]Ansteel Beijing Research Institute Co., Ltd, Beijing 102200, China
[3]College of Ocean Science and Engineering, Shanghai Maritime University, Shanghai, 201306, China

[*]Corresponding author's e-mail: wchy_huayun@163.com

Abstract: For ships sailing in polar ice regions, the marine steel not only has to be safely served at low temperatures but also may be abraded by hard sea ice. However, marine steel is inevitably affected by corrosion in seawater whose surface state and oxide type have a great influence on wear performance. To study the influence of the surface state on the wear performance of low alloy high-strength marine steel, the 460 MPa class high-strength marine steel was used as the research object to test the strength and plastic of the material. Further, γ-FeOOH oxide layer steel sample, untreated steel sample, and Fe_3O_4 oxide layer steel sample were prepared. The light interferometer and scanning electron microscope were used to characterize the microstructure morphology and wear scar morphology of the different steel samples. The test results show that different surface states have a great influence on the wear properties of Marine steel. The Fe_3O_4 steel sample with a dense and complete oxide layer has the lowest wear amount, its wear marks are the shiniest and narrowest, and the surface is mainly adhesive abrasion. The largest amount of wear is the relatively loose structure γ-FeOOH oxide steel sample, which has the smallest friction coefficient, its wear mechanism is mainly abrasive wear, and the surface has more furrows and pits. The wear resistance of untreated steel samples was between the above two.

1. Introduction

Low alloy high strength steel has been widely used in bridges, ships, railways, construction, and other major projects [1-4]. With the development of the world economy, especially the continuous expansion of polar shipping routes, the low-temperature steel used in polar ships has attracted more and more attention [5]. Ships sailing in polar sea ice environments can peel surface coatings due to ice loads. The marine steel is corroded by seawater and then forms oxides at the steel surface, which are then rubbed and worn by ice loads. Some studies have reported the importance of forming rust layers on low alloy high strength steel [6-7]. Adding some elements (Ni, Cr) into low-alloy steels can effectively improve the resistance of rust layers to Cl^-, thus the stability of rust layers is improved [8]. Although polar ships are generally coated with impact-resistant coatings with excellent wear resistance, the impact of ice loads on ships sailing in sea ice environments often results in surface coating peeling, as shown in Figure 1 (a). The contact between steel materials and seawater will accelerate the corrosion behavior. With the increase of corrosion immersion time, the corrosion oxidation products attached to the surface of the steel sample will change with the composition change of low alloy steel. The existing studies have shown that the rust layer on the surface of polar sailing ship materials is mainly composed of hydroxyl oxides such as α-, β- and γ-FeOOH and iron oxides such as Fe_2O_3 and Fe_3O_4, as shown in Figure 1 (b). Different oxides have different structural characteristics and binding forces with metal surfaces, and their interaction behavior and mechanism with environmental media are also different.

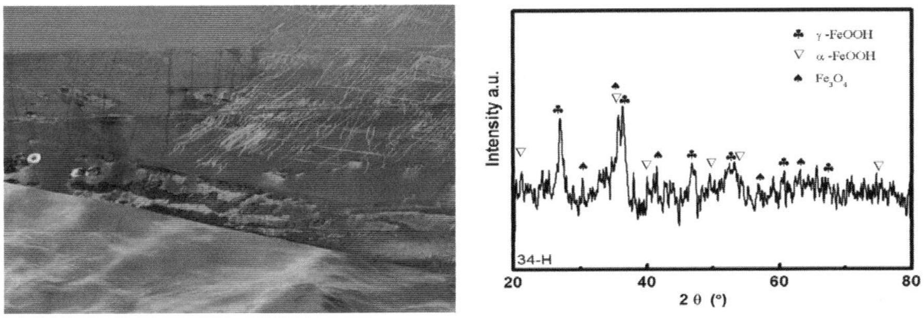

(a) (b)

Figure 1. Photos of the bow of the polar ice breaker and its composition.

In this paper, a 460 MPa class high-strength marine steel was corroded to form an γ-FeOOH layer and a Fe_3O_4 layer on the surface. The influence of different rust layer compositions on the abrasion performance of steel samples was studied by using friction and wear testing machines. The research work of this paper is helpful for the shipbuilding industry and steel research industry to evaluate the reliability of ship materials for navigation in polar seas, provide development ideas for the design of Marine steel materials, and help to promote the development of the marine economy.

2. Experiment method and result

2.1. Material and experiment method

The material used in this experiment is a new low-temperature steel for polar ships. The steel adopts a low-carbon micro-alloy composition, appropriately adding Cu, Ni, Cr, and other alloying elements. The TMCP technology is adopted to promote phase transformation strengthening and structure control as the core. The fine multi-phase microstructure is obtained to improve the strength of the material and ensure that the material has good toughness at low temperatures. The chemical composition and mechanical properties of tested steel are shown in Table 1 and Table 2 respectively.

Table 1. Chemical composition (weight fraction) of the tested steel.

C	Si	Mn	P	S	Cu+Cr+Ni	Nb+V+Ti
0.06	0.258	1.47	< 0.02	< 0.005	Trace	Trace

Table 2. The Mechanical properties of the tested steel.

ReH[a]/MPa	Rm[b]/MPa	A[c]/%	KV_2[d] (-60°C)/J		
465	552	29.0	234	220	242

a The upper yield strength of the test material.
b The tensile strength of the material.
c The elongation after fracture of the material.
d The Charpy impacts the toughness of materials.

Wire cutting equipment is used to prepare low-temperature steel for new ships to be tested to a size of 10 mm × 10 mm × 3 mm, and 180-mesh, 400-mesh, 600-mesh, 1,000-mesh and 1,200-mesh sandpaper is used to grind the friction surface of the steel sample and clean. After drying, it is wrapped with plastic wrap to reduce the surface oxidation of the friction surface. Three parallel samples are prepared for each group of tests to ensure the reliability and objectivity of the experiment. The steel sample is vertically put into a beaker filled with deionized water, the mouth of the beaker is sealed with a sealing film to prevent the external environment from affecting the experiment, taken out after 7 days of static corrosion, and dried in a vacuum drying box at room temperature to obtain a γ-FeOOH film on the steel surface. By dissolving 1.251 g $FeSO_4 \cdot 7H_2O$ in 30 ml deionized water, adding 5 ml polyethylene glycol 20000 (50 g/L) to the $FeSO_4$ aqueous solution, stirring well, and mixing with 5 ml of 2.5% $NH_3 \cdot H_2O$, 0.15 ml H_2O_2 is slowly added to react. The reaction mixture is stirred for 5 min to obtain a homogeneous solution and the etching solution is transferred to a 50 ml reactor liner. Then the steel sample is vertically put into the lining solution of the reactor and sealed. After oxidizing it at a high temperature of 160 °C for 8 hours, it is taken and cooled to room temperature. The blackened steel sample is taken and washed with deionized water. Then, after drying at 80°C, a surface Fe_3O_4 film layer is obtained. Then, the corrosion products were analyzed by the Bruker D8-advanced XRD equipment.

The Bruker UMT-2 multifunctional friction and wear testing machine is used to conduct experiments to analyze the influence of different rust layer components on the wear resistance and protection performance of the substrate. Al_2O_3 grinding balls are selected with a diameter of 8 mm and the balls face the grinding method. The experimental parameters include that the experimental load is 15 N, the grinding time is 30 minutes, the reciprocating friction distance is 5 mm and the sliding frequency is 2 Hz.

According to the ASTMD1141-98 standard, simulated artificial seawater with 3.5% NaCl is prepared as the environmental medium, at room temperature and standard atmospheric pressure. The above-mentioned untreated steel samples, γ-FeOOH layer steel samples, and Fe_3O_4 layer steel samples are respectively subjected to linear reciprocating friction experiments, and the friction coefficient is measured. After the friction experiment, the surface of the sample is ultrasonically cleaned with deionized water. The Bruker Contour GT-II white light interferometer is used to observe the wear scar morphology of the steel sample after friction corrosion and draw the wear scar cross-section curve. The Zeiss GeminiSEM-500 scanning electron microscope is used to observe the surface structure and the wear scar morphology of the steel samples. The auto-lab elemental composition is determined by an energy-dispersive spectrometer on the micro-area abrasion scratches of the steel samples.

2.2. Composition and Friction coefficient of surface oxidation products

The steel sample before and after surface oxidation treatment is shown in Figure 2. As can be seen from the figure, the surface of steel samples without surface treatment presents metallic luster, the surface corrosion products of steel samples with frequent warm oxidation are brown-yellow, and the surface corrosion products of steel samples with high-temperature oxidation are brown-black, and the two corrosion products are closely bonded with the surface of steel samples.

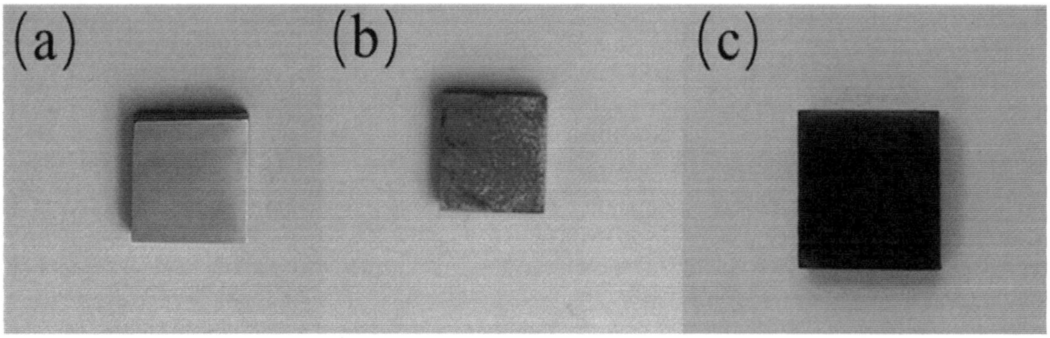

Figure 2. Photographs of steel samples before and after surface oxidation treatment.
(a) Untreated steel sample (b) γ-FeOOH layer steel sample (c) Fe_3O_4 layer steel sample.

The XRD spectra of these two kinds of corrosion products are shown in Figure 3. According to the calibration of diffraction peaks, it can be determined that the products obtained by pre-oxidation at room temperature and high temperature are γ-FeOOH and Fe_3O_4 respectively.

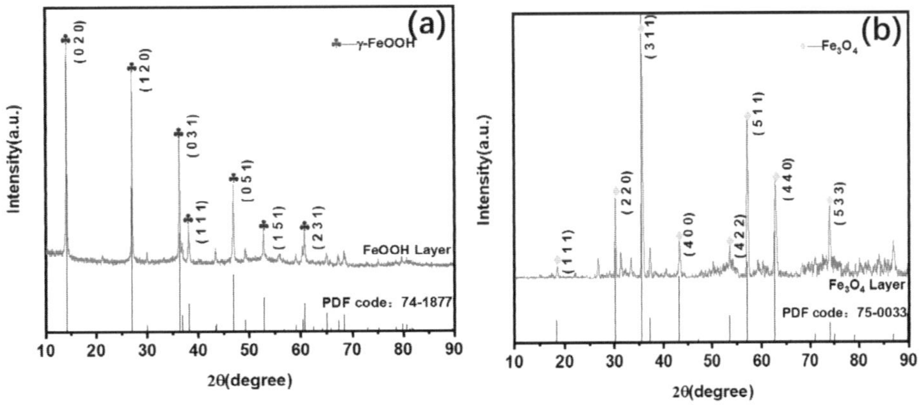

Figure 3. XRD spectrum of the oxidation product (a) γ-FeOOH, (b) Fe_3O_4.

Combined with SEM images (Figure 4), γ-FeOOH film is coniferous and Fe_3O_4 film is layered. According to the experimental results, these two kinds of corrosion products exhibit distinct tribological behaviors. The needle-like layer may act as a lubricant to resist wear, while the denser particles in the layer-like structure may facilitate adhesive wear.

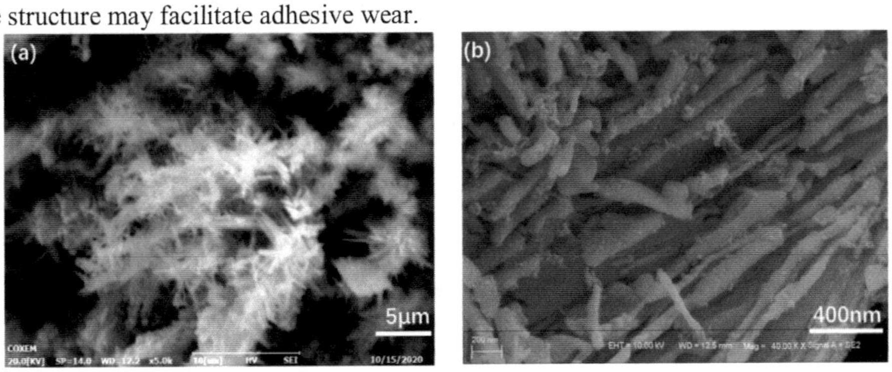

Figure 4. SEM image of oxidation product (a) γ-FeOOH, (b) Fe_3O_4.

The friction coefficient is an important feature to characterize the friction characteristics of low-carbon alloy steel. Figure 5 (a) shows the variation in the friction coefficient of steel samples with different iron oxide layers over time. The friction coefficient of the Fe_3O_4 layer steel sample quickly reaches the steady-state value at the beginning of friction, and then the amplitude fluctuates regularly within a certain time interval, which can be attributed to the formation of wear debris [9]. The γ-FeOOH layer sample has a lesser friction coefficient than other samples.

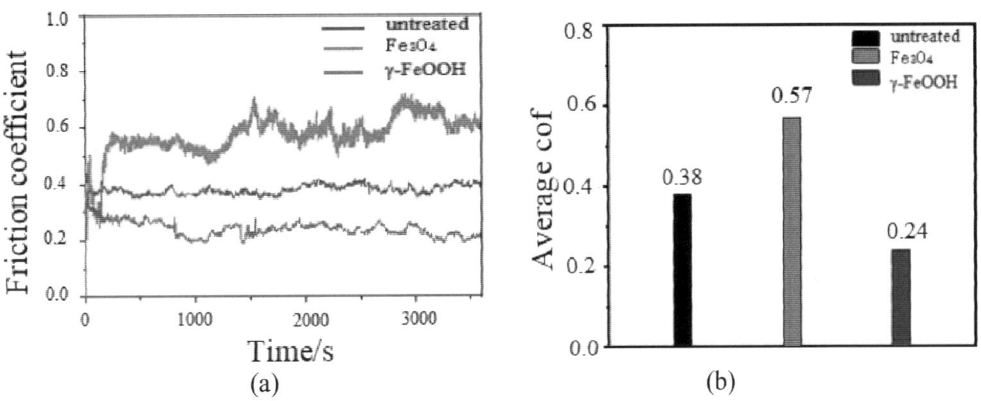

Figure 5. Friction coefficient (a) and the average friction coefficient (b) of different samples.

It can be seen from Figure 5 (b) that the average friction coefficient of the experimental steel sample was 0.57 (Fe_3O_4 layer) > 0.38 (untreated) > 0.24 (γ-FeOOH layer). It may be because γ-FeOOH has a smooth contact surface, which can reduce the friction coefficient. However, when the dense surface of the Fe_3O_4 oxide layer appears between the friction pair of the steel sample and the grinding ball, the glazed surface will be formed during the friction process and the main mechanism of wear may become adhesion wear, leading to the increase of friction coefficient [10].

2.3 Analysis of wear trace profile and wear quantity

Wear amount refers to the amount of material loss by friction, which can be obtained by measuring the mass or volume change before and after the wear which is called mass wear amount or volume wear amount. Figure 6 (a) shows the cross-section of the wear trace profile of the steel sample after abrasion in the salt solution. It can be seen from the figure that the wear trace depth of the γ-FeOOH layer steel sample is the largest, reaching 5 μm, followed by 4.5 μm of untreated steel sample and the shallowest wear trace profile of the Fe_3O_4 layer steel sample is 2.8 μm. It is noteworthy that friction corrosion products accumulate on both sides of the wear marks. This is mainly because the steel sample is not only affected by friction and wear but also by corrosion behavior. After the surface oxidation layer is worn, the exposed steel sample contacts with the active Cl^- ion in seawater, which increases the sensitivity of local corrosion and makes the surface more vulnerable to pitting corrosion. Figure 6 (b) shows that the wear degree of the steel sample is similar to that in Figure 6 (a). The order of wear quantity from heavy to light is 10.5×10^{-3}g (γ-FeOOH layer) > 8.4×10^{-3}g(untreated) > 2.5×10^{-3}g (Fe_3O_4 layer). Generally, the loss of wear mass is proportional to the coefficient of friction, but it is different in this case. The friction coefficient increases with the increase of the adhesion of the steel oxide layer, while the wear volume loss decreases with the increase of the contact area between the substrate and the wear debris [11].

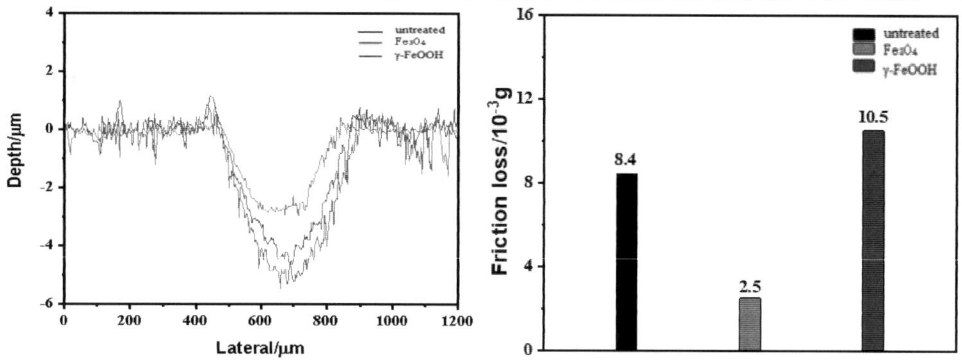

(a) (b)

Figure 6. Cross section of wear scar profile (a) and Abrasion (b) of steel sample.

3. Analysis and discussion

Figure 7 shows the friction and wear morphology of steel samples with different treatments under the same load and load time. For the three steel samples, the larger material loss occurred in the central part of the wear track, which was consistent with the wear trace profile shown in Figure 7 (a).

By observing Figure 7 (a) and (b), it is found that the wear marks of the steel samples without pre-corrosion treatment are of large width and have obvious characteristics of furrows and pits, which are important characteristics of plastic deformation. Furrow is generated because the grinding ball continuously generates heat energy in the friction process and the grinding debris will adhere to the spherical surface and form a micro-convex body, resulting in stress concentration. The micro-convex body is hardened in the process of the reciprocating friction on the surface of the steel sample [12], thus leaving furrow-like wear on the surface of the steel sample. Pits are generated because steel samples are not only subjected to mechanical wear but also corroded by seawater. Corrosion pits may appear under the influence of friction-corrosion coupling [13].

It can be seen from Figure 7 (c) that the wear mark width of the steel sample of the Fe_3O_4 layer is small, which is consistent with the wear data shown in Figure 6 above. Meanwhile, the surface and surrounding finish of the wear mark are high, indicating that the oxidation layer of Fe_3O_4 has a certain corrosion resistance to seawater. It can be seen from Figure 7 (d) that micro-cutting wear exists and the analysis shows that the wear mechanism is adhesive wear. Fe_3O_4 acts as an oxide layer, which increases friction adhesion and shear force and laminates the surface of steel samples during grinding. The exfoliated Fe_3O_4 film is transferred to the Al_2O_3 ball and the steel sample surface is continuously rolled along the sliding direction. The micro-convex peak of the steel sample surface is changed into grinding debris by cutting action, and the surface roughness is reduced, which is also the reason for the regular fluctuation of the amplitude of the friction coefficient in the middle and late periods. Some scholars reported a similar conclusion that the appearance of an oxidation layer on the surface of a steel sample reduces wear [14].

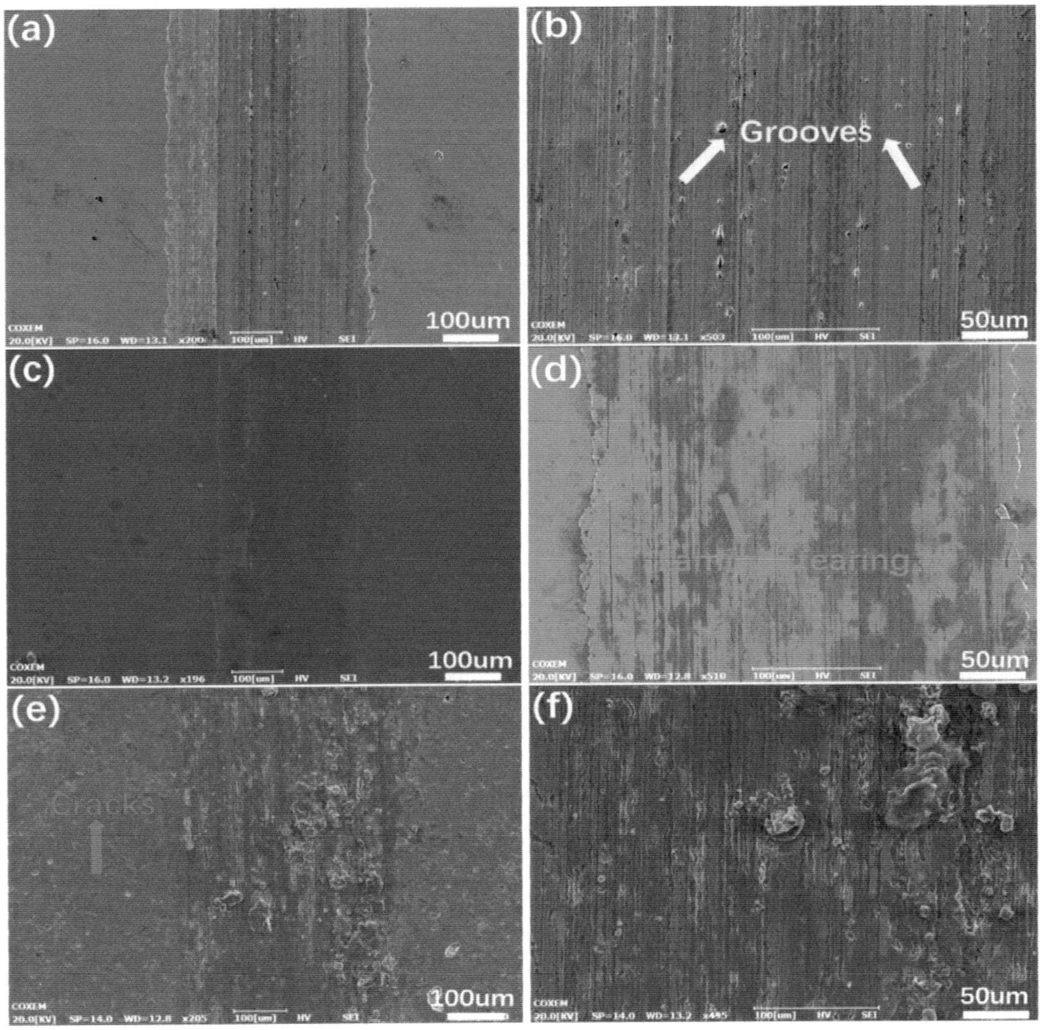

Figure 7. Friction and wear morphology of steel samples before and after surface oxidation treatment Untreated: (a) 200 ×, (b) 500 ×; Fe_3O_4: (c) 200 ×, (d) 500 ×; γ-FeOOH: (e) 200 ×, (f) 500 ×.

According to Figure 7 (e) and (f), it is found that the wear degree of the steel sample in the γ-FeOOH layer is the largest and there are obvious cracks in the wear marks. This is because micro-cracks form on the surface of the steel sample after pre-corrosion treatment. When the Al_2O_3 ball and the steel sample are ground against each other, the micro-cracks will extend from the surface to the inside of the steel sample at a certain angle with the load direction. This also provides a channel for the O element and Cl element to participate in corrosion diffusion. Cl^- in the environmental solute penetrates the matrix crack through diffusion, forming a local potential difference, which has the primary influence on galvanic corrosion and greatly promotes the occurrence of electrochemical corrosion. At the same time, water-soluble O diffuses into the matrix through microcracks, forming differential corrosion batteries for oxygen supply, which further aggravates the occurrence of corrosion. Because the γ-FeOOH layer is easy to flake off due to its small adhesion force, a large number of corrosion products and wear chips will re-enter the wear marks under the reciprocating sliding of the friction pair and gradually fill the furrow, reducing the friction coefficient. This conclusion is also in line with the changing trend of the friction coefficient curve shown in Figure 5.

4. Conclusion

The wear resistance of the polar marine steel after surface oxidation treatment was studied, exploring the effects of two main components of rust on the surface of the polar marine low-temperature steel, Fe_3O_4, and γ-FeOOH, on the abrasive properties of the experimental steel samples, investigating the technical requirements of surface protection for the polar marine steel. The results of this study are helpful to the development of the high corrosion resistance polar marine steel, to establish the corrosion evaluation standards for polar marine steel, and to provide technical support for long-term safe service of polar marine steel. The following conclusions are drawn:

(1) The Fe_3O_4 oxide layer formed on the surface of low-temperature steel has good corrosion resistance, dense microstructure, and can protect the matrix; γ-FeOOH has no passivation performance due to its loose and porous structure. The corrosion behavior and mechanism are the same as those of bare steel.

(2) The abrasion resistance of γ-FeOOH oxide layer steel sample, untreated steel sample, and Fe_3O_4 oxide layer steel sample increased successively. Among them, the friction coefficient of the γ-FeOOH oxide layer steel sample is the smallest, but its wear amount is the largest. The abrasion resistance of the untreated steel sample is better than that of the γ-FeOOH oxide layer steel sample, while the Fe_3O_4 oxide layer steel sample has the lowest wear amount and the best wear resistance, and the wear mark profile is also the shallowest and narrowest among the three, and the wear mechanism is mainly adhesive abrasion.

(3) For the actual use conditions of low-temperature polar ship steel, the surface rust layer structure should be fully considered. When the coating falls off due to ice load, the steel plate should have strong passivation ability to ensure its excellent corrosion resistance and wear resistance and meet the requirements of safe use of ships. In the process of ship design, the passivation ability of the steel plate should also be fully considered to better estimate its corrosion margin and achieve the purpose of energy saving.

Acknowledgments

National Key Research and Development Program (2022YFB3705303): Study on service performance and damage failure mechanism of special alloys under ultra-low temperature and alternating load environment

References

[1] Yang Y, Cheng X Q, Zhao J B, et al. A study of rust layer of low alloy structural steel containing 0.1 % Sb in the atmospheric environment of the Yellow Sea in China [J]. Corrosion Science, 2021, 188:109549.
[2] Wang Y F, Li J G, Wang Q F et al. Some new discoveries on the structure of the rust layer of weathering steel in a simulated industrial atmosphere by STEM-EDS and HRTEM [J]. Corrosion Science, 2021,183:109322.
[3] Zhou L J, Yang S W. Investigation on crack propagation in band-like rust layers on weathering steel [J]. Construction and Building Materials, 2021, 281:12564.

[4] Gong K, Wu M, Xie F, et al. Effect of Cl− and rust layer on stress corrosion cracking behavior of X100 steel base metal and heat-affected zone in marine alternating wet/dry environment [J]. Materials Chemistry and Physics, 2021, 270:124826.
[5] Li Z, Zhang D C, Wu H Y, et al. Fatigue properties of welded Q420 high-strength steel at room and low temperatures[J]. Construction and Building Materials, 2018, 189:955-966.
[6] Wang Y F, Li J G, Zhang L, et al. Structure of the rust layer of weathering steel in A high chloride environment: A detailed characterization via HRTEM, STEM-EDS, and FIB-SEM [J]. Corrosion Science, 2020, 177, 108997.
[7] Roman V, Iryna I, Olena B, et al. Segmentation of rust defects on painted steel surfaces by intelligent image analysis [J]. Automation in Construction, 2021, 123: 103515.
[8] Zhang T Y, Xu X X, Li Y, et al. The function of Cr on the rust formed on weathering steel performed in a simulated tropical marine atmosphere environment [J]. Construction and Building Materials, 2021, 277: 122298.
[9] Lee L, Behera P, Sriraman K R, et al. Effects of humidity on the sliding wear properties of Zn–Ni alloy coatings [J]. RSC Advances, 2017, 2017 (37): 22662-22671.
[10] Tian X, Wang C B, Zhu Y. Finite element analysis of fatigue wear of SiC and ZrO2 friction pairs [J]. Lubrication and sealing, 2009 (08): 70-73
[11] Qian A, Yang X H, Jin P, et al. Micro electrochemical behavior of AerMet100 steel in salt spray environment under the action of Cl - [J]. Equipment environmental engineering, 2019, 016 (010): 88-94
[12] Cao C N. Principles of corrosion electrochemistry (Third Edition) [M]. Chemical Industry Press, 2008
[13] Tan L, Wang Z, Ma Y. Tribocorrosion Behavior and Degradation Mechanism of 316L Stainless Steel in Typical Corrosive Media [J]. Acta Metallurgica Sinica (English Letters), 2021: 813-824.
[14] Lee L, Behera P, Sriraman K R, et al. Effects of humidity on the sliding wear properties of Zn–Ni alloy coatings [J]. RSC Advances, 2017, 2017 (37): 22662-22671.

Lifetime prediction of EPDM sealing materials with thermal aging under constant compression

Zhaonan Xie, Junhong Chen *, and Kai Zhang

Institute of Systems Engineering, China Academy of Engineering Physics, Mianyang 621990, China

* Corresponding author's e-mail: chenjh@lnm.imech.ac.cn

Abstract. EPDM (Ethylene Propylene Diene Monomer) is an important engineering sealing material. Thermal aging leads to a decline in its service life. In this study, the static sealing lifetime of EPDM sealing materials is investigated by thermal aging experiments under constant compression. Based on the variation pattern of compression set over aging time at different aging temperatures (80 °C, 100 °C, and 120 °C), a lifetime prediction model for EPDM sealing materials is derived using chemical reaction kinetics and the Arrhenius equation. The model is compared with the experimental results. The results show that the prediction model is highly reliable. The theoretical life of EPDM after thermal aging at 25 °C is 30.6 years, 53.1 years, and 89.1 years, respectively, when the compression set is 40%, 45%, and 50 %.

1. Introduction

EPDM is a synthetic rubber composed of ethylene, propylene, and diene monomers. Due to its stable chemical structure, EPDM exhibits excellent qualities of aging resistance, radiation resistance, corrosion resistance, insulation, and hydrolysis resistance. It has widespread application in engineering fields such as aerospace, shipbuilding, vehicle engineering, nuclear power generation, and tunnel construction [1-4]. In these domains, EPDM is fabricated into various functional components for a variety of purposes, such as sealing, vibration isolation, insulation, and transportation [5-6]. Among these applications, EPDM sealing components play a vital role in ensuring the safety and reliability of engineering systems. Their primary functions include preventing the leakage of internal substances and protecting critical components from corrosion caused by external factors [7].

EPDM serves as sealing materials primarily due to their hyperelasticity, which utilizes elastic deformation to provide the necessary rebound stress for achieving sealing functionality [8]. However, long-term use of EPDM is inevitably affected by environmental factors (thermal oxygen, irradiation, and chemical media), which lead to chemical aging [9-10]. This results in the degrading of its mechanical performance and consequently reduces its service life. In order to ensure its performance and improve its reliability, it is necessary to conduct a reasonable evaluation of the service life of EDPM.

Static seal is the most common seal state of EPDM sealing products, so the static seal life prediction of EPDM sealing materials has always been an important issue and research focus. At present, the establishment of rubber static seal life prediction model is usually divided into two steps [11-13]: 1) The relationship between rubber compression set and aging time is established under different aging temperatures; 2) Then, the compression set is determined, and the service life model of rubber at different temperatures is established in combination with the Arrhenius relation. The model established by this method can only be used to predict the service life of rubber at different temperatures under a single compression set. To solve this problem, the relationship between compression set, aging time, and aging temperature was directly derived based on constant compression thermal aging experiment, first-order chemical reaction kinetics, and Arrhenius relation. The service life prediction model of EPDM under different compression sets and different temperatures is given by conversion.

2. Experimental

2.1. Materials
EPDM cylindrical specimens with a diameter of 12.5 mm and a height of 29 mm were prepared, which were provided by the Institute of Systems Engineering of the China Academy of Engineering Physics.

2.2. Experimental devices.
The compression device consisted of compression plates, height limiters, and a clamping device, as shown in Figure 1. The EPDM samples are placed in a compression device to achieve constant compression. The thermal aging test of EPDM samples was carried out in an air circulation oven, as shown in Figure 2. The temperature of the oven could be adjusted and maintained, and the temperature deviation was ±2 °C. The air-circulated oven is provided by Chongqing Sida Experimentaling Equipment Company Limited, with the model LR-02.

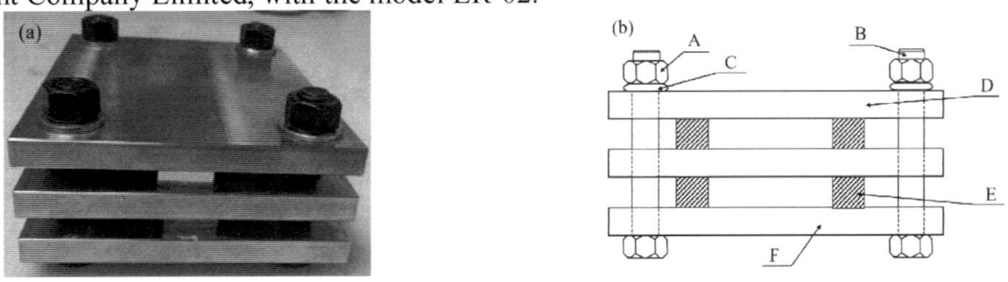

Figure 1. The compression device. (a) The actual diagram. (b) The schematic diagram: A, nut; B, screw; C, gasket; D, upper plate; E, height limiter; and F, lower plate.

Figure 2. The air-circulated oven.

2.3. Aging methods
The EPDM sample was compressed to 25% and held constant using two compression plates and a height limiter. Then, samples in the compressed state were put into an air-circulated oven for the experimental thermal oxygen aging. Twelve groups of thermal oxygen aging experiments were carried out on the samples. The experimental aging temperatures were set at 80 °C, 100 °C, and 120°C, and the experimental aging times were selected as 9 d, 21 d, and 65 d (d stands for day). To avoid the effects of the occasional factors on the experimental results, each group of experiments contained three parallel samples.

2.4. Compression set
After constant compression-thermal oxygen aging, the EPDM sample was taken out from the air-circulated oven and placed at room temperature for recovery. When the sample height did not change, the normalized compression set (CS) was calculated by the following formula:

$$CS = \frac{h_0 - h_1}{h_0 - h_s} \times 100\% \tag{1}$$

where h_0 refers to the length of a fresh sample along the direction of the central axis before the experiment, h_1 refers to the recovered length of an aged sample along the direction of the central axis after the experiment, and h_s refers to the length of a fresh sample after compression, h_0-h_s represents the initial amount of compression, while h_0-h_1 represents the amount of compression set after the aging experiment. Accordingly, $1\text{-}CS=(h_1\text{-}h_s)/(h_0\text{-}h_s)$, where h_1-h_s represents the residual amount of compression after aging, and 1-CS measures the proportion of residual compression.

2.5. Experimental results
The compression set of EPDM sealing materials is an important criterion to evaluate its service life. As can be seen from Table 1, the compression set of EPDM increases with the aging time, and the higher the aging temperature, the more obvious this phenomenon is. This data will be used in the subsequent modeling of the service life of EPDM sealing materials

Table 1. Compression set of EPDM cylindrical sample under different thermal oxygen environments.

Aging temperature (°C)	Aging time (d)	Compression set
80	9	0.168
	21	0.293
	65	0.425
100	9	0.332
	21	0.406
	65	0.523
120	9	0.484
	21	0.543
	65	0.594

3. Mathematical modeling
EPDM is a kind of high molecular polymer, and its excellent recovery capability owes to its unique macromolecular chain network structure. However, EPDM cannot always maintain its recovery ability in the thermal oxygen environment, because its macromolecules chemically react with oxygen, and high temperature accelerates the reaction, which destroys the structure of its molecular chain network and therefore causes its recovery ability to decrease [14-15].

From the analysis above, it can be seen that the compression set of EPDM in a thermal oxygen environment is essentially caused by a chemical reaction, so the method of chemical kinetics can be used for analysis. However, chemical kinetics describes the relationship between reactant concentration and reaction rate. Since 1-CS is continuously decreased by chemical reactions, similar to the change of reactant concentration with time, it is assumed that 1-CS and $\mathrm{d}(1\text{-}CS)/\mathrm{d}t$ follow chemical kinetics. According to the thermal-oxygen aging mechanism of rubber, the macromolecular reactants of EPDM undergo complex chemical reactions. Generally, they are described using a first-order chemical reaction formula [16]. The degradation relationship of EPDM residual compression under a thermal-oxygen environment can be expressed as follows:

$$\frac{\mathrm{d}(1\text{-}CS)}{\mathrm{d}t} = k(T)(1 - CS) \tag{2}$$

where t is the thermal aging time, $k(T)$ is the chemical reaction rate satisfying the Arrhenius equation: $k(T) = A\exp(-E_a/RT)$, where T is the absolute temperature, E_a is the reaction activation energy, R is the universal gas constant 8.314 J/(mol·K), and A is the pre-exponential factor.

According to the initial conditions $t = 0$ and $CS = 0$, Formula (2) can be solved to obtain:

$$CS(t,T) = 1 - \exp\left(At\exp\left(-\frac{E_a}{RT}\right)\right) \tag{3}$$

The least square method was used to fit the experimental data, as shown in Figure 3. The correlation coefficient between the model value and the experimental value is 83.2%. The fitting results show that Formula (3) cannot precisely describe the thermal aging trend of the EPDM compression set. To resolve this problem, the time term is modified into exponential form to improve the predictive performance of the Formula (3). The modified model can be expressed as follows:

$$CS(t,T) = 1 - \exp\left(At^n\exp\left(-\frac{E_a}{RT}\right)\right) \tag{4}$$

where n is the fitting constant. The least square method was also used to fit the experimental data, and the fitting effect of the modified model is shown in Figure 4. The correlation coefficient between the model value and the experimental value is significantly improved, increasing to 96.1%. The fitting results show that the modified model can effectively capture the thermal aging trend of the compression set. The model parameters A, n, and E_a are obtained by fitting, which are -292.9, 0.3, and 22333.1 J/mol, respectively.

By adjusting Formula (4) and substituting the fitting parameter values, the lifetime prediction model of EPDM can be expressed as follows:

$$t = \left(\frac{\ln(1-CS)}{-292.9\exp(-\frac{2646.1}{T})}\right)^{3.3} \tag{5}$$

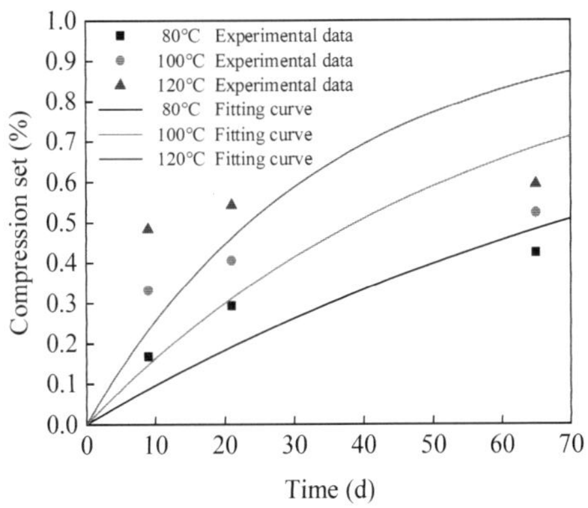

Figure 3. Comparison of EPDM unmodified model results and experimental results under different thermal aging conditions.

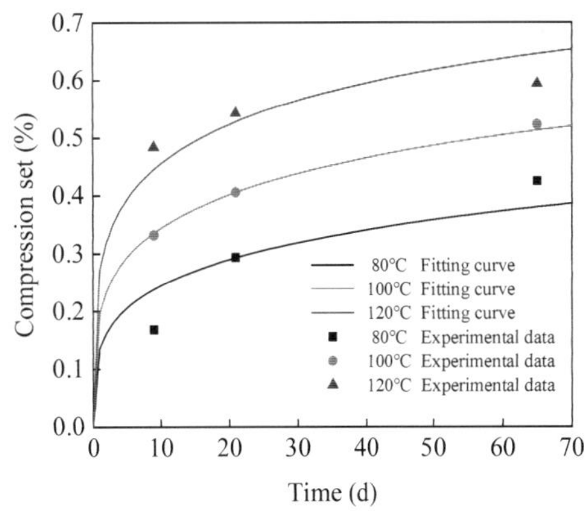

Figure 4. Comparison of EPDM modified model results and experimental results under different thermal aging conditions.

4. Model validation

In order to verify the reliability of the proposed model, it is necessary to compare the prediction results of the model with non-fitting experimental data. Through the experimental method proposed in this study, the compression sets at 80 °C-189 d, 100 °C-4 d, 120 °C-3 d, and 55 °C-9 d, 21 d, 65 d, 189 d have been measured.

According to the data in Table 2, the average relative error of the model can be obtained as 6.7%. From the calculation of the data in Figure 5, the following conclusions can be drawn: 1) The relative errors between the predicted value and the experimental value at 9 d, 21 d, 65 d, and 189 d are 29.4%, 16.0%, 11.9%, and 2.0%, respectively; 2) The average relative error between the predicted value and the experimental value is 14.8%. These results show a trend in which the relative deviation between the predicted value and the measured value decreases with the extension of the aging time. These results also show that the model predicted value is higher than the experimental value, so the use of the predicted value to judge the state of use provides a certain safety factor, which is acceptable in engineering.

After the validity of the model is proved, the service life of EPDM at different aging temperatures is predicted. Assuming a compression set of 40%, 45%, and 50%, EPDM lost its sealing performance. The lifetime of EPDM under different failure criteria can be calculated by Formula (5), as shown in Figure 6. The following conclusions can be drawn from the data in Figure 6: 1) When the compression set value is 40%, 45%, and 50%, the service lives of EPDM at 25 °C are 30.6 a, 53.1 a, and 89.1 a (a is year), respectively. 2) At 25 °C to 45 °C, the service life of EPDM decreased sharply from 30.6 a, 53.1 a, and 89.1 a to 4.2 a, 7.29 a, and 12.2 a.

Table 2. Comparison between experimental and predicted results of compression set of EPDM cylindrical samples.

Aging temperature (°C)	Aging time (d)	Experimental value	Predicted value
80	189	0.449	0.477
100	4	0.251	0.277
120	3	0.363	0.350

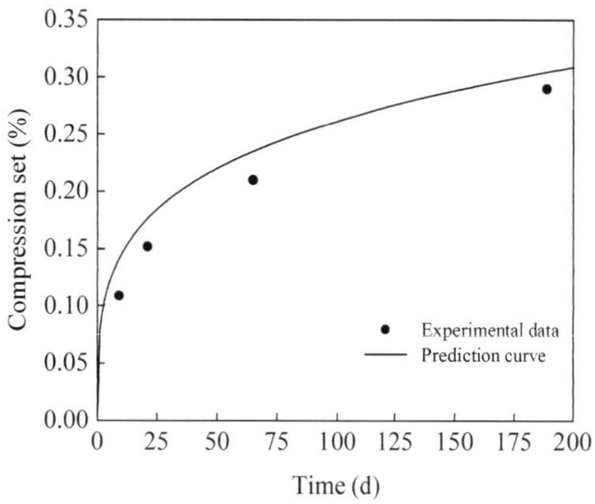

Figure 5. Comparison of model predicted and experimental values of the compression set at 55 °C.

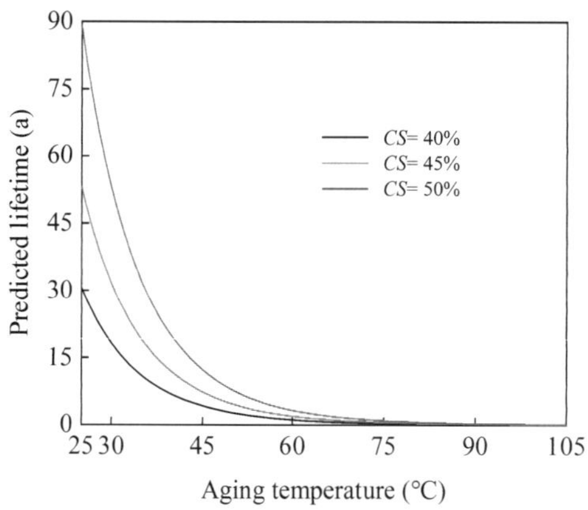

Figure 6. EPDM lifetime prediction curves at different temperatures when the compression set is 40%, 45%, and 50%.

5. Conclusions

In this study, an EPDM service life prediction model that considers different compression sets and aging temperatures is proposed. By comparing with the experimental data, the reliability of the EPDM prediction model is evaluated, and the service life under different compression set criteria is predicted. The main conclusions are as follows:

(1) At 80 °C-189 d, 100 °C-4 d, and 120 °C-3 d, the average relative error between predicted and experimental values of the EPDM compression set is 6.7%; The average relative error between the predicted and experimental values of the EPDM compression set at 55°C is 14.8%.

(2) After aging at 55 °C for 9 d, 21 d, 65 d, and 189 d, the relative errors between the predicted and experimental values of the EPDM compression set decreased gradually, which were 29.4%, 16.0%, 11.9%, and 2.0%, respectively.

(3) When the compression set of EPDM is 40%, 45%, and 50%, the service life of EPDM at 25 °C is 30.6 a, 53.1 a, and 89.1 a.

Acknowledgments
This work was supported by the National Natural Science Foundation of China, Nos. 12302494. The authors wish to thank Dr. Ren He for his work in constructing the compression device. Valuable suggestions provided by Dr. Xi-Cheng Huang, Dr. Wen-Jun Hu, and Dr. Wei-Zhou Zhong are gratefully acknowledged. The permission of the Institute of Systems Engineering of China Academy of Engineering Physics, who sponsored this work, to publish these results is appreciated.

References

[1] Bouguedad D, Mekhaldi A, Jbara O, et al. (2015) Physicochemical study of thermally aged EPDM used in power cables insulation. *IEEE Trans. Dielect. Electr. Insul.*, 22(6): 3207-3215.
[2] Liu Y K, Zhang Q S, Liu R T, et al. (2022) Compressive stress-hydrothermal aging behavior and constitutive model of shield tunnel EPDM rubber material. *Const. Build. Mater.*, 320:126298.
[3] Wei B X, Yi X T, Xiong Y J, et al. (2020) The preparation and characterization of a carbon fiber-reinforced epoxy resin and EPDM composite using the co-curing method. *RSC Adv.*,10(35):20588-20594.
[4] Delor-Jestin F, Lacoste J, Barrois-Oudin N, et al. (2000) Photo-, thermal and natural ageing of ethylene–propylene–diene monomer (EPDM) rubber used in automotive applications. Influence of carbon black, crosslinking and stabilizing agents. *Polym. Degrad. Stab.*, 67(3): 469-477.
[5] Huang X M, Gu J M, Li Ming, et al. (2023) A leakage prediction model for sealing performance assessment of EPDM o-rings under irradiation conditions. *Polymers*, 15(14):3073.
[6] Li X, Gong Z B. (2024) Investigation of Long-Term Waterproof Performance and Hardness Change of EPDM Rubber Gasket Used for Shield Tunnels. *J. Mater. Civ. Eng.*, 36(2): 04023577.
[7] Zhang Z X, Jia X H, Guo F, et al. (2021) Effect of vulcanization on the tribological behavior and sealing performance of EPDM seals: Experimentation and simulation. *Tribol. Int.*, 164: 107219.
[8] Zaghdoudi M, Kömmling A, Jaunich M, et al. (2023) Understanding the recovery behaviour and the degradative processes of EPDM during ageing. *Polym. Test.*, 121: 107987.
[9] Šarac T, Quiévy N, Gusarov A, et al. (2016) Influence of γ-irradiation and temperature on the mechanical properties of EPDM cable insulation. *Radiat. Phys. Chem.*, 7(125): 151-155.
[10] Li C J, Ding Y Q, Yang Z, et al. (2020) Compressive stress-thermo oxidative ageing behaviour and mechanism of EPDM rubber gaskets for sealing resilience assessment. *Polym. Test.*, 84:106366.
[11] Woo C S, Kim H S, Lee S B, et al. (2010) Useful lifetime prediction of rubber components using accelerated testing. *IEEE Trans. Rel.*, 59(1): 11-17.
[12] Li K H, Chen Z Y, Shi W K. (2022) Rubber aging life prediction based on interpolation and improved time-temperature superposition principle. *Mater. Res. Express.*, 9(1): 015301.
[13] Bensalem K, Easaee M, Hassanipour M, et al. (2024) Lifetime estimation models and degradation mechanisms of elastomeric materials: A critical review. *Polym. Degrad. Stab.*, 220: 110644.
[14] Colin X, Hassine M B, Nait-Abelaziz M. (2019) Chemo-Mechanical model for predicting the lifetime of EPDM rubbers. *Rubber Chem. Technol.*, 92(4): 722-748.
[15] Wang Z N, Shen S L, Zhou A N, et al. (2020) Experimental Evaluation of Aging Characteristics of EPDM as a Sealant for Undersea Shield Tunnels. *J. Mater. Civ. Eng.*, 32(7): 04020182.
[16] Hu Q C, Chen Q, Song P R, et al. (2023) Performance of thermal-oxidative aging on the structure and properties of ethylene propylene diene monomer (EPDM) vulcanizates. *Polymers*, 15(10): 2329.

Effect of surface micro-texture morphology on the wettability of aluminum alloy on B_4C surface

Yuchi Dai [1, 2], Jia Liu [1, 2, *] and Yan Shi [1, 2]

[1] School of Electromechanical Engineering, Changchun University of Science and Technology, Changchun, Jilin, 130022, China
[2] National Base of International Science and Technology Cooperation in Optics, Changchun, Jilin, 130022, China

[*] Corresponding author's email: liujia@cust.edu.cn (Jia Liu)

Abstract: To enhance the wettability of AlSi10Mg on the B_4C surface, a 2D laminar two-phase flow, phase-field coupled multi-field model is utilized to investigate the effect of surface micro-texture morphology on the wettability of AlSi10Mg on B_4C surfaces and optimize the morphology of the micro-texture with the best improvement of the interfacial wetting effect. The results of the study show that compared to the square texture, the conical texture has a better effect on the improvement of the wettability of the interface. The best wettability of AlSi10Mg on the B_4C surface was obtained at a 1:1 depth-to-width ratio of the conical texture, and the contact angle of the Al droplet is 47.9, which is improved by 21.22% compared to that of the smooth interface. The interfacial wetting state is transformed from the Wenzel state to the Cassie state with the increasing depth-to-width ratio of the textures, and the wettability of the interface deteriorates.

1. Introduction

Boron carbide ceramic has high-temperature resistance, low density, high modulus of elasticity, and excellent wear resistance [1, 2]. These excellent physical properties make B_4C widely used in engineering applications such as abrasives, nozzles, reactor control rods, and lightweight armor [3, 4]. However, B_4C ceramic has high hardness and low fracture toughness [5]. Thus, it is difficult to be machined, which limits the range of applications of B_4C ceramic. The combination of ceramic and metal materials can expand the application range of ceramic materials [6]. Aluminum alloys are the most widely used metal with a low density, high strength-to-weight ratio, and good corrosion resistance. Therefore, aluminum alloys have many applications in the automotive industry and aerospace [7]. However, the wettability of aluminum alloys on boron carbide surfaces is poor due to the large differences in the physical and chemical properties between the metal and ceramic materials. Surface texturing treatment can achieve surface modification by preparing a series of regular geometrical structures on the surface of the material, which can uniformly improve the surface roughness and surface energy [8].

Surface texturing treatment can achieve surface modification by preparing a series of regular geometrical structures on the surface of the material, which can uniformly improve the surface roughness and surface energy. Therefore, surface texture can improve the specific surface area of ceramic, which promotes the diffusion of Al melt droplets and improves the wettability of aluminum alloy on the ceramic surface [9]. Laser processing technology has the advantages of high precision, controllable texture shape, and processing depth [10]. Thus, it has been widely used in the surface processing of ceramics.

The study of surface texture in terms of hydrophilic/hydrophobicity shows that the wetting state of a liquid on a solid surface transforms between the hydrophilic Wenzel state and the hydrophobic Cassie state with the change in the surface texture morphology [11]. Therefore, the wetting state of a metal on the surface of a textured ceramic will also transition with the change of surface texture morphology. In the present studies, the wettability of Al droplets on the textured ceramic surfaces are all in the Wenzel state, and there are few reports on the transition of the wetting state due to the change of the surface texture morphology.

In this paper, the 2D surface texture models with different shapes and sizes are established to study the effect of the geometrical shape and size of the surface micro-texture on the wetting and spreading state of Al droplets on the boron carbide surface. The theoretical guidance is provided to improve the wettability of the metal/ceramic interface by ceramic surface texturing treatment.

2. Description of the wetting model

2.1. Physical field model

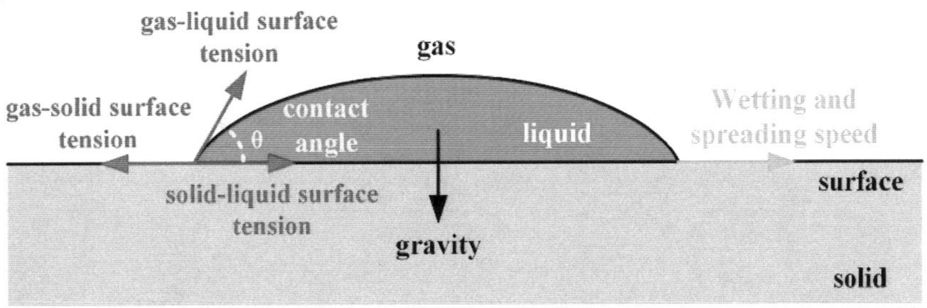

Figure 1. The schematic diagram of the physical field model for surface wetting.

A laminar-two-phase flow, phase-field coupled multi-field model with a 2D spatial dimension is established in COMSOL Multiphysics software. A schematic representation of the surface-wetting physical field model is shown in Figure 1. The model satisfies the following basic assumptions: The fluid flow pattern is assumed to be laminar. The flow mode of the fluid is assumed to be the laminar flow mode, and the fluid is also assumed to be an incompressible, viscous Newtonian fluid. The contribution of the chemical reactions occurring between Al and B_4C in driving the wettability is ignored, and the effect of surface micro-structure morphology and dimensional changes on the wetting spreading condition is considered exclusively. During the simulations, the velocity was used as the mobility adjustment parameter, and an open boundary containing hydrostatic pressure compensation was used as the simulation boundary condition.

2.2. Geometric model

A 2D computational domain is created by defining the attributes of the geometric nodes in the geometric modeling module of the software. A rectangular area with 1.2 mm×0.9 mm dimensions is defined as the air domain, and the bottom surface is defined as the wetting wall. As shown in Figure 2, a circle with a radius of 0.09 mm is created at the central position of the bottom edge, and the inner region of the circle is defined as the initial region of the AlSi10Mg melt droplet. The surface micro-textures with various geometries are created by changing the coordinates of the geometric nodes at the bottom edge. The geometric parameters of the surface textures are listed in Table 1.

A free triangular mesh divides the computational domain, the mesh of the regions near the Al melt droplet. The wetting wall is refined to improve the computational accuracy. The results of the model mesh division are shown in Figure 1. The quality of the smallest cell is 0.35, which is higher than the technical standard, and the quality of the smallest cell of the two-dimensional grid is not lower than

0.3. The average cell quality of the grid is 0.931, which can effectively guarantee the accuracy of the calculation.

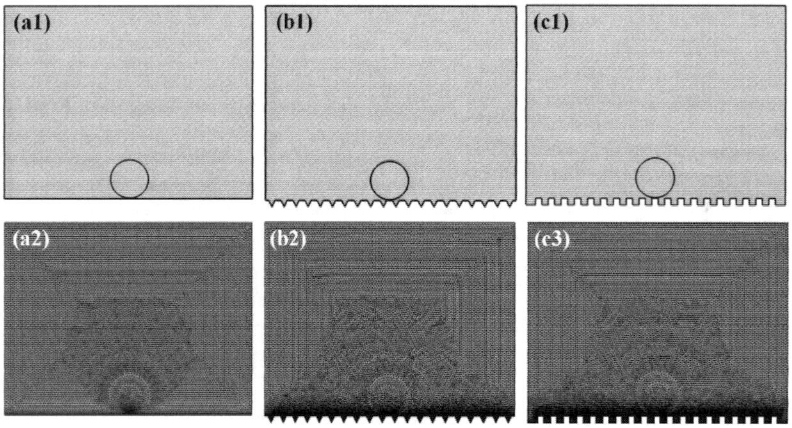

Figure 2. The schematic diagrams of surface micro-texture morphologies and meshing results: (a) untextured surface, (b) conical texture, and (c) square texture.

Table 1. The geometric parameters of the surface textures.

Groove shape	Grooves pacing (mm)	Groove width (mm)	Groove depth (mm)	Depth-to-width ratio
	0.03	0.03	0.03	1: 1
Conical	0.03	0.03	0.06	2: 1
	0.03	0.03	0.09	3: 1
	0.03	0.03	0.015	1: 2
Square	0.03	0.03	0.03	1: 1
	0.03	0.03	0.06	2: 1
	0.03	0.03	0.09	3: 1

2.3. Material properties

The material "Air" is selected from the material database of the software, and the scope of air is set to all fields of the geometric entity. Then, "Aluminum" is selected in the material database and applied to the Al melt droplet region. The wetting and spreading condition of AlSi10Mg single track with process parameters of 200W laser power and 2000mm/s scanning speed is adapted to provide a reference for the simulation of wetting of Al melt on the smooth surface. The temperature of the AlSi10Mg melt track is 1458K at this process parameter, as monitored by the Infrared inspection system of the melt pool. The surface tension coefficient and kinetic viscosity of AlSi10Mg at 1458K are 0.664N/m and 6.7×10^{-4}Pa•s, respectively, calculated using JMat Pro software.

3. Experiment

Based on the simulation results of Al melt droplet wetting on the surfaces with different textures, the corresponding microtextures were processed on the B_4C surface according to the geometrical parameters shown in Table 1. As shown in Figure 3, a picosecond laser with a wavelength of 1030 nm (Zhonghui Laser Technology Co., Ltd.) was used to texture the surface of B_4C ceramic. The laser process parameters are listed in Table 2.

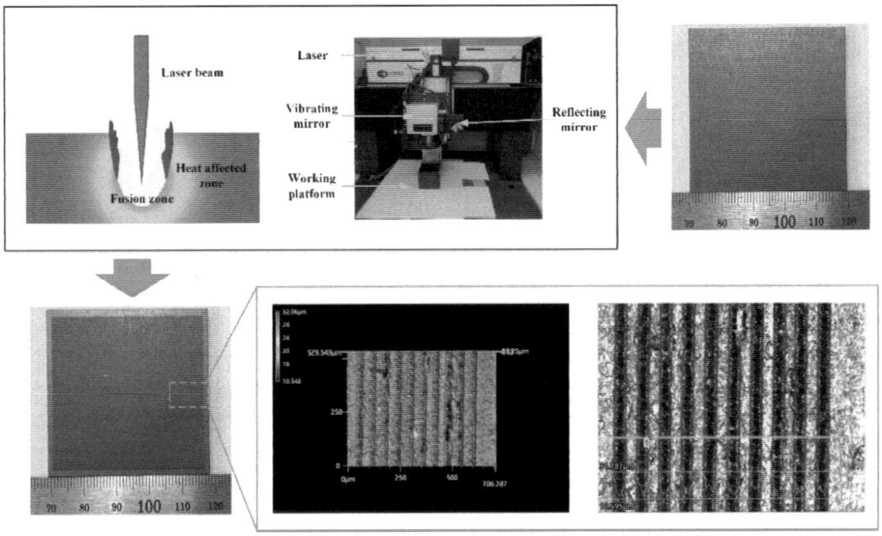

Figure 3. Schematic diagram of the processing of B_4C surface micro-texture.

Table 2. Surface micro-texture processing parameters.

Process parameter	Value
Laser power (W)	80
Scanning speed (m/s)	10
Laser pulse frequency (kHz)	800
Scan times	400

As shown in Figure 4, the textured ceramics were placed into a Truprint 3000 additive facility (TRUMPF, Germany), and the single-track spreading experiments of the AlSi10Mg were carried out according to the laser process parameters shown in Table 3. The three-dimensional morphologies of the AlSi10Mg melt tracks were constructed using ZEISS LSM 700 confocal laser scanning microscopy to obtain information on melt height, melt width, and contact angles.

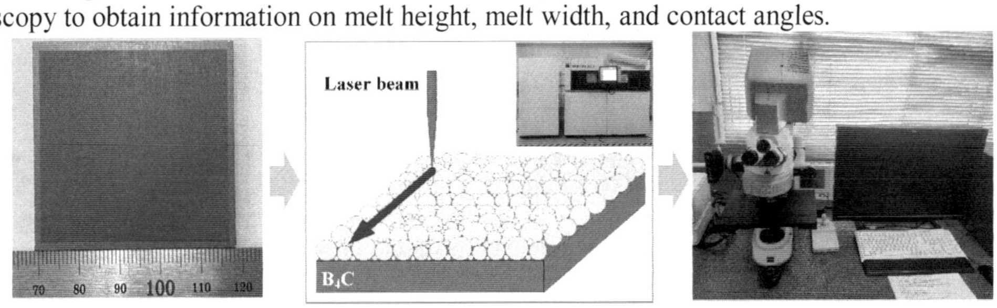

Figure 4. Schematic diagram of AlSi10Mg wetting and spreading experiments on the textured B_4C surfaces.

Table 3. The laser parameters of AlSi10Mg wetting and spreading experiments on the textured B_4C surfaces.

Process parameter	Value
Laser power (W)	200, 240, 280
Scanning speed (mm/s)	2000
Thickness of powder layer (μm)	60
Oxygen content (ppm)	100

4. Result and discussion

For conical textures, the wetting spreading situations of Al melt droplets on the surface of B_4C with various sizes of micro-textures are shown in Figure 5. It can be observed that the contact angles of Al melt droplets on the B_4C surface tend to increase with the increase of the depth-to-width ratio of the micro-textures. When the depth-to-width ratio of the texture is 1:1, the Al droplet can completely infiltrate into the grooves of the micro-texture, and the interface wetting state is the Wenzel state at this time [12]. While the depth-to-width ratio of the texture is 2:1, the Al droplets cannot fully infiltrate the texture grooves, and the wetting state of the interface is at the transition stage from the Wenzel state to the Cassie state [13].

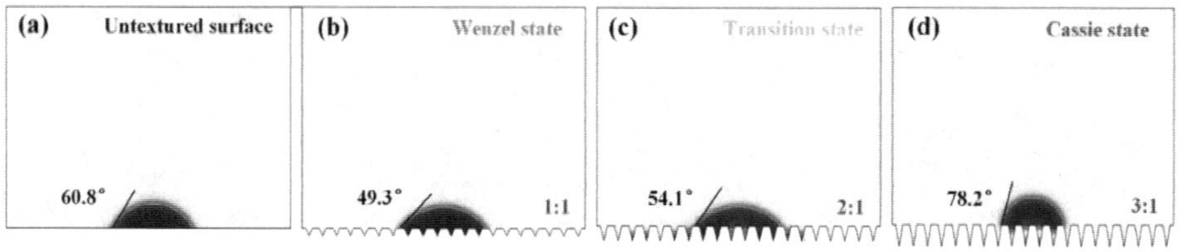

Figure 5. The wetting spreading situations of Al droplets with different depth-to-width ratios of the conical surface textures: (a) smooth surface; (b) 1:1; (c) 2:1 and (d) 3:1.

Once the depth-to-width ratio of the texture reaches 3:1, the Al droplet is unable to infiltrate the texture grooves, and the wetting state of the interface is completely transferred to the Cassie state, where the contact angle of the interface is larger than that of the Al droplet on the smooth surface, and the wetting situation gets deteriorated. The reason for this phenomenon is due to the relatively large depth to width of the groove; the metal melt droplet is unable to expel the residual air inside the groove completely, and the contact line force overcomes the gravitational force of the metal droplet on the side wall of the groove. Thus, the solid-liquid surface tension, the gas-liquid surface tension, and the gas-solid surface tension reach equilibrium on the side wall of the groove, and the wettability state of the Al droplet is completely transformed from the Wenzel state to the Cassie state, which results in the poor interfacial interface wettability.

For square textures, the wetting spreading situations of Al melt droplets on the surface with different sizes of micro-textures are shown in Figure 6. The contact angle of the Al melt droplet on the B_4C surface shows a tendency to increase gradually with the increase of the depth-to-width ratio of the textures. When the depth-to-width ratio of the texture is 1:2, the Al droplet can infiltrate into the grooves completely, the wetting state of the interface is the Wenzel state, the wettability of the interface gets improved, and the contact angle of the Al droplet is smaller than that of its contact angle on the smooth surface. With the further increase of the depth-to-width ratio of the square textures, the Al droplet could not infiltrate the grooves completely, and the interfacial wetting state was transformed to the Cassie state, which deteriorated the wetting state and the interfacial contact angle increased.

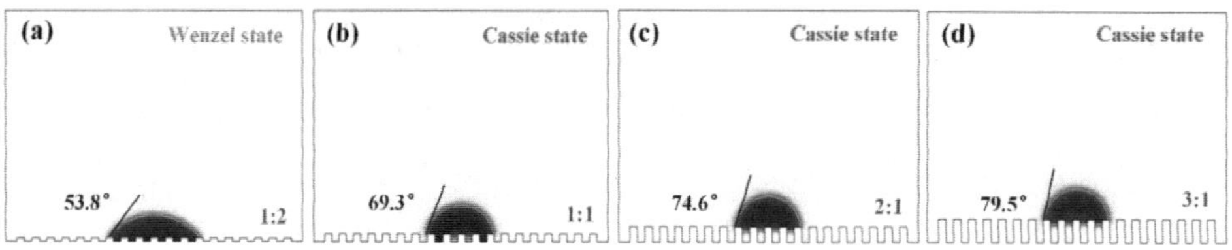

Figure 6. The wetting spreading situations of Al droplets with different depth-to-width ratios of the square surface textures (a) 1:2; (b) 1:1; (c) 2:1 and (d) 3:1.

The simulation results show that the wettability of Al melt droplets on the conical textured surfaces is better than on the square textured surfaces. The optimum wettability of the Al melt droplet is achieved when the depth-to-width ratio of the conical texture is 1:1. The contact angle at this point is

Journal of Physics: Conference Series **2842** (2024) 012009 doi:10.1088/1742-6596/2842/1/012009

49.3°, which is enhanced by 18.91% compared to the smooth surface. Therefore, the conical texture with a 1:1 depth-to-width ratio was processed on the surface of the B_4C ceramic to perform the validation experiments.

Figure 7 demonstrates the wetting spreading situations of the AlSi10Mg melt tracks under different laser process parameters. It can be found that the contact angle of the AlSi10Mg melt track on the textured surface is close to the value obtained by the simulation at the laser power of 200W and scanning speed of 2000mm/s. The value of the contact angle measured by the experiment is slightly smaller than that of the simulation, which is because the interfacial reaction between the aluminum alloy and the B_4C ceramic is neglected in the numerical simulation calculations, and this interfacial reaction can promote the wettability of the AlSi10Mg melt droplet on the B_4C surface.

In addition, it can be seen that the contact angles of the AlSi10Mg melt tracks both on the smooth surfaces and the textured surfaces become smaller with the increase of the laser power. This is because the increasing laser power leads to a corresponding increase in the temperature of the AlSi10Mg melt tracks. On the one hand, the increase in temperature can reduce the surface tension of the metal droplet, and on the other hand, the increasing temperature can promote the interfacial reaction [14]. Therefore, the wettability of the AlSi10Mg melt tracks on the textured B_4C surfaces becomes better with the increase of laser power.

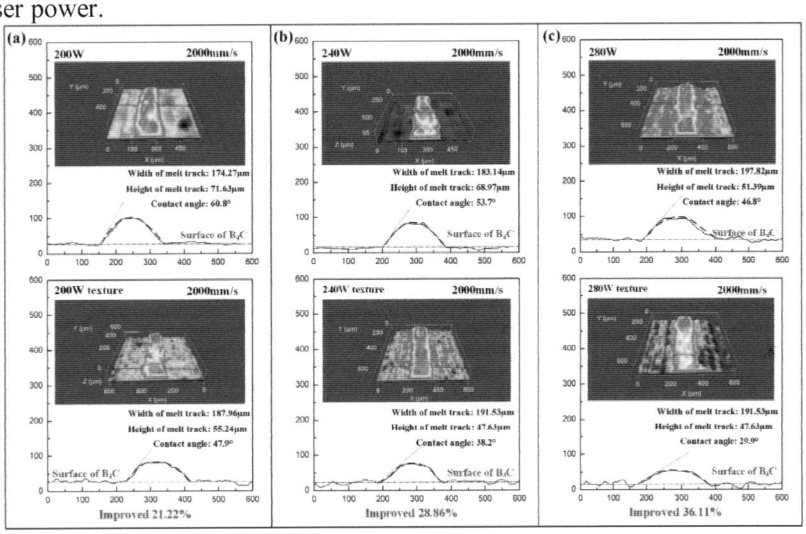

Figure 7. Wetting spreading situations of AlSi10Mg melt tracks under different laser powers (a) 200W, (b) 240W, and (c) 280W.

5. Conclusions

The effect of surface micro-texture morphology on the wettability of AlSi10Mg melt tracks on textured B_4C surfaces was investigated based on the phase field model. The micro-texture morphology with the best effect on wettability improvement was optimized, and experiments verified the simulation results. The principal research findings are generalized as follows:

(1) The best wettability of AlSi10Mg on the B_4C surface was obtained at a conical texture with a 1:1 depth-to-width ratio, which improved by 18.91% compared to the smooth surface. With the increasing depth-to-width ratio of the texture, the wetting state of the interface transforms from the Wenzel state to the Cassie state, and the interfacial wettability deteriorates.

(2) The improvement in the wettability of AlSi10Mg on the B_4C surface with square textures is not as good as with conical textures. The best wettability of AlSi10Mg on the B_4C surface was achieved at a depth-to-width ratio of 1:2 for the square texture, which is improved by 11.51% compared to the smooth surface. Similarly, the interfacial wetting state transformed from the Wenzel state to the Cassie state with the increased depth-to-width ratio of the textures, and the interfacial wettability deteriorated.

(3) The experimental results show that the surface micro-texture can effectively improve the wettability of AlSi10Mg on the B_4C surface. With the laser power of 200 and scanning speed of 2000mm/s, the contact angle of the AlSi10Mg melt track on the textured surface is 47.9°, which is improved by 21.22% compared with that on the smooth surface, which verifies the accuracy of numerical simulation calculation. The wettability of AlSi10Mg on the B_4C surface is improved with increased laser power.

Acknowledgments
This work was supported by the Equipment Pre-Research Common Technology Project of China [50923030512].

References

[1] Lakhdar Y, Tuck C, Terry A, Spadaccini C and R Goodridge (2021) Direct ink writing of boron carbide monoliths. J. Eur. Ceram. Soc., 41: 76-92.
[2] Zhang W (2021) A review of tribological properties for boron carbide cerami. Prog. Mater. Sci., 116: 100718.
[3] Xie K Y, Kuwelkar K, Haber R A, LaSalvia J C and Hemker K J (2016) Microstructural Characterization of a Commercial Hot-Pressed Boron Carbide Armor Plate. J. Am. Ceram. Soc., 99: 2834-2841.
[4] Lu R, Chandrasekaran S, Frane W L D, Landingham R L, Worsley M A and Kuntz J D (2018) Complex shaped boron carbides from negative additive manufacturin. Mater. Des., 148: 8-12.
[5] Yue X Y, Wang J J, Li Y S and Ru H Q (2013) Properties of B4C/Al-B4C composite with a two-layer structure. Mater. Res. Bull., 48: 1720-1724.
[6] Xia J B, Ding T, Ren W, Zhao X, Ren K and Wang Y G (2023) Ceramics/metals joining under the influence of electric field: A review. J. Eur. Ceram. Soc., 43: 5061-5077.
[7] Gorjan L, Boretius M, Blugan G, Gili F, Mangherini D, Lizarralde X, Ferraris M, Graule T, Igartua A, Mendoza G, and Kuebler J (2016) Ceramic protection plates brazed to aluminum brake discs. Ceram. Int., 42: 15739-15746.
[8] Wang R, Bai X S, (2015) Effect of droplet size on wetting behavior on laser textured SiC surface. Appl. Surf. Sci., 353: 564-567.
[9] Xu J Y, Ji M, Li L F, Wu Y, Yu Q and Chen M (2022) Improving wettability, antibacterial and tribological behaviors of zirconia ceramics through surface texturing. Ceram. Int., 48: 3702-3710.
[10] Chen X, Wang C Y, Jiang J Y, Ji C H, Feng S H, Yang C, Xiang Z Y and Long Y (2023) Effect of picosecond laser texturing on the friction behavior of silicon carbide in hybrid ceramic bearings under dry and water lubrication. Ceram. Int., 49: 29857-29869.
[11] Jiang Y, Lian J S, Jiang Z H, Li Y C, and Wen C E (2020) Thermodynamic analysis on wetting states and wetting state transitions of rough surfaces. Adv. Colloid Interface Sci., 278: 102136.
[12] Chen B D, Chai N Y, Zhang J Y, Ren L, Wang W M, Zhang F, Wang X W and Fu Z Y (2023) Wetting tuning of Al/B4C interface via femtosecond laser irradiation. Appl. Surf. Sci., 636: 157802.
[13] Ruiz S, Yoo Y, Wadle L, Chen X, Li N, Lu Y F, Wohl C J, Wiesner V and Cui B (2023) Laser ablative patterning of B4C and MoAlB ceramics for hydrophobic surfaces. J. Am. Ceram. Soc., 106: 5712-5722.
[14] Guo R F, Chen S M and Shen P (2023) Influence of Si, Ti, and Cu as alloying elements on the wettability and reactivity of an Al/B4C system. Appl. Surf. Sci., 136: 157802.

Research on hydrogen distribution of electroless nickel plating on aluminum-based silicon carbide composites

Nan Wang*, **Qingxin Cui, Sizhen Li, Jingrun Wang, Jiafeng Li, Xuguang Wang, Junjie Xu, Xiaobo Tong and Jun Wang**

China Academy of Space Technology, Beijing Spacecrafts, Beijing, China, 100094

*Corresponding author's e-mail: wnan1215@163.com

Abstract. Aluminum-based silicon carbide composite materials have the characteristics of high thermal conductivity, low thermal expansion, and good dimensional stability. Therefore, the composites are widely used in space microwave component shell products. In order to meet the functional requirements of conductivity, welding, and heat dissipation when applied to antenna microwave components, aluminum-based silicon carbide composite materials need to be chemically plated with nickel, which introduces hydrogen, thus requiring research on hydrogen release and control. Here, SEM, XRD, and other methods are used to characterize the microstructure of aluminum-based silicon carbide substrate and nickel plating layer. There are certain pore structures in both the substrate and plating layer, providing a place for hydrogen storage. The experimental method of hydrogen microprinting has been used to characterize the distribution of hydrogen. The results showed that the hydrogen distribution pattern in aluminum-based silicon carbide nickel plating is as follows: after nickel plating on aluminum-based silicon carbide, hydrogen is distributed in the nickel plating layer and the shallow substrate under the coating; with the extension of time, hydrogen spontaneously diffuses to the deep substrate; hydrogen is mainly distributed in the Al matrix and less on SiC particles.

1. Introduction

Aluminum-based silicon carbide (SiCp/Al) composite materials have the characteristics of high thermal conductivity, low thermal expansion, and good dimensi Aluminum-based silicon carbide (SiCp/Al) composite materials have the characteristics of onal stability, finding wide utilizations in space microwave component shell products [1-3]. In practical applications, especially, when applied to antenna microwave components, it is necessary to prepare surface coatings of aluminum-based silicon carbide composite materials to meet the functional requirements of conductivity, welding, heat dissipation, etc. Generally, chemical nickel plating is used to prepare the metal layer on SiCp/Al composites [4,5].

According to Gutzeit G's theory of primary hydrogen, during the process of electroless nickel plating, primary hydrogen is generated when the hypophosphite ions are oxidized, and nickel ions are reduced to metallic nickel by primary hydrogen. Some of the hypophosphite ions are reduced to elemental phosphorus by primary hydrogen, and phosphorus co-deposits with nickel to form the electroless nickel plating layer. Only about one-third of the hypophosphite ions are adsorbed on the nickel surface in the form of H-bonds during the entire reaction process, while the rest are consumed to produce hydrogen [6-8]. The hydrogen generated during the nickel plating process will diffuse to the substrate, the interface between the coating and the substrate, and can desorb from the surface of the coating, then slowly release, causing hydrogen poisoning of hydrogen-sensitive chips in microwave components [9], leading to device performance degradation, functional failure, and other problems. This has an important impact on the quality reliability of the load product, even the performance of the entire satellite, and the stable operation in orbit. Therefore, it is necessary to study the hydrogen release and control of electroless nickel plating on aluminum-based silicon carbide composite materials. This article studies the distribution of hydrogen in aluminum-based silicon carbide nickel-plated materials, analyzes the distribution pattern of hydrogen in the materials, and provides support for hydrogen control in the shell-plating layer.

2. Experimentation

2.1 Materials

The experiment used a high volume fraction aluminum-based silicon carbide composite material (North China University of Technology), in which the volume fraction of SiC particles is about 65%, and the specimen size is 10mm x 10mm x 2mm. The surface of the specimen is ground [10].

2.2 The process of chemical nickel plating

The main process of chemical nickel plating on specimens is as follows[4,11]: roughening → zinc immersion → zinc removal → secondary zinc immersion → chemical nickel plating.

The roughening solution is a mixture of nitric acid (65%), hydrogen peroxide, and deionized water, with a volume ratio of 1.5:1:1, at room temperature and a roughening time of 5 minutes. The formula of the zinc immersion solution is 500g/L sodium hydroxide, 90g/L zinc oxide, 1g/L iron chloride, and 8g/L potassium sodium tartrate. Here, the activation process is treated at room temperature for 20 s. The zinc removal treatment is carried out with a mixture of nitric acid (65%) and water in a volume ratio of 1:10 for 5-10 seconds at room temperature. The chemical nickel plating solution formula is mixed with 25g/L nickel sulfate, 20g/L sodium hypophosphite, 15g/L citric acid, 4g/L succinic acid, 1mg/L thiourea, and 5g/L ammonium chloride, then used at pH 4.5 and 80 ℃.

2.3 Hydrogen microprinting experimental process

After nickel plating on the sample, hydrogen microprinting experiments were conducted immediately to directly observe the distribution of hydrogen after nickel plating.

Hydrogen microprinting experiment process: silver bromide solution is used to react with aluminum-based silicon carbide nickel-plated samples to generate silver particles. Excess silver bromide is washed off with a cleaning solution, then washed off with water, and the surface of the sample is cleaned with alcohol and finally blown dry. The silver bromide solution used is to fully dissolve silver bromide in a 5% aqueous solution of sodium nitrite, reaching saturation. The cleaning solution is a mixture of 15w% sodium thiosulfate and 10w% sodium nitrite aqueous solution.

2.4 Coating characterization

Scanning electron microscopy (SEM, ZEISS SUPER 55VP) with EDS energy spectrum has been used to characterize the microstructure and elemental composition of the substrate material and coating of aluminum-based silicon carbide nickel-plated samples, and to characterize the distribution of Ag particles on the surface of the samples after hydrogen microprinting experiments.

X-ray diffraction (XRD) has been used to characterize the matrix material and coating crystal structure of aluminum-based silicon carbide nickel plating.

The porosity of aluminum-based silicon carbide materials was tested.

3. Results and Discussion

3.1 Microstructure of aluminum-based silicon carbide materials and nickel plating

SEM testing has been conducted on aluminum-based silicon carbide materials to observe the surface microstructure and composition distribution, as shown in Figure 1. It is evident that silicon carbide particles are intermingled in the aluminum matrix, forming Al-Si interfaces and pores of different phases.

The density test has been conducted on aluminum-based silicon carbide, and the actual measured density is 2.31. Compared with the theoretical density (3.02), the resulting density is 76.5%, which indicates that there are certain pores in the aluminum-based silicon carbide substrate.

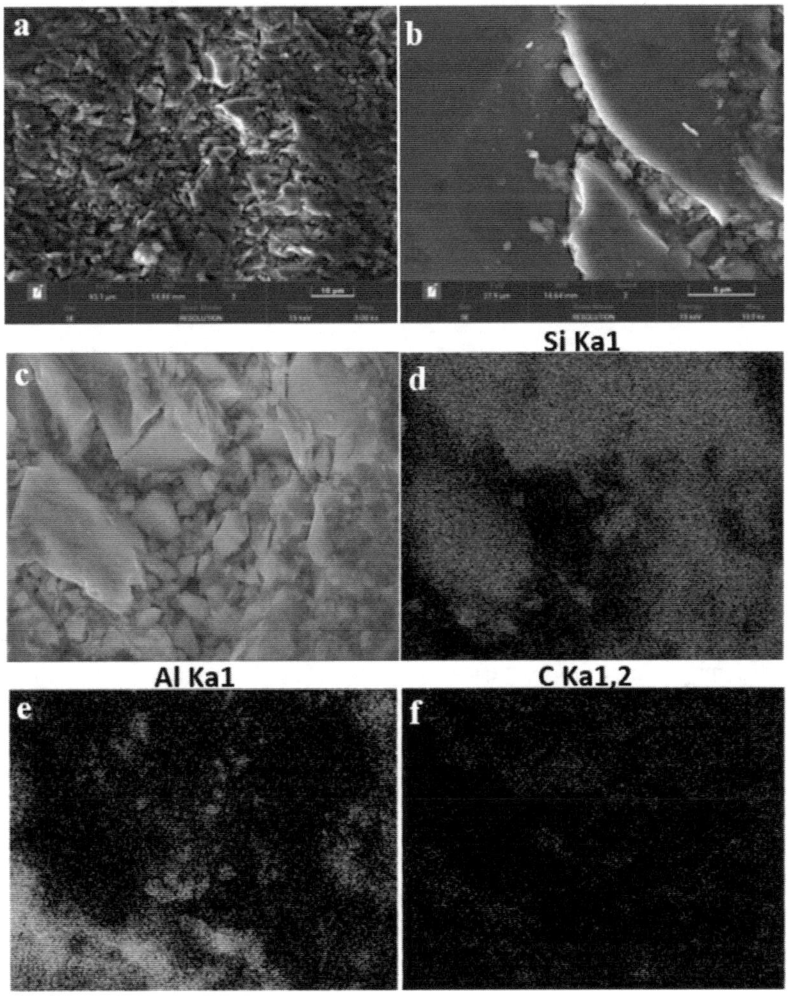

Figure 1. SEM images and EDS test results of aluminum-based silicon carbide materials.

Microscopic morphology tests have been conducted on the nickel plating layer at different magnifications (Figure 2a~d), and it was found that the nickel plating layer is porous and porous, which can affect the adsorption and diffusion of hydrogen.

We cut small pieces of sample using the wire cutting method and added a layer of protection on the side (to prevent nickel plating). After nickel plating on the upper and lower surfaces, we polished the side with 600 # to 5000 # sandpaper. After that, we observed the surface morphology under electron microscopy and performed composition analysis using EDS. Figures 2e~h show the SEM image of the surface morphology of aluminum-based silicon carbide nickel plating and the distribution of elements in this area. It can be seen that the material is composed of an aluminum matrix and irregular silicon carbide particles, among which the size difference of silicon carbide particles is significant, with 30% μ large particles larger than 30 μ m, as well as medium particles of about 10 μ m, and small particles of a few microns as well. Besides, there is a good and uniform combination of aluminum matrix and silicon carbide particles.

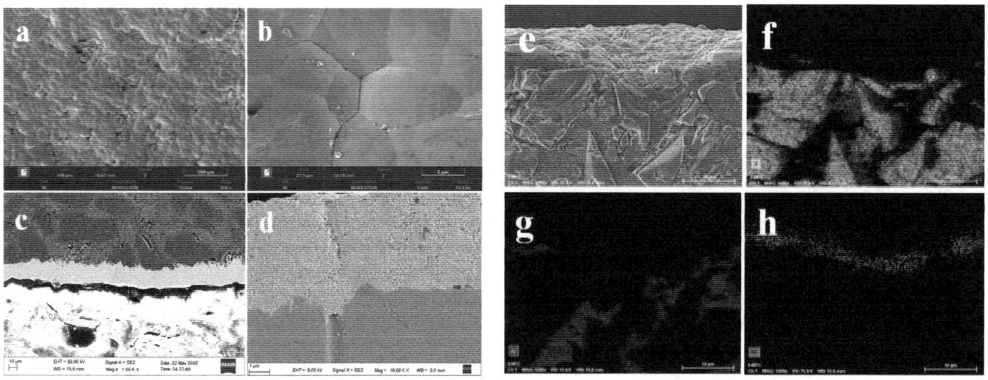

Figure 2. SEM images and EDS test results of Ni layer on aluminum-based silicon carbide materials.

XRD testing has been conducted on aluminum-based silicon carbide and surface nickel-plated materials, as shown in Figure 3. In the XRD spectrum of aluminum-based silicon carbide materials, there are three obvious strong peaks at 35.6 °, 60.1 °, and 75.4 °, corresponding to the (111), (220), and (311) crystal planes of SiC, and there are also obvious peaks at 38.4 °, 44.5 °, 65.0 °, and 78.3 °, corresponding to the aluminum phase in the aluminum-based silicon carbide substrate. The XRD results indicate that only SiC and aluminum phases exist in the aluminum-based silicon carbide substrate, and no intermediate phases, such as A14C3, have been found. XRD testing was conducted on aluminum-based silicon carbide nickel plating, and a clear peak appeared at 44.8 °, reflective of a diffuse scattering diffraction peak in the (111) crystal plane diffraction direction of nickel, indicating that the nickel-phosphorus coating is indeed amorphous. The nickel phosphorus components in spherical particles combine in an amorphous form, while phosphorus exists in a solid solution form in the nickel lattice, forming a nickel-based supersaturated solid solution. The amorphous coating has no defects and high-stress zones.

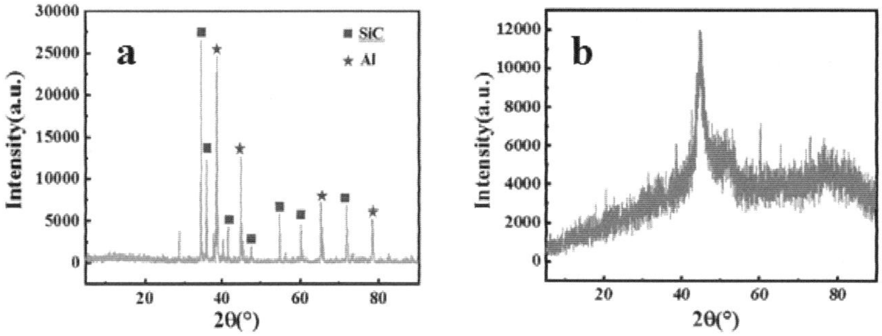

Figure 3. XRD patterns of aluminum-based silicon carbide and surface nickel-plated materials.

3.2 Aluminum-based silicon carbide material and hydrogen distribution in nickel plating layer

The position of hydrogen in the material is difficult to observe directly. The hydrogen microprinting test is conducted on aluminum-based silicon carbide nickel plating. The process involves immediately immersing the aluminum-based silicon carbide nickel plating into a solution containing silver bromide, reducing the silver ions in the solution to silver particles using the hydrogen in the nickel-plated part, and observing the distribution of silver particles in the nickel-plated part through SEM, thereby characterizing indirectly the distribution of hydrogen.

Silver ions can undergo displacement reactions with hydrogen atoms in materials, generating silver particles that settle on the surface of the material, thereby reflecting the position of hydrogen atoms in the material based on the position of the silver particles. The chemical reaction equation that occurs during the reaction is Equation (1)

$$Ag^{+}+H=Ag+H^{+} \tag{1}$$

After nickel plating on the surface of aluminum-based silicon carbide, hydrogen microprinting tests were immediately conducted, and the distribution of hydrogen in the nickel plating on aluminum-based silicon carbide was tested using SEM.

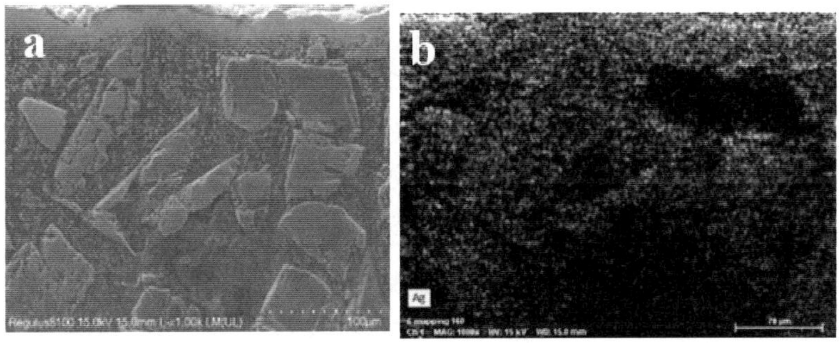

Figure 4. Hydrogen microprinting reaction results, (a) SEM image of the lateral morphology, (b) the distribution map of the silver element in EDS analysis.

Figure 4 shows the SEM image after 4 hours of hydrogen microprinting reaction and the distribution of silver elements analyzed by EDS in this region. It can be seen that the Ag particles in the aluminum matrix show a gradually decreasing trend from the surface to the deep layer. There are fewer silver particles on silicon carbide without an obvious pattern. This distribution may be related to the slower diffusion rate of hydrogen in aluminum. According to the basic principle of hydrogen diffusion [12,13], an approximate estimation is made using the diffusion model of a one-dimensional infinite object. The diffusion coefficient of hydrogen in aluminum is calculated as about 10^{-10} cm^2/s [14,15], and the distance of hydrogen diffusion corresponding to one hour of nickel plating time is 6 μ m. After processing time and reaction time (totaling about 6 hours), the diffusion distance should be in the range of tens of micrometers, which is consistent with the experimental results of 4 hours. In other words, at room temperature, the diffusion coefficient of hydrogen in silicon carbide is extremely low [16,17], which can be considered as non-diffusion. Only a layer of adsorption layer can be formed on the surface of silicon carbide. Both the content and adsorption positions of adsorbed hydrogens are related to the atomic surface of silicon carbide, so there is no obvious distribution pattern of silver particles on silicon carbide.

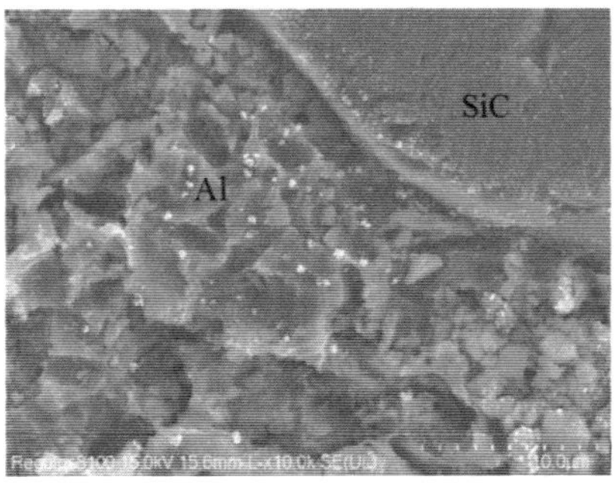

Figure 5. SEM morphology of hydrogen micro imprinting reaction after 5 hours of hydrogen charging, 1 hour of hydrogen release, and 2 hours of hydrogen micro imprinting reaction.

We removed and smoothed the nickel plating layer on aluminum-based silicon carbide, charged hydrogen at a current density of 0.5 mA/cm^2 for 5 hours in the first solution, and cleaned the surface after hydrogen charging. After releasing hydrogen for 1 hour, we then performed a hydrogen micro imprinting reaction for 2 hours to observe the surface morphology, as shown in Figure 5. It can be clearly seen that there are relatively uniform silver particles on the aluminum matrix, and the distribution of silver on silicon carbide is uneven.

The distribution morphology of Ag particles near the surface of aluminum-based silicon carbide nickel plating after 4 hours of hydrogen micro-printing reaction is shown in Figure 6. It can be seen that in the matrix near the surface of aluminum-based silicon carbide nickel plating, Ag particles are distributed on both the Al matrix and SiC particles, especially around some small SiC particles, as indicated by the blue arrow in Figure 6b. On large SiC particles, as shown in Figure 6d, the distribution of Ag particles is clearly related to the material structure.

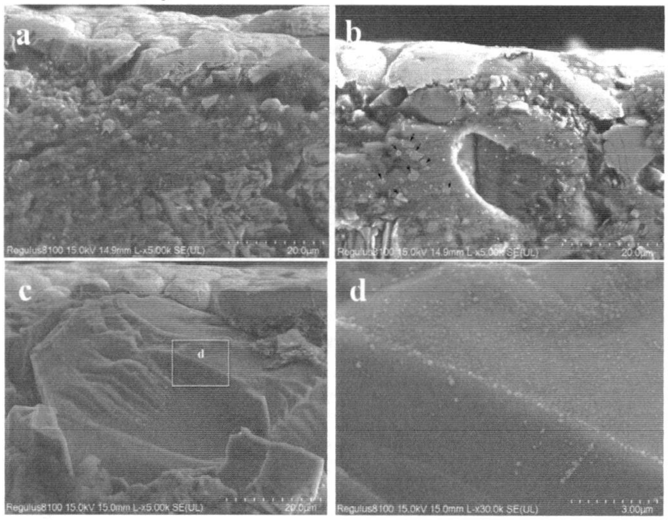

Figure 6. The cross-sectional morphology of the aluminum-based silicon carbide nickel plated sample near the surface of the aluminum-based silicon carbide nickel plated sample after 4 hours of hydrogen microprinting reaction.

Figure 7 shows the SEM morphology near the center of the cross-section of the aluminum-based silicon carbide nickel-plated sample. After zooming in step by step, it can be seen that Ag particles are distributed uniformly in both SiC particles (Figure 7d) and Al matrix (Figure 7e).

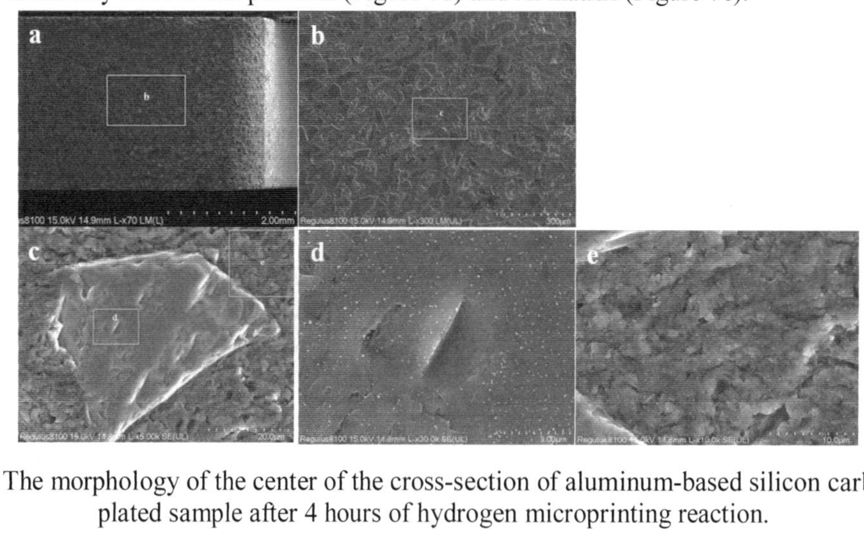

Figure 7. The morphology of the center of the cross-section of aluminum-based silicon carbide nickel plated sample after 4 hours of hydrogen microprinting reaction.

3.3 Hydrogen release testing and analysis of nickel plating on aluminum-based silicon carbide

Firstly, TDS testing was conducted on aluminum-based silicon carbide nickel-plated parts. Figure 8 shows the test results of the aluminum-based silicon carbide substrate and the nickel-plated parts on it after vacuum dehydrogenation at 250 ℃ and 300 ℃ for 96 hours, respectively. From the perspective of hydrogen release rate, aluminum-based silicon carbide nickel-plated parts have a significant hydrogen release effect compared with aluminum-based silicon carbide substrates. At the same time, compared with 250 ℃ dehydrogenation treatment, the hydrogen release rate of the test piece after 300 ℃ dehydrogenation treatment is reduced significantly. Besides, in contrast to vacuum dehydrogenation at 250 ℃, the peak area of hydrogen trap peaks significantly decreases after vacuum dehydrogenation at 300 ℃, indicating that higher dehydrogenation temperatures excite more hydrogen from the material and reduce the residual hydrogen in the material significantly. From the hydrogen trap peak, the sample treated with 250 ℃ dehydrogenation began to show peak fluctuations around 310 ℃, indicating a significant hydrogen release rate when heated to 310 ℃. Peak fluctuations appeared around 340 ℃ for the sample after dehydrogenation treatment at 300 ℃.

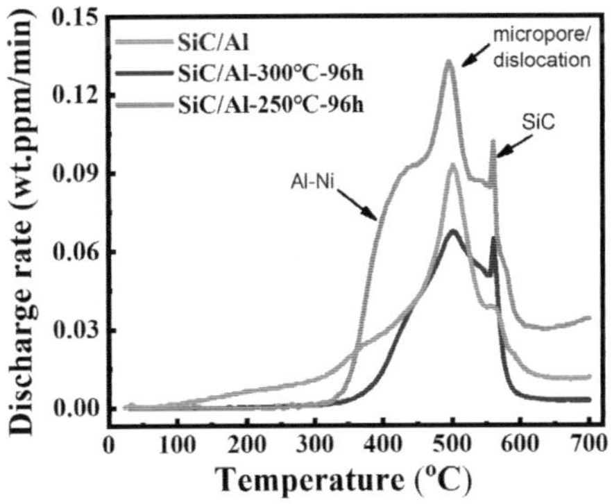

Figure 8 TDS testing of aluminum-based silicon carbide and nickel-plated samples.

Further peak fitting was performed on the TDS test curve (Figure 9), and three distinct hydrogen trap peaks were obtained at 401.88 ± 1.58 ℃, 493.64 ± 1.96 ℃, and 564.86 ± 0.85 ℃, each accounting for 10.6%, 85.9%, and 3.5%, respectively, corresponding to the adsorbed hydrogen at the Al-Ni interface, micropores/dislocations, and SiC particles. This indicates that hydrogen in aluminum-based silicon carbide nickel plating is more distributed at the micropores and dislocation defects in the substrate and coating, which is consistent with the microstructure of aluminum-based silicon carbide and nickel plating layers in SEM testing, where the aluminum-based silicon carbide substrate behaves a certain porosity (aluminum-based silicon carbide porosity of 98.7%).

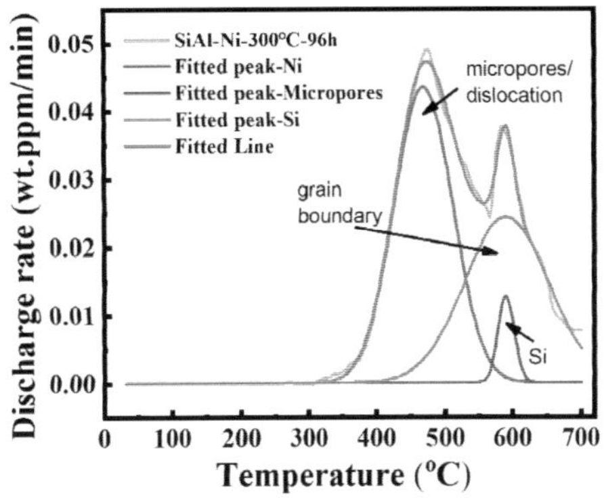

Figure 9. TDS testing peak fitting diagram of aluminum-based silicon carbide nickel plated samples.

4. Conclusion

SEM, XRD, and other methods have been used to characterize the microstructure of aluminum-based silicon carbide substrate and nickel plating layer. There are certain pore structures in both the substrate and plating layer, providing a place for hydrogen storage. The experimental method of hydrogen microprinting has been used to characterize the distribution of hydrogen here. The results showed that after nickel plating on aluminum-based silicon carbide, hydrogen distributes in the nickel plating layer and the shallow substrate under the coating. With the extension of time, hydrogen diffuses into the deep substrate spontaneously. Herein, hydrogen is mainly distributed in the Al matrix and less on SiC particles.

References

[1] Guoqiang Gan, Zhen Han, Jianhua Bao, et al. Research status of particle surface modification technology for silicon carbide particle reinforced aluminum matrix composites [J] Precision Forming Engineering, 2023,15 (12): 58-67.
[2] A. N. Nyafkin, A. A. Shavnev, V. M. Serpova, et al. Study of the Effect of the Fractional Composition of a SiC Reinforcing Phase on Mechanical and Thermophysical Characteristics of a Metallic Composite Material Based on an Aluminum Al-Si Alloy[J]. Inorganic Materials: Applied Research, 2024,14(1):37-41.
[3] Hongying Ren, Preparation and Thermal Expansion Performance Study of Al-50Si/6063 Composite Structure [D]. Heilongjiang: Harbin Institute of Technology, 2022.
[4] Beijing Spacecrafts Co., Ltd. A method of electroless nickel plating on the surface of high volume aluminum-based silicon carbide composite materials: CN202111271969.9 [P] March 8, 2022.
[5] Yifeng Jiang, Research on the process, mechanism, and application of electroless nickel phosphorus alloy plating [D] Fujian: Xiamen University, 2015.
[6] Jing Li, Yeqin Feng, Fangyu Fu, et al. Polyoxometalate-organic cage with {Ni4SiW9} node for photocatalytic hydrogen evolution [J]. Chinese Chemical Letters, 2024,35(3):177-181.
[7] Tonghe Liu, Xiaoting Yuan, Hai Huang. Primary irradiation damage in Ni-graphene nanocomposites with pre-existing hydrogen: insights from atomistic simulations[J]. The European Physical Journal Plus, 2024, 139(1), 22.
[8] Yan Qi, Xin Zhang, Jun Li, et al. Hydrogen storage thermodynamics and kinetics of as-cast Ce-Mg-Ni-based alloy [J]. Journal of Iron and steel research international. 2024, 31(3): 752-766.
[9] Cui Xihui, Fang Jie, Xu Bing, et al. Microwave product dehydrogenation process method [J] Electronic Process Technology, 2020,41 (5): 264-266.
[10] Lan Jianfei Experimental study on aluminum-based silicon carbide polishing solution [J] Agricultural Equipment and Vehicle Engineering, 2023,61 (6): 159-162.
[11] Xing Lehong, Shi Xinting, Meng Fanxu, et al. The effect of plating solution temperature on the electrocatalytic hydrogen evolution performance of nickel electrodes [J] Electroplating and Coating, 2022, 41 (11): 774-776.
[12] Oger L, Odemer G, Blanc C, et al. Influence of dislocations on hydrogen diffusion and trapping in an Al-Zn-Mg aluminum alloy. Materials & Design, 2019, 180: 107901.
[13] Takano N, Hydrogen diffusion and embrittlement in 7075 aluminum alloy, Materials Science and Engineering A, 2008, 483, 336-339.
[14] Matsuo S, Hirata T. Diffusion of hydrogen in aluminum. Journal of the Japan Institute of Metals, 1967, 31: 590.
[15] Hashimoto E, Kino T. Hydrogen diffusion in aluminum at high temperatures. Journal of Physics F: Metal Physics, 1983, 13: 1157.
[16] Doremus R H, Diffusion of hydrogen in silicon: Diffusion-reaction model, Materials Research Innovations, 2000, 4:49-59.
[17] Causey R A, Fowler J D, Ravanbakht C, et al. Hydrogen Diffusion and Solubility in Silicon Carbide. Journal of the American Ceramic Society, 2010, 61: 221-225.

Study on the continuous cooling transformation behavior of X65 hydrogen transportation pipeline steel

Tianyi Li[1,2], Wenyue Liu[1,2*], Tao An[1,2], Liu Liu[2], Chunduo Dai [1,2] and Meihui Sun[1,2]

[1]State Key Laboratory of Metal Materials for Marine Equipment and Application, Anshan, Liaoning, 114009, China
[2]Ansteel Beijing Research Institute Co. LTD, Beijing, 102200, China

[*]Corresponding author's email: liuwenyue@ansteel.com

Abstract: In this work, the continuous cooling transformation (CCT) behavior of X65 hydrogen transportation pipeline steel was studied by Gleeble-3800 machine. The CCT curve was constructed by temperature-dilation curves and microstructures of samples at different cooling speeds. The A_{c1} and A_{c3} temperatures of the tested steel are 715 °C and 901 °C. Polygonal ferrite and pearlite can be recognized when the cooling speed is between 0.2 °C/s ~ 5 °C/s, bainite and polygonal ferrite can be recognized when the cooling speed is higher than 10 °C/s. The hot-rolling process was designed according to CCT. After finishing rolling, the air cooling steel plates possesses higher hydrogen diffusion coefficient, and the steel plates coiled at 650 °C possesses lower hydrogen diffusion coefficient.

1. Introduction

Hydrogen has enormous potential as a clean fuel for the global energy transition, which has high combustion value and can completely replace fossil fuels without emitting carbon dioxide and causing air pollution [1]. Pipeline transportation is an important method for gaseous hydrogen delivery [2]. However, exposure of steel to hydrogen will lead to embrittlement of the material and cracking of the pipeline [3]. If the pipeline ruptures during hydrogen transportation, hydrogen leaks can lead to deflagrations and explosions, posing a major threat to public safety.

Most of the existing pipelines were built to transport crude oil and natural gas, and the problem of hydrogen embrittlement (HE) was not considered in their original material design [4-5]. New pipeline steels with lower susceptibility to hydrogen embrittlement need to be developed for hydrogen transportation. Materials that have a low hydrogen diffusion coefficient are also less susceptible to hydrogen embrittlement [6]. In addition, different microstructures also have different susceptibility to hydrogen embrittlement [7]. Polygonal ferrite and acicular ferrite are less sensitive to HE; pearlite, granular bainite and martensite are more sensitive to HE [8]. The formation of microstructures with high susceptibility to HE, such as pearlite and granular bainite, must be prevented in the production of pipeline steel for hydrogen transportation. Therefore, for pipeline steel used in hydrogen environments, both the chemical composition and the manufacturing process must be redesigned to ensure that the final product has a low susceptibility to hydrogen embrittlement.

To support the research and development of hydrogen transport pipeline steel, the CCT behaviour of a newly developed X65 pipeline steel and its hydrogen permeation behaviour after thermo-mechanical control process (TMCP) are investigated.

2. Materials and experiments

The material investigated is a newly developed X65 pipeline steel billet, which is manufactured by a 50 kg electric arc furnace. The chemical composition of the tested steel is shown in Table 1. Thermodynamic calculations of this steel were performed using JMatPro 12.1® software to determine the phase transformation temperature and to construct CCT diagram.

Table 1. Chemical composition in wt. %.

C	Si	Mn	Ni	Mo	Ti	Nb	V	Fe
0.03	0.2	1.35	0.4	0.15	0.02	0.035	0.045	Bal.

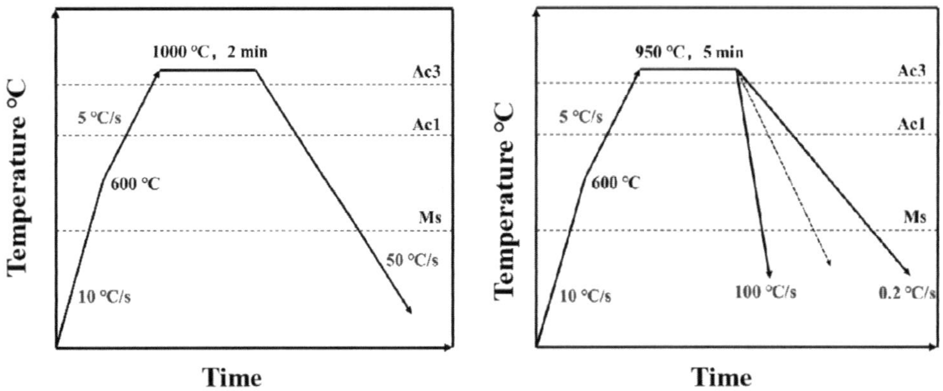

Figure 1. Experimental process chart for determination of (a) A_{c1} and A_{c3}, (b) static CCT curve.

The controlled cooling with different cooling speed was conducted on tested steel to obtain its phase transformation behaviour. The size of the used samples is round bar with diameter of Φ 8 x15 mm.

Glbble-3800 machine was utilized to obtain the phase transformation temperature of the tested steel. The thermocouple was spotwelded on the sample to measure and record the temperature data.

During phase transformation from austenite to ferrite, bainite, martensite and pearlite, the expansion of the sample changes under different cooling speeds. And the exact phase transformation temperature can be revealed according to temperature-dilation curve.

Through the process shown in Fig.1 (a), The A_{c1} and A_{c3} temperatures of the tested steel were calculated by using the temperature-dilation data.

The tangential method was used to decide the temperature at inflection point and the phase transition temperature under each cooling speed was further determined, as shown in Fig.1 (b). The exact phases were determined later by microstructure analysis.

The samples after cooling with different cooling speeds were characterized by optical microscope (OM) and scanning electron microscope (SEM). Before microstructure characterization, the samples were firstly ground by sand paper until 1000 #, then polished with a 2.5 μm diamond suspension. Finally, the 4 % nital solution were used for sample etching. OM investigation was performed on the ZEISS Axio Observer 7. The used SEM is a Zeiss Sigma field emission SEM.

After the microstructure characterization and the CCT diagram construction, the TMCP strategy was designed accordingly and the steel billet was hot-rolled into steel plate, the thickness of the steel plate is 14 mm. The hydrogen permeation test was carried out for hydrogen embrittlement sensitivity assessment. The hydrogen permeation sample size was 15×15×1.5 mm. The sample was firstly ground by sand paper until 2000 # and then plated by nickel on one side by vacuum magnetron sputtering method to ensure the accuracy of discharge current measurement, which comes from the hydrogen atoms. The double compartment cell was used for hydrogen permeation test [9], which is originally developed by Devanathan, the experiment process is illustrated in Fig.2. The hydrogen was charged in the left side by a DC power supply, the current density was set to 5 mA/cm^2. The solution used is 0.2 mol/L NaOH + 0.2 g/L CH_4N_2S. The hydrogen atoms were measured in the right side, which is a three-electrode system for anodic current measurement, consisting of a platinum electrode, a saturated glycerol electrode, and the hydrogen permeation sample. The solution used was 0.2 mol/L NaOH.

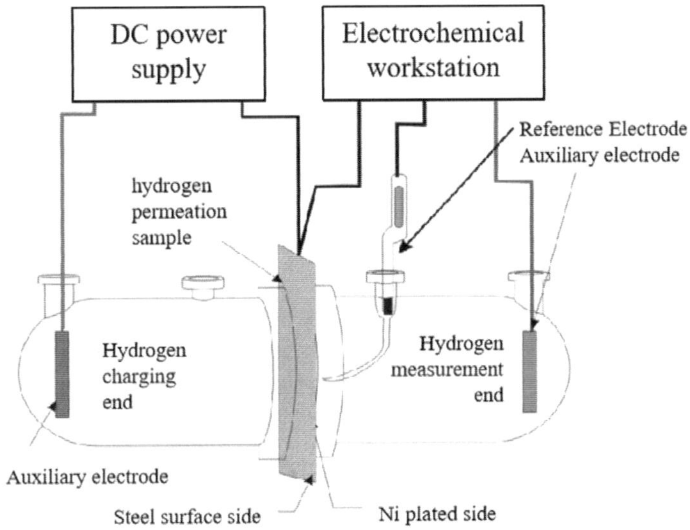

Figure 2. Illustration of double compartment cell used for hydrogen permeation test.

Once the hydrogen permeation curve was plotted, the hydrogen diffusion coefficient D was defined by Eq.1 [10]:

$$D = \frac{L^2}{6t_{0.63}} \quad (1)$$

where L is sample thickness, $t_{0.63}$ is the lag time when the raito between the momentary current density value and the final steady-state current density value, which can be directly determined from hydrogen permeation curve.

3. Results and discussion

3.1 Thermodynamic calculations

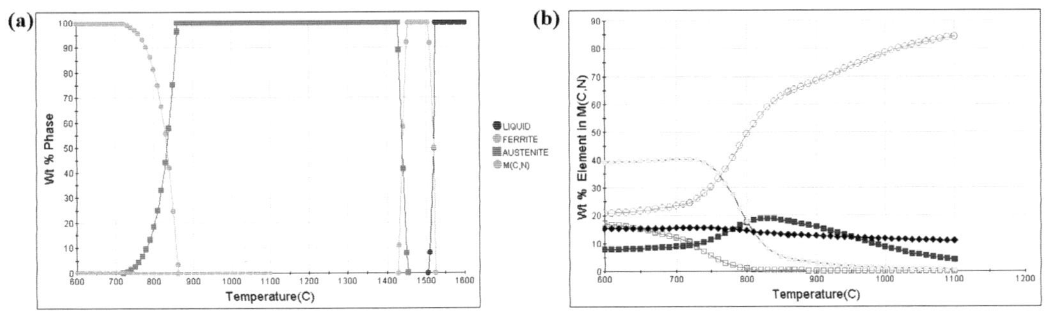

Figure 3. (a) Equilibrium phases of tested steel, (b) chemical composition of M (C, N) phase.

The phase transformation behaviour of the tested steel during cooling is shown in Fig.3 (a). Under equilibrium state, M (C, N) phase will firstly precipitate from austenite at 1100 °C, ferrite starts to form at 861°C, and the austenite will fully transform to ferrite at 725 °C. The chemical composition of M (C, N) phase is shown in Fig.3 (b), it mainly contains C, Nb, Ti, V, the precipitation phase is (Ti, Nb, V) C.

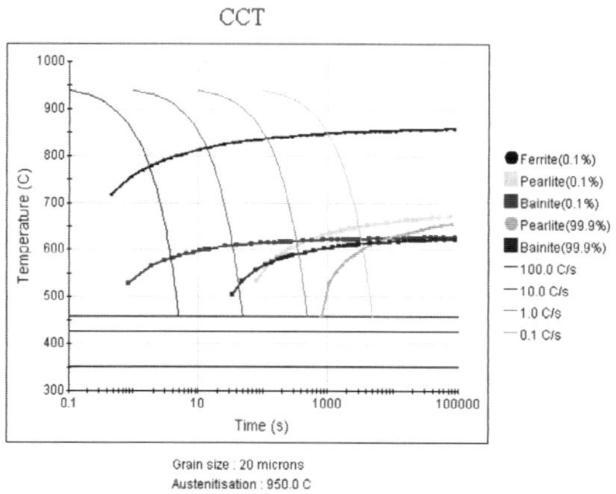

Figure 4. CCT diagram constructed using JMatPro software.

The CCT diagram constructed by JMatPro software is shown in Fig.4. The setting austenization temperature before cooling is 950 °C. The microstructure is fully bainite and ferrite with the 10 °C/s cooling speed. Increasing cooling speed will decrease the amount of ferrite. At the cooling speed about 5 °C/s, ferrite, pearlite and bainite will exist. There is no bainite left when the cooling speed is 1 °C/s. Further decreasing cooling speed leads to more pearlite and less ferrite.

3.2 CCT curve construction

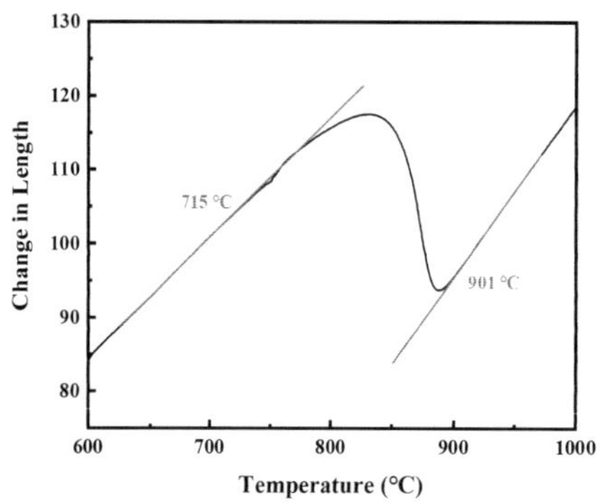

Figure 5. Change of radial thermal expansion of tested steel during heating.

The dilation curve for determination of A_{c1} and A_{c3} point temperatures during heating was shown in Fig.5, which are 715 °C and 901 °C.

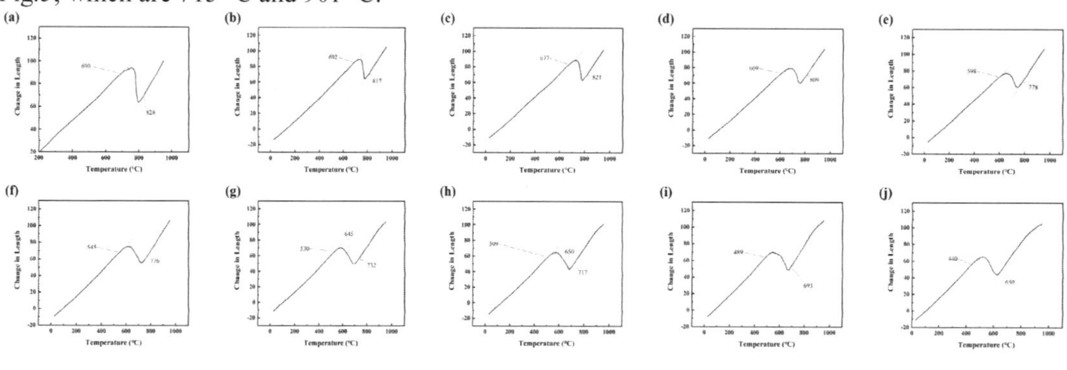

Figure 6. Temperature-dilation curves of tested steel with different cooling speeds: (a) 0.2 °C/s, (b) 0.5 °C/s, (c) 1 °C/s, (d) 3 °C/s, (e) 5 °C/s, (f) 10 °C/s, (g) 20 °C/s, (h) 30 °C/s, (i) 50 °C/s, (j) 100 °C/s.

The dilation curves of different cooling speeds were shown in Fig.6. The phase transformation temperatures were listed in Table 2. With faster cooling speed, the phase transformation temperature will decrease. Besides, when the cooling speed is slower than 10 °C/s, higher cooling speed leads to larger divergence between the austenite transition temperature range, which means that that the austenite phase transformation was blocked, and expanded the temperature ranges. The cooling speed has no obvious influence on the austenite transition range if it is faster than 10 °C/s.

Table 2. The austenite transition temperature with different cooling speeds.

Phase transformation temperature °C	Cooling speed °C/s									
	0.2	0.5	1	3	5	10	20	30	50	100
Start temperature	828	817	821	809	778	776	732	717	693	650
Finish temperature	690	692	677	609	598	545	530	509	489	440
Temperature range	138	125	144	200	180	231	202	208	204	210

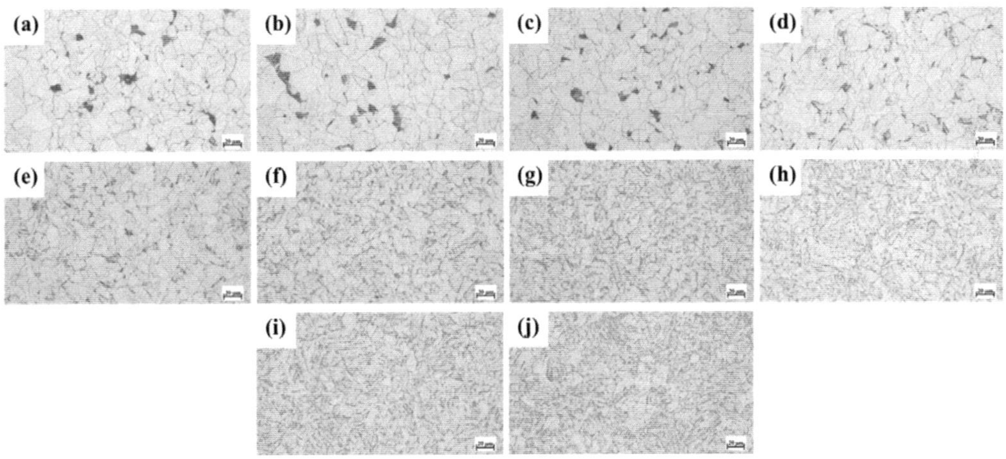

Figure 7. Microstructures of the tested steel with different cooling speeds under OM: (a) 0.2 °C/s, (b) 0.5 °C/s, (c) 1 °C/s, (d) 3 °C/s, (e) 5 °C/s, (f) 10 °C/s, (g) 20 °C/s, (h) 30 °C/s, (i) 50 °C/s, (j) 100 °C/s.

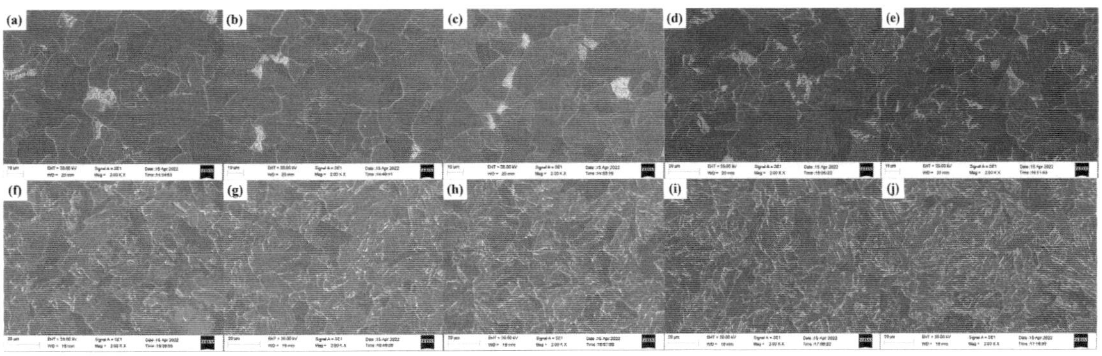

Figure 8. Microstructures of the tested steel with different cooling speeds under SEM: (a) 0.2 °C/s, (b) 0.5 °C/s, (c) 1 °C/s, (d) 3 °C/s, (e) 5 °C/s, (f) 10 °C/s, (g) 20 °C/s, (h) 30 °C/s, (i) 50 °C/s, (j) 100 °C/s.

The microstructures with different cooling speeds are shown in Fig.7 (OM) and Fig.8 (SEM). Both OM and SEM analysis results at same cooling speed are similiar. When the cooling speed was 0.2°C/s, polygonal ferrite can be observed, and some pearlite can be recognized on the boundaries of ferrite grains. With increasing cooling speed, the amount of pearlite starts to decrease, and the ferrite grain size becomes finer. Only a little pearlite can be distinguished on the ferrite grain boundary with 5 °C/s cooling speed. With 0.2 °C/s cooling speed, the temperature range of austenite transition is from

Journal of Physics: Conference Series **2842** (2024) 012011 doi:10.1088/1742-6596/2842/1/012011

828-690 °C. Increasing the cooling speed to 5 °C/s leads to the decreasing of corresponding transition range to 828-690 °C. At this time, the diffusional transition such as austenite to ferrite transformation and austenite to pearlite transformation will be delayed due to the inhibition of carbon diffusion. The expansion of the temperature ranges could also be explained due to inhibition of carbon diffusion. The growth of ferrite grains is also restrained, the ferrite grains become finer with higher cooling speed. Only bainite, quasi polygonal ferrite and can be recognized with 10 °C/s cooling speed, pearlite cannot form under this cooling speed. With higher cooling speed, the ferrite fraction will reduce. Almost 100 % bainite can be observed when the cooling speed reached 100 °C/s, and a little quasi polygonal ferrite exist.

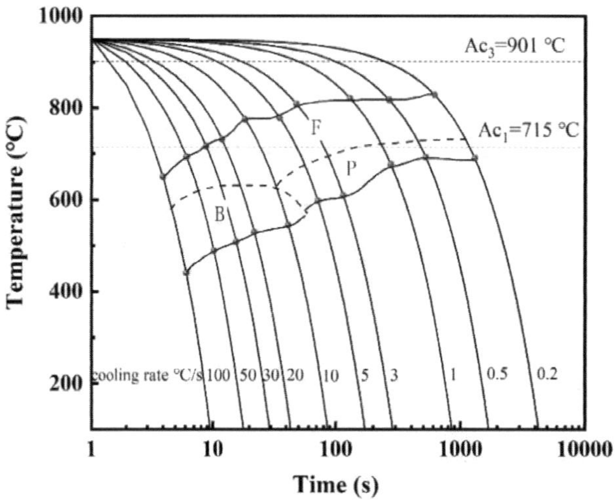

Figure 9. The static CCT curve of the tested steels.

The static CCT curve with the cooling speed ranging from 0.2-100 °C/s was plot according to the phase transformation temperatures and microstructures, as shown in Fig.9. The austenite can transformed to ferrite, bainite and pearlite at different cooling speeds. With decreasing cooling speed, the transformation temperature from austenite to ferrite increased, and the transformation temperature range of austenite to ferrite and pearlite will decrease. However, the transition range of austenite to bainite is not influenced by cooling speed. There is no pearlite formation at 10 °C/s cooling speed, only bainite and ferrite exist. Besides, the static CCT curve constructed by experiment result is similar to the CCT curve calculated by JMatPro software.

3.3 Hydrogen permeation result

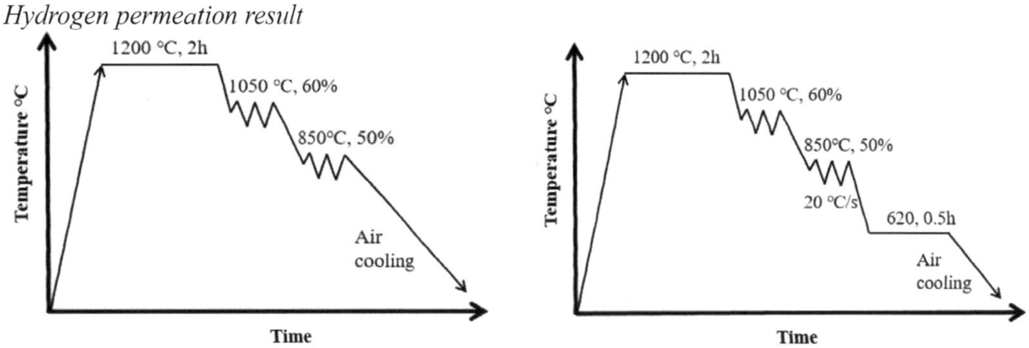

Figure 10. Schematic diagram of hot rolling process: (a) Air cooling, (b) 620 °C coiling.

The hot rolling technique was designed according to the CCT curve of the tested steel. The tested steel was firstly heated to 1200 °C for fully austenitization, the holding time was set to 2 hours.

Then the tested steel was rolled to 28 mm at 1050 °C (rough rolling) for thickness reduction, finishing rolled to 14 mm at 850 °C (finishing rolling) for grain size refinement, air cooled to room temperature, or laminar flow cooled with cooling speed of 20 °C/s to 620 °C and keep warm for 0.5 hour to simulate coiling processes, as shown in Fig.10. The cooling speed of 20 °C/s was specifically chosen to avoid the formation of pearlite. After rolling, two samples underwent hydrogen permeation test, and the results of hydrogen diffusion coefficient are listed in Table 3. The air-cooled steel plates possesses higher hydrogen diffusion coefficient, because the air-cooled steel plate contains the least amount of precipitates and its microstructure consists pearlite, and hydrogen can not be trapped effectively. The steel plates coiled at 620 °C possesses lower hydrogen diffusion ability.

Table 3. Hydrogen diffusion coefficients under different conditions.

Hot rolling technique	620 °C Coiling	Air cooling
$t_{0.63}$ (s)	1202	558
Hydrogen diffusion coefficient (cm^2/s)	3.11×10^{-6}	6.72×10^{-6}

4. Conclusion

In this work, the CCT behavior and hydrogen permeation behaviour of a newly developed X65 hydrogen transportation pipeline steel is studied, the conclusions are as follows:

1. The A_{c1} temperature and A_{c3} temperature are 715 °C and 901 °C. During the continuous cooling process, polygonal ferrite and pearlite can be recognized when the cooling speed is between 0.2 °C/s ~ 5 °C/s, bainite and polygonal ferrite can be recognized when the cooling speed is higher than 10 °C/s.
2. The CCT diagram constructed by temperature-dilation curve and microstructure is similar to the CCT diagram calculated by JMatPro software.
3. After hot rolling, the air cooling steel has higher hydrogen diffusion coefficient compared with 620 °C coiling steel.

Acknowledgments

The authors acknowledge the support of the National Natural Science Foundation of China (No. 52101068)

References

[1] Rong Y, Chen S, Li C, Chen X, Xie L, Chen J and Long R 2023 Techno-economic analysis of hydrogen storage and transportation from hydrogen plant to terminal refueling station. *Int J Hydrogen Energ* Published Online.
[2] Bouledroua O, Hafsi Z, Djukic M B and Elaoud S 2020 The synergistic effects of hydrogen embrittlement and transient gas flow conditions on integrity assessment of a precracked steel pipeline. *Int. J. Hydrogen. Energ* **45** 18010-20.
[3] Lang F, Huang F, Yue J, Li L, Xu J and Liu J 2023 Hydrogen trapping and hydrogen embrittlement (HE) susceptibility of X70 grade high-strength, acid-resistant, submarine pipeline steel with Mg treatment. *JMR&T* **24** 623-38.
[4] Li H, Niu R, Li W, Lu H, Cairney J and Chen YS 2022 Hydrogen in pipeline steels: Recent advances in characterization and embrittlement mitigation. *J. Nat. Gas. Sci. Eng* **105** 104709.
[5] Hardie D, Charles E A and Lopez AH 2006 Hydrogen embrittlement of high strength pipeline steels. *Corros. Sci* **48** 4378-85.
[6] Cheng X Y, Zhang H X 2020 A new perspective on hydrogen diffusion and hydrogen embrittlement in low-alloy high strength steel. *Corros. Sci* **174** 108800.
[7] Luppo M I, Ovejero-Garcia J 1991 The influence of microstructure on the trapping and diffusion of hydrogen in a low carbon steel. *Corros. Sci* **32** 1125-36.
[8] Hui W, Xu Z, Zhang Y, Zhao X, Shao C and Weng Y 2017 Hydrogen embrittlement behavior of high strength rail steels: A comparison between pearlitic and bainitic microstructures. *Mater. Sci. Eng. A* **704** 199-206.

[9] Devanathan M A, Stachurski Z 1962 The adsorption and diffusion of electrolytic hydrogen in palladium. *Proceedings of the Royal Society of London. Series A Mathematical and Physical Sciences* **270** 90-102.
[10] Ifergane S, David R B, Sabatani E, Carmeli B, Beeri O and Eliaz N 2018 Hydrogen diffusivity and trapping in custom 465 stainless steel. *J. Electrochem. Soc* **165** C107.

Study on the dilution mechanism of laser-cladding AlCoCrFeNi high-entropy alloy coatings

Ying Li, and Yongjun Shi*

College of Electromechanical Engineering, China University of Petroleum (East China), Qingdao, Shandong, 266580, China

*Corresponding author's e-mail: shiyj_upc@126.com

Abstract. High-performance AlCoCrFeNi high-entropy alloy (HEA) coating is manufactured by employing laser cladding to enhance the wear-resistant properties of 45# steel. Finite element simulations and experiments are combined to examine the effects of dilution rate on coating properties. The findings demonstrate that the rate of dilution increases as the speed of the laser scan increases, leading to a noticeably greater Fe concentration in the coating. Due to the low solidification rate at 4 mm/s, a moderate convection can be formed, which leaves the Fe content comparable to that of the other elements. Convection of the melt pool is exacerbated by the faster scanning speed, which allows more Fe to flow from the substrate into the coating and raises Fe concentration. The average values of microhardness are 581.54 HV, 601.86 HV, and 276.75 HV at scanning speeds of 4, 6, and 8 mm/s, respectively. Moderate dilution of the substrate can increase the microhardness, while excessive dilution can lead to a higher Fe concentration in the coating, which seriously deteriorates the properties. This study provides a guide to fabricating high-performance AlCoCrFeNi HEA coatings for 45# steel.

1. Introduction

45# steel (carbon content is 0.45 wt.%, Chinese standard) is a common material for machine parts, with good plasticity and toughness, high strength, and better cutting and processing. However, the low microhardness makes its wear resistance unsatisfactory. The problem can be solved by surface modification techniques, such as laser cladding (LC), chemical vapor deposition, and magnetron sputtering technology. During the LC process, the powders are pre-positioned or synchronously fed onto the surface of a substrate, which is melted through a focused high-energy laser beam and solidified to form a coating with better wear resistance, corrosion, oxidation, etc. [1]. In addition, high-performance surface-modified materials are crucial for enhancing the performance of the substrate. In the new coating materials systems, HEAs comprising multiple elements in identical or nearly identical molar ratios have attracted a great deal of interest. Due to the high mixing entropy, HEAs are also made up of simple solid solutions [2]. In particular, AlCoCrFeNi HEA has high compressive strength, microhardness, and frictional properties attributed to the body-centered cubic (BCC)/B2 structure [3]. Previous studies have reported the improvement in abrasion and wear resistance of 45# steel by cladding with AlCoCrFeNi HEA coating [2, 3]. However, during the LC process, substrate dilution can result in an excessive amount of Fe content in the coating due to the fact that the main element of 45# is Fe [2]. The dilution mechanism of Fe and the effect on the coating properties are still unclear. To further optimize the elemental composition and fabricate the AlCoCrFeNi HEA coatings on the surface of 45# steel, the effect of high Fe concentration from substrate dilution on the coating properties is investigated at various scanning speeds. In addition, the relationship between melt pool flow and elemental distribution is analyzed by means of experiments and finite element simulations.

2. Materials and methods

Atomized AlCoCrFeNi powder is utilized as a cladding material in the size range of 53-105 μm. Besides, the 45# steel, measuring 55 × 15 × 10 mm^3, serves as substrate. To examine the impact of various scanning speeds on the dilution rate, the laser system is equipped with a 2000 W continuous wave fiber laser, a powder feeder with argon as the powder carrier gas, a KUKA six-axis robot, and a cladding head. Meanwhile, the whole process is carried out under argon protection. Table 1 contains a list of the process parameters.

Table 1. Process Parameters of LC.

Sample Number	Laser Power / W	Feeding Rate / g•min^{-1}	Scanning Speed / mm•s^{-1}
E1			4
E2	1000	10.75	6
E3			8

After cladding, a wire-cutting machine is used to cut the sample to a size of 7 x 7 x 8 mm^3 transverse to the direction of the cladding. The specimens are ground with 360-3000 mesh sandpaper and then polished with a 1 μm diamond suspension. The polished samples are directly analyzed by using Scanning Electron Microscopy (SEM) to evaluate the microstructure and associated elemental distribution. A Vickers hardness tester (HVT-1000) is employed to conduct microhardness examinations, applying a 500 g load for 15 seconds. At intervals of 0.1 mm, points have been determined from the substrate to the surface of the coating.

The laser-powder-substrate interaction leads to complex multi-field coupling in the LC process, which is a dynamic physical metallurgical process. It is difficult to directly observe the melt pool evolution during LC by experiment. Therefore, a finite element model created based on the COMSOL Multiphysics® software has been used to investigate the melt pool flow. As shown in Figure 1(a), only half of the geometry is modeled due to the symmetry. Before solving the model, reasonable assumptions are also established according to the features of LC. First, the molten metal satisfies the continuity assumption. Second, the molten metal flow pattern is laminar. Then, the effect of the evaporation heat and the viscosity loss are neglected. Finally, the substrate and powder are isotropic materials. The governing equations can be expressed in the following form based on the above assumptions [4]:

$$\nabla \cdot \vec{u} = 0 \tag{1}$$

$$\rho \frac{\partial \vec{u}}{\partial t} + \rho \vec{u} \cdot \nabla \vec{u} = -\nabla p + \nabla \cdot (\mu(\nabla \vec{u} + \nabla \vec{u}^{T})) + \vec{F} \tag{2}$$

$$\rho C_p \frac{\partial T}{\partial t} + \rho C_p \vec{u} \cdot \nabla T = \nabla \cdot (k \nabla T) - \frac{\partial H}{\partial t} - \rho \vec{u} \cdot \nabla H \tag{3}$$

where μ is the hydrodynamic viscosity; k indicates the thermal conductivity; C_p refers to the heat capacity; H represents the latent heat of the fusion; $\vec{F}$ is the force from the outside acting on the fluid. The starting and ambient temperatures are set to 25°C, representing the actual occurrence of cladding. Other boundary conditions include the heat exchange occurring within the coating and the environment, the surface tension of the melt pool, and thermal buoyancy [4]. Besides, the laser heat source is described through the double ellipsoid model, as shown in Figure 1(b), and its heat flux distribution can be established as:

$$q_1(x,y,z) = \frac{6\sqrt{3} f_1 Q_l}{\pi a_1 bc\sqrt{\pi}} \exp(-3\frac{(x-vt)^2}{a_1^2}) \exp(-3\frac{y^2}{b^2}) \exp(-3\frac{z^2}{c^2}) \quad x > 0 \tag{4}$$

$$q_1(x,y,z)=\frac{6\sqrt{3}f_2Q_l}{\pi a_2bc\sqrt{\pi}}\exp(-3\frac{(x-vt)^2}{a_2^2})\exp(-3\frac{y^2}{b^2})\exp(-3\frac{z^2}{c^2})\quad x\le 0 \tag{5}$$

where a_1, a_2, b and c are the ellipsoidal geometry parameters; f_1 is the heat distribution function of the anterior and f_2 represents the posterior ellipsoids; Q_l is the laser power. Considering the temperature dependence of the thermophysical parameter, the variation curves are obtained through the JmatPro software. The results are shown in Figure 2. The melt pool flow is dependent on the temperature field. Therefore, a laser power of 800 W as well as a scanning speed of 6 mm/s is considered to validate the model. As illustrated in Figure 1(d), the melt pool shape can be adjusted to match the actual cladding shape by modifying the parameters of the double ellipsoid model. Besides, Figure 1(c) displays that the highest temperature during the LC process is 2020 K, which is consistent with the temperature tested by the infrared camera pyrometer.

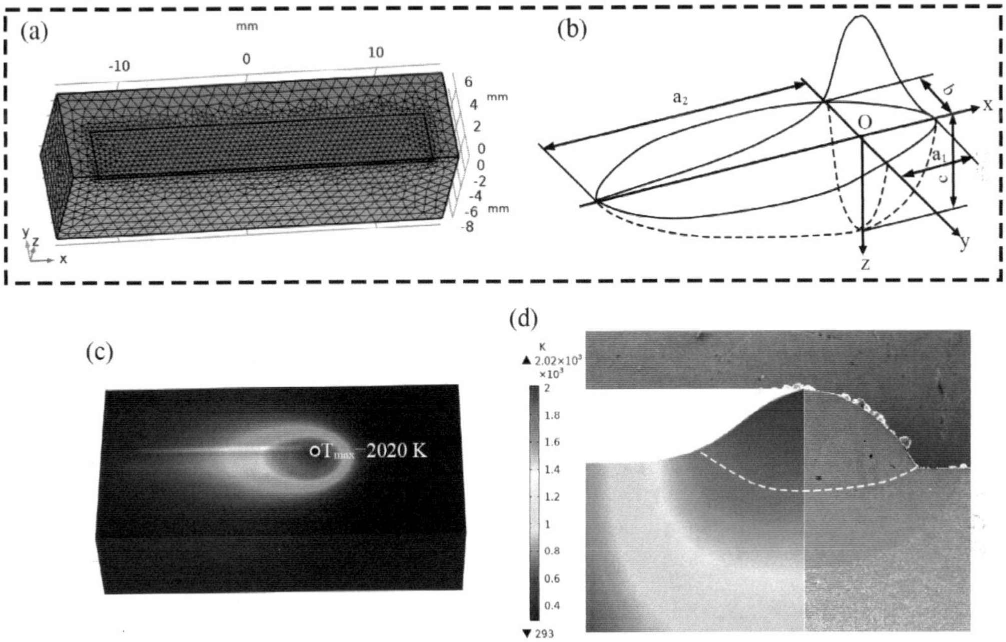

Figure 1. (a) The size and meshing of the computational domain; (b) Double ellipsoid heat source model; (c) Evolution of the temperature field; (d) Comparison of the cross-sectional morphology from simulation and experimental result.

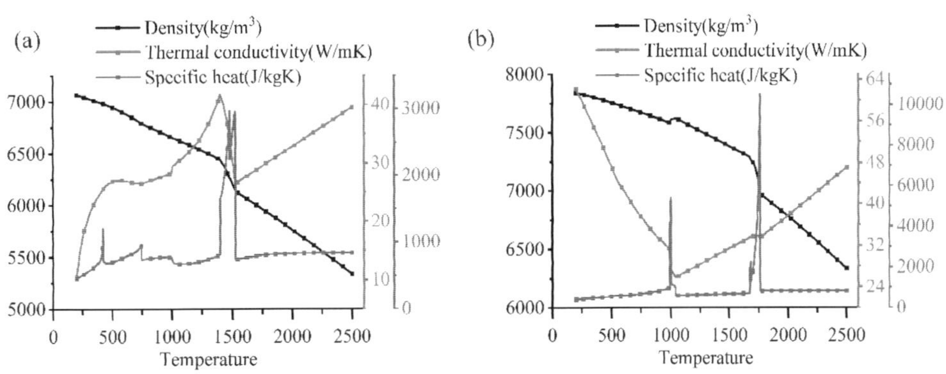

Figure 2. Variations in the physical properties with temperature. (a) AlCoCrFeNi HEA; (b) 45# steel.

Journal of Physics: Conference Series **2842** (2024) 012012 doi:10.1088/1742-6596/2842/1/012012

3. Results and discussions

3.1. Dilution rate

Figure 3 illustrates the cross-sectional morphologies, showing that the scanning speeds do not have a considerable effect on the width of the coating, but have a highly significant effect on the height. Some unmelted powders can be seen on each coating surface. In addition, the cross-sections are dense and crack-free, indicating that AlCoCrFeNi HEA has better laser formability. However, some micro-pores appear for E3, which can be attributed to a higher solidification rate under faster scanning speed, making it too late for the gas to escape. As shown in Figure 3, the substrate absorbs the heat of the laser to form a melt pool, marked by the red dotted line. The molten pool depth is positively correlated with the scanning speed, but the converse is true for the coating height. The primary reason for this is that greater energy absorption by the substrate and less powder deposition per unit of time occur at high scanning speeds. According to Figure 3(d), The depth of the melt pool and the coating height are denoted by h and H, respectively. The dilution rate can be defined by Equation (6).

$$\eta = \frac{h}{H+h} \tag{6}$$

The dilution rates for the E1, E2, and E3 coatings are 15.46%, 27.00%, and 38.37%, respectively, demonstrating a rise in dilution rate with increasing scanning speed.

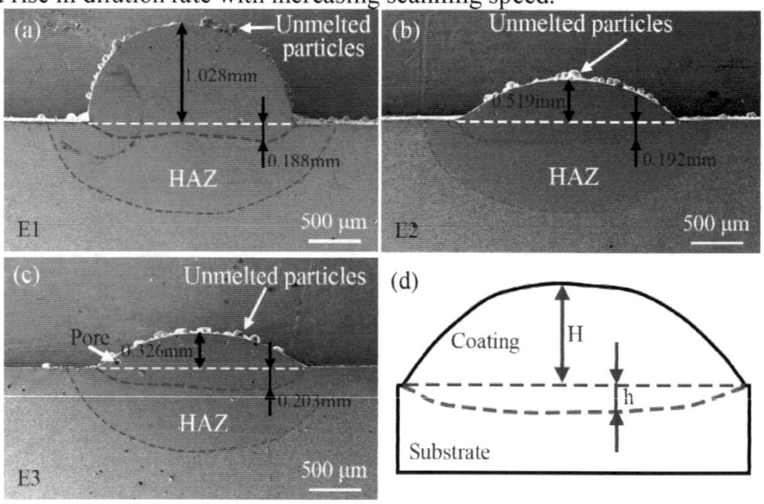

Figure 3. Cross-sections of coating with different scanning speeds.

3.2. Elemental distribution

Figure 4 shows an elemental composition SEM line scan of the coating, where Figure 4(d) represents the scanned area. It can be noticed that the Fe concentration rises abruptly from the coating to the substrate, but other elements exhibit an inverse tendency. In addition, it can be observed that the levels of Al, Cr, Co, and Ni contained in the coating are comparable, while the Fe concentration is high due to substrate dilution. As is demonstrated in Figure 4, E1 has the lowest dilution rate, while E3 is the highest, which agrees with the results of Figure 3. Therefore, the Fe content of E1 coating is close to other elements. High scanning speed can aggravate the dilution of Fe, resulting in higher Fe content relative to other elements. A critical factor influencing the distribution of elements is the convection of the melt pool. To investigate the mechanisms of dilution at different scanning speeds, the flow in the melt pool is analyzed by finite element simulation. Figure 5 compares the velocity fields at different scanning speeds. The surface of the substrate is irradiated through a laser, causing the temperature to rise rapidly and create an uneven temperature gradient. A surface tension gradient can result from an uneven temperature gradient. The molten mass in the melt pool starts to flow under surface tension and melt buoyancy to form complete convection, which causes the elements to be uniformly distributed in the melt pool. Meanwhile, elemental exchange between the substrate and the coating occurs, resulting in a metallurgical bond in the interfacial bonding zone. However, it will result in a high content of that element in the coating when the coating material contains the same element as the substrate, which can affect the performance of the coating.

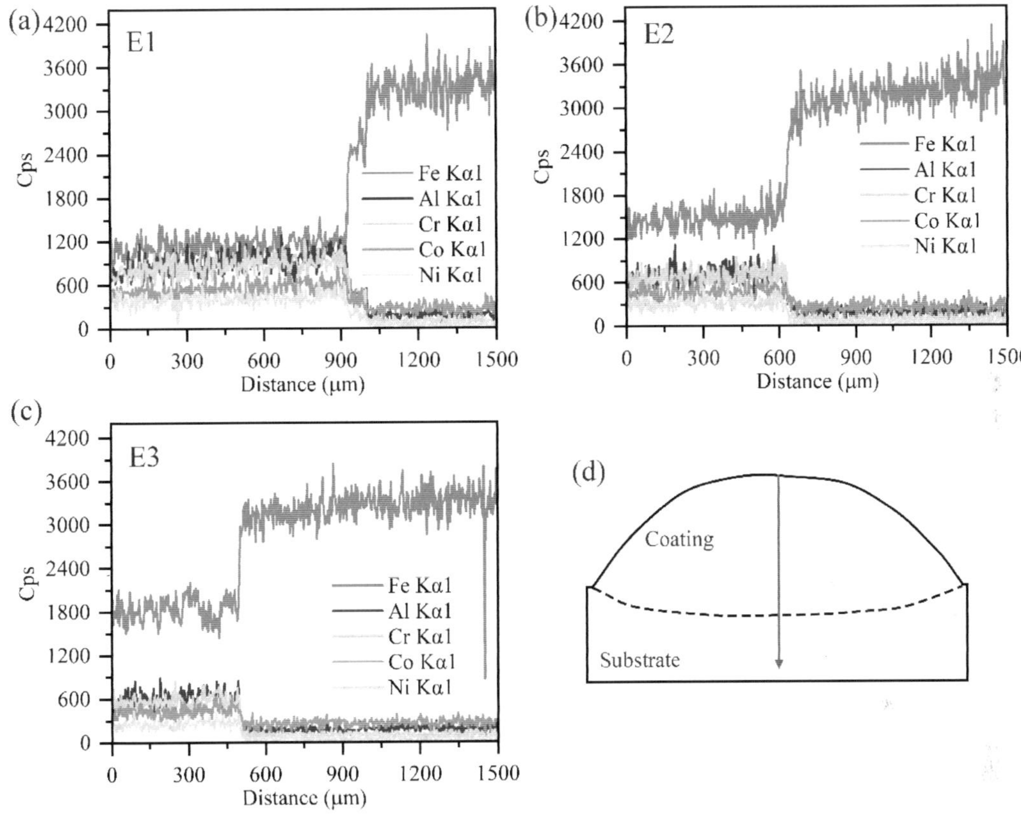

Figure 4. SEM-EDS line scan of the elemental composition.

Figure 5 depicts that the melt on the free surface flows from high to low temperatures due to surface tension. The typical exterior convection is displayed, flowing from the center of the melt pool outward. The highest melt velocities are concentrated at melt margins, while central melt velocities are the lowest. The maximum flow velocity corresponding to E1, E2, and E3 coatings is 0.15, 0.17, and 0.14 mm/s, respectively, reflecting an increase followed by a decrease. The high-temperature melt in a central melt pool is driven to a surface by surface tension and along a central interface by buoyant and gravitational forces. The melt flows rapidly close to the border of the melt pool because of the continuity of the low-temperature melt movement. When the liquid reaches the surface, it flows along the solid-liquid interface to the bottom of the melt pool, then converges and rises in the central region, forming a melt circulation flow. The segregation of elements is associated with the strength of convection in the molten pool. An even distribution of the elements and a high density of the coating can be gained under a moderate amount of convection in the molten pool. However, both inadequate and excessive convection can lead to segregation of cladding composition, forming porosity and non-fusion defects. The majority of energy from laser irradiation is captured by powder as it is scanned at 4 mm/s speed, increasing the thickness of the coating. This slow solidification rate allows sufficient time for complete melt pool convection, thus there is little change in coating element content and no significant Fe dilution. There is a positive association between the temperature gradient and the scanning speed. Convection in the melt pool is aided by increased temperature and surface tension gradients that occur with faster scanning speeds. Furthermore, more energy is captured through the substrate with increased scanning speed, which intensifies the laser-substrate interaction intensity to raise the melting depth of the substrate. Strong convection increases the Fe concentration in the coating. By further increasing the scanning speed, convection becomes insufficient for E3 coatings due to the high solidification speed. The substrate absorbs energy to melt and then flows to the coating under buoyance, which increases the Fe content in the coating.

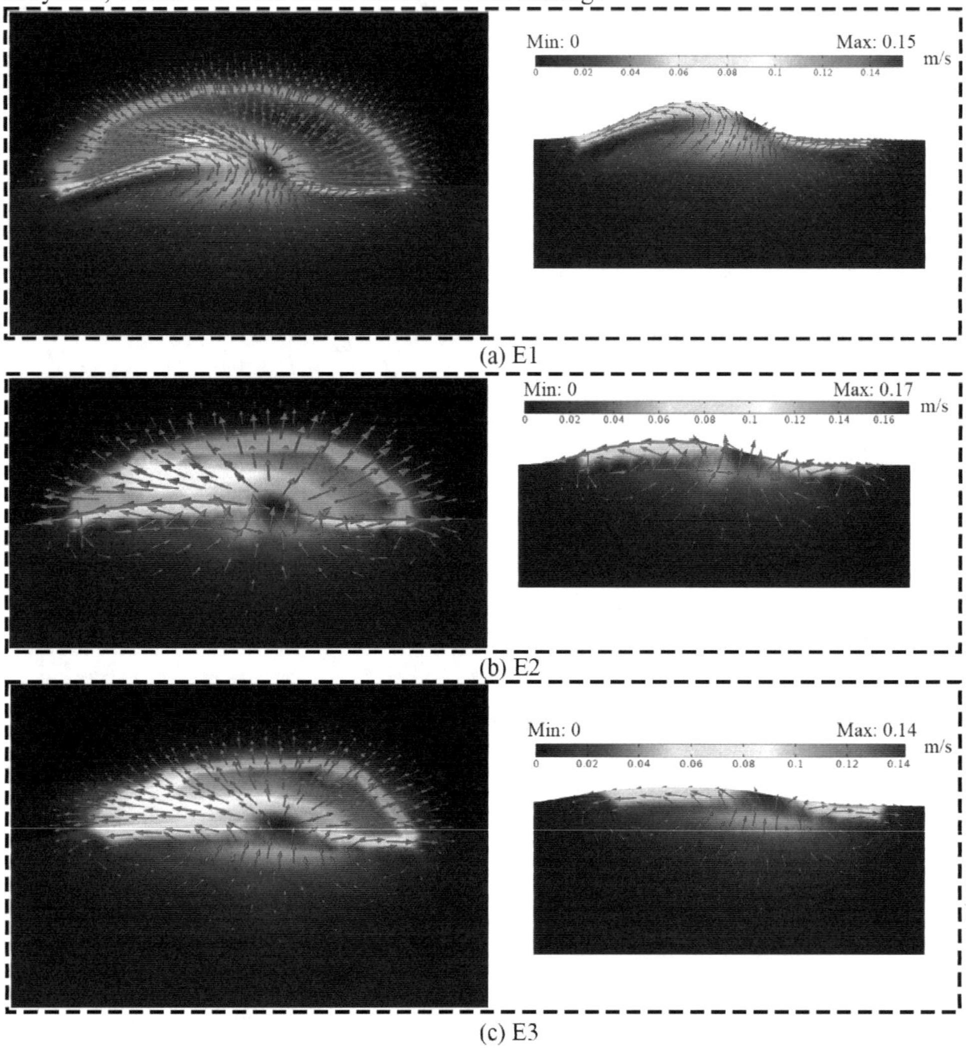

Figure 5. Velocity fields of melt pool at different scanning speeds.

3.3. Microhardness

The distribution of microhardness at varying scanning speeds from the substrate to the top of the coating is depicted in Figure 6. Coatings E1, E2, and E3 have average microhardness values of approximately 581.54 HV, 601.86 HV, and 276.75 HV, respectively. The microhardness of the coatings shows an evolutionary trend of increase and subsequent decrease. In a previous study, AlCoCrFeNi HEA coatings with a larger Fe concentration than the equimolar ratios have the highest microhardness due to the change of the solidification order of the main phases A2 and B2 [5]. However, excessive Fe content can accelerate the formation of a softer face-centered cubic (FCC) phase, which greatly reduces the microhardness of the coating [6]. Therefore, moderate dilution can increase the hardness, but excessive dilution can lead to a high content of Fe in the coating, which can seriously deteriorate the coating properties.

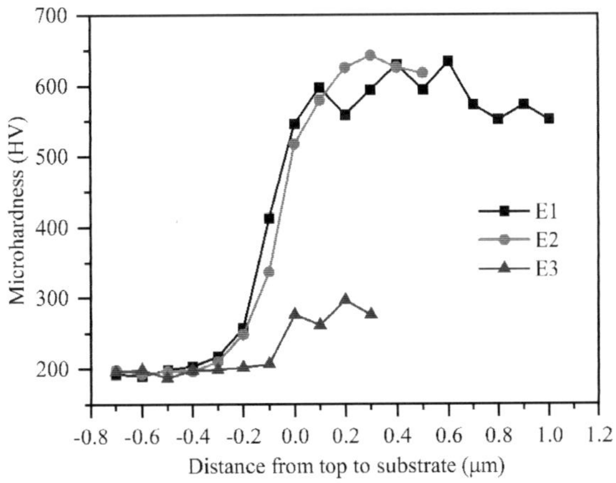

Figure 6. Evolution of microhardness.

4. Conclusion

The employing of laser cladding to produce high-performing AlCoCrFeNi HEA coatings is explored in this paper. The effects of substrate dilution on coating properties are analyzed in combination with experiments and finite element simulations. The following are the primary conclusions.

(1) The dilution rate of the coating gradually increases with the rise of scanning speed, and the dilution rates of E1, E2, and E3 are 15.46%, 27.00%, and 38.37%, respectively.

(2) The elemental distribution of coating is closely correlated with the melt flow. The convection of the molten pool accelerates at a higher scanning speed and more Fe on the substrate surface flows to the coating, resulting in Fe content in the coating much higher than the theoretical value.

(3) The average microhardness of E1, E2, and E3 is 581.54 HV, 601.86 HV, and 276.75 HV, respectively. Moderate dilution can enhance the microhardness of the coating, but excessive dilution can lead to a high content of Fe in the coating, which can seriously deteriorate the coating properties.

References

[1] Zhou Z-Y, Liu X-B, Zhuang S-G, Yang X-H, Wang M, and Sun C-F 2019 Preparation and high-temperature tribological properties of laser in-situ synthesized self-lubricating composite coatings containing metal sulfides on Ti6Al4V alloy. Appl. Surf. Sci., 481 209-18.

[2] Haftlang F, Kim E S, Kwon J, Heo Y-U and Kim H S 2023 Extraordinary combination of strength and ductility in an additively manufactured Fe-based medium entropy alloy through in situ formed η-nanoprecipitate and heterogeneous microstructure, Addit. Manuf., 63 103421.

[3] Günen A, Lindner T, Karakas M S, Kanca E, Toberling G, Vogt S and Gok M S 2022 Effect of the boriding environment on the wear response of laser-clad AlCoCrFeNi high entropy alloy coatings, Surf. Coat. Technol., 447 128830.

[4] Gao Q, Ling W, Zhou Y, Zhan X and Shen H 2023 Correlation between special bud-shaped area and molten pool flow in laser cladding high-entropy alloy on TC4 surface. Int. J. Therm. Sci., 183 107825.

[5] Zhang G J, Tian Q W, Yin K X, Niu S Q, Wu M H, Wang W W, Wang Y N and Huang J C 2020 Effect of Fe on microstructure and properties of AlCoCrFexNi (x=1.5, 2.5) high entropy alloy coatings prepared by laser cladding. Intermetallics, 119 106722.

[6] Guo L, Xiao D, Wu W, Ni S and Song M 2018 Effect of Fe on microstructure, phase evolution and mechanical properties of (AlCoCrFeNi)100-xFex high entropy alloys processed by spark plasma sintering. Intermetallics, 103 1-11.

Effects of shapes on the deformation of thin-walled artifacts manufactured by selective laser melting

Xiaona Tong and Weishi Li*

School of Instrument Science and Opto-electronics Engineering, Hefei University of Technology, Hefei 230009, China

*Corresponding author's e-mail: Weishili@hfut.edu.cn

Abstract. In selective laser melting technology, the metal powder is subjected to a temperature gradient effect during the melting process, leading to uneven distribution of internal stress, and thus forming residual stress after cooling. At the same time, the high temperature gradient will cause complex changes in the shape of the parts. In this paper, the selective laser melting process is simulated for several typical thin-walled test artifacts to predict the residual stress, and the fabricated artifacts are measured to evaluate the shape deformation. The results show that the sharper the local corners of a part, the greater the residual stress within it, making it more prone to significant deformation. We show that we can suppress the deformation by changing the local shape of the artifact, such as substituting sharp corners with circular fillets, or replacing straight shapes with curved ones.

1. Introduction

As a crucial metal processing technology in the field of Additive Manufacturing (AM), selective laser melting (SLM) has received much attention in recent years [1]. Over the last decade, SLM technology has been used in the manufacturing of medical devices, complex molds and aerospace parts [2].

During the SLM process, the large temperature gradient results in compressive and tensile stress at different locations of the part [3-5]. Consequently, this phenomenon leads to significant shape deformation, which poses a significant challenge to the dimensional accuracy of the SLM technology. Both scanning strategies [6] and process parameters [7] can have an impact on the residual stress and shape distortion. However, once the material is determined, the process parameters usually remain constant. This implies that the geometric dimension and shape of the parts play a decisive role in the impact on residual stress [8-10]. To address this problem, it's essential to delve into how the geometric size and structural characteristics of thin-walled parts influence the distribution of residual stress and deformation rule.

In this paper, we employed Simufact Additive, a commercial software for simulating metal-based additive manufacturing processes, to investigate the residual stress distribution in thin-walled test artifacts fabricated with Renishaw AM 250, and measured the deformation of the printed artifact with a 3D scanner in order to analyze the impact of shape and dimension on the deformation of the part.

2. Artifact design and simulation results

Several thin-walled artifacts, including L-shaped artifacts, semi-tubes, thin plates and square tubes, are designed to evaluate the residual stress.

We use Simufact Additive software to simulate the residual stress of the artifacts. In Simufact Additive, a hexahedral mesh is used to improve the accuracy of the simulation results, and the relevant material characteristic curves for the 316 L material have been preset. Considering the computational efficiency of the simulation, the element size is set to 1 mm. The remaining parameters are the same as the process parameters and the optimum processing parameters recommended by Renishaw for 316 L stainless steel powder, i.e. laser power of 200 W, powder layer thickness of 60 μm, hatch spacing of 100 μm and laser scanning speed of 500 mm/s, and a stripe scanning strategy with rotation angle between adjacent layers set to 67° was employed.

For the solution of the final residual stress of the parts, the software uses the inherent strain method to simulate the stress field. When the laser heat source melts the powder material, plastic strain ε_p, phase change strain ε_χ, elastic strain ε_e, and thermal strain ε_T are generated inside the sample. The relationship between the total strain ε and the four strains is as follows:

$$\varepsilon = \varepsilon_e + \varepsilon_{in} = \varepsilon_e + \varepsilon_p + \varepsilon_T + \varepsilon_\chi \tag{1}$$

When the sample is detached from the stress state, the elastic strain $\varepsilon_e = 0$, the remaining strain is the inherent strain of the sample in the forming state, and the inherent strain ε_{in} is expressed as follows:

$$\varepsilon_{in} = \varepsilon_p + \varepsilon_T + \varepsilon_\chi \tag{2}$$

The stress field of an object during the forming process can be expressed as follows:

$$[\sigma] = [D]([\varepsilon] - [\varepsilon_{in}]) \tag{3}$$

where $[\sigma]$ is the stress state matrix of the object, $[D]$ is the elastic-plastic stiffness matrix, and $[\varepsilon]$ is the strain state matrix. When simulating, as long as the inherent strain state matrix $[\varepsilon_{in}]$ is calculated. Therefore, the inherent strain method only considers thermal strain and plastic strain, without considering the impact of phase change on the forming process, thereby improving computational efficiency.

2.1. L-shaped thin-walled artifacts

In the design and manufacturing process of a part, the sharp corner is a position that needs special attention. The stress level in this area may significantly exceed the stress level in the surrounding areas, resulting in stress concentration. Therefore, we designed L-shaped thin-walled artifacts with different angles, and L-shaped thin-walled artifacts with circular fillets are also designed for comparison as shown in Table 1. All artifacts have a side length and height of 35 mm and a wall thickness of 1 mm.

Figure 1 shows the effective stress distribution of L-shaped artifacts. The maximum stress is mainly concentrated at the corner and the side edge of the artifact, and the sharper the corner is, the higher the stress is as given in Table 1. The circular fillet is effective in reducing the stress concentration. The stress level at the top corner of the artifacts is lowest, since the cooling shrinkage at this position is subject to little restriction from the solidified region.

Table 1. L-shaped artifacts.

	Angle/°	Circular fillet	Maximum Stress/MPa
L1	30	No	1323.6
L2	30	Yes	1293.1
L3	60	No	1307.2
L4	60	Yes	1272.6
L5	90	No	1299.1
L6	90	Yes	1232.3

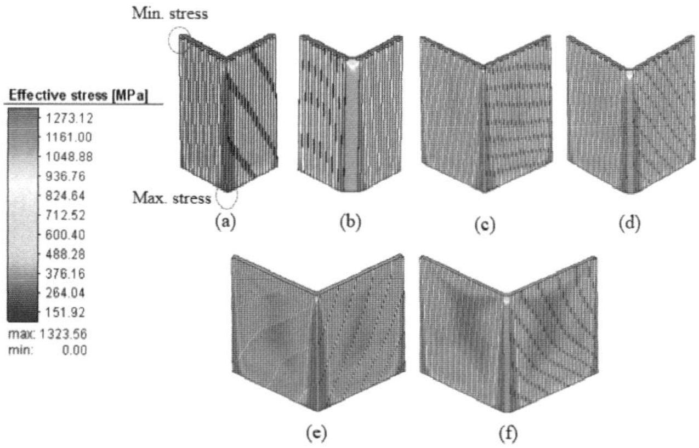

Figure 1. Effective stress distribution for L-shaped thin-walled artifacts: (a) L1; (b) L2; (c) L3; (d) L4; (e) L5; (f) L6.

2.2. Semi-tubes

We used two inscribed regular polylines to approximate a semi-circle of 30 mm diameter as shown in Table 2. All semi-tubes have a height and wall thickness of 30 mm and 1 mm, respectively.

Table 2. Semi-tubes.

	Cross-section	Maximum Stress/MPa
T1	Semi-regular hexagon	1273.5
T2	Semi- regular decagon	1212.1
T3	Semi-circle	1189.1

The effective stress distribution of all semi-tubes is shown in Figure 2. We can observe that the stress level at the corner of the semi-tubes is the highest. The maximum stress of T1 is higher than that of the other two, and the maximum stress of T3 is the lowest as given in Table 2. Similar to the L-shaped artifacts, the stress level at the top corner of the semi-tubes is the lowest.

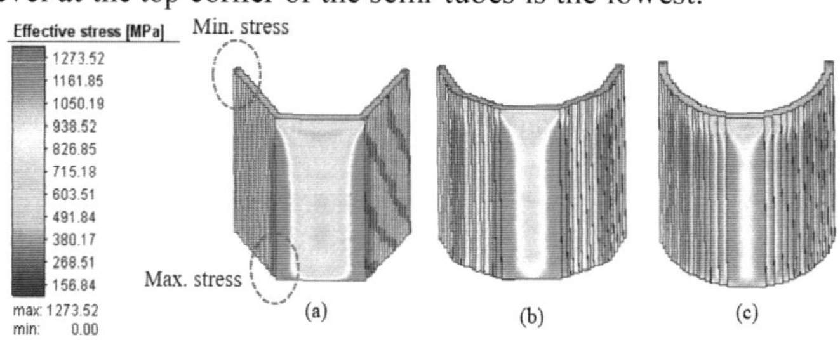

Figure 2. Effective stress distribution for semi-tubes: (a) T1; (b) T2; (c) T3.

2.3. Thin plates

The sine function $y = A\sin x$ was used to improve the stress distribution of the flat plate. All artifacts have a side length of 37.7 mm, a height of 40 mm, and a wall thickness of 1 mm, and the sine curves with different amplitudes are shown in Table 3.

Table 3. Single-walled artifacts.

	Cross-section is $y = A\sin x$	Maximum Stress/MPa
P1	A=0	1271.9
P2	A=0.5	1212.6
P3	A=1	1200.1

As shown in Figure 3, the effective stress distribution of all single-walled artifacts. We can observe from the figure, that the stress at the side edges of the artifacts is the smallest. The maximum stress of P1 is higher than that of the other two, and the maximum stress of P3 is the lowest as given in Table 3.

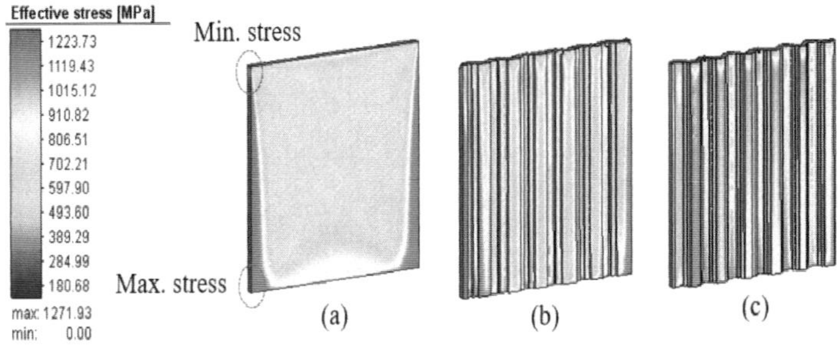

Figure 3. The effective stress distribution for single-walled artifacts: (a) P1; (b) P2; (c) P3.

3. Experimental results

3.1. SLM Machine and materials

The SLM process was carried out using Renishaw AM 250. During the process, all parameters and simulation parameters are set the same to ensure the reliability of the experimental results. The build chamber volume inside the equipment is 250×250×300 mm³, which has a Ti-6Al-4V substrate (250×250×30 mm³) and a pure argon atmosphere. 316L stainless steel powder is used as the material for the artifacts, with a diameter distribution of 30 μm to 60 μm. The chemical composition of 316L stainless steel powder is given in Table 4.

Table 4. Chemical composition of 316L stainless steel powder [11].

Element	Cr	Ni	Mo	Mn	Si	N	Fe
Mass/%	18	14	3	<2	<1	<0.1	Balance

Figure 4 shows all the manufactured thin-walled artifacts. It should be noted that a reference plate with a thickness of 3 mm is designed at the bottom of the thin-walled semi-tubes and thin plates. This reference plate serves as a datum for measuring the dimensional deviation.

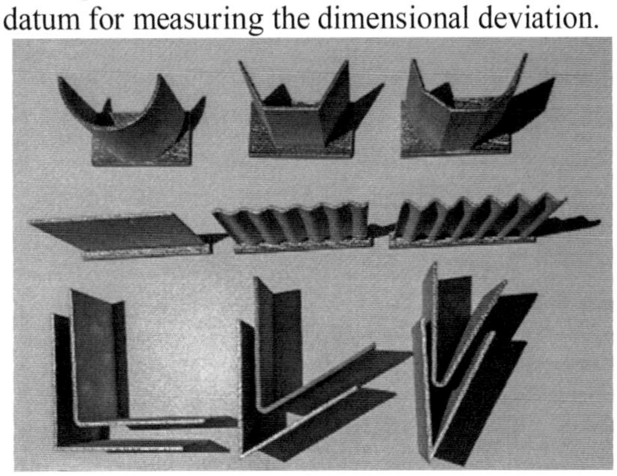

Figure 4. Thin-walled artifacts manufactured by SLM.

3.2. Measurement results

Figure 5 shows the AICON StereoScan neo 3D scanner used to measure the actual shapes of all artifacts, and the point clouds were imported into Geomagic Studio to evaluate the form deviation, respectively.

Journal of Physics: Conference Series **2842** (2024) 012013 doi:10.1088/1742-6596/2842/1/012013

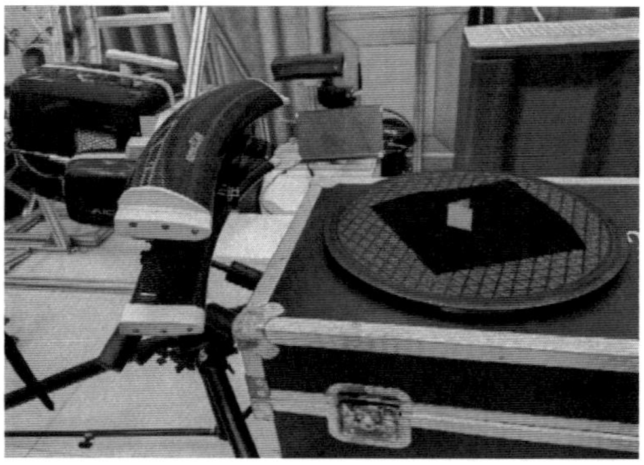

Figure 5. AICON StereoScan neo 3D scanner.

As shown in Figure 6, the angle between the two outer surfaces of the L-shaped thin-walled artifacts was measured to evaluate and compare the dimensional accuracy of different designs. Then we can see from Figure 7, that the angle deviation gradually decreases as the angle of the L-shaped part increases, indicating that increasing the angle can disperse the residual stress at the corner by comparing L1, L3, and L5. In addition, by comparing L1 and L2, L3 and L4, and L5 and L6, respectively, we can conclude that we can improve the dimensional accuracy of a part by adding a circular fillet at the corner.

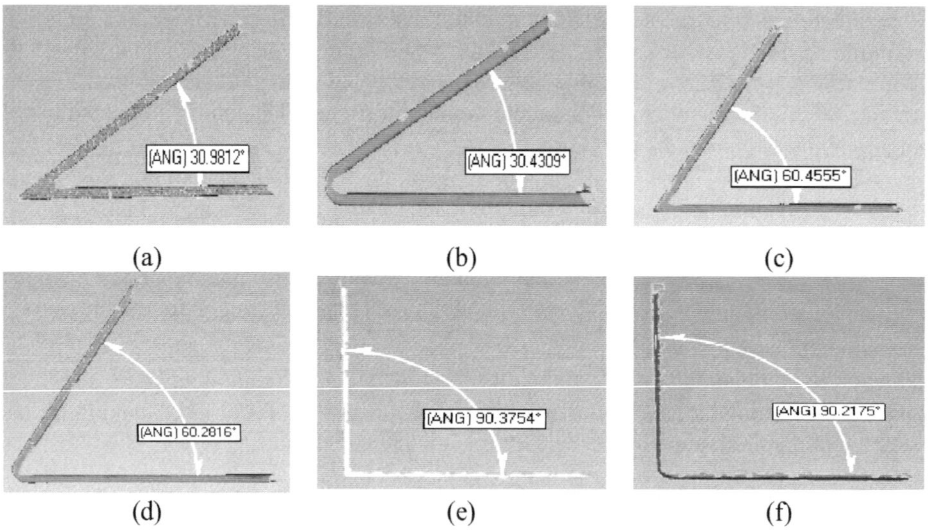

Figure 6. Measurement results for L-shaped artifacts: (a) L1; (b) L2; (c) L3; (d) L4; (e) L5; (f) L6.

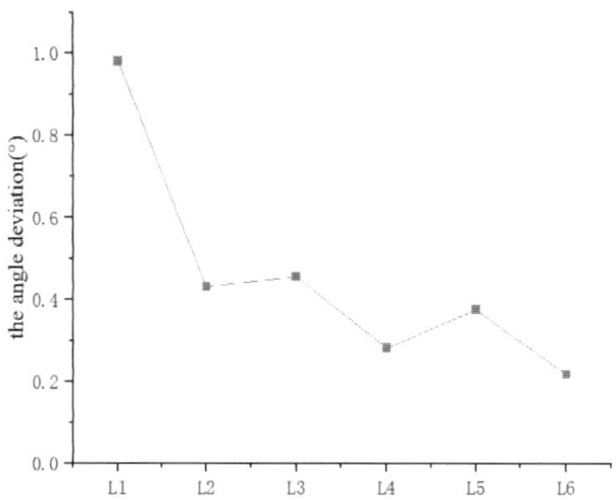

Figure 7. Angle deviation for L-shaped artifacts

Journal of Physics: Conference Series **2842** (2024) 012013 doi:10.1088/1742-6596/2842/1/012013

The surface profile error of the outer surface of each semi-tube was measured, the measurements are shown in Table 5. Then we can see from Figure 8, that as the cross-section of the semi-tube approximates a circular shape, the surface profile error decreases gradually, indicating that we can optimise a shape with corners by increasing the corner angle to suppress the deformation.

Table 5. Semi-tubes.

	Cross-section	The surface profile error
T1	Semi-regular hexagon	0.2807
T2	Semi-regular decagon	0.1882
T3	Semi-circle	0.1837

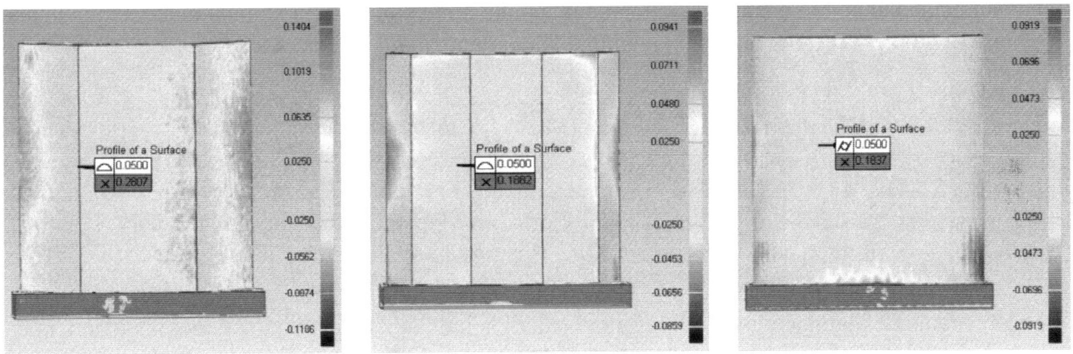

(a) (b) (c)

Figure 8. Measurement results for semi-tubes: (a) T1; (b) T2; (c) T3.

The surface profile of each single-walled artifact is measured, the measurements are given in Table 6. As shown in Figure 9, as the amplitude of the sine curve increases, the deformation is effectively suppressed. This means that we can optimise a thin plate to a corrugated plate when it is acceptable.

Table 6. Single-walled artifacts.

	Cross-section is y=Asin(x)	The surface profile error
P1	A=0	0.2582
P2	A=0.5	0.2431
P3	A=1	0.1747

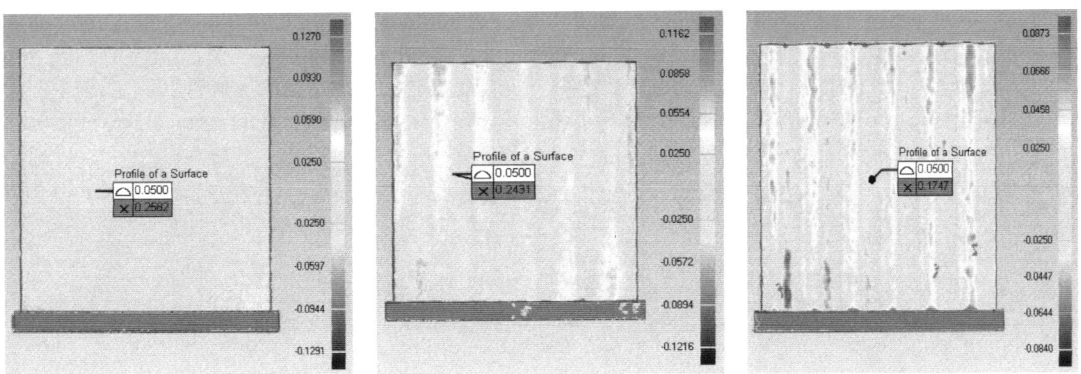

(a) (b) (c)

Figure 9. Measurement results for semi-tubes: (a) P1; (b) P2; (c) P3.

4. Conclusion

In this study, we found that the geometry shape will affect the residual stress and deformation of several additively manufactured thin-walled artifacts. The simulation results of residual stress coincide well with the shape deformation of the manufactured artifacts. According to the results of both simulation and shape measurement, the deformation of a part can be suppressed by reducing sharp corners, and replacing straight shapes with sine curves is also helpful for suppressing deformation. The results can serve as design rules for parts to be manufactured with the SLM process.

Acknowledgments

This work was financially supported by the National Natural Science Foundation of China (Nos. 51975177).

References

[1] DebRoy T, Wei H L and Zuback J S (2018) Additive manufacturing of metallic components–process, structure and properties. Progress in Materials Science, 92: 112-224.
[2] Li Z, Xu R and Zhang Z (2018) The influence of scan length on fabricating thin-walled components in selective laser melting. International Journal of Machine Tools and Manufacture, 126: 1-12.
[3] Miranda G, Faria S and Bartolomeu F (2019) A study on the production of thin-walled Ti6Al4V parts by selective laser melting. Journal of Manufacturing Processes, 39: 346-355.
[4] Bartlett J L (2018) Revealing mechanisms of residual stress development in additive manufacturing via digital image correlation. Additive Manufacturing, 22: 1-12.
[5] Ahmed A, Majeed A, Atta Z and Jia G (2019) Dimensional quality and distortion analysis of thin-walled alloy parts of AlSi10Mg manufactured by selective laser melting. Journal of Manufacturing and Materials Processing, 3(2): 51.
[6] Buchbinder D, Meiners W, Pirch N, Wissenbach K and Schrage J (2014) Investigation on reducing distortion by preheating during manufacture of aluminum components using selective laser melting. Journal of laser applications, 26(1).
[7] Strantza M (2021) Effect of the scanning strategy on the formation of residual stresses in additively manufactured Ti-6Al-4V. Additive Manufacturing, 45: 102003.
[8] Lu X, Cervera M, Chiumenti M and Lin X (2021) Residual stresses control in additive manufacturing. Journal of Manufacturing and Materials Processing,5(4): 138.
[9] Wu A S, Brown D W, Kumar M, Gallegos G F and King W E (2014) An experimental investigation into additive manufacturing-induced residual stresses in 316L stainless steel. Metallurgical and Materials Transactions A, 45: 6260-6270.
[10] Qi H, Lee B, Gao Y and Azer M (2008) Studies of thermal distortion and residual stresses of an Inconel 718 part fabricated by the laser powder deposition process. In: International Congress on Applications of Lasers & Electro-Optics. New York. pp. 782-788.
[11] Strantza M, Ganeriwala R K, Clausen B, Phan T Q, Levine L E, Pagan D and Brown D W (2018) Coupled experimental and computational study of residual stresses in additively manufactured Ti-6Al-4V components. Materials Letters, 231: 221-224.

The sealing performance of PTFE O-rings at low temperature

Jing Wang* **and Haijun Hao**

North China University of Technology, Beijing 100144, China

*Corresponding author's e-mail: 3459956122@qq.com

Abstract. Polytetrafluoroethylene (PTFE) has the advantages of high chemical stability, wide operating temperature range, excellent lubrication, etc. It is widely used in aerospace, automotive industry and other engineering fields and plays an important role in sealing applications of low-temperature media. However, in practical engineering applications, the design of PTFE O-rings in low-temperature variable-temperature working conditions lacks a reference basis. Based on the sealing structure of low-temperature pressure vessel, this paper establishes a two-dimensional axisymmetric model using the finite element analysis software Abaqus to study the sealing performance of PTFE O-rings in the static sealing system, and analyzes the effects of temperature, compression rate, cold shrinkage on the deformation of the PTFE O-rings, the Mises stresses, and the contact pressure.

1. Introduction

With the development of science and technology, the working conditions of pressure vessels are becoming more and more demanding, such as working under high pressure, low temperature, and strong corrosive conditions. Especially as hydrogen energy is gradually receiving high global attention, more stringent requirements have been placed on the sealing performance of cryogenic pressure vessels. The essence of sealing is to prevent mass exchange between the medium inside the equipment and the external environment. Leakage of the sealing medium may lead to energy loss, causing environmental pollution, the equipment can not be used normally to affect the continuous production, etc. Leakage of liquid hydrogen medium may also cause combustion or even deflagration, explosion, frostbite and other hazards to the human body [1-2].

Rubber O-Ring is the more widely used self-tightening sealing ring, in the lower preload can obtain good sealing performance. However, some researchers have studied the low-temperature characteristics of NBR materials, and the results show that at low temperatures, the seals usually suffer from cold shrinkage and embrittlement, and the sealing performance of NBR decreases significantly at low temperatures of -55°C [3-5]. Therefore, conventional rubber materials are not suitable for ultra-low-temperature seals. PTFE has a wide range of operating temperatures and is favored for sealing low-temperature pressure vessels [6-8]. Liu [9] applied PTFE O-ring seals to the sealing of throttle valves at 105 MPa, and verified the relay of PTFE seals in high-pressure occasions through simulation and test. However, the factors affecting the sealing performance of PTFE O-rings under low-temperature conditions, especially the effect of low-temperature cold shrinkage on the sealing performance, have not been fully investigated yet.

Therefore, based on the sealing structure of a low-temperature pressure vessel, this paper establishes a two-dimensional axisymmetric model to study the sealing performance of PTFE O-ring in the low-temperature static sealing system, analyzes the effects of temperature, compression rate, and shrinkage in cold on the PTFE O-ring, and then determines a reasonable range of values of the relevant parameters, in order to provide references for the sealing structure of low-temperature pressure vessel design.

2. Finite element model

The O-ring seal of a cryogenic pressure vessel with an inner diameter of 156.5 mm and a cross-section diameter of 7 mm was selected for this study. In order to solve the problem rationally and efficiently, the following assumptions and simplifications are made to the model during finite element simulation: (1) The seal material is homogeneous and continuous, isotropic, and has a fixed Poisson's ratio; (2) The sealing structure is weightless and the geometry and loading of the structure is perfectly axisymmetric; (3) The modulus of elasticity of the seal seat and cover material is much greater than the seal material, which is assumed to be rigid.

2.1. Geometric model and meshing

Based on the above assumption that the geometry and forces of the sealing structure are axisymmetric, the study of three-dimensional sealing structures is simplified into a two-dimensional problem for analysis. A geometric model is developed for finite element analysis including the O-ring, seal seat, and cover. The maximum Mises stress change is used as an indicator to verify the mesh-independence of the model as well as the meshing. The results are shown in Figure 1 and Figure 2. There are 5181 cells and 6197 nodes in the model grid.

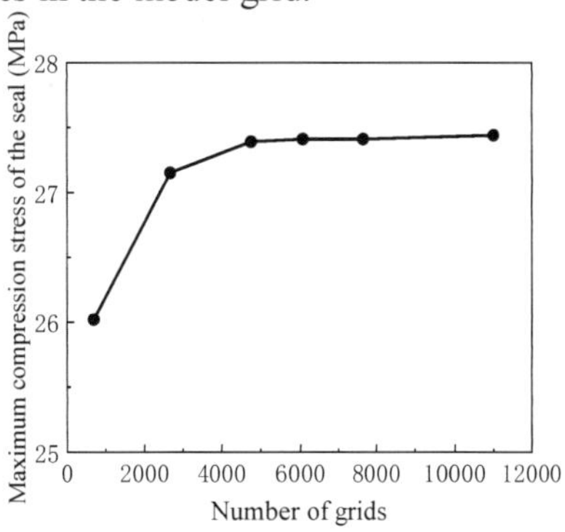

Figure 1. Grid-independent verification.

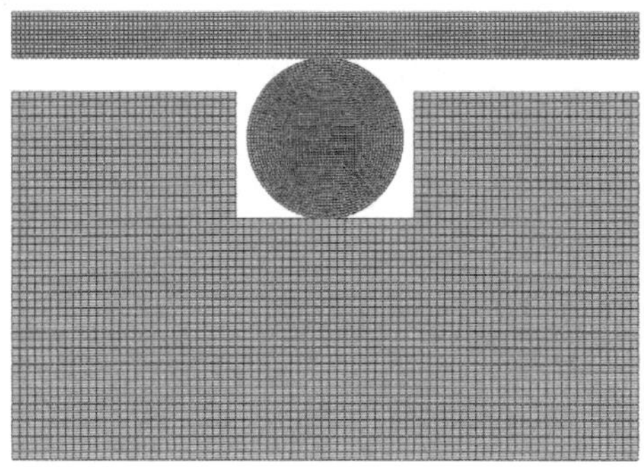

Figure 2. Seal finite element model.

2.2. Material model and parameters

According to the functional requirements of the sealing structure and the actual working conditions, the material of the seal seat and cover is 304 stainless steel, with a density of 7930 kg/m^3 and a Poisson's ratio of 0.3, and the material of the O-ring is PTFE, with a density of 2210 kg/m^3 and a Poisson's ratio of 0.45. PTFE is a hard sealing member, its small deformation can produce large contact stresses, and PTFE exhibits linear elastic behavior in small deformation, so the linear elastic model is used in this model to characterize the mechanical behavior of PTFE. PTFE shows linear elastic behavior, so the linear elastic model is used in the model to characterize the mechanical behavior of PTFE. The mechanical parameters of the model materials at different temperatures are shown in Table 1 [10-14].

Table 1. Material model parameters.

Temperature (K)	Modulus (MPa)		Specific Heat (J/K·g)		Thermal expansion (10^{-5}·K^{-1})		Heat conductivity (W/m·K)	
	PTFE	304 steel	PTFE	304 steel	PTFE	304 steel	PTFE	304 steel
20	7555	204	0.10	13.45	0.49	1.08	0.11	2.17
77	6728	204.81	0.48	136.92	4.27	1.25	0.45	6.65
100	6468	203.84	0.55	275.5	4.82	1.35	0.49	9.22
150	4460	201.55	0.68	349.1	6.63	1.42	0.50	11.08

200	1825	198.92	0.77	396.07	7.93	1.48	0.45	12.07
250	924	193.01	0.84	463.41	8.57	1.56	0.42	14.25
300	494	189.72	0.88	470.54	9.29	1.59	0.39	15.31
350	263	187.65	0.90	478.25	9.56	1.60	0.37	16.24

2.3. Contact settings and boundary conditions
In the model, two contact pairs are defined: (1) the contact between the seal and the cover plate; and (2) the contact between the seal and the seal seat. The contact problem of the seal with the seal seat and cover plate belongs to the rigid-flexible contact. We set the sealing surface of the seal seat and cover plate as the target surface, the surface of the O-ring as the contact surface, and the contact type as the friction contact. The algorithm of the normal contact is chosen as the augmented Lagrangian algorithm, the algorithm of the tangential contact is the penalized function method, and the friction coefficient is set to 0.15.

In simulating the compression and cooling process of the seal, the full degrees of freedom of the seal seat are restricted. Two loading steps are set up. The first step applies a downward displacement to the cover plate to bring the seal to a specified compression rate; the second step applies a temperature boundary condition to the seal seat as well as to a portion of the O-ring surface, which is used to simulate the cooling process.

3. Results and discussion

3.1. Effect of temperature on sealing performance
Figure 3 demonstrates the compressive stress cloud of the O-ring at 77 K and 300 K temperatures when the compression rate is 3.5%, and the maximum Mises stress of the O-ring is mainly distributed between the two contact surfaces subjected to extrusion. This part of the seal undergoes the largest amount of extrusion deformation during compression, resulting in a large combined equivalent stress. The maximum Mises stress of O-rings increased gradually with the decrease of temperature, as shown in Figure 4.

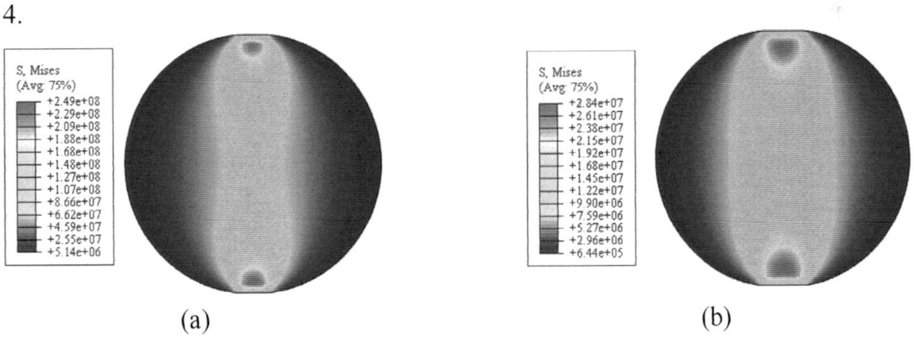

(a) (b)

Figure 3. Stress clouds of O-ring seals at different temperatures. (a) 77 K and (b) 300 K.

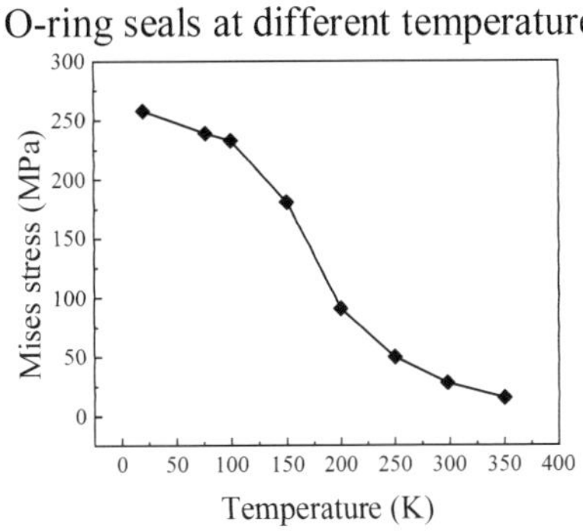

Figure 4. Variation of maximum Mises stress with temperature.

In order to facilitate the analysis of the contact stress situation at different contact locations of the O-rings, the boundary nodes on the contact surface of the O-rings are numbered as shown in Figure 5. Figure 6 shows the comparison of the contact stress values at different temperatures at a 3.5% compression rate, which is consistent with the variation of the maximum Mises stress value with temperature, and the contact stress decreases gradually with the increase in temperature. The number of mesh nodes where contact occurs becomes more as the temperature increases and the contact area between the seal and the sealing surface increases. The reason for this phenomenon is that as the temperature increases, the seal gradually becomes softer and the deformation of the material under the same compression ratio is greater, resulting in an increase in the contact area.

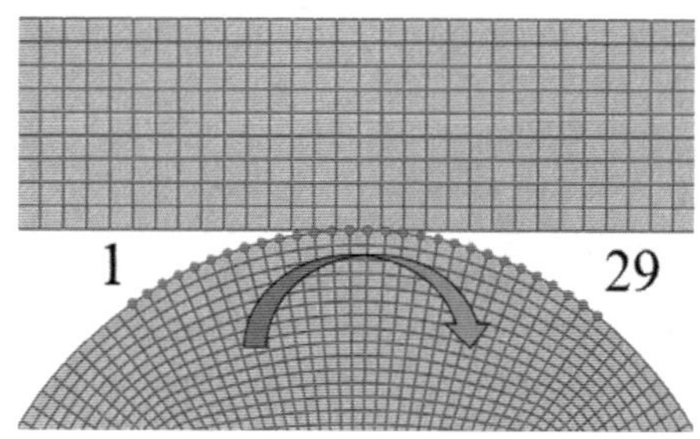

Figure 5. Mesh contact node numbering.

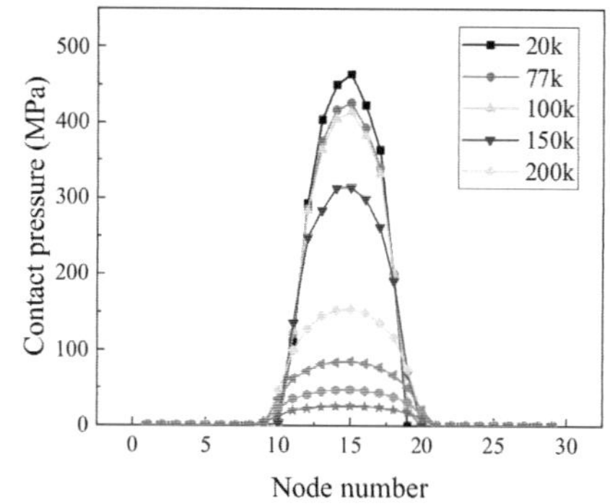

Figure 6. Contact pressure at each contact node.

3.2. Effect of compression ratio on sealing performance

Figure 7 illustrates the stress distribution cloud of the O-ring at room temperature for compression rates of 2% and 10%. As the compression rate increases, the Mises stress inside the O-ring gradually increases and extends from the vicinity of the contact surface to the inside of the O-ring. The maximum Mises stress increases with increasing compression rate, as shown in Figure 8.

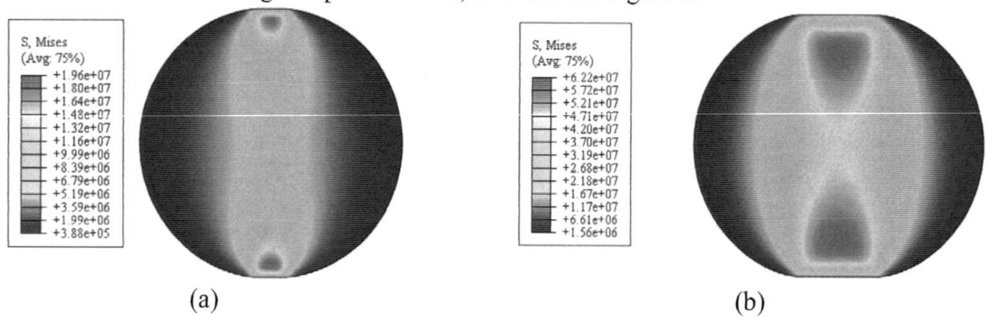

Figure 7. Stress clouds of seals with different compression ratios. (a) $\varepsilon = 0.02$, and (b) $\varepsilon = 0.1$.

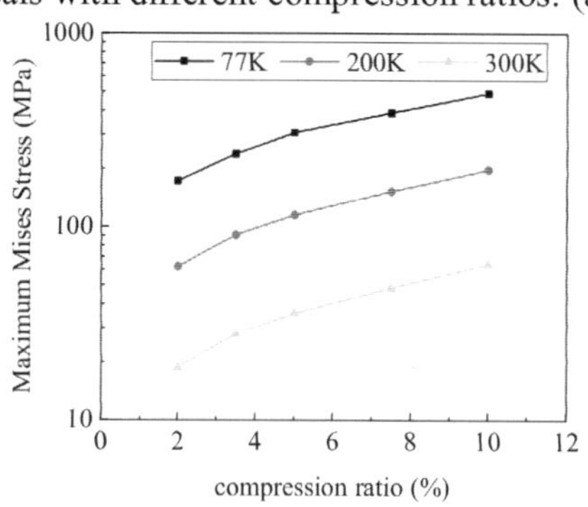

Figure 8. Maximum Mises stress vs. compression ratio for O-ring.

Figure 9 illustrates the values of contact stresses at each contact node for different compression rates at temperatures of 77 K and 300 K. As the compression rate increases, the number of nodes producing contact stress on the contact surface increases, and the contact area increases. The contact stresses on the contact surfaces of O-rings with different compression ratios are parabolic, and the maximum contact stresses appear in the center of the contact surfaces. 10% compression ratio of the maximum contact stresses is about 2.4-2.7 times the maximum contact stresses of 2% compression ratio. For the sealing pressure less than 25 MPa medium and low-pressure sealing, room temperature and low temperature can ensure that there is a 2% compression rate can ensure the sealing effect.

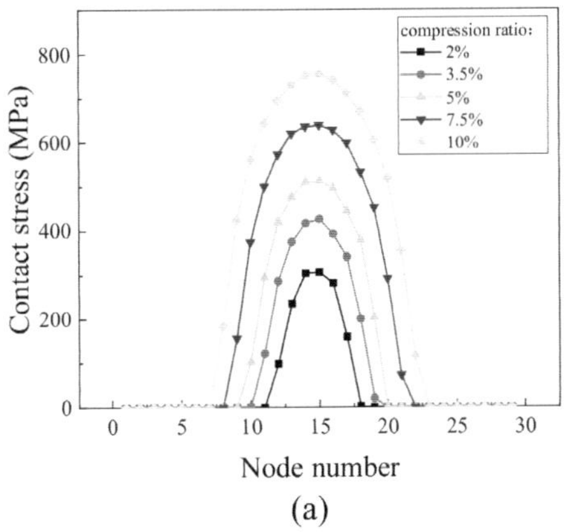

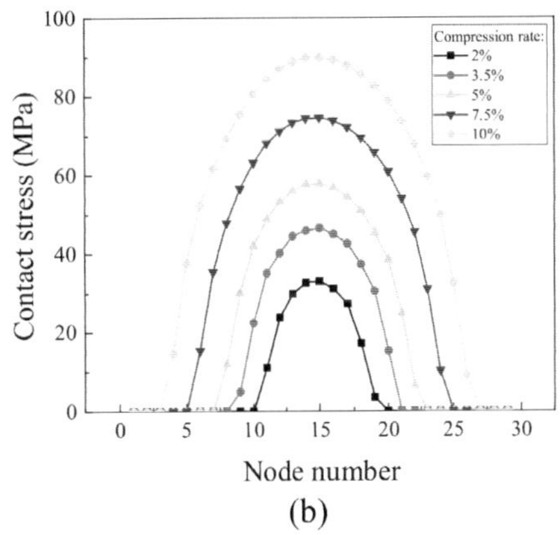

Figure 9. Contact stress values of O-rings with different compression rates. (a) 77 K and (b) 300 K.

3.3. *Influence of cold shrinkage on sealing performance*

PTFE O-ring will shrink when it is cold. If the seal's room temperature compression ratio is small, in the cooling process, the seal will gradually shrink and seal the gap between the surface, losing sealing ability. When the sealing medium is liquid nitrogen, the room temperature compression rate of the seal is 3.5%, and the PTFE seal maintains good sealing performance from room temperature down to low temperature. However, when the sealing medium is liquid hydrogen, the temperature drops to a lower temperature, at which time PTFE seals produce a greater contraction and a gap between the seal and the cover plate, and no stress is generated. Its stress cloud diagram is shown in Figure 10. When the sealing medium is liquid hydrogen, the PTFE seals room temperature compression rate of 3.5% does not meet the sealing conditions. The room temperature compression ratio of the seal should be increased.

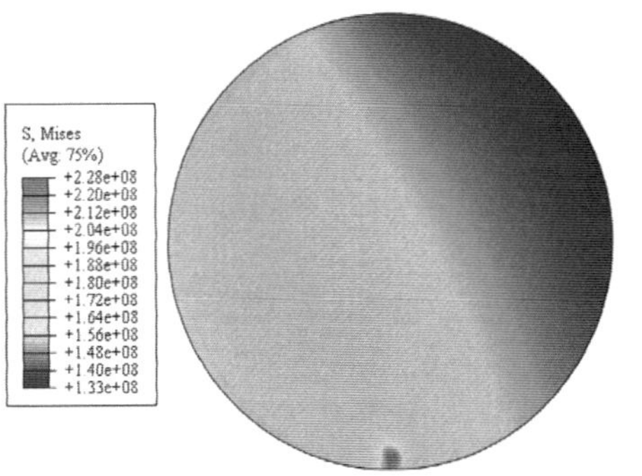

Figure 10. Stress cloud at room temperature compression rate of 3.5% liquid hydrogen temperature.

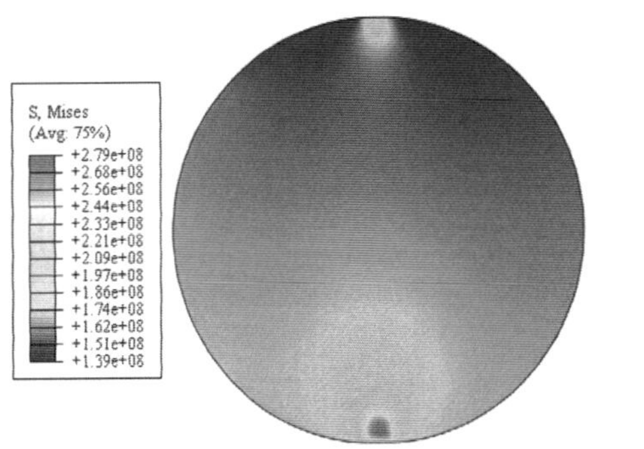

Figure 11. Stress cloud at 5% liquid hydrogen temperature for room temperature compressibility.

When the room temperature compression rate is 5%, the Mises stress value of the O-ring at the liquid hydrogen temperature is between 139 MPa and 279 MPa, and the maximum Mises stress of 279 MPa occurs at the position of the inner diameter of the seal, as shown in Figure 11. In the process of temperature reduction, the inner diameter of the seal is cold shrinkage. The value of the inner diameter cold shrinkage is larger than the value of the sealing groove cold shrinkage, and the seal and the sealing groove are more tightly fitted to each other. The seal generates a larger Mises stress inside the seal and the contact stress between the seal and the sealing groove is larger than the contact stress between the seal and the cover plate. As shown in Figure 12, the maximum contact stress of the O-ring and cover contact is 252 MPa, and the maximum contact stress in contact with the seal seat is 362 MPa, which is much larger than the fluid pressure of the sealing medium, and can ensure the sealing performance.

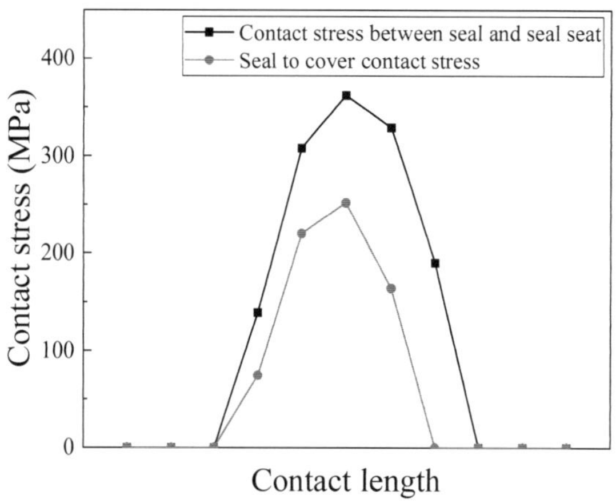

Figure 12. Contact stress distribution of O-ring at liquid hydrogen temperature.

4. Conclusion

Based on the sealing structure of the low-temperature pressure vessel, this paper establishes a two-dimensional axisymmetric model using the finite element analysis software ABAQUS, and researches the sealing performance of the PTFE O-ring in the static sealing system by analyzing the effects of temperature, compression rate, and shrinkage in case of cold on the deformation of the PTFE O-ring, the Mises stress, and the contact pressure. The following conclusions are obtained.

(1) The maximum Mises stress of the O-ring is mainly distributed between the two contact surfaces subject to extrusion, and gradually extends to the inside of the seal as the compression rate increases.

(2) O-ring Mises stress and contact stress gradually decrease with the increase of temperature and increase with the increase of compression rate. The contact stress on the contact surface is parabolic distribution, and the maximum value appears in the center of the contact surface. At room temperature and low temperature, it can maintain 2% of the compression rate, and can be used for sealing at media pressures of less than 25 MPa.

(3) The PTFE O-ring seal will shrink when it is cold. When the sealing medium is liquid nitrogen, its room temperature compression rate should be not less than 3.5%. When the sealing medium is liquid hydrogen, its room temperature compression rate should meet the requirement of not less than 5%.

References

[1] Mahankar P S, Dhoble A S. Review of hydraulic seal failures due to effect of medium to high temperature [J]. Engineering Failure Analysis, 2021, 127: 105552.
[2] Guo L, Su J, Wang Z, et al. Hydrogen safety: An obstacle that must be overcome on the road towards future hydrogen economy [J]. International Journal of Hydrogen Energy, 2024, 51: 1055-1078.

[3] Jiang J, Li X, Su C, et al. The performance of double-layer rubber sealing ring under the action of low temperature environment [J]. Vibroengineering Procedia, 2023, 48: 120-125.
[4] Wang J, Zhang Y. Influence of High and Low Temperature Environment on Sealing Performance of Nitrile Butadiene Rubber O-ring [J]. Lubrication Engineering, 2018, 43(08): 148-152.
[5] Tang Z H. Cryogenic Property of NBR Sealing Materials [J]. Internal Combustion Engine & Parts, 2020(04): 220-221.
[6] Run J, Huang J. Development of Ultra-High-Pressure, High-and Low-Temperature Seals for Cage Throttle Valves [J]. Journal of Failure Analysis and Prevention, 2023, 23(6): 2351-2361.
[7] Feng K, Shi Z Y. Numerical Study on the Reliability of PTFE Hose in Aeroengine [J]. Noise and Vibration Control, 2024, 44 (02): 208-213.
[8] Li L, Song M, Shi Z J. Polarity effect of impact surface flashover of polytetrafluoroethylene in liquid nitrogen [J]. Cryogenics, 2023(06):58-62.
[9] Liu P Z. Development and performance test of ultra-high pressure non-metallic seals [D]. Southwest Petroleum University, 2021. DOI:1 0.27420/d.cnki.gxsyc.2018.000228.
[10] Baccaredda M, Butta E. Transitions in polytetrafluoroethylene [J]. Journal of Polymer Science, 1958, 31(122): 189-191.
[11] Hartwig G. Polymer properties at room and cryogenic temperatures [M]. Springer Science & Business Media, 1995.
[12] McKeen L W, Ebnesajjad S. Handbook of Thermoplastic Fluoropolymers: Properties, Characteristics and Data [M]. William Andrew, 2023.
[13] Ledbetter H M, Frederick N V, Austin M W. Elastic-constant variability in stainless-steel 304 [J]. Journal of Applied Physics, 1980, 51(1): 305-309.
[14] Qu F, Liu L, Tao G, et al. Effect of thermal oxidative ageing on the molecular structure and tribological properties of polytetrafluoroethylene [J]. Tribology International, 2023, 188: 108850.

Anti-icing performance and influencing factors of super-lubricated surface in simulated glaze icing

Bo Li [1], Jie Bai [1], Lusong Zhang [1], Huiying Xiang [2], and Yuan Yuan [2*]

[1] Environmental Protection and Materials Technology Research Center, Institute of Electric Power Science of Guizhou Power Grid Co., Ltd., Guiyang, China
[2] College of Materials Science and Engineering, Chongqing University, Chongqing, China

[*]Corresponding author's email: yuany@cqu.edu.cn

Abstract. The icing of the overhead aluminum conductor has caused the conductor load to be too heavy or uneven, which results in huge accidents. The super-lubricated surface (SLIPS) shows great potential in anti-icing. Glaze ice is a common type of icing, and its damage to transmission lines is the most serious. However, the properties and influencing factors of anti-icing lubricated surfaces at glaze ice surroundings are rarely reported. In this study, the super-lubricated surface was prepared by the anodic oxidation method, and the anti-icing properties and influencing factors of surfaces were investigated in glaze ice surroundings. The results demonstrate that the SLIPS can decrease the icing amount, which is only 41% of the original aluminum surface. SLIPS exhibits excellent anti-icing performance in different environments temperatures and rainfalls. This study provides experimental guidance for the practice application of SLIPS in anti-icing aluminum conductors.

1. Introduction

The icing of overhead aluminum conductors has caused the conductor load to be too heavy or uneven, which leads to accidents such as wire galloping, wire breakage, and tower (rod) collapse. According to relevant statistics, there have been thousands of transmission line accidents caused by icing since 1950. In particular, in 2008, 12 provinces and autonomous regions in southern China experienced more than 20 days of snow coagulation and disastrous weather. Transmission lines and equipment were severely iced, and a wide range of power systems were paralyzed, causing significant social impacts. Super-lubricated surface (SLIPS) is a bionic surface based on the predation effect of Nepenthes, which injects a specific lubricant into the pore structure to produce a smooth liquid layer on the solid surface [1]. The porous structure and stable lubricant layer of the super-lubricated surface make it show great potential in anti-icing.

It was found that the lubricant film of SLIPS significantly prolonged the freezing time of a single water droplet by static freezing experiments of a single water droplet. In addition, the dynamic icing experiment was simulated by the way of supercooled water droplets falling, and the water droplets adhered to the untreated surface to form a large area of ice. On the contrary, supercooled water droplets can quickly slide away from SLIPS, and the icing area and icing weight on the surface are smaller [2-3]. To further approach the real environment, researchers installed nozzles in the climate chamber to produce micro-droplets in the form of spray [4], and then simulated wind tunnel icing. The results show that the SLIPS constructed on the wind turbine blade model can greatly inhibit the accumulation of impact ice. Due to the influence of micro-topography and micro-meteorology, the icing types of aluminum conductors are mainly divided into three categories [5]: the first category is atmospheric icing, including rime (soft rime), mixed rime (hard rime), and glaze (usually related to precipitation). The second category is precipitation icing, including freezing rain and wet snow. This kind of icing is similar to glaze icing. The third type is frosting, that is, water vapor condenses on the cold surface. Among them, glaze icing is the most serious harm to conductors and towers. In this study, the anti-icing properties and influencing factors of SLIPS were investigated under glaze ice surroundings. Firstly, SLIPS with the nanoporous substrate was fabricated on the aluminum surface by the oxidization method. Secondly, the properties of icephobic SLIPS were studied in glaze ice surroundings. Finally, the effect of environmental factors on the properties of SLIPS was explored.

2. Methodology

2.1 Materials

Aluminum plate (50×50×1 mm^3) was purchased from Shandong Dijian Aluminum Co., Ltd. Oxalic acid and phosphoric acid were supplied by Jinan Century Tongda Chemical Co., Ltd. Sodium hydrate and anhydrous ethanol were supplied by Chengdu Cologne Chemical Co., Ltd. Octadecyltrimethoxysilane (OTS) and dimethyl silicon oil were purchased from Macklin Co., Ltd.

2.2 Fabrication

The preparation process of SLIPS includes porous structure construction, low surface energy modification, and lubricant injection. The preparation process of anodic oxidation is mainly divided into cleaning, anodic oxidation, and drying. Firstly, the aluminum plate was alkaline and washed for 2 min. Secondly, the oxidization was conducted on the cleaned aluminum plate, and then anodic aluminum oxide was formed on the surface. The electrolyte of anodic oxidation was oxalic acid (0.3 mol/L). The anodized current was 160 mA/cm^2 and the anodized time was 16 min. The prepared oxide was placed in phosphoric acid (5 wt. %, 1.0 h). Then the AAO was modified by OTS. The oxide was modified in 2 wt. % OTS-ethanol for 0.5 h and the modified oxide was then dried at 110°C (1.0 h). The lubricant was injected into the modified oxide to prepare a super-lubricated surface. First, the modified oxide was placed in lubricant and vacuumed for 12 h; Secondly, the surface was placed vertically. The SLIPS with nanoporous structure was completed.

2.3 Characterization

The micrographs were obtained by Zeiss Auriga scanning electron microscopy (SEM) produced by Zeiss Company. The CA and CAH were obtained by an SDC-300 contact angle measuring instrument, which was purchased from Dongguan Shengding Precision Instrument Co., Ltd.

The glaze ice simulation experiment was conducted in the icing test box, which was mainly composed of a 1×1×1 m^3 chamber, a nozzle, a built-in circulating water freezer, and a control panel. The top nozzle was used to produce spray water drops. When spraying, the humidity in the chamber was about 99%, and the temperature was in the range of -4 ~ -8°C. The icing time was 5.0 h, and the icing macrographs were obtained. The size of the sample was suspended on the shelf of the sample room, tilted 20°. The rainfall range was 40~200 L/h/m^2. The icing weight was calculated by weighting the samples during the glaze ice test.

3. Methodology

3.1 Micrographs

The micrographs and wettability of oxides are demonstrated in Figure 1 (a). A large number of nanopores are evenly distributed on the surface of AAO, and the pore diameter is 120 nm. The cross-sectional micrographs show a large number of nanopore tubes with a high aspect ratio, and the pore depth was 31 μm. The CA and CAH of AAO modified by OTS were 156.3° and 8.0°, respectively.

OTS is a straight-chain alkane with-CH_3, which can functionalize porous surfaces to make them hydrophobic. SLIPS was prepared after lubricant infusion. The SLIPS is covered by lubricant, showing a uniform film. From the cross section, the nanotubes of SLIPS are filled with lubricant, which indicates that the lubricant is completely injected into the pore structure. The CA and CAH of SLIPS are 103.9° and 1.5°, respectively. The CAH of SLIPS is low (< 2.5°), which is consistent with the characteristics of the super-lubricated surface [6]. The results indicate that the oil layer gives the super-lubricated surface excellent water droplet movement performance.

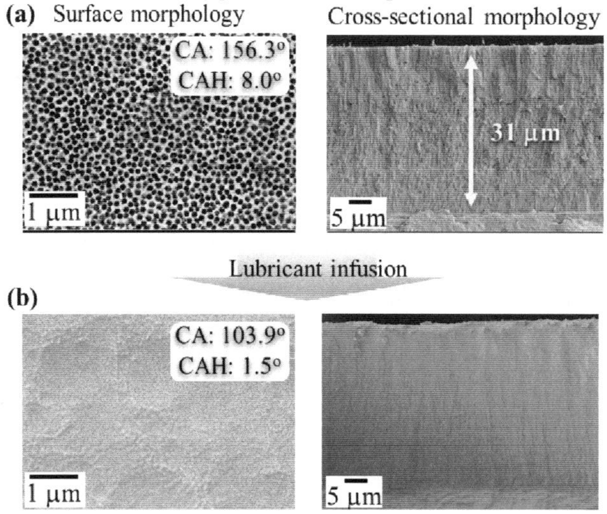

Figure 1. Morphology and wettability of (a) AAO and (b) SLIPS.

3.2 Anti-icing properties in glaze ice surroundings

The icing morphology of SLIPS and untreated aluminum surfaces in simulated glaze icing is shown in Figure 2. The temperature was -8°C. Most of the untreated surface area is covered by a thick icing layer. At 5.0 h, a long icicle appeared on the untreated surface, and the ice weight (W) reached 20.0 g. A small amount of ice particles at 1.0 h were observed on the SLIPS. The formed ice tends to slide down and leave the surface. SLIPS has good water droplet sliding performance. As the time continues to increase, the ice particles staying on the surface gradually grow. The icing weight (W) of SLIPS was 8.3 g at 5.0 h, and this is 41% of the original aluminum. Therefore, SLIPS demonstrates excellent anti-icing properties in simulated glaze ice.

Figure 2. Icing morphology of SLIPS and untreated aluminum surface in glaze ice.

The factors affecting glaze ice mainly include temperature, rainfall, etc. The properties of SLIPS under different influencing factors were investigated here. Figure 3 shows the freezing micrographs and icing amount of SLIPS at different temperatures, and the rainfall was controlled at 40 L/h/m^2. When the temperature was -4°C, discontinuous ice particles were observed on the sample surfaces at 5.0 h, and the ice amount was 4.0 g, which was 27% of the untreated. The icing amount of SLIPS increases with the decrease in temperature. The lower the temperature is, the faster the heat exchange between the water droplets and the outside is. The anti-icing effect of the super-lubricated surface was characterized by the weight ratio of SLIPS to untreated aluminum. Although the weight ratio decreases with the increase in temperature, SLIPS exhibits excellent anti-icing performance at different temperatures.

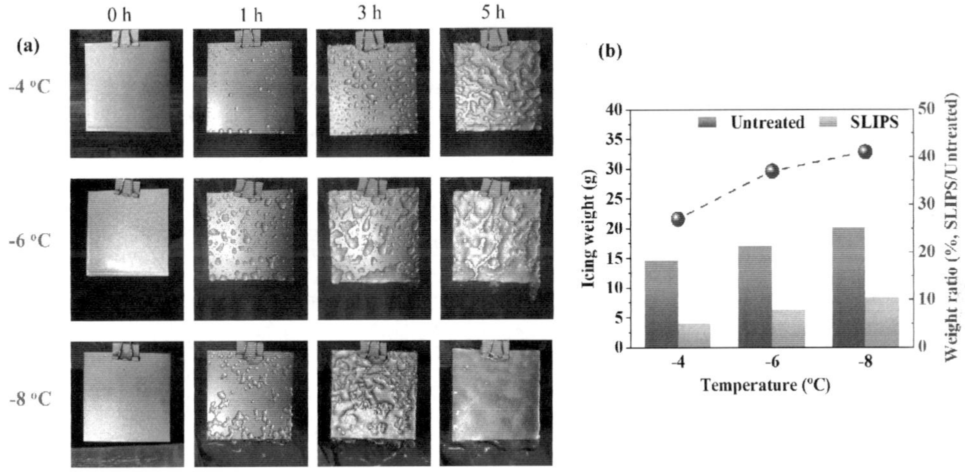

Figure 3. (a) Icing morphology; (b) Icing amount at 5.0 h in simulated glaze ice surroundings under different temperatures.

The icing morphology and weight of SLIPS under different rainfalls are shown in Figure 4, and the temperature is -8°C. With the increase in rainfall, the icing amount of SLIPS increases first and then decreases. When the rainfall increases from 40 L/h/m^2 to 100 L/h/m^2, the icing is aggravated. At 100 L/h/m^2, the ice amount of SLIPS reached the maximum of 11.7 g at 5.0 h, but it was still only 36% of the untreated aluminum. As the rainfall continues to increase, the water droplets on the sample surface are difficult to freeze due to the accelerated update speed, and the amount of ice on the surface of the sample is significantly reduced. When the rainfall was 200 L/h/m^2, there were almost no large ice particles on SLIPS, and the ice weight was only 0.5 g, which was 10% of that of the original surface. The weight ratio decreases with the rainfall, which demonstrates that the SLIPS has better properties under heavy rainfall.

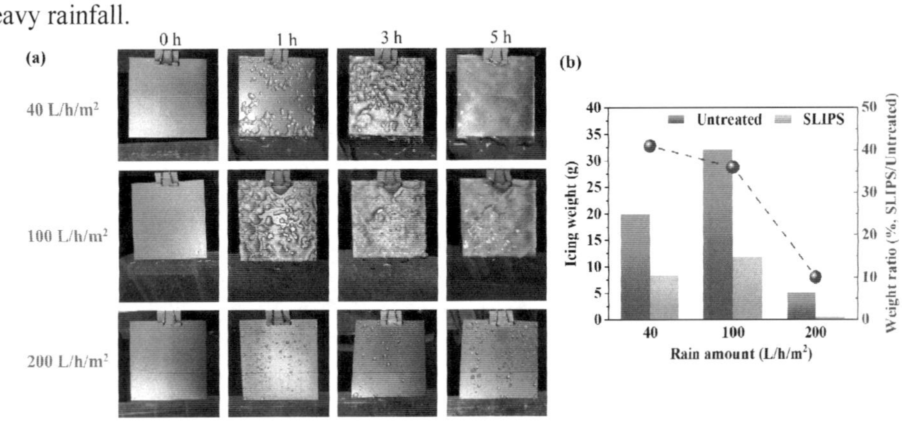

Figure 4. (a) Icing morphology; (b) Icing amount at 5.0 h in simulated glaze ice surroundings under different rainfalls.

4. Conclusion

This work systematically investigated the properties and influencing factors of icephobic lubricated surfaces in simulated glaze ice surroundings. The key conclusions include: (1) SLIPS significantly reduces the accumulation of icing due to its excellent water droplet sliding ability, and the icing amount is 41% of that of the original aluminum; (2) The decrease in temperature accelerates the freezing rate of raindrops, and SLIPS exhibits excellent anti-icing performance at different temperatures; (3) The icing amount of SLIPS increases first and then decreases with the increase of rainfall, and it has better properties under heavy rainfall. This paper demonstrates the excellent properties of anti-icing SLIPS in glaze ice surrounding, and the influence law of temperature and rainfall on the properties of SLIPS was obtained.

Acknowledgment

This research work was funded by the National Natural Science Foundation of China (U23B20133). Additionally, the Electric Power Research Institute of Guizhou Power Grid Co., Ltd., China (Contract number, 066600202203010lHX00001) and Key Laboratory of Ice Prevention & Disaster Reducing of China Southern Power Grid Co., Ltd. (Meihuashan Base) also supported this work.

References

[1] Wong T. S., S. H. Kang, S. K. Tang, E. J. Smythe, B. D. Hatton, A. Grinthal, and J. Aizenberg. Bioinspired self-repairing super-lubricated surfaces with pressure-stable omniphobicity, Nature, 477 (7, 365), 443-447 (2011).

[2] Luo H., S. Yin, S. Huang, F. Chen, Q. Tang, and X. Li. Fabrication of slippery Zn surface with improved water-impellent, condensation, and anti-icing properties, Appl. Surf. Sci., 470 (1, 139-1, 147 (2019).

[3] Pan W., J. Ma, D. Yan, W. Xu, Y. Chen, L. Huang, and J. Song. A facile and highly efficient method to fabricate slippery liquid-infused porous surfaces with enhanced functionality, Surf. Coat. Technol., 472 (129, 897 (2023).

[4] Uranai S., K. Fukudome, H. Mamori, N. Fukushima, and M. Yamamoto. Numerical Simulation of the Anti-Icing Performance of Electric Heaters for Icing on the NACA 0012 Airfoil, Aerospace, 7 (9), 123 (2020).

[5] Parent O. and A. Ilinca. Anti-icing and de-icing techniques for wind turbines: Critical review, Cold Reg. Sci. Technol., 65 (1), 88-96 (2011).

[6] Heydarian S., R. Jafari, and G. Momen. Recent progress in the anti-icing performance of slippery liquid-infused surfaces, Prog. Org. Coat., 151 (106, 096 (2021).

Effect of Ag coatings prepared by magnetron sputtering on the wear properties of Ti6Al4V alloy

Xinyu Huang [1,2], Wenjuan Rong [1,2*] and Jialin Wang [1,2]

[1]Beijing National Innovation Institute of Lightweight. Ltd, Beijing 10140, China
[2]State Key Laboratory of Advanced Forming Technology and Equipment, China Academy of Machinery Science and Technology, Beijing 101400, China

[*]Corresponding author's e-mail: hxy12034@163.com

Abstract. Ag coatings were prepared on the surface of Ti6Al4V alloy by magnetron sputtering technique, and the effects of different Ag target power and sputtering time on the coating properties were investigated. The coating thickness was firstly measured by a step meter, and then the coatings obtained under different parameters were characterized by Vickers hardness tester, scratch tester, and friction wear tester. The variation in the furnace temperature had a significant effect on the hardness of the coatings: the coating thickness increased slightly with increasing temperature, but the increase was not dramatical. The Vickers hardness of the coatings showed a tendency to increase and then decrease with increasing temperature, reaching a peak at 150 °C. The hardness of the coatings was found to increase with increasing temperature and then decrease with increasing temperature. Scratch tests further showed that the coatings prepared in the range of 150 °C to 180 °C exhibited superior bond strength with the substrate. The wear morphology of the coatings showed that the coatings prepared at 150 °C had the best wear resistance. At this temperature, the flaking of the coating due to adhesive wear was significantly reduced, and the depth of scratches due to abrasive wear was also reduced. The variation of the Ag target power also had a significant effect on the hardness of the coatings. As the target power increases, the coating thickness increases significantly, and the Vickers hardness increases and then decreases, reaching a maximum at 12 W. The coatings prepared at 10 W show better performance in terms of scratch and wear patterns. These experimental results provide strong data support for the subsequent preparation of Ti6Al4V joint prostheses.

1. Introduction

Titanium alloy materials have the highest percentage of clinical applications due to their excellent mechanical strength, good machinability, and high-quality biocompatibility [1-4]. Among them, Ti6Al4V alloy occupies about 80% of the medical titanium alloy application market and has been widely used in the field of artificial joint prostheses [5-7].Ti6Al4V alloy plays a vital role in the manufacture of load-bearing components due to its excellent mechanical properties, such as tensile strength, yield strength, and fatigue strength. It shows great potential in the fabrication of artificial joint prostheses. However, the direct implantation of Ti6Al4V alloy joint prosthesis in the human body also has some problems, such as the poor friction and wear performance of Ti6Al4V alloy [8,9]. After the implantation of the prosthesis in the human body, the Al element and the V element of the surface of the Ti6Al4V alloy will continue to be precipitated in the human body's tissue fluids, which will lead to the human body's occurrence of the poisoning of the Al and V elements, resulting in the infection of the medical device, causing the failure of the postoperative period and causing great harm to the patient. This can lead to medical device infections, resulting in postoperative failures and great harm to patients.

The use of surface modification to improve the surface wear properties of titanium alloys has become a research hotspot [10]. For instance, Tian et al. [11] studied titanium alloy substrates with different power (60, 110, 150 W) to prepare Ag-TiN coating. Through the friction and wear test, it was found that titanium alloys prepared for Ag-TiN composite coatings have lower coefficients of friction than uncoated titanium alloys and less wear, where the sputtering power is lower than the uncoated titanium alloys. The composite coating with a sputtering power of 110 W has a lower coefficient of friction and better friction and wear resistance than the other groups of samples.

Liu [12] sputter deposited Ni-Ti coating on the surface of Ti6Al4V to study the wear properties of the coating at different temperatures (25, 120, 260, 400, and 540 °C). The results showed the maximum microhardness of Ni-Ti film at 400°C, the best film base bonding at 260 °C, and the lowest friction coefficient.

Guo [13] investigated the wear mechanism of magnetron-sputtered Ti-Ag-(N) thin films, and the results showed that the friction coefficients of the sample films were all slightly lower than that of the substrate, indicating that the modified layer had little or no friction reduction effect. The wear mechanism is based on abrasive wear and adhesive wear.

2. Experimental equipment

2.1. Ag coating preparation equipment

The equipment uses an AS500MTXB magnetron sputtering coating machine (North China - Damp) to deposit Ag coating. The substrate is Ti6Al4V alloy, and the composition is shown in Table 1. The specification is φ = 20mm, the thickness of r = 2 mm, and the target material is selected as the purity of 4 N (99.99 %) Ag as the sputtering use of planar target material. The rotational speed of the carrier table was 1r/min; the argon flow rate was 240 sccm, and the hydrogen flow rate was 30 sccm; the loading bias voltage was 81 V/0.3 A; and the pressure in the furnace was 0.5Pa.

2.2. Coating characterization equipment

Characterization of coating thickness is reflected by Bruker's DektakXT Step Meter.

The HR-1500T Electric Rockwell hardness tester was used to characterize the Vickers hardness pattern of the coating, using a diamond positive prismatic indenter with a load of 1000 gf, holding the load for 10 S. The substrate was subjected to three puncture tests, and the average value was obtained.

The MFT-4000 multifunctional scratch tester was used to characterize the bond strength between the coating and the substrate, setting a loading rate of 100 N/min, an end load of 100 N, and the total length of the scratches of 8 mm.

The MS-T3100 rotary friction and wear tester, with Al_2O_3 ceramic ball as friction partner, was used to conduct friction and wear tests on the substrate after deposition of the coating with a test time of 30 min, a rotational speed of 200 r/min, and a load of 1000 g. The test was carried out using an Olympus orthotropic metallographic system.

In the Olympus orthogonal metallographic microscope, the surface scratch morphology of the deposited Ag coating was photographed and spliced after the work of a scratch tester, and the scratch morphology was analyzed to get the comparison of the bonding strength between the coating and the substrate. It also aids in observing the morphology of the friction and wear of low magnification of the organization characteristics and analyzing the abrasion mechanism.

3. Discussion of results

3.1. *Influence of coating thickness on wear resistance*

3.1.1 Effect of Furnace Temperature on Coating Thickness and Vickers Hardness. The center part of the substrate was selected for the Vickers hardness hitting test, and three points were hit for each specimen to obtain the Vickers hardness value of the coating at different furnace temperatures. In general, the higher the Vickers hardness of the coating, the better its wear resistance. For magnetron sputtering, the higher the density of Ag coating, the higher the hardness. This is based on Archard's law [14].

$$x = k \cdot H \quad (1)$$

where K is the wear coefficient, L is the sliding stroke, N is the load, and H is the hardness. The wear rate x is inversely proportional to the hardness of the alloy and increasing the hardness of the coating can effectively improve the wear resistance of the coating.

A proper substrate temperature can promote the bonding of the coating with the substrate and reduce the defects, thus increasing the hardness. Table 1 shows the Ag coating thickness and average Vickers hardness of coatings at different furnace temperatures. In order to further analyze the pattern of change of Vickers hardness of coatings at different furnace temperatures, Figure 1 demonstrates the variation between different furnace temperatures and coating thicknesses with respect to Vickers hardness.

Table 1. Ag coating thickness and average Vickers hardness values for different furnace temperatures.

Experimental sequence	Sputtering time/min	Ag target power/W	Furnace temperature /°C	Thicknesses /μm	Average Vickers hardness/HV25gF
A_1	10	10	90	0.764	192
A_2	10	10	120	0.791	203
A_3	10	10	150	0.833	256
A_4	10	10	180	0.925	227
A_5	10	10	210	0.973	213

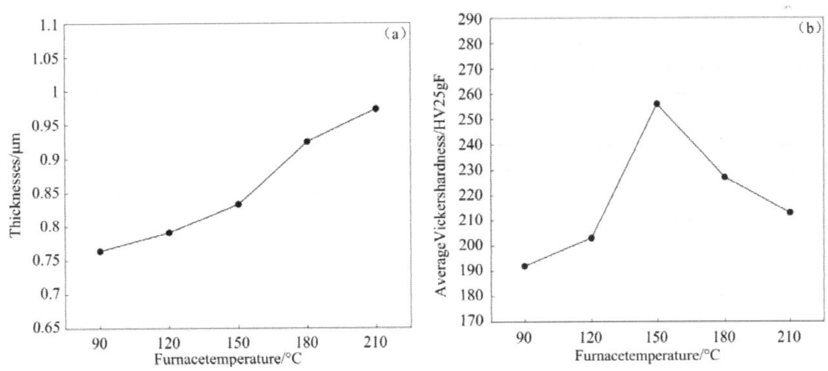

Figure 1. Variation of coating thickness and Vickers hardness with furnace temperature.

As can be seen from Figure 1(a), the thickness of the Ag coating shows a slow increase with increasing temperature, from 0.764 μm at 90 °C to 0.973 μm at 210 °C. There is not much change at 90 °C-120 °C versus 180-210 °C, which is due to the fact that the mobility and diffusion rate of the Ag atoms are lower at lower temperatures (90-120 °C), which results in a slower growth rate of the coatings slower. As the temperature increases, the mobility and diffusion rate of Ag atoms increases, which promotes the growth of the coating. Secondly, the change in Ag coating thickness is affected by the interaction between the coating and the substrate. At high temperatures, the bonding between the coating and the substrate is enhanced, thus favoring the uniform deposition and growth of the coating. However, when the temperature continues to increase (180-210 °C), a different coefficient of thermal expansion between the coating and the substrate occurs, leading to stress or cracks in the coating, thus limiting further growth of the coating.

As can be seen in Figure 1(b), the Vickers hardness showed a tendency of increasing and then decreasing, reaching its highest value when the temperature in the furnace was 150 °C. The corresponding Vickers hardness values show an increasing trend when the temperatures are 90, 120, and 150 °C, and the Vickers hardness of the coating decreases when the temperature increases to 180 and 210 °C. In the lower temperature range (90-120 °C), the coating hardness gradually increases, which is due to the fact that the coating material gradually achieves homogeneous grain growth and densification at relatively low temperatures, thus increasing the hardness of the coating. The Vickers hardness of the coating reaches its highest value when the temperature in the furnace reaches 150 °C. At this temperature, the coating material achieves optimum conditions for crystal growth and phase transformation, resulting in the highest hardness of the coating. However, when the temperature continued to increase to 180 and 210 °C, the Vickers hardness of the coating showed a decreasing trend. This is due to the fact that too high temperatures lead to changes in the growth mode in the coating material, the densification of the coating is reduced, and there are a large number of Ag ion particles that accumulate into clusters, which makes the hardness of the coating decrease. In summary, a temperature of 150 °C can be selected for the deposition of Ag coatings.

3.1.2 Effect of Ag target power on coating thickness and Vickers hardness. Table 2 demonstrates the thickness and average Vickers hardness values of the coatings under different Ag target power. In order to further analyze the change rule of coating Vickers hardness under Ag target power, the coating thickness and Vickers hardness changes under different Ag target power were generated, as shown in Figure 2.

Table 2. Silver coating thickness and average Vickers hardness values for different Ag target powers.

Experimental sequence	Sputtering time/min	Furnace temperature/°C	Ag target power/W	Thicknesses /μm	Average Vickers hardness /HV25gF
B_1	10	150	6	0.373	167
B_2	10	150	8	0.526	230
B_3	10	150	10	0.837	251
B_4	10	150	12	1.563	289
B_5	10	150	14	1.740	263

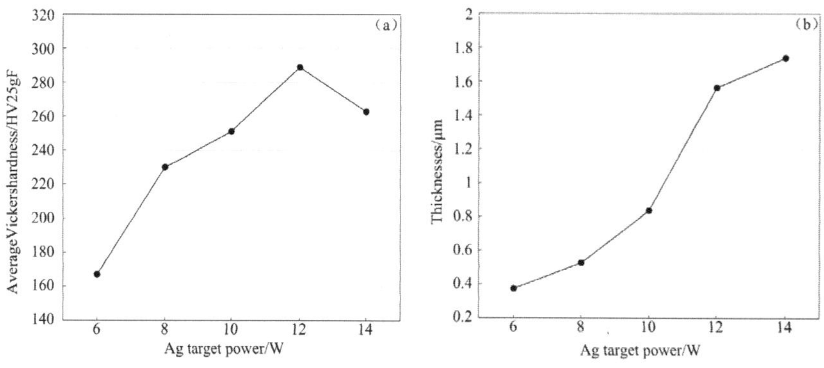

Figure 2. Variation of coating thickness and Vickers hardness with Ag target power.

As can be seen in Figure 2(a), the coating thickness changes significantly with increasing power. The coating thickness is very small at lower power (6 and 8 W), only 0.373 and 0.526 μm, and it increases significantly to 1.563 and 1.740 μm as the power is increased to 12 and 14 W. Firstly, higher power means that more energy is transferred to the coating material, which promotes the Ag ionization of the target, thus increasing the number of particles deposited onto the substrate surface. Secondly, the higher power resulted in the particles acquiring higher kinetic energy in flight, which contributed to the formation of a denser and thicker coating on the substrate. When the power was further increased to 14W, the coating thickness continued to increase, but the rate of increase seemed to slow down. This is because, with the increase in power, more collisions and scattering may occur during the flight of the particles, resulting in some of the particles failing to reach the surface of the substrate.

As seen in Figure 2(b), the Vickers hardness shows a trend of increasing and then decreasing with the increase of Ag target power, reaching a maximum of 12W. The Vickers hardness of the coating increases when the Ag target power is 6, 8, 10, or 12 W, and it decreases when the Ag target power is 14 W. The Vickers hardness value of the coating decreases with the increase of Ag target power. As the target power increases, the energy of sputtered atoms also increases. The sputtered atoms with high energy are able to interact more effectively with the substrate surface atoms when deposited onto the substrate surface, forming a denser coating structure. This dense coating structure can reduce defects and holes in the coating and improve the hardness and wear resistance of the coating. However, when the Ag target power was further increased to 14 W, the Vickers hardness of the coating showed a decreasing trend. This is due to the fact that the excessively high Ag target power leads to the secondary etching phenomenon of the sputtered Ag ions, resulting in an increase in the grain size and a decrease in the number of grain boundaries in the coating, which reduces the hardness of the coating.

3.2. Characterization of coating-to-substrate bond strength

3.2.1 Effect of Furnace Temperature on Coating-Substrate Bonding Strength. We combined microscope observation and scratch morphology diagram, according to GB/T 30707-2014 scratch morphology based on the comparison of the effect of different furnace temperature coating bonding strength. The main judgment is the pitting rupture perpendicular to the direction of the scratch (the position of the bold arrow in the figure) and the Ag coating stacking phenomenon at the end of the scratch (the thin arrow in the figure), as shown in Figure 3.

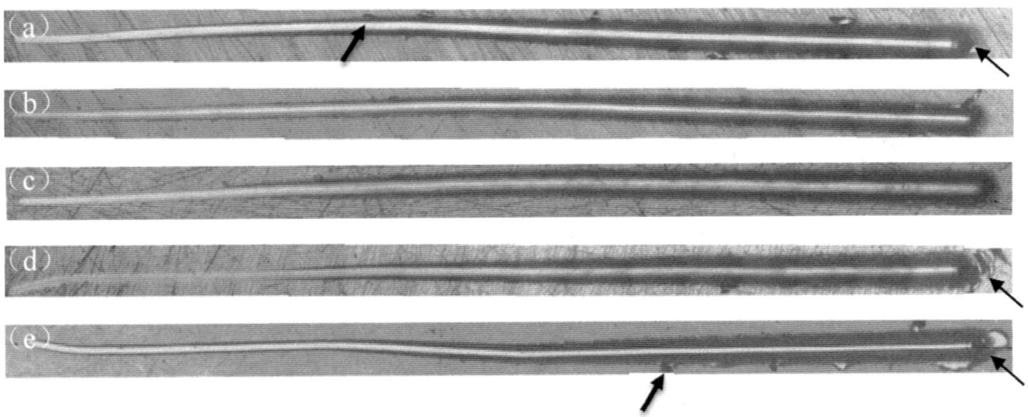

Figure 3. Scratch morphology of coatings at different furnace temperatures.

Different furnace temperatures have little effect on the coating and the substrate, and there is pitting rupture perpendicular to the direction of the scratch in Figs. 4 (a) and (e), and Ag coating stacking in (a), (d), and (e). The coating-substrate bonding is better at furnace temperatures of 120-150 °C.

3.2.2 Effect of Ag target power on the bond strength between coating and substrate. Figure 4 shows the morphology of Ag coating scratches under different Ag target powers. Different Ag target powers have a significant effect on the coating morphology, and when the power is lower (6 and 8 W), there is a rupture at the initial end of the scratch, and the bonding strength between the coating and the substrate is poor. This is due to the fact that at lower power, only fewer Ag atoms are uniformly distributed on the substrate surface, resulting in a weaker bond between the coating and the substrate. As the power of the Ag target increases, the energy and kinetic energy of the Ag ions increase accordingly, resulting in a uniform distribution of Ag ions on the surface of the substrate. As a result, the bonding strength between the coating and the substrate is also enhanced. When the power of the Ag target is increased to

14 W, the initial section is again ruptured. The excessive energy of the sputtered Ag ions at high power leads to excessive stress in the formation of the coating and secondary etching. When the coating was cooled, this led to cracks and defects within the coating, which reduced the bond strength of the coating to the substrate.

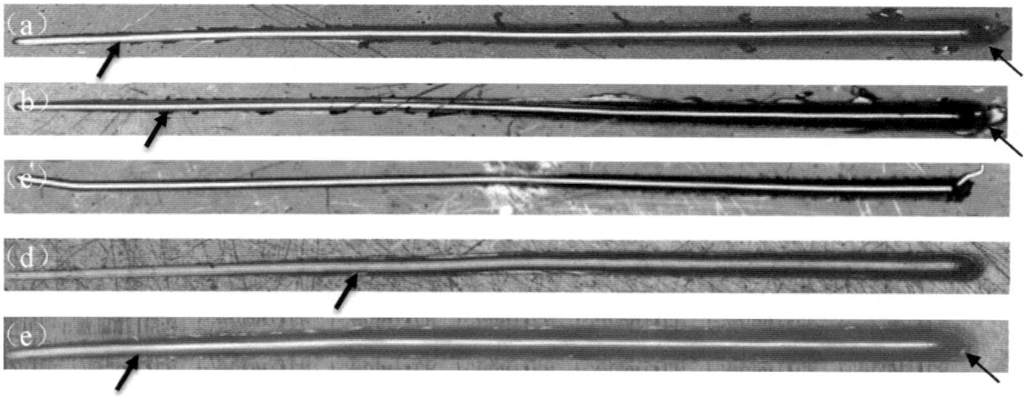

Figure 4. Scratch morphology of coatings at different Ag target power.

3.3. Morphological analysis of low-fold wear of Ag coatings after abrasion

3.3.1 Wear patterns of Ag coatings at different furnace temperatures. The power of the Ag target was set at 10 W and sputtering time at 10 min to study the change of wear morphology of the coating low-fold organization under the furnace temperatures of 90, 120, 150, 180, and 210 °C, and the wear morphology of the low-fold center of the Ag coating at 500 times is shown in Figure 5:

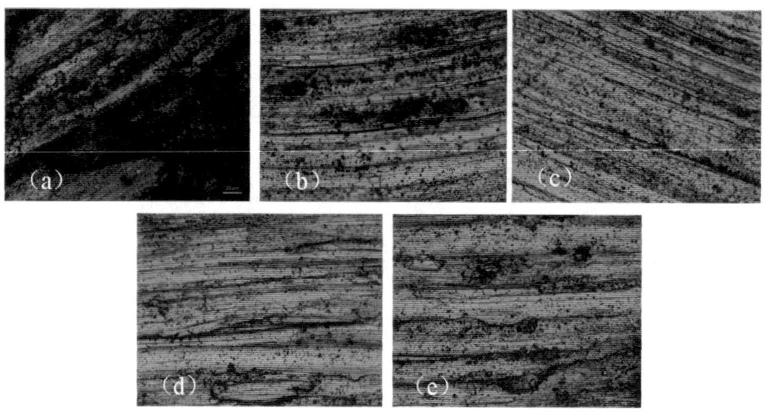

Figure 5. Wear morphology of Ag-coated low-profile centers at different furnace temperatures (a) 90 °C; (b) 120 °C; (c) 150 °C; (d) 180 °C; (e) 210 °C.

The wear mechanism of the coating is adhesive wear, and at low temperatures of 90 °C and 120°C, the growth rate of the coating grains is relatively slow, which leads to irregularities in the morphology of the coating surface, accompanied by a large number of coating ruptures. This is due to the increased brittleness of the coating material at low temperatures, making the coating prone to fracture when subjected to external forces. With the gradual increase in temperature to 150 °C, the grain growth rate in the coating gradually accelerates, the surface morphology gradually becomes regular, and the rupture phenomenon gradually disappears. At the same time, the holes and cracks in the coating will also be reduced with the increase in temperature, and the high temperature will help the release of internal stress and repair of defects in the coating. When the temperature continues to rise to 180-210 °C, the surface of the coating will again appear as large particles of holes. This is because the high temperature causes the coating material to become soft, and the hardness is reduced, making the coating easy to deform and plastic flow when subjected to external forces. This plastic flow causes the granular material on the surface of the coating to be extruded, resulting in the formation of large-grained holes.

The low magnification edge wear morphology of the Ag coating at 500× is shown in Figure 6:

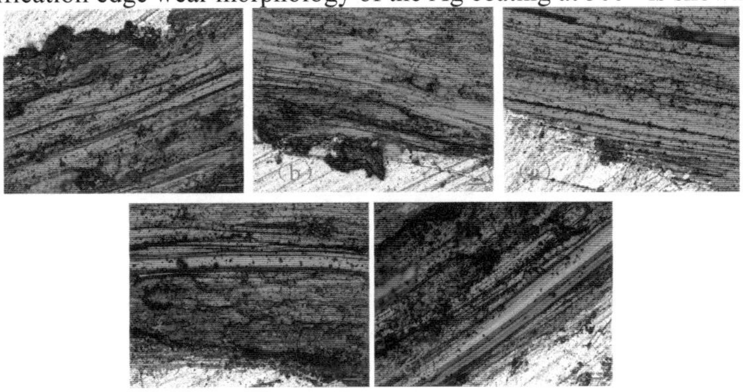

Figure 6. Low times edge wear morphology of Ag coatings at different furnace temperatures (a) 90 °C (b) 120 °C (c) 150 °C (d) 180 °C (e) 210 °C.

3.3.2 Wear patterns of Ag coatings at different Ag target power. The wear morphology of the coating edges changed significantly with the change in temperature in the furnace. At 150 °C, no significant width-wise wear scratches appeared on the coating edges, indicating that the coating has good wear resistance at low temperatures. However, as the furnace temperature increased, the wear of the coating edges gradually intensified. When the temperature in the furnace increased to 180 °C and 210 °C, the coating edge showed wear scratches with larger widths and deeper depths. These patterns are typical of abrasive wear patterns. The change in morphology indicates that the coating material is subjected to more severe abrasion at high temperatures, and the substrate material is also worn away by the friction side.

Setting the furnace temperature at 150°C and the sputtering time at 10 minutes, we investigate the Ag target powers of 6, 8, 10, 12, and 14 W under the coating low times to observe changes in the wear morphology. The edge wear morphology after 500 cycles of Ag coating low times is shown in Figure 7:

Figure 7. Low times center wear morphology of Ag coating at different Ag target power (a) 6W (b) 8W (c) 10W (d) 12W (e) 14W.

The effect of Ag target power on the wear morphology of the coating is significant. When the target power is 6 W, the surface of the coating shows wide abrasion marks and relatively heavy wear, resulting in increased plastic damage on the surface of the coating. As the target power was gradually increased to 8 W, 10 W, and 12 W, the width of the abrasion marks on the coating gradually decreased, and the degree of abrasion was obviously reduced, which showed the gradual improvement of the wear resistance of the coating. When the power of the Ag target was increased to 14 W, the coating surface began to show a flaking phenomenon, and the wear performance was obviously weakened, which was due to the uneven distribution of stress inside the coating caused by too high power.

The low magnification edge wear morphology of the Ag coating at 500× is shown in Figure 8:

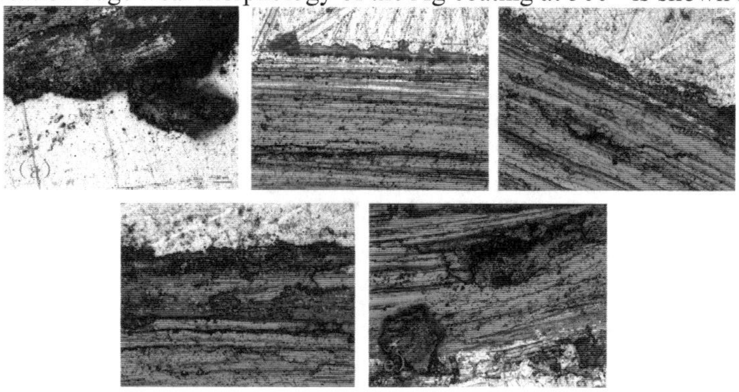

Figure 8. Low times edge wear morphology of Ag coating at different Ag target power (a) 6W (b) 8W (c) 10W (d) 12W (e) 14W.

When the Ag target power was set to 6 W, 8 W, 12 W, and 14 W, respectively, significant flaking of the coating edges was observed, resulting in the exposure of the substrate layer. At 12 W and 14 W power conditions, the width and depth of the wear scratches at the coating edges are increased, indicating that the coating has suffered severe damage. The appearance of this wear morphology reduces the coating lifetime. When the power of the Ag target was set at 10 W, the wear pattern of the coating edges performed relatively well. At 12 W, the hardness, structure, and wear resistance of the coating reached a relatively balanced state, thus effectively resisting the action of wear forces.

4. Conclusion

In this experiment, Ag coatings with different wear-resistant properties were prepared by magnetron sputtering with deposition time of 10 min and pressure of 0.5 Pa, furnace temperatures of 90, 120, 150, 180, 210 °C, and Ag target power of 6, 8, 10, 12, 14 W. The conclusions are as follows:

(1) The furnace temperature is an essential factor affecting the wear-resistant properties of the coating, and the coating thickness increases slowly with the increase of the furnace temperature. The Vickers hardness value of the coating is the largest at the furnace temperature of 150 °C, which is 256HV25gF. At this temperature, the obtained coatings exhibit improved bonding morphology with the substrate, better wear morphology between the coating edges and the center, reduced rupture of the coating morphology at the center, and decreased presence of pores and cracks in the coatings as the temperature increases.

(2) The power of the Ag target has a significant effect on the coating performance, and the coating thickness increases significantly with the increase of power. When the power of the Ag target is set to 12 W, the Vickers hardness value of the coating is the largest, which is 289HV25gF. The scratch morphology and wear morphology of the coating are better when the power of the Ag target is 10 W, the width of the abrasion mark of the coating is gradually reduced, and the degree of wear is also reduced significantly.

Acknowledgments

This work was financially supported by Beifang Huachuang-Beijing Danpu Surface Technology Co., Ltd. and the China General Research Institute of Mechanical Sciences (CGIMS), for which we are grateful.

References

[1] Han Sheng, Wang Yuan, Lin Hualin, et al. Research progress on Surface modification and Biotribological Properties of medical titanium alloys [J]. Chinese Journal of Applied Technology, 2019, 24(01): 35-42.
[2] YAVARISA, STOK JVD, CHAI Y C, et al. Bone re-generation performance of surface treated porous titani-um[J]. Biomaterials, 2014, 35(24): 6172-6181.
[3] FAN Jingyi, Ma Xun, Li Wei, et al. Research progress of surface modification technology of medical titanium alloys [J]. Journal of Functional Materials, 2022, 53(07): 7027-7039.
[4] MCCARTY C P, PARKSH, HON C, et al. Taper material loss in total hip replacements: is it affected by jointfriction[J]. The Journal of Bone Joint Surgery, 2022, 104(9): 796-804.
[5] SHIW, DONG H. Improvement in the tribological properties of UHMWPE sliding against Ti-6Al-4V by surfacemodification[J]. Journal of Shanghai University, 2005(9): 164-171.
[6] PAN Y, XIONG D, MA R. A study on the friction properties of poly(vinyl alcohol) hydrogel as articular cartilage against titanium alloy [J]. Wear, 2007, 262(7-8): 1021-1025.
[7] Wei F R, Wang H, Jin X D, et al. Biomedical titanium alloy materials and their applications[J]. World Nonferrous Metals.2018, (2): 260-262.
[8] Sousa V F C, Silva FJG, Alexandre R, et al. Study of the wear behavior of TiAlSiN and TiAlN PVD coated tools on milling operations of pre-hardened tool steel[J]. Wear, 2021, 476(1): 203695.
[9] Sd A, Sg B, Rg A, et al. A comparative analysis of different properties of TiN, TiAlN and TiAlSiN thin film coatings grown in nitrogen gas atmosphere-Science Di-rect[J]. Materials Chemistry and Physics, 2020, 258: 123866.
[10] Yu S, YuZT, Niu JL, et al. Research progress of surface/interface interaction between titanium alloy implants and human tissue[J]. Hot Working Technology, 2017, 46(22): 18-22.
[11] Tian Qing-Qing, XU Jin-yong, Li Tong, et al. Effect of sputtering power on microstructure and wear resistance of Ag-TiN coating [J]. Journal of Guilin university of electronic science and technology, 2022, and (3): 194-198.
[12] LIU Shuai. Study on mechanical properties of Ni-Ti films deposited by Ti-6A1-4V surface sputtering [D]. Shandong University, 2019.
[13] Guo Yangyang. Preparation and properties of magnetron sputtering Ti-Ag-(N) films [D]. Taiyuan University of Technology, 2015.
[14] Wu L L. Dry sliding tribological behavior of Zr-based bulk metallic glass[J]. Transactions of Nonferrous Metals Society of China, 2012, 22(3): 585-589.

Impact of adding Na_2SiF_6 on the crystal phase and copper valence state in glass ceramics made from leftover granite for use as architectural ornamentation

Minghong Ma and Yongqian Huang *

School of Materials Science & Engineering, Sichuan University, Chengdu 610065, China

* Corresponding author's e-mail: huangyq@scu.edu.cn

Abstract: The accumulation of granite waste can result in the occupation of a considerable amount of land resources. Using granite waste as glass ceramics for architectural decoration represents a potential solution to this issue. The microstructure, valence state, and crystalline phase of copper ions were examined by using X-ray photoelectron spectroscopy (XPS), scanning electron microscopy (SEM), differential scanning calorimetry (DSC), and X-ray diffraction (XRD). The findings demonstrated that the concentration of Na_2SiF_6 affects the valence condition of Cu ions in crystallized glass. The crystalline phases of the sample containing 1.84 wt% Na_2SiF_6 consisted of forsterite ($MgSiO_3$) and diopside ($CaMgSi_2O_6$), and the Cu^+ accounted for 39.9% of the total Cu ions in the sample which showed a grey-black color. For the sample containing 3.67 wt% Na_2SiF_6, the crystalline phases consisted of richterite (Na, F) syn (Na (Na, Ca) $Mg_5Si_8O_{22}F_2$), cuprite (Cu_2O), and forsterite, its Cu^+ accounted for 39.9%, and the sample appeared orange-red. With the increase of Na_2SiF_6 content, the percentage of Cu^+ in the total Cu ions showed an increasing trend, and the corresponding red color of the samples gradually deepened.

1. Introduction

Large quantities of waste were generated during the granite mining process. The accumulation of granite waste not only represents a land resource utilization issue but also poses an environmental contamination risk. Granite wastes are primarily composed of SiO_2 and Al_2O_3, which act as glass network formers. Additionally, they contain a small amount of oxides, including Na_2O, K_2O, CaO, and MgO. These oxides are a crucial part of glass ceramics. Therefore, granite wastes can be used as a raw material for the preparation of glass ceramics.

Glass ceramics are a type of material containing both crystalline and glass phases prepared by heat treatment of glass with a specific chemical composition, which has the characteristics of both glass and ceramics. Glass ceramics are widely utilized in a range of industries, including aerospace, electronics and communications, biomedicine, and architectural ornamentation, because of their many advantages, which include outstanding mechanical qualities, great resistance to acid and alkali corrosion, and good thermal stability.

Red is a fascinating color. Red ceramic glazes and glass ceramics for decorative materials can be prepared by the method of adding coloring agents. Among them, the red coloring agents include metal monomers or oxides such as gold, tin, cadmium, and copper. Copper red is an economical and environmentally friendly red color, so many researchers have used copper oxide as a coloring agent to prepare red decorative materials [1].

In our experimental studies, it was discovered that the Na_2SiF_6 significantly affected the hue of glass ceramics made from granite refuse that included copper oxide. According to the mechanism of copper coloring, Cu^+ or Cu^0 makes glass ceramics red. The higher the content of Cu^+ or Cu^0 glass ceramics is, the redder the glass ceramics are. Therefore, it is important to increase the proportion of Cu^+ or Cu^0 in the total copper ions for the preparation of red glass ceramics with the same amount of CuO. In this paper, it is firstly reported that the proportion of Cu^+ or Cu^0 in the total copper ions was regulated by adding Na_2SiF_6, and the red glass ceramics were successfully prepared by using granite waste as the main raw material.

2. Materials and methods

The granite waste's chemical components are shown in Table 1.

Table 1. Chemical composition of the granite waste (wt%).

Composition	SiO_2	Al_2O_3	K_2O	Na_2O	CaO	MgO	P_2O_5	TiO_2	Fe_2O_3	others
wt%	76.61	11.96	4.49	3.67	0.20	0.12	0.05	0.06	1.05	2.844

In accordance with Table 2, $CaCO_3$, $Mg_2(OH)_2CO_3 \cdot 5H_2O$, Na_2SiF_6, $CaHPO_4 \cdot 2H_2O$, TiO_2, and CuO were incorporated into the stone wastes and well blended. After melting at 1, 450°C for 2 h, the mixture was cooled to 1, 400°C and held for half an hour. An hour was spent in annealing the glass melting in a warmed steel frame at 600°C. The samples with 1.84 wt%, 3.67 wt%, 5.51 wt%, and 7.35 wt% Na_2SiF_6 were named A_1, A_2, A_3, and A_4 respectively.

Table 2. The chemical composition of basic glass samples (wt%).

Samples	SiO_2	Al_2O_3	MgO	CaO	K_2O	Na_2O	P_2O_5	TiO_2	Na_2SiF_6	CuO
A_1	45.74	7.14	17.82	6.23	2.68	2.20	2.17	4.31	1.84	7.35
A_2	45.74	7.14	17.82	6.23	2.68	2.20	1.56	3.10	3.67	7.35
A_3	45.74	7.14	17.82	6.23	2.68	2.20	0.95	1.88	5.51	7.35
A_4	45.74	7.14	17.82	6.23	2.68	2.20	0.33	0.66	7.35	7.35

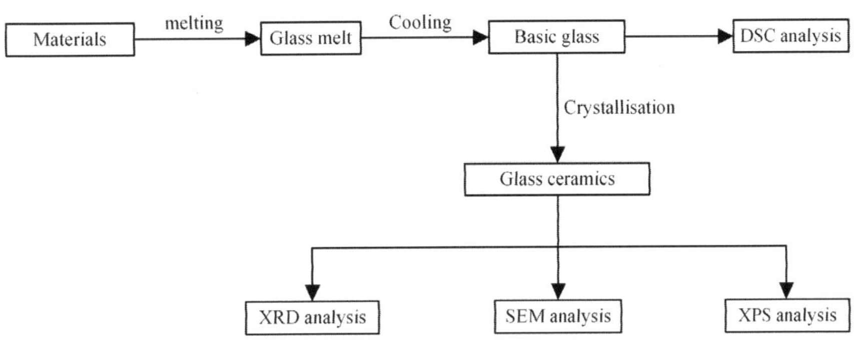

Figure 1. The experience process.

Phase evolution was detected in the sample powders by evaluation utilizing DSC (NETZSCH STA 449C, Germany). Samples were heated from 40°C to 1, 200°C at a rate of 10°C per minute. In the DSC experiences, alumina was the reference material. XRD (XRD-6100, Japan) was utilized to identify the materials' crystalline phases. SEM (Apreo S, USA) was used to investigate the samples to look at their microstructure. After fifty seconds of etching with three-volume hydrofluoric acid (HF) water, the samples were cleaned with distilled water. After drying at room temperature, the etched samples were coated with gold. The surface elements of the samples were studied qualitatively and quantitatively by XPS (AXIS Supra, Britain), and the data was assessed by using CasaXPS (Casa Software). The experience process is shown in Figure 1.

3. Results and discussion

3.1 Thermal analysis

Figure 2 shows the DSC information. The two producing heat peaks and one needing heat peak represented the temperature at which crystals formed (Tp1, Tp2) and the temperature at which glass transitioned (Tg). The temperature range for the sample nucleation was above Tg but below Tg + 50°C. The DSC curve and the outcomes of earlier tests indicated that the basic glass nucleated at 640°C for 120 minutes and subsequently crystallized at 986.6°C for 120 minutes.

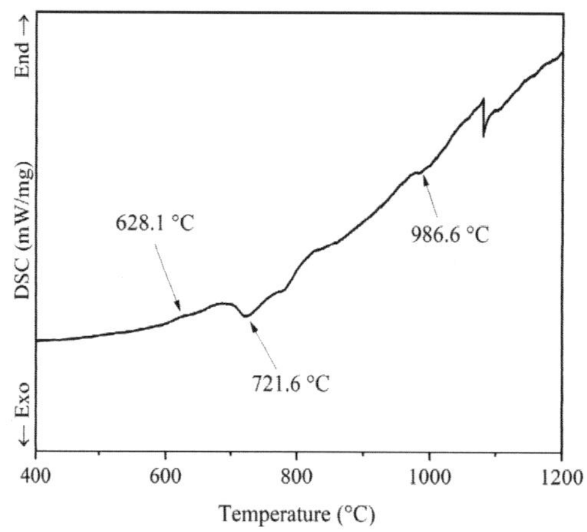

Figure 2. The DSC curve of basic glass.

3.2 XRD analysis

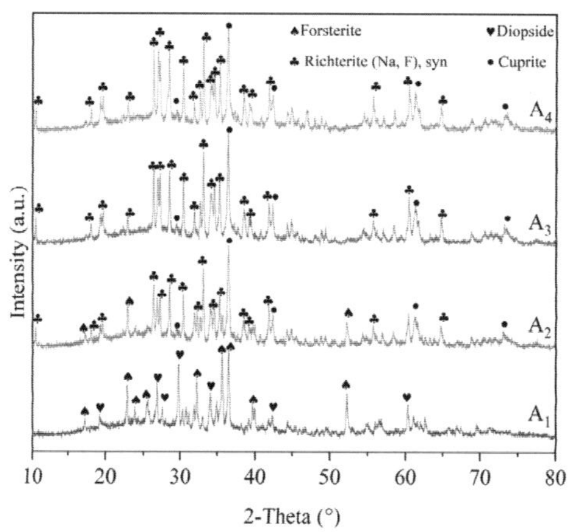

Figure 3. The XRD pattern of samples with different Na_2SiF_6 content.

Figure 3 displays the crystal forms in the samples determined by XRD. There were forsterite ($MgSiO_3$, JCPDS 71-1080) and diopside ($CaMgSi_2O_6$, JCPDS 78-1390) in the A_1 sample. The crystalline phase of the A_2 sample differed from that of the A_1 sample. In addition to forsterite, it also contained richterite (Na, F) syn (Na (Na, Ca) $Mg_5Si_8O_{22}F_2$), JCPDS 75-1666, and cuprite (Cu_2O, JCPDS 78-

2076). The forsterite diffraction peak of the A_2 sample was weaker than that in the A_1 sample, which suggested that the A_2 sample contained less forsterite than the A_1 sample. Both the A_3 and A_4 samples have Na (Na, Ca) $Mg_5Si_8O_{22}F_2$ and Cu_2O as their crystalline phases. Na (Na, Ca) $Mg_5Si_8O_{22}F_2$ diffraction peaks in the A_3 sample were more intense than in the A_2 sample, suggesting that more Na (Na, Ca) $Mg_5Si_8O_{22}F_2$ precipitated from the A_3 sample. The XRD pattern of the A_4 sample did not show any significant change compared to the A_3 sample.

3.3 Microstructure

Figure 4 displays typical SEM pictures of every sample's Na_2SiF_6 composition. Granular forsterite crystals and short columnar diopside crystals were distributed in the matrix of the A_1 sample [2]. Granular Cu_2O and lamellar crystals of Na (Na, Ca) $Mg_5Si_8O_{22}F_2$ were seen in the A_2 sample [3]. The A_3 sample contained more Na (Na, Ca) $Mg_5Si_8O_{22}F_2$ than the A_2 sample did. The Na (Na, Ca) $Mg_5Si_8O_{22}F_2$ found within the A_4 sample was bigger than in the A_3 sample, and both samples shared the same crystalline phase. Since fluorine is the important constituent of richterite (Na, F) syn, an increase in fluorine content can help richterite (Na, F) syn precipitation.

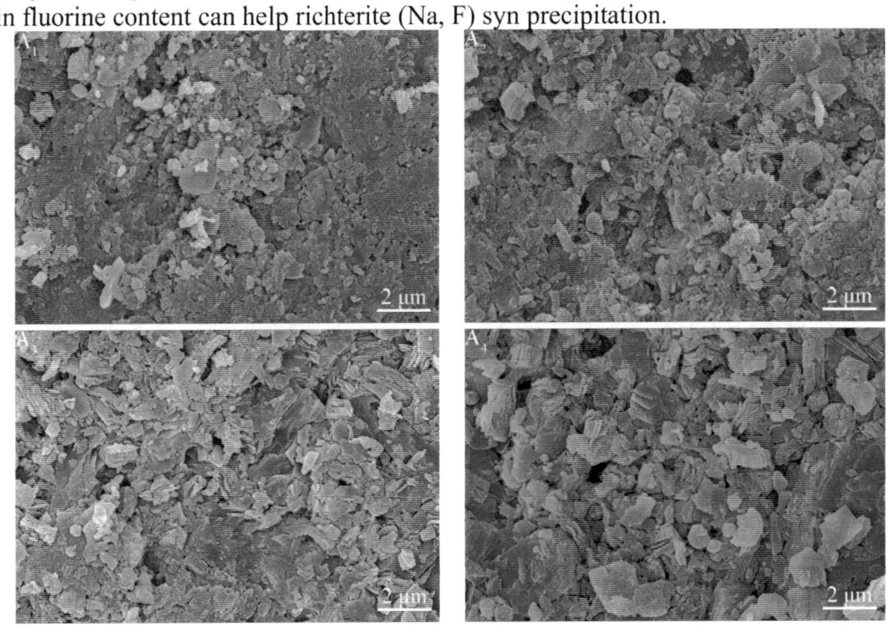

Figure 4. The typical SEM pictures of every sample's Na_2SiF_6 composition.

3.4 XPS analysis

Glass-ceramic materials have three distinct types of copper ions: Cu^0, Cu^+, and Cu^{2+}. According to the mechanism of copper coloring, Cu^+ or Cu^0 makes glass ceramics red. The higher the content of Cu^+ or Cu^0 glass ceramics is, the redder the glass ceramics are [4]. In general, the origin of Cu^+ can be explained by the reduction of CuO in fluoride ion-rich melts.

$$2Cu^{2+}+2F^- \leftrightarrow 2Cu^+ + F_2 \quad (1)$$

$$Cu^+ + F^- \leftrightarrow Cu^0 + 0.5F_2 \quad (2)$$

Cu_2O is a kind of colloid produced by Cu^+ aggregation in matrixes, as shown in Equation (3).

$$2nCu^+ + 0.5nO_2 \rightarrow (Cu_2O)_n \quad (3)$$

The surface characteristics of samples containing different concentrations of Na_2SiF_6 are shown in Figure 5. The A_1, A_2, A_3, and A_4 samples all displayed different colors.

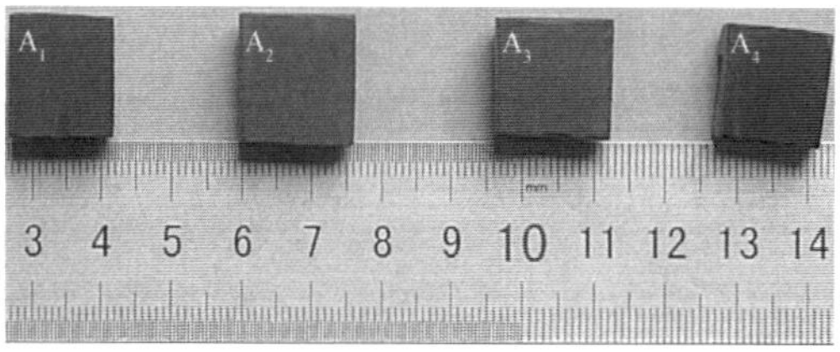

Figure 5. The surface color of the samples.

The exterior chemical makeup and electrical condition of the components were analyzed by using XPS. In Figure 6, we demonstrated that the binding energies of Cu $^2P_{3/2}$ and Cu $^2P_{1/2}$ in the samples were 932 eV and 953 eV, respectively.

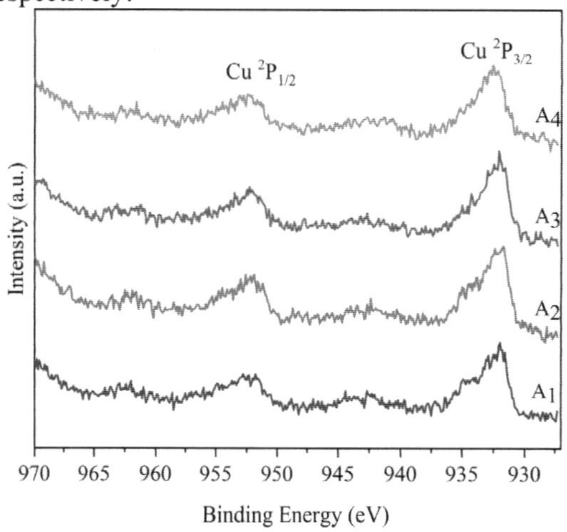

Figure 6. The XPS spectra of Cu $^2P_{3/2}$ and Cu $^2P_{1/2}$ of samples.

To investigate the valence changes of copper ions in glass ceramics during the change of F content, the satellite peaks of Cu $^2P_{3/2}$ of samples with different F contents were split and fitted. As shown in Figures 7 (a)-7 (d), Cu $^2P_{3/2}$ has two characteristic peaks at 932 ev and 933 ev, which correspond to the characteristic peaks of Cu^+ and Cu^{2+}, respectively [5-6]. The area of the characteristic peaks represents the percentage of the corresponding copper ions in the total copper ions. Equations (4) and (5) were used to compute the specific content of Cu^+ in the samples [7]:

$$\text{Cu sum number=CuO number} \times MR_{Cu}/MR_{CuO} \quad (4)$$

$$Cu^+ \text{ number=Cu sum number} \times A_{Cu^+} \quad (5)$$

where Cu's atomic weight in relative terms is denoted by MR_{Cu}, CuO's molecular weight in relative terms is denoted by MR_{CuO}, and the percentage of the Cu^+ number to the overall Cu number, which includes Cu^{2+}, Cu^+, and Cu^0, is known as $A_{Cu}+$.

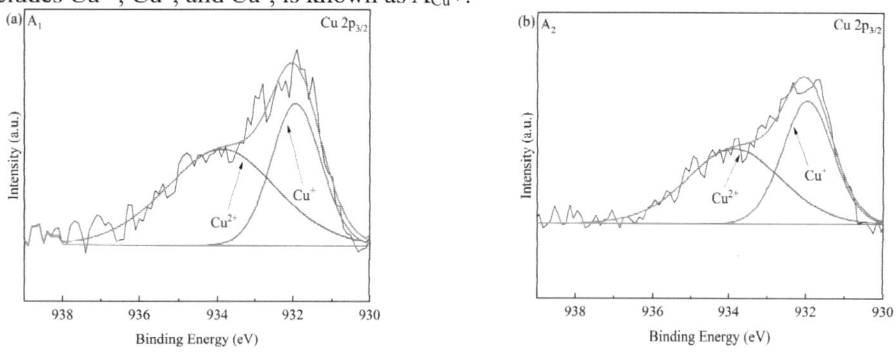

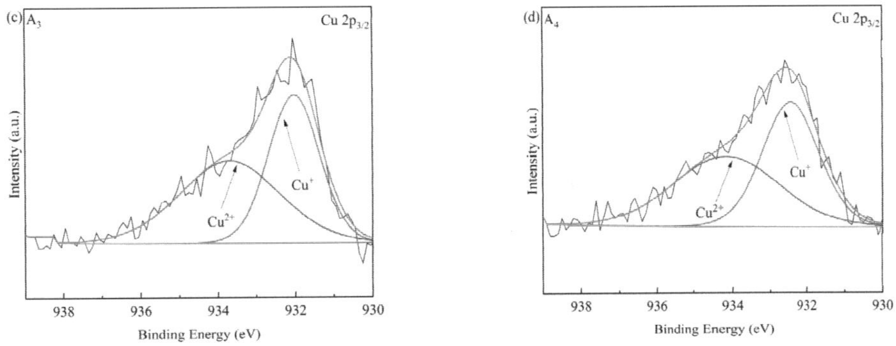

Figure 7. (a-d) The characteristic peaks in satellite peaks of A_1-A_4 samples.

Table 3 presents the XPS data. The A_{Cu}^{+} of the A_1, A_2, A_3, and A_4 were 39.9 %, 46.5%, 47.8%, and 51.4% respectively. The A_1 sample crystallized contained 2.4 wt% Cu^+, having been grey-black. The A_2 sample contained 2.7 wt% Cu^+, having been red-orange color. The A_3 sample contained 2.8 wt% Cu^+, having been red-orange color. A_4 contained 3.0 wt% Cu^+, having appeared red. The primary cause of the hue shift in the samples containing CuO derived from granite wastes is the alteration in the Cu^+ concentration of the samples during crystallization. The Na_2SiF_6 content affected the content of Cu_2O in the samples, which made the samples show different colors.

Table 3. The data of XPS for glass ceramics varying in the concentration of Na_2SiF_6.

Samples	The binding energy of Cu $^2P_{3/2}$ (eV)		FWHM of Cu $^2P_{3/2}$ (eV)		Cu total content (wt%)	Cu^+ ratio (%)	Cu^+ content (wt%)
	Cu^+	Cu^{2+}	Cu^+	Cu^{2+}			
A_1	931.9	933.9	1.3	3.4	5.9	39.9	2.4
A_2	932.0	933.9	1.5	2.9	5.9	46.5	2.7
A_3	932.0	933.7	1.5	3.0	5.9	47.8	2.8
A_4	932.4	934.2	1.8	3.3	5.9	51.4	3.0

4. Conclusion

This study looked at how the valence condition of copper ions in glass ceramics made from granite waste, which was the primary raw material, was affected by the Na_2SiF_6 content. The findings were as follows:

(1) According to the XRD findings, the A_1 sample's crystal phases, which contained 1.84 weight percent Na_2SiF_6, were made up of forsterite and diopside; the A_2 sample's crystalline phases contained 3.67 weight percent Na_2SiF_6, precipitated forsterite, Cu_2O, and Na (Na, Ca) $Mg_5Si_8O_{22}F_2$; the A_3 and A_4 samples' crystal phases, which contained 5.51 and 7.35 weight percent Na_2SiF_6, respectively, were all Na (Na, Ca) $Mg_5Si_8O_{22}F_2$ and Cu_2O.

(2) Although the samples with different Na_2SiF_6 contents all contained 7.35 wt% copper oxide, they had different percentages of Cu^+ in the total copper ions, which in turn showed different physical appearances: the A_1 sample had 39.9 percent Cu^+ of the total copper ions, and the sample with the Cu^+ content of 2.4 wt% was grey-black. The A_2 sample with 3.67 wt% Na_2SiF_6 had 46.5% Cu^+ of total copper ions, containing 2.7 wt% Cu^+, and the sample appeared orange-red. The proportion and content of Cu^+ in the A_4 sample with 7.35 wt% Na_2SiF_6 were 51.4% and 3.0 wt%, respectively, and the sample appeared red.

Acknowledgment

The business EMIN Nano Glass provides funding for the project.

Journal of Physics: Conference Series **2842** (2024) 012017 doi:10.1088/1742-6596/2842/1/012017

References

[1] M. Ma, Y. Yu, and Y. Huang (2024). Effect of heating rate on crystallization and properties of R_2O-CaO-MgO-CuO-Al_2O_3-SiO_2 glass-ceramics. Ceramics International, 50: 12, 326. Doi: 10.1016/j.ceramint.2024.01.137.

[2] H. Li, S. Liu, W. Xu, et al. (2021). The mechanism of the crystalline characteristics of spinel-induced epitaxial growth of diopside in CMAS glass-ceramics. Journal of the European Ceramic Society, 41: 1, 603. Doi: https://doi.org/10.1016/j.jeurceramsoc.2020.09.037.

[3] J. Xue, J. Zhong, Y. Mao, C. Xu, W. Liu, and Y. Huang (2020). Effect of CuO on crystallization and properties of red R_2O-CaO-MgO-Al_2O_3-SiO_2 glass ceramics from granite wastes. Ceramics International, 46: 23, 186. Doi: 10.1016/j.ceramint.2020.06.099.

[4] G. K. Inwati, P. Kumar, W. D. Roos, H. C. Swart, and M. Singh (2020). UV-irradiation effects on tuning LSPR of Cu/Ag nanoclusters in an ion-exchanged glass matrix and its thermodynamic behavior. Journal of Alloys and Compounds, 823: 153, 820. Doi: https://doi.org/10.1016/j.jallcom.2020.153820.

[5] C. Pischetola, S. M. Francis, F. Grillo, C. J. Baddeley, and F. Cardenas Lizana (2021). Phenylacetylene hydrogenation coupled with benzyl alcohol dehydrogenation over Cu/CeO_2: A consideration of Cu oxidation state. Journal of Catalysis, 394: 316. Doi: 10.1016/j.jcat.2020.11.002.

[6] N. Sondezi, Z. Njengele Tetyana, K. P. Matabola, and T. A. Makhetha (2024). Sol-Gel-Derived TiO_2 and TiO_2/Cu Nanoparticles: Synthesis, Characterization, and Antibacterial Efficacy. ACS Omega, 9: 15, 959. Doi: 10.1021/acsomega.3c09308.

[7] J. Wu, X. Huang, H. Chen, M. Gou, and T. Zhang (2021). Fabrication of Cu-Al_2O_3/ceramic particles by using brick particles as supports for highly efficient selenium adsorption. Journal of Environmental Chemical Engineering, 9: 105, 008. Doi: https://doi.org/10.1016/j.jece.2020.105008.

Journal of Physics: Conference Series **2842** (2024) 012018 doi:10.1088/1742-6596/2842/1/012018

Enhanced hydrolysis of spiramycin in aqueous solution using SiO_2/SO_3H

Hongyu Lyu [1], Shuai Chen [1], Rong Zhang [2], Chunliang Zhang [2], Ruiruo Jiang [1*], and Wen Yang [1*]

[1] Institute of NBC Defence PLA ARMY, Beijing 102205, China
[2] PLA Strategic Support Force's Characteristic Medical Center, No. 9 Anxiang North Lane, Chaoyang District, Beijing 100101, China

*Corresponding author's e-mail: yangwen_bnu@163.com

Abstract: Sulfonated silicon dioxide (SiO_2/SO_3H) solid acids were synthesized and they were characterized and used for spiramycin hydrolysis pretreatment of antibiotic wastewater. Results show that the spiramycin removal effect rankings are as follows: $SiO_2/SO_3H > SiO_2$ (pH = 3 or so) > control group. A first-order model was used to reflect the hydrolytic kinetics. The hydrolysis rate of the SiO_2/SO_3H solid acid calcined at 600°C is the highest, up to 3.59×10^{-2} (k). The performance of SiO_2/SO_3H is in positive correlation with the total acidity, which was determined by using the n-butylamine titration method. This hydrolytic method eliminates the need for the on-site handling of hazardous and corrosive mineral acids. As an attractive selective catalytic pretreatment method, it is targeted at removing related functional groups from antibiotics to reduce the inhibitory effect and allow antibiotic wastewater to further be easily biologically treated.

1. Introduction

Antibiotics are unique medicine because of their selective actions on bacteria and pathogens without affecting human cells and tissues [1]. Currently, over 250 registered antibiotics are applied in human and veterinary medicine [2]. Macrolide antibiotics have been frequently used to treat infectious diseases that gram-negative and gram-positive bacteria cause in human beings and animals [4]. Recent studies show that mass antibiotic production and consumption have produced large quantities of antibiotics in the environment. The findings of Watkinson et al. [3] show that the low-concentration (ng/L scale, up to 2 μg/L) detection rate of macrolide, quinolone, and sulphonamide antibiotics within the investigated urban watershed ranges from 15% to 83%. Oxytetracycline was found in both the surface water and groundwater of Wangyang River in China at 13, 640.9 ng/L and 8, 325.8 ng/L respectively, with its detection rate up to 100% [4]. The global consumption of antibiotics has rapidly increased, thus resulting in a considerable rise in the amount of wastewater produced during antibiotic production, about which there is widespread concern. In recent years, concentrations of antibiotics in wastewater have been increasing. How to reduce their concentrations has become a big challenge. Thus, there is an urgent need to develop new methods of degrading or pretreating antibiotics.

As a low-cost green alternative to liquid acids, solid acids have attracted much attention. With the growing interest in studies on the effects of solid acids across the world, remarkable advancements have been achieved in solid acid development. Such solid acids include zeolite, heteropolyacid, acid-functionalized carbonaceous materials, silicon-based materials, inorganic/mixed oxides, and acid resins. Silicon-based materials have been widely applied to prepare solid acids by introducing sulfonic groups (-SO_3H). It is reported that sulfonated silicon dioxide has been widely applied in reactions, such as hydrolysis, dehydration, and alkylation, with good results.

In this study, multiple sulfonated silicon dioxide solid acids were prepared and the enhanced spiramycin hydrolysis effects (depending on the concentration) of different preparation methods were evaluated. In addition, the hydrolytic activity differences of the SiO_2/SO_3H solid acids based on different preparation programs were also evaluated. This paper can present a useful approach to pretreating antibiotic wastewater before further biotreatment.

2. Procedure

2.1 Preparation of solid acids

Sulfonated silicon dioxide was prepared from TEOS and $ClSO_3H$ on-site [5]. Ten milliliters of TEOS and a certain amount of $ClSO_3H$ were put into a beaker. The beaker was magnetically stirred till a white gel-like solid was formed. The gel-like solid was put into a crucible. The crucible was placed in a furnace for calcination at 100~800°C for three hours (Thermolyne F48000 w/ a temperature decrease rate of 4°C/min under atmospheric pressure). The obtained solid was cooled to room temperature, pulverized, and washed by using hot water until a negative $BaCl_2$ test could be given. The test confirmed that the adherent $ClSO_3H$ was completely removed. Each catalyst program is expressed as ISS-*T*, where *T* is the calcination temperature (°C).

2.2 Characterization of solid acids

An XRD-7000 diffractometer (Shimadzu, Japan), together with a Cu *K*α-radiation source, was utilized to characterize the XRD patterns of the samples. These samples were scanned from 10° to 80° at the speed of 2°/min at 40 kV and 40 mA. A Spectrum GX spectrophotometer with a resolution of 2 cm^{-1} (Perkin Elmer, USA) was utilized to record the FTIR spectra of these samples at room temperature. A JEOL-2011 high-resolution scanning electron microscope (JEOL, Japan) was applied to observe their surface morphologies.

2.3 Hydrolysis experiments

Spiramycin hydrolyses took place in a series of 500 mL beakers. Some of them contained SiO_2/SO_3H and the others did not contain SiO_2/SO_3H. The initial spiramycin concentration in each beaker was 20 mg/L and the catalyst usage was 1 g/L. The solutions were incubated at 35°C in the darkness within a water bath for 360 minutes and were shaken at the speed of 170 rpm. These solutions were sampled at intervals of 30 minutes, filtered with 0.22 μm membranes, and stored at the temperature of -20°C. The analyses of the samples were completed within 2 days. Each experiment was performed twice and the average values were reported.

2.4 Analytic procedure

2.4.1 Spiramycin concentration analysis. An Agilent 1200 high-performance liquid chromatograph (HPLC) and a C-18 reverse phase column (4.6×250 mm, 5 cm; Agilent, USA) were used to determine the quantity of spiramycin. Micro-syringe filters (0.22 μm) were used to filter the samples initially. The measured mobile phase consisted of 50 mM ammonium acetate water solution and acetonitrile (vol/vol: 63:37). The injected sample volume was 10 μL each time. A photodiode array detector (Agilent, USA) was used to determine the spiramycin grain size. The detected spiramycin size was 232 nm.

3. Results and discussions

3.1 Characterization of SiO_2/SO_3H solid acids

The scanning electron microscope images of the SiO_2/SO_3H solid acids prepared under different preparation conditions are shown in Figure 1. We can see that the grains of the SiO_2/SO_3H solid acids are spherical.

XRD analyses were made to characterize the SiO_2/SO_3H solid acids' crystalline structures and compositions. As shown in Figure 2 (a), all the samples were similar in XRD pattern. The characteristic peaks at 22.0° and 23.3° were attributed to cristobalite high according to PDF 39-1425; those at 20.9° and 22.1° to CIT according to PDF 51-1382; those at 21.6° and 35.6° to cristobalite syn according to PDF 27-0605.

The samples' FTIR spectra are shown in Figure 2 (b). The peaks at 806^{-1} and 952 cm^{-1} were considered to originate from the vibrations of the terminal Si-O-Si groups; those at 1, 170^{-1} and 610 cm^{-1} were considered to originate from the O=S=O and S-O stretch vibrations of the sulfonic groups ($-SO_3H$). A striking increase in peak strength at 1, 170 cm^{-1} corresponds to an increase in the $ClSO_3H$ loaded ratio. This indicates that an increase in the $ClSO_3H$ loaded ratio can enhance the acidity of the SiO_2/SO_3H solid acid.

Figure 1. SiO_2/SO_3H SEM images.

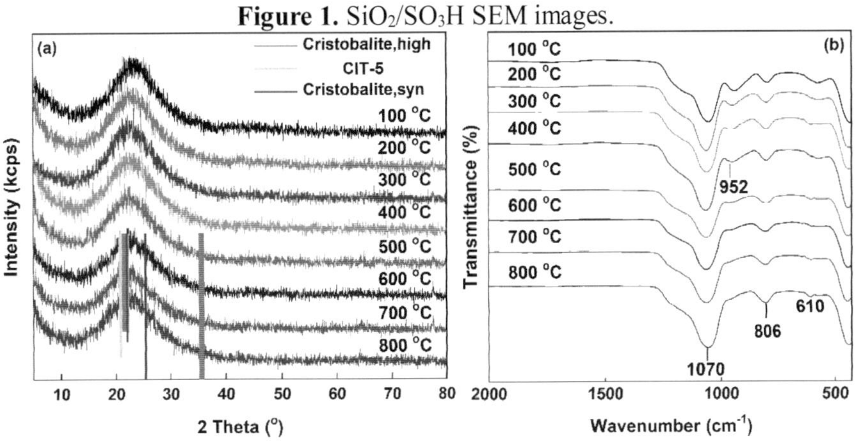

Figure 2. (a) XRD patterns; (b) SiO_2/SO_3H FTIR spectra at different temperatures.

3.2 Study on hydrolytic activity

The calcination temperature has a strong impact on the catalytic activity of SiO_2/SO_3H [6]. Increasing the calcination temperature appropriately during preparation can heighten the catalytic activity of this solid acid. This section deals with the effect of the calcination temperature.

The preparation programs for different temperatures including 100°C, 200°C, 300°C, 400°C, 500°C, 600°C, 700°C, and 800°C, which are expressed as ISS-100, ISS-200, ISS-300, ISS-400, ISS-500, ISS-600,

ISS-700, and ISS-800, were examined, and the results were as shown in Figure 3. All removal processes follow the pseudo-first-order kinetic law and all their R^2 values are higher than 0.80. Both the removal rate and the reaction rate constant increase as the calcination temperature increases below 600°C. When the calcination temperature is 600°C, the 20 mg/L spiramycin in the solution can be entirely degraded within 360 minutes. As the calcination temperature continues to increase, both the removal rate and the reaction rate constant decrease. The estimated K_h values along with the observation data are plotted in Figure 3 (a). 600°C was regarded as the optimal calcination temperature, and ISS-600 was considered as the best SiO_2/SO_3H preparation program.

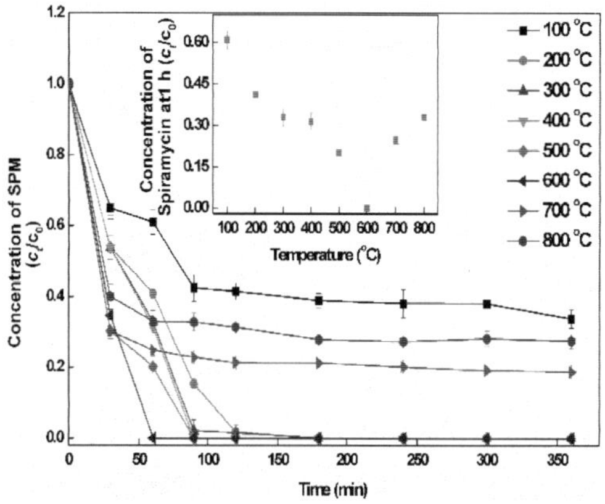

Figure 3. Spiramycin concentration decline curves of SiO_2/SO_3H solid acids calcined under different calcination temperatures (initial spiramycin concentration: 20.0 mg/L; SiO_2/SO_3H: 1.0 g/L; solution volume: 500 ml; temperature: 35±1°C; shaking time: 360 minutes).

3.3 In-depth study on SiO_2/SO_3H-enhanced spiramycin hydrolysis

The *n*-butylamine titration method was used to determine the total acidity values of the SiO_2/SO_3H solid acids prepared at different temperatures. The results are shown in Figure 4. Their total acidity values at 100°C~600°C range from 1.12 mmol/g to 1.22 mmol/g and their total acidity value differences are small. At 700°C and 800°C, however, the total acidity values drop sharply to 0.43 mmol/g and 0.24 mmol/g respectively. This might be because SO_3H was decomposed at high temperatures. The spiramycin hydrolysis experiments show that a SiO_2/SO_3H solid acid calcined at a very high temperature produces much lower hydrolytic efficiency. This indicates that a dramatic reduction in acidity can significantly reduce the efficiency of the SiO_2/SO_3H-enhanced hydrolysis.

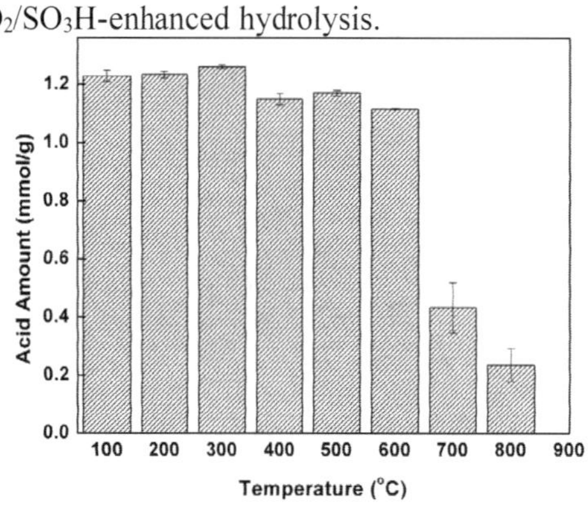

Figure 4. Acidity values of SiO_2/SO_3H solid acids calcined at different temperatures.

4. Conclusion

At present, the problem of antibiotic pollution has become a significant problem in water treatment. In this study, SiO_2/SO_3H solid acids were prepared by calcinating the reaction products of $ClSO_3H$ and tetraethyl orthosilicate. The catalysis activity is in a positive correlation with the calcination temperature. The optimal calcination temperature was 600°C. This is the first time that SiO_2/SO_3H was used for enhanced hydrolysis in a mild environment (35°C/2 h). A pseudo-first-order model was used to describe the spiramycin hydrolysis kinetics. Its findings show that the SiO_2/SO_3H-enhanced hydrolysis method can be regarded as a feasible and promising pretreatment method for spiramycin-containing antibiotic wastewater. Thus, this paper presents a useful approach to pretreating antibiotic wastewater before further biological treatment.

References

[1] Tian Y., Li J., Liu J., Meng J., and Li J. (2024). Insights into the microbiological mechanism of antibiotic removal in an aerobic-microaerobic process treating manure-free piggery wastewater. Chemical Engineering Journal, 484: 149, 634.

[2] Menkem Z. O. E., Ngangom B. L., Tamunjoh S. S. A., and Boyom F. F. (2019). Antibiotic residues in food animals: Public health concern. Acta Ecologica Sinica, 39 (5): 411-415.

[3] Watkinson A. J., Murby E. J., Kolpin D. W., and Costanzo S. D. (2009). The occurrence of antibiotics in an urban watershed: From wastewater to drinking water. Science of the Total Environment, 407 (8): 2, 711-2, 723.

[4] Jiang Y., Li M., Guo C., An D., Xu J., Zhang Y., and Xi B. (2014). Distribution and ecological risk of antibiotics in a typical effluent – receiving river (Wangyang River) in north China. Chemosphere, 112: 267-274.

[5] Jorge E. Y. C., Lima C. G. S., Lima T. M., Marchini L., Gawande M. B., Tomanec O., Varma R. S., and Paixão M. W. (2020). Sulfonated dendritic mesoporous silica nanospheres: a metal-free Lewis acid catalyst for the upgrading of carbohydrates. Green Chemistry, 22 (5): 1, 754-1, 762.

[6] Dumpalapally M., Boggala S., Nappuni R. C., Varimalla S., and Akula V. (2023). Cerium sulfate doped ZrO2 as a solid acid catalyst for the synthesis of polyoxymethylene dimethyl ethers: Influence of Brønsted acidity on the catalytic activity and product selectivity. Chemical Engineering Science, 278, 118, 922.

Different temperature-dependent fluorescence properties of $CsPbBr_3$ nanorods and nanowires

Fengjiao Chen, Liting Li, Xiong Shen, Jing Zhang, Siyu Ma, Yue Wang and Limin An*

College of Physics Science and Technology, Heilongjiang University, Harbin, Heilongjiang Province, 150080, China

*Corresponding author's e-mail: anlimin@hlju.edu.cn

Abstract. The morphology of totally inorganic metal halide perovskites has a certain influence on their optoelectronic properties. In this study, one-dimensional $CsPbBr_3$ nanorods (NRs) and nanowires (NWs) were arranged utilizing a boiling injection strategy. By controlling the processing time, NRs with a length of roughly 63 nm and NWs with a length of 300 nm were obtained, and it was found that the nanowires were grown from the nanorods. Temperature-dependent photoluminescence (PL) measurements were conducted on $CsPbBr_3$ NRs and NWs, and NRs exhibit a more stable excitonic state at high temperatures (>240 K). This study provides a method to change the morphology of one-dimensional metal halide perovskite nanomaterials without doping, further promoting the application of one-dimensional all-inorganic perovskite materials in the field of optoelectronics.

1. Introduction

With the advancement of science and technology, the demand for semiconductor materials is increasing. At the same time, the issue of energy scarcity is gradually emerging. Therefore, finding new materials with lower cost, higher performance, and energy utilization efficiency than traditional materials has become a primary task for researchers. Halide-based hybrid perovskite semiconductor nanocrystals (NCs) have accumulated meaningful interest owing to their customizable PL within the light spectrum, considerable photoluminescence quantum yield (PLQY), broad assimilation, and slight emission capacity [1-3]. Traditional metal halide perovskite materials are 3D crystals. In recent years, with the continuous development of perovskites, colloidal quantum dots, One-dimensional nanowires or nanorods, and 2D nanobelts or nanosheets have been continuously studied and developed by scientists [4]. Research has found that the morphology of metal halide perovskite materials is an important factor affecting their performance. 1D metal halide perovskites, owing to their distinctive electromagnetic and light properties, including quantum size effects and surface effects, can modify the energy spectrum and behavior of electrons, thereby altering the optical characteristics and making them novel optoelectronic materials [5-7]. Calcium titanium oxide nanowires have been shown to possess low threshold laser operation, and their physical structure can act as a Fabry-Perot resonant cavity [8-11]. Moreover, they can serve as the gain medium for semiconductor lasers.

In this study, 1D $CsPbBr_3$ NRs and NWs were prepared by a facile hot-injection method. XRD patterns show that the crystal structure of $CsPbBr_3$ is not changed by the change in morphology and size. The fluorescence lifetime of $CsPbBr_3$ NRs is 49 ns, while the fluorescence lifetime of $CsPbBr_3$ NWs increases to 107 ns, indicating that NWs have fewer surface defects than the NRs, so the lifespan increases. Fitting the fluorescence intensity of NRs and NWs showed that the exciton binding energy of NRs was greater than that of NWs, indicating that the structure of NRs was more stable.

2. Experimental procedures

Reagents and chemicals, including cesium carbonate (Cs_2CO_3, 99.99%), lead bromide ($PbBr_2$, 98%), oleic acid (OA, 90%), 1-octadecene (ODE, 90%), n-hexane (99%), oleamine (OAm, 90%) and n-octylamine (OTA, 99%) were procured from Aladdin company and utilized without any additional purification steps.

To prepare the cesium-oleate solution, a blend containing 0.2 g of Cs_2CO_3, 7 ml of ODE, and 0.7 ml of OA was incorporated into a 30 ml triple-neck bottle. The chemical process mixture was gradually warmed from 20°C to 120°C and sustained at 120°C for 60 minutes. Subsequently, the temperature was further increased to 150°C, resulting in the transformation of the transparent solution into a deep yellow color, indicating the successful synthesis of the precursor.

To synthesize $CsPbBr_3$ perovskite NRs, a mixture containing 7 ml of ODE, 1.4 ml of OAm, and 1 ml of OTA was incorporated into a 50 ml triple-neck bottle. The chemical process solution was gradually heated from 20°C to 120°C and maintained at this temperature for 60 minutes. 0.13 g of $PbBr_2$ was incorporated into the reaction mixture and proceeded for another 60 minutes. The temperature was elevated to 125°C, at which point 1.4 ml of cesium oleate was swiftly infused into the blend. After the solution blended was allowed to react for 10 seconds, the bottle was promptly placed in a frozen water basin for rapid cooling.

To synthesize $CsPbBr_3$ perovskite NWs, a mixture containing 7 ml ODE, 1.3 ml of OAm, and 1 ml of OTA was incorporated into a 50 ml triple-neck bottle. The steps are consistent with the preparation process of $CsPbBr_3$ perovskite NRs. After the mixed solvent was allowed to react for 3 minutes, the bottle was promptly placed in a frozen water basin for rapid cooling.

The purification procedure for $CsPbBr_3$ perovskite NRs and NWs involved centrifugation at 9500 rpm for 5 minutes to separate the solution. Subsequently, the NRs solution was subjected to further centrifugation at 8500 rpm, while the NWs mixture was rotated at 3000 rpm for 1 minute. Thereafter, the clear liquid of NRs was centrifuged again at a centrifuge speed of 9000 rpm, and the supernatant of NWs was centrifuged again at a centrifuge speed of 6000 rpm for 1 minute, then the supernatant was discarded, retained by precipitation, and dispersed into n-hexane.

The optical properties of $CsPbBr_3$ NRs and NWs were tested and characterized. The morphology of $CsPbBr_3$ NRs and NWs was analyzed utilizing transmission electron microscopy (JEM-F200), while their visualized attributes were analyzed utilizing ultraviolet spectroscopy (UV-2550), fluorescence spectrometry (F-4600), and time-resolved spectroscopy (SP-2500).

3. Results and discussions

$CsPbBr_3$ NRs and NWs were successfully synthesized through the thermal infusion approach. The transmission electron microscopy (TEM) images of $CsPbBr_3$ NRs and NWs are depicted in Figure 1 (a, b). From the captured images, it is evident that the size distribution of the NRs and NWs exhibits uniformity, and their morphology remains stable, indicating minimal external interference during the nucleation process. The length of the NRs is approximately 63 nm, with a dimension of about 30 nm. The measurement of the NWs is around 300 nm, with a width of roughly 5.7 nm. The main difference between the growth processes of the NRs and NWs lies in the growth time.

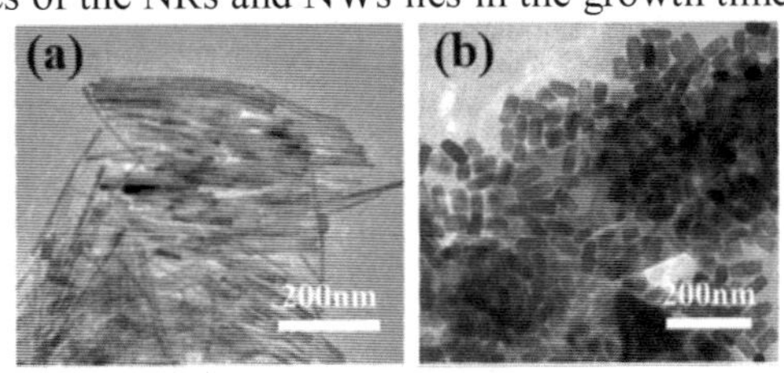

Figure 1. (a, b) TEM images illustrating the shape of $CsPbBr_3$ NRs and NWs.

Journal of Physics: Conference Series **2842** (2024) 012019 doi:10.1088/1742-6596/2842/1/012019

The equilibrium assimilation spectrums and PL spectra of $CsPbBr_3$ NRs and NWs are presented in Figure 2a and exhibit distinct characteristic peaks. The ingestion spectra of the NRs and NWs exhibit similar trends, indicating that the change in morphology does not affect the structure of $CsPbBr_3$ perovskite. A restricted and radiant PL radiation peak is observed at a wavelength of 522 nm in the $CsPbBr_3$ NWs, with a width at half maximum (FWHM) of 16 nm. Compared to the NRs, the radiation summit of the NWs shows a significant redshift, which is attributed to the quantum size effect. As the magnitude of the material augments, the bandgap width decreases, indicating that the phenomenon of energy level splitting is not significant during the growth process from 60 nm to 300 nm. Furthermore, the Stokes shift increases from 6 nm to 10 nm, reducing the overlap between absorption and emission spectra and avoiding energy transfer that may lead to a decrease in fluorescence efficiency. The XRD spectra of $CsPbBr_3$ NRs and NWs are presented in Figure 2b. Their diffraction peaks are consistent with standard $CsPbBr_3$ (PDF No.54-0752), indicating that the change in size does not affect the crystal phase of $CsPbBr_3$. All the samples were one-dimensional $CsPbBr_3$ materials.

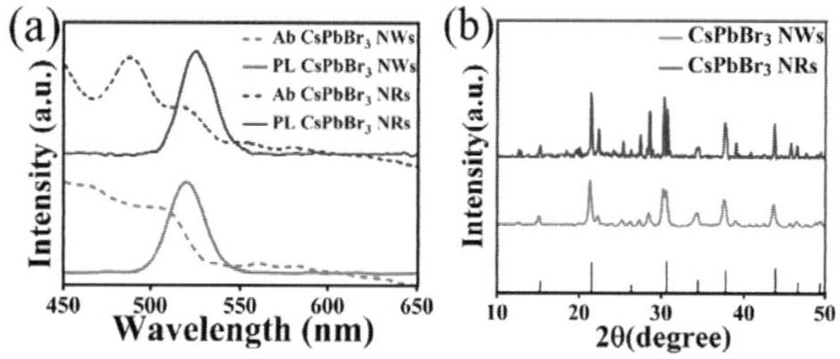

Figure 2. (a) PL spectra and Absorption of $CsPbBr_3$ NRs and NWs. (b) XRD patterns

Figure 3 (a, b) displays the PL lifetimes of $CsPbBr_3$ NRs and NWs, which were analyzed using a three-exponential model described by the subsequent mathematical expression [12]:

$$I(t) = A_1 e^{\frac{-t}{t_1}} + A_2 e^{\frac{-t}{t_2}} + A_3 e^{\frac{-t}{t_3}} \tag{1}$$

Based on the equation, the lifetimes of $CsPbBr_3$ NRs and NWs are approximately 49 ns and 107 ns, respectively. The increase in nanowire size reduces the density of internal crystal surface defect states, thereby reducing the probability of carrier recombination with surface defects and allowing more carriers to participate in radiative transitions [13]. Therefore, the lifetime of $CsPbBr_3$ NWs is significantly greater than that of NRs.

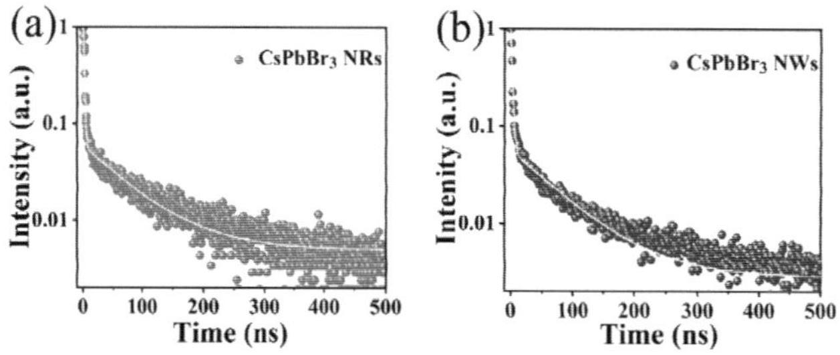

Figure 3. (a, b) Normalized time-resolved PL spectra from $CsPbBr_3$ NRs and NWs.

To further investigate the microstructural photoluminescent characteristics, specifically the engagement between excitonic complexes and vibrational quanta within the samples, based on the peak positions observed in Figure 2a, the PL spectral profiles of $CsPbBr_3$ NRs and NWs with varying temperatures are depicted in Figure 4(a, b). Within the temperature interval of 80-230 K, $CsPbBr_3$ NRs and NWs exhibit similar properties, with a certain degree of blue shift observed as the temperature increases. In the temperature interval of 240 K to 340 K, there is little change in the fluorescence peak. This is due to a phase transition occurring in the crystal at this temperature. The association between the PL spectrum intensity of $CsPbBr_3$ NRs and NWs and temperature is fitted using the Arrhenius equation [14]:

Journal of Physics: Conference Series **2842** (2024) 012019 doi:10.1088/1742-6596/2842/1/012019

$$I(T) = \frac{I_0}{1 + A\exp\left(-E_b / K_B T\right)} \tag{2}$$

Here, $I(T)$ and I_0 represent the combined strengths of PL at T K and 0 K, individually. A denotes the adjustment fixed value, K_B represents the Boltzmann constant, and E_b is the exciton binding power. The exciton binding power E_b for $CsPbBr_3$ NRs and NWs is determined to be 291.85 meV and 44.48 meV, respectively. The higher exciton binding power of the NRs compared to the NWs indicates stronger coulomb interactions and greater structural stability within the NRs at high temperatures (>240 K). The higher exciton binding energy also suggests that $CsPbBr_3$ NRs exhibit better thermal stability. Additionally, it can be inferred that $CsPbBr_3$ NRs are in a more stable electron-hole interaction regime. The smaller shift of the luminescence peak of the NRs compared to the NWs at higher temperatures in Figure 4(a, b) further validates this conclusion. Figure 4c illustrates the temperature-dependent evolution of the PL spectral intensity for $CsPbBr_3$ NRs and NWs. Within the temperature interval of 180 K to 240 K, a significant increase in fluorescence intensity can be observed, and within the decreasing trend of fluorescence intensity, the decrease rate of the NRs is higher than that of the NWs. This is attributed to the changes in non-radiative transitions caused by surface defect states. The slower decrease rate of the NWs indicates a lower density of surface defects and a smaller contribution of non-radiative transitions during carrier recombination.

Figure 4d shows the variation of the bandgap width of $CsPbBr_3$ NRs and NWs with temperature. From the graph, it can be observed that the bandgap width of both the NRs and NWs increases with increasing temperature. This phenomenon can be clarified by contemplating the electron-vibration engagement and lattice heat-related enlargement taking place within the $CsPbBr_3$ nanomaterial. As the temperature continues to rise, the motion of phonons within the crystal becomes more prominent, resulting in a relatively smaller thermal expansion. This enhanced electron-phonon coupling reduces the bandgap width. Figure 4d shows the fitting of the correlation between peak energy and temperature for $CsPbBr_3$ NRs and NWs using the following equation[15]:

$$E_g(T) = E_0 + A_{EP}\left[\frac{2}{\exp(\hbar\omega / K_B T)\text{-}1} + 1\right] \tag{3}$$

where $E_g(T)$ denotes the bandgap of the material at temperature T, A_{EP} is the coefficient for electron-phonon coupling, $\hbar\omega$ denotes the average energy of optical phonons, K_B is the Boltzmann constant, and E_0 is the bandgap width at 0 K. The fitting results for $\hbar\omega$ in $CsPbBr_3$ NRs and NWs are found to be 10.7 meV and 1.54 meV, respectively. This observation suggests that $CsPbBr_3$ NRs demonstrate a more pronounced interaction impact with high-velocity optical phonons than $CsPbBr_3$ NWs.

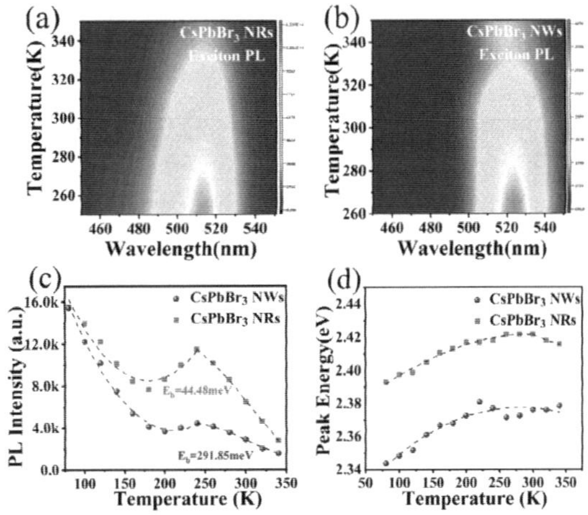

Figure 4. Temperature-dependent spectra of $CsPbBr_3$ NRs and NWs.

4. Conclusions
In summary, $CsPbBr_3$ perovskite NRs and NWs were successfully synthesized. Analysis of the fluorescence lifetime spectra demonstrated that the NWs exhibit longer lifetimes compared to the NRs, indicating a lower density of surface defects in the NWs. Furthermore, investigation of the temperature-responsive PL spectrums revealed that $CsPbBr_3$ NRs possess elevated longitudinal phonon levels and excitonic binding strength, suggesting enhanced photoluminescent performance and thermal stability. Consequently, one-dimensional $CsPbBr_3$ perovskite nanomaterials are deemed more suitable for optoelectronic device applications.

Acknowledgment
This work was supported by the Natural Science Foundation of Heilongjiang Province (LH2020A021).

References
[1] Wang, Y., Liu, X., He, Q., Chen, G. (2021) Reversible transformation between CsPbBr3 perovskite nanowires and nanorods with polarized optoelectronic properties. Advanced Functional Materials, 31: 2011251.
[2] Hua, Z., Tara, S., Zhang, J, Y., Berkinsky, D, B. (2022) One-dimensional highly-confined CsPbBr3 nanorods with enhanced stability: synthesis and spectroscopy. Nano Letters, 22: 8355-8362.
[3] Shamsi, J., Urban, A., S., Imran, M. (2019) Metal halide perovskite nanocrystals: synthesis, postsynthesis modifications, and their optical properties. Chemical Reviews, 119: 3296-3348.
[4] Akkerman, Q, A., Rainò, G., Kovalenk, M, V. (2018) Genesis, challenges and opportunities for colloidal lead halide perovskite nanocrystals. Nature Materials, 17: 394-405.
[5] Peters, J., Liu, Z., Bulgin, O., He, Y., Klepov, V., Siena, M, D. (2021) Excitons in CsPbBr3 halide perovskite. The Journal of Physical Chemistry Letters, 12: 9301-9307.
[6] Zhao, Q, Q., Chen, F, T., Huang, Q. (2024) Shape-controlled synthesis of CsPbBr3 nanorods with bright pure blue emission and high stability. Journal of Materials Chemistry C, 12: 4234-4242.
[7] Shi, J, D., Wang, M, Q., Da, Z, Y. (2023) Studies on the optical stability of CsPbBr3 with different dimensions (0D, 1D, 2D, 3D) under thermal environments. Nanoscale, 15: 11190-11198.
[8] Liu, B, H., Wang, S., Lai, X, F. (2024) CsPbBr3 Nanorod Luminescent Property Study. The Journal of Physical Chemistry C, 128: 642-650.
[9] Zhang, Y, H., Sun, R, J., Ou, X, Y. (2019) Metal Halide Perovskite Nanosheet for X-ray High-Resolution Scintillation Imaging Screens. American Chemical Society Nano, 13: 2520-2525.
[10] Suhail, A., Saini, A., Beniwal, S. (2023) Tunable Optoelectronic Properties of CsPbBr3 Perovskite Nanocrystals for Photodetectors Applications. Journal of Physical Chemistry C, 127: 17298-17306.
[11] Huang, K., Liu, J., Chang, X. (2018) Integrating Properties Modification in the Synthesis of Metal Halide Perovskites. Advanced Materials Technologies, 4: 1800321.
[12] Sun, Y, L., Zhu, H., Jin, N. (2021) Fast Lifetime Blinking in Compact CdSe/CdSCore/Shell Quantum Dots. Journal of Physical Chemistry C, 125: 15433-15440.
[13] Shu, B., Chang, Y., Dong, L. (2021) Highly stable CsPbBr3 perovskite quantum dots incorporated in aluminum stearate. Journal of Luminescence, 234: 117962.
[14] Wei, K., Xu, Z., Chen, R. (2016) Temperature-dependent excitonic photoluminescence excited by two-photon absorption in perovskite CsPbBr3 quantum dots. Opt Lett, 41: 3821-3824.
[15] Monreal, R., C. (2024) Electron–Electron and Electron–Phonon Interactions in the Dynamics of Trap-Filling in Charged Quantum Dots. Journal of Physical Chemistry C, 128: 2506-2517

Journal of Physics: Conference Series **2842** (2024) 012020 doi:10.1088/1742-6596/2842/1/012020

Influence of replacement ratios of spontaneous-combustion coal gangue aggregate on the compressive strength of concrete

Bo Gao [1,*], Dadong Xia [1], Xiuqing Liang [2], Tirui Zhang [3], and Qinghe Wang [3]

[1] China State Construction Railway Investment & Engineering Group Co., Ltd., Beijing 100855, China

[2] China Construction Third Engineering Bureau Group Co., Ltd., Wuhan 430070, Hubei, China

[3] School of Civil Engineering, Shenyang Jianzhu University, Shenyang 110168, Liaoning, China

[*] Corresponding author's e-mail: 329751777@qq.com

Abstract: To quantify the influence of coarse spontaneous-combustion coal gangue aggregate (SCGA) replacement ratio (r_c) on the compressive strength of concrete, we reveal the deterioration mechanism of the compressive performance of spontaneous-combustion coal gangue aggregate concrete (SCGAC). 5 groups of SCGAC with different aggregate replacement ratios were designed to measure compressive strength by test, and the structures of concrete was acquired by Scanning Electron Microscope. Data reveal that as the r_c rises, there is a corresponding decrease in the compressive strength of SCGAC. Compared with natural aggregate concrete (NAC), the 28 d axial compressive strength of SCGAC with the r_c of 50% and 100% decreases by 14.61% and 19.40%, respectively. The Weibull distribution of SCGAC compressive strength is closer to the actual compressive strength distribution characteristics, which is more suitable for predicting the compressive strength of SCGAC.

1. Introduction

Coal refuse refers to the residual solid material produced as a byproduct of coal extraction and processing operations [1]. If left piled up under natural conditions for an extended period, it can spontaneously ignite once it reaches its ignition temperature, thereby forming spontaneous-combustion coal gangue [2-3]. This poses a significant threat to the natural environment. Consequently, substituting SCGA for natural aggregates in the production of concrete not only achieves efficient utilization of coal gangue but also reduces the consumption of sand and stone resources, thus mitigating the issue of natural resource scarcity [4].

Previous research has explored the mechanical characteristics of SCGAC based on equal mix proportions. Recent studies show that with an increase in the r_c of SCGA, the porosity of SCGAC gradually increases [5], and the mechanical performance gradually reduces [5-7]. However, high water absorption of the coarse SCGA can also cause a loss of concrete slump.

Based on this, and assuming that the workability of concrete remains consistent, the research conducted here evaluates the effect that replacement ratios of SCGA have on the concrete's compressive strength. SEM is employed to examine the concrete's meso-structures at a microscopic level, which aids in understanding the underlying reasons for the decline in compressive strength within SCGAC.

2. Experimental procedures

2.1. Raw materials

The cementitious materials employed were Class I fly ash and standard P·O 42.5 Portland cement. The aggregates were naturally sourced, with the natural fine aggregate (NFA) being river sand featuring a size from 0 to 5.0 mm, and the natural coarse aggregate (NCA) being crushed stone, sized between 5.0 to 20 mm. Coal gangue from spontaneous combustion initially sized between 40.0 to 100.0 mm, was processed by crushing and sieving to achieve a size range of 5.0 to 20 mm, acting as an alternative to the NCA. To ensure compatibility, the grain size of the SCGA was manually calibrated to equate the NCA distribution.

The density and water absorption of the aggregates, along with the crush index for the coarse aggregates, were determined following the JGJ 52-2006 standard [8]. The spontaneous-combustion gangue aggregates (SCGA) have comparatively inferior mechanical attributes, with a dry density 7.7% less than that of the NCA. The crush index for SCGA is notably higher, being 5.6 times the index of NCA. Moreover, the water absorption characteristic of SCGA is considerably greater than natural aggregate, at 9.80%, 1.35%, and 3.45%, respectively. The variance in the physical characteristics between the coarse SCGA and NCA is predominantly attributed to the disparities in their meso-structures. The SCGA has a more porous and less compact surface than the NCA. This porous nature results in increased water absorption and diminished mechanical properties, which can notably impact the concrete's workability and strength. As a result, a water-reducing admixture, with a density of 1, 210 kg/m³, is incorporated into the mix to guarantee adequate workability.

2.2. Concrete preparation and testing methodology

Five groups of specimens with spontaneous-combustion gangue aggregate concrete (SCGAC) were created, varying the replacement ratios from 0% to 100% in increments of 25%. These were labeled as r_c-0, r_c-25, r_c-50, r_c-75, and r_c-100. For each mix, six concrete prisms measuring 100 mm by 100 mm by 300 mm were fabricated for uniaxial compression testing. For the mixes with 0%, 50%, and 100% replacement ratios, an additional 30 cubic concrete specimens, each 100 mm^3, were also made to measure the changes in cubic compressive strength. To mitigate the impact of the water absorption of the coarse SCGA on the concrete's workability, these aggregates were pre-soaked for one hour. The specific mix ratios for the concrete are detailed in Table 1. Post casting, all samples were removed from their molds after 24 h and then stored in a controlled curing environment with a humidity level of 98% and a temperature maintained at 20°C until reached the appropriate age for testing.

Table 1. Concrete mix design (kg/m^3).

Specimen ID	Water	Cement	Fly ash	NFA	NCA	SCGA	AE [a]
r_c-0	180	320	80	950	870	0	3.2
r_c-25	180	330	82	905	624	208	3.3
r_c-50	180	339	85	860	395	395	3.4
r_c-75	180	349	87	814	188	564	3.5
r_c-100	180	358	90	769	0	710	3.6

AE [a] stands for a type of water-reducing admixture.

During testing, a 1, 000 kN capacity testing machine was used, and the loading rate was adjusted to 5.0 kN/s according to the specifications of EN 12390-3 [9]. The strength of the SCGAC was subsequently evaluated in accordance with the GB/T 50081-2019 standard [10].

3. Experimental results and discussion

By measuring the slump of the SCGAC with different r_c, it was found that the slump for r_c-0, r_c-25, and r_c-50 was 70 mm; for concrete with r_c of 75% and 100%, the slump slightly increased to about 75 mm.

Overall, this indicates that all the fresh concrete mixtures exhibited workability close to the expected levels.

3.1. Compressive strength

The axial compressive strengths (f_{cp}) of the SCGAC of 28 d and 90 d are detailed in Table 2. It is evident that as the r_c of the aggregate increases, the f_{cp} of the SCGAC diminishes. To be more precise, when comparing each mix to the r_c-0, the reductions in the 28 d strengths for r_c-25 , r_c-50 , r_c-75 , and r_c-100 were 9.82%, 14.61%, 15.11%, and 19.40% respectively. At 90 d, the strength reductions were 7.34%, 9.64%, 12.87%, and 14.68% respectively. This trend indicates a direct relationship between the proportion of SCGA used and the resulting compressive performance of the concrete.

Table 2. Axial compressive strength (f_{cp}) and elastic modulus (E) of SCGAC

	f_{cp} (MPa)		E (GPa)	
Specimen ID	28-day	90-day	28-day	90-day
r_c-0	39.7	47.7	29.7	34.1
r_c-25	35.8	44.2	24.4	29.9
r_c-50	33.9	43.1	21.2	27.3
r_c-75	33.7	41.6	21.0	24.7
r_c-100	32.0	40.7	20.3	23.6

3.2. Stress-strain curves

Figure 1 presents the entire stress-strain curves for SCGAC, considering diverse r_c. It should be noted that in the 28 d uniaxial compression test for SCGAC with a 100% replacement ratio, the measured data were incorrect due to external factors, so it was not considered. It can be observed that the stress-strain curves for concrete have a generally similar shape. The complete stress-strain curves reveal that the SCGAC exhibits significantly lower mechanical properties when juxtaposed with NAC. Specifically, as the r_c of SCGA increases, the peak stresses at 28 d and 90 d decrease by 19.40% and 14.68%, respectively, with corresponding peak strains ranging from 0.0015 to 0.0022 and from 0.0017 to 0.0026, respectively.

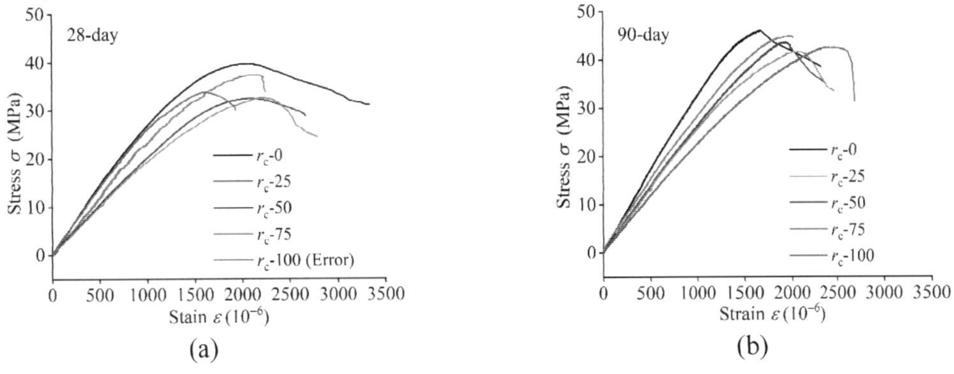

Figure 1. Influence of r_c on the stress-strain relationships of SCGAC

3.3. Characteristics of ITZ of concrete

The meso-structures of the 28-day cured r_c-0 and r_c-50 concrete specimens were observed by using SEM, as shown in Figure 2. It is clear that considerable variations exist in the hydration products and the intermediate structures of the ITZ across various types of concrete. Specifically, in the ITZ of NAC, needle-like C-S-H gel is formed, resulting in a very tight bond between the NCA and the cement mortar

(Figures 2a and 2b). In the SCGAC with a 50% replacement ratio, amorphous C-S-H gel is monitored (Figure 2c).

The porous nature of the SCGA enhances the absorptive capacity of the aggregates, which in turn lowers the hydration ratio in the ITZ of the concrete. This results in a reduced level of hydration, particularly in the concrete with a 50% replacement ratio. The diminished hydration reaction weakens the ITZ, making it the most vulnerable part of the SCGAC. As the testing continues, the initial signs of cracking manifest in this area, as depicted in Figure 2c. To encapsulate, the heightened water absorption and inferior mechanical attributes of the SCGA contribute to the ITZ being the most vulnerable area in the concrete, which in turn causes a relatively diminished compressive strength in SCGAC when set against NAC.

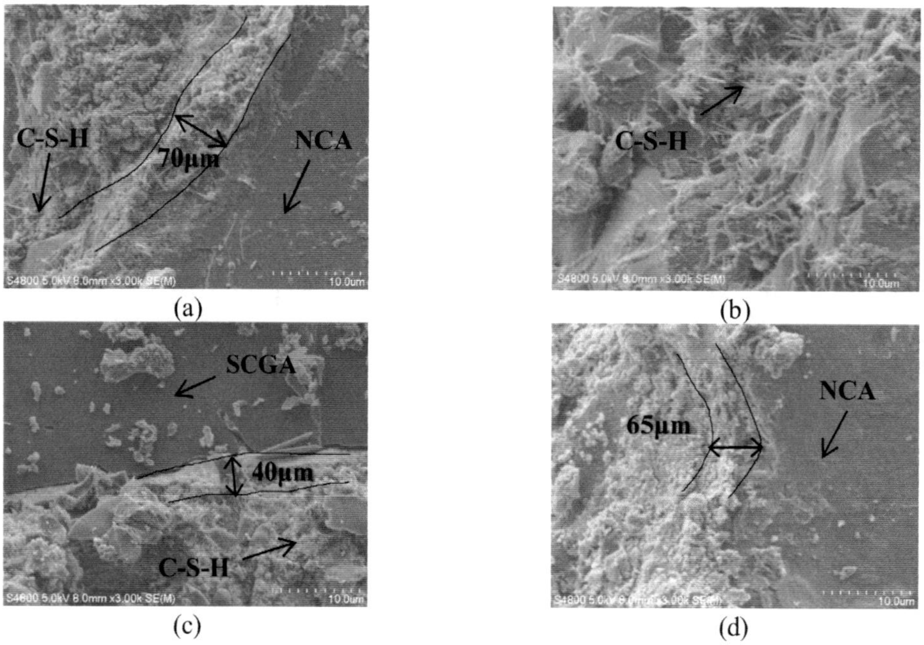

Figure 2. Meso-structures of concrete ITZs

3.4. Characteristics of compressive strength distribution

The strength of concrete is typically modeled by a normal distribution, and its probability density function is shown in Equation (1). However, the compressive strength of brittle materials is always analyzed by using the Weibull distribution, with its probability density function shown in Equations (2) to (4). Thus, to gain a deeper insight into the distribution traits of the compressive strength of SCGAC, a statistical examination employing both Weibull and normal distributions was conducted on the concrete's compressive strength figures.

$$f(\sigma)=\frac{1}{\sqrt{2\pi}s}e^{\frac{[-(\sigma-\mu)^2]}{2s^2}} \tag{1}$$

where μ is the mean strength, s is the standard deviation, and σ is the stress, which is the independent variable of the probability density function $f(\sigma)$.

$$f(\sigma)=\frac{m}{\sigma_0^m}(\sigma-\sigma_{\text{th}})^{m-1}\exp[-(\frac{\sigma-\sigma_{\text{th}}}{\sigma_0})^m] \tag{2}$$

$$P(\sigma)=1-\exp[-(\frac{\sigma-\sigma_{\text{th}}}{\sigma_0})^m] \tag{3}$$

$$\ln\ln(\frac{1}{1-P})=m\ln\frac{\sigma}{\sigma_0} \tag{4}$$

where σ is the stress; $P(\sigma)$ is the probability of failure at stress σ; $f(\sigma)$ is the probability density function; m is the Weibull modulus, delineating the variability of failure strength, with a higher value indicating a reduced dispersion; σ_0 and σ_{th} are the characteristic strength and scale parameter (minimum compressive strength), respectively.

The experimental data, Weibull distribution, and normal distribution curves for r_c-0, r_c-50, and r_c-100 are shown in Figure 3. The results indicate that the Weibull distribution and the normal distribution share a geometric similarity in shape and can both be used to predict the strength of SCGAC. However, there are significant differences in the peak probabilities and corresponding compressive strengths obtained from the two distributions.

Specifically, taking r_c-50 as an example, the peak probabilities from the experimental data, Weibull distribution, and normal distribution are 0.20, 0.17, and 0.26, respectively, with corresponding compressive strengths of 43.5 MPa, 43.9 MPa, and 43.1 MPa, respectively. It is noticeable that the Weibull distribution of the SCGAC strength aligns more closely with the actual pattern of the strength distribution. Consequently, the Weibull distribution proves to be a more fitting model for forecasting the distribution traits of the SCGAC strength.

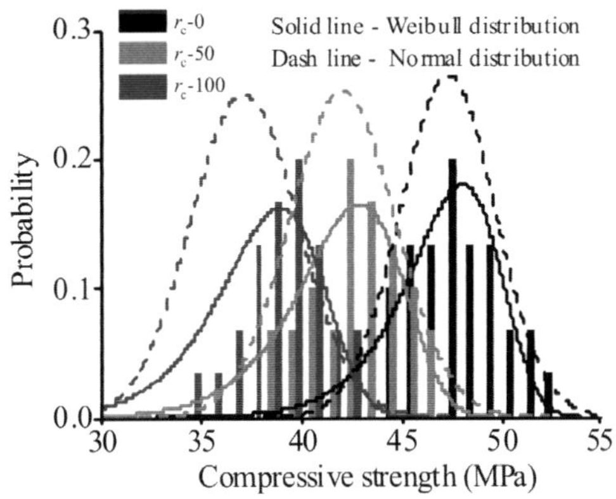

Figure 3. Comparison between Weibull and normal distribution of SCGAC compressive strength

4. Conclusion

(1) The strength of concrete that incorporates SCGA diminishes as the proportion of this aggregate replacing the traditional one increases. When contrasted with concrete made from natural aggregates, the spontaneous-combustion coal gangue concretes at 28 d intervals see a reduction in axial compressive strength by 9.82%, 14.61%, 15.11%, and 19.40% for r_c of 25%, 50%, 75%, and 100% respectively. Similarly, at 90 d, the decrease in axial compressive strength is observed to be 7.34%, 9.64%, 12.87%, and 14.68% for the same replacement ratios.

(2) Unlike natural aggregates, coal gangue from spontaneous combustion exhibits greater water absorption and a relatively reduced hydration level. As a consequence, the amorphous C-S-H gel is more evident in the ITZ of concrete incorporating this type of aggregate. This feature renders the ITZ the most vulnerable area within the concrete's microstructure. Thus, the concrete's compressive strength, when made with coal gangue from spontaneous combustion, is typically lower in comparison to that of concrete using natural aggregates.

(3) Comparing the Weibull to the normal distribution shows that the Weibull distribution more accurately reflects the actual distribution of this concrete's compressive strength.

Acknowledgments
The research conducted here was financially supported by China's Liaoning Province Applied Basic Research Program, designated with the grant identifier 2022JH2/101300130.

References

[1] M. Zhu, J. Qiu, and J. Chen (2022). Effect and mechanism of coal gangue concrete modification by basalt fiber. Construction and Building Materials, 328 (1): 126, 601. https://doi.org/10.1016/j.conbuildmat.2022.126601.

[2] A. Moussadik, F. Ouzoun, H. Ez-zaki, et al. (2023). Mineralogical study of a binder based on alkali-activated coal gangue [J]. Materials Today: Proceedings. https://doi.org/10.1016/j.matpr. 2023.07.091.

[3] M. Murtaza, J. Zhang, C. Yang, et al. (2024). Performance analysis of self-compacting concrete by incorporating fly ash, coal gangue powder, cement kiln dust, and recycled concrete powder by absolute volume method [J]. Construction and Building Materials, 431: 136, 601. https://doi.org/10.1016/j.conbuildmat.2024.136601.

[4] A. Moussadik, H. Fadili, M. Saadi, et al. (2024). Lightweight aerated concrete based on activated powders of coal gangue and fly ash. Construction and Building Materials, 417: 135, 333. https://doi.org/10.1016/j.conbuildmat.2024.135333.

[5] L. Yu, J. Xia, Z. Xia, et al. (2022). Study on the mechanical behavior and micro-mechanism of concrete with coal gangue fine and coarse aggregate. Construction and Building Materials, 338: 127, 626. https://doi.org/10.1016/j.conbuildmat.2022.127626.

[6] L. Yu, A. Wang, and X. Zhong (2023). Progress in the study of mechanical and durability performance of concrete with coal gangue aggregates. Materials Report, 38 (20): 23, 080, 244. https://doi.org/10.11896/cldb.23080244.

[7] Y. She, S. Chen, and G. Qian (2023). Research progress in mechanical properties of spontaneous-combustion coal gangue aggregate concrete. Shanxi Architecture, 49 (6): 133-135+139. https://doi.org/10.13719/j.cnki.1009-6825.2023.06.035.

[8] China Standard. Standard for technical requirements and test method of sand and crushed stone (or gravel) for ordinary concrete: JGJ 52-2006.

[9] British Standard. Testing hardened concrete. Part 3: Compressive strength of test specimens: EN 12, 390-3.

[10] China Standard. Standard for test methods of concrete physical and mechanical properties: GB/T 50081-2019.

Physically crosslinked polyvinyl alcohol hydrogel for electromagnetic shielding

Zian Liu, Yufeng Wu, Zheng Zhang, Xinyu Xie, Wanting Cao, Jianchun Xu and Kai Huang*

State Key Laboratory of Information Photonics and Optical Communications & School of Science, Beijing University of Posts and Telecommunications, Beijing, 100876, China

*Corresponding author's e-mail: huang-kai@bupt.edu.cn

Abstract. As military and civilian electronic communication technologies advance rapidly, there is a growing interest in the design and study of high-performance materials for electromagnetic shielding. In marine environments especially, the use of devices like wireless communication, radar, and satellite navigation is common. Hydrogels, a type of porous material, possess characteristics such as low density, high surface area, and exceptional flexibility. Conductive hydrogel materials, which combine the benefits of conductive and hydrogel materials, exhibit conductivity, adjustable mechanical flexibility, and ease of processing. They also have strong electromagnetic wave absorption capabilities. To enhance the effectiveness of electromagnetic shielding in the broadband range of 5-18 GHz, we propose a method to create polyvinyl alcohol hydrogel through physical cross-linking. This involves preparing the hydrogel using the freeze/thaw technique and then physically cross-linking it by submerging it in various salt solutions. Our method significantly improves the electromagnetic shielding effectiveness, with the hydrogel in magnesium chloride solution showing the greatest enhancement in performance, achieving a ~7 dB improvement across different frequencies. The physical cross-linking approach for hydrogel preparation offers the benefits of simplicity and affordability, showcasing a noteworthy electromagnetic shielding effect with promising applications in the field.

1. Introduction

As the economy continues to grow and science and technology advance, the rapid expansion of the digital and information sectors has led to an improvement in people's quality of life. The increasing interconnectedness of various electronic devices with our daily routines has provided us with convenience, but it has also introduced problems related to electromagnetic interference. In our daily routines and professional endeavors, we are increasingly utilizing electronic devices that feature low power consumption, high speed, and advanced integration circuits, all of which enhance our daily experiences significantly. The trend toward the miniaturization of electronic components within highly integrated systems has resulted in smaller devices operating at faster speeds with closely spaced electronic parts [1]. Additionally, the demand for wearable electronic products has surged, leading to closer proximity between electronic devices and the human body. These devices, which consume and distribute electrical energy, are capable of producing electromagnetic interference (EMI), which can disrupt highly sensitive neighboring electronic systems by introducing noise into the input signals, potentially diminishing the longevity and proper functioning of electronic devices [2-4]. Therefore, it is imperative to implement appropriate shielding measures to mitigate this electromagnetic pollution. Doing so not only safeguards electronic components from malfunction but also protects the surrounding environment and individuals from harmful electromagnetic radiation [5].

The primary function of electromagnetic shielding is to either deflect or absorb electromagnetic waves by interacting with free electrons or holes in the conductor with the electromagnetic field. Whether the system deflects or absorbs electromagnetic waves is dependent on its unique properties. The capacity to absorb electromagnetic waves is influenced by the collective dielectric and magnetic losses, which are correlated with the dielectric constant and magnetic permeability, while the capacity to deflect is primarily determined by conductivity. When electromagnetic waves reach the surface of the shielding material, they are initially deflected. The remaining waves that are not deflected penetrate the interior of the shielding material and experience attenuation and loss during propagation. Upon reaching the opposite side of the shielding material, a portion of the electromagnetic waves will be deflected once more. These waves undergo numerous reflections and transmissions within the shielding material [6-8]. The shielding effectiveness (SE) is commonly used to quantitatively describe the shielding performance of a uniform shielding system:

$$SE = SE_A + SE_R + SE_M \quad (1)$$

where SE denotes the electromagnetic shielding effectiveness, SE_A denotes the absorption loss of the shielding material, SE_R denotes the reflection loss of the shielding material, and SE_M indicates the multiple reflection loss of the shielding material [9].

EMI shield materials must not only possess effective EMI shield performance but also adhere to standards of cost-effectiveness, lightweight, pliability, and thinness to broaden their range of applications, particularly in sectors requiring compact packaging and/or adaptable/wearable platforms. While the use of conductive metals like copper and silver conventionally yields superior EMI shielding efficiency, it proves difficult to sustain such high levels while accommodating these criteria. Conductive hydrogels, serving as versatile and adaptable conductive substances, merge sensory, driving, and information-processing capabilities and find widespread application in wearable electronic gadgets. Owing to their exceptional conductivity, porous composition, and aqueous nature, conductive hydrogels have garnered significant interest in the realm of electromagnetic shielding [10]. Optimal conductive hydrogel materials possess the ability to combine the benefits of both conductive and hydrogel materials, showcasing excellent conductivity, customizable mechanical flexibility, and straightforward processing attributes. The abundant aqueous environment within hydrogels, along with their porous structure and heightened electrolyte concentration, contribute to the enhancement of ion water transport and conductivity improvement. The consistently interconnected 3D network structure found in hydrogels grants flexibility and superior mechanical properties. Conductive self-repairing hydrogels can mend themselves through external stimuli (such as heat or the addition of self-repairing substances) or autonomous interactions within the hydrogel itself (such as dynamic chemical bonds and non-covalent interactions). In this study, we present a technique for fabricating polyvinyl alcohol (PVA) hydrogel for electromagnetic shielding using physical cross-linking.

2. Materials and methods

2.1. Materials and reagents

Polyvinyl alcohol (PVA, 99% hydrolyzed, DP=1750±50) was purchased from Sinopharm Chemical Reagent Co., Ltd. Calcium chloride ($CaCl_2$), aluminum chloride ($AlCl_3$), potassium chloride (KCl), magnesium chloride ($MgCl_2$), zinc chloride ($ZnCl_2$), sodium chloride (NaCl), sodium carbonate (Na_2CO_3), and sodium sulfate (Na_2SO_4) were purchased from Macklin. Deionized water was purified in the laboratory. All chemical agents were directly used without any purification.

Journal of Physics: Conference Series **2842** (2024) 012021 doi:10.1088/1742-6596/2842/1/012021

2.2. Fabrication of PVA hydrogel

We prepared physically cross-linked PVA hydrogel using the well-known freeze/thaw method. PVA powder (1 g) was dissolved in distilled water (7 g) and vigorously stirred at around 95°C. The gel solution obtained in the previous step was poured into a mold with dimensions of 5 cm×3 cm×0.2 cm. Subsequently, the gel was freeze-dried at 20°C and then thawed. This process was repeated three times to obtain the hydrogel. The hydrogel was then immersed in 10 wt.% solutions of calcium chloride, aluminum chloride, potassium chloride, magnesium chloride, zinc chloride, sodium chloride, sodium carbonate, and sodium sulfate for 3 days, thus getting $CaCl_2$/PVA hydrogel, $AlCl_3$/PVA hydrogel, KCl/PVA hydrogel, $MgCl_2$/PVA hydrogel, $ZnCl_2$/PVA hydrogel, NaCl/PVA hydrogel, Na_2CO_3/PVA hydrogel, and Na_2SO_4/PVA hydrogel with the comparison of deionized water (H_2O/PVA hydrogel).

3. Results and discussion

3.1. EMI SE measurement

As shown in Figure 1, the performance of PVA as an electromagnetic shielding material does not improve significantly after immersion in water. The slight enhancement is due to its water absorption, but the increase is minimal. In contrast, the hydrogels immersed in solutions of aluminum chloride, calcium chloride, magnesium chloride, and zinc chloride showed more significant performance improvements. Among them, the hydrogel in the magnesium chloride solution shows the highest enhancement in performance, with a ~7 dB improvement of SE_T (the sum of SE_A and SE_R) across different frequency bands. Additionally, as can be seen from Figure 2 and Figure 3, the improvement of SE_T is mainly due to the enhancement of SE_A, whose change is significantly greater than that of SE_R. Particularly in the 8-12 GHz frequency band, SE_A increases while SE_R decreases.

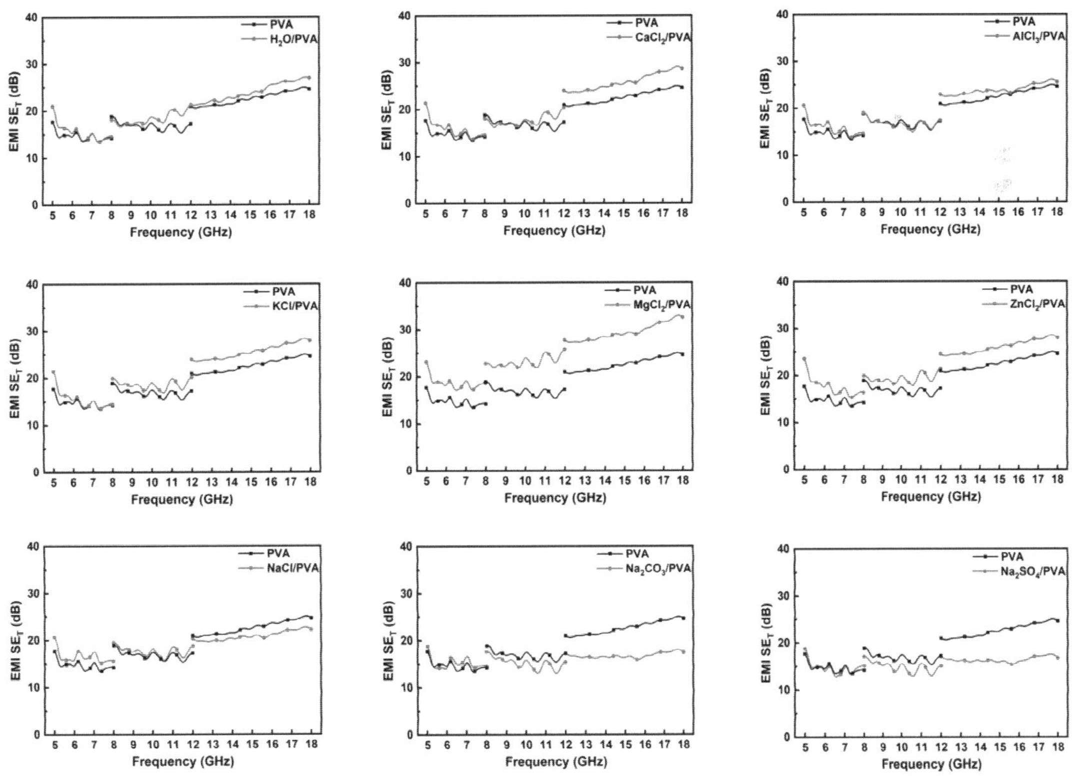

Figure 1. SE_T of H_2O, $CaCl_2$, $AlCl_3$, KCl, $MgCl_2$, $ZnCl_2$, NaCl, Na_2CO_3, and Na_2SO_4/PVA hydrogel.

Journal of Physics: Conference Series **2842** (2024) 012021 doi:10.1088/1742-6596/2842/1/012021

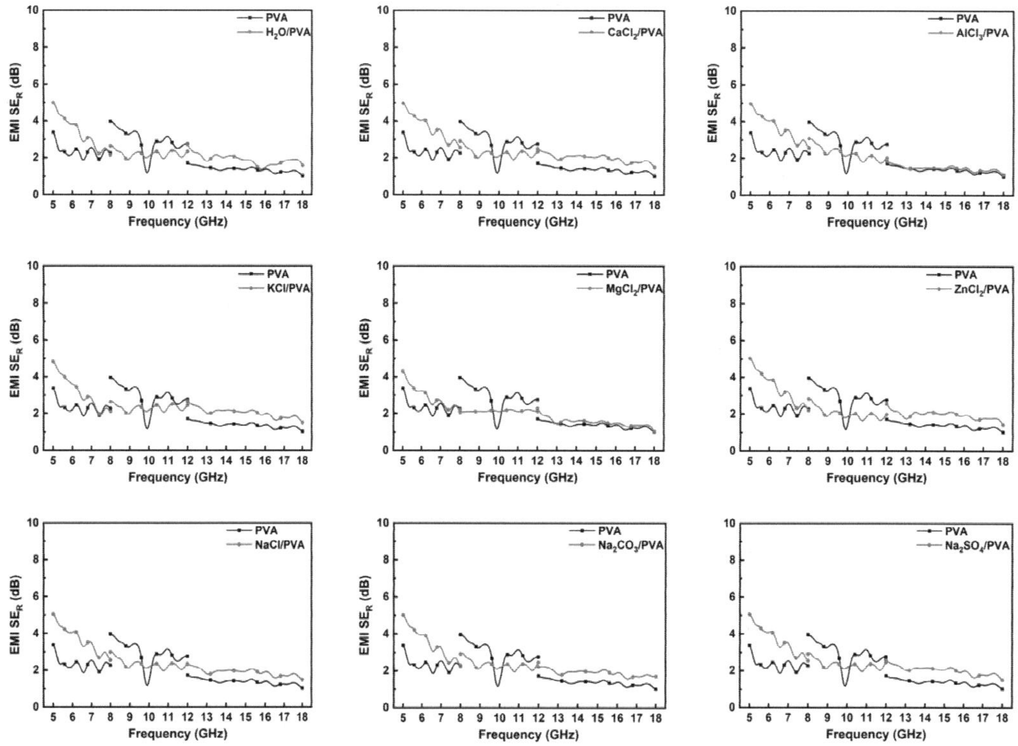

Figure 2. SE_R of H_2O, $CaCl_2$, $AlCl_3$, KCl, $MgCl_2$, $ZnCl_2$, NaCl, Na_2CO_3, and Na_2SO_4/PVA hydrogel.

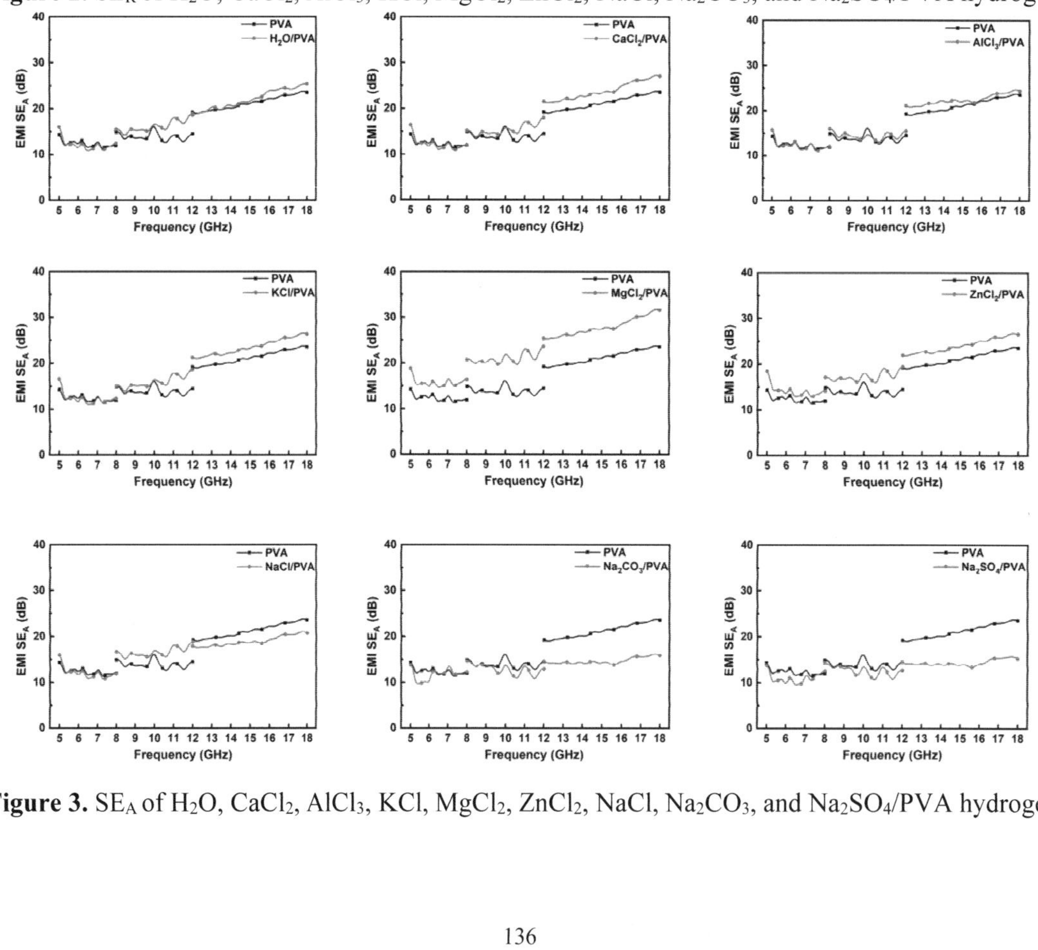

Figure 3. SE_A of H_2O, $CaCl_2$, $AlCl_3$, KCl, $MgCl_2$, $ZnCl_2$, NaCl, Na_2CO_3, and Na_2SO_4/PVA hydrogel.

Journal of Physics: Conference Series **2842** (2024) 012021 doi:10.1088/1742-6596/2842/1/012021

Reflection rate (R) and absorption rate (A) are important data for judging the shielding method of electromagnetic shielding materials, which can be calculated through Formulas (2) and (3):

$$R(\%) = \left(1 - 10^{-\frac{SE_R}{10}}\right) \times 100 \tag{2}$$

$$A(\%) = \left(1 - 10^{-\frac{SE_A}{10}}\right) \times (100 - R) \tag{3}$$

Table 1. Reflectance and absorbance of hydrogel before and after immersion.

Samples	Reflectance (%)	*Absorbance* (%)
PVA	48.84	51.16
H_2O/PVA	39.74	60.26
$CaCl_2$/PVA	40.43	59.57
$AlCl_3$/PVA	40.57	59.43
KCl/PVA	37.66	62.34
$MgCl_2$/PVA	39.49	60.51
$ZnCl_2$/PVA	37.05	62.95
NaCl/PVA	39.51	60.49
Na_2CO_3/PVA	37.19	62.81
Na_2SO_4/PVA	37.10	62.90

As shown in Table 1, the reflection rate of the hydrogel after immersion decreased, and the absorption rate increased. Additionally, the $MgCL_2$/PVA shows the greatest performance improvement across the range of 8-12 GHz, with the reflection rate decreasing from 48.84% to 39.49% and the absorption rate increasing from 51.16% to 60.51%. This demonstrates that it is an absorption-dominated electromagnetic shielding material as expected by the experiment. The magnesium ions introduced by physical crosslinking successfully enhanced the polarization of the hydrogel, thereby improving the absorption of electromagnetic waves.

3.2. Mass change of hydrogels

As shown in Figure 4(b), the mass of the hydrogels soaked in different types of solutions undergoes changes corresponding to their change in electromagnetic shielding performance. When immersed in sodium chloride, sodium carbonate, and sodium sulfate solutions, the hydrogel will undergo significant volume shrinkage and toughness enhancement, but its structure will be damaged, as evidenced by the white flocculent matter shown in Figure 4(a), and there will be a noticeable decline in mass and electromagnetic shielding performance. On the other hand, the hydrogels immersed in water, calcium chloride, aluminum chloride, magnesium chloride, and zinc chloride solutions exhibit an increase in electromagnetic shielding performance. This confirms that water, calcium chloride, aluminum chloride, magnesium chloride, and zinc chloride have successfully undergone physical cross-linking with polyvinyl alcohol, resulting in improved electromagnetic shielding performance.

Figure 4. (a) Physical image of NaCl, Na_2CO_3, and Na_2SO_4/PVA hydrogel. (b) Weight analysis of hydrogel before and after immersion.

4. Conclusion

The present study reports on the preparation of polyvinyl alcohol hydrogel as an absorption-dominated electromagnetic shielding material through the physical cross-linking method of freeze-drying and dissolution. The immersion of the hydrogel in salt solutions such as magnesium chloride effectively enhances the SMI SE value of the material, with a ~7 dB improvement across a wide frequency band (5-18 GHz). This research demonstrates the significant potential of hydrogels in the field of electromagnetic shielding in marine environments. The proposed method offers advantages such as simplicity of steps and low cost, which holds great potential for further development in the field of electromagnetic shielding.

Acknowledgments

This work was supported by the Basic Science Center Program of the National Natural Science Foundation of China (NSFC) Grant No. 51788104, Fundamental Research Funds for the Central Universities (2023ZCJH03, 2021XD-A041), Fund of State Key Laboratory of Information Photonics and Optical Communications (Beijing University of Posts and Telecommunications, P.R. China), National Natural Science Foundation of China Grant No. 12174154, Teaching Reform Projects at BUPT (2022CXCY-B03), and BUPT Excellent Ph.D. Students Foundation (CX2022237).

References

[1] He C et al. Research progress of electromagnetic shielding materials. Earth and Environmental Science, 2018, 170(3): 1315-1362.
[2] Yan D X, Pang H, Li B, et al. Structured reduced graphene oxide/polymer composites for ultra-efficient electromagnetic interference shielding [J]. Advanced Functional Materials, 2015, 25(4): 559-566.
[3] Chen Z, Xu C, Ma C, et al. Lightweight and flexible graphene foam composites for high-performance electromagnetic interference shielding [J]. Advanced materials, 2013, 25(9): 1296-1300.
[4] Yousefi N, Sun X, Lin X, et al. Highly aligned graphene/polymer nanocomposites with excellent dielectric properties for high-performance electromagnetic interference shielding [J]. Advanced Materials, 2014, 26(31): 5480-5487.
[5] Shahzad F, Alhabeb M, Hatter C B, et al. Electromagnetic interference shielding with 2D transition metal carbides (MXenes) [J]. Science, 2016, 353(6304): 1137-1140.
[6] Lin S, Liu J, Wang Q, et al. Highly robust, flexible, and large-scale 3D-metallized sponge for high-performance electromagnetic interference shielding [J]. Advanced Materials Technologies, 2020, 5(2): 1900761.
[7] Das N C, Liu Y, Yang K, et al. Single-walled carbon nanotube/poly (methyl methacrylate) composites for electromagnetic interference shielding [J]. Polymer Engineering & Science, 2009, 49(8): 1627-1634.
[8] Li Q, Zhang Z, Qi L, et al. Toward the application of high frequency electromagnetic wave absorption by carbon nanostructures [J]. Advanced Science, 2019, 6(8): 1801057.
[9] Song Q, Ye F, Kong L, et al. Graphene and MXene nanomaterials: toward high-performance electromagnetic wave absorption in gigahertz band range [J]. Advanced Functional Materials, 2020, 30(31): 2000475.
[10] Shu J C, Cao W Q, Cao M S. Diverse metal-organic framework architectures for electromagnetic absorbers and shielding [J]. Advanced Functional Materials, 2021, 31(23): 2100470.

Formation mechanism of Hf-based ceramic particles in tungsten alloys

Jincheng Li, Xiang Wei, Lin Gao, Longhui Lou, Benzhe Sun and Yongli Liu*

College of Materials Science and Engineering, Northeastern University, Shenyang, Liaoning, 110819, China

*Corresponding author's e-mail: ylliu@imp.neu.edu.cn

Abstract: To meet the requirements of high strength of tungsten (W) based alloys at high temperature applied in the aerospace field, the preparation and formation of Hf-related ceramic particles in W were studied. It is found that the HfO_2 phase with a size of about 500 nm was observed at the grain boundary in both W-0.26 wt.% HfC and W-0.25 wt.% HfH_2-0.125 wt.% C systems. Further study of phase stability and thermodynamic properties of Hf(W)-O(C) compounds reveals that the HfO_2 phase possesses the most stability of all the Hf-related compounds, and the oxidation of Hf is preferred to the carbonization. It is predicted that more stability of W base metal, higher temperature, and higher carbon concentration can reduce the formation of HfO_2 while promoting the formation of the HfC particle phase.

1. Introduction

Tungsten (W) possesses the highest melting point (3695 K) and extreme hardness of all metals, which makes it widely used in electronics, military, aerospace, etc. [1-2]. In the nuclear fusion reactors, according to the research and demonstrations over the years, it can be determined that W is the significant material for plasma-facing materials (PFMs) [3]. When utilized in non-cooled nozzles of solid fuel rocket engines with working temperatures exceeding 2000 °C, W will encounter high-temperature, high-velocity gas erosion, as well as intense mechanical impact and thermal stress [4-6]. However, severe drawbacks like high ductile-to-brittle transition temperature (DBTT), insufficient high-temperature strength, low-temperature brittleness, and low toughness prevent W from its wide use in most applications.

A range of studies have shown that appropriate annealing temperature and high plastic deformation can effectively reduce the DBTT of tungsten and improve its toughness [7]. Meanwhile, the solid solution strengthening method can improve the brittleness of the material, such as adding Re to W to improve the low-temperature plasticity and tensile strength of W [8-10]. Additionally, hard and dispersed second-phase particles have been adopted in W base metal to enhance its strength and toughness [11-15]. The present work mainly focuses on the dispersed strengthening of the ceramic particle phase in W. The hard particles can effectively impede the movement of dislocations and increase the strength and hardness of the alloy with little decrease in plasticity and toughness. The dispersed phase is generally oxide, carbide, or nitride particles with high thermal stabilities so that the strengthening effect can still be sustained at higher temperatures.

In W alloys, the preferred particle phase is hafnium carbide (HfC), which possesses the highest melting point among carbides (3890°C). A small quantity of HfC particles can be transformed into monoclinic HfO_2 particles by actively adsorbing oxygen impurities, which can effectively purify the

Journal of Physics: Conference Series **2842** (2024) 012022 doi:10.1088/1742-6596/2842/1/012022

grain boundaries, lower the DBTT of W, and improve its overall mechanical properties [7, 16].

The conventional preparation way of the particle phase is to add the corresponding particle material into the W matrix by powder metallurgy method [11-12, 17]. For example, Vikhman et al. [11] obtained HfC particle phase with a diameter of 2 um by keeping the raw material containing 5-30 Vol% HfC at 5-1000 MPa, 1850 °C for one hour. However, the particle phase produced by this method tends to be oversized (1-4 um). When the particles are distributed on the grain boundaries, it is prone to induce cracks and lower the material properties. In recent years, researchers have proposed several improved methods for the sintering process such as ultra-high pressure sintering [18] and spark plasma sintering (SPS) [12, 19-22], which can effectively enhance the mechanical properties of materials. Xu et al. [18] prepared nanocrystalline HfC ceramics with a grain size of 130 nm by sintering at 1500°C and 15 GPa.

In-situ reaction synthesis involves the formation of particle phases through reactions between raw materials and can result in uniformly distributed particles with smaller sizes (< 0.5 μm), improving the effect of dispersion strengthening. Lee et al. [12] adopted carbothermal reduction (CTR) and SPS method based on W and 30 Vol% Hf, and obtained the particle size of 0.5 um. More studies are needed to explore the effect of preparation methodology and raw materials for the formation of Hf-O(C) particle phase. In this work, two different kinds of raw materials (W+HfC/W+HfH_2+C) were designed to prepare the dispersed particles. The formation and existence state of Hf were analyzed by XRD method and TEM characterization method. Further, the formation mechanism of the Hf related particle phase was analyzed by combining with calculations of the phase stability and thermodynamic properties.

2. Materials and methods

2.1. Materials preparation and microstructure characterization

Samples were prepared based on two kinds of ingredients: W-0.26 wt.% HfC and W-0.25 wt.% HfH_2-0.125 wt.% C. HfH_2 is expected to react with the carbon (C) from raw materials or the O_2 in the atmosphere to form carbides or oxides in the sintering process due to its low melting point [23]. The raw material was prepared by mixing, cold isostatic pressing, and intermediate frequency furnace sintering. The synthesis process is as follows. Firstly, the alloying powder and W ball grinder were put into the ball milling jar, and then the ball was milled for 24 hours with the appropriate amount of alcohol in the turning device. Secondly, the mixed raw materials were dried in the nitrogen-protected oven for 24 hours. Then, the cold isostatic pressing was applied to the ball-milled powder in the rubber sleeve to complete the molding of the powder. Finally, the cold isostatic pressed bar was sintered in the intermediate frequency furnace in a hydrogen atmosphere for one hour at 1200 °C, 1600 °C, and 1850 °C, respectively. The sintered billets were taken out after natural cooling. Microstructure characterization was conducted using X-ray diffraction and transmission electron microscopy (TEM, FEI Tecnai G2 T20) tool with a point resolution of up to 2.3 nm. The X-ray was produced by Cu-Kα radiation (λ =1.541 nm) operated at 30 mA and 30 kV.

2.2. Computational models and methods

2.2.1. Calculation of ground state properties. For the purpose of investigating the formation mechanism of the particle phase, the present work studied the phase stability of eight compounds related to W and Hf that may be produced during the precipitation process of Hf-O(C) compounds. The calculations were conducted based on density-functional theory, where the interactions between the ion core and the electrons were described by projected augmented wave (PAW) [24-26]. The stress and Hellman-Feynman forces were minimized by the conjugate gradient method with a stress convergence of less than 10^{-3} eV/Å to determine the ground state geometry. The plane-wave cut-off energy was set to 450 eV to extend the Kohn-Sham orbital. We used a 13 × 13 × 13 Monkhorst-Pack k-point mesh [27] for both structure optimization and total energy calculations. The valence electron configurations of W, Hf, C, O, and H are 5p6s5d, 4p5s4d, 2s2p2, 2s2p4, and 1s1, respectively.

2.2.2. Calculation of Gibbs free energy. During the sintering process, the W alloy reacts with the added HfC, HfH_2, carbon powder, and residual oxygen. Therefore, the change in standard free energy ΔG^{θ} is adopted to evaluate the reaction direction and precipitation tendency of the chemical compounds [28]. ΔG^{θ} is calculated according to the following formula:

$$\Delta G^{\theta} = \sum v_i G_i \tag{1}$$

where v_i is the fraction of substance i. G_i is the relative molar free energy value of each substance i at different temperatures, which can be found in Table 2-1 in the literature [29].

3. Results analysis and discussion

3.1. Phase analysis

The XRD results of the above two W-Hf systems are shown in Figure 1. It can be seen that the 2θ angle is between 60° and 65°, and two strong diffraction peaks come from the standard Si single-crystals, which are adopted for the calibration of the overall XRD patterns. The sample mainly consists of the W matrix and a small amount of the HfO_2 phase. Identification of XRD results shows that the HfO_2 phase (JCPDS No. 78-0049) possesses a monoclinic structure with crystal lattice constants of a = 5.117 Å, b = 5.175 Å, c = 5.292 Å, α = 90.00°, β = 99.22°, and γ = 90.00°. Comparison of the diffraction peaks of W with the standard PDF card (JCPDS No. 04-0806) reveals that each peak is shifted to a low-angle direction, indicating that the lattice parameter of the W matrix observed in the present experiment is slightly larger than that of the perfect W crystal. Two reasons are related to this result. Firstly, a very small amount of Hf is dissolved in the W matrix. The atomic radius of Hf (2.16 Å) is slightly larger than that of W (2.02 Å), thus causing the enhancement of atomic volume. Secondly, the C atoms in the raw material may diffuse into the W matrix under high-temperature conditions and occupy the octahedral interstitial in the body-centered cubic W matrix, which leads to a slight increase in the lattice parameter.

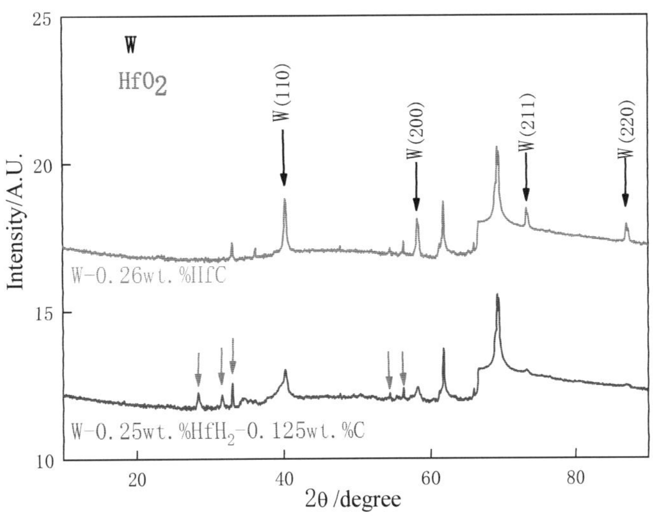

Figure 1. XRD results of two kinds of sintered W-Hf samples.

Figure 2(a) shows the TEM morphology of the W-0.25 wt.% HfH_2-0.125 wt.% C sample. It can be seen that a spheroid particle with about 500 nm is located on the grain boundary between two W grains. Figure 2(b) shows the selected electron diffraction pattern of this grain, which clearly shows two sets of diffraction patterns, one of which belongs to the W matrix with body-centered cubic structure, and the other set with denser diffraction spots corresponds to the HfO_2 phase. From the morphology (Figure 2a), it can be clearly seen that the boundary between the HfO_2 particles and the W matrix is relatively dense and flat, indicating a strong interfacial bonding strength between the two phases. Meanwhile, from the diffraction pattern (Figure 2b), it can be seen that although there is a certain mismatch between the HfO_2 particles and the W matrix, the two-phase still possess the following crystal orientation relationship: $[01\bar{1}]_{HfO2}//[210]_W$ and $(333)_{HfO2}//(1\bar{2}1)_W$.

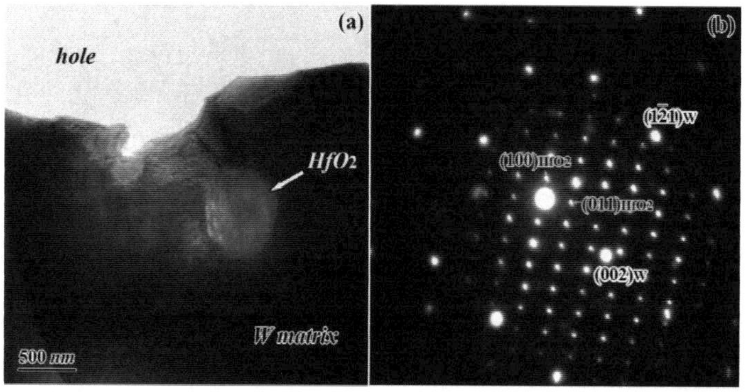

Figure 2. TEM morphology of W-0.25 wt.% HfH_2-0.125 wt.% C sample and the corresponding selected area electronic diffraction pattern.

In summary, during the preparation process, in addition to being dissolved in the matrix, Hf is also oxidized to form HfO_2. HfC may also exist, but it is not detected due to the low content. The formation of the HfO_2 particle phase is related to two reasons. From one perspective, residual oxygen exists even in the hydrogen protective condition, which provides the source of O for oxidation. From the other perspective, Hf and HfC in the raw material can be easily oxidized at the temperature of 400 °C-500 °C, and the high-temperature preparation environment of W makes the oxidation reaction more sufficient. HfO_2 has a variety of crystal structures due to different growth conditions. The study of HfC thermodynamic oxidation by Li Hui et al. [29] shows that Hf in HfC is oxidized to produce monoclinic HfO_2 at the low-temperature region (T < 1699.85 °C) and tetragonal HfO_2 at the high-temperature region (1699.85 °C ≤ T < 2899.85 °C). According to this analysis, at the early stage of sintering, the oxidation of HfC causes the formation of the monoclinic HfO_2 phase. Considering that no C content is left, the oxidation reaction formula of HfC should be as follows:

$$HfC+2O_2=HfO_2+CO_2 \quad (2)$$

For the raw materials of W-0.25 wt.% HfH_2-0.125 wt.% C, as HfH_2 is easily decomposed into Hf and H_2, it is possible that the decomposed Hf reacts with carbon powder and residual oxygen to form HfC and HfO_2 phases. The reason why Hf forms HfO_2 in W instead of HfC remains to be further explored.

3.2. Formation mechanism of HfO_2 particle phase

The first prerequisite for the phase formation of W(Hf)-C(O,H) compounds is that the phase is stable under specific conditions. The ultimate existence of a compound depends mainly on the ΔG^{θ} of the reaction that forms the compound. The more negative the value is, the higher the reaction driving force is, and the compound will form preferentially. Therefore, the present work further studys the stability of Hf-related carbides and oxides (Figure 3) that may be formed during the preparation of the W-HfC sample based on the change in standard Gibbs free energy between the raw ingredients and the components. The results are shown in Table 1 and Figure 4.

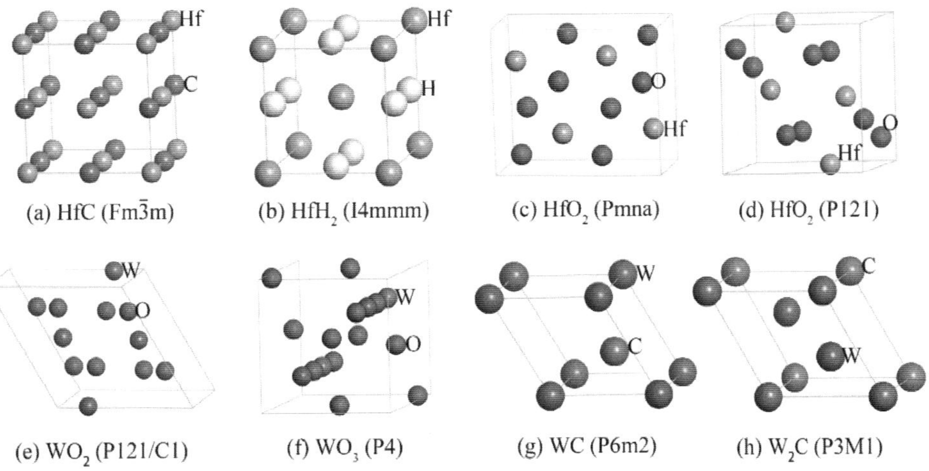

Figure 3. Crystal structures of W(Hf)-C(O,H) compounds.

Table 1. Lattice parameters and formation energy of W(Hf)-C(O,H) phases.

Structure/ space group	Lattice constant (Å)		Formation energy (eV/atom)
	Calculated value	Experimental value	
WC (P6m2)	a=2.9188, c=2.8446	a=2.906, c=2.838 [30]	-0.1667
W_2C (P3M1)	a=3.0613, c=4.6600	---------------	0.0160
HfC (Fm$\bar{3}$m)	a=4.6470, c=4.6470	a=4.6500, c=4.6500 [31]	-0.9596
HfH_2 (I4mmm)	a=3.4868, c=4.2862	a=3.4780, c=4.3610 [32]	-0.5825
HfO_2 (Pmna)	a=3.3198, b=5.5442, c=6.4978	a=3.3020, b=5.5690, c=6.4530 [33]	-3.5842
HfO_2 (P121)	a=5.1423, b=5.1951, c=5.3260	a=5.1170, b=5.1754, c=5.2915 [34]	-3.5849
WO_3 (P4)	a=5.3236, b=5.3236, c=7.9842	a=5.2759, b=5.2759, c=7.8462 [35]	-2.1398
WO_2 (P121/C1)	a=5.5850, b=4.9442, c=5.70830	a=5.5630, b=4.8960, c=5.5630 [36]	-1.9290

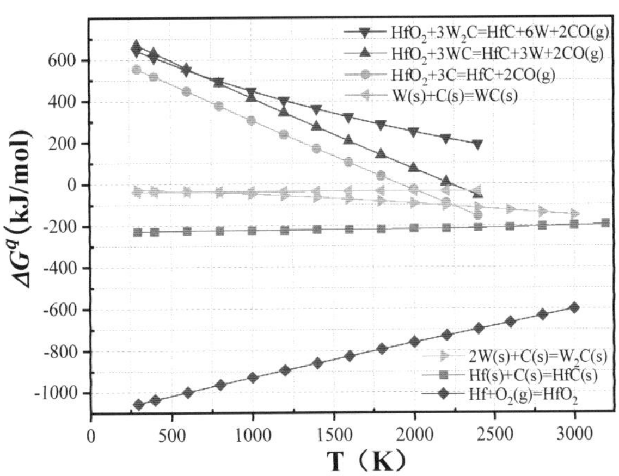

Figure 4. Variations of ΔG^{θ} change with the temperature for W(Hf)-C(O,H) compounds.

3.2.1. Ground-state stability of W(Hf)-C(O,H) compounds. Figure 3 shows the crystal structures of W(Hf)-C(O,H) compounds that may be formed during the experimental preparation process. The comparison of the lattice parameters of these structures between the calculated values and experiment results is shown in Table 1. According to Table 1, the predicted lattice parameters conform to the experimental results, which proves the reliability of the model and the setups adopted for the calculations. Negative values of the formation energies indicate that the compounds can be generated spontaneously and keep stable under equilibrium conditions; positive values indicate that the stability needs to be further investigated or that they are metastable phases; and positive values of the formation energy of W_2C indicate that it is an unstable phase or metastable phase that may exist under special conditions. The formation energies of other W(Hf)-C(O,H) phases are negative, suggesting that these structures are stable under equilibrium conditions. The formation energies of the considered compounds change in the following order: $E_{HfO2} < E_{WO3} < E_{WO2} < E_{HfC} < E_{HfH2} < E_{WC} < E_{W2C}$, and accordingly, the stability of the phases descends in the order of $E_{HfO2} > E_{WO3} > E_{WO2} > E_{HfC} > E_{HfH2} > E_{WC} > E_{W2C}$. Based on the above calculations, it can be seen that HfO_2 is the most stable Hf-related ceramic compound, followed by WO_3 and WO_2. The oxides are more stable than the carbide. Meanwhile, the formation energies of the Hf-related compounds are lower than that of W-related ones, indicating that the Hf-related compounds are easier to form than the W-related ones under the same situation.

3.2.2. Thermodynamic states of W(Hf)-C(O) compounds. During the preparation process of W-0.25 wt.% HfH_2-0.125 wt.% C, the decomposition reaction of HfH_2 occurs at the early stage of sintering because the sintering temperature is considerably higher than the melting point of HfH_2 [23]:

$$HfH_2=Hf+H_2 \tag{3}$$

The generated Hf can subsequently react with carbon in the raw material and oxygen in the gaseous condition as follows:

$$Hf+O_2=HfO_2 \tag{4}$$

$$Hf+C=HfC \tag{5}$$

$$C+O_2 \rightarrow CO_2 \tag{6}$$

To further explore the formation tendency of W(Hf)-C(O,H) compounds at high temperatures, the variations of ΔG^{θ} change with the temperature for W(Hf)-C(O,H) compound are shown in Figure 4. Since the products of reactions (2) and (4) are both HfO_2, and the driving force of the corresponding reactions is much larger than that of (5), oxygen is mainly consumed to form HfO_2 in the sintering stage. For the W-0.26 wt.% HfC system, the driving force of reaction (2) is larger than that of (5), which suggests that the oxidation reaction of HfC occurs more easily than the oxidation of Hf. Meanwhile, because defect locations such as grain boundaries are more likely to absorb O atoms and become channels for rapid diffusion of O atoms, a large number of HfO_2 particles are formed at the Hf grain boundaries. Therefore, the HfO_2 phase is reasonably observed in our experiment study rather than HfC.

3.3. Formation of HfC particle phase

The key to reduce HfO_2 and retain more HfC is to control the oxidation of Hf. The existence state of Hf in W alloys is closely related to the stability of the matrix, temperature, atmosphere, and ingredients. On the one hand, the oxidation of Hf in W alloys is accomplished through the outward diffusion of Hf and the inward diffusion of O. It is expected that the oxidation can be effectively suppressed by the addition of matrix stabilizing elements so that Hf is dissolved in the matrix. On the other hand, it can also be seen from Figure 4 that when the temperature is higher than 2000 K, and the C source is sufficient, the ΔG^{θ} of reaction $HfO_2+3C=HfC+2CO$ will drop to the negative value. When the temperature is higher than 2500 K, the ΔG^{θ} is lower than that of the reaction Hf+C=HfC, predicting that the carbonation of HfO_2 into HfC may occur.

Among above mentioned factors, both the temperature and content are important. For the two present raw ingredients, though the sintering temperature (1850 °C, 2123 K) is slightly higher than

2000 K, the HfO_2+3C=HfC+2CO reaction may occur to form HfC, and HfC has not yet been detected by the XRD method due to its low concentration of Hf or C. The Hf content of both two raw materials preparation systems is low. After preferential oxidation to form HfO_2, the content of Hf is further decreased. Therefore, little HfC remains finally to be detected in XRD. In all, sufficient C is required in addition to the high temperature. It is predicted that higher temperature, sufficient C, and longer reaction time can not only promote the formation of HfC particles but also help to transform the HfO_2 into HfC by reacting with C. The researchers believe that the HfC particles observed in heat treatment samples are likely to originate from the reduction reaction of HfO_2 particles during sintering [37].

4. Conclusion

In this work, experiments and calculations were combined to study the formation of Hf-related ceramic particles in W. The main results are as follows,

(1) Sintering of two kinds of raw ingredients (W-0.26 wt.% HfC and W-0.25 wt.% HfH_2-0.125 wt.% C) produces HfO_2 particle phase with size of about 0.5 um. No HfC particle phase was detected, which may be related to its low concentration. It is found that the oxidation reaction of Hf in W is preferred to the carbonization reaction.

(2) The oxidation mechanism of Hf in the sintering process was studied. It is found that the HfO_2 phase is the most stable of all Hf-related ceramic compounds, and the Hf oxides are more stable than the Hf carbide. The ΔG^{θ} of Hf or HfC oxidation reaction is much more negative than the carbonization reaction of Hf. Therefore, the oxidation of Hf preferentially occurs than the carbonization of Hf.

(3) The formation conditions of the HfC phase were explored. The state of Hf in W alloys is related to the lattice stability of the W base metal, O and C concentration, and sintering conditions. More stable W lattice, higher temperature (>2500 K), longer preparation time, and higher carbon concentration may promote the formation of HfC.

References

[1] Shen D N (2020) Preparation and investigation on microstructure and properties of ultra-fine grained tungsten alloys. Nanjing University of Science & Technology, Nanjing.
[2] Li Y C, et al. (2021) Microstructure and high temperature mechanical properties of advanced W-3Re alloy reinforced with HfC particles. Mater. Sci. Eng.A, 814:141198.
[3] Wang Q (2021) Theoretical Study on thermodynamic and interfacial properties of Tungsten and Tungsten alloys for fusion reactors. Nanchang University, Nanchang.
[4] Zhao Y L and Lei M (2021) Research progress of Tungsten-based plasma materials in fusion reactors. Rare Met. Mater. Eng., 50(09):3399-3407.
[5] Xu Z and Zheng J W (2021) Molding technology of solid rocket motor C/C throat liner. New Chem. Mater., 49(09):222-225.
[6] Wang F Q, et al. (2020) Overview of large solid rocket motor throat development. Aerosp. Mater. Technol., 50(06):8-14.
[7] Wang Y K (2020) Fabrication and properties of Hafnium Carbide dispersion strengthened Tungsten based alloys. University of Science and Technology of China, Hefei.
[8] Zhao B L (2021) Fabrication and properties of fine-grain Tungsten alloy. University of Science and Technology of China, Hefei.
[9] Wang M M (2021) Microstructure and properties control of W-ZrC alloys with multi-scaled interface structure. University of Science and Technology of China, Hefei.
[10] Li Y C, et al (2021) Microstructural evolution and hot deformation behavior of W-3Re-5HfC alloy. Int. J. Refract. Met. Hard Mater., 98:105535.
[11] Vikhman S V, et al. (2011) Activated sintering of W-HfC composite materials. Russ. J. Non-Ferrous Met., 52:285-289.
[12] Lee J M, et al. (2013) Advanced W-HfC cermet using in-situ powder and spark plasma sintering. J. Alloys Compd., 552:14-19.

[13] Yan Z S (2020) Properties improvement of Tungsten alloy based on plasma facing material by Y_2O_3 and TiC. Hefei University of Technology, Hefei.
[14] Liu W L (2020) Study on microstructure of tungsten-based ultra-high temperature refractory metal composites. Northeastern University, Shenyang.
[15] Gu M (2019) Study on recrystallization behavior of rolled yttrium oxide dispersion strengthen tungsten. Hefei University of Technology, Hefei.
[16] Liu X S, et al. (2021) Investigations on microstructure and mechanical properties of HfC particle-reinforced W-Mo composites. Int. J. Refract. Met. Hard Mater., 100:105634.
[17] Huang B Y, et al. (2019) Development of modern powder metallurgy materials and technology. Chin. J. Nonferrous Met., 29(09):1917-1933.
[18] Xu H Y, et al. (2024) Microstructure evolution of ultra-fine grained HfC ceramics with high hardness sintered under ultrahigh pressure by plastic deformation densification mechanism. Int. J. Refract. Met. Hard Mater., 118:106425.
[19] Xiao P, et al. (2021) Improved toughness of HfC with tantalum fibre net and tantalum foil. Ceram. Int., 47(13):19260-19266.
[20] Qian Y J (2021) Preparation of multi-element WC-based composites by spark plasma sintering. Henan University of Technology, Zhengzhou.
[21] Peng T S (2021) Study on SPS sintering of Hafnium Carbide and Tungsten Carbide. Hefei University of Technology, Hefei.
[22] Yang S (2021) Study on fabrication and modification of Ta_2O_5 ceramics by spark plasma sintering. Chongqing University, Chongqing.
[23] Yura K, et al. (1985) Isotherms of the hydrogen-hafnium system. J. Chem. Soc., Faraday Trans., 82:101-114.
[24] Kresse G and Furthmuller J (1996) Efficient iterative schemes for ab initio total-energy calculations using a plane-wave basis set. Phys. Rev. B, 54:11169-11186.
[25] Kresse G and Joubert D (1999) From ultrasoft pseudopotentials to the projector augmented-wave method. Phys. Rev. B, 59:1758-1775.
[26] Blöchl P E (1994) Projector augmented-wave method. Phys. Rev. B, 50:17953-17979.
[27] Monkhorst H J and Pack J D (1976) Special points for Brillouin-zone integrations. Phys. Rev. B, 13:5188-5192.
[28] Li H, et al. (2015) A thermodynamic study on the oxidation of HfC. Materials Reports, 29:136-151.
[29] Liang Y J and Che Y C (1993) Inorganic Thermodynamic Data Manual. Northeastern University Press, Shenyang.
[30] Litasov K D, et al. (2010) Pressure-volume-temperature equation of state of tungsten carbide to 32 GPa and 1673 K. J. Appl. Phys, 108(5):053513.
[31] Aigner K, et al. (1994) Lattice parameters and thermal expansion of $Ti(C_xN_1\text{-}x)$, $Zr(C_xN_{1-x})$, $Hf(C_xN_{1-x})$ and TiN_{1-x} from 298 to 1473 K as investigated by high-temperature X-ray diffraction. J. Alloys Compd., 215:121-126.
[32] Sidhu S S (1954) The effect on metal-metal bonds of increased concentration of hydrogen in hafnium dihydride. Acta Crystallogr., 7:447-449.
[33] Tang J, et al. (1998) A high-pressure high-temperature X ray study of phase relations and polymorphism of HfO_2. Washington DC American Geophysical Union Geophysical Monograph Series, Washington.
[34] Hann R E, et al (1985) Monoclinic crystal structures of ZrO_2 and HfO_2 refined from X-ray. J. Am. Ceram. Soc., 68:285-286.
[35] Vogt T, et al. (1999) The high-temperature phases of WO_3. J. Solid State Chem., 144:209-215.
[36] Palmer D J and Dickens P G (1979) Tungsten dioxide: structure refinement by powder neutron diffraction. Acta Crystallogr., 35:2199-2201.
[37] Liu M, et al. (2022) Microstructural characterization of the second-phase particles in W-Re-HfC alloys. Journal of Shenyang University of Technology, 44(04):403-408.

Study on porous carbon materials derived from Chitosan

Shasha Wang[1], Zhi Tian[2], Jianlin Zhao[2], Yuelong Xu[1], Yangguang Zhang[2], and Lihui Zhang[1*]

[1] Institute of Energy Resources, Hebei Academy of Sciences, Shijiazhuang, 050081, China
[2] Hebei Baoli Engineering Equipment Corporation Limited, Hengshui, 053000, China

[*] Corresponding author's email: zlhkxy@sohu.com

Abstract. The materials with high specific surface area, uniform pore size distribution, and high nitrogen content were prepared by using chitosan as raw material and a low co-soluble mixture (lithium nitrate, potassium nitrate, and potassium sulfate) as a pore-forming agent and salt template agent. The CO_2 adsorption capacity of the material was tested at 25°C (2.15 mmol/g) and 0°C (2.88 mmol/g). It was found that more pores and higher nitrogen content were conducive to the adsorption of CO_2 by the material.

1. Introduction

A low eutectic mixture refers to the phenomenon of two or more solid substances being added together to reduce the melting point of the mixture, which generally has the characteristics of low cost, environmental friendliness, and design[1]. In recent years, there have been many reports on the application of low co-soluble mixtures as salt template agents to the synthesis of carbon materials. In the process of calcination of carbon materials, low co-soluble mixtures can melt at a lower temperature, and can also flow in the structure of carbon materials raw polymer, which plays the role of supporting structure, that is, the role of pore-making agents[2]. Porous carbon material with a high specific surface area and high porosity is a good CO_2 adsorption material. Large specific surface area can provide more CO_2 adsorption sites, and large pore volume can store more CO_2[3-4]. In recent decades, scientists have studied more carbon materials and CO_2 adsorption[5-7]. In this paper, nitrogen-rich chitosan was used as raw material to prepare carbon materials, and the same amount of lithium nitrate, potassium nitrate, and potassium sulfate mixture was used as template agent and nitrogen source, which could play a role in increasing the content of pores and nitrogen.

2. Methods

10 grams of chitosan and 6 grams of low eutectic salt mixture (2 grams of lithium nitrate, 2 grams of potassium nitrate, and 2 grams of potassium sulfate) are accurately weighed into the ball mill, at the speed of 500 revolutions per minute, a total of 1.5 hours of ball milling, each ball milling for half an hour to rest for 10 minutes, to get intermediate products. Finally, the intermediate product is calcined at a high temperature, and the intermediate product is calcined in a tubular carbonization furnace under a nitrogen atmosphere, and the heating procedure is as follows: From room temperature heating rate of 3°C/min to 240°C, after maintaining 240°C for one hour, then heating at 5°C/min rate to 900°C, maintaining 900°C for 2 hours, and finally cooling at 4°C/min rate to room temperature, black solid product is obtained. The product carbon material can be obtained by washing the black solid product

Journal of Physics: Conference Series **2842** (2024) 012023 doi:10.1088/1742-6596/2842/1/012023

in 1 M hydrochloric acid aqueous solution until it is neutral.

3. Results and discussion

To test the porosity of the obtained carbon material, we tested the specific surface area and pore size distribution of the carbon material, as shown in Figure 1. It can be seen from Figure 1 (a) that the nitrogen absorption and desorption curves have a large hysteresis loop, indicating that the material structure is rich in mesoporous structures, which can be confirmed by Figure 1 (b). Figure 1 (b) shows that the pore size distribution of carbon materials is mainly concentrated in the range of 18-22 nm, which belongs to a typical mesoporous structure. The specific surface area of the material is high, 890 m^2/g, indicating that the added low co-soluble salt mixture (lithium nitrate, potassium nitrate, and potassium sulfate) plays the role of pore-making agent and template agent.

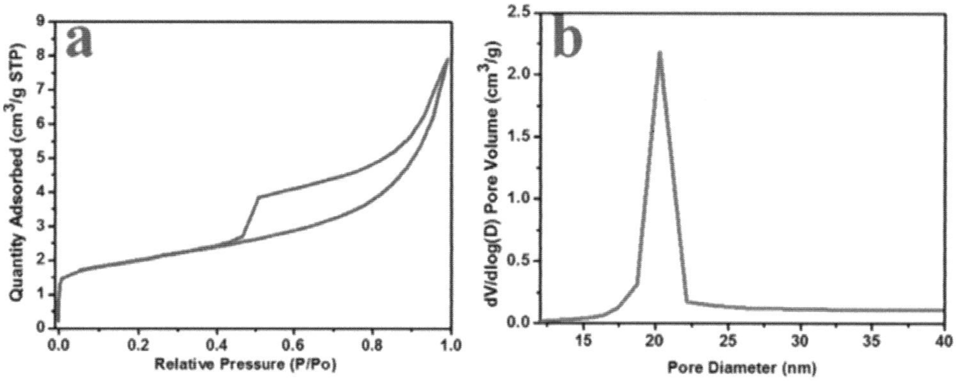

Figure 1. Absorption and desorption curve (a) and pore size distribution diagram (b).

The microstructure of the material was observed by scanning electron microscopy. As shown in Figure 2, the material is a coral-coated structure with abundant pores, indicating that chitosan is one of the excellent raw materials for the preparation of porous carbon materials, and lithium nitrate, potassium nitrate, and potassium sulfate also play an active role in the structure of the material.

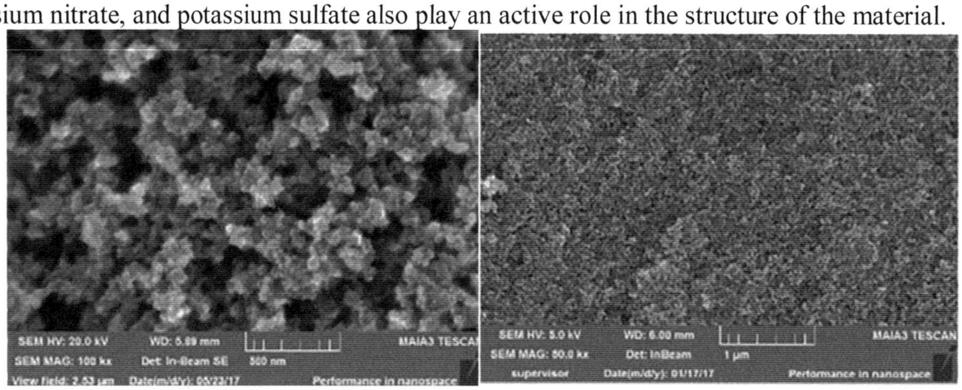

Figure 2. SEM images at different magnification.

To detect whether the nitrogen in the raw material and the nitrogen source was successfully doped into the carbon material, an XPS test was performed on the carbon material, and sub-maps of the total yield (Figure 3a), carbon (Figure 3b), nitrogen (Figure 3c), and oxygen (Figure 3d) were obtained. From Figure 3, we can see that nitrogen and oxygen are successfully doped into the carbon material. The atomic content of this material of carbon, nitrogen, and oxygen are 83.83at%, 0.82at%, and 15.35at%.

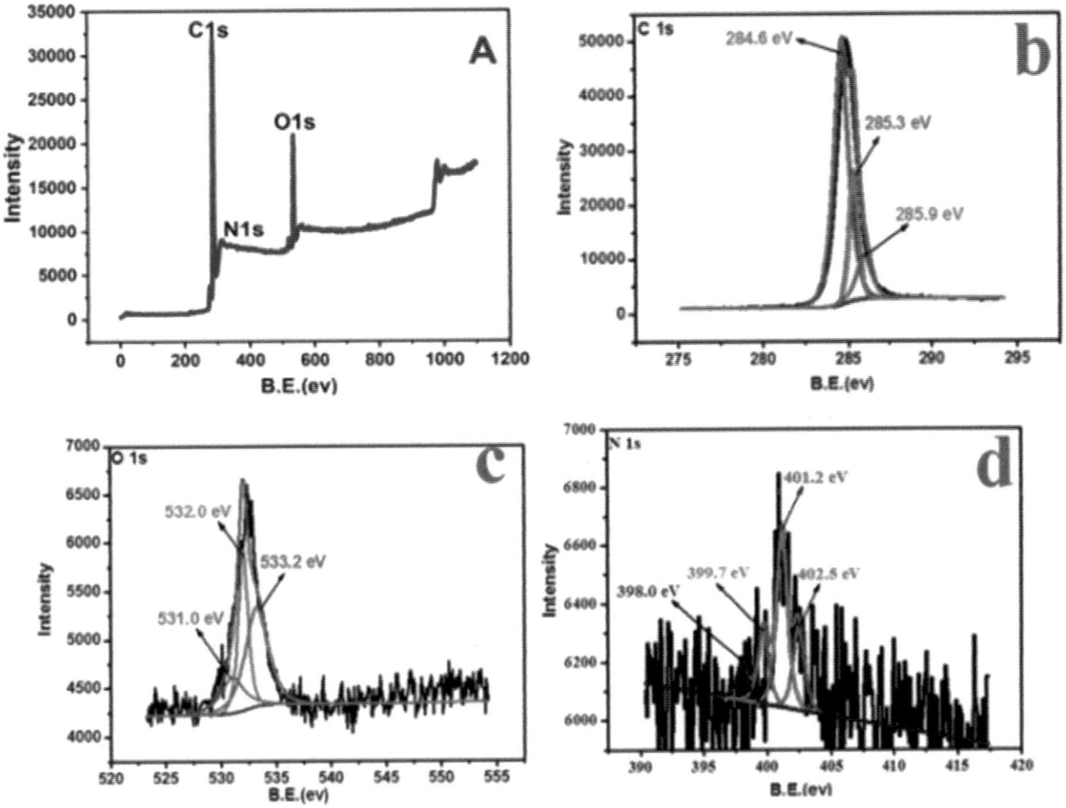

Figure 3. XPS atlas of materials.

Carbon materials rich in nitrogen and oxygen are potential materials for CO_2 adsorption and are also one of the research hotspots in recent years[8-9]. We conducted CO_2 adsorption experiments on carbon materials at 0°C and 25°C to study the adsorption capacity of the materials. As shown in Figure 4, the adsorption capacities of CO_2 materials at 0°C and 25°C are 2.88 mmol/g and 2.15 mmol/g, respectively. We found that the lower the temperature, the greater the adsorption capacity of CO_2 materials because the kinetic energy of CO_2 is small at low temperatures. Strong adsorption force, large adsorption capacity. The alkaline impurity atoms rich in carbon materials can effectively adsorb acidic CO_2 molecules, so the higher the nitrogen content of the material, the greater the adsorption capacity of the material for CO_2.

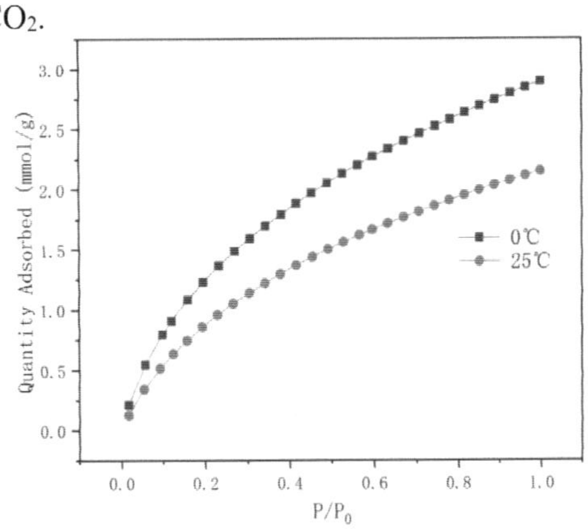

Figure 4. CO_2 adsorption curve.

Understanding the affinity of the material for CO_2 adsorption is also important, as these properties determine the ability of the carbon material to capture and store CO_2. The ability of a material to trap CO_2 can be determined by measuring the heat of adsorption (enthalpy) of the material for carbon dioxide. The isometric heat of adsorption is usually obtained by using the Clausius-Clapeyron equation to analyze the gas adsorption isotherms collected at different temperatures[10]. When the surface coverage is very low, a value of adsorption heat can be obtained by selecting a common adsorption quantity from the adsorption isotherms at different temperatures.

In the experiment, the CO_2 adsorption isotherm of activated carbon samples was tested by Mack Instrument's gas adsorption instrument. During the test, the sample constant temperature was controlled by iso-controller low-temperature thermoelectric refrigeration Dewar. At the end of the test, the equivalent heat of adsorption was calculated using Mack Instruments' Micro Active software for all surface coverage ranges from zero to saturation (Figure 5). The relationship between CO_2 adsorption capacity and adsorption heat is 0°C and 25°C. As shown in Figure 6, the initial adsorption heat value of the material (41 KJ mol^{-1}) is lower than the typical value of chemisorption (>60 KJ/mol)[11], indicating that carbon materials mainly adsorb CO_2 through physical processes. Therefore, when applied to gas adsorption, they have the advantages of rapid adsorption-desorption and reduced regeneration costs.

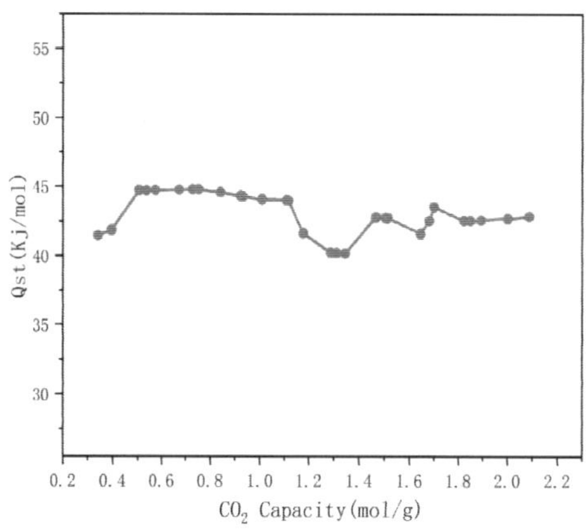

Figure 5. CO_2 adsorption heat.

Figure 6 shows the stability and cyclicity of the sample. After saturated CO_2 at 0°C and 1 bar, the adsorbent was degassed at 100°C for 6 h, and the sample was placed at 0°C after cooling, and then CO_2 capture was restarted to obtain the adsorption capacity. This process is repeated five times at a time. As shown in Figure 6, the sample hardly observed the attenuation of adsorption capacity. After 5 consecutive cycles, the retention rate of adsorption capacity of adsorbed CO_2 was still 97.2%.

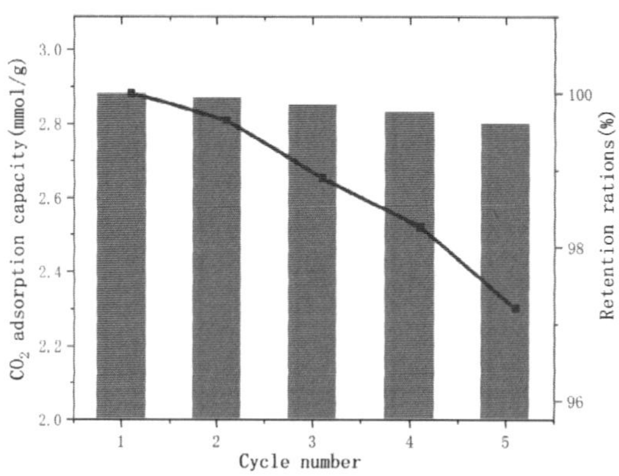

Figure 6. Regeneration performance for CO_2 adsorption at 0°C.

4. Conclusion

This paper innovatively developed a carbon material preparation method, using chitosan as raw material and nitrogen source, 2 grams of lithium nitrate, 2 grams of potassium nitrate, and 2 grams of potassium sulfate as template agent and nitrogen source. Through the one-pot ball milling method to prepare the material, this material has large adsorption of CO_2, which benefits from the material's high specific surface area and nitrogen content, and it is a good adsorption material. The material as prepared has good cyclic stability, and the adsorption capacity of CO_2 adsorbed after five consecutive cycles is still 97.2%, which is a potential adsorption material.

Acknowledgments

This work was financially supported by the Hebei Academy of Sciences high-level talent training and funding project (2024G03).

References

[1] Mu, D H. (2024) Quercetin-crosslinked chitosan nanoparticles: a potential treatment for allergic rhinitis. Scientific reports, 14:4021-4028.

[2] Esenturk, E. (2018) Activated carbon-supported CuO nanoparticles: a hybrid material for carbon dioxide adsorption. Journal of Nanoparticle Research, 20: 59-64.

[3] Zhou, C. (2024) Carbon edge engineering of nitrogen-doped carbon nanosheets via a self-sacrifice template strategy for enhanced hydrogen evolution reaction. Journal of Nanoparticle Research, 26: 43-51.

[4] Ye, H. (2024) Construction of mannose-grafting Fe-doped carbon dots for CpG delivery and photothermal ablation of tumor. Journal of Nanoparticle Research, 26:25-30.

[5] Mohamed, T. (2024) Fabricating a 3D floating porous PDMS − Ag/AgBr decorated g-C3N4 nanocomposite sponge as a re-usable visible light photocatalyst. Scientific reports, 14: 4184-4190.

[6] Mc Muratic (Shanghai) Instrument Co., LTD, 2020. The determination of CO2 is sorption heat for microporous carbon materials. http://mic.cnpowder.com.cn/news_27973.html.

[7] Riebeek H. 2011. The Carbon Cycle. https://www.earthobservatory.nasa.gov/features/CarbonCycle.

[8] North M. (2015) Chapter 1-What is CO2? Thermodynamics, Basic Reactions, and Physical Chemistry. Carbon Dioxide Utilisation. 3-17. https://doi.org/10.1016/B978-0-444-62746-9.00001-3.

[9] Shivangini B. (2023) Tungsten and molybdenum based polyoxometalates for photo and electrocatalytic carbon dioxide conversion - A critical review. 72:100430.

[10] Alberto A. (2022). Carbon dioxide removal potential from decentralized bioenergy with carbon capture and storage (BECCS) and the relevance of operational choices. Biomass and Bioenergy. 159: 106406-106410.

[11] Renato P. (2019) A synergistic approach for the simultaneous decarbonization of power and industry via bioenergy with carbon capture and storage (BECCS). International Journal of Greenhouse Gas Control, 87; 221-230.

Spreading modeling of prepreg tape used for manufacturing of variable angle tow composites

Siyu He[1] *, Zhao Feng[2] and Qiang Zhang[1]

[1] School of Intelligent Manufacturing, Jianghan University, Wuhan 430056, China
[2] School of Power and Mechanical Engineering, Wuhan University, Wuhan 430072, China

*Corresponding author's e-mail: syhe@jhun.edu.cn

Abstract. Variable angle tow (VAT) laminates are gaining popularity for enhancing composite structure performance. Automated fiber placement (AFP) is an effective method for tailoring VAT laminates to boost manufacturing efficiency. However, traditional AFP equipment faces limitations, such as the inability to eliminate defects like gaps and overlaps due to a minimum steering radius. This study addresses these defects by examining the spreading behavior of prepreg tapes. A simplified Navier-Stokes equation-based model is developed to analyze the impact of pressure and temperature on tape spreading width. Experimental validation, conducted under conditions that simulate automated layup operations, demonstrated that the tapes can be significantly broadened by up to 30% with increased temperature and pressure. The experimental data is curve-fitted using the least square method, and the model shows a good fit with the results.

1. Introduction

Advanced composite materials, renowned for their lightweight and high-strength attributes, have seen widespread adoption in sectors such as automotive, aeronautics, and astronautics[1-2]. Structures incorporating Variable Angle Tow (VAT) laminates have garnered considerable interest due to their potential to enhance compressive buckling and first-ply failure performance through tailored in-plane stiffness without additional weight. The design of in-plane stiffness distributions is facilitated by precise fiber path arrangements, making the efficient manufacturing of VAT laminates a burgeoning research focus[3-7].

The Automated Fiber Placement (AFP) technique holds promise for VAT laminate manufacturing. However, traditional AFP machinery faces limitations, particularly when navigating complex surfaces requiring tight turning radii, leading to defects such as gaps, overlaps, and fiber discontinuities[8-10]. These defects can be mitigated by enhancing the deformability of fibers, with methods such as Tailored Fiber Placement (TFP) and Continuous Tow Shearing (CTS) showing promise[11]. Despite advancements, challenges remain, including maintaining mechanical properties and improving process efficiency.

Innovative approaches, including the spreading of prepreg tape widths during layup, have demonstrated potential in eliminating gaps and improving mechanical performance. The integration of a spreading device with existing AFP equipment can maintain high layup efficiency. The relationship between spreading width and parameters such as pressure and temperature requires further investigation to refine the control of spreading widths.

This paper introduces a deformation model founded on the Navier-Stokes equation, accounting for variables like pressure, temperature, and geometric constraints within the operating environment of automated layup equipment. The model, tailored for single-layer prepreg spreading, is validated through experimental results, indicating that temperature and pressure significantly influence prepreg spreading width. The model's accuracy is confirmed through the least squares method, with width errors typically under 0.1mm, suggesting its utility for closed-loop control in the layup process to enhance spreading width accuracy.

2. Model analysis

Integration of Spreading Device: As illustrated in Fig. 1(a), the automated layup equipment integrates a spreading device positioned ahead of the main roller, ensuring controlled spreading of the prepreg tape prior to layup. The focus of this study is on the tape's spreading behavior within this device.

In Fig. 1(b), the prepreg tape, subjected to pressure between the roller and heater plate, exhibits an increase in width with descending height due to resin flow. This is explained by the squeezing flow concept, applicable to viscoelastic fluids between parallel planes. The spreading is considered a transverse squeezing flow, occurring perpendicular to the fiber direction in an isotropic plane.

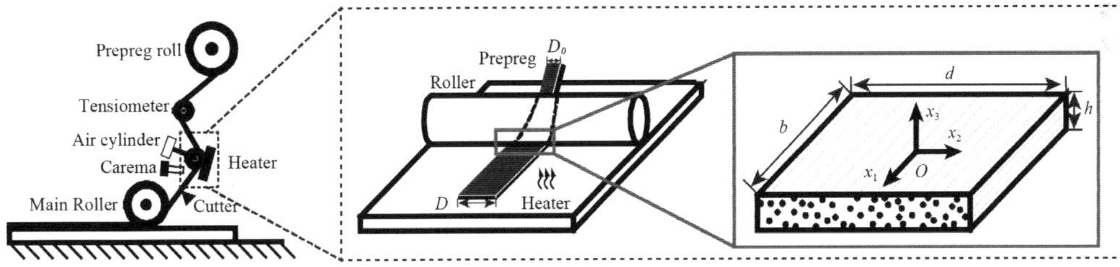

Figure 1. (a) Automated layup machine, (b) Spreading device, (c) Coordinate system.

Coordinate System and Assumptions: Fig. 1(c) shows the prepreg tape under roller pressure, modeled as a thin element with negligible thickness compared to width and length (*d*, *b*, and *h*, respectively). The fiber direction is x_1, width x_2, and thickness x_3, with the coordinate system's origin at the element's center. External pressure in the x_3 direction affects the area $d * b$, with the transverse flow in the $x_2 - x_3$ plane. This section details the prepreg squeezing flow process based on several assumptions, leading to a deformation model that explains the tape's spreading behavior.

Model Assumptions: The prepreg element is treated as a Newtonian-incompressible fluid, with viscosity and shear rate defined by Eq. (1). The volume $V = d * b * h$ is constant, and the deformation tensor rate's trace is zero, leading to boundary Eq. (2):

$$\tau_{23} = \mu \frac{\partial v_2}{\partial x_3} \tag{1}$$

$$\frac{\partial v_2}{\partial x_2} + \frac{\partial v_3}{\partial x_3} = 0 \tag{2}$$

where τ_{23} is the shearing stress in the $x_2 - x_3$ plane, v_2 is the velocity in the x_2 direction, v_3 is the velocity in the x_3 direction, and μ is the Newtonian viscosity coefficient (unit $Pa \cdot s$).

Given the fiber's strength exceeds the resin's, the flow in the fiber direction is negligible, considering length b as constant and smaller than width d. Additional assumptions are:

$$\frac{\partial v_2}{\partial x_1} = \frac{\partial v_3}{\partial x_1} = 0 \tag{3}$$

Spreading Width Expression: The Navier-Stokes equation governs the dynamic squeezing flow process, analyzing stress for the prepreg element. The balanced equation in the x_2 direction simplifies to:

$$-\frac{dP}{dx_2} + \mu \frac{\partial v_2^2}{\partial x_3^2} = 0 \tag{4}$$

By integrating the Eq. (4) of the x_3 motion twice, another expression of velocity v_2 can be written as Eq. (5):

$$v_2 = \frac{1}{\mu}\frac{dp}{dx_2}\left(\frac{x_3^2}{2} - \frac{h^2}{8}\right) \tag{5}$$

The following boundary conditions of the integrating process are given in the Eq. (6):

$$\begin{aligned} \frac{\partial v_3}{\partial x_1} &= 0 \quad \text{for} \quad x_3 = 0 \\ v_2 &= 0 \quad \text{for} \quad x_3 = \pm h/2 \end{aligned} \tag{6}$$

The final spreading width expression (Eq. 7) depends on parameters, including initial tape width D_0, external force F, viscosity coefficient μ, and compression duration t. The viscosity coefficient μ, primarily temperature-dependent, influences these parameters, which will be discussed further in the next section.

$$\frac{D^5}{5} - \frac{D_0^2 D^3}{3} + \frac{2D_0^5}{15} = \frac{V^2 F t}{6h^3\mu} \tag{7}$$

3. Experimental validation

3.1 Experimental setup and procedure

Platform Overview: The experimental platform, as detailed in Fig. 2, is divided into three main components: the spreading machine, vision system, and control system. The spreading machine is designed to adjust the prepreg width under specific pressure and temperature conditions. The vision system, equipped with an IP camera and a single-axis robot, captures images of the prepreg tape in its initial and spread states. The control system integrates a computer and sbRIO-9637 for executing data recording and image processing algorithms, with motion and parameter control managed by the sbRIO-9637.

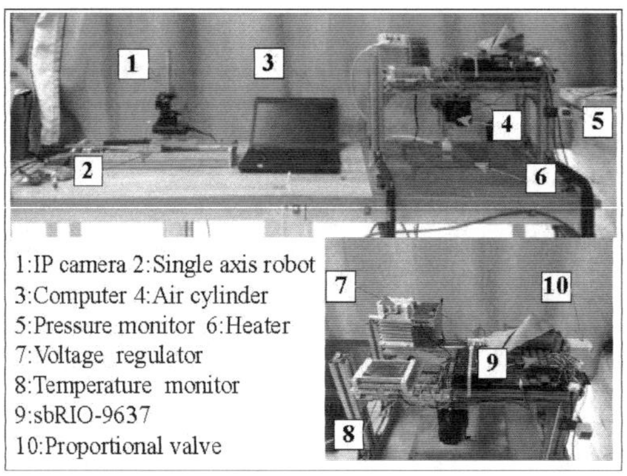

Figure 2: The platform for the spreading prepreg width experiment.

Material Specifications: The undirectional thermoset prepreg, model EM710 from Hengshen Jiangsu, China, features an epoxy resin with a curing temperature range of 85 to 120 °C. Key material parameters are summarized in Table 1, highlighting typical values for width and thickness.

Table 1: Undirectional prepreg parameters.

Model	Resin content	Density	Width	Thickness
EM710	33%	125 gsm	6 mm	0.20 mm

Temperature and Viscosity Control: Resin viscosity (μ) is temperature-dependent, with a substantial decrease from 210 to 8 $Pa \cdot s$ as temperature rises from 50 to 75 °C. Experiments are conducted over a temperature range of 40 to 80 °C at 10 °C intervals, tailored to the material's specifications.

Pressure Application and Effects: The external force (F) drives the squeezing flow of the prepreg tape, with spreading width varying by force duration at a set temperature. While higher pressures increase spreading width, they are limited by fiber obstruction and the controllability of the automated equipment. Experimental pressures are set from 0.1 to 0.7 MPa, incremented by 0.1 MPa, with a minimum pressure of 0.15 MPa due to valve constraints.

3.2 Experimental method

The original equipment (Fig. 2) is utilized to examine the spreading behavior of unidirectional thermoset prepreg tapes with epoxy resin. The prepreg tapes are subjected to pressures ranging from 0.15 to 0.7 MPa., maintained by a proportional valve for 8 seconds. Temperature is regulated between 40 to 80 °C using a voltage regulator. The experiment is structured into thirty-five groups with varying pressure and temperature, totaling eighty sample points per group. Spreading width measurements are taken using an IP camera with a telecentric lens for precision.

3.3 Results discussion

Geometrical results (Fig. 3) indicate the variation in the spreading width of prepreg tapes under different temperature and pressure conditions. The experimental results indicate that the width of the prepreg tapes increases significantly with the rise in temperature and pressure. Specifically: **1)** At 40°C and 0.2 MPa, the width increases by 10%; **2)** At 70°C and 0.15 MPa, the width increases by 20%; **3)** At 70°C and 0.6 MPa, the width increases by 30%. These results demonstrate that by adjusting the temperature and pressure, the spreading width of the prepreg tapes can be effectively controlled, thereby eliminating gaps during the layup process. Parameter settings for different spreading widths are detailed in Table 2.

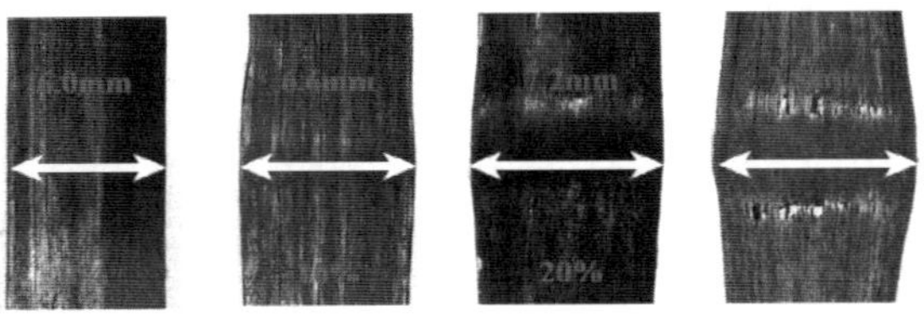

Figure 3: Difference in spreading width of the prepreg tapes.

Table 2: The parameter settings of the various spreading width.

Width increase (%)	Temperature (°C)	Pressure (MPa)
10	40	0.2
20	70	0.15
30	70	0.60

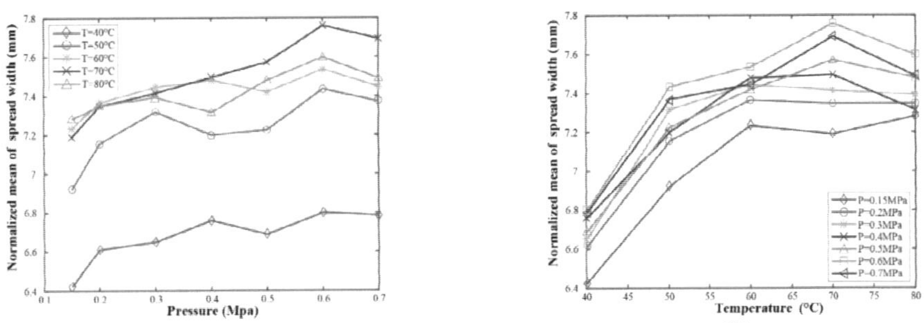

(a) (b)

Figure 4: Experimental results: (a) The normalized mean of prepreg spreading width against the pressure P for various values of temperature T, (b) The normalized mean of prepreg spreading width against the temperature T for various values of pressure P.

Influence of Pressure and Temperature: Figures 4(a) and 4(b) illustrate the normalized mean spreading width of prepreg tapes against pressure and temperature, respectively. The results show that the spreading width increases non-linearly with the increase in pressure and temperature. Specifically, at 70°C and 0.6 MPa, the spreading width reaches its maximum increase of nearly 30%. The non-linear relationship indicates that as the pressure increases, the resistance of the fibers to the resin flow intensifies, limiting further increases in spreading width. Similarly, with the rise in temperature, the spreading width increases significantly due to the decrease in resin viscosity, which enhances resin flow and promotes the spreading of the prepreg tapes. At a pressure of 0.6 MPa, the spreading width increases by about 20% as the temperature rises from 40°C to 70°C.

Model Discussion and Analysis: Figure 5 compares the experimental results with the predictions of the Navier-Stokes equation-based model under different temperature conditions. The experimental data were fitted using the least squares method, and the results indicate that the model has high predictive accuracy. In most cases, the error in spreading width is less than 0.1 mm. Figure 5(f) further shows the variation of model parameter k with temperature. As the temperature increases from 50°C to 60°C, the value of k decreases significantly, indicating that temperature has a substantial impact on the model parameters.

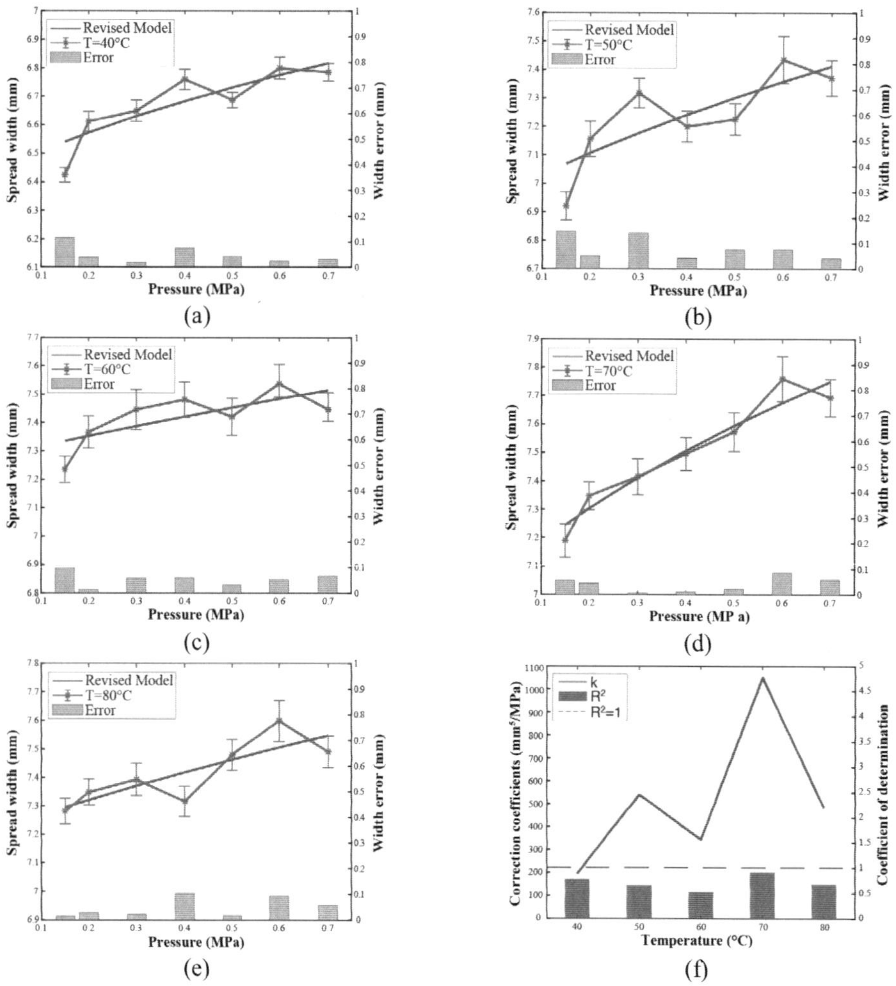

Figure 5: The experiment results compared to the developed model at various values of temperature. (a) T = 40 °C, (b) T = 50 °C, (c) T = 60 °C, (d) T = 70 °C, (e) T = 80 °C, (f) various coefficient k and coefficient of determination R^2

4. Conclusions

In this study, a comprehensive investigation was conducted to understand the spreading behavior of prepreg tapes under varying conditions of temperature and pressure. The main conclusions are as follows:

(1) Model Development and Validation: This study developed a deformation model based on the Navier-Stokes equations and Newtonian fluid dynamics, predicting the spreading behavior of prepreg tapes under various conditions. The model was validated through extensive experiments, demonstrating high accuracy with an error margin of less than 0.1 mm in most cases.

(2) Impact of Temperature and Pressure: Experimental results showed that temperature and pressure significantly affect the spreading width of prepreg tapes. At optimal conditions of 70°C and 0.6 MPa, the spreading width increased by up to 30%. The experimental data fit the model well, with a high determination coefficient (R^2), highlighting the importance of the viscosity coefficient, which is primarily influenced by temperature.

(3) Practical Applications: The integration of a spreading device with automated layup equipment has the potential to significantly enhance the efficiency and quality of VAT laminate manufacturing.

References

[1] Zaghloul, Mahmoud Yousry Mahmoud, Moustafa Mahmoud Yousry Zaghloul, and Mai Mahmoud Yousry Zaghloul. (2021). Developments in polyester composite materials–An in-depth review on natural fibres and nano fillers. Composite Structures, 278, 114698.

[2] Brasington, A., Sacco, C., Halbritter, J., Wehbe, R. and Harik, R. (2021). Automated fiber placement: A review of history, current technologies, and future paths forward. Composites Part C: Open Access, 6, 100182.

[3] Zhang K, Liu D and Wang Q (2023) Multi-level intelligent design of variable angle tow laminates via image-driven method[J]. Composite Structures, 303: 116354.

[4] A. M., Ribeiro, P., Rodrigues, J. D. and Akhavan, H. (2020). Modal analysis of a variable stiffness composite laminated plate with diverse boundary conditions: experiments and modeling. Composite Structures, 239, 111974.

[5] Zhang Y, Kong B, and Gu J. (2023) Experimental investigation on the buckling and post-buckling behavior of variable stiffness laminates[J]. Thin-Walled Structures, 184: 110450.

[6] Sheikhi, M., Anamagh, M. R., Bediz, B. and Tunc, L. T. (2024). Design of manufacturable variable stiffness composite laminates using spectral Chebyshev and normalized cut segmentation methods. Composite Structures, 330, 117836.

[7] Izzi, M. I., Montemurro, M. and Catapano, A. (2023). Variable-stiffness composites optimization under multiple design requirements and loads. International Journal of Mechanical Sciences, 258, 108537.

[8] Rajasekaran. (2022). Steering of Thermoplastic Composite Tapes using Hot Gas Torch-assisted Automated Fiber Placement (Doctoral dissertation, Concordia University).

[9] Qian, S., Liu, X., Ye, Y., Xu, Q., Zhang, T. and Li, X. (2021). Effect of gap and overlap fiber placement defects on the delamination behavior of L-shaped composite laminates. Composite Structures, 268, 113963.

[10] Yadav N. and Schledjewski, R. (2023). Review of in-process defect monitoring for automated tape laying. Composites Part A: Applied Science and Manufacturing, 107654.

[11] Simon, J., Hamila, N., Binetruy, C., Comas-Cardona, S. and Masseteau, B. (2022). Design and numerical modeling strategy to form Tailored Fibre Placement preforms: Application to the tetrahedral part with orthotropic final configuration. Composites Part A: Applied Science and Manufacturing, 158, 106952.

Study on preparation of ettringite-based fireproofing material for cable trench and its fireproofing mechanism

Jianguo Yuan[1], Jun Ji[2,*], Youchun Li[1], Jingkai Nie[2], Xuming Fan[1], Hui Liu[2] and Yihang Zhang[1]

[1]Jinhua Power Supply Company, State Grid Zhejiang Electric Power Co., Ltd., Jinhua, 321000, China
[2]State Grid Smart Grid Research Institute Co., Ltd., Beijing, 102209, China

[*]Corresponding author's e-mail: jijun2006@163.com

Abstract: The construction of a new power system puts forward higher requirements for the fire safety of power equipment. Fiber cement board is a kind of fire-proof material for cable trenches in electric power fields, but it cannot meet the requirement of electric fire protection. In order to meet the new requirement of high fire-proof grade for cable trench in substation, Ettringite-rich fire-proof separating material was prepared by using a metallurgical slag-sulphoaluminate cementitious system. The high temperature behavior of samples was studied by high temperature calcination test, thermal analysis and microstructure observation. The results show that the temperature rise of ettringite cementitious material is 5.2°C and 22.1°C lower than that of fiber cement board at 500°C and 1000°C respectively, and it can be used in the field of fire protection of substation cable trench. The reversible decomposition of ettringite is the main reason why ettringite-based cementitious materials can be reused.

1. Introduction

With the construction of new power systems, wind energy, nuclear energy, solar energy, and other new energy access to a large proportion of ultra-high voltage power projects, the power equipment is overloaded [1-2]. Long-term overload heating of power equipment will lead to power switch short circuits, cable fire and other phenomena. When "Control cable" and "Power cable" are arranged in the cable trench of a substation, once the power cable catches fire, the control cable will be damaged, resulting in out-of-control transmission [3]. Cable trench fire not only affects the operation safety of UHV equipment, but also causes casualties and serious social impact [4]. In order to solve this problem, electric power designers will use different types of cable layering, the key position of the fire partition used to separate the support. Many research results show that adding a fire barrier between the power cable and the control cable can effectively prevent cable trench fire spreading and ensure the normal operation of the control cable [5,6]. With the increase of the UHV voltage level, the heating capacity of the cable is also increasing, which puts forward higher requirements for the fire performance of the fire barrier [7].

At present, there are two kinds of fire-proof partitions: assembled type and L-type fire-proof partition. The advantage of the assembled fire barrier is that it can be prepared in the factory first, and then directly spliced on the spot, which greatly saves labor cost and time cost [8]. L-type fireproof partition board is made of fiber cement board, which has the advantage of low cost. The research shows that the cable protective sheath above the fire barrier should not be damaged when the fire source is ablated at 400 kW for the fire barrier with a voltage of 10 kV, and the thickness of the L-type fire barrier should not be less than 0.125m respectively [9]. The increase in the thickness of the fire barrier can prolong the fire resistance time of the cable trench fire barrier, but it brings great inconvenience to the operation and maintenance of the converter station (substation). Therefore, it is urgent to carry out research on the preparation of light and thin reusable cable trench fire-proof partitions suitable for substations.

Ettringite is a kind of calcium sulphoaluminate hydrate (AFt), which is produced by the combination of C-A-H (calcium aluminate hydrate) and sulfate ions. At present, the chemical composition and crystal structure of pure ettringite prepared by experimental means are basically the same as that of the mineral ettringite. Ettringite is a mineral containing 32 crystals of water, and it is produced in large quantities in sulphoaluminate cementitious materials. When the ettringite-based fireproofing materials are fully calcined by open fire, they will dehydrate, absorb heat, and decompose into other substances, making them have good fire performance. The decomposition and reforming of ettringite are reversible, but there is a lag between 30-400 Torr and 55-95°C. An important result of this study is to determine the reversibility of the decomposition process, but it occurs with significant hysteresis [10-11]. Therefore, it is very important to study the decomposition and reversibility of ettringite from steel slag, slag and desulphurization gypsum at high temperatures.

In this study, the thermal decomposition principle of ettringite was used to prepare fire-resistant partition material for the power ditch, and the fire protection mechanism of ettringite cementitious materials was revealed by IR and SEM analysis.

2. Test process

2.1. Test material

The slag is a glassy substance with a specific surface area of 560 m^2/kg. The mass fraction of CaO, Al_2O_3 and SiO_2 is 34.14%, 13.80% and 31.06% respectively. The water-hardness coefficient is 1.76, the activity coefficient is 0.45 and the alkalinity coefficient is 0.88. The steel slag was obtained by grinding hot muggy steel slag after iron removal. The specific surface area was 450 m^2/kg. The mass fraction of CaO, Al_2O_3 and SiO_2 was 38.96%, 4.31% and 14.14% respectively. The main phases were C_3S, β-C_2S, CA $(OH)_2$ and Ro. The specific surface area of FGD gypsum is 420 m^2/kg, and the main phase is dihydrate gypsum. The sulphoaluminate cement clinker is 72.5 grade calcium sulphoaluminate cement produced by Tangshan Polar Bear Company.

2.2. Sample preparation and testing

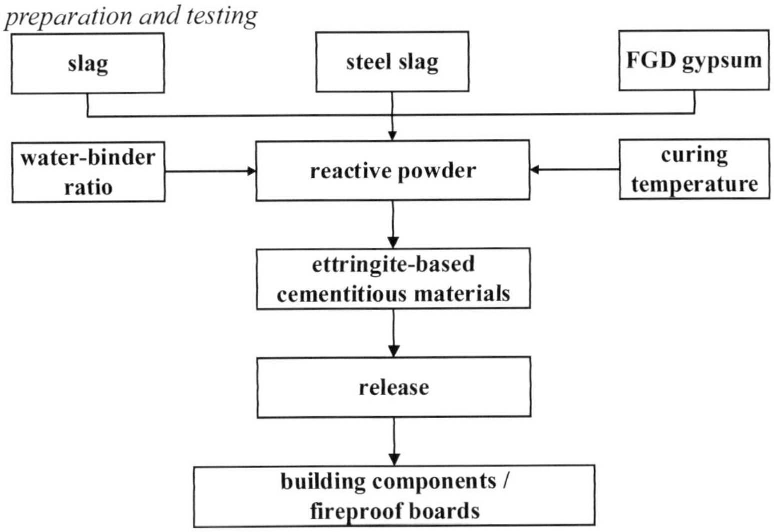

Figure 1. Flow chart of test process.

The metallurgical slag cementitious material is made of slag, steel slag and desulphurization gypsum with a mass ratio of 16:4:3, the water-binder ratio is set at 0.32, and the amount of polycarboxylate water reducer is 1% of the quality of the cementitious material. The raw materials were mixed and poured into a mortar mixer, placed in a curing box of 20°C and relative humidity over 95% for 1 day, then demolded and maintained to a total age of 14 days. The sample was dried in a 50°C oven for 48 h to obtain the product number M1. The contrast sample is the fibre cement board purchased in the market, product number M2.

In order to study the change of the sample at high temperatures, the sample with the size of 80 mm×20 mm×20 mm was calcined in a muffle furnace at 500°C and 1000°C for 2 h, and the backfire surface temperature, phase composition, and microstructure of calcined samples were analyzed and compared with those of pre-calcined samples. The test process is shown in Figure 1.

3. Results and analysis

3.1. Back surface temperature analysis

Sample M1 and sample M2 were calcined at 500°C and 1000°C, respectively, and the temperature rise of the backfire surface of the sample was collected. The test results are shown in Table 1. Table 1 shows that ettringite cementitious material has 5.2°C and 22.1°C lower preheating surface temperatures than fiber cement board at 500°C and 1000°C, respectively. Ettringite has 32 crystal water, and the theoretical dehydration weight loss rate is 45.93%. The dehydrating temperature of Ettringite is about 100°C, the dehydrating temperature of gypsum is about 150°C, the endothermic valley of ettringite is partly duplicated with C-S-H, and the dehydrating temperature of $Al(OH)_3$ is 200°C~350°C. The thermal decomposition of crystal water in the ettringite-based fireproof board during the heating process will lead to lower temperatures on the backfire surface of the material.

Table 1. Statistical Table of temperature rise of samples at different temperatures (°C).

Sample number	Temperature rising on the face of fire/°C	Temperature rises on the back side of the fire/°C
M1	500	67.9
M2	500	73.1
M1	1000	99.3
M2	1000	121.4

3.2. IR analysis

The samples M1 and M2 were calcined at 500°C and 1000°C, and the spectra of M1-500°C and M1-1000°C were compared with those of M1, and the spectra of M2-500°C and M2-1000°C are compared with those of M2, as shown in Figure 2 and Figure 3, respectively. The position of 3428 cm^{-1} in Figure 2 and 3414 cm^{-1} in Figure 3 correspond to ettringite, partially reactive dihydrate gypsum, as well as trace amounts of crystal water in calcium dihydroxide crystals produced by O-H lattice stretching vibration, respectively, which can be seen with increasing temperature. Whether in the M1 system or the M2 system, when the calcination temperature reaches 1000°C, the peak of the functional group almost disappears completely, indicating that the hydration products with structural water are no longer present in the cementitious material system, and the hydration products have been calcined and decomposed.

The peak of HCO_3^- ion at 1683 cm^{-1} in Figure 1 and at 1676 cm^{-1} in Figure 3 disappeared at 500°C with the increase in temperature. The peak of the strong stretching vibration of CO_3^{2-} ion with wave number 1414 cm^{-1} in Figure 2 and 1435 cm^{-1} in Figure 3 was enhanced at 500°C, but disappeared at 1000°C. Combined with the change of HCO3$^-$ ion and CO_3^{2-} ion, it can be concluded that the HCO_3^- ion in the sample decomposes into CO_3^{2-} ion by calcination at 500°C, and CO_3^{2-} ion decomposes into CO_2 when the temperature rises to 1000°C. Therefore, the strong stretching vibration peak does not

Journal of Physics: Conference Series **2842** (2024) 012025 doi:10.1088/1742-6596/2842/1/012025

exist. The wave number 1123 cm^{-} in M1 and M2 is the strong stretching vibration peak of SO_4^{2-} ion. With the increase of temperature, the absorption peak of SO_4^{2-}ion increases continuously, which shows that with the increase of temperature, ettringite and gypsum continue to decompose to form $CaSO_4$. The wave numbers of 868 cm^{-1} and 621 cm^{-1} in M1 are weak deformation vibration peaks of CO_3^{2-} ion and SO_4^{2-} ion, respectively. The wave numbers of 985 cm^{-1} and 673 cm^{-1} in M2 are weak deformation vibration peaks of SO_4^{2-} ion and CO_3^{2-} ion, respectively.

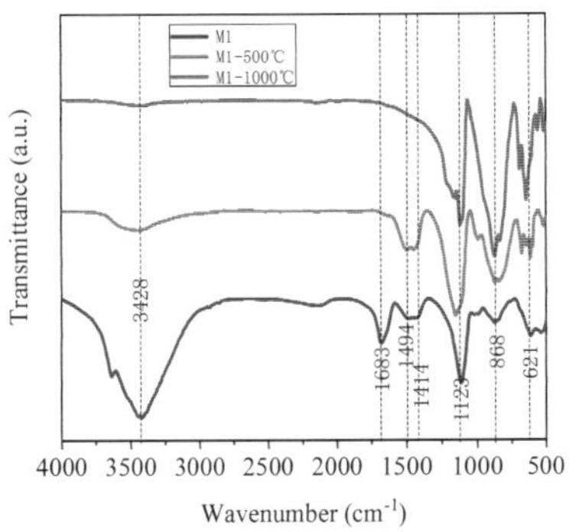

Figure 2. IR analysis of sample M1 at different temperatures.

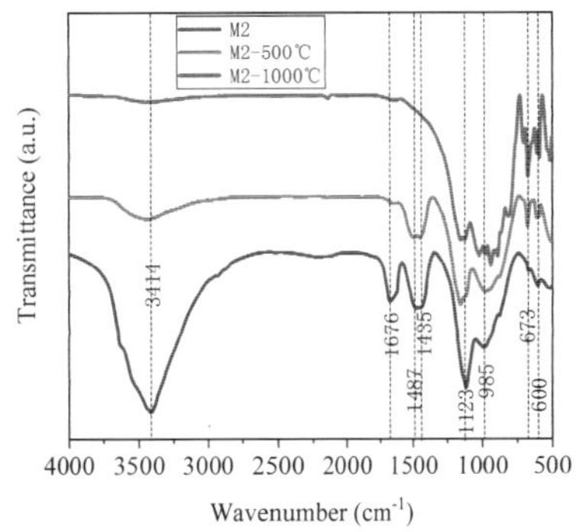

Figure 3. IR analysis of sample M2 at different temperatures.

3.3. SEM analysis

After soaking the M1 fireproof board sprayed by open fire at 1000°C for 3 days, a small portion of the M1 fireproof board calcined in the face of the fire was taken and observed by SEM. The test results are shown in Figure 4, it was found that a large amount of ettringite whisker was regenerated in its internal structure. This shows that the decomposition of ettringite is reversible. The decomposed ettringite can be recombined with crystalline water in a humid environment.

Figure 4. SEM of ettringite whisker.

4. Conclusion

Based on the results and discussions presented above, the conclusions are obtained as below:

(1) Compared with 500°C and 1000°C, ettringite cementitious material has a lower temperature rise of 5.2°C and 22.1°C, which can be used in the field of cable trench fire protection in substation.

(2) The results of IR analysis of ettringite cementitious material and fiber cement board show that the O-H lattice stretching vibration of ettringite cementitious material crystal water is produced, which disappears completely at 1000°C. The weak deformation vibration peaks of $CO3^{2-}$ ion and $SO4^{2-}$ ion in the fiber-cement plate increase with the increase in temperature.

(3) The reversible decomposition of ettringite is the main reason why ettringite-based cementitious materials can be reused.

Acknowledgments
This work was financially supported by the Science and Technology Project of State Grid Company Ltd., (5500-202119246A-0-0-00).

References
[1] Xue G, Liu S, Gong D, et al (2019) Environmental intelligent control of underground integrated pipe gallery based on equipment failure rate. J. IEEE Access, 7: 160062-160071.
[2] Cao Q C (2021) Analysis on safety of fire protection system of urban underground pipe gallery. J. Building Technology Development, 48(10): 92-94.
[3] Chen D P, Hou K Y, Wang L J, et al (2017) Status and development of fire protection materials based on super thermal insulator and their application prospect in urban underground space. J. Chinese Journal of Engineering, 39(6): 811-822.
[4] State GRID Corporation of China. Technical specification for fire prevention of HV power cables and channels. S. Q/GDW 12067-2020, 2021-04-30.
[5] Chen Y D, Zhang H L, Baosen B and et al (2023) Preparation of hydromagnesite-vemiculite based fire barrier and its fire protection performance. J. Non-Metallic Mines, 46(1): 6-10.
[6] Chen B, Johannes K, Horgnies M, et al (2021) Characterization of an ettringite-based thermochemical energy storage material in an open-mode reactor. J. The Journal of Energy Storage, 33: 102159.
[7] Chen B, Kuznik F, Horgnies M, et al (2019) Physicochemical properties of ettringite/meta-ettringite for thermal energy storage: Review. J. Solar Energy Materials and Solar Cells, 193: 320-334.
[8] Baquerizo L G, Matschei T and Scrivener K L (2016) Impact of water activity on the stability of ettringite. J. Cement and Concrete Research, 79: 31-44.
[9] Li J R (2023) Preparation of sulphoaluminate-metallurgical slag cementation system materials and its fire performance research. D. Beijing: University of Science and Technology Beijing.
[10] López B, Suesta F (2012) Ettringite- and gypsum-based binder with high fire-resistance and method for the production thereof. P. Wo 2012020152 A1, 2012-02-16
[11] Jia X S, Chen D P, Yu X W, et al (2022) Experimental study on total solid waste cementitious materials socked in seawater. J. Bulletin of the Chinese Ceramic Society, 41(3): 913-921.

Ionic gels with high mechanical strength enabled by supramolecular interactions between small molecular solvents and polymer chains

Zhaoqiang He, Chenyang Xing and Zhengchun Peng*

College of Physics and Optoelectronic Engineering, Shenzhen University, Shenzhen, 518060, China

*Corresponding author's e-mail: zcpeng@szu.edu.cn

Abstract. The limited mechanical properties of hydrogels significantly impede their practical applications. In this research, we have synthesized supramolecular ionic gels using betaine, citric acid, and acrylamide as the main components. These specimens exhibit exceptional tensile properties (exceeding 1000% elongation at break) and toughness (approximately 3 MJ/m^3), as well as recovery capabilities under substantial compressive strains. Furthermore, these transparent and conductive ionic gels have been employed in the fabrication of alternating current electroluminescent devices with single-electrode configurations. The outcomes of this study indicate that the high-strength ionic gel preparation method outlined herein shows great potential for expanding the application spectrum of ionic gels.

1. Introduction

Hydrogels exhibit immense potential in the field of biomedical applications due to their flexibility and compatibility with biological tissues. They have found widespread use in areas such as tissue engineering[1], wound dressings[2, 3], and drug delivery[4], benefiting from their unique characteristics. However, the high water content and swelling behavior of hydrogels have presented challenges, particularly regarding their strength and toughness, which have limited their broader application[5]. One strategy involves incorporating nanomaterials as crosslinkers during the formation of hydrogels[6]. However, concerns regarding nanomaterial toxicity, biocompatibility, and biodegradation need to be addressed[5]. Another significant advancement in hydrogel technology is the development of toughened hydrogels with a double network (DN) structure[7]. Although chemically crosslinked DN hydrogels provide high stiffness, their irreversible covalent bonds can result in permanent damage[8]. In contrast, physically crosslinked DN hydrogels, formed through reversible non-covalent interactions, offer a promising solution. These physical DN hydrogels employ various non-covalent interactions, such as metal-coordination bonds[9], hydrogen bonds[10], hydrophobic interaction[11], electrostatic interaction[12], and host-guest interaction[13]. Supramolecular polymer hydrogels, constructed via these non-covalent intermolecular interactions, display unique properties, including excellent mechanical properties and stimulus responsiveness[14, 15]. This opens new opportunities for designing hydrogels with customized functionalities suitable for a wide range of applications, including flexible electronic devices.

In this study, we successfully fabricated a series of supramolecular ionic gels with superior mechanical properties, transparency, and ionic conductivity. The method involves a simple process of mixing aqueous solutions followed by UV-initiated polymerization, using acrylamide (AAm), citric acid (CA), and zwitterionic betaine (Bet) as the main components. The resulting PAAm/Bet/CA gels exhibit exceptional characteristics, including high strength, toughness, stretchability, and outstanding recoverability. Notably, the tensile strength and elongation at break can be adjusted within wide ranges, providing versatility in material properties. Furthermore, these ionic gels maintain their structural integrity even under significant compressive strain, demonstrating impressive compressive strength.

2. Experimental section

2.1 Synthesis of gels

2.1.1 Preparation of supramolecular ionic gels. Firstly, we dissolved the α-KGA photoinitiator in deionized water at a mass ratio of 1:10. Then, betaine (Bet) was mixed with citric acid (CA) at a 1:1 mass ratio. Afterward, we added varying amounts of deionized water to this mixture to create different ratios, specifically 90/80/50/25 wt% solutions. Consequently, the mass of water in each solution accounted for 10/20/50/75%, respectively. Next, we combined acrylamide (AAm) with the Bet/CA solution in various mass ratios (1:9, 1:4, 3:7, 2:3), corresponding to AAm making up 10/20/30/40% of the total mass. The mixture was thoroughly dissolved, followed by the addition of an α-KGA solution at 1wt% of the mixture. It is vital to ensure that the components were evenly dispersed, and that any air bubbles were removed through water bath sonication. Finally, the solution was exposed to UV light for 30 minutes to polymerize AAm. The resulting ionic gels were designated as PAAm/Bet/CAα-β, where α and β represent the water content in the Bet/CA solution and the quantity of AAm added in wt%, respectively.

2.1.2 Preparation of hydrogels based on other solvent systems. To compare the effect of different solvents on the mechanical properties of the gels, various solvent systems were utilized instead of the Bet/CA/water solution. The solvent systems included ethylene glycol (EG)/water, glycerol (Gly)/water, choline chloride (ChCl)/water, Bet/water, and CA/water, all formulated with 50 wt% water content. Mix AAm monomers with each solvent system in a 1:4 ratio (AAm accounting for 20% of the total mass). The subsequent steps of initiator addition, bubble removal, and UV polymerization are consistent with the previous text.

2.2 Preparation of the ACEL device.

The ACEL devices were prepared by sequentially spin-coating the precursor solution of PAAm/Bet/CA-20-20 and the mixture of Zinc-copper sulfide (ZnS:Cu) phosphor, Eclflex00-30 and barium titanate ($BaTiO_3$). The mass ratio of ZnS:Cu, $BaTiO_3$, and Ecoflex is 1:1:1.

3. Result and discussion

3.1 Results of gel preparations with different solvent systems

The hydrogels prepared using H_2O, EG/H_2O, Gly/H_2O, ChCl/H_2O, Bet/H_2O, CA/H_2O, and Bet/CA/H_2O were extracted from columnar molds following equivalent durations of UV irradiation. The photos of these preparations are depicted in Figure 1. We can observe an obvious result, namely, the introduction of small molecules such as Bet, and CA would enhance the structural stability of obtained supramolecular gels. Gels without chemically crosslinked polymer networks cannot form stable structures derived from the solvent systems EG/H_2O, Gly/H_2O, and ChCl/H_2O. In contrast, gels prepared from Bet/H_2O, CA/H_2O, and Bet/CA/H_2O exhibited resilience, preserving their original structure following a 2-hour incubation period (Figure 1(a)).

Additionally, we compressed the PAAm/Bet/CA ionic gel to 90% in compressing strain, resulting in the gel reverting to nearly its original state within 1 minute (see Figure 1(b)(i)). Conversely, gels derived from pure water collapsed due to their excessive softness (refer to Figure 1(b)(iii)). Subsequently, a minute quantity of N,N'-methylene diacrylamide (MBAA) was incorporated as a crosslinking agent during the preparation of gels using pure water. The findings indicated that the inclusion of MBAA augmented the robustness of the PAAm hydrogels, albeit the covalently crosslinked hydrogels were susceptible to fracturing owing to their increased brittleness (see Figure 1(b)(ii)).

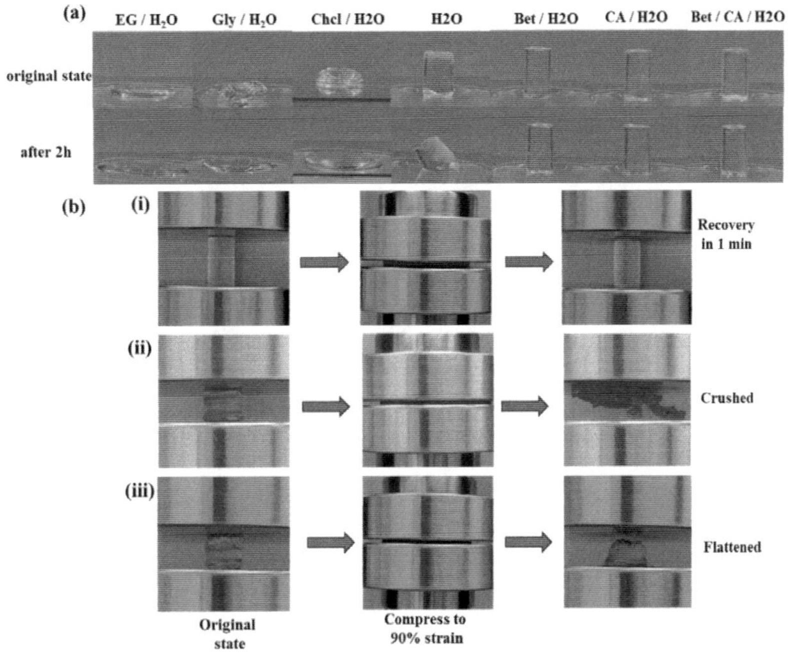

Figure 1. (a)Results of gel preparations with different solvent systems.(b) Compression characterization of (i)PAAm/Bet/CA-50-20, (ii) PAAm hydrogel with MBAA, and (iii) PAAm hydrogel without MBAA.

3.2 Mechanical properties of PAAm/Bet/CA ionic gels

To study the influence of water content in supramolecular solvents on the mechanical properties of the gels, Bet/CA solutions were prepared with varying water concentrations (pure water, 10%, 20%, 50%, and 75%). The resulting tensile stress-strain curves, depicted in Fig. 2(a), revealed notable trends. As the water content in the Bet/CA solvent decreased, Young's modulus of the gels exhibited a pronounced increase, rising from 158 kPa to 472 kPa. Similarly, the tensile stress escalated from 64.8 kPa to 1.59 MPa with diminishing water content in the solvent (Fig. 2(b)). Interestingly, the elongation at the break of the gels peaked at 50% water content, reaching 1101.9%(Fig. 2(c)). Conversely, gels prepared from solutions with 75% water content and pure water displayed plastic deformation under increased stretching, unable to revert to their original state. These findings present the delicate balance between water content and mechanical properties in supramolecular gels, with implications for optimizing gel formulations for specific applications.

The influence of AAm content on the mechanical properties of the gel was further investigated by controlling the water content of the Bet/CA solution at 20% by adjusting the AAm-to-Bet/CA ratio. The resulting tensile stress-strain curves of the gels are depicted in Fig. 2(d). As the AAm content increased, both the modulus and tensile stress of the gels exhibited a gradual increase, as illustrated in Fig. 2(e). According to the compressive stress-strain curve, shown in Fig. 2(g), by increasing the AAm percentage, the compressive modulus of the gel experienced a notable increase from 19.9 kPa to 207 kPa. Additionally, the compressive strength gradually increased, with the gel containing 30% AAm displaying a compressive stress of 8.92 MPa at 90% strain (Fig. 2(h)). Remarkably, none of the samples experienced crushing during the test. Figure 2(i) presents the loading-unloading measurements of PAAm/Bet/CA-20-20 at various strains, revealing the minimal tensile hysteresis of the gels.

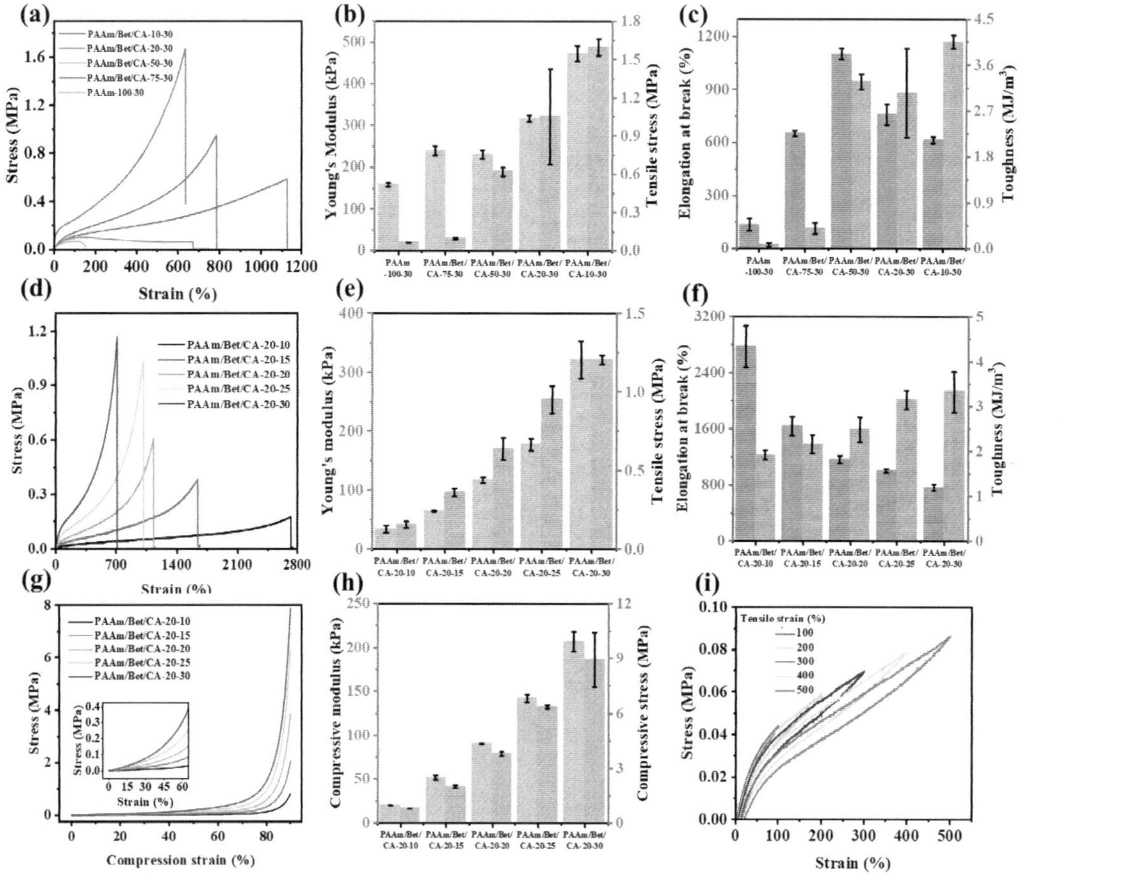

Figure 2. Mechanical properties of PAAm/Bet/CA ionic gels. (a)~(f) Tensile mechanical properties of the gels. (g)~(h) Compressive mechanical properties of the gels. (i)Loading-unloading measurements of PAAm/Bet/CA-20-20.

3.3 Application of PAAm/Bet/CA ionic gels in ACEL device

The application of PAAm/Bet/CA ion gel in the development of flexible alternating current electroluminescent (ACEL) devices demonstrates its promising potential in the field. In the construction of ACEL devices with a single-electrode structure and interactive display functionality, PAAm/Bet/CA ionic gel serves as an electrode material. Figure 3(a) elucidates the schematic equivalent circuit of the prepared ACEL device. When subjected to an AC electric field, the interaction between the human epidermis and grounded polar droplets serves to create floating electrodes above the ACEL device, thereby exciting the fluorescent layer to emit light. The luminescence display pattern, showcased in Figure 3(b), exemplifies the device's capability to render clear letter patterns, such as "SZU" upon the introduction of deionized water onto the designated template and grounding. Furthermore, Figures 3(c) and 3(d) present luminescent images of templates depicting "SZU" and a "cat," respectively, demonstrating the device's versatility in facilitating customized interactive pattern displays, even under the pressure of a finger. Remarkably, Figure 3(e) underscores an additional intriguing aspect: when the PAAm/Bet/CA ion gel atop the fluorescent layer is touched by hand, the entire gel emits light, thus illustrating the potential of ACEL devices to achieve out-of-plane luminescent displays through the judicious selection of conductor materials. These findings not only highlight the adaptability and functionality of PAAm/Bet/CA ionic gels in flexible electronic devices but also open avenues for innovative applications in interactive displays.

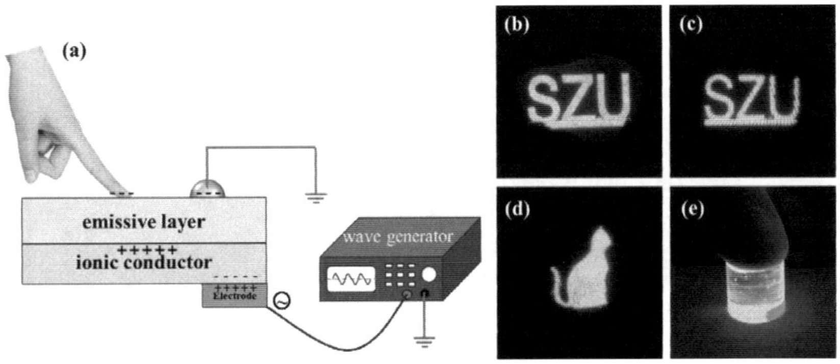

Figure 3. ACEL device based on ionic gel and the application to interactive luminescent displays.

4. Conclusion

In summary, we introduce herein high-strength and high-toughness, transparent, and conductive ionic gels bolstered by supramolecular interactions. These ionic gels surmount the challenge posed by the inadequate mechanical properties of traditional hydrogels and hold promise for utilization in the realm of flexible electronics. An ACEL device integrating PAAm/Bet/CA ionic gels as pliable electrodes has effectively realized interactive light-emitting displays. Collectively, this investigation underscores the promise of supramolecular ionic gels within the domain of flexible electronics, thus paving the way for further advancements and breakthroughs in functional materials research.

Acknowledgments

This work was financially supported by the Science and Technology Innovation Council of Shenzhen under grants KQTD20170810105439418 and JCYJ20200109114237902.

References

[1] Wang W, et al. (2023) Extracellular matrix mimicking dynamic interpenetrating network hydrogel for skin tissue engineering. Chemical Engineering Journal, 457, 1-12.
[2] Wang L, et al. (2022) Multifunctional hydrogel as wound dressing for intelligent wound monitoring. Chemical Engineering Journal, 433, 1-7.
[3] Koehler J, et al. (2018) Hydrogel wound dressings for bioactive treatment of acute and chronic wounds. European Polymer Journal, 100, 1-11.
[4] Goh M, et al. (2024) Advancing burn wound treatment: exploring hydrogel as a transdermal drug delivery system. Drug Deliv, 31 (1), 1-15.
[5] Huang H, et al. (2023) High-strength hydrogels: Fabrication, reinforcement mechanisms, and applications. Nano Research, 16 (2), 3475-3515.
[6] Yao C, et al. (2015) Poly (N-isopropylacrylamide)-Clay Nanocomposite Hydrogels with Responsive Bending Property as Temperature-Controlled Manipulators. Advanced Functional Materials, 25 (20), 2980-2991.
[7] Gong J P, et al. (2003) Double - Network Hydrogels with Extremely High Mechanical Strength. Advanced Materials, 15 (14), 1155-1158.
[8] Lin X, et al. (2022) Progress in the mechanical enhancement of hydrogels: Fabrication strategies and underlying mechanisms. Journal of Polymer Science, 60 (17), 2525-2542.
[9] High-Strength and Injectable Supramolecular Hydrogel Self-Assembled by Monomeric Nucleoside for Tooth-Extraction Wound HealingWang X-H, et al. (2018) Strong and tough fully physically crosslinked double network hydrogels with tunable mechanics and high self-healing performance. Chemical Engineering Journal, 349, 588-594.
[10] Wang Z, et al. (2022) High-Strength and Injectable Supramolecular Hydrogel Self-Assembled by Monomeric Nucleoside for Tooth-Extraction Wound Healing. Adv Mater, 34 (13), 1-20.

[11] Liu X, et al. (2020) Dual Physically Crosslinked Hydrogels Incorporating Hydrophobic Interactions with Promising Repairability and Ultrahigh Elongation. Advanced Functional Materials, 31 (3), 1-9.
[12] Cai Y, et al. (2023) Anti-swelling hydrogels based on surfactant–polymer interactions for underwater sensing with excellent mechanical properties. Journal of Materials Chemistry C, 11 (38), 12981-12991.
[13] Cai T, et al. (2024) Double physical network hydrogels with rapid self-recovery and multistimuli-responsive shape memory effects based on low-methoxyl pectin and host-guest interactions. Sustainable Materials and Technologies, 40, 1-10.
[14] Zhang W, et al. (2021) Skin-like mechanoresponsive self-healing ionic elastomer from supramolecular zwitterionic network. Nat Commun, 12 (1), 1-12.
[15] Chen R, et al. (2023) Reinforcing supramolecular hyaluronan hydrogels via kinetically interlocking multiple-units strategy. Carbohydr Polym, 310, 1-12.

Nucleation promotion for the atomic layer deposition of HfO_2 onto carbon nanotubes for device applications

Bingxian Li*

Hunan Institute of Advanced Sensing and Information Technology, Xiangtan University, Xiangtan, Hunan, 411105, China

*Corresponding author's e-mail: 202121521241@smail.xtu.edu.cn

Abstract. The potential of semiconducting single-walled carbon nanotubes (CNTs) as the next-generation semiconductor information material as a successor to silicon-based technology is promising. However, a significant challenge for CNT based electronics at device level lies in the obtainment of high-quality gate dielectrics such as HfO_2 on the dangling bond-free and chemically inert surface of CNTs. To address this issue, this paper studies the atomic layer deposition (ALD) of HfO_2 onto the surface of carbon-based materials, and focuses on how to promote the nucleation via two methods, i.e. low-temperature Nanofog method and high-temperature variable parameter method. Both methods are confirmed to be effective in improving the continuity and uniformity of ALD HfO_2 films.

1. Introduction

As silicon-based complementary metal-oxide-semiconductor (CMOS) technology is approaching the physical limits [1], novel channel materials are suggested for next-generation nanoelectronics. Among the various candidates, semiconducting single-walled carbon nanotubes (CNTs) have garnered significant attention due to their unique structures and exceptional electrical properties such as high carrier mobility, current carrying capability, and injection velocity [2].

In the past three decades, tremendous progress has been achieved in exploring the potentials of CNTs-based field-effect transistors for application in electronic devices, ranging from individual FETs [3] to sensors [4], flexible [5] and wearable devices [6], thin-film devices [7], radiofrequency devices [8], digital devices [9] and circuits [10]. Challenges also arise for CNT-based CMOS and in particular the obtainment of high-quality and high-k dielectrics such as HfO_2 onto CNTs seems to be difficult, due to the dangling bond-free and chemically inner surface of CNT [11], a feature different from silicon. To promote the nucleation of HfO_2 during processes like atomic layer deposition (ALD), the CNT's surface is usually modified with additional seed layers and functional groups such as DNA molecules [12]. The drawback of these approaches is also apparent, as they may either introduce surface defects and thus additional carrier scattering centers, or decrease the overall dielectric constant due to the introduction of relatively low-κ seeding layers. As a result, they are likely hard to meet the request for scaling and thus novel approaches are on demand [13-15].

In this study, we explored the application potential of two ALD recipes onto arrayed CNTs, i.e. the low-temperature Nanofog method [16-18] and parameter-tuning high-temperature ALD [19], which were originally developed for depositing high-quality high-k dielectrics onto low-dimensional materials such as graphene, MoS_2, and nanocarbon materials. In specific, the low-temperature Nanofog process is similar to chemical vapor deposition (CVD), allowing the precursors to either physically adsorb onto the substrate or undergo reactions similar to CVD in the gas phase and then deposit onto the substrate, serving as the seeding layer for subsequent growth of HfO_2 under normal ALD conditions. Compared with traditional ALD recipes, the Nanofog process adopts a lower deposition temperature, a shorter purge time, and a larger dosage of precursor pulses [20]. The variable parameter high-temperature ALD method originally reported by I. Adrianus et al. was proposed due to the consideration that high-temperature processes prefers the formation of form uniform and dense HfO_2 film on substrate due to their self-limiting nature. The following three parameters are mainly controlled: (i) the doses of precursors used in the reaction; (ii) the doses of precursors deposited on the substrate surface; and (iii) the slow adsorption process related to the precursor reaction on the substrate surface. Three effective routes are suggested to promote the nucleation of ALD HfO_2 on arrayed CNTs, that are Pretreatment Mode (PM), Multipulse Mode (MM), and Stop-flow Mode (SM). Note that this work is mainly focused on the nucleation behaviors (early-stage in ALD) of ALD HfO_2 on carbon-based materials, as this is a key step to achieving HfO_2 films with good uniformity and conformality on carbon-based materials.

2. Materials and methods

For ease of characterization and comparison, relatively clean two-dimensional planar carbon-based materials with sp^2 surface such as highly oriented pyrolytic graphite (HOPG, purchased from Nanjing Beige Electronic Technology Co., Ltd., model ZYB_DS) and the CVD grown on SiO_2/Si substrate (via transfer) were used for the optimization of ALD parameters. Another consideration for such a choice is related to the complex morphology of the arrayed CNTs with polymer residues [21], making the direct comparison under different growth conditions difficult. The SiO_2/Si-supported high-density arrayed CNTs (A-CNTs) are purchased from Suzhou Xijing Semiconductor Technology Co., Ltd.

A commercial ALD system (T200Q from Wuhan Kenadi Vacuum Technology Co., Ltd) was used for the ALD deposition. Solid tetra (dimethylamino), hafnium (TDMAHf, purity 99.9999% and heating temperature of 75°C), and liquid H_2O (purity 99.9999%, at room temperature) are used respectively as the metal Hf precursor (denoted as A in Figure 1) and oxygen precursor (denoted as B in Figure 1). Nitrogen gas (N_2) with a purity of 99.9999% was used as the purge and carrier gas, both under a flow rate of 10 sccm. The load-in and load-out of all the samples were conducted when the chamber reached the setting deposition temperature (T_{dep}). Prior to the ALD, N_2 gas with pressure (P_{base}) of ~ 0.19 Torr was introduced to clean the chamber for at least six minutes (t_{purin}). Different ALD deposition conditions were achieved by varying the deposition recipes as explained later in detail. Unless otherwise specified, only ten ALD cycles were used to highlight the nucleation process.

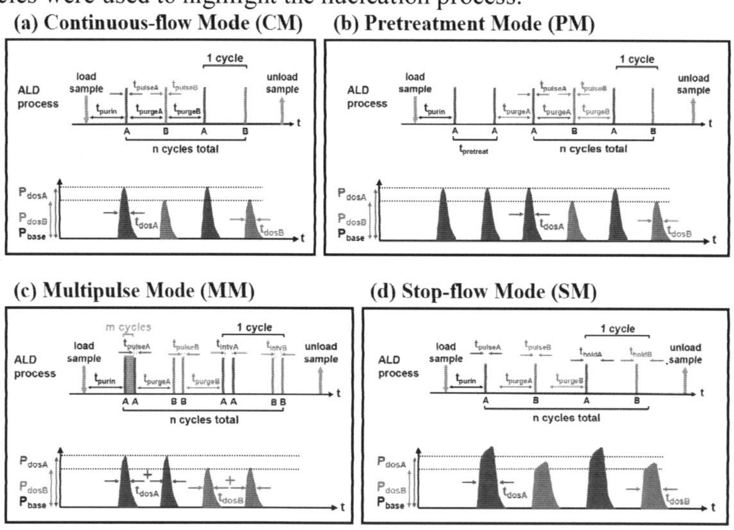

Figure 1. Schematic diagrams of (a) CM, (b) PM, (c) MM, and (d) SM.

Generally, the ALD process is a continuous alternating supply of two precursor sources (Hf-, and O-containing precursors). Following that proposed by I. Adrianus et al. [19], four different ALD modes were conducted: (i) continuous-flow mode (CM mode), controlled by alternating pulses of each precursor; (ii) pretreatment mode (PM mode), samples are pretreated with an exposure to a series of oxidant pulses before the CM-ALD; (iii) multipulse mode (MM), controlled by the number of consecutive alternating pulses of each precursor; (iv) stop-flow mode (SM mode), the stop time is introduced after the precursor pulse. Details for each mode are illustrated in Figure 1a-1d. The T_{dep} was maintained at 200 °C in all these ALD modes.

The purge time (t_{purge}) between the pulses of the two precursors at high temperatures is 20 s. To maintain consistency, the product of P_{dos} (the full width at half maximum of the chamber pressure profile during the pulse) and t_{dos} (the time the precursor resides on the substrate surface during the pulse) was used to describe the amount of precursor more accurately, instead of directly using pulse time (t_{pulse}) as a measurement metric [19]. It was based on the following considerations: if the carrier gas flow rate, source bottle vapor pressure, and the amount of precursor available for growth are different, the same pulse time may result in different pulse pressures and residence times. Using the variable control method and orthogonal experimental method, the effects of key parameters such as deposition temperature, precursor pulse dosage, and purge time on the ALD deposition of HfO_2 films on carbon-based materials were determined.

This work uses scanning electron microscopy (SEM, Sigma 500) to characterize the growth of HfO_2 on the substrate surface, and atomic force microscopy (AFM, Dimension icon) is used to characterize the surface morphology, roughness, and other properties of the samples.

3. Results and discussion

3.1. The low-temperature Nanofog method

At low temperatures, precursor molecules are mainly adsorbed physically onto the substrate rather than chemically adsorption, thus meaning that desorption of precursor molecules also occurs at a considerable rate when being subject to external disturbances such as heating and prolonged purging [22]. Hence, we first investigate the effect of different deposition temperatures (75 °C, 90 °C, 110 °C, 120 °C, 140 °C, 150 °C, and 200 °C) on the ALD growth of HfO_2 films and present the results in Figure 2. The top (Figure 2a) and bottom (Figure 2b) panels represent the AFM images and secondary-electron SEM images of as-deposit HfO_2 clusters with the associated surface coverage (the percentage of as-deposit HfO_2 clusters occupying the total area of the substrate, denoted as θ) were quantified and marked in each Figure 2. The reason for using SEM instead of AFM is that CVD-G/SiO_2 is not easily locatable in the optical microscope. As seen, the coverage of HfO_2 on both HOPG and CVD-G/SiO_2 substrates decreases with increased deposition temperature. This result is consistent with the growth mechanism of ALD at low temperatures. As the deposition temperature decreases, the impact of physical adsorption becomes more profound, favoring the deposition of HfO_2 onto the surface of carbon-based materials. In contrast, elevating the deposition temperature disfavors the physical adsorption of precursors, and thus decreases the surface coverage of HfO_2.

Based on Figure 2, by comparing the topographical images of the two substrates, it is evident that when the deposition temperature is 120 °C, the clustering distribution of HfO_2 is the most dense and uniform, with the smallest cluster particles. This is conducive to the compactness of the subsequently deposited films. Therefore 120 °C can be selected as the optimal temperature for ALD HfO_2 on the surface of carbon-based materials and this temperature will be used as the standard in subsequent experiments. When the temperature is too high or too low, the as-deposit HfO_2 films exhibit fractal structures, rather than continuous films. The presumed cause is related to the fact that the too-low deposition temperature will lead to the precursor molecules diffusing slower and thus clusters, which in turn causes the formation of fractal structures of HfO_2 particles. When the deposition temperature is too high, physical adsorption hardly occurs and transitions to chemical adsorption, but due to the inertness of the substrate surface, nucleation occurs at places such as steps with dangling bonds.

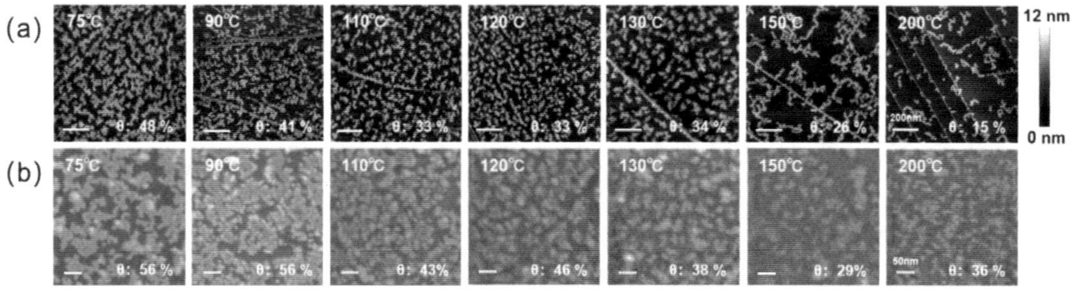

Figure 2. (a) HOPG and (b) Morphology of HfO_2 growing on CVD-G/SiO_2 substrate at different temperatures.

After the optimization of the deposition temperature, we explore the effects of two important factors for the low-temperature ALD process, i.e. the precursor quantity and purge time. Following the orthogonal experimental design, we selected the pulse pressure values of ALD precursors H_2O and TDMAHf, as well as the pulse holding time, as the three factors for analysis, under which nice comparative experiments were conducted with the setting parameters summarized in Table 1 and the results were displayed in Figure 3.

Table 1. A table with headings spanning two columns and containing notes[a].

No.	1	2	3	4	5	6	7	8	
Exp.								9	
Dose H_2O (Torr·ms)	3	16	143	3	16	143	3	16	143
Dose Hf (Torr·ms)	12	12	12	170	170	170	340	340	340
t_{purge} (s)	7	12	20	12	20	7	20	7	12

The experimental results indicate that shortening the purge time can improve the coverage of HfO_2, regardless of the doses of H_2O or TDMAHf. For instance, this phenomenon was observed under the conditions of 3/12 Torr·ms, 143/170 Torr·ms, and 46/340 Torr·ms. Specifically, when using a long pulse of 20 s (as shown in groups 3/5/7), the coverage of HfO_2 is relatively low, suggesting that a shorter purge time is more conducive to the physical adsorption of HfO_2 on the substrate. This is because, within a shorter t_{purge}, unreacted physically adsorbed precursor molecules lack time to desorb from the substrate and chamber walls. Conversely, using a longer t_{purge} can lead to more precursors desorbing from the substrate surface, thus reducing the amount of material deposited on the substrate. This aligns with the growth mechanism of nanofog-ALD, where nanoparticles may nucleate within the ALD chamber before migrating to the substrate surface or directly forming nuclei on the substrate surface. Therefore, precise control of t_{purge} is crucial for optimizing the deposition process of HfO_2 on carbon-based surfaces.

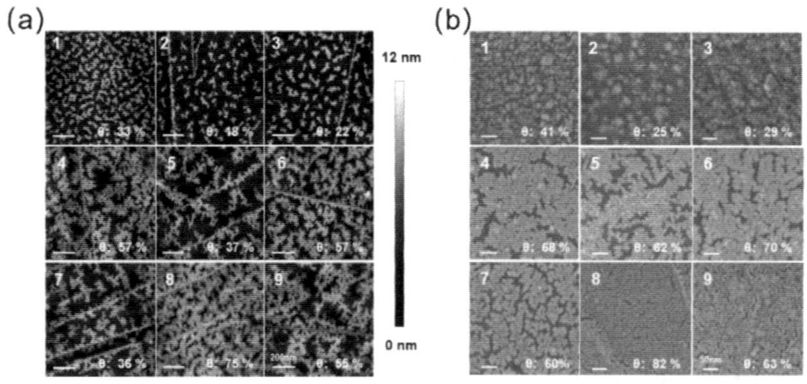

Figure 3. (a) HOPG, (b) CVD-G/SiO_2 substrate surface HfO_2 morphologies grown under different precursor pulse time and purge time.

When the purge time and the doses of H_2O are fixed, it is observed that the higher the doses of TDMAHf the higher the coverage of HfO_2 and the more prominent the fractal structure (as shown in 1/6/8, 2/4/9, or 3/5/7 in Figure 3). This phenomenon may be caused by the fact that TDMAHf is more prone to physical adsorption on the substrate than H_2O, thus acting as the nucleation sites for further growth of HfO_2. Additionally, it is observed that the coverage of HfO_2 deposited on the CVD-G/SiO_2 is generally higher than that on the HOPG substrate under the same condition, which may be related to the specific substrate condition. For example, the as-exfoliate HOPG should be much cleaner than CVD-G/SiO_2.

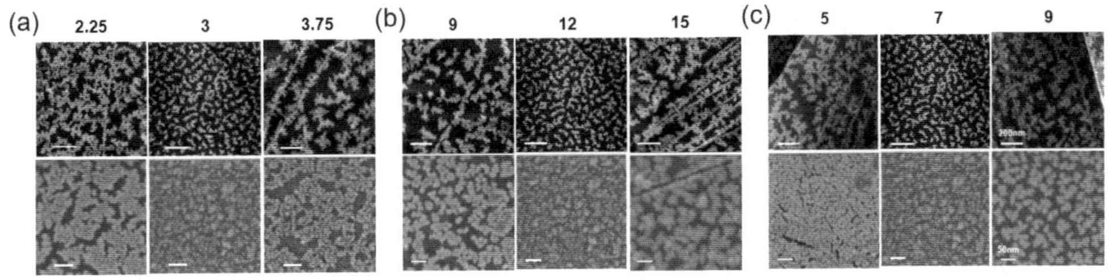

Figure 4. HOPG and CVD-G/SiO_2 substrate surface HfO_2 morphologies grown under different precursor pulse time and purge time. (a) Dose H_2O (Torr·ms), (b) Dose Hf (Torr·ms), (c) t_{purge} (s).

As shown in Figure 4, note that under the low-temperature ALD condition, a clear and wide "optimal window" that found in true ALD condition (where the chemical adsorption induced self-limiting surface reaction occurs) was hardly noticed. For the physical adsorption, due to the sensitivity of the ALD growth mechanism, even minor parameter adjustments can have a significant impact on the morphology, making it difficult to find a clear one. Therefore, we directly selected the five processes of 1 (with the smallest and most uniform growth particles) and 4, 6, 8, 9 (with higher films coverage) to prepare five samples using ALD to grow 50 cycles of HfO_2, namely #1, #2, #3, #4, and #5, for subsequent roughness (RMS) characterization. Through this approach, we hope to evaluate and optimize these process conditions more accurately, thereby achieving precise control over the growth process of HfO_2 films.

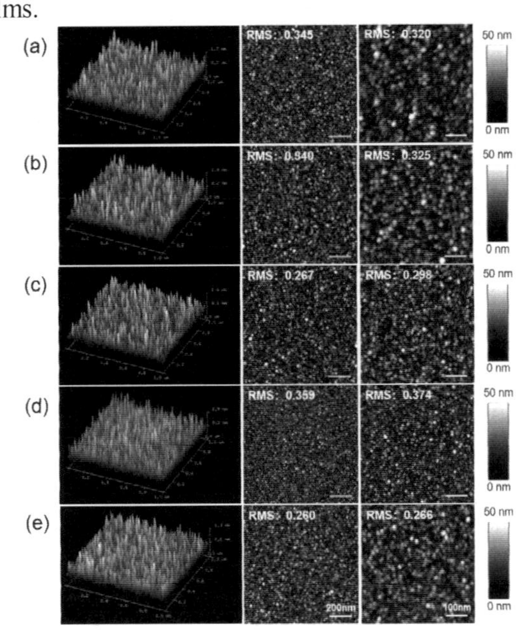

Figure 5. AFM 3D images and surface morphology diagrams of HfO_2 grown on SiO_2 substrates through 50 cycles using ALD process (a) #1; (b) #2; (c) #3; (d) #4; (e) #5.

Observing Figures 5a, 5b and 5e, we found that the ALD HfO_2 particles grown on samples #1, #2, and #5 were relatively large, and the roughness of samples #1 and #2 was high, with RBS values of 0.345 and 0.340 within the 1 μm^2, and 0.320 and 0.325 within the 0.25 μm^2, respectively. Meanwhile, the 3D images revealed that the deposition of HfO_2 on the surfaces of samples #1 and #2 was not flat, with obvious concave-convex phenomena corresponding to the locations of the holes in the plan view, which contributed to the higher roughness of the samples. However, when the ratio of H_2O to TDMAHf was larger, the deposition of the HfO_2 films was more uniform, as shown in Figures 5c and 5e. This indicates that increasing the amount of H_2O precursor helps reduce the roughness of the film.

To illustrate the similarity between A-CNTs and CVD-G/SiO_2 substrates, we attempted to directly apply these five processes to arrayed carbon nanotubes to observe whether similar results can be obtained. First, a two nm nucleation layer was deposited on the carbon nanotubes using these five processes, followed by the growth of a six nm HfO_2 film. The experimental results are shown in Figure 6. For the traditional high-temperature process, the SEM image of the six nm HfO_2 film deposited directly on the surface of the original A-CNTs showed large particles and poor film continuity, as shown in Figure 6a. Subsequently, we conducted ALD growth experiments on A-CNTs using processes 1, 4, 6, 8, and 9 based on the pulse and purge time parameters. As shown in Figures 6b and 6c, similar to the results in Figures 4a and 4b, although the continuity of the HfO_2 film was improved, the particles were large, and the film coverage was continuous but unevenly distributed. Although the HfO_2 particles grown using process 9 decreased in size, there was still discontinuity. However, when using ALD processes 6 and 8, the grown HfO_2 particles were more uniform, and the film continuity was stronger, as shown in Figures 6d and 6e. This result is similar to the morphology structures in Figures 5c and 5d, indicating that the film morphology is more uniform due to the small and dense HfO_2 particles grown using these processes.

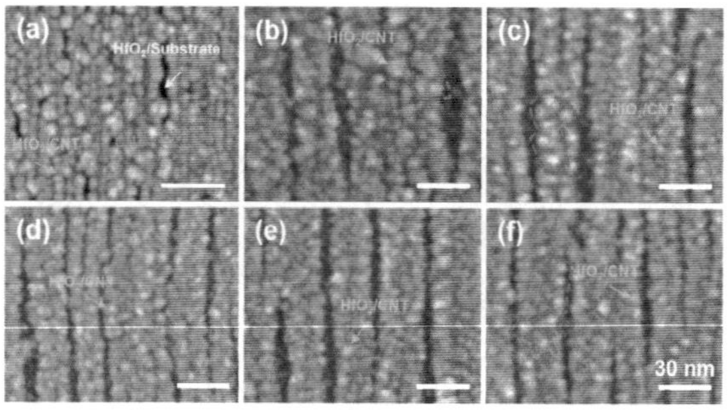

Figure 6. (a) Original A-CNTs; (b) 1; (c) 4; (d) 6; (e) 8; (f) 9. Growth of a six nm HfO_2 film after growing a nucleation layer on the surface of original A-CNTs using ALD processes.

In terms of morphology, compared to the other four ALD processes, process 6 produces HfO_2 films with a relatively small roughness of 0.267, and the HfO_2 particles grown on A-CNTs are more uniform with strong film continuity. Process 8 deposits the smallest and densest HfO_2 particles. Although process 9 deposits larger HfO_2 particles compared to process 8, it achieves the smallest HfO_2 films roughness of 0.260. In contrast to processes 6, 8, and 9, processes 1 and 4 result in HfO_2 films with greater roughness (0.345, 0.340) and larger deposited particles, indicating poorer morphological outcomes for these two processes. The other three processes each have their own advantages, and further testing may be required to verify performance differences.

3.2. Tuning of parameters at high-temperatures

Figure 7 shows the SEM images of 10 cycles HfO_2 on the surface of CVD-graphene/SiO_2 substrates, where the doses of H_2O in Figures 7a, 7b, and 7c are 1.52, 4.18, and 6.51 Torr/ms, respectively, and the doses of TDMAHf are 2.8, 10.01, and 18.1 Torr/ms, respectively. The resulting HfO_2 coverage θ is 26.97%, 23.53%, and 21.41%, respectively. The results indicate that with the increase of precursor quantity, the coverage of HfO_2 clusters on the surface of the graphene substrate only fluctuates within a range of 5.56%, showing no significant change.

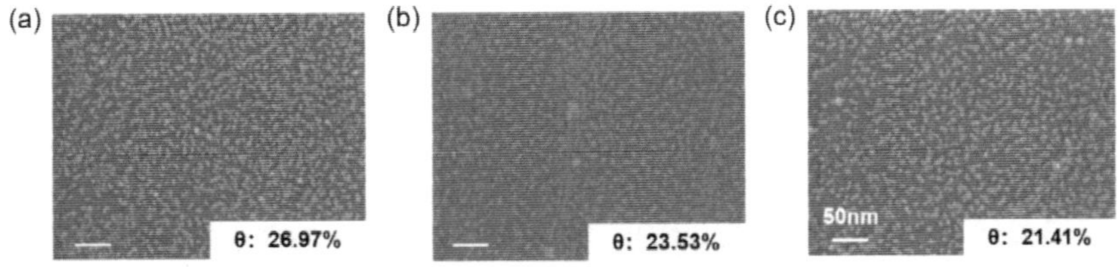

Figure 7. SEM images of HfO_2 ALD with different precursor amounts on CVD-G/SiO_2 at a T_{dep} of 200 °C. The deposition of HfO_2 on CVD-G/SiO_2 is only 10 ALD cycles.

This suggests that for the CVD-graphene substrate used in this experiment, the doses of H_2O (1.52 Torr·ms) and TDMAHf (2.80 Torr·ms) corresponding to Figure 7a is already sufficient for saturated reactions. Due to the self-limiting mechanism of ALD, supplying more precursor molecules cannot increase the nucleation rate, which is consistent with our expectations. To further confirm this, we grew 50 cycles of HfO_2 films and measured their thicknesses using a film thickness gauge. The results showed that the thicknesses of the three samples were 13.45 nm, 13.09 nm, and 13.16 nm, respectively, further indicating that the quantity of oxygen precursor (1.52 Torr/ms) and metal precursor (2.80 Torr/ms) corresponds to the precursor quantity for saturated reactions.

Based on the determination of the CM-ALD process, we attempted to deposit HfO_2 on the CVD-graphene/SiO_2 substrate using the PM-ALD process. It is worth noting that we have subdivided the PM-ALD process into two types, as shown in the yellow part of Figure 8:

- HPM process: In the pretreatment stage, 1.52 Torr·ms of H_2O is pulsed first, followed by a 20-second purge, which constitutes the first complete cycle; subsequently, 1.52 Torr·ms of H_2O is pulsed again to enter the next pretreatment, as shown in Figure 8a.
- NPM process: Pulsing 1.52 Torr·ms of H_2O constitutes one cycle, followed by another pulse of 1.52 Torr·ms of H_2O, which is the second cycle; after all cycles are completed, a 20-second purge process is performed, as shown in Figure 8b.

After the completion of the pretreatment process, we immediately proceeded to the deposition process while maintaining the vacuum state unchanged. The deposition parameters were consistent with the aforementioned CM-ALD, and namely, the doses of H_2O and TDMAHf were 1.52 Torr·ms and 2.80 Torr·s, respectively.

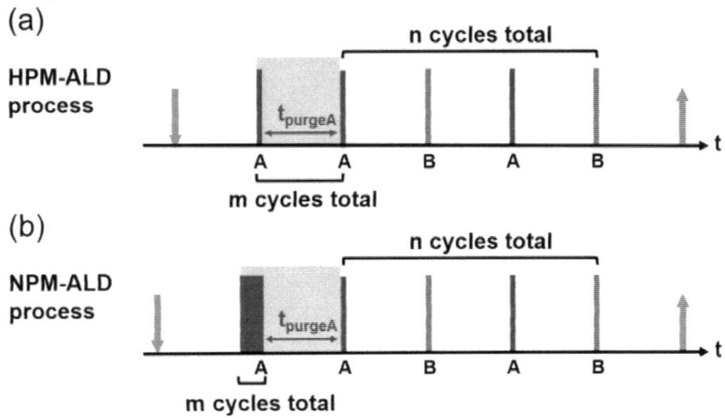

Figure 8. Schematic diagrams of (a) HPM-ALD and (b) NPM-ALD.

To explore the PM-ALD process and cycles corresponding to the highest HfO_2 coverage, we designed a series of pretreatment cycles for both HPM and NPM pretreatment processes, including 5, 20, 50, 100, 150, and 300 cycles. Subsequently, based on the pretreatment processes, CM-HfO_2 is grown for 10 cycles, and the corresponding growth results are shown in Figure 9. Figure 9a shows the results of the HPM-ALD process. It can be seen that as the number of pretreatments gradually increases, the coverage of CM-HfO_2 on the substrate gradually increases and reaches 40% at 150 pretreatments, which is much higher than the CM-ALD process without pretreatment (θ = 26.41%, as shown in Figure 7a). However, as the number of pretreatments continues to increase, the HfO_2 coverage no longer increases, indicating that in the HPM-ALD process, the number of pretreatments corresponding to the highest HfO_2 coverage is 150 times. Figure 9b shows the results of the NPM-ALD process. Its growth trend is similar to the HPM-ALD process, both showing that the HfO_2 coverage increases with the increase in the number of pretreatments and then decreases. The difference is that in the NPM-ALD process, the maximum HfO_2 coverage is achieved when the number of pretreatments is 50 times, and it is as high as 50%, almost twice that of the pretreatment-free process. These series of results indicate that both the HPM-ALD and NPM-ALD processes can effectively increase the HfO_2 coverage. Among them, the NPM-ALD process with 50 pretreatments has a short time consumption and the highest coverage, making it the best-optimized process. In addition, we notice an interesting phenomenon: whether it is HPM-ALD or NPM-ALD when the number of pretreatments is excessive, the HfO_2 coverage will decrease. This may be due to the agglomeration of excessive H_2O, resulting in the selective growth of subsequent TDMAHf on the agglomerated H_2O, thereby reducing the HfO_2 coverage.

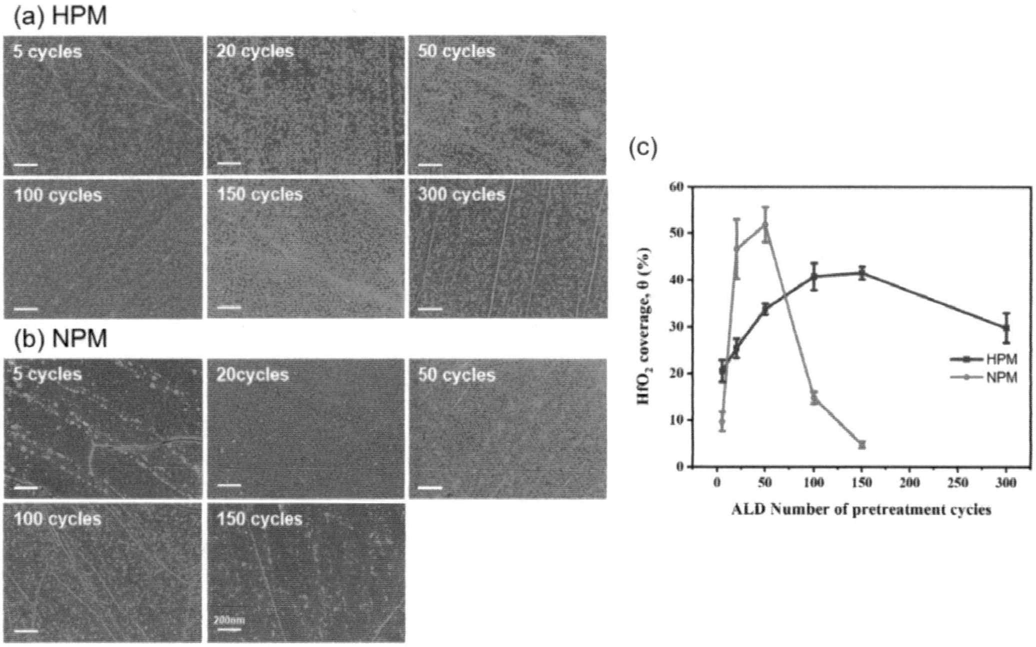

Figure 9. (a) Morphology of HfO_2 grown on CVD-G/SiO_2 using CM process after 5~300 cycles of HPM process; (b) Morphology of HfO_2 grown on CVD-G/SiO_2 using CM process after 5~150 cycles of NPM process; (c) Coverage of HfO_2 deposited on CVD-G/SiO_2 using CM process after pretreatment cycles of HPM and NPM.

Under the condition of a deposition temperature of 200 °C, using the PM method on CVD-G/SiO_2 does not allow for conformal HfO_2 nucleation, meaning that the film cannot be uniformly deposited on every exposed surface of the substrate. This prompts us to delve deeper into the limiting parameters of the ALD process under such high-temperature conditions. Since the ALD process highly depends on surface saturation to achieve self-limiting reactions, we speculate that the previously observed selective HfO_2 nucleation on graphene may be due to the surface not reaching saturation conditions. However, it is unclear in the current literature whether these conditions can always be met. Therefore, we decide to try using more doses of precursors than conventional ones, aiming to saturate the surface and thus achieve conformal HfO_2 nucleation on graphene.

Given that the doses of precursors is determined by both the P_{dos} and t_{dos}, we can obtain a sufficiently high pulse volume to achieve the desired nucleation by increasing P_{dos} or extending t_{dos}. However, due to the interdependent relationship between P_{dos} and t_{dos}, it becomes quite complex to independently explore the influence of each parameter in CM. This is because they both are controlled by a single parameter, the pulse time (t_{pulse}) of precursors. To address this issue, we introduce improvements to the ALD process, here referred to as MM and SM. These improvements allow us to independently adjust P_{dos} and t_{dos}, thus decoupling their relationship.

In the MM-ALD process, each dose of H_2O/TDMAHf is rapidly delivered through multiple consecutive pulse sequences. That is, when the precursor is continuously pulsed m times, its pulse residence time t_{dos} is equal to the sum of m pulse times (t_{pluse}). Therefore, MM-ALD can independently regulate t_{dos} without changing P_{dos}. To explore the optimal number of consecutive pulses, in the MM-ALD process, when H_2O/TDMAHf is continuously pulsed four times, as shown in Figure 10b, uniform nucleation can be easily obtained with a pulse volume of 3.04/5.60 Torr·s without preferential nucleation behavior; further increasing to 0.76/14.00 Torr·s, i.e., continuously pulsing five times, can achieve highly uniform and fully surface-covered HfO_2 nucleation.

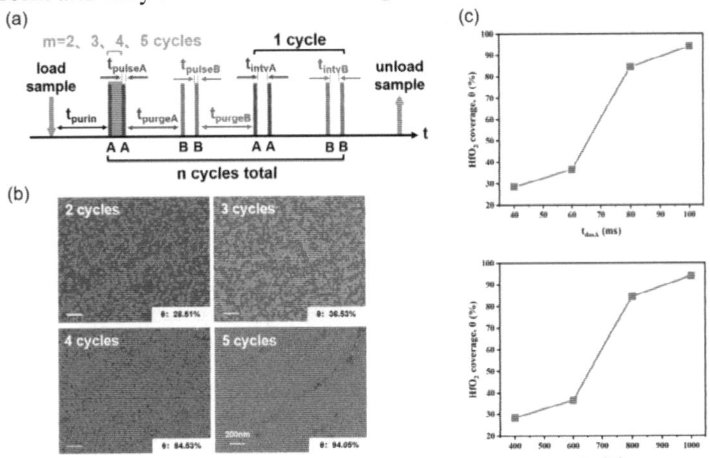

Figure 10. (a) Schematic diagram of the MM-ALD process flow; (b) Nucleation morphology of MM-ALD HfO_2 on CVD-G/SiO_2 at m=2, 3, 4, 5 cycles; (c) Coverage of HfO_2 on CVD-G/SiO_2 with varying precursor residence time under the conditions of t_{dosA} and t_{dosB}.

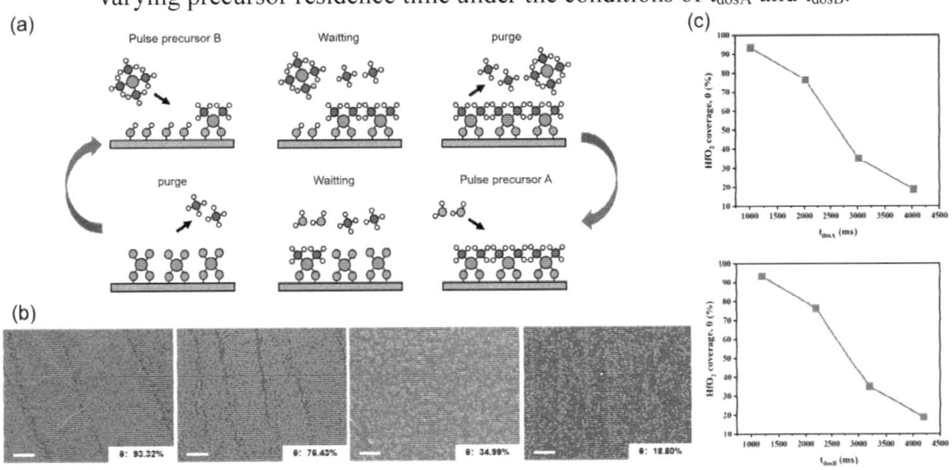

Figure 11. (a) Schematic diagram of the SM-ALD HfO_2 growth process; (b) Morphology of ALD HfO_2 grown on CVD-G/SiO_2 using SM-ALD for 10 cycles with t_{dos} of 1020, 2020, 3020, and 4020 ms, respectively, while keeping P_{dos} constant; (c) Coverage of HfO_2 on CVD-G/SiO_2 with varying precursor residence time under the conditions of t_{dosA} and t_{dosB}.

For SM-ALD, the sample is soaked (held) in the H_2O/TDMAHf pulse for several seconds, allowing t_{dos} to be controlled by the residence time rather than t_{pulse}. Therefore, SM allows t_{dos} to be completely independent of P_{dos} and can be almost indefinitely extended, as shown in Figure 11a. Similarly, in SM, only using the doses of H_2O/TDMAHf of 77.52/16.8 Torr·s, a highly uniform nucleation can also be achieved through ALD when soaking for one second. A strong linear correlation can be observed between θ and t_{dos} in all ALD modes, as a higher t_{dos} usually leads to a higher θ. However, in SM, when the doses of H_2O/TDMAHf exceed 77.52/16.8 Torr·s, θ actually decreases, which may be due to the precursor staying on the substrate surface for too long, leading to particle aggregation. Finally, using MM-ALD and SM-ALD at a deposition temperature of 200°C, we successfully achieve conformal HfO_2 nucleation on CVD-G/SiO_2, as shown in Figure 10b and Figure 11b.

We conducted ALD growth experiments on A-CNTs using the parameters of HPM = 150 cycles, NPM = 50 cycles, MM = 0.76/14.00 Torr·s, and SM = 77.52/16.8 Torr·s, and the results are shown in Figure 12. Figure 12a shows the SEM image of a six nm HfO_2 thin film deposited directly on the surface of the original A-CNTs. When using the PM process, whether it is HPM or NPM, although the continuity of the obtained HfO_2 thin film is improved due to the pretreatment effect of H_2O, the HfO_2 particles are significantly smaller, and the coverage of the thin film is continuous, the distribution is not uniform and there are still a few larger particles. Additionally, the MM process can also improve the continuity of the HfO_2 thin film and enhance the growth of HfO_2 by adjusting t_{dos}. In contrast, although the continuity of HfO_2 grown using the SM process is improved, the particles are generally larger, which may be due to the results obtained on graphene are not completely applicable to carbon nanotubes with more arrangement defects, so it is not possible to directly apply the same process parameters to obtain the ideal thin film morphology.

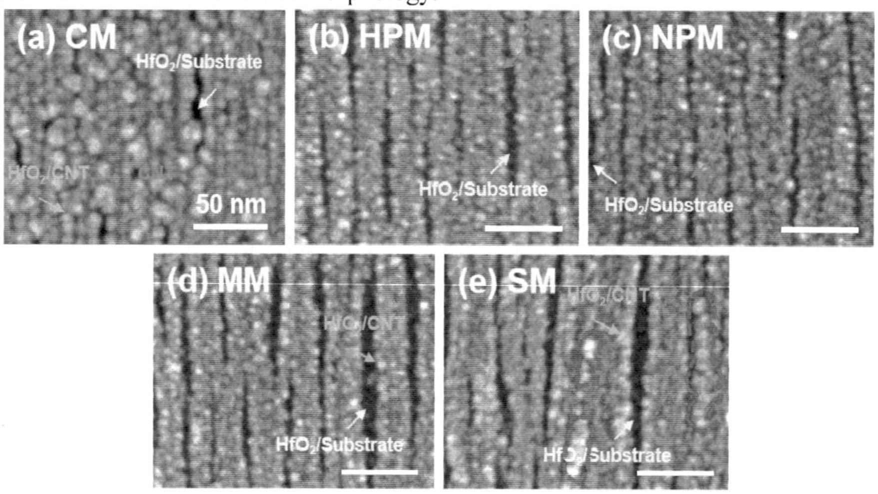

Figure 12. Using the optimal high-temperature process of (a) CM; (b) HPM; (c) NPM; (d) MM; and (e) SM, a two nm nucleation layer and a six nm HfO_2 thin film on the surface of the original ACNTs are grown.

4. Conclusion

The research results of this paper focus on two methods proposed to promote the ALD nucleation of gate dielectrics on carbon-based material surfaces, which indicate that

By utilizing low-temperature NF-ALD technology, the ALD of HfO_2 directly on graphene using standard H_2O/TDMAHf precursors can result in different nucleation behaviors across the entire surface depending on the deposition conditions. This nucleation behavior can be highly selective or completely uniform. Experimental results show that the low-temperature process on carbon nanotubes with more arrangement defects has a similar effect to that on graphene. Deposition temperature, the amount of precursor pulses, and purge time, significantly affect the growth of HfO_2 films on arrayed carbon nanotubes. Through systematic control of ALD parameters, we have successfully identified the optimal process conditions: with H_2O/TDMAHf pulse amount of 143/170 Torr·ms and a purge time between the two precursors of 7 s, a more continuous and uniform HfO_2 films can be prepared compared to traditional high-temperature conventional processes.

Under variable parameter high-temperature processes, PM, MM, and SM were studied to improve nucleation issues. At the same time, we noticed an interesting phenomenon: although all three processes improved the nucleation of HfO_2 on carbon-based materials, it was difficult to achieve full coverage of HfO_2 for PM, regardless of whether it was HPM or NPM, and regardless of whether the pretreatment was excessive or insufficient. This may be due to excessive H_2O molecule clustering, which affects the selective growth of subsequent TDMAHf on the clustered H_2O molecules, resulting in reduced HfO_2 coverage. As shown in this paper, even though most of the process parameters on graphene cannot be directly applied to A-CNTs, in SM-ALD, using only 77.52/16.8 Torr·s of H_2O/TDMAHf pulse amount and soaking for one second, highly uniform nucleation can be achieved through ALD, enabling the preparation of sub-two-nm-thin continuous HfO_2 films directly on the surface of graphene. Moreover, the SM-ALD results can be directly applied to carbon nanotubes.

This ability to directly integrate thin and continuous HfO_2 films with carbon-based materials will allow for further development and exploitation of carbon nanotubes as channel materials for a wide range of applications. This achievement provides us with a new approach for the preparation of high-quality HfO_2 films on different types of carbon-based materials.

References

[1] Robertson J, Wallace R M. High-K materials and metal gates for CMOS applications[J]. Materials Science and Engineering: R: Reports, 2015, 88: 1-41.
[2] Charlier J C, Blase X, and Roche S. Electronic and transport properties of nanotubes [J]. Reviews of modern physics, 2007, 79(2): 677-732.
[3] Avouris P. Carbon nanotube electronics and optoelectronics[J]. Mrs Bulletin, 2004, 29(6): 403-410.
[4] Hierold C, Helbling T, Roman C, et al. CNT based sensors[J]. Advances in Science and Technology, 2009, 54: 343-349.
[5] Park S, Vosguerichian M, and Bao Z. A review of fabrication and applications of carbon nanotube film-based flexible electronics[J]. Nanoscale, 2013, 5(5): 1727-1752.
[6] Di J, Zhang X, Yong Z, et al. Carbon-nanotube fibers for wearable devices and smart textiles [J]. Advanced materials, 2016, 28(47): 10529-10538.
[7] Hu L, Hecht D S, and Gruner G. Carbon nanotube thin films: fabrication, properties, and applications[J]. Chemical reviews, 2010, 110(10): 5790-5844.
[8] Zhong D, Zhang Z, and Peng L M. Carbon nanotube radio-frequency electronics[J]. Nanotechnology, 2017, 28(21): p. 212001.
[9] Peng L M, Zhang Z, and Qiu C. Carbon nanotube digital electronics[J]. Nature Electronics, 2019, 2(11): 499-505.
[10] Wei H, Shulaker M, Hills G, et al. Carbon nanotube circuits: Opportunities and challenges[C]//2013 Design, Automation & Test in Europe Conference & Exhibition (DATE). IEEE, 2013: 619-624.
[11] Geim A K, Novoselov K s. The rise of graphene [M]. Nanoscience and Technology, 2009: 11-19.
[12] Lu Y, Bangsaruntip S, Wang X, et al. DNA functionalization of carbon nanotubes for ultrathin atomic layer deposition of high κ dielectrics for nanotube transistors with 60 mV/decade switching[J]. Journal of the American Chemical Society, 2006, 128(11): 3518-3519.
[13] McDonnell S, Azcatl A, Mordi G, et al. Scaling of HfO2 dielectric on CVD graphene[J]. Applied Surface Science, 2014, 294(3): 95-99.
[14] Pitner G, Zhang Z, Lin Q, et al. Sub-0.5 nm interfacial dielectric enables superior electrostatics: 65 mV/dec top-gated carbon nanotube FETs at 15 nm gate length[C]//2020 IEEE International Electron Devices Meeting (IEDM). IEEE, 2020: 3.5.1-3.5.4.

[15] Wang X, Tabakman S M, and Dai H. Atomic layer deposition of metal oxides on pristine and functionalized graphene[J]. Journal of the American Chemical Society, 2008, 130(26): 8152-8153.
[16] Kwak I, Kavrik M, Park J H, et al. Low interface trap density in scaled bilayer gate oxides on 2D materials via nanofog low temperature atomic layer deposition[J]. Applied Surface Science, 2019, 463: 758-766.
[17] Lin Y S, Hoo J Y, Chung T F, et al. Low-Temperature Physical Adsorption for the Nucleation of Sub-10 nm Al2O3 Gate Stack on Top-Gated WS2 Transistors[J]. ACS Applied Electronic Materials, 2020, 2(5): 1289-1294.
[18] Ko J S, Zhang Z, Lee S, et al. Ultrathin Gate Dielectric Enabled by Nanofog Aluminum Oxide on Monolayer MoS 2[C]//ESSDERC 2023-IEEE 53rd European Solid-State Device Research Conference (ESSDERC). IEEE, 2023: 1-4.
[19] Aria A I, Nakanishi K, Xiao L, et al. Parameter space of atomic layer deposition of ultrathin oxides on graphene[J]. ACS Applied Materials & Interfaces, 2016, 8(44): 30564-30575.
[20] Knoops H C M, Potts S E, Bol A A, et al. Atomic layer deposition[M]//Handbook of Crystal Growth. North-Holland, 2015: 1101-1134.
[21] Liu L, Han J, Xu L, et al. Aligned, high-density semiconducting carbon nanotube arrays for high-performance electronics[J]. Science, 2020, 368(6493): 850-856.
[22] Nam T, Seo S, and Kim H. Atomic layer deposition of a uniform thin film on two-dimensional transition metal dichalcogenides[J]. Journal of Vacuum Science & Technology A, 2020, 38(3): 1-17.

Journal of Physics: Conference Series **2842** (2024) 012028 doi:10.1088/1742-6596/2842/1/012028

The research of electrochemical performance of Mg-doped $NaNi_{1/3}Mn_{1/3}Fe_{1/3}O_2$ positive electrode materials

Xue Bai*, Aijia Wei, Rui He, Xiaohui Li, Lihui Zhang, and Zhenfa Liu

Institute of Energy Resources, Hebei Academy of Sciences, Shijiazhuang, 050081, China

*Corresponding author's e-mail: bx_ier@yeah.net

Abstract. The demand for Li-ion batteries is increasing year by year, which poses a great challenge to the scarce Li and Co resources. Na-ion batteries are considered to be one of the potential energy storage batteries because of their low raw material cost, wide source of sodium resources, and prices that are not susceptible to market fluctuations. The layered oxide positive electrode (such as $NaNi_{1/3}Mn_{1/3}Fe_{1/3}O_2$, NMF) is a potential replacement due to its high-capacity characteristics and low-cost advantages. However, the rate performance of NMF is compromised by the slow migration kinetics of Na^+ ions. Here, we used the Mg doping strategy to realize the enhancement of the rates and cycling properties of NMF. Our research shows that the 2% Mg-doped cathode material has superior rate performance and cycling performance. The discharge-specific capacities are 120.5 mAh/g at 0.1C and 88.9 mAh/g for 100 cycles at 0.5C with a high retention rate (85.2%).

1. Introduction

The demand for Li-ion batteries (LIBs) is gradually increasing, especially in the field of power batteries and energy storage power stations, while the shortage of Li resources has led to significant price fluctuations and a gradual increase in the contradiction between high demand. In addition to Li resources, the scarcity and uneven distribution of cobalt, nickel, and other metals also contribute to the high cost of LIBs, hindering large-scale expansion in application areas. Na-ion batteries have a similar internal structure to LIBs, where the positive and negative electrodes are embedded compounds of sodium ions with different potentials, and the electrolyte is an organic electrolyte containing sodium salt (e.g., $NaPF_6$) [1-3]. At the same time, the production process and equipment of SIBs are similar to those of LIBs, and most importantly, sodium resources are very abundant and widely distributed. At the same time, these features offer several benefits, including low cost, rapid charging capabilities, superior low-temperature performance, and environmental safety [4]. Within the framework of the national dual-carbon strategy, SIBs are emerging as an optimal power source for future distributed energy storage, centralized energy storage systems, and low-speed electric vehicle power supplies. Consequently, the advancement of SIB technology holds significant implications for alleviating the supply pressure of LIBs and promoting the development of the energy storage industry [5, 6].

In SIBs, the electrochemical performance is largely limited by the cathode material, as it directly influences safety, energy density, and cycling performance. Layered oxide materials assume an important role in the industrial manufacturing of SIBs due to their high theoretical specific capacity and production ease [7]. However, these materials suffer from complex phase transitions and poor cycling performance. Among them, iron-based transition metal oxides (α-$NaFeO_2$), which belong to the tripartite crystal system and have a space group of R3m, is a typical O3-structured layered material with low price. However, when the cut-off voltage for charging exceeds 3.5 V, it leads to an irreversible phase transition, resulting in reversible capacity reduction. To enhance the stability of $NaFeO_2$ cathode materials under high operating voltages, researchers frequently incorporate elements such as Mn, Ni, and Co to partially substitute Fe. This substitution not only augments the specific capacity but also bolsters the cycling stability [8, 9]. However, the doping of the Co element is uneconomical due to cost considerations, so more studies use Mn and Ni elements for doping [10, 11].

In this work, Mn and Ni elements were employed to dope the $NaFeO_2$ cathode material, while a small amount of Mg element was doped to further enhance the reversible specific capacity and cycling stability. SEM and XRD characterization of the synthesized materials revealed that the cathode materials have spherical-like morphology and layered structure belonging to α-$NaFeO_2$. The XRD diffraction pattern revealed no diffraction peaks of other impurity phases, suggesting the successful synthesis of pure-phase Mg-doped NMF cathode material. Electrochemical performance tests demonstrated that the first discharge-specific capacities of cathode material are 120.5 and 104.3 mAh/g at 0.1C and 0.5C, respectively. Furthermore, when the number of cycles reaches 100 at 0.5C, its capacity retention is 85.2. The use of MgO as a dopant does not have an impact on the environment and does not produce CO_2 emissions.

2. Material and methods

2.1 Synthesis and characterization

The NNMF cathodes with different Mg doping doses were prepared by the solid-state method. The precursor (P-NNMF) was synthesized by the co-precipitation method at 60 °C for 15 h by adding $NiSO_4$, $MnSO_4$, $FeSO_4$, NaOH as a precipitant agent, $NH_3 \cdot H_2O$ as a complexing agent, and MgO as a doping agent. The molar ratios of $NiSO_4$, $MnSO_4$, and $FeSO_4$ were 1:1:1, and the doping amounts of Mg were 1%, 2%, and 3%. The precursors were homogeneously mixed with Na_2CO_3. After calcination at 750 °C for 15 h, we obtain different Mg-doped NNMF materials labeled as Mg-1, Mg-2, and Mg-3, respectively.

The phase composition of the synthesized cathode materials was analyzed by XRD (Ultima IV, Riguku) with diffraction angles in the range of 10-90° and tested at a rate of 5°/min. The morphology and surface structural information were observed by SEM (Inspect S50, FEI).

2.2 Electrochemical performance test

The electrochemical properties of the materials were evaluated by constructing a button cell. The electrode slurry consists of cathode material, SP, and PVDF were combined in an NMP solution at a mass ratio of 8:1:1. Aluminum foil was selected as the collector of the electrode slurry and dried at 85°C for 45min. The positive electrode piece was made into a 12 mm circle. The coin cell was assembled in a glove box with lithium metal for the negative electrode, polypropylene diaphragm for the separator, and 1 mol/L $NaPF_6$ solution for the electrolyte. The rate and cycle performance of the button cell were tested using a constant current charge and discharge mode, with a voltage range of 2.0-4.0 V.

2.3 Calculation section

The formula for calculating initial efficiency:

$$\eta = \frac{C_d}{C_c} \times 100\% \tag{1}$$

where η is the initial charge-discharge efficiency, C_d is the initial discharge-specific capacity, and C_c is the initial charge-specific capacity.

The formula for calculating capacity retention rate:

$$\omega = \frac{C_{100}}{C_1} \times 100\% \tag{2}$$

where ω is the capacity retention rate, C_{100} is the DSC after 100 cycles at 0.5C, and C_1 is the initial DSC at 0.5C.

3. Results and discussion

3.1 Material characterization

The crystal structures of the precursor and the cathode material with different Mg doping amounts were tested by XRD. As shown in Figure 1a, the XRD pattern of the precursor obtained by co-precipitation shows that there are no obvious characteristic peaks of crystal structure, which indicates that the precursor, without high-temperature treatment, is not well crystallized. The material, post-high-temperature annealing treatment, exhibited characteristic peaks consistent with the R-3m space group, indicative of an O3-$NaFeO_2$ structure. The absence of any diffraction peaks corresponding to impurity substances in the XRD patterns confirms the successful doping of the Mg element into the NNMF materials. With the increase of Mg doping, the diffraction peaks on the (104) crystal plane of the NNMF cathode material gradually weakened, and the (104) crystal plane possessed the strongest diffraction peaks when the Mg doping was 2%, which implied that the material had better crystallinity. Figure 1b compares the greenhouse gas (GHG) emissions of different Mg-containing chemicals during thermal decomposition, and there are no GHG emissions when MgO is used as a Mg dopant, which minimizes the impact on the environment.

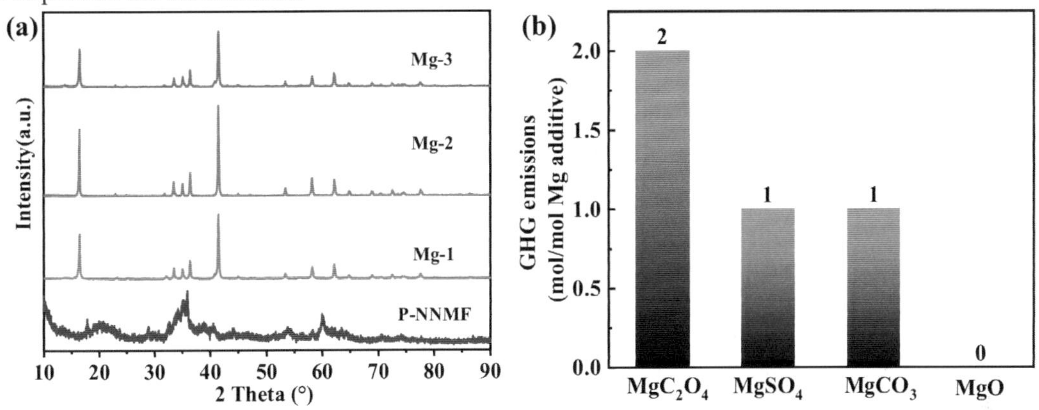

Figure 1. XRD results of P-NNMF and cathode materials with different Mg-doped content.

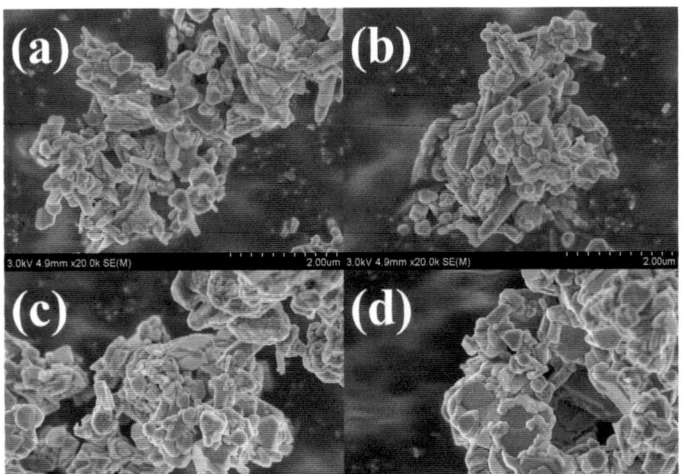

Figure 2. SEM images of (a) P-NNMF, (b) 1wt% Mg-doped, (c) 2wt% Mg-doped, and (d) 3wt% Mg-doped.

Figure 2 presents the SEM images of both the precursor and cathode material, each with varying degrees of Mg doping. As shown in Figure 2a, the precursor presents a mixture of spherical and rod shapes. After high-temperature calcination, the 1% Mg-doped cathode material still retains part of the rod-like structure (Figure 2b). As the concentration of Mg doping increased, the rod-like structure in the 2% Mg-doped material progressively dissipated (Figure 2c), giving way to a more homogenous granular structure with a consistent particle size. However, when the Mg doping level reached 3% (Figure 2d), a relatively large lamellar structure emerged within the Mg-doped material.

3.2 Electrochemical performance

Figure 3 presents the initial charge-discharge curve, discharge-specific capacity at varying rates, and cycling performance curves at 0.5C. As illustrated in Figures 3a and b, the specific charge capacities for 1wt%, 1.5wt%, 2wt%, 2.5wt%, and 3wt% Mg-doped are recorded as 127.3, 135.5, 141.6, 138.4, and 134.4 mAh/g at 0.1C, respectively. Conversely, the specific discharge capacities for these cathodes are noted as 107.5, 110.5, 120.5, 113.2, and 109.2 mAh/g. This results in an initial efficiency of 84.4%, 81.5%, 85.1%, 81.8%, and 81.3% respectively. Figure 3c illustrates that the specific capacities of discharge of 2% Mg-doped cathode material at different rates are 120.5, 111.0, 104.3, and 93.7 mAh/g, respectively. These values surpass those recorded for other materials. Upon reverting the discharge rate back to 0.2C, the specific discharge capacity of the 2% Mg-doped material was found to be 104.1 mAh/g, representing approximately 93.8% of its initial discharge capacity at the same rate, higher than 92.2% and 89.4% for 1% Mg and 3% Mg doping, which indicated that the 2% Mg-doped cathode material had a better rate performance. This phenomenon may be ascribed to the more uniform particle size, which facilitated the migration of Na^+ between the materials. The cycling performance of the cathode material is also an important index for evaluating the material. Figure 3d illustrates the discharge-specific capacity curves for cathode materials with varying degrees of Mg doping, evaluated over 100 cycles at a rate of 0.5 C. After 100 charge/discharge cycles, the capacities recorded for Mg-1, Mg-2, and Mg-3 were 63.6, 88.9, and 61.2 mAh/g, respectively, with retention rates of 66.2%, 85.2%, and 70.3%, which indicate that the 2% Mg-doped cathode material has better cycling stability. Based on the above results, the optimal doping ratio of Mg is 2%, which can significantly improve the performance of SIB cathode materials.

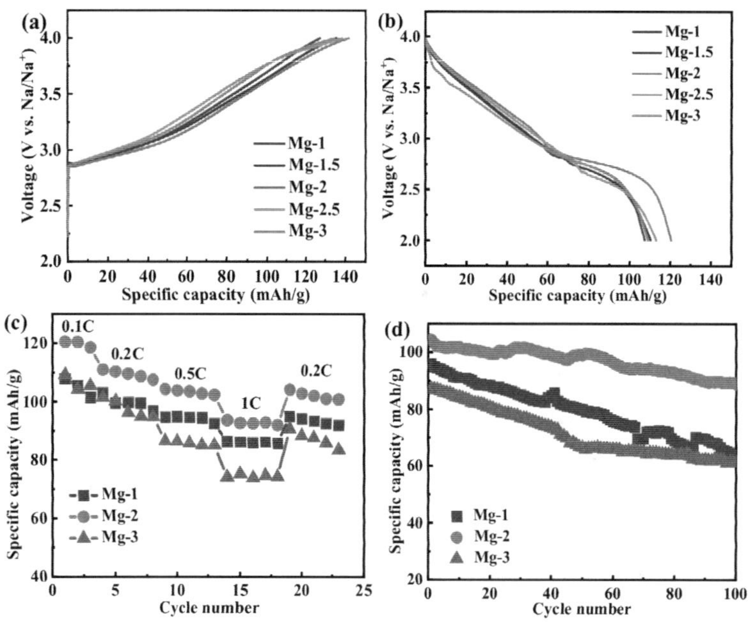

Figure 3. (a) Specific capacity of discharge stages at different rates, (b) cycling performance curves at 0.5C rate.

4. Conclusion.

In this study, the synthesis of precursors for the Na-ion battery cathode was carried out using the co-precipitation method. Subsequently, various Mg-doped $NaNi_{1/3}Mn_{1/3}Fe_{1/3}O_2$ were prepared through a high-temperature solid-phase method. The XRD test results showed that the Mg-doped cathode materials belonged to the R-3m space group, which is an O3-$NaFeO_2$ structure. The morphology characterization of the SEM results showed that the 2% Mg-doped material had a more homogeneous particle size. The electrochemical performance evaluations revealed that the 2% Mg-doped cathode material exhibited superior rate and cycling capabilities. Specifically, the initial discharge capacity and the capacity after 100 cycles are 120.5 and 88.9 mAh/g, respectively, with a retention rate of 85.2%. Therefore, proper Mg doping has a positive effect on improving the O3-$NaNi_{1/3}Mn_{1/3}Fe_{1/3}O_2$ cathode materials.

Acknowledgments

This work was supported by the Project of Hebei Academy of Sciences (24709).

References

[1] Nagmani, Pahari D, Verma P and Puravankara S (2023) Are Na-ion batteries nearing the energy storage tipping point? – Current status of non-aqueous, aqueous, and solid-sate Na-ion battery technologies for sustainable energy storage. J. Energy Storage, 56: 105961.

[2] Li X *et al.* (2024) O3-type cathodes for sodium-ion batteries: recent advancements and future perspectives. Batteries & Supercaps, 7: e202300618.

[3] Xu S, Chen H, Zhang X, Zhou M and Zhou H (2023) NASICON-type $NaTi_2(PO_4)_3$ surface modified O3-type $NaNi_{0.3}Fe_{0.2}Mn_{0.5}O_2$ for high-performance cathode material for sodium-ion batteries. ACS Appl. Mater. Interfaces, 15: 47764–47778.

[4] Jeong M, Lee H, Yoon J and Yoon W S (2020) Structural evolution and redox mechanism of O3-$NaNi_{1/3}Fe_{1/3}Mn_{1/3}O_2$ layered cathode for Na rechargeable batteries. Meet. Abstr. MA2020-01: 567.

[5] Zhao X, Hou L, Liu Q, Zhao Y, Mu D, Zhao Z, Li L, Chen R and Wu F (2024) Insights into the capacity fading and failure mechanism of an O3-$NaNi_{1/3}Fe_{1/3}Mn_{1/3}O_2$ layered oxide cathode material for sodium-ion batteries. J. Mater. Chem. A, Advance article.

[6] Jiang C *et al.* (2024) Toward high stability of O3-type $NaNi_{1/3}Fe_{1/3}Mn_{1/3}O_2$ cathode material with zirconium substitution for advanced sodium-ion batteries. Carbon Neutralization, 3: 233-244.

[7] Cai H, Yan L, Yang X and Pan H (2023) Structural evolution and sodium-storage performance of O3/P2-$Na_xNi_{1/3}Co_{1/3}Mn_{1/3}O_2$ multiphasic cathode materials. Energy Storage Science and Technology, 12: 2707-2714.

[8] Gao X *et al.* (2023) Post-substitution modulated robust sodium layered oxides. Small Methods, 7: 2300635.

[9] Hong N *et al.* (2023) An in situ dual-modification strategy for O3-$NaNi_{1/3}Fe_{1/3}Mn_{1/3}O_2$ towards high-performance sodium-ion batteries. J. Mater. Chem. A,11: 18872-18880.

[10] Liu G, Xu W, Wu J, Li Y, Chen L, Li S, Ren Q and Wang J (2023) Unlocking high-rate O3 layered oxide cathode for Na-ion batteries via ion migration path modulation. J. Energy Chem. 83: 53-61.

[11] Samigullin R, Drozhzhin O and Antipov E (2022) Comparative study of the thermal stability of electrode materials for Li-ion and Na-ion batteries. ACS Appl. Energy Mater. 5: 14-19.

Experimental study on activation of coal gangue and preparation of cementing materials

Guangliang Wang[1], Xingmei He[1], Youliang Yang[1], Mingzhou Xiong[1], Yonghua Ji[1,*], and Bingke Qin[2]

[1] School of Mining and Mechanical Engineering, Liupanshui Normal University, Liupanshui 553004, China

[2] School of Chemistry and Materials Engineering, Liupanshui Normal University, Liupanshui 553004, China

[*]Corresponding author's e-mail: haizhigong@126.com

Abstract. The production of coal gangue in China is constantly increasing, and it has become a large amount of solid waste. Open-air storage not only leads to a waste of land resources but also the harmful elements in coal gangue can cause pollution to the environment. This article studies the use of activated coal gangue in the preparation of cementitious materials. The research results show that the optimal activation conditions are an activation temperature of 650°C and an activation time of 120 minutes. When the amount of coal gangue mixed under these activation conditions is 25%, the cementitious material prepared has the maximum flexural and compressive strengths at 3d, 7d, and 28d.

1. Introduction

Coal gangue is a blackish-gray rock associated with coal seams during the coal formation process. Currently, China's stockpile of coal gangue exceeds 7 billion tons, and it continues to increase by approximately 10% annually. The accumulation of large amounts of coal gangue has formed coal gangue mountains, which cause pollution to the natural environment and are prone to natural disasters such as landslides and mudslides [1-2]. Given the current shortage of sandstone resources for civil construction, if the sorted and processed coal gangue can be used in cementitious materials, it will not only greatly reduce the amount of coal gangue but also broaden the sources of concrete admixtures and sandstone for engineering applications [3].

The activity level of coal gangue depends on the content of soluble SiO2 and Al2O3, as well as the ability of glass phase depolymerization. When coal gangue is calcined, its thermal decomposition and vitrification are the main sources of its activity, resulting in the production of burned coal gangue with pozzolanic activity. This burned coal gangue can then be used to prepare cementing materials [4-5].

This article takes the coal gangue from a coal preparation plant in Liupanshui City, Guizhou Province as the research object. After crushing and screening, the coal gangue is placed in a muffle furnace for calcination and activation. The activated coal gangue is finely ground and then mixed with ordinary Portland cement, polycarboxylate-based water reducer, and water to prepare cementitious materials.

2. Experiment

2.1. Sample preparation
After sampling coal gangue, it is first dried in the air and then crushed using a jaw crusher. Part of it is made into analytical samples for testing and analysis of coal gangue, while the majority is crushed to below 1 mm for activation experiments.

2.2. Sample analysis
The coal gangue was subjected to full industrial analysis, sulfur content testing, calorific value testing, and XRF testing analysis. The test results are shown in Tables 1 and 2, and the XRD testing analysis of coal gangue is shown in Figure 1.

Table 1. Analysis and Testing of Coal Gangue.

Sample	M_{ad}/%	A_d/%	V_{daf}/%	FC_{daf}/%	$S_{t,d}$/%	$Q_{gr,d}$/(MJ/kg)
Coal gangue	2.96	83.94	88.72	11.28	0.63	0.03

Table 2. XRF test results of coal gangue/ %.

Sample	SiO_2	Al_2O_3	Fe_2O_3	CaO	TiO_2	K_2O	SO_3	MgO	P_2O_5	other
Coal gangue	44.40	19.70	18.90	4.53	7.24	2.10	1.01	0.30	0.30	1.52

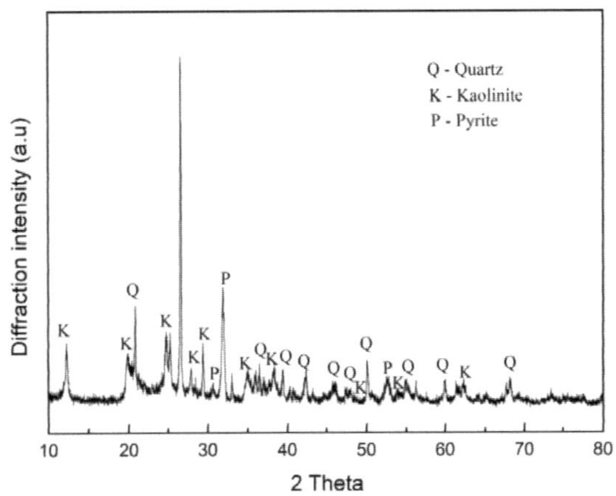

Figure 1. XRD spectrum of coal gangue.

According to the XRD spectrum in Figure 1, the minerals contained in coal gangue mainly include quartz, kaolinite, and pyrite.

3. Results and discussion
The screened material after crushing and screening is activated using the method of thermal activation. The calcination of coal gangue can cause lattice distortion of coal gangue minerals, as well as the destruction of saturated Si-O and Al-O bonds, resulting in more glassy and amorphous substances in coal gangue and higher activity.

3.1. Effect of activation temperature on the activity of coal gangue
The XRF test results of coal gangue at different activation temperatures are presented in Table 3, while the XRD test results are shown in Figure 2. According to Table 5, when the calcination temperature is 650°C, the Al2O3 content decreases from 18.8% before calcination to 18.50%, and the chemical component content of SiO2 decreases from 42.8% to 41.90%. From the data, it can be concluded that the activation effect of coal gangue is significant when the calcination temperature is 650°C.

Table 3. XRF test results of coal gangue at different calcination temperatures/%.

Temperature/°C	MgO	Al_2O_3	SiO_2	Fe_2O_3	SO_3	K_2O	CaO	TiO_2	P_2O_5	Other
850	0.53	21.40	46.50	15.20	2.49	2.11	4.354	6.43	0.30	0.78
750	0.43	21.00	46.30	15.60	2.71	2.14	.08	6.22	0.29	1.23
650	0.82	18.502	41.90	21.00	2.34	2.10	5.66	6.03	0.30	1.39
550	0.52	0.70	44.60	17.20	2.98	2.03	4.23	6.47	0.36	0.86
450	0.48	20.90	45.80	15.70	2.56	2.13	4.35	6.78	0.34	0.80

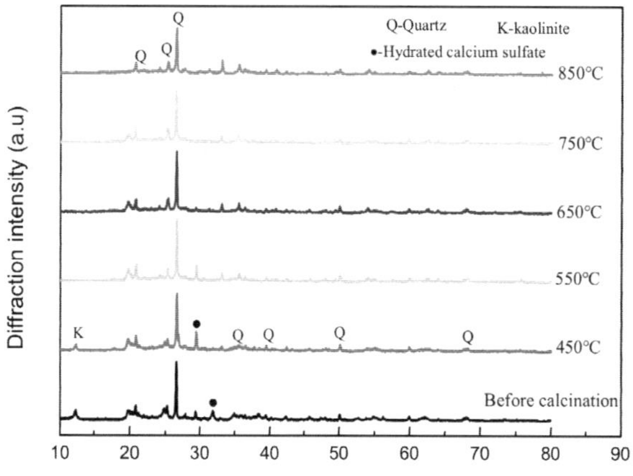

Figure 2. The XRD patterns of coal gangue calcined at different temperatures.

As can be seen from Figure 2, the intensity of the quartz peak in coal gangue decreases with increasing activation temperature. After the temperature exceeds 450°C, the diffraction peak of kaolinite disappears, and the diffraction peak of hydrated calcium sulfate disappears when the temperature rises above 550°C. When the temperature reaches 650°C, the quartz diffraction peak tends to stabilize. Therefore, the optimal calcination temperature is 650°C.

3.2. Analysis of the effect of time on the activity of coal gangue

With a fixed calcination temperature of 650°C, the chemical composition and XRD patterns of coal gangue under different activation times are presented in Table 4 and Figure 3, respectively.

Table 4. The chemical composition of coal gangue/%.

Time/min	MgO	Al_2O_3	SiO_2	Fe_2O_3	SO_3	K_2O	CaO	TiO_2	P_2O_5	Other
240	0.66	19.30	44.80	18.00	2.64	1.99	5.776.	5.69	0.28	0.90
180	0.59	18.90	44.30	18.20	2.49	1.98	27	6.11	0.37	0.85
120	0.63	18.50	43.60	19.10	2.70	1.85	6.46	6.04	0.33	0.79
90	0.55	19.20	45.50	17.40	2.33	2.05	5.14	6.66	0.29	0.85
60	0.56	19.40	46.00	17.10	2.56	2.00	5.22	5.87	0.36	0.86
30	0.86	19.00	44.40	18.20	2.65	2.02	5.28	6.44	0.27	0.82

The diffraction peak of kaolinite disappears when the activation time exceeds 30 minutes (Figure 3). At a scanning angle of 27°, the intensity of the quartz diffraction peak decreases as the time increases and stabilizes after 120 minutes. Compared to the XRD pattern of coal gangue before activation, the intensity of the diffraction peaks of quartz and kaolinite at a scanning angle of 27° decreases sharply under different activation times and temperatures. Therefore, the optimal activation time is 120 minutes.

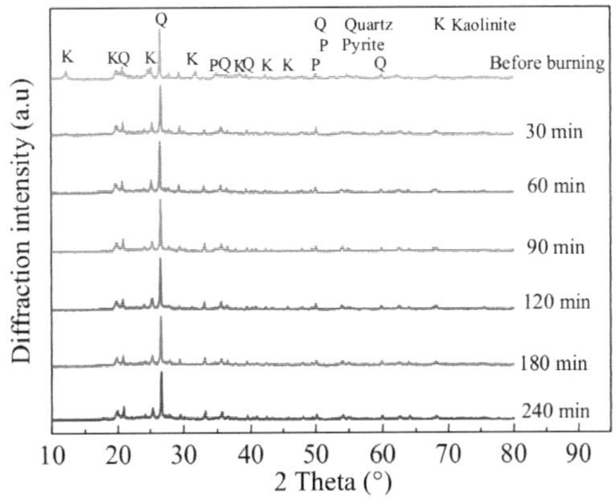

Figure 3. The XRD patterns of coal gangue were calcined for different durations.

3.3. Research on mechanical properties of coal gangue cementitious materials

Using calcined and activated coal gangue as an admixture, a cementitious material is prepared by adding a 1% water-reducing agent, a water-binder ratio of 0.44, and a sand ratio of 0.5. First, coal gangue is calcined at 650°C to prepare the first group of cementitious materials with admixture ratios of 15%, 25%, 35%, and 45%. Then, coal gangue calcined at different temperatures is used to prepare the second group of cementitious materials with a coal gangue admixture ratio of 25% and a cement content of 75%. Samples with different curing times of 3d, 7d, and 28d are maintained and cared for according to the national standard GB/T17671-1999, and the flexural and compressive strengths of the cementitious materials are tested using a WAW-600C universal testing machine.

3.3.1. Analysis of the impact of coal gangue admixture ratio on the mechanical properties of cementitious materials. The flexural strength and compressive strength of cementitious materials prepared with different admixture ratios of coal gangue were measured (Table 5, Figure 4, and Figure 5).

Table 5. Strength of Cementitious Materials under Different Coal Gangue Content.

Addition amount/%	Flexural strength/MPa			Compressive strength/MPa		
	3d	7d	28d	3d	7d	28d
0	10.0	13.0	33.0	20.0	31.0	39.0
15	9.0	11.0	16.0	25.5	33.0	67.0
25	11.0	13.0	26.0	28.5	36.0	80.0
35	8.0	10.0	16.0	21.0	24.0	47.0
45	6.0	7.0	12.0	14.5	21.0	22.0

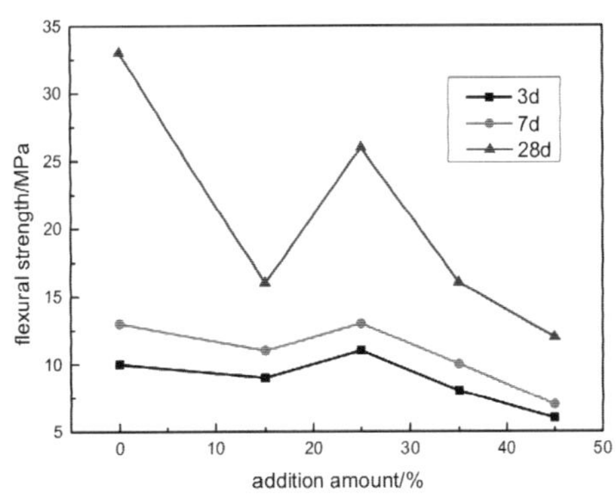

Figure 4. Bending strength of cementitious materials with different dosages.

According to Table 5 and Figure 4, the flexural strength of cementitious materials with different dosages first decreases and then increases, reaches its maximum value, and then decreases again, reaching its maximum value at a dosage of 25%. At this time, the flexural strength of the cementitious material at 3d, 7d, and 28d is 11 MPa, 13 MPa, and 26 MPa, respectively.

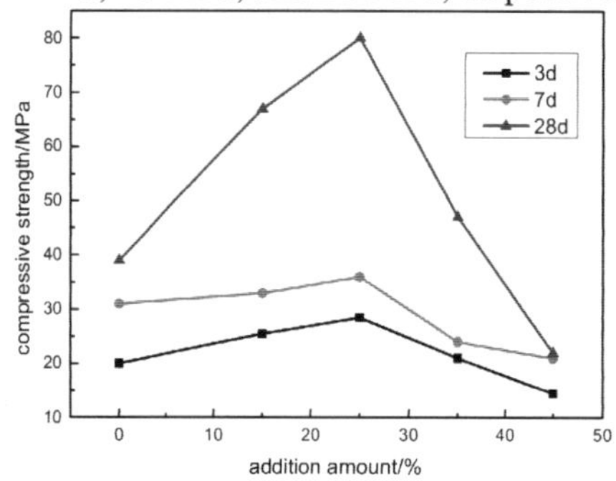

Figure 5. Effect of admixture ratio on the compressive strength of cementitious materials.

When the admixture ratio of coal gangue is below 25%, the compressive strength of the cementitious material increases with the increase in admixture ratio, reaching a maximum value at an admixture ratio of 25%. At this point, the compressive strengths at 3d, 7d, and 28d are 28.5 MPa, 36.0 MPa, and 80.0 MPa, respectively. When the admixture ratio exceeds 25%, the compressive strength begins to decrease (Table 5 and Figure 5). Therefore, the optimal admixture ratio of coal gangue is 25%.

3.3.2. Analysis of the impact of activation temperature of coal gangue on the mechanical properties of cementitious materials. Mechanical properties of cementitious materials prepared using coal gangue with different activation temperatures were analyzed (Table 6, Figure 6, and Figure 7).

Table 6. Strength of coal gangue cementitious materials at different activation temperatures.

Temperature/°C	Flexural strength/MPa			Compressive strength/MPa		
	3d	7d	28d	3d	7d	28d
450	6.0	9.0	20.0	14.6	19.3	35.0
550	7.0	11.0	22.0	15.0	19.5	39.0
650	10.0	12.0	26.0	28.5	36.0	80.0
750	6.5	8.0	18.0	12.0	21.0	38.0
850	5.0	7.0	9.0	11.0	15.0	28.0

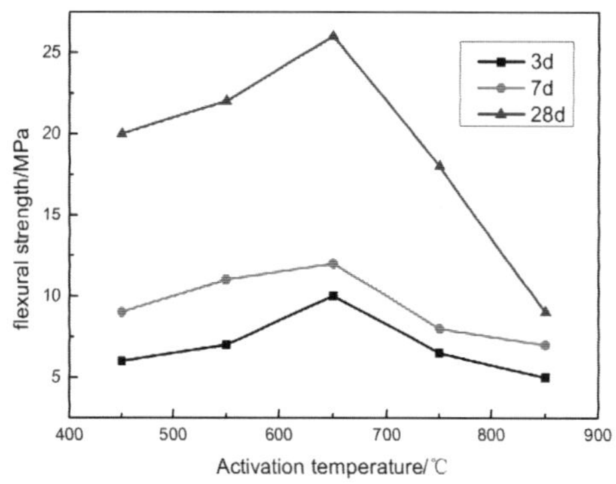

Figure 6. Bending strength of cementitious materials with different activation temperatures.

As can be seen from Table 6 and Figure 6, when the temperature is below 650°C, the flexural strength continues to increase, reaching a maximum value at an activation temperature of 650°C. At this point, the flexural strengths at 3d, 7d, and 28d are 10 MPa, 12 MPa, and 26 MPa, respectively. When the temperature exceeds 650°C, the flexural strength begins to decrease.

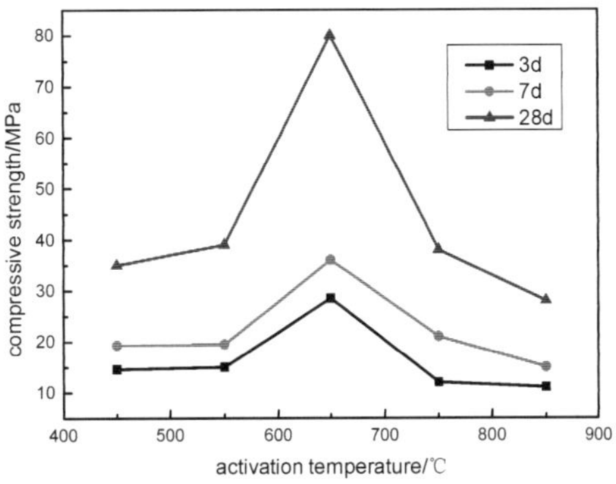

Figure 7. Relationship curve between compressive strength and activation temperature.

As can be seen from Table 6 and Figure 7, when the temperature is below 650°C, the compressive strength continues to increase, reaching a maximum value at an activation temperature of 650°C. At this point, the compressive strengths at 3d, 7d, and 28d are 28.5 MPa, 36 MPa, and 80 MPa, respectively. When the temperature exceeds 650°C, the compressive strength begins to decrease.

4. Conclusion

After careful analysis, the following conclusions can be drawn from this article:

(1) Analysis of the XRD spectra of coal gangue under different activation temperatures and durations reveals that when the activation temperature is 650°C (with an activation duration of 120 minutes) or when the activation duration is 120 minutes (at an activation temperature of 650°C), the quartz diffraction peak at a scanning angle of 27° decreases significantly, and the kaolinite diffraction peak disappears. This indicates that the coal gangue exhibits good pozzolanic activity under these conditions.

(2) When the mixing amount of coal gangue is 25% and the activation time is 120 minutes, the flexural and compressive strengths of the cementing material first increase and then decrease under different activation temperature conditions. When the activation temperature is 650°C, the flexural and compressive strengths of the cementing material reach their maximum values at 3d, 7d, and 28d, with flexural strengths of 10 MPa, 12 MPa, and 26 MPa, and compressive strengths of 28.5 MPa, 36 MPa, and 80 MPa, respectively.

(3) The activation of coal gangue can be used for the preparation of cementing materials, providing a utilization pathway for the large-scale utilization of coal gangue.

Acknowledgments

This paper was supported by the following project: the Innovation Training Program for University Students(202210977020) and the Carbon Neutral Engineering Research Center of Guizhou Colleges and Universities in the Coal Industry (Qian Jiao Ji [2023] No. 044).

References

[1] Zesheng Xu, Jiarui Chen, Senbiao Wang, et al. (2021) Research on graded and quality-based processing and utilization of coal gangue. Chain Coal., 47 (11): 61-68.

[2] Tao Zhu, Xinjuan Wu, Cheng Xing, et al. (2024) Current Status and Progress of Resource Utilization of Coal Gangue. Coal Science and Technology., 52 (01): 380-390.

[3] Laihong Feng, Liangfa Chen, Yichao Li, et al. (2023) Status and progress of coal gangue resource utilization under dual carbon background in China. Multipurpose Utilization of Mineral Resources., pp. 1-14.

[4] Kezhou Yan, Li Cui, Fangqin Cheng, Helen H. Lou, Yanxia Guo (2014) Effect of Na2CO3 additive on the activation of coal gangue for alumina extraction.

[5] Wang Fuke, Gong Jinwei, Hu Yanshao, Sun Zuozheng, Jin Peng, Han Xiaoyan, Li Keliang (2023) Study on the Reactivity Activation of Coal Gangue for Efficient Utilization.

Synthesis of amino silicone oil for use in oil agent in carbon fiber

Junsha Wang[1], Xianwen Huang[2*], and Cheng Zhang[1]

[1]School of Materials Science and Engineering, Shanghai Institute of Technology, Shanghai, 201418, China
[2]Department of high polymer material, Ningbo Institute of Materials Technology & Engineering, Ningbo 315201, China

[*]Corresponding author's email: huangxw@nimte.ac.cn

Abstract: An oil with uniform particle size distribution, good heat storage, and good wettability for carbon fiber was prepared. In other words, by mixing with non-ionic emulsifiers E1303 and E1309, the most suitable synthesis temperature, solid content, and the proportion of silicone oil emulsifiers were explored by using the control variable method. The results show that carbon fiber oil with uniform particle size distribution and good heat-resistant storage can be obtained when the synthesis temperature is 50°C, the solid content is 20% and the emulsifier ratio is 1:3. The particle size of the carbon fiber oil is 200 nm, the particle size distribution index is less than 0.3, and the particle size distribution is uniform.

1. Introduction

Carbon fiber is the main reinforcement for the preparation of high-performance, lightweight, and functional composite materials, and is the strategic basic material that is highly valued by various countries [1]. In the actual production process, raw silk oil is one of the core chemical additives for the preparation of high-performance raw silk and carbon fiber [2] and plays an extremely important role in the oil, pre-oxidation, and carbonization of raw silk. A suitable oil agent can avoid the occurrence of fiber surface defects and thus determine the yield of carbon fiber [3]. Polyacrylonitrile (PAN) based preoxidation and low-temperature carbonization require no melting bonding, fuelling, and breaking, and high-quality carbon fiber oil can effectively solve the above process problems [4].

The composition of traditional PAN raw silk oil is complex, and most of them are a combination of modified organic silicone oil, such as epoxy-modified silicone oil, amino-modified silicone oil, polyether-modified silicone oil, surfactants, antistatic agents, antioxidants, and stabilizers [5]. At present, the research work of carbon fiber filament oil at home and abroad mainly focuses on the preparation of modified silicone oil and the control of technological parameters of silicone oil microemulsion [6]. Most enterprises focus on the influence of the oiling process on the quality of carbon fiber, while the research direction of universities mostly focuses on the analysis of the composition of imported oil, the formulation of functional silicone oil microemulsion, and the structural design and modification of polysiloxanes are rarely studied.

In this study, a modified polysiloxane structure containing an epoxy group chain end was designed and prepared from the basic raw material of organosilicon through a one-step organic synthesis reaction. A PAN profilamentous oil with a simple process flow could be obtained by compounding it with a certain proportion of commercial surfactants. The O-PDMS was characterized by Fourier transform infrared spectroscopy (FT-IR) and hydrogen nuclear magnetic resonance spectroscopy (1H-NMR). The oil agent (P-OA) was prepared by combining epoxy-modified silicone oil with a commercial non-ionic surfactant and was compared with Japanese Takemoto oil (J-OA). P-OA was characterized by a laser particle size analyzer, thermogravimetric analyzer, and surface tensiometer. The results show that the emulsion properties of P-OA oil can meet the needs of PAN filament production. Water was taken from the PAN filament production line for washing, and P-OA and J-OA were used to produce the oiled polyacrylonitrile-based carbon filament P-PAN and J-PAN, respectively. The corresponding carbon fibers P-CF and J-CF were prepared by carbonization of P-PAN and J-PAN. The PAN filament and carbon fiber were tested and characterized by a monofilament tensile instrument and contact Angle measuring instrument. After oiling, the tensile strength of P-PAN monofilament is not significantly higher than that of J-PAN. Other performance indexes are at the same level as J-OA, which provides a new idea for the research and development of carbon fiber filament oil.

2. Experiment

2.1 materials

Octamethy lcyclotetrasiloxane (D4): Analytically pure, Sinopharm Group Co., LTD.; 1, 1, 3, 3-tetramethyldisiloxane (HMM), allyl glycidyl ether (AGE): Analytically pure, Sinopharm Group; Isopropyl alcohol: Analytically pure, Aladdin Reagent Co., LTD. Chloroplatinic acid: reagent grade, Sinopharm Group; Concentrated sulfuric acid: Analytically pure, Sinopharm Group Co., LTD.; Sodium bicarbonate, potassium bromide: Analytically pure, Sinopath Group Co., LTD.; Glacial acetic acid: Analytically pure, Sinopharm Group Co., LTD.; Deionized water, laboratory made; Imported PAN carbon fiber filament oil is produced in Japan.

2.2 Synthesis of epoxy-modified cationic silicone oil

A one-step method was used to design and synthesize hydroxy-amino-modified silicone oil (O-PDMS). Fig. 1 is the schematic diagram of the process.

0.3 g chloroplatinic acid was dissolved in 29.7 g isopropyl alcohol dehydrated by anhydrous sodium sulfate and reacted under 45°C N_2 atmosphere for 3 ~ 4 hours to prepare chloroplatinate-isopropyl alcohol catalyst, which was stored at 4°C for use.

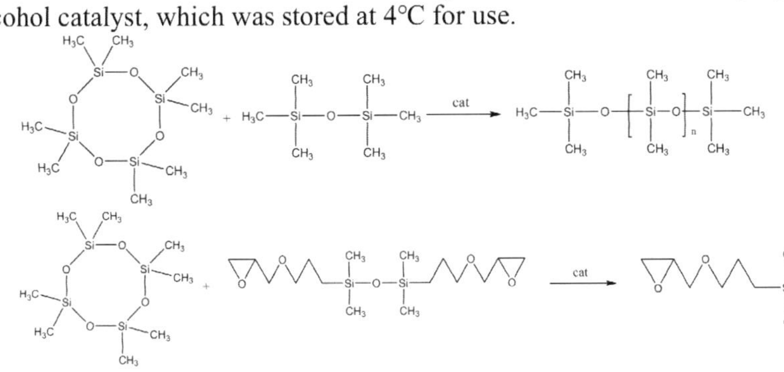

Figure 1. Synthesis route of O-PDMS.

Chloroplatinic acid was dissolved in isopropyl alcohol dehydrated with anhydrous sodium sulfate, the chloroplatinic acid content was 1%, and the catalyst was prepared by stirring continuously at an N_2 atmosphere of 45°C for 3~4 h. The catalyst was stored in a refrigerator at 4°C for use. Excess AGE was placed in a four-neck reactor equipped with a thermometer, agitator, and constant pressure drip funnel, dehumidified for 2 h (N_2 atmosphere, 110°C), activated with chloroplatinate-isopropanol catalyst at 60°C for 1~2 h, followed by PDMS at 70~110°C for 3~7 h. Until the product was tested by infrared spectroscopy to be free of silicon hydrogen bonding (Si-H). At the end of the reaction, the product was decolorized by activated carbon powder, and then the decolorized activated carbon and catalyst were separated by a diatomite column. Finally, the residual AGE was removed by vacuum distillation at 120°C, and the clear transparent liquid was O-PDMS.

2.3 Compounding and emulsification

After mixing O-PDMS with polyoxyethylene ether emulsifier (provided by Guangzhou Xumei Chemical Co., LTD.), according to the mass ratio of 1:4, adding a certain amount of glacial acetic acid, we stir the mixture evenly, slowly add deionized water, and stir the solution at a fixed stirring speed for 0.15 h to obtain a milky emulsion, that is, P-OA oil. The properties of P-OA emulsion were measured and compared with J-OA. The results are shown in Table 1.

Table 1. Emulsion properties of P-OA and J-OA.

Sample	Particle diameter/nm	Surface Tension mN m^{-1}	pH
P-OA	178	22.60	5.5
J-OA	257	24.31	4.5

As can be seen from Table 1, the surface tension of the P-OA emulsion is 22.60 mN/m, and that of the J-OA emulsion is 24.31 mN /m.

It can be seen that the addition of O-PDMS can reduce the surface tension of P-OA, which is lower than that of J-OA. The lower the surface tension of the oil agent, the better its spreading property and the easier it is to form a film on the surface of the polyacrylonitrile-based carbon fiber filament

2.4 Production and oiling process of polyacrylonitrile-based carbon fiber filament

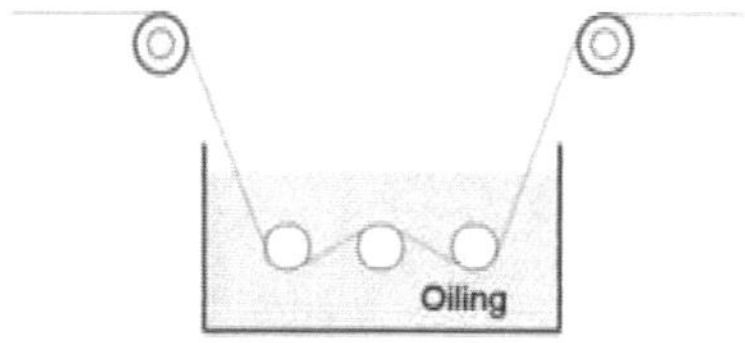

Figure 2. Oiling process of PANs.

As shown in Figure 2 that the sample is a PAN precursor with a deoiled surface, which is impregnated with oil. Firstly, an oil with a mass concentration of 1% was prepared, and then the weighed sample was placed in 1% oil solution and impregnated for 5min. Then the impregnated PAN filament was placed on the weighing paper and excess water was extruded to obtain the sample.

2.5 Preoxidized carbonization of PAN filament and its process

As shown in Figure 3 that Carbonization and sizing of PAN precursor P-PAN and J-PAN are completed. The carbon fibers P-CF and J-CF were prepared by pre-oxidation for 60 min at 180~260°C air atmosphere, low-temperature carbonization for 3 min at 400~ 700°C nitrogen atmosphere.

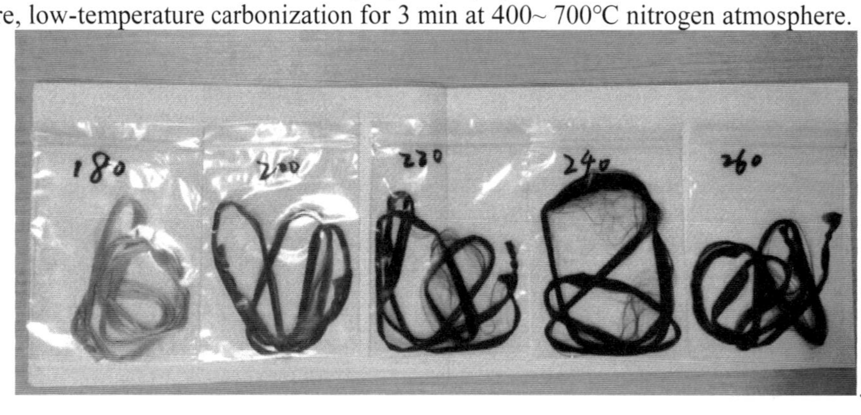

Figure 3. Pre-oxidized fiber.

2.6 Test and characterization

Japan Shimadzu IR-100 Fourier Transform infrared spectroscopy (FI-TR) instrument, resolution 2 cm[1], scan times 20; AdvanceII500 MHz nuclear magnetic resonance spectrometer (1H-NMR) of Bruker Company, Switzerland, using deuterium chloroform as a solvent, TMS as internal standard, 600 MHz nuclear magnetic resonance spectrometer determination; German SNano laser particle size meter, we take 2 drops of sample into the reagent bottle, dilute it with distilled water until the infrared light path is clear and transparent, and test repeatedly for 3 times to avoid sampling error; TG209F1 thermogravimetric analyzer, test temperature range 0~500°C, using N_2 as purge gas and protection gas, a flow rate of 60 mL/min, a heating rate of 10 °C/min; German company K100SF surface tensiometer, emulsion test solid content 3%, each sample tested 3 times to take the average. Germany automatic single fiber physical properties analyzer, according to GB/T14337-2008, tests the tensile strength of the filament.

3. Results and discussion

As shown in Figure 4 that the infrared spectrum of the product. As can be seen from Fig. in 962 cm^{-1} is the stretching vibration peak with an absorption peak of $-CH_3$, 1141 cm^{-1} is the deformation vibration peak of $-CH_2-$ in the epoxy group 9, and the stretching vibration peak of -C-O-C- appears at 1107 cm. It can be seen that this structure is consistent with the designed O-PDMS structure. The hydrosilylation reaction has been completed and O-PDMS has been synthesized.

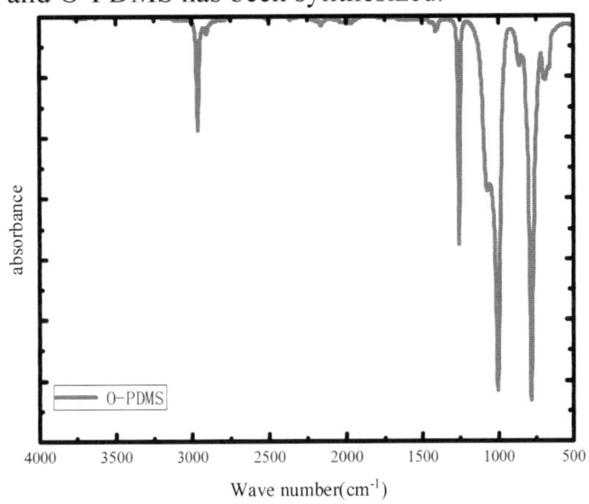

Figure 4. FT-IR spectra of O-PDMS.

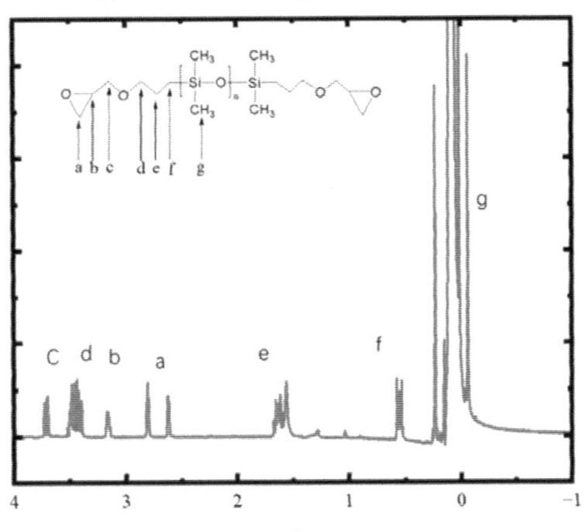

Figure 5. NMR spectra of O-PDM.

As shown in Figure 5 that The formation of O-PDMS was characterized by 'H-NMR.The displacement at a (4.57) was attributed to hydrogen in Si-H, and the displacement at b (0.08) was attributed to hydrogen in Si-CH_3, indicating that H-PDMS had been synthesized. As shown in the figure, the displacements at c and d (2.61 ~ 2.80, 3.15) belong to the hydrogen on the epoxy group, the displacements at e and f (3.70, 3.36) belong to the hydrogen on both ends of the ether bond, and the displacements at g and h (1.62, 0.57) belong to the hydrogen on -CH_2-. The displacement (0.08) at i is attributed to the hydrogen in Si-CH_3, which proves that O-PDMS has been synthesized.

In the wet spinning line of polyacrylonitrile, wash silk was taken, and P-OA and J-OA were used to oil PAN wash silk, and the running stability of P-PAN raw silk was observed. During the winding process, P-PAN fiber runs smoothly, there is no hair, broken yarn, or sticky roller.

4. Conclusion

Through molecular structure design, O-PDMS was synthesized by a one-step method, and polyacrylonitrile-based carbon fiber precursor oil (O-OA) was prepared by combining it with commercial polyoxyethylene ether surfactants. The results show that the particle size of the O-OA emulsion is 178 nm, the surface tension is lower than that of J-OA, and the centrifugal stability is better, which makes it easy to resist shaking during transportation. P-PAN and J-PAN were prepared by oiling the PAN filament with P-OA and J-OA respectively. Two kinds of carbon fibers, P-CF and J-CF, were prepared by carbonization. FT-IR, elemental analyzer, XRD, XPS, EDS, surface wetting Angle, etc., PAN filament, and carbon fiber were tested and characterized by a fiber strength meter. Compared with the Japanese oil, the wetting Angle is reduced from 33° to 31°, and the wettability is improved, which is conducive to the oil penetrating the fiber, and the tensile strength of P-PAN monofilament is increased by 4% compared with that of J-PAN. Compared with J-CF after carbonization, the tensile strength and tensile modulus of P-CF increased by 8% and 6%, which provided a new idea for the research and development of domestic carbon fiber precursor oil.

References

[1] Frank E, Hermanutz F, Buchmeiser M R. (2022) Carbon Fibers: Pre- cursors, manufacturing and properties . Macromolecular Materials and Engineering, 297(6): 493-501.
[2] Q.Gao, and C.G. Wang, New insight into structure-property correlation of polyacrylonitrile precursor fibers and resultant carbon fibers. Journal of Polymer Research, 2022. 29(6)254-261.
[3] Aldosari M, Khan M, and Rahatekar S, Manufacturing carbon fibres from pitch and polyethylene blend precursors: a review. Journal of Materials Research and Technology-Jmr&T, 2020. 9(4): p. 7786-7806.
[4] M.Q.Jian, Y.Y. Zhang, and Z.F. Liu, Graphene Fibers: Preparation, Properties, and Applications. Acta Physico-Chimica Sinica, 2022. 38(2) 214-233.
[5] Y.D.Liu, and Kumar S, Recent Progress in Fabrication, Structure, and Properties of Carbon Fibers. Polymer Reviews, 2012. 52(3-4): p. 234-258.
[6] Poursorkhabi V, et al., Processing, Carbonization, and Characterization of Lignin Based Electrospun Carbon Fibers: A Review. Frontiers in Energy Research,2020.8(2): p. 413-429.

Research on moisture absorption characterization of wound composites based on surface concentration modified model

Zilong Yang[1, 2, *], Yuxia Hu[1, 2], Wusheng Li[1, 2] and Fujiang Yang[1, 2]

[1] National Key Laboratory of Particle Transport and Separation Technology, Tianjin, 300180, China

[2] Research Institute of Physical and Chemical Engineering of Nuclear Industry, Tianjin, 300180, China

[*] Corresponding author's e-mail: yangzilong1214@163.com

Abstract. Four groups of hygroscopic experiments of T700/epoxy composite wound rings at temperatures and humidities of 45°C/60%, 45°C/90%, 70°C/60%, and 70°C/90% were completed in order to study the moisture absorption characterization. Surface concentration correction models characterized by different types of functions such as linear, quadratic, sinusoidal, exponential, and others were proposed to establish a "three-stage" moisture absorption model, in which the initial stage was smooth, the middle stage grew rapidly, and the later stage tended to equilibrium. The theoretically predicted hygroscopic curves derived from quadratic and exponential functions were in good agreement with the experimental data, which revealed the difference in the moisture absorption characterization between the wound rings with high fiber volume and that of the laminates and fabrics and provided a basis for the prediction of the mechanical degradation and life of the wound composites in hygrothermal environments.

1. Introduction

Hygrothermal environments have a significant effect on the mechanical properties of resin matrix composites [1~2]. Predicting the hygroscopic dynamical curves of composites during transportation, storage, and service is the basis for material aging analysis and life assessment [3]. It is generally accepted that the moisture absorption characterization of carbon fiber-reinforced resin matrix composites (CFRP) conforms to Fick's second law [4]. A "two-stage" model is generally used to describe the hygroscopic dynamical curve of composites [5]. In the first stage, due to the rapid water absorption by defects and cracks in the composite material itself, the moisture absorption rate is linearly related to the 1/2 power of time. In the second stage, due to the slow absorption of water by the polar hydrophilic groups in the resin polymer, the rate of moisture absorption slows down until it reaches equilibrium [6]. The ASTM standard provides an equation for estimating the rate of moisture absorption [7], but the accuracy of this Equation is insufficient, and it is generally only used for predicting the initial moisture absorption process.

In order to improve the accuracy of the theoretical prediction of the late process of the hygroscopic dynamical curves, many scholars have carried out related studies. Cao [8] studied the moisture absorption characterization of T700/9916 composite laminates in a constant temperature water bath and high humidity environment and obtained a highly accurate curve of hygroscopic progress by finite element simulation with the mass diffusion module of ABAQUS. Yu [9] researched moisture absorption characterization of carbon fiber-reinforced polymer using Fickian and non-Fickian models. Souza, Wang, and Qi [10~13] calculated the hygroscopic dynamical curves of the composite single-cell model with transient thermal analysis according to the principle of formal equivalence between the mass diffusion equation and the heat conduction equation. Based on the single-cell model, the moisture absorption behavior of the laminate as a whole was predicted, and the results were more consistent with the experimental data. Hui [14] tested the hygroscopic dynamical curve of composites with angled ply, which conformed to Fick's second law. Feng [15] observed the moisture absorption behavior at the interface of composites by SEM, analyzed the process of water molecule damage at the interface from the fine view, and gave the reasons for the gradual slowing down of the moisture absorption rate. In summary, the "two-stage" moisture absorption model of composite laminates or fabrics has been well studied, but the moisture absorption model of wound composites has been less studied. At present, some experiments show that the initial and later moisture absorption rates of wound composites are small and do not conform to the "two-stage" curve.

In order to study the moisture absorption law of wound composites more clearly and to establish a hygroscopic dynamical model more accurately to provide design inputs for the moisture absorption aging study and life prediction of wound composites during transportation, storage, and use, four groups of hygroscopic experiments of T700/epoxy composite wound rings at different temperatures and humidities were completed in the paper. By fitting the experimental curves, the accuracy and validity of the established multiple surface concentration modified models were compared and verified, and the dynamic relationship of moisture absorption of wound composites with time was grasped to realize a more precise prediction of the hygroscopic curves of wound composites.

2. Hygroscopic dynamical model

2.1. General hygroscopic dynamical model

The general hygroscopic dynamical model of composites follows Fick's second law.

$$\frac{\partial C(z,t)}{\partial t} = D\frac{\partial^2 C(z,t)}{\partial^2 Z} \tag{1}$$

In the above Equation, C is the concentration of the diffusing substance, D is the diffusion coefficient, and Z is the direction of diffusion of the substance.

For the case of water diffusion into the plane of the plate for a thickness of h, the surface concentration is assumed to reach the equilibrium concentration instantaneously, which is modeled as a transient equilibrium model (TEM). The boundary condition is,

$$\begin{cases} C(z,0) = C_0 \\ C(-h/2,t) = C(h/2,t) = C_\infty \end{cases} \tag{2}$$

where C_0 is the initial concentration, and C_∞ is the equilibrium concentration.

Combining Equations (1) and (2) and integrating over Z from $-h/2$ to $h/2$ yields the moisture absorption rate $M(t)$ at any instant as,

$$M(t) = M_0 + (M_\infty - M_0)\{1 - \frac{8}{\pi^2}\sum_{n=0}^{\infty}\frac{1}{(2n+1)^2}\exp[-D(2n+1)^2\pi^2 t/h^2]\} \tag{3}$$

where M_0 is the initial moisture absorption rate, and M_∞ is the equilibrium moisture absorption rate.

The hygroscopic dynamical curves obtained by the surface concentration boundary condition in Equation (2) show an obvious "two-stage" characteristic. In the early stage of hygroscopicity, there is linear and rapid growth, and in the late stage of hygroscopicity, it gradually flattens out.

In particular, for general composite laminates or woven fabrics, the moisture absorption rate increases linearly by approximately 1/2 power of time during the initial period of moisture absorption. The diffusion coefficient can be estimated using the following Equation.

$$D = \pi\lambda^2 (\frac{h}{4M_\infty})^2 \tag{4}$$

In the above Equation, λ is the slope of the linear segment of the $M - t^{1/2}$ plot, reflecting the rate of moisture absorption.

$$\lambda = \frac{M(t_2) - M(t_1)}{\sqrt{t_2} - \sqrt{t_1}} \tag{5}$$

2.2. Surface concentration modified model

The fiber volume of wound composites is usually high. The water molecules diffuse slowly in the early stage of moisture absorption, and the surface concentration cannot reach the equilibrium instantaneously. Therefore, the surface concentration condition in Equation (2) needs to be corrected, where C(±h/2) is not a constant value but a variable parameter related to moisture absorption and time. The corrected boundary condition is called the surface concentration modified model.

In this paper, the following hypothetical model with a simpler form of proportional coefficients was established. The equation form and solution process of the model are easier, and its accuracy and validity will be verified by the experiments in Chapter 3.

Assuming that the moisture diffusion rate on the surface of composites is related to the difference between the actual concentration on the surface and the equilibrium concentration, which differs by a scaling factor α, and the boundary condition is,

$$-D\frac{\partial C(\pm h/2,t)}{\partial z} = \alpha[C_\infty - C(\pm h/2,t)] \tag{6}$$

It can be derived that,

$$M(t) = M_0 + (M_\infty - M_0)\sum_{n=0}^{\infty}\{\frac{2\sin^2\beta_n}{\beta_n(\sin\beta_n\cos\beta_n + \beta_n)^2}\exp(-4\beta_n^2 Dt/h^2)\} \tag{7}$$

In the above Equation, $\beta_n \in [n\pi, \frac{\pi}{2} + n\pi], (n = 0,1,2...)$ is the solution to the following Equation.

$$\beta\tan\beta = \frac{h}{2}k \tag{8}$$

In the above Equation, $k = \alpha/D$ is the relative scale factor. When the value of k is large, it converges to the boundary condition of instantaneous equilibrium (TEM).

In order to better fit the high fiber volume of the wound composites and the slow growth of the moisture absorption rate in the early and last stages of moisture absorption, a non-constant scaling factor $\alpha = \alpha(t)$ can also be used. When α has different functional forms, the representation means how fast or slow the moisture absorption rate of the wound composites grows at different stages.

Considering several common functional forms, including linear approximation model(LAM), quadratic approximation model(QAM), sinusoidal approximation model(SAM), and exponential approximation model(EAM), α can be written as,

$$\begin{aligned} \alpha_1(t) &= a_1 + b_1 t \\ \alpha_2(t) &= a_2 + b_2 t^2 \\ \alpha_3(t) &= \alpha_\infty \sin(a_3 t + b_3) \\ \alpha_4(t) &= a_4 e^{t/b_4} \end{aligned} \tag{9}$$

Using Equations (6) and (9) as the boundary conditions, the hygroscopic dynamical curve with a "three-stage" shape can be acquired, which can better fit the characteristics of wound composites with lower moisture absorption rates in the early and late stages of moisture absorption than the "two-stage" curve obtained according to Equation (2). In the following, it will be further observed through experiments which functional form of Equation (9) matches the experimental results better.

3. Examples and discussion

3.1. Experimental conditions

(1) Sample preparation

The experimental samples were T700/epoxy composite rings wound with an angle-90° purely. The thickness of the rings was 3 mm, and they were divided into four groups to carry out the experiments with different temperatures and humidity.

Table 1. Experimental groups.

Group	#1	#2	#3	#4
Temperature (°C)	45	45	70	70
Relative humidity (%)	60	90	60	90

(2) Experimental setup

The experiments were carried out in a temperature-humidity climatic chamber (Figure 1).

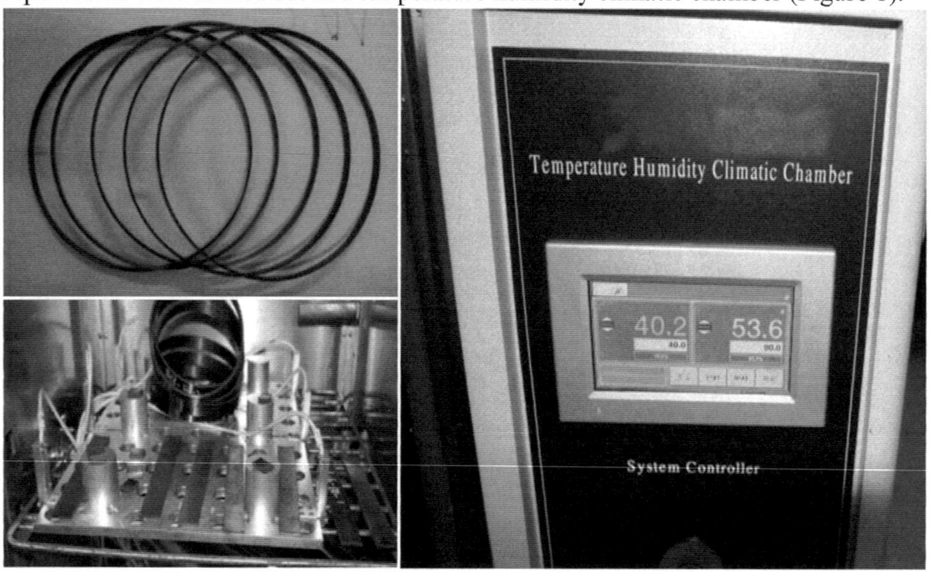

Figure 1. Experimental pieces and setup.

The following formula is used to calculate the moisture absorption rate.

$$M(t) = \frac{W(t) - W_0}{W_0} \times 100\% \tag{10}$$

where $M(t)$ is the moisture absorption rate at moment t, $W(t)$ is the weight of the experimental piece at moment t, and W_0 is the initial experimental piece weight.

(3) Experimental steps

a. All the experimental pieces were dried in a vacuum environment at 80°C for 15 days so that the initial moisture absorption of the composite winding circle was approximately equal to 0% and weighed to obtain W_0. Then, the experimental pieces were put into the temperature-humidity climatic chamber, and the temperature and humidity were set according to Table 1.

b. Remove the experimental pieces every three days and weigh and record $W(t)$, after which the experimental pieces are put back into the temperature humidity climatic chamber.

c. Step *b* continued for 105 days.

d. Four groups of experiments were carried out separately, according to Table 1.

3.2. Results and discussion

The results of the four groups of experiments and the theoretical prediction curves with different surface concentration-modified models are shown in Figure 2.

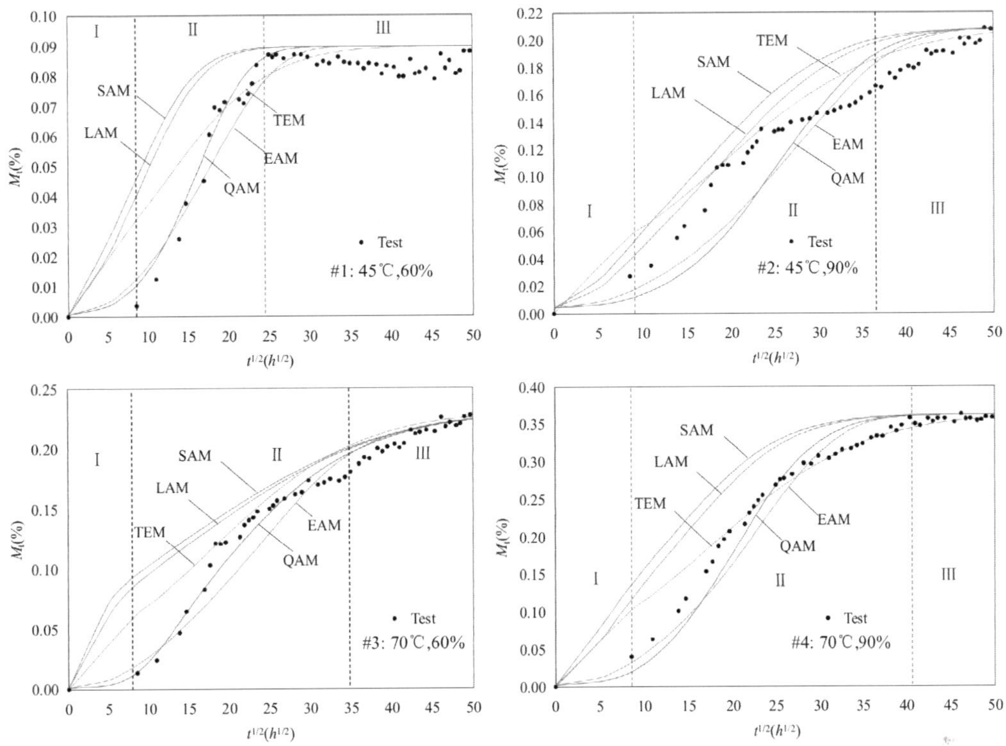

Figure 2. Experimental results and theoretical prediction curves.

As can be seen in Figure 2:

(1) The hygroscopic dynamical curve of wound composites is relatively smooth in the early and late stages, and the middle stage shows linear and rapid growth, which can be classified as "three-stage," and there is a significant difference with the "two-stage" curves of general laminates and woven fabrics. The higher the temperature and the higher the humidity, the longer the middle stage will last.

(2) The two-stage instantaneous equilibrium model has a large deviation from the experimental data at the early stage of moisture absorption, reflecting that the moisture absorption rate of wound composites is slower at the early stage of moisture absorption, and the more obvious it is at lower temperature and lower humidity. Among the various types of surface concentration-modified models, the exponential function and quadratic function models are in good agreement with the experimental curves.

(3) The higher the temperature and humidity, the greater the equilibrium moisture absorption rate and the slope of growth of the linear segment (Table 2).

Table 2. Equilibrium moisture absorption rate and slope of linear section.

Group	#1	#2	#3	#4
M_∞ (%)	0.08958	0.20806	0.22949	0.36147
λ ($t^{-1/2}$)	0.00499	0.00504	0.00650	0.00951

The surface concentration and diffusion rate are plotted in Figure 3.

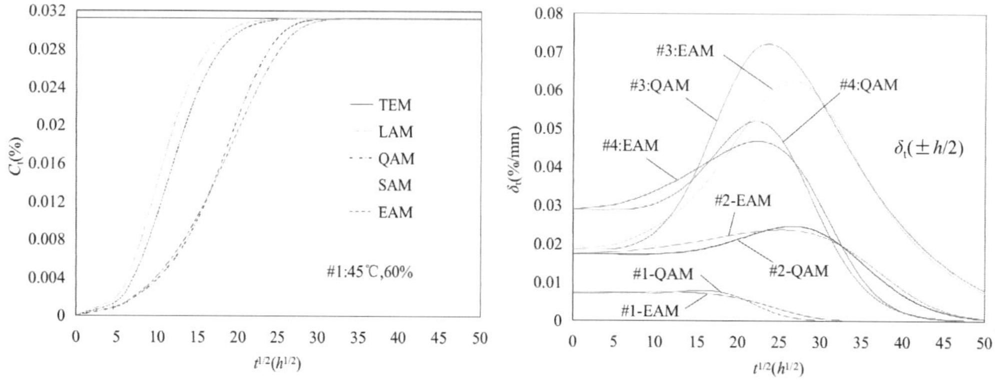

Figure 3. Surface concentration and diffusion rate.

The rule of change of surface concentration under different temperature and humidity environments is basically the same, so Figure 3 only gives the #1 group of experimental curves as a representative. It can be seen from Figure 3:

(1) Except for the transient equilibrium model in which the surface concentration reaches the equilibrium concentration instantaneously, the surface concentration of the composites in the rest of the models grows gradually, and the shape of the curves is very similar to the process of the growth of moisture absorption rate.

(2) The surface water diffusion rate is first fast and then slow, which means a maximum diffusion rate at an extreme point can be fined. The phenomenon is more significant in high temperature and high humidity environments. The reason for this phenomenon is that the general fiber volume of wound composites is high, and the porosity is small. The resistance of water molecules diffusion in the resin in the early stage of moisture absorption is large, and the diffusion channel is fully formed only after a faster diffusion rate.

In addition, the variation curves of the β values in Equation (8) are given in Figure 4. Similarly, since the shapes of the β curves with time are similar for different groups, only the experimental curves of group #1 are given as representative.

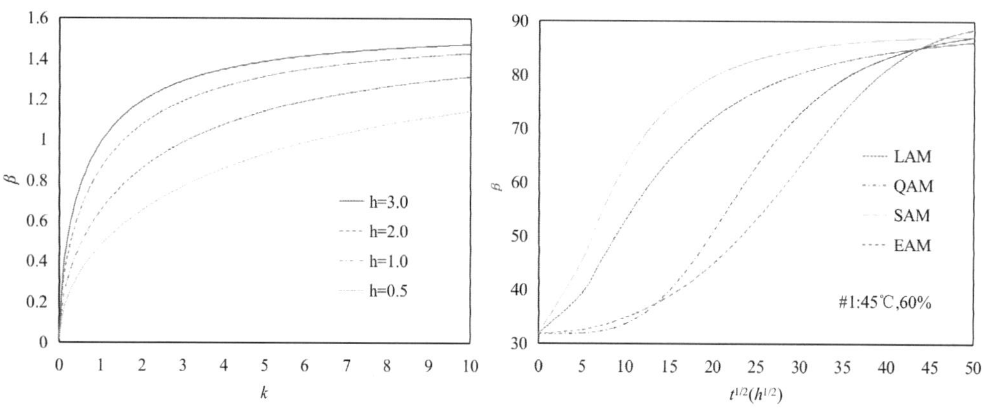

Figure 4. Variation curve of β.

As can be seen in Figure 4:

(1) The thicker the composite winding ring is, the larger the β-value is, and the β-value curve tends to be horizontal (TEM) as the relative scale factor k increases.

(2) The β-value grows gradually with time, and the β-value of the exponential model and the quadratic model grows relatively flat. Especially in the early stage of hygroscopicity, the slope of the β curve is very small and close to the horizon, which is consistent with the change rule of the

hygroscopic dynamical curve and surface concentration curve. Therefore, it can also be approximated to predict the hygroscopic dynamical curve from the change of β-value.

4. Conclusion

In the paper, the theoretical study and experimental validation of the modeling of moisture absorption dynamics of wound composites were carried out, and the main conclusions were obtained.

(1) The hygroscopic dynamical curve of wound composites can be described by the "three-stage" model, with the initial and later stages being flat and the middle stage being linear and fast growth, which is significantly different from the "two-stage" model for general laminates and woven fabrics. The phenomenon is mainly related to the slow diffusion rate of water molecules between resins at the early stage of moisture absorption under conditions of high fiber volume of wound composites.

(2) The higher the temperature and the higher the humidity, the larger the equilibrium moisture absorption rate and the slope of growth of the linear segment of wound composites.

(3) The exponential model and quadratic model can better describe the scale factor of surface concentration growth. Different from the laminates or woven fabrics, the water molecule diffusion rate of which is largest in the early stage of moisture absorption, the surface moisture diffusion rate of wound composites increases first and then decreases, tending to an extreme value in the hygroscopic dynamical curve.

References

[1] Huang Z B, Luo W Q, Cong J, et al. (2022) Moisture absorption characteristics and interface properties of Glass Fiber/Nylon 6 composite[J]. China Plastic Industry. 50(10):106-112. DOI:10.3969/j.issn.1005-5770.2022.10.018.
[2] Zhang Y W, Ge D D, Li S Y, et al. (2021) Preparation and characterization of low water absorbent copolyamide resin[J]. China Plastic Industry. 49(8):162-166. DOI:10.3969/j.issn.1005-5770.2021.08.033.
[3] Lin G J, Zhang S G, Li H, et al. (2024) Study of moisture absorption and aging behavior of composite interface riveted structures[J/OL]. Composites Science and Engineering. http://link.cnki.net/urlid/10.1683.tu.20230802.1732.018.html.
[4] Akbar Sohaib, Zhang T, et al. (2008) Moisture diffusion in carbon/epoxy composite and the effect of cyclic hygrothermal fluctuations: characterization by dynamic mechanical analysis (DMA) and interlaminar shear strength (ILSS) [J]. Adhesion. 84(7): 590. doi: 10.1080/00218460 802255434.
[5] Zheng L, Chang X L, Zhao F, et al. (2007) Research on moisture absorption of composites in the hydrothermal environment [J]. Fiber Composites. 2:37-39. DOI: 10.3969/j.issn.1003-6423.2007.02.012.
[6] Lv X J, Zhang Q, Ma Z Q, et al. (2005) Study on hydrothermal aging effect on mechanical properties of carbon fiber/epoxy resin composites[J]. Journal of Materials Engineering. 11:50-57. DOI: 10.3969/j.issn.1001-4381.2005.11.014.
[7] American Society for Testing and Materials. (1992) ASTM D5229/D 5229M Standard test method for moisture absorption properties and equilibrium conditioning of polymer matrix composite materials[S]. PA: ASTM International. DOI:10.1007/springerreference_3468.
[8] Cao S, Wang B, Jiao G Q. (2012) Research on water-sorption behaviors of T700/9916 Composite Laminates [J]. Journal of aeronautical materials. 32(2):59-64. DOI: 10.3969/j.issn.1005-5053.2012.2.011.
[9] Yu H Y, Xing P. (2022) Moisture absorption characterization of carbon fiber-reinforced polymer using Fickian and non-Fickian models[J]. Polymer Composites. DOI:10.1002/pc.27074.
[10] Wang D, Zhang T F, Gao Q, et al. (2021) Simulation analysis of moisture absorption process of CFRP composites in the hygrothermal environment[J]. Computer Integrated Manufacturing Systems. 38(7):236-240. DOI:10.3969/j.issn.1006-9348.2021.07.051.

Journal of Physics: Conference Series **2842** (2024) 012031 doi:10.1088/1742-6596/2842/1/012031

[11] Qi S J, Zhang J K, Cheng X Q. (2016) Numerical simulation of rapid heating induced delamination growth of composite laminates agter moisture absorption [J]. Acta Materiae Compositae Sinica. 33(8):1687-1693. DOI:10.13801/j.cnki.fhclxb.20151120.002.
[12] Sun Z G, Zhao L, Chen L, et al. (2013) Research on failure criterion of composite based on unified macro and micro-mechanical model[J]. Chinese Journal of Aeronautics(S1000-9361), 26(1):122-129. DOI:10.1016/j.cja.2012.12.015.
[13] M L de Souza, J I Marde, B L Fox. (2013) Hygrothermal effects on painted carbon fiber composite surfaces [J]. Journal of Composite Materials (S0021-9983), 47(5): 539-547. DOI: 10.1177/0021998312442555.
[14] Hui L, Xu L, Xu L, et al. (2014) Research on water-sorption behavior of carbon fiber reinforced epoxy composite laminates[J]. New Chemical Materials. 42(7):90-92. DOI: CNKI: SUN: HGXC.0.2014-07-030.
[15] Feng T B, Zhao Y, Luo Y F, et al. (2011) Moisture absorption and desorption behavior CCF300/5405 composites in the cyclic hygrothermal environment [J]. Acta Materiae Compositae Sinica. 28(5):77-82. DOI: 10.3724/SP.J.1146.2010.00242.

Influence of secondary peak temperature on simulated ICCGHAZ microstructure and toughness of high-strength low-alloy steels

Liangliang Bao*, Yanhong Xu, Xinming Zhang and Jun Jiang

China Construction Industrial & Energy Engineering Group Co., Ltd, Nanjing 210000, China

*Corresponding author's e-mail: baoliangfighting@163.com

Abstract. The ICCGHAZ specimens of laser-arc hybrid welding under various secondary peak temperatures were prepared using welding thermal simulation technology. Investigations were conducted into how the toughness and microstructure of the simulated ICCGHAZ specimens were correlated with the secondary peak temperature. Results showed that the size, distribution, and morphology of the M-A constituents in the ICCGHAZ specimens were primarily influenced by the secondary peak temperature. With this temperature at 760 °C, the blocky M-A constituents mostly aggregated into chains along the grain boundaries. As the secondary peak temperature reached 800 °C, while the blocky M-A constituents typically stayed close to the grain boundaries, the necklace M-A constituents began dispersing. The former austenite grain boundaries vanished as the M-A constituents migrated outward when the temperature further rose to 840 °C. At secondary peak temperatures of 760 °C and 800 °C, the volume proportion of the M-A constituents was 2.9% and 3.2%, respectively. In comparison, the volume fraction rapidly decreased to 0.9% at 840 °C. The size of the M-A constituents shrank as the secondary peak temperature rose, whereas the crack initiation energy Ei, crack propagation energy Ep, and impact energy Et increased. The toughness of the ICCGHAZ specimen was diminished because of the grain boundary necklace distribution characterization and the large size of the M-A constituents.

1. Introduction

Due to the good strength and toughness matching, high-strength low-alloy (HSLA) steels are extensively employed in offshore platform construction, shipbuilding, steel structure bridges, fabrication of pressure vessels, and other industries. The superior mechanical properties of HSLA steels may be destroyed by the quick heating and cooling of the welding process. For HSLA steels, particularly the ultra-high-strength steels, embrittlement in the heat affected zone (HAZ) during welding has long been a major concern. Numerous studies show that the local embrittlement zone of the HSLA steel welding HAZ is the inter-critically reheated coarse-grained HAZ (ICCGHAZ) [1]. Li et al. [2] discovered that the combination of microstructure coarsening and the existence of martensite austenite (M-A) elements resulted in ICCGHAZ having the lowest toughness in submerged arc welding of X100 pipeline steels. They also found that using shielded metal arc welding, ICCGHAZ was the zone of embrittlement in the simulated HAZ of ASTM4130 steels.

Laser-arc hybrid welding (LAHW) has been extensively investigated by academia as well as industry in recent years, because of its high quality and efficiency. LAHW has been applied in thick-plated HSLA steel welding and has obvious advantages in improving welding efficiency and controlling welding deformation [3]. The HAZ embrittlement is also a problem that needs special attention in HSLA steels LAHW. Previous studies [4] have found that coarse-grained HAZ (CGHAZ) and ICCGHAZ are the embrittlement zones, but there are few reports on the systematic study of the ICCGHAZ embrittlement in HSLA steels LAHW. Hao et al. [5] studied how heat produced by LAHW influenced the toughness and microstructure of CGHAZ in HSLA.

This study prepared ICCGHAZ specimens of laser-arc hybrid welding under various secondary peak temperatures using welding thermal simulation technology. The instrumented Charpy impact test was performed to evaluate how tough the ICCGHAZ specimens were, and OM, SEM, TEM, and EBSD were used to analyze the microstructure and fracture surface morphology. We looked into how secondary peak temperatures affected the toughness and microstructure of the ICCGHAZ specimens.

2. Materials and methods

The steel under investigation has the following chemical composition: Fe-0.12C-1.09Mn-0.25Si-0.58Gr-0.53Mo-0.28Cu-0.04V-0.004P-0.0039N-0.0009B-0.074Al (wt pct). Table 1 shows its mechanical properties. The A_{c1} and A_{c3} of the steel, as measured by thermodilatometry, are 720 °C and 850 °C, respectively.

Table 1. Physical properties of the steel under investigation.

Yield strength (MPa)	Tensile strength (MPa)	Reduction of area (%)	-40 °C Absorbed energy (J)			
833	890	19	Testing value	188	177	153
			Average value		173	

The Gleeble-3800 thermal-mechanical testing bench was used to prepare the specimens. Previous research has examined the welding thermal cycle and temperature distribution of the laser-arc hybrid welding process [6]. It has been discovered that from 800 °C to 500°C the cooling time $t_{8/5}$ was 4 S~6 S, and the dwelling time t_{1100} above the temperature of 1100 °C was 0.8 S~1.3 S with an average heating rate of 400 °C/s. The simulated parameters of the thermal cycle were determined according to the aforementioned results, as shown in Figure 1. Firstly, the primary peak temperature was set as 1300 °C to obtain CGHAZ specimens. The second peak temperatures were set as 760 °C, 800 °C, and 840 °C to obtain different ICCGHAZ specimens.

In accordance with GB/T 229–2007, an Instrumented Charpy impact test (55mm×10mm×10mm) was conducted at 233K (−40 °C). SEM (JEOL-JSM7200F) and OM (Leica DM2500M) were then used to examine the fractured surfaces of the specimens.

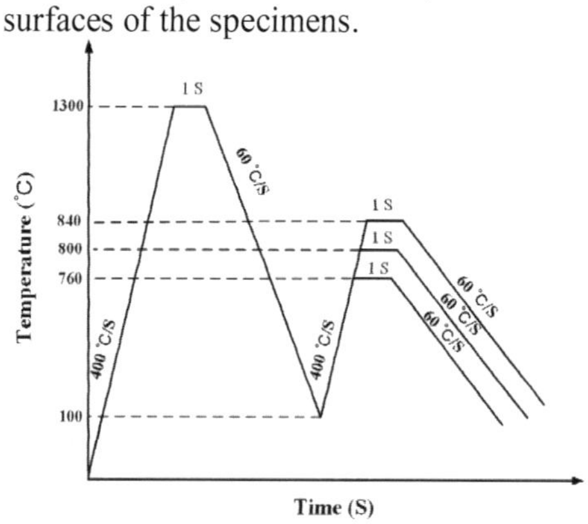

Figure 1. The welding thermal cycles of the simulated ICCGHAZ specimens.

Having been etched with the Lepera etchant to reveal the M-A constituents, metallurgical specimens were then observed using TEM (FEI-Talos-F200), SEM (JEOL-JSM7200F), and OM (Leica DM2500M). Applying a field emission SEM fitted with an Oxford Symmetry EBSD detector, we performed the EBSD characterization.

3. Results

3.1. Impact of secondary peak temperature on microstructure

Figure 2 demonstrates the microstructures of the simulated ICCGHAZ specimens under various secondary peak temperatures. Lath martensite with M-A constituents dispersed at grain and sub-grain boundaries made up the ICCGHAZ specimens. The form, distribution, and quantity of the M-A constituents varied with the secondary peak temperature. According to Figures 2a and 2b, when this temperature was 760 °C, the M-A constituents were blocky and mostly gathered close to the previous austenite grain boundaries, forming the necklace M-A constituents. As this temperature rose to 800 °C, the necklace M-A constituents began to unlink, with some dispersed at the sub-grain boundaries, as shown in Figures 2c and 2d. When the temperature increased further to 840 °C, which was close to A_{c3}, the majority of the original martensite was reversed into austenite. After that, during the fast cooling process, this austenite was converted into fine lath martensite. The former boundaries surrounding austenite grains vanished when the grain boundary of M-A constituents continued to spread.

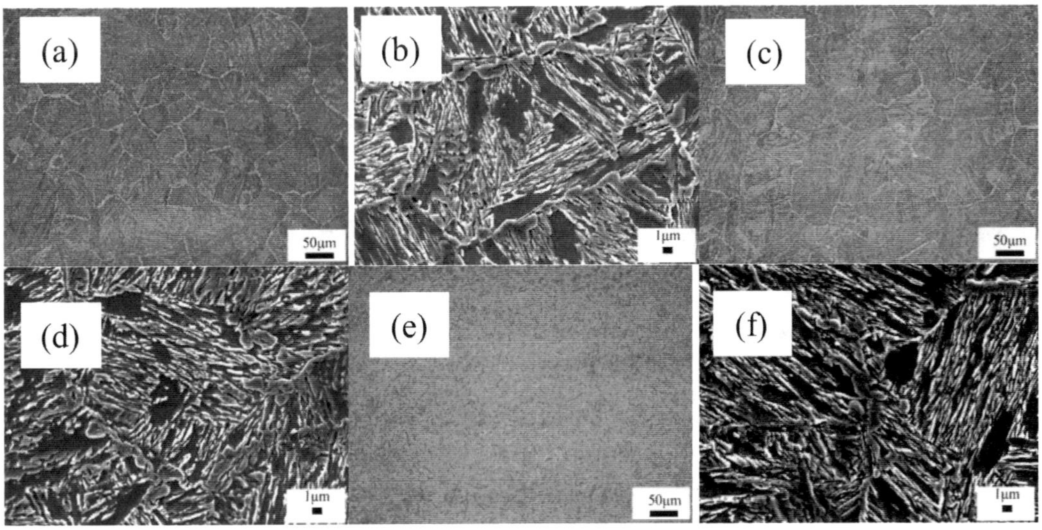

Figure 2. The microstructures of the simulated ICCGHAZ specimens
(a) OM micrograph of the specimen at 760 °C; (b) SEM micrograph of the specimen at 760 °C; (c) OM micrograph of the specimen at 800 °C; (d) SEM micrograph of the specimen at 800 °C; (e) OM micrograph of the specimen at 840 °C; (f) SEM micrograph of the specimen at 840 °C.

The metallographic micrographs were processed by Image-Pro Plus software, and the volume proportion of M-A constituents was counted to look into how the quantity and size of M-A constituents were influenced by the secondary peak temperature. The area fraction of two-dimensional metallographic images was used to approximately represent the volume fraction, and at least five metallographic images were counted at each specimen. Figure 3 depicts the processed metallographic images of the simulated ICCGHAZ specimens. The M-A constituent size was calculated by scanning electron microscope images; at least 30 M-A constituents were counted and the average value was taken. Figure 4 shows the effects of the secondary peak temperature on the volume proportion and size of M-A constituents. With the increase of the secondary peak temperature, the M-A constituent size diminished from 2.5 μm at 760 °C, to 2.1 μm at 800 °C, and further to 1.8 μm at 840 °C. However, the influence of secondary peak temperature on the volume proportion of M-A elements was not straightforward. The volume fraction of M-A constituents of the specimen at 840 °C was 0.9%, while this figure was significantly higher at 760 °C and 800 °C, which were 2.9 % and 3.2 %, respectively.

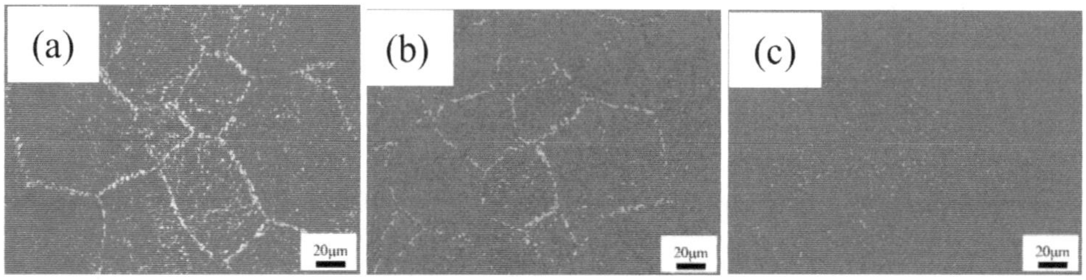

Figure 3. The processed metallographic images of the simulated ICCGHAZ specimens at (a) 760 °C; (b) 800 °C; (c) 840 °C.

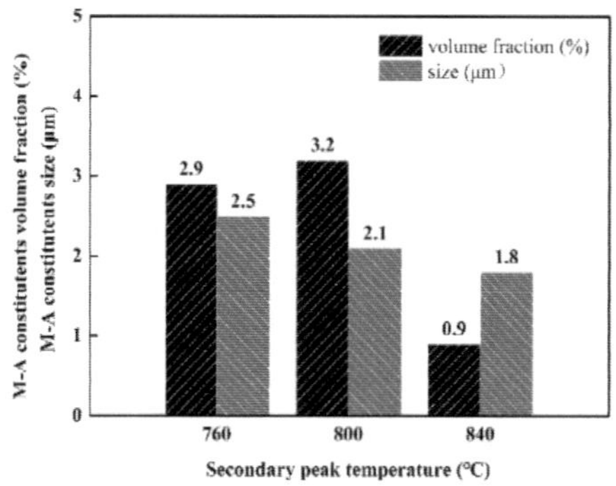

Figure 4. The volume fraction and size of the M-A constituents at different secondary peak temperatures.

3.2. Impact of secondary peak temperature on toughness

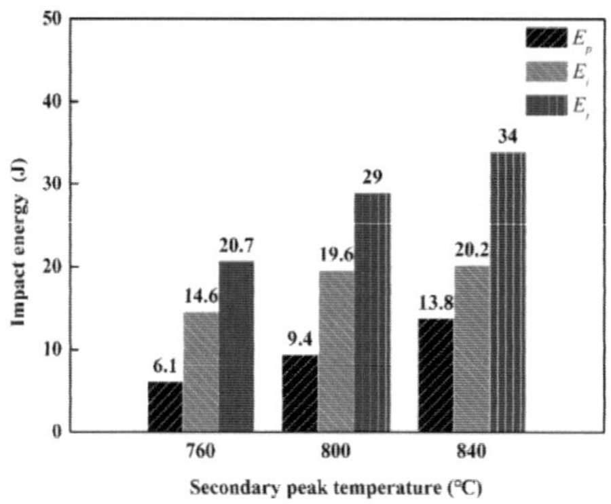

Figure 5. The instrumented impact test values at various secondary peak temperatures.

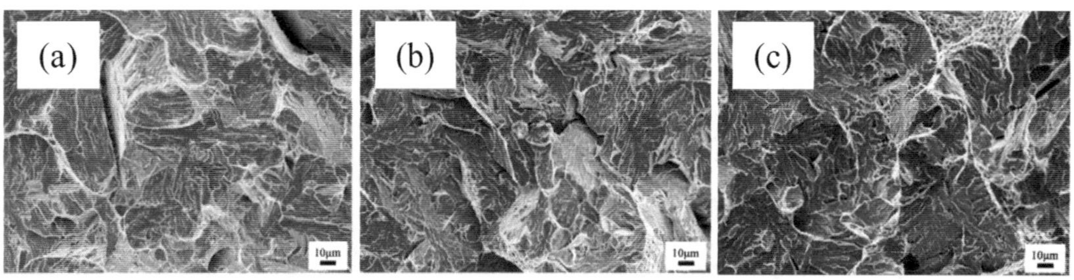

Figure 6. Fracture morphologies of the simulated ICCGHAZ specimens (a) 760 °C specimen; (b) 800 °C specimen; (c) 840 °C specimen.

Figure 5 demonstrates the instrumented impact test values of the ICCGHAZ specimens under various secondary peak temperatures. The components of the impact energy Et were the crack propagation energy E_p and the crack initiation energy E_i. The instrumented impact test values were explained in detail in reference [4]. Figure 5 shows increased instrumented impact values at higher secondary peak temperatures. When this temperature was 760 °C, E_i and E_p were both at low levels. When the temperature rose to 800 °C, E_i increased, while E_p barely changed. When the secondary peak temperature increased further to 840 °C, E_i and E_p both improved.

Figure 6 displays the microscopic fracture morphologies of the crack propagation zone of ICCGHAZ specimens under several secondary peak temperatures. All specimens showed brittle quasi-cleavage fracture characteristics. The fracture surfaces were mainly composed of cleavage facets and tearing edges at secondary peak temperatures of 760 °C and 800 °C, while at 840 °C, the local dimple bands increased significantly, indicating an improved hindering effect on crack propagation.

The above results showed that the impact toughness of the simulated ICCGHAZ specimens was closely correlated with secondary peak temperatures. In particular, E_t, E_i, and E_p were at low levels when this temperature was at the low critical temperature, but they all improved when this temperature was at the middle and upper critical temperatures.

4. Discussions

It was found that the secondary peak temperature had a significant influence on the E_i of the simulated ICCGHAZ specimens. Figure 5 demonstrates increased crack initiation energy E_i at higher secondary peak temperatures. Figure 4 indicates that the change of crack formation work had little correlation with the M-A constituents' volume fraction. In particular, at secondary peak temperatures of 760 °C and 800 °C, the M-A constituents' volume fractions were close to each other at 2.9% and 3.2%, respectively. Nevertheless, this figure dropped to 0.9% at 840 °C.

TEM and EBSD characterizations were used to analyze the specimen microstructures at a number of secondary peak temperatures, as shown in Figure 7, Figure 8, and Figure 9. The morphology and size of the M-A constituents of the simulated ICCGHAZ specimens depended on the secondary peak temperature. As depicted in Figure 7, the continuous dispersion of the blocky M-A constituents at the preceding austenite grain boundary formed the necklace M-A constituents, which was found to reduce the ability to resist crack formation [7]. As shown in Figure 8(a), these necklace M-A constituents started to scatter at a secondary peak temperature of 800 °C, yet they were still widely distributed. Austenite films were found between the intracrystalline laths, and M-A constituents in short rod-like and acicular shapes were distributed between the laths, as shown in Figures 8(c) and 8(d). Studies have found that austenitic films can effectively improve toughness [8]. The average grain dimension of the M-A constituents at 760 °C was 2.5 μm, and it dropped to 2.1 μm at 800 °C. As the M-A constituent began to disperse and its size decreased, the crack initiation energy Ei increased. Figure 9 shows the EBSD characterizations of the 840 °C specimen. The majority of the microstructure transformed into austenite when the secondary peak temperature reached 840 °C, which was close to Ac3. Lath martensite and M-A constituents were then created after cooling. The grain boundaries of the previous austenite were invisible, and the grains shrank in size. The M-A constituents saw a further decrease in average grain size to 1.8 μm and an increase in crack initiation energy Ei. In conclusion, the variance in E_i was highly correlated with the size and distribution of the M-A constituents. The larger these constituents were in size, the lower E_i was. The E_i of the specimen was low with M-A constituent distribution in necklace close to grain boundaries; however, it was improved when these constituents scattered discretely along grain boundaries or within grains.

The secondary peak temperature also significantly influenced the crack propagation energy E_p of the ICCGHAZ specimens. Figure 5 shows that increased secondary peak temperature resulted in increased crack propagation energy E_p. It was found that high-angle grain boundaries had a beneficial effect on preventing crack propagation [7]. Li et al. [4] demonstrated that grain boundaries at an angle greater than 45° can effectively stop cracks from spreading and thereby enhance the toughness of the HAZ. The high-angle grain boundaries above 45° were represented by the yellow lines in Figures 7, 8, and 9. The proportions of the high-angle grain boundaries over 45° of the ICCGHAZ specimens under various secondary peak temperatures obtained by software statistics were 12.6%, 13.9%, and 16.3%, respectively. As the secondary peak temperature rose, the fraction of high-angle grain boundary climbed as the effective grain size shrank. This enhanced the capability of blocking crack propagation, so crack propagation energy E_p kept increasing. The high-angle grain boundary ratio of the 800 °C specimen was similar to that of the 760 °C specimen, but the crack propagation energy E_p was increased, and the toughness was improved due to the existence of film reversal structure between the laths.

In conclusion, the low critical temperature ICCGHAZ specimen (760 °C specimen) had the lowest toughness because of the existence of the necklace M-A constituents close to the grain boundary and their enormous size. The crack initiation energy E_i of the intermediate critical temperature ICCGHAZ specimen (800 °C specimen) was increased compared with that of the 760 °C specimen due to the dispersion of M-A constituents at grain boundaries. The percentage of the high-angle grain boundary of the 800 °C specimen was similar to that of the 760 °C specimen, but the crack propagation energy E_p and the toughness were improved due to the existence of the film reversal structure between the laths. The upper critical temperature ICCGHAZ specimen (840 °C specimen) had a relatively good toughness due to the increased percentage of high-angle grain boundary, grain refinement, and the small dimension of the M-A constituents, which was able to scatter within the grains.

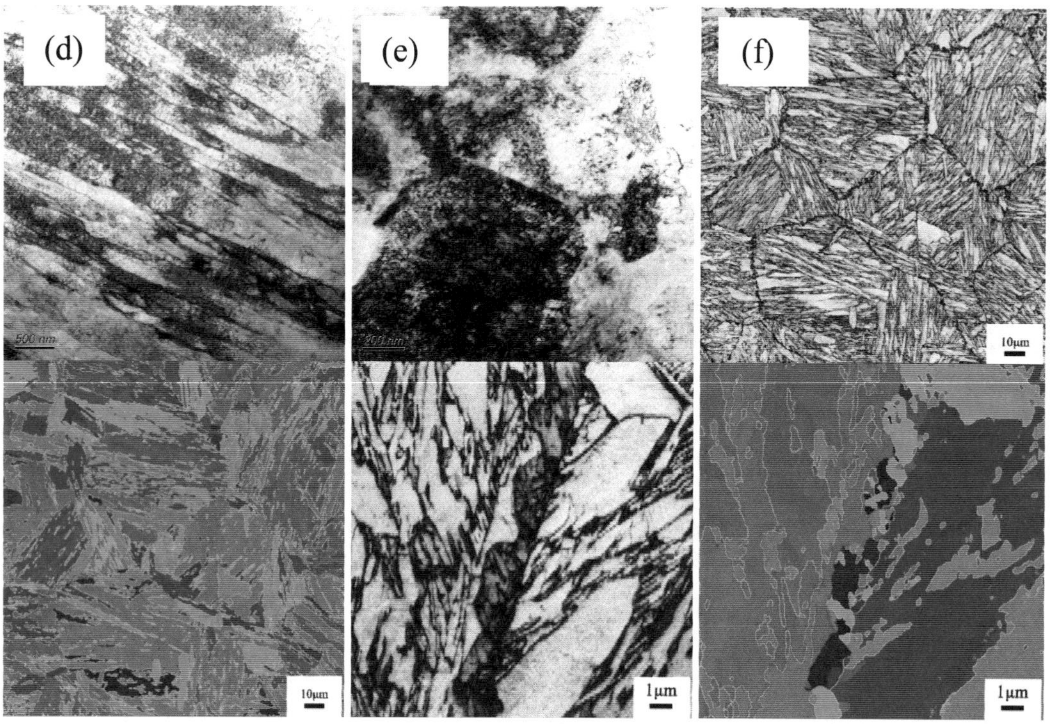

Figure 7. TEM pictures and EBSD characterization of the 760 °C specimen
(a) Internal microstructure TEM picture; (b) grain boundary blocky M-A constituent TEM picture; (c) BC map; (d) euler map; (e) grain boundary M-A constituent BC map; (f) grain boundary M-A constituent euler map.

Journal of Physics: Conference Series **2842** (2024) 012032 doi:10.1088/1742-6596/2842/1/012032

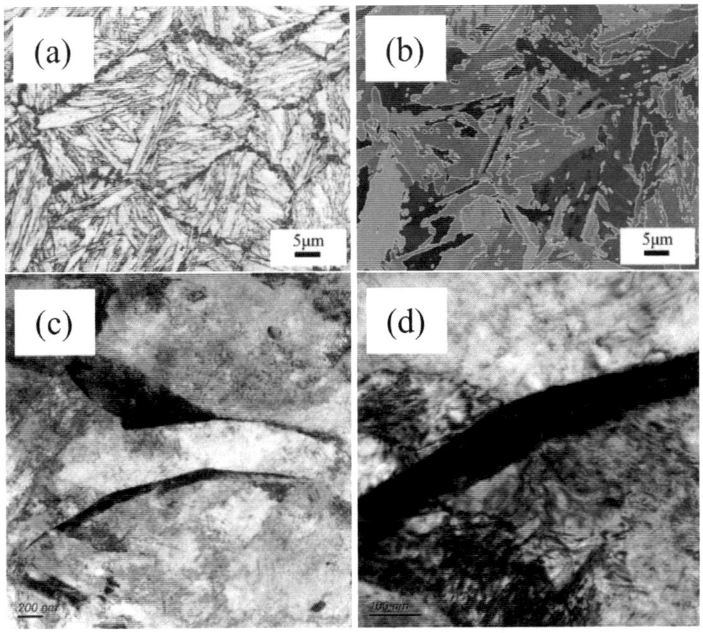

Figure 8. TEM pictures and EBSD characterizations of the 800 °C specimen
(a) BC map; (b) euler map; (c) inter lath reversed microstructure; (d) inter lath reversed microstructure.

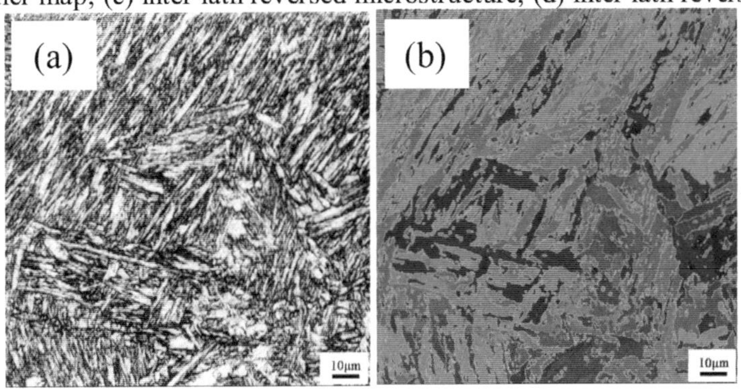

Figure 9. EBSD characterization of the 840 °C specimen
(a) BC map; (b) eulur map.

5. Conclusions

(1) The simulated ICCGHAZ specimens were composed of lath martensite with M-A constituents distribution at sub-grain and grain boundaries. With a secondary peak temperature of 760 °C, the M-A constituents were blocky and primarily scattered close to the previous austenite grain boundaries, which formed the necklace M-A constituents. As this temperature reached 800 °C, the necklace M-A constituents began to unlink, with some distributed near the sub-grain boundaries. When this temperature further went up to 840 °C, the previous austenite grain boundaries vanished, with the increased diffusion of grain boundary M-A constituents. The volume percentage of these constituents reached as much as 2.9% and 3.2%, respectively, at secondary peak temperatures of 760 °C and 800 °C. In comparison, the volume fraction rapidly decreased to 0.9% when the temperature was 840 °C. As the secondary peak temperature rose, the M-A constituents shrank in size.

(2) High secondary peak temperature resulted in increased E_i, E_p, and E_t. The 760 °C specimen lost some of its toughness because of the existence of large-size M-A constituents in the necklace at the grain boundaries. The E_i of the 800 °C specimen was increased compared with that of the 760 °C

specimen because of the gradual dispersion of the grain boundary M-A constituents. The toughness and crack propagation energy E_p of the 800 °C specimen were improved due to the existence of the film reversal structure between the laths. The 840 °C specimen possessed a rather acceptable toughness owing to the high percentage of the high-angle grain boundaries, the grain refinement, as well as the tiny size of the M-A constituents that were diffused within the grains.

Acknowledgments
This work was supported by the China State Construction Technology Research and Development Program (CSCEC-2022-Q-42).

References

[1] Bao L, Pan C, Liu F, et al. Research on the HAZ microstructures and toughness of the laser-arc hybrid welding of high strength low alloy steels. Transactions of the China Welding Institution, 2022, 5: 90 – 97.
[2] Li X, Fan Y, Ma X, et al. Influence of Martensite-Austenite constituents formed at different intercritical temperatures on toughness. Materials & Design, 2015,67:457-463.
[3] Bunaziv I, Frostevarg J, Akselsen O, et al. Process stability during fiber laser-arc hybrid welding of thick steel plates. Optics & Lasers in Engineering, 2018,102:34-44.
[4] Bao L, Liu F, Xu Y, et al. Investigation on microstructure and impact toughness of double-pass laser-arc hybrid welding heat affected zone. Transactions of the China Welding Institution, 2022, 12: 90 – 99.
[5] Hao K, Zhang C, Zeng X, et al. Effect of heat input on weld microstructure and toughness of laser-arc hybrid welding of martensitic stainless steel. Journal of Materials Processing Technology, 2017, 245:7-14.
[6] Bao L, Wang Y, Zhang H, et al. Welding thermal cycle of the laser-arc hybrid welding of the EQ70 steel and its effects on the microstructure evolution of the heat affected zone. Transactions of the China Welding Institution, 2021, 3: 26 – 33.
[7] Davis L, King E. Cleavage initiation in the intercritically reheated coarse-grained heat-affected zone: Part I. fractographic evidence. Metallurgical & Materials Transactions A, 1994, 3:563-573.
[8] Deng W, Zhang X, Zhang T, et al. Microstructure transformation and toughness of coarse grain heat affect zone of 10Ni8CrMoV steel during second welding thermal cycle. Hot working technology, 2006, 15:5-8.

Analysis of thermal stability and enthalpy in SSD systems with various thickening agents

Rui Duan*, Jinning Wang, and Ao Zhang

College of Energy and Mechanical Engineering, Shanghai University of Electric Power, Shanghai, 201306, China

*Corresponding author's e-mail: duanrui@shiep.edu.cn

Abstract. Phase change materials (PCMs) play a vital role in the improvement of efficiency of cold storage air conditioning for demand side management of electrical power. To enhance the thermal stability and the enthalpy of the system, the separation of Sodium Sulfate Decahydrate (SSD) hydrate crystals from their aqueous solution is a significant factor that needs to be addressed. The melt blending technique has been used to explore the effect of various thickening agents, namely Carboxymethyl Cellulose (CMC), Xanthan Gum (XG), and Sodium Alginate (SA) on the thermal stability and enthalpy of the system. The choice of the best thickening agent and its optimum concentration led to the enhancement in the thermal stability and enthalpy of the system. Experimentally, among the three thickeners, CMC proved to be the best thickening agent that remarkably improved the thermal physical properties of the SSD formulation. The paper described the detailed experimental procedures, including melt blending set-up, thermal characterization unit, and cyclic stability test unit. The experimental results showed that the formulation, SSD-CMC at 2 wt% CMC and 50% SSD significant, improved in their thermal physical performance, wherein the enthalpy decreased from 104.23 kJ/kg to 83.71 kJ/kg over fifty cycles, which are the 19.69% of the decay for such a long cycling. It was then concluded that the results would improve the effectiveness of thermal energy storage materials and that phase change materials should be optimized to be used in sustainable and efficient cooling applications and useful for the improvement of the storage capacity of thermal energy storage solutions.

1. Introduction

The technology of Thermal Energy Storage (TES) for air conditioning, makes it feasible to produce and store a stack of cooling, during off-peak periods, that can be used at the time of the peak demand, raises the energy-use efficiency considerably, and therefore also reduces energy consumption significantly [1, 2, 3]. This is achieved, because the electric power necessary to generate and store cooling during completely off-peak periods is substantially less costly than the power required to produce cooling at times when its need may, together with that of other electric loads, cause the demand for power to exceed the supply [4]. TES is also part of a group of technologies that are known as "load management" because their application is also a means for "smoothing" the load curves, which represents the hourly demand upon the grid, and minimizing or perhaps eliminating the demand humps, that is, the transient deviations from these demand curves, which result from the starting of the various types of equipment that are electric motor driven [5]. Among the numerous consequences of its imposition, the effect of the first of these is to enhance the stability of the electrical grid's operation, and that of the second is to assure increased reliability. The ultimate benefit is that the extent of the reduction of electricity consumption, achieved indirectly through diminished carbon dioxide and other major pollutants from electricity production, keeps the air quality improvement equivalent of these efforts aligned with the most ambitious of environmental conservation efforts [6]. In their application to air conditioning systems, TES also plays a central role in enabling the architecture, which executes green building design, that is, the design of a building such that it reduces the use of non-renewable and non-recovered resources minimizes overall building impact on the environment, and creates a sustainable building, as it reduces the building's overall energy consumption and then, it tremendously, to the realization of the most important objective of this new type of architecture, high reductions of multiple emissions, of pollutants and greenhouse gases [7]. Therefore, all of TES, and not just their applications to air conditioning, are among the comprehensive solutions, that are required to the reduction in energy usage and its environmental impact, to permit the realization, so quickly as can be, that sustainable building practices, will collectively become a compelling successor of the hitherto socially dominant ones, crumbling more rapidly by the day, which they will replace [8].

Latent heat storage, compared to sensible heat storage, has a higher energy storage capacity during the process of phase change. It has broad prospects as a seasonal or long-term energy storage technology. Phase change materials (PCMs) act as the storage medium in latent heat storage systems, absorbing and releasing heat during the phase change process. PCMs have property advantages such as high energy storage density, isothermal operation, and small temperature fluctuation with the receiving medium. According to the type of phase transition, PCMs can be divided into liquid-gas, solid-liquid, solid-gas, and solid-solid. Since containers require higher volume and pressure due to the presence of gas, they are not fit to be used as a phase change cooling material for either liquid-gas or solid-gas phase transition [9]. Solid-liquid phase change materials currently make up the majority of latent heat storage materials and can be divided into organic, inorganic, and composite phase change materials based on their chemical composition. Each of these has different application characteristics and advantages in energy storage [10].

Compared with phase change temperature ranges, inorganic hydrated salts have a much broader range of applications. It is also very important to develop composite phase change materials for diverse temperature ranges [11]. The materials can provide significant references for the selection and optimization of the formula. In current years, the inorganic hydrated salts which are used to phase change materials include Sodium Sulfate Decahydrate ($Na_2SO_4 \cdot 10H_2O$) [12, 13, 14, 15], Calcium Chloride Hexahydrate ($CaCl_2 \cdot 6H_2O$), and Sodium Acetate Trihydrate ($C_2H_3O_2Na \cdot 3H_2O$) [16, 17] for phase change purposes. Inorganic hydrated salts phase change materials have phase separation, the degree of supercooling is larger [18, 19, 20], which uses additives to change the materials, thereby enhancing the thermal performance and stabilizing, to realize stores energy and the efficient release of it [2, 21]. Methods that are already in use to resolve phase separation of phase change materials include the employment of thickening agents to increase their viscosity [14, 22], and nano-microencapsulation treatment for the materials [23]. After the thickening of the materials, the phase separation problem is resolved due to (the) solid particles being evenly distributed in the increased viscosity material instead of settling at the bottom of the material [15]. The phase separation problem of inorganic hydrated salts phase change materials was resolved easily due to its (the material's) phase separation mechanism in nano-microcapsules [19, 24, 25]. These methods are very critical to the stabilization and effective use of inorganic hydrated salt phase change materials for thermal energy storage.

The key to phase change material formulation is the choice of a suitable thickening agent. Muller et al. [26] recently demonstrated the feasibility of using Xanthan Gum (XG) as a thickening agent. In their work, a mixture was formulated that contained 2wt% Ammonium bi-borate tetrahydrate and 1wt% Xanthan Gum. The composite achieved a melting temperature of 25.7 °C and an enthalpy of 120.2 kJ/kg. This formulation validated the viability of Xanthan Gum (XG) as a thickening agent in phase change material formulations. In the work by Meng et al. [27], Sodium Alginate (SA) was formulated as the thickening agent with Sodium Sulfate Decahydrate ($Na_2SO_4 \cdot 10H_2O$, SSD) for thermal energy storage. The resultant composite phase change material (PCM) made of Sodium Alginate (SA) and

SSD achieved a high phase change enthalpy of 160 kJ/kg and improved cyclic stability of the phase change material. The stabilization mechanism is based on the ionic and hydrogen bonding interactions between Sodium Alginate (SA) and SSD. The work by Lin et al. [28] successfully developed an advanced phase change refrigeration material from Sodium Sulfate Decahydrate (SSD, $Na_2SO_4\cdot 10H_2O$) by blending Carboxymethyl Cellulose (CMC) to suppress phase separation. It exhibited a phase change temperature of 6.8°C, a melting latent heat of 97.05 kJ/kg, a degree of supercooling of 0.7°C, and a thermal conductivity of 0.264 W/m·K. A novel sodium sulfate decahydrate (SSD) based composite phase change material, which can be used for thermal energy storage was synthesized to reduce phase separation problems of the SSD/hygroscopic salt/organic additive composite materials. The optimized ratio of SSD: Borax: CMC in the present study (29) was proposed to be 100: 5: 3, and the new composite phase change material was made using solidification and melting methods. The degree of subcooling of the 100:5:3=SSD:Borax: CMC composite PCM was 1.5°C, and the phase separation was eliminated. Another phase change cooling material mainly composed of Sodium Sulfate Decahydrate (SSD) was developed by Jin et al. [29], and the impact of different kinds and contents of thickeners (Sodium Polyacrylate (PAAS), Polyacrylamide(PAM), Carboxymethyl Cellulose (CMC), Poly Anionic Cellulose (PAC), Xanthan Gum (XG)) on the phase separation, phase change enthalpy of SSD was investigated. The phase separation was eliminated by adding PAAS at 1.0 wt%-2.0 wt%. The phase change temperature and latent heat of SSD of the optimized composite phase change cooling material were 7.4°C and 117.4 kJ/kg, respectively.

This study aims to research thickening agents suitable for the Sodium Sulfate Decahydrate ($Na_2SO_4\cdot 10H_2O$, SSD) system to solve the issue of phase separation. The agents include Carboxymethyl Cellulose (CMC), Xanthan Gum (XG), Sodium Alginate (SA), Super Absorbent Polymer (SAP), Poly Anionic Cellulose (PAC), Guar Gum (GG), Polyacry amide (PAM), and Polyacrylic Acid Sodium (PAAS). The composite PCM was prepared through melt blending, and the optimal ratio of different agents was determined based on the suitability of the PCM as a phase change material using thermal property characterization. After that, the optimal thickening agent was selected through ten cycles of stability tests, and the optimal ratios of each thickening agent and SSD were determined. Finally, fifty cycles of stability tests were used to verify the best thickening agent and to measure the thermal conductivity of the PCM. The results showed that the composite phase change material (CPCM) exhibited excellent thermal physical properties, which made the material a good prospect for them in terms of cold storage air conditioning systems to improve its application for cold storage air conditioning technology.

2. Experimental

2.1 Materials

All chemical reagents used in this research, as shown in Table 1, namely Sodium sulfate decahydrate ($Na_2SO_4\cdot 10H_2O$, SSD), sodium tetraborate decahydrate ($Na_2B_4O_7\cdot 10H_2O$, Borax), ammonium chloride (NH_4Cl), sodium hexametaphosphate ($(NaPO_3)_6$, SHMP), Carboxymethyl Cellulose (CMC), Xanthan Gum (XG), Sodium Alginate (SA), Super Absorbent Polymer (SAP), Poly Anionic Cellulose (PAC), Guar Gum (GG), Polyacrylamide (PAM), and Polyacrylic Acid Sodium (PAAS), were in analytical grade and obtained from Shanghai Macklin Biochemical Technology Co., Ltd. of China. Such reagents were considered to have high purity and good reliability, thereby having minimum disturbance on the experimental results. For this study, they were used as received without any extra treatment and were not further purged. This was aimed to ensure the fairness of comparison and the true result regarding the capabilities of reagents in their normal condition.

Table 1. Material name and abbreviation.

Material	Abbreviation	Material	Abbreviation
Sodium sulfate decahydrate	SSD	Sodium Alginate	SA
Sodium tetraborate decahydrate	Borax	Super absorbent polymer	SAP
Ammonium chloride	NH_4Cl	Poly anionic cellulose	PAC

Sodium hexametaphosphate	SHMP	Guar gum	GG
Carboxymethyl cellulose	CMC	Polyacrylamide	PAM
Xanthate gumand	XG	Polyacrylic acid sodium	PAAS

2.2 Preparation of the composites

To address the phase separation challenge associated with the original formulation, as discussed in detail in Table 2, a systematic approach was adopted, incorporating the inclusion of various thickening agents. These agents, specifically Carboxymethyl Cellulose (CMC), Xanthan Gum (XG), Sodium Alginate (SA), Super Absorbent Polymer (SAP), Poly Anionic Cellulose (PAC), Guar Gum (GG), Polyacrylamide (PAM), and Polyacrylic Acid Sodium (PAAS), were integrated into the formulation incrementally. The process was initiated at a concentration of 0.0 wt%, with the concentration of thickening agents being increased by 0.5 wt% after each successive addition. Following each addition, the mixture was vigorously stirred at a temperature of 50°C to facilitate blending and homogenous dispersion across the combination of thickening agents. This procedure was used consistently until a visually homogenous thickened consistency was achieved without any phase separation for at least 24 hours beyond mixing. This progressive addition and thorough mixing approach provided the ability to determine the concentration levels necessary for each thickening agent and minimized the tendency of the formulation to phase separate, thereby ensuring the stability of the formulation. As these tests were carried out, it was found that the systematic addition and complete mixing of the combination of thickening agents allowed for stepwise knowledge of the most productive concentration of each thickening agent, while identifying key players in the prevention of phase separation for these types of complex formulations.

In the previous studies noted in Table 2, a notable work was presented, optimizing phase change materials for thermal storage applications, particularly within air conditioning systems. 82.00 wt% Sodium Sulfate Decahydrate (SSD), primary phase change material (PCM) was reinforced with 3.00 wt% Borax that reduced the supercooling to 0.50°C. While the Borax was instrumental in enhancing the phase change experience, it was also critical for the transformation of the viscous SSD to a free-flowing liquid. 14.00 wt% NH_4Cl was added to adjust the phase change temperature to 9.50°C/13.90°C (T_c/T_m) for the air conditioning thermal storage application. When the composition was slightly modified with 1.00 wt% Sodium Hexametaphosphate (SHMP), the enthalpy increased to 110.85 kJ/kg, certifying the material would be ideally tailored for the intended air conditioning application. Despite these advances, the PCM with all three salt hydrates suffered the same problem noted in the use of SSD as a PCM; phase separation was observed and was persistent. Given the potential for loss of efficiency and long-term stability of material with a phase separation issue, as for the ternary chloride system as well, this recent work recommended, and sought out, solutions to the problem of phase separation. A key recommendation of this recent work was the use of thickeners to make a slurry that might not phase separate, as suggested and employed in this report; the solution to the phase separation, also improved the homogenization of the phase change materials formulations, allowing another avenue to a better PCM formulation and one that might be robust and reliable.

Table 2. Basic formulation.

SSD	Borax	NH_4Cl	SHMP	T_c(°C)	T_m(°C)	T_s(°C)	H (kJ/kg)
82.00%	3.00%	14.00%	1.00%	9.50	13.90	0.50	110.85

2.3 Characterization

Enthalpy measurements were conducted employing a METTLER TOLEDO DSC3 differential scanning calorimeter, Switzerland, over a temperature range of -10°C to 30°C, at a scan rate of 5°C per minute. The chamber was supplied with nitrogen gas at a flow rate of 50 ml/min, thus maintaining a protective atmosphere. K-type thermocouple ensured that the temperature within the apparatus was continuously and accurately monitored, with a precision reaching ±0.5°C, and which was recorded in real-time by a compatible data logging system. The trials were performed within a temperature-regulated cycling chamber employing the same temperature range of -10°C to 50°C, as indicated by its thermostat that precisely balanced temperature fluctuations that were essential for the examination of material properties under inquiry. The thermal conductivity of samples was evaluated with a Hot disk TPS 2500S instrument, in Sweden, using a heating power of 30 mW, with measurements being acquired over 10 seconds.

3. Results and discussion

3.1 Optimum proportion

Given in Section 2.3 and documented in Table 3 concerning the various agents, the minimum concentration thresholds necessary for achieving a requisite efficient thickening are critical determinants for maintaining their consistency and stability for future applicability. Carboxymethyl Cellulose (CMC), at a concentration of 2.00 wt%, was able to effectively thicken a mixture, underlining the strong thickening potential of the agent. In contrast, it is noteworthy that Xanthan Gum (XG) thickened effectively, at a much higher concentration of 7.50 wt%, consequently exposing a tremendous difference between the various agents in terms of their efficiency. Sodium Alginate (SA) thickened at a concentration of 3.50 wt%, revealing a very moderate requirement for efficient thickness among the agent spectrum evaluated. Furthermore, at a concentration of 2.00 wt%, both Super Absorbent Polymer (SAP) and Polyacrylic Acid Sodium (PAAS) were able to effectively thicken the mixture, exhibiting strong efficiency among these valid thickening agents. Finally, at a nearly identical concentration of 1.50 wt%, Poly Anionic Cellulose (PAC) was the most effective of all the above-noted agents, while Guar Gum (GG) and Polyacrylamide (PAM), respectively, required 6.00 wt% and 5.00 wt%, to also effectively thicken the mixture. Herein, these concentrations delineate at what minimal levels each of those agents can exponentially contribute to the improved mixture's viscosity; this is central in securing these improvements in stability/consistency, and accordingly their optimal performance for future application scenarios.

Table 3. Optimal ratio of thickening agent.

Thickening agent	Optimum proportion (wt%)	Thickening agent	Optimum proportion (wt%)
CMC	2.00	PAC	1.50
XG	7.50	GG	6.00
SA	3.50	PAM	5.00
SAP	2.00	PAAS	2.00

The calculations were meticulously executed to address the challenges of phase separation. These revealed that Xanthan Gum (XG), Guar Gum (GG), and Polyacrylamide (PAM) necessitate a minimum concentration of 5.00 wt% to demonstrate their thickening capabilities effectively. Irrevocably, the substantially high mass fraction these agents demand introduces a significant potential to radically alter the composition of the primary phase change materials — with duly consequential impacts on the formulation's enthalpy values as well. It is this balance between enthalpy and compositional integrity that must be maintained, however, as they are inextricably linked to the thermal performance and stability of the formulation. Ultimately, they are indicative of the formulation's fitness for thermal energy storage applications.

Given the profound influence of the aforementioned thickening agents on both the enthalpy values and the compositional balance, the inclusion of XG, GG, and PAM in the present formulation was very carefully re-evaluated. As an optimized balance between the material's functional performance and its thermal energy storage capacity remained paramount, the conscious choice was made to exclude XG, GG, and PAM from the present formulation. This strategic exclusion ensures the efficacy and stability of the formulation while maintaining its performance integrity across its intended application scenarios. It underscores the necessity of selecting thickening agents that rather than compromise essential thermal properties, or unduly alter the phase change material's composition-contribute positively to the formulation's overall performance while simultaneously preserving its practicality and effectiveness in real-world applications.

3.2 Thermal properties

The thermal behavior of the considered thickening agents is comprehensively illustrated in Figure 1 and Figure 2, allowing a detailed visualization of the heating and cooling profiles of these agents based on thermocouple measurements, and the results of Differential Scanning Calorimetry (DSC) analysis, respectively. These instrumental evaluations are pivotal for determining with a high degree of precision the key thermal parameters (crystallization temperature (T_c), melting temperature (T_m), degree of supercooling (T_s), and enthalpy change (H)) that are necessary to interpret the phase transition behavior of these agents. The detailed data obtained from these analytical procedures are methodically collated and listed in Table 4.

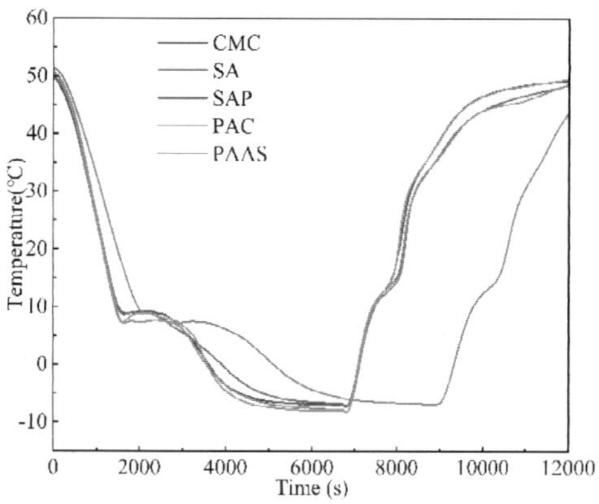

Figure 1. Cooling and heating curves.

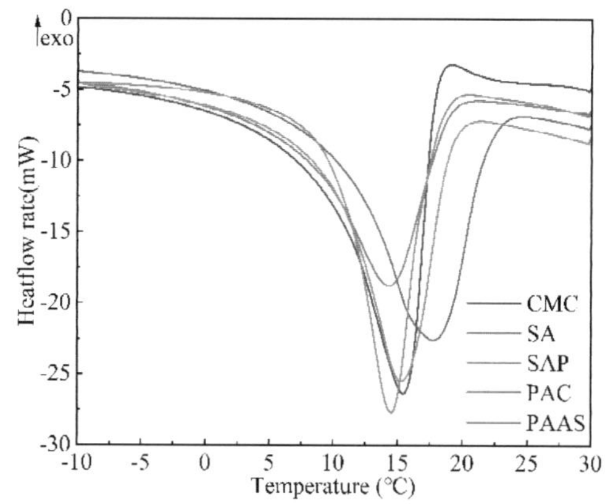

Figure 2. DSC curves.

Table 4. Thermal properties.

	T_c(°C)	T_m(°C)	T_s(°C)	H (kJ/kg)
Basic formulation	9.50	13.90	0.50	110.85
CMC	9.40	11.70	0.50	104.23
SA	7.50	13.00	0.30	53.50
SAP	9.10	11.60	0.40	81.34
PAC	8.90	11.60	1.70	58.05
PAAS	7.50	11.70	0.40	58.78

This encompassing examination of thermal properties, succinctly depicted in Table 4, provides a nuanced comparison of the varying thermal stability impacts of different thickening agents on the studied formulation. In particular, Carboxymethyl Cellulose (CMC) serves to illustrate one extreme; its inclusion in the formulation appears to have an almost negligible influence on the thermal characteristics of the formulation, as evidenced by the fact that CMC mirrors the baseline sample enthalpy values. This behavior indicates its remarkable compatibility with the formulation and the tacit suggestion that its inclusion alters the overall thermal stability of the formulation negligibly. As such, the appropriateness of the use of CMC as a thickening agent is unquestioned in the present formulation.

Conversely, it is visually evident that the enthalpy values are significantly affected by the incorporation of Sodium Alginate (SA), Super Absorbent Polymer (SAP), Poly Anionic Cellulose (PAC), and Polyacrylic Acid Sodium (PAAS) into the formulation attesting to the significant alteration of the thermal characteristics of the formulation by these agents. Nonetheless, the close adherence of these changes to the crystallization temperature (T_c), melting temperature (T_m), and degree of supercooling (T_s) indicate that these modifications do not fall significantly beyond the well-defined experimental requirements that limit these temperature changes, and as such compliance as paralleled by a certain level of thermal performance is still maintained of the agents. In contrast, the effect of the same was quite dramatic on enthalpy (H) (a crucial parameter for thermal energy storage capacity). More specifically, concerning enthalpy (H), the CMC reduced it by 5.97%, while for SA, the reduction was quite substantial (51.74%). In comparison, the SAP and PAC were reduced by 26.62% and 47.63%, respectively, and the PAAS was reduced by 46.97%.

3.3 Thermal reliability

Stability is critical for air conditioning thermal storage applications. Rigorous evaluation of the formulations, with different thickening agents, was carried out for ten thermal cycles to gauge their endurance over time. A visual representation of the durability and resistance of the formulation over many cycles is presented below in Figures 3 and 4. The quantitative data is compiled and presented below in Table 5. These are of utmost importance in charting the long-term performance and reliability of the thickening agents in the thermal storage matrix. The qualitative analysis was performed by carefully analyzing at the end of each cycle, and it was evident that all of the materials' products have good separation stability. Even for the SAP case, no visible signs of product degradation or separation were observed.

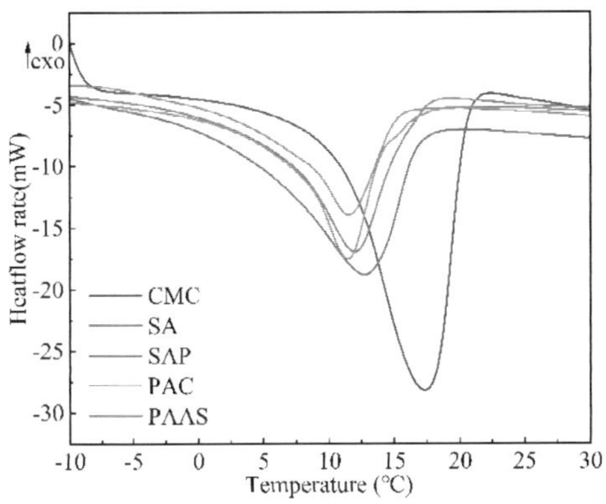

Figure 3. DSC curves after ten cycles.

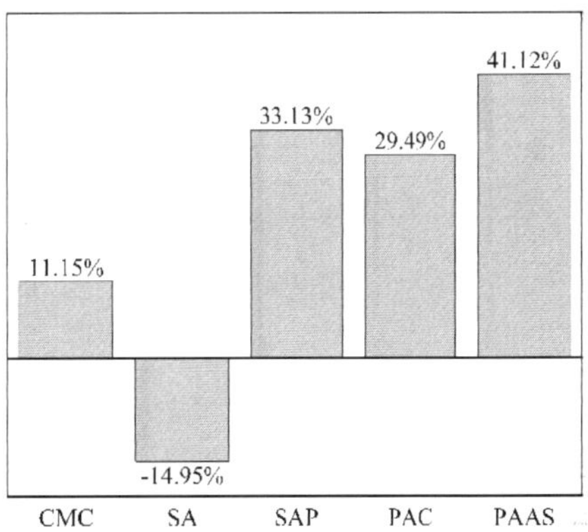

Figure 4 .Heat flow change percentage.

Table 5. Thermal reliability.

	H after ten cycles (kJ/kg)	Attenuation rates (%)
CMC	92.61	11.15
SA	61.50	-14.95
SAP	54.39	33.13
PAC	40.93	29.49
PAAS	34.61	41.12

The detailed analysis of thermal performance, as illustrated by the data in Table 5 over ten cycles, shows the consequences of cyclical thermal loading on formulations that have been enhanced with different thickening agents. The formulation with Carboxymethyl Cellulose (CMC) was observed to experience a reduction in its enthalpy from 104.23 kJ/kg to 92.61 kJ/kg from the 1st to the 10th cycle for a decay rate of 11.15%. These suggest a modest level of thermal degradation over the cycles. Meanwhile, the formulation with Sodium Alginate (SA) experienced an odd trend, with its enthalpy increasing from 53.50 kJ/kg to 61.50 kJ/kg for a negative decay rate of -14.95%, indicating an increase in thermal storage capacity. The Super Absorbent Polymer (SAP) formulation was able to bring about a significant decrease in enthalpy with a decay rate of 33.13% from 81.34 kJ/kg to 54.39 kJ/kg from Cycle 1 to Cycle 10. In the case of the Poly Anionic Cellulose (PAC) formulation, a substantial decrease in enthalpy from 58.05 kJ/kg to 29.49 kJ/kg (decay rate of 49.29%) was brought about from Cycle 1 to Cycle 10. Finally, the Polyacrylic Acid Sodium (PAAS) based formulation also observed a decrease in its enthalpy with a decay rate of 30.03% from 58.78 kJ/kg to 41.12 kJ/kg from cycle 1 to cycle 10. Following the cycling process, it might be reasonable to presume that the pronounced diminution in enthalpy values seen for the Carboxymethyl Cellulose (CMC), Super Absorbent Polymer (SAP), Poly Anionic Cellulose (PAC), and Polyacrylic Acid Sodium (PAAS) formulations signifies the deterioration or shutdown of the thickening agents due to the high ionic concentrations and inherent system instability. The formulation with CMC, however, expressed an enthalpy reduction of just 11.15% by far the smallest observed rate of enthalpy decay and thus is seen as the eminently suitable candidate that meets the rigorous standards that air conditioning thermal storage systems demand.

Conversely, the enthalpy for the saturated Sodium Alginate (SA) formulation rose 14.95%, which implies that this microsystem was unstable in its appearance post-formulation, but gradually stabilized throughout the ten cycles, leading up to its increased enthalpy value. The SA formulation's ultimate enthalpy value, however, was 61.50 kJ/kg, less than the CMC formulation. Its lesser enthalpy valuation illustrates plain and simple that it falls short of meeting the thermal storage requirements that define air conditioning systems. As such, from a comparison perspective of thermal performance and stability post cycling, the Carboxymethyl Cellulose (CMC) formulation was the ultimate selection and it should be used to optimize the performance and reliability of air conditioning thermal storage solutions.

3.4 Optimal formulation

Carboxymethyl Cellulose (CMC) is a thickening agent, which after being studied in the comprehensive study outlined before, is the better one to include in the formulation analyzing Table 6. This formulation is studied to forecast the behavior and compare results in the thermal storage system to use in air conditioning. An optimal formulation is further determined by a balance of the results predicted in this comprehensive analysis, along with environmental compatibility and manufacturing efficacy. This balance is very challenging and can result in a significant number of additives found in various studies. The effort here has been made to limit these extraneous additives to optimize the formulation and minimize environmental impact. The primary phase change of the final formulation, required based on the above comprehensive analysis, is 80.00 wt% Sodium Sulfate Decahydrate (SSD). In addition to 3.00 wt% Borax used as a nucleating agent to assist with crystallization efficiency, 14.00 wt% NH_4Cl is added as a temperature-adjusting additive to ensure the required temperature range of the process. 1.00 wt% Sodium Hexametaphosphate (SHMP) is added as a crystal modifier to ensure the sodium sulfate is held in the proper phase and promotes better crystal formation for improved thermal conductivity. 2.00 wt% CMC is a thickening agent to increase the viscosity and stability of the mixture. This formulation results in a crystallization temperature (T_c) of 9.40°C, a melting temperature (T_m) of 11.70°C, a minimal (T_s) of 0.50°C and a enthalpy (H) of 104.23 kJ/kg, ensuring a precise temperature range required in air conditioning and process to process thermal storage ability.

Table 6. Optimal formulation.

SSD	Borax	NH_4Cl	SHMP	CMC
80.00wt%	3.00wt%	14.00wt%	1.00wt%	2.00wt%

To expand the scope and breadth of the cycling test beyond the initial ten cycles and ensure the enthalpy (H) values of the optimally selected Carboxymethyl Cellulose (CMC) formulation were robust and consistent, each cycling test was extended to a comprehensive fifty cycles to completely validate thermal stability and performance of the formulation. The findings of the expanded analytical testing are documented and depicted in Figure 5, where the extended cycling of enthalpy is visualized, as load cycles continue beyond the original ten cycles at a consistent 100% load.

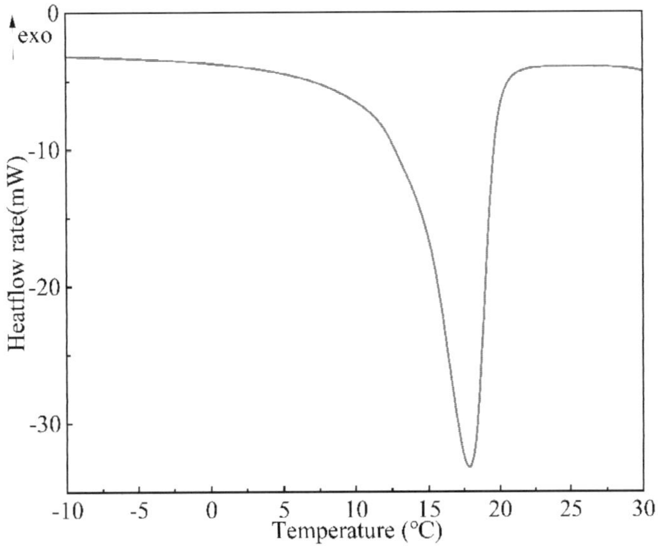

Figure 5. DSC curve after 50 cycles.

Through the detailed and analytical lenses of Differential Scanning Calorimetry (DSC) testing, a compelling and data-driven investigation of the enthalpy (H) of a Carboxymethyl Cellulose (CMC) enhanced formulation was conducted under rigorous thermal cycling conditions. The testing involved fifty (50) thermal cycling processes to expose the formulation to the repeated thermal stresses necessary to gauge the impact of repeatedly cycling through the thermal stress, as indicated in the results established for the enthalpy of Table 7 and presented in full technological disclosure for complete scholarly review.

Table 7. The enthalpy test after cycles.

H(kJ/kg)	H after ten cycles (kJ/kg)	Attenuation rates (%)	H after 50 cycles (kJ/kg)	Attenuation rates (%)
104.23	92.61	11.15	83.71	19.69

The initial enthalpy measured for the formulation was found to be 104.23 kJ/kg, and after we applied the thermally induced ten cycles, we recorded a reduced enthalpy of 92.61 kJ/kg, an 11.15% enthalpy decay rate. The decay continued after the test reached 50 cycles and resulted in an enthalpy of 83.71 kJ/kg, resulting in a cumulative decay of 19.69%, demonstrating the superior thermal stability, enhanced performance, and unrivaled thermal resilience of the formulation.

The observed reduced enthalpy values, though indicative of potential thermal decomposition, remained well within the expected ranges for actual thermal storage applications, particularly air conditioning systems. These results not only successfully validate the stringent standards that must be met for these applications, but also defensibly support the applicability and efficacy of CMC as a formidable thickening agent within the Sodium Sulfate Decahydrate (SSD) based formulation herein. Empirical data collected from this extensive battery of tests ultimately corroborates this formulation's capacity to withstand critical thermal properties after being subjected to repetitive thermal stresses. In so doing, this study further compounds the already overwhelming body of data confirming the suitability of CMC-enhanced formulations for deployment within the new generation of advanced thermal storage devices.

3.5 Thermal conductivity

To extensively characterize the material's thermal conductivity, the Hotdisk TPS 2500S system, which is known for its unrivaled precision, was employed. The instrument was configured to apply a heating power of 30 mW with a measurement duration of 10 s to yield the most accurate assessment. After three consecutive trials, the initial measurement was made at 0.6410 W/m·K, the second at 0.6239 W/m·K, and the final trial produced a measurement of 0.6356 W/m·K. The aggregated results of these experiments yielded an average thermal conductivity value of 0.6335 W/m·K, which was comprised of the results of three consecutive measurements. Detailed numerical data about this intact experiment is systematically amalgamated within Table 8.

Table 8. Tests of thermal conductivity.

Test 1 (W/m·K)	Test 2 (W/m·K)	Test 3 (W/m·K)	Average value (W/m·K)
0.6410	0.6239	0.6420	0.6356

4. Conclusion

In this research, we performed intensive and systematic research to address the critical issue of phase separation in Sodium Sulfate Decahydrate (SSD) based systems that have been a significant challenge for thermal energy storage. Specifically, from a carefully selected list, six thickening agents: Carboxymethyl Cellulose (CMC), Xanthan Gum (XG), Sodium Alginate (SA), Super Absorbent Polymer (SAP), Poly Anionic Cellulose (PAC), Guar Gum (GG), Polyacrylamide (PAM), and Polyacrylic Acid Sodium (PAAS) were studied for their effectiveness. Through a structured experimental plan including the melt blending technique for material preparation, comprehensive thermal property characterization, and extensive cyclic stability testing, we identified the concentration of the most effective of these agents that virtually eliminated phase separation in the SSD systems. Among these, Carboxymethyl Cellulose (CMC) rose as the primary thickening agent, as it negligibly impacted the thermal stability and enthalpy of the SSD system, virtually eliminating phase separation. With the selection process, this formulation with CMC was evaluated extensively. This included evaluation of stability over fifty thermal cycles and detailed thermal conductivity measurement. This confirmed beyond doubt the improvement in the thermal physical properties of the formulation, making it viable as a candidate for use in cold storage air conditioning systems.

Future research must intensify our understanding of the cycle stability of phase change materials, focusing on comprehensive studies of thermal stability challenges of Sodium Sulfate Decahydrate, Na2SO4·10H2O, systems. Given the commercial & industrial transformations that phase change thermal storage technologies are likely to initiate, future work must concentrate on refining formulations that will minimize production costs and enhance economic viability. This may involve a wide range of permutations of ingredient combinations that will hit the optimized balance between high performance and low cost. Such advancements would have the potential to revolutionize the feasibility and utility of phase change thermal storage solutions that could then become mainstays of an enormous range of commercial and industrial applications.

References

[1] Oh K, Shen Z, Kwon S, Toivakka M. 2021. Thermal properties of graphite/salt hydrate phase change material stabilized by nano-fibrillated cellulose. Cellulose. 28, 6845-56.
[2] Cong L, Zou B, Palacios A, Navarro ME, Qiao G, Ding Y. 2022. Thickening and gelling agents for the formulation of thermal energy storage materials - A critical review. Renew Sust Energy Rev.155.
[3] Li YX, Li CC, Lin NZ, Xie BS, Zhang DY, Chen J. 2021. Review on tailored phase change behavior of hydrated salt as phase change materials for energy storage. Mater Today Energy. 22, 26.
[4] Feng G, Xu X, Li G, Li H, Huang K. 2015. Application of Na2SO4 center dot 10 H (2) O thermal storage in air-conditioning cooling water system. MATERIALS RESEARCH INNOVATIONS.19, 983-7.
[5] Zhai XQ, Wang XL, Wang T, Wang RZ. 2013. A review on phase change cold storage in air-conditioning system: Materials and applications. Renew Sust Energy Rev. 22, 108-20.

[6] Xu XF, Zhang XL, Munyalo JM. 2019. Key technologies and research progress on enhanced characteristics of cold thermal energy storage. J Mol Liq. 278, 428-37.
[7] Hirschey J, Gluesenkamp KR, Mallow A, Graham S, editors. Review of Inorganic Salt Hydrates with Phase Change Temperature in Range of 5°C to 60°C and Material Cost Comparison with Common Waxes 2018 2018-07-01; United States. Research Org.: Oak Ridge National Lab. (ORNL), Oak Ridge, TN (United States) Sponsor Org.: USDOE Office of Energy Efficiency and Renewable Energy (EERE).
[8] Oro E, de Gracia A, Castell A, Farid MM, Cabeza LF. 2012. Review on phase change materials (PCMs) for cold thermal energy storage applications. Appl Energy. 99, 513-33.
[9] Veerakumar C, Sreekumar A. 2016. Phase change material based cold thermal energy storage: Materials, techniques, and applications - A review. Int J Refrig-Rev Int Froid. 67, 271-89.
[10] Li Y, Kumar N, Hirschey J, Akamo DO, Li K, Tugba T, et al. 2022. Stable salt hydrate-based thermal energy storage materials*. Composites Part B-Engineering. 233.
[11] Qureshi ZA, Ali HM, Khushnood S. 2018. Recent advances on thermal conductivity enhancement of phase change materials for energy storage system: A review. Int J Heat Mass Transf. 127, 838-56.
[12] Li C, Zhang B, Xie B, Zhao X, Chen J. 2020. Tailored phase change behavior of Na2SO4 center dot 10H(2)O/expanded graphite composite for thermal energy storage. Energy Conv Manag. 208.
[13] Zheng M, Xie C, Liu J, Jin H. 2019. Composite hydrate salt Na2HPO4 center dot 12H (2) O-Na2SO4 center dot 10H (2) O and its thermal storage properties. Emerging Materials Research. 8, 68-76.
[14] Xu X, Zhang X, Zhou S, Wang Y, Lu L. 2019. Experimental and application study of Na2SO4 center dot 10H (2)O with additives for cold storage. J Therm Anal Calorim. 136, 505-12.
[15] Jiang Z, Tie S. 2017. Preparation and thermal properties of Glauber's salt-based phase-change materials for Qinghai-Tibet Plateau solar greenhouses. International Journal of Modern Physics B. 31.
[16] Xing D, Dong C, Li Z, Liu Y, Wang B, Hua Q, et al. 2022. Preparation and Characterization of Sodium Sulfate Decahydrate-Sodium Acetate Trihydrate Shape-Stabilized Phase Change Materials. Energy & Fuels. 36, 4947-56.
[17] Salunkhe PB, D JK. 2017. Investigations on latent heat storage materials for solar water and space heating applications. Journal of Energy Storage. 12, 243-60.
[18] Feng G, Xu X, He N, Li H, Huang K. 2015. Testing research of energy storage system during Na2SO4 center dot 10H (2)O phase change. Materials Research Innovations. 19, 972-7.
[19] Kumar N, Hirschey J, LaClair TJ, Gluesenkamp KR, Graham S. 2019. Review of stability and thermal conductivity enhancements for salt hydrates. Journal of Energy Storage. 24, 100794.
[20] Luo Y, Zou L, Qiao J, Zhang J, Liu K, Wu H, et al. 2022. Boron Nitride-Doped Inorganic Hydrated Salt Gels Demonstrating Superior Thermal Energy Storage and Wearability Toward High-Performance Personal Thermal Management. ACS Appl Energy Mater. 5, 11591-603.
[21] Lin NZ, Li CC, Zhang DY, Li YX, Chen J. 2021. Enhanced cold storage performance of Na2SO4 center dot 10H (2)O/expanded graphite composite phase change materials. Sustain Energy Technol Assess.48.
[22] Hou P, Mao J, Chen F, Li Y, Dong X. 2018. Preparation and Thermal Performance Enhancement of Low Temperature Eutectic Composite Phase Change Materials Based on Na(2) SO(4).10H(2)O. Materials (Basel). 11.
[23] Tie J, Liu X, Tie SN, Zhang JQ, Jiang SY, Tao RX, et al. 2020. Packing and properties of composite phase change energy storage materials based on SiC nanowires and Na2SO4 center dot 10H (2) O. J Therm Anal Calorim. 139, 855-62.

[24] Liu X, Tie J, Zhang J, Jiang S, Tie S. 2019. Preparation and properties of porous diatomite-supported multi-walled carbon nanotubes with $Na_2SO_4 \cdot 10H_2O$-based phase change energy storage composites.SN Applied Sciences. 1, 710.
[25] Li G, Hwang YH, Radermacher R. 2012. Review of cold storage materials for air conditioning application. Int J Refrig-Rev Int Froid. 35, 2053-77.
[26] Muller L, Guarda D, Leiggener J, Muller A, Maranda S, Stamatiou A, et al., editors. Findings on Thickening Glauber's Salt. 6th IIR Conference on Thermophysical Properties and Transfer Processes of Refrigerants (TPTPR) / 13th IIR Conference on Phase Change Materials and Slurries for Refrigeration and Air Conditioning (PCM); 2021 2021Sep 01-03; Electr Network 2021.
[27] Meng L, Ivanov AS, Kim S, Zhao X, Kumar N, Young-Gonzales A, et al. 2022. Alginate-Sodium Sulfate Decahydrate Phase Change Composite with Extended Stability. ACS Appl Polym Mater. 4, 6563-71.
[28] Lin N, Li C, Zhang D, Li Y, Chen J. 2022. Emerging phase change cold storage materials derived from sodium sulfate decahydrate. Energy. 245.
[29] Jin Y, Yonggao Y, Wanhe C, Jingyuan W, Jiufa C. 2022. Preparation and performance optimization of phase change cold storage materials with sodium sulfate hydrate salt. Chemical Industry and Engineering Progress. 41, 5977-85.

Comparison of high temperature oxidation behaviour of high speed steel and Hi-Cr steel

Wanli Li and Yujing Ou*

Institute of Petrochemical Engineering, Lanzhou University of Technology, Lanzhou, 730050, PR China

*Corresponding author`s e-mail: 732302148@qq.com

Abstract. High speed steel and Hi-Cr steel are widely used as one of the important materials for rolls. The oxide layer during rolling plays a key role in the service life of the rolls and the quality of the rolled material. Therefore, it is very important to study the oxidation behaviour of high speed steel and Hi-Cr steel. In this paper, the oxidation behaviour of HSS and Hi-Cr steel is compared in the range of 500-700°C. The oxidation kinetic curve of HSS is dominated by a parabolic law, while the oxidation kinetic curve of Hi-Cr steel is dominated by a logarithmic law. To analyse the effect of their oxidation products on the oxidation behaviour, V_2O_5 in high speed steel plays a destructive role in the oxide layer, while Cr_2O_3 in Hi-Cr steel can effectively reduce the thickness of the oxide layer.

1. Introduction

Rolls have always played an important role in steel production. If 1 kg of rolls is required to roll 1 ton of steel, the consumption of rolls is huge. Therefore, rolls with the development of steel rolling technology and rapid development. To adapt to the high quality of rolled material processing and harsh processing conditions, enhancing the performance and service cycle of rolls is the focus of attention of the rolling industry. The high-temperature oxidation behaviour of the roll material is one of the important research directions of rolls. Commonly used roll materials are HSS (high speed steel) and Hi-Cr steel. Due to the use of work rolls in the process of high temperature, high stress, and the existence of rolling pieces of relative sliding and other special working conditions, the work roll will inevitably cause oxidation and oxide film to influence the rolling efficiency, rolling material quality, production costs and so on. The roll surface oxide layer in the formation can be greater to slow down the wear and tear of the slab on the roll, prolonging the use of the roll cycle, improving the quality of the material to be tied and the wear and tear of the roll will also be reduced. If the thickness of the oxide film is not enough, it will cause thermal adhesion to affect the quality of the product. When the oxide film reaches a certain critical thickness, the increase of thermal stress and growth stress to the oxide film will crack and spalling, because the oxide film is difficult to relax the stress through plastic deformation. To affect the quality of the rolled products in serious cases, even production accidents.

Many studies have investigated the effect of temperature on the oxidation behaviour of high carbon steels. The oxidation of HSS (high speed steel) is not obvious in the oxidation temperature below 400°C. When the oxidation temperature of HSS is 700°C, the change in the surface oxide layer is obvious, and the oxidation products mainly Fe_2O_3 can be detected on the surface of HSS [1-3]. Hao et al. [4] studied the oxidation behavior of high speed steel under dry and wet atmospheres at 650°C and 700°C, and the results showed that the oxide layer was divided into two phases, hematite and (Fe,

$Cr)_3O_4$, under dry conditions, and three phases, magnetite, hematite and (Fe, $Cr)_3O_4$, under wet conditions. Garza-Montes-de-Oca et al. [5] investigated the oxidation behaviour of working roll grade HSS and found that there was little difference between the two oxidation layers in dry and humid atmospheres at 550 and 615°C. Flakes of iron oxides can be observed after 10 min of oxidation and the iron phase can still be observed by XRD. Gaona-Martinez et al. [6] concluded that changing the morphology of carbides affects the rate of oxidation and the state of oxides in HSS, and that tempering HSS reduces its antioxidant properties. The oxidation ability of Hi-Cr steel depends on the oxidation atmosphere and the metallurgical conditions of the sample. Ramire et al. [7] heated Hi-Cr steel at 650°C for 240 min, and the tempering process can strip the Cr element from the carbide into the matrix, thus improving the overall oxidation resistance. Within a water vapour atmosphere, then the oxidation is exacerbated by the water providing O. The products are dominated by Fe/Cr oxides. Zhang et al. [8] suggested that during the oxidation of Hi-Cr steel, there is a state of constant oxidative weight gain, with Fe oxides preferentially emerging from grain boundaries. Drobne et al. [9] suggested that the reduction of Cr content at grain boundaries around eutectic carbides by 4.5-5.5 wt% with respect to the substrate results in the preferential formation of Fe oxides. By comparing the oxidation properties of its HSS and high chrome cast iron, Delaunois et al. [10] believed that its initial oxidation rate of HSS is lower than that of high chrome cast iron at 575°C, and that the oxidation rate of high chrome cast iron will be lower than that of HSS as the reaction proceeds. At present, the oxidation research of HSS and high chromium cast iron on relatively few at different high temperatures. In this paper, the oxidation of HSS and high chromium cast iron at 500, 600 and 700°C for comparison, to explore the composition of its oxide film, microstructure change law and its mechanism, for the follow-up roll material oxide film and friction relationship between the study to lay the foundation, and the production of practical guidance.

2. Experimental methods

The constant temperature oxidation experiments were conducted by discontinuous weighing method. The samples for the oxidation experiments were Φ15 mm×11 mm cylinders of HSS and Hi-Cr steel, whose main chemical compositions are shown in Table 1. The surface of the samples was polished with 60, 400 and 1200 mesh sandpaper using a metallographic grinding and polishing machine (LAP-2MV, Nebel Testing Technology (Shanghai) Co., Ltd., China). The polished specimens were immersed in an anhydrous ethanol solution and ultrasonicated for 15 min using an ultrasonic cleaner (KQ-500, Kunshan Ultrasonic Instrument Co., Ltd., China). The cleaned samples were dried and placed in a chamber heating furnace (KSL-1100, Heifei Kejing Material Technology Co., Ltd., China) for oxidation treatment at temperatures of 500, 600 and 700°C, respectively. The constant temperature oxidative heating time was 100 h. The samples were retained and observed and analysed at oxidation times of 0.5, 1, 2, 5, 10, 20, 40, 60, 80 and 100 h, respectively. The samples were retained, observed and analysed. The samples were weighed using an electronic analytical balance (AUY120, Shimadzu Corp., Japan.) with an accuracy of 0.1 mg, and the microstructures of the surface and cross-section of the samples were observed using an X-ray diffractometer (PANalytical B.V, Holland) and a scanning electron microscope (Thermo Fisher Scientific, Waltham, MA, USA) to observe the microstructure of the sample surface.

Mathematical fitting by $y = a + bx + cx^2$ [11] and $y = alg(bx + c)$ was also performed using the formula fitting software that comes with the origin, where a, b and c are constants.

Table 1. Main chemical composition of HSS samples (wt%).

Sample	Fe	V	Cr	Mo	Mn	C	O
HSS	64.48	6.24	4.48	4.18	0.27	20.36	-
Hi-Cr steel	59.33	0.40	14.49	1.06	0.29	21.12	3.32

3. Experimental results and discussion

3.1 Comparison of oxidation kinetics curves

Figure 1 shows the summary diagram of high-temperature oxidation kinetics of Hi-Cr steel and HSS. Combined with Table 2, the oxidation rate of HSS changes greatly at the oxidation temperature of 500-700°C, while the oxidation rate of Hi-Cr steel changes little and the overall oxidation rate is lower than that of HSS. Figures 2 and 3 show the oxidation kinetics curves for HSS and Hi-Cr steel at 500, 600 and 700°C, respectively. Combined with Table 1 and Table 2, Hi-Cr steel contains a large number of Cr elements that are conducive to slowing down the oxidation process, so that its oxidation rate is significantly lower than the HSS. Based on their oxidation kinetic curves, they were fitted using Origin formula fitting software to obtain the oxidation kinetic curve equations as shown in Figures 2 and 3, both of which have R^2 in the range of 0.88-1. At the same time, the oxidation kinetics of the HSS curve was fitted, and it was found that in the 600°C, the HSS oxidation kinetics curve was parabolic law, while in 500 and 700°C, respectively, it changed from a parabolic law to a linear law. In contrast, the oxidation kinetic curve of Hi-Cr steel could not be fitted at 500°C, and the oxidation rate was essentially unchanged after the initial oxidation weight gain. And the oxidation kinetic curve at 600 and 700°C shows a logarithmic law. When the oxidation kinetic curve is a straight line, it indicates that the oxide layer basically does not have a protective effect, and cannot prevent the external O elements from contacting the metal matrix inside the oxide layer, and the oxidation rate is faster at this time. When the oxidation kinetic curve is parabolic law, the oxide layer has a certain protective effect, and can effectively impede the contact between the external O elements and the metal matrix inside the oxide layer, slowing down a certain oxidation rate. At this time, the factors controlling the oxidation rate are determined by the thermal diffusion of ions. When the oxidation kinetic curve is logarithmic law, the oxidation reaction rate is slower, and the contact between the metal and the external O elements is lower. At this time, the denseness of the oxide layer of Hi-Cr steel is much higher than that of the oxide layer of HSS, so the oxidation rate is significantly reduced.

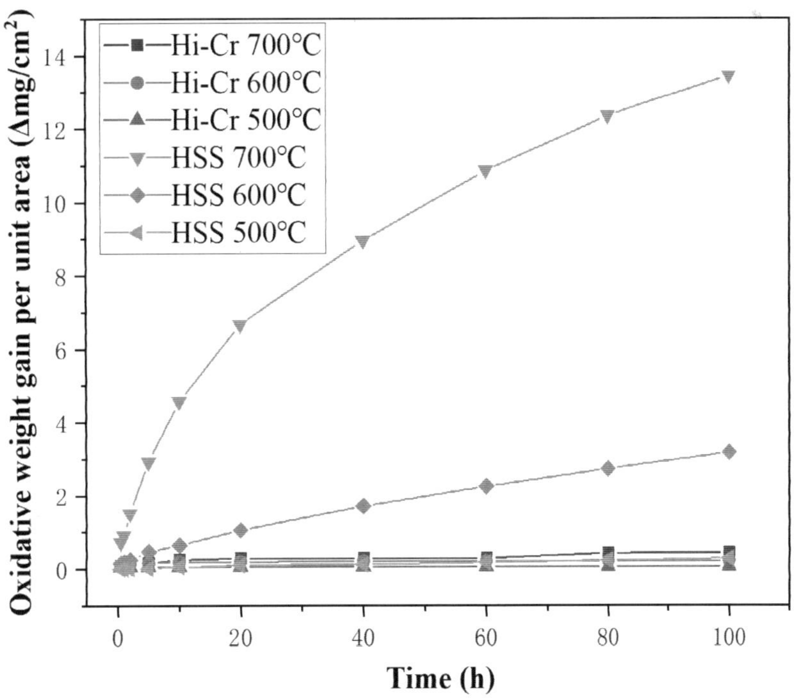

Figure 1. Summary diagram of high-temperature oxidation kinetics of Hi-Cr steel and HSS

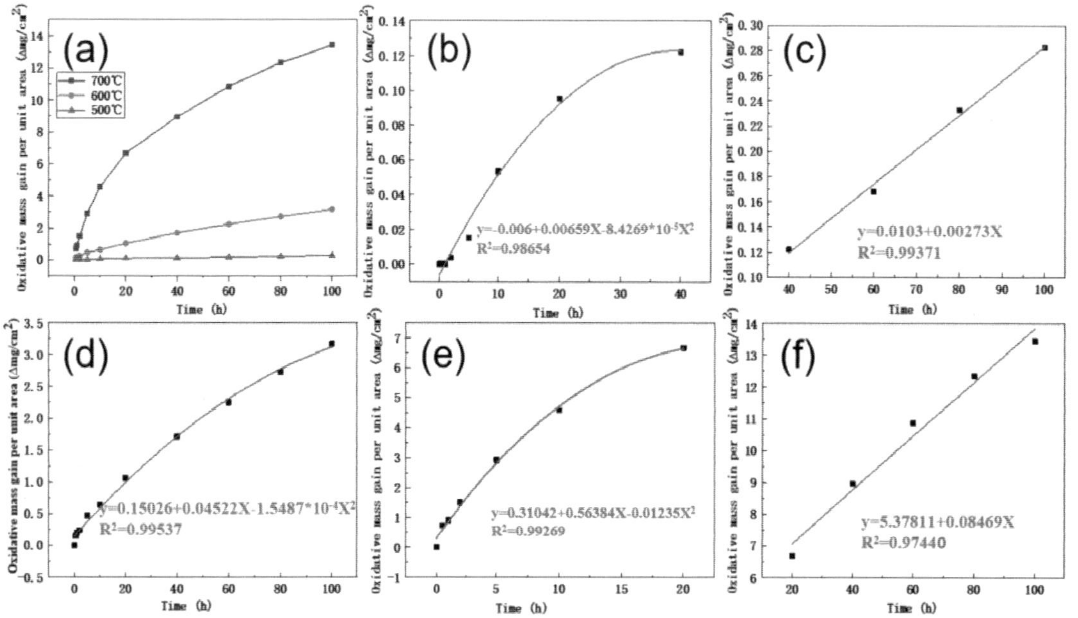

Figure 2. High-temperature oxidation kinetic curves of HSS (a) Summary diagram; (b) 500°C 0-40 h; (c) 500°C 40-100 h; (d) 600°C 0-100 h; (e) 700°C 0-20 h; (f) 700°C 20-100 h;

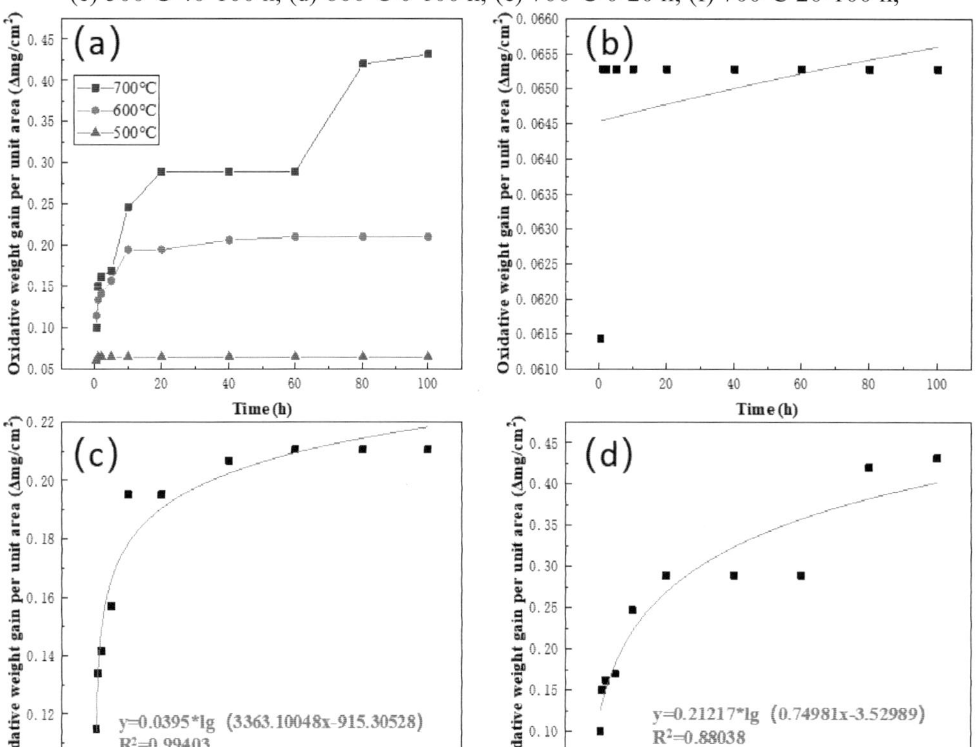

Figure 3. High-temperature oxidation kinetic curves of Hi-Cr steel (a) Summary diagram; (b) 500°C 0-100 h; (c) 600°C 0-100 h; (d) 700°C 0-100 h;

Table 2. Oxidation rates of HSS and Hi-Cr steel.

Sample name	Temperature (°C)	Time (h)	Oxidation rates ($g \cdot m^{-2} \cdot h^{-1}$)			
			1	2	3	Average
HSS	500	100	0.0422	0.0425	0.0442	0.0430
	600	100	0.3364	0.3031	0.3110	0.3169
	700	100	-	1.3022	1.3868	1.3446
Hi-Cr steel	500	100	0.0058	0.0081	0.0058	0.0065
	600	100	0.0287	0.0173	0.0172	0.0210
	700	100	0.0406	0.0403	0.0486	0.0432

3.2 Comparison of oxidation morphology

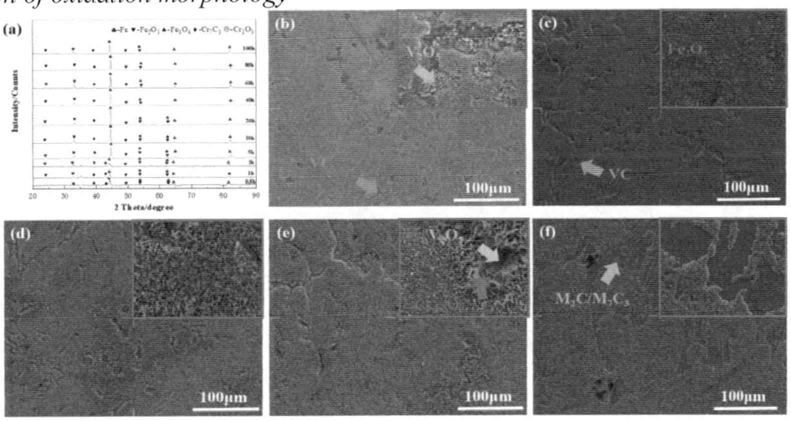

Figure 4. Oxidation surface of Hi-Cr steel at 500°C (a) XRD; (b) 1 h; (c) 10 h; (d) 20 h; (e) 60 h; (f) 100 h.

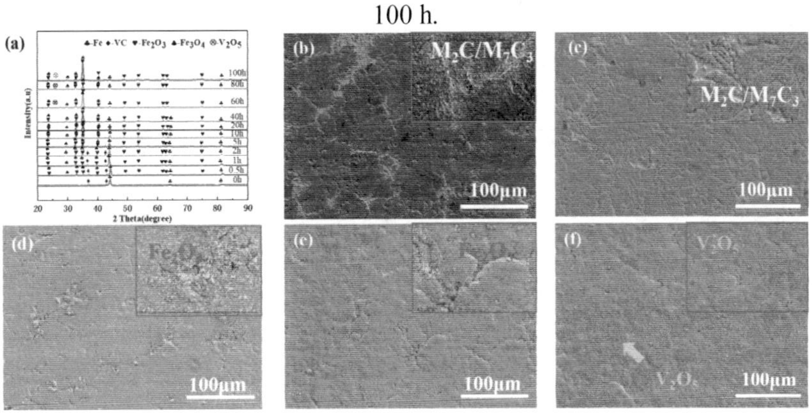

Figure 5. Oxidation surface of HSS at 500°C (a) XRD; (b) 1 h; (c) 10 h; (d) 40 h; (e) 80 h; (f) 100 h.

Comparing the oxidation surface of Hi-Cr steel and HSS at 500°C, the biggest difference in their products by XRD comparison (Figures 4(a) and 5(a)) is Cr_2O_3, which is present in Hi-Cr steel in large amounts and in high speed steel in smaller amounts. These Cr_2O_3 are mainly concentrated on M_7C_3-type carbides, as shown in Figure 4(f). Whereas V oxides are observed in Hi-Cr steel with V_2O_5 concentrated on VC at the beginning of the oxidation (Figure 4(b)), in HSS it is not until 60 h of oxidation that fine V_2O_5 is observed on the surface face of the oxide layer. For the oxide of Fe, the matrix part is covered by a thin layer of α-Fe_2O_3. In the junction of carbide and matrix, the growth degree of α-Fe_2O_3 is significantly higher than of the carbide in other parts of the matrix, which is due to the junction of carbide and matrix is at the grain boundary, where the energy is high, which is conducive to the adsorption and reaction of the O element and accelerates the formation of the oxide.

When the oxidation time is 100 h, Cr_7C_3 carbides can still be clearly observed in Hi-Cr steel, the surface of HSS has basically been covered by Fe oxides.

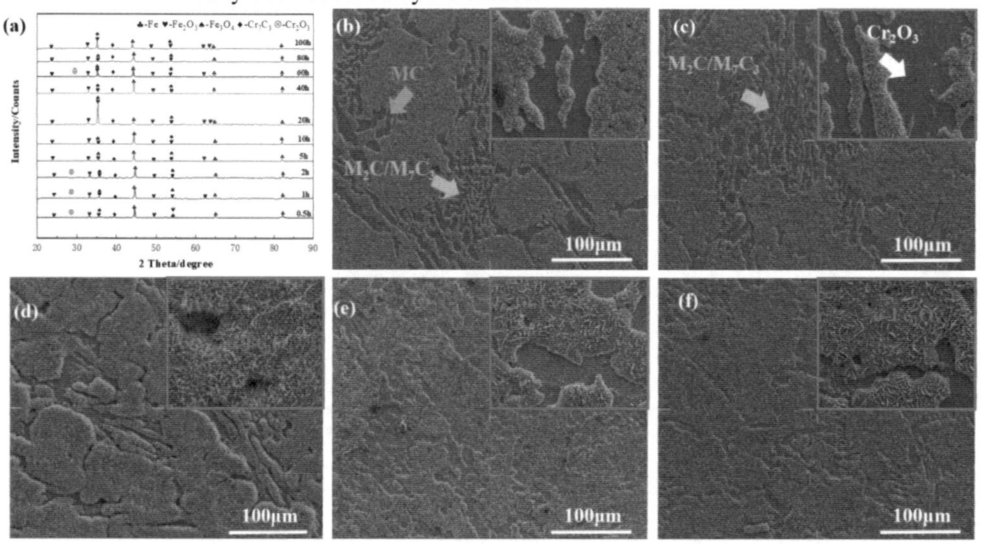

Figure 6. Oxidation surface of Hi-Cr steel at 600°C (a) XRD; (b) 1 h; (c) 10 h; (d) 20 h; (e) 60 h; (f) 100 h.

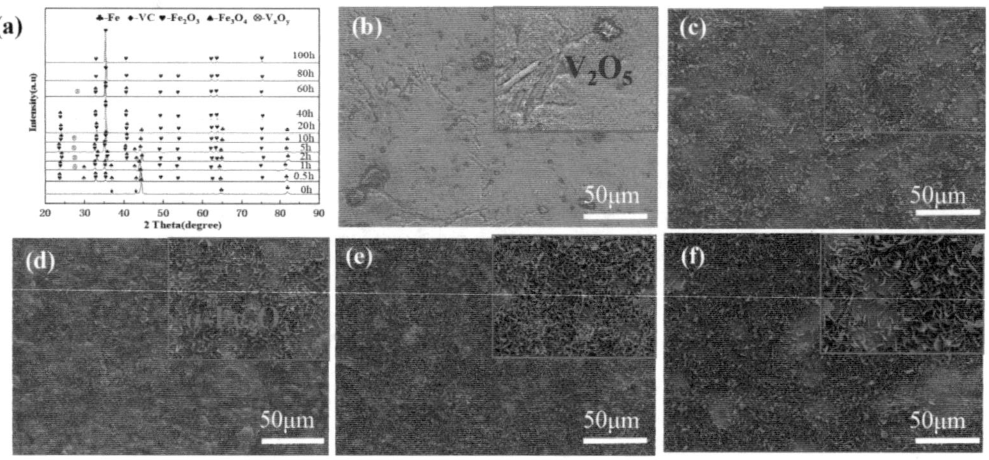

Figure 7. Oxidation surface of HSS at 600°C (a) XRD; (b) 1 h; (c) 10 h; (d) 20 h; (e) 60 h; (f) 100 h.

At 600°C, the degree of oxidation of Hi-Cr steel and HSS was significantly higher than at 500°C. The presence of Cr_2O_3 was found by XRD throughout the Hi-Cr steel. The Cr_7C_3 carbides are clearly visible throughout the oxidation process at 600°C (Figure 6(b-f)), and the Cr_2O_3 on top of them is also clearly observed. In HSS, V_2O_5 can be found in the early stage of oxidation, and V_2O_5 disappears after a period of time, and subsequently disappears after V_2O_5 is found at 60 h because of the progress of oxidation. Combined with Figure 7(b-f), it was found that the oxide layer was gradually covered by flakes of α-Fe_2O_3 as the oxidation proceeded, and it is possible that cracks appeared on the surface of the oxide layer at 60 h leading to V_2O_5 being observed again. Overall the oxide layer in Hi-Cr steel is thin but whisker-like α-Fe_2O_3 is observed, whereas flaky α-Fe_2O_3 is observed in HSS. This is due to the different degrees of oxidation of the two samples, with the flaky α-Fe_2O_3 oxidation to a greater extent and the Hi-Cr steel oxidation to a lesser extent, with needle-and-whisker shaped α-Fe_2O_3 on its surface.

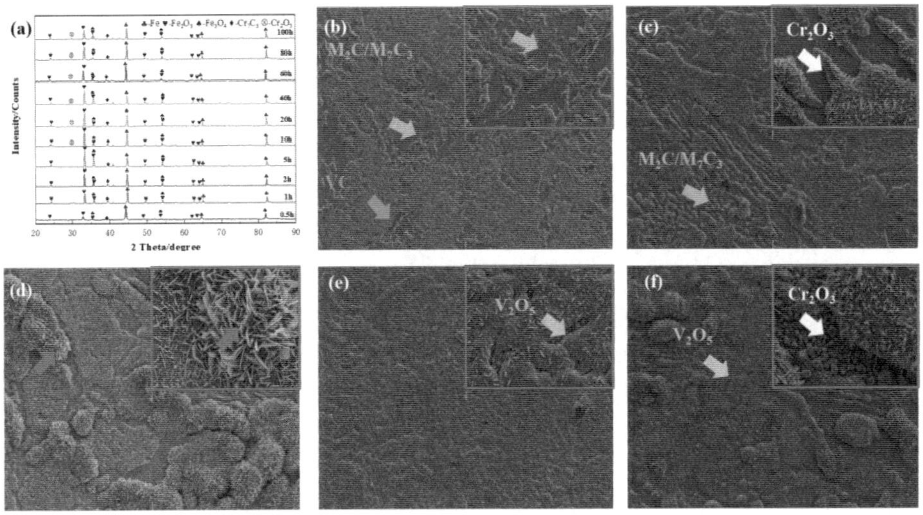

Figure 8. Oxidation surface of Hi-Cr steel at 700°C (a) XRD; (b) 1 h; (c) 10 h; (d) 20 h; (e) 40 h; (f) 100 h.

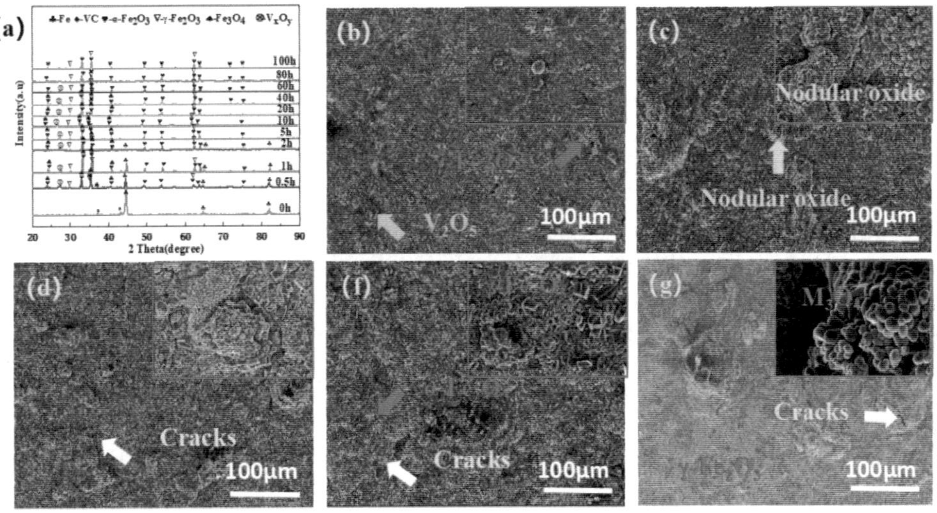

Figure 9. Oxidation surface of HHS at 700°C (a) XRD; (b) 1 h; (c) 10 h; (d) 20 h; (e) 40 h; (f) 100 h.

At the oxidation temperature of 700°C, the surface of the matrix of its Hi-Cr steel and HSS is basically completely covered by the oxide layer. At the time, it can be found by XRD patterns that the oxidation products of Hi-Cr steel are basically the same as those at 500°C and 600°C (Figure 8(a)), but the oxidation products Fe_2O_3 in HSS are divided into γ-Fe_2O_3 and α-Fe_2O_3 (Figure 9(a)). Through Figure 9(f), γ-Fe_2O_3 is loose and porous. On the oxidation surface of the Hi-Cr steel, the iron oxide at the grain boundaries is much higher than that in other parts of the oxide layer (Figure 8(d)), and the Cr_2O_3 gradually grows from dense scales to lumps (Figure 8(f)). Compared with the oxide surface of Hi-Cr steel, the oxide surface of HSS has more oxide defects, such as nodular V_2O_5 oxides (Figure 9(c)), cracks (Figure 9(d)), and loose and porous γ-Fe_2O_3 (Figure 9(f)), leading to a decrease in the protective properties of the oxide layer, accelerating the contact between the external O elements and the internal metal, and making its oxidation rate higher. Eventually, the oxide layer of the HSS cracked, and M_3O_4 (M for Fe, Cr, V, etc) inside the oxide cracks could be observed through the cracks, as in Figure 9(g).

Overall, the oxidation resistance of Hi-Cr steel is better than that of HHS at 500-700°C. By elemental comparison, Hi-Cr steel has high Cr element and low V element characteristics. Cr element in Hi-Cr steel in addition to the distribution in the carbide Cr_7C_3, will be diffusely distributed in the matrix to form a Fe-Cr solid solution, the oxidation process to form a dense scaly oxide Cr_2O_3, so that the overall oxidation rate of Hi-Cr steel is reduced. Meanwhile, the lower V element at 700°C can effectively reduce the defects of the oxide layer, thus maintaining a better oxide layer. However, the lower oxidation rate makes the thickness of the oxide layer lower, which affects its friction mechanism in the friction wear state thus increasing the wear rate. For HHS, it has the characteristics of low Cr element and high V element, so its hardness is high, and it can form a relatively dense oxide layer at 600°C. However, at 700°C, because the presence of V elements will lead to oxidation defects, the formation of the oxide layer is not dense and easy to break or even fall off.

4. Conclusions

(1) In 500-700°C, Hi-Cr steel oxidation resistance is stronger than HSS. The oxidation kinetic curve of Hi-Cr steel is dominated by logarithmic form, while HSS is dominated by parabolic law.

(2) The oxidation products of high-chromium steel and high-speed steel are basically the same, both of which are mainly Fe oxides. However, Cr_2O_3 in Hi-Cr steel will reduce the overall oxidation rate, and V_2O_5 in HSS will form oxidation defects to improve the overall oxidation rate.

References

[1] Wang Y, Mao B, Chu S, et al. (2023) Revealing the Oxidation mechanism of High vanadium High-speed Steel Using Multi-scale Characterization. Corrosion Science, 225:11592.
[2] Yin Y, Sun J, Teng S, et al. (2016) Oxidation behavior of high-speed steel used for hot roll. Oxidation of Metals, 86:45-57.
[3] Deng G.Y, Tieu A.K, Su L.H, et al. (2019) Characterizing deformation behaviour of an oxidized high speed steel: Effects of nanoindentation depth, friction and oxide scale porosity. International Journal of Mechanical Sciences, 155:267-285.
[4] Hao L, Li T, Z Xie, et al. (2020) The Oxidation Behaviors of Indefinite Chill Roll and High Speed Steel Materials. Metal, 10(8), 1095.
[5] Garza-Montes-de-Oca N.F, Colas R, Rainforth W.M. (2011) High temperature oxidation of a work roll grade high speed steel. Oxidation of Metals, 76:451-468.
[6] Gaona-Martinez M. J, Guajardo-Lopez L. N, Perez-Patino J. A, et al. (2019) Effect of Temperature on the Oxidation Behaviour of a Roll-Grade High-speed Steel During Thermal Cycling. Oxidation of Metals, 91:641-656.
[7] Ramirez-Ramirez. J.H, Colas. R, Garza-Montes-de-Oca N.F. (2013) High Temperature Oxidation of a Work Roll Grade High-chromium White Cast Iron. Journal of Iron and Steel Research, International, 20(10):122-129.
[8] Zhang B, Li H, Yuan N, et al. (2023) A Pre-oxidation Scheme Based on the Evolution Difference of Oxides on the Surface of High-chromium Steel under Different Heating Conditions. Corrosion Science, 213:11103.
[9] Drobne M, Klancnik U, Farjfar P, et al. (2021) Roll of Oxidation and Microstructure Constituents on high chromium steel work roll surface degradation. Journal of Materials Research and Technology, 12:186-182.
[10] Delaunois F, Stanciu V. I, Megret A, et al. (2022) Oxidation and Wear Behavior of High-speed Steel and Semi-high-speed Steel Used in Hot Strip Mill. The International Journal of Advanced Manufacturing Technology, 119:677-689.
[11] Li T. (2003) High Temperature Oxidation and Thermal Corrosion of Metals. Chemical Industry Press. The Beijing.

Evaluation of the mechanical properties of stiffened composite panels under consideration of water load

Junchao Yang[1,*], Xiangming Chen[1], Zhe Wang[1], Peng Zou[1] and Qiang Zhu[2]

[1]National Key Laboratory of Strength and Structural Integrity, Aircraft Strength Research Institute of China, Xi'an Shaanxi 710065, China
[2]China Special Vehicle Research Institute, Hubei Jingmen 448035, China

[*]Corresponding author's e-mail: yangjc008@avic.com

Abstract: To meet the strength evaluation requirements of a typical stiffened composite panel of the lower fuselage of a new generation amphibious aircraft underwater load, the experimental testing technique and finite element analysis method were investigated. Firstly, according to the loading and support requirements of the typical stiffened composite lower fuselage panel, a testing device for combined axial and pneumatic loading was designed and developed. Then, a finite element model of the stiffened composite plate and the combined load testing device were created. Based on the experiments and finite element analysis, the static strength of the stiffened composite plate was verified to meet the load-bearing capacity requirements, and the strain behavior during loading was determined. Compared with the experimental results, the maximum strain error at the skin is 6.16%, and the maximum strain error at the stiffener is -11.6%. This confirms the effectiveness of the finite element model and also confirms that the test technique and equipment can accurately simulate in-plane axial loading and out-of-plane water loading. Finally, based on the validated finite element model, the load-bearing capacity of the stiffened composite panel under combined loads was predicted.

1. Introduction

An amphibious aircraft is a fixed-wing aircraft that can glide, take off, and land on both land and water [1]. When testing the structural strength of the fuselage, in addition to conventional loads such as aerodynamic loads, propulsion system loads, and inertial loads, the surface reaction force carried by the fuselage structure during takeoff and landing on water must also be taken into account [2,3]. Due to their excellent weight reduction properties, unique design capability, and good technology, composite materials are increasingly used for the structure of aircraft panels [4,5], such as fuselage and wing panels. The new generation of amphibious aircraft and other surface aircraft have higher weight requirements and are considering using composite materials to replace part of the fuselage structure made of metal to further reduce the structure's weight and improve the load-bearing capacity. However, the performance of the composite material between the layers is weak [6]. When applied to the fuselage of amphibious aircraft, there is a risk that the interlayers will be damaged by the external water load, resulting in a significant reduction in structural load-bearing capacity and a serious threat to flight safety. Therefore, it is important for designers to investigate and verify the mechanical properties of typical composite panel structures of the fuselage under water loading.

Researchers have done a lot of research on direct water testing of amphibious aircraft. In 2015, Panciroli [7] conducted a study on water impact tests in free fall with wedge components. The results showed that most of the kinetic energy is converted into kinetic beam energy and potential energy required for the buoyancy of the liquid surface when the structure is immersed in water. In 2019, Dong [8] conducted experimental research and numerical simulations on the impact of a free-falling wedge profile after immersion in water, discussed the influence of gravity, impact velocity, oblique angle of buoyancy, and the effect of structural mass, and pointed out that the stiffness of the wedge structure has a great influence on the water load.

Direct water stress tests have a long implementation period, high costs, and poor repeatability and reliability of the results of water stress tests under wave conditions. Therefore, the researchers proposed a quasi-static pressure loading test method that simulates dynamic water loading [9]. Shang [10] developed an elastomer sandwich water loading method based on the characteristics and loading requirements of water loading of amphibious aircraft fuselage structures and verified the feasibility of the mechanical model and loading method through finite element analysis and experiments. However, the water load simulation method of the elastomer sandwich leads to a large out-of-plane stiffness of the skin, and the water load simulation used in the testing of the panel may increase the out-of-plane performance of the composite panel, thereby overestimating the strength of the panel and creating design risks.

In this work, a combined axial load and airtightness testing device was designed and developed according to the strength verification requirements of typical stiffened composite panels. An independently supported airbag was used to simulate the effect of water load, and lateral support components were used to support the panels laterally to ensure the accuracy of load and boundary conditions. Then, the finite element model of the test is created, and the results of the test and finite element analysis are used to check whether the structure meets the limit load requirements. Finally, the load-bearing capacity of the stiffened slab is predicted using the verified finite element model.

2. Test pieces & load

2.1. Test pieces

The composition and size of the test piece are shown in Figure 1, including the skin, the stiffener, the horn rod, the corner piece, the frame, and the aluminum box.

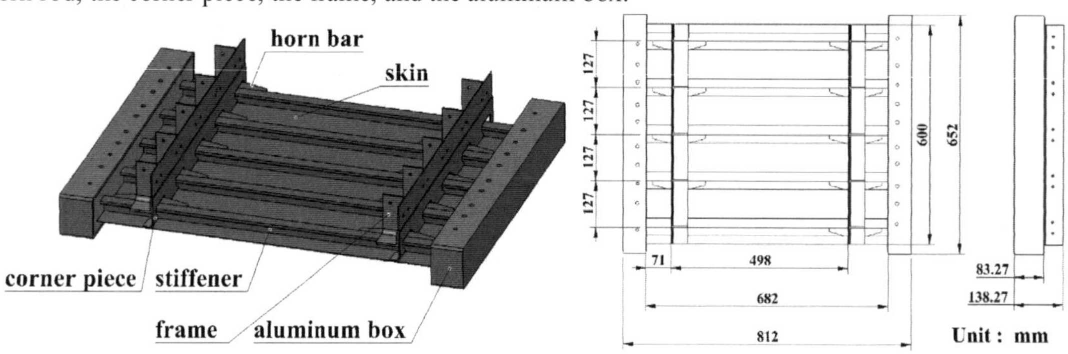

Figure 1. Composition and configuration of the stiffened panel.

The material of the corner piece, frame, and aluminum box is 7075-T73 aluminum alloy. The skin, stiffener, and horn rod are made of unidirectional CCF800H/AC531 carbon fiber tape, and the nominal thickness of a single layer is 0.14 mm. The mechanical properties are listed in Table 1, and the layer sequences of the composite structures are listed in Table 2.

2.2. Test loads

The load includes the axial load F inside the plane and the uniform pressure p outside the plane, as shown in Figure 2. The magnitude of the base load is p = 50 kPa, F = 15 kN. It must be checked whether the test body made of stiffened panels can withstand the design limit load (420 % base load, p = 210 kPa, F = 63 kN).

Table 1. The mechanical property parameters of CCF800H/AC531.

Property	Value	Property	Value
Longitudinal elastic modulus, E_1/ GPa	154.5	Transverse tensile strength, Y_T/MPa	64.7
Transverse elastic modulus, E_2/ GPa	8.89	Transverse compressive strength, Y_C/MPa	214
Poisson's ratio, v_{12}	0.336	Shear strength, S_L/MPa	149
Shear modulus, G_{12}/ GPa	4.33	Mode I fracture toughness, G_{IC}/(kJ·m^{-2})	0.194
Longitudinal tensile strength, X_T/MPa	2538	Mode II fracture toughness, G_{IIC}/(kJ·m^{-2})	2.447
Longitudinal compressive strength, X_C/MPa	1378	Fitting parameter, η	0.921

Table 2. Layer sequences of composite structures.

Position	Layer sequences
Skin	1.54 mm: $[\pm45/0_2/90/0_{1/2}]_s$
Stiffener	2.38 mm: $[(\pm45)_2/0/90/0/90/0_{1/2}]_s$
	3.50 mm: $[(\pm45)_2/0_2/90_2/0_2/90_2/0_{1/2}]_s$
Horn bar	2.94 mm: $[(\pm45)_2/0_2/90_2/0_2/90_{1/2}]_s$

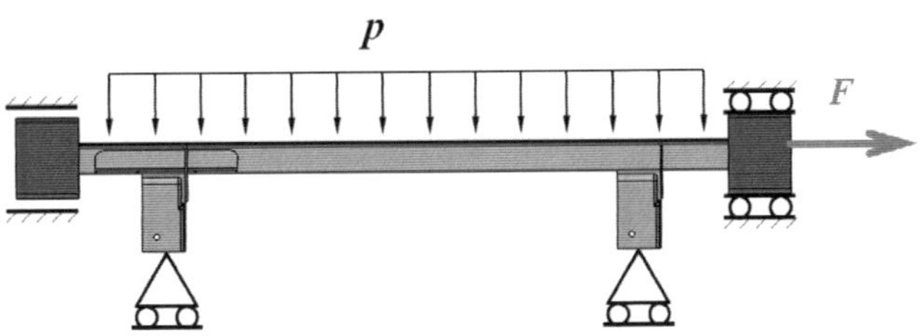

Figure 2. Schematic diagram of the load state.

3. Test load application method

The combined axial/outer plane load testing device was designed according to the load condition of the test specimen. According to the structure of the equipment, it can be divided into test specimen components, axial load components, airtight load components, and lateral support components, as shown in Figure 3.

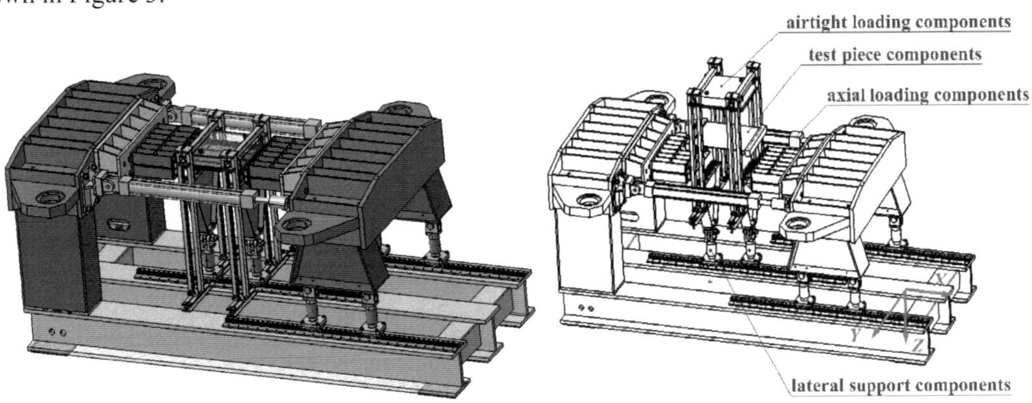

Figure 3. Exploded diagram of the test facility.

3.1. Axial load loading method

The axial loading components, as shown in Figure 4, consist of the bottom beam, fixed end column, fixed end beam, extension piece, corner box, actuator cylinder, active end beam, active end column, and adjustment support, which can slide axially (X-axis) on the guide rail of the bottom beam. The axial load unit is a self-balancing system. Both ends of the test specimen are bolted to the corner box, and the movable end beam is driven to slide axially on the guide rail by controlling the retraction and extension of the actuator cylinder, thereby exerting an axial tension/compression load on the test specimen.

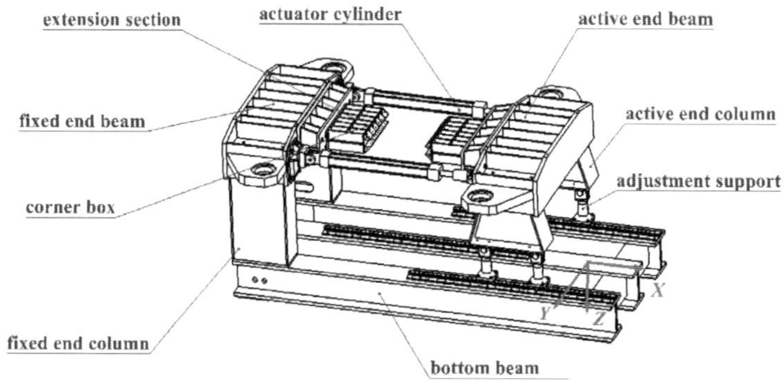

Figure 4. Axial loading components.

3.2. Water load loading method

The water load corresponded to the evenly distributed pressure on the skin. The components of the airtight load are shown in Figure 5. They consist of an airtight box, four support columns, and four lower support beams.

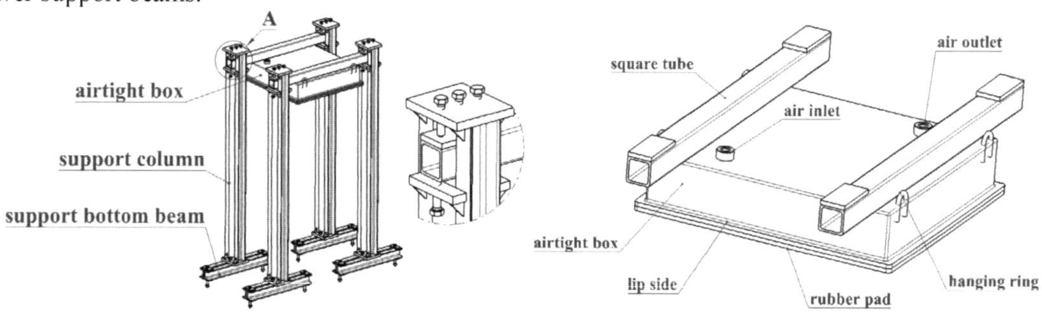

Figure 5. Airtight loading components.

As shown in Figure 5, two square tubes, two standard air inlets and outlets, and four hanger rings are welded to the airtight box, which is used for height adjustment, airway connection, and lifting, respectively. A rubber pad is glued to the lip side to prevent scratches on the skin surface during loading. A balloon cloth is glued to the inside of the airtight box.

4. Finite element model

The finite element model of the test system is shown in Figure 6. Rivet and bolt connections are simulated by the Fasteners unit. A COHESIVE unit is inserted between the stiffener and the skin to simulate the interface debonding behavior. A TIE connection is placed between the stiffener and the horn rod.

The grid of the test component is set to 5 mm, the grid of the connected device to 10 mm, and the grid of the other test devices to 40 mm. The load and boundary conditions are set as follows: the Z-displacement (U3 = 0), the y-displacement (U2 = 0) in the center line of the lower beam, and the X-displacement (U1 = 0) in the sideline near the fixed end of the tension and compression platform are restricted. Apply a uniform pressure of 0.05 MPa to the skin surface. Two loading points are placed at the fixed end and coupled with the beam at the fixed end, and a concentration force of -7.5 kN is applied in the X-axis direction. Two load points are placed at the active end and coupled to the active end beam, and a concentration force of 7.5 kN is applied in the direction of the X-axis.

Journal of Physics: Conference Series **2842** (2024) 012035 doi:10.1088/1742-6596/2842/1/012035

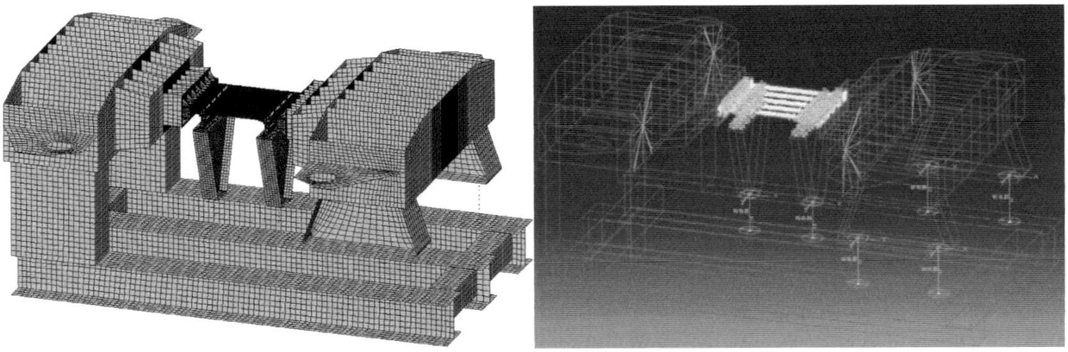

Figure 6. Finite element model of the experimental system.

5. Results & discussion

5.1. Check the fixture strength before the test

The strength of the testing device is checked using the finite element model of the test system created in Chapter 4. The stress distribution (p = 210 kPa, F = 63 kN) of the test rig is shown in Figure 7, and the maximum Mises stress was 104 MPa. The worst-performing material in the test fixture is Q345 (yield stress 345 MPa), so the strength of the fixture meets the test requirements.

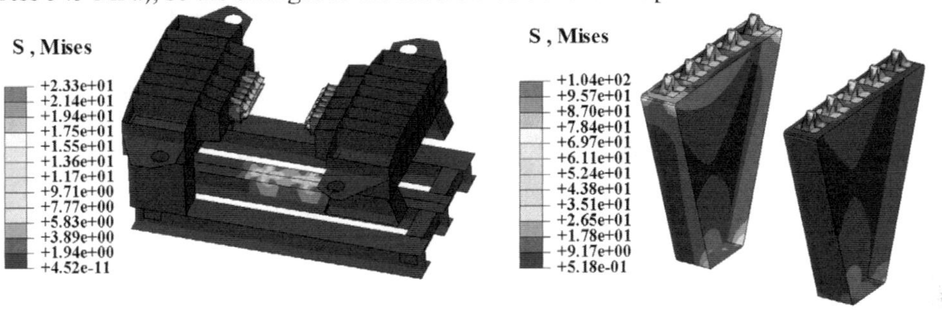

Figure 7. Stress cloud map of the test facility.

5.2. Strain analysis

Before the test, the finite element analysis showed that the main deformation of the stiffened plate was the bending deformation of the skin and the beam of the center section (Figure 8). Therefore, 11 strain gauges were glued on the skin and beam (the strain gauge number of the smooth side of the skin was in the back-to-back glued transverse (y-direction) brackets, and the beam was only glued axially (x-direction) on the outside of the upper edge strip, as shown in Figure 9) to measure the strain of the test specimen.

Figure 8. Displacement cloud map.

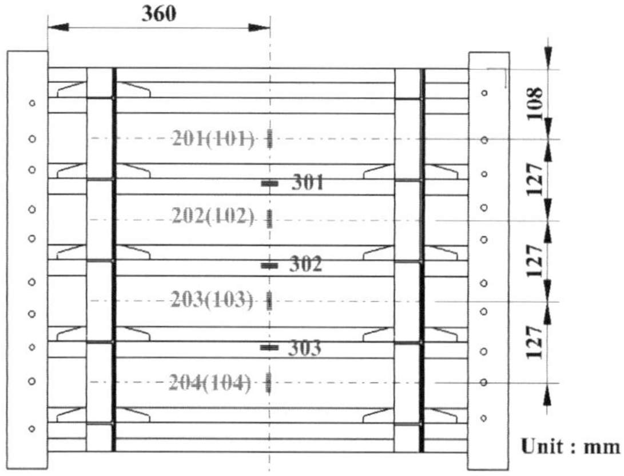

Figure 9. Strain gauges distributions and numbers.

The photo of the test site can be seen in Figure 10. The 100% benchmark load test was carried out first. The strain curve is shown in Figure 11. The strain is linear and symmetrical.

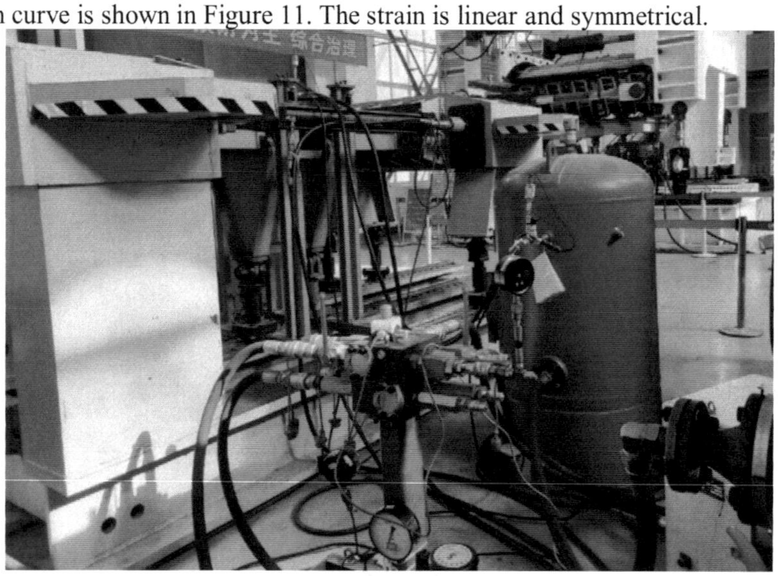

Figure 10. Photo of the test site.

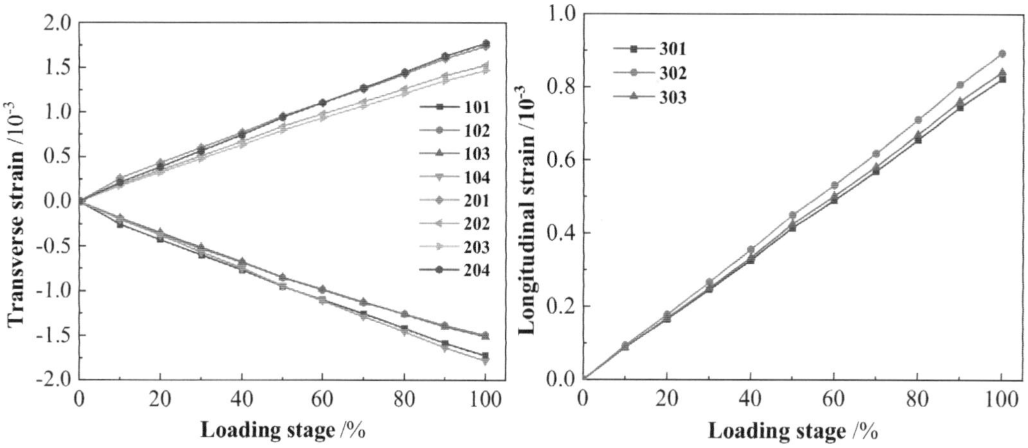

Figure 11. Strain-load curves of critical measurement points (p = 50 kPa, F = 15 kN).

Then, the 420% benchmark load test was performed. The noise was heard when the load reached 380% and 400% of the benchmark load. After the test, non-destructive testing showed that there were no abnormalities in the stringer, skin, and stringer-skin interface in the test area. The sound heard during the test could be the sound of the end bolt crushing the filler resin. The comparison of the test and finite element results is shown in Figure 12 and Table 3.

The strain of the skin determined by the test agrees well with the results of the finite element analysis, with a maximum error of 6.16%. The finite element analysis results of the strains on the stringer are all lower than the test values, with a maximum error of 11.6%. The possible reasons are that the stringer is small, the cross-section is semi-closed, the manufacturing process is difficult, the fiber laying angle is difficult to control, and the R zone of the upper edge strip is prone to wrinkling and other manufacturing defects, resulting in the stiffness of the stringer of the test piece being lower than the theoretical stiffness. Therefore, the result of the test is slightly larger than that of the finite element analysis.

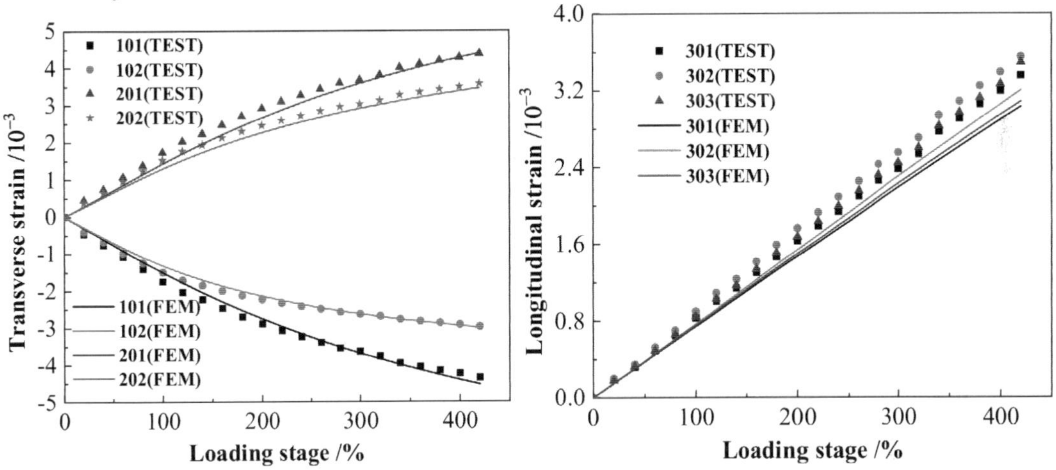

Figure 12. Strain-load curves of critical measurement points (p = 210 kPa, F = 63 kN).

Table 3. Comparison of strain values (p = 210 kPa, F = 63 kN).

Position	Gauge number	Experimental value/με	Analytical value/με	Error/%
Skin	101	-4335	-4526	4.41%
	102	-2967	-2999	1.08%
	103	-3056	-2957	-3.24%
	104	-4336	-4603	6.16%
	201	4380	4398	0.41%
	202	3576	3460	-3.24%
	203	3517	3421	-2.73%
	204	4317	4505	4.35%
Stiffener	301	3357	3032	-9.68%
	302	3549	3200	-9.83%
	303	3491	3086	-11.60%

As can be seen in Figure 12, the slope of the strain curve of the stiffener did not change significantly during the loading process because the stiffener resisted bending deformation by the upper and lower edge strips, while the distance between the upper and lower edge strips was large, and the influence of deformation on the distance between the upper and lower edge strips was almost negligible during the loading process, so the bending stiffness of the stiffener (i.e., the slope of the strain curve) hardly changed. However, when the skin is loaded with 150% base load, the slope of the strain curve becomes smaller and smaller as the load increases. This is because as the skin is loaded, the curvature of the neutral surface gradually increases, resulting in an increase in bending stiffness, which causes the slope of the strain curve of the skin to decrease gradually as the load increases.

The above finite element and test results show that the typical stiffened composite panel structure can meet the limit load requirements. At the same time, the test results confirm the validity of the finite element model.

5.3. Carrying capacity prediction

The finite element model was used to predict the load-bearing capacity of the stiffened composite panel. The results show that the stiffened panel will fail at a load of p = 368 kPa and F = 110.4 kN. The failure mode is shown in Figure 13. Due to the sudden stiffness change at the web of the stiffener, the local stress concentration of the skin and the sudden stress change at the interface between the stiffener and the skin lead to damage to the skin and tearing of the interface between the stiffener and the skin after the load is high.

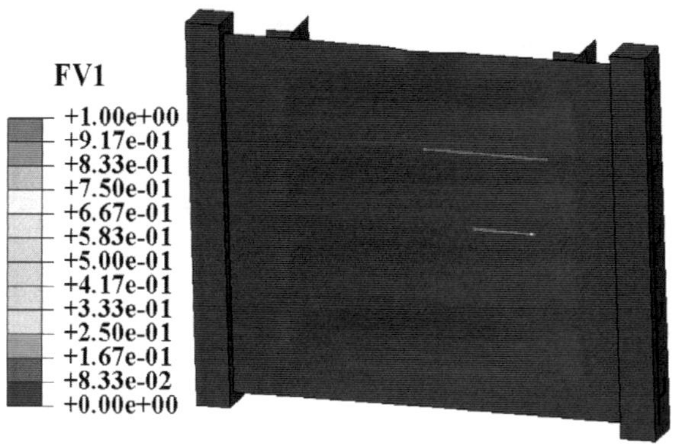

Figure 13. Failure mode (p = 368 kPa, F = 110.4 kN).

6. Conclusions

A series of loading and support devices have been developed to evaluate the strength of a typical stiffened composite panel under water loading. The fixture can realize the combined loading by in-plane axial load and out-of-plane uniform pressure and support the frame laterally.

The finite element model of the test system is created, and the strength of the test fixture is checked to ensure the test's safety. Compared with the test results, the maximum strain error of the skin is 6.16%, and the maximum strain error of the stiffener is -11.6%, which confirms the validity of the model and predicts the load-bearing capacity of the stiffened plate on this basis.

The finite element analysis method and the test prove that the typical composite plate structure of the hull under the hull meets the limit load requirements.

Acknowledgments

The research was supported by the National Key R&D Program of China (2019YFA0706800), the Aviation Science Foundation of China (2020Z055023002), and Special scientific research and technology research project for civil aircraft (MJZ5-3N21).

References

[1] ZHANG Y J, CUI B, WANG M Z, et al. (2023) Research progress of amphibious aircraft water landing test and theoretical analysis methods. Acta Aeronautica et Astronautica Sinica, 44:528665
[2] Campbell J C, Vignjevic R. (2012) Simulating structural response to water impact. International Journal of Impact Engineering, 49: 1-10.
[3] Qiu L J, Song W B. (2013) Efficient decoupled hydrodynamic and aerodynamic analysis of amphibious aircraft water takeoff process. Journal of Aircraft, 50: 1369-1379.

[4] DU S Y, GUAN Z D. (2008) Strategic considerations for development of advanced composite technology for large commercial aircraft in China. Acta Materiae Compositae Sinica, 25: 1-10.
[5] ZHAO T, LI Y, ZHANG C, et al. (2022) Fundamental mechanical problems in high-performance aerospace composite structures: State-of-art review. Acta Aeronautica et Astronautica Sinica, 43: 526851.
[6] CAO Y, ZHANG C. (2022) Impact damage behavior of thin-ply composites: A review. Acta Aeronautica et Astronautica Sinica, 43: 525323.
[7] PANCIROLI R, SHAMS A, and PORFIRI M. (2015) Experiments on the water entry of curved wedges: High-speed imaging and particle image velocimetry. Ocean Engineering, 94: 213-222.
[8] DONG C R, SUN S L, SONG H X, et al. (2019) Numerical and experimental study on the impact of a free-falling wedge on water. International Journal of Naval Architecture and Ocean Engineering, 11: 233-243.
[9] SEDDON C M, MOATAMEDI M. (2006) Review of water entry with applications to aerospace structures. International Journal of Impact Engineering, 32:1045-1067.
[10] SHANG H X, WANG H, HE Y Z, et al. (2019) Simulation method of water load in structural strength test of amphibious aircraft. Science Technology and Engineering, 19: 371-376.

Journal of Physics: Conference Series **2842** (2024) 012036 doi:10.1088/1742-6596/2842/1/012036

Experimental and simulation study on mechanical properties of ZK60 magnesium alloy after accumulative roll bonding

Shaojie Li[1], Jianfeng Jin[1,2*], Yongbo Wang[1], Hucheng Pan[1] and Gaowu Qin[1]

[1]School of Materials Science and Engineering, Northeastern University, Shenyang, Liaoning, 110819, China

[2]State Key Laboratory of Special Rare Metal Materials, Northwest Rare Metal Materials Research, Shizuishan, Institute Ningxia Co. Ltd., 753000, China

[*]Corresponding author's e-mail: jinjf@atm.neu.edu.cn

Abstract: The mechanical properties of two- and four-layered wrought ZK60 Mg alloy sheets after two passes of accumulative roll bonding (ARB) with different reduction were studied and the microstructure were analyzed. The four-layered ZK60 sheets after ARB with 50% reduction demonstrate better strength-ductility combinations with yield strength of 223 MPa, ultimate tensile strength of 327 MPa, and total elongation of 23.7%, compared with the initial ZK60 of 222 MPa, 291 MPa and 10.0%, respectively. The average grain size of four-layered ARB ZK60 sheet is reduced from 21.8 μm of the initial sample to 4.4 μm. Using a micromechanical finite element model, the strain hardening behavior of ZK60 sheet was successfully predicted. The predictions were in good agreement with the experimental results. The simulation results reveals that the high performance of ZK60 sheets may be contributed by the lamella structure formed by the fine and coarse grains, which can avoid the local stress concentration and inhibit initiation and propagation of the microcracks. These findings provide a theoretical basis for the structural optimization design of ZK60 Mg alloys.

1. Introduction

With the increasing sustainable and energy-saving concerns, lightweight becomes a crucial issue in material designing for productions in automobiles, aerospace and other industries. As the lightest metallic structural material available today, Mg alloy is the most prosperous materials candidate in these fields, which has a high specific strength, good electromagnetic shielding effect, shock absorption, and recyclability. However, due to the HCP structure of Mg, the limited slip systems at room temperature makes plastic deformation and forming process difficult, which limits its applications [1]. Thus, alloying and severe plastic deformation are commonly employed to enhance the ductility of Mg alloys by grain refinement and texture weakening. However, it is difficult to avoid the strength-ductility trade-off dilemma caused by grain refinement.[2]. Recently, some experimental results [3, 4] show that heterostructure can facilitate the synergy of strength-ductility in Mg alloys. Heterogenous bimodal structure Mg alloys can be prepared from the hot extrusion [5] and hard plate rolling [3] processes by controlling the severe plastic deformation process parameters, in which the bimodal structure is composed by dynamic recrystallized fine grains (FGs) and the remaining coarse grains (CGs). In this work, the two- and four-layered wrought ZK60 Mg alloy sheets are prepared by two passes of accumulative roll bonding (ARB) process with different reduction and the mechanical properties of ZK60 after ARB are studied. The lamella microstructure (LS) is observed and the deformation behaviors of the ZK60 are analyzed by a micromechanical finite element model. These finds may provide a guidance on optimize ZK60 alloy through the microstructural designing.

2. Experimental procedure and simulation setting

The composition of the commercial Mg-Zn-Zr alloy (ZK60) was 6 wt.% Zn, 0.5 wt.% Zr and balanced Mg respectively and the thickness of initial wrought sheet was 2 mm. The sheet was cut into the rectangular pieces with the size of 120 × 100 mm^2. After surface treatment, two layers of the pieces were stacked together and accumulatively rolled twice at 623K with 30%, 50% and 70% reduction and the samples were rotated 180° between each rolling passes. Four layers of the pieces underwent the same ARB processing with 50% reduction. After the rolling, the sample was further annealed by holding at 573K for 10 minutes.

To investigate the microstructural characteristics, optical microscopy (OM) was employed. Prior to the OM analysis, the specimens were subjected to mounting and polishing processes, followed by etching using a picric acid solution. Make specimens from the rolling direction of the ARB sheets for tensile testing. These tensile specimens featured a gauge length of 25 mm and a cross-sectional area of 6 × 2 mm^2. The tensile tests were carried out at room temperature conditions utilizing an AG-X-PLUS100KN testing machine with a crosshead speed of 1 mm/min.

On the other hand, a micromechanical 2D finite element (FE) model is implement by Simdroid Multiphysics Software, Beijing Internet Based Engineering Co., Ltd. [6] and the work hardening behavior of ZK60 alloys after ARB is predicted, associated with local stress/strain distribution. In the FE model, the microstructure of ZK60 alloy is abstracted into three parts: fine grain (FG) zone, coarse grain (CG) zone and the interface affected (IA) zone between the CG and FG with the interface width $\Gamma = (\mu/\sigma_y)^2 b$ [7], where μ is the shear modulus, σ_y is the yield strength of CG and b is the Burgers vector, with the values of 16.3 GPa, 220 MPa and 0.25 nm [8], respectively, corresponding to the Γ of 1.37 μm. The constitutive stress-strain input of the FG zone (σ_{FG}) derives from the experimental data of ZK60 with an average grain size of 2.8 μm [9]. For the CG zone, the constitutive input (σ_{CG}) comes from the experimental data of the ZK60 sample before ARB process. Incorporating the back-stress effect (σ_{BS}) due to the pileups of geometrically necessary dislocations [10] and the σ_{CG}, the constitutive input of the interface affected zone (σ_{IA}) is given by $\sigma_{IA} = \sigma_{CG} + \sigma_{BS}$ with the $\sigma_{BS} = M\mu bN/d_{CG}$ [10], in which the Taylor constant (M) is 3.06 [8] and the CG size (d_{CG}) is 23 μm. The N represents the number of pile-up dislocations within the interface affected zone, described by $N = N_0\ [1 - \exp(-\xi\varepsilon_p/bN_0)]$ [10], where N_0 is the saturation dislocation number and ξ is the spacing between the neighboring pileups with the values of 200 and 200 nm by fitting the experimental data of necking point, respectively. Further details about the FE model setup can be found in our previous work [8].

3. Results and discussion

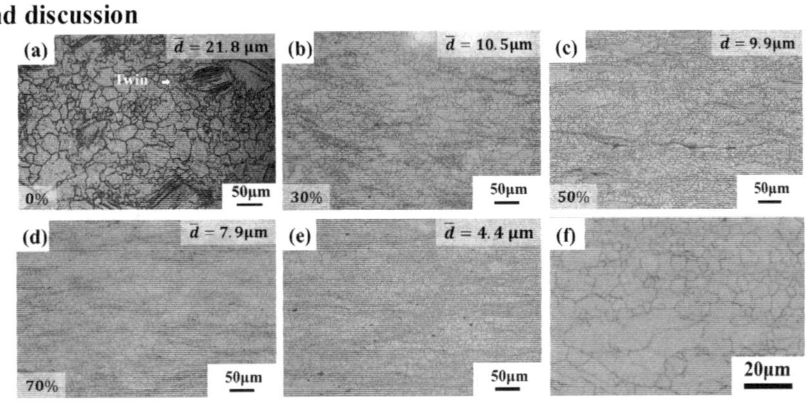

Figure 1. Optical microstructure of two-layered ZK60 samples after ARB process with the reduction of (a) 0%, (b) 30%, (c) 50% and (d) 70% per pass and (e, f) optical microstructure of four-layered ZK60 samples after ARB process with the 50% reduction.

The microstructural characteristics of the ZK60 alloy before ARB process were investigated on the longitudinal section, as depicted in Figure 1a. The microstructure contains equiaxed grains and some twins. The average grain size estimated by intercept procedure is about 21.8 μm. Figure 1b~d shows typical lamella structure (LS) from incomplete recrystallization in the two-layered ZK60 sample after ARB process with 30%, 50% and 70% reduction per pass, respectively. The microstructure is mainly composed of two parts: the elliptical CG stretched along the rolling direction caused by the ARB and recrystallized FG distributed around CG. After ARB processing, the grain size of two-layered ZK60 samples significantly decreases with the initial size of 21.8 μm reducing to 10.5 μm, 9.9 μm and 7.9 μm at 30%, 50% and 70% reduction, respectively. At a reduction of 70%, the elliptical shape coarse grains almost completely disappear, transforming into medium-sized equiaxed grains.

Figure 2a and 2b shows that the yield strength of the initial and two-layered samples after ARB with 30%, 50% and 70% reductions is 222 MPa, 190 MPa, 235 MPa and 200 MPa, respectively, the ultimate tensile strength is 291 MPa, 300 MPa, 325 MPa and 318 MPa, and the total elongation is 10.0%, 20.3%, 21.6% and 23.0%, respectively. It is found that the two-layered ARB sample with the 50% reduction has the better synergy of strength and ductility, associated with the average grain size of 9.9 μm and CG volume fraction of 17%, in which the length of elliptical CG is over 80 μm and the length-width ratio is over 4.

Afterward, four-layer sample under ARB with the 50% reduction is carried out to further optimize mechanical properties. Figure 1e and 1f show that the average grain size of four-layered sample after ARB is 4.4 μm with the CG size of ~23 μm and volume fraction of 41%. Compared to the two-layered one, the FGs in the four-layered sample are finer with the grain size ranging from 1μm to 4 μm.

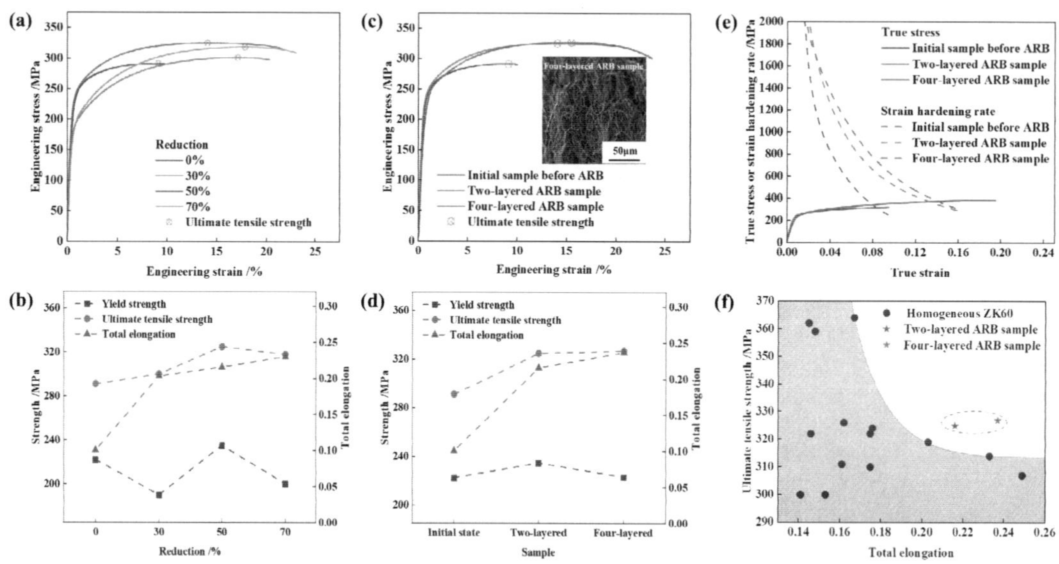

Figure 2. Mechanical properties of two-layered ZK60 samples after ARB with different reduction (a) the engineering stress-strain curves and (b) the strength and elongation comparison; mechanical properties of two- and four-layered ZK60 samples with 50% reduction: (c) the engineering stress-strain curves, (d) the strength and elongation comparison, (e) true stress-strain curves and strain hardening rate and (f) ultimate tensile strength vs. total elongation of the ZK60 samples after ARB and the published data of homogeneous structure ZK60 alloys.

The experimental stress-strain curves of different samples indicate that the ARB process significantly enhances the mechanical properties of the ZK60 alloys, as shown in Figure 2c. Figure 2d shows that the yield stress of the initial, two-layered and four-layered ARB samples with 50% reduction per pass are 222 MPa, 235 MPa and 223 MPa, respectively, the ultimate tensile strength are

291 MPa, 325 MPa and 327 MPa, and the total elongation is 10.0%, 21.6% and 23.7%, respectively. It is found that the four-layered ARB sample has the best synergy of strength and ductility. The strain hardening rate of the four-layered sample also exhibits the highest one in Figure 2e. Furthermore, it is shown that the fracture morphology in the four-layered sample is a typical ductile fracture, characterized by the dimples and tear ridges in Figure 2c inset. In Figure 2f, the mechanical properties of the ARB ZK60 are compared with the published data of homogeneous structure ZK60 alloys collected by our group in the home-made database of Mg alloys (http://39.101.169.161:8865/#/home) and it proves that the excellent synergy of strength and ductility of Mg alloys can be achieved by multiple passes of accumulative roll bonding process.

To further investigate the reasons behind the excellent strength-ductility synergy of ARB samples, a 2D finite element analysis is carried out. In Figure 3a and 3b, the 2D model dimensions are 250×250 μm^2, in which the gray, blue and black area represents the FG, CG and IA zone, respectively. In the four-layered ARB sample, the area fraction of FG, CG and IA zones is 53%, 31%, and 16%, respectively, compared with the initial one of 7.5%, 86%, and 6.5%. Figure 3c shows that the predicted stress-strain curve of the initial and four-layered ARB ZK60 samples under uniaxial tension agrees well with the experimental one with the stress error less than 5% in the whole curve. The overall mechanical response in Figure 3c indicates that the four-layered ARB sample has better strain hardening capability than the initial one. In Figure 3d~3f, the microstructure shows that the four-layered ARB sample contains more FGs and most of CGs are isolated by these FGs. According to the Considère criterion [11], the high strain concentration area is defined as the area with the local Y-component strain (ε_y) higher than the necking strain (*i.e.*, 10% of initial and 14% of four-layered ARB sample), marked in white area in Figure 3d~3f and the contours of the Y-component strain (ε_y) / stress (σ_y) show that during deformation, the local strain concentration firstly appears in the CG zone near the IA zone and then extended along the 45° direction rapidly in the four-layered ARB sample. As the strain concentration area is expanded, discrete shear bands are observed and gradually becomes thicker in the CG zones. In Figure 3g, it is shown that the total proportion of high strain concentration area in the four-layered ARB sample is 41.9% with 15.9%, 23.1% and 2.9% in the FG, CG, and IA zone, respectively. For comparison, the total proportion of high strain area in the initial one is 84.3% with 2.3%, 81.3% and 0.7% in the FG, CG, and IA zone, respectively and there is no obvious shear band in the initial sample and the overall strain distribution is uniform before failure. This indicates that the CG zones in lamella structure can accumulate more strain and balance strain partitioning during plastic deformation. Moreover, the lamella structure fully utilizes the coordinated deformation ability of CG, and the higher low strain area reserves sufficient coordination space for possible local stress/strain concentration.

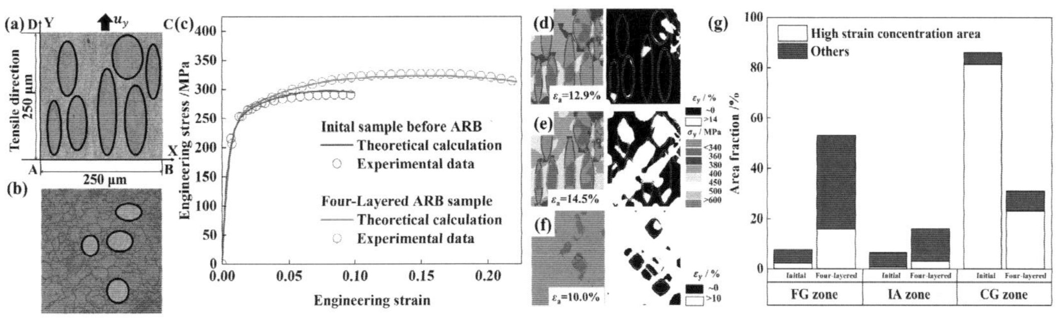

Figure 3. Schematic of 2D finite element model for the (a) four-layered ARB sample and (b) the initial sample under uniaxial tension in the Y direction, where the gray, blue and black area represents the fine-grain (FG), coarse-grain (CG) and interface-affected (IA) zone, respectively, (c) comparison of the predicted stress-strain curve and experimental one, and the contour of the Y-component stress (σ_y) / strain (ε_y) at the applied strain (ε_a) of (d) 12.9% and (e) 14.5%, respectively; (f) the σ_y and ε_y contours of the initial sample at the ε_a of 10.0%.

The strain partitioning between the CG and FG zones can also lead to the pileups of geometrically necessary dislocations in the IA zone, generating back stress that enhances the strength of the IA zone. As the area fraction of the IA zone decreases from 16% in the initial sample to 6.5% in four-layered ARB sample, the effect of back stress strengthening is weakened. What's more, some studies [12] indicate that the complex stress state induced in the IA zone facilitates the activation of non-basal slip in the neighboring CG zones, which can compromise stress concentration and ensure continuous strain hardening by increasing dislocation density propagated from Frank-Read sources. This may explain that the four-layered ARB sample can still undergo about 10% strain before failure occurs after necking. In summary, the lamella structure plays a crucial role in enhancing the mechanical properties of the ZK60 alloys by releasing stress concentration and inhibiting the initiation and propagation of microcracks.

4. Conclusions

In this study, combining the experimental and simulation method, the mechanical properties of two- and four-layered wrought ZK60 Mg alloy sheets after two passes of accumulative roll bonding (ARB) with different were studied and the microstructure were analyzed. The conclusions are following,

1. The yield strength of the initial-state and two-layered samples after ARB with 30%, 50% and 70% reduction is 222 MPa, 190 MPa, 235 MPa and 200 MPa, the ultimate tensile strength is 291 MPa, 300 MPa, 325 MPa and 318 MPa, and the total elongation is 10.0%, 20.3%, 21.6% and 23.0%, respectively.
2. The microstructure of wrought ZK60 sheet can be significantly optimized by the ARB process, in which the average grain size of two-layered and four-layered ZK60 after ARB with 50% reduction decreases to 9.9 μm and 4.4 μm, respectively, after compared with 21.8 μm of the initial one.
3. The four-layered ZK60 sheet after ARB with 50% reduction demonstrates the best strength-ductility combination with yield strength, ultimate tensile strength and total elongation of 223 MPa, 327 MPa and 23.7%, respectively.
4. The finite element analysis indicates that the high performance of ZK60 sheets may be contributed by the lamella structure formed by the fine and coarse grains, which can release the local stress concentration more effectively and inhibit initiation and propagation of the microcracks.

Acknowledgement

We thank the National Key Research and Development Program of China for funding the present research (No. 2023YFB3812900) and the State Key Laboratory of Special Rare Materials (Contract No. SKL2018K003), Northwest Rare Metal Materials Research Institute Ningxia Co. Ltd.

References

[1] Kocich R, Kunčická L, Král P, et al. 2016 Mater. Design 90 1092-1099. [CrossRef]
[2] Kou H, Lu J, Li Y. 2014 Adv. Mater. 26 5518-5524. [CrossRef]
[3] Wang H, Yu Z, Zhang L, et al. 2015 Sci. Rep. 5 1-9. [CrossRef]
[4] Xu Y, Li J, Qi M, et al. 2020 J. Mater. Res. Technol. 9 9856-9867. [CrossRef]
[5] Luo Y, Cheng W, Li H, et al. 2022 Mat. Sci. Eng. A 834 142623. [CrossRef]
[6] Beijing Internet Based Engineering Co. L I. (2024) Simdroid Multiphysics Software. https://www.ibe.cn/en/about
[7] Huang C, Wang Y, Ma X, et al. 2018 Mater. Today 21 713-719. [CrossRef]
[8] Li S, Jin J, Song Y, et al. 2022 Acta Metall. Sin. 58 114-128. [CrossRef]
[9] Peng P, Zhang K, She J, et al. 2021 J. Alloy Compd. 861 157958. [CrossRef]
[10] Zhu L, Lu J. 2012 Int. J. Plasticity 30-31 166-184. [CrossRef]
[11] Yasnikov I S, Kaneko Y, Uchida M, et al. 2022 Mat. Sci. Eng. A 831 142330. [CrossRef]
[12] Li Y, Zha M, Jia H, et al. 2021 J. Magnes. Alloy 9 1556-1566. [CrossRef]

Effect of thermal deformation on the microstructure and tensile property of the TiZrAl medium entropy alloy

Rong Guo[1], Jiaxiang Cui[1], Bang Dou[1], Qian Zhao[1], Shihai Sun[1, 2, *], Liang Wang[1] and Yunfei Xue [1, 2]

[1] School of Materials Science and Engineering, Beijing Institute of Technology, Beijing, 100081, China
[2] Tangshan Research Institute, Beijing Institute of Technology, Tangshan, Hebei, 063099, China

[*] Corresponding author's e-mail: shihaisun@bit.edu.cn

Abstract: Light high/medium entropy alloys with HCP structure are getting attention because of their high strength and good work hardening ability. However, low ductility is an urgent problem to be solved in them. In this paper, a lightweight medium entropy alloy TiZrAl with an HCP structure was used as the model alloy. The layered structure was transformed into a spherical structure through a thermal deformation process, which made the tensile strength up to 918 MPa, plasticity up to 19.36%, and showed good work hardening ability. This study provides a useful reference for the development of lightweight medium entropy alloys with high performance.

1. Introduction

Lightweight high/medium entropy alloys (LHEA/LMEA) have attracted great attention because of their wide composition designing space and good physical and mechanical properties [1]. They are promising candidates in the fields such as aerospace, military, energy, and transport [2].

Among the LHEA/LMEA, BCC lightweight high-entropy alloys consisting of elements such as Al, Ti, and Cr are investigated most owing to their high specific strength and good high-temperature strength [3]. However, these alloys generally lack work-hardening capability and have insufficient uniform deformation capacity, which can easily lead to component failure due to overloading during practical services [4]. Compared with the body-centered cubic (BCC) structure, the alloys of the HCP structure are less prone to cross-slip migration, and the L-C immobile dislocations can be an obstacle to dislocation motion, bringing about significant work-hardening [5]. Therefore, lightweight high/medium entropy alloys with HCP structures are beginning to attract the interest of researchers. However, because of the small number of slip systems in the HCP structure, the plasticity of the alloy is usually poor, which limits the engineering applications of the alloy [6]. For this reason, there is an urgent need to enhance the plasticity of lightweight high/medium entropy alloys with HCP structures.

In this paper, TiZrAl LMEA with HCP structure was selected as a model alloy, and the effect of thermal deformation on the microstructure and mechanical properties of the alloy was investigated. The good strength-plasticity match of the alloy was achieved by transforming the lamellar microstructure to the globular microstructure, and the good work hardening ability was maintained.

Journal of Physics: Conference Series **2842** (2024) 012037 doi:10.1088/1742-6596/2842/1/012037

2. Experimental methods

The TiZrAl LMEA ingot with a weight of ~500 g was prepared in a cold crucible induction melting furnace using the high purity (> 99.9%) Ti, Zr, and Al as the raw material. To ensure the homogeneity of the composition, the ingot was flipped and remelted three times. The ingot was held at 1000 °C for 1 h to make the microstructure become uniform.

The blocks were cut from the ingot by avoiding the central shrinkage cavity, and coated with 314 stainless steels to prevent oxidation during hot rolling. The blocks were hot rolled at 750 °C, 800 °C, and 850 °C with a deformation of 50%, respectively, followed by air cooling. Before hot rolling, the blocks were heated at the target temperature for 30 min to ensure uniform heating. To regulate the microstructure further, the samples rolled at 800 °C were annealed at 650 °C, 700 °C, and 750 °C for 1 h, respectively.

The microstructure observation was carried out on a scanning electron microscopy (SEM, Hitachi Regulus 8230) and a transmission electron microscope (TEM, FEI Tecnai G2 F20). The dog bone sample was used for the tensile test, whose gauge part was 1 mm in thickness, 3 mm in width, and 10 mm in length. The tensile test was carried out on the universal electronic testing machine (CMT4305) with a strain rate of 10^{-3} s^{-1}. Each sample was tested three times.

3. Results and discussion

3.1. Microstructure characterizations

The as-cast TiZrAl LMEA shows the lamellar microstructure with a thickness of 6.16 μm, as shown in Figure 1. According to the TEM images, the TiZrAl LMEA is an HCP structure grain with lamellar microstructure. This microstructure is easy to cause stress concentration at the grain boundary and intergranular facture, and usually results in poor plasticity of the alloy. Therefore, it is very important to optimize the microstructure of TiZrAl LMEA by a thermal deformation process to improve the mechanical properties.

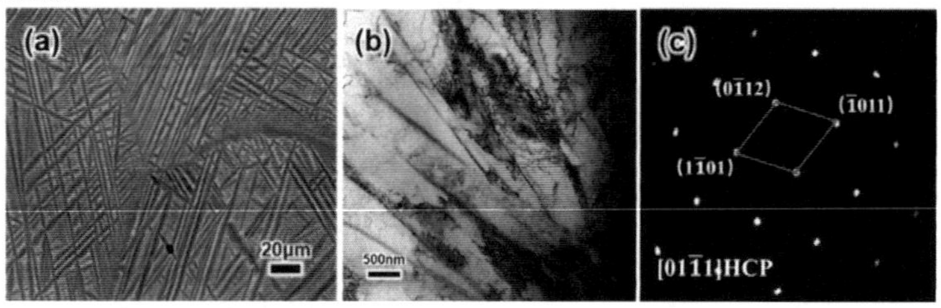

Figure 1. Microstructure and phase structure of as-cast TiZrAl LMEA (a) the SEM image; (b) the bright field image under TEM; (c) the SAED image.

Figure 2 shows the SEM images of TiZrAl LMEA under different hot-rolled temperatures. As shown in Figure 2(a), the original lamellar structure of TiZrAl LMEA was severely distorted at 750 °C, exhibiting a kinked shape. When the temperature was increased to 800 °C (Figure 2(b)), the alloy exhibited mixed microstructure consisting of short rod-like and spherical-like grains and the spheroidization rate was about 29%. When the temperature was further increased to 850 °C as shown in Figure 2(c), the sample still exhibited the lamellar microstructure, similar to the original state. This is because the transformation temperature from the BCC phase to the HCP phase is around 830 °C in the TiZrAl LMEA. When the deformation temperature is higher than the phase transition temperature of the alloy (Figure 2(c)), the deformation of the alloy will mainly occur in the easily deformed BCC phase, and the lamellar microstructure will form by the phase transformation during the cooling process after rolling.

At a temperature a little lower than the phase transformation temperature, the lamellar HCP grain will be broken into a short rod-like grain by the stress during hot deformation, and the short rod-like grains will become spherical shape gradually by the thermal diffusion. Because the hot deformation time is short, only partly spheroidization occurs in the 800 °C rolled sample.

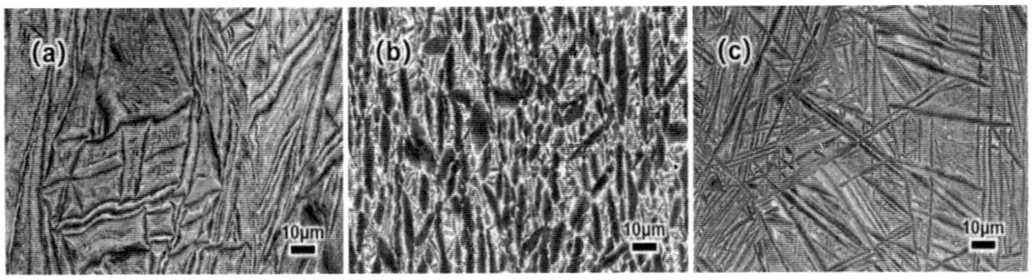

Figure 2. SEM images of hot-rolled TiZrAl LMEA at (a) 750 °C; (b) 800 °C; (c) 850 °C.

From the above experimental results, it can be seen that TiZrAl LMEA did not completely recrystallize after hot rolling, and part of it was transformed into a spherical microstructure after hot rolling at 800 °C. In order to obtain the spherical microstructure of TiZrAl LMEA, the alloy after hot rolling at 800 °C was heat treated at different temperatures.

Figure 3(a) shows that only short rod-like microstructure and partially spherical-like microstructure occur in TiZrAl LMEA at 650 °C for 1 h, indicating that the alloy is still incomplete recrystallization at 650 °C. The recrystallization temperature of TiZrAl LMEA is not reached at 650 °C. When the annealing temperature is increased to 750 °C, the degree of spheroidization of the alloy is significantly increased, and the microstructure distribution is more uniform with fine grains. According to the statistics of the percentage of spheroidization of the alloy, it is concluded that the percentage of spheroidization of TiZrAl LMEA is greater than 80%, which basically realizes the complete globalization. As shown in Figure 3(c), when the annealing temperature continues to rise to 800 °C, TiZrAl LMEA presents lamellar microstructure. This is because when the alloy is annealed at a temperature higher than the phase transition temperature, the phase transformation will occur during the heat preservation process, and the single HCP phase in the alloy will be transformed into the single BCC phase. The BCC phase will be transformed again during the subsequent cooling process, forming the lamellar HCP phase.

It can be seen from the above analysis that the spherical microstructure of TiZrAl LMEA can be obtained after hot rolling at 800 °C and annealing at 700 °C for 1 h.

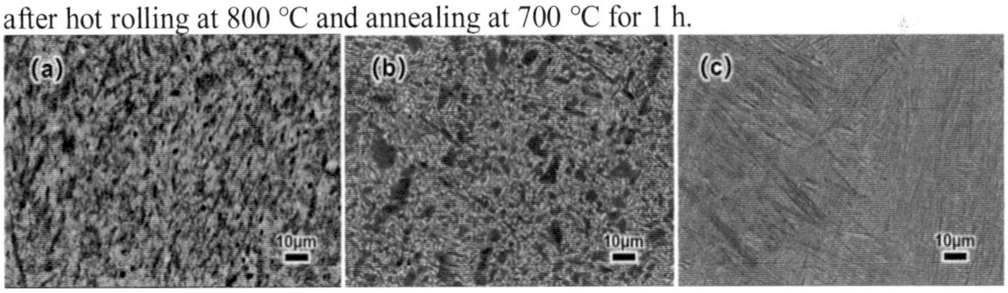

Figure 3. SEM images of TiZrAl LMEA after heat treatment at (a)650°C-1 h; (b) 700°C-1 h; (c) 750°C-1 h.

3.2. Mechanical properties

The quasi-static tensile mechanical properties of TiZrAl LMEA with spherical microstructure after annealing, with as-cast lamellar microstructure, and with partial spherical microstructure after hot rolling at 800 °C were compared, and the mechanical properties curves obtained were shown in Figure 4(a). It can be seen that after recrystallization annealing, the tensile strength change of TiZrAl LMEA is not obvious and can reach ~918 MPa, but the plasticity is significantly improved. The plasticity of the annealed alloy is greatly improved, and the elongation can reach 19.36%. Figure 4(b) shows the work hardening rate curve of the TiZrAl LMEA. It can be seen that the annealed alloy in the steady-state plastic deformation stage is greater than 2 GPa, and it has a good work hardening ability.

At the initial stage of quasi-static tensile deformation, the work hardening rate linearly decreases with the increase of strain. In the deformation process of hot-rolled and annealed alloys, a strain hardening rate plateau appeared, and the work hardening rate was greater than 2 GPa.

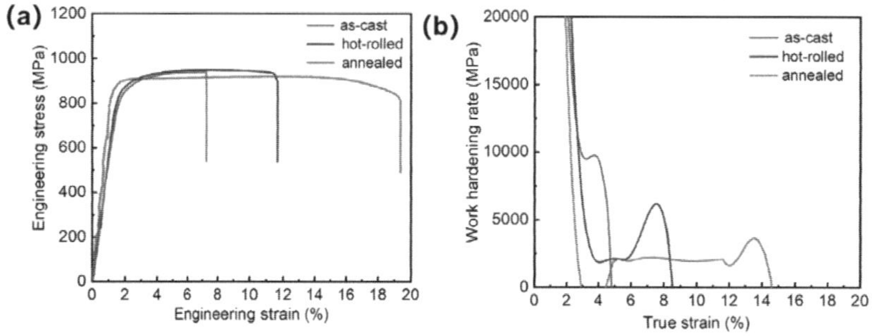

Figure 4. (a) Tensile stress-strain curves; (b) work hardening rate curves of TiZrAl LMEA in different states.

Figure 5 shows the tensile fracture morphology of TiZrAl LMEA as-cast, hot-rolled at 800 °C and annealed at 750 °C for 1 h, respectively. The fractures of the samples show the characteristic of dimples, which are ductile fractures. The number of dimples in the annealed alloy is increased, and the dimples are larger and deeper, which indicates that the alloy has a strong plastic deformation ability.

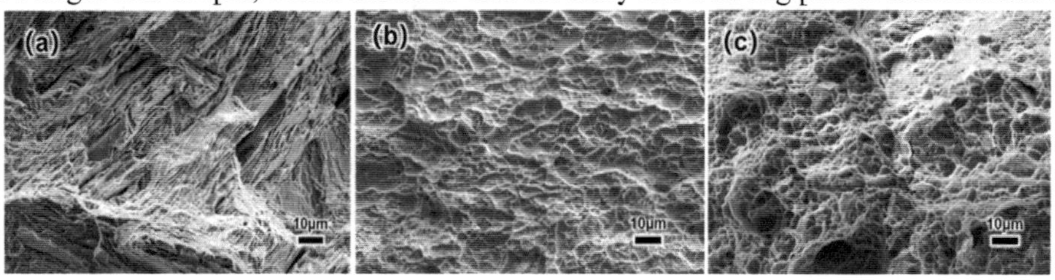

Figure 5. The fracture surface of TiZrAl LMEA in different states after the tensile test (a) as-cast; (b) hot-rolled at 800 °C; (c) annealed at 700 °C for 1 h.

4. Conclusions

In this paper, the microstructure and mechanical properties of TiZrAl LMEA during thermal deformation and annealing are studied. The results show that after hot rolling at 800 °C and annealing at 700 °C for 1 h, the TiZrAl LMEA changes from lamellar microstructure to spherical microstructure, and the ductility improves obviously. The TiZrAl LMEA with good strength-plasticity matching is obtained, with tensile strength up to 918 MPa and elongation up to 19.36%.

Acknowledgments

This work was financially supported by the National Natural Science Foundation of China under Grant number 52171028.

References

[1] Diao G, He A, Wu M, Tang Y, Chen W G, Li Q Y, and Li D. (2024) Lightweight Al_3Ti-based medium-entropy alloys with well-balanced strength and ductility. Scripta Materialia, 242: 115928. [https://doi.org/10.1016/j.scriptamat.2023.115928]

[2] Kumar D, Seetharam R, and Ponappa K. (2024) A review on microstructures, mechanical properties and processing of high entropy alloys reinforced composite materials. Journal of Alloys and Compounds, 972: 172732. [https://doi.org/10.1016/j.jallcom.2023.172732]

[3] Yan X, Liaw P K, and Zhang Y. (2022). Ultrastrong and ductile BCC high-entropy alloys with low-density via dislocation regulation and nanoprecipitates. Journal of Materials Science & Technology, 110: 109-116. [https://doi.org/10.1016/j.jmst.2021.08.034]

[4] Xiong W, Guo A X Y, Zhan S, Liu C T, and Cao S C. (2023) Refractory high-entropy alloys: A focused review of preparation methods and properties. Journal of Materials Science & Technology, 142: 196-215. [https://doi.org/10.1016/j.jmst.2022.08.046]

[5] Zhang Y, Yang X, and Liaw P K. (2012) Alloy Design and Properties Optimization of High-Entropy Alloys. JOM, 64: 830-838. [https://doi.org/10.1007/s11837-012-0366-5]
[6] Youssef K M, Zaddach A J, Niu C, Irving D L, and Koch C C. (2015). A novel low-density, high-hardness, high-entropy alloy with close-packed single-phase nanocrystalline structures. Materials Research Letters, 3(2): 95-99. [https://doi.org/10.1080/21663831.2014.985855]

Journal of Physics: Conference Series **2842** (2024) 012038 doi:10.1088/1742-6596/2842/1/012038

Application of instrument indentation test for residual stress characterization of 5083 aluminium alloy

Zhuotong Du [1], Rutao Zhao [1], Chen Gao [1], Pei Chen [1, 2*] and Kuilong Xu [1]

[1] Luoyang Ship Material Research Institute, Luoyang, Henan, 471023, China
[2] National Key Laboratory of Marine Corrosion and Protection, Luoyang, Henan, 471023, China

[*] Corresponding author's e-mail: 15978638526@163.com

Abstract. An abnormal cracking phenomenon occurred at the corner of a ventilation window on the 5083 aluminum alloy wallboard used in certain transportation equipment. The residual stresses of the wallboard were measured using an instrumented indentation test. Initially, stress loading was applied to the 5083 specimens using a residual stress loading device to verify the effectiveness of the instrument indentation test (IIT) for testing the residual stresses of 5083 aluminum alloy. Subsequently, IIT was conducted under static conditions at the serving site of the equipment, and the working stresses under dynamic conditions were detected using a strain monitor. The results indicate that the total tensile stress at the uncracked corner of the ventilation window has exceeded the 5083 yield strength, so excessive tensile stress is the main cause of the abnormal cracking of the wallboard.

1. Introduction

The 5083 aluminum alloy is a high-magnesium five-series aluminum alloy with good corrosion resistance and welding properties, widely used in lightweight equipment such as aircraft, ships, and automobiles [1-3]. In order to reduce structural weight, certain transportation equipment utilizes 5083 aluminum alloy material for its internal wallboards, which have a rectangular through-hole ventilation window in the middle. During the operation of the equipment, cracking occurred at the corner of some ventilation windows, as shown in Fig. 1. It was preliminarily speculated that this could be due to excessive local residual stresses. The user of the equipment requested that testing methods that would compromise the structural integrity, such as drilling or cutting, should not be used. Additionally, the X-ray testing equipment was hard to be installed on the non-magnetic vertical wallboard. Therefore, it is proposed to use the IIT to determine the residual stress.

IIT is a high-throughput non-destructive mechanical performance testing method developed in recent years. It involves pressing a material surface with a spherical, conical, or composite indenter and collecting load-depth curves during the indentation process. By analyzing the relationship between the load and depth information, the mechanical constitutive relationship of the material could be established. Various mechanical properties can be characterized, such as residual stresses, yield strength, tensile strength, elastic modulus, fracture toughness, and so on [4-7]. It is currently essentially a unique technology to simultaneously detect mechanical property and residual stress. Due to its typical indentation depth not exceeding 200 μm, it is particularly suitable for on-site engineering testing and has broad application prospects in the non-destructive testing of structural materials in in-service equipment.

The principle of commonly used instrumented indentation test for residual stresses is comparing the indentation response parameters between samples with and without residual stress, and inferring residual stress based on the changes in these response parameters. Suresh et al. first proposed a way to detect equal biaxial residual stresses according to the ratio of thc action arcas with and without residual stresses [9]. Lee et al. further improved the method by estimating the contact area using the indentation load [10]. Lu et al. established an equal biaxial residual stress detection method based on the phenomenon of residual compressive stress and residual tensile stress causing changes in the curvature of the indentation force-depth curve during the conical indentation loading segment [11]. Peng et al. developed a spherical indentation test for detecting equal biaxial residual stresses by the change of the indentation load at the same depth [12]. However, the aforementioned studies mainly focus on constructing theoretical models for steel materials, and further research is needed to assess their applicability to aluminum alloy materials.

In this study, the instrumented indentation test with Vickers indenter was employed to characterize the equal biaxial residual stresses of 5083 aluminum alloy. Initially, stress loading was applied using a residual stress loading device in the laboratory to verify the relevant calculation parameters for residual stresses. Subsequently, equal biaxial residual stress measurement and working stress detection were conducted at the in-service equipment to investigate the cause of the abnormal cracking of the wallboard.

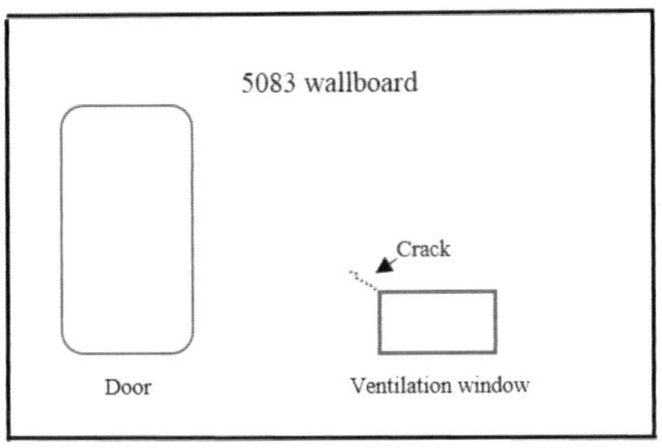

Figure 1. Schematic of 5083 wallboard cracking.

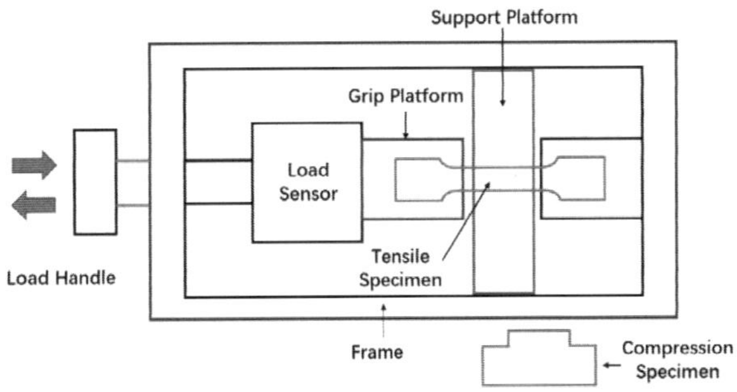

Figure 2. Residual stress loading device.

2. Specimen and test method

The cracked wallboard is made of 10mm thick H116 state 5083 aluminum alloy. In the laboratory, experiments were conducted on the 5083 plates from the same manufacturer, with the same specifications, similar chemical composition, and similar mechanical properties. The chemical composition is in Table 1. The mechanical properties of the laboratory test materials, the cracked wallboard from the test report before the construction, and the in-service wallboard characterized by IIT under spherical indenters on-site are shown in Table 2 [13].

Table1. Chemical composition of 5083 alloy (wt.%).

Material	Si	Fe	Cu	Mn	Mg	Cr	Ni	Zn	Ti
In Lab	0.062	0.175	0.047	0.896	4.35	0.151	0.019	0.031	0.044
In Service	0.062	0.175	0.043	0.903	4.33	0.152	0.020	0.030	0.044

Table 2. Mechanical properties of 5083 alloy.

Material	Yield strength (MPa)	Tensile strength (MPa)	A (%)	E (Gpa)	n
In Lab	216	338	13.0	73.6	0.174
In Service (from the historical report)	223	335	14.0	74.2	/
In Service (by IIT)	209	329	/	75.7	0.166

Firstly, residual stress validation tests were conducted in the laboratory. The TIMEH1110IITP instrumented indentation tester used in this paper has a load range of 1 kN, a depth range of 2 mm, and a resolution of 0.01μm. Depth control was utilized with a speed of 0.2mm/min, and the maximum depth was 100μm in the indentation test. A residual stress loading device, as shown in Fig. 2, is used to apply the corresponding residual stresses to the tensile and compressive specimens. The residual stress loading device is mounted below the instrumented indentation tester to test the stress after applying the residual stresses. The applied stresses were firstly adjusted to zero to obtain the zero-stress reference point. Then the applied stresses were adjusted to ±50 MPa, ±100 MPa, ±150 MPa, and ±200 MPa, and instrumented indentation tests were conducted under these stress levels.

After completing the laboratory validation, on-site testing of the equipment was carried out. The test area was polished using 800-grit sandpaper to remove surface paint and oxide layers. Subsequently, the instrumented indentation tester was fixed at the test position using a suction cup device, and tests were conducted in a static condition. Next, a 120Ω strain gauge was attached next to the test area, and a DH3816N dynamic strain monitor was used to monitor the work stresses dynamically to obtain the maximum working stress. Finally, the overall stress levels of in-service wallboards were acquired.

3. Results and analysis

For elastoplastic materials with conical indentation, the loading segment curve satisfies Kick's law [14]:

$$F = ch^2 \tag{1}$$

F is the indentation force, h is the depth of indentation, and c is the curvature of the loading segment curve. When the loading curvature c_0 of the indentation under zero stress is obtained, the residual stress can be characterized according to the measured changes in loading curvature. For the same depths, the tensile residual stresses make the indentation force decrease and the corresponding c value increase; the compressive residual stresses make the force increase and the corresponding c value decrease, as shown in Fig. 3.

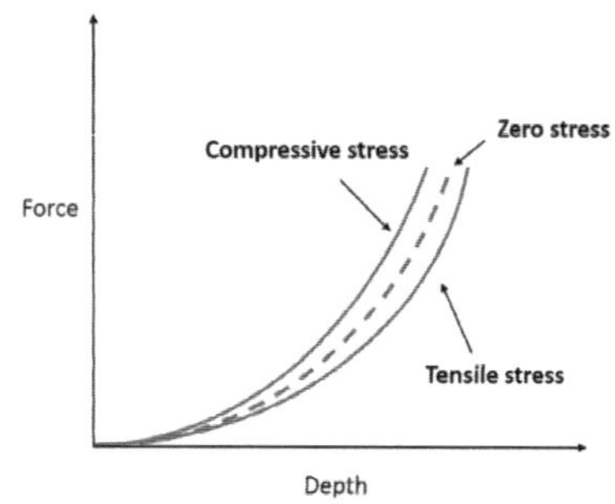

Figure 3. Indentation force-depth curve under different residual stresses.

A functional relationship between material parameters, indentation parameters, and residual stress was constructed in Eq. 2:

$$F = f(E, \upsilon, n, h, \theta, \sigma_y, \sigma_r) \tag{2}$$

where E is the material's modulus of elasticity, υ is Poisson's ratio, taken as 0.30, n is the strain hardening index, θ is the indenter taper angle, taken as 70.3°, σ_y is the yield strength, and σ_r is the equal biaxial residual stress.

Through the FE simulation of a quantity of different material mechanical parameters with different residual stress levels, statistical analysis found that the residual stresses and the variations of the loading curvature have an approximately linear relationship. The dimensionless functional relationship is established in Eq. 3 [15]:

$$\frac{c-c_0}{\sigma_y} = \prod\left(\frac{\sigma_y}{E}, n, \frac{\sigma_r}{\sigma_y}\right) \tag{3}$$

For the mechanical parameter characteristics of aluminum alloy material, different σ_r/σ_y, n and E values are input to carry out the finite element analysis, and finally the equal biaxial residual stress calculation equation for aluminum alloy material is established, as shown in Eq. 4:

$$\frac{c-c_0}{\sigma_y} = -1.09\left(22.14\left(\frac{\sigma_r}{\sigma_y}\right)^2 + 103.96\frac{\sigma_r}{\sigma_y} + 1\right)\left(-49376\left(\frac{\sigma_y}{E}\right)^2 + 388.25\frac{\sigma_y}{E} + 1\right)(-0.142n^2 + 1.483n + 1) \tag{4}$$

The σ_y, E, and n of 5083 aluminum alloy in Table 1, and the loading curvature c from the indentation curve were used to calculate the equal biaxial residual stresses by Eq. 4. The results of residual stresses tests in the laboratory are shown in Table 3. The value of unidirectional residual stresses applied to the 5083 samples can be expressed in terms of the equal biaxial residual stress, which is two times the equal biaxial residual stress [9].

Table 3. Residual stress results in laboratory.

Applied residual stress (MPa)	Calculated equal biaxia residual stress (MPa)	Calculated single biaxia residual stress (MPa)	Relative error (%)
-200.4	-115.7	-231.4	15.47
-150.6	-85.5	-171.0	13.55
-99.6	-53.9	-107.8	8.23
-50.2	-26.8	-53.6	6.77
+49.8	+23.9	+47.8	4.02
+100.6	+54.1	+108.2	7.55
+150.8	+84.7	+169.4	12.33
+199.4	+108.9	+217.8	9.23

In the table, "+" stands for tensile stress and "-" stands for compressive stress. The maximum error of the residual stress of 5083 material calculated by this model is 15.47%, and the minimum error is 4.02%. The study shows that in the traditional residual stress test methods, due to the influence of human operation errors and other factors, the measurement accuracy of the residual stress is generally between 10 and 30%. Therefore, the IIT method in this paper basically meets the requirements of residual stress testing of materials in engineering applications.

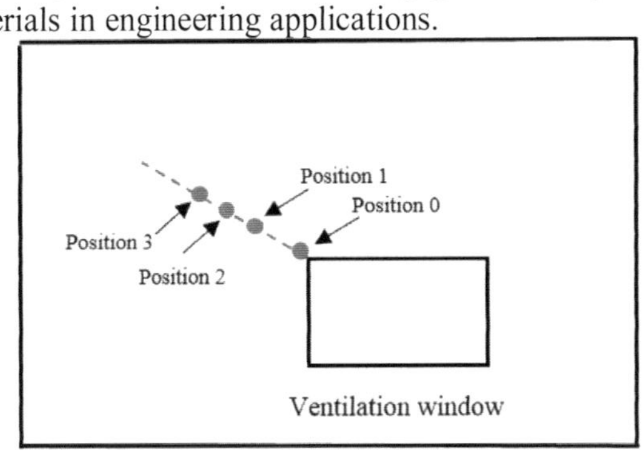

Figure 4. Schematic of in-service equipment test points.

Instrumented indentation tests were first conducted in the corner region of the vent window on the in-service aluminum alloy wallboard where no cracking had occurred to obtain the residual stresses under static conditions. Subsequently, the strain gauges were pasted close to the instrumented indentation test points to detect the working stresses under dynamic conditions. The schematic of the test points is shown in Fig. 4: Position 0 is the window corner, but due to size limitations of the instrumented indentation tester, it is not possible to directly measure residual stress at this location; only strain gauges are attached to monitor working stress. The distance between position 0 and position 1 is 20mm along the 45°direction, and the interval from position 1 is 10mm for position 2 and position 3 in turn. The results are in Table 4.

Table 4. Residual stress results in service.

Test position	Equi-biaxia residual stress (MPa)	Working stress (MPa)	Total stress (MPa)
0	/	+19.0	/
1	+188.9	+15.9	+204.8
2	+157.4	+11.6	+169.0
3	+103.2	+7.4	+110.6

The total stress at Position 1 is 204.8 MPa, reaching 91.8% of the yield strength. Based on the trend of working stress, the actual total stress at Position 0 at the window corner probably exceeded the yield strength of 5083. Working under such high tensile stress conditions, coupled with the low-cycle fatigue effect of superimposed alternating loads, is highly likely to induce cracking. Therefore, the cause of cracking at the window corner of the 5083 aluminum alloy wallboard is attributed to excessively high tensile stress. The sharp angle of the window corner leads to local stress concentration, which is the direct cause of the high tensile stress. If the rounded corner design of the door is used, as shown in Fig. 1, the local stress concentration can be effectively reduced.

4. Conclusion

(1) The method of instrumented indentation for residual stresses measurement in 5083 aluminum alloy is established, and the maximum error of the method is verified to be 15.47%, which has the capability of engineering application.

(2) The residual stresses at the corner of ventilation windows on 5083 aluminum alloy wallboard of in-service transportation equipment are tested, and combined with the working stresses, the total maximum tensile stress at the window corners has probably exceeded the yield strength of 5083 alloy, as the main cause of abnormal cracking.

References

[1] Sarvesha, R., Garcia, D., Dolmetsch, T., Wang, T., Eberheim, R., Champagne, V. K., ... & Agarwal, A. (2024) In Situ Study on Fracture Behavior of Thick - Section Friction Stir - Welded AA 5083 Alloy. Advanced Engineering Materials, 26(9), 2301574.

[2] Chen, H., Tianli, Z., Zhu, Z., Xu, L., Lin, S., Yang, S., & Kou, S. (2024) Study on the effect of Ar–He shielding gas on the weld formation, microstructure and mechanical properties of 5083 aluminum alloy weld. Journal of Materials Research and Technology, 28, 683-694.

[3] Guoyuan, L., Yue, W. (2009) Research on Stress Corrosion Cracking and Exfoliation Corrosion Resistance of 5083 Alloy Integral Panel. Development and Application of Materials, 24(4):5-7.

[4] VanLandingham, M. R. (2003) Review of instrumented indentation. Journal of research of the National Institute of Standards and Technology, 108(4), 249.

[5] Ma, Q. P., Basterrech, S., Halama, R., Omacht, D., Měsíček, J., Hajnyš, J., ... & Petrů, J. (2024) Application of Instrumented Indentation Test and Neural Networks to determine the constitutive model of in-situ austenitic stainless steel components. Archives of Civil and Mechanical Engineering, 24(2), 129.

[6] Moon, S., Choi, M., Hong, S., Kim, S. W., & Yoon, M. (2024). Non-equibiaxial residual stress evaluation methodology using simulated indentation behavior and machine learning. Nuclear Engineering and Technology, 56(4), 1347-1356.

[7] Dou, G., Jiang, H., Xu, F., Chen, J., Zhang, T., & Peng, G. (2024) A Correction Function to Improve the Accuracy of Measuring Elastic Modulus by Instrumented Spherical Indentation. Journal of Testing and Evaluation, 52(4).

[8] Peng G, Zhang T. (2022) Progress in instrumented indentation methods for determination of surface residual stress, Chinese Joumal of Theoretical and Applied Mechanics, 54(8): 2287-2303.
[9] Suresh S, Giannakopoulos AE. (1998) A new method for estimating residual stresses by instrumented sharp indentation. Acta Materialia, 46(16): 5755—5767.
[10] Lee Y, Kwon D. (2004) Estimation of biaxial surface stress by instrumented indentation with sharp indenters. Acta Materialia, 52: 1555-1563.
[11] Lu Z, Feng Y, Peng G. (2014) Estimation of surface equi-biaxial residual stress by using instrumented sharp indentation. Materials Science and Engineering A, 614: 264—272.
[12] Peng G, Lu Z, Ma Y. (2018) Spherical indentation method for estimating equibiaxial residual stress and elastic—plastic properties of metals simultaneously. Journal of Material Research, 33(8): 884-897.
[13] Du, Z., Xu, K., Wang, L., Zhang, X., & Zhang, X. (2022) Characterization of the local mechanical properties of 5083-5383 MIG welded joints. In Journal of Physics: Conference Series. 2390, 012013.
[14] Jang, J. I. (2009) Estimation of residual stress by instrumented indentation: A review. J. Ceram. Process. Res, 10(3), 391-400.
[15] Geng Zhao. (2022) Research on the evaluation technique of mechanical properties and residual stresses of ductile metals based on micrometer indentation. Guizhou University, Guiyang.

Study on the effect of heat treatments on microstructure and mechanical properties of nano SiC reinforced AlSi7Mg alloy

Tianchun Zou[1], Zhe Song[1,*], and Minying Chen[2]

[1]College of Safety Science and Engineering, Civil Aviation University of China, Tianjin, 300300, China
[2]Jiangxi Hongdu Aviation Ind Co. Ltd, Nanchang, Jiangxi, 330000, China

[*]Correspondence: zhe13848465220@163.com (Zhe Song)

Abstract: This paper investigated the influence of heat treatment on the microstructure evolution and mechanical performance of nano SiC-enhanced AlSi7Mg alloy fabricated using the laser powder bed fusion technique. After heat treatment, the grain size of the specimen increases significantly, the reticular structure is gradually broken, and Si is gradually precipitated after annealing. With increasing annealing temperature, elongation increases, while the yield strength decreases. After solution-aging heat treatment, the grain distribution is more uniform, the grain size increases slightly, and larger-sized Si and Mg2Si particles precipitate. As the solution temperature increases, the yield strength and the tensile strength of the samples show an increase followed by a decrease, while the trend of elongation is opposite. Both heat treatments exhibit ductile fracture as the predominant damage mode.

1. Introduction

Laser powder bed fusion (LPBF) has shown wide application in the aerospace field as a metal additive manufacturing (AM) technology, meeting the demands of complex structures and individual customization [1, 2]. However, the short liquid lifetime and high thermal gradient under preparation could cause high residual stresses in LPBF-formed specimens, leading to the forming of cracking and distortion [3, 4]. Consequently, post-treatment techniques are required to improve the relevant properties. Generally, heat treatment of LPBF-formed parts can eliminate stress and reduce potential deformation and cracking risks [5, 6].

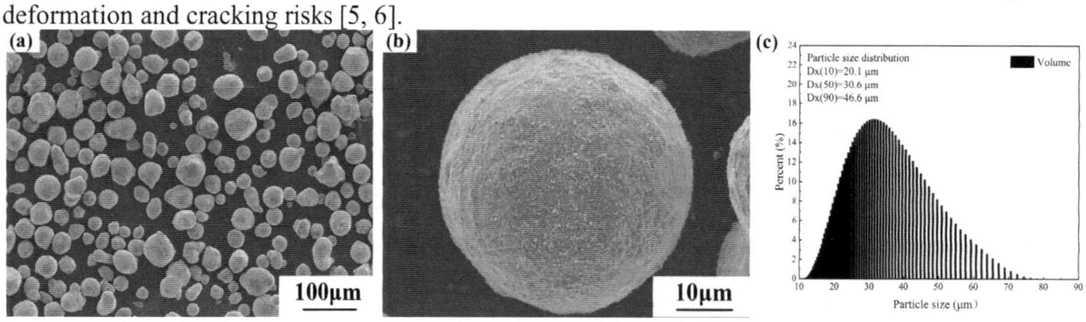

Figure 1. SEM images of nano SiC/AlSi7Mg mixed powder (a, b) and powder size distribution diagram (c).

Journal of Physics: Conference Series **2842** (2024) 012039 doi:10.1088/1742-6596/2842/1/012039

2. Experimental procedure

2.1. Powder preparation
The SiC (2 wt.%) nanoparticles and AlSi7Mg powders were mixed by a mechanical mixer, and the microstructure is displayed in Figures 1(a) and 1(b). After mixing, the powder still retained a certain sphericity. Figure 1(c) shows the particle size distribution.

2.2. Heat treatment
Some specimens were annealed at various temperatures (220°C, 250°C, 280°C, 310°C, 340°C, 2 hours). Others were treated under solution conditions at various temperatures (440°C, 470°C, 500°C, 530°C, 560°C, 2 hours), water quenching and finally, artificial aging was taken (150°C, 6 hours). The treated specimens were cooled in a room-temperature environment.

2.3. Experimental methods
The vertical sections of specimens were polished and etched by a chemical reagent (Keller) for 30 seconds and finally rinsed with alcohol. Scanning electron microscopy equipment (SEM) was used to examine the microstructure of specimens. The electron backscatter diffraction analysis (EBSD) was analyzed for the grain morphology of the samples. Tensile tests were conducted by an INSTRON 5982 testing machine with a 1 mm/min rate at room temperature.

3. Results

3.1. The effect of heat treatment on microstructure
After annealing at 220°C for 2 hours, Al and eutectic Si boundaries are observed in Figure 2, which indicates that the lower temperatures have not changed the cellular eutectic Si sub-structures formed in the process of rapid solidification. With increasing annealing temperature, the network structure gradually breaks down, leading to the dissolution and coarsening of Si. When the temperature is higher than 310°C, the fine secondary Si particles almost disappear and produce significantly larger particle sizes.

The microstructure of the solution specimens is shown in Figure 3. In contrast to the annealed specimens, Si particles precipitate after solution treatment. As the solution temperature increases, the size of precipitated Si particles shows an increase followed by a decrease, the number of particles decreases gradually, and the spacing between particles increases, as diagrammed in Figures 3 (b) and 3(f). When the temperature increases from 440°C to 500°C, the size of the precipitated particles increases from 0.5-1.5 μm to 1.5-3 μm, and the size is virtually unchanged at 530°C. Further increasing the temperature to 560°C results in a slight decrease in particle size, with particles primarily distributed around 1-2 μm, while the distance between particles increases significantly.

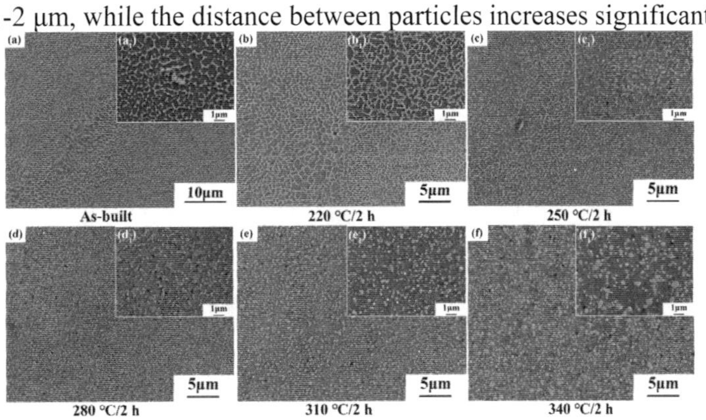

Figure 2. SEM images of as-built (a) and different annealing temperature nano SiC/AlSi7Mg composites (b-f), and the enlarged SEM images inserted (a_1-f_1).

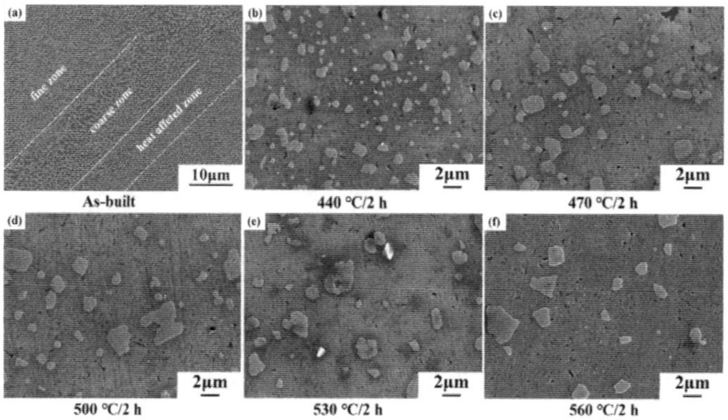

Figure 3. SEM images of as-built (a) and different solution temperature nano SiC/AlSi7Mg composites (b-f).

Furthermore, EDS mapping was conducted to investigate the phase evolution under annealed (280°C) and solution (500°C) conditions using Al, Si, Mg, and C as the primary tracking elements, as illustrated in Figure 4. For the annealed sample, the reticular Si structure is not completely broken, but the distribution of Si is relatively uniform. The distribution of Mg is no longer uniform, indicating the intermetallic compounds of Mg contained in the precipitated particles [7]. For the solution samples, most precipitates consist of Si particles (Points 1 and 2), while some particles are Mg_2Si phase, precipitated in the aging process (Point 3), with an obvious element segregation phenomenon.

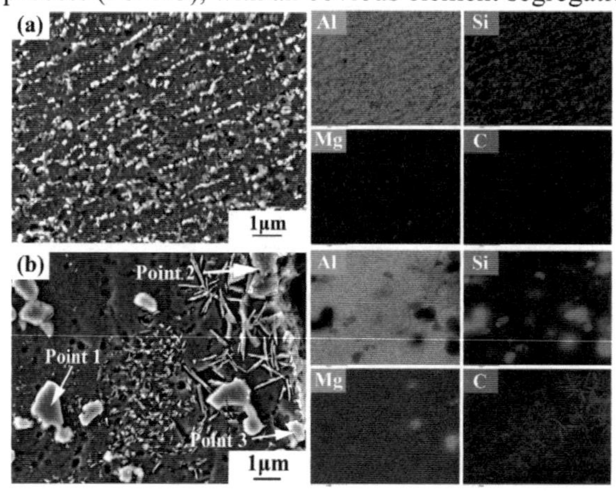

Figure 4. EDS phase maps of elements Al, Si, Mg, and C for annealing (a) and solution (b) conditions.

The EBSD results of specimens under as-built, annealed (280°C 2 hours), and solution-aging conditions (500°C 2 h + 150°C 6 hours) are shown in Figure 5. Most grains in Figures 5 (a), (b), and (c) exhibit crystal orientations of (001), (110), and (111) when referring to the inverse pole figure (IPF). However, red and green color areas are more than blue color areas in Figure 5(a), indicating that the intensities of (001) and (101) orientations are stronger than that of the (111) orientation in the as-built sample. After annealing, the red area decreases and the intensities of the three orientations are almost equal in Figure 5(b). After solution heat treatment, the blue area almost disappears in Figure 5(c), indicating a decrease in the intensity of the (111) orientation. The local crystallographic texture intensity of the (100) pole figure is shown in Figures 5 (d)-(f). The texture index of as-built, annealing, and solution specimens was determined to be 19.80, 11.39, and 12.93, respectively. Figures 5 (h)-(j) show that the annealing treatment increases average grain size and solution treatment decreases average grain size. Figure 6 shows that the average dislocation density of the as-built sample is 1.78 × 1014 /m^2. The average dislocation densities of the annealed and solution samples are 0.83 × 1014 /m^2 and 2.00 × 1014 /m^2, respectively.

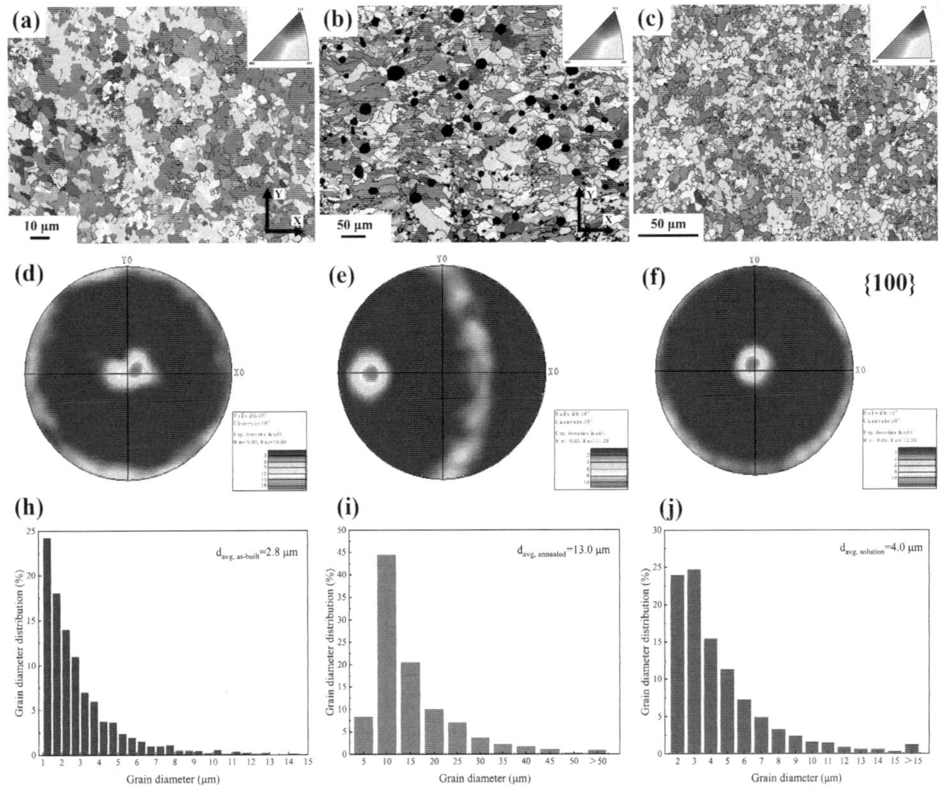

Figure 5. (a-c) EBSD maps, (d-f) pole figures, and (g-i) grain size distribution of specimens with different conditions: as-built (a, d, h) annealing (b, e, i) and solution (c, f, i) conditions.

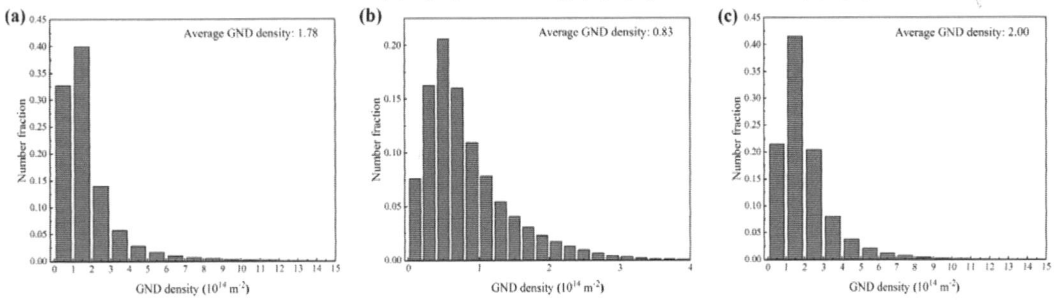

Figure 6. GND density of as-built (a) annealed (b) and solution (c) specimens.

3.2. The effect of heat treatment on mechanical properties

Figure 7 illustrates the tensile characteristics of specimens treated with annealing and solution processes at different temperatures. As the annealing temperature increases, the yield strength and the tensile strength both decrease (196.61 MPa, 120.21 MPa), while the elongation increases (18.64%). As the solution temperature increases from 440°C to 560°C, both yield strength and the tensile strength increase first and then decrease, while the trend of elongation is opposite.

As the annealing temperature increases, the diameter of the holes and cracks increases in Figure 8. When the temperature is above 280°C, equiaxed dimples become visible on the samples. The fracture morphology is characterized by ductile fracture. Figure 9 shows equiaxed dimples are present on the surface of the fracture. When the solution temperature increases from 470°C to 530°C, the plastic deformation ability of the dimple accumulation region decreases relatively, leading to a lower elongation compared to other specimens.

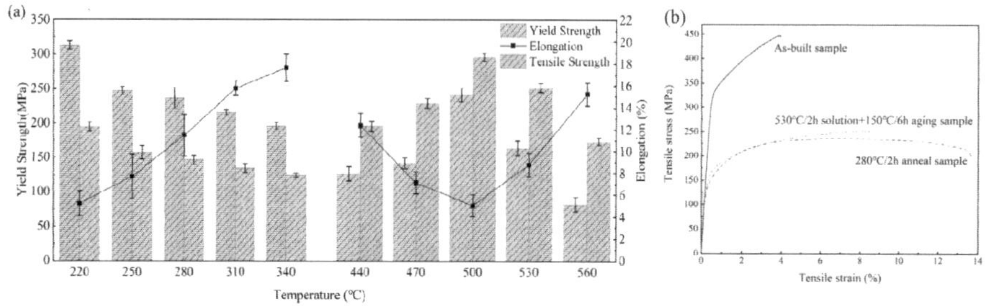

Figure 7. (a) Tensile characteristics of different heat-treated nano SiC/AlSi7Mg composites, (b) tensile curves of specimens under different conditions.

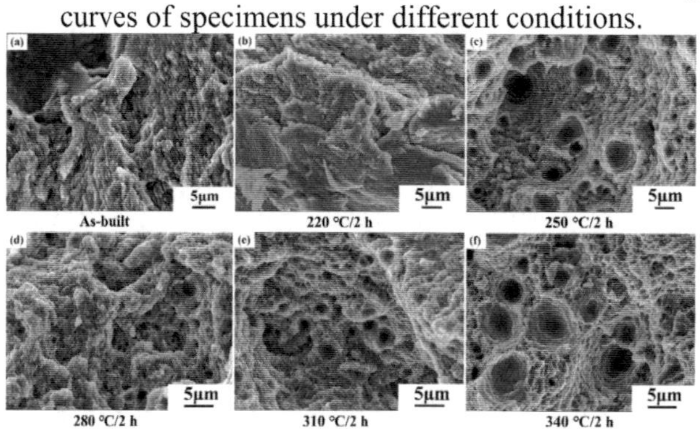

Figure 8. Fracture diagrams of as-built (a) and different annealing temperature (b-f) specimens.

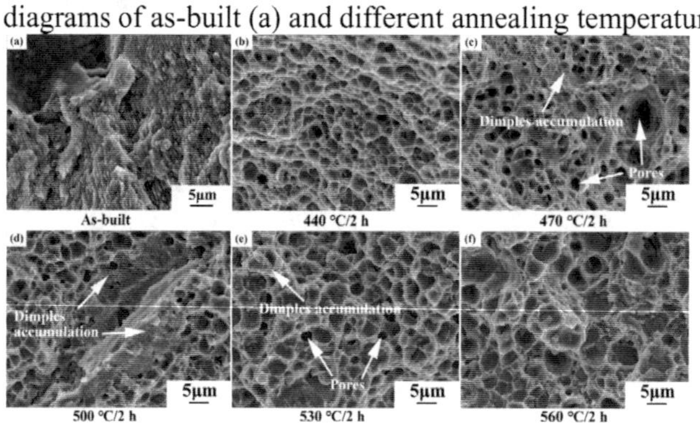

Figure 9. Fracture diagrams of as-built (a) and different solution temperatures (b-f) specimens.

4. Conclusions

This study investigated the effect of heat treatment on the microstructure and mechanical properties of nano SiC/AlSi7Mg specimens fabricated by LPBF.

(1) As the annealing temperature increased, the network structure of the microstructure in the LPBF-formed nano SiC/AlSi7Mg sample was progressively disrupted, leading to a gradual increase in the size of precipitated particles. This resulted in a significant increase in grain size, negatively impacting strength. Consequently, both yield strength and tensile strength gradually decreased, while elongation produced an increase. When the temperature is above 280°C, equiaxed dimples become visible on the samples. The damage mode shows ductile fracture.

(2) As the solution temperature increased, the diameter of the precipitated particles initially increased before decreasing, the number of particles decreased gradually, and the spacing of the particles increased. When the solution temperature increases, the plastic deformation ability of the dimple accumulation region decreases relatively, both yield strength and the tensile strength show an increase followed by a decrease, and the elongation decreases first and then increases.

Acknowledgment
Funding and other support was provided by the National Natural Science Foundation of China [No. 52071069].

References

[1] Yu, W.H., Sing, S.L. (2019) Particle-reinforced metal matrix nanocomposites fabricated by selective laser melting: A state of the art review, Progress in Materials Science, 104, 330-379.
[2] Gu, D.D. (2011) Nanocrystalline TiC reinforced Ti matrix bulk-form nanocomposites by Selective Laser Melting (SLM): densification, growth mechanism and wear behavior, Composites Science and Technology, 71, 1612-1620.
[3] Cabrini, M. (2016) Effect of heat treatment on corrosion resistance of DMLS AlSi10Mg alloy, Electrochim Acta, 206, 346-355.
[4] Wang, M., Song, B. (2019) Effects of annealing on the microstructure and mechanical properties of selective laser melted AlSi7Mg alloy, Materials Science and Engineering: A., 739, 463-472.
[5] Lakshmikanthan, A., Prabhu, T.R. (2020) The effect of heat treatment on the mechanical and tribological properties of dual size Sic reinforced A357 matrix composites, Journal of Materials Research and Technology, 69, 6434-6452.
[6] Chen, K.J. (2020) Improving the applicability of wear-resistant Al–10Si–0.5Mg alloy obtained through selective laser melting with T6 treatment in high-temperature and high-wear environments, Journal of Materials Research, 9, 9242.
[7] Takata, N., Kodairal, H. (2017) Change in the microstructure of selective laser melted AlSi10Mg alloy with heat treatments, Materials Science and Engineering: A., 705, 218-228.

Continuous cooling transformation behavior of V microalloyed hot-bent X80 pipeline steel

Tao An[1,2*], Wenyue Liu[1,2], Tianyi Li[1,2], Yan Zang[1,2] and Jiangwen Li[1,2]

[1]State Key Laboratory of Metal Materials for Marine Equipment and Application, Anshan, Liaoning, 114009, China
[2]Ansteel Beijing Research Institute Co. LTD, Beijing, 102200, China

*Corresponding author's e-mail: AnTao202372@outlook.com

Abstract: Continuous cooling transformation (CCT) behavior of a vanadium-microalloyed hot bent pipeline steel was studied by using Gleeble 3800 thermal simulation machine, and the hot-rolling technique of the tested steel was designed. Acicular ferrite and granular bainite can be observed with the cooling speeds ranging from 1~ 40 °C/s. Higher cooling speed leads to grain refinement. The yield strength is 570 MPa. The ultimate tensile strength (UTS) is 757 MPa. The fracture elongation is 22.50 %. The low temperature toughness at -40 °C is 325 J. The tested steel has both high strength and good low temperature toughness, and meets the mechanical performance requirement of API Spec 5L specification for X80 pipeline steel.

1. Introduction

Pipeline transportation is the fastest and most economical method for the transportation of natural gas [1]. The microalloying of niobium and titanium is often used in pipeline steel [2-4], which can enhance the strength and low temperature toughness of the steel through grain refinement, solid solution strengthening and precipitation hardening. However, the microalloying of vanadium in pipeline steel is usually not utilized, because during thermomechanically controlled process, vanadium carbide and nitride will not precipitate, and have little influence on mechanical properties.

During laying a long transmission pipeline, a large number of bends are required, which must be adapted to the main pipeline [5]. Existing techniques for changing the direction of pipelines include elastic laying, cold bending and hot bending. Hot bending pipelines are widely used in construction due to their advantages such as reliable quality, good applicability and high production efficiency [6]. During the hot pipe bending process, the pipeline is inductively heated to soften and bent, and then tempered. But the strength of the pipeline steel with Nb and Ti microalloyed will decrease during the induction heating and tempering process. Recent studies have shown that the loss of strength of pipeline steel during hot bending process can be offset by vanadium-microalloying, because of the precipitation of vanadium carbide and vanadium nitride during tempering process. [7]. Besides, the addition of V can improve the low temperature toughness of heat affected zone during welding [8].

CCT curves is very important for the designation of steel production process [9,10]. In this work, the influence of vanadium element on the CCT behivour was studied. The CCT diagram was plotted to support the production of vanadium-microalloyed hot-bent pipeline steel. The microstructure of the investigated steel after hot-rolling was characterized. And its mechanical properties were measured.

2. Material and methods

The tested material is a V-microalloyed, hot-bent X80 pipeline steel. Table 1 shows the chemical composition. Besides, the phase transition temperature were calculated by thermodynamic calculations, the used software was Thermo-Calc 2020b (TC) software with the TCFE10 database.

Table 1. Chemical composition in wt. %.

C	Si	Mn	Ni	Cr	Cu	Ti	Nb	V	Mo	Al	Fe
0.06	0.28	1.7	0.2	0.25	0.18	0.025	0.055	0.06	0.15	0.025	Bal.

The billet was firstly manufactured by a vacuum induction furnace. Cylindrical samples were produced from the billet for controlled cooling test. The diameter and length of the sample is 6 mm and 80 mm. The investigation of the phase transition behavior of the tested steel was conducted on a Gleeble 3800 machine. The samples were controlled heated and cooled for the construction of CCT diagram, the heating history is shown in Figure 1.

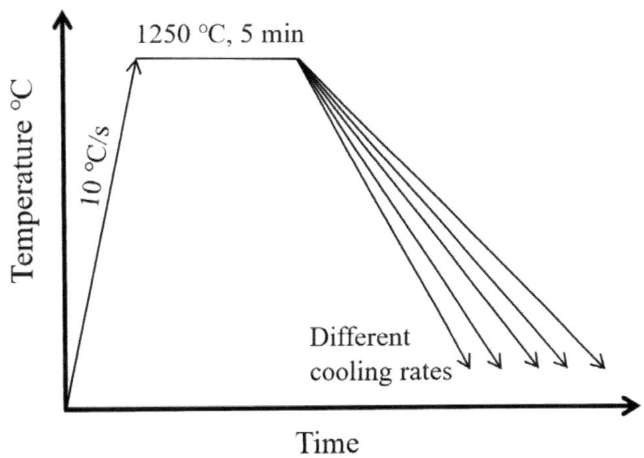

Figure 1. Experimental process chart for determination of CCT curve.

Optical microscope (OM) and X-ray diffractometer (XRD) were used for microstructure analysis and matrix determination of samples after different cooling speeds. Prior to OM characterization, OM samples were ground and polished by sandpaper. 4 % nital solution was used for the sample etching. The Vikers hardness was then measured on the OM samples using HV10. The used OM is a ZEISS Axio Observer 7 OM.The used XRD is bruker D8 XRD using Cu-Kα radiation.

After the construction of the CCT, the hot-rolling process of the tested steel was designed. The mechanical properties was tested, which include yield strength, UTS, fracture elongation and low temperature toughness at -40 °C. Beside, the tested steel was characterized by transmission electron microscope (TEM), the used TEM is Talos F200i TEM.

3. Results and discussion

3.1 Thermodynamic calculation

The temperature dependence of the equilibrium phases in tested steel is shown in Figure 2 (a) and (b). At 500 °C, the matrix phase in the steel is the ferrite (BCC_A2) phase, while the precipitate phases in the steel are the MC_ETA, FCC_A1#2 and FCC_A1#3. As can be seen in Figure 2 (b) and (c), the FCC_A1#2 is niobium-rich carbide and titanium-rich carbides, which are supposed to be stable below 1220°C. In other words, niobium, titanium-rich carbides precipitate from the austenite phase. The equilibrium calculation of FCC_A1#3 is shown in Figure 2 (b) and (d). The FCC_A1#3 is vanadium-rich carbides which are stable below 780°C. According to the equilibrium calculation, in addition to the element V, the MC_ETA also consists of a high proportion of the element Mo, which is stable below 690°C. vanadium-rich carbides and molybdenum, vanadium-rich carbides begin to precipitate when both austenite and ferrite are present.

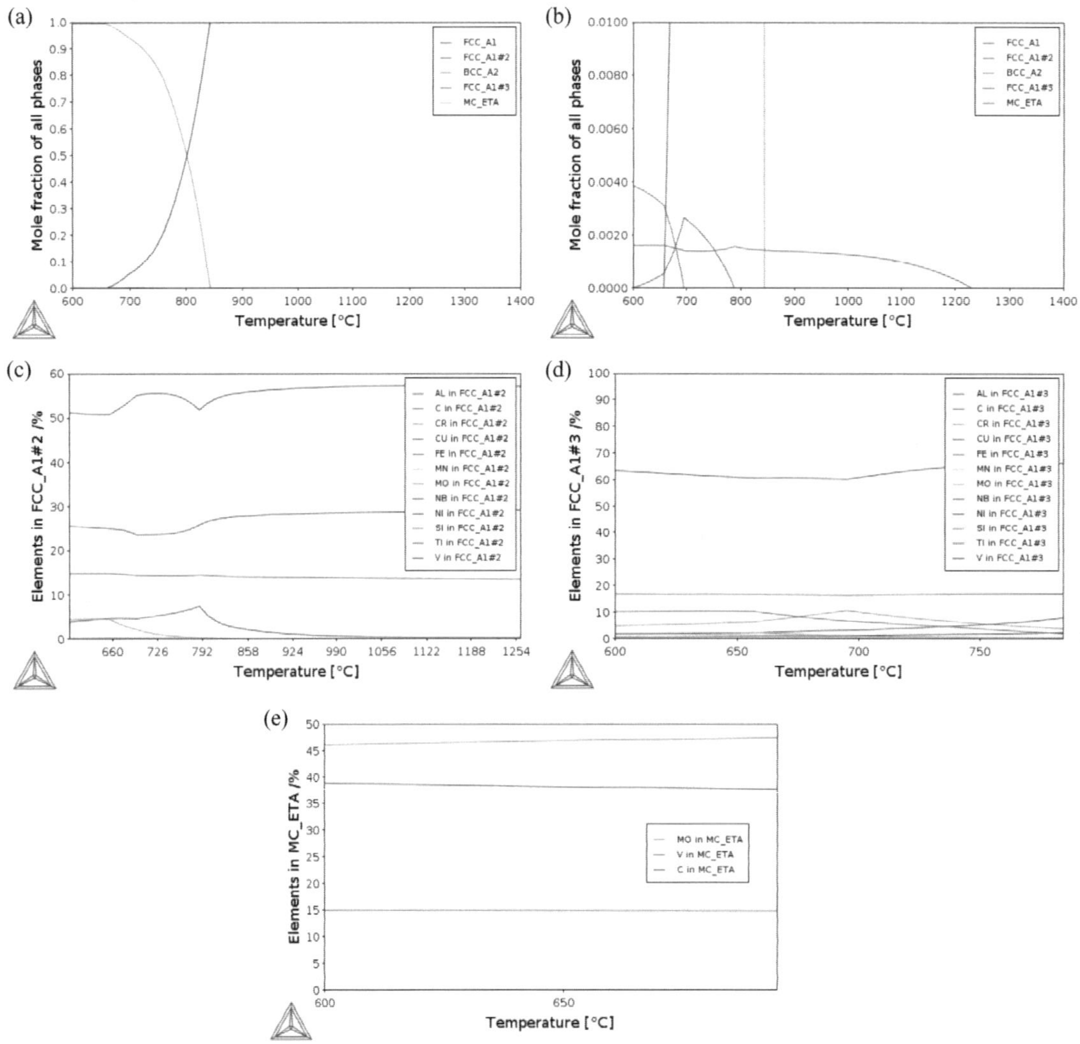

Figure 2. Phase fraction of equilibrium phases (a) and precipitates (b) in the tested steel, chemical composition of FCC_A1#2 (c), FCC_A1#3 (b) and MC_ETA (c) precipitates.

3.2 Continuous cooling transformation curves

Figure 3 shows the microstructures with different cooling speeds. Faster cooling speed leads to finer grain size. When the cooling speed is 0.3°C/s, polygonal ferrite and granular bainite can be recognized. The microstructure of the tested steel gradually changes from polygonal ferrite + granular bainite to granular bainite + acicular ferrite when the cooling speed is increasing. If the cooling speed is faster than 1 °C/s, the formation of polygonal ferrite is prevented. At this time, only granular bainite and acicular ferrite can be recognized.

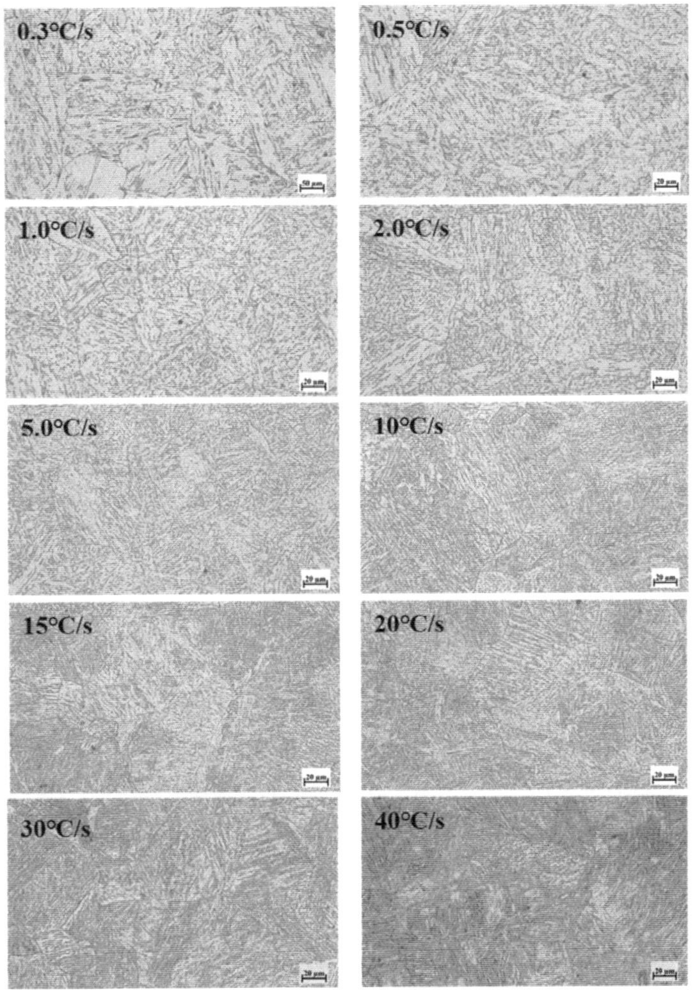

Figure 3. Microstructure evolution of the tested steel during controlled cooling process.

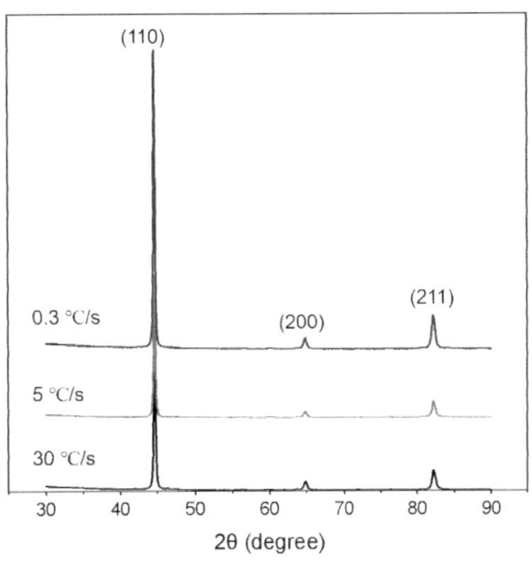

Figure 4. XRD patterns of the tested steel.

The XRD spectra with the cooling speeds of 0.3, 5, 30 °C/s is shown in Figure 4. The samples with different cooling speeds are all BCC structure and there is no trace of FCC structure.

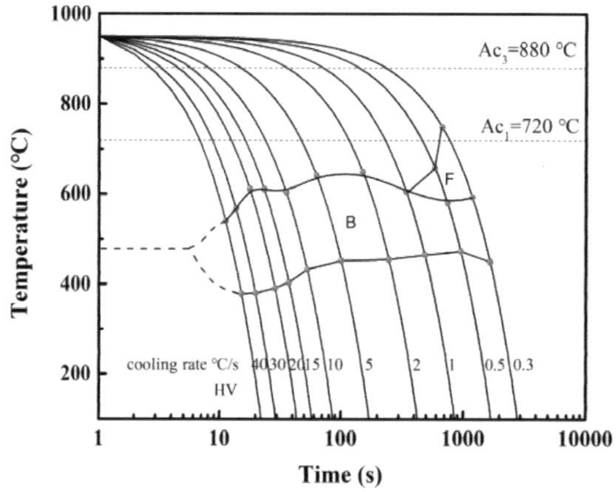

Figure 5. The constructed CCT curve.

The dilation of the sample against time was recorded by thermal simulation with different cooling speeds. The tangential method based on the inflection point of the temperature-dilation curve was utilized for the determination of phase transition temperature of austenite, which was used for the construction of CCT diagram in combined with microstructure analysis. Figure 5 plots the CCT diagram of the tested steel. The measured A_{c1} and A_{c3} temperatures are 720 °C and 880 °C, respectively. At a cooling speed of 0.3 ~ 1 °C/s, only ferrite and bainite exit. With higher cooling speed, the beginning of phase transition will be delayed and the proportion of ferrite decreases. At a cooling speed of 1 ~ 20 °C/s, only bainite exits and the starting temperature of bainite formation rises slowly. At a cooling speed of 20 ~ 40 °C/s, still only bainite will form, and the start temperature of the bainite formation decreases slowly.

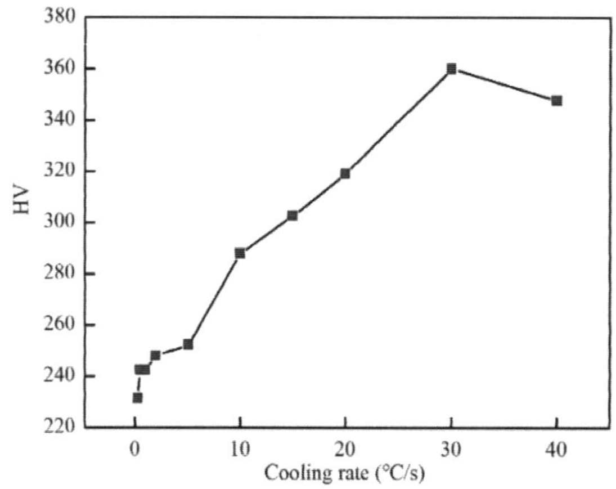

Figure 6. The Vickers hardness.

The hardness of the samples with different cooling speeds was measured by the Vickers hardness machine with HV10, as shown in Figure 6. The hardness can be influenced by various factors, such as grain size, phase, dislocation density, etc. And the hardness can also reflect the changes in the UTS of the steel. As can be seen in Figure 6, increasing the cooling speed leads to an increase in hardness. When the cooling speed increases from 0.3 to 1 °C/s, the hardness value improves significantly from

231.6HV to 242.5HV; when the cooling speed increases from 1 to 30 °C/s, the hardness improves from 242.5HV to 360.1HV; when the cooling speed exceeds 30 °C/s, the hardness value decreases a little.

3.3 Production practice of the tested steel

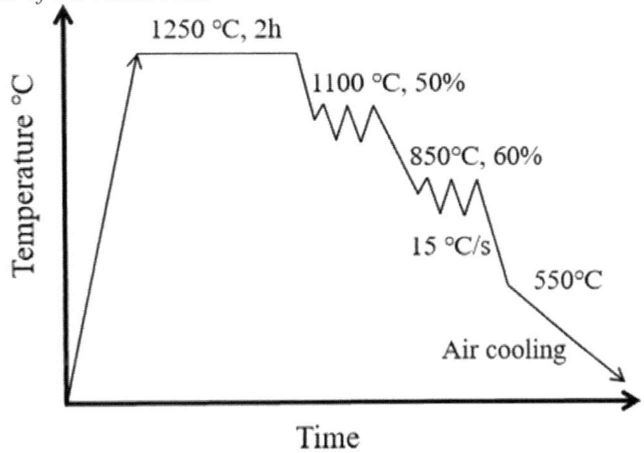

Figure 7. Hot-rolling technique for tested steel.

The hot-rolling process was designed according to the CCT curve for a high strength and good toughness. The billet was heated to 1250 °C for 2 hours to be fully austenitized, and then rough rolled at 1100 °C with a reduction of 50%, finish rolled at 850 °C with a rolling reduction of 60%, then fast cooled to 550 °C with a cooling speed of 15 °C/s and air-cooled, as shown in Figure 7.

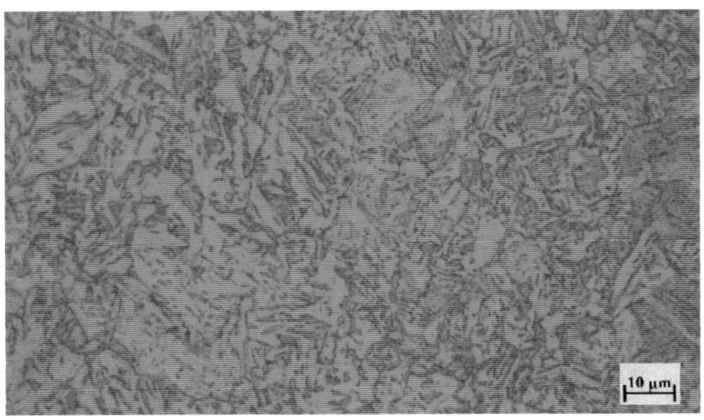

Figure 8. Microstructure of the tested steel under OM.

The microstructure of the tested steel after hot-rolling is shown in Figure 8 and Figure 9. The OM microstructure consists of acicular bainite and granular bainite. TEM examination shows an obvious lath-like microstructure and massive niobium, vanadium-rich carbides. The presence of these carbides is consistent with the results of thermodynamic calculations. The grain size of the steel after hot-rolling is finer than in CCT experiments at the same cooling speed. This phenomenon is due to the facts that 1) under the conditions of applied deformation, the grains are more broken and the dislocation density increases, which hinders grain growth; 2) the deformation before the phase transition generates a large amount of residual strain, so that additional strain energy can provide nucleation energy for the formation of a grain core, causing more grains to nucleate.

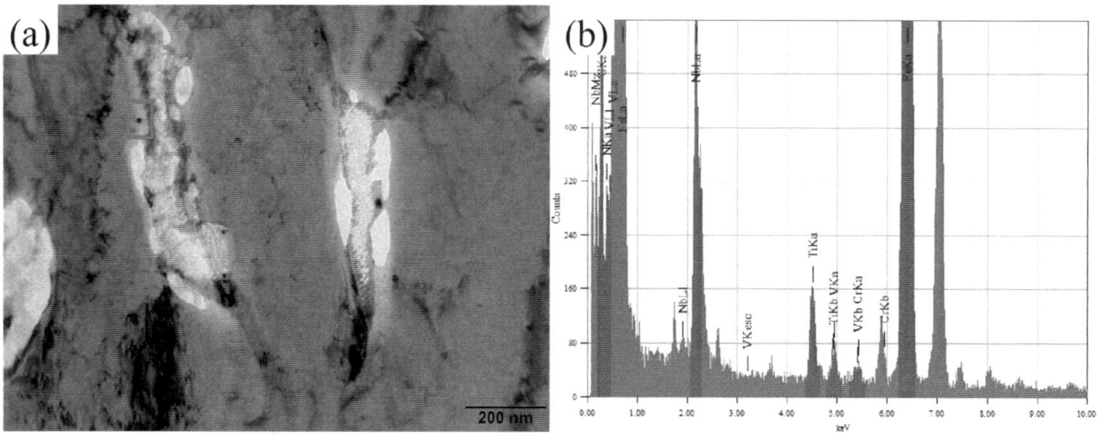

Figure 9. Microstructure of the tested steel under TEM.

Table 2 shows the detailed mechanical properties of the tested steel, which meets the mechanical performance requirements of API Spec 5L specification for X80 pipeline steel.

Table 2. Mechanical properties.

UTS /MPa	Yield Stength /MPa	Fracture Elongation /%	Charpy impact absorbed energy at -40 °C /J
757	570	22.50	325

4. Conclusions

In this paper, the CCT curve of a vanadium-microalloyed steel for hot-bent pipelines are investigated and the conclusions are as followed:

1) The measured A_{c1}, A_{c3} temperature of the tested steel are 720 °C and 880 °C, respectively.
2) When the cooling speed is 0.3 ~ 1 °C/s, polygonal ferrite and granular bainite can be recognized; when the cooling speed is faster than 1 °C/s, only granular bainite and acicular ferrite can be observed.
3) After hot-rolling, the properties of the investigated steel are in accordance with API Spec 5L specification for X80 pipeline steel, whose yield strength, UTS and fracture elongation are 570 MPa, 757 MPa and 22.50 %, respectively, and the Charpy impact absorbed energy at -40 °C is 325 J. Acicular bainite and granular bainite was found under OM.

References

[1]Saniere A, Hénaut I and Argillier J F 2004 Pipeline transportation of heavy oils, a strategic, economic and technological challenge. Oil & Gas Sci. and Tec.- Rev. IFP 59 455-66.

[2]Niu T, Kang Y L, Gu H W, Yin Y Q and Qiao M L 2010 Precipitation behavior and its strengthening effect of X100 pipeline steel. J. Iron Steel Res. Int. 17 73-8.

[3]Zhao A M, Wang Y, Chen Y L, Tang D, Gao X T and Zuo B Q 2011 Precipitation behaviors of X80 acicular ferrite pipeline steel. Int. J. Miner. Metall. Mater. 18 309-13.

[4]Zhang J M, Huo C Y, Ma Q R and Feng Y R 2018 NbC-TiN co-precipitation behavior and mechanical properties of X90 pipeline steels by critical-temperature rolling process. Int. J. Press. Vessel. Pip. 165 29-33.

[5]Karamanos S A 2016 Mechanical behavior of steel pipe bends: an overview. Journal of Pressure Vessel Technology 138 041203.

[6]Wang L, Zhou P S, Hu Y W and Wang B 2022 Effect of microstructure and texture on the mechanical properties in high strength pipeline bend. Int. J. Press. Vessel. Pip. 195 104604.

Journal of Physics: Conference Series **2842** (2024) 012040 doi:10.1088/1742-6596/2842/1/012040

[7]Silva R A, Pinto A L, Kuznetsov A and Bott I S 2018 Precipitation and grain size effects on the tensile strain-hardening exponents of an API X80 steel pipe after high-frequency hot-induction bending. Metals 8 168.

[8]Stornelli G, Tselikova A, Mirabile G D, Mortello M, Schmidt R, Sgambetterra M, Testani C, Zucca G and Di S A Influence of Vanadium Micro-Alloying on the Microstructure of Structural High Strength Steels Welded Joints. 2023 Materials 16 2897.

[9]Huda N, Midawi A, Gianetto J A and Gerlich A P 2021 Continuous cooling transformation behaviour and toughness of heat-affected zones in an X80 line pipe steel. J. Mater. Res. Technol. 12 613-28.

[10]Chu F, Zhang J, Zhang Y 2020 Continuous Cooling Phase Transformation Behavior of Sino-Russian East Line Large Wall Thickness X80 Pipeline Steel. Hot Working Technology 49 146-8.

Study on the mechanism of pressure-induced erosion in molten pool of EDM metal

Chao Chen, Shoudong Ni* and Fouad Ben Mhahe

School of Mechanical and Power Engineering, Nanjing Tech University, Nanjing, 211800, China

*Corresponding author's e-mail: nsd@njtech.edu.cn

Abstract: This study employs molecular dynamics methods to explore the mechanism of pressure-induced erosion due to the pressure differential within and outside the molten pool during the EDM process. Through simulations of the EDM process for metals such as copper and tungsten, it was found that high melting point materials exhibit slow melting and evaporation rates within the molten pool, resulting in lower internal pressures and reduced efficacy in material erosion, thereby exhibiting high erosion resistance. Experimental validation has demonstrated that the machining efficiency of metals increases with the decrease in their melting points, thus confirming the accuracy of the simulation results. This research not only unveils the micro-dynamic mechanisms of material erosion but also provides a theoretical foundation and experimental support for the advancement of EDM technology.

1. Introduction

Electrical discharge machining (EDM) employs a sophisticated approach to the precision shaping of metals by harnessing thermal forces. During the machining operation, the plasma channel's elevated temperatures cause the electrode material to transition into molten and vapor states, facilitating the erosion of the material [1]. Jiang significantly enhanced the machining efficiency of Polycrystalline diamonds (PCD) by utilizing Cu-Ni electrodes and graphene suspended in dielectric fluid [2]. Peng, building on the tri-phase flow model, identified that material removal in EDM predominantly occurs through molten splashing [3]. During the machining operation, the plasma channel's elevated temperatures cause the electrode material to transition into molten and vapor states, facilitating the erosion of the material [4]. Piyar has enhanced the surface quality of nickel-based superalloys by integrating powder-mixed dielectric fluid into the EDM methodology [5]. Ji discovered that surfaces textured using oil-in-water working fluids exhibited high wear resistance [6]. However, a comprehensive explanation of the specific mechanism behind molten material erosion during EDM is still lacking.

In the course of spark discharge, the plasma channel's temperature significantly surpasses the boiling thresholds of conventional materials, inducing localized melting and vaporization within the molten pool, thereby influencing erosion dynamics [7]. As the electrode material melts and vaporizes due to high temperatures, the evaporated material separates from the surface in the form of atomic clusters, and the material at the bottom of the molten pool also reaches its boiling point and begins to evaporate [8]. This evaporation process occurs within a very small area and over a brief period, causing a rapid increase in pressure within that area. The rapid increase in internal pressure leads to the expansion of the material volume, forming a protruding convex surface. When internal pressure disparities exceed the surface tension threshold of the molten pool, its convex surface breaches, leading to the expelling of molten constituents.

This investigation employs molecular dynamics to explore the erosion phenomena at the atomic scale during discharge events, corroborated by empirical studies, with the goal of elucidating the micro-dynamic erosion mechanisms. This provides both theoretical insights and experimental validation crucial for advancing EDM applications. It should be noted that due to limitations in computational resources, the simulation model used in this study differs in time and spatial scales from the actual EDM process. Nevertheless, the qualitative study of the small-scale discharge process through molecular dynamics methods offers an important perspective for understanding the spark discharge process and the mechanisms of material erosion.

2. Molecular dynamics simulation of single-pulse spark discharge

2.1. Basic simulation theory
Molecular Dynamics Simulation (MDS) is a simulation technique that employs numerical methods to simulate the dynamic behavior of particles based on the principles of Newtonian classical mechanics. This technique, by integrating the equations of motion of particles, can predict the position and velocity of particles over time, thereby providing a deep understanding of the connection between the microscopic states of particles (such as coordinates, velocity, acceleration, etc.) and macroscopic physical properties (such as temperature, pressure, elastic modulus, etc.), offering a powerful tool for analyzing the physical and chemical properties of materials. Utilizing LAMMPS, a computational tool developed by Sandia National Laboratories, this study examines the pressure-driven erosion mechanisms in the molten pool under spark discharge conditions, specifically analyzing metals Cu and W due to their contrasting melting points.

2.2. Potential function selection
In MDS, the potential function plays a crucial role as it defines the interactions between particles within the simulation system. These interactions not only influence the motion patterns of the particles but are also directly related to the overall properties of the material. Thus, the accuracy of the simulation largely depends on the precision with which the potential function describes the interparticle forces. In this chapter, the Embedded Atom Method (EAM) potential function proposed by Daw and Baskes was used, and it was developed based on the density functional theory and effective medium theory. This method, by considering the impact of local electron density on atomic interactions, enhances the computational accuracy in simulating the structure of metals and their thermodynamic properties [9]. Therefore, this study employed the EAM potential function to simulate the interactions between metal atoms in the simulation, thereby ensuring the accuracy of the simulation results and the feasibility of practical applications.

2.3. Heat source model
The spark discharge plasma channel, comprising ions, electrons, and neutral atoms, rapidly expands following the dielectric breakdown. In the simulation model, the heat source can be simplified into two states: expanding and constant diameter states. According to phenomena observed by Kojima using high-speed photography, the plasma channel rapidly expands within 2 μs, after which its diameter remains constant [10]. Consequently, this analysis does not delve into the arc column's formation or expansion, nor the plasma particles it contains, but rather models it as a thermal source following a Gaussian heat flux distribution.

2.4. Simulation condition
Given the realities of EDM, where the working gap, although filled with dielectric fluid, predominantly contains bubbles, the model simplistically assumes the working gap is in a vacuum state for ease of analysis. To enhance the accuracy of the simulation and ensure that the system can

quickly balance while maintaining the initial total momentum at zero, the initialization of particle velocities was conducted using the Maxwell-Boltzmann distribution method, which facilitates the rapid equilibration of the system.

2.5 Simulation process of material erosion

MDS was utilized to examine the impact of internal pressure variations within the molten pool during a single-pulse discharge on molten material erosion. The simulation revealed that the sustained application of a Gaussian heat source incrementally elevated the material's temperature in the discharge zone. When the electrode material's temperature surpasses its melting and boiling points, melting and evaporation ensue, as depicted in Figures 1, 2, and 3. During the single-pulse discharge simulation, a large part of the processed copper material reached its melting point at the beginning of the discharge, but material erosion was not observed until 80 ps later. Prior to 80 ps, the pressure in the molten zone was insufficient to disrupt interatomic bonds significantly, resulting in minimal material erosion beyond a modest amount of vaporization. For the high-melting-point refractory metal tungsten, the rate of material erosion was even lower, with only a few atoms showing activity after 200 ps, and no significant bulging of the molten pool was observed, indicating that the erosion of high melting point materials requires a longer discharge time.

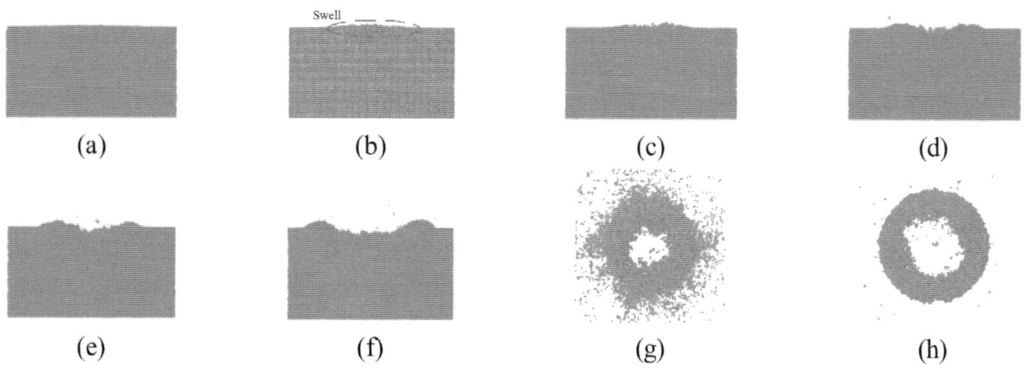

Figure 1. Cross-section of simulation process of single-pulse discharge corrosion Cu : (a) 20 ps, (b) 40 ps, (c) 60 ps, (d) 80 ps, (e) 100 ps, (f) 200 ps, (g) 100 ps, (h) 200 ps.

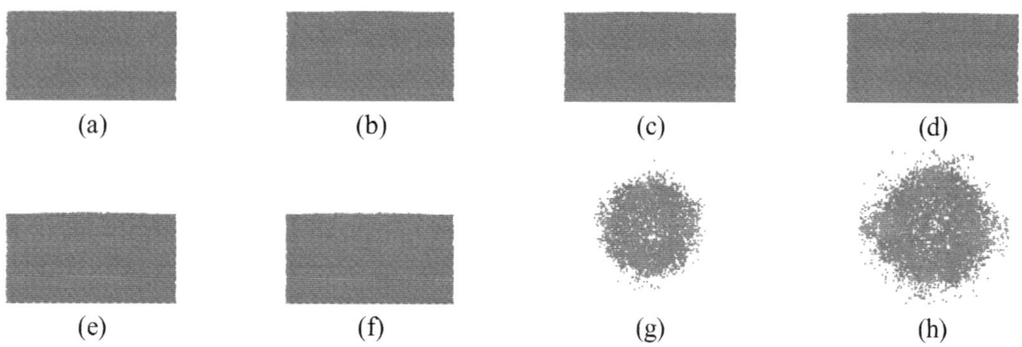

Figure 2. Cross-section of simulation process of single-pulse discharge corrosion W : (a) 20 ps, (b) 40 ps, (c) 60 ps, (d) 80 ps, (e) 100 ps, (f) 200 ps, (g) 100 ps, (h) 200 ps.

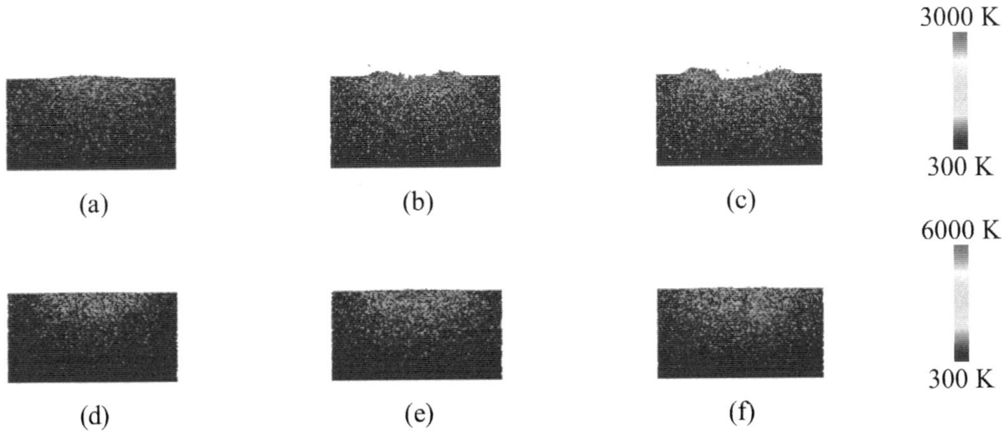

Figure 3. Cross-section of material temperature field distribution during simulated discharge. Cu: (a) 40 ps, (b) 80 ps, and (c) 120 ps. W: (d) 40 ps, (e) 80 ps, and (f) 120 ps.

Thus, the pressure differential across the molten pool's surface exceeding its surface tension contributes to the expulsion of molten material, a phenomenon described as the internal pressure erosion mechanism. Simulation data demonstrate that in practical discharges, the extraction of molten material is independent of the presence of bubbles within the pool. Significant material removal was noted in both gaseous and vacuum conditions, with vacuum simulations particularly showing that bulges in the molten pool could expand and burst outward, leading to substantial material removal and the formation of larger discharge pits. However, in actual discharges, the external gap pressure of the molten pool suppresses the expansion and rupture of the bulge, so the intense expansion and rupture of the bulge, as shown in the simulations, do not occur. These findings underscore that the rise in internal pressure within the molten pool plays a crucial role in driving material removal.

3. Comparative experiment of metal holes in EDM

3.1. Experimental condition

The experimental platform chosen was a precision EDM machine. Before starting the machining process, the metal workpiece was placed in an ultrasonic cleaner and cleaned for 10 minutes using anhydrous ethanol. The corresponding parameters of the machine tool were set according to Table 1, followed by the execution of the machining process. Throughout the process, the mass of both the workpiece and the tool electrode before and after the machining, as well as the machining duration, were documented to assess material removal efficiency and electrode wear rate.

Table 1. Experimental processing conditions.

Machining parameter	Value
Workpiece electrode	W/Mo/Ni/SUS304/Cu
Workpiece electrode size (mm)	40×15×2
Tool electrode	Brass
Tool electrode diameter (mm)	0.5
Tube electrode rotation (r/min)	100
Electrode feed depth (mm)	30
Working fluid	H_2O
Working fluid pressure (MPa)	5

3.2. Comparison of processing efficiency of metals with different melting points

This experiment considered the internal pressure erosion mechanism within the molten pool during machining, aiming to verify the impact of different melting points of metal materials on machining efficiency. EDM was performed on five different materials using deionized water as the working fluid. The experimental results, as shown in Figure 4, indicate that the machining efficiency of metals increases with the decrease in melting point. The reason for this is that materials with higher melting points possess stronger atomic bonds, which slow down their melting and evaporation during the discharge process, resulting in a lower internal pressure within the molten metal and, consequently, reduced material removal efficiency. These observations align with the internal pressure difference erosion mechanism identified in the molten pool and corroborate the findings from the single-pulse discharge simulations.

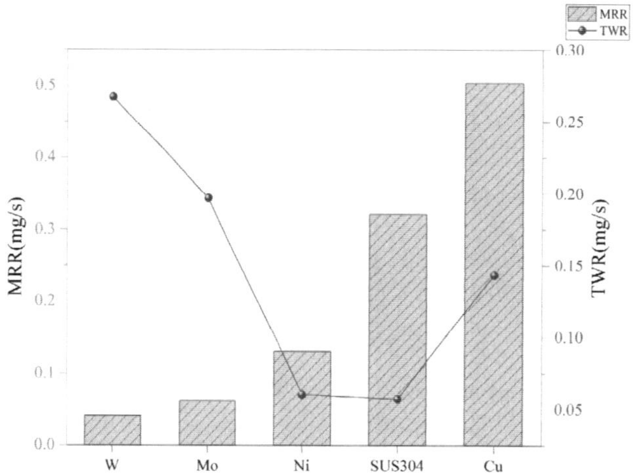

Figure 4. MRR and TWR of different metals in EDM.

4. Conclusion

This paper conducted molecular dynamics simulations of single-pulse discharge in EDM of metals, particularly focusing on the erosion mechanism caused by the pressure differences within the molten pool. The principal conclusions of this research are outlined below:

(1) The simulations demonstrated that the erosion mechanism is primarily driven by the pressure difference inside and outside the molten pool. This differential pressure influences the material removal process during EDM.

(2) Using copper and tungsten as examples, significant differences in the internal bonding forces of metals were observed. These differences affect the erosion resistance of materials.

(3) It was found that high-melting-point materials, such as tungsten, exhibit slow melting and evaporation rates within the molten pool. This results in lower internal pressures, which make it challenging to erode the electrode material and lead to high discharge erosion resistance.

(4) Experimental results indicate that the machining efficiency of metals increases with decreasing melting points. High-melting-point materials have stronger atomic bonds that slow down the melting and evaporation processes during discharge, consequently reducing material removal efficiency.

(5) The experimental findings confirm the theoretical predictions about the internal pressure erosion mechanism within the molten pool, as suggested by the results of the single-pulse discharge simulations.

Acknowledgments

This work was supported by the School of Mechanical and Power Engineering at Nanjing Tech University. We extend our gratitude to the editor and reviewers whose insightful comments have significantly enhanced this manuscript.

References

[1] Wang X, Liu Z, Hui Z, Qiu M, Tian Z, and Cao Z (2016) Microcosmic mechanism of electrical discharge machining ablation on titanium alloy. J. Int J Adv Manuf Technol, 83(5-8): 929-935. [CrossRef]
[2] Jiang M, Li L, Sun X, Wang W, and Liu Z (2023) Research on the mechanism and process of polycrystalline diamond by EDM. J. Int J Adv Manuf Technol, 125, 819-830. [CrossRef]
[3] Peng W, Tang J, Li Z, Qiu Z, and Yin M (2024) Analysis of mechanisms of bubble characteristics and molten pool dynamics in EDM based on a three-phase flow model. J. Journal of Physics D: Applied Physics, 57(25): 255303. [CrossRef]
[4] Panda D, Sahu S K and Das S R (2024) Comparative performance evaluation and economic assessment between traditional and vibration assisted EDM during machining of Inconel 718. P I Mech Eng C-j Mec, 0(0). [CrossRef]
[5] Piyar M U, Deb J B and Arindam M (2023) Effect of process parameters on the performance of powder mixed EDM of Ni-based superalloys: a review. J. Surf Rev Lett, 30(11). [CrossRef]
[6] Ji R, Zhao Q, Zhao L, Liu Y, Jin H, Wang L, Wu L, and Xu Z (2023) Study on high wear resistance surface texture of electrical discharge machining based on a new water-in-oil working fluid. J. Tribology International, 180: 108218. [CrossRef]
[7] Kitamura T and Kunieda M (2014) Clarification of EDM gap phenomena using transparent electrodes. J. Cirp Ann-manuf Techn, 63(1): 213-216. [CrossRef]
[8] Yue X and Yang X (2017) Molecular dynamics simulation of single pulse discharge process: clarifying the function of pressure generated inside the melting area in EDM. J. Molecular Simulation, 43(12): 935-944. [CrossRef]
[9] Sanchez A J, Plaza S, Lacalle D L N L and Lamikiz A (2016) Computer simulation of wire-EDM taper-cutting. J. Int J Comput Integ M, 19(7): 727-735. [CrossRef]
[10] Kojima A, Natsu W and Kunieda M (2008) Spectroscopic measurement of arc plasma diameter in EDM. J. Cirp Ann-manuf Techn, 57(1): 203-207. [CrossRef]

Prediction of flow stresses for a typical Nickel-Based superalloy during hot deformation based on constitutive equation

Jianxiao Liu[1], Chao Jiang[1], Yuewen Zhai[1*], Leyu Zhou[1], Haiyu Zong[1], Huizhen Wang[1] and Yuheng Zhan[2]

[1]Beijing Research Institute of Mechanical& Electrical Technology Co., Ltd. CAM, Beijing City, 100083, China
[2] Department of Aerospace Engineering, Faculty of Aerospace Engineering, Beijing Institute of Technology, Beijing 100081, China

[*]Corresponding author's e-mail: zhaiyuewen@163.com

Abstract: Utilizing the Gleeble-3500 thermal simulation testing machine, a series of isothermal constant strain rate thermal compression tests were executed to study the high-temperature rheological properties of the GH4169 alloy. These tests were carried out under deformation temperatures between 900°C and 1100°C, with strain rates from 0.001 to 1 s^{-1}, and involved a deformation of 60%. The results reveal that the peak stress of the GH4169 alloy decreases as the temperature increases, and the strain rate decreases during hot deformation. The activation energy for thermal deformation is calculated to be 806.39 kJ/mol. Through regression analysis and polynomial fitting of true stress-strain curves from thermal compression tests, an Arrhenius constitutive equation integrating strain compensation for high-temperature deformation was established. The model exhibits an average relative error of 8.86% between predicted and experimental values, indicating the model's good accuracy in predicting the alloy's rheological behavior during thermal deformation. By adjusting the strain rate and deformation temperature, this model can be utilized to control the stress level in the hot working process.

1. Introduction

GH4169 alloy has oxidation resistance, high strength, radiation resistance, hot processing performance, and does not contain scarce resources Co., making it a crucial material for applications in aviation, aerospace, nuclear energy, and the petroleum industry. The greatest advantage of this material is that by adjusting the thermal deformation process parameters, a variety of metallurgical products and forgings with different grain sizes and different levels of performance can be obtained, meeting the requirements of different applications in the engine of the various types of parts, which makes the application of this material continues to expand, and the amount of use is also increasing[1].

The rheological behavior of metals during hot deformation is intricate, with the parameters of the hot working process impacting the hardening and softening mechanisms of metals. Generally, an intrinsic relationship is employed to describe the plastic rheological properties of metals. In recent years, numerous scholars have delved into the deformation behaviors and organizational evolution of the GH4169 alloy. Chen et al. [2] examined the hot compressive deformation behavior of δ-phase-aged GH4169 alloy, predicting its rheological stresses and establishing a set of fifth-order polynomials of material constants for the intrinsic equation of the alloy, with strain variables as the independent variable. Si et al. [3]investigated the thermoforming behavior of GH4169 alloy treated by the DP process, establishing a high-temperature rheological constitutive equation in the stages of work hardening-dynamic restitution and dynamic rheological softening, respectively. Various methods, including hot rolling, ring forging, vacuum induction melting, and different billet-opening techniques for ingots, were used to prepare the GH4169 high-temperature alloy, whose hot deformation behavior was studied by Azarbarmas[4], Gauri [5], Ren [6] and Liu [7]. Their studies demonstrated that the initial state of GH4169 influences the hot deformation process under different microstructural conditions resulting from various thermal processing methods.

Thus, this paper explores the thermal deformation behavior of the GH4169 high-temperature alloy using isothermal constant strain rate compression tests. A constitutive equation based on strain compensation is developed to describe the rheological stresses of the metal under varying strain rates, deformation temperatures, and levels of strain. The study aims to provide data support for GH4169 hot forming processing and numerical simulation.

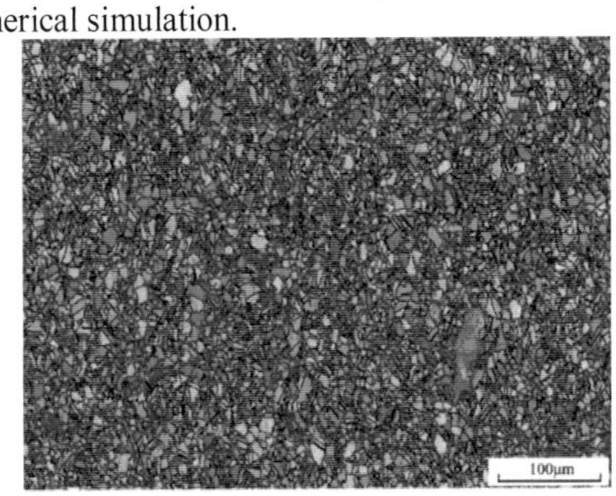

Figure 1. Microstructure of GH4169 superalloy sample.

2. Experimental materials and research methods

The material utilized in the test is the treated GH4169 high-temperature alloy bar, from which several specimens measuring φ8mm×12mm were extracted. Figure 1 illustrates the initial organization. The specimens were heated using a Gleeble-3500 thermal simulation compressor at a rate of 10°C/s to deformation temperatures ranging from 900°C, 950°C, 1000°C, 1050°C, to 1100°C. These compression temperatures were selected to encompass the forging temperature range for this alloy (900 to 1050°C). Subsequently, the specimens were allowed to dwell for 5 minutes after reaching the desired temperature to ensure uniform internal and external temperatures. Thermoforming tests were then conducted under compression at strain rates of 0.001 s^{-1},0.01 s^{-1}, 0.1 s^{-1}, and 1 s^{-1}, respectively, with a deformation of 60%. The chosen strain rates were representative of those typically encountered in conventional die-forging processes. Following deformation, they were halved along their central axis using wire-cutting equipment. After grinding, polishing, and immersion in an alcohol solution for ultrasonic cleaning for 30 seconds, electrolytic polishing was performed using a solution composed of H_2SO_4 and CH_3OH at room temperature and 20°C for 10-15 seconds, applying a voltage of 25-30 V. The microstructure of the specimens, including morphology, size, and distribution, was observed using a Zeiss Gemini 500 microscope.

3. Results and discussions

3.1 Rheological behavior of GH4169 alloy during hot forming process

True stress-true strain curves for GH4169 alloy were derived from thermal simulation compression tests. **Figure 2** demonstrates the changes in rheological stress with true strain at different deformation process parameters.

The curves of the alloys under the above process conditions exhibit similar characteristics: at the onset of deformation, the number of dislocations increases rapidly with strain, resulting in a rapid accumulation of rheological stress. As deformation continues, the applied pressure persists until a critical strain is reached, triggering dynamic recrystallization. Despite ongoing work hardening, this is gradually counteracted by restitution or recrystallization, though work hardening remains dominant, causing the curve to rise with a decreasing slope. Subsequently, stress reaches a peak, signifying the acceleration and dominance of dynamic recrystallization, leading to a gradual decrease in rheological stress with increasing strain until it stabilizes, indicative of steady-state rheology.

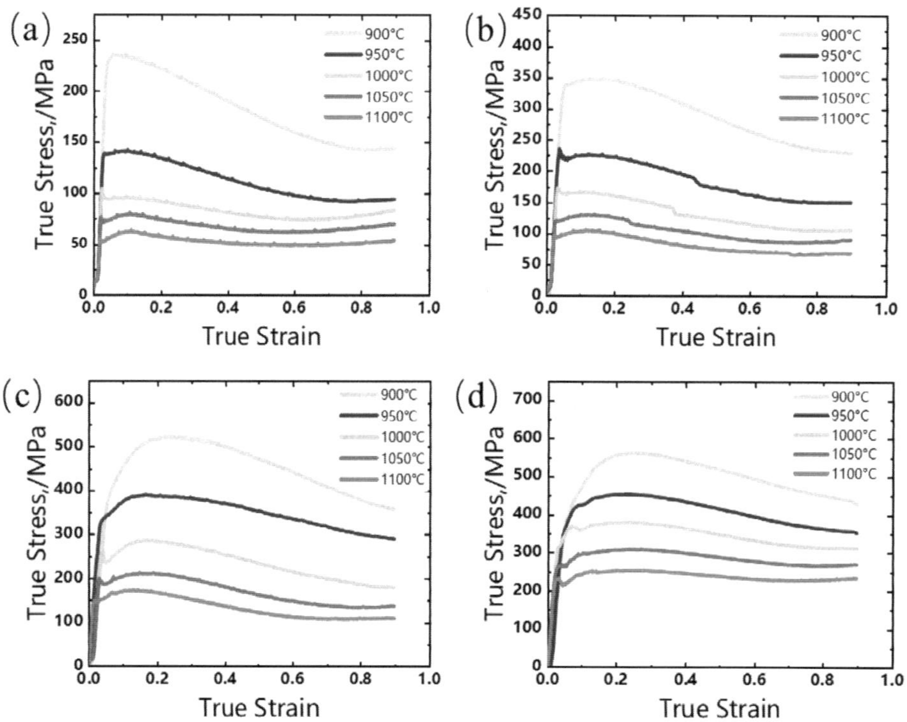

Figure 2. True stress-true strain curve of GH4169 high temperature alloy in hot compression in HIP state: (a)0.001s^{-1}; (b) 0.01 s^{-1}; (c) 0.1 s^{-1}; (d) 1 s^{-1}.

The peak stress values for each deformation condition are extracted from the $\varepsilon-\sigma$ curves, as depicted in Figure 3. Rheological stress decreases as the deformation temperature increases at a given strain rate. This phenomenon is mainly due to the higher temperature, which enhances the mobility and frequency of metal atom movements.

Rheological stress decreases with increasing deformation temperature at a specific strain rate. This is primarily attributed to the elevated temperature, which accelerates the movement and frequency of metal atoms and weakens the interaction forces between atoms. Simultaneously, higher temperatures facilitate crystal nucleus growth, thereby expediting dynamic recrystallization and mitigating work hardening induced during deformation. Consequently, higher temperatures result in shorter durations for completing the dynamic recrystallization process and facilitating plastic deformation.

At a fixed deformation temperature, rheological stress increases as the strain rate rises. Elevated deformation rates lead to relatively shorter durations for completing plastic deformation, limiting the time for dislocation motion during deformation and consequently increasing deformation resistance. Moreover, the deformation rate determines whether dynamic recrystallization occurs completely. Higher deformation rates may not afford sufficient time for dynamic recrystallization to take place, weakening the softening effect and prolonging the retention of work hardening induced during deformation.

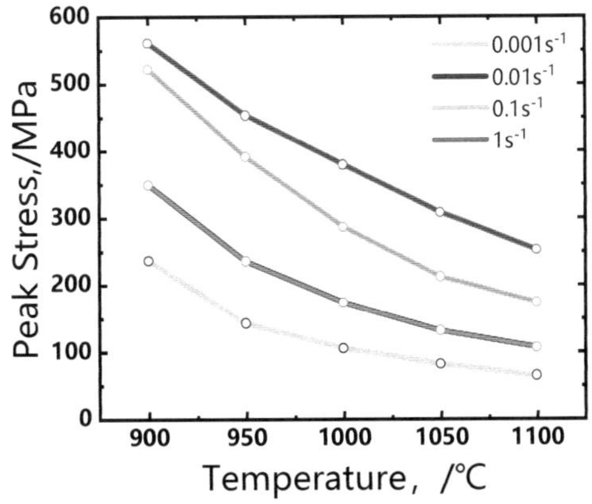

Figure 3. Peak stress of GH4169 alloy with different process parameters.

3.2 Constitutive modeling of hot forming of GH4169 alloy

A constitutive equation serves as a mathematical model derived by fitting experimental data to the material. It predicts the stress-strain relationship of the material under various processing conditions, effectively characterizing its rheological behavior during thermal processing. Constructing a constitutive model aids in optimizing processing parameters, enhancing material properties, and improving processing efficiency[8].

The Arrhenius constitutive equation is often used to describe the intricate relationship between deformation rate, deformation temperature, and the stress-strain response during high-temperature deformation of materials. To encompass the combined roles of temperature and strain rate on thermal deformation behavior, the Zener-Hollomon parameter (Z) is utilized. The formula for the Z parameter is given as follows:

$$Z = \dot{\varepsilon} \exp(Q / RT) \tag{1}$$

Q denotes the activation energy for deformation (KJ/mol), T stands for the absolute temperature(K), R represents the gas constant, generally taken, R=8.314J/(mol·K), and $\dot{\varepsilon}$ signifies the strain rate (s^{-1});.

The Arrhenius equation has 3 forms as follows:

$$\dot{\varepsilon} = A_1 \sigma^{n_1} \exp(-Q / RT) \tag{2}$$

$$\dot{\varepsilon} = A_2 \exp(\beta\sigma) \exp(-Q / RT) \tag{3}$$

$$\dot{\varepsilon} = A[\sinh(\alpha\sigma)]^n \exp[-Q / (RT)] \tag{4}$$

There, n and n_1 represent the stress exponential constant; A_1 , A_2 and β are constants, $A_1 = A / 2^{n_1}$, $A_2 = A\alpha^{n_1}$, $\alpha = \beta / n_1$.

Taking logarithms of Equation (2) simultaneously and collapsing, we obtain:

$$\ln \dot{\varepsilon} = \ln A_1 + n_1 \ln \sigma - Q / (RT) \tag{5}$$

With a constant temperature, a linear relationship is obtained by plotting $\ln \dot{\varepsilon}$ and $\ln \sigma$, with the slope represented by n_1. By inputting each data point into Equation (5) and creating the $\ln \dot{\varepsilon} - \ln \sigma$ plot shown in Figure 4, the average slope across different temperatures can be obtained through linear fitting, n1=5.5868.

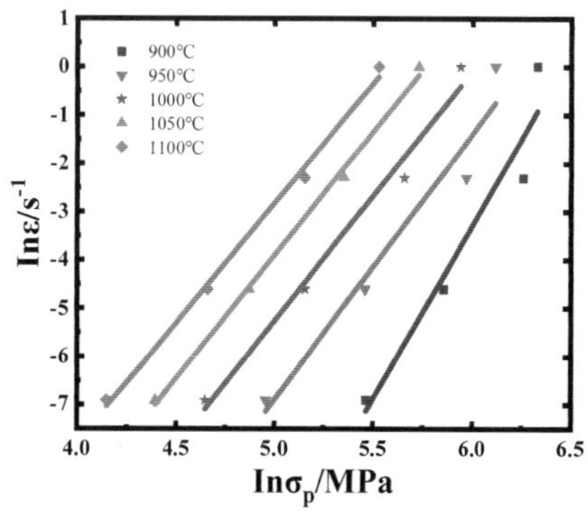

Figure 4. Relation curves of $\ln\dot{\varepsilon}-\ln\sigma$ at different temperatures.

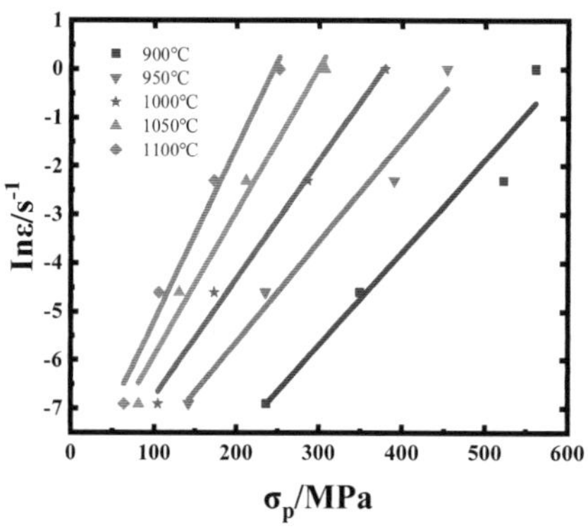

Figure 5. Relation curves of $\ln\dot{\varepsilon}-\sigma$ under different temperatures.

Taking logarithms on both sides of Equation (3)simultaneously and simplifying, we get:

$$\ln\dot{\varepsilon}=\ln A_2+\beta\sigma-Q/(RT) \tag{6}$$

Substituting into Equation (6) and plotting the $\ln\dot{\varepsilon}-\sigma$ relationship curve, in **Figure 5**, we obtain $\beta=0.0259MPa^{-1}$, then $\alpha=\beta/n_1=0.0046MPa^{-1}$.

Taking logarithms of Equation(4) yields:

$$\ln\dot{\varepsilon}=\ln A+n\ln[\sinh(\alpha\sigma)]-Q/(RT) \tag{7}$$

Substituting α into Equation (7) and plotting $\ln[\sinh(\alpha\sigma)]-\ln\dot{\varepsilon}$ against $\ln\left[\sinh\left(\alpha\sigma\right)\right]-1/\mathrm{T}$ relationship curve as shown in Figure 6 and Figure 7, we obtain that n is 4.0050 and Q=806.39kJ/mol.

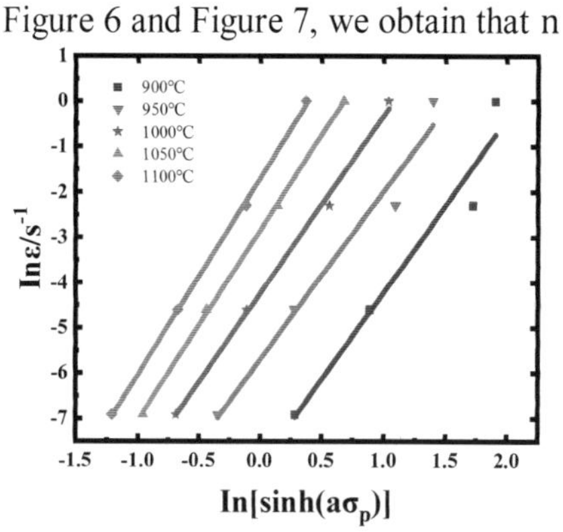

Figure 6. Relation curves of $\ln\dot{\varepsilon}-\ln[\sinh(\alpha\sigma)]$ under different temperatures.

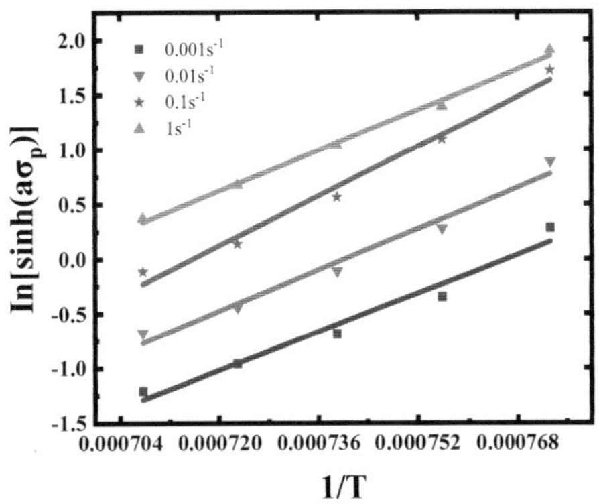

Figure 7. Relation curves of $\ln[\sinh(\alpha\sigma)]-1/T$ under different strain rates.

Combining equations (1) and(4), we have:

$$Z = A[\sinh(\alpha\sigma)]^n \tag{8}$$

The Z value can be determined by substituting into equation (1). Taking the logarithms of equation (8), we get:

$$\ln Z = \ln A + n\ln[\sinh(\alpha\sigma)] \tag{9}$$

Based on Equation (9), we plot $\ln[\sinh(\alpha\sigma)]$against $\ln Z$, as depicted in Figure 8, the ln*A* is determined. Then, A=1.4470×10^{29}.

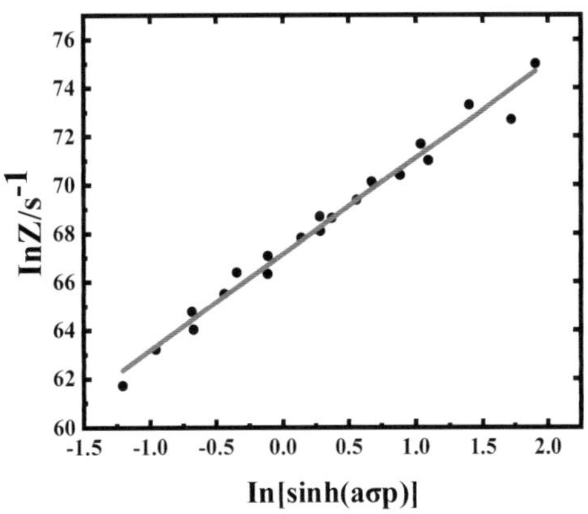

Figure 8. $\ln[\sinh(\alpha\sigma)]-\ln Z$ Relationship curve.

Substituting the obtained parameters into Equation (4), the constitutive equation for the GH4169 alloy is derived as follows:

$$\dot{\varepsilon} = 1.4470\times10^{29}[\sinh(0.0046\sigma)]^{4.0050}\times\exp(806.39/RT) \tag{10}$$

$$\sigma = \frac{1}{0.0046}\ln\left\{\left(\frac{Z}{1.4470\times10^{29}}\right)^{1/4.0057} + \left[\left(\frac{Z}{1.4470\times10^{29}}\right)^{2/4.0050} + 1\right]^{1/2}\right\} \tag{11}$$

3.3 Modification of the intrinsic model of GH4169 alloy

Equation (11) represents the material constant obtained by considering only the peak stress, without accounting for the effect of strain on the flow stress. To establish a more accurate constitutive model covering the full strain range, it is necessary to introduce a strain term to compensate for the strain dependence of material constants such as thermal deformation parameters *Q*, stress exponential constants *n*, structural factors *A,* and stress level parameters *α*.

Journal of Physics: Conference Series **2842** (2024) 012042 doi:10.1088/1742-6596/2842/1/012042

The related parameters (α, n, Q, InA) corresponding to 18 groups ranging from 0.05 to 0.9 intervals were solved respectively.

Scatter plots were drawn for each material parameter against the strain value. Polynomial fittings of degrees 2 to 9 were performed for each material parameter and strain value. Upon comparison, it was found that the 8th-degree polynomial fitting value closely matched the experimental values. As the fitting curve is shown in Figure 9, the solid line represents the fitted values, while the dots represent the experimental values.

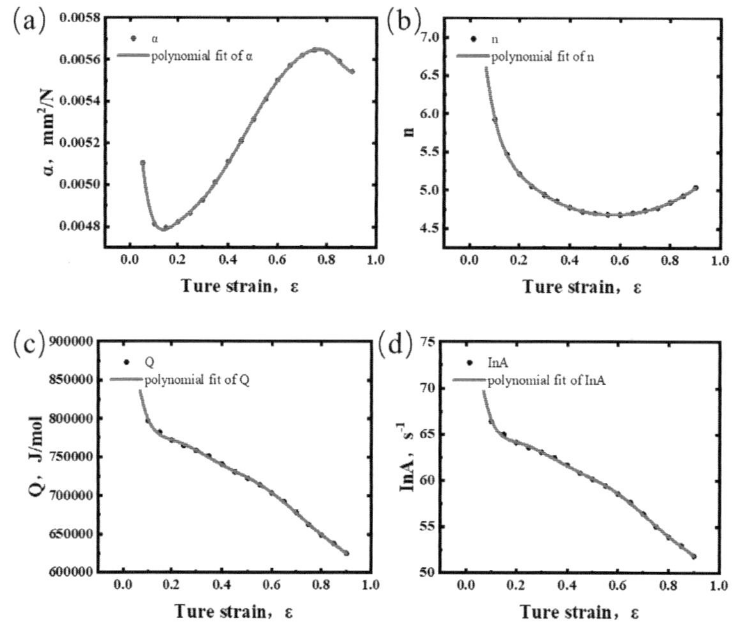

Figure 9. Relationship between the relevant parameters and strain.

The coefficients of the material constant and strain fitting equations are shown in Table 1.

Table 1. Fitting coefficients for material constants.

a (ε)/(mm^2•N^{-1})		n (ε)		Q (ε)/(kJ·mol^{-1})		lnA(ε)/s^{-1}	
X_{10}	1.4319	X_{20}	544.0910	X_{30}	8.1138E7	X_{40}	6702.2497
X_{11}	-5.7617	X_{21}	-2438.6047	X_{31}	-3.924E8	X_{41}	-32899.9175
X_{12}	9.7259	X_{22}	4667.7665	X_{32}	7.9179E8	X_{42}	67239.0377
X_{13}	-8.9360	X_{23}	-4969.9305	X_{33}	-8.6517E8	X_{43}	-74284.7583
X_{14}	4.8480	X_{24}	3212.07780	X_{34}	5.5574E8	X_{44}	48202.6077
X_{15}	-1.5818	X_{25}	-1289.5691	X_{35}	-2.1324E8	X_{45}	-18684.0268
X_{16}	0.3053	X_{26}	317.5152	X_{36}	4.7278E7	X_{46}	4188.3041
X_{17}	-0.0313	X_{27}	-45.8960	X_{37}	-5.6339E6	X_{47}	-504.1309
X_{18}	0.0061	X_{28}	6.7046	X_{38}	1.0556E6	X_{48}	89.6582

The average absolute relative error between the predicted stresses obtained from the calculation and the stresses obtained from the tests was 8.86%, and the correlation coefficient was 0.98. The rheological stresses obtained from the modeling calculation conformed to the experimental values to a high degree, indicating a good correlation.

Journal of Physics: Conference Series **2842** (2024) 012042 doi:10.1088/1742-6596/2842/1/012042

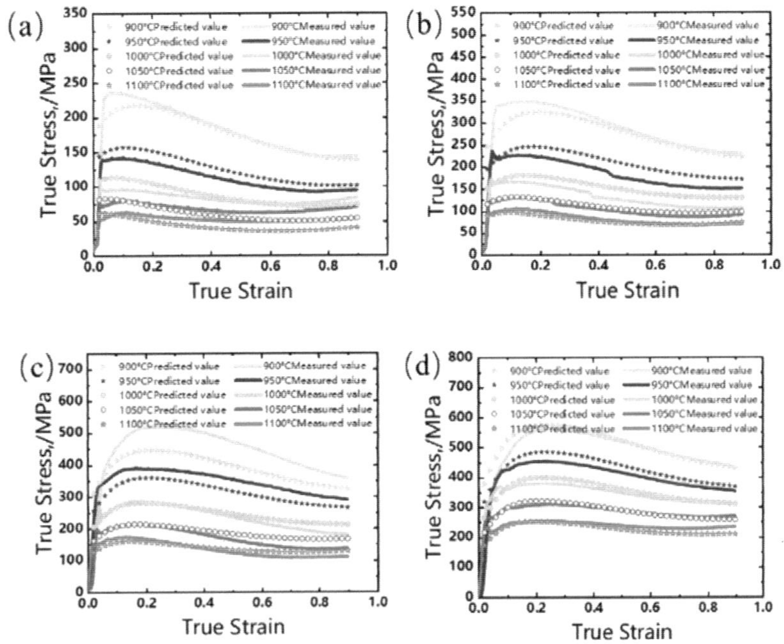

Figure 10. Comparison of simulated curves (solid lines) with predicted values (dots): (a) 0.001 s^{-1}; (b)0.01 s^{-1}; (c)0.1 s^{-1};(d)1s^{-1}.

Comparing the rheological stress predicted by the constructed intrinsic model with the data obtained from the tests, as shown in **Figure 10**, where the solid lines indicate the simulated values and the dots indicate the predicted values.

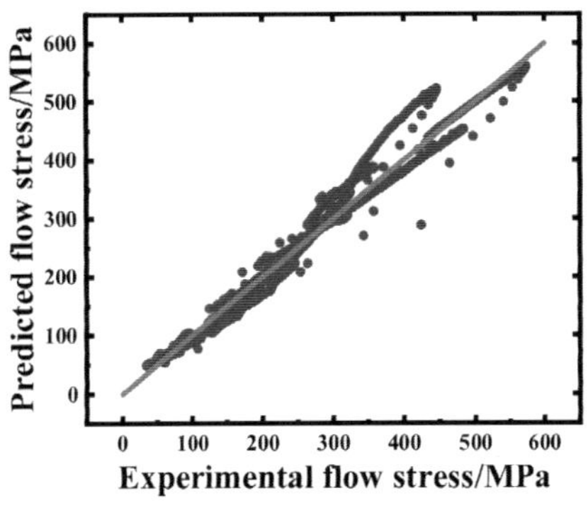

Figure 11. Correlation of modeling results.

A plot of predicted stress versus test stress is conducted to further verify the model accuracy, as depicted in Figure 11.

Under the experimental conditions outlined in this paper, the strain-compensated and coupling-based rheological stress intrinsic model proves to be highly accurate. This model accurately estimates the rheological stress of GH4169 high-temperature alloy under various hot deformation scenarios.

4. Conclusions

Based on the discussions and findings presented above, the conclusions obtained are as follows:

(1) During thermal deformation, the peak stress of the GH4169 high-temperature alloy diminishes as the temperature rises and the strain rate falls. This suggests that dynamic recrystallization is more

likely to occur at higher temperatures and lower strain rates, reducing the rheological stress.

(2) An Arrhenius hyperbolic sinusoidal constitutive equation model of GH4169 high-temperature alloy, incorporating strain compensation, is established through regression analysis. The correlation coefficient between its predicted value and experimental value is 0.98, with an average relative error of 8.86%. These results indicate that the constructed constitutive equation is highly accurate. It holds significant guidance for the numerical simulation of plastic deformation and optimization of process parameters for GH4169 alloy. These conclusions provide valuable insights for understanding the thermal deformation behavior of GH4169 high-temperature alloy and contribute to enhancing the efficiency and precision of its plastic deformation processes.

References:

[1] Cheng M, Ye N Y, Zhang S H. Research progress of main plastic processing technologies for GH4169 alloy[J]. China Materials Progress, 2016, 35(4): 241-250+260-261.

[2] Chen Q, Wang Y. Hot working behavior of δ phase aged state GH4169 alloy[J/OL]. Journal of the Chinese Society of Nonferrous Metals, 2015, 25(10): 2727-2737.

[3] Si J Y, Chen L, Liao X H, et al. New high temperature constitutive model and quantitative study of microstructure for GH4169 alloy treated by DP process[J/OL]. Forging & Stamping Technology, 2017, 42(10): 180-188.

[4] AZARBARMAS M, AGHAIE-KHAFRI M, CABRERA J M, et al. Microstructural evolution and constitutive equations of Inconel 718 alloy under quasi-static and quasi-dynamic conditions[J/OL]. Materials & Design, 2016, 94: 28-38.

[5] MAHALLE G, KOTKUNDE N, GUPTA A K, et al. Microstructure Characteristics and Comparative Analysis of Constitutive Models for Flow Stress Prediction of Inconel 718 Alloy[J/OL]. Journal of Materials Engineering and Performance, 2019, 28(6): 3320-3331.

[6] Ren Y H, Cheng Z, Wang L X, et al. Constitutive equation and hot processing map of forged GH4169 alloy[J]. Journal of Precision Forging Engineering, 2023, 15(5): 148-155.

[7] Liu Z, Dai P C. Influence of two different hot deformation methods on microstructure of GH4169 alloy forged bar[J/OL]. Metal Functional Materials, 2020, 27(3): 65-68.

[8] LIU J X, JIANG Ch, ZHAI Y W, et al. Dynamic recrystallization behavior of FGH4096 high temperature alloy[J/OL]. Forging and pressing technology, 2023, 48(7): 242-248.

Analysis of electrothermal damage in a radome exposed to electromagnetic radiation

Zhongxiang Gui[1*], Ruoxuan Xia[2], and Shengwen Wei[1]

[1]The 38th Research Institute of China Electrons Technology Groups Corporation, Hefei, Anhui, 230088, China

[2]Department of Military Representative Bureau of Aerospace System, Nanjing, Jiangsu, 210000, China

[*]Corresponding author's e-mail: gxiang518@163.com

Abstract. After nearly 9 months of electromagnetic irradiation with a power density of 10 W/cm^2, the unit blocks of an A-sandwich radome in the electromagnetic beam scanning area experienced electrothermal damage. Both the inner skin and core layer foam of the A-sandwich unit blocks in that area present different levels of damage. The inner skin is wrinkled and cracked, and the core layer foam is locally charred, carbonized, and ablated. The temperature images recorded through the infrared thermal beam method show that the temperature distribution of the unit blocks in the beam irradiation area is abnormal and strongly related to the discontinuous structure distribution of the core foam and resin. The non-uniformity of core material is the main cause of the electrothermal damage to unit blocks. By process improvements, the shaping of the polyurethane foam core layer is changed from cutting, resin bonding, and splicing to one-step foaming without resin binder, and the continuity of the core layer material is ensured. The usage results of unit blocks made by the new process show that the temperature field distribution is uniform, with a maximum temperature of only 31.3°C, which is 71.62% lower than the maximum temperature of 110.3°C in the damaged block.

1. Introduction

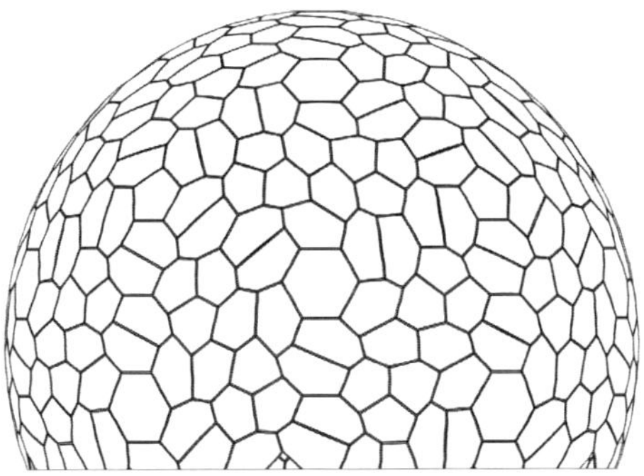

(a) Division of the radome

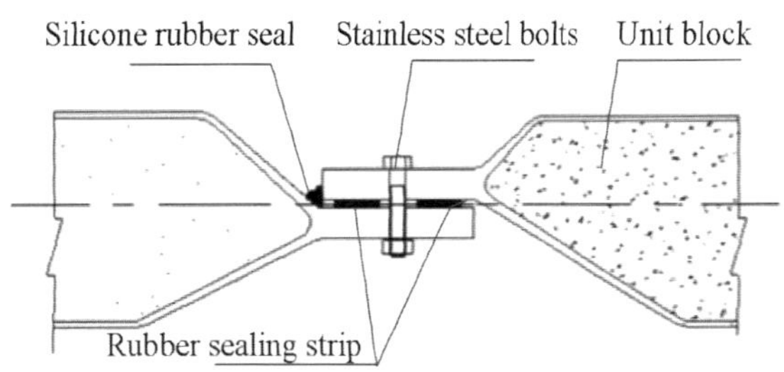

(b) Connection of unit blocks

Figure 1. Structure design of the radome.

The radome is a functional and structural part integrating electrical performance, structural strength, stiffness, and special functional requirements. Its role is to prevent the impact and damage of strong wind, salt fog, rainstorms, hail, and other environments on the antenna [1-3], extend the service life of the antenna, improve reliability, and ensure all-weather operation on the premise of ensuring the antenna performance indicators. In this article, the radome is a large truncated spherical cover that cannot be formed and installed as a whole due to its large size. It is divided into several unit blocks for manufacturing and then assembled into a complete cover. The blocks are fastened by bolts in their lapped rib edges made of pure fiberglass (Fig. 1). The unit block adopts an A-layer structure design; moreover, its skin is made of polyester fiberglass reinforced with E-glass fiber as the reinforcement material and unsaturated polyester resin as the binder, while the core material is polyurethane foam.

2. Experimental

The phenomena of the electrothermal damage in unit blocks are shown in Fig. 2. After installation, the radome began to work by scanning by electromagnetic beams with a power density of 10 W/cm^2, with a daily time of 22 hours. Almost 9 months later, on-site inspection revealed that there were obvious black streaks on the inner skin surface of some unit blocks in the electromagnetic beam scanning area, further, the streaks exhibited a grid-like distribution (Fig. 2a). Disassembling the corresponding blocks, it is observed that:

(1) Block A (Figs. 2b and 2e) had a high degree of damage to the inner skin, with severe wrinkling and obvious grid characteristics. Opening the inner skin, it could be seen that the core layer foam had been locally severely ablated and carbonized, with a trend of spreading from the rib edge of the block to the interior.

(2) The rib edge of block B overlapping block A and the transition area to the block B interior were severely ablated, furthermore, the inner skin of the transition area cracked (Fig. 2f). In addition, the integrity of the inner skin in the middle of the block B was also lost (Fig. 2c), local ablation and carbonization of the internal sandwich foam could also be observed from the crack opening.

(3) The connecting ribs of block C (Fig. 2d) maintained their integrity, but there was significant damage inside with visible grid marks. After separating its skin from block C (Fig. 2g), it was found that its internal sandwich foam had also been carbonized (Fig. 2h). Compared with blocks A and B, the overall appearance of the inner skin of block C remained milky white, while the inner skin of the other two blocks appeared noticeably burnt yellow.

(4) The degree of damage to the radome blocks showed a decrease from A to C.

3. Results and discussion

3.1. Analysis of the electrothermal damage

The infrared thermal image monitoring results show that the core layer of the radome unit block is cemented with small pieces of foam and the adhesive distribution at the joints is uneven. Under electromagnetic beam irradiation, the temperature distribution of unit blocks is abnormal, which is strongly correlated with the discontinuous structural distribution of the core layer foam and resin. The temperature of block connecting side ribs, core layer foam joints, and joints between foam is significantly higher than that of other areas, with the highest temperature of 110.3°C (Fig. 3). In addition, the influence of electromagnetic beam thermal effects only occurs in the beam irradiation area. Due to the absence of open flames in the usage environment of the radome, based on the damage situation of the block and the 10 W/cm^2 power density, the cause of the damage is identified as electric thermal damage caused by long-term (nearly 9 months) continuous electromagnetic beam irradiation. In addition, the shortcomings of the infrared thermal imaging monitoring method, electromagnetic interference, and spatial height limitations make it impossible to directly test and obtain the internal temperature of the blocks, but it does not affect the analysis because the actual internal temperature will be higher.

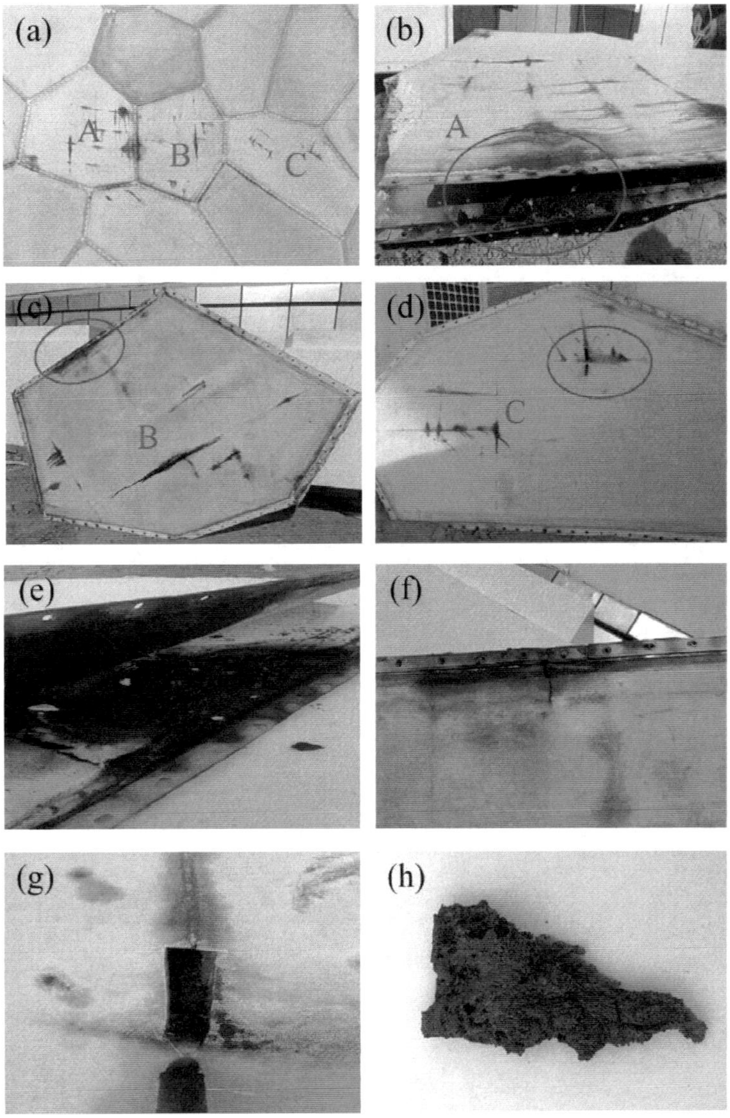

Figure 2. Electrothermal damage to the unit blocks.

The radome operates under electromagnetic beam irradiation, where electromagnetic beams interact with the material of the unit block and are affected by power density and irradiation time. For the A-sandwich unit block inner skin, when electromagnetic beams act on the E-fiberglass composite material skin, some of the electromagnetic beam energy is reflected, some is transmitted, and most of the energy is absorbed by the composite material. The deposition of electromagnetic beam energy can create a certain depth of bulk heat source in composite materials [4-6]. Under the action of a body heat source, skin composite materials undergo a series of physical and chemical changes, and the basic process is [4-6]: wrinkling along the fiber direction - resin matrix volatilization and decomposition - resin and fiber coking - carbonization and cracking in a transverse and longitudinal direction. The heating simulation test results of the unit block also showed the same phenomenon (Fig. 4). The key to thermal damage lies in chemical changes [4-6]. Coking and carbonization cause severe shrinkage of resin and fibers, and the stress generated by shrinkage promotes damage to composite materials with deteriorated performance, resulting in cracks in Fig. 2c and Fig. 2f. The unsaturated polyester resin in the E-glass fiber composite skin has fusible properties. When it is irradiated by electromagnetic beams to a certain temperature, the resin is plasticized, melted, and flows, which is the fundamental reason for the appearance of black hanging objects on the unit blocks. During the plasticization process, the resin matrix becomes hard and brittle, causing the skin to be unable to effectively transmit stress concentration caused by uneven heating of the core material. The combined effect of the two causes the shrinkage and wrinkling of the skin to appear in a grid shape at the splicing joints of the core foam.

Figure 3. Temperature monitoring by thermal infrared imager.

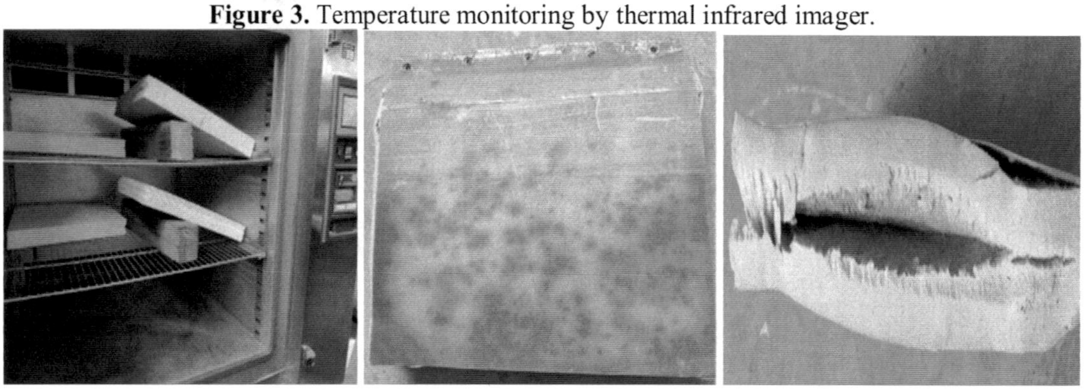

Figure 4. Simulated heating test of the unit blocks.

Table 1. Thermophysical properties of unit block materials [7-9].

Materials	Dielectric constant /10 GHz	Loss tangent /10 GHz	Heat resistance /°C	Thermal conductivity /W/(m·K)	Density /g·cm^{-3}
E-glass fiber	6.13	0.0038	-	0.035	2.54
Unsaturated polyester resin	3.0	0.018	130	0.3324	1.3
E-glass fiber reinforced plastics	3.92	0.018	140	0.40	1.78
Polyurethane foam	1.08	0.003	130	0.023	0.08

For the foam core layer of A-sandwich unit block, on the one hand, since the core layer foam is composed of polyurethane foam cut and bonded with unsaturated polyester resin, the resin bonding joint can be regarded as a material design defect for the entire core layer (Table 1), resulting in the non-uniformity of the material properties of the core layer; On the other hand, in the manufacturing process of the block, the process document did not put forward quantitative requirements for the bonding process of the core layer foam, resulting in a large difference in the amount of resin filling at the splice joints, further strengthening the non-uniformity of the core layer foam material characteristics.

The non-uniformity of this material affects the transmission characteristics of the unit block (Fig. 5). The transmission loss of the block with joints and the word without joints differ by more than 10 times and expand with the increase of frequency band. The electromagnetic beam is converted into heat more during the process of passing through the antenna cover. Of course, the contribution to the body heat source is outstanding, especially in the area with large resin deposition, the temperature rises after continuous irradiation of high power density electromagnetic beam for a long time, and when the heat resistance temperature of the connected foam is exceeded, foam coking and carbonization will be caused. In addition, due to the decrease in power transmission coefficient with increasing temperature [10], the transmission performance of the unit block decreases, the deposition of electromagnetic beam energy increases, and the body heat source effect intensifies.

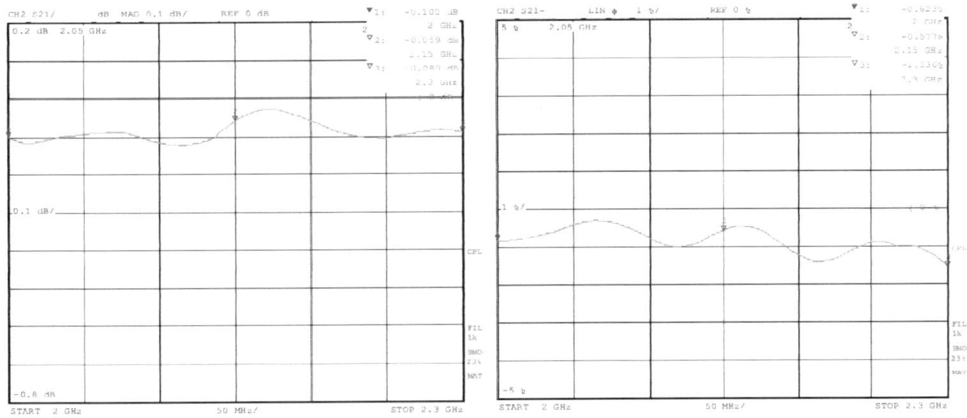

Figure 5. Electrical performance testing of the unit block without (a) and with (b) adhesive joints.

3.2. Process improvement and its validation

Through process improvements, the shaping of the polyurethane foam in the core layer of the radome unit block is changed from cutting, resin bonding, and splicing to one-step foaming without resin binder, to ensure the continuity of the core material. New unit blocks were made according to the new process, and temperature observations were conducted using an infrared thermal imager after installation. The results showed that the temperature field distribution of the new unit blocks was uniform, with only high-temperature areas appearing at the joints of the blocks, reaching a maximum of only 31.3°C. Compared with the highest temperature of 110.3°C in the damaged block, the decrease reaches 71.6%.

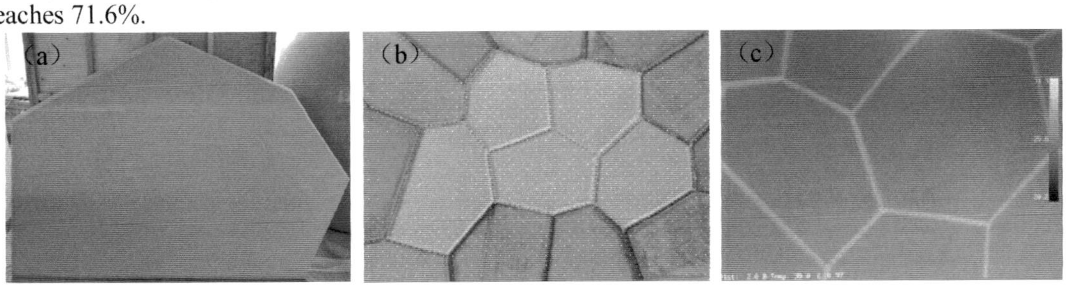

Figure 6. (a) unit block by one-step foaming; (b)validation test; (c) Temperature observation.

4. Conclusions

(1) The non-uniformity of the core material of the radome unit blocks leads to thermal damage caused by long-term irradiation of high-power density electromagnetic beams;

(2) By process improvements, the shaping of the polyurethane foam in the core layer of the radome unit block is changed from cutting, resin bonding, and splicing to one-step foaming without resin binder, effectively improving the continuity of the core material. The verification results show that the temperature field distribution of the newly made unit blocks is uniform, with a maximum temperature of only 31.3°C, which is 71.62% lower than the maximum temperature of the damaged block.

References

[1] Aamir M. T., Nasir M. A., Iqbal Z., et al. (2023) Multi-disciplinary optimization of hybrid composite radomes for enhanced performance. Results in Engineering.20(101547):1-13.
[2] Özdemir B., Salamci E., Kuloğlu M., et al. (2021) Comparison of radome sandwich composite structures with finite element method, Mater. Today: Proc., 34(1): 297-303.
[3] Wenwen Z., SONG Liyuan S., WANG Xiaojun W., et al. (2016) Influence of acetone on the combination between epoxy resin and E-glass fiber under the effect of DMP-30. Thermosetting Resin, 26(1):39-43.
[4] Huan L., Jun L., Jiayu X., et al. (2012) An Overview on the Technology and Electromagnetic Performance of Radome. Mater. Rev., 26(8):48-52.
[5] Xin J., Zhenguo Y., Hansheng J., et al. (2006) Progress of materials applied to ground radomes. PTCA (PART: A PHYS. TEST.), 42(5):217-223.
[6] Chuanyang L., Jianghua H. (2018) Effect of Different Fillers on the Thermal Properties of UFRP. J. Ord. Equip. Eng., 39(10):144-148.
[7] Taylor K., Ni Y., Chengying X.(2022) Dielectric and mechanical properties of hypersonic radome materials and metamaterial design: A review.J. EUR. CERAM. SOC.,42(1):1-17.
[8] Seok-Min Y., Seung Jun L., Jinwoo P., et al. (2021) Fiber-reinforced plastic material with de-icing capability for radome application. MATER. LETT., 284(2):128943-128953.
[9] Pavan Chandra M., Abdul Ayaz M., Nikhilesh Goud P., et al. (2022) Modelling and analysis of aircraft radome using different materials. Mater. Today: Proc., 62(6): 4492-4497.
[10] Hengqing Z. (2002) Power Transmission Character of Radome Which Temperature Variety Impacts. GUIDANCE&FUZE, 23(3):45-48.

Journal of Physics: Conference Series **2842** (2024) 012044 doi:10.1088/1742-6596/2842/1/012044

Molecular dynamics simulation of irradiation damage accumulation in α-Fe and model alloy $FeMn_{1.4}Ni_{0.7}$

Wenqing Zhang[†], Jiaxin Wu[†], Feng Zhou, Jinhong Chen, Fei Zhu[*], Ligang Song[*] and Xianfeng Ma[*]

Sino-French Institute of Nuclear Engineering and Technology, Sun Yat-Sen University, Zhuhai, Guangdong, 519082, China
[†]Authors contribute equally

[*]Corresponding author's e-mail: zhuf56@mail.sysu.edu.cn (Fei Zhu)
songlg6@mail.sysu.edu.cn (Ligang Song)
maxf6@mail.sysu.edu.cn (Xianfeng Ma)

Abstract: The neutron irradiation of RPV steel leads to the generation of defects and further degrades the mechanical properties. MD simulation was used to study the accumulated damage of α-Fe and model alloy $FeMn_{1.4}Ni_{0.7}$ at room temperature. Two methods of introducing PKA, i.e., concentrated irradiation and uniform irradiation, were performed. The simulation results showed that concentrated irradiation is beneficial for the clustering of vacancies and the formation of vacancy DLs <100>. All the types of interstitial DLs are $1/2 < 111 >$, and all the types of vacancy DLs are $< 100 >$. The addition of Mn and Ni restrained the clustering of interstitials under uniform irradiation conditions.

1. Introduction

The reactor pressure vessel (RPV), serving as the vital component that cannot be replaced in a light water reactor (LWR), faces a harsh service environment during operation. The neutron irradiation of RPV steels induces various changes in the microstructure, i.e., matrix damage, copper-rich precipitates, and so on [1]. These defects can impede the movement of dislocations and further lead to mechanical properties deterioration, e.g., irradiation-induced hardening and embrittlement, which seriously threaten the lifespan of reactors. In the A508-3 RPV steel, a steel with a low content of copper, the matrix damage, including interstitial and vacancy clusters, dislocations, and vacancy-solute complexes, are the main causes of the irradiation-induced hardening and embrittlement [2]. In addition, the presence of other solutes also has an influence on the damage in low copper content alloys [3]. However, the fundamental understanding of the matrix damage and the influence of other solutes on the damaged microstructures are still limited.

The formation of irradiation-induced defects is a result of the evolution of interstitials and vacancies produced by the displacement cascade under high-energy particles. Molecular dynamics (MD) simulation can study the primary damage and the evolution of defects under successive recoils at the atomic level and has been used extensively to investigate the radiation damage of materials [4, 5]. In this study, the accumulation of defects, the size distribution of clusters, and the evolution of dislocations of pure metal Fe and model alloy $FeMn_{1.4}Ni_{0.7}$ under successive recoils were studied by MD simulation. Besides, the impact of the method of introducing PKA, i.e., concentrated irradiation and uniform irradiation, on the microstructural damage was also examined. The study of point defects, clusters, and dislocations generated under neutron irradiation contributes to comprehending the changes in mechanical properties of RPV steels caused by irradiation and provides a research basis for the subsequent prediction of the lifespan of RPV steels.

2. Method

The Fe-Mn-Ni-Cu EAM potential [6, 7] was employed. First, a pure Fe model was established. Then, Fe atoms were randomly replaced by Mn and Ni atoms with a Fe:Mn:Ni atomic ratio of 1:1.4:0.7 to obtain the desired $FeMn_{1.4}Ni_{0.7}$ model. A cubic box of dimensions 114.8Å×114.8Å×114.8Å was constructed, containing 128000 atoms, with periodic boundary conditions and canonical (NVT) ensemble employed to relax the system for 50ps to ensure equilibrium. This obtained a fully relaxed structure.

Two approaches were used for cumulative irradiation: concentrated irradiation, where a random atom from a spherical shell centered in the simulation cell was chosen as the primary knocked-atom (PKA) and imparted with 5 keV of recoil energy, with the velocity directed towards the center of the box. The radius of the shell was 23 Å, based on the calculation of the distance between the cascade explosion region and the initial PKA of 5 keV cascade in Fe, ensuring that all the cascades occurred at the center of the box. The second approach was uniform irradiation, where an atom at the box center was randomly selected as PKA and endowed with an energy of 5 keV, with a randomly oriented velocity. After each recoil, the box was moved along the x, y, and z directions randomly to ensure uniform cascade distribution. The first cascade cooled down to 300K for 50ps, and then another PKA was selected to initiate the next cascade. We have conducted 300 cascades in each cell. OVITO [8] was used to show the evolution of defects during the cascade overlap. Employing the Wigner-Seitz method in OVITO to identify defects, sites occupied by no atom were identified as vacancies, while sites occupied by more than one atom were identified as interstitials. The DXA was used to identify the dislocation line. The point defects within the second nearest neighbor range (2.87 Å) were identified as one cluster.

3. Results and discussion

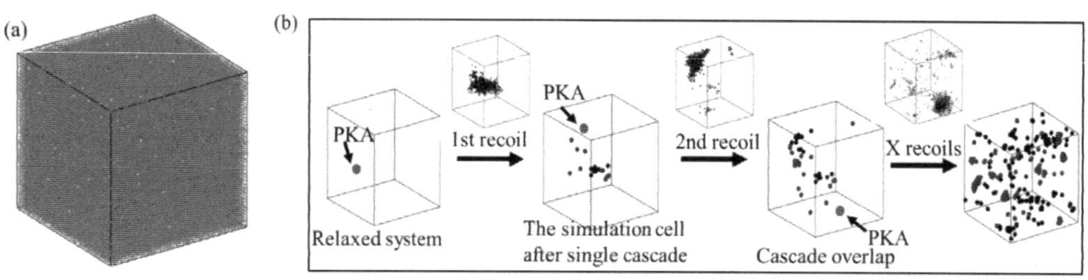

Figure 1. (a) The model alloy $FeMn_{1.4}Ni_{0.7}$ and (b) The process of cascade overlap under uniform irradiation. The red, green, and yellow spheres represent Fe, Mn, and Ni atoms, respectively. The blue and black spheres represent interstitials and vacancies, respectively.

Figure 1 (a) depicts the model alloy $FeMn_{1.4}Ni_{0.7}$, and We can observe that Mn and Ni atoms are randomly distributed in corresponding proportions. Figure 1(b) visualizes the process of cascade overlap under uniform irradiation. The first recoil represents the primary irradiation damage process of the alloy. Randomly select a PKA atom to start the cascade collision. The atoms in the box undergo the processes of collision, melting, and cooling during the cascade, generating a certain number of vacancies and interstitials. Then, in the second recoil, a new PKA atom is selected to carry out the cascade collision and generate a certain number of defects. Repeating this cycle 300 times yields the final cumulative collision defects morphology. Therefore, the defects in the early stage of the uniform irradiation cascade increased linearly, as there was no cascade overlap at this point, and the number of defects was purely additive. However, the growth of defects in concentrated irradiation was initially slow because all the cascades occurred at the center of the box, leading to cascade overlap. Overlapping recoils can enhance the interstitial-vacancy recombination [9], so the number of defects does not linearly accumulate. However, the number of defects in the two forms of irradiation eventually tended to saturate (about 300 FPs) as the cascade progressed. Comparing the two materials, the final number of FPs shows no significant difference under concentrated irradiation and uniform irradiation, but the fluctuation of the number of FPs in α-Fe is larger.

The snapshots of defect distribution for the α-Fe and $FeMn_{1.4}Ni_{0.7}$ under concentrated irradiation are exhibited in Figure 2 (a1-a3) and Figure 2 (b1-b3). After 100 recoils, both interstitials and vacancies in Fe form a predominated large cluster, respectively. As the number of cascades increases, the sizes of the predominated interstitial clusters and vacancy clusters increase after 200 and 300 recoils. Similar to Fe, $FeMn_{1.4}Ni_{0.7}$ also has predominated interstitial and vacant clusters after 100 recoils, and the cluster sizes further increase after 200 recoils, but then the predominated t cluster sizes in $FeMn_{1.4}Ni_{0.7}$ seem to decrease after 300 recoils, which could be due to the fluctuation in the number of defects during the cumulative recoils. For both materials, the predominated vacancy cluster was always in the box center, while the predominated interstitial cluster was off center, which is due to the high mobility of interstitials and the low mobility of vacancies. Furthermore, all the interstitial-type dislocation loops (DLs) $1/2 < 111 >$ and all the vacancy DLs $< 100 >$ in both materials can be observed. A mixture of 1/2<111> and <100> DLs in the irradiated α-Fe and its alloys have also been reported by our previous work [10] and other studies [11]. Comparing the two materials, no noticeable difference in the clusters and dislocations was observed under the concentrated irradiation.

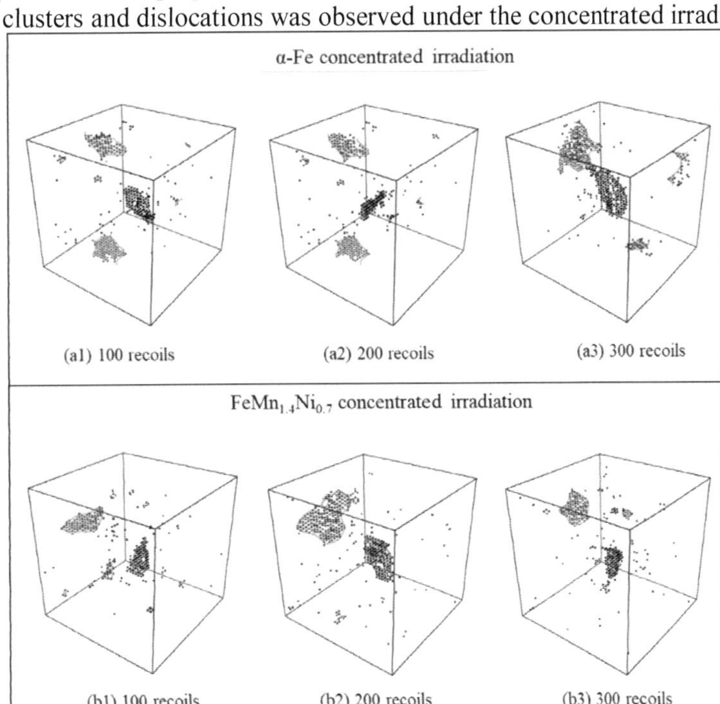

Figure 2. Snapshots of the spatial distribution of defects in α-Fe and $FeMn_{1.4}Ni_{0.7}$ under concentrated irradiation at 100, 200, and 300 recoils, respectively.

Figure 3 shows the snapshots of defect distribution after 100, 200, and 300 recoils for the α-Fe and $FeMn_{1.4}Ni_{0.7}$ under uniform irradiation. After 100 recoils, a predominated interstitial cluster forms in Fe, while the vacancies are randomly distributed throughout the simulation cell without aggregating into large vacancy clusters. After 200 recoils, the size of the predominated interstitial cluster further increases, and some small vacancy clusters also emerge. After 300 recoils, more and larger vacancy clusters can be observed. The uniform irradiation behavior of $FeMn_{1.4}Ni_{0.7}$ is similar to that of Fe. After 200 cascades, $FeMn_{1.4}Ni_{0.7}$ forms a predominated large interstitial cluster, while the vacancies are randomly distributed in the simulation box. As the number of recoils increases to 300, the size of the predominated large interstitial cluster in $FeMn_{1.4}Ni_{0.7}$ further increases, but no large-sized vacancy clusters are formed. In summary, in contrast to the concentrated irradiation, only the interstitials aggregated to a predominated large cluster, but the vacancy clusters were almost evenly distributed in the simulation box, and no vacancy-type dislocation loops were observed in both materials. This phenomenon can also be explained by the defect diffusion. Due to the high mobility of interstitials, the generated interstitials under subsequent recoils can migrate rapidly and then be absorbed by a large interstitial cluster, resulting in the formation of the predominated large cluster. For vacancies, they are almost immobile at room temperature, so the clustering of vacancies would be suppressed.

By comparing the defects morphology of the two materials in Figure 3, the interstitials in α-Fe favored clustering more than that in $FeMn_{1.4}Ni_{0.7}$. For example, after 100 recoils, the predominated interstitial clusters have appeared in Fe, but there are only small and scattered interstitial clusters in $FeMn_{1.4}Ni_{0.7}$, which are not gathered in a large amount. After 200 recoils, it can be clearly observed that the size of interstitial clusters in Fe is larger than that in $FeMn_{1.4}Ni_{0.7}$. After 300 recoils, it can be clearly observed that the size of vacancy clusters in Fe is larger than that in $FeMn_{1.4}Ni_{0.7}$. In other words, the addition of Mn and Ni suppressed the clustering of interstitials, which is due to the slow diffusion of interstitials or the facilitated nucleation by Mn and Ni [6]. By comparing the two methods of introducing PKA, it could be found that concentrated irradiation is conducive to the clustering of vacancies in both materials and the clustering of interstitials in $FeMn_{1.4}Ni_{0.7}$. For example, after 100 recoils, the formation of predominated interstitial clusters has been observed in both materials under concentrated irradiation. It can be found that the size of interstitial clusters formed under concentrated irradiation in Fe is significantly larger than that under uniform irradiation. In comparison, no large interstitial clusters have formed in $FeMn_{1.4}Ni_{0.7}$ under uniform irradiation after 100 recoils. This can be explained by the higher degree of aggregation of the generated point defect under concentrated irradiation. In addition, concentrated irradiation also means a higher damage degree than uniform irradiation.

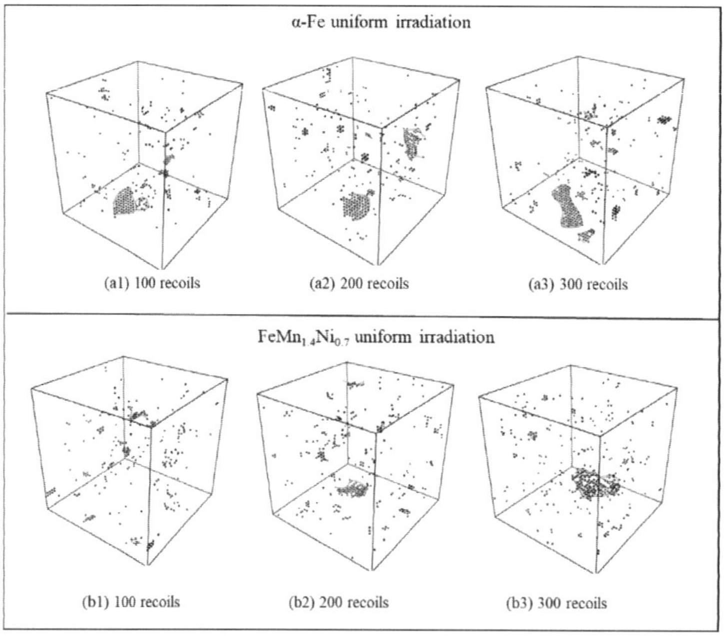

Figure 3. Snapshots of the spatial distribution of defects in α-Fe and $FeMn_{1.4}Ni_{0.7}$ under uniform irradiation at 100, 200, and 300 recoils, respectively.

4. Conclusion

In this study, the accumulated damage of α-Fe and $FeMn_{1.4}Ni_{0.7}$ under two irradiation conditions, i.e., concentrated irradiation and uniform irradiation, was simulated using MD to understand the matrix damage of RPV steel and the influence of solutes. It was found that the method of introducing PKA has a significant influence on the damaged microstructure, suggesting that the concentrated irradiation promotes the clustering of vacancies and the formation of vacancy <100> DLs and propels the clustering of interstitials in $FeMn_{1.4}Ni_{0.7}$. All the types of interstitial DLs are $1/2 < 111 >$, and all the types of vacancy DLs are $< 100 >$. Under uniform irradiation, the addition of Mn and Ni suppresses the clustering of interstitials, which may be due to the reduced mobility of interstitials or the facilitated nucleation by Mn and Ni.

Acknowledgments

The authors acknowledge the support of Guangdong Major Project of Basic and Applied Basic Research (2019B030302011), National Natural Science Foundation of China (12305325, 1220050457), Guangdong Provincial Major Youth Talents Project (2021TQ06L123), Guangdong Basic and Applied Basic Research (2022A1515110266, 2023A1515012504), Guangdong Provincial International Sci & Tech Cooperation Program (2020A0505020005), and STRFML-2022–11 Fund of the Science and Technology on Reactor Fuel and Materials Laboratory, Nuclear Power Institute of China.

References

[1] Margolin B, Fomenko V, Shvetsova V, et al. (2022) Engineering Fracture Mechanics, 267: 108400. DOI: 10.1016/j.engfracmech.2022.108400.
[2] Ding Z N, Han X X, Yang Y T, et al. (2021) Nuclear Instruments and Methods in Physics Research Section B: Beam Interactions with Materials and Atoms, 504: 14-20. DOI: 10.1016/j.nimb.2021.06.021.
[3] Castin N, Bonny G, Bakaev A, et al. (2020) Materials Today Energy, 17: 100472. DOI: 10.1016/j.mtener.2020.100472.
[4] Granberg F, Byggmästar J, Nordlund K (2020) Journal of Nuclear Materials, 528: 151843. DOI: 10.1016/j.jnucmat.2019.151843.
[5] Li R, Guo L, Liu Y, Xu Q S, et al. (2023) Acta Metallurgica Sinica-English Letters, 36: 1482-1492. DOI: 10.1007/s40195-023-01577-w.
[6] Terentyev D, Zinovev A, Bonny G (2016) Journal of Nuclear Materials, 475: 132-139. DOI: 10.1016/j.jnucmat.2016.04.005.
[7] Terentyev D, He X, Bonny G, et al. (2015) Journal of Nuclear Materials, 457: 173-181. DOI: 10.1016/j.jnucmat.2014.11.023.
[8] Stukowski A (2010) Modelling and Simulation in Materials Science and Engineering, 18: 015012. DOI: 10.1088/0965-0393/18/1/015012.
[9] Ullah M W, Aidhy D S, Zhang Y, et al. (2016) Acta Materialia, 109: 17-22. DOI: 10.1016/j.actamat.2016.02.048.
[10] Ma X, Zhang Q, Song L, et al. (2022) Crystals, 12. DOI: 10.3390/cryst12081091.
[11] Ding Z, Zhang C, Zhang X, et al. (2020) Nuclear Materials and Energy, 22. DOI: 10.1016/j.nme.2020.100727.

Journal of Physics: Conference Series **2842** (2024) 012045 doi:10.1088/1742-6596/2842/1/012045

Outer spiral drill rod lunar soil sampling resistance moment simulation study

Dehui Li [1,*], Haitao Luo [2], Yuxin Li [3], and Yuming Dong [1]

[1] School of Mechanical Engineering, Shenyang University of Technology, Shenyang, 110870, China.
[2] State Key Laboratory of Robotics, Shenyang Institute of Automation, Chinese Academy of Sciences (CAS), Shenyang, 110016, China.
[3] University of Chinese Academy of Sciences, Beijing, 100049, China.

[*] Corresponding author's e-mail: lidehui@sia.cn

Abstract: The task of unmanned automated sampling of China's lunar exploration is to bring back lunar soil samples with a certain depth, and sampling of external auger drill pipe is the best way. The power and moment size of the lunar surface is limited, so a special design is required. In this paper, a unique structural model of a double-auger drilling tool is designed. It is a movement analysis of lunar soil and drill pipe. We use simulated lunar soil simulations and experiments. The Hertz-Mindlin slip-free model in the discrete element software EDEM was used for numerical simulation. Given the disadvantage of the large amount of computation of discrete elements, the "embedding simulation method" proposed in this paper reduces the amount of computation. When drilling 0.4m and 1m under lunar and earth gravity, the magnitude and influence of drill pipe rotation and vertical velocity on the drag moment were analyzed. We provide the relevant technical parameters of the real unmanned drilling activities on the lunar surface. Through simulation, the rationality parameters of several groups of drill pipe drilling are obtained to ensure the smooth progress of drilling.

1. Introduction

The landmark activity of lunar exploration is lunar soil sampling, and the third phase of China's lunar exploration project has encountered some problems in the implementation of lunar soil drilling, so it is still necessary to carry out some basic research and optimize and improve [1]. The landmark activity of lunar exploration is lunar soil sampling. China has encountered some problems in implementing lunar soil drilling, so it is still necessary to carry out some basic research and optimize and improve [2]. Sampling is divided into surface sampling and deep sampling, and hollow external auger pipe drilling sampling is an effective way for deep sampling [3]. First, the sampling of deep soil helps clarify the properties of different layers of lunar soil, which is conducive to further research on lunar soil resources and the formation of the universe [4]. Secondly, it is conducive to studying the law of lunar soil action and lays a theoretical foundation for constructing lunar bases and mineral exploitation. Third, basic research on lunar soil drilling and sampling could pave the way for drilling on other planets [5].

Since lunar soil is an ultra-rare substance, we use simulated lunar soil for related experiments [6]. The drilling tools are made of titanium alloy, which can not only meet the requirements of use in terms of strength but also reduce weight and load for rocket launches. The power of real lunar soil drilling is limited, and the resistance moment of drilling tools is one of the important factors that cannot be ignored [7], so it is necessary to pay attention to the interaction between lunar soil and drilling tools, such as load law and drilling power consumption. For the limitation of environmental conditions for lunar drilling and production, the auger drill pipe can meet the design requirements of low-power unmanned automatic sampling to a certain extent [8]. The coring tube is located inside the drill pipe and is used to mothball the transferred lunar soil. Faced with the difficulty of a large computational volume of discrete elements, a pre-embedded simulation method is adopted. It can not only ensure the accuracy and authenticity of the working conditions but also reduce the amount of calculation [9].

2. Drilling tool structure design and drilling mechanics chip removal model

Lunar drilling gives drilling tools low energy consumption and small weight [10]. The best way to automate unmanned sampling is to drill and sample the hollow outer auger drill pipe, design its structure, and establish a mechanical chip removal model.

2.1 Structural parameters of Auger drilling tools

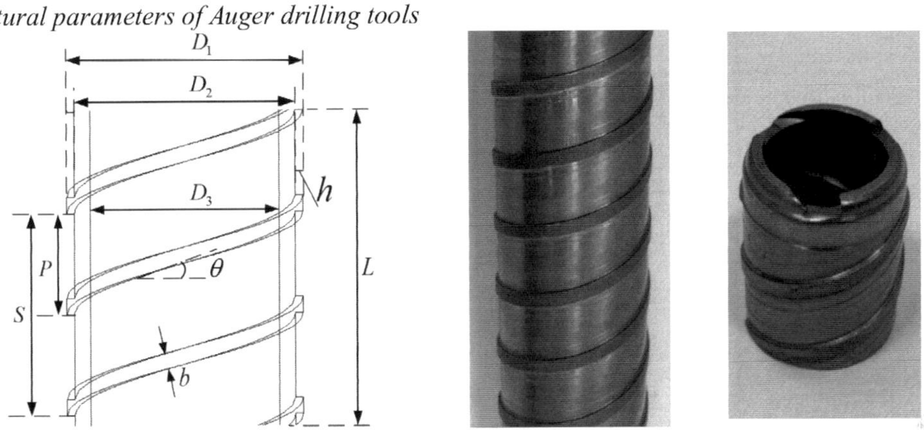

(a) (b) (c)

Figure 1. Drilling tool structural parameters and physical objects.

Figure 1 shows the structural parameters and physical drawings of drilling tools. Figure 1(a), (b), and (c) are respectively drill pipe structural parameters, test drill pipe, and test drill bit. The main function of the drill bit is to cut and crush, and the main role of the auger drill pipe is to transport the soil to the outside of the borehole. The drill pipe adopts a double helix structure with high conveying efficiency, and the material is selected as a high-quality titanium alloy material for aerospace, which has good strength and rigidity. The coring tube is located inside the auger drill pipe and is used for soil sampling. Due to the design of the drilling depth of 1m, considering the convenience of carrying it to the lunar surface, a segmented drill pipe design is adopted. Each drill pipe is 500mm long, and the key connection transmits torque. The parameters are listed in Table 1 below.

Table 1. Drill pipe parameters.

Structural parameters	D_1 / mm	D_2 / mm	D_3 / mm	θ /°	b / mm	P / mm	S / mm	H / mm	L / mm
Valid values	30	28	23	8.33	2	12.5	2	1	500

D_1 is the rod's outside diameter; D_2 is the shaft's outer diameter; D_3 is the inner diameter; θ is the spiral rising angle; b is the thickness; P is the pitch; S is the leads; h is the spiral height; L is the rod length.

2.2 Kinematics and mechanical analysis of chip removal process

Through the analysis of soil stress, there is a motion relationship between soil chip removal and the motion parameters of the auger drilling tool, as shown in Figure 2.

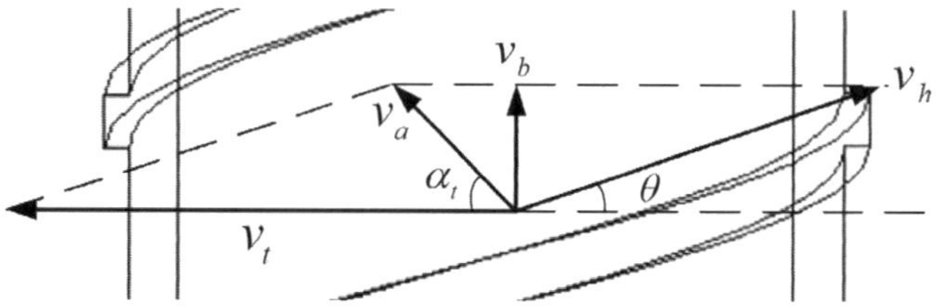

Figure 2. Parameters of lunar soil chip removal movement.

v_a is the chip removal speed; v_h is the relative speed of movement of the chips; v_t is the speed of the rotation line; v_b is the vertical speed; θ is the spiral rising angle; α_t is the chip removal angle [11]. The following relationships exist between the speeds:

$$\cot\alpha_t \cdot v_b + \cot\theta \cdot v_b = v_t \tag{1}$$

$$v_b = \frac{\tan\alpha_t \cdot \tan\theta}{\tan\alpha_t + \tan\theta} v_t \tag{2}$$

The ratio of the upward velocity of the debris to the linear velocity of rotation is:

$$k = \frac{\tan\alpha_t \cdot \tan\theta}{\tan\alpha_t + \tan\theta} \tag{3}$$

In the stable chip removal area, the soil is subjected to the external force of sliding friction, as shown in Figure 3.

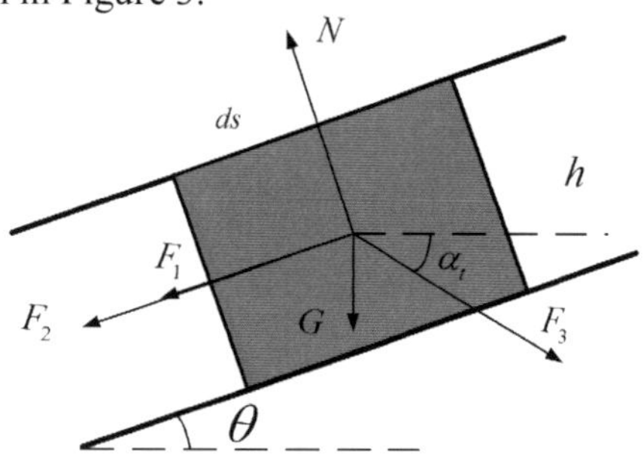

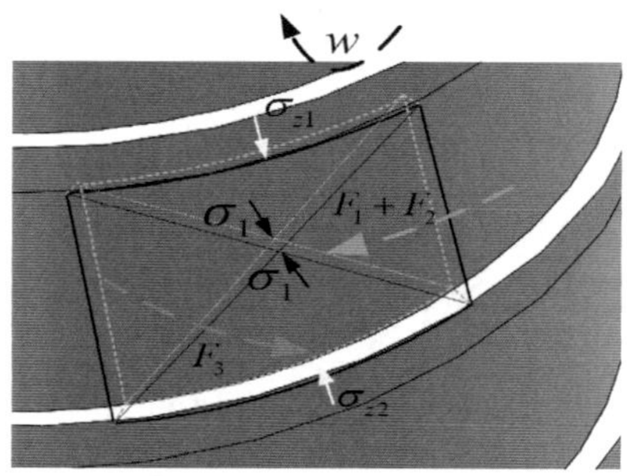

Figure 3. Stress state of lunar soil in a spiral. **Figure 4.** Spiral inner lunar soil stress.

The spiral support force is N, and the direction is perpendicular to the surface upward. The sliding friction of the spiral against the soil is F_1. The self-weight of the soil unit is G. The friction of the inner surface against the soil is F_2. The shear friction force of the borehole wall on the lunar soil element is F_3, which is opposite to the velocity of the lunar soil element relative to the stationary coordinate system. The forces have the following equilibrium relationship in the horizontal direction:

$$\cos\theta(F_1 + F_2) + N\sin\theta = F_3\cos\alpha_t \tag{4}$$

The unit length is dS. Differentiating it and bringing it into the equilibrium equation gives the relationship between θ and α_t. σ_1 is the lateral compressive stress in the spiral groove space; σ_{z1} and σ_{z2} are the top surface compressive stress and the bottom surface compressive stress in the spiral groove, respectively. As shown in the Figure 4. The spiral flute chip evacuation equation in equilibrium can be written as:

$$\int_{S_1} \sigma_1 f_1 \cos(\alpha_t + \theta) dS_1 = \int_{S_1} \sigma_1 f_2 dS_1 + \int_{S_2} (\sigma_{z1} + \sigma_{z2}) f_2 dS_2 \tag{5}$$

f_1 is the Equivalent coefficient of friction between lunar soils; f_2 is the equivalent coefficient of friction between the lunar soil and the drill pipe surface. The condition for the existence of a solution to this equation is $f_1 > f_2$.

Journal of Physics: Conference Series **2842** (2024) 012045 doi:10.1088/1742-6596/2842/1/012045

3. Calibration of macroscopic parameters of simulated lunar soil

EDEM modeling accuracy largely depends on reasonable parameters [12]. To make the mechanical parameters of the lunar soil discrete element model consistent with the real simulated lunar soil, in-situ experiments must be carried out to determine the specific material properties, such as elastic recovery coefficient, sliding friction coefficient, accumulation angle, etc.

3.1 Elastic recovery factor

$$e = \frac{v_2}{v_1} = \sqrt{\frac{2gh}{2gH}} = \sqrt{\frac{h}{H}} \tag{6}$$

v_1 is the contact velocity; v_2 is the bounce speed; g is the gravitational acceleration; H is the initial height; h is the rebound height. The particle-to-particle recovery coefficient is 0.1. The recovery coefficient of the particles and the geometry is 0.15.

3.2 Static and dynamic friction coefficients

Figure 5. Coefficient of friction determination test.

This is shown in Figure 5. According to Newton's second law, $mg\sin\alpha = \mu mg\cos\alpha$, and the static friction between the soil and the drill pipe is 0.2; the rolling friction coefficient is 0.1. The coefficient of static friction of the particles is 0.53; the rolling friction coefficient is 0.2.

3.3 Bulk density

Figure 6. Bulk density testing.

The density of the soil involved in the simulation software EDEM is the density of solids. Model stacking a cylinder in EDEM so that particles with a radius of 0.5 mm fill the cylinder. We adjust the particle density so that the simulation and actual experiment quality are consistent. The measured bulk density is $2280\,kg/m^3$. This is shown in Figure 6.

3.4 Stacking corners

The contact parameters between the particles can be obtained by visually indexing the stacking angle [13]. This is shown in Figure 7. The accumulation angle is the maximum bottom angle the soil produces in its general state. The recovery coefficient and friction coefficient of the particles together affect the magnitude of the accumulation angle [14].

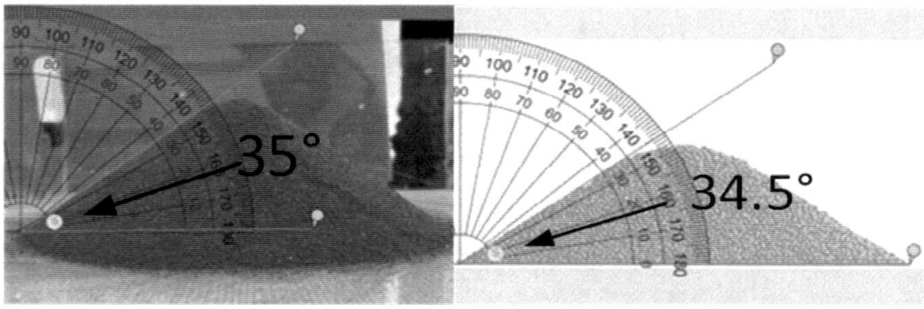

Figure 7. Stacking angle test and simulation.

4. Lunar soil discrete element drilling and production simulation

4.1 Lunar soil drilling and production modeling and working condition setting

Table 2. Condition parameters.

Gravity (g)	1		1/6	
Feed rate (mm/min)	240		360	
Rotate speed (rpm)	120	180	120	180

This is shown in Table 2. The result parameters obtained by calibration are input into the EDEM pre-processing using the pre-embedded simulation method. The drill geometry is imported first, and then the pellet plant is generated [15]. After the pellet generation, the drill pipe is given a constant rotational speed and feed rate. The magnitude of the drag moment when drilling to 0.4m and 1m under the gravity conditions of the earth and the moon are simulated, respectively. Figure 8 illustrates the drilling process.

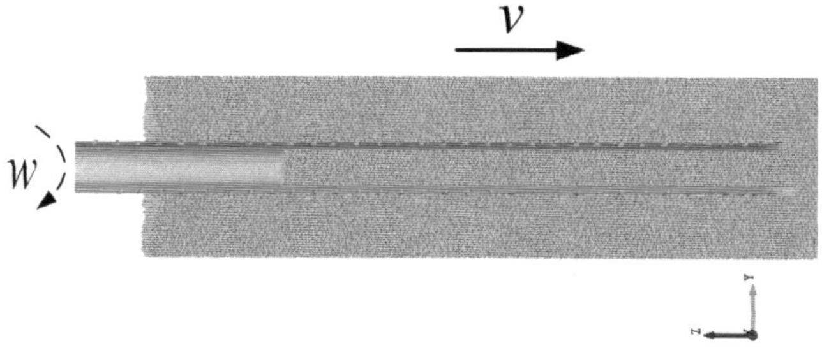

Figure 8. Drilling process.

4.2 Analysis of lunar soil drilling simulation results

This is shown in Figure 9, Figure 10, Figure 11 and Figure 12. Under the ground gravity acceleration, the magnitude of resistance moment of drill pipe at different depths, feed speeds, and rotational speeds is obtained by simulation. When the rotational speed is the same, the drag moment positively correlates with the feed rate. At the same feed rate, the drag torque is negatively correlated with the rotational speed. At a depth of 1m, a rotation speed of 120 rpm, and a feed rate of 360 mm/min, the drag moment of surface drilling is 2.75 Nm, while the drag moment of moon drilling is 1.55 Nm.

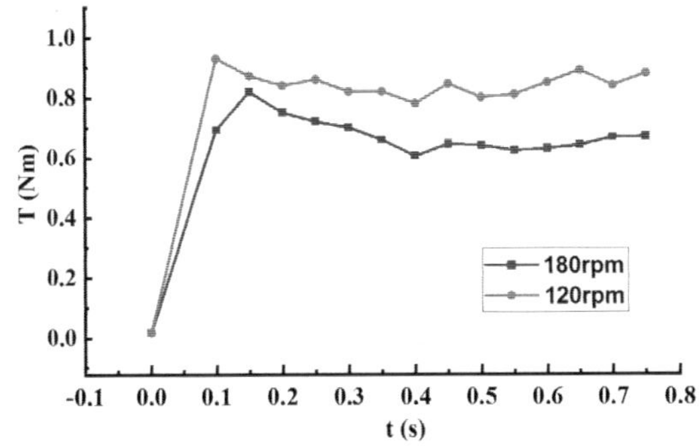

Figure 9. Depth 0.4m and feed rate 240mm/min.

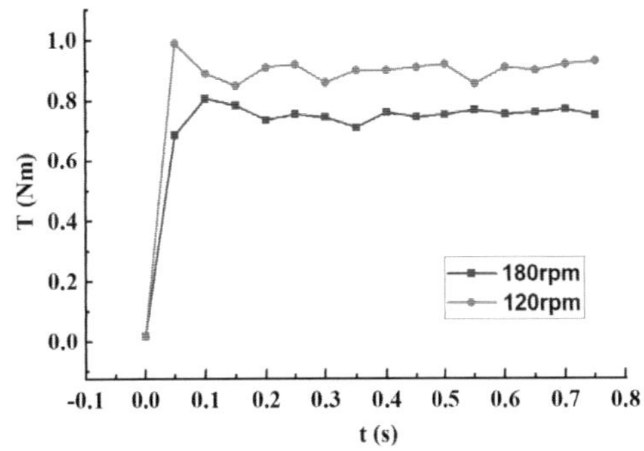

Figure 10. Depth 0.4m and feed rate 360mm/min.

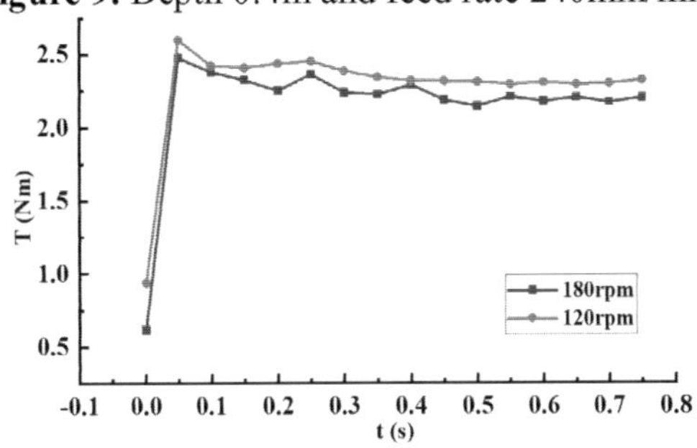

Figure 11. Depth 1m and feed rate 240mm/min.

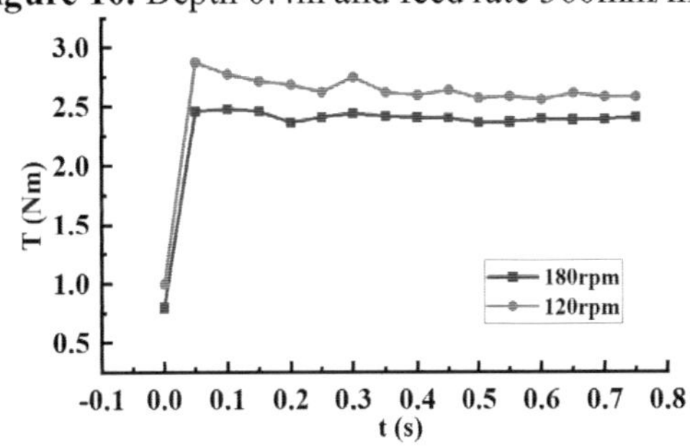

Figure 12. Depth 1m and feed rate 360mm/min.

We adjust the acceleration to the gravitational acceleration of the lunar surface, and the other parameters remain unchanged. The results are shown in Figure 13, Figure 14, Figure 15 and Figure 16.

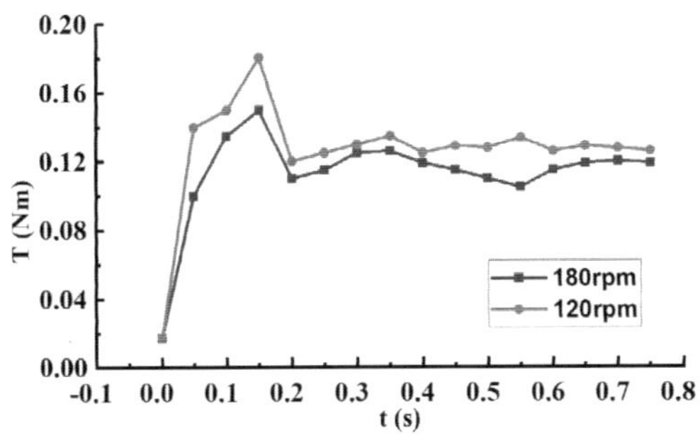

Figure 13. Depth 0.4m and feed rate 240mm/min.

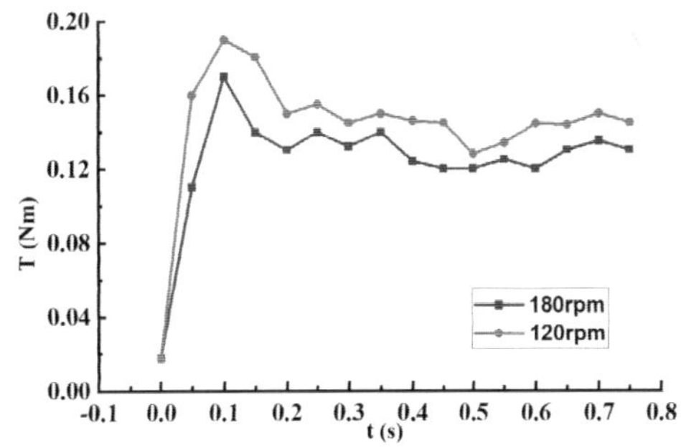

Figure 14. Depth 0.4m and feed rate 360mm/min.

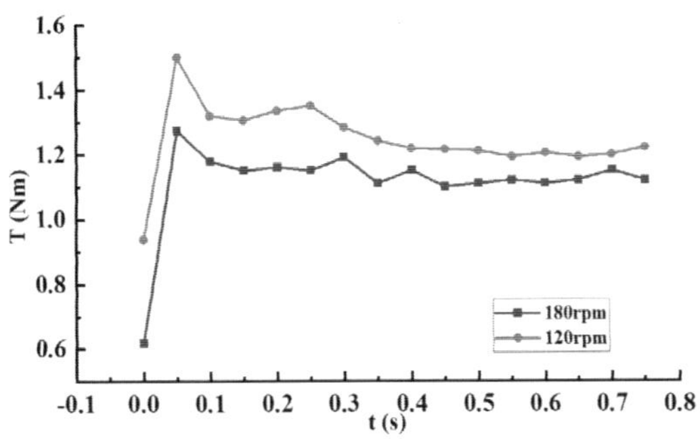

Figure 15. Depth 1m and feed rate 240mm/min.

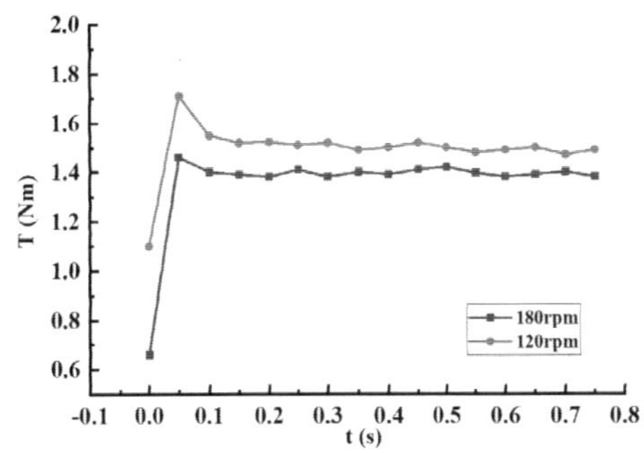

Figure 16. Depth 1m and feed rate 360mm/min.

5. Conclusion

The structural design and dynamic analysis of the auger drill pipe are carried out, and the parameter calibration test and simulation of the simulated lunar soil are carried out. Given the inherent defects of the discrete element method, the "embedded simulation method" was proposed for the first time. The particles are stacked to the target depth, and the working conditions are simulated for simulation. In the case of earth gravity and lunar gravity, the rotation speed and feed rate are used as variable parameters, and the magnitude of the drag moment of the drill pipe at 0.4m and 1m is obtained through simulation analysis. When the depth is 1m, the ground gravity, the rotation speed is 120rpm, and the feed speed is 360mm/min, the resistance torque of the drill pipe in the steady state is 2.75Nm. When the acceleration is changed to the acceleration of the moon's gravity, the drag moment of the drill pipe is 1.55Nm. Therefore, the size of this resistance torque should be considered when selecting the drilling motor to prevent problems such as stuck drilling. The analysis of the simulation post-processing results shows that the lunar soil is stored in the coring tube, and the sampling efficiency is 80%. It meets the requirements of use and has good sampling efficiency.

References

[1] Khademian Z, Kim E, Nakagawa M. Simulation of Lunar Soil with Irregularly Shaped, Crushable Grains: Effects of Grain Shapes on the Mechanical Behaviors [J]. Journal of Geophysical Research: Planets, 2019, 124(5). DOI: 10.1029/2018JE005889.
[2] Da-Fo L I, Yan L, Shao-Ning X U. Study on Particular Coring Bit for Lunar Soil Drilling [J]. Exploration Engineering (Rock & Soil Drilling and Tunneling), 2013.
[3] Zhang K, Hou X, Pan W, et al. The study of the drilling core features of a multi-pipe deep lunar soil sampling driller for manned lunar exploration based on the discrete element technology [C] //IEEE International Conference on Robotics & Biomimetics. IEEE, 2015.
[4] Liu Tx. Research on drilling-sampling performance based on the interaction model between drill and lunar soil [D]. Harbin Institute of Technology, 2015.
[5] Wei C, Wang H, Liu T. Mechanical model of hollow-external-screw drill rod for lunar soil particle vertical conveying [J]. IEEE, 2013.DOI: 10.1109/ICCA.2013.6565063.
[6] Xu-Yan H, Zong-Quan D, Hong-Hao Y, et al. Research on the Impact of a Drill Bit on the Original State of Artificial Lunar Soil using DEM Simulation [J]. Transactions of the Japan Society for Aeronautical and Space Sciences, 2016, 59(4): 236-242. DOI: 10.2322/tjsass.59.236.
[7] J. Liang, L. Tao, W. Zhang, J. Tang, Y. Pang, S. Jiang, Analysis of the lunar regolith sample obstruction in the Chang'E-5 drill and its improvement, Adv. Space Res. 69 (5) (2022) 2248–2258, https://doi.org/10.1016/j.asr.2021.12.004.
[8] Chen Hw, Ran Hq, Wang Yl, etc. Lunar drilling simulation experiments and technical prospects [J]. Drilling Engineering, 2023, 50(S1): 192-199.
[9] Liu Tx, Jun Z, Lei L, et al. A systematic calibration and validating method for lunar soil DEM model [J]. Advances in Space Research, 2021, 68(9): 3925-3942.
[10] Liu Tx, Ma L, Liang L, etc. Analysis on Influence of Drill Parameters in the Sampling of Deep Lunar Soil [J]. Journal of Mechanical Engineering, 2018, 54(09): 27-36.
[11] Liu Tx, Jun Z, Lei L, et al. Effect of drill bit structure on sample collecting of lunar soil drilling [J]. Advances in Space Research, 2021, 68(1): 134-152.
[12] Shen Zq, Li S, Liu Dy, etc. Research on development of extraterrestrial object detection and drilling technology represented [J]. Spacecraft Engineering, 2023, 32(01): 97-104.
[13] Tao C, Zhen Z, Qi W, et al. Modeling and experimental investigation of drilling into lunar soils [J]. Applied Mathematics and Mechanics, 2019, 40(1): 153-166.
[14] Liu F. Research on drilling-sampling performance based on interaction model between drill and lunar soil [D]. Harbin Institute of Technology, 2011.
[15] Zhou J, Liu Tx, Liang L, etc. Formation mechanism analysis and predictive modeling of footage force in lunar soil drilling and sampling [J]. Journal of Astronautics, 2023, 44(01): 34-42.

Journal of Physics: Conference Series **2842** (2024) 012046 doi:10.1088/1742-6596/2842/1/012046

The impact of induction heating time on the temperature distribution of high-temperature alloy discs

Tuermengke*, **Weidong Zhu, Yuewen Zhai**

China Beijing Research Institute of Mechanical and Electrical Technology Co., Ltd. CAM, Beijing 100083

*Corresponding author's e-mail: tuermengke201809@163.com

Abstract: The induction heating time is directly related to the temperature uniformity of high-temperature alloy disc components. Therefore, investigating the influence of induction heating time on the temperature field distribution of disc components provides guidance for temperature control strategies in the forming process. In this study, the finite element analysis software COMSOL Multiphysics was employed to examine the impact of induction heating time on the radial and thickness-wise temperature fields of GH4169 discs. The results indicate that the heating time affects the thickness of the heating layer, and appropriately extending the heating time can improve the temperature uniformity in the thickness direction, achieving a temperature difference of ±5 °C in the disc component thickness direction under certain conditions. When the heating time ranges from 1229 s to 1239 s, the temperature in the R=30 mm to R=100 mm range reaches the suitable forging temperature.

1. Introduction

The high-temperature alloy disc, specifically the turbine disc, serves as a crucial rotating component in engines, necessitating precise control for optimal performance. This study is based on the second-generation non-die rolling forming machine independently developed by the Beijing Institute of Mechanical and Electrical Research, China Academy of Machinery Science and Technology. The research employs induction heating technology to achieve localized heating of high-temperature alloy discs, investigating the influence of induction heating time on the temperature field. [1]

Choosing localized induction heating for disc deformation is motivated by the precise heating of specific regions using electromagnetic induction [2] [3]. This approach, compared to uniform heating methods, mitigates uneven temperature gradients, enhancing temperature uniformity. To meet forging requirements, maintaining disc rotation introduces radial temperature gradients through rotation. This ensures higher temperatures at the disc edge and lower temperatures at the center, aligning with the process requirements for forging temperature and thickness-wise temperature differentials.

During the rolling process, the disc undergoes plastic deformation through rolling head units. Employing rotation facilitates flexible adjustment of the heating area, ensuring the induction coil is unaffected by rolling head units. This helps guarantee uniform heating of the disc, avoiding temperature non-uniformities caused by localized obstructions.

Temperature control is crucial for forging nickel-based high-temperature alloy discs as the appropriate temperature range directly impacts mechanical performance and microstructure formation. The rolling zone near the disc center should not fall below 1000 °C, and the temperature range near the outer edge should be between 1100 °C and 1150 °C. The higher temperature requirement at the disc's outer edge ensures sufficient plasticity on the surface during the rolling process, preventing defects such as cracks. The induction heating time directly influences the temperature uniformity of high-temperature alloy discs. Under constant conditions, suitable heating time helps avoid local overheating or cooling, improving overall temperature uniformity. Excessive or insufficient induction heating time may result in excessively large or small grain sizes, affecting the mechanical properties and high-temperature stability of the disc. [4] [5]

In this study, a symmetrically distributed induction coil configuration is employed for localized induction heating of GH4169. Through numerical simulations, the research explores the influence of induction heating time on the radial and thickness-wise temperature field distribution of the disc. This lays the foundation for temperature control strategies in subsequent induction heating systems for disc plastic forming processes.

2. Model establishment

2.1 Establishment of the geometric model

In accordance with the process requirements for rolling discs, a scaled-down induction heating model is established. Considering the need to avoid rolling head units and achieve a gradient temperature distribution on the disc, this study employs a symmetrical localized induction heating configuration for the disc. A hollow copper tube with dimensions of 14×14×1 mm is wound around the disc in an equidistant manner. Cooling water is circulated through the copper tube during the experimental process.

Based on the proportionally scaled induction heater, a circular disc with a diameter of 200 mm and a thickness of 30 mm is created. To accommodate the gripping mechanism of the rotating device, a central hole with a diameter of 20 mm is machined at the center of the disc. Since the electromagnetic field is not localized, a suitable space is required, which can be either air or vacuum. A sufficiently large air domain is essential to ensure accurate calculations of the electromagnetic field in the small local space. Thus, an air domain, as illustrated in Figure 1, is established for the induction heating model.

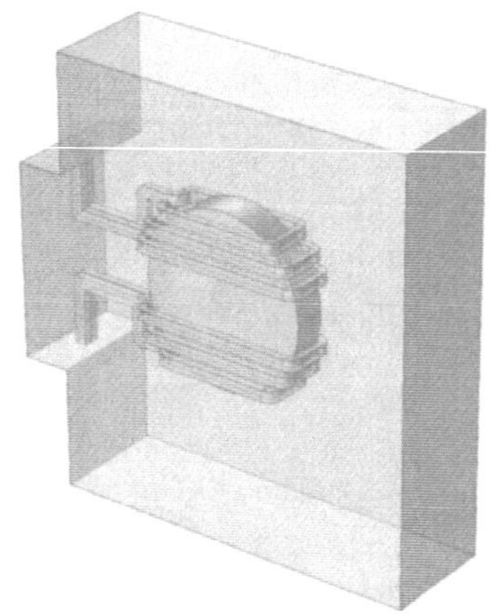

Figure 1. Induction heating model.

2.2 Preprocessing

The localized induction heating process of the disc under investigation in this study is simulated using COMSOL Multiphysics software. The software's built-in user-controlled meshing capabilities are employed for the overall meshing of the model. Custom meshing is applied to the smallest elements and narrow regions that trigger errors to enhance computational accuracy, utilizing a free triangular mesh. Subsequently, the remaining sections are meshed with free tetrahedral elements. In total, the model is divided into 152, 863 elements. Figure 2 illustrates the meshing results of the three-dimensional model for localized induction heating.

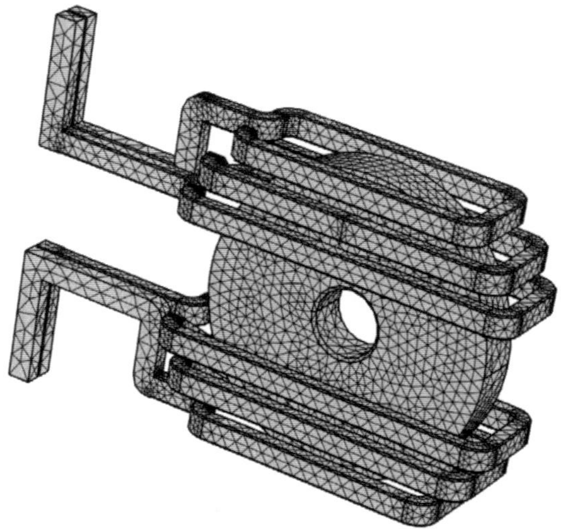

Figure 2. Induction heating meshing.

The present study necessitates the definition of material properties for both the disc and the induction coil. Additionally, the air domain needs to be characterized, with the relative magnetic permeability of air set to 1. The chosen material for the disc is GH4169, and its chemical composition is outlined in Table 1. GH4169 is a precipitation-strengthened nickel-chromium-iron-based high-temperature alloy. Relevant material properties such as electrical conductivity, resistivity, thermal conductivity, and specific heat capacity for GH4169 can be calculated using JMatPro software. Apply a rotating domain to the disc, specifying the rotation type as rotational speed. Define the angular velocity formula as V0*step1(t/10[s]) rad/s.

Table 1. Chemical composition of high-temperature alloy GH4169 (Mass fraction, wt%).

	Ni	C	Mn	Si	Cu	Nb	Cr	Al	Mo	Ti	Co
GH4169	53.70	0.026	0.03	0.05	0.02	5.37	18.10	0.44	2.95	0.87	0.30

3. Simulation result analysis

3.1 The impact of induction heating time on the thickness-wise temperature field of the disc

Setting the excitation current to I = 990 A and the frequency to 13500 Hz, a simulation is conducted to study the localized electromagnetic induction heating process of the high-temperature alloy GH4169 within the inductor for a duration of 1300 s. The focus is on investigating the temperature uniformity in the thickness direction. Figure 3 illustrates the disc's thickness-wise temperature distribution at different time points under these operating conditions. The four sets of images present sectional views of the disc after induction heating at a current of 990 A and a rotational speed of 2 revolutions. They depict the temperature distribution in the thickness direction at different radii. The horizontal axis represents the thickness of the disc, with the reference point at the disc's center and a thickness of 30 mm. Seven three-dimensional intersection points with different thicknesses at the same radius are selected and interpolated as probe points. The vertical axis represents the temperature of the disc's material points. From the graphs, it can be observed that, at the same moment, there is a difference in thickness-wise temperature at different radii. The results indicate that the temperature is relatively uniform in the thickness direction near the disc's edge, while there is a relatively larger temperature difference near the disc's center.

Journal of Physics: Conference Series **2842** (2024) 012046 doi:10.1088/1742-6596/2842/1/012046

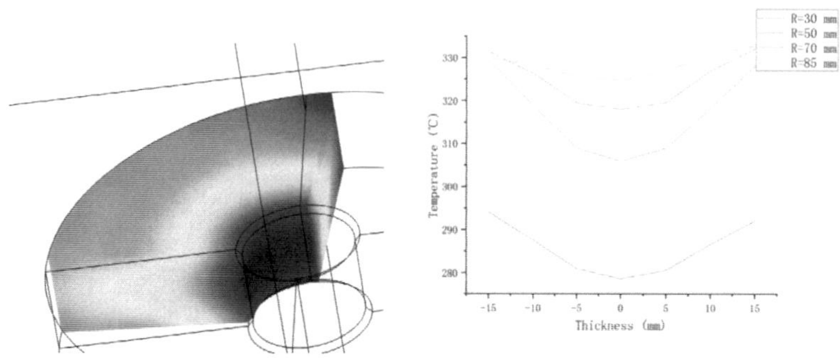

(a) 300 s

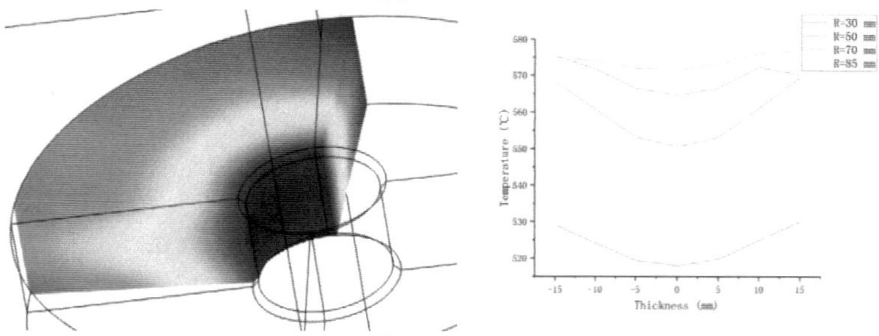

(b) 600 s

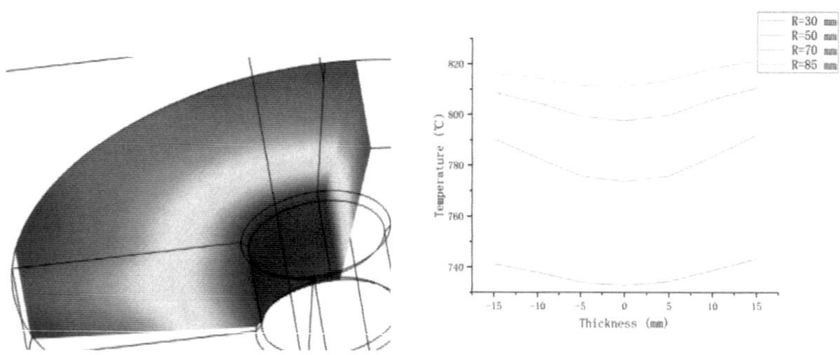

(c) 900 s

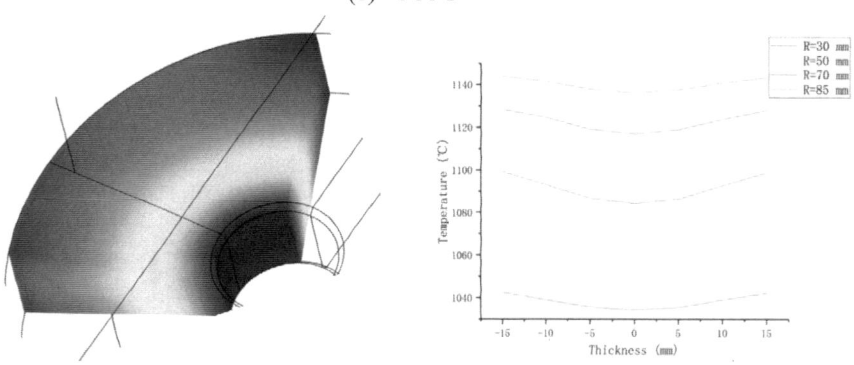

(d) 1300 s

Figure 3. Thickness-wise Temperature Distribution of the Disc at Different Time Points.

In addition, Table 2 provides the point temperatures at four different radii in the thickness direction of the disc after 1300 s of induction heating with a current of 990 A and a rotational speed of 2 revolutions. Combining this analysis with the previous figure, it can be concluded that with the increase in time, the thickness-wise temperature difference gradually decreases, reaching a temperature difference of ±5 °C.

Table 2. Thickness-wise Temperature Distribution at Different Radii (1300 s).

Thickness mm	Temperature (R = 30 mm) degC	Temperature (R = 50 mm) degC	Temperature (R = 70 mm) degC	Temperature (R = 85 mm) degC
15	1042.25	1098.8	1127.95	1143.31
10	1039.98	1092.7	1123.82	1140.87
5	1035.45	1086.39	1118.88	1137.6
0	1034.38	1084.46	1117.01	1136.33
-5	1035.62	1086.66	1119.08	1137.87
-10	1039.08	1093.18	1124.7	1141.6
-15	1042.58	1099.09	1128.27	1143.89

3.2 The impact of induction heating time on the radial temperature field of the disc.

Based on heating with a current of 990 A for 1300 s, the temperature variation over time at two points near the disc's center and edge was observed. Initially, as observed in Figure 4, the temperature increase is faster near the disc's edge. By 1239 s, the temperature at the disc's center reaches 1000 °C, while at 1229 s, the temperature at the disc's edge reaches 1100 °C. Therefore, the conclusion can be drawn that when the heating time is between 1229 s and 1239 s, the temperature range from R=30 mm to R=100 mm reaches the suitable forging temperature.

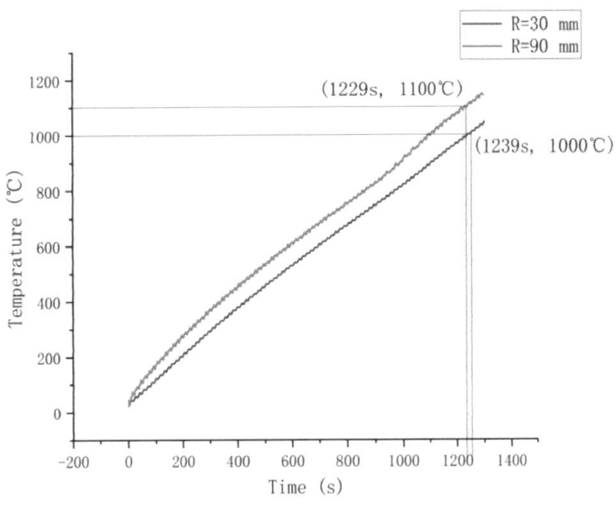

Figure 4. Temperature variation with time at different radii.

4. Conclusion

(1) The longer the heating time, the larger the region reaching the Curie point at the surface and, correspondingly, the larger the demagnetized area. The heat conduction effect becomes more pronounced, resulting in a larger heated region. Longer heating times are advantageous for achieving uniform temperature distribution across the thickness.

(2) When the excitation current I set to 990 A with a frequency of 13500 Hz, as time increases, the thickness-wise temperature difference gradually decreases, reaching a temperature difference within ±5 °C. In the heating duration of 1229 s-1239 s, the temperature ranging from R=30 mm to R=100 mm reaches the suitable forging temperature.

References

[1] Fan L, Liu X, and Zhai Y, et al. Technical Progress and Industrialization Analysis of Dieless Rolling for Turbine Disk [J]. Journal of Mechanical Engineering, 2020,56(14):44-56.
[2] Kovac D and Melnykov V. Simulation and Laboratory Model of Induction Heating in COMSOL Multiphysics [C] //2021 IEEE International Conference on Modern Electrical and Energy Systems (MEES). IEEE, 2021: 1-6.
[3] Smol'yanov I A, Tarasov F E and Shvydkii E L. Numerical Calculation of the Induction Heating of Large Objects [J]. Russian Metallurgy (Metally), 2021, (12): 1510-1516.
[4] Yermekova M and Galunin S A . Numerical simulation and automatic optimization of the disk induction heating system [C] // 2017 IEEE Conference of Russian Young Researchers in Electrical and Electronic Engineering (EIConRus). IEEE, 2017: 1085-1090.
[5] Mannanov E , Galunin S and Yermekova M , et al. Numerical Simulation, Investigation, and Development of Induction Systems for Heating of Disks with Complex Profile [C] // 2019 IEEE Conference of Russian Young Researchers in Electrical and Electronic Engineering (EIConRus). IEEE,2019: 606-608.

Study on quasi-static tensile mechanical properties of biomimetic helicoidal composite laminates

Shutong Huang[1], Xiaoli Jia[1,*], Hengxiang Liu[1], Jianfeng He[1], Qingyu Yang[1], Liaoliang Ke[2], Jie Yang[3] and Sritawat Kitipornchai[4]

[1]College of Mechanical and Transportation Engineering, China University of Petroleum (Beijing), Beijing, 102249, China
[2]School of Mechanical Engineering, Tianjin University, Tianjin, 300072, China
[3]School of Aerospace, Mechanical and Manufacturing Engineering, RMIT University, PO Box 71, Bundoora, VIC 3083 Australia
[4]School of Civil Engineering, Guangzhou University, Guangzhou,510006, China

[*]Corresponding author's e-mail: xljia@cup.edu.cn

Abstract: Numerous biomaterials exhibit intricate helicoidal structures renowned for their exceptional toughness. This study drew inspiration from the collagen fiber structure found in the scales of coelacanths to design composite laminates featuring double helicoidal layup configurations with varying pitch angles. A damage evolution model with damage initiation and linear stiffness degradation based on the Hashin failure criterion was used to describe the damage behavior of the materials. The impact of varying pitch angles on the quasi-static tensile properties of double helicoidal composite laminates is investigated. The results show that the magnitude of the pitch angle has a significant effect on the tensile properties of the laminates, and the tensile modulus of double composite laminates is highest when the pitch angle is 90°.

1. Introduction

Composite materials are usually made of a combination of two or more materials of different properties, with the advantages of light weight, high strength, corrosion resistance, and flexible design. Compared with traditional metals, ceramics, and other materials, fiber-reinforced composites have the advantages of high specific modulus, high specific strength, strong designability, and easy processing, and researchers have carried out a large number of studies on fiber-reinforced composites [1-2].

Fiber composites, as a material with excellent performance and a wide range of application prospects, have been further optimized and expanded through the application of bionics [3-4]. As the key to defending fish against external enemies, scales have evolved a subtle spiral structure through years of natural selection. For example, the microstructure of carp scales shows a single spiral arrangement, and the coelacanth, as a "living fossil", has evolved a double spiral arrangement of microstructure scales. It has been shown that the smaller pitch angle has a toughening effect on the material [5]. At present, there are more studies on single helical structures and fewer studies on double helical structures, especially the lack of research on the effect of different pitch angles on the tensile properties of helical structures and the damage failure behavior.

In this paper, a bionic double helicoidal laminate (BDHL) inspired by coelacanth scales, with epoxy resins and carbon fibers as matrix and reinforcement materials, respectively, have been designed. A combination of numerical simulations is used to investigate the mechanical properties and damage failure behavior of double helicoidal laminates with different pitch angles under static tensile loading.

2. Numerical simulation

2.1 Finite element model

In this paper, the finite element model of carbon fiber composite laminates with bionic helical structure is established by using the display integral solver in ABAQUS. The mechanical properties and failure behaviors of double helical composite laminates with different pitch angles under static tensile action are investigated. The stacking order of different helical composite laminates is shown in Table 1. Among them, the finite element model and layup structure of the double helicoidal carbon fiber composite laminate with a pitch angle of 10° is shown in Figure 1. The reference point RP1 is created at 50 mm from the left end face, coupling constraints are formed with the left end face, and fixed constraints are applied at the reference point. The reference point RP2 is created at 50 mm from the right end face, coupling constraints are formed with the right end face, and displacement load is applied at the reference point to simulate the tensile load.

Table 1. Stacking sequence of double helicoidal composite laminates.

Sample	Pitch angle	No. of plies	Stacking sequence
BDHL-10	10°	20	[0°/90°/10°/100°/..../90°/180°]
BDHL-20	20°	20	[0°/90°/20°/110°/..../180°/270°]
BDHL-30	30°	20	[0°/90°/30°/120°/..../270°/360°]
BDHL-40	40°	20	[0°/90°/40°/130°/..../360°/90°]
BDHL-50	50°	20	[0°/90°/50°/140°/..../90°/180°]
BDHL-60	60°	20	[0°/90°/60°/150°/..../180°/270°]
BDHL-70	70°	20	[0°/90°/70°/160°/..../270°/360°]
BDHL-80	80°	20	[0°/90°/80°/170°/..../360°/90°]
BDHL-90	90°	20	[0°/90°/90°/180°/..../90°/180°]

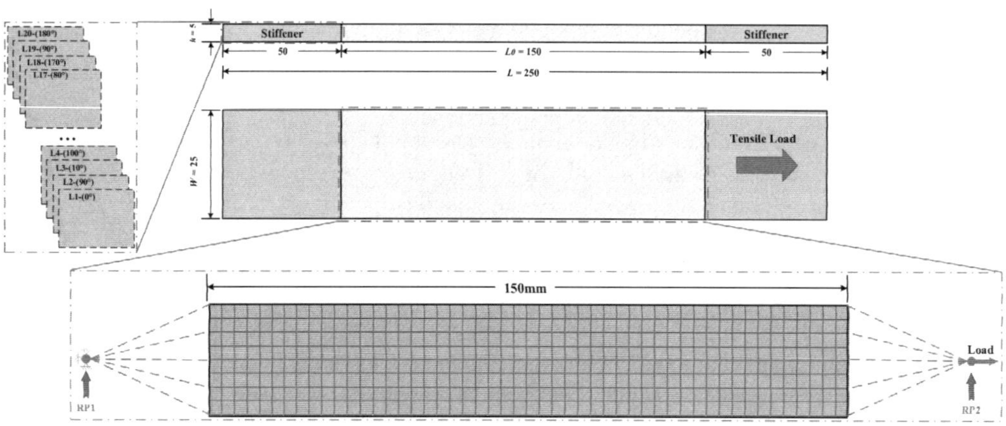

Figure 1. Schematic of the finite element model and layup structure of the composite laminate.

2.2 Damage failure and evolution simulation of composite layers

2.2.1 Damage initiation based on hashin failure criterion. The Hashin failure criterion, as one of the most widely used failure mode discrimination criteria, is mainly used for the evaluation of the mechanical properties of fiber-reinforced composites. The Hashin failure criterion is not only more accurate in predicting the gradual damage process inside the composites but also visualizes the damage process in the form of cloud diagrams, which helps analyze the damage state of the composites more clearly.

The damage initiation criteria for the 2D Hashin failure criterion are as follows [6-7]:
Fiber tensile damage ($\hat{\sigma}_{11} \geq 0$):

$$F_f^t = \left(\frac{\hat{\sigma}_{11}}{X_t}\right)^2 + \alpha\left(\frac{\hat{\sigma}_{12}}{S_l}\right)^2 \tag{1}$$

Fiber compression damage ($\hat{\sigma}_{11} < 0$):

$$F_f^c = \left(\frac{\hat{\sigma}_{11}}{X_c}\right)^2 \tag{2}$$

Matrix tensile damage ($\hat{\sigma}_{22} \geq 0$):

$$F_m^t = \left(\frac{\hat{\sigma}_{22}}{Y_t}\right)^2 + \alpha\left(\frac{\hat{\sigma}_{12}}{S_l}\right)^2 \tag{3}$$

Matrix compression damage ($\hat{\sigma}_{22} < 0$):

$$F_m^c = \left(\frac{\hat{\sigma}_{22}}{2S_t}\right)^2 + \left[\left(\frac{Y_c}{2S_t}\right)^2 - 1\right]\frac{\hat{\sigma}_{22}}{Y_c} + \left(\frac{\hat{\sigma}_{12}}{S_l}\right)^2 \tag{4}$$

where $\hat{\sigma}_{11}$, $\hat{\sigma}_{12}$ and $\hat{\sigma}_{22}$ are the effective stress components, X_t, X_c, Y_t, Y_c, S_l and S_t are the fiber longitudinal tensile strength, fiber longitudinal compression strength, fiber transverse tensile strength, fiber transverse compression strength, longitudinal shear strength, and transverse shear strength, and the calculation process of $\hat{\sigma}$ and σ is as follows:

$$\hat{\sigma} = M\sigma \tag{5}$$

where M is the damage operator matrix [8]:

$$M = \begin{bmatrix} \frac{1}{(1-d_f)} & 0 & 0 \\ 0 & \frac{1}{(1-d_m)} & 0 \\ 0 & 0 & \frac{1}{(1-d_s)} \end{bmatrix} \tag{6}$$

where d_f, d_m and d_s are the internal fiber damage variables, the internal matrix damage variables, and the internal shear damage variables.

$$d_f = \begin{cases} d_f^t\,(\hat{\sigma}_{11} \geq 0) \\ d_f^c\,(\hat{\sigma}_{11} < 0) \end{cases} \tag{7}$$

$$d_m = \begin{cases} d_m^t\,(\hat{\sigma}_{22} \geq 0) \\ d_m^c\,(\hat{\sigma}_{22} < 0) \end{cases} \tag{8}$$

$$d_s = 1 - (1-d_f^t)(1-d_f^c)(1-d_m^t)(1-d_m^c) \tag{9}$$

Before any damage occurs, the damage operator matrix M is a unitary matrix when $\hat{\sigma} = \sigma$. Once damage occurs, the damage operator matrix M is no longer a unitary matrix, and the post-damage stresses are redistributed, with damage in any mode having a significant effect on other damage modes. The parameters of composite materials are shown in Table 2.

Table 2. Parameters of composite materials.

Performance parameter	Value	Performance parameter	Value
$\rho / (\text{g}\cdot\text{cm}^{-3})$	1.7	X_c / MPa	555.2
E_1 / GPa	110.5	Y_t / MPa	24.6
E_2 / GPa	8.9	Y_c / MPa	104.2
ν	0.28	S / MPa	41.7
G_{12} / GPa	4.5	G_{ft} / (N/mm)	100

G_{13} / GPa	4.5	G_{fc} / (N/mm)	100
G_{23} / GPa	2.43	G_{mt} / (N/mm)	0.22
X_t / MPa	1001.58	G_{mc} / (N/mm)	1.1

3. Results and discussion

3.1 Verification of model

To verify the correctness and reliability of the finite element model, a finite element model with the same size as the experimental specimen in [9] was established, and the material parameters were all from [9]. The force-displacement curve obtained by numerical simulation was compared with the experimental data, as shown in Figure 2. The variation trend of the experimental and simulated force-displacement curves is consistent. Compared with the experimental data, the error of the tensile limit load in the numerical simulation results is only 0.82%, the error of the failure displacement is about 8%, and the failure displacement error of simulation and experiment in [9] is 6%, both of which are within the acceptable range. Therefore, the finite element model in this paper is correct and has high reliability. The reason why the slope of force-displacement curves of simulation and experiment is not completely consistent may be that the loading method is inconsistent. The loading method of uniform speed is adopted in the experiment, and the loading method of smooth analysis step is adopted in the simulation to make the results smoother.

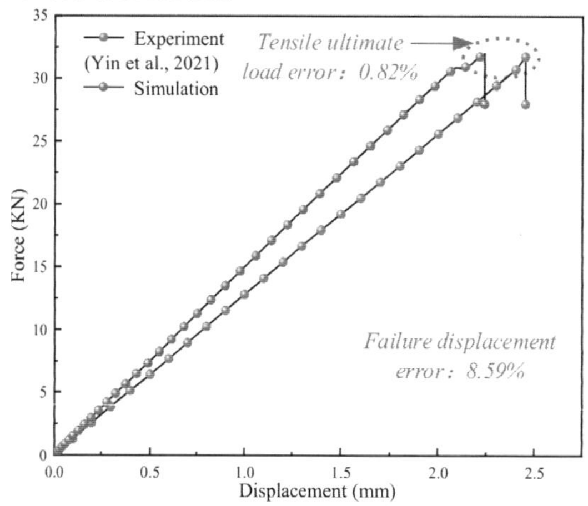

Figure 2. Validation of the numerical model.

3.2 Tensile mechanical properties of double helicoidal composites

From Figure 3(b), it can be seen that the failure displacements of different pitch angle laminates can be roughly categorized into three classes, in which the failure displacements of BDHL-10, BDHL-30, BDHL-60, and BDHL-80 are roughly equal to about 1.93 mm, the failure displacements of BDHL-20, BDHL-40, BDHL-50, and BDHL-70 are also roughly equal to about 2.62 mm, and BDHL-90 has the shortest failure displacement of 1.33 mm. From Figure 3(c), it can be seen that the failure strength of BDHL-90 is significantly higher than that of other laminates, which reaches 490.59 MPa, and the failure strengths of laminates with pitch angle increasing from 20° to 70° are almost equal, and the failure strengths of BDHL-10 and BDHL-80 are the lowest, which are about 375 MPa. From Figure 3(d), it can be seen that the tensile modulus follows the same pattern as the failure displacement, and the tensile modulus of laminates with different pitch angles can be roughly classified into three classes: the tensile modulus of BDHL-10, BDHL-30, BDHL-60, and BDHL-80 are roughly equal, about 30 GPa; the tensile modulus of BDHL-20, BDHL-40, BDHL-50, and BDHL-70 are also roughly equal; and the tensile modulus of BDHL-10 and BDHL-70 are about 375 MPa.

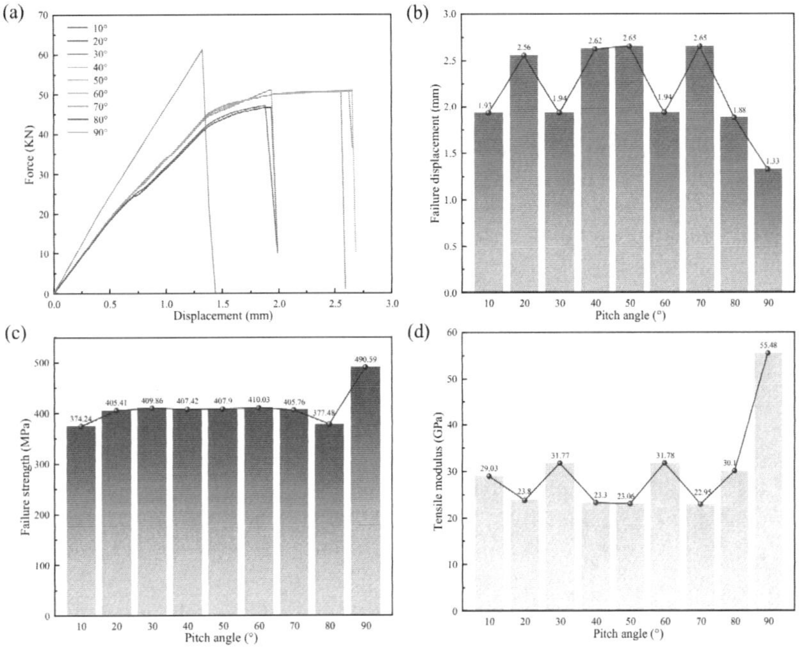

Figure 3. (a) Force-displacement curves, (b) Failure displacement, (c) Failure strength, and (d) Tensile modulus of double helicoidal composite laminates.

Since the tensile modulus at the pitch angle of 90° is significantly higher than that of the laminates with other pitch angles, the focus is placed on the analysis of the damage failure behavior of the laminates with the pitch angle of 90° next.

3.3 Damage failure process of BDHL-90

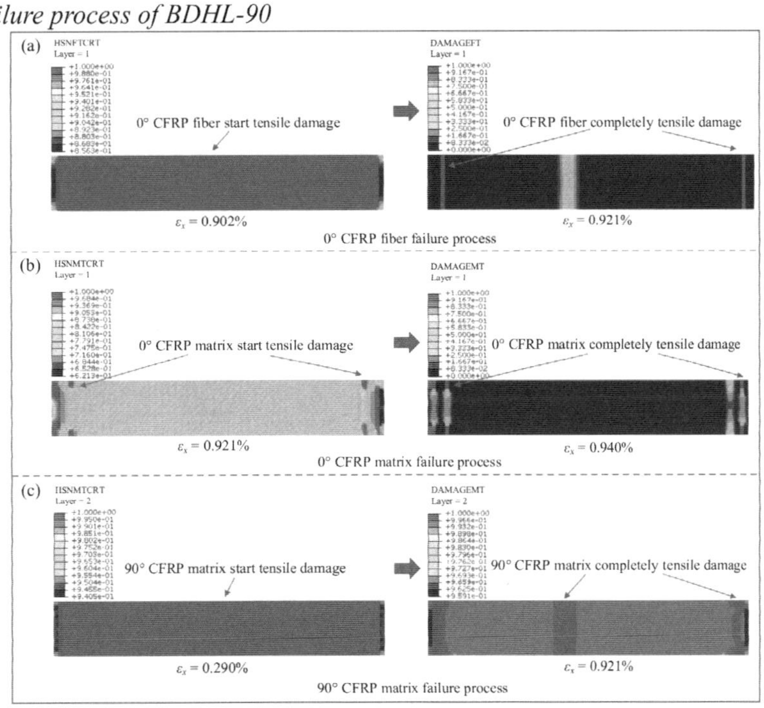

Figure 4. Damage evolution of BDHL-90.

The damage evolution of BDHL-90 is shown in Figure 4. Figure 4(a) shows that when the strain reaches 0.902%, the fiber damage state variable of the 0° layer reaches 1 from the legend, indicating that the fibers begin to show damage. The fibers are completely damaged when the strain reaches 0.921%, and cracks run completely through the specimen at both ends of the specimen. From Figure 4(b), it can be seen that the matrix starts to show damage when the strain reaches 0.921%, the matrix fails when the strain reaches 0.940%, and the failure location is located at both ends of the specimen. Figure 4(c) shows the matrix failure process of the 90° layer. The damage starts to appear when the strain is 0.29%, and the matrix completely fails when the strain reaches 0.921%. Overall, the 0° layer showed earlier damage to the matrix of the 90° layer than the 0° layer due to the higher tensile strength of the fibers.

4. Conclusion

In this paper, the mechanical properties and damage failure modes of laminates with different pitch angles under static tensile loading are investigated by numerical simulations, based on the damage initiation of the Hashin failure criterion and the damage evolution of the linear stiffness degradation model. The main conclusions are as follows:

(1) When the pitch angle increases from 10° to 80°, the pitch angle of double helicoidal plywood has less effect on the failure strength and tensile modulus. The failure strength and tensile modulus are the highest when the pitch angle is 90°, but the toughness is the worst, and the failure displacements of double helicoidal plywood can be roughly divided into three grades. The toughness is the best when the pitch angle is 20°, 40°, 50° and 70°.

(2) The strength and toughness of fiber-reinforced composites are a contradiction. Composites with high strength have poor toughness, while composites with good toughness have low strength, so hybrid fiber composites can be considered to improve this situation.

References

[1] Prasanthi P P, Kumar M S R N, Chowdary M S, et al. (2023) Mechanical properties of carbon fiber reinforced with carbon nanotubes and graphene filled epoxy composites: experimental and numerical investigations. Materials Research Express, 10(2): 025308.

[2] Wang C, Guo L, Xia Y, et al. (2024) Flexural performance and damage evolution of multiple fiberglass-reinforced UV-CIPP composite materials--A view from mechanics and energy release. Journal of Materials Research and Technology, 29: 3317-3339.

[3] Liu Y X, Li A H, Lin S Y, et al. (2023) Research on biomimetic design and impact characteristics of periodic multilayer helical structures. Frontiers in Bioengineering and Biotechnology, 11: 999137.

[4] Wang H, Wu Z, Tao J, et al. (2024) Bamboo-Inspired Crack-Face Bridging Fiber Reinforced Composites Simultaneously Attain High Strength and Toughness. Advanced Science, 11(10): 2308070.

[5] Liu J L, Lee H P and Tan V B C. (2018) Effects of inter-ply angles on the failure mechanisms in bioinspired helicoidal laminates. Composites Science and Technology, 165: 282-289.

[6] Hashin Z and Rotem A. (1973) A fatigue failure criterion for fiber reinforced materials. Journal of composite materials, 7(4): 448-464.

[7] Hashin Z. (1981) Fatigue failure criteria for unidirectional fiber composites. Journal of Applied Mechanics, 47(2): 329-334.

[8] Zheng Y, Hu K, Zhang M, et al. (2024) Characterizing damage patterns and evolution in Multi-Hole GLARE laminates under tensile load via integrated AE and DIC techniques. Composite Structures, 331: 117911.

[9] Yin D, Zheng J, Xiong C, et al. (2021) Experimental and numerical studies of perforated CFRP laminates under quasi-static tensile load. In: New Material and Chemical Industry. Xiamen. 012011.

Effects of Co and Mo on the microstructure properties of GH4282 and the mechanical properties of metal bellows

Zhuoliang Li[1,*], Dalin Zhang[1], Linhong Song[1], Xianglin Yu[1], Songhan Wang[1] and Zhaoxia Shi[2]

[1]Shenyang Academy of Instrumentation Sciences Co., Ltd, Shenyang 110043, China
[2]Beijing Key Laboratory of Advanced High Temperature Materials, Beijing 100081, China

[*]Corresponding author's e-mail: zhuoliang_li@foxmail.com

Abstract. In the present study, a new type of Ni-Cr-Co-Mo (GH4282) precipitation-hardening superalloy was prepared on the basis of GH4202 alloy. By adding Co and Mo elements to enhance the solution-strengthening effect, good creep resistance and toughness were obtained. The results show that the strength and microstructure stability of GH4282 alloy are significantly higher than those of traditional GH4202 alloy under the condition of 700~900°C. The precipitated phase of GH4282 alloy is dominated by γ' phase, which contains $M_{23}C_6$, MC, and a small amount of M_6C carbide. The precipitation is mainly distributed along the grain boundaries, and has good stability at temperatures below 900°C, which can effectively pin the grain boundaries. The thermal stability of the alloy is increased. The metal bellows made of GH4282 alloy have significantly higher fatigue life and high-temperature stability, the increase of Co and Mo element content leads to the enhancement of precipitation, and the combined effect of solution strengthening and precipitation strengthening leads to the above results.

1. Introduction

As an important elastic component, metal bellows can realize the functions of measurement, connection, conversion, compensation, isolation, sealing, and shock absorption, and have a wide range of applications in petrochemical, instrumentation, electric power, metallurgy, and other industries [1-2]. At present, the main manufacturing materials of domestic metal bellows are stainless steel and carbon steel. Among them, austenitic stainless steel is the most widely used. Austenitic stainless steels have good plasticity, but their strength (especially at high temperatures) is low, so they can only be used at lower temperatures and cannot be strengthened by heat treatment.

With the continuous development of cutting-edge technology fields, more and more important occasions have added the application of metal bellows, and the performance of metal bellows components used in these fields (especially high-temperature performance) has been put forward higher and higher requirements. In order to meet the elastic properties, sealing properties, pressure-bearing capacity, and fatigue properties in high-temperature environments, more and more special alloys have been used in the manufacture of metal bellows in recent years [3-5]. For example, corrugated pipes made of Hastelloy-C corrosion-resistant alloy, and high-temperature alloy GH4169 and GH3625 materials have been widely used in various special working conditions and have become common high-temperature materials for bellows products.

As a material specially designed for high-temperature environments, GH4202 alloy has become a common choice for high-temperature metal bellows manufacturing because of its excellent mechanical properties and corrosion resistance. GH4202 alloy has an elongation at break of more than 30% at room temperature and the tensile strength is about 1100 MPa [6]. GH4202 superalloy can maintain good mechanical properties and stability at high temperatures, and the tensile strength of the material can reach more than 900 MPa when the ambient temperature reaches 700°C [7-8]. However, with the continuous improvement of the working temperature of precision metal bellows products for aerospace, the requirements for the thermal properties are getting higher and higher, and the mechanical properties of GH4202 are becoming more and more difficult to meet the requirements.

In this study, a new type of Ni-Cr-Co-Mo (GH4282) precipitation-hardening superalloy was prepared on the basis of GH4202 alloy. By adding Co and Mo elements, the solution strengthening effect is enhanced, so as to prevent grain boundary sliding to obtain good creep resistance and toughness.

2. Materials and experiments

In this research, the hot-rolled GH4282 alloy sheet with a thickness of 0.3 mm was selected for the manufacture of metal bellows. The chemical composition of the material and the GH4202 material is shown in Table 1.

Table 1. Chemical composition of the GH4282 alloy.

Element		Cr	Al	Ti	Mo	Co	Ni
wt%	GH4282	17.9	1.5	2.5	8.7	11.5	Balance
	GH4202	17.6	1.4	2.4	4.5	0.04	

The tensile tests were carried out on a SANSCMT5000 electronic universal material testing machine at room temperature, 700°C, 800°C, and 900°C, respectively, with an equal chuck moving speed. The initial strain rate was $1.0\times10^{-3}s^{-1}$ and a tensile specimen size was shown in Figure 1. More than 3 specimens were tested under each condition to ensure the reliability of the test results. The specimen was surface ground and then polished with argon ions at -30°C. Specimens were characterized using the Ultra Plus Field Emission Scanning Electron Microscope (SEM).

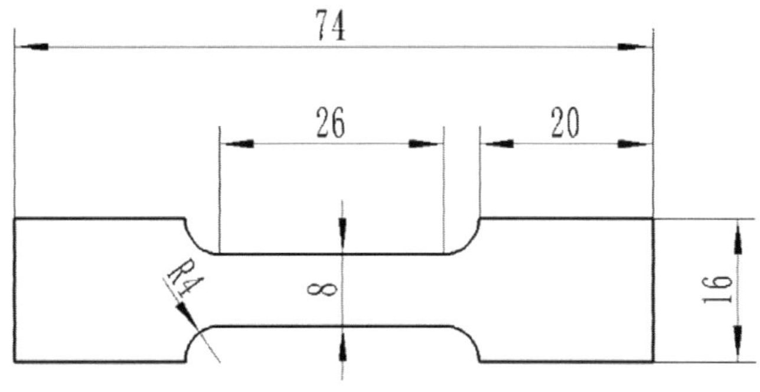

Figure 1. Schematic diagram of the tensile test specimen.

The GH4282 alloy material was used to prepare the multilayer metal bellows. The inner diameter of the bellows was 64mm, the outer diameter was 92mm, and the wave depth ratio of the bellow was 1.44. The layers of bellows were 2. The formed bellows were tested for fatigue life under the condition of internal pressure of 0.55MPa, the stroke was ±2mm, and the frequency was 1Hz. The fatigue life test was carried out using the SDZ-200 electro-hydraulic servo fatigue test machine. The bellows stiffness test was carried out using DF13.504T high-precision axial stiffness tester with a compression of 1.2mm. The high-temperature reliability test was carried out in the WZJQ-60G vacuum air quenching furnace. After the bellows were stretched to 80% displacement and limited, they were heated to 800°C in a vacuum furnace and then cooled to room temperature naturally after holding for 10h. After repeating this cycle 20 times, the fixture was removed, and the residual deformation was tested using the DF13.504T high-precision axial stiffness tester.

3. Results and discussion

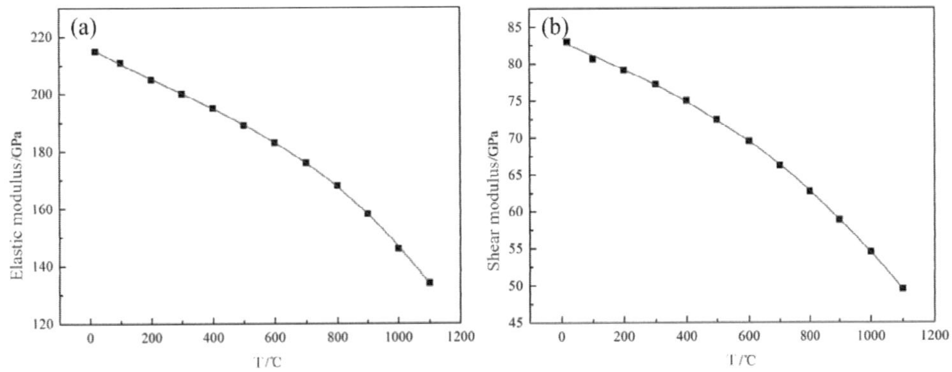

Figure 2. Modulus of elasticity-temperature curve (a) and shear modulus-temperature curve (b) of GH4282 alloy.

As shown in Figure 2, with the gradual enhancement of deformation temperature, the elastic modulus and shear modulus of GH4282 alloy show a downward trend, which is in line with the law of general metal materials. At room temperature, 700°C and 900°C, the elastic modulus of GH4282 alloy is 215 GPa, 176 GPa, and 158 GPa, respectively. Compared with GH4202 alloy (216GPa, 173GPa and 142GPa) at the same temperature, GH4282 material has a higher modulus of elasticity at high temperatures. The data in Table 2 shows the compare of the mechanical properties of GH4282 alloy with that of conventional GH4202 alloy under different temperature conditions. The GH4282 material obviously has higher high-temperature strength when the material density and plastic deformation ability are basically the same, even if the strength difference between the two materials gradually decreases with the increase in temperature.

Table 2. Mechanical properties of GH4282 and GH4202 alloys.

	RT		700°C		800°C		900°C	
	Rm/MPa	A/%	Rm/MPa	A/%	Rm/MPa	A/%	Rm/MPa	A/%
GH4282	1042	27	900	35	745	26.5	515	26
GH4202	1060	35	879	30	714	25	490	25

Figure 3 shows the microstructure of the GH4282 alloy. The microstructure of GH4282 alloy is relatively homogeneous. The grain is basically equiaxed, with an average grain size of ~35μm. There is obvious precipitation inside the alloy, and these precipitation phases are mainly distributed along the grain boundary position. The results of energy spectroscopy analysis showed that the second phase was a carbide with Ti and Mo, and its main components (wt.%) were 47.88% Ti, 1.77% Co, 4.88% Cr, 31.86% Mo, and 6.59% Ni. The results of high-power microstructure observation showed that the precipitated phase was fine and equiaxed, with an average size of about 24~61nm. The presence of these precipitated phases could effectively hinder the grain boundary sliding during high-temperature deformation, and thus GH4282 alloy showed higher strength above 700°C.

Figure 4 shows the XRD pattern of the precipitated phase of the GH4282 alloy. The results show that the alloy mainly contains γ' phase strengthening phase. In addition, $M_{23}C_6$, MC, and a very small amount of M_6C carbide are also contained in the material.

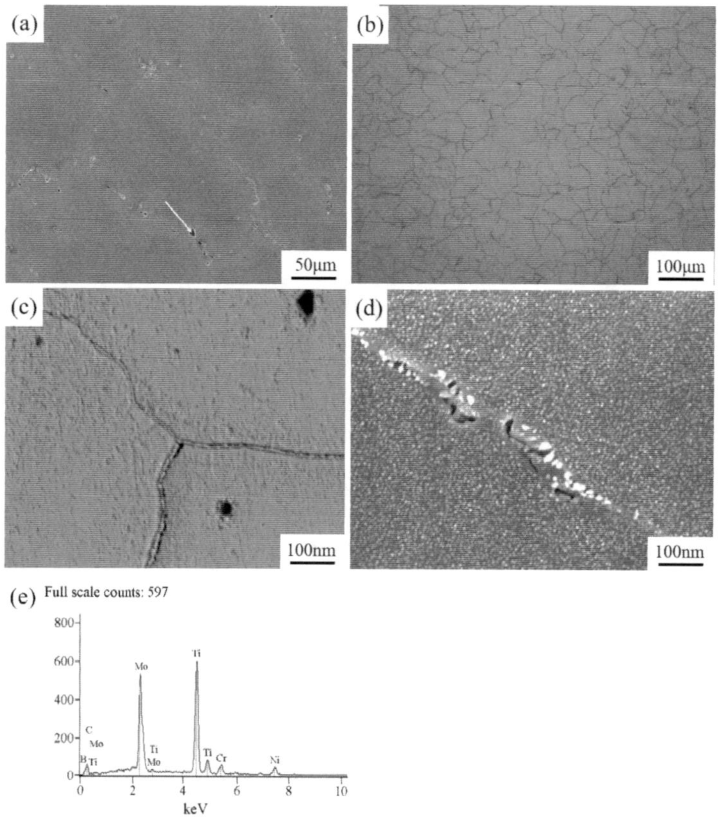

Figure 3. Microstructure of GH4282 alloy.
(a)(b) Optical microstructure; (c)(d) SEM tissue; (e) EDS composition analysis

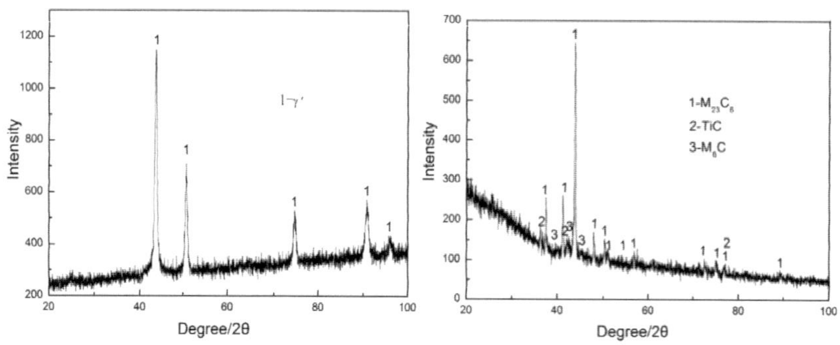

Figure 4. XRD pattern of the precipitated phase of GH4282 alloy.
(a) γ′ phase; (b) Carbide

Figure 5 shows the DSC curves of the heating and cooling process of the GH4282 alloy. It can be seen from the curve over the full temperature range (Figure 5(a)) that the γ matrix begins to melt significantly at 1327 °C, with a significant endothermic peak. At 1338 °C, the MC carbide begins to melt. At 1364 °C, the melting of the alloy ends, indicating that the melting temperature of the alloy is 1327~1364 °C. Zooming in on the heating DSC curve in the temperature range of 950~1150 °C, two small endothermic peaks can be observed, as shown in Figure 5(b). As shown in the picture, the peak temperature of the first small endothermic peak is 1003 °C, and the peak temperature of the second small endothermic peak is 1092 °C. Combined with the thermodynamically calculated precipitation temperature of the γ' phase (1007 °C) and the precipitation temperature range of M6C (808~1210 °C), it is considered that the first and second small endothermic peaks indicate the dissolution of the γ' phase M_6C, respectively. It can also be seen from the curve of the cooling process (Figure 5(c)) that the alloy begins to solidify at 1369 °C, γ dendrites form rapidly, and by 1365 °C the dendrite framework has been formed, which releases a large amount of latent heat of crystallization and forms a very steep exothermic peak. The remaining liquid phase between the dendrites slowly solidifies over a wide range of temperatures until the exothermic reaction is completed at 1304°C, indicating the end of solidification. Therefore, the solidification temperature range of the alloy is 1304~1369 °C. From the above results, it can be seen that the precipitated phase of GH4282 alloy can be basically stable under the condition of 700~900°C, which can effectively improve the strengthening effect [9-11]. Under the combined action of solution strengthening and precipitated phase, the high-temperature strength and stability of the studied GH4282 alloy are significantly higher than those of the traditional GH4202 alloy.

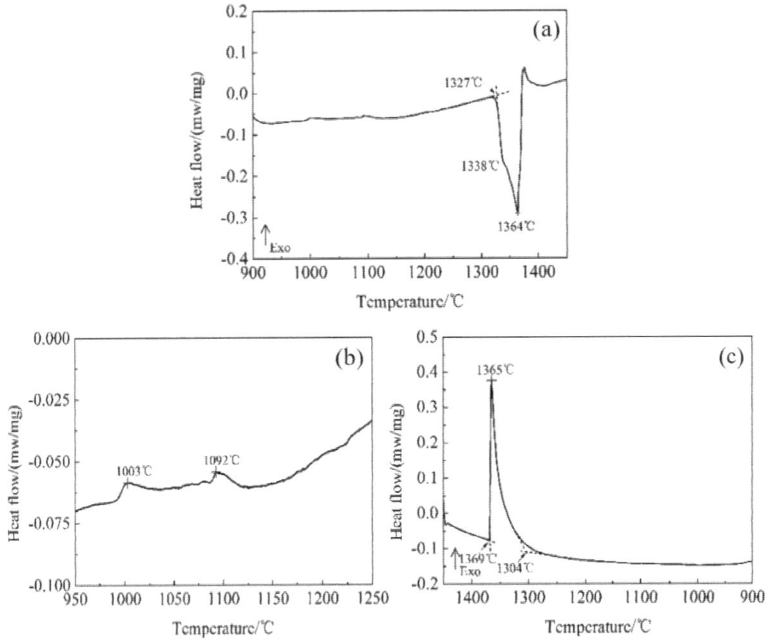

Figure 5. DSC curves of GH4282 alloy heating and cooling processes.
(a) Heating curves; (b) Partial enlargement of the heating curve; (c) Cooling curve

Figure 6. GH4282 alloy bellow.

Figure 6 shows the metal bellows fabricated using the studied GH4282 material. The mechanical properties test results showed that the average stiffness of GH4282 metal bellows was 120.4N/mm under the condition of compression of 1.2mm. The internal pressure was applied to 1.0MPa, the pressure was maintained for 5min, and there was no leakage in the metal bellows. Pressurized to 2.0MPa for 10min, no leakage of the bellows was observed, without obvious deformation of the bellows. Compared with the GH4202 metal bellows (117N/mm) with the same specifications, there was no significant change in the axial stiffness of the products of the studied material. The elastic modulus of GH4282 material is close to that of GH4202, and the product performance is basically the same under the same design and manufacturing parameters. According to the results obtained from the fatigue test, under the conditions of displacement ±2mm, internal pressure 0.55MPa, and frequency of 1.5Hz, the GH4282 metal bellows did not leak after 85,500 times of fatigue reciprocation process, which was much higher than the fatigue life of GH4202 bellows (76,950 leaks). The analysis results show that the addition of Co and Mo elements significantly improves the solution-strengthening effect of the material, and the generated γ' phase, $M_{23}C_6$, MC, and a small amount of M_4C carbides could effectively pin the grain boundaries. According to Orowan's relation [12], the presence of the precipitated phase could significantly improve the yield strength of GH4282 alloy. In the present study, with the increase of the yield strength, the elastic segment was significantly elongated, the plastic deformation generated by each cycle decreased during the fatigue test, the strain accumulation at the peak and/or trough of the bellows was slower, the stress concentration was not easy to form, and the initiation and propagation of cracks were relatively slow [13-15]. As a result, the fatigue life of the metal bellows had been significantly improved.

The high-temperature reliability tests of the products were carried out in the furnace. After the bellows were stretched to 80% displacement and limited, it was kept at 800°C in the furnace, and then cooled to room temperature naturally after holding for 10h. After repeating this cycle 20 times, the fixture was removed, and the residual deformation was tested. The results show that the residual deformation of GH4282 metal bellows is 0.75%, which meets the design requirements and is significantly higher than that of GH4202 bellows (0.95%), indicating that GH4282 material has better high-temperature stability and high-temperature strength. At 800°C, there are more precipitated phases in GH4282 material, which can effectively limit grain boundary sliding, inhibit grain growth, and improve the strength and microstructure thermal stability of the material [16-17].

4. Conclusion

In this study, the effects of the addition of Mo and Co elements on the microstructure properties of GH4842 alloy were revealed, and the fatigue properties and high-temperature reliability of GH4282 bellows were studied. The following main conclusions were obtained:

(1) At room temperature, 700°C and 900°C, the elastic modulus of GH4282 alloy is 215 GPa, 176 GPa, and 158 GPa, respectively, which is higher than that of traditional GH4204 alloy.

(2) GH4282 alloy mainly contains γ' phase strengthening phase; in addition, $M_{23}C_6$, MC and a very small amount of M_6C carbide are also contains. These precipitates can effectively pin grain boundaries and improve the strength and high-temperature stability of the material.

(3) The presence of the precipitation significantly improved the yield strength of GH4282 material, and the elastic segment of the material was elongated. During the fatigue test, the plastic deformation generated by each cycle decreased, the strain accumulation at the peak and/or trough of the bellows was slower, the stress concentration was not easy to form, and the crack initiation and propagation process was relatively slow. As a result, the fatigue life of the metal bellows had been significantly improved.

(4) There are more precipitated phases in GH4282 material, which can effectively limit grain boundary sliding, inhibit the grain growth process, and improve the strength and microstructure stability of the material at high temperatures.

Journal of Physics: Conference Series **2842** (2024) 012048 doi:10.1088/1742-6596/2842/1/012048

References

[1] W. Zhang. Technical problem analysis and prospect of nuclear power valve, Valve, 2020(05):41-44.
[2] C. Feng. Research on key points of quality control of nuclear power valve, China Equipment Engineering, 2022(19):258-260.
[3] S. H. Huo, W. Z. Yan, X. J. Xu, et al. Bending characteristics of the reinforced S-shaped bellows under internal pressure, The International journal of pressure vessels and piping, 2021(192):104412.
[4] G. Faraji, M. Mashhadi, V. Norouzifard, et al. Evaluation of effective parameters in metal bellows forming process, Journal of materials processing technology, 2009 (7):3431-3437.
[5] L. F. Jiang, Y. He, Y. C. Lin. Influence of process parameters on thinning ratio and fittability of bellows hydroforming, International journal of advanced manufacturing technology, 2020(107):3371-3387.
[6] Z. H. Wang, Y. G. Wang, D. J. Lv, et al. Performance and wear mechanisms of coated PCBN tools in high-speed turning superalloy GH4202, International journal of advanced manufacturing technology, 2023(127):5083-5097.
[7] X. Yu, Y. G. Wang, D. J. Lv, et al. Study on machining characteristics with variable distribution density micro-texture tools in turning superalloy GH4202, International journal of advanced manufacturing technology, 2022(123):187-197.
[8] X. Yu, D. J. Lv, Y. Gao, et al. The design and application of variable distribution density micro-textured tool, International journal of advanced manufacturing technology, 2023(128):315-327.
[9] H. Zhang, Y. Li, T. Ma, et al. Tailoring of nanoscale γ' precipitates and unveiling their strengthening mechanisms in multimodal nickel-based superalloy GH4720Li, Materials Characterization, 2022(188):111918.
[10] X. Zheng, C. Shi, X. Zhu, et al. Effect of Solution Treatment on Grain Growth and Precipitates in Electroslag Remelted 15Cr-22Ni Iron-Base Superalloy, Metallurgical and Materials Transactions B, 2022(53): 877-894.
[11] S. Taller, T. Austin. Using post-processing heat treatments to elucidate precipitate strengthening of additively manufactured superalloy 718, Additive Manufacturing, 2022(60):103280.
[12] Srivatsan T S. A review of: "Dispersion Strengthened Aluminium Alloys" edited by Y. W. Kim and W. M. Griffith TMS, Warrendale, Pennsylvania 795 pages, hardcover, 1988[J]. Materials and Manufacturing Processes, 1991, 6(3): 565-568.
[13] Z. L. Li, X. L. Yu, Q. H. Zhang, et al. Failure mechanism of the Inconel 625 bellows for nuclear control valves, Ninth International Conference on Mechanical Engineering, Materials, and Automation Technology, 2023: 1280146.
[14] A. M. A. Mohamed, F. H. Samuel, Saleh Alkahtani. Microstructure, tensile properties and fracture behavior of high temperature Al–Si–Mg–Cu cast alloys, Materials Science and Engineering A, 2013(577): 64-72.
[15] D. M. Kim, C. Kim, C. H. Yang, et al. Heat treatment design of Inconel 740H superalloy for microstructure stability and enhanced creep properties, Journal of Alloys and Compounds, 2023(946): 169341.
[16] Z. Tang, X. Dong, Y. Geng, et al. The effect of warm laser shock peening on the thermal stability of compressive residual stress and the hot corrosion resistance of Ni-based single-crystal superalloy, Optics & Laser Technology, 2022(146): 107556.
[17] H. Z. Niu, X. J. Chen, Y. F. Chen, et al. Microstructural stability, phase transformation and mechanical properties of a fully-lamellar microstructure of a Mo-modified high-Nb γ-TiAl alloy, Materials Science & Engineering. A, 2020(784):139313.

Study on spring-back characteristics of magnesium alloy plate in multi-point forming

Mengjun Zhao*, Dongxiong Lv, skib, Pengcheng Zhu, Zan Liu

The School of Mechanical Engineering, Jiangsu University of Science and Technology

* Corresponding author's email: z_mj_2005@just.edu.cn

Abstract: Multi-point forming (MPF) is an advanced method of plate forming with its characteristics of dieless, rapid, and flexible. In this paper, the springback of Magnesium alloy (Mg alloy) plates was researched under different geometrical conditions, mechanical properties, and constitutive models with the MPF method. The studies showed that the springback of Mg alloy plates reduced with the thickness of the plates, increased with curvature radius, and enlarged with the initial yield strength. Meanwhile, the simulated results indicated that the constitutive model had a significant influence on the springback of the Mg alloy plate forming.

1. Introduction

Magnesium alloys (Mg alloys) are ideal structural materials with their mechanical properties, and their use has been increasing in the last century [1-6]. However, the special crystal structure of magnesium alloys leads to plasticity at room temperature.

The method of multi-point forming (MPF) is a new method of plate forming along with its characteristics of dieless, rapid, and flexible, and it has been successfully employed in complex three-dimensional plate metal, such as sail-shaped and saddle-shaped plates [7-9]. The MPF has been studied and used widely in high-strength steel plates, but currently, there are few application studies on the MPF method in Mg alloy plates.

In this paper, springback characteristics of the commonly-used AZ80 plate in MPF are investigated. The finite element method was used to calculate the springback of AZ80 plates under four different plate thicknesses (2.0, 3.0, 4.0, 5.0 mm), curvature radius (200, 300, 400, 500 mm), and the initial yield strengths. Additionally, three different material constitutive models (elastic-perfectly plastic model, bilinear isotropic hardening, and bilinear kinematic hardening) were also discussed in this work.

2. Forming process of MPF for Mg alloy plate

The forming processes of MPF are described in the following steps [10]:

Step 1: Adjusting. We adjust the shapes of the lower die and upper punch, as shown in Figure 1(a), where the profiles of the lower die and the upper punch are in accordance with the target surface of the blank (Mg alloy plate).

Step 2: Placing. We place the blank (Mg alloy plate) between the lower die and the upper punch, as shown in Figure 1(b).

Step 3: Forming. The lower die is immobilized, the upper punch moves along vertically downward, and the blank is pressed into the profile of the lower die and the upper punch, as shown in Figure 1(c).

Journal of Physics: Conference Series **2842** (2024) 012049 doi:10.1088/1742-6596/2842/1/012049

Step 4: Opening. The lower die is still immobilized, and the upper punch moves vertically upward until the upper punch keeps a certain distance from the formed Mg alloy plate, as shown in Figure 1(d).

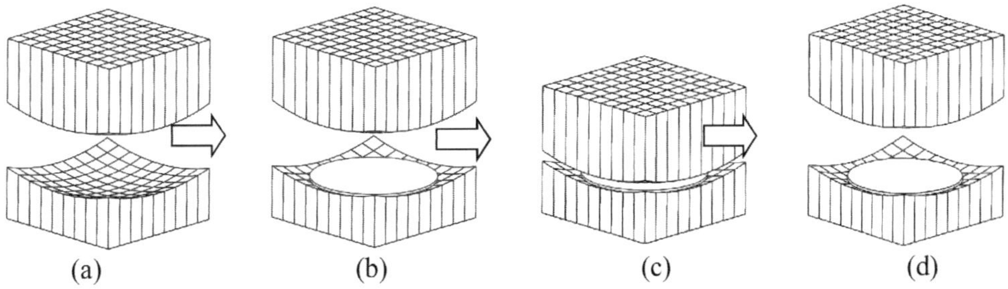

Figure 1. Forming processes of MPF: (a) Step 1; (b) Step 2; (c) Step 3; (d) Step 4.

Springback of the formed Mg alloy plate will occur after open mold as shown in Figure 1(d) and Figure 2, and the schematic process of the springback may be shortly described as follows:

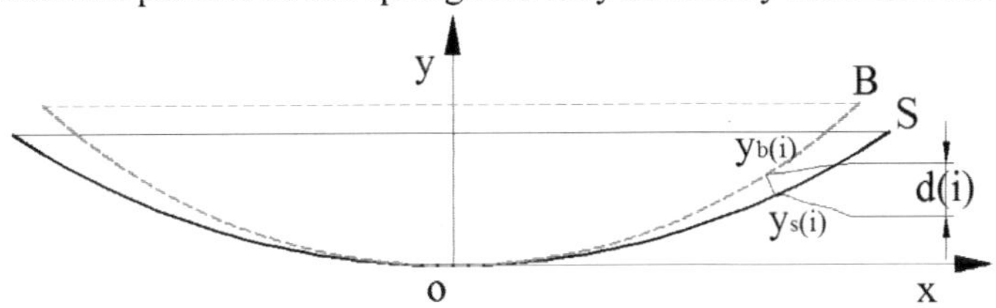

Figure 2. Springback calculation for the formed Mg alloy plate.

The initial profile of the AZ80 plate (before springback) is displayed with a line dashed line OB and springback profile OS with the solid line. The springback is calculated as follows:

$$d(i) = y_s(i) - y_b(i) \quad (1)$$

where d(i) is the difference value of the corresponding node(i) before and after springback; $y_b(i)$ is the vertical coordinate of the corresponding node(i) before springback; $y_s(i)$ is the vertical coordinate of the corresponding node(i) after springback.

3. Finite element model of MPF for AZ80 plate

The present paper mainly researches the springback of spherical surface parts in the MPF, and a circle blank plate with a diameter of 200 mm was chosen. Due to the symmetry of the bank plate and the target profile, the geometrical model of the simulation was represented by a 1/4 structure to reduce the computation amount, as shown in Figure 3.

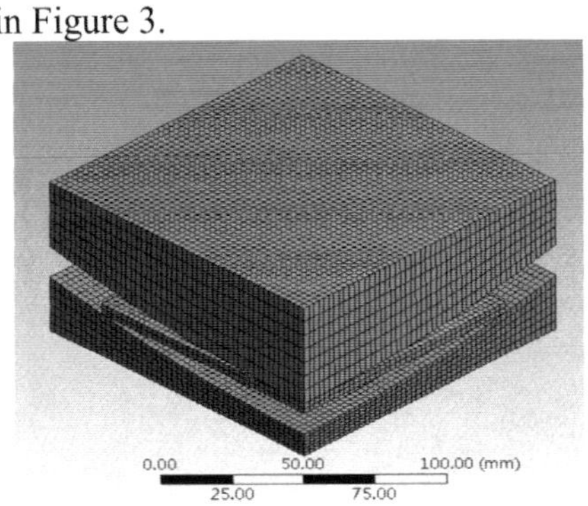

Figure 3. Finite element model with 1/4 structure.

In this simulation, the elastic-perfectly plastic model, bilinear isotropic hardening, and bilinear kinematic hardening were considered, and materials of the blank and its mechanical properties [11] were shown in Table 1.

Table 1. Mechanical properties of the AZ80 plate.

Mg alloy number	Yield strength /MPa	Ultimate strength /MPa	Tangent modulus /MPa	Elastic modulus /MPa	Poisson ratio λ
AZ80	205 [12]	279.5 [12]	2.0×10^3 [13]	4.5×10^4 [14]	0.35 [14]

In the finite element model, the bottom dies and upper punch were rigid, the AZ80 plate was flexible, the element type was a 3D solid element, and friction coefficients were 0.125 between the AZ80 plate, bottom die, and upper punch.

Since the implicit method yields accurate results for simulation and is unconditionally stable [15-16], the finite element model was solved by using the implicit method.

In this manuscript, four different plate thicknesses, including 2.0, 3.0, 4.0, and 5.0 mm, four curvature radii, including 200, 300, 400, and 500 mm, and four initial yield strengths, including 155, 205, 255, and 305 MPa, were used to study the springback characteristics of the AZ80 plate. Additionally, three different material constitutive models including the elastic-perfectly plastic model, bilinear isotropic hardening, and bilinear kinematic hardening were discussed in this work.

4. Results and discussion

It can be observed in Figure 4 that the springback of the AZ80 plate reduces obviously with an increase of the plate thickness under the given curvature radius, such as R=500 mm. When the curvature radius of the AZ80 plate increases, the springback will increase apparently under the given thickness, such as T=5 mm, as shown in Figure 5. Meanwhile, from Figure 4, it can also be found that the material constitutive models have little impact on the springback of the AZ80 plates.

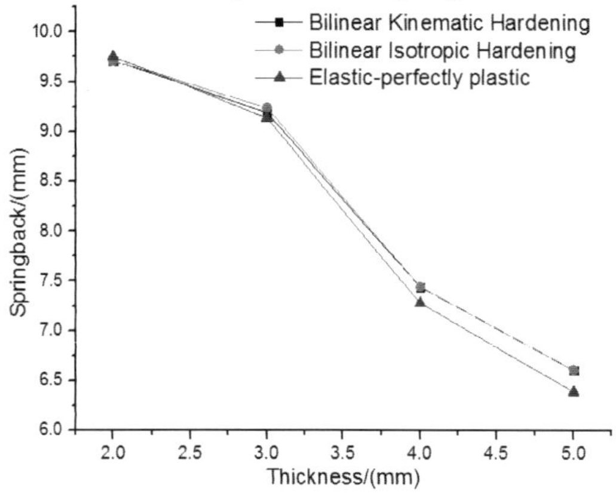

Figure 4. Springback vs thickness under three constitutive models (R=500 mm).

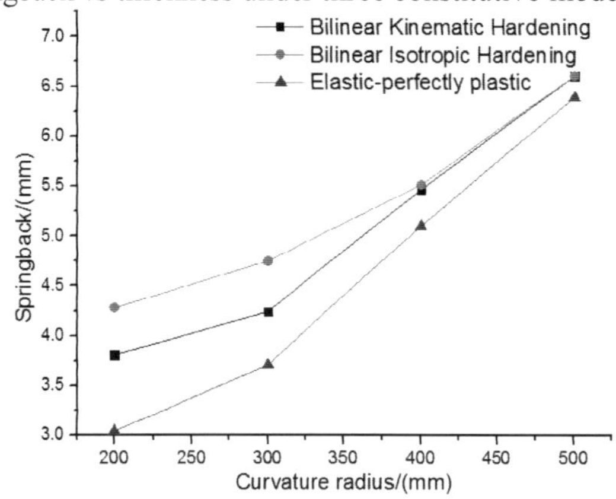

Figure 5. Springback vs. curvature under three constitutive models (T=5 mm).

However, Figure 5 shows that when the curvature radius dwindles, there are big differences in the springbacks under the three different material constitutive models, and when the curvature radius reaches 200 mm, the difference between the springbacks under the three different material constitutive models is maximum. The reason for the above difference may be the degree of plastic deformation for AZ80 plate under different curvature radii, and it can be illustrated in Figure 6 that the maximum plastic strain increases with the decrease of curvature radii, and the plastic deformation range increases gradually from R=500 mm to R=200 mm, as shown in Figure 6 (a)-(d).

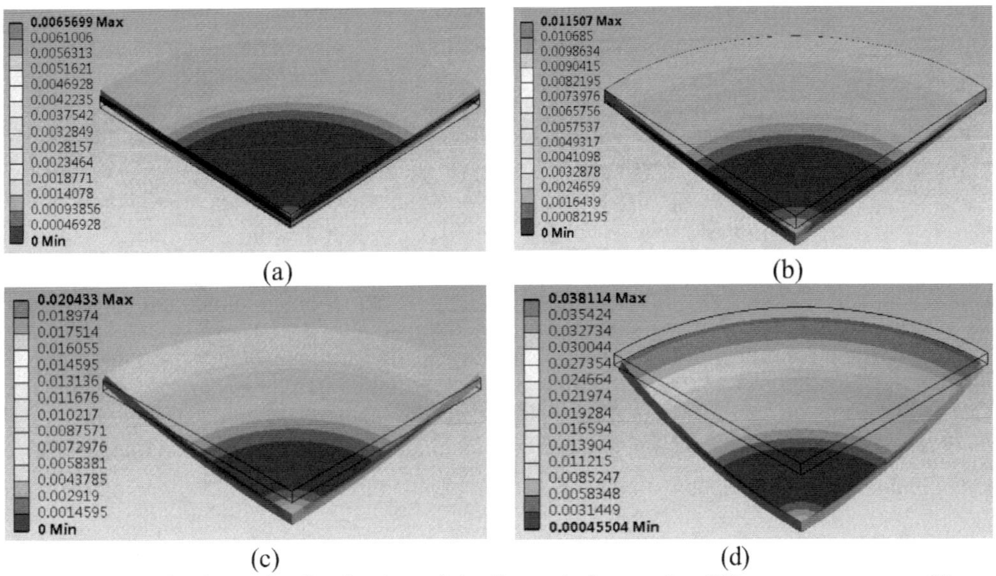

Figure 6. Plastic strain distribution of the formed plate under different curvature radii: (a) R=500 mm; (b) R=400 mm; (c) R=300 mm; (d) R=200 mm.

Additionally, assuming that the initial yield strength of AZ80 happens to change, the springback of the formed plate will happen to change too, and it can be seen in Figure 7 that the springback will increase noticeably with the initial yield strength. Like Figure 6, Figure 8 shows that the maximum plastic strain decreases with the increase of the initial yield strengths and the plastic deformation range diminishes gradually from Y1=155 MPa to Y4=305 MPa, as shown in Figure 8 (a)-(d).

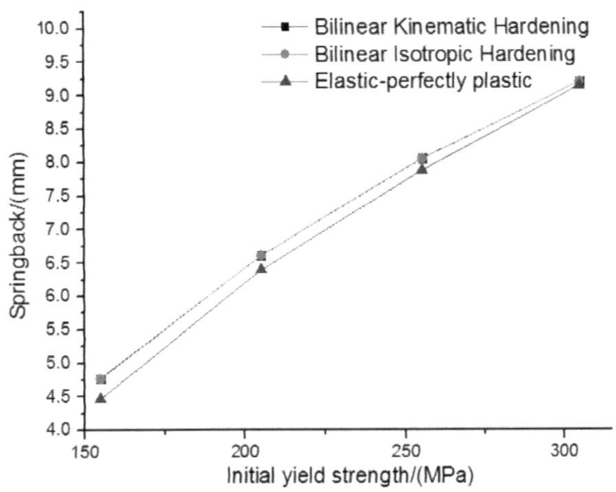

Figure 7. Springback vs initial yield strength under three constitutive models (T=5 mm).

Journal of Physics: Conference Series **2842** (2024) 012049 doi:10.1088/1742-6596/2842/1/012049

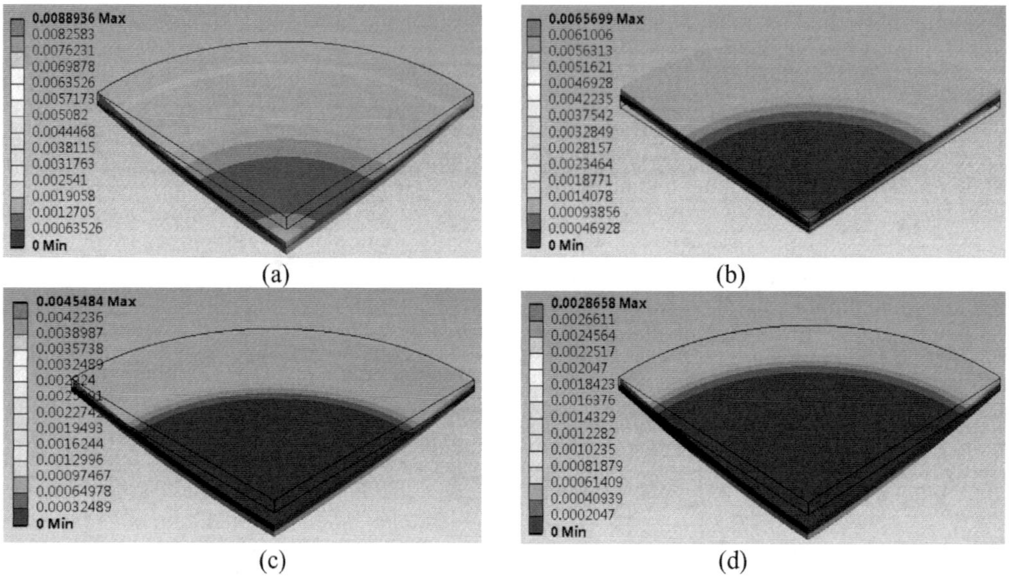

Figure 8. Stress distribution on the plate under different initial yield strengths:
(a) Y1=155 MPa; (b) Y2=205 MPa; (c) Y3=255 MPa; (d) Y4=305 MPa, after processing by the MPF method.

5. Conclusions

It is difficult to deform plastically for Mg alloys; however, the MPF is one of the most effective ways to achieve AZ80 plate forming. This is our first study to investigate the springback characteristics of the AZ80 plate based on MPF technology, and some important conclusions are drawn as follows:

(1) The thickness, curvature radius, and initial yield strength have obvious effects on the springback of the AZ80 plates, and the springback increases with the thickness, curvature radius, and initial yield strength.

(2) The springback is also affected by the different material constitutive models under large plastic deformation of the AZ80 plate.

(3) The analysis shows that the essential factor deciding the springback of the AZ80 plate is its plastic deformation.

References

[1] S. Jayasathyakawin, M. Ravichandran, N. Baskar, C.A. Chairman, and R. Balasundaram, Mechanical properties and applications of Magnesium alloy-Review, *Materials Today: Proceedings*, 2020, **27**(5).

[2] A.H. Musfirah and A.G. Jaharah, Magnesium and aluminum alloys in automotive industry, *Journal of Applied Sciences Research*, 2012, 8(10), p 4868-4875.

[3] M.K. Kulekei, Magnesium and its alloys applications in automotive industry, *The International Journal of Advanced Manufacturing Technology*, 2008, 39(9), p 851-865.

[4] A. Lg, A. Jy, C. Jpa, Z.A. Ying, Z.B. Kai, and B. Jj, Study of the microstructure, bonding evolution and mechanical properties of continuously extruded magnesium AZ31 sheet, *Materials Science and Engineering: A*, 2021.

[5] W.J. Joost and P.E. Krajewski, Towards magnesium alloys for high-volume automotive applications, *SCRIPTA MATERIALIA*, 2016, **128**, p 107-112.

[6] F. Kelen, Importance and applications of magnesium and its alloys in automotive industry, *Journal of the Institute of Science and Technology*, 2021, p 548-562.

[7] M.Z. Li, J.G. Yao, Z.Y. Cai, and S.H. Li, The research on multi-point alternate forming of sheet metal to minimize springback, *Journal of Plasticity Engineering*, 2000, **7**(1), p 22-25.

[8] S.H. Li, M.Z. Li, Z.Y. Cai, and J.G. Yao, Numerical simulation of multi-point bending and springback processes of sheet metal parts, *Journal of Agricultural Machinery*, 2000, **31**(1), p 112-115.
[9] S.J. Su, Y. Hu, and C.F. Wang, Research on prediction model of hull plate bending springback based on SVM, *Ship Science and Technology*, **2015**, 37(5), p 104-108.
[10] M.Z. Li, S.H. Li, J.J. Chen, and J.G. Yao, Numerical simulation on different multi-point forming techniques for sheet metal parts, *China Mechanical Engineering*, 2000, 11(9), P 1042-1047.
[11] L.J. Qu, H.Q. Li, and Y.Q. Wang, Stress-strain relationship and modulus of elasticity of 16 Mn steel under loading and constant temperature, *Fire Safety Science*, 2009, **18**(4), p 187-191.
[12] W.D. Li, H.T. Zhou, H.X. Chen, W.J. Qi, and K.H. Zheng, Effect of rolling deformation on microstructure and properties of AZ80 alloy, *Material & Heat Treatment*, 2012, 41(4), p 22-28.
[13] X.Y. Zhang, J.J. Li, F.C. Lang, X.H. Hou, S.T. Yang, and Y.M. Xing, Effects of annealing process on microstructure and mechanical properties of AZ31B magnesium alloy, *Hot Working Technology*, 2021, **50**(22), p 121-131.
[14] M.J. Zhao, Z.L. Wu, and H.M. Cai, Effects of mandrel angle on hydrostatic extrusion process of magnesium alloy tube. *Rare Metal Materials & Engineering*, 2017, **46**(5), p 1214-1218.
[15] A.A. Muhammad and Q. Ala, Numerical simulation of sheet metal forming: a review, *The International Journal of Advanced Manufacturing Technology*, 2017.
[16] H. En, S. Yilmaz, and R.A. Yildiz, Springback behavior of dp600 steel: an implicit finite element simulation. *Journal of Engineering Research*, 2020, **8**(2), p 252-264.

Simulation and study of the flow field in a kettle-type reactor for ethylene polymerization

Hou Ji[1,2,*], Zhang Hong[1,2], Ren Xinhua[1,2], Zhang Jia[1,2] and Peng Chen[1,2]

[1]College of Mechanical Engineering, Dalian University of Technology, Dalian 116024, Liaoning, China
[2]National Key Laboratory of High Performance Precision Manufacturing, Dalian 116024, Liaoning, China

[*]Corresponding author's e-mail: houji@mail.dlut.edu.cn

Abstract. Addressing the complexities inherent in the polymerization process for solution method polyethylene production, particularly the challenges in describing the turbulent three-dimensional flow field within the reactor and the precise design of reactor structural parameters, this research leverages computational fluid dynamics (CFD) methods integrated with polymerization reaction kinetics equations. We have formulated User-Defined Functions (UDFs) for a quantitative representation of the polymerization process. Employing the Realizable k-epsilon turbulence model and Sliding Grid (SG) technique, we analyzed the flow dynamics and field distribution within the reactor, elucidating the distribution patterns of temperature, pressure, and concentration fields. A comparative assessment of simulation outcomes against experimental data revealed that the average deviation of the simulated temperature field from actual operating conditions falls within the permissible error range, thereby affirming the simulation model's validity. Utilizing local sensitivity analysis, we investigated the impact of impeller structural parameters on mixing time and stirring power. Our analysis of the most sensitive structural parameters indicated that impeller diameter and the number of impellers are pivotal in reducing mixing time and stirring power, respectively. Notably, the stirring power escalates with an increase in the number of impellers, while the mixing time exhibits a trend of initial decrease followed by an increase with varying impeller diameters. The methodologies and patterns discerned in this study provide insightful design references for the design and optimization of reactors across different tonnages.

1. Introduction

The kettle-type stirred reactor, integral for the solution method of ethylene polymerization, operates by solubilizing feedstocks such as ethylene and butene in a solvent, followed by catalysis-driven polymerization under conditions ranging from 160-200°C and 15-20 MPa. Within this system, the copolymerization of ethylene and butene exhibits intricate behaviors, heavily influenced by temperature, viscosity, shear rates, polymer concentration, molecular weight, and other factors. Furthermore, the reactor typically incorporates internal components like stirrers and baffles, complicating the medium's flow dynamics. The construction of experimental setups is both time-consuming and labor-intensive, and the precision in measuring temperature and concentration distributions is severely limited by current measurement methodologies [1]. Consequently, the intricate interplay of chemical reactions, rheology, and fluid dynamics in the ethylene and butene copolymerization process presents significant challenges in quantitatively characterizing the flow properties within the reactor.

Beginning in the 1990s, researchers globally have incrementally engaged in simulation studies of flow fields within kettle-type reactors utilizing Computational Fluid Dynamics (CFD). In Yang's work [2], an innovative approach was adopted to design critical components for a fluidized bed reactor within a 50 kg/h gas-phase polyethylene facility, focusing on the reactor's structure and fluidization dynamics. The research utilized optimized key components from pilot-scale setups and a newly developed catalyst system to investigate the copolymerization behavior of 1-hexene and ethylene using gas-phase technology. This investigation led to the production of medium-density polyethylene suitable for rotational molding applications and polyolefin elastomer (POE) products. Zhang and He [3] delved into the operational aspects of various polymerization reactors, emphasizing the significance of gas-liquid mixing, mass and heat transfer processes. By analyzing the inherent properties of the materials and making informed choices regarding the reactor's design and the polymerization process, the study highlighted the pivotal role such decisions play in minimizing investment and operational costs as well as maintenance expenses. Li and Wu [4] focused on the design aspects of conical capshaped porous distributors, elaborating on the methodology for determining crucial parameters such as the open area ratio, pressure drop, and inter-hole spacing. The paper outlined the design steps for these distributors and demonstrated their application in the design of distributors for fluidized bed reactors at an industrial scale, showcasing the practical implications of such design considerations.

In this study, we investigate the complex polymerization process in solution method polyethylene production and the challenge of quantitatively characterizing the reactor's internal three-dimensional flow field. The research focuses on the complex rheological phenomena within the reactor. Initially, the study integrates polymerization reaction kinetics in Computational Fluid Dynamics (CFD) simulations, revealing the transient distributions of temperature, velocity-pressure, and concentration fields within the reaction vessel. The fidelity of the simulation model is corroborated by juxtaposing simulation outcomes with empirical data. Based on this model, the objective is to minimize mixing duration and stirring energy. To this end, a local parameter sensitivity analysis and an impact study are conducted on five structural parameters, including the quantity and diameter of stirrer blades. The goal is to elucidate the impact of variations in stirrer structural parameters on the internal flow dynamics of the reactor and to suggest novel strategies for the refinement of stirrer design.

2. Reactor calculation model and structural parameters

2.1. Governing equations and modeling of polymerization reaction processes

In this study, we utilize the core principles of fluid dynamics, encompassing the equations of mass conservation, momentum conservation, and energy conservation, to model and analyze the fluid flow dynamics inside the reactor. These governing equations are articulated through a system of partial differential equations, providing a comprehensive mathematical framework to capture the intricate dynamics of fluid behavior within the reactor environment. This approach enables a detailed and accurate representation of the fluid flow, essential for understanding and optimizing reactor performance in various industrial applications.

Continuity equation, momentum equation, and species transport equation:

$$\frac{\partial u_i}{\partial x_i} = 0 \tag{1}$$

$$\frac{\partial u_i}{\partial t} + \frac{\partial u_i u_j}{\partial x_j} = -\frac{1}{\rho}\frac{\partial p}{\partial x_i} + \nu \frac{\partial^2 u_i}{\partial {x_j}^2} \tag{2}$$

$$\frac{\partial}{\partial t}(\rho Y_m) + \frac{\partial}{\partial x_i}(\rho \cdot u_i \cdot Y_m) = -\frac{\partial}{\partial x_i} J_{m,i} + S_m \tag{3}$$

where Y_m denotes the mass fraction of species m, J_{ml} signifies the diffusive flux of species m, and S_m corresponds to the source term arising from chemical reactions. The equation is structured such that the initial term on the left delineates the temporal variation of species m within the system, followed by the convective term. On the right, the equation is bifurcated into a term representing diffusion and another term encapsulating the influence of chemical reactions on species m, thereby providing a comprehensive description of the transport phenomena of different species in a reactive flow context.
Energy equation:

$$\frac{\partial}{\partial t}(\rho E)+\frac{\partial}{\partial x_i}\left[u_i(\rho E+p)\right]=\frac{\partial}{\partial x_i}\left[k_{eff}\frac{\partial T}{\partial x_i}-\sum_m h_m J_m+u_j\left(\tau_{ij}\right)_{eff}\right]+S_h \tag{4}$$

The equation delineates various thermodynamic and transport properties, where E signifies the total energy in Joules, h the specific enthalpy per kilomole, k_{eff} the effective thermal conductivity, and k_t the turbulent heat conductivity, both in units of meters squared per second. J_m represents the mass diffusive flux of species m(kg·m^{-2}·s^{-1}), and h_m denotes the enthalpy of species m. $(\tau_{ij})_{eff}$ corresponds to the effective stress tensor, and S_h represents the heat source term from chemical reactions, quantified in kilojoules.

This study utilizes the polymerization kinetics model advanced by Kiparissides and Verros [5], noted for its robust suitability and precision in the design, optimization, and control of industrial-scale high-pressure ethylene copolymerization reactors.

The thermal energy generated in polymerization predominantly arises from the chain elongation mechanism inherent in the polymerization process. The exothermic nature of the reaction, denoted as ΔQ, can be mathematically represented as:

$$\Delta Q=\Delta H\cdot R_p \tag{5}$$

where ΔH represents the heat capacity of polymerization, measured in kilojoules per mole (kJ·mol^{-1}). The rate of the polymerization reaction is expressed in terms of kilomoles per cubic meter per second (kmol·m^{-3}·s^{-1}). The rate of elementary reactions in this process adheres to the Arrhenius equation, which is articulated as follows:

$$R_p=k_p\left[M\right]\left[C^*\right] \tag{6}$$

$$k_p=k_0\exp\left[\frac{-\left(E_a+V_a\cdot P\right)}{R_g\cdot T}\right] \tag{7}$$

where [M] represents the monomer concentration(kmol·m^{-3}); [C*] denotes the concentration of active catalyst sites in kmol·m^{-3}; k_p is the kinetic constant; k_0 is the pre-exponential factor (s^{-1}); E_a is the activation energy (kJ·kmol^{-1}); V_a is the activation volume (m^3·kmol^{-1}); P is the pressure in kilopascals gauge (kPaG); R_g is the molar gas constant (J·mol^{-1}·K^{-1}); and T is the temperature in Kelvin.

Regarding the impact of viscosity on the reaction process, Zhao [6], using polyolefin elastomers (POE) as raw materials, systematically studied the effects of molecular weight, concentration, temperature, shear rate, and other factors on the viscosity of polymer and monomer mixtures. Their findings are particularly accurate for high-viscosity systems under medium to high shear rates, aligning with the actual conditions of the reactor in this study. The predictive model is as follows:

$$\eta=B_1\cdot(e^{\mu\cdot c\cdot\overline{M}_\omega{}^\theta}\cdot e^{\frac{E_\eta}{RT}})^\beta\times\dot{\gamma}^m \tag{8}$$

The equation specifies viscosity η (Pa·s), temperature T (K), shear rate y (s^{-1}), and weight-average molecular weight $\bar{M}_W$(g·mol^{-1}). The molar gas constant R is set at 8.314 J·(mol·K)$^{-1}$, and E_η denotes the viscous flow activation energy (J·mol-1). Coefficients B_1, B_2, μ, β, and m are integral to the equation. By integrating Equation (8) into the User-Defined Function (UDF), the study accounts for the impact of viscosity variations on the reaction system. In modeling the polymerization reaction process, both the mass and heat source terms of the reaction are embedded into the conservation equations via UDFs, employing the DEFINE_SOURCE macro for execution, thus ensuring a comprehensive representation of the polymerization dynamics within the reactor system.

2.2 Physical model and grid division

2.2.1 Physical model. This study focuses on a kettle-type stirred reactor with an annual production capacity of 80,000 tons. Figure 1(a) illustrates the three-dimensional model of the reactor, constructed based on real-world parameters from an operational plant. The reactor is comprised of five layers of stirrer blades. Raw materials such as ethylene and 1-butene are fed from the bottom inlet, while catalysts and toluene enter through the catalyst inlet. The combined action of the rotating stirrer blades and baffles ensures a thorough mixing of polyethylene and other products, which subsequently exit through the discharge outlet. Detailed specifications are provided in Figure 1(b), highlighting dimensions such as the reactor diameter (D=1016 mm), spacing between layers (T_d=635 mm), and the heights of the bottom and top stirrer blades from their respective ends (T_b=436 mm and T_f=494 mm, respectively), with the overall reactor height being H=3470 mm.

Figure 1(c) depicts an inclined-blade disk turbine stirrer, distinct from conventional straight-blade designs, noted for its higher power usage and increased shear stress [7]. Inclined blades offer a larger contact surface with the fluid and possess some axial mixing capabilities. Key dimensions include the stirring shaft diameter (r=190 mm), blade thickness (b=9 mm), blade inclination angle (α=45°), vertical blade height (h=102 mm), and blade diameter (l=635 mm).

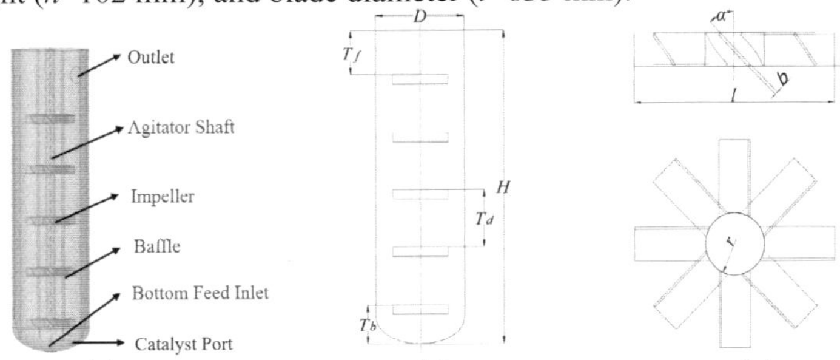

Figure 1. Structure of reactor and stirring paddle. (a) Reactor component diagram. (b) Reactor structural parameters. (c) Impeller structural parameters.

2.2.2 Grid division and independence verification. The Sliding Grid (SG) methodology is adept at tracking the fluid dynamics at incremental time steps, providing a high degree of accuracy in computational predictions, and aligning closely with experimental data. The SG approach delineates the reactor's internal computational domain into a rotational zone (proximal to the stirrer blades) and a stationary zone (comprising the remainder of the reactor). The interface between these zones, where grid sliding occurs, necessitates pertinent interpolation to ensure congruence in momentum and energy transfer.

Table 1. Grid measurement standard.

Maximum Grid Skewness	Minimum Orthogonality Quality	Quality Results
0.8-0.9	0.2-0.5	Excellent Quality
0.6-0.8	0.1-0.2	Good Quality
0.4-0.6	0.01-0.1	Average Quality

The reactor's internal flow domain is discretized using a tetrahedral unstructured grid system. This system employs Boundary-Oriented Inflation (BOI) for enhanced mesh density near the impeller blades and incorporates boundary layers at wall surfaces. As shown in Table 1, the grid exhibits a

Journal of Physics: Conference Series **2842** (2024) 012050 doi:10.1088/1742-6596/2842/1/012050

maximal skewness value of 0.856 and a minimal orthogonality quality of 0.24, indicative of superior grid integrity and suitability for the computational fluid dynamics analysis of the reactor's internal flow characteristics.

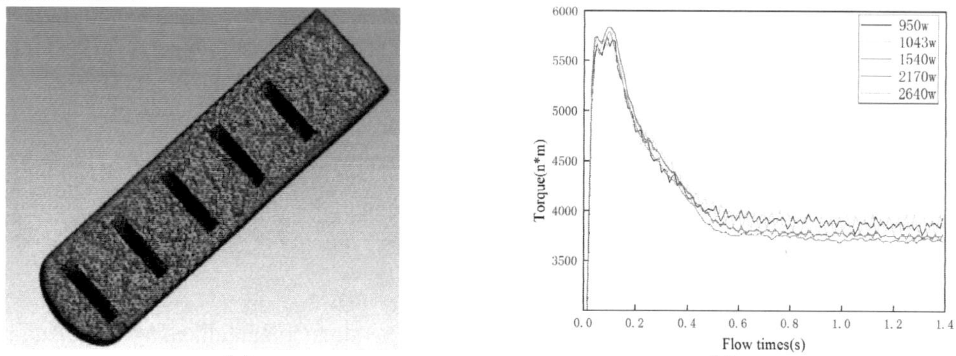

(a) (b)

Figure 2. Structure of reactor and stirring paddle. (a) Reactor component diagram. (b) Reactor structural parameters. (c) Impeller structural parameters.

To examine the impact of grid density on simulation accuracy, we monitored the torque produced by the reactor's bottom impeller, adjusting the grid count in the stirring and non-stirring regions while keeping all other variables constant. Figure 2 reveals that when grid counts reached 9.5 million and 10.43 million, there were noticeable changes in torque. However, between 15.4 million and 26.4 million grids, the torque fluctuations stabilized, with the average variance staying below 2%, which satisfies the requirements for grid independence. Considering the trade-off between computational precision and expense, the decision was made to proceed with a model comprising 10.43 million grids for the simulation tasks.

2.2.3 CFD simulation conditions and model selection. In this simulation, the Component Transport Model is utilized to replicate the interaction and amalgamation of reactants and catalysts inside the reactor. The Realizable k-epsilon model, known for its efficacy in handling turbulence, is chosen for this purpose [8]. The agitation zone, characterized by the stirrer's operation, is set to a rotational speed of 230 r/min, and the computational interval is calibrated at 0.01 seconds per time step. To maintain thermal isolation within the model, the reactor walls are designed as adiabatic. For the simulation's purpose, liquid ethylene and 1-butene are selected as the primary reactants, with toluene serving as a surrogate for the actual catalyst, offering a pragmatic approach to studying the reactor's dynamics under these simulated conditions.

2.3. Temperature field simulation and experimental validation

Utilizing the physical modeling and grid stratification detailed in Section 2.2, along with integrating reaction kinetics, the study simulates the internal temperature distribution within the reactor. Temperature monitoring points are established congruent with those used in an industrial setting, facilitating a comparative analysis between the simulation outcomes and actual empirical data.

As depicted in Figure 3(a), liquid ethylene and 1-butene, entering from the reactor's bottom feed inlet, react exothermically upon contact with the catalyst. This reaction induces a dynamic where incoming ambient temperature materials continuously feed into the reactor, causing the already heated materials to ascend, thereby establishing a temperature profile with a cooler base and warmer middle and top sections. The temperature at the reactor's apex is recorded at 242.9°C, while the base measures 213.6°C, yielding a substantial 30°C variance between the highest and lowest temperature points. Within each impeller blade zone, the temperature disparity is relatively marginal. However, a noticeable temperature gradient of 5-10°C exists between each impeller blade layer, underscoring the substantial thermal stratification within the reactor.

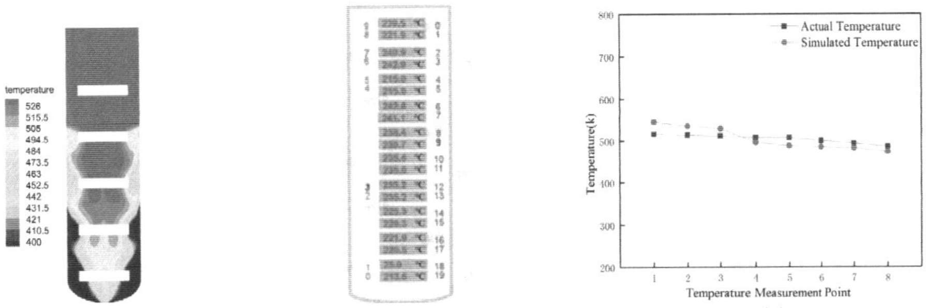

(a) (b) (c)

Figure 3. Comparison of Simulated Temperature Distribution with Actual Temperature Measurement Data. (a) Simulated Temperature Distribution. (b) Actual Temperature Measurement Data. (c) Comparison Between CFD Simulation and Actual Operating Conditions.

A comparative analysis is conducted between the outcomes of the temperature field simulation and the temperature data acquired from thermocouples during the actual production process of a specified enterprise, as illustrated in Figure 3(b). Figure 3(c) presents the juxtaposition of the simulation results with the real-world measurements. The mean deviation between the simulated and actual production temperature measurement points stands at 3.41%, falling within an acceptable error threshold. This degree of congruence suggests the simulation model's relevance as a reference tool and affirms its validity in replicating the temperature dynamics within the actual production setting.

2.4. Calculation of stirring power

In the context of industrial production, the stirring power directly influences the overall energy expenditure of the plant. Thus, it's essential to monitor the torque exerted by the stirring blades and calculate the resultant stirring power, a vital aspect of designing and operating a stirring reactor. The equation for calculating stirring power is given as:

$$P = 2\pi MN \tag{9}$$

where P signifies the power generated by the stirring process in Watts, M denotes the torque applied by the stirrer blades in Newton meters, and N is the rotational speed of the blades in revolutions per second. The torque data obtained from Fluent simulations are used to calculate the maximum power exerted by the stirring shaft, providing a foundational metric for in-depth studies on the sensitivity of parameters pertaining to the reactor's stirrer blades.

3. Reactor flow field simulation

In Section 2.1, we developed a comprehensive model for the reactor's internal flow field, utilizing Equation (1) through Equation (4). Additionally, Equations (5) to (8) were employed to construct a detailed kinetic model for the polymerization reaction. These models enable a thorough analysis of the reactor's internal fluid dynamics, including the examination of velocity, pressure, and concentration distributions. This approach is particularly focused on understanding the flow mechanics of the polymerization reaction taking place within the reactor

3.1. Velocity and pressure field distribution

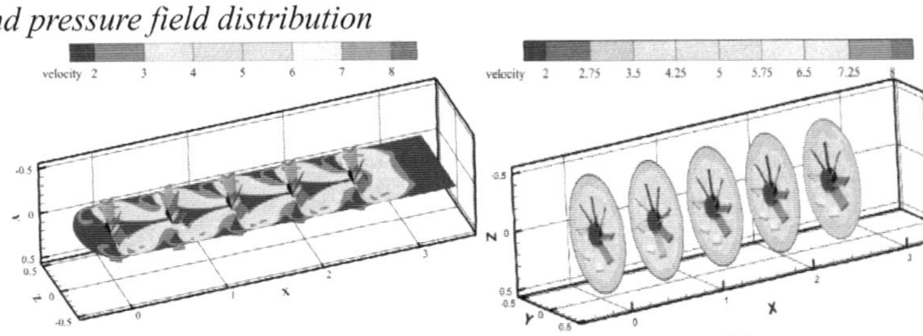

(a) (b)

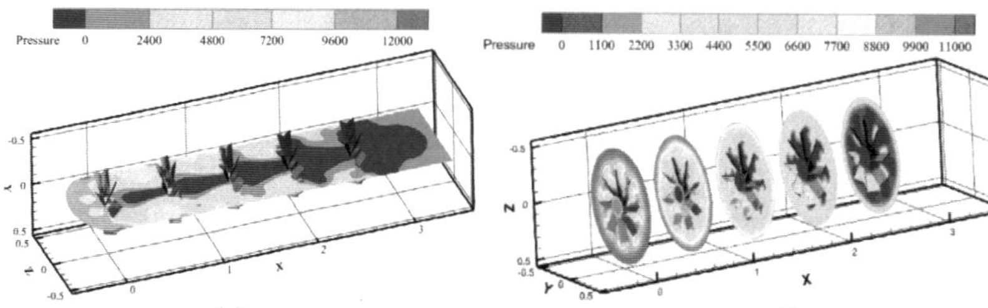

(c) (d)

Figure 4. Velocity and pressure Distribution in the YZ and XZ Cross-Sections of the Reactor. (a) Velocity Distribution in the YZ Cross-Sections of the Reactor (b) Velocity Distribution in the XZ Cross-Sections of the Reactor (c)Pressure Distribution in the YZ Cross-Sections of the Reactor (b) Pressure Distribution in the XZ Cross-Sections of the Reactor.

In Figure 4(a), the yz cross-sectional view of the stirring paddle reveals that the velocity field within the reactor is distinctly segmented into five layers, each forming a relatively independent circulating flow area. Similar patterns of velocity field distribution are observed across these layers, with the maximum velocity, reaching 8.56 m/s, occurring near the left tip of the stirring paddle. The velocity distribution within the paddle's rotating area is higher on both the left and right sides, but lower in the middle. This pattern results from the greater linear velocities at the sides, intensely influenced by the paddle's rotational motion. However, as the fluid progressively moves towards the positive x-direction, it comes increasingly under the influence of the above layer's stirring paddle, leading to a gradual decrease in velocity. At the upper layer's stirring paddle, the velocity field displays a symmetric, droplet-shaped distribution.

Figure 4(d) shows that the pressure distribution along the x-axis initially decreases before increasing. The highest relative pressure, reaching 12000 Pa, is observed near the bottom of the lower stirring paddle. In the feeding area, fluid entering through the outlet is impeded by the first layer's stirring paddle. The downward-moving material is swept along by the high-speed fluid from the inlet, forming a symmetric swirling area with higher flow velocity but lower pressure. In the main flow area, the pressure is exerted primarily near the bottom stirring paddle as the material enters, resulting in lower pressure near the second layer's stirring paddle. As the fluid moves upwards to the upper three layers of stirring paddles, the pressure gradually increases.

3.2 Concentration field distribution

A catalyst concentration of 1 was set at the catalyst inlet. Through iterative calculations, the concentration value of the catalyst was determined after it had fully diffused and mixed uniformly within the reactor. If the concentration at all monitoring points reaches 95% of the uniformly mixed concentration, the time elapsed from the start of the calculation to this point is considered as the mixing time for that monitoring point, defined as θ^{95} [9]. Figure 5 illustrates the concentration distribution of the tracer within the reactor at 2 s, 5 s, 10 s, 15 s, 25 s, and 27 s during the mixing process. This visually demonstrates the dispersion of the tracer in the stirred tank.

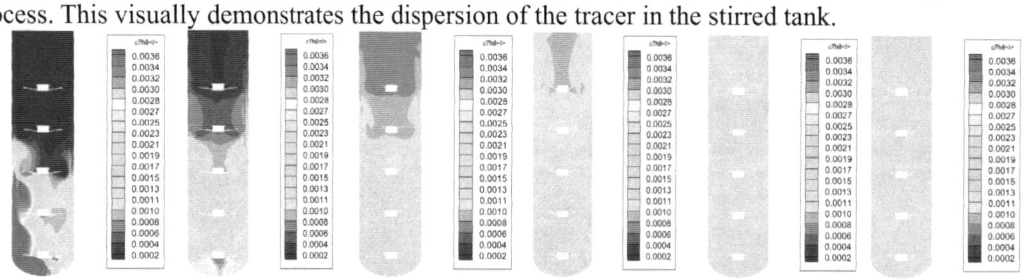

(a) (b) (c) (d) (e) (f)

Figure 5. The concentration field in the reactor varies with time. (a) 2 s; (b) 5 s; (c) 10 s; (d) 15 s; (e) 25 s; (f) 27 s.

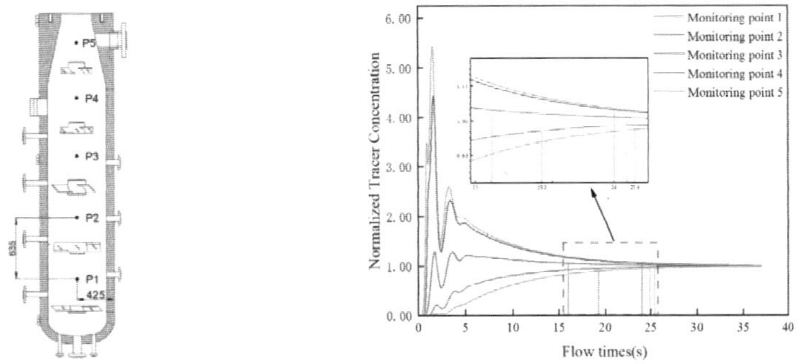

(a) (b)

Figure 6. Location distribution and concentration-response curve of concentration monitoring points. (a) Location distribution; (b) Concentration-response curve of concentration monitoring.

Figure 6(a) displays the reactor's interior where five monitoring points are evenly distributed among the five stirring layers, serving to gauge the catalyst-material mixing time. Analysis of Figure 6(b) reveals varying mixing durations at these points: 24.9 s for Point 1, 24 s for Point 2, 16 s for Point 3, 19.3 s for Point 4, and 25.4 s for Point 5, illustrating an initial reduction followed by an increase in overall mixing time. Notably, Points 1 and 2, situated near the bottom feed, experience marked fluctuations in concentration. Conversely, Points 3, 4, and 5, progressively distanced from the feed point, demonstrate a steady increase in concentration with less volatility and extended mixing periods

4. Sensitivity study of reactor agitator parameters

4.1 Fundamentals of local sensitivity analysis

This study aims to evaluate the influence of impeller structural variables on mixing time and stirring power. A detailed local sensitivity analysis was carried out on five key impeller parameters: quantity, diameter, vertical height, tilt angle, and thickness. The purpose is to inform and guide refinements in impeller design.

A prevalent method for this type of analysis is the modified Morris screening technique. It involves altering parameters in consistent increments and conducting a series of simulations. The parameter's sensitivity is then deduced from the average variation observed in the model's outputs [10]. The method is encapsulated in the following equation:

$$S=\sum_{i=1}^{n}\frac{\left(Y_{i+1}-Y_i\right)/Y_0}{\left(P_{i+1}-P_i\right)/100}/i \tag{10}$$

where 'S' denotes sensitivity. The terms 'Y_i' and 'Y_{i+1}' refer to the output values of the i-th and i+1-th runs, respectively, while 'Y_0' is the baseline value of the model. 'P_i' and 'P_{i+1}' represent the relative change rates of the parameter in the i-th and i+1-th runs compared to the baseline. The index 'i' indicates the number of runs executed.

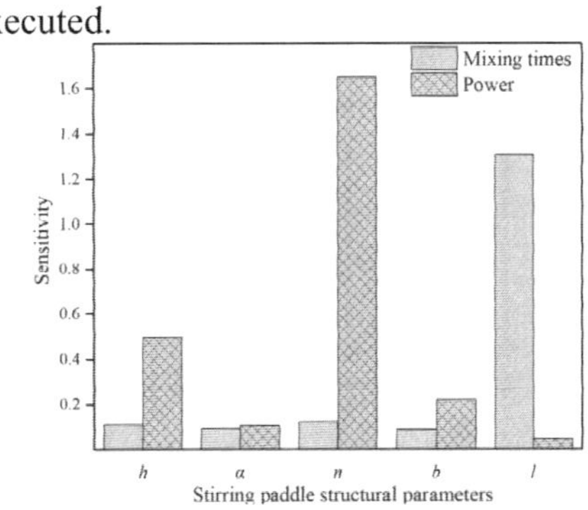

Figure 7. Sensitivity of each parameter of the impeller.

From the insights of Figure 7, it's evident that the parameter 'l' exhibits a high sensitivity of 1.31 for mixing time, classifying it as highly sensitive and underscoring the significant role impeller diameter plays in mixing efficiency. The sensitivity of other parameters, '*n*', '*h*', '*α*', and '*b*', follows a descending order, all falling into the moderately sensitive category, thus similarly influencing mixing time. Notably, '*n*' shows a high sensitivity of 1.65 in relation to stirring power, highlighting its substantial impact. The vertical height and thickness of the impeller, represented by '*h*' and '*b*', are sensitive parameters, that significantly affect stirring power. The sensitivity of '*l*' is moderate at 0.11 for stirring power, whereas it is low at 0.048, suggesting a minor effect on stirring power.

To summarize, the parameters affecting mixing time θ^{95} are ranked as follows: $l > n > h > \alpha > b$, where an increase in impeller diameter notably enhances the stirred volume and intensifies fluid agitation within the reactor, significantly shortening mixing time. The influence on stirring power P is ranked as $n > h > b > \alpha > l$, with the number of impellers '*n*' being a major determinant. Increasing the number of impellers escalates the cumulative torque and consequently the stirring power.

4.2 Parameter influence analysis

Section 4.1's sensitivity analysis reveals that the number of impellers and their diameter, '*l*', are crucial factors affecting stirring power, P, and mixing time, θ^{95}, respectively. This necessitates a focused examination of these parameters' influences.

From Figure 8(a), it's clear that an increase in the number of impellers, '*n*', leads to a corresponding rise in stirring power. This is attributed to the increased torque on the stirring shaft, enhancing the stirring power. In practical production settings, balancing energy consumption with process demands makes opting for lower stirring power a cost-effective strategy.

Figure 8(b) depicts how changes in impeller diameter 'l' affect mixing times within a reactor. The mixing time initially decreases and then increases, peaking at a diameter of 675 mm (24.29 s). Below 675 mm, the impeller diameter is insufficient for effectively stirring the reactor's internal flow, resulting in suboptimal material-catalyst mixing. Above this diameter, the proximity of the impeller to the reactor's internal walls causes high-speed fluid impacts, disrupting inter-layer mixing. Since mixing time directly impacts the residence time inside the reactor, selecting impeller parameters that reduce mixing time can significantly boost reaction efficiency. Impeller diameters ranging from 635-715 mm lead to shorter mixing times and higher process efficiency, making this range a valuable reference for optimizing impeller design.

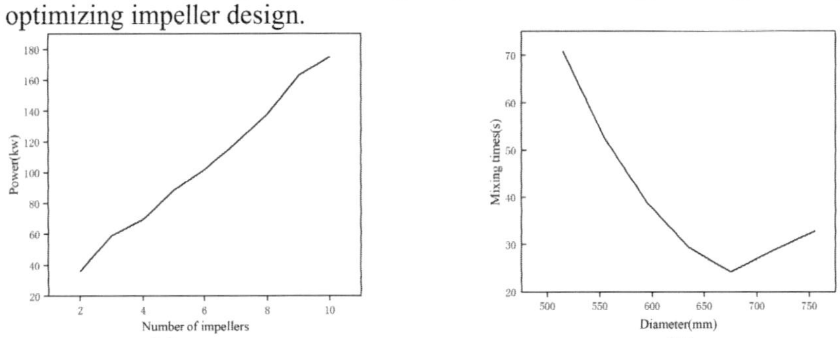

(a) (b)

Figure 8. Impeller Parameter Influence Analysis. (a) Effect of different numbers of impellers on the stirring power; (b) Effect of different impeller diameters on mixing time.

5. Conclusion

Based on the results and discussions presented above, the conclusions are obtained as below:

(1) The investigation of the temperature field within the reactor reveals a gradient decrease from the top to the bottom. The bottom part registers lower temperatures, contrasting with the higher temperatures in the middle and upper sections. When the simulation was conducted with temperature monitors set as in the actual plant, the results deviated by an average of 3.41% from the real factory

data, falling within the acceptable error margin. This confirms the simulation model's reliability.

(2) An examination of the internal flow dynamics of the reactor shows that the velocities are highest in the inlet and outlet areas, with a gradual decrease along the jet area moving in the positive x-axis direction. Pressure distribution follows a pattern of initial decrease and subsequent increase along the x-axis. Concentration-response curve analysis indicates that the mixing times at the monitoring points decrease and then increase from the bottom to the top of the reactor. Points 1 and 2 experience more intense concentration shifts, whereas Points 3, 4, and 5 maintain a steady upward trend, with less fluctuation and longer mixing durations.

(3) Results from the local parameter sensitivity analysis highlight that the number of impellers and their diameter are critical factors influencing stirring power and mixing time θ^{95}, respectively. An increase in the number of impellers leads to a rise in stirring power, while variations in impeller diameter cause the mixing time to first decrease and then increase

References

[1] Ma X Y and Liu K Q and HU J J (2023) CFD study on mixing and reaction process in POE solution polymerization reactor. CIESC Journal., 1-27.

[2] Yang G X (2021) Design of key components for gas phase fluidized bed reactor and research on polymerization process. PhD thesis. Harbin Institute of Technology.

[3] Zhang F L and He J (2022) X Research Progress in the Application of Polymerization Reactor. Contemporary Chemical Industry., 51(08):1871-1875.

[4] Li X X and Wu L T (2021) Design of sparger for fluidized bed polymerization reactor [J]. Petrochemical Equipment Technology., 42(03):11-13

[5] Kiparissides G and Verros J F (1993) Macgregor. Mathematical modeling, optimization, and quality control of High-Pressure ethylene polymerization reactors. Journal of Macromolecular Science Reviews in Macromolecular Chemistry and Physics., C33(4): 437-527.

[6] Zhao G G (2018) Study on the rheological properties of polyolefin elastomer subconcentrated solution and melt. PhD thesis. Zhejiang University.

[7] Hoseini S S and Najafi G and Ghobadian B (2021) Impeller shape-optimization of stirred-tank reactor: CFD and fluid structure interaction analyses. Chemical Engineering Journal., 413: 127497.

[8] Lin Z Q and Yu Y (2021) Numerical simulation of the flow field of centrifugal blood pump by SRF method and slip grid method. Journal of Medical Biomechanics

[9] Luo S and Zhou Q X and Zhou Q Q (2022) Study on Mixing Characteristics of 6-Pitched-Blade Turbine Stirred Tank Mechanical & Electrical Engineering Technology, 51(11):178-180

[10] Kang D D and Qiu F J and Wen R J (2023) Analysis on Local and Global Sensitivity of LID Parameters Based on SWMM. China Water & Wastewater., 39(17):131-138.

Theoretical investigation of Cu influence on the stability characteristics of T-phase in crossover alloys

Boyu Xue[1,2,3,4], Wei Xiao[2,3,4*], Xiwu Li[2,3,4*], Guanjun Gao[2,3,4], Xiaowu Li[1], Hongwei Yan[2,3,4], Lizhen Yan[2,3,4], Yongan Zhang[2,3,4], and Baiqing Xiong[2,4]

[1] Department of Materials Physics and Chemistry, School of Materials Science and Engineering, Key Laboratory for Anisotropy and Texture of Materials, Ministry of Education, Northeastern University, Shenyang 110819, China
[2] State Key Laboratory of Nonferrous Metals and Processes, China GRINM Group Co., Ltd., Beijing 100088, China
[3] GRIMAT Engineering Institute Co., Ltd., Beijing 101407, China
[4] General Research Institute for Nonferrous Metals, Beijing 100088, China

[*]Corresponding author's email: xiaowei@grinm.com (Wei Xiao); lixiwu@grinm.com (Xiwu Li)

Abstract. In crossover Al-Mg-Zn alloys, the T-phase ($Mg_{32}(Al, Zn)_{49}$) is crucial for reinforcement. The critical nucleation size of the precipitated T-phase is notably reduced by Cu addition, as well as the activation energy. There is a lack of theoretical research on the influence of Cu elements on the stability of T-phases and mechanical properties. In this work, we systematically investigate the influence of varying Cu concentration on the T-phases by examining their stability, mechanical properties, and electronic structure characteristics through first-principles calculations. The findings reveal that the T-phase maintains lattice stability within the 0 to 5% Cu content range, with a linear decrease in the lattice parameter and formation energy (–0.16 ~ –0.15 eV/atom). Additionally, there is an observable positive correlation trend between the elastic moduli (*B*, *E*, and *G*) of the T-type phase and the Cu content. It is shown that the T-phase exhibits inherent ductility. Analysis based on Pugh's criterion and Poisson's ratio indicates a reduction in the ductility with increasing Cu content in the T-phase. This work theoretically validates the previous experimental observations and reveals the impacts of Cu concentration on the T-phase.

1. Introduction

Crossover Al-Mg-Zn alloys have emerged as novel aluminum alloys in recent years, and its innovative design strategy of "cross-alloying" [1] [2] is gradually evolving into an avenue for aluminum alloy innovation applicable in the commercial sector. This newly developed alloy is formed by combining different commercial alloys (2xxx/5xxx/7xxx) [1], integrating multiple advantages, including solid solution strengthening, precipitation hardening, and dynamic aging [3] [4]. Among these advantages, the T-phase and its precursors, crucial for aging-strengthening, significantly enhance the mechanical performance of the alloy. However, due to its comparatively elevated activation energy and considerable critical nucleation size [5], its formation typically needs a longer aging time. In recent years, researchers have accelerated the alloy's aging response or enhanced the comprehensive mechanical performance through strategies involving increasing the Zn content [6], introducing microalloying elements such as Sc, Zr, Ag, and Cu [6-9], and adjusting the heat treatment process [10] [11]. For instance, Zhao et al. [12] investigated Al-xMg-(7−x) Zn-1Cu (wt.%) alloys and found that by raising the Zn/Mg ratio and employing a two-step artificial aging method, the aging response was substantially enhanced and accelerated. The pre-aging process resulted in a rise in the UTS and YS of the alloys by at least 40 MPa. Besides heat treatment processes, the incorporation of alloying elements like Sc, Zr, Cu, and Ag has significantly enhanced the alloy's performance. Among them, Cu, as a common alloying element, has been more widely investigated and reported in numerous experiments given economic considerations. Zhang et al. [13] demonstrated an outstanding balance of ductility and strength in an Al-Mg-Zn alloy containing 1.5 wt.% Cu. This is attributed to the accelerated formation kinetics of nano-sized precipitates at high Cu concentrations, leading to an 80.5% increase in yield strength by augmenting the solid solution strengthening contribution. Cu alloying effectively narrows the alloy's PFZ, enhances the contribution of forest dislocation strengthening, and introduces coarsen precipitates. These actions collectively enhance the strain-hardening capacity of the alloy while maintaining its ductility. Incorporating minor Cu elements can effectively boost and expedite the alloy's hardening process. Stemper et al. [14] enhanced T-phase precursor precipitation in the Al-4.7Mg-3.5Zn (wt.%) alloy by incorporating Cu (0.15 and 0.5 wt.%), thereby shifting the hardening onset to earlier times.

Also, the previous DFT theoretical calculations [15] have energetically predicted the possible occupation of individual Cu atoms in the T-phase. However, the interaction between Cu elements and the T-phase is still not well understood, and the impacts of varying Cu contents on its stability, electronic structure, and mechanical characteristics remain unknown. Therefore, this work strives to systematically explore the strengthening mechanisms of various Cu contents on the T-phase, providing a theoretical framework for the advancement of advanced aluminum alloys.

2. Computational methodology

In Figure 1, the crystal structure of the T-phase is categorized under the cubic structure, characterized by the space group $Im\bar{3}$. It relies upon the regular alignment of Bergman clusters within a unit cell consisting of eight sites (A-H). In this arrangement, Al/Zn atoms demonstrate chemical disorder by randomly occupying the B, C, and F sites. Meanwhile, Mg atoms are situated at sites other than those mentioned earlier, as well as A sites [16]. As mentioned in the previous studies [14, 15], it is known that Cu tends to substitute Al in the T-phase, particularly at the C-site. Therefore, to simplify the model, we characterize the T-phase structure by varying the Cu content at the C-site in this work. Specific modeling details can be referred to in [15]. In the present work, crystal structure models were constructed for different Cu contents based on the $Mg_{64}Al_{42}Zn_{54}$ composition using the special quasi-random structure method [17].

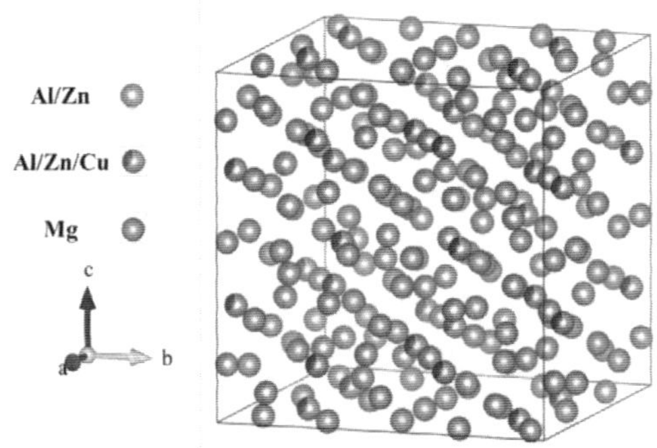

Figure 1. The T-phase's structure.

All calculations were performed using the density functional theory (DFT) method. These analyses were performed within the Vienna ab initio simulation package [18]. We employed the projector augmented wave (PAW) method to characterize electron-ion interactions. The exchange-correlation potential was determined using the Perdew-Burke-Ernzerhof general function of the generalized gradient approximation [19]. For the plane-wave basis set, the cutoff energy was set to 500 eV. During Brillouin-zone sampling, a 5 × 5 × 5 Gamma-centered k-mesh was employed for structure optimization. Additionally, a 7 × 7 × 7 k-mesh was employed for property calculations. The energy convergence threshold was set to be smaller than 10^{-5} eV.

3. The calculation and discussion

3.1. Structural stability

To investigate the influence of Cu concentration on the T-phase's crystal structure, Figure 1 illustrates the crystal structure of the Cu-containing T-phase. Within this structure context, Cu replaces the Al atom in the C-site, leading to the occupation of this site by Al, Zn, and Cu atoms. Figure 2 illustrates the T-phase crystals' lattice constants with varying Cu concentrations (ranging from 0 to 5 at. %), obtained by structural optimization and energy calculations for the Cu-containing T-phase. Its lattice parameter spans from 1.413 to 1.424 nm, comparable to a = 1.425 ± 0.003 nm for T-phase (Al_6CuMg_4), and a BCC-structured homogeneous phase coexisting in Al-Mg-Cu alloys [20]. Given that the atomic diameter of Cu is 0.256 nm [21], less than that of Al (0.286 nm), Mg (0.320 nm) [22], and Zn (0.274 nm) [15], the lattice constant diminishes linearly as the Cu concentration rises in Cu-containing T-phase. The atomic bond lengths of Cu near-neighbors are shorter than those of the relevant Al bonds. This is ascribed to the size effect, where the smaller atomic diameter of Cu leads to shorter Cu-Al (0.256 nm), Cu-Zn (0.252 nm), and Cu-Mg (0.301 nm) bonds compared to Al-Al (0.267 nm), Cu-Zn (0.254 nm), and Al-Mg (0.307 nm) bonds, potentially resulting in stronger bonding interactions. This trend suggests that higher Cu content reduces inter-atomic distance and enhances interactions, consequently increasing the stiffness of elastic deformation. This observation is further supported by subsequent mechanical analysis.

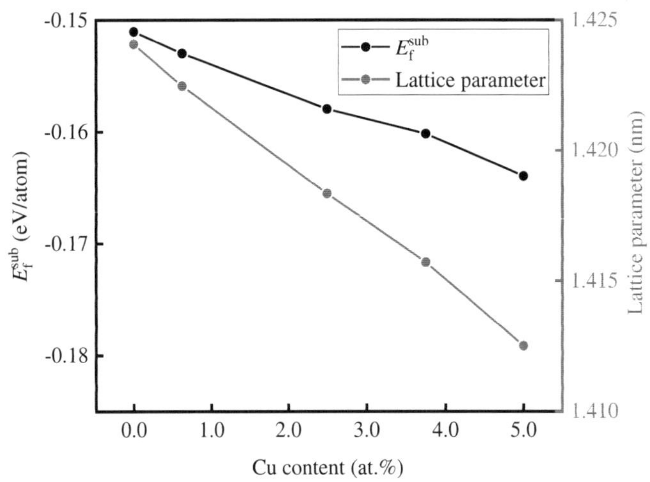

Figure 2. The lattice parameters and formation energies for T-phases containing varying Cu content.

Additionally, we evaluate the structural stability of Cu-containing T-phase by calculating the solid solution formation energy at 0 K according to Equation (1):

$$\Delta E_f^{sub}(Mg_{x_1}Al_{x_2}Zn_{x_3}Cu_{x_4}) = \frac{E_0 - [x_1\mu_{Mg}^{sub} + x_2\mu_{Al}^{bulk} + x_3\mu_{Zn}^{sub} + x_4\mu_{Cu}^{sub}]}{\sum x_i}, \quad (1)$$

where E_0 corresponds to the energy of the T-phase ($Mg_{x_1}Al_{x_2}Zn_{x_3}Cu_{x_4}$). x_i is the atoms' number of different elements in the T-phase. μ_i represents the chemical potential of the respective element. The total energy difference can be determined by comparing the $Al_{107}X$ dilute solution (where X represents Mg, Zn, or Cu) with Al, utilizing identical 3 × 3 × 3 supercell configurations. Additional information can be found in the cited [15]. According to Figure 2, the Cu-containing T-phase's formation energy ranges from –0.16 to –0.15 eV/atom approximately as the Cu content decreases from 5% to 0. In essence, the formation energy exhibits a decreasing trend as the Cu content increases. Hence, it can be expected that the inclusion of Cu contributes to the stabilization of the T-phase and facilitates its precipitation, consistent with the experimental observation that Cu can enhance the aging-strengthening effect of alloys [14].

3.2. Mechanical properties

The elastic constants and resulting elastic moduli for the Cu-containing T-phases were determined based on the strain-stress method [23] for varying Cu concentrations, as depicted in Table 1. The elastic modulus of the cubic system was

$$\begin{cases} B=\frac{1}{2}(B_V+B_R)=\frac{1}{2}\left[\frac{C_{11}+2C_{12}}{3}+\frac{1}{3(S_{11}+2S_{12})}\right] \\ G=\frac{1}{2}(G_V+G_R)=\frac{1}{2}\left[\frac{(C_{11}-2C_{12})+3C_{44}}{5}+\frac{5}{4(S_{11}-S_{12})+3S_{44}}\right] \end{cases} \Rightarrow \begin{cases} E=\frac{9BG}{3B+G} \\ v=\frac{3B-2G}{2(3B+G)} \end{cases} . \quad (2)$$

where C_{ij} is the stiffness matrix, and S_{ij} represents the compliance tensor [24]. All structures of the T-phase exhibit mechanical stability, satisfying the stability conditions of BCC structures: C_{11} and C_{44} are positive, and both $C_{11} + 2\,C_{12}$ and $C_{11}–C_{12}$ are greater than zero [25], which is also deduced to be $C_{12} < B < C_{11}$ [26]. C_{11} represents the unidirectional compression induced along the a-axis, with a larger value indicating stronger corresponding resistance. In the structures of the T-phase, C_{44} is comparatively smaller than C_{11}, indicating a higher susceptibility to shear strain [26].

Typically, the mechanical properties of materials reflect the ability to withstand deformation when subjected to external pressure. A larger bulk modulus (*B*) indicates greater resilience to volume changes under pressure. The shear modulus (*G*) indicates the capacity of a material to withstand in shape caused by shear deformation. Besides, Young's modulus (*E*) serves as an indicator of material rigidity, indicating the relationship between stress and strain [27, 28]. The calculated elastic modulus demonstrates that the values of *B*, *E,* and *G* for T-phases elevate as the Cu content rises. In Table 1, it is evident that the $Mg_{64}Al_{34}Zn_{54}Cu_8$ structure exhibits superior elastic modulus (*B*, *E,* and *G*) values of 61, 77, and 30 GPa, surpassing the T-phase structure without Cu, which has elastic modulus values of 59, 70, and 27 GPa, respectively.

Furthermore, the elastic constants, as shown in Table 1, can serve to evaluate *G/B* and Poisson's ratio. According to the materials' ductility and brittleness criterion: crystals having $G/B < 0.57$ and $v > 0.26$ are classified as ductile [27,28]. Therefore, the considered T-phase structures are all ductile, as they exhibit Poisson's ratio values higher than 0.26 (0.29 ~ 0.30) and *G/B* ratios lower than 0.57 (ranging from 0.45 to 0.49). The Poisson's ratio of the T-phase crystals aligns closely with the typical value of 0.33 for metals, surpassing 0.25 for ionic crystals, and 0.1 for covalently bonded crystals [26], indicating the metallic character of T-phases. As the Cu concentration of the T-phase increases, Poisson's ratio decreases, but the *G/B* ratio and material hardness tend to increase. The machinability index μ_M (=B/C_{44}) serves as another measure of material hardness [29], which evaluates the ease of cutting or machining a material. It is crucial for determining the lubricity and plasticity of a solid. As observed in Table 1, the machinability index is lower due to the higher hardness of the T-phase (e.g. $Mg_{64}Al_{34}Zn_{54}Cu_8$) with increased Cu content (2.05).

Table 1. The C_{ij} values and resulting elastic moduli (GPa), along with the *G/B* ratio, *v,* and μ_M of varying Cu content T-phases.

Cu content (at. %)	$Mg_{64}Al_{42}Zn_{54}$ (0)	$Mg_{64}Al_{41}Zn_{54}Cu_1$ (0.6%)	$Mg_{64}Al_{38}Zn_{54}Cu_4$ (2.5%)	$Mg_{64}Al_{36}Zn_{54}Cu_6$ (3.8%)	$Mg_{64}Al_{34}Zn_{54}Cu_8$ (5.0%)
C_{11}	94.7	95.1	98.4	99.7	101.2
C_{12}	41.1	41.3	40.8	41.1	41.5

C_{44}	26.7	26.7	27.9	28.8	30
B	58.9	59.2	60	60.7	61.4
E	69.7	69.8	73.3	75	77.2
G	26.8	26.8	28.3	29	29.9
G/B	0.45	0.45	0.47	0.48	0.49
v	0.303	0.303	0.296	0.294	0.29
μ_M	2.21	2.22	2.15	2.11	2.05

To gain insights into the Cu content's influence, we employ electronic structure analysis. Figure 3 illustrates the density of states of T-phases with varying Cu concentrations ($Mg_{64}Al_{41}Zn_{54}Cu_1$ and $Mg_{64}Al_{34}Zn_{54}Cu_8$). Setting the Fermi energy level (E_F) at 0 eV, it's evident that the density of states at E_F for T-phases is non-zero, indicating the metallic nature of T-phases with varying Cu contents. Furthermore, in the $Mg_{64}Al_{34}Zn_{54}Cu_8$ structure, the E_F shifts to the left, resulting in a more pronounced decrease in the density of states for the elements Cu, Mg, and Al in $Mg_{64}Al_{34}Zn_{54}Cu_8$ at the E_F compared to their counterparts in the $Mg_{64}Al_{41}Zn_{54}Cu_1$. This results in $Mg_{64}Al_{34}Zn_{54}Cu_8$ exhibiting lower metallicity and ductility.

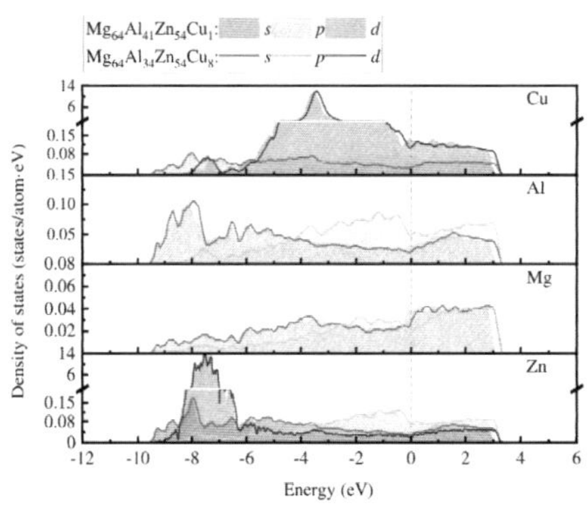

Figure 3. The PDOS of different Cu substitutions for Al in $Mg_{64}Al_{41}Zn_{54}Cu_1$ and $Mg_{64}Al_{34}Zn_{54}Cu_8$. The green, blue, and red lines depict the *s*, *p*, and *d* electron orbitals, correspondingly.

4. Conclusion

DFT calculations were employed to explore the stability and mechanical characteristics of T-phases with varying Cu content in this work. The findings indicate that Cu-containing T-phases maintain lattice stability within 0-5% Cu content, with concurrent reductions in lattice parameters and formation energy (–0.16 ~ –0.15 eV/atom) as Cu content increases. Significantly, a positive correlation emerges between elastic moduli (*B*, *E*, and *G*) and Cu content in T-phases. Poisson's ratio analyses indicate a reduction in T-phase ductility as Cu concentration increases, revealing the inherent ductility of T-phases.

Acknowledgments

This work was funded by the National Key R&D Program of China (No.2023YFB3710403).

References

[1] Stemper L, Tunes M A, Tosone R, Uggowitzer P J, and Pogatscher S (2022) Prog. Mater. Sci., 124: 100873. [CrossRef]
[2] Tunes M A, Stemper L, Greaves G, Uggowitzer P J, and Pogatscher S (2020) Adv. Sci., 7: 2002397. [CrossRef]
[3] Yang X B, Chen J H, Liu J Z, Qin F, Xie J, and Wu C L (2014) J. Alloys Compd., 610: 69-73. [CrossRef]

[4] Zhang X Z, Zhou D S, Li Y, and Zhang D L (2022) Mater. Lett., 854: 143800. [CrossRef]
[5] Zhang Z R, Li Y, Li H X, Zhang D, Zhao Z Y, Du Q, and Zhang J S (2021) Calphad, 74: 102298. [CrossRef]
[6] Guo C, Zhang H T, and Li J H (2022) J. Alloys Compd., 904: 163998. [CrossRef]
[7] Jiang L, Zhang Z F, Bai Y L, Wang Y B, and Mao W M (2023) J. Alloys Compd., 969: 172425. [CrossRef]
[8] Huang Y S, Sun P, Sun L L, Li Y, Zhang X Y, Li X W, Yan H W, Li B, Liu Y L, and Du Y (2024) J. Alloys Compd., 976: 173368. [CrossRef]
[9] Zhang H T, Nan Y, Guo C, and Cui J Z (2022) J. Alloys Compd., 910: 164767. [CrossRef]
[10] Ma Q B, Zhang D, Zhuang L Z, and Zhang J S (2018) Mat. Charact., 144: 264-273. [CrossRef]
[11] Yao J J, Zhang D, Geng Y X, Pan Y L, and Zhang J S (2023) Mat. Charact., 202: 113000. [CrossRef]
[12]Zhao Y Q, Tian T, Jia H L, Ma P K, Yang Z Z, Xu J, Zha M, and Wang H Y (2023) J. Mater. Res. Technol., 27: 1874-1885. [CrossRef]
[13]Zhang Z R, Li Y, Li H X, Zhang D, and Zhang J S (2022) J. Mater. Res. Technol., 20: 4585-4596. [CrossRef]
[14] Stemper L, Tunes M A, Dumitraschkewitz P, Mendez-Martin F, Tosone R, Marchand D, Curtin W A, Uggowitzer P J, and Pogatscher S (2021) Acta Mater., 206: 116617. [CrossRef]
[15] Xue B Y, Xiao W, Li X W, Gao G J, Li X W, Zhang Y A, Wang L G, and Xiong B Q (2024) J. Mater. Sci. Technol., 173: 237-246. [CrossRef]
[16] Bergman G, Waugh J L T and Pauling L (1957) Acta Crystallogr., 10: 254-259. [CrossRef]
[17] Jiang C (2009) Acta Mater., 57: 4716-4726. [CrossRef]
[18] Kresse G and Furth Müller J (1996) Phys. Rev. B, 54: 11169. [CrossRef]
[19] Perdew J P, Burke K, and Ernzerhof M (1996) Phys. Rev. Lett., 77: 3865. [CrossRef]
[20] Zhang L, Li X Y, Nie Z R, Huang H, and Niu L Q (2016) Mater. Des., 92: 880-887. [CrossRef]
[21] Ostermann F (2014) Springer-Verlag Berlin Heidelberg, Berlin Heidelberg. [CrossRef]
[22] Lee C S and Miller G J (2000) J. Am. Chem. Soc., 122: 4937-4947. [CrossRef]
[23] Shao H B, Huang Y C, Liu Y, and Xiao Z B (2022) Solid State Commun., 59: 87-93. [CrossRef]
[24] Mouhat F and Coudert F X (2014) Phys. Rev. B, 90: 224104. [CrossRef]
[25] Hou Z F (2010) Solid State Commun., 150: 1874-1879. [CrossRef]
[26] Iyigör A, Al S, and Arikan N (2022) Chem. Phys. Lett., 206: 140052. [CrossRef]
[27] Pettifor D G (1992) Mater. Sci. Tech., 8: 345-349. [CrossRef]
[28] Bao W Z, Liu D and Duan Y H (2018) Ceram. Int., 44: 14053-14062. [CrossRef]
[29] Islam M and Abdur R S M (2023) Physica B, 668: 415244. [CrossRef]

Investigation of high field breakdown characteristics in CNTFETs

Weihao Zeng and Zijie Zhou*

Hunan Institute of Advanced Sensing and Information Technology, Xiangtan University, Xiangtan, Hunan, 411105, China

*Corresponding author's e-mail: zzj@xtu.edu.cn

Abstract. The enhancement of stability in carbon nanotube field effect transistors (CNTFETs) is crucial for mitigating high field failures, thereby influencing device reliability through modifications in channel dimensions (W and L), as well as the incorporation of dielectric layers and gate metals during fabrication. The present study demonstrates the correlation between device withstand voltage and channel size, as well as the impact of gate oxide layer and gate metal introduction on device reliability. The findings reveal that both the breakdown voltage and breakdown current of the device are influenced by W, while only the breakdown voltage is affected by L. Furthermore, a gradual decrease in device robustness is observed with continuous integration of the dielectric layer and gate metal. The increase in device index simultaneously leads to an effective augmentation of the device failure current. These results provide valuable insights and guidance for optimizing the structural design of devices.

1. Introduction

The semiconductor industry is transitioning into the post-Moore era as advancements in processes persist and device dimensions continue to diminish. Carbon nanotubes are widely recognized as optimal materials for nanoelectronics and optoelectronics owing to their exceptional electrical, optical, thermal, and mechanical properties [1,2]. Recent studies have also demonstrated the advantages of carbon nanotubes (CNTs) as a two-dimensional material for device trenching, including applications in flexible devices, gas-sensitive sensing, biosensing, and RF devices [3,4,5]. However, due to the relentless pursuit of device performance enhancement, there has been a gradual reduction in device size, thereby exposing the device to an increased risk of breakdown. Consequently, it becomes crucial to strike a balance between the channel size of the device and minimizing the impact of the dielectric layer and gate metal on the breakdown voltage of carbon nanotube (CNT) channels.

The detrimental impact of high voltage electric fields on semiconductor devices primarily manifests in the occurrence of semiconductor breakdown phenomena. When the electric field strength surpasses a certain threshold, there is an exponential rise in carriers within the semiconductor, leading to a significant surge in conductivity and resulting in a phenomenon known as breakdown. In high-pressure environments, both electrons and holes experience heightened kinetic energy, thereby elevating the risk of semiconductor breakdown.

In the case of semiconductor devices, this breakdown phenomenon can result in device damage or failure. For instance, certain semiconductor components such as diode lasers, field-effect transistors, and light-emitting diodes have been found to sustain damage when subjected to reverse voltages ranging from 5 V to 30 V. Additionally, shrinking CNTFETs also face potential risks.

The current research focus on CNTFETs primarily lies in investigating the high-field characteristics and robustness of back-gate devices, as well as exploring the voltage breakdown capabilities of down-row array tubes [6,7]. However, the advantages of back-gate devices in circuit integration are not as significant as those of top-gate devices. Moreover, the inclusion of a dielectric layer and gate metal in top-gate devices complicates the device failure situation. To compare the failure difference between back-gate and top-gate devices under a high-voltage field and investigate the effect of introducing a dielectric layer and gate metal on channel stability, this paper conducts high-voltage field failure tests for back-gate and top-gate devices with different channel scales.

This paper will be divided into four sections: Part II - CNTFET Preparation Flow, where we will provide a detailed description of the preparation flow for CNTFETs and characterize the transfer characteristic curve to ensure device performance reliability; Part III - high-voltage Test Results, in which we will conduct high-voltage field failure tests on the devices and analyze their failure behavior; Part IV - Conclusion, where we summarize the failure behavior of CNTFETs based on current experimental results and provide recommendations for improving device reliability.

2. CNTFET fabrication

The device preparation process is illustrated in Figure 1. Figure 1(a) depicts the electron beam evaporation of Pd/Au as the source and drain metal, facilitating a contact formation between Pd and CNT without any potential barrier. Figure 1(b) showcases the device channel created through plasma etching. Figure 1(c) exhibits the oxide layer generated by thermal oxidation, employing Y_2O_3 as the medium layer for this particular device. Figure 1(d) demonstrates the electron beam evaporation of Ti/Au as the gate metal.

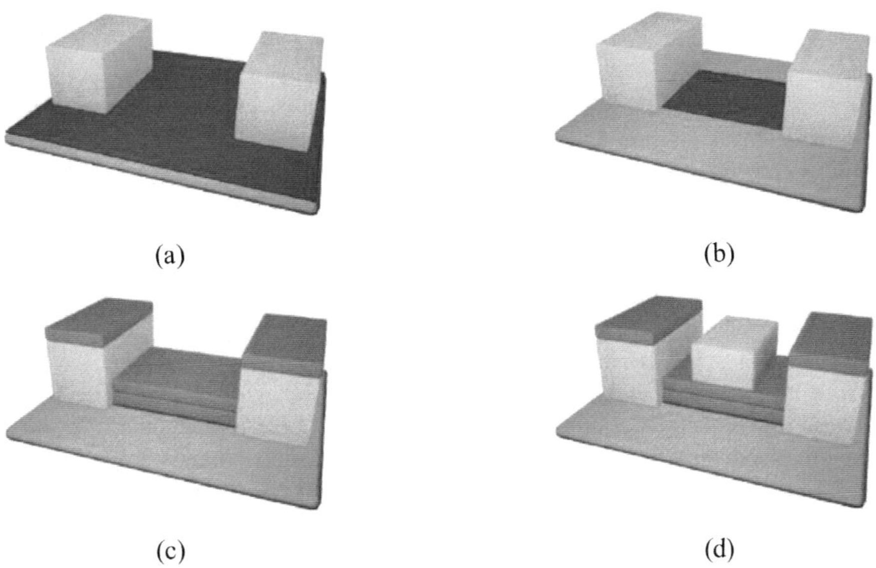

Figure 1. (a) Evaporation source-drain metal. (b) Etching channel CNT. (c) Growth dielectric layer. (d) Evaporation gate metal.

In order to ensure the device performance reliability, we conducted a test on the transfer characteristic curve of a device with a gate length of 11 um and a channel width of 108 um. The results are presented in Figure 2 below. Among them, Figure 2(a) illustrates the OM characterization results of the device, while Figure 2(b) displays the test results for the device's transfer characteristic curve at IDS=-0.1 V, with gate voltage ranging from -3 V to 1 V in steps of 0.05 V. The obtained subthreshold swing is measured as 118 mV/dec with a switching ratio of 5. These test outcomes demonstrate that CNTFET exhibits excellent performance and can serve as an effective reliability testing device.

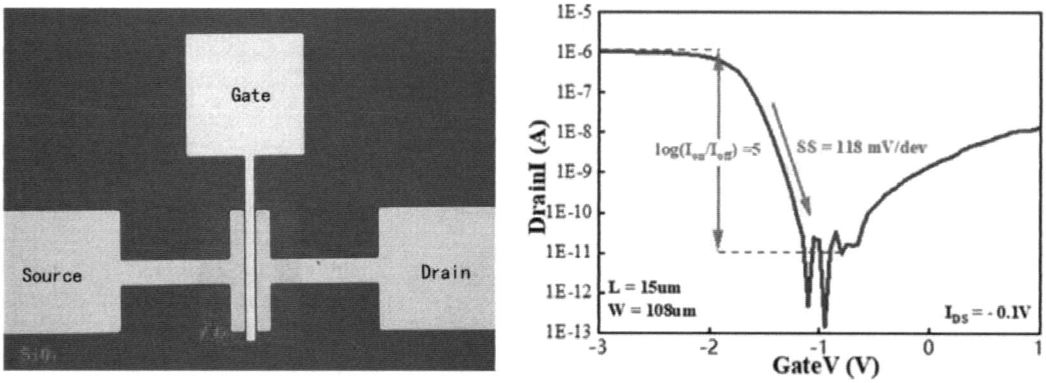

(a) (b)

Figure 2. (a) Device light mirror (OM) diagram. (b) Transfer characteristic curve test results.

3. Results and analysis

The test is divided into two parts: one part focuses on the failure analysis of back-gate CNTFET to investigate the impact of channel size on device failure behavior, while the other part examines the failure behavior of top-gate CNTFET and discusses how both channel size and gate finger influence it. The influence of the dielectric layer and gate metal on the reliability of channel CNT is concurrently examined by comparing experimental results from both components.

3.1. High-voltage testing of back-gate CNTFET

3.1.1. Relevance of channel width W. The device characteristic parameters encompass a series of key parameters that reflect the physical and structural characteristics of the device, exerting a direct influence on its performance. The channel width (W) and length (L) of CNTFET serve as fundamental elements for device design and optimization, holding significant implications for enhancing device reliability. The simplest back-gate CNTFET is employed to investigate the influence of channel width (W) on reliability under a high-pressure field, while eliminating the effects of dielectric layer and gate metal. The test device has a fixed channel length (Lch) of 15 um, with varying W values of 3, 6, 10, 15, 20, 25 and 30 um. Figure 3(a) presents the DC failure I-V curves for these seven groups of devices subjected to a high-voltage field. The results demonstrate an increase in both DC breakdown voltage and breakdown current as the channel width increases. The normalized failure voltage, as depicted in Figure 3(b), reveals that the channel stress of the device is 18.3 V/um when W=3 um, whereas it decreases to only 5.8V/um as W increases to 30 um. Notably, there exists an inverse relationship between W and the voltage borne by the device per unit length. This result shows the following: 1) When the channel width increases, the electric field distribution in the channel becomes more uniform, thereby enhancing the breakdown voltage of the device. However, as W continues to increase, the impact of improved electric field distribution diminishes, resulting in a reduced increment in DC breakdown voltage. Therefore, in device design, it is crucial to identify the appropriate channel width (W) based on the specific circumstances while ensuring a uniform distribution of electric field, thereby avoiding diminishing returns resulting from excessive W incrementation. 2) The increase in channel width (W) significantly influences the migration and diffusion of the carrier (hole), as it enlarges the flow cross-section for carriers, thereby mitigating their accumulation at the gold semi-contact. 3) Moreover, an increased channel width (W) also impacts both channel resistance and thermal effects during device operation. A larger W reduces channel resistance, consequently minimizing heat generation.

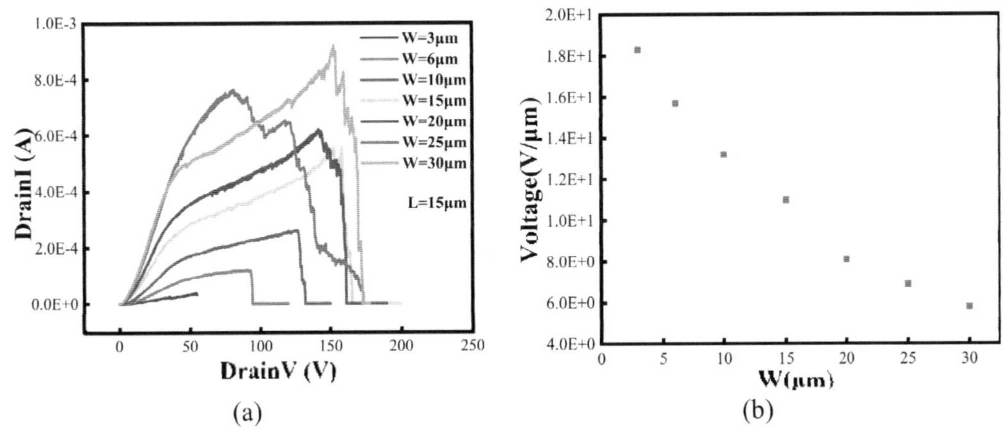

Figure 3. (a) I-V curves of high-voltage failure test with different channel widths W. (b) Failure voltage normalization results.

3.1.2. Relevance of channel length L. In semiconductor devices, the electrical performance is directly influenced by the channel length. To investigate the impact of high-voltage electric field failure on CNTFETs, devices with a fixed width (W) of 15 um and varying channel lengths (L) of 3, 6, and 10 um were employed to establish their relationship. The experimental results are presented in Figure 4 below. The breakdown voltage in Figure 4(a) increases with the increase of L due to the enhanced uniform distribution of the channel electric field and the improved heat dissipation resulting from the increased length, while maintaining a constant current flow cross section with fixed W, which ensures that the device failure current remains unchanged regardless of variations in L. The relationship between the breakdown voltage of the device and L is depicted in Figure 4(b), demonstrating a clear linear correlation. This observation can be attributed to several underlying factors: 1) With the increase in channel length L, the electric field distribution within the channel becomes more uniform, thereby reducing the maximum intensity of the electric field and enhancing mitigation of electric field concentration phenomena; 2) A longer channel facilitates increased carrier transmission distance, which aids in lowering current density within the channel and consequently mitigating thermal failure risks associated with excessive current. However, elongation of the channel also presents additional challenges, such as elevated channel resistance.

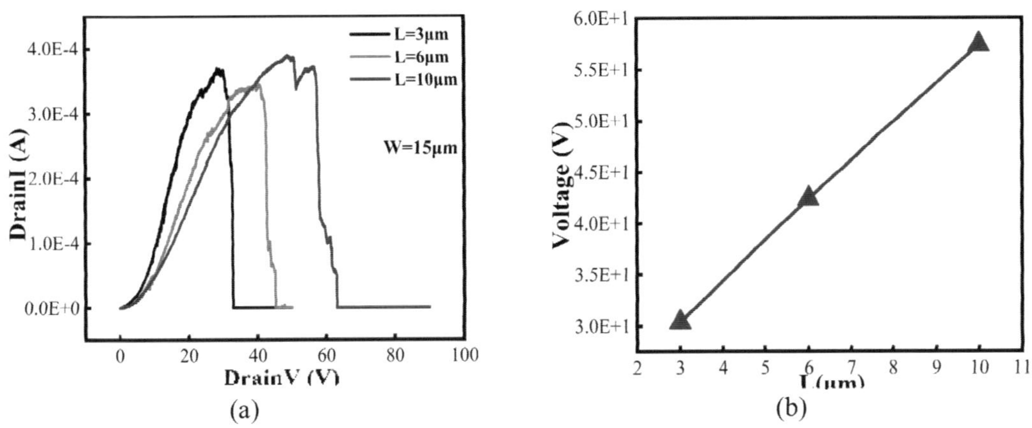

Figure 4. (a) I-V characteristics of high-pressure failure test for various channel lengths L. (b) Correlation between groove length and failure voltage.

3.2. Top-gate CNTFET high-voltage test

Due to its distinct structure and operational principles, the top-gate device exhibits different failure mechanisms and performance characteristics compared to the back-gate device. Therefore, it is imperative to conduct similar tests on the top-gate device after comprehending the impact of channel width and channel length on the back gate device in order to unveil its failure behavior under a high-voltage field. In the experiment, devices with a channel length (L) of 15 um and varying channel widths (W) of 3, 6, 10, 20, and 25 um were employed. The experimental results are presented in Figure 5 below. As shown in Figure 5(a), the failure voltage and failure current of the top-gate device exhibit different degrees of decrease compared to the corresponding data for the back-gate device. In Figure 5(b), a comparison is made between the small-voltage I-V curves of both devices with L=15 um and W=35, 40, and 45 um, where the dashed line represents the top-gate device. The test outcomes reveal that due to electrostatic doping from the dielectric layer and metal gate on the channel region, there are two orders of magnitude reduction in channel conductance for top-gate devices as compared to back-gate devices. The channel carrier mobility and carrier concentration of top-gate devices is observed to be lower in the off state, indicating reduced device performance. Furthermore, the high initial resistance value of the top-gate CNTFET channel leads to increased carrier scattering and collisions during migration, rendering the device more susceptible to damage. Additionally, in top-gate CNTFET, the decrease in failure voltage is closely associated with variations in electric field distribution. Specifically, due to the introduction of gate metal, the electric field intensity becomes more complex within localized regions of the device under high-voltage conditions. Consequently, these areas experience higher electric field intensities and are prone to breakdown.

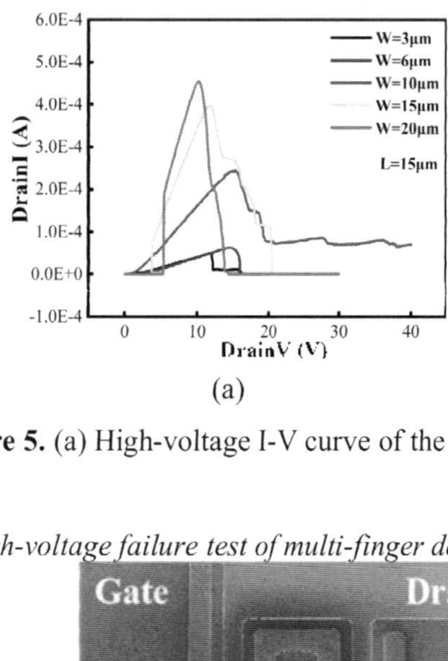

(a)

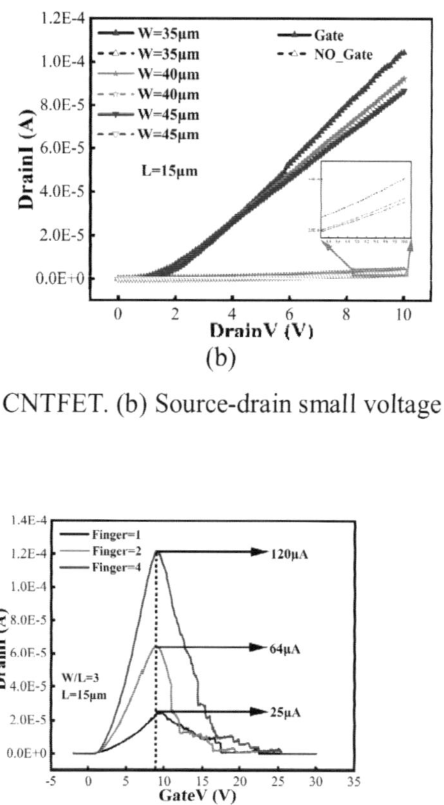

(b)

Figure 5. (a) High-voltage I-V curve of the top-gate CNTFET. (b) Source-drain small voltage I-V curve.

3.3. High-voltage failure test of multi-finger devices

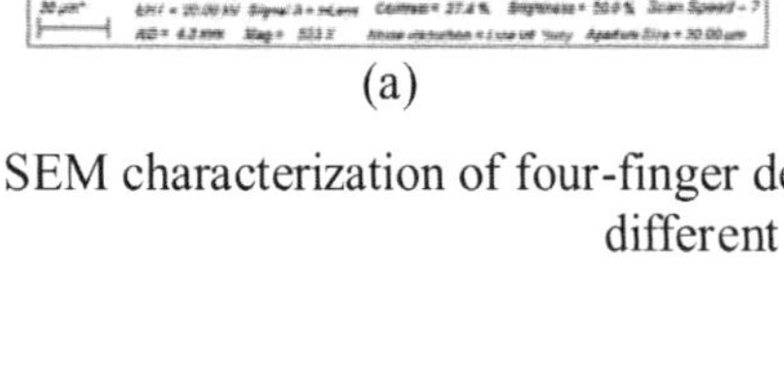

(a)

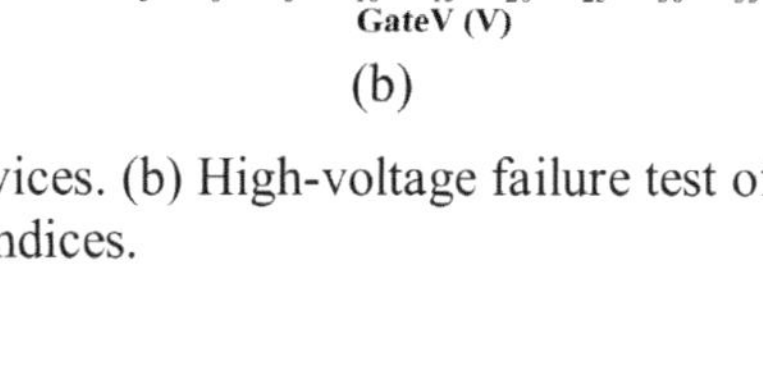

(b)

Figure 6. (a) SEM characterization of four-finger devices. (b) High-voltage failure test of devices with different indices.

To further investigate the factors influencing high voltage failure of top gate devices, we conducted high voltage failure tests on devices with varying parameters. We conducted experiments using single-finger devices with a width-to-length ratio (W/L) of 3 and a length (L) of 15 um, as well as multi-finger devices with indices of 2 and 4. The scanning electron microscopy (SEM) analysis in Figure 6(a) illustrates the characterization of the device with an index of 4, while Figure 6(b) presents the test results for all three devices. The test results demonstrate a linear correlation between the device failure current and the finger, with maximum single-finger currents of 25 uA, 32 uA, and 30 uA for devices with Fingers 1, 2, and 4 respectively. The results demonstrate that augmenting the finger can effectively enhance the failure current of the device.

4. Conclusion

The failure characteristics of back-gate CNTFET and top-gate CNTFET in the high-voltage field are discussed in this paper, with emphasis on the influence of channel size on device stability. The results demonstrate that by increasing the width and length of the channel, the breakdown voltage of the device can be adjusted within a certain range, highlighting the tradeoff between device performance and reliability during the design process. Additionally, when introducing a dielectric layer and gate metal, it is observed that the high field failure voltage of top-gate devices is significantly reduced. Furthermore, for multi-finger devices, an increase in device finger leads to a substantial rise in failure current which exhibits a linear relationship with the finger. These findings contribute to a deeper understanding of CNTFET's failure behavior while providing valuable insights for optimizing device design and improving reliability.

References

[1] Charlier J C, Blase X and Roche S, (2007) Electronic and transport properties of nanotubes. Reviews of modern physics, 79(2): p.677.
[2] Neto A C, Guinea F, Peres N M, Novoselov K S and Geim A K, (2009) The electronic properties of graphene. Reviews of modern physics, 81(1): p.109.
[3] Vijay M M, Kumar O P, Francis S A J, Stalin A D and Vincent S, (2024) Enhancing high-speed digital systems: MVL circuit design with CNTFET and RRAM. Journal of King Saud University-Computer and Information Sciences, p.102033.
[4] Anjanashree M R, Raja T and Reena Monica P, (2023) CNTFET-Based Gas Sensor with a Novel and Safe Testing Chamber Design. Smart Grids for Smart Cities Volume 2, pp.311-322.
[5] Valdez-Sandoval L M, Cid-Delgado A, Pacheco-Sánchez A, Aguilar M A E, Annamalai M, Schröter M and Ramírez-García E, (2023) A 5.2 GHz inductorless CNTFET-based amplifier design feasible for on-chip implementation. IEEE Transactions on Nanotechnology.
[6] Gupta M P, Behnam A, Lian F, Estrada D, Pop E and Kumar S, (2013) High field breakdown characteristics of carbon nanotube thin film transistors. Nanotechnology, 24(40): p.405204.
[7] Fikri M A, Pandey A K, Samykano M, Kadirgama K, George M, Saidur R, Selvaraj J, Abd Rahim N, Sharma K and Tyagi V V, (2022) Thermal conductivity, reliability, and stability assessment of phase change material (PCM) doped with functionalized multi-wall carbon nanotubes (FMWCNTs). Journal of Energy Storage, 50: p.104676.

Nonlinear damage dynamics modeling and characteristics simulation of bevel gear set

Jiao Hu[1], Lun Tang[2*], Xingshu Liu[1] and Jian Shen[3]

[1]Innovation Collaboration Center Hunan Automotive Engineering Vocational University Zhuzhou, China
[2]College of Vehicle Engineering Hunan Automotive Engineering Vocational University Zhuzhou, China
[3]School of Intelligent Manufacturing Engineering Changsha Vocational and Technical College Changsha, China

[*] Corresponding author's email: yvhygatb@163.com

Abstract: As a key core device for power transmission in the main reducer of a helicopter, the bevel gear set affects the flight performance of the helicopter directly. A nonlinear damage dynamic model of bevel gear set was constructed using the centralized parameter method. The inherent characteristics and nonlinear behavior characteristics of bevel gear sets were studied, and the time-frequency domain characteristics under different health states were simulated. The fault response transmission law and fault mechanism of the system were deeply revealed, which provides guidance and support for the optimization design of helicopter transmission system and the diagnosis algorithms development.

1. Introduction

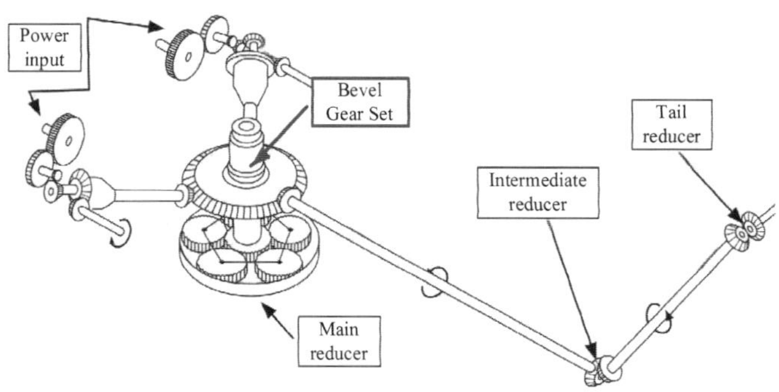

Figure 1. Structural diagram of the main transmission.

As a typical dual-use product, helicopters are widely used in fields such as medical rescue, maintenance, pest control, tourism, and sightseeing. The bevel gear set is widely used as the power input end of helicopter main reducers due to its advantages of high load-bearing capacity, high reliability, smooth operation, and long service life, as shown in Figure 1. The service environment of helicopters determines that the bevel gear set often operates under complex and variable working conditions, and the gear system is prone to fatigue faults. Once faults occur, they cannot be eliminated timely, which will seriously affect the stability and safety of helicopter flights. For revealing the faults mechanism of the bevel gear set, diagnosing early damage and faults, and improving the security and reliability of the helicopter transmission system, it is necessary to carry out research and diagnostic analysis on the damage dynamics modeling of the bevel gear set.

Many scholars have carried out research on the dynamic modeling methods of the bevel gear set. Liu et al. established a 12-degree-of-freedom nonlinear dynamic model including speed fluctuation and rotor unbalance excitation [1]; Gou et al. established a bending torsion-axial coupling model considering time-varying parameters [2]. Xu et al. introduced nonlinear support characteristics into the bevel gear set and analyzed the bifurcation and chaotic behavior by constructing an 8-degree-of-freedom dynamic model [3]; Li et al. believed that tooth side clearance could affect the consistency of impact on both sides of the tooth surface, resulting in asymmetric meshing stiffness on both sides of the bevel gear [4]. Molaie et al. established a nonlinear dynamic model for bevel gear sets that considered manufacturing error, tooth elasticity, and backlash [5].

The literature on dynamic modeling of bevel gear sets mentioned above mostly focuses on simulating the response characteristics, providing guidance for optimal design. There is little research on damage dynamics modeling for fault diagnosis needs. This article takes the bevel gear set of a helicopter as the research object, establishes a nonlinear damage dynamics model using the centralized parameter method, and conducts research on the inherent characteristics, nonlinear behavior characteristics, and fault characteristics.

2. Damage dynamic modeling of bevel gear set

Figure 2 shows the mechanical diagram of the spur bevel gear set. The driving gear and driven gear both contain four degrees of freedom for lateral, longitudinal, axial displacement, and axial rotation, namely x, y, z, θ. δ is cone angle, and k_m and c_m denote the meshing stiffness and damping, respectively. b is tooth-side clearance, and $e(t)$ is comprehensive transmission error.

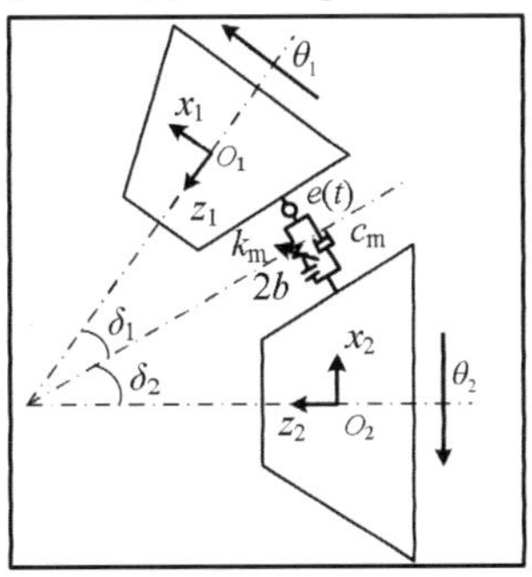

Figure 2. Model of bevel gear.

According to the geometric motion relationship, the dynamic transmission error can be written as

$$\begin{aligned}\lambda &= x_1 \sin\alpha \cos\delta_1 + y_1 \cos\alpha + z_1 \sin\alpha \sin\delta_1 \\ &+ R_1\theta_1 - x_2 \sin\alpha \cos\delta_2 - y_2 \cos\alpha \\ &+ z_2 \sin\alpha \sin\delta_2 - R_2\theta_2 - e(t)\end{aligned} \tag{1}$$

where R is the base circle radius.

So, the meshing force F_m can be written as [6]

$$F_m = k_m(t) f(\lambda) + c_m \dot{\lambda} \tag{2}$$

$$f(\lambda) = \begin{cases} \lambda - b, & \lambda > b \\ 0, & |\lambda| \le b \\ \lambda + b, & \lambda < -b \end{cases} \tag{3}$$

The dynamic equation of the bevel gear set could be obtained from Newton's theorem as follows

$$\begin{aligned} & m_1\ddot{x}_1 + c_{\mathrm{m}}\dot{\lambda}\sin\alpha\cos\delta_1 + k_{\mathrm{m}}(t)f(\lambda)\sin\alpha\cos\delta_1 \\ & \quad + c_{\mathrm{bx1}}\dot{x}_1 + k_{\mathrm{bx1}}x_1 = 0 \end{aligned} \tag{4}$$

$$m_1\ddot{y}_1 + c_{\mathrm{m}}\dot{\lambda}\cos\alpha + k_{\mathrm{m}}(t)f(\lambda)\cos\alpha + c_{\mathrm{by1}}\dot{y}_1 + k_{\mathrm{by1}}y_1 = 0 \tag{5}$$

$$\begin{aligned} & m_1\ddot{z}_1 + c_{\mathrm{m}}\dot{\lambda}\sin\alpha\sin\delta_1 + k_{\mathrm{m}}(t)f(\lambda)\sin\alpha\sin\delta_1 \\ & \quad + c_{\mathrm{bz1}}\dot{z}_1 + k_{\mathrm{bz1}}z_1 = 0 \end{aligned} \tag{6}$$

$$J_1\ddot{\theta}_{z1} + c_{\mathrm{m}}\dot{\lambda}R_1 + k_{\mathrm{m}}(t)f(\lambda)R_1 = T_1 \tag{7}$$

$$\begin{aligned} & m_2\ddot{x}_2 - c_{\mathrm{m}}\dot{\lambda}\sin\alpha\cos\delta_2 - k_{\mathrm{m}}(t)f(\lambda)\sin\alpha\cos\delta_2 \\ & \quad + c_{\mathrm{bx2}}\dot{x}_2 + k_{\mathrm{bx2}}x_2 = 0 \end{aligned} \tag{8}$$

$$m_2\ddot{y}_2 - c_{\mathrm{m}}\dot{\lambda}\cos\alpha - k_{\mathrm{m}}(t)f(\lambda)\cos\alpha + c_{\mathrm{by2}}\dot{y}_2 + k_{\mathrm{by2}}y_2 = 0 \tag{9}$$

$$\begin{aligned} & m_2\ddot{z}_2 + c_{\mathrm{m}}\dot{\lambda}\sin\alpha\sin\delta_2 + k_{\mathrm{m}}(t)f(\lambda)\sin\alpha\sin\delta_2 \\ & \quad + c_{\mathrm{bz2}}\dot{z}_2 + k_{\mathrm{bz2}}z_2 = 0 \end{aligned} \tag{10}$$

$$J_2\ddot{\theta}_{z2} - c_{\mathrm{m}}\dot{\lambda}R_2 - k_{\mathrm{m}}(t)f(\lambda)R_2 = -T_2 \tag{11}$$

where m and J represent the gear mass and moment of inertia, respectively.

3. Simulation analysis of response characteristics of bevel gear set

3.1 Natural frequency analysis of linear system

Mass and stiffness are crucial dynamic parameters in solving the natural frequency of gear systems, which will greatly affect the dynamic response characteristics. The basic parameters of the bevel gear set are listed in Table 1, and the quality parameters are obtained by constructing a three-dimensional model, as shown in Table 2. Based on the previous research fruits [7], the stiffness curves of gear pairs under different health states are evaluated as shown in Figure 3, where q represents crack size.

Table 1. Basic parameters of bevel gear set.

Parameters	Driving gear	Driven gear
Number of teeth	18	36
Module m_{d} (mm)	4.25	4.25
Pressure angle α (°)	20	20
Tooth width L (mm)	27.66	27.66
Bearing support stiffness (N·m^{-1})	$k_{\mathrm{bx1}} = k_{\mathrm{by1}} = k_{\mathrm{bx2}} = k_{\mathrm{by2}} = 1.75\times10^{8}$ $k_{\mathrm{bz1}} = k_{\mathrm{bz2}} = 2.27\times10^{8}$	
Bearing support damping (N·s·m^{-1})	$c_{\mathrm{bx1}} = c_{\mathrm{by1}} = c_{\mathrm{bx2}} = c_{\mathrm{by2}} = 933$ $c_{\mathrm{bz1}} = c_{\mathrm{bz2}} = 2615$	

Table 2. Mass parameters of bevel gear set.

Parameters	Driving gear	Driven gear
Mass (kg)	1.313	3.679
Moment of inertia J_{d} (kg·m^2)	7.672×10^{-4}	4.677×10^{-3}
Polar moment of inertia J_{p}/(kg·m^2)	8.270×10^{-4}	7.340×10^{-3}

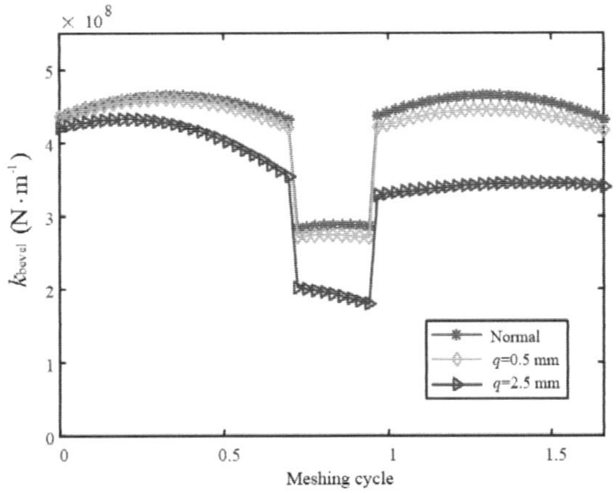

Figure 3. Mesh stiffness curves.

When the nonlinear time-varying system of the bevel gear set is degraded into a linear system, the natural frequencies under normal conditions are solved, which are shown in Table 3. Obviously, the first-order natural frequency is 0 due to the rigid body vibration in the direction of torsion.

Table 3. Natural frequencies of linear system.

Order	Natural frequencies ω_i /(rad·s^{-1})	Dimensionless natural frequencies ω_i^r
1	0.00	0.000
2	17783.82	0.354
3	19488.79	0.388
4	22101.89	0.440
5	27487.34	0.547
6	32650.42	0.650
7	37170.24	0.739
8	66007.03	1.313

3.2 Nonlinear dynamic characteristics analysis

Sweeping frequency excitation is used in the bevel gear set model using a wide input speed. Assuming that the comprehensive transmission error follows a sine function, and $e(t) = E_b \sin(2\pi f_{bm} t + \varphi_0)$ [7], E_b, f_{bm}, φ_0 are the error amplitude, meshing frequency, and initial phase, respectively. The tooth side clearance is 20 μm and E_b is 50 μm. The sampling frequency is set to 500 calculation steps in a meshing cycle and the simulation duration is a total of 600 meshing cycles. The RMS of y_1 is extracted to plot the amplitude frequency response curve, which is shown in Figure 4.

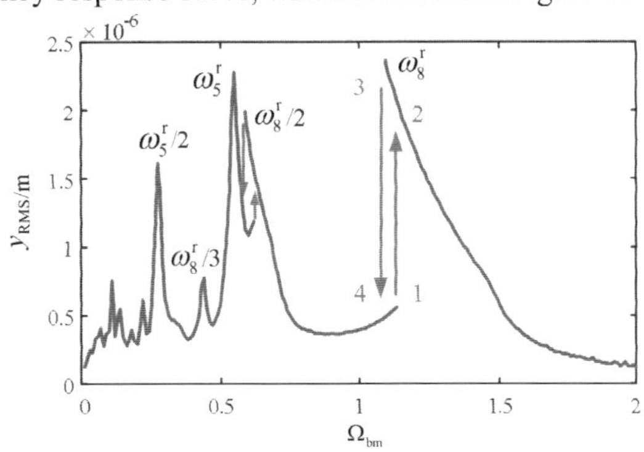

Figure 4. Amplitude frequency response characteristics.

As shown in Figure 4, the bevel gear set exhibits obvious nonlinear behaviors such as multi-value, jumping, and superharmonic resonance. Due to the characteristics of superharmonic resonance and parametric excitation, the amplitude-frequency response curve exhibits multiple resonance peaks in the range of $\Omega_{bm} > 0.25$, and there are dense little amplitude resonance peaks in the low-speed region ($\Omega_{bm} < 0.25$). Among them, the important resonance peaks appear near $\Omega_{bm} = 1.2$ and $\Omega_{bm} = 0.55$, corresponding to the excitation of the 8th and 5th natural frequency resonances, namely ω_8^r and ω_5^r, respectively. When the excitation frequency $\Omega_{bm} = 1.2$ which is near the torsional natural frequency of the linear system, it is most affected by the tooth side clearance, resulting in the obvious jumping phenomenon. In addition, superharmonic resonances of the 8th and 5th natural frequencies are excited at $\Omega_{bm} = 0.65$ and $\Omega_m = 0.27$.

To further explore the dynamic response details under sweep frequency conditions, the time-domain response, phase diagram, Poincaré mapping, frequency spectrum at the resonance peak near $\Omega_{bm} = 0.65$ are plotted as follows.

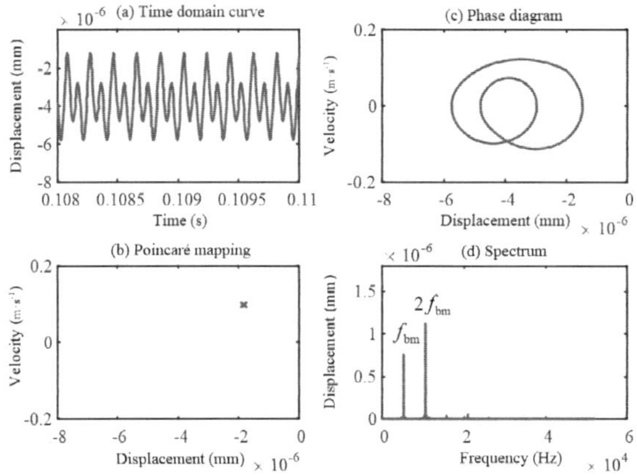

Figure 5. Nonlinear behavioral characteristics of the driving gear when $\Omega_{bm} = 0.65$.

Figure 5 shows that the time curve is clearly not a single periodic motion, and the Poincaré mapping shows a single point. However, there are two loops in the phase diagram, which means that it moves back and forth twice within a cycle and $2f_{bm}$ is the main component of the spectrum. So, it can be concluded that nonlinear factors and time-varying parameters have caused the system to exhibit superharmonic resonance. From the Figure 6, it can be seen that the resonance peak is a superharmonic resonance near $\Omega_{bm} = 0.27$.

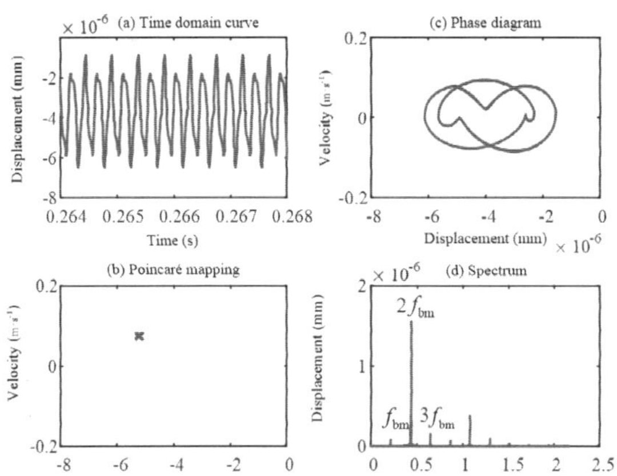

Figure 6. Nonlinear behavioral characteristics of the driving gear when $\Omega_{bm} = 0.27$.

3.3 Characteristics analysis of fault response

The meshing stiffnesses of three different healthy states are substituted into the dynamic equation, and the dynamic responses are calculated. Assuming the driving gear speed is 6000 r/min, and the external load of the driven gear is 715 $\mathrm{N \cdot m}$, the operating states of the bevel gear set are listed in Table 4.

Table 4. Operating parameters of bevel gear set.

Parameter	Meshing frequency f_{bm}	Driving gear frequency f_{b1}	Driven gear frequency f_{b2}
Value (Hz)	1800	100	50

The time-domain curves and frequency spectrums of the driving gear acceleration response $\ddot{y}_1$ are shown in Figure 7. The time-domain curve under normal conditions presents a simple periodic motion, and the frequency spectrum is composed of the meshing frequency and higher-order harmonics of the bevel gear set. When the crack size is 0.5 mm, the effect of the crack on time-domain vibration is not significant, and the frequency spectrum is very similar to the normal state. Therefore, when the crack is in the early stage, it is hard to reflect the damage level from the time-frequency domain characteristics.

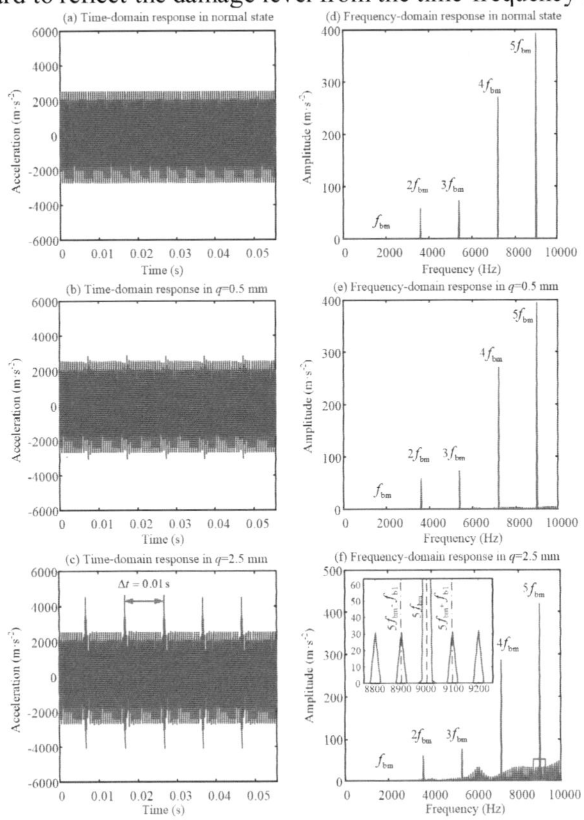

Figure 7. Effect of crack propagation on $\ddot{y}_1$.

When the crack extends to 2.5 mm, there are obvious periodic pulses in the time-domain curve, and there are dense sidebands with larger amplitudes near the meshing frequencies. The interval Δt of the time-domain fault pulse is 0.01 s, corresponding to a pulse frequency 1/Δt=100 Hz, equal to the driving gear frequency. Figure 7(f) shows a locally enlarged frequency band within a rectangular box. The modulation sidebands ($5f_{bm} - f_{b1}$, $5f_{bm} + f_{b1}$) labeled on both sides of $5f_{bm}$ have an 100 Hz interval, which is consistent with the characteristic reflected by the time-domain response.

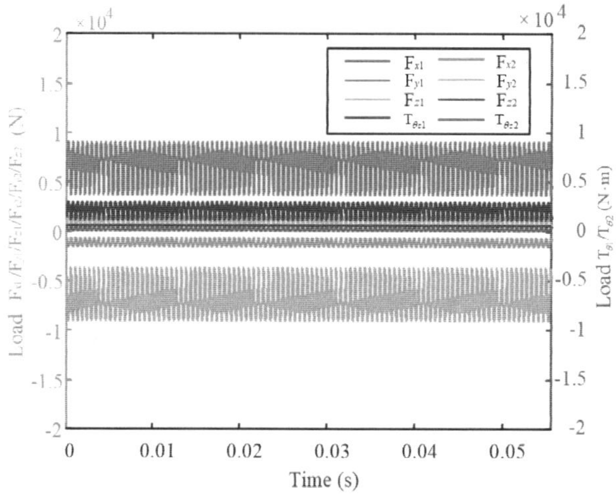

Figure 8. The meshing force of bevel gear set in normal state.

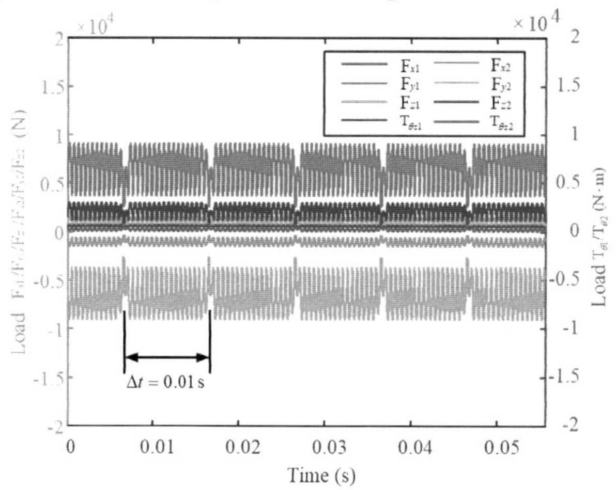

Figure 9. Meshing force of bevel gear set when q=2.5 mm.

Figure 8 and Figure 9 show the meshing forces of the bevel gear set under different states.

In summary, the damage dynamics model of the bevel gear set can reflect the influence of crack propagation on the vibration responses. The time-frequency domain characteristics of the simulated acceleration responses and meshing forces indicate the occurrence of tooth fault in the driving gear.

4. Conclusion

Aiming at the bevel gear set of a helicopter main reducer, this paper establishes a bending-torsion-axial coupling damage dynamics model using the lumped-parameter method.

1) The intrinsic characteristics and nonlinear characteristics are analyzed, and the nonlinear system may exhibit multi-value, jumping, superharmonic response, and parametric excitation phenomena under different operating states.

2) The influence of crack expansion on the failure response characteristic is investigated. Local tooth faults will cause periodic impacts of vibration responses, and the impact frequency is equal to the rotation frequency of the faulty gear.

3) The simulation analyses show that the damage model can effectively reflect the health state, reveal the failure mechanism and response conduction mechanism of the bevel gear set, and provide theoretical and simulation data support for the development of helicopter fault diagnosis algorithms.

Acknowledgment
This work is financially supported by the Natural Science Foundation of Hunan Province of China (Grant No. 2024JJ8096), the School Level Scientific Research and Innovation Team, and the Education Department of Hunan Province of China (Grant No. 23B1141).

References

[1] Liu J, Zhang H, Zhai J, et al. Nonlinear dynamic analysis of the central bevel gear transmission system in aero engine subjected to internal and external excitation [J]. Proceedings of the Institution of Mechanical Engineers, Part C: Journal of Mechanical Engineering Science, 2022, 236(11): 5940-5953.

[2] Gou Xiangfeng, Li Guyu, Zhu Lingyun. Dynamic characteristics of a straight bevel gear drive system considering multi-state meshing and time-varying parameters [J]. Mechanism and Machine Theory, 2022, 171: 104779.

[3] Xu J, Zeng F, Su X. Coupled bending-torsional nonlinear vibration and bifurcation characteristics of spiral bevel gear system [J]. Shock and Vibration, 2017: 6835301.

[4] Li Y, Li G, Zheng L. Influence of asymmetric mesh stiffness on dynamics of spiral bevel gear transmission system [J]. Mathematical Problems in Engineering, 2010: 124148.

[5] Molaie M, Samani F S, Zippo A, et al. Spiral bevel gears: nonlinear dynamic model based on accurate static stiffness evaluation [J]. Journal of Sound and Vibration, 2023, 544: 117395.

[6] Liang X, Zuo M, Hoseini M R. Vibration signal modeling of a planetary gear set for tooth crack detection [J]. Engineering Failure Analysis, 2015, 48: 185-200.

[7] Jiao Hu, Niaoqing Hu, Yi Yang, Lun Zhang, Guoji Shen. Nonlinear dynamic modeling and analysis of a helicopter planetary gear set for tooth crack diagnosis [J]. Measurement, 2022. 198.

Multipeak ultraviolet lasing based on F-P mode from in-situ ZnO nanowire

Ying Zhang and Ruibin Liu*

School of Physics, Beijing Institute of Technology, Beijing 100081, PR China

*Corresponding author's e-mail: liusir@bit.edu.cn

Abstract. High-quality ultraviolet singlemode and multimode lasing from zinc oxide nanowires are highly desirable. In this paper, we achieve high-Q lasing oscillation in vertically grown Zinc oxide (ZnO) microwires. The broad gain range reaches 13 lasing modes under high-energy excitation. We observe multiple peak emissions from the Fabry-Perot (F-P) lasing mode. In contrast, nanorod arrays with low aspect ratios exhibit amplified spontaneous emission (ASE) due to significant diffraction losses. Our work demonstrates that high-quality ZnO nanowires with an appropriate aspect ratio realize high-Q ultraviolet multipeak lasing based on F-P mode. These findings will benefit the design and fabrication of nanolasers based on nanowires.

1. Introduction

Strong coherent light-emitting nanostructures are always a focus for their wide applications, including optoelectronic devices and micro lasers [1-3]. ZnO nanostructures are potential gain media for ultraviolet lasers with large exciton combination energy (60 meV), providing a new route to generate intense nanoscale ultraviolet coherent light sources. Lasing behaviors differ with morphology due to microcavity-determined emission mechanisms under in-situ excitation. Under higher excitation energy, lasing is dominated by electron-hole plasma (EHP), widely reported in many structures, and leading to wide linewidth emission. EHP is easily formed in small nanowires due to 2-D confinement [4]. Meanwhile, EHP lasing with broader linewidth in larger, higher-quality nanowire cavities results in multimode lasing from high-Q oscillation. Optical micro-photoluminescence (PL) studies reveal emission peaks correlated with F-P modes achieve high Q-factor ($Q \approx 10^3$) monocrystalline nanowires, inducing much narrower and sharper emission. A redshift and broadening of emission under higher pumping fluence demonstrate EHP dominates the gain at room temperature. EHP-induced lasing has been documented across a range of ZnO structures, encompassing microcolumns and nanowires [5]. Research into the lasing properties of densely arranged ZnO nanowires has highlighted the potential of high-quality single-crystal ZnO nanowires as optimal nanolaser cavities, fulfilling the roles of both gain medium and resonator. The inaugural demonstration of laser emission from an individual nanowire occurred in 2001, with subsequent studies deepening the comprehension of the fundamental mechanisms involved [6].

In the majority of studies, nanowires have been extracted from native sapphire substrates and then repositioned onto empty Si/SiO_2 wafers for individual characterization through mechanical or ultrasonic methods. Given the geometric arrangement, the nanowires interface with the substrate along their entire length, causing a reduced refractive index differential and elevated cavity losses. This scenario makes the observation of F-P modes less likely. Conversely, it provides favorable conditions for the easy attainment of WGM resonances.

In this paper, we utilized vertically grown ZnO microwires to achieve high-Q lasing oscillation. The broad gain range allowed up to 13 lasing peaks to occur under high-energy excitation. For the first time, we observed multiple peak emissions from the F-P lasing mode rather than the WGM. In contrast, the lasing behavior from low aspect-ratio nanorod arrays exhibited only ASE due to significant diffraction losses. Our work demonstrates that high-Q ultraviolet multipeak lasing can be realized from high-quality ZnO nanowires with an appropriate aspect ratio, based on the F-P mode. These findings will benefit the design and fabrication of nanolasers based on nanowires.

2. Experimental section

Vertical ZnO nanowires were prepared using low-density Au nanoparticles on *a*-plane sapphire substrate. Initially, the sapphire substrate is immersed in a 10% aminopropyltriethoxysilane (APTES) solution for 10 s and in a colloidal solution containing gold (Au) nanoparticles as catalysts (Ted Pella, with a diameter of 80 nm). Subsequently, the substrate is rinsed with deionized (DI) water, placed in a mixture with graphite to ZnO ratio of 1:1, and heated to 900 ° C at a constant Ar flow rate for 30 min. This process results in the growth of nanowires with a density ranging from 4×10^4 to 2×10^7 nanowires/cm^2, which correspond to inter-nanowire distances between 50 and 2 μm, respectively. The gain in nanowire is proportional to the length of the nanowire, so we get a longer nanowire up to 50 μm to get a large gain. In contrast, low aspect ratio ZnO nanorods were synthesized through metal-organic chemical vapor deposition (MOCVD). The ZnO layers were epitaxially deposited via MOCVD, employing iso-propanol, i-PrOH ($(CH_3)_2CHOH$) and diethylzinc, DEZn ($(C_2H_5)_2Zn$), as precursors for oxygen and zinc. C-plane sapphire substrates served as the foundation for the epitaxial growth. T_{growth} was established at 380°C. T_{DEZn} and T_{PrOH} precursors were kept at 17°C and 20°C, respectively. The combined gas flow rate was 5.5 slm, and the flow rate specifically for the DEZn gas was 50 umol/min at a pressure of 400 mbar and a V/II ratio of 36. The layers were grown at a rate of 0.95 um/h, achieving a thickness of 1.4 um.

The scanning electron microscope (SEM) with the model Hitachi S-5200 is used to observe the morphology of ZnO. Micro PL and PL dynamics of ZnO were obtained using a mode-locked Ti-sapphire laser. As shown in Figure 1, the UV beam was divided into two parts after passing through the attenuator and beam splitter. One beam served as an energy reference measured by the energy meter (0~1000 μJ/cm^2). The other beam of light acted on a point with a diameter of about 50 μm in a 45° direction through the X10 microscope objective and the PL excited by it was obtained using the objective lens in a confocal microscopy system, which was collected and fed into a monochromator using a CCD detector analysis, showing the spectral resolution of 0.065 nm.

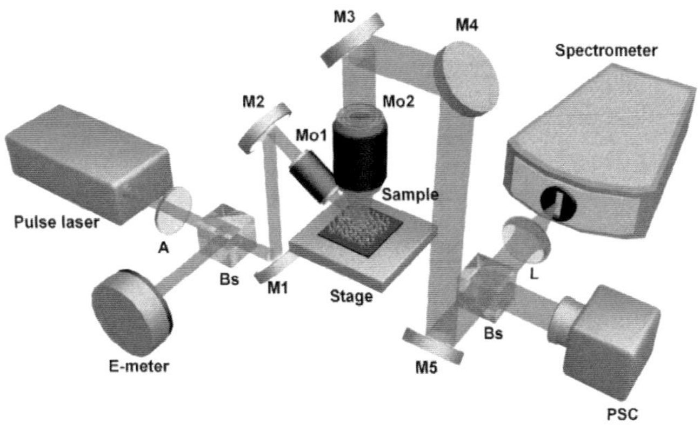

Figure 1. Photoluminescence (PL) spectra acquisition system of ZnO nanowires. A: Attenuator; Bs: Beam splitter; M1-M5: Mirror; Mo1-Mo2: Microscopic objective; L: Convex lens; E-meter: Energy meter; PSC: Photon-counting Streak Camera; Sample: ZnO nanowires.

3. Results and discussion

Here, we optimized the growth conditions of ZnO and obtained a vertical ZnO nanowire array with a spacing between 20-50 μm. This arrangement is beneficial for the optical characterization of a single vertical nanocavity by focusing light spots smaller than the nanowire spacing and confocal microscopy with UV laser scanning.

At ambient temperature, a study was conducted to examine the optical behavior of ZnO nanorods when subjected to the impact of intense fs pulsed laser irradiation. The SEM image inset in Figure 2(a) depicts regular ZnO nanorods, which measure approximately 2 μm in length and nearly 2 μm in diameter. The corresponding emission spectra, illustrated in Figure 2(a), vary with different excitation fluences. Notably, a broad UV spontaneous emission spectrum is observed at a relatively low excitation level, featuring a full width at half maximum (FWHM) of approximately 14 nm. The relationship between pumping fluence and the intensity of the emission spectra is depicted in Figure 2(b). Beyond the threshold pumping fluence of 60 μJ/cm², the emission spectra in Figure 2(a) exhibit sharp peaks with linewidths less than 0.4 nm, and the linewidths of the emission spectra become even sharper as the pumping fluence increases. Moreover, the emission peak wavelength does not change with pumping fluence but remains constant at 370 nm, indicating the occurrence of near-ultraviolet band-edge amplified spontaneous emission of ZnO nanorods. The emission peak at 370 nm also demonstrates the quality of nanowires is high. If there are many impurities in ZnO nanowires, more trap-bound states-related bands will appear near the band gap, making the gap narrower. Therefore, the band edge emission peaks will redshift to around 380 nm, even 390 nm. Figure 2(c) illustrates the decay curve of the ZnO nanorods under different pumping fluence. Each instance of blue emission is observed at a pumping fluence of 31 μJ/cm², a value that is below the threshold required for stimulated emission. A single exponential decay model provides an excellent fit, with a characteristic time constant of approximately 1 ns. For the red curves, the pumping fluence is just at around the threshold at 59 μJ/cm², and the lifetime becomes a little fast to around 0.8 ns, illustrating some up-conversion carriers recombine at the same time and coherent light appears. As the pumping fluence achieves 109 μJ/cm², the decay curve can be fitted well with two componential functions with the time constant of 0.2 ns and 0.7 ns, respectively. The fast part is the obvious indicator of the coherent light emission occurrence. The emission without multipeak appearance also illustrates that the multiple oscillation is not formed. Only ASE occurs due to the small size cavity-induced large diffraction loss. Limited 2 μm length also means large mode spacing, thus can only accommodate a limited number of longitudinal modes inner such nanowire cavities.

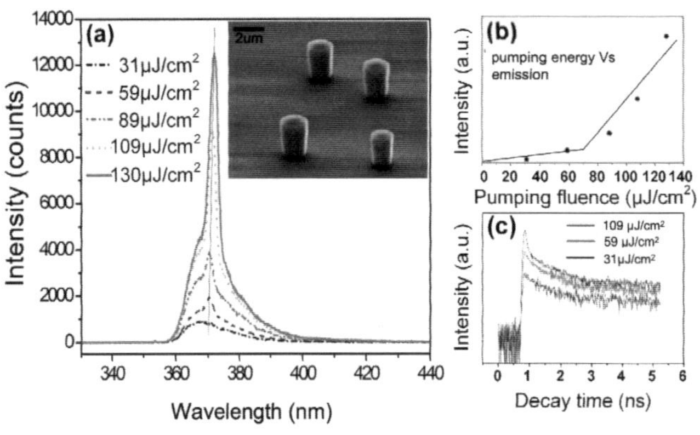

Figure 2. (a) The emission spectra for a ZnO nanorod are captured across a spectrum of excitation fluences, with the pump fluence varying from 31 to 130 μJ/cm². The inset presents an enlarged SEM image of the ZnO nanorods; (b) The graph displays light-intensity curves, which are optically excited at room temperature, corresponding to the near-band-edge emission of ZnO nanorods; (c) The photoluminescence decay time of the ZnO nanorods is measured at excitation intensities of 122, 160 and 391 μJ/cm².

Optical gain and lasing behavior are important to realize the application of ZnO nanowires for laser diodes and nanolasers. The supplementary SEM image inset in Figure 3(a) depicts ZnO nanowires, which exhibit a length of approximately 50 μm and a diameter close to 5 μm. Figure 3(a) displays the emission spectra as a function of power, which corresponds to the excitation energy respectively. The PL spectra from the nanowire under low excitation are characterized by a broad spontaneous emission band (the left inset in Figure 3(a)). Upon further increasing the pumping fluence, the concentration of charge carriers rises to a more elevated state, promoting the creation of electron-hole pairs EHP. The amplified optical emission within an optically confined microcavity possesses sufficient gain to compensate for the optical losses [7]. Subsequently, multiple lasing modes were observed within these nanowires, as depicted in Figure 3(a). Initially, the dominant mode corresponds to the peak position of ASE, where the optical gain is maximized, overcoming the optical losses along the c-axis of the ZnO nanowires. It is widely recognized that the emission band widen, and the emission peak are the principal features associated with the recombination of EHP that occurs in redshift [8]. As a result, an increasing number of modes gain the ability to counteract the optical losses and initiate lasing action between the two end facets by increasing the UV-ASE band and the pumping fluence. As shown in Figure 3, there exist more modes with up to 13 peaks under higher excitation energy in Figure 3(a) (iii). Similarly, the ASE wave packet takes place redshift, causing the most intense mode to move towards the lower energy side. Comparing scenarios depicted in Figure 3(a) (ii) with those in (i), the observed behavior aligns with the expectation. However, the case presented in Figure 3(a) (iii) exhibits diversity. The variations in the intensity of distinct lasing modes can be ascribed to the pronounced susceptibility of F-P modes to the surface characteristics and the purity level inside the microcavity [9]. Each mode in the nanowire has its own route along the nanowire. Different route has different losses, which will influence the emission intensity of each mode. These will induce the intensity fluctuation of each mode. The observation of multiple lasing modes indicates that the nanowire end facets possess high quality. Mastery of the optical gain and stimulated emission mechanisms is essential for the effective design and enhancement of photonic materials and devices [8].

Figure 3(b) shows the intensity of integrated spontaneous emission and lasing peaks from the edge-emission of ZnO nanorods as a function of the pumping fluence. From the light-light curve of stimulated emission, a superliner increase occurs at a pump intensity of about 60 μJ/cm^2, which is the threshold pumping fluence.

As depicted in Figure 3(c), upon exceeding the threshold energy, single-mode lasing emission is initially observed. FWHM of UV spectra rapidly narrows down to a minimum of 0.2 nm. The lasing peak in the central position of the spontaneous emission band, which is maximum gain, only in which the mode can exceed the loss, thus allowing the lasing to firstly occur only in this zone of gain, while other modes are still lower than loss. The output intensity increases rapidly in only one peak, and with the increase of the pumping fluence, generally, other peaks gradually appear.

As is known, vertically oriented nanowires utilize the end-facets as a F-P cavity, with boundary conditions requiring an integer number of half-wavelength standing waves. Consequently, the spacing of the F-P modes is approximately given by the expression in Equation (1):

$$\Delta\lambda = \frac{\lambda^2}{\left[2L\left(n - \frac{\lambda dn}{d\lambda}\right)\right]} \tag{1}$$

where $\Delta\lambda$ is mode spacing. n (2.4) and dn/dλ (-0.015 nm^{-1}) at λ=396.436 nm and $\Delta\lambda$ are estimated at 0.164 nm L =57μm, which aligns with the dimensions ascertained through SEM. The phenomenon of bandgap renormalization within semiconductor nanowires precipitates a redshift in the lasing modes and an enhanced amplification of wavelengths in the longer spectrum under conditions of elevated excitation pump dosage [10]. The modal quality factor is a pivotal parameter in the design of nanowire lasers, reflecting the confinement of the guided mode within the nanowire cavity. High Q means the quality of the nanowire is perfect, including very smooth end-facets. Utilizing the quality factor Q=λ /Δλ, Q of the cavity under this excitation can get 396.436/ (396.524-396.36) =2417. This indicates that the vertically aligned ZnO nanorods can achieve high optical quality.

Journal of Physics: Conference Series **2842** (2024) 012054 doi:10.1088/1742-6596/2842/1/012054

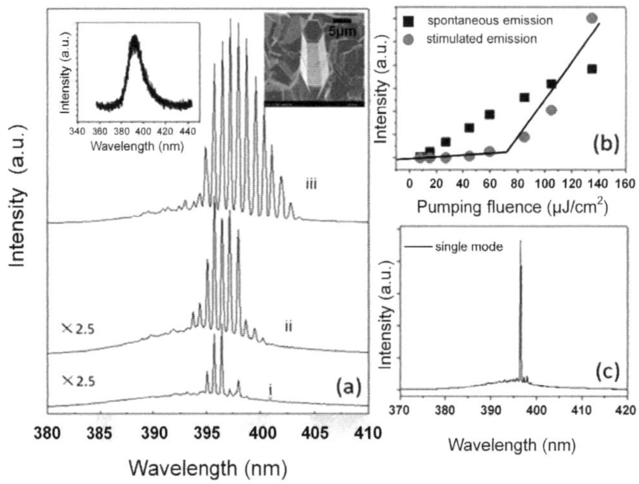

Figure 3. (a) The emission spectra of ZnO nanowires at excitation intensities of 225, 285, and 375 $\mu J/cm^2$ are presented. The left inset shows the PL spectrum of ZnO nanorods at low pumping fluence, while the right inset displays SEM image of single ZnO nanorod; (b) Intensity of integrated and stimulated spontaneous emission from the edge-emission of ZnO nanowires versus the optical excitation intensity; (c) Single mode lasing emission spectrum from the ZnO nanowires under excitation at 80 $\mu J/cm^2$.

4. Conclusion

In summary, the aligned ZnO nanowires successfully synthesized vertically. With the crystalline facets of the nanowires acting as laser cavity mirrors, 13-fold multimode room-temperature lasing behavior with Q factor of 2417 from F-P mode upon EHP mechanism was realized in such in-situ ZnO nanowire for the first time, which was proved by the relation between mode spacings and cavity lengths in photoluminescence spectra. While the lasing behavior from low aspect-ratio nanorod arrays only exhibits ASE due to significant diffraction losses. These results suggest that the individual in-situ ZnO nanowires can act as high Q-factor optical microcavity to realize F-P mode nanolaser, which is the working mode of general solid-state laser, such Nd:YAG laser. It demonstrates under proper cavity conditions, the nanolaser can be easily fabricated and widely applied in miniaturized light sources and optoelectronic devices.

References

[1] Consoli A, Caselli N, Lopez C. (2022) Electrically driven random lasing from a modified Fabry-Perot laser diode. Nat. Photonics, 3:16-22.

[2] Hjort F, Enslin J, Cobet M, et al. (2021) A 310 nm optically pumped AlGaN vertical-cavity surface-emitting laser. ACS Photonics, 2021, 8: 135-141.

[3] Schmiedeke P, Doganlar C, Jeong H W, et al. (2024) Low-threshold single ternary GaAsSb nanowire lasers emitting at silicon transparent wavelengths. Appl. Phys. Lett.,7:124-128.

[4] Daniel J G, Michael C M, Adrian N, et al. (2010) Whispering Gallery Mode lasing from zinc oxide hexagonal nanodisks. ACS Nano,4: 3270-3276.

[5] Sekar K, Doineau R, Mayarambakam S, et al. (2024) Control of ZnO nanowires growth in flexible perovskite solar cells: A mini-review. Heliyon,10:1-8.

[6] Yin L, Chun L, Huang Y, et al. (2020) Plasmonic Nanolasers in On-Chip Light Sources: Prospects and Challenges. ACS Nano,14: 14375-14390.

[7] Liu Z, Hu M, Du J, et al. (2021) Subwavelength-polarized quasi-two-dimensional perovskite single-mode nanolaser. ACS Nano, 2021, 15(4): 6900-6908.

[8] Sun L, Dong H, Xie W, et al. (2010) Quasi-whispering gallery modes of exciton-polaritons in a ZnO microrod. Opt Express, 18: 15371-15376.
[9] Zhang C, Zhu G, He S, et al. (2020) Side-mode suppression in ultraviolet quasi-semicircle microlaser cavity. Mod. Phys. Lett. B, 34:1-6.
[10] Gargas D J, Toimil-Molars M E, Yang P. (2008) Imaging single ZnO vertical nanowire laser cavities using UV-laser scanning confocal microscopy. J. Am. Chem. Soc.131:1-8.

Journal of Physics: Conference Series **2842** (2024) 012055 doi:10.1088/1742-6596/2842/1/012055

Preparation of carbon nanotube composite solder and analysis of the thermal properties

Shizhong Cui, Wenfan Miao, Yang Ding and Bin Ju*[*]

College of Electrical Engineering and Automation, Anhui University, Anhui Hefei, 230601, China

[*] Corresponding author's email: jubin2016@ahu.edu.cn

Abstract. The solder layer constitutes a crucial component of IGBT power modules. It endures significant thermal stress during operation, and the magnitude of this stress correlates with the coefficient of thermal expansion. To prevent fatigue damage to the solder layer during operation, it is necessary to reduce the solder's coefficient of thermal expansion, thereby alleviating the impact of thermal stress. In this paper, the method of adding carbon nanotube powder to solder is proposed to solve this problem. Firstly, the relationship between thermal stress and the coefficient of thermal expansion is analyzed, and then the mechanical ball milling method is employed to prepare composite solder with a 0.05% carbon nanotube content. Scanning electron microscopy observation of the prepared composite solder reveals homogeneous mixing and uniform distribution of carbon nanotubes on the solder particle surfaces. Furthermore, thermal expansion coefficient tests were conducted on the sintered carbon nanotube composite solder, revealing a 16.2% reduction in the coefficient due to the incorporation of carbon nanotubes.

1. Introduction

The solder layer is a crucial component of the IGBT power modules, providing support and electrical connections to the chips while serving as a significant heat dissipation channel [1]. Compared to other components, the solder layer is particularly vulnerable during high-temperature operation. Due to the differences in thermal expansion coefficients, various layers of materials in the module undergo deformation and compression, generating thermal stress that leads to fatigue damage in the solder layer [2]. The primary forms of fatigue damage are solder cracking and warping, with severe warping potentially resulting in chip failure. When solder cracks occur, air fills the interior of the solder layer. However, the air has a lower thermal conductivity, increasing the module's thermal resistance, affecting heat transfer, and causing a further rise in chip junction temperature, thus creating a vicious cycle [3,4].

The addition of carbon nanotubes can effectively enhance the ultimate tensile strength, shear strength, and creep resistance of the solder [5,6]. Additionally, carbon nanotubes can mitigate chip warping issues caused by excessive thermal stress between the chip and solder [7]. However, with the increase in carbon nanotube content, the mechanical properties of the solder initially increase and then decrease [8]. Therefore, controlling the carbon nanotube content is crucial during the preparation of carbon nanotube composite solder. Currently, the common method for preparing the composite solder is mechanical mixing [9]. Therefore, this paper utilizes the mechanical ball milling method to incorporate carbon nanotube powder into the solder, aiming to reduce the coefficient of thermal expansion (CTE). Furthermore, the paper compares the effects of solder layers with different coefficients of thermal expansion on the maximum thermal stress of the IGBT encapsulation layers through simulation.

2. IGBT thermal stress analysis and simulation

Each layer of the materials inside the IGBT experiences temperature changes during operation, and the intrinsic parameters of the materials also affect the magnitude of thermal stress. According to the strain-stress theory, the temperature and stress jointly produce the strain, with the thermal stress formula as follows [10]:

$$\sigma = E\alpha\Delta T \tag{1}$$

where α represents the CTE, T denotes the temperature, E stands for Young's modulus, and σ represents the thermal stress. Equation 1 reveals that the magnitude of the thermal stress is proportional to the CTE. When the CTE and ΔT increase, the thermal stress also increases.

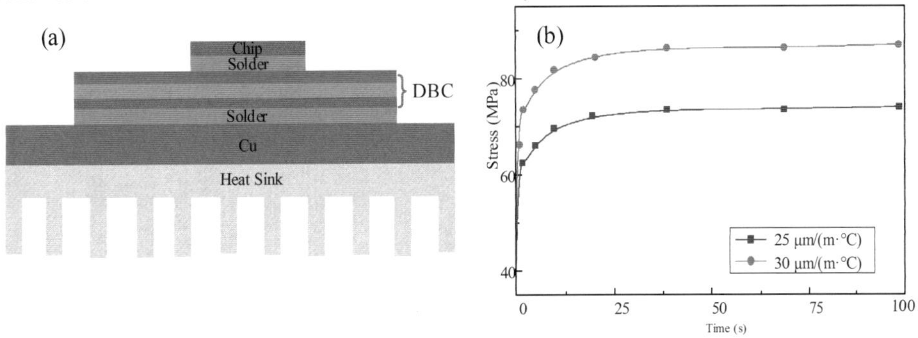

Figure 1. (a) The simulation structure of IGBT. (b) Maximum thermal stress in the solder layers of IGBT with two different coefficients of thermal expansion.

During the simulation process, the chip thermal source is set to 100 W. The solder layer material is set as Sn-Ag. Due to the IGBT's heat transfer occurring from top to bottom, dissipating through the bottom substrate, the thermal boundary conditions are set as three-sided thermal insulation, with a bottom heat transfer coefficient set to 2000 $W \cdot m^{-1} \cdot K^{-1}$. Under high power density operating conditions, the IGBT chip generates considerable heat, causing its junction temperature to continuously rise. Moreover, the temperature experienced by the solder layer in contact with the chip also increases continuously. The tensile strength of Sn-Ag is 41 MPa. When subjected to prolonged thermal stress exceeding this strength, it is prone to damage.

Figure 1 (a) is the simulation structure of IGBT. Figure 1 (b) depicts the maximum thermal stress of two solder layers with different coefficients of thermal expansion calculated through simulation. It can be observed that the maximum thermal stress of the solder layer with a CTE of 25 μm/(m·°C) is 74.03 MPa after a period of operation, which is 16% lower than that of the solder layer with a CTE of 30 μm/(m·°C).

3. Preparation of carbon nanotube composite solder

In this paper, we prepared composite solder with a carbon nanotube content of 0.05%. The preparation process is shown in Table 1. Initially, carbon nanotube powder was dispersed in ethanol at a ratio of 1:500, and the mixture was treated with an ultrasonic device. Ultrasonic treatment for 30 minutes ensures a more uniform distribution of carbon nanotubes in the solder, preventing clustering of carbon nanotube powder during mixing with the solder, thus enhancing the wettability of the solder. Subsequently, ethanol in the mixed solution was evaporated using a drying oven to obtain pure carbon nanotube powder. This experiment selected lead-free solder Sn-Ag3.0-Cu0.5 and mixed it with carbon nanotube powder at a ratio of 100:0.05. Finally, the mixture was placed in a planetary ball mill and ball-milled at a speed of 300 r/min for 2.5 hours to ensure thorough mixing of carbon nanotube powder with the solder. The prepared carbon nanotube composite solder is shown in Figure 2 (a). Lastly, solder particles were placed in a cylindrical mold and uniaxially pressed at 14 MPa, then the extruded composite solder was cut into elongated strips measuring 3 × 5 × 10 mm. Compaction improves the shape and density of the solder, enhancing the sintering quality of the solder during subsequent sintering processes.

Table 1. The preparation process for the carbon nanotube composite solder.

Procedure	Preparation Equipment	Process	Materials
1	/	Carbon nanotubes Mixed with C_2H_5OH	Carbon nanotubes, C_2H_5OH
2	Ultrasonic Disperser	Ultrasonic Dispersion 30 min	/
3	Drying Oven	35°C Heat Treatment 1 hour	Carbon nanotubes-C_2H_5OH Mixed Solution
4	Planetary Ball Mill	Rotational Speed: 300 r/min	Carbon nanotubes, Sn-Ag3.0-Cu0.5
5	Mold	Extrusion Pressure: 140 MPa	Carbon nanotube composite solder
6	Reaction Vessel	Sintering Temperature: 210°C	Carbon nanotube composite solder

The solder was sintered in a reaction vessel at 110°C and atmospheric pressure for 2 hours to remove excess flux. After removing the surface flux with ethanol, the solder was dried in an oven at 35°C for 30 minutes. Then, the solder was sintered again at 160°C and atmospheric pressure for 6 hours to initially fix the shape of the solder. Finally, the solder was sintered at 3 MPa pressure and 210°C for 8 hours. After sintering, the solder was polished using sandpaper with grit sizes of P400, P1000, and P2000 to remove surface oxides and achieve a smooth and bright surface, as shown in Figure 2 (b). This process reduces the presence of oxides, ensuring the purity of the solder surface and providing more accurate shape changes for subsequent thermal expansion coefficient tests.

Figure 2. (a) Prepared carbon nanotube composite solder. (b) Carbon nanotube composite solder after sintering.

4. Performance Testing

The left side of Figure 3 shows a scanning electron microscope (SEM) micrograph of the solder at a scale of 100 μm. The solder is evenly distributed and exhibits intact morphology. This indicates that the solder particles remained intact after undergoing high-energy ball milling in the planetary ball mill, without experiencing breakage or other issues. The right side of Figure 3 shows an enlarged image of the solder particles at a scale of 2 μm. The white areas in the image represent carbon nanotubes. Hence, the carbon nanotubes were uniformly mixed with the solder after milling, and the carbon nanotubes were embedded into the surface of the solder particles.

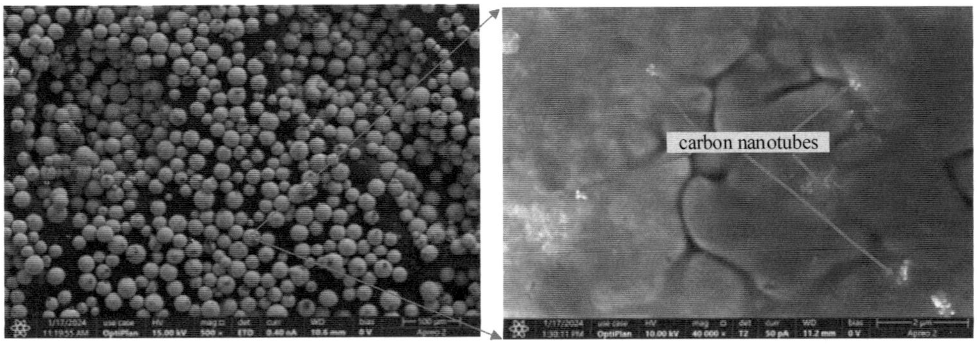

Figure 3. The SEM image of the carbon nanotube composite solder particles.

The CTE is determined by measuring the dimensional changes of a material as a function of temperature. This experiment employs a Thermomechanical Analyzer (TMA) to conduct coefficient of thermal expansion tests on composite solder materials. The composite solder materials are mounted onto the TMA using fixtures. The experimental temperature range is set from 20 to 180°C using a heating device, with a heating rate of 5°C/min and temperature control accuracy of 0.5°C. A mechanical stress of 0.02 N is applied to the samples using pressure rods and heads. As the temperature changes, the material's volume also varies accordingly. Displacement sensors are employed to record the volume changes of the solder material. The equation for calculating the coefficient of thermal expansion is as follows:

$$CTE = \frac{\Delta L}{L_0 \cdot \Delta T} \tag{2}$$

where ΔL represents the dimensional change under temperature variation, L_0 denotes the initial dimension, and ΔT indicates the temperature change.

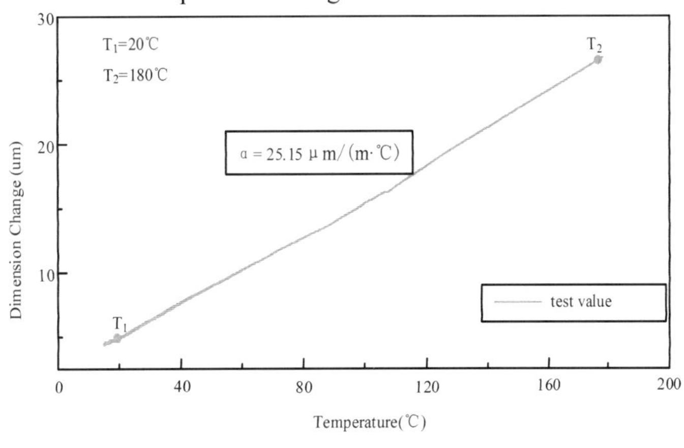

Figure 4. TMA testing curve of the carbon nanotube composite solder.

Figure 4 depicts the temperature and deformation curves obtained for the carbon nanotube composite solder under TMA testing. As shown in the figure, within the temperature range of 20-180°C, the CTE for the composite solder is 25.15 μm/(m·°C). This represents a decrease of 16.2% compared to the coefficient of the original solder, which was 30 μm/(m·°C).

5. Conclusions

This study simulated and analyzed the maximum thermal stress in the IGBT solder layer when using solders with different coefficients of thermal expansion. As the coefficient of thermal expansion decreases, the maximum thermal stress in the solder layer decreases accordingly. In this study, the mechanical ball milling method was employed to prepare carbon nanotube composite solder to reduce the coefficient of thermal expansion of the solder. Through scanning electron microscopy observation, it was found that carbon nanotubes were uniformly mixed with solder particles, and carbon nanotubes were evenly distributed on the surface of solder particles. Subsequently, the solder after sintering was tested using a coefficient of thermal expansion tester. After testing, the CTE of the carbon nanotube composite solder was determined to be 25.15 μm/(m·°C). This represents a decrease of 16.2% compared to the CTE of the original solder, which was 30 μm/(m·°C).

Acknowledgments
This work was supported in part by the National Natural Science Foundation of China (52175529), the University Outstanding Youth Project of Anhui Province (2023AH030003), and the University Synergy Innovation Program of Anhui Province (GXXT-2022-048).

References

[1] Choi U-M, Blaabjerg F, and Lee K-B (2014) Study and handling methods of power IGBT module failures in power electronic converter systems[J]. IEEE Transactions on Power Electronics, 30(5): 2517-2533. [CrossRef]
[2] Li J, Yaqub I, Corfield M, et al (2017) Interconnect materials enabling IGBT modules to achieve stable short-circuit failure behavior[J]. IEEE Transactions on Components, Packaging and Manufacturing Technology, 7(5): 734-744. [CrossRef]
[3] Jian Z, Jiesen L, Zongbei D, et al (2022) Failure Analysis of Dielectric Material Defects Inside IGBT Chips[C]. 2022 4th International Conference on System Reliability and Safety Engineering (SRSE): 421-424. [CrossRef]
[4] Ciappa M (2002) Selected failure mechanisms of modern power modules[J]. Microelectronics Reliability, 42(4-5): 653-667. [CrossRef]
[5] Li L, Qin W, Mai B, et al (2023) Effect of Carbon Nanotubes on the Mechanical, Thermal, and Electrical Properties of Tin-Based Lead-Free Solders: A Review[J]. Crystals, 13(5): 789. [CrossRef]
[6] Yang Z, Zhou W, Wu P (2014) Effects of Ni-coated carbon nanotubes addition on the microstructure and mechanical properties of Sn–Ag–Cu solder alloys[J]. Materials Science and Engineering: A, 590: 295-300. [CrossRef]
[7] Ginga N J, Sitaraman S K (2022) Thermomechanical Reliability Investigation of Carbon Nanotube Off-Chip Interconnects for Electronic Packages[J]. IEEE Transactions on Components, Packaging and Manufacturing Technology, 12(8): 1282-1292. [CrossRef]
[8] Mao J, Yang W, Song Q, et al (2021) The effects of the addition of CNT@ Ni on the hardness, density, wettability, and mechanical properties of Sn-0.7 Cu lead-free solder[J]. Journal of Materials Science: Materials in Electronics, 32: 10843-10854. [CrossRef]
[9] Chen B, Wang H, Zou M, et al (2022) Evolution of interfacial IMCs and mechanical properties of Sn–Ag–Cu solder joints with Cu-modified carbon nanotube[J]. Journal of Materials Science: Materials in Electronics, 33(24): 19160-19173. [CrossRef]
[10] Zhang C, Ge X, Lin C, et al (2023) Optimization and Analysis of Thermal Conductivity Structure for IGBT Module Embedded with the Vapor Chamber[J]. IEEE Transactions on Components, Packaging and Manufacturing Technology, 13(6):798-807. [CrossRef]

Journal of Physics: Conference Series **2842** (2024) 012056 doi:10.1088/1742-6596/2842/1/012056

Low cycle fatigue behavior and failure mechanism of 34CrNiMo6 steel used as connecting rod for diesel engine

Zengquan Wang [1], Guangshuo Zhao [2, 3], Guoxi Jing [2, 3 *], Liqiang Zhang [1], Yexing Hou [1], and Jianchao Pang [4]

[1] China North Engine Research Institute, Tianjin 300400, China
[2] Tianjin Key Laboratory of New Energy Automobile Power Transmission and Safety Technology, School of Mechanical Engineering, Hebei University of Technology, Tianjin 300130, China
[3] School of Mechanical Engineering, Hebei University of Technology, Tianjin 300400, China
[4] Shi-changxu Innovation Center for Advanced Materials, Institute of Metal Research, Chinese Academy of Sciences, Shenyang 110016, China

[*] Corresponding author's email: okjgx@163.com

Abstract: In this paper, for the 34CrNiMo6 steel material used in diesel engine connecting rods, the low-cycle fatigue properties and failure mechanism were studied. The metallographic structure of 34CrNiMo6 has been investigated by using optical microscopes, scanning electron microscopes, and transmission electron microscopes. Tensile tests and low-cycle fatigue experiments were performed at different levels of strain amplitudes, and the microstructural changes were monitored. The examination indicates that the structure of 34CrNiMo6 steel consisted primarily of a ferrite base and small spherical and rod-shaped carbides, with a mean particle size measuring 2 μm. It had exceptional mechanical characteristics of high strength and moderate plasticity. The total strain amplitude (TSA) of 34CrNiMo6 steel affected the cycling behavior, which was stable at low TSA (0.2% ~ 0.3%). At high TSA (0.4% ~ 0.7%), a large number of dislocations were annihilated through the dynamic recovery process, showing a typical cyclic softening phenomenon. The low-cycle fatigue failure behavior was mainly a combined mode of quasi-cleavage fracture and cross-grain fracture.

1. Introduction

As an important connecting component between the piston and crankshaft in the diesel engine, the connecting rod usually endures fatigue failure under the harsh working environment of alternating stress and strain [1]. Therefore, suitable selection for the connecting rod material is of vital importance to ensure a satisfactory fatigue life [2-3]. The most commonly applied materials to fabricate connecting rods are classified into hardened and tempered steels, non-hardened and tempered steels, free-cutting steels, light metal matrix composites, and titanium alloys [4-10]. Among them, the 34CrNiMo6 steel (AISI 4340) with medium carbon and low alloying elements, as a representative hardened and tempered steel, possesses excellent comprehensive mechanical properties, good processability, satisfactory tempering, and nitriding performance [11-12]. The 34CrNiMo6 steel has become an ideal connecting rod material [13]: (1) Its heat treatment procedures are maturely established. The microhardness, tensile strength, elongation, and impact toughness can reach 308.9 HV,

1, 025 MPa, 15%, and 8 kg/cm^2, respectively via heat treatment; (2) It has relatively low alloying elements compared with other structural alloy steels, which makes it much easier for smelting, rolling, machining, etc.; (3) Low alloying elements and simple processing processes make it cheaper for large-scale applications.

Currently, the primary focus of research on 34CrNiMo6 steel is the improvement of its static mechanical properties through heat treatment or alternative processing techniques [14-18]. Hou et al. enhanced the mechanical characteristics of 34CrNiMo6 steel by implementing a pre-heat treatment involving normalizing and high-temperature tempering prior to quenching and tempering, resulting in an advantageous balance of strength and ductility [17]. Wu et al. successfully repaired the damaged 34CrNiMo6 steel with no observable defects by the friction stir processing (FSP) and achieved a maximum ultimate strength of 900 MPa [18]. In addition, Avilés et al. enhanced the dynamic mechanical properties of the 34CrNiMo6 steel by various surface modification methods [19-20]. Avilés et al, for instance, contrasted the effects of low-plasticity burnishing (LPB) and shot-peening (SP) on the high cycle fatigue of 34CrNiMo6 steel. The results showed that its fatigue limit was increased by 39% and 52% owing to the more uniform residual stress after SP treatment and LPB treatment, respectively [19]. Branco et al. studied the room-temperature fully-reversed strain-controlled low cycle fatigue (LCF) behavior of 34CrNiMo6 steel. They noticed that the material displayed a highly excellent Masing-type behavior in the form of hysteresis loops and that it gradually softened under tension until failure fracture [20].

However, the static and dynamic mechanical behavior of the 34CrNiMo6 steel is closely related to the testing conditions. The correlation between the microstructure and mechanical characteristics of 34CrNiMo6 steel has been thoroughly examined. However, the operational efficiency of 34CrNiMo6 connecting rods with real structure and dimensions remains undisclosed. The service life of a diesel engine is affected by the fatigue behavior of the connecting rod. Failure usually occurs at the connecting rods under large and low cycle loads. In this regard, determining the 34CrNiMo6 steel connecting rod's failure mechanism is essential. Therefore, the tensile mechanical properties and LCF behavior of the 34CrNiMo6 steel in association with the microstructure evolution were innovatively and systematically investigated.

2. Materials and methods

2.1 Sample preparation

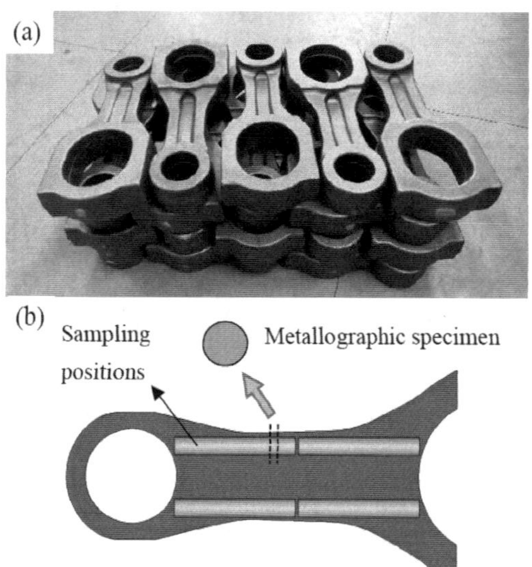

Figure 1. (a) Appearances; (b) Sampling positions of the connecting rods made of 34CrNiMo6 steel.

Journal of Physics: Conference Series **2842** (2024) 012056 doi:10.1088/1742-6596/2842/1/012056

Figure 1 (a) depicts the actual appearances of the connecting rods, and Figure 1 (b) shows the sampling positions of tensile test samples and low-cycle fatigue samples on the connecting rod. The connecting rods were firstly kept at 850°C for 1 hour and oil quenched, and then it was kept at 560°C for 1.5 hours and air cooled. The purpose of all these heat treatments was to reduce stress concentration in the production process of connecting rods. A plasma atomic emission spectrometer with an inductive coupling (ICP-6300, Thermo Fisher Scientific, USA) was used to analyze the chemical makeups of the 34CrNiMo6 according to the Chinese GB/T20125-2006 standard. The corresponding chemical compositions of the 34CrNiMo6 steel used as a connecting rod are listed in Table 1, whose element constitutions conform to the AISI 4030 steel.

Table 1. The chemical makeup of the material used as a connecting rod.

Elements	C	Si	Mn	Cr	Mo	Ni	Fe
Content (wt.%)	0.33	0.25	0.54	1.43	0.21	1.62	Bal.

2.2 Microstructure characterization

The metallographic microstructure was carefully inspected at different magnification times from 100X to 1000X by an optical microscope (OM, OLYMPUS DP73, Japan) according to the GB/T13298-2015 standard (Metal microstructure examination method). As shown in Figure 1 (b), the metallographic sample of 34CrNiMo6 was processed on the bar material cut from the connecting rod. Before observation, the specimens were successively mechanically buffed and chemically scratched with 5% nitrate ethanol reagent (95 ml ethanol + 5 ml HNO_3) for 30 seconds. The grain orientations and size distributions were analyzed by the electron backscattering diffraction (EBSD) attachment on the scanning electron microscope (SEM, Zeiss Supra 55, Germany). For EBSD analysis, these samples were mechanically polished by using waterproof abrasive papers from 400# to 2000# and then electrochemically polished with 10% perchlorate alcohol solution at 17 V for 20 seconds.

2.3 Tensile tests

The static mechanical properties were measured via tensile tests at ambient temperature on the electronic universal testing machine (Instron 5982, USA). The samples were obtained from the center part of the connecting rods and processed into rod-like shapes with specified sizes. Figure 2 shows the three-dimensional sample size for the tensile tests. The working part measured 30 mm in parallel length and 5 mm in diameter. The transition section between the working section and the clamping section had an arc radius of R = 5 mm. For the tensile tests, the deformation rate was set as 1×10^{-3} s^{-1} under the displacement control mode and the extensometer with a measurement distance of 25 mm was removed once the deformation degree reached 2%. Following the tensile testing, the fracture morphology was examined by using the Quanta 600 SEM (FEI, USA).

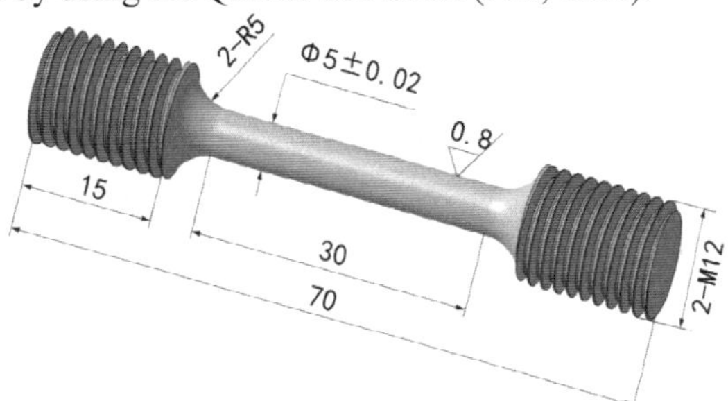

Figure 2. Three-dimensional sample size for the tensile tests (units: mm).

2.4 LCF tests

The dynamic mechanical properties were assessed by LCF tests at ambient temperature by using mechanical fatigue testing equipment (Instron 8862, USA). For the LCF testing, comparable sample and processing techniques were used in the tensile tests. The three-dimensional sample size for the

LCF tests is indicated in Figure 3. The working segment measured 15 mm in parallel length and 6 mm in diameter. The transition section between the working section and the clamping section had an arc radius of R = 15 mm. The LCF tests were carried out under the complete strain control mode based on the Chinese GB/T26077-2010 standard. The extensometer had a gauge distance of 12.5 mm and the total strain was set within a range of 0.2%-0.7% at a strain rate of 0.6% s^{-1}. The samples were defined as failure once the cyclic stress dropped to 60% of the median life. After the LCF tests, the fracture morphology was observed by Quanta 600 SEM (FEI, USA). The internal flaws near the fractured region were examined by using a field emission transmission electron microscope (TEM, FEI Tecnai F20, USA) to elucidate the cyclic stress response during testing. For the preparation of the TEM specimen, the samples were mechanically grounded to 50 μm with sandpapers. After that, the samples were electrolytically double-sprayed with perchlorate alcohol solution by the automatic twin-jet electropolisher (TenUPOL-5, Struers, Denmark) to prepare a thin region. The corresponding parameters were as follows: a voltage of 20 V, a temperature of -20°C, a luminousness of 100%, a flow rate of 29 m/s, a current of 74 mA, and a time of 2 min.

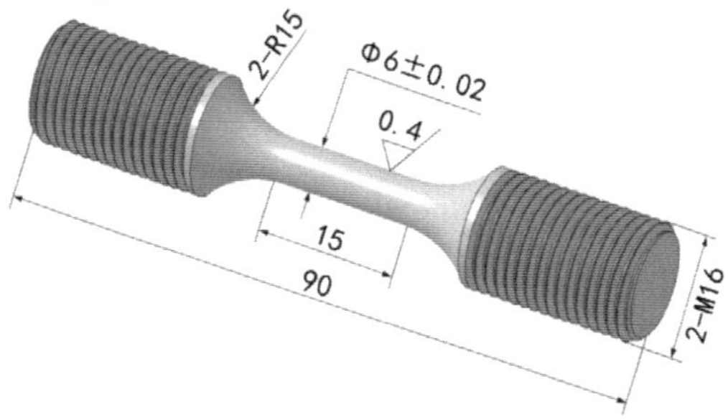

Figure 3. Three-dimensional sample size for the LCF tests (units: mm).

3. Results and discussion

3.1 Microstructure

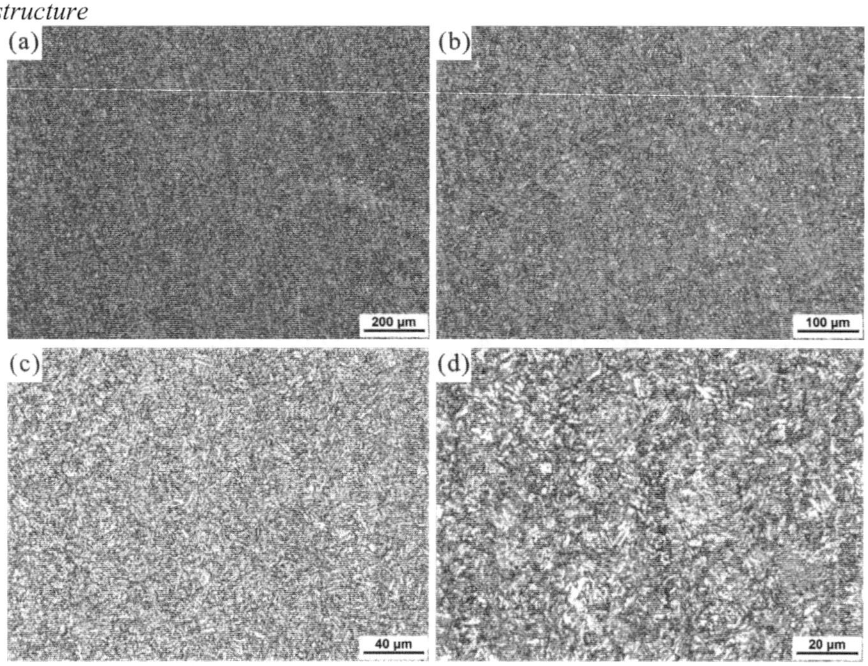

Figure 4. Metallographic microstructure of the 34CrNiMo6 steel at different enlargements: (a) 100×; (b) 200×; (3) 500×; (d) 1000×.

Figure 4 indicates the metallographic structure of the 34CrNiMo6 steel at different magnifications. From a macro perspective, as shown in Figures 4 (a)-4 (c), the 34CrNiMo6 steel exhibits uniform and fine grain structure at relatively low magnification from 100X to 500X. It is seen in Figure 4 (d) that the 34CrNiMo6 steel is mainly composed of ferrite matrix (with white contrast) and plenty of granular carbides (with dark contrast), which are inferred as the typical tempered sorbites microstructure formed after quenching and tempering.

Figure 5 displays the EBSD orientation mapping results and grain size distributions of the 34CrNiMo6. Figure 5 (a) shows that the structure and shape of the grains are consistent with the OM results shown in Figure 4. The grains of the 34CrNiMo6 steel are deformed by the thermal-mechanical processing procedures to exhibit various orientations. Furthermore, the grain sizes are calculated by the intercept method and their size distributions are displayed in Figure 5 (b). The grain size of the ferrite matrix of the 34CrNiMo6 steel exhibits a Gaussian-like distribution shape and its average grain size is approximately 2 μm. Comparable research findings about the microstructural development of a 34CrNiMo6 alloy tempering steel during heating processes can also be found in [21].

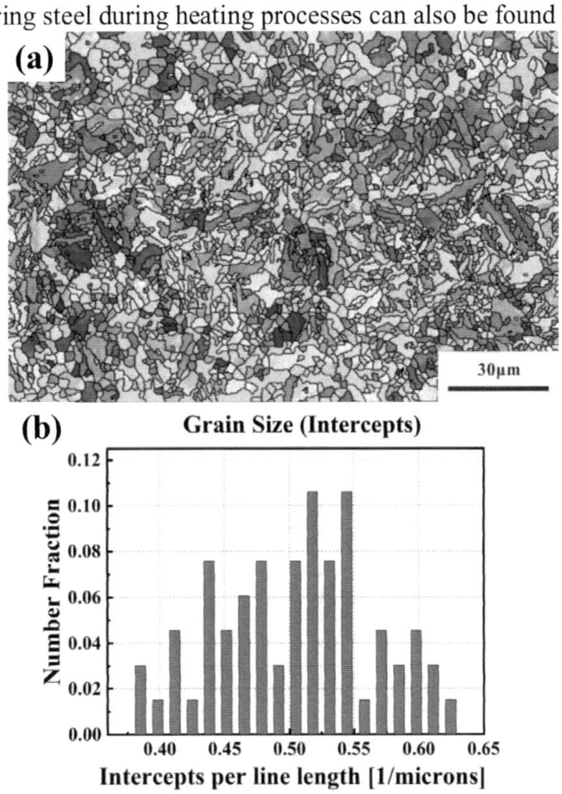

Figure 5. (a) EBSD orientation mapping results; (b) Grain size distributions of the 34CrNiMo6 steel.

3.2 Tensile mechanical properties

Figure 6 indicates the tensile stress-strain diagrams of the 34CrNiMo6 steel and the corresponding appearances of the fractured samples. This material has uniform mechanical properties, as its stress-strain curves obey almost the same tensile deformation behavior. An obvious necking phenomenon can be observed for the fractured samples, indicating that the ductile fracture happens at the 34CrNiMo6 steel to some degree. Moreover, the tensile mechanical properties of yield strength (YS), ultimate tensile strength (UTS), elongation (δ), and area reduction (A) are presented in Table 2. The mean values of YS, UTS, and δ are (898±9) MPa, (1, 025±7) MPa, and (14.6±0.5)%, respectively, which suggests that the 34CrNiMo6 steel possesses outstanding mechanical qualities of high strength and moderate plasticity after heat treatment.

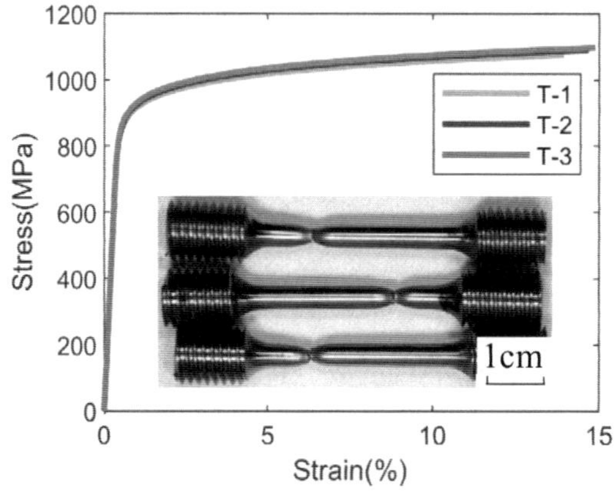

Figure 6. Tensile stress-strain diagrams of the 34CrNiMo6 and corresponding appearances of fractured samples (Note: T-1, T-2, and T-3 are three parallel samples.).

Table 2. Tensile mechanical properties of the 34CrNiMo6 steel.

Samples	YS (MPa)	UTS (MPa)	δ (%)	A (%)
T-1	889	1, 020	14.1	41.5
T-2	901	1, 032	14.7	44.0
T-3	905	1, 023	14.9	42.5
Averages	898±9	1, 025±7	14.6±0.5	42.7±1.3

Figure 7 presents the fracture morphology of the 34CrNiMo6 steel after tests. It is seen in Figure 7 (a) that there are three characteristic regions with different macroscopic morphologies on the fracture surface: (I) the shear lip zone at the edge; (II) the radical zone along the diameter direction; (III) the fibrous zone in the center. The shear lip zone is formed, accompanied by the necking phenomenon. The radical zone, occupying a large area proportion of the fracture surface, is an indication of the high strength of the 34CrNiMo6 steel. The radical zone is also filled with several coarse radioactive cracks, meaning that the material has good toughness. In addition, the fibrous zone at high magnification, as shown in Figure 7 (b), has limited dimples with small size and shallow depth, suggesting that the plasticity of 34CrNiMo6 steel is relatively low.

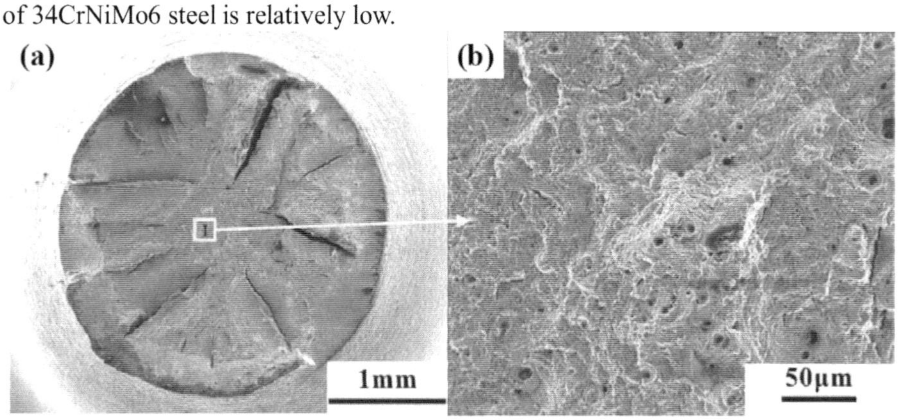

Figure 7. Images from SEM displaying the 34CrNiMo6 steel's fracture surface following tensile tests at different magnifications: (a) 1 mm; (b) 50 μm.

The tensile tests indicate that the 34CrNiMo6 steel has a favorable balance of high strength and mild plasticity, making it appropriate for manufacturing connecting rods.

3.3 LCF behavior

Figure 8 depicts the relationship in the double logarithmic coordinate between the cycle number and total strain magnitude of the 34CrNiMo6 steel. The overall strain amplitude increases with a significant decrease in the number of cycles to failure (Nf). The least square method is adopted to analyze the experimental data. The fitted curve (in red) indicates that the total strain amplitude (TSA) at a cycle number of 1×10^5 (which was specified in the LCF tests) is 0.242%, which is in good consistency with the measured data (0.2%). Therefore, it is considered that no failure will occur, provided that the overall strain amplitude is less than 0.242%.

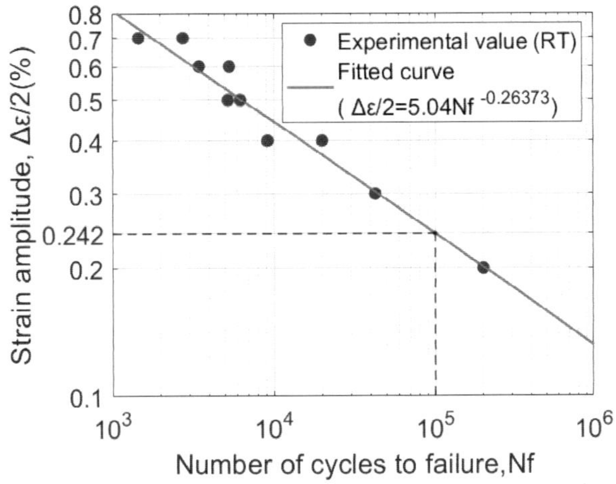

Figure 8. Correlation between overall strain amplitude and cycle amount of 34CrNiMo6 steel plotted on a double logarithmic scale.

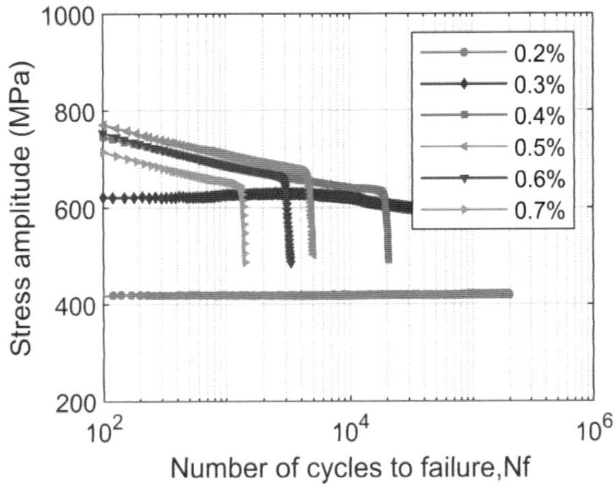

Figure 9. Cyclical stress amplitude response curves of the 34CrNiMo6 steel under different total strain amplitudes.

The cyclical stress amplitude response curves of 34CrNiMo6 steel under different total strain amplitudes are exhibited in Figure 9. For the smaller TSA of 0.2%-0.3%, the 34CrNiMo6 steel presents a cyclic stabilization behavior. The stress amplitudes tend to be stable as the amount of cycles rises. While for larger TSA of 0.4%-0.7%, an obvious cyclic softening behavior is observed with increasing cycle number. This phenomenon is consistent with the empirical rule proposed by Bannantine et al. in 1990 [22]: the cyclic softening behavior could happen on materials with a UTS/YS ratio lower than 1.2. The UTS/YS ratio of the 34CrNiMo6 steel in this work is 1.14, which explains the reason for the cyclic softening behavior under large total strain amplitudes (0.4%-0.7%). Furthermore, it is noted that the cyclic stress amplitudes (< 800 MPa), especially at the initial stage, are lower than the YS value (898±9 MPa), which may be due to the different loading rates (1×10^{-3} s^{-1} and 0.6% s^{-1} for tensile tests and LCF tests, respectively).

Figure 10 displays the 34CrNiMo6 steel's plastic strain amplitude and cyclic hysteretic energy at various total strain amplitudes. It can be easily seen from Figure 10 (a) that the plastic strain amplitude enlarges with increasing TSA value. When the TSA is at a low level of 0.2%, there is almost no plastic strain amplitude observed throughout the entire LCF tests. When the TSA increases to 0.3%, the plastic strain amplitude remains negligible until reaching approximately 3×104 cycles, after which it starts to slightly increase. With a continuous increase in total strain amplitude up to 0.7%, the plastic strain amplitude shows a linear growth pattern as the cycle number increases, eventually leading to fracture. To evaluate the accumulated fatigue damage inside the materials, the cyclic hysteretic energy (W_a, MJ/m^3) is determined by calculating the corresponding area of hysteresis curves, and the results are shown in Figure 10 (b). Similar change trends can also be found for the cyclic hysteretic energies with those of the plastic strain amplitudes, as seen in Figure 10 (a). The irreversible cyclic plastic work represents the changes of internal friction inside the materials, which is consistent with the fatigue life shown in Figure 8.

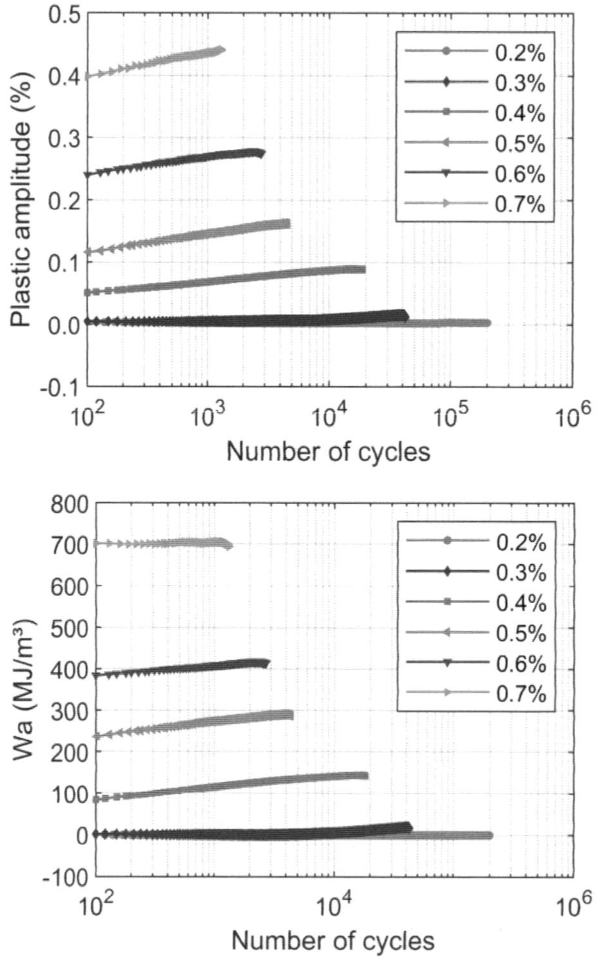

Figure 10. (a) Plastic strain amplitude; (b) Cyclic hysteretic energy of the 34CrNiMo6 steel under various total strain amplitudes.

3.4 Fatigue failure mechanism

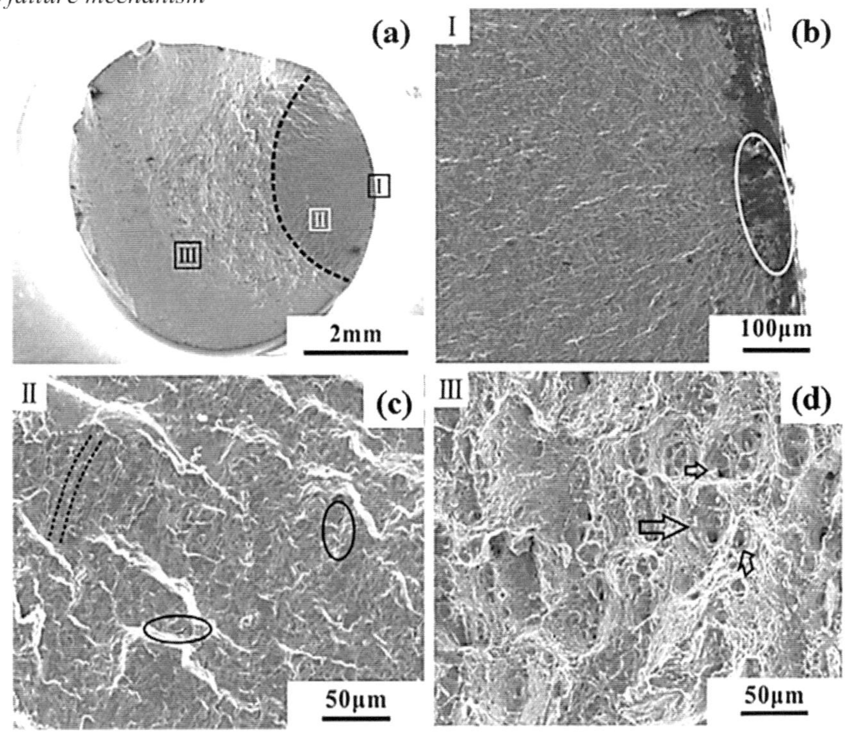

Figure 11. Fatigue fracture morphologies of the 34CrNiMo6 steel under 0.3% total strain amplitude: (a) overall appearance; (b, c, d) magnified views of Regions I, II, and III.

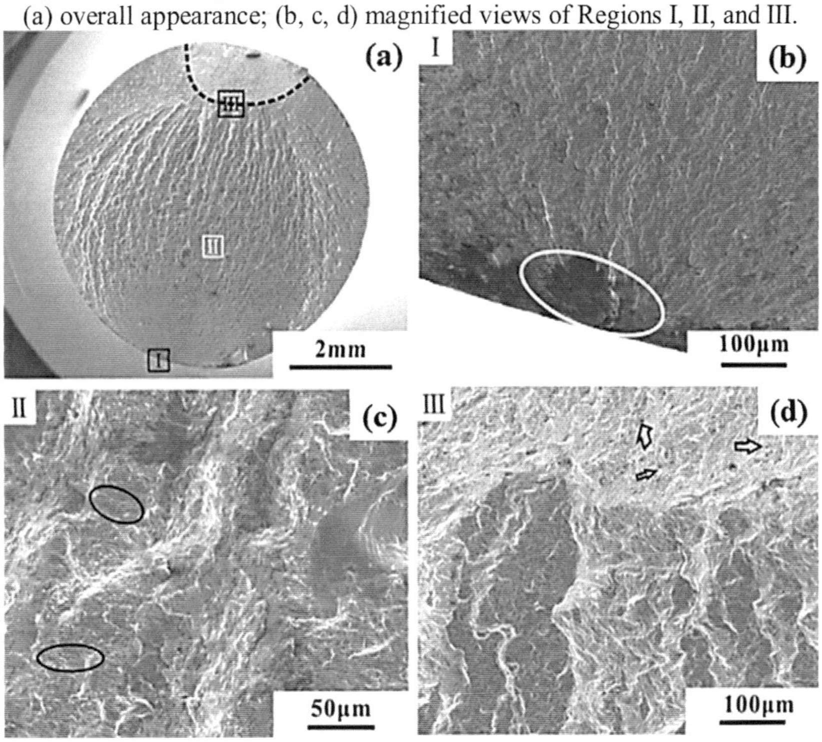

Figure 12. Fatigue fracture morphologies of the 34CrNiMo6 steel under 0.4% total strain amplitude: (a) overall appearance; (b, c, d) magnified views of Regions I, II, and III.

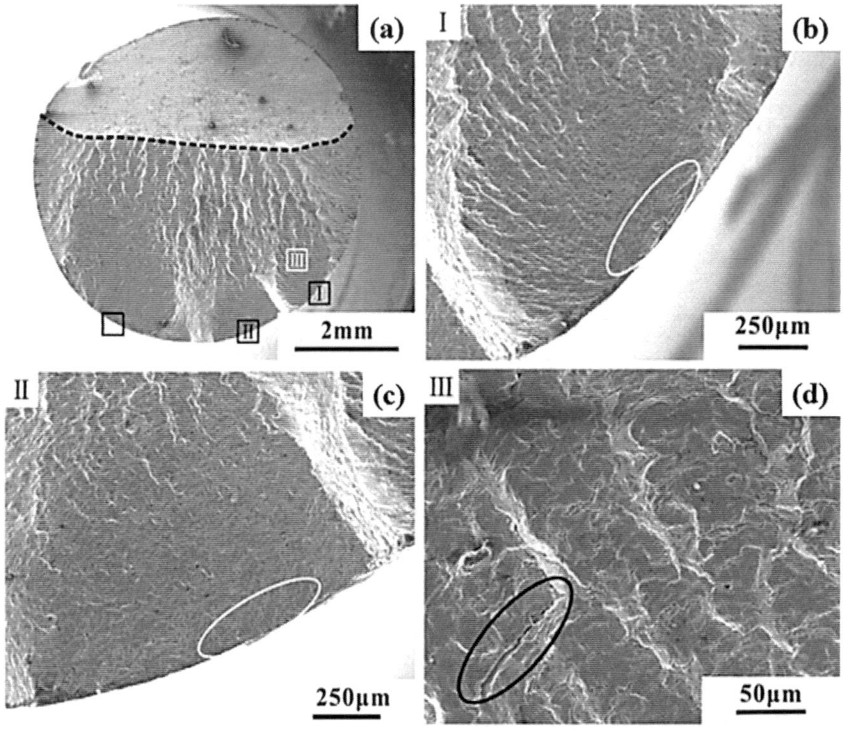

Figure 13. Fatigue fracture morphologies of the 34CrNiMo6 steel under 0.6% total strain amplitude: (a) overall appearance; (b, c, d) magnified views of Regions I, II, and III.

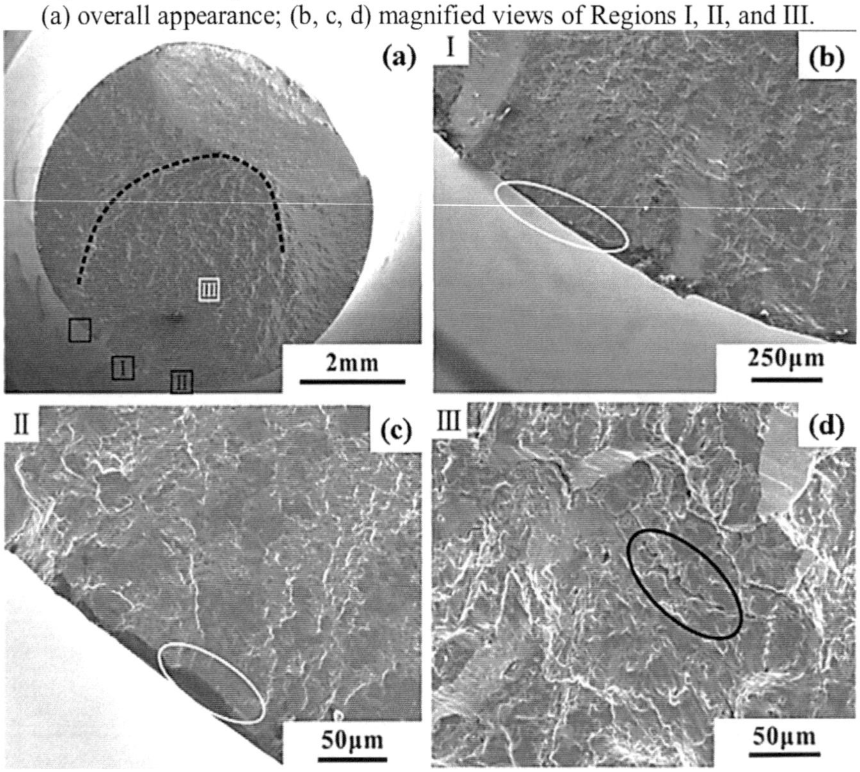

Figure 14. Fatigue fracture morphologies of the 34CrNiMo6 steel under 0.7% total strain amplitude: (a) overall appearance, (b, c, d) magnified views of Regions I, II, and III.

Figures 11, 12, 13, and 14 show the fatigue fracture morphologies of the 34CrNiMo6 steel under varying amplitudes of the total strain of 0.3%, 0.4%, 0.6%, and 0.7%, respectively. These fractured surfaces under different TSA values are featured with three typical regions: (I) crack initiation zone; (II) crack propagation zone; (III) transient fracture zone. Figures 11-14 demonstrate crack initiation zones identified by a black box and white ellipse, with the boundaries between the crack propagation zone and the transient fracture zone highlighted by a blue dotted line. The above three regions exhibit different fatigue failure behaviors:

In the crack initiation zone, only a single crack origin site is observed under low TSA values of 0.3% and 0.4% (see Figures 11 (a)-11 (b) and Figures 12 (a)- 12 (b)), but multiple crack origin sites are found under high TSA values of 0.6% and 0.7% (see Figures 13 (a)-13 (c) and Figures 14 (a)-14 (c)). In the crack propagation zone, there exist some randomly distributed secondary cracks (indicated by black ellipses in Figure 11 (c), Figure 12 (c), Figure 13 (d), and Figure 14 (d)). The size of secondary cracks enlarges with increasing TSA value, which may be attributed to the higher cyclic stress under higher TSA. Moreover, there are a few fatigue striations (see black dotted lines in Figure 11 (c)) under 0.3% TSA, suggesting that the cracks propagate via trans-granular mode. Within the transitory fracture zone, the failure behavior is primarily influenced by a combination of quasi-cleavage fracture and trans-granular fracture (see Figures 11 (d), 12 (d), 13 (d), and 14 (d)). A slightly ductile fracture feature is found under low TSA values of 0.3% and 0.4%, as evidenced by the tiny and shallow dimples (see the blue arrows in Figures 11 (d) and 12 (d)). However, no distinct dimples are observed under high TSA values of 0.6% and 0.7%, which means that the 34CrNiMo6 steel is broken down via brittle fracture under high cyclic stress.

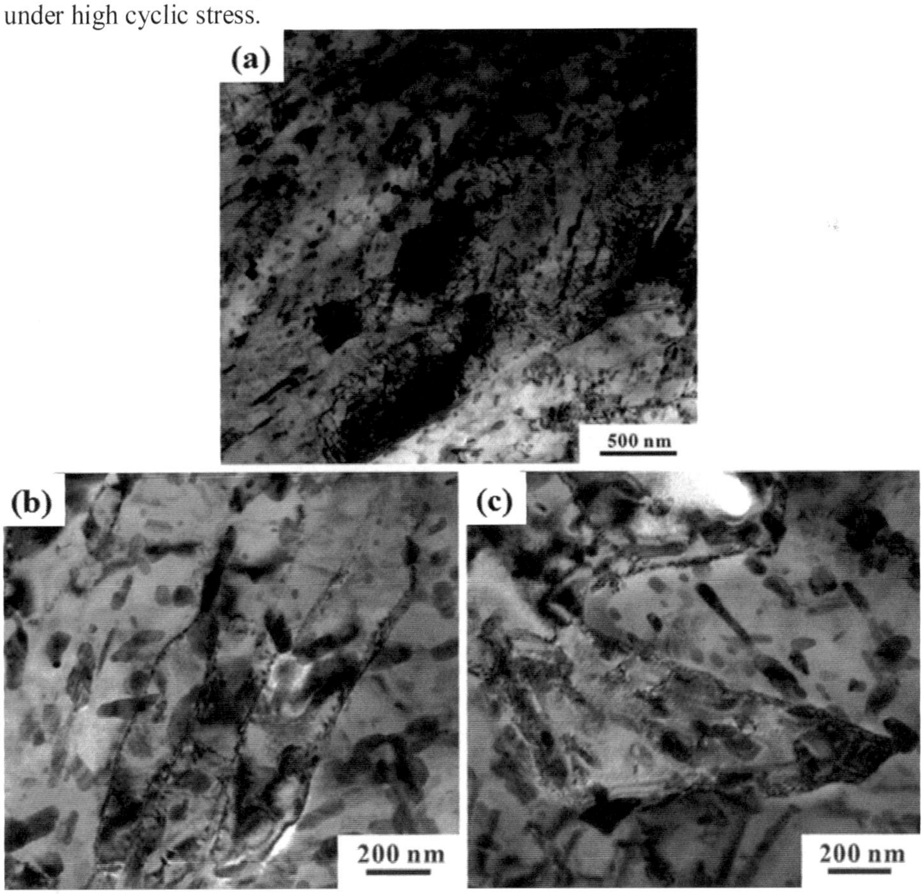

Figure 15. TEM images showing the internal microstructure of the 34CrNiMo6 steel (a) before and (b, c) after LCF tests under 0.7% total strain amplitude.

Figure 15 demonstrates the internal microstructure of the 34CrNiMo6 steel before and thereafter LCF tests under 0.7% total strain amplitude. It is seen in Figure 15 (a) that there are many high-density dislocation entanglements and several parallel deformation twins formed during the thermo-mechanical treatment. Meanwhile, there also exist lots of globular and rod-like carbides, which could enhance the 34CrNiMo6 steel via precipitation strengthening. After the LCF tests, as shown in Figures 15 (b) and 15 (c) at two different positions, the dislocation density is remarkably decreased. Plenty of dislocations annihilate via the dynamic recovery process that happens during the cyclic deformation, which reduces the slippage resistance and thus leads to the cyclic softening behavior (see Figure 9). Furthermore, more deformation twins are generated after LCF tests under 0.7% TSA.

4. Conclusions

The microstructural evolution, tensile mechanical characteristics, and LCF behavior of the 34CrMoNi6 steel used as the connecting rod for diesel engines were investigated in this work. The key conclusions are drawn as follows:

(1) The 34CrNiMo6 steel exhibited a mean grain size of 2 μm and was characterized by typical tempered sorbites after quenching and tempering heat treatment. There exist lots of tiny carbides along and within the grains, which serve as strengthening precipitates for the ferrite matrix.

(2) The 34CrNiMo6 steel possessed excellent mechanical properties of high strength and moderate plasticity. Its YS, UTS, and δ values reached (898±9) MPa, (1, 025±7) MPa, and (14.6±0.5)%, respectively. Its fractured surface featured three different regions of fibrous zone in the center, radical zone along the diameter direction, and shear lip zone at the edge.

(3) The 34CrNiMo6 steel's LCF life was seen to diminish as the TSA rose from 0.2% to 0.7%. It exhibited cyclic stabilization behavior under 0.2% and 0.3% TSA but presented cyclic softening behavior under 0.4%-0.7% TSA. Its fatigue failure behavior was dominated by a combined mode of quasi-cleavage fracture and trans-granular fracture. The high-density dislocation entanglements are annihilated via a dynamic recovery process during cyclic deformations.

Acknowledgment

This work was supported by the Tianjin Municipal Natural Science Foundation (19JCQNJC02800).

References

[1] M. M. Topaç, A. Tanriverdi, O. Çolak, L. Bilal, and M. Maviş. Analysis of the failure modes and design improvement of an eccentrically loaded connecting rod for a double front axle steering linkage prototype, Eng. Fail. Anal., 2021, 122, p. 105-204.

[2] G. He, C. Zhang, B. Jiang, X. Y. Ma, and J. Li. Effect of Anisotropy on the Manufacturing Process of Mg-Zn-Zr Alloy Connecting Rod, J. of Mater. Eng. Perform, 2022, 31 (3), p. 2, 159-2, 167.

[3] S. Khare, O. P. Singh, K. B. Dora, and C. Sasun. Spalling investigation of connecting rod, Eng. Fail. Anal., 2012, 19, p. 77-86.

[4] J. W. Qiu, Y. Liu, Y. B. Liu, B. Liu, B. Wang, E. Ryba, and H. P. Tang. Microstructures and mechanical properties of titanium alloy connecting rod made by powder forging process, Mate. Design, 2012, 33, p. 213-219.

[5] Y. Basavaraj, R. Joshi, G.R. Setty, M. Samiullah, M. Museb, M. Tayab, and H. Banu. FEA of NX-11 Unigraphics modeled connecting rod using different materials, Mater. Today, 2021, 46, p. 2, 807-2, 813.

[6] O. K. Ajayi, B. O. Malomo, S. D. Paul, A. A. Adeleye, and S.A. Babalola. Failure modeling for titanium alloy used in special-purpose connecting rods. Mater. Today, 2021, 45, p. 4, 390-4, 397.

[7] S. Mohammed, D. J. Li, X. Q. Zeng, and D. L. Chen. Low-cycle fatigue behavior of a newly developed cast aluminum alloy for automotive applications, Fatigue Fract. Eng. M., 2019, 42, 1, 912-1, 926.
[8] İ. M. Özkara and M. Baydoğan. Optimization of Thixoforging Parameters for C70S6 Steel Connecting Rods, J. of Mater. Eng. Perform., 2016, 25 (11), p. 5, 020-5, 029.
[9] T. Sathish, S. D. Kumar, and S. Karthick. Modeling and analysis of different connecting rod material through finite element route, Mater. Today, 2020, 21, p. 971-975.
[10] J. W. Qiu, Y. Liu, Y. B. Liu, B. Liu, B. Wang, E. Ryba, and H. P. Tang. Microstructures and mechanical properties of titanium alloy connecting rod made by powder forging process [J]. Mater. Design, 2012, 33, p. 213-219.
[11] C. P. Huang, X. Lin, H. O. Yang, F. C. Liu, and W. D. Huang. Microstructure and tribological properties of laser forming repaired 34CrNiMo6 steel, Materials, 2018, 11 (9), p. 1, 722.
[12] Kumar A., Jayakumar P., and Sharma V. K. Microstructure, and properties of thermomechanically treated and bake hardened AISI 4340 steel, Mater. Today, 2019, 18, p. 848-860.
[13] R. Liang, C. Huang, H. Hao, F. G. Liu, F. C. Liu, M. H. Song, and L. M. Ke. Microstructure and mechanical properties of 34CrNiMo6 steel repaired by laser remelting, J. Mater. Res. Technol., 2020, 9 (6), p. 13, 870-13, 878.
[14] J. Cochet, S. Thuillier, T. Loulou, N. Decultot, P. Carré, and P.Y. Manach. Heat treatment of 34CrNiMo6 steel used for mooring shackles, Int. J. Adv. Manuf. Tech., 2017, 91 (5-8), p. 2, 329-2, 346.
[15] Q. Fang, G. Chen, Y. B. Wu, Y. Huang, Z. Wang, and Y. L. Li. Influence of heat treatment on microstructure and mechanical properties of 34CrNiMo6 steel. Foundry Tech., 2017, 38 (08), p. 1, 866-1, 867+1, 871 (in Chinese).
[16] Z. Si, N. Yuan, and H. Fu. Effect of Quenching and Partitioning Process on Microstructure and Properties of Mn-Si-Cr Steel, J. of Mater. Eng. Perform., 2022, p. 1-13.
[17] Z. L. Hou, Z. F. Wang, X. D. Zhang, J. S. Guo, and F. Kang. Effects of heat treatment on the performance of 34CrNiMo6 steel, J. NetShape Form. Eng., 2018, 10 (6), p. 38-42 (in Chinese).
[18] Z. W. Wu, C. P. Huang, and F. C. Liu. Microstructure and mechanical properties of 34CrNiMo6 steel repaired by friction stir processing, Materials, 2019, 12 (2), p. 279-284.
[19] A. Avilés, R. Avilés, J. Albizuri, L. Pallarés Santasmartas, and A. Rodríguez. Effect of shot-peening and low-plasticity burnishing on the high-cycle fatigue strength of DIN 34CrNiMo6 alloy steel, Int. J. Fatigue, 2019, 119, p. 338-354.
[20] R. Branco, J. D. Costa, and F. V. Antunes. Low-cycle fatigue behavior of 34CrNiMo6 high strength steel, Theor. Appl. Fract. Mec., 2012, 58 (1), p. 28-34.
[21] S. Herbst, H. B. Besserer, O. Grydin, A. Milenin, H. J. Maier, and F. Nürnberger. Holistic consideration of grain growth behavior of tempering steel 34CrNiMo6 during heating processes, J. Mater. Process. Tech., 2016, 229, p. 61-71.
[22] J. A. Bannantine, J. J. Comer, and J. L. Handrock. Fundamentals of metal fatigue analysis, Englewood Cliffs, New Jersey, Prentice Hall, 1990, p. 49-50.

Analysis of the temperature rise process of carbon ceramic axle-mounted brake disc

Qingfeng Qiao[1*], and Mingchen Xia[2]

[1]CRRC QINGDAO SIFANG CO., LTD., Qingdao 266111, China
[2]Institute of Rail Transit, Tongji University, Shanghai 201804, China

[*]Corresponding author's email: sf-qiaoqingfeng@cqsf.com

Abstract: At present, high-speed train braking modules gradually adopt carbon ceramic brake discs. Based on the calculation results, the temperature simulation calculation of carbon ceramic brake disc for high-speed trains is of important engineering reference value for design and application. This paper established a finite element thermal simulation model based on the material properties of carbon-ceramic composite materials and the size and structural characteristics of the brake disc, and conducted a simulation analysis of the 400km/h emergency braking condition. After calculation, the carbon ceramic disc surface temperature reaches about 1250°C during the braking process, which does not exceed the material limit of 1350°C, and can adapt to the braking requirements of high-speed trains. The evolution rules of the temperature field and temperature value changes of the carbon ceramic shaft-mounted brake disc were further studied, and the general rules of disc temperature changes during emergency braking at 400km/h were summarized. The relevant content can provide a reference for the thermal analysis research of carbon ceramic axle-mounted brake disc.

1. Introduction

As high-speed trains further increase their speed to 400km/h, their braking heat load has approached the service limit of traditional steel brake disc. Carbon ceramic axle-mounted brake disc shows good application prospects with their excellent heat-carrying capacity and thermal stability [1-4]. Therefore, the analysis and research on the temperature rise process of carbon ceramic axle-mounted brake disc in high-speed trains have important engineering reference value for the design and application of carbon ceramic axle-mounted brake disc.

At present, research on carbon ceramic axle-mounted brake disc at home and abroad mainly focuses on material performance analysis. Researchers usually use experimental methods to measure the friction and wear performance and heat-carrying capacity of carbon ceramic materials [5-7]. In terms of simulation analysis, some scholars conduct thermal simulation research on carbon-ceramic disc by constructing finite element models. Their research focus usually focuses on fatigue damage or structural optimization of brake disc [8-10]. However, the analysis content of the detailed evolution process of the temperature field of the carbon ceramic disc is insufficient.

Aiming at solving the problem of insufficient research on the temperature rise process of carbon ceramic axle-mounted brake disc, this paper establishes a thermal simulation model of carbon ceramic axle-mounted brake disc based on the ANSYS software. Based on the parameters of 400km/h emergency braking conditions, the model is used for simulation calculations. Based on the simulation results, the temperature change process and evolution rules of the disk are analyzed.

2. Brake disc finite element model

2.1 Thermal analysis theory

This paper calculates the heat flux load input to the brake disc surface based on the friction power method. As shown in Fig. 1. Assuming that friction heat is evenly distributed in the circumferential direction of the brake disc, the heat flow load calculated based on the radial contact length of the friction pair is:

$$q(t,r) = \eta \frac{mar\varphi(v_0 + at)}{2\pi n r_{fri} S} \tag{1}$$

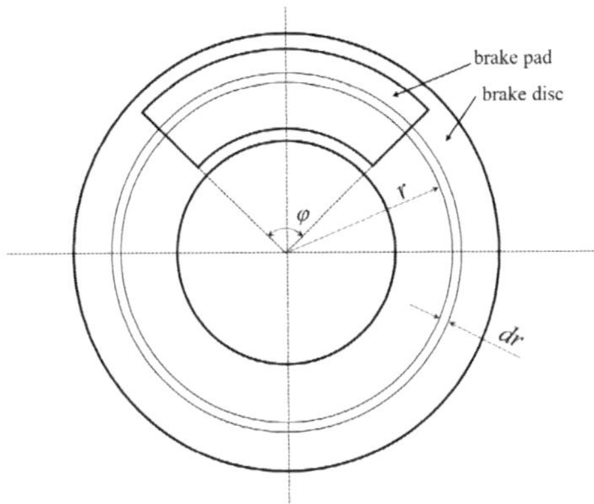

Figure 1. Schematic diagram of friction pair.

where, η is the heat absorption ratio of the brake disc. m is the axle load. a is the braking deceleration. r is the radius of any point on the disc. φ is the brake pad radian. v_0 is the braking initial speed. t is the braking duration. r_{fri} is the brake disc friction radius. n is the friction surface number of the brake disc. S is the pad area.

According to the analysis of heat transfer theory, the expression of the convection heat dissipation coefficient on the surface of the brake disc during the heat transfer process of the carbon ceramic disc during the braking process is:

$$h = \left(0.037\,\mathrm{Re}^{0.5} - 871\right)\mathrm{Pr}^{\frac{1}{3}} \frac{\lambda}{L} \tag{2}$$

where, L is the disc diameter. Re is the Reynolds number. h is the convective cooling coefficient. λ is the thermal conductivity. Pr is the Prandtl number.

Since there are many influencing factors of thermal radiation and it is difficult to determine accurately, this paper converts the thermal radiation load into a convection heat transfer coefficient and applies it. The conversion formula is as follows:

$$h_0 = \varepsilon\sigma\left(T^2 + T_0^2\right)\left(T + T_0\right) \tag{3}$$

where, T is the brake disc surface temperature. $\varepsilon = 0.3$ is the blackness of the surface of the object. $\sigma = 5.67\times10^{-8}\ \mathrm{W/(m^2\cdot K^4)}$ is the blackbody radiation constant. T_0 is the ambient temperature.

2.2 Calculation model

This paper conducts computational research on shaft-mounted brake disc. Because the brake disc structure is cyclically symmetrical. Therefore, this article selects the 1/6 model of the disk body in the calculation process. Tetrahedral elements are used to mesh the brake disc model. In terms of calculation units, considering the complex structure of the disk model, it is impossible to use plate and shell units commonly used in composite material analysis. Therefore, the Solid278 laminated solid unit is selected. The divided brake disc mesh model is shown in Fig. 2, and the number of mesh model units is 157051.

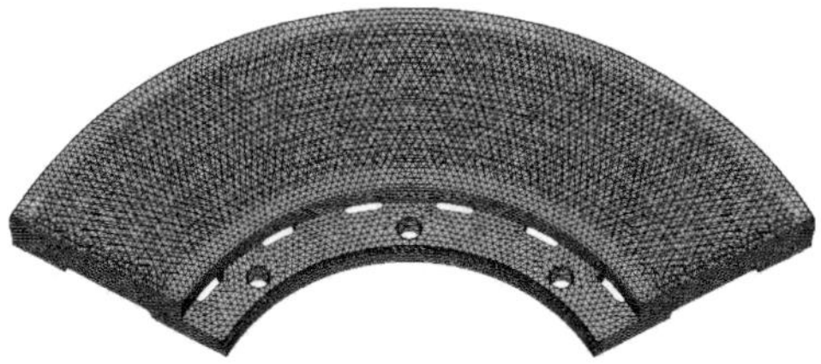

Figure 2. Calculation model of brake disc.

2.3 Materials and loads

The thermophysical performance parameters of the carbon ceramic axle-mounted brake disc are shown in Table 1. Analysis shows that the carbon ceramic material has a small density and a high specific heat capacity, so it has a strong heat-bearing capacity. In addition, its thermal conductivity in the direction perpendicular to the disc surface is low, which may cause more heat to accumulate on the disc surface during braking.

Table 1. C/C-SiC performance parameters.

Temperature (°C)	Density (kg/m^3)	Specific heat capacity (J/(kg·°C)$^{-1}$)	Thermal conductivity (W·(m·°C)$^{-1}$)
100	2100	800	43(//) 20(⊥)
300		1300	
500		1550	
700		1700	
900		1600	

This paper calculates the heat flux load input to the disc surface during emergency braking according to Formula 1, and applies it to the corresponding area of the brake disc surface. We set the brake body cutting surface as an insulating surface, and then apply corresponding convection heat transfer load to each heat dissipation surface.

3. Working condition parameters and heat input analysis

3.1 Working condition parameters

The emergency braking conditions of high-speed trains with a speed of 400km/h are analyzed. The specific braking conditions parameters are shown in Table 2. The equivalent load of the axle plate corresponds to the working condition where the train axle load is 17t and the influence of the train's idling and air resistance is ignored.

Table 2. Value range of factors.

Parameter	Factor value range
Ambient temperature (°C)	20
Brake disc equivalent load (t)	5.7
Braking initial velocity (km/h)	400
Deceleration (m/s^2)	$a=\begin{cases}0.6680 & 360\text{km/h}<v\le 400\text{km/h}\\ 0.9121 & 315\text{km/h}<v\le 360\text{km/h}\\ 1.1560 & 250\text{km/h}<v\le 315\text{km/h}\\ 1.3400 & 0<v\le 250\text{km/h}\end{cases}$

Journal of Physics: Conference Series **2842** (2024) 012057 doi:10.1088/1742-6596/2842/1/012057

3.2 Heat input analysis

Since the decisive part of the temperature rise simulation for a brake disc is the heat input load on the disc surface, the changes in the heat flux density load input on the disc surface during braking were analyzed. The heat flux density at different radii of the disc surface is calculated according to Formula 1 and input into Formula 4. The corresponding three-dimensional change curve is shown in Fig. 3.

$$h=\begin{cases}(3930053.59-23627.48t)r & 0<t\le 16.63\text{s}\\(5545812.24-44050.36t)r & 16.63\text{s}<t\le 30.34\text{s}\\(7455994.56-70758.76t)r & 30.34\text{s}<t\le 46\text{s}\\(9205726.95-95076.70t)r & 46\text{s}<t\le 97.78\text{s}\end{cases}\quad (\text{W/m}^2) \tag{4}$$

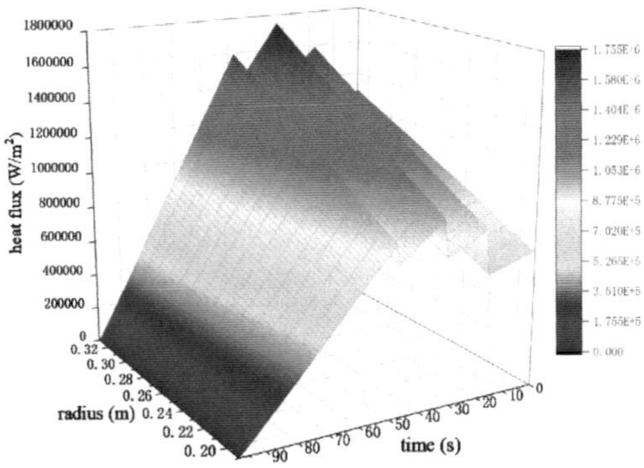

Figure 3. Brake disc surface heat flow density input curve.

The analysis shows that as the radius increases, the heat flux density input to the disc surface continues to increase at the same braking moment, and there is a step increase at the deceleration mutation point. Further analysis will be conducted based on the simulation calculation results.

4. Simulation result analysis

4.1 Temperature field evolution

Fig. 4 shows the carbon ceramic axle-mounted brake disc temperature cloud diagram at different times under emergency braking conditions. It can be observed that the high-temperature area is mainly distributed on the friction surface of the disk. In addition, the temperature band distribution on the disk shows a wavy distribution trend. This is because its internal structure is not uniform, resulting in uneven heat dissipation on the disk. The low-temperature areas are mainly distributed inside the disc body and around the disc hub. As time goes by, the proportion of areas with higher temperatures on the surface of the brake disc gradually increases. As the braking time continues to increase, the temperature in the area near the disc hub is affected by heat conduction. Therefore, it shows an increasing trend. The overall temperature change of the plate shows a consistent development trend.

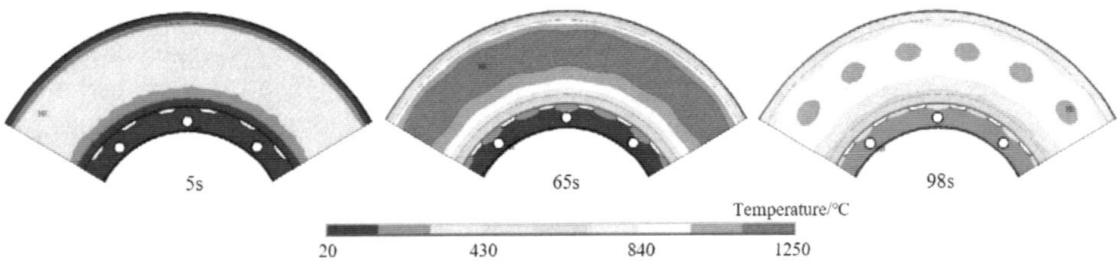

Figure 4. Brake disc temperature cloud chart at different times.

4.2 Brake disc temperature changes

As shown in Fig. 5, this paper takes several equidistant measuring points in the radial direction and thickness direction of the disc body surface. Based on the calculated change trends and numerical differences of multiple nodes, the temperature change rules of the carbon ceramic shaft-mounted brake disc are summarized.

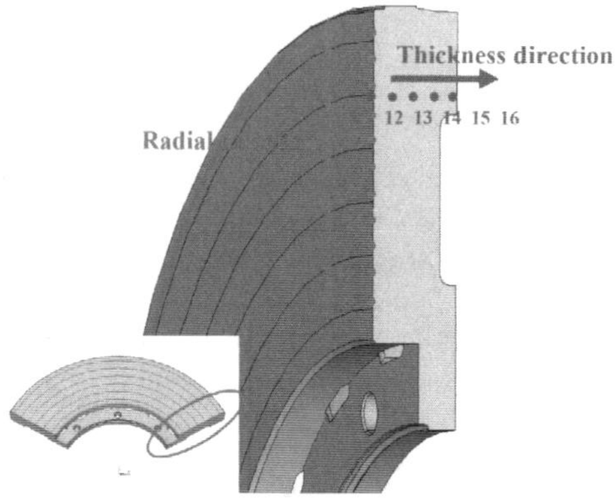

Figure 5. Brake disc value point.

As shown in Fig. 6, it can be found that the temperatures from nodes 1 to 11 all show a similar change pattern. The temperature rise rate changes suddenly at the same time point, and it first rises and then falls. The peaks of the curves are mainly concentrated at nodes 6 to 8. The temperature drop trend of nodes 1 and 11 located at the edge of the disc is not significant. This also reflects that a large amount of frictional heat energy in the edge area of the disc is quickly transferred to the surrounding area. As shown in Fig. 7, the temperature rise curves of nodes 13 to 16 are relatively smooth, which is due to the relatively gentle heat input process of the internal structure of the disc. In addition, the temperatures of each node gradually approach each other as time goes by, and their temperature rise curves form an intersection trend, which is similar to the temperature trend of each node in the radial direction.

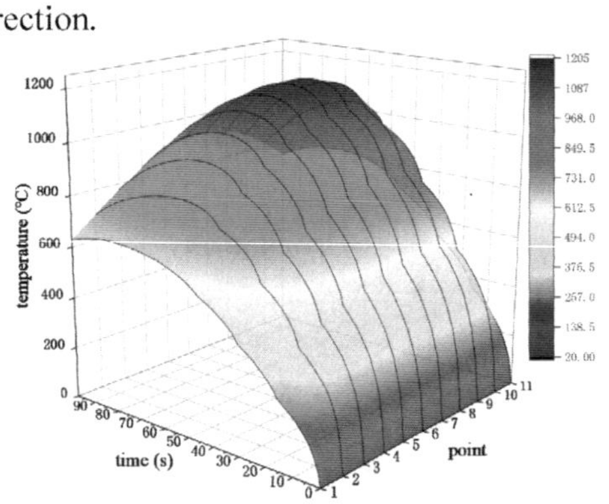

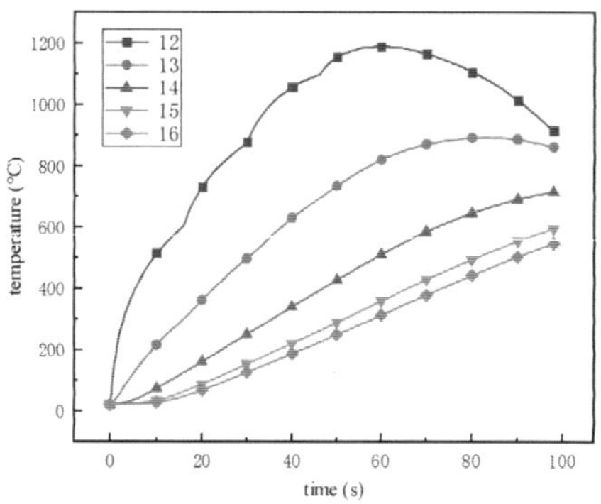

Figure 6. Temperature change in radial direction. **Figure 7.** Temperature change through thickness.

In addition, the surface temperature of the carbon ceramic brake disc reaches a maximum of about 1250°C during emergency braking at 400km/h, and it does not exceed the upper limit of 1350°C. Therefore, the heat-carrying capacity of carbon ceramic axle-mounted brake disc can adapt to the needs of 400km/h high-speed trains, showing good prospects for use.

5. Conclusion

1. When the train brakes in an emergency, the temperature in the middle surface area of the disc body rises rapidly, which is the main high-temperature accumulation area. The middle part is the main high-temperature area, and the temperature around the edge of the disk and the disk hub is lower.

2. The temperature change trends of different nodes along the radial direction of the carbon ceramic disk surface are similar. The overall trend is a rapid increase in the early stage and a slow decrease in the later stage. The internal node temperature changes slowly, showing a gradual upward trend.

3. In the case of emergency braking at 400km/h, the maximum temperature of the disc surface reaches about 1250°C, which is less than the material's heat capacity limit of 1350°C. Therefore, carbon ceramic axle-mounted brake disc can adapt to the emergency braking needs of 400km/h high-speed trains.

Acknowledgment
This work was supported by the National Key R&D Program of China (2021YFB3703805) and Science and Technology Research and Development Programme Topics of China State Railway Group Co., Ltd. (Grant No. N2022J019).

References

[1] Li Y, Xiao P, and Shi Y. (2017) Mechanical behavior of LSI based C/C-SiC composites subjected to flexural loadings. Composites Part A, 95: 315-324. DOI: 10.1016/j.compositesa. 2017.01.024.
[2] Xie J, Fang G, and Chen Z. (2017) High-temperature constitutive model for three-dimensional needled C/C-SiC composite under tensile loading. Journal of Composite Materials, 51(18): 2619-2629. DOI: 10.1177/0021998316674350.
[3] Zhu Y, Zhang Q, and Meng X. (2020) Adhesive joint properties of advanced carbon/ceramic composite and tungsten–copper alloy for the hybrid rocket nozzle. International Journal of Adhesion and Adhesives, 102. DOI: 10.1016/j.ijadhadh.2020.102670.
[4] Yang B, Wang D, and Lin H. (2006) Synthesis of dimethyl carbonate from urea and methanol catalyzed by the metallic compounds at atmospheric pressure. Catalysis Communications, 7(7): 472-477. DOI: 10.1016/j.catcom.2006.01.009.
[5] Lyu B. (2019) Friction-wear properties of BC Contents to C/C-SiC brake materials in high-speed EMU. Railway locomotive car. 39(05): 5-9. DOI: 10.3969/j.issn.1008-7842.2019.05.02.
[6] Zhao S, Zhang X, Zhong W, et al. (2021) The wet braking and recovery behaviors of the P/M pad mated with C/C–SiC disc for high-speed trains. Wear, 468-469. DOI: 10.1016/J.WEAR. 2020.203609.
[7] Yang C, Qiao F, and Yao F. (2023) Characteristics of C/Ceramic brake discs and their application in rail vehicles. Materials for Rail Transportation System, 2(06): 10-13. DOI: 10.3969/j.issn.2097-1923.2023.06.003.
[8] Qu J, Wang W, Shangguan Y, et al. (2023) Thermomechanical modelling and strength assessment of C/C-SiC composite brake disc. Applied Composite Materials, 30(5): 1547-1568. DOI:10.1007/S10443-023-10136-6.
[9] Deng P, Li P, Xiao P, et al. (2021) Oxidation behaviour of C/C–SiC brake discs tested on full-scale bench rig. Ceramics International, 47(24): 34783-34793. DOI: 10.1016/J.CERAMINT. 2021.09.017.
[10] Xin J, Chen H, and Nan L. (2023) Numerical study on air pumping effect and convective heat dissipation performance of carbon ceramic axle-mounted brake discs. Journal of Wuhan University of Technology, 45(11): 132-138+156. DOI: 10.3963/i.issn.1671-4431.2023.11. 019.

Journal of Physics: Conference Series **2842** (2024) 012058 doi:10.1088/1742-6596/2842/1/012058

Preliminary study on smelting micro-carbon ferromanganese alloy by electro silicon thermal one-step process

Yaoxin Song[1, 2, *], Jianming Pang[1, 2], Lingen Luo[1, 2] and Chongfeng Yue[1, 2]

[1]Resource Application and Alloy Materials Division, China Iron & Steel Research Institute Group, Beijing, 100081, China

[2]Beijing Cisri-nmt Engineering Technology Co., Ltd., Beijing, 100081, China

[*]Corresponding author's e-mail: syx68@126.com

Abstract: In this paper, the related thermodynamic analysis of the one-step electro silicon thermal smelting process of micro-C-ferroalloy was carried out, and the effects of smelting temperature (T), slag alkalinity (R), silicon ratio (K_{Si}) and holding time (t) on the appearance quality of the alloy, the Mn content, the Si content of the alloy and the Mn element recovery rate were studied. It is concluded that under the conditions of a smelting temperature of 1400°C, the basicity of slag R is 1.2, the silicon addition coefficient K_{Si} is 1.0, the holding time is 60 min, the alloy appearance quality is good, Mn content of the alloy is high, Si content is low, and Mn yield is relatively high. Under laboratory conditions, the Mn yield of the alloy obtained by this method is low, and the thermodynamic equilibrium and reaction kinetics of the alloy melt and slag system of this method still need to be further studied to provide a theoretical basis for the expansion test.

1. Introduction

Ferro manganese alloy is an alloy of manganese and iron [1], pure ferro manganese alloy (such as micro-carbon ferro manganese, etc.) is an alloy additive for smelting high-quality pure steel, suitable for low-carbon alloy structural steel and high-quality varieties of steel, and in the preparation process, there is no need to change the original steelmaking process, can optimize the alloy, improve the internal quality of steel, reduce the cost of steelmaking alloy, with obvious economic benefits. At present, the conventional preparation method of micro-carbon ferromanganese alloy is a two-step method [2], firstly, the production of silicon-manganese alloy by carbothermal reduction process in the submerged arc furnace, and the second step is a desilication method, which means that the silicon in the silicon-manganese alloy is oxidized by oxides in the manganese ore to produce different brands of micro-carbon ferromanganese alloy. The desilication production methods include an electric silicon-thermal method and a shake ladle method. Due to the use of ore smelting, and the need for submerged arc furnace and electric furnace/shaking ladle double production, the process is complex, and to meet the requirements of micro-carbon ferromanganese alloy, it is necessary to strictly control the carbon content of silicon manganese alloy, the process requires that the silicon manganese alloy as an intermediate alloy has a low carbon content and a high silicon content.

The traditional two-step process makes it difficult to control the carbon content of silicon-manganese alloy because its raw materials are manganese ore, silica, coke, and iron filings. The reducing agent is coke, carbon is used as a reducing agent for carbothermal reduction, and carbon emission is high. Song et al. [3] introduced a practical study on the production of micro-carbon ferromanganese by melting method. The raw material used in this method is manganese metal. When the price of manganese metal is high, this method has no economic benefit. Baisanov et al. [4] proposed to use aluminum-silicomanganese ASM instead of silicomanganese alloy to react with manganese ore to produce refined ferromanganese, which is also difficult to use for the production of micro-carbon ferromanganese alloys. Yessengaliyev et al. [5] studied the feasibility of using high-alkalinity manganese ore to produce refined ferromanganese. This method mainly reduced the amount of flux lime, and the product could not meet the requirements of micro-carbon ferromanganese. Therefore, a new method of smelting micro-carbon ferromanganese alloy by electro silicon thermal one-step process is proposed, and preliminary experimental research is carried out.

2. Experimental principle and thermodynamic analysis

The metallurgical principle of the silicothermal method is the principle of metal thermal reduction reaction, which is a mature metallurgical principle widely used in ferroalloy production. The silicothermal method is a method that uses siliceous materials as a reducing agent to reduce oxides in ores. The reaction process of this method can be expressed by the following equation [6].

$$[Me,\ (1+n)\ Si]_{Fe} + (2MeO) \xrightarrow{2CaO} [3Fe,\ nSi]_{Fe} + (2CaO,\ SiO_2) \quad (1)$$

Silico thermic reduction reaction is an exothermic reaction, but the heat is not enough to meet the continuous progress of the above reaction, so additional heat needs to be added, which is generally supplemented by electric energy, it is called electro silicon thermic method. In addition, it can be seen from the reaction Equation (1) that to make the reaction balance proceed to the right and reduce the metal oxide to metal, a certain flux needs to be added. Then we increase the activity of MeO in slag or reduce the activity of SiO_2.

In the experiment of one-step electro silicon thermal smelting of micro-carbon ferromanganese alloy, the reducing agent is industrial silicon, and the manganese ore is semi-carbonate manganese ore. During the heating process, manganese carbonate in manganese ore is decomposed into manganese oxide, the best decomposition temperature is about 500°C~530°C, and silicon reduces manganese oxide and iron oxide to produce micro-carbon ferromanganese. In the process of silicon thermal reaction, silicon is oxidized to SiO_2, because SiO_2 and MnO are easy to form manganese silicate, thus inhibiting the silicon reduction of manganese oxide reaction. Compared with the formation of manganese silicate, the reaction between CaO and SiO_2 to calcium silicate is more likely to occur, therefore, it is necessary to add flux such as lime to promote the silicon reduction of manganese oxide and iron oxide reaction balance to the right. The main reactions in this process are as follows [7].

$$2MnOS+Sis=2MnS+SiO_2S,\ \Delta G^{\Theta}=-90052+3.36T \quad (2)$$

$$2MnOS+Si_l=2Mn_l+SiO_2S,\ \Delta G^{\Theta}=-61396+39.98T \quad (3)$$

$$2/3Fe_2O_3S+Si_s=4/3FeS+SiO_2S,\ \Delta G^{\Theta}=-403000+30.23T \quad (4)$$

$$2/3Fe_2O_3S+Si_l=4/3Fe_l+SiO_2S,\ \Delta G^{\Theta}=-471940+70.37T \quad (5)$$

$$2MnOS+SiO_2=2MnO\bullet SiO_2,\ \Delta G^{\Theta}=-53600+24.73T \quad (6)$$

$$2CaOS+SiO_2=2CaO\bullet SiO_2,\ \Delta G^{\Theta}=-118800+6.7T \quad (7)$$

According to the diagram of Reaction (2) - Reaction (5), it can be seen from Figure 1 that silicon reduction of manganese oxide and iron oxide is an exothermic reaction, and the solid phase reduction reaction occurs at the temperature below the melting point. The metals Mn and Fe are miscible metals, and after solid phase reduction, they are melted into an alloying solution. According to the Gibbs free energy formula of the above reactions, it can be seen that at the same temperature, the reaction of reducing iron oxide by silicon is easier to carry out than that of reducing manganese oxide. For example, at 1573 K, the ΔG^{Θ} of the reaction of reducing manganese oxide by silicon is -84.8 kJ/mol, while that of reducing iron oxide is -355.4 kJ/mol. It shows that the reduction of iron oxide by silicon is more thorough. With the increase of temperature, the value of ΔG^{Θ} decreases, so the high temperature is not conducive to the occurrence of silicon thermal reduction reaction, in addition, due to the high temperature, the Mn element is easy to volatilize, resulting in element loss. In order to obtain a better yield of alloy, it is necessary to determine the appropriate smelting temperature, appropriate reducing agent dosage, and flux addition amount.

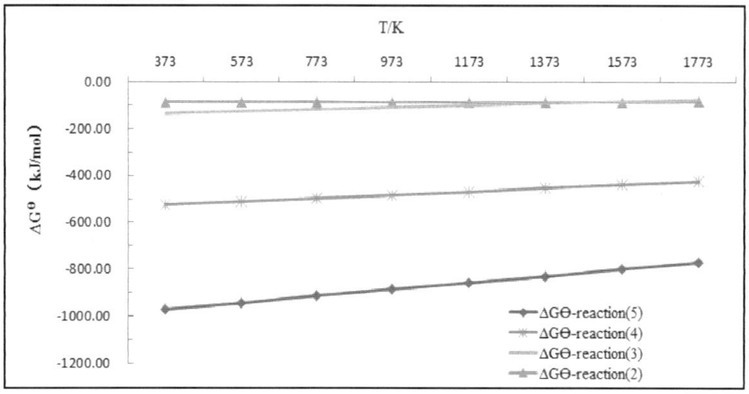

Figure 1. Values of ΔG^{Θ} for reactions.

3. Experimental equipment and raw materials

The equipment used in the experiment of smelting micro-carbon ferromanganese by electric silicon thermal one-step process is high-temperature electric heating experimental furnace and the material of the crucible is MgO. The manganese ore selected as raw material for the experiment is South African manganese semi-carbonate ore, and its composition is shown in Table 1. The siliceous reducing agent is industrial silicon, and its composition is shown in Table 2. The flux used is lime.

Table 1. Main chemical composition of manganese ore %.

Mn	Fe	SiO_2	P	CaO	MgO	Al_2O_3	H_2O
38.2	3.5	7.2	0.05	17.0	1.8	1.1	1.5

Table 2. Main chemical composition of industrial silicon%.

Si	Fe	Al	Ca
≥99.6	≤0.37	≤0.03	≤0.03

4. Experimental procedures

The experiment was carried out in a high-temperature experimental furnace without dynamic stirring means. To make the materials react evenly, all the experimental materials were made of powder and mixed. Firstly, the manganese ore powder, industrial silicon, and lime are crushed and ground to about 80 meshes, and the materials are mixed evenly according to the calculated data to make pellets, which are dried for 1 hour at 120°C in a drying oven until the moisture is completely removed. The dried pellets were placed in a magnesia crucible and then heated in a high-temperature experimental furnace. The maximum heating temperature and holding time were the temperature and time set for the experiment. After the heat preservation was completed, the experimental samples were taken out after cooling to 1200°C in the furnace and cooled by water cooling. After cooling, the material in the crucible is separated from the slag iron, and the obtained ferromanganese alloy is weighed and sent to the testing department for chemical composition analysis.

5. The experiment design

In this experiment, the influence of multiple factors on the product quality and Mn recovery rate of micro-carbon ferromanganese alloy smelting by the one-step silicon thermal reduction method was investigated. The experimental scheme was designed according to the smelting temperature (T), slag basicity (R), silicon addition coefficient (K_{Si}), holding time (t), and other process parameters and the influence of various factors on the Mn and Si content of the obtained alloy and the Mn recovery rate was investigated.

According to the Fe-Mn phase diagram, manganese and iron are completely miscible in liquid and solid-state and do not form compounds. When the Mn content is about 65%~85%, the melting temperature is about 1240°C~1350°C. According to the above analysis, it is necessary to add a certain amount of lime flux to make the reduction reaction proceed to the right. The alkalinity of the slag is considered alkaline slag, and the basicity of the slag is proportional to the melting temperature of the slag. The higher the basicity value of the slag is, the higher the melting temperature of the slag will be. Because the Si content of the micro-carbon ferromanganese is required to be ≤1.0%, the amount of silicon needs to be strictly limited. Therefore, considering the good slag fluidity, the characteristics of Mn element volatilization at high temperatures, and research experience, the range of smelting temperature T is 1250°C~1550°C, the silicon addition coefficient K_{Si} is 0.8~1.4, the slag basicity R is 1.2-1.5, and the range of holding time t is 30 min~90 min. A total of 14 furnace tests were designed, and the scheme is shown in Table 3.

Table 3. Experiment scheme.

No.	T/°C	K_{Si}	R	t/min
1-1	1250	1.0	1.2	60
1-2	1350	1.0	1.2	60
1-3	1450	1.0	1.2	60
1-4	1550	1.0	1.2	60
2-1	1400	1.0	1.0	60
2-2	1400	1.0	1.2	60
2-3	1400	1.0	1.3	60
2-4	1400	1.0	1.5	60
3-1	1400	0.8	1.2	60
3-2	1400	1.0	1.2	60
3-3	1400	1.2	1.2	60
3-4	1400	1.4	1.2	60
4-1	1400	1.0	1.2	20
4-2	1400	1.0	1.2	40
4-3	1400	1.0	1.2	60
4-4	1400	1.0	1.2	90

Note: No. 2-2, 3-2, and 4-3 are for the same furnace.

6. Experimental results and analysis

6.1 Experimental result

The experimental results are shown in Table 4.

Table 4. Experiment results.

No.	Alloy weight/g	Mn%	Si%	C%	yield rate of Mn /%
1-1	9.5				-
1-2	21.7	67.77	0.78	0.047	38.50
1-3	28.2	80.76	0.42	0.064	59.62
1-4	20.5	65.93	2.11	0.043	35.38
2-1	19.0	70.87	1.89		35.25
2-2	30.0	83.42	0.42	0.039	65.51
2-3	28.9	84.19	0.454		63.69
2-4	25.8	81.39	0.716		54.97

3-1	24.5	76.79	0.22		49.25
3-2	30.0	83.42	0.42	0.039	65.51
3-3	29.8	77.78	1.35		60.68
3-4	31.2	67.17	3.7		54.86
4-1	24.5	57.11	3.2		36.63
4-2	28.9	76.28	1.03		57.71
4-3	30.0	83.42	0.42	0.039	65.51
4-4	29.8	82.12	0.73		64.06

Note: 1) Because the weight of the product obtained from the 1-1 furnace is too little, the composition analysis was not done. 2) The C content of some products has been analyzed.

6.2 Analysis of experimental results

6.2.1 Effect of smelting temperature on product quality. In Table 3, the four furnace tests numbered 1-1; 1-2, 1-3, and 1-4 mainly studied the effect of smelting temperature on the quality of ferromanganese alloy products (Mn and Si content) and the yield rate of Mn. The selected test conditions are: silicon addition coefficient is 1.0, slag basicity is 1.2, holding time is 60 min, and smelting temperature is 1250°C, 1350°C, 1450°C, 1550°Crespectively. Figure 2 and Figure 3 show the appearance of test products and the Mn and Si contents and yield rate of Mn element at different smelting temperatures respectively.

(1) The influence of different smelting temperatures on the appearance quality of alloy

In Figure 2, 1-1, 1-2, 2-2, 1-3, and 1-4 show the appearance of the product at the smelting temperature of 1250°C, 1350°C, 400°C, 1450°C and 1550°C respectively. According to Figure 2, when the smelting temperature is 1250°C, due to the poor fluidity of the slag, the slag and iron separation effect is not good, only a small amount of metal is collected and the texture is hard. At the same time, there are small iron beads scattered in the slag, and the reduction degree of manganese ore is low. When the smelting temperature is 1350°C, a small amount of slag is coated in the obtained alloy, and a few small iron beads are scattered in the smelting slag. When the smelting temperature is 1400°C and 1450°C, the slag iron separation is good, and the obtained alloy has a good metallic luster, and the smelting temperature is 1550°C, the obtained alloy color brightness is slightly worse.

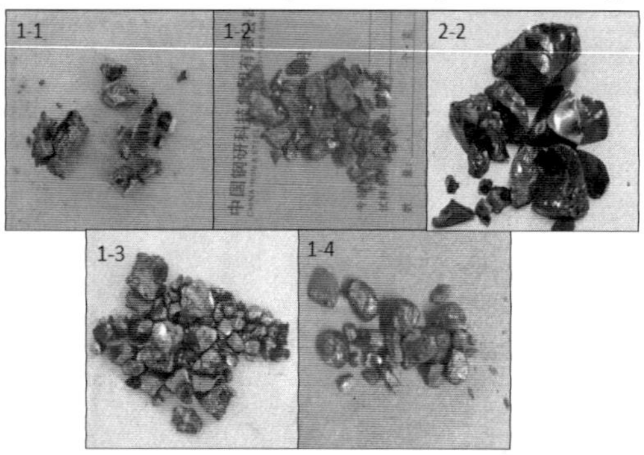

Figure 2. Appearance of ferromanganese alloy at 1250°C, 1350°C, 1400°C, 1450°C, and 1550°C.

(2) Effect of different smelting temperatures on Mn content and Mn yield rate of alloy

It can be seen from Table 4 that under the test conditions of silicon addition coefficient of 1.0, slag basicity of 1.2, and holding time of 60 min, among the five smelting temperatures, the Mn% is higher at 1400°C and 1450°C, and the Mn yield rate is higher. As can be seen from Figure 3, when the smelting temperature is 1350°C~1550°C, the Mn content and Mn yield rate of the alloy show an obvious trend of first increasing and then decreasing. At 1400°C, the Mn content and Mn yield rate of the alloy are the highest values, which are 83.42% and 65.51% respectively. Combined with the alloy composition, it is preliminarily believed that 1350°C to 1450°C is the appropriate temperature range for alloy smelting, especially around 1400°C, beyond the temperature range, because the volatilization of Mn element leads to the decline of alloy yield.

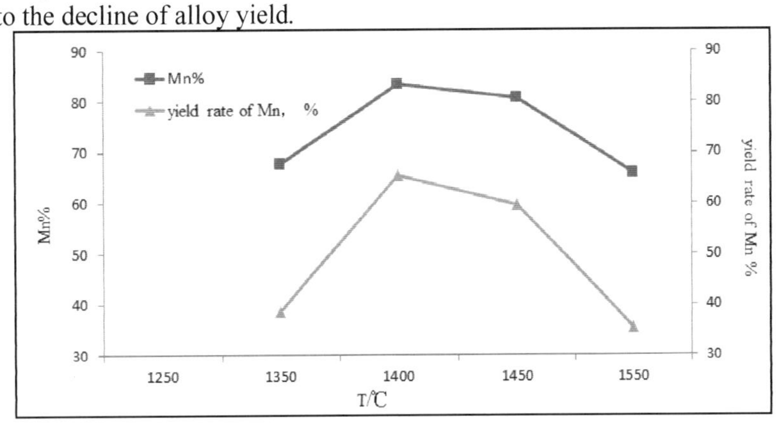

Figure 3. Mn% and Mn yield of ferromanganese alloys at different temperatures.

6.2.2 Effect of slag basicity on product quality. Slag basicity mainly affects the mobility and the mass transfer effect of slag. To preliminarily study the influence of slag basicity on Mn content and Mn yield rate of alloy, exploration tests were conducted under different basicity of slag, and the test conditions were selected as follows: The smelting temperature is 1400°C, the silicon addition coefficient is 1.0, the holding time is 60 min, and the basicity of the slag is 1.0, 1.2, 1.3, and 1.5 respectively. The weight of the alloy obtained under the condition of four kinds of slag basicity is shown in Table 4. The main difference is that the weight and composition of the alloy are different, and the appearance is not very different.

It can be seen from Figure 4 that when the slag basicity is 1.0~1.5, the Mn content and Mn yield rate of the alloy show a trend of first increasing and then slowly decreasing. When the basicity of slag increases from 1.0 to 1.2, the Mn content and yield of the alloy increase significantly. When the basicity is 1.2, the Mn content of the alloy is 83.42%, and the Mn yield rate is 65.51%. When the basicity of slag increases from 1.2 to 1.5, the Mn content and Mn yield rate of the alloy decrease slowly. The preliminary analysis shows that at 1400°C, when the basicity of slag is lower than 1.2, the high SiO_2 content in slag is not conducive to the silicon thermal reduction reaction to the right, resulting in low Mn content and low Mn yield rate of the alloy. Increasing the basicity of slag appropriately (R=1.2-1.3) can reduce the activity of SiO_2, and promote the reaction to the right, and the Mn content and Mn yield rate of the alloy are relatively high. When the basicity of slag is 1.5, the melting point of slag increases, and the mass transfer effect of slag decreases, which leads to the decrease of Mn content and Mn yield rate of the alloy.

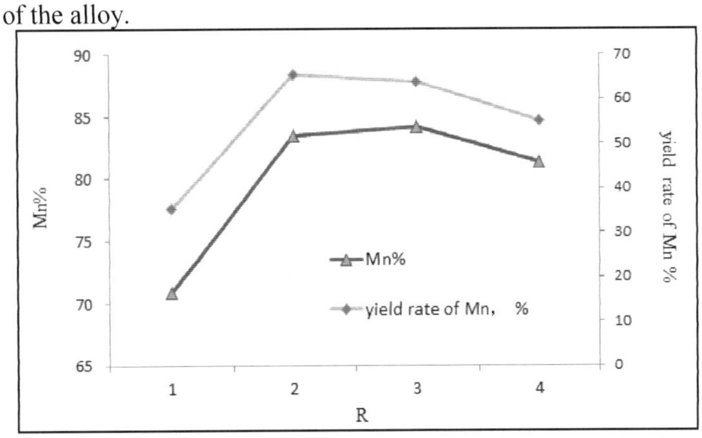

Figure 4. Test results of slag basicity R of 1.0, 1.2, 1.3, and 1.5.

Journal of Physics: Conference Series **2842** (2024) 012058 doi:10.1088/1742-6596/2842/1/012058

6.2.3 Effect of silicon addition coefficient on product quality. The reducing agent used in this study is industrial silicon, and the silicon addition coefficient directly affects the reduction effect of manganese ore and also has a great influence on the Si content of the obtained alloy. Under the conditions of smelting temperature of 1400°C, slag basicity of 1.2, and holding time of 60 min, the experiments of silicon addition coefficient K_{Si} of 4 furnace times are 0.8, 1.0, 1.2, and 1.4 respectively. The obtained alloy with higher Si content has a dark gray color. For example, the alloy obtained from furnace No. 3-4 has a Si content of 3.7%, and its appearance is shown in Figure 5. In addition, the relationship between Mn content and Si content of the alloy with different silicon addition coefficients K_{Si} is shown in Figure 6.

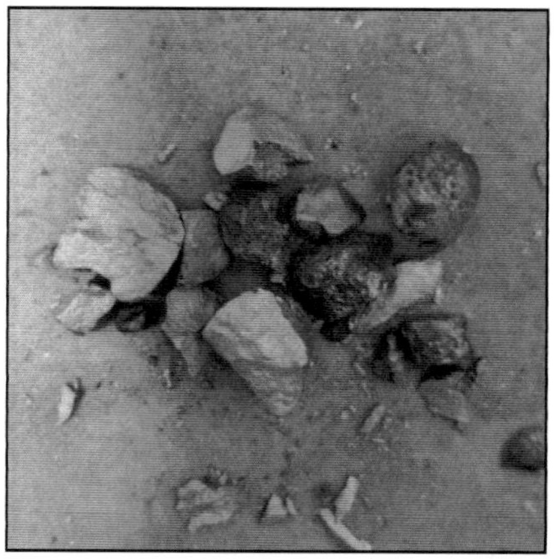

Figure 5. Test furnace (3-4) alloy appearance.

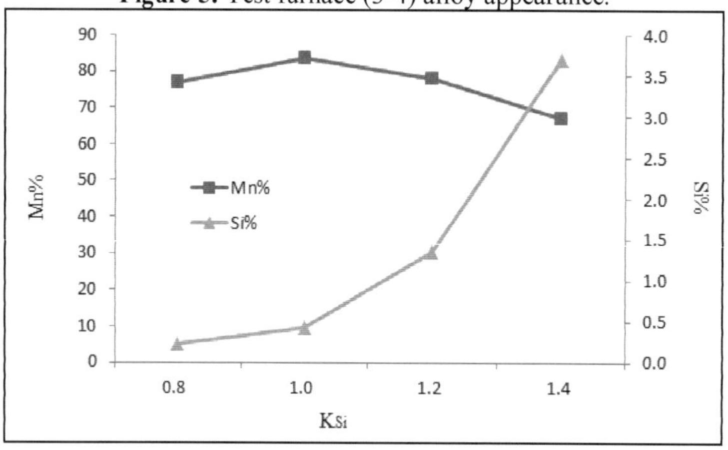

Figure 6. Test results at silicon ratios of 0.8, 1.0, 1.2, and 1.4.

It can be seen from Figure 6 that when the silicon addition coefficient K_{Si} rises from 0.8 to 1.0, the Mn% and Si% of the alloy rise slightly; when it rises from 1.0 to 1.4, the Mn% of the alloy slowly decreases, but the Si% increases obviously. When the silicon addition coefficient is low, it indicates that the amount of reducing agent is insufficient, so the Mn% of the alloy is low. When the silicon addition coefficient is increased, the amount of Si added will increase, and the amount of Si in the obtained alloy will increase, resulting in a decrease in the Mn content of the alloy.

6.2.4 The effect of holding time on product quality. Under the conditions of smelting temperature of 1400°C, silicon addition coefficient K_{Si} of 1.2, slag basicity of 1.0, and holding time of 20, 40, 60, and 90 min respectively, the test was carried out (corresponding to numbers 4-1, 4-2, 4-3, and 4-4 in Table 3). The effects of different holding times on the reduction of manganese ore, Mn content, and Mn yield rate of the alloy were studied.

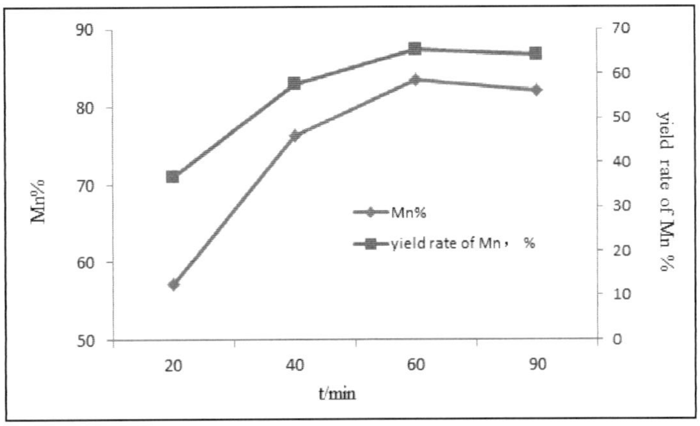

Figure 7. Mn% and the Mn yield rate of the alloy under different holding times.

According to Figure 7, when the holding time is 20 min, the Mn content and Mn yield rate of the obtained alloy are the lowest. It is analyzed that due to the short time, the silicon thermal reduction reaction cannot be fully carried out, and the generated ferromanganese alloy does not have enough time to settle and agglomerate. With the extension of the holding time, the two tend to increase respectively. From 60 min to 90 min, the Mn% and the Mn yield rate of the alloy are relatively stable.

7. Conclusions

(1) The micro-carbon ferromanganese alloy with qualified composition can be smelted in one step by electro silicon thermal method. Under laboratory conditions, due to the lack of kinetic conditions, the reaction process is not sufficient, and the overall yield rate of the Mn element is low.

(2) Smelting temperature, slag basicity, silicon addition coefficient K_{Si} and holding time have important effects on the test index. Under the conditions of 1400°C, slag basicity of 1.2, silicon addition coefficient of 1.0, and holding time of 60 min, the metal luster of the alloy is better, and the Mn content and Mn yield rate of the alloy are relatively high.

(3) Under laboratory conditions, the Mn yield rate of the alloy obtained by this method is lower than that of the traditional process. The thermodynamic balance between alloy melt and slag system and the reaction kinetics of this method still need to be further studied to improve the Mn yield rate. At the same time, it provides a theoretical basis for the expansion test.

References

[1] Zhou Jinhua. (1991) Ferroalloy production technology. Science Press, Beijing.

[2] Cui Xianyun et al. (2022) Study on the technology of producing low silicon and low micro carbon ferromanganese in multifunctional furnace. Ferroalloys. 53(03):1-5.

[3] Song Yaoxin, Pang Jianming, et al. (2022) Production practice of melting micro-carbon ferromanganese alloy in non-vacuum induction furnace. Ferroalloys. 53(05): 15-17.

[4] Baisanov S.O., Baisanov A.S., et al.(2018)Development of smelting technology of refined ferromanganese with special complex reducing agents. Izvestiya Visshikh Uchebnykh Zavedenii [J]. 61(9):689-694.

[5]Yessengaliyev D.; Kainenova T.et al. (2022)Feasibility study of using with high basicity manganese ore for smelting refined ferromanganese. Metalurgija [J]. 61(2):371-373.

[6] Li Hongxia. (2020) Advanced Technology of Energy Saving and Emission Reduction in Raw and Auxiliary Materials Production of Iron and Steel [M]. Metallurgical Industry Press, Beijing.

[7] Liang Yingjiao, Che Yichang. (1993) Data manual of inorganic thermodynamics. Northeastern University Press, Shenyang.

Journal of Physics: Conference Series **2842** (2024) 012059 doi:10.1088/1742-6596/2842/1/012059

Research on the influence of applying counterweight on the deflection and stability of cantilever beam of machine tool

Xinguo Jiang[1], Xifeng Fang[1*], Zhiduo Li[2,3], Shengwen Zhang[1]

[1]School of Mechanical Engineering, Jiangsu University of Science and Technology, Zhenjiang 212100, China
[2]Spallation Neutron Source Science Center, Guangdong Dongguan 110142, China
[3]Institute of High Energy Physics, Chinese Academy of Sciences, Beijing 100049, China

[*]Corresponding author's e-mail: xgjiang1997@163.com

Abstract. The deflection and stability of the cantilever beam of the machine tool are important factors that affect the machining accuracy of the machine tool. Because of the space size, it needs to be weighted within a limited size. By designing different weighting schemes, the static analysis and modal analysis are carried out by finite element analysis software, and the influence law of the deflection and stress of the beam is obtained, so the appropriate weighting scheme is selected. Through the analysis of different schemes, it is known that the deflection of the beam is the smallest under the action of the arm of 900mm and the force of 1000N, and the machine tool structure is more compact, which is the key component of the subsequent machine tool.

1. Introduction

Vertical machine tools are widely used because of their compact structure and high precision. However, it is challenging to ensure the theoretical analysis of the deflection and stability of the cantilever beam. For the cantilever beam under normal working conditions, the problem of excessive cantilever distance can be improved by increasing the auxiliary guide rail, and the counterweight block can be used to achieve force balance. However, the working conditions are special, and the length and width of the machine tool are limited. In this paper, the problem is improved by applying different counterweights to select the better counterweight scheme under this special working condition. The simulation test is carried out to test the stability of the deflection machine of the cantilever beam of the machine tool under different counterweight conditions, so as to select the appropriate counterweight scheme. The size design of the machine tool counterweight scheme is controlled within a reasonable range, which has important theoretical significance and practical value for improving the stability of the machine tool operation [1-3].

2. Analysis of influencing factors of the cantilever beam

2.1. Vertical machine tool cantilever beam

The machine tool bed is placed horizontally on the ground, and the cantilever beam moves in the XY direction under the action of the cross slide. Because the maximum distance of the cantilever beam is 900 mm and the maximum size of the beam is 1150 mm, in order to ensure the stability of the machine tool beam, it is necessary to apply a counterweight to reduce the deflection and improve the stability. The machine tool beam shutdown and operating conditions are shown in Figure 1.

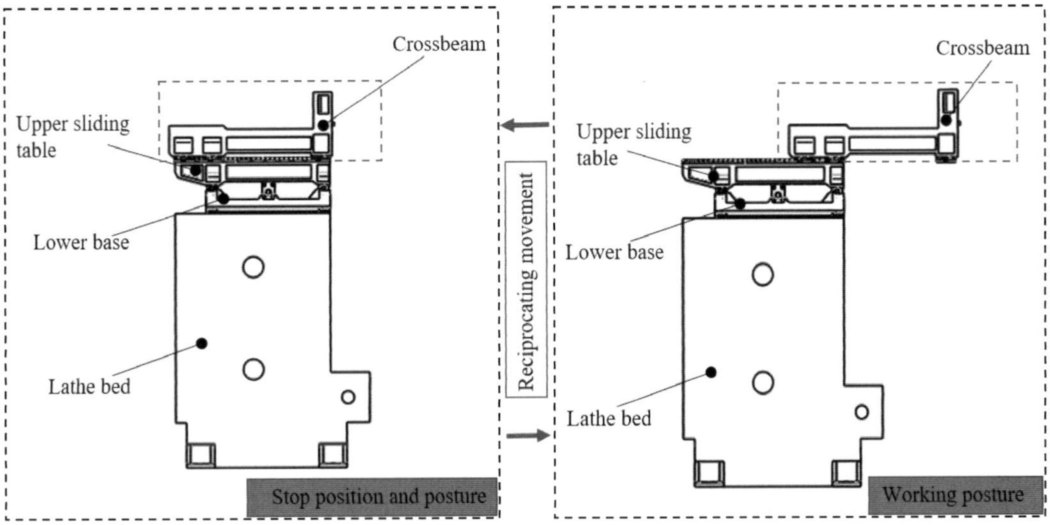

Figure 1. Working diagram of different positions and postures of machine tool crossbeam.

2.2. Deflection analysis of cantilever beam

In the project, the span is longer than the deflection of the bending member so that the beam axis presents a flat bending curve under the action of the bearing force [4]. The simplified model of the cantilever beam expansion deformation is shown in Figure 2.

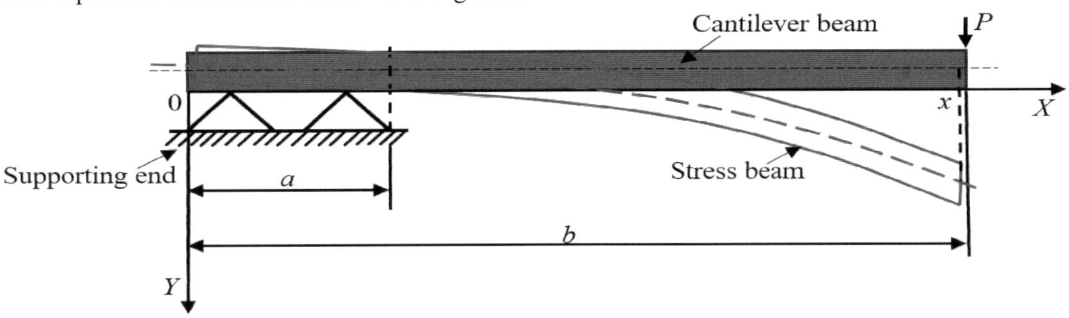

Figure 2. Schematic diagram of a cantilever beam of machine tool.

In the figure, a is the contact support surface between the cantilever beam and the slider; b is the overall length of the cantilever beam, x is the abscissa of any point before the deformation of the cantilever beam, and P is the normal stress on the end face of the cantilever beam.

Therefore, the coordinate system is set, and the relationship between the abscissa and the axis deflection after the beam deformation is:

$$\omega = f(x) \tag{1}$$

In the formula: ω is the deflection at this time. The axis of the cantilever beam after deformation is called the deflection curve, and the expression described is called the deflection curve equation. The deflection curve equation of the beam is obtained, and the relationship between curvature k and bending moment M is established:

$$k(x) = \frac{1}{p(x)} = \frac{M(x)}{EI} \tag{2}$$

In the formula, EI represents the bending strength, and the curvature of the plane curve can be expressed as:

$$\frac{1}{p(x)} = \pm\frac{\omega''}{(1+\omega'^2)^{\frac{3}{2}}} \quad (3)$$

Bringing (2) into (1), taking into account the opposite of M and ω, we can deduce that:

$$\frac{\omega''}{(1+\omega'^2)^{\frac{3}{2}}} = -\frac{M(x)}{EI} \quad (4)$$

Here, ω^2 is very small compared to 1, so $(1+\omega^2) \approx 1$, and (4) is approximately expressed as follows :

$$\omega'' = -\frac{M(x)}{EI} \quad (5)$$

Since the bending stiffness EI of the beam is a constant, the above equation can be rewritten as:

$$EI\omega'' = -M(x) \quad (6)$$

After integration, we can get:

$$EI\omega'' = -\int\left[\int M(x) + dx\right]dx + c_1 x + c_2 \quad (7)$$

The line load set p is derived from the load set P:

$$p = \frac{P}{b} \quad (8)$$

Based on the above theoretical analysis and calculation, it can be concluded that:

$$\omega = \frac{px^2}{24EI}(x^2 + 6l^2 + 4lx) \quad (9)$$

In the formula l = b-a, it is the overhang distance of the beam.

2.3. Deflection analysis of cantilever beam

The dynamic stiffness of the cantilever beam of the machine tool has an important influence on the machining accuracy and stability of the machine tool. When the machine tool is working, different excitations may cause the excitation frequency to coincide with the natural frequency of the cantilever beam, causing resonance, which affects the machining accuracy of the machine tool. Through modal analysis, we can understand the dynamic characteristics of the cantilever beam of the machine tool and thus improve the dynamic stiffness of the whole machine tool [5-6].

According to the theory of modal analysis, the response of the structure can be expressed by different modal parameters such as natural frequency, damping ratio, and vibration mode. These parameters constitute the response of each point of the structure under excitation. By superimposing these modal parameters, the modal shapes of different orders of the structure can be obtained [7]. The basic dynamic equation of the structure can be expressed as:

$$M\ddot{x} + C\ddot{x} + Kx = F(t) \quad (10)$$

In the formula, M is the mass matrix, C is the damping matrix, K is the stiffness matrix, x is the displacement vector, and F (t) is the excitation force vector.

When $F(t) = 0$, ignoring the effect of damping, the equation is simplified to the undamped free vibration equation:

$$M\ddot{x} + Kx = 0 \quad (11)$$

During free vibration, simple harmonic vibration occurs at each point of the structure, and the displacement of each point is:

$$x = \emptyset \cdot e^{-i\omega} \quad (12)$$

Bringing Formula (12) into Formula (11), we can get:

$$(\dot{K} - \omega^2 M) \cdot \emptyset = 0 \quad (13)$$

The characteristic equation of (11) is:

$$|K - \omega^2 M| = 0 \tag{14}$$

where ∅ represents the feature vector, which is used to describe the vibration mode of the structure; ω^2 represents the eigenvalue, and the arithmetic square root ω_i of the i-th eigenvalue is called the i-th-order natural frequency. The higher the natural frequency of the structure is, the higher the stiffness of the unit mass is.

3. Design of different counterweight schemes

According to the above mechanical analysis model, it is concluded that the cantilever beam needs to avoid the deflection and stability of the cantilever beam under working conditions. According to the principle of force balance, it is necessary to apply a counterweight to the end of the cantilever beam to adjust the balance [8-9]. The counterweight can be driven by a power source or mechanically apply different forces along the guide rod. The force balance model is shown in Figure 3.

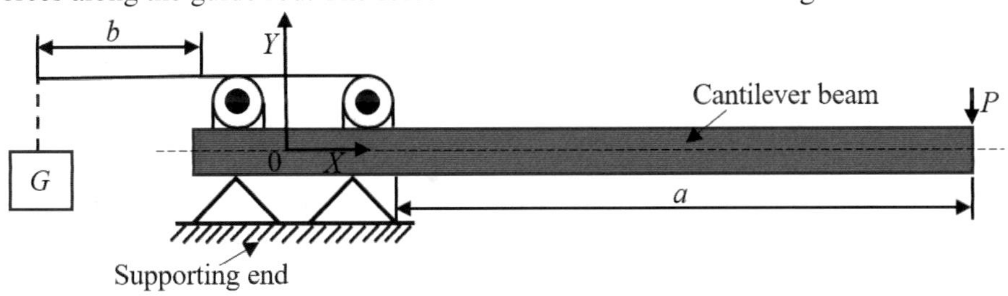

Figure 3. Schematic diagram of force balance of cantilever beam of machine tool.

In Figure 3, a means that the cantilever distance of the beam during operation is 750mm; P means that the normal stress on the end face of the cantilever beam is 1000N; B is the distance from the object to the supporting end; G is the weight of the counterweight.

Due to the limited working space of the machine tool and other factors, the distance from the object to the supporting end is $0 < b < 700$ mm, and the equation that can satisfy the force balance condition is:

$$G \cdot b \geq P \cdot a = 900000 N \cdot mm$$

According to its bending moment value, different combinations of counterweight and arm can be selected so as to determine the best counterweight scheme. Thus, the cantilever beam of the machine tool can achieve the minimum deflection and improve stability [10-11]. The selected counterweight and arm schemes are shown in Table 1.

Table 1. Reference Points for Machine Tool Movement at Different Positions.

Scheme type	Arm of force, b (mm)	Balance weight, G(N)
1	100	9000
2	200	4500
3	300	3000
4	400	2250
5	500	1800
6	600	1500
7	700	1285
8	800	1125
9	900	1000

4. Finite element analysis

In the actual installation process, the lathe bed is fixed on the ground by its own weight, the lower base is connected with the lathe bed by bolts, and the upper slide table and the lower base and the upper slide table and the cross beam are connected with the slider by guide rails. First, in order to improve the speed

Journal of Physics: Conference Series **2842** (2024) 012059 doi:10.1088/1742-6596/2842/1/012059

of finite element analysis, the analysis model is simplified, and some unnecessary features, such as small parts, holes, and round corners, are removed [12]. Then, the simplified model is imported into ANSYS Workbench software for analysis, and the types of bound contact and non-separated contact are set. Then, the grid is divided, and a total of 168,682 cells and 307,544 nodes are obtained in this grid division. After that, a 1000N normal load is applied to the front end of the beam, and the bottom of the bed is fixed. When no counterweight is applied, the deformation and stress changes of the beam during the movement are obtained by statics, as shown in Figure 4.

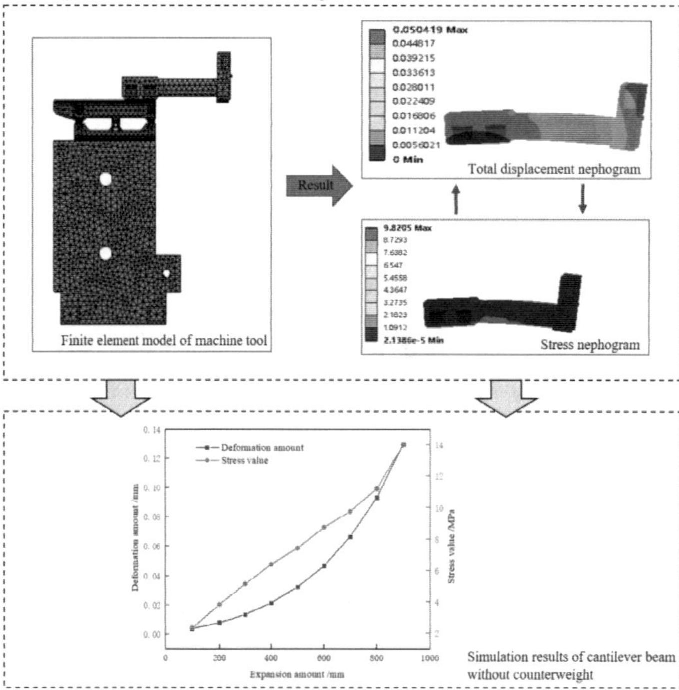

Figure 4. Simulation of cantilever beam without counterweight.

In order to clearly understand the influence of applying different counterweights on the whole process of beam movement, the positions where the beam of the machine tool moves to different positions are taken as reference points. By selecting different position points, the influence of different counterweight on the beam structure is analyzed, and the improvement of the deformation and stress value of the beam under all working conditions by different counterweight schemes is shown in Figure 5.

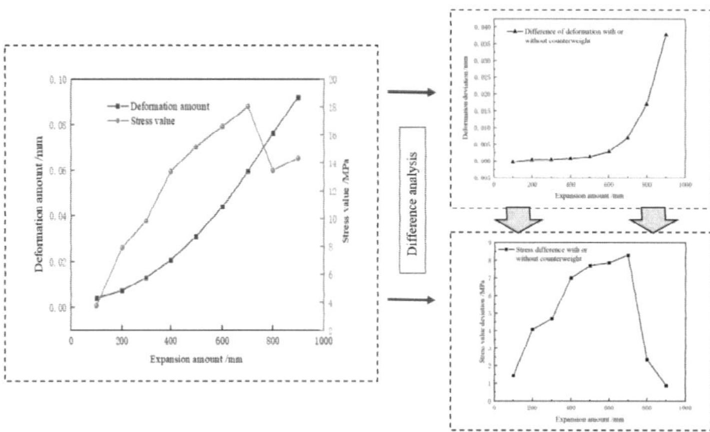

Figure 5. Simulation difference analysis of cantilever beam with counterweight.

Through the simulation analysis of the working motion stroke of the cantilever beam with a counterweight leveling device, the result is shown in Figure 5. With the increase of the cantilever distance of the beam, its deflection and deformation also increase, but the balancing weight leveling device has an obvious effect when the cantilever working distance of the beam is large, which can improve the deflection and deformation of the beam. The stress values of the cantilever beam under these two working conditions are numerically compared, and the deviation is obtained. The stress values increase with the increase of the cantilever distance of the beam, but the stress values change little when the cantilever beam is in the limit position. The overall stress values are less than the allowable stress, which meets the requirements of machine tool operators and is stable in operation. The analysis shows that the effect is the best when the counterweight is located at 900mm, and the mass is 1000 N.

5. Conclusion

Through the static and dynamic analysis of the cantilever beam with or without counterweight during machine tool operation, the deflection deformation and stress change of the beam are obtained, and the deflection of the beam is reduced by applying different counterweight schemes. According to the analysis results, the deflection of the beam is reduced. The results show that the scheme of applying counterweight under the action of arm 900mm and counterweight 10000N is reasonable and reliable, which provides a useful reference for the design of other key parts of the machine tool.

Acknowledgments

This work was financially supported by the Reconstruction of the thermal chamber of the spallation neutron source target station in China (DSS-WXGZ-2021-0010) and the Jiangsu Postgraduate Innovation Program Project (KYCX23_3858).

References

[1] Li Rongrong, Zhu Shuangkai. Deflection analysis of reinforced concrete cantilever beam based on ABAQUS [J]. Henan Science and Technology, 2018(07):111-113.

[2] Guo Tieneng, Cao Jinxuan, Cai Ligang, et al. Identification method of weak stiffness link of a cantilever beam using curve angle [J]. Journal of Beijing University of Technology, 2020,46(12):1323-1334.

[3] Zhang Dongsheng, Ning Wei, Song Chunming, et al. Finite element modal analysis of beam structure of gantry machine tool [J]. Mechanical Design and Manufacturing Engineering, 2014,43(05):21-23.

[4] Zhang W H, YAN J Y, HONG Y L, et al. Numerical analysis of large deflection of the cantilever beam subjected to a force pointing at a fixed point [J]. Applied Mathematical Modelling, 2021(92): 719-730.

[5] Wu Kepeng, Zhang Bang, Miao Fuxing. Finite element analysis of vibration response of flexible base/silicon laminated cantilever beam [J]. Journal of Applied Mechanics, 2021,38(04):1523-1530.

[6] Xue Chuang, Zeng Qingsheng. Finite element analysis and optimal design of telescopic arm of dredging pump lifting equipment [J]. Journal of South China University (Natural Science Edition), 2019,33(05):41-45.

[7] V. ONDRA, B. TITURUS. Free vibration and stability analysis of a cantilever beam axially loaded by an intermittently attached tendon[J]. Mechanical Systems and Signal Processing, 2021(158): 1-18.

[8] Xie Zhikun, Shi Keke, Liu Bocong, et al. Dynamic simulation of a cantilever beam of a laser cutting machine [J]. Modular Machine Tool and Automatic Machining Technology, 2013(07):19-21.

[9] Dai Yalan. Research on modal analysis method of machine tool structure based on particle swarm optimization [D]. Wuhan: Huazhong University of Science and Technology, 2019.

[10] Li Xiaolei. Research on reliability design and test technology of high-grade CNC lathe [D]. Harbin: Harbin Institute of Technology, 2020.

[11] Yang Ze, Li Hongzhou, Li Jianyong, et al. Simulation analysis of cantilever beam reliability based on ANSYS software [J]. Journal of Jilin Institute of Chemical Technology, 2017,34(05):34-37.
[12] Liu Jianying, Wang Xiaoyue, Gong Jinliang. Modal study of a cantilever beam with different boundary conditions [J].2017,36(19):221-226.

The influence of heat treatment on AlSi-polymer abradable coating

Caiyan Liu[1], Zhengping Mao[2] and Xiaojie Xu[1*]

[1]School of Artificial Intelligence, Shanghai Normal University Tianhua College, Shanghai 201815, China
[2]Oerlikon Metco Surface Technology (Shanghai) Co. Ltd, Shanghai, 201814, China

[*]Corresponding author's email. hitment@163.com

Abstract. Abradable coatings, as one of the most important technologies for aero and industry gas turbine engines, have an extremely significant influence on engine performance. The macro-hardness of AlSi-Polyester for low-pressure compressor and as-sprayed coating degradation mechanism has been investigated by heat treatment process in imitation service condition. SEM (Scanning Electron Microscope) and OM (Optical Microscope) were used to observe the morphology of powder and coating microstructure in cross-section, respectively. Rockwell hardness tester was used to measure the macrohardness of coatings before and after heat treatment. It shows that AlSi-Polyester powder has an approximately spherical shape and uniform microstructure with alloy phase and polyester phase distribution in the coating prepared by air plasma spraying. Oxidation of the alloy phase and reduced polyester, as well as macrohardness in the as-sprayed coating, occur after the heat treatment process.

1. Introduction

Due to the continuous advancement of technology and fast development of the aviation industry, the requirements for aeroengines are getting higher and higher. Economic fuel consumption, great efficiency, and high thrust have proven to be the prime targets of engine manufacturing and design. Abradable sealing coatings can improve the sealing performance of the engine's air path, enhance engine performance, and reduce fuel consumption and greenhouse gas emissions. They have been widely recognized by various engine manufacturers and successfully applied to commercial engines [1-3]. According to research data, a 1% improvement in the efficiency of the globally popular 2500 GW engine can save about 100 million tons of fuel per year and reduce approximately 300 million tons of carbon dioxide greenhouse gas emissions annually, effectively slowing down the trend of global warming [4]. Reducing the tip clearance is one of the most effective methods to improve engine efficiency and reduce fuel consumption [5]. Different materials in the engine have different thermal expansion coefficients, coupled with processing deviations in parts, tolerances in equipment, displacement deformation caused by vibration, and other factors. Inevitably, this will cause scraping between the rotor and stator in the engine and lead to tip wear. In severe cases, engine failure may occur, which limits the gap design between the blade and the casing [6]. As a sacrificial coating, the abradable sealing coating can be gently abraded away when the rotor contacts it, avoiding tip wear and effectively reducing the gap between the tip and the casing, thereby improving engine efficiency [7]. The main material systems for abradable coatings include aluminum-silicon-polymer, nickel-graphite, NiCrAl-bentonite, YSZ-polyphenylene ester, etc., which are applied to different parts of the engine according to temperature requirements [8-10]. In this study, thermal spraying technology was used to prepare aluminum-silicon-polymer abradable coatings for low-pressure compressor sections, revealing their microstructure and comparing performance changes before and after heat treatment.

2. Experimental process

2.1 Powder for abradable seal coatings

In this study, the Metco 601NS aluminum-silicon polyester powder from Oerlikon Metco was employed. With a composition of Al7Si40Polyester, this powder has been certified by numerous aviation OEMs. Prepared using a mixing method, it exhibits a particle size distribution ranging from 11 to 125 μm and a bulk density of approximately 0.9 g/cm³. The scanning electron microscope was used to inspect the powder shape.

2.2 Sample and coating preparation

The stainless steel test piece of 50 mm × 50 mm size and 3 mm thick was used in this experiment. The coating material was Metco 601NS powder. This coating was prepared using SinplexPro cascade plasma spray with a Multicoat system from Oerlikon Metco. Prior to adding the powder to the hopper, it was baked in an oven at approximately 70°C for 2 hours to remove atmospheric water vapor trapped in the powder and enhance its flowability. In order to increase the coating bond strength, the stainless steel test piece was sandblasted with 60 mesh white alumina sand. To investigate coating performance changes in an actual service environment, 24-hour heat treatment in air at 350°C was done for as-sprayed coating. The coating hardness was checked and expressed using HR15Y Rockwell hardness before and after heat treatment.

3. Discussion of results

3.1 Powder and coating microstructure

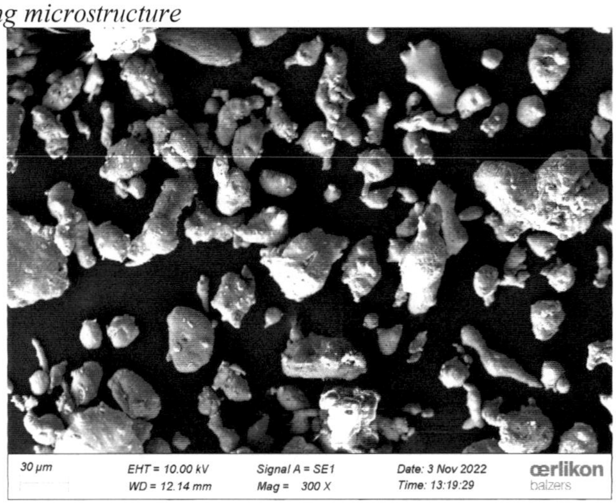

Figure 1. Morphology of Metco 601NS Powder.

As can be seen from the SEM image in Figure 1, the aluminum-silicon polyester powder exhibits an approximately spherical morphology, ensuring good flowability. The distribution of metal alloy phases and polyester is relatively uniform, meeting the consistency requirements for the composition of the subsequent coating. To control the macro-hardness and abradable performance of the coating, the particle size of the polyester is significantly larger than that of the metal phase. Using the optimized spraying process detailed in Table 1, the coating was prepared using the SinplexPro cascade spray gun. The microstructure of the coating is shown in Figure 2.

Table 1. Coating parameters of Metco 601NS sprayed by SinplexPro gun.

Parameters	Ar flow (NLPM)	H2 flow (NLPM)	Carrier gas flow (NLPM)	Nozzle diameter (mm)	Current (A)	Powder feeder rate (g/min)	Surface speed (m/min)	Spray distance (mm)
Metco601NS	90-150	1-4	2-4	9	300-500	60-100	60-100	100-150

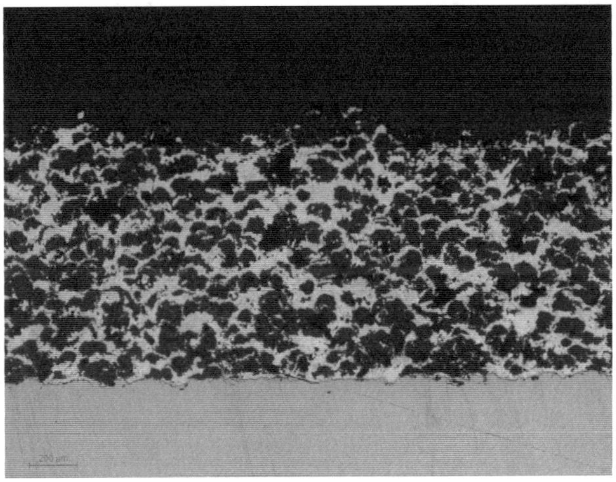

Figure 2. Cross-section morphology of Metco 601NS coating.

As can be observed from Figure 2, the metal phase (gray areas) and the abradable phase (polyester, black areas) are distributed relatively uniformly within the coating. The coating exhibits a high degree of density, ensuring the required abradable performance and corrosion resistance performance when the coating is in service in an engine. This microstructure ensures that the coating can effectively fulfill its function in high-temperature and high-pressure environments while maintaining its durability and performance during service.

3.2 Heat treatment for the coating

To investigate the changes in coating performance before and after exposure to actual operating conditions, the as-sprayed samples were heat treated in an atmospheric environment. 24-hour heat treatment in an ambient atmosphere was conducted at 350°C. Figure 3 compares surface topography for as-sprayed coating and the coating after heat treatment, where (a) shows the appearance of the as-sprayed coating and (b) the appearance of the heat-treated coating. The heat-treated coating photomicrograph is presented in Figure 4.

(a) (b)

Figure 3. Surface morphology of Metco 601NS coating before and after heat treatment.

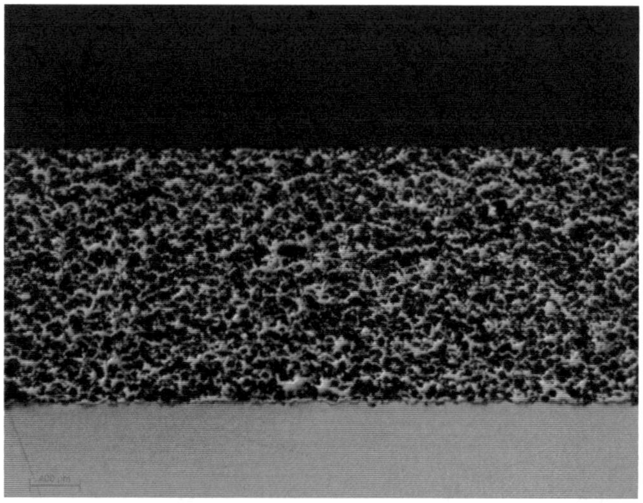

Figure 4. Cross-section morphology of heat-treated Metco 601NS coating.

Afterward, heat-treated for these as-sprayed samples, a noticeable change in coating appearance color was shown. The surface color of the coating transformed from a light gray color in the as-sprayed state to a darker gray after heat treatment. This change is primarily attributed to the burnout and volatilization of the polyester component during the heat treatment process, as well as the oxidation of the surface AlSi metal phase, resulting in the formation of corresponding oxides. The coating's micro-structure remained largely unchanged after heat treatment, maintaining a uniform distribution of metal (gray areas) and abradable polyester (black areas). However, the content of the abradable polyester phase decreased from 56% in the as-sprayed state to 46% after heat treatment. Additionally, the surface Rockwell hardness decreased from 53.3 HR15Y in the as-sprayed state to 23.6 HR15Y after heat treatment; see Figure 5 for details.

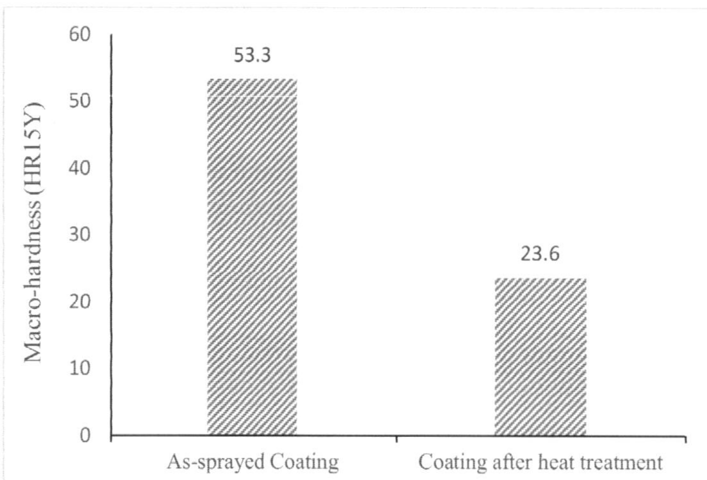

Figure 5. The macro-hardness for as-sprayed and heat-treated Metco 601NS coatings.

According to the specifications of aerospace engine OEMs for aluminum-silicon polyester coatings, the maximum operating temperature of this coating should not exceed 325°C. At the 350°C heat treatment condition, some polyester underwent burnout, creating pores within the coating. This led to a reduction in the abradable polyester phase and a decrease in macro-hardness. These observations align well with the change in coating surface color.

4. Conclusion

In this study, the microstructure of atmospheric plasma sprayed AlSi polyester coatings and the effects of heat treatment on both the micro-structure and macro-structure of the coatings were investigated. The specific conclusions are as follows:

(1) Using an optimized plasma spraying process and the SinplexPro cascade spray gun, a microstructure with a uniform distribution of metal and abradable phases can be achieved, resulting in a coating with excellent density.

(2) After atmospheric heat treatment at 350°C for 24 hours, the surface AlSi metal phase of the as-sprayed coating underwent oxidation, and the polyester underwent burnout. The polyester content in the as-sprayed coating decreased from 56% to 46%, and the surface Rockwell hardness decreased from 53.3 HR15Y to 23.6 HR15Y.

References

[1] Rajendran R. Gas turbine coatings- an overview[J]. Engineering Failure Analysis, 2012, 26, 355-369.

[2] Cui Y, Guo M, Wang C, et al. Zirconia-based abradable coatings fabricated on single-crystal superalloy: Microstructure, residual stress and mechanical properties[J]. Journal of Alloys and Compounds, 2022, 907:164537.

[3] Rahimov E, Watson M, Hadjisoteriou A, et al. Investigation of wear mechanisms in AlSi-polyester abradable - Ti(6Al4V) blade contacts using stroboscopic imaging[J]. Wear, 2022(494-495):204207.

[4] Siddiqui M S, Joshi P, Nayak N S, et al. Thermal spraying, optimization, and characterization of abradable seal coating for a gas turbine for service temperature up to 750 C[J]. Advanced Materials Letters, 2014, 5(9):506-510.

[5] Yang Guiming, Wang Mingfu, Wang Lianshi. The control technology of tip clearance on turboprop launched 6 engines [J]. Aviation Maintenance & Engineering,1997,(9):23-24.

[6] Goldspiegel F, Portemont G, Berthe J. On the Divergence mechanisms analysis of a vibrating blade subjected to repetitive rubs with an abradable coating[J]. Journal of Turbomachinery, 2022, 144(5):051008.

[7] Ali R, Huang T, Song P, et al. Tribological performance and phase transition of MAX-phase/YSZ abradable seal coating produced by air plasma spraying [J]. Ceramics International, 2022, 48(3):4188-4199.

[8] Chupp R E, Ghasripoor F, Turnquist N A, et al. Advanced seals for industrial turbine applications: dynamic seal development[J]. Journal of Propulsion and Power, 2002, 18(6): 1260-1266.

[9] Cui Y, Guo M, Wang C, et al. Zirconia-based abradable coatings fabricated on single-crystal superalloy: Microstructure, residual stress and mechanical properties[J]. Journal of Alloys and Compounds, 2022, 907:164537.

[10] Huang J, Liu R, Hu Q, et al. High-temperature abradable sealing coating for SiCf/SiC ceramic matrix composites [J]. Ceramics International, 2023,49(2):1779-1790.

Journal of Physics: Conference Series **2842** (2024) 012061 doi:10.1088/1742-6596/2842/1/012061

Fast determination of thermal conductivity of aluminum alloy by laser-induced breakdown spectroscopy

Yuheng Shan and Ruibin Liu*

Key Laboratory of Precision Spectroscopy and Optoelectronics Technology, School of Physics, Beijing Institute of Technology, China

*Corresponding author's email: liusir@bit.edu.cn

Abstract: The matrix thermal properties are closely linked to laser-induced plasma, because it is the heat effect predominantly governs the process when the nanosecond-pulsed laser acting on the material, particularly in metallic materials. In the study using a series of pure metal samples, We detected a substantial inverse linear relationship linking the matrix's thermal storage coefficient of the material to the temperature of the plasma. This discovery reveals that metals exhibiting reduced thermal conductivity or lower specific heat capacity necessitate a smaller amount of laser energy to achieve thermal spreading and to facilitate the transitions to the melted and vaporized states, which consequently results in a higher rate of material removal and higher plasma temperatures. Based on this correlation, a prediction model for the thermal conductivity of aluminum alloys has been developed, employing LIBS technique as analysis method, alongside PLS regression, with a relative error of below 1.5%. It presents a pioneering technique for the swift evaluation of thermal conductivity in aluminum alloys.

1. Introduction

As a laser-based spectroscopic analysis technique, Laser-induced Breakdown Spectroscopy (LIBS) enables identification and measurement of a sample's constituents and properties. Owing to its distinctive benefits of swift analysis and minimal sample impairment, it has gained extensive application across diverse domains, including metallurgy[1-3], coal sector[4], environmental studies[5], and realm of energetic materials[6]. Most studies on LIBS quantitative analysis of metals focus on measuring material composition content through stoichiometric methods. Rifai et al. [7] utilized two analytical methods for assessing the elemental composition within cooper-nickel ore samples, evaluating the performance of the Partial Least Squares Regression (PLSR) analysis in comparison to univariate calibration technique. Univariate calibration technique was moderately successful in predicting the levels of major elements, whereas PLSR was capable of quantifying every element and significantly enhancing the accuracy of the predictions for the major elements. The physical properties of the sample itself often have a close correlation with plasma induced by laser, offering a basis for the rapid determination of related physical parameters. For example, Sattar et al. [8] explored hardness impacts on the temperature within plasma generated by a laser focused on tungsten-heavy alloys, as well as electron density. The discussion delved into mechanism of influence of the ablation rate and plasma produced by the laser, taking into account the band gap traits, as well as grain size. A method for detecting surface hardness in alloys using LIBS has been put forward. The thermal properties are notably significant in the matrix effects within LIBS, given that the thermal impact holds a leading role throughout the interaction process between the nanosecond-pulsed laser and the material. In essence, laser advances the ionization via thermal transfer, thereby offering a potential way to gauge the thermal performance of samples via LIBS

In this paper, we concentrate on examining how the thermal attributes of the matrix impact ns-pulsed LIBS analysis with a selection of pure metals. Herein, temperature of plasma and ablation rate were measured using Saha Boltzmann plot method and a step profiler, respectively. By integrating the cumulative impacts of thermal characteristics, a robust correlation was established between the thermal retention factor of the pure metals and both plasma temperature and the rate of ablation, with the underlying mechanisms being elucidated. Based on this, we detected the thermal conductivity of aluminum alloy samples using the PLSR method, and finally completed model training based on 5 types of aluminum alloy samples. Therefore, the proposed method offers an innovative pathway to achieve swift and precise determination of the thermal conductivity of metallic substances.

2. Experiments and methods

2.1 Samples

During the experimental phase, a selection of 8 distinct grades of high-purity metals (exceeding 99% purity) were designated as S1 through S8. Five types aluminum alloys of 5000 series, were designated as A1 through A5. All these samples were accompanied by certification from JiNanZhongBiao Company. A compilation of key physical attributes for each sample is presented in Table 1, with their visual aspects depicted in Figure 1.

Table 1. Physical characteristics of all metals.

No.	Name	Density (g/cm^3)	Thermal Conductivity (W/m·K)	Specific Heat Capacity (J/g·°C)
S1	Pure Cd	8.64	92.00	0.230
S2	Pure Co	8.80	69.21	0.440
S3	Pure Fe	7.87	76.20	0.440
S4	Pure Mg	1.74	159.00	1.025
S5	Pure Ni	8.88	60.70	0.460
S6	Pure Sn	7.29	63.20	0.213
S7	Pure Ti	4.50	17.00	0.528
S8	Pure Zn	7.10	112.20	0.390
A1	Aluminum 5005	2.70	200.00	0.9
A2	Aluminum 5050	2.69	193.00	0.9
A3	Aluminum 5052	2.68	138.00	0.88
A4	Aluminum 5154	2.68	125.00	0.9
A5	Aluminum 5083	2.66	117.00	0.9

2.2 Devices

As depicted in Figure 1, a traditonal LIBS device was built to accomplish experimental measurements. The ablation process employs a Nd: YAG laser, operating at 1064 nm wavelength, featuring pulse duration from 3 to 5 ns, as well as an adjustable repetition under 10 Hz. Upon reflection by reflective mirror (R), laser pulse is concentrated onto sample's surface by lens (L1), with 50 mm focal length. High laser energy with 80 μm diameter focal spot, leading to an extraordinarily high density of laser power that surpasses 95 $GW\cdot cm^{-2}$. The lateral acquisition of plasma radiation signals is based on a lens module (L2 and L3) and linked to spectrometer through a fiber. The analysis of the spectrum is carried out using a four-channel spectrometer, covering the wavelength range of 180nm-690nm with a resolution of 0.07-0.1 nm. The overall timing control is achieved through a delay generator. In actual experiments, the laser energy is selected as 24 mJ, and the time for signal integration and the spectral acquisition delay are both set, correspondingly, to 30 μs and 1 μs.

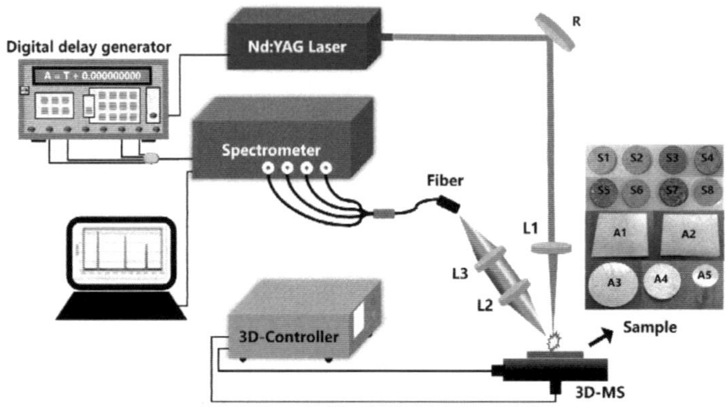

Figure 1. Schematic of self-built LIBS system.

For each sample, 25 positions are selected on the surface for ablation. For each position, 15 pulses are executed. The first 5 pulses serve as pre-ablation, with the aim of reducing the abnormal effects of surface oxide layers. The last 10 pulses are collected as effective spectral data. No oxygen atomic lines are observed in the 250 spectra collected from each sample, confirming the effectiveness of pre-ablation.

2.3 Analysis methods

In a state of complete or partial local thermodynamic equilibrium, the allocation of particles across different energy levels within the plasma adheres to the Boltzmann distribution. The transition probability between these levels significantly affects the intensity of spectral lines emitted from atoms and ions during plasma radiation. In such a scenario, a sequence of spectral lines from the same atomic species along with a corresponding Boltzmann plot[9], as depicted in Eq. (1), the plasma temperature is able to be determined.

$$\ln\frac{\lambda I_\lambda}{g_k A_{ki}} = -\frac{1}{k_B T}E_k + c \quad (1)$$

Herein, T represents the temperature of plasma, k_B denotes Boltzmann constant, λ signifies wavelength and I_λ corresponds to the intensity of the spectral line. The spectroscopic data A_{ki}, g_k, E_k are transition probability, degeneracy, and energy level's energy, respectively, which can be accessed through the spectral database provided by NIST[10].

If distinct spectra lines from atoms and ions species are discernible within spectra, plasma temperature can be deduced through the Saha-Boltzmann graph [11], as illustrated in Eq. (2).

$$\left(\ln\frac{\lambda I_\lambda}{g_k A_{ki}}\right)^{\wedge} = -\frac{1}{k_B T}E_k{}^{\wedge} + c \quad (2)$$

Here, the spectral lines of ions must be adjusted in accordance with Eqs. (3) and (4).

$$\left(\ln\frac{\lambda I_\lambda}{g_k A_{ki}}\right)^{\wedge} = \ln\frac{\lambda I_\lambda}{g_k A_{ki}} - \ln[2\left(\frac{2\pi m k_B}{h^2}\right)^{\frac{3}{2}}\frac{T^{\frac{3}{2}}}{n_e}] \quad (3)$$

$$E_k{}^{\wedge} = E_k + E_{IP} + \Delta E \quad (4)$$

where m symbolizes the mass of an electron and h stands for the Planck constant. The terms E_{IP} and ΔE refer to ionization potential of corresponding element and the associated plasma interaction correction, respectively. These values can likewise be sourced from the NIST database [10].

Considering S1 as an illustrative instance, the temperature of the plasma produced form pure metals is ascertained by employing the Saha-Boltzmann graph as shown in Figure 2(a). Compared to Boltzmann plot, Saha-Boltzmann graph improves temperature measurements' accuracy by utilizing more spectra lines generated by both atoms and ions species of identical element. This method allows for consideration of upper energy levels within a much broader range, which enhances fitting accuracy.

For instance, the goodness of fit for S1 is 0.999. Therefore, the plasma temperatures determined through the Saha-Boltzmann plot are selected for the subsequent studies of the pure metal specimens.

Surface morphology of metals after ablation is measured using a step profiler, as shown in Figure 2(b), and the ablation volume is calculated as representative of the ablation rate.

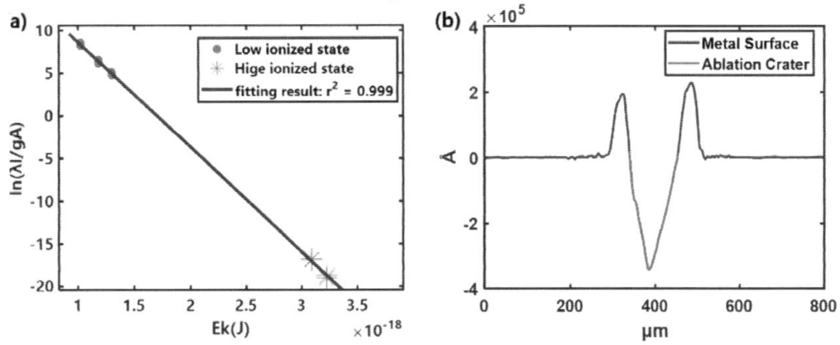

Figure 2. (a) Saha-Boltzmann plot of S1; (b) Cross-sectional view of ablation crater of S1.

3. Results and discussions

3.1 Thermal properties in metals

During the engagement of the laser with the material, the thermal conductivity and specific heat capacity collectively influence the thermal dispersion across the material's surface, consequently impacting laser energy's efficacy in the key reaction region, as shown in properties of plasma induced by laser. Recognizing that laser energy dissipation extends beyond thermal dispersion, the residual energy on localized surface exerts more pronounced effect on plasma's temperature. In this context, S, the heat storage coefficient, is delineated to quantify the material's capacity to retain heat in discrete areas, encompassing both the thermal diffusivity and the heat storage capacity of the substrate, as illustrated in Eq. (5) [12].

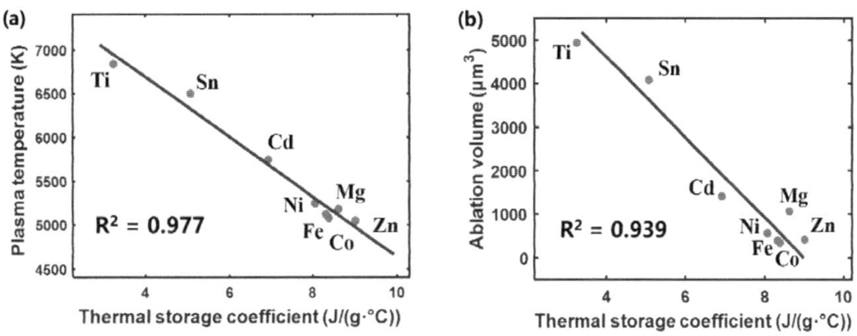

Figure 3. Relationship between the thermal storage coefficient of S1-S8 and (a) the temperature of plasma, as well as (b) ablation volume.

$$S = \sqrt{\frac{2\pi\rho C_p \kappa}{\tau}} \tag{5}$$

where τ represents the period of harmony, which is commonly considered to be 24 hours..

Figure 3(a) illustrates plasma's temperature fluctuations related to the heat storage coefficient, which is observable there is a invese linear relation between plasma's temperature induced by the laser and matrix's heat storage coefficient, with an R-squared value exceeding 0.95. This correlation indicates that plasma's temperature inversely relates to thermal conductivity and specific heat capacity of samples matrix. When irradiating sample surface by a ns-pulsed laser, lower thermal conductivity

Journal of Physics: Conference Series **2842** (2024) 012061 doi:10.1088/1742-6596/2842/1/012061

metals may lose less energy via thermal transfer, resulting in a greater concentration of laser energy in confined areas and consequently more pronounced ablation. Lower specific heat capacity metals demand less laser energy for melting process and vaporization process over a localized surface area. As a result, this scenario enhances ablation efficiency and accelerates temperature increase. In light of these observations, metals with a low heat storage coefficient expend less laser energy on thermal diffusion and phase change processes, allowing for a greater amount of energy to be absorbed by the ensuing plasma, culminating in elevated plasma temperatures."The correlation between ablation volume and the heat storage coefficient, as illustrated in Fig.3(b), substantiates the mechanism explanation detailed previously.

Based on this, it can be considered that the laser-induced breakdown spectra contain hidden information about the thermal properties of metals, including thermal conductivity. This provides us with opportunity to employ the LIBS technique to quickly measure thermal conductivity of aluminum alloys.

3.2 Thermal conductivity prediction models for aluminum alloys

The thermal conductivity prediction models for aluminum alloys have been developed utilizing the PLSR method. Before the training of the model, among the five, one aluminum alloy sample's LIBS spectrum is randomly selected as the data of the prediction set. The remaining quartet is employed as calibration datasets for modeling. During the training of the model, the quantity of principal components, which is the key hyperparameter in the PLSR method, is determined to be two. This decision comes from that cumulative interpretation rate of the initial two key components have achieved 98.04%, which is above the threshold of 95%. This surpassing suggests that the initial pair of principal components encompasses the majority of the information inherent in the spectral data pertaining to aluminum alloys.

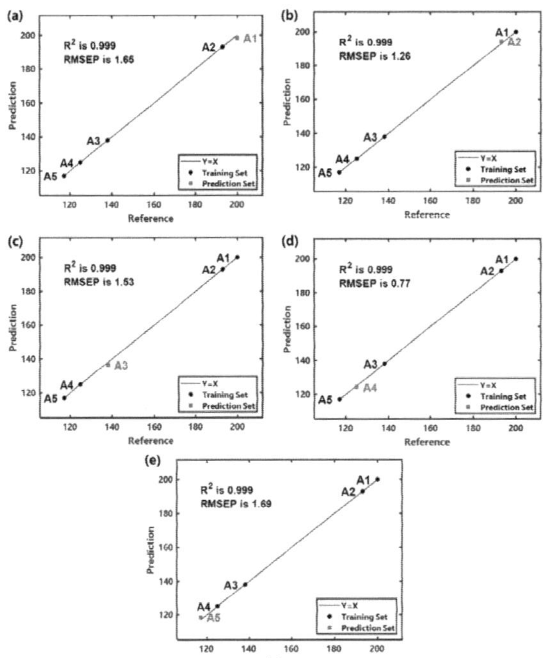

Figure 4. The thermal conductivity prediction model for aluminum alloys. The prediction set is (a) A1, (b) A2, (c) A3, (d) A4, and (e) A5.

The predictive model of the thermal conductivity of aluminum alloys derived from different calibration sets is shown in Figs. 4 (a)-(e), respectively. The standard values are provided by the manufacturer using the traditional method, as shown in Table 1, whereas the predicted values are obtained from our thermal conductivity prediction model. The principal assessment criteria of each model are also displayed in Figure 4, with R^2 signifying the determination coefficient and RMSEP indicating root mean square error of prediction part. Other important model evaluation indicators of prediction part are displayed within Table 2, including relative error (REP) and absolute error (AEP). Each of the five prediction models boasts an R^2 of 0.999 and a REP below 1.5%. The favorable outcomes of the predictions substantiate the efficacy of employing the LIBS-based technique for assessing thermal conductivities of aluminum alloys.

Table 2. Model evaluation indices of the thermal conductivity prediction model.

Prediction Set	R^2	REP	AEP
A1	0.999	0.83%	1.65 W/m·K
A2	0.999	0.65%	1.26 W/m·K
A3	0.999	1.11%	1.53 W/m·K
A4	0.999	0.62%	0.77 W/m·K
A5	0.999	1.44%	1.69 W/m·K

4. Conclusion

In our research, we employed a series of pure metal specimens to examine the impact of the matrix's thermal characteristics on plasma induced by ns-pulsed lasers. The temperature of the plasma, ascertained through spectroscopic analysis, was found to have a correlation with the matrix's thermal attributes, notably thermal diffusivity. Consequently, we introduced heat storage coefficients, revealing a substantial inverse relationship with both plasma temperature and the volume of ablation. This phenomenon is understandable when considering that metallic materials with lower thermal conductivity or specific heat capacity demand less heat in processes of thermal diffusion and phase change, leading to increased ablation efficiency and elevated plasma temperatures. Consequently, we developed an efficacious PLSR predictive model for the thermal diffusivity of aluminum alloys, utilizing the spectra from laser-induced breakdown. The thermal conductivity prediction model has an R^2 of 0.999 and a REP below 1.5%, thereby substantiating the precise and rapid means to determine the thermal conductivity of aluminum alloy is feasible.

Acknowledgments.

The authors express deep gratitude for the sponsorship provided by the National Key Research and Development Project.

References

[1] D. Díaz-Romero, S. Van den Eynde, I. Zaplana, and etc. "Classification of aluminum scrap by laser induced breakdown spectroscopy (LIBS) and RGB+ D image fusion using deep learning approaches," Resources, Conservation and Recycling **190**, 106865 (2023).

[2] J. Yao, Q. Yang, X. He, and etc. "Spectral filtering method for improvement of detection accuracy of Mg, Cu, Mn and Cr elements in aluminum alloys using femtosecond LIBS," RSC advances **12**, 32230-32236 (2022).

[3] S. Grünberger, V. Ehrentraut, S. Eschlböck-Fuchs, J. Hofstadler, and etc. "Overcoming the matrix effect in the element analysis of steel: Laser ablation-spark discharge-optical emission spectroscopy (LA-SD-OES) and Laser-induced breakdown spectroscopy (LIBS)," Analytica Chimica Acta **1251**, 341005 (2023).

[4] X. Xu, A. Li, X. Wang, and etc., "The high-accuracy prediction of carbon content in semi-coke by laser-induced breakdown spectroscopy," Journal of Analytical Atomic Spectrometry **35**, 984-992 (2020).

[5] Y. Zhang, T. Zhang, and H. Li, "Application of laser-induced breakdown spectroscopy (LIBS) in environmental monitoring," Spectrochimica Acta Part B: Atomic Spectroscopy **181**, 106218 (2021).

[6] A. Li, X. Zhang, Y. Yin, and etc."High-accuracy measurement of the heat of detonation with good robustness by laser-induced breakdown spectroscopy of energetic materials," Journal of Analytical Atomic Spectrometry **38**, 810-817 (2023).
[7] Kheireddine Rifai, Lütfü Özcan, François Doucet, and etc. "Quantification of copper, nickel and other elements in copper-nickel ore samples using laser-induced breakdown spectroscopy," Spectrochimica Acta Part B: Atomic Spectroscopy **165**.
[8] H. Sattar, H. Ran, W. Ding, M. Imran, and etc. "An approach of stand-off measuring hardness of tungsten heavy alloys using LIBS," Applied Physics B **126**, 5 (2019).
[9] B. Bousquet, V. Gardette, V. M. Ros, and etc. "Plasma excitation temperature obtained with Boltzmann plot method: Significance, precision, trueness and accuracy," Spectrochimica Acta Part B: Atomic Spectroscopy **204**, 106686 (2023).
[10] A. Kramida, Y. Ralchenko, and N. A. Team, "NIST Atomic Spectra Database (version 5.9)," (National Institute of Standards and Technology, 2021).
[11] J. A. Aguilera, and C. Aragón, "Multi-element Saha–Boltzmann and Boltzmann plots in laser-induced plasmas," Spectrochimica Acta Part B: Atomic Spectroscopy **62**, 378-385 (2007).
[12] R. Singh, and M. Dabas, "Neural Network-based Prediction of Effective Heat Storage Coefficient of Building Materials," in *Defect and Diffusion Forum*(Trans Tech Publ2014), pp. 73-78.

Journal of Physics: Conference Series **2842** (2024) 012062 doi:10.1088/1742-6596/2842/1/012062

Determination of stress triaxiality by experiment-simulation and obtaining of johnson-cook damage criterion for Al-Mn alloy

Wei Wang [1], Yongbo Hu [1,*], Dashuai Zhang [1], Dao Chi [2] and Guobin Wang [2]

[1] School of Automotive Engineering, Wuhan University of Technology, Wuhan, Hubei, 430000, China

[2] Valeo Automotive Air Conditioner Hubei Co., Ltd, Jingzhou, Hubei, 434000, China

[*]Corresponding author's e-mail: yongbo.hu@whut.edu.cn

Abstract: Al-Mn alloy is extensively implemented in new energy vehicles. Damage prediction is the key to its safety application. The Johnson-Cook (JC) constitutive relationship parameters of Al-Mn alloy were first determined by standard tensile tests to predict the damage behavior of Al-Mn alloy. Subsequently, different notch specimens were designed to study the influence of stress state on damage behavior. By combining tensile tests and simulations, the stress triaxiality of Al-Mn alloy under different stress states was accurately obtained. On this basis, the JC damage criterion parameters were obtained. Finally, the obtained JC constitutive relationship and damage criterion parameters were imported into the tensile simulation models. The results showed that the force-displacement curves of the tensile experiments and simulations matched well, which can effectively simulate the deformation and fracture behavior of the Al-Mn alloy.

1. Introduction

Lightweight materials are extensively implemented in new energy vehicles to improve driving range. Aluminum alloy is one of the preferred materials for automotive lightweighting. While considering vehicle lightweighting, it is also necessary to consider the reliability of the vehicle and the safety of drivers and passengers. When a car experiences bumps or collisions, the body and its components need sufficient stiffness and strength to resist deformation or fracture. Therefore, studying the plastic and fracture behavior of aluminum alloys is crucial for their safe application in new energy vehicles [1].

Due to the widespread application of aluminum alloys, many scholars have studied the plastic properties of aluminum alloys. Some studies have shown that the plastic properties of aluminum alloys are sensitive to strain rates. Guo et al. [2] studied the stress-strain relationship of 3003 aluminum alloy through compression experiments from $10^{-3}s^{-1}$ to $10^{4}s^{-1}$ rates and found that 3003 aluminum alloy has significant strain rate sensitivity. Some studies have also shown that the plastic properties of aluminum alloys are temperature-sensitive. Wang et al. [3] investigated the plastic property of 7050-T7451 aluminum alloy. The results showed that all the flow stresses of tested specimens presented decreasing trends when temperature increased from 25°C to 400°C. Based on the research on the plastic property of aluminum alloys, some scholars have established a series of constitutive relationship models to characterize the plastic property of aluminum alloys on the influence of work hardening, strain rate,

Journal of Physics: Conference Series **2842** (2024) 012062 doi:10.1088/1742-6596/2842/1/012062

and temperature, like the Johnson-Cook (JC), the Khan-Huang-Liang and the Hansel-Spittel constitutive model. The JC constitutive model is extensively applied in various fields because of its agreement between simulation and experiments [4]. Seshadri et al. [5] used the JC constitutive relationship to successfully predict the deformation behavior of 2024 aluminum alloy at various rates.

Damage prediction is also crucial for the safe application of aluminum alloys. Some studies have confirmed that the material damage behavior is affected by its stress state, which affects its failure behavior [6]. Johnson and Cook proposed the JC damage criterion by studying material fracture under different stress states. Mehra et al. [7] effectively predicted the fracture behavior of 6061 aluminum alloy during impact using the JC fracture criterion.

To precisely fit the JC damage criterion, it is crucial to obtain the stress triaxiality of the material under different stress states. Bridgman [8] designed circular bar tensile specimens with different notch radii and provided a stress triaxiality calculation formula for these specimens. However, during the actual tensile process, the stress triaxiality of the specimen will change with the increase of strain. In addition, for notched specimens, the stress triaxiality in cross-section varies. Therefore, the stress triaxiality calculated through theoretical formulas is not applicable. To address this issue, some scholars have used a combination of simulation and experimental methods to determine the stress triaxiality of materials during the tensile process. Cao et al. [9] obtained the stress triaxiality of the maximum local strain position during the loading process of the tensile specimen by simulation. Then, the JC damage criterion was established and combined with experimental results, and the fracture behavior of the pipeline steel was successfully predicted.

Among numerous aluminum alloys, the Al-Mn alloy is mainly used in heat exchangers and power battery casings of new energy vehicles. When the casings undergo severe deformation or even rupture, the battery pack will experience thermal runaway, seriously affecting the safety of the driver and passengers. Therefore, studying the plastic and fracture behavior of Al-Mn alloys is crucial. However, it is hard to process round bar-notched specimens for aluminum alloy sheets commonly used in industrial applications. Thus, the method to get stress triaxiality by bar-notched specimens is not applicable. In response to this issue, some scholars have studied the stress state of flat groove specimens. Bai et al. [10] investigated the stress triaxiality of flat tensile specimens with grooves and provided a calculation formula. But, for sheets with smaller thicknesses, it is difficult to process the grooves above. In addition, the range of groove radius is also limited, which leads to a smaller range of stress triaxiality that can be studied. In response to this issue, the flat notch tensile tests were designed in this paper to obtain different stress states for Al-Mn alloy sheets. Subsequently, the fracture strain-stress triaxiality curve of Al-Mn alloy was obtained by combining experiments and simulation. Finally, the JC damage criterion of Al-Mn alloy sheets was established and validated by comparing the experiment results with those of the numerical calculation results.

2. Experiment

The plastic and damage behavior of Al-Mn alloys has been studied through tensile tests. The test specimens were cut from the Al-Mn alloy sheets with a thickness of 0.8mm. Four specimens were designed, including standard tensile specimens, R5, R10, and R20 notched tensile specimens. The geometric dimensions of these specimens are shown in Figure 1. A standard tensile specimen was used to obtain the constitutive relationship of the Al-Mn alloy. Both standard and three different notched specimens were designed to obtain the failure strain under various stress states, thereby obtaining the JC damage criterion of the Al-Mn alloy. All samples were tested on an MTS microcomputer-controlled electronic universal testing machine with 2.5mm/min, 25mm/min, and 100mm/min as the loading rates. To accurately obtain the strain during the tensile process of the specimen, a 50mm gauge length extensometer was applied to obtain the deformation. For each strain rate, three specimens were tested to reduce testing errors. The experimental results are shown in Figure 2. The tensile strength and fracture strain did not show significant rate dependency. The fracture strain was mainly affected by the specimen notch morphology, which was also because the stress state was related to the notch morphology.

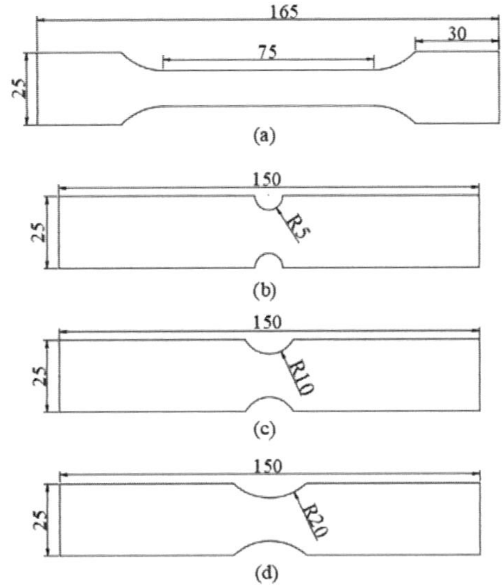

Figure 1. Geometric dimensions of specimens (a) Standard (b) R5 notched (c) R10 notched (d) R20 notched.

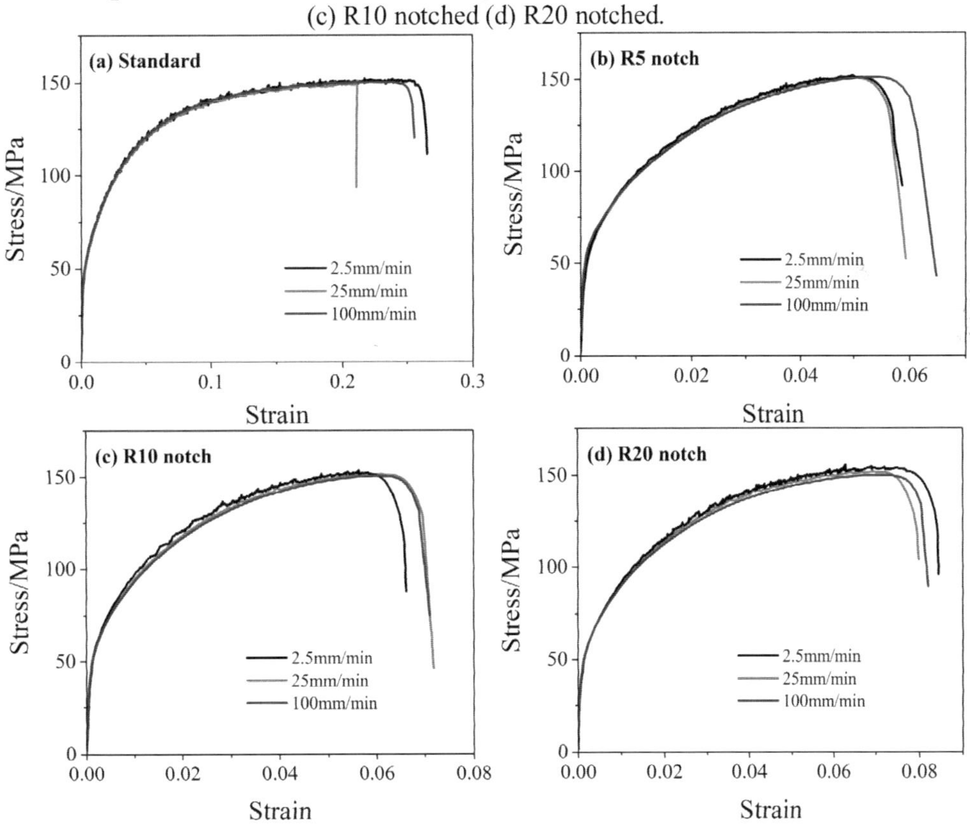

Figure 2. Stress-strain curves of different specimens.

3. Results and discussion

3.1. JC constitutive model

The mathematical model of the JC constitutive relationship takes the form of the following:

$$\sigma_{eq} = (A + B\varepsilon_p^n)(1 + C\ln\dot{\varepsilon}^*)(1 - T^{*m}) \quad (1)$$

In the formula, $A, B, n, C,$ and m are the constitutive model parameters. σ_{eq} represents equivalent flow stress, ε_p represents plastic strain. $\dot{\varepsilon}^* = \dot{\varepsilon}/\dot{\varepsilon}_0$, where $\dot{\varepsilon}$ is the strain rate and $\dot{\varepsilon}_0$ is the reference strain rate. $T^* = (T - T_r)/(T_m - T_r)$, where $T, T_r,$ and T_m represent the working temperature, room temperature, and melting point of the material, respectively.

The Al-Mn alloy tensile experiments in this study were implemented in a room-temperature environment, so the effect of temperature on the plastic behavior of this alloy was not considered. As shown in Figure 3, the alloy hardening effect does not exhibit rate sensitivity within the loading range of 2.5mm/min-100mm/min. So, Equation (1) can be simplified as:

$$\sigma_{eq} = A + B\varepsilon_p^n \quad (2)$$

According to the results of standard tensile tests, the elastic modulus of the material is 51.23 GPa. Model parameter A can be obtained from the true stress-strain curve start point. The model parameters B and n were acquired by fitting the true stress-strain curve. Finally, the JC constitutive model parameters for Al-Mn alloy are summarized in Table 1.

Table 1. JC constitutive model parameters for Al-Mn alloy.

Parameter	A (MPa)	B (MPa)	n
Value	30.3	287.7	0.37

3.2. JC damage criterion

The JC damage criterion assumes that material damage accumulates during plastic strain. The mathematical model of JC damage criterion takes the form of:

$$\varepsilon_f = (D_1 + D_2\exp(-D_3\eta))(1 + D_4\ln\dot{\varepsilon}^*)(1 + D_5T^*) \quad (3)$$

In the formula, D_1- D_5 are damage criterion parameters. η is stress triaxiality.

Like the study of constitutive relationships, the temperature factor is not considered. So, Equation (3) can be converted to:

$$\varepsilon_f = (D_1 + D_2\exp(-D_3\eta))(1 + D_4\ln\dot{\varepsilon}^*) \quad (4)$$

After simplifying the JC damage criterion, the obtaining of the parameters from D_1 to D_4 is carried out in two steps:

Step 1: We determine parameters D_1- D_3.

We select 2.5mm/min as the reference strain rate. At this rate, the value of $\dot{\varepsilon}^* = \dot{\varepsilon}/\dot{\varepsilon}_0$ is 1, so Equation (4) can be converted to:

$$\varepsilon_f = D_1 + D_2\exp(-D_3\eta) \quad (5)$$

To obtain D_1- D_3, it is necessary to establish the failure strain-stress triaxiality relationship. This paper investigates the effect of stress triaxiality on the failure strain of Al-Mn alloy by designing notched specimens for tensile testing. Due to the variation of the stress triaxiality of the notched specimen over time during the tensile process, this paper obtains the stress triaxiality of fracture moment through a combination of experiments and simulation analysis.

The standard and notched specimens finite element models established in ABAQUS software are shown in Figure 3. The element size at both ends is 2mm, and the elements in the middle section are refined with a size of 1mm. A velocity load was performed on the top of the model. The bottom of the model was fixed. C3D8R solid element was selected as the element type. The material property was set according to the JC constitutive model obtained above.

Firstly, for each type of tensile specimen, the PEEQ (equivalent plastic strain) and stress triaxiality curves of the central element at the fracture section of the finite element model over simulation time can be obtained. Based on the intermediate variable of simulation time, the stress triaxiality and PEEQ curves were established, as shown in Figure 4.

Secondly, the PEEQ can be acquired at the fracture moment based on experimental results. Then, according to Figure 4, the stress triaxiality at the corresponding moment is read. As shown in Table 2, the fracture strain decreases when stress triaxiality increases.

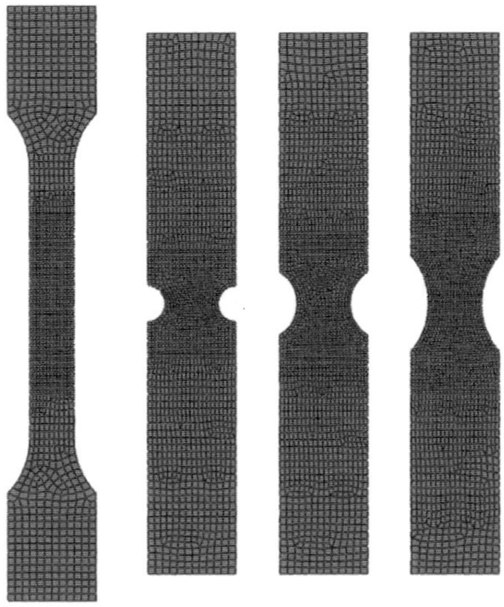

Figure 3. FE model of a tensile specimen.

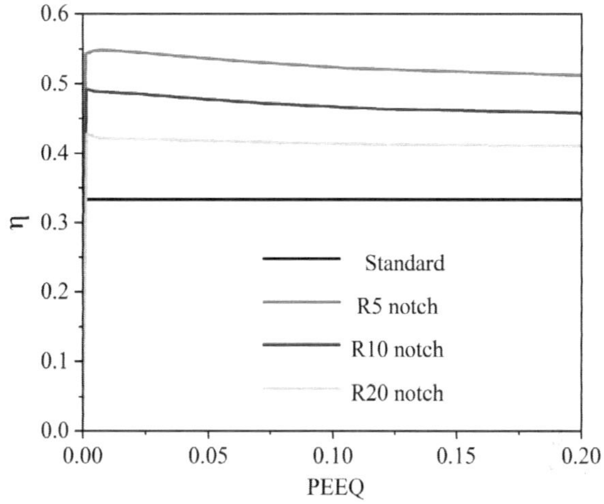

Figure 4. Stress triaxiality-equivalent plastic strain the curve of different specimens.

Table 2. Fracture strain and stress triaxiality of different specimens.

	Standard	R5	R10	R20
ε_f	0.226	0.085	0.119	0.170
η	0.333	0.518	0.461	0.409

According to Table 2, the JC damage criterion was fitted, as shown in Figure 5.

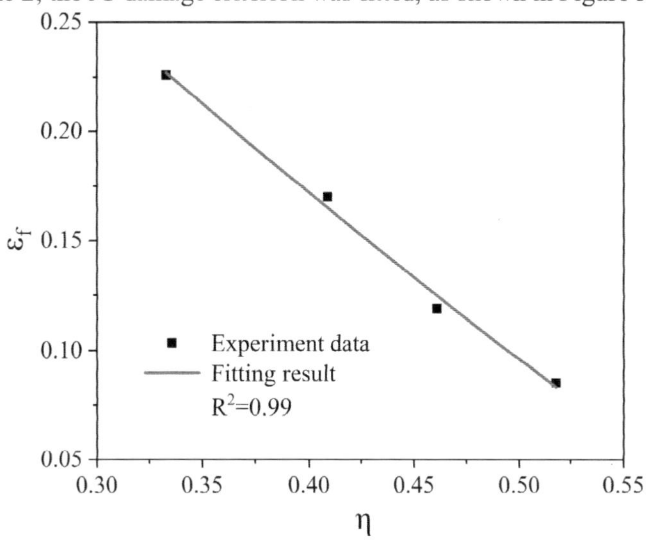

Figure 5. Fitting of JC damage criterion.

Step 2: We determine parameters D_4.

According to the fracture strain of tensile tests with 25mm/min and 100mm/min as tensile rate, damage parameter D_4 was obtained by fitting Equation (4).

In summary, the Johnson-Cook damage criterion parameters for the Al-Mn alloy were obtained, as shown in Table 3.

Journal of Physics: Conference Series **2842** (2024) 012062 doi:10.1088/1742-6596/2842/1/012062

Table 3. JC damage criterion parameters for Al-Mn alloy.

Parameter	D_1	D_2	D_3	D_4
Value	-0.708	1.264	0.905	-0.009

3.3. Verification of JC constitutive model and damage criterion

Finite element models were used to verify the effectiveness of the JC constitutive relationship and damage criterion. The models are consistent with the previous models in Section 3.2, except for including the damage criterion. The fracture location of simulations and experiments with different tensile specimens are shown in Figure 6. It is observed that the notched specimens all fractured from the root of the notch, which is consistent with the experiment and verifies the effectiveness of the JC damage criterion in simulating the damage behavior of the notched specimens. Furthermore, the load-displacement curves of finite models under various stress triaxiality were compared with experiment results. The load-displacement curves of finite models match well with the experimental results, which further verifies the accuracy of the JC damage criterion. The fracture displacement errors of standard, R5, R10, and R20 specimens are 2.7%, 16.9%, 14.4%, and 6.9%, respectively. The peak loads errors of standard, R5, R10, and R20 specimens are 2.8%, 1.1%, 3.2%, and 4.3%, respectively.

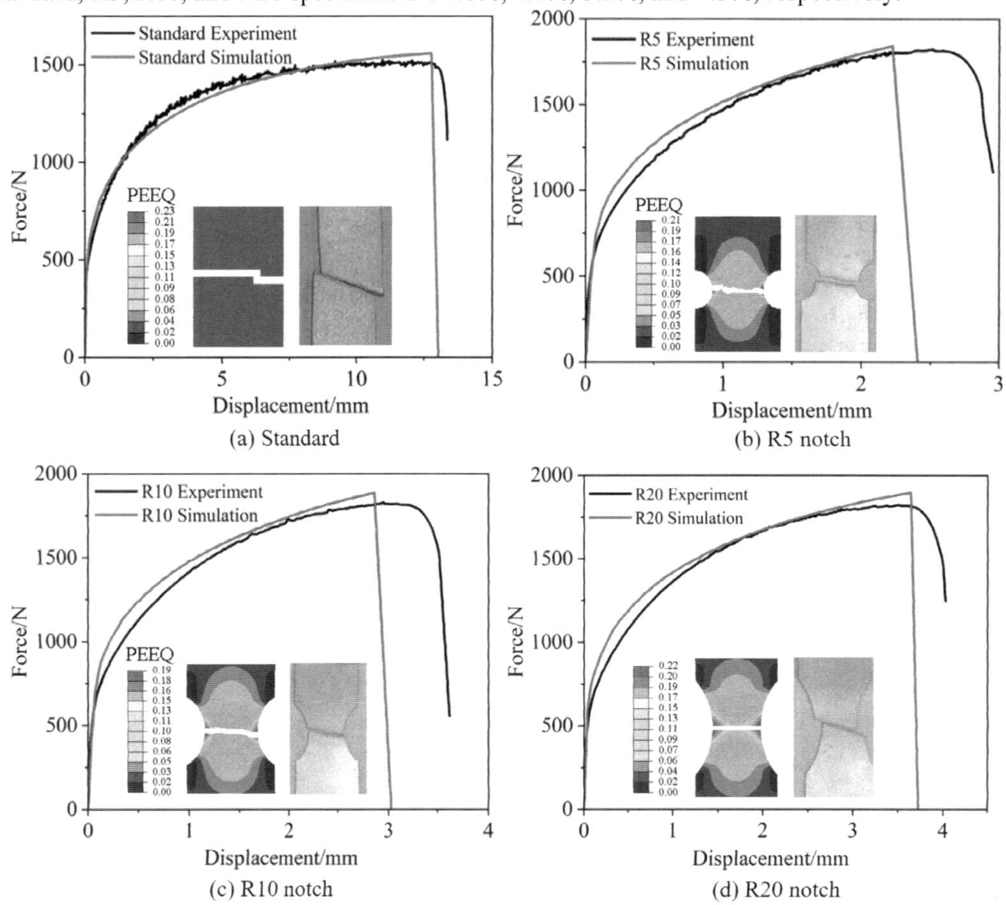

Figure 6. Experiments and simulations force-displacement curves of different stress triaxiality.

The SEM results of the Al-Mn alloy tensile fracture section are shown in Figure 7. Aluminum alloy is generally a ductile material. However, as shown in Figure 7, the plastic deformation of this aluminum alloy in the necking stage is small due to the size effect of thin plate materials. The dimples are shallow and not particularly obvious. This can also be seen from the stress-strain curve in Figure 2, where the average plastic strain between the starting point of necking and fracture for all samples was only 0.7%. The simulation result showed that the average plastic strain between the starting point of necking and fracture was 0.4%. This also demonstrated the effectiveness of the simulation results.

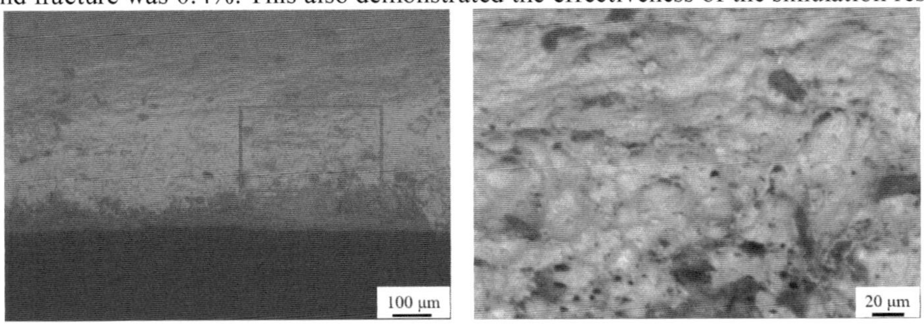

(a) (b)

Figure 7. (a) Tensile fracture section of Al-Mn alloy (b) Enlarged view of the red box area.

4. Conclusion

This paper investigated the plastic and fracture behavior of Al-Mn alloy, obtained the JC constitutive relationship and damage criterion of Al-Mn alloy, and validated the effectiveness of the model by a combination of numerical simulation and experiments. The conclusion is as follows:

(1) As the strain increases, the stress of Al-Mn alloy first rapidly grows, then slows down. Finally, the flow stress sharply decreases after reaching the tensile strength, with a short necking. Within the rate range (2.5 mm/min-100 mm/min), the plastic behavior of Al-Mn alloy does not show significant rate dependence. JC constitutive relationship parameters for Al-Mn alloy were obtained based on standard tensile test results.

(2) The fracture strain of Al-Mn alloy decreases when stress triaxiality increases. JC damage criterion parameters for Al-Mn alloy were obtained by combining tensile experiments and simulations.

(3) The obtained JC constitutive relationship and damage criterion parameters were imported into the tensile simulation model. The results showed that the force-displacement curves of tensile experiments and simulations matched well, which can effectively simulate the plastic and failure behavior of the Al-Mn alloy.

References

[1] Binze W, Zhengyang Z, Guochang X, et al. (2023) Wrought and cast aluminum flows in China in the context of electric vehicle diffusion and automotive lightweighting. J. Resources, Conservation and Recycling, Vol.191: 106877.

[2] Guo W G, Zhang X Q, Su J, et al. (2011) The characteristics of plastic flow and a physically-based model for 3003 Al-Mn alloy upon a wide range of strain rates and temperatures. J. European Journal of Mechanics A/Solids, Vol. 30 (No. 1): 54-62.

[3] Wang B, Liu Z Q, Song Q H, et al. (2019) A Modified Johnson-Cook Constitutive Model and Its Application to High Speed Machining of 7050-T7451 Aluminum Alloy. J. Journal of Manufacturing Science & Engineering, Vol. 141 (No. 1): 1-15.

[4] O'Toole L, Haridas R S, Mishra R S, et al. (2023) Determination of Johnson-Cook plasticity model parameters for CoCrMo alloy. J. Materials Today Communications, Vol.34.

[5] Seshadri R, Naveen I, Sharan S, et al. (2013) Finite Element Simulation of the Orthogonal Machining Process with Al 2024 T351 Aerospace Alloy. J. Procedia Engineering, Vol. 64 (No. 0): 1454-63.

[6] Fanlei M, Huiping L, Dejing Z. (2024) Ductile fracture prediction of 7A62 high-strength aluminum alloy under a wide range of stress states. J. Engineering Research Express, Vol. 6 (No. 1): 015525.

[7] Vishal M, Sambaran P, Shashank C. (2014) Possibility of Shear Fracture in Low-Velocity Impact of Al-6061 Projectiles on Al-6061Plates. J. Procedia Engineering, Vol. 86 (No.0): 295-301.
[8] Bridgman PW. (1952) Studies in large plastic flow and fracture. New York: McGraw-Hill.
[9] Cao Y, Zhen Y, Song M, et al. (2020) Determination of Johnson–Cook parameters and evaluation of Charpy impact test performance for X80 pipeline steel. J. International Journal of Mechanical Sciences, Vol. 179 (No. 0): 105627.
[10] Yuanli B, Xiaoqing T, Tomasz W. (2009) On the Application of Stress Triaxiality Formula for Plane Strain Fracture Testing. J. Journal of Engineering Materials & Technology, Vol. 131 (No. 2): 021002.

Journal of Physics: Conference Series **2842** (2024) 012063 doi:10.1088/1742-6596/2842/1/012063

Influence of heat treatment process on the microstructure and hardness of Q345B steel

Yan Wang *

College of Mechanical & Power Engineering, China Three Gorges University, Yichang 443002, Hubei, China

*Corresponding author's e-mail: wychinasxdx@ctgu.edu.cn

Abstract: This paper studied and determined the heat treatment processing technology of Q345B steel, and analyzed its metallographic structure and hardness after different quenching and tempering methods. The results showed that the salt quenching cooling rate was the fastest, resulting in the most martensite and the highest hardness. The hardness of salt-quenched, water-quenched, and oil-quenched steel decreased sequentially, but all of them are significantly higher than the initial hardness of Q345B steel. 950°C was more suitable for quenching Q345B steel than 900°C, with better microstructure and hardness. After quenching, tempering is conducted, and under the same quenching process, three different types of tempering are conducted. The hardness of low-temperature tempering, medium-temperature tempering, and high-temperature tempering decreased sequentially, but they were not lower than the initial hardness of Q345B steel.

1. Introduction

Low alloy high-strength steel is a type of engineering structural steel developed by adding a small amount of alloying elements based on carbon steel to improve its strength and usability. Q345 steel belongs to low alloy high-strength steel and is the most commonly used type of steel in China. With the increasingly widespread application of steel, the requirements for the performance of Q345 steel are also increasing, and in-depth research is also needed [1-7]. Q345 steel can be divided into five types according to grade: Q345A, Q345B, Q345C, Q345D, and Q345E. This paper investigated the influence of the heat treatment process (quenching+tempering) on the microstructure and hardness of Q345B steel. The research in this paper had both theoretical and practical significance.

2. Experimental materials and parameters

Q345B is a low alloy high-strength steel, in which Q represents the yield strength, 345 represents the yield strength value in MPa, and B represents the qualified impact at 20°C. The chemical composition of Q345B is C≤0.20%, Si≤0.50%, Mn≤1.70%, P≤0.035%, S≤0.035%, Nb≤0.07%, V≤0.15%, Ti≤0.20%, Cr≤0.30%, Ni≤0.50%, Cu≤0.30%, N≤0.012%, and Mo≤0.10%. Under a 500x microscope, the initial microstructure of Q345B steel was observed, as shown in Figure 1. The initial microstructure of Q345B steel was ferrite and pearlite. The initial hardness of Q345B steel was 182.11 HV. This paper mainly studied the influence of the heat treatment process on the microstructure and hardness of Q345B, to obtain Q345B steel with better performance through heat treatment. Finally, the heat treatment method of quenching and tempering was chosen. Quenching is followed by tempering. The main purpose of quenching was to obtain martensite and improve mechanical properties.

Journal of Physics: Conference Series **2842** (2024) 012063 doi:10.1088/1742-6596/2842/1/012063

Tempering mainly adjusted the hardness to facilitate cutting and eliminate residual internal stress. The final heat treatment process parameters were determined as shown in Table 1, with a heating rate of 8.83 °C/min. In the text, salt cooling referred to the cooling of 10% NaCl solution, water cooling referred to the cooling of clean water, and oil cooling referred to the cooling of No. 10 engine oil.

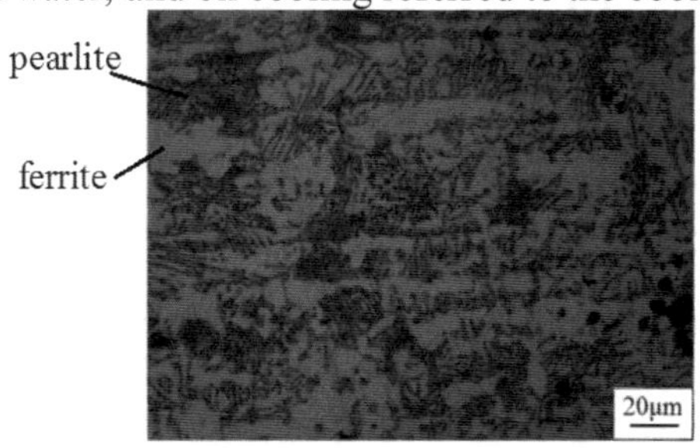

Figure 1. Initial microstructure of Q345B.

Table 1. Q345B steel heat treatment process parameters.

Sample number	Quenching	Tempering
Sample 1	900°C holding time 60 min (salt cooling)	200°C holding time 60 min (air cooling)
Sample 2	900°C holding time 60 min (salt cooling)	400°C holding time 60 min (air cooling)
Sample 3	900°C holding time 60 min (salt cooling)	650°C holding time 60 min (air cooling)
Sample 4	900°C holding time 60 min (water cooling)	200°C holding time 60 min (air cooling
Sample 5	900°C holding time 60 min (water cooling)	400°C holding time 60 min (air cooling)
Sample 6	900°C holding time 60 min (water cooling)	650°C holding time 60 min (air cooling)
Sample 7	900°C holding time 60 min (oil cooling)	200°C holding time 60 min (air cooling)
Sample 8	900°C holding time 60 min (oil cooling)	400°C holding time 60 min (air cooling)
Sample 9	900°C holding time 60 min (oil cooling)	650°C holding time 60 min (air cooling)
Sample 10	950°C holding time 60 min (salt cooling)	200°C holding time 60 min (air cooling)
Sample 11	950°C holding time 60 min (salt cooling)	400°C holding time 60 min (air cooling)
Sample 12	950°C holding time 60 min (salt cooling)	650°C holding time 60 min (air cooling)
Sample 13	950°C holding time 60 min (water cooling)	200°C holding time 60 min (air cooling)
Sample 14	950°C holding time 60 min (water cooling)	400°C holding time 60 min (air cooling)
Sample 15	950°C holding time 60 min (water cooling)	650°C holding time 60 min (air cooling)
Sample 16	950°C holding time 60 min (oil cooling)	200°C holding time 60 min (air cooling)
Sample 17	950°C holding time 60 min (oil cooling)	400°C holding time 60 min (air cooling)
Sample 18	950°C holding time 60 min (oil cooling)	650°C holding time 60 min (air cooling)

3. Metallographic analysis of Q345B steel after quenching

3.1. Microstructure after quenching at 900°C

We place Samples 1-9 in a heat treatment furnace, heat them up to 900°C, and then set the holding time for 60 minutes. We waited until the end of the heat preservation. Samples 1-3 were cooled in salt water, Samples 4-6 were cooled in water, and Samples 7-9 were cooled in oil. The microstructure obtained after quenching at 900°C is shown in Figure 2.

The microstructure of Sample 1 in Figure 2 (a) after salt quenching was martensite, with relatively fine grains. The microstructure of Sample 4 in Figure 2 (b) after water quenching was martensite, with larger grains than those after salt quenching. The microstructure of Sample 7 in Figure 2 (c) after oil quenching was bainite and pearlite.

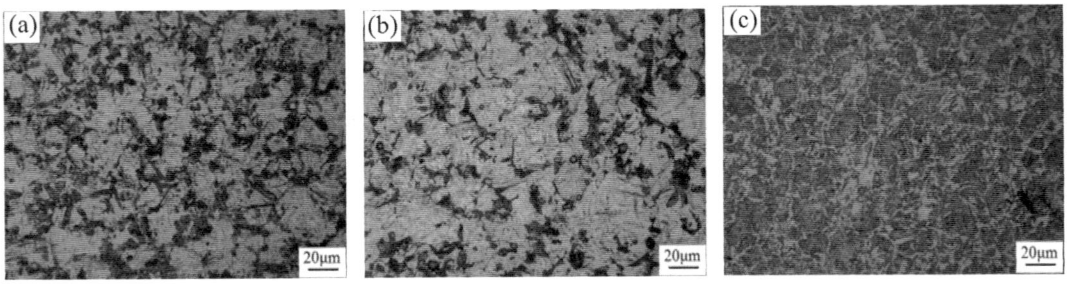

Figure 2. Microstructure after quenching at 900°C: (a) Sample 1 salt quenching; (b) Sample 4 water quenching; (c) Sample 7 oil quenching.

After observation and analysis, the cooling rate of Sample 1 quenched with salt water at 900°C was the highest, so there was the most martensite in the microstructure and the grains were relatively small. The cooling rate of Sample 4 after water quenching was lower than that of salt quenching and higher than that of oil quenching, so there was less martensite in the microstructure than that of salt quenching, and the grain size was coarser than that of salt quenching. The cooling rate of oil-quenched Sample 7 was lower than that of salt-quenched and water-quenched samples, and there was no martensite obtained, only bainite and pearlite.

3.2. Microstructure after quenching at 950°C

We place Samples 10-18 in a heat treatment furnace, heat them up to 950°C, and then set the holding time for 60 minutes. After heat preservation was completed, Samples 10-12 were cooled in salt water, Samples 13-15 were cooled in water, and Samples 16-18 were cooled in oil. The microstructure obtained after quenching at 950°C is shown in Figure 3.

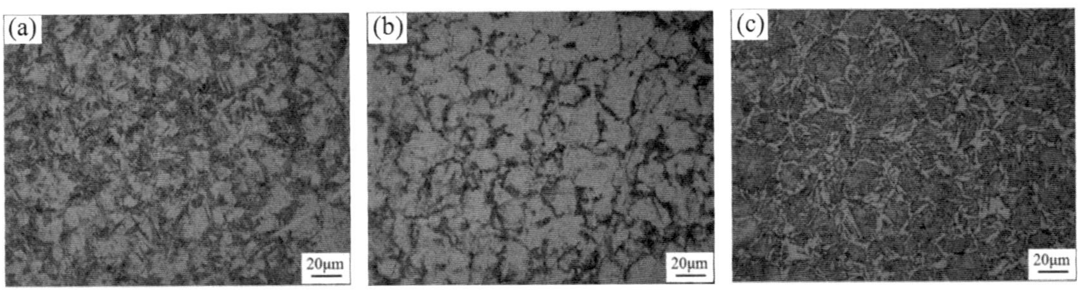

Figure 3. Microstructure after quenching at 950°C: (a) Sample 10 salt quenching; (b) Sample 13 water quenching; (c) Sample 16 oil quenching.

The microstructure of the salt-quenched Sample 10 in Figure 3 (a) was martensite, with very fine grains, and there was more martensite quenched in salt than quenched in water. The reason was that the cooling rate of salt-quenched Sample 10 was the highest. The microstructure of water-quenched Sample 13 in Figure 3 (b) was martensite, with larger grains than salt quenched. The martensite obtained by water cooling was less than that obtained by salt cooling. The reason was that the cooling rate of water quenching was lower than that of salt quenching. The microstructure of the oil-quenched Sample 16 in Figure 3 (c) was bainite and pearlite. The cooling rate of oil-quenched Sample 16 was the lowest, indicating that there was almost no martensite in the metallographic structure.

3.3. Comparison of metallographic structures between 900°C quenching and 950°C quenching

The microstructure after salt quenching at 950°C had more martensite and smaller grains compared to 900°C. After water quenching, the grain size at 950°C was smaller than that at 900°C. After oil quenching, the grain size at 950°C was coarser than that at 900°C, but the bainite content at 950°C was higher than that at 900°C. The reason was that appropriately increasing the austenitizing temperature can improve the stability of austenite, shift the C curve to the right, and thus improve the hardenability of the steel.

4. Metallographic analysis of Q345B steel after quenching and tempering

After quenching treatment and different tempering treatments, the microstructure of Samples 1-18 will undergo varying degrees of change.

4.1. Microstructure after quenching at 900°C and tempering at different temperatures

(1) 900°C quenching + low temperature tempering

We place salt-quenched Sample 1, water-quenched Sample 4, and oil-quenched Sample 7 into a heat treatment furnace for heating at 200°C, holding for 60 minutes, and then conduct air cooling. We observe the metallographic structure after low-temperature tempering as shown in Figure 4.

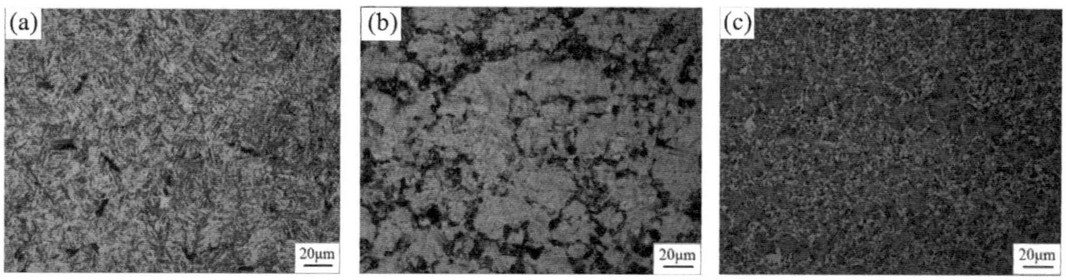

Figure 4. Microstructure after 900°C quenching (salt-cooled, water-cooled, and oil-cooled) + low-temperature tempering: (a) Sample 1 salt quenching; (b) Sample 4 water quenching; (c) Sample 7 oil quenching.

The microstructure of salt-quenched and low-temperature tempered Sample 1 contained more tempered martensite, with relatively small grains. The microstructure of water-quenched and low-temperature tempered Sample 4 also contained tempered martensite, with obvious grain growth. The microstructure of Sample 7 after oil quenching and low-temperature tempering was bainite and pearlite. By comparative analysis, it can be found that salt quenching and low-temperature tempering had the most tempered martensite, which means that the retained strength and hardness were the most.

(2) 900°C quenching + medium temperature tempering

We place salt-quenched Sample 2, water-quenched Sample 5, and oil-quenched Sample 8 into a heat treatment furnace for heating at 400°C, holding for 60 minutes, and then conduct air cooling. We observe the microstructure after medium temperature tempering as shown in Figure 5.

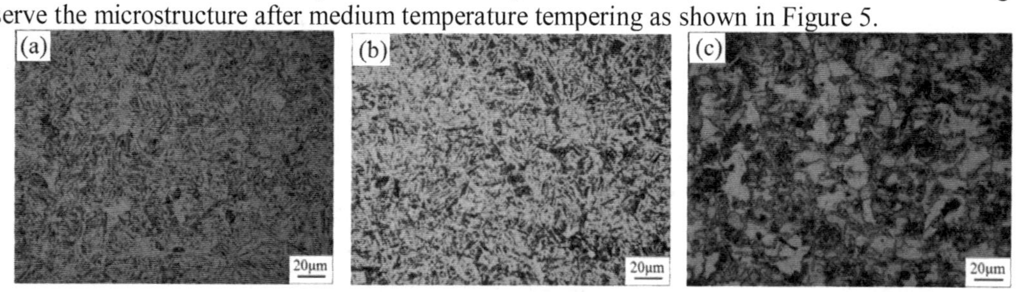

Figure 5. Microstructure after 900°C quenching (salt-cooled, water-cooled, and oil-cooled) + medium temperature tempering: (a) Sample 2 salt quenching; (b) Sample 5 water quenching; (c) Sample 8 oil quenching.

The microstructure of Sample 2 tempered at medium temperature in salt quenching was mainly tempered troostite. The microstructure of the water-quenched medium temperature tempered Sample 5 was also tempered troostite, but the grain size was slightly larger than that of salt quenched. The microstructure of Sample 8 tempered at medium temperature in oil quenching was ferrite and pearlite. The distribution of metallographic structure in the above three figures was relatively uniform. Comparing the analysis of three figures, it was found that after salt quenching and medium temperature tempering, the tempered troostite was the most abundant and relatively small, which meant that the strength and hardness were the highest.

(3) 900°C quenching + high temperature tempering

We place salt-quenched Sample 3, water-quenched Sample 6, and oil-quenched Sample 9 into a heat treatment furnace for heating at 650°C, holding for 60 minutes, and then conduct air cooling. We observe the metallographic structure after high-temperature tempering as shown in Figure 6.

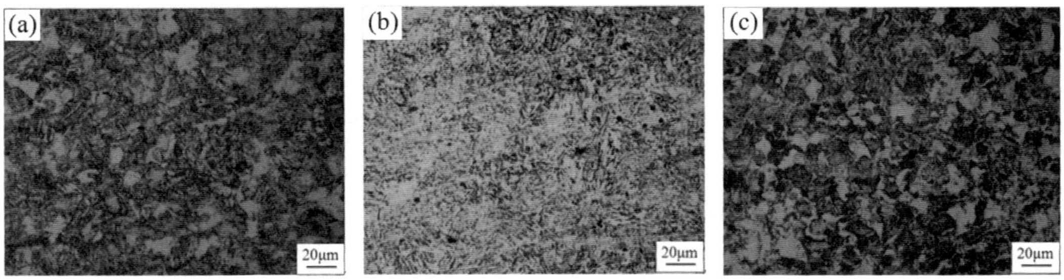

Figure 6. Microstructure after 900°C quenching (salt cooled, water cooled, and oil cooled) high-temperature tempering: (a) Sample 3 salt quenching; (b) Sample 6 water quenching; (c) Sample 9 oil quenching.

The microstructure of Sample 3 after salt quenching and high temperature tempering was mainly tempered sorbite, with some pearlite and ferrite also present. The microstructure of the water-quenched and high-temperature tempered Sample 6 was mainly composed of tempered sorbite, pearlite, and ferrite. The microstructure of Sample 9 after oil quenching and high-temperature tempering was pearlite and ferrite. From these three figures, the tempered sorbite composition was the highest in the microstructure after salt quenching and high-temperature tempering.

4.2. Microstructure after quenching at 950°C and tempering at different temperatures

(1) 950°C quenching + low temperature tempering

We place salt-quenched Sample 10, water-quenched Sample 13, and oil-quenched Sample 16 into a heat treatment furnace for heating at 200°C, holding for 60 minutes, and then conduct air cooling. We observe the metallographic structure after low-temperature tempering as shown in Figure 7.

Figure 7. Microstructure after 950°C quenching (salt cooled, water cooled, and oil cooled) +low temperature tempering: (a) Sample 10 salt quenching; (b) Sample 13 water quenching; (c) Sample 16 oil quenching.

Journal of Physics: Conference Series **2842** (2024) 012063 doi:10.1088/1742-6596/2842/1/012063

The microstructure of the salt-quenched and low-temperature tempered Sample 10 contained more tempered martensite. The microstructure of the water-quenched and low-temperature tempered Sample 13 was also tempered martensite. The microstructure of Sample 16 subjected to oil quenching and low-temperature tempering was pearlite and ferrite. By comparing and analyzing the above three figures, the tempered martensite was the most after 950°C salt quenching and low-temperature tempering. Due to the excellent comprehensive mechanical properties of tempered martensite structure, the microstructure with 950°C salt quenching and low-temperature tempering was the best.

(2) 950°C quenching + medium temperature tempering

We place salt-quenched Sample 11, water-quenched Sample 14, and oil-quenched Sample 17 into a heat treatment furnace for 400°C heating, holding for 60 minutes, and then conduct air cooling. We observe the microstructure after medium temperature tempering as shown in Figure 8.

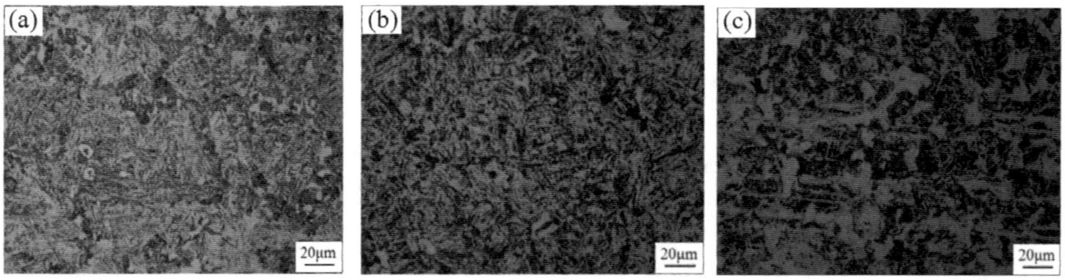

Figure 8. Microstructure after 950°C quenching (salt-cooled, water-cooled, and oil-cooled) + medium temperature tempering: (a) Sample 11 salt quenching; (b) Sample 14 water quenching; (c) Sample 17 oil quenching.

The microstructure of Sample 11 at 950°C salt quenching + medium temperature tempering mainly consisted of tempered troostite. The microstructure of the water-quenched medium-temperature tempered Sample 14 was also tempered troostite. The microstructure of Sample 17 at 950°C oil quenching + medium temperature tempering only consisted of pearlite and ferrite, which are uniformly distributed. By comparison, the tempered troostite was the most abundant in the microstructure of salt-quenched medium-temperature tempering.

(3) 950°C quenching + high temperature tempering

We place salt-quenched Sample 12, water-quenched Sample 15, and oil-quenched Sample 18 into a heat treatment furnace for heating at 650°C, holding for 60 minutes, and then conduct air cooling. We observe the microstructure after high temperature tempering as shown in Figure 9.

Figure 9. Microstructure after 950°C quenching (salt cooled, water cooled, and oil cooled) + high-temperature tempering: (a) Sample 12 salt quenching; (b) Sample 15 water quenching; (c) Sample 18 oil quenching.

The microstructure of Sample 12 at 950°C salt quenching and high-temperature tempering was mainly composed of tempered sorbite, with less pearlite and ferrite. The microstructure of Sample 15 at 950°C water quenching and high-temperature tempering was also tempered sorbite, pearlite, and ferrite. The microstructure of Sample 18 at 950°C oil quenching and high-temperature tempering only consisted of pearlite and ferrite, with coarse grains.

5. Hardness analysis of Q345B steel after heat treatment

5.1. Hardness after quenching

The initial hardness of Q345B was 182.11 HV, and the changes in hardness after quenching are shown in Figure 10.

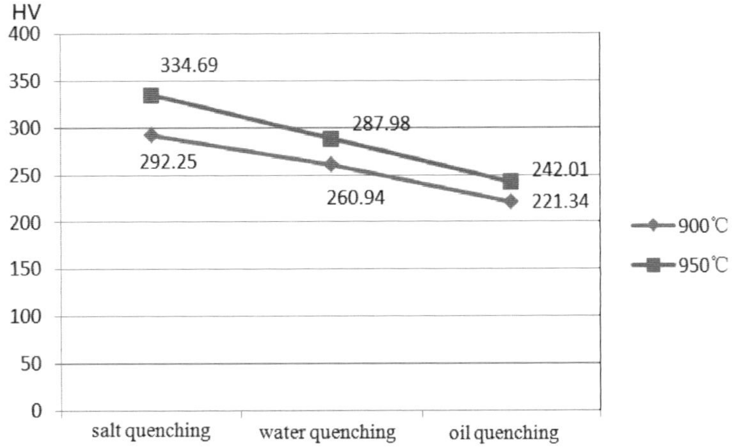

Figure 10. Changes in hardness under different quenching processes.

From Figure 10, the hardness of salt-cooled, water-cooled, and oil-cooled steel decreased sequentially, but all were higher than the initial hardness of Q345B steel. The hardness of different cooling methods quenched at 950°C was higher than that of 900°C, indicating that the austenitization at 950°C was sufficient, the hardenability of the steel was improved, and there were more martensites and bainite in the microstructure, resulting in higher hardness. Q345 steel was not easy to quench, and it was necessary to increase the quenching heating temperature appropriately to improve the hardenability of Q345 steel.

5.2. Hardness after quenching and tempering

The hardness after quenching at 900°C and then tempering is shown in Figure 11. From Figure 11, the hardness of salt quenching, water quenching, and oil quenching after quenching and tempering decreased sequentially. The hardness of low-temperature tempering, medium-temperature tempering, and high-temperature tempering also decreased sequentially, but they were all higher than the initial hardness of Q345B steel.

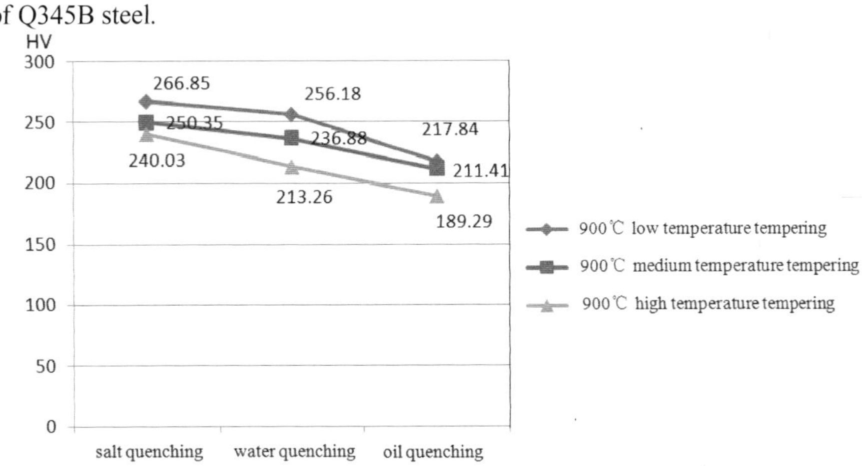

Figure 11. Hardness changes after 900°C quenching and different temperatures tempering.

The hardness after quenching at 950°C and then tempering is shown in Figure 12. From Figure 12, the hardness of salt cooling, water cooling, and oil cooling after quenching and tempering decreased sequentially. The hardness of low-temperature tempering, medium-temperature tempering, and high-temperature tempering also decreased sequentially, but all were higher than the initial hardness of Q345B steel.

From Figures 10-12, tempering reduced the hardness of quenching, which meant improving the toughness of the steel.

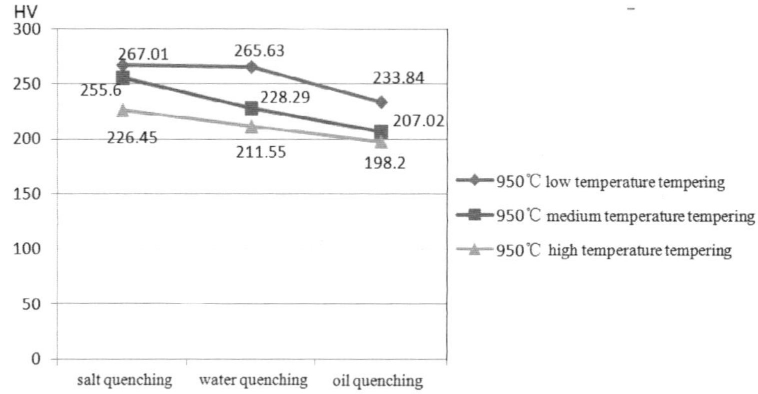

Figure 12. Hardness changes after 950°C quenching and different temperatures tempering.

6. Conclusion

The effects of different heat treatment processes on the microstructure and hardness of Q345B steel were studied, and the following conclusions were drawn:

(1) In metallographic observation and comparison, the microstructure quenched at 950°C had more martensite than that quenched at 900°C. It indicated that austenitization at 950°C was more thorough and had better hardenability than at 900°C. Therefore, the hardness after quenching at 950°C was significantly higher than that after quenching at 900°C.

(2) After heat treatment, the hardness of salt quenching, water quenching, and oil quenching decreased sequentially. However, the quenching hardness was significantly higher than the initial hardness of Q345B steel.

(3) Quenching followed by tempering reduced the hardness of the previous quenching. The hardness decreased sequentially during low-temperature tempering, medium-temperature tempering, and high-temperature tempering. However, they were not lower than the initial hardness of Q345B steel.

References

[1] Ivanov K. D., Gavze A. L., and Pavlov A. A. (2023). Influence of final heat treatment modes on structure and properties of high strength sheet steel. *Steel in Translation* **53** 701-706.

[2] Badaruddin M., Sugiyanto, Sumardi S., and Asmi D. (2024). Improvement of the fatigue crack growth resistance in AISI 4140 steel under single- and multi-austempering heat treatments. *Results in Engineering* **21** 101, 814.

[3] Omar S. M. T. and Plucknett K. P. (2022). The influence of DED process parameters and heat-treatment cycle on the microstructure and hardness of AISI D2 tool steel. *Journal of Manufacturing Processes* **81** 655-671.

[4] Zhang W. W., Cui J. X., Wang Z. M., and Peng Y. (2018). Carbonitride precipitation behavior of Q345D steel during rapid tempering. *Heat Treatment of Metals* **43** 13-19.

[5] Wang H. F. (2018). Effect of different quenching and tempering treatments on impact toughness of Q345 steel. *Foundry Technology* **39** 186-188.

[6] Sun Y. W. and Wang L. S. (2021). Effects of heat treatment on microstructure and properties of Q345E steel for unseating prevention. *Hot Working Technology* **50** 115-119.
[7] Yang B., Liu B. X., Fan K. Y., Zhang B. Y., Ji P. G., Feng J. H., Tong H. C., and Yin F. X. (2023). Obvious different formability and mechanical properties of warm calibre rolling Q345 steel with normalized and tempered states. *Materials Science & Engineering A* **887** 145, 765.

Analysis of conductive filament heat transfer in TiO_2-based RRAM device

Jing Li*, **Jinyan Pan, Qiao Huang, Ruotong He and Lunbo Xie**

School of Ocean Information Engineering, Jimei University, Xiamen, Fujian, 361000, China

*Corresponding author's e-mail: 2571815196@qq.com

Abstract: Resistive random access memory (RRAM) has developed into a new type of non-volatile memory that has attracted much attention for its high density and low power; TiO_2 has been studied to make RRAM because of its high conductivity. Based on the COMSOL Multiphysics finite element method and domain decomposition method, an electrically-thermally coupled model of TiO_2-based RRAM with oxygen vacancy (Vo) conduction mechanism is constructed, the formation and cutting process of RESET and SET conductive filament (CF) under different voltages is simulated. Moreover, the characteristics of the internal temperature distribution of the CF are explored. The results show that the RESET process is more sensitive to heat changes, the transverse direction thermal value of the CF is more obvious than the longitudinal direction, and the rate of change is faster. Additionally, the electric field induces the migration of the Vo in filament, which affects the enthalpy change of the device's heat transfer and resistive properties. This is an important reference for a deeper understanding of the switching behavior of RRAM devices and the control mechanisms for thermal studies.

1. Introduction

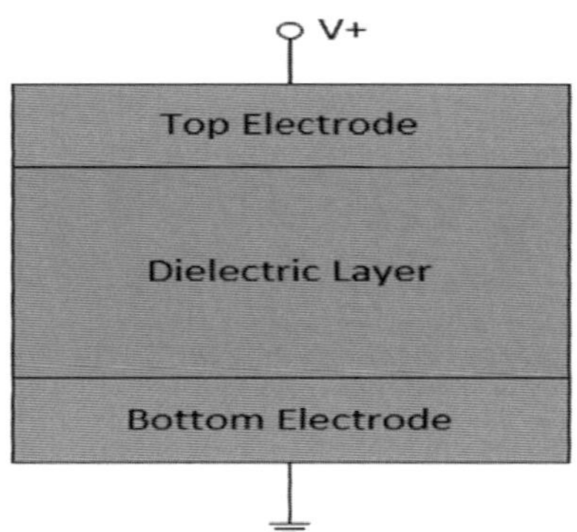

Figure 1. RRAM basic structure.

The RRAM [1] device is a metal/insulator/metal (MIM) structure that is like a sandwich, as in Figure 1, various insulating layer materials were used to confirm the existence of resistive behavior, such as organic substance, electrolyte material, metal oxides and so on. RRAM based on binary metal oxides are the most promising for application owing to their simple structure and CMOS process compatible features. Among them, TiO_2 material is getting widespread attention and presents better resistive characteristics, which have high permittivity and good thermal stability. It is considered one of the eligible binary oxides for studying RRAM devices [2]. TiO_2 has been widely studied for its excellent photocatalytic properties and unique crystal structure, but its properties are affected by different physical fields in practical applications, and temperature is a factor that is bound to be present. In this paper, we will use COMSOL Multiphysics to study the internal heat conduction properties of CFs formed based on oxygen vacancies.

2. Modelling

Firstly, the Ni/TiO_2/Pt model, which belongs to the typical MIM structure, is built. The dielectric layer is TiO_2, the top electrode is Ni, and the bottom electrode is Pt. As Figure 2, it is constructed using a two-dimensional (2D) axisymmetric component as Figure 2(a) and a three-dimensional (3D) rotational model as Figure 2(b). In order to simplify the analysis, it is assumed that just one CF [3] comes into the orange part of the picture, and the CF is part of the dielectric layer that generates the connection channel. Secondly, according to Zhao[4] et al. research found that the main component inside the CF is Ti_4O_7 when TiO_2 film is used as the switching layer of the device, to simulate the on-off of the filament, CF is defined as composed of Ti_4O_7, the top electrode Ni is usually used in the device's Vo reservoir [5], and the bottom electrode using the inert electrode Pt, which plays a role of connecting to the ground in the device work, and Table 1 gives the parameters of the materials.

Finally, the initial conditions are as follows: top electrode connected to DC scanning voltage, positive pressure range (0~1 V), negative pressure range (-1.5~0 V), bottom electrode is grounded, and initial temperature of 300 K, Vo concentration is set to 0 for the region of non-CF, and Vo concentration of $1e^{21}/cm^3$ in the region of CF.

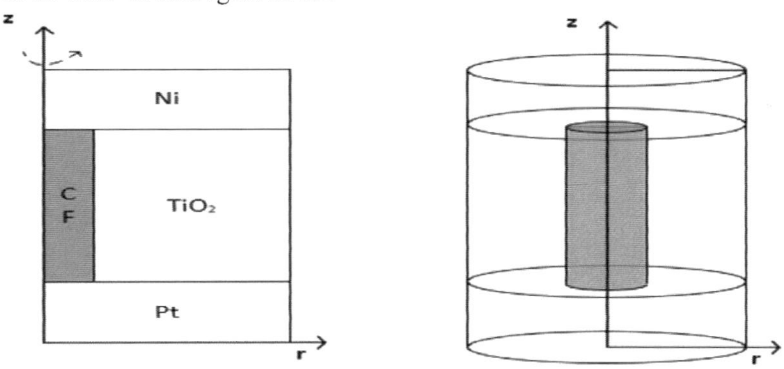

(a) (b)

Figure 2. Ni/TiO_2 /Pt RRAM (a) 2D and (b) 3D structure model.

Table 1. Material parameters of each part in the model.

	Pt	Ni	TiO_2
K_{th} (W/(m*K))	71.6	75.5	5.18
σ (S/m)	$9.6*10^6$	$14.3*10^6$	$1.2*10^{-12}$
ρ (Kg/m^3)	2,1450	8902	4260
C_p (J/(Kg*K))	130	512	700

3. Results and discussion

As in Figure 3(a), during the RESET, with the increase in potential, the highest temperature point of the CFs gradually moves upward, and the rest of the position of the temperature is relatively lower; the high-temperature point is more prominent, to the 19 nm near the increase to -0.85 V, due to the Joule heating [6] effect of the highest temperature occurs in the smaller jump, at this time, the CFs have been completely broken, the dielectric layer is almost in the insulating state, the temperature distribution of each region of the filaments. The temperature distribution of each region of the filament is obvious; the highest temperature is located in the upper half of the edge of the region, and in the sparsest oxygen vacancies, the original high-temperature zone into a medium-temperature zone.

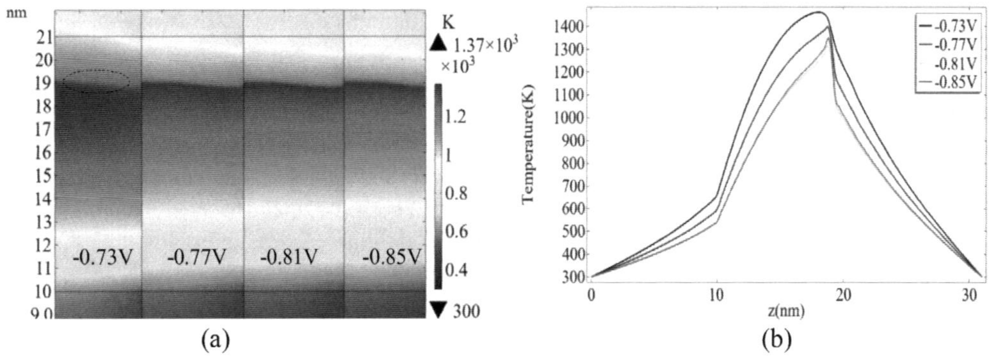

Figure 3. RESET process (a) 2D and (b) temperature distribution of CFs.

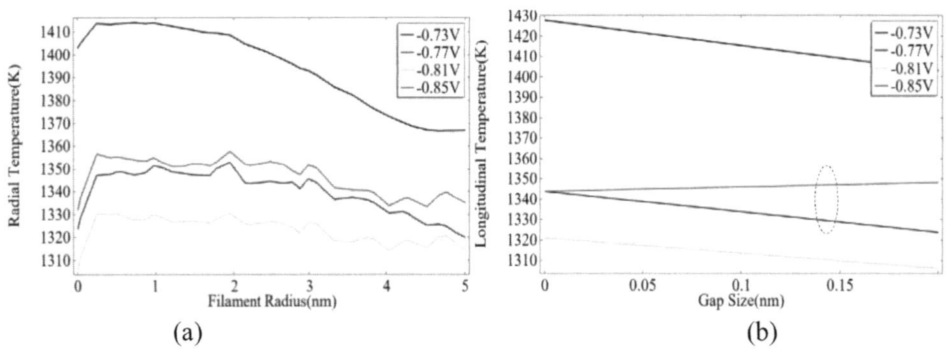

Figure 4. Distribution in (a) transverse and (b) longitudinal temperature at the breakage of RESET.

Figure 4 gives the temperature distribution of CFs in the transverse and longitudinal directions of the breaks during the fracture process. In the transverse direction, due to the existence of space charge [7] and charge distribution is not uniform, the temperature at the starting point will be slightly lower, the temperature is wave distribution, longitudinal due to the emergence of tiny fracture, the charge can be seen as a uniform distribution, so the temperature is a linear distribution, and all with the voltage increases and decreases (except when -0.85 V), the longitudinal and transverse temperature attenuation is almost the same when the voltage reaches -0.85 V, the temperature is instead greater than at -0.81 V, indicating that the filament breakage occurs after the highest temperature and the oxygen vacancies are RESET by continuous Joule heating, heat generation, and dissipation is a dynamic equilibrium process. Figure 5 shows the SET at different voltages; the temperature of the CF region is higher than the other regions, and the temperature of the upper part is higher than the lower part. This is because, excited by an electric field, the current flows through the dielectric layer to the bottom electrode, and the non-CF part of the Vo concentration is low, which can be regarded as titanium dioxide with high purity and conductivity [8] is very low.

As in Figure 6, CF has a certain capacitance and charge under the action of the linear electric field; during SET, the free energy of the ohmic heating system increases, which leads to an increase in temperature, and the capacitor also conducts under the action of current tunneling [9], as Equation (1) the lattice oxygen in the dielectric layer of TiO_2 loses two electrons, forming Vo and 1/2 mol oxidation.

Journal of Physics: Conference Series **2842** (2024) 012064 doi:10.1088/1742-6596/2842/1/012064

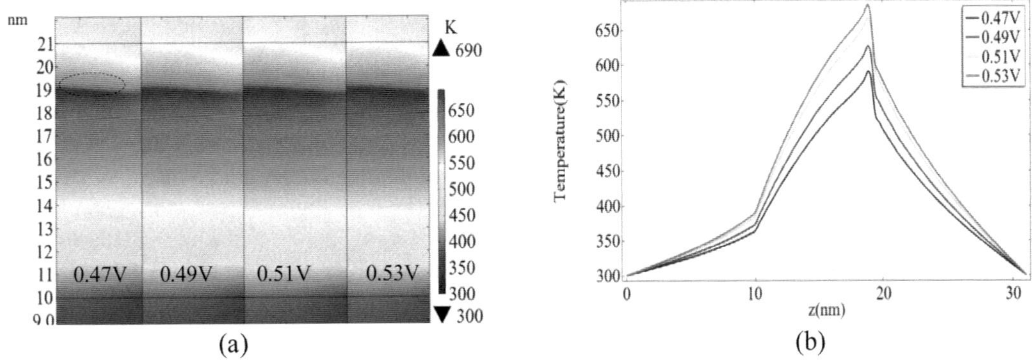

Figure 5. SET process (a) 2D and (b) CF temperature distribution.

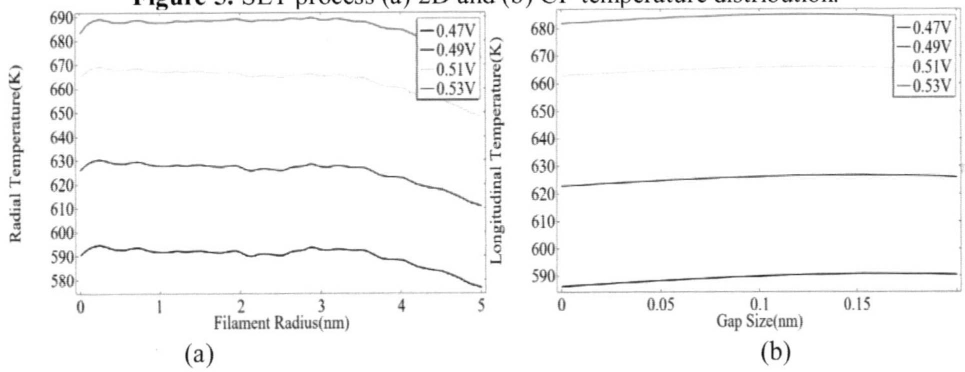

Figure 6. Distribution in (a) transverse and (b) longitudinal temperature at the breakage of SET.

$$O_o^x - 2e^- \rightarrow V_o + \frac{1}{2}O_2 \quad (1)$$

$$N_V = N_C e^{\left(-\frac{E_V}{kT}\right)} \quad (2)$$

Equation (2) [10] describes the effect of temperature on the diffusion of Vo in the dielectric layer of an RRAM device. N_V denotes the amount or concentration of defects (cm^{-3}), the sum of the energies required for vacancy generation is denoted by E_V in (J/mol), Boltzmann's constant is usually denoted by (8.62×10^{-5} eVK^{-1}) and is expressed as k, and T(K) is the optimum temperature for CF. N_C is the number of lattice sites per cubic centimeter number of lattice sites. However, we investigated several values for the average temperature of the filament during SET operation, which is obtained in Figure (6).

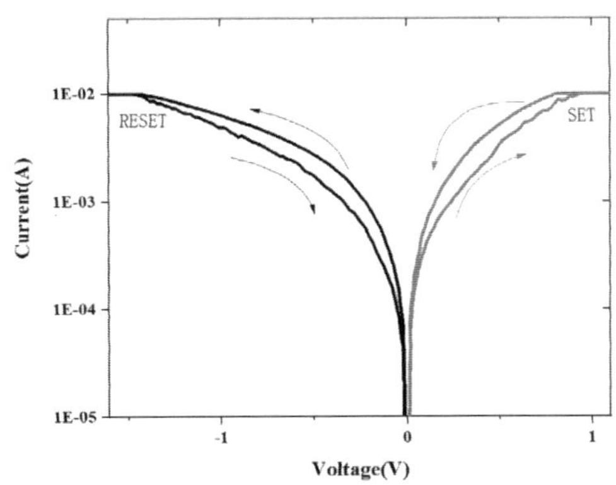

Figure 7. I-V curve of Ni/TiO_2/Pt the experiment.

As in Figure 7, the experimental I-V characteristic curve of the Ni/TiO_2/Pt structure of the device shows a slow-varying characteristic, with insignificant SET and RESET processes, almost no jumps, and a small window value, V_{SET} and V_{RESET} are larger than the simulated results, probably due to the lack of purity caused by impurities, air pressure, and RF power during the film preparation process, which is consistent with the valence change memory (VCM) [11] of type conductivity mechanism based on the formation of CFs from oxygen vacancies. This conductive mechanism usually involves the migration and redistribution [12] of oxygen vacancies in the oxide, which leads to a change in the resistive switch of the device. The coincide between experimental and simulation results shows that the COMSOL Multiphysics model successfully captures the key physical processes of Vo migration and CF formation in the device. This further validates the applicability of the VCM-type conductive mechanism based on the formation of CFs from oxygen vacancies in Ni/TiO_2/Pt structured devices.

4. Conclusion

In summary, a 2D axisymmetric structural model of a TiO_2-based RRAM device is developed using COMSOL Multiphysics to analyze the thermal conduction phenomenon of the internal CFs during operation. The dynamic influence of Joule thermal effect on the resistance switching during the electric-field driven Vo ion migration is investigated, and the effect of temperature on the formation and fracture of CF, as well as the change of temperature distribution of the system by the growth of fracture, are explored, it is concluded that the migration of the internal oxygen vacancies will be increased with the increase of temperature. The transverse thermal effect of the CFs is larger, the rate of change is faster, and the longitudinal temperature distribution is more uniform and stable. It is valuable for realizing the controllable preparation and characterization of the device.

Acknowledgments

Gratitude is given to Yunlong Gao, school of Pen-Tung Sah Institute of Micro-Nano Science and Technology, Xiamen University, Fujian, 361102 (e-mail: gaoyl@xmu.edu.cn), Special Foundation of Fujian Province to Promote High-quality Development of Marine and Fishery Industries, under Grant FJHYF-ZH-2023-05 and Natural Science Foundation of Fujian Province 2022J01061. This project was supported by the Natural Science Foundation of Fujian Province, China (Grant No. 2020J01713).

References

[1] Pujari R, Gasasira A, Kim Y, et al. Introduction to Analog Testing of Resistive Random Access Memory (RRAM) Devices Towards Scalable Analog Compute Technology for Deep Learning[C]//2021 32nd Annual SEMI Advanced Semiconductor Manufacturing Conference (ASMC).2021.

[2] Wang C, Peng X, Fang W, et al. TiO_2-based cathode with modest oxygen vacancies and defective Ti^{3+} for long-life lithium-oxygen batteries[J]. Applied Surface Science, 2023, 614:156262-.

[3] Donglan Z, Jiong W, Qing W, et al. Exploring the direction-dependency of conductive filament formation and oxygen vacancy migration behaviors in HfO_2-based RRAM[J]. Physical chemistry chemical physics: PCCP, 2023.

[4] Zhao L, Zhang J, He Y, et al. Dynamic Modeling and Atomistic Simulations of SET and RESET Operations in Based Unipolar Resistive Memory[J]. IEEE Electron Device Letters, 2011, 32(5):677-679.

[5] Lv Q, Wu S, Lu J, et al. Conducting nanofilaments formed by oxygen vacancy migration in Ti/TiO_2/TiN/MgO memristive device[J]. Journal of Applied Physics, 2011, 110(10):28.

[6] Zhang Z, Liu Y, Su X, et al. Electro-triggered Joule heating method to synthesize single-phase CuNi nano-alloy catalyst for efficient electrocatalytic nitrate reduction toward ammonia[J]. Nano Research: English Edition, 2023, 16(5):6632-6641.

[7] Tsang-Tse Fang, Yu-Chi Lu and Hsing-I Hsiang. Reassessment of the effect of space charge potential on the electrical conductivity of polycrystalline ceramics[J]. Journal of the American Ceramic Society, 2023.

[8] Ishkov A V, Kunitsyn R A, and Malikov V N. Superminiature Eddy Current Probe for Measuring the Electrical Conductivity of Copper Thin Films[J]. Key Engineering Materials, 2022, 910.
[9] Ahn Y, Son J Y . Conductive filament nanostructures and their magnetization reversal in NiO resistive switching memory cells[J]. Materials Science & Engineering, B. Solid-State Materials for Advanced Technology, 2022:283.
[10] Chen S, Doolen G D. Lattice Boltzmann Method for Fluid Flows[J]. Annual Review of Fluid Mechanics, 2003, 30(1):329-364.
[11] Kaniselvan M, Luisier M, Mladenovic M. An atomistic modeling framework for valence change memory cells[J]. Solid-State Electronics, 2023.
[12] Wu B , Guo X , Liu B, et al. Insight into hydrogen migration and redistribution characteristics during co-pyrolysis of coal and polystyrene[J]. Journal of analytical & applied pyrolysis, 2023(Aug.):173.

Synthesis and thermophysical properties of rare-earth tantalate ZrO_2-Dy_3TaO_7 ceramics

Dongbo Li*, Ming Zhang, Ge Deng, Yuan Yang, Rong Gui, and Guizhong Yang

State Key Laboratory of Complex Nonferrous Metal Resources Clean Utilization, Kunming University of Science and Technology, Kunming, 650093

YUNNAN COPPER CO., LTD. SOUTHWEST COPPER BRANCH, Kunming 650102, China

*Corresponding author's email: dbliwill@163.com

Abstract: Thermal Barrier Coatings (TBCs) are of major concern to researchers because of their outstanding properties, such as high-temperature resistance, compressive resistance, hardness, and toughness. Its applications are wide-ranging and critical. Examples include hypersonic vehicles, aero-engines, and service in high-temperature and high-impact extreme environments. This work focuses on the preparation, thermal properties (coefficient of thermal expansion, thermal conductivity), and performance evaluation (thermal shock resistance, thermal shock resistance, and bonding) of ZrO_2-Dy_3TaO_7 rare earth tantalate ceramic coating materials. The results show that ZrO_2-Dy_3TaO_7 possesses a low thermal conductivity (1.25-1.50 W K^{-1} m^{-1}, 900°C), a high thermal expansion coefficient (10.80-11.14×10-6 K^{-1}, 1200°C), which is about half of the value of the thermal conductivity of 7YSZ (2.8-2.2 W $K^{-1}m^{-1}$). It is better than that of the 7YSZ (10.04×10^{-6} K^{-1}, 1200°C), which meets the performance requirements of the new thermal barrier coating. In addition, ZrO_2-Dy_3TaO_7 can achieve more than 6, 000 thermal fatigue cycles at 1200°C, and the ideal number of thermal shock cycles is 4, 040 cycles at the same temperature. These outstanding test results demonstrate that the ceramic coating can effectively protect the substrate material and provide long-term service in high-temperature copper liquid. This research lays the theoretical foundation for the development of next-generation coating materials, particularly for thermal barrier/environmental barrier integrated coating materials.

1. Introduction

Along with the rapid development of modern society in the field of industry, science, and technology, hot-end components require higher and higher temperatures, which have reached the limiting conditions of high-temperature alloys and single-crystal materials [1-2]. Taking the heated parts of the lance tube of the ESAB furnace as an example, they exist within a severe environment characterized by conditions of high-temperature oxidation and elevated copper erosion, withstanding temperatures as high as 1,200°C, which has exceeded the limiting temperature for the use of high-temperature nickel alloys (1,075°C). Therefore, we urgently need to develop new materials to meet these harsh service conditions. Over the years, researchers have continued to innovate and explore the advantages of combining the high strength and toughness of metals with the high-temperature resistance of ceramics to prepare thermal barrier coatings that can solve the above problems, which can exhibit significant contributions to thermal insulation, oxidation and corrosion resistance, and have found application in steam turbines, diesel generators, jet engines, and other high-temperature environments, effectively extending the operational lifespan of critical components [3].

In recent years, domestic and foreign academics have paid great attention to the research and development of thermal barrier coatings. The main research on oxide ceramics as candidate materials for thermal barrier coatings has focused on rare earth zirconates ($RE_2Zr_2O_7$), rare earth phosphates ($REPO_4$), lanthanum cerate (La_2Ce_2O7), and rare earth silicates [RE_2SiO_5, $RE_2Si_2O_7$, and $RE_{9.33}(SiO4)_6O_2$], calcite-structured oxides (ABO_3), magnetite-structured compounds ($REMgAl_{11}O_9$), garnet-structured compounds ($Y_3Al_5O_{12}$), and aluminate (Ba_2REAlO_5, $Ba_6RE_2Al_4O_{15}$) are among the material systems [4-7]. However, these thermal barrier coatings also have some drawbacks, La_2Ce_2O7 thermal barrier coatings cannot work stably for a long period above 1350°C, the coefficient of thermal expansion between room temperature and 1350°C is not stable, and it is easy to react with the thermal oxidation growth layer, which leads to structural instability and low fracture toughness, etc. [8]. Oxide coatings with chalcogenide structure have much lower coefficients of thermal expansion than rare earth thermal barrier coatings and have severe crack defects, etc. [9]. Rare earth tantalate materials have extremely low thermal conductivity, matched thermal expansion coefficient, high hardness, low modulus, ferroelastic phase transition toughening, excellent resistance to CMAS corrosion, high-temperature water-oxygen corrosion, sintering resistance, high-temperature phase stability, excellent chemical compatibility, and high deposition rate, etc. [10-12], and the use of the temperature can be reached to 1500°C and above. It is an insulation and corrosion resistance of the ultra-high temperature. It has great potential for an ultra-high temperature thermal barrier coating material with great potential for thermal insulation and corrosion resistance. In this paper, we will carry out the rare earth tantalate spray powder preparation technology and its performance assessment, evaluate and analyze the key service performance of the new thermal barrier coating, and provide key materials and technical support for developing a new generation of thermal barrier coating materials.

2. Experimental methodology

2.1 Preparation of the bonding layer

The supersonic flame coating alloy bonding layer [13-14] is achieved through the utilization of propane, propylene, and other hydrocarbon gas or hydrogen alongside high-pressure oxygen. This creates a high-temperature, high-speed combustion flame flow that can reach speeds of up to Mach 5 (1500 m/s) or higher. Often referred to as HVOF, the powder is fed axially into the flame, which results in the sprayed particles reaching a molten or semi-molten state and being accelerated to velocities of up to 300-500 m/s or even higher, resulting in a high-quality coating with high bonding strength and denseness. The spraying process parameters are shown in the table 1 below:

Table 1. Spraying parameters of the binder layer.

Powder	Replacement	Air duct		White line		Airbrush		
	Gun barrel	Oxidation SCFH	Diesel gpm	Powder delivery g/min	Carrier gas SCFH	Spray distance mm	Tread mm	Linear velocity mm/s
NiCoCrAlY	4 inch	1850	7	45	23	350	5	500

2.2 Preparation of ceramic layers

Ceramic coatings were prepared by plasma spraying [15-16], a plasma spraying method operated in an atmospheric atmosphere with particle velocities below 300 m/s. The ceramic powder sprayed was ZrO_2-Dy_3TaO_7 (YTa-1). The sprayed ceramic powder was ZrO_2-Dy_3TaO_7 (YTa-1). The spray gun's operational power was augmented to 80 kW, and the sprayed particle velocity was increased from 0.2 Ma ~ 0.5 Ma to 1 Ma or even higher. The enthalpy increased significantly from 2093.4 kJ/kg to 9211 kJ/kg, while the bonding strength of the coating improved from below 50 MPa to a range of 50 MPa to

Journal of Physics: Conference Series **2842** (2024) 012065 doi:10.1088/1742-6596/2842/1/012065

70 MPa. Moreover, the coating's porosity was reduced to below 10%. The spraying process parameters are shown in the table 2 below:

Table 2. Ceramic layer coating parameters.

Spraying Process-Powder	Power supply			Air duct		Powder Line (4-Way Powder Feed)		Cool off
	Power kw	Amps A	Input voltage V	Nitrogen L/h	Hydrogen L/h	Turbulence L/h	Rpm r/s	Cooling water temperature °C
APS-YTa-1	45.3	533	85	2403	534	240	0.5	12

3. Results and discussion

3.1 Topographic characterization of coatings

As shown in Figure 1, the experiments were carried out by supersonic plasma spraying to prepare the binder layer, and the ceramic layer was prepared on the surface of the binder layer using atmospheric plasma spraying technology. Figure 1(a) is the surface topography of the thermal barrier coating, and we can observe that the ceramic powder on the surface is uniformly distributed on the binder layer, which is very dense. Figure 1(b) is the cross-section morphology of the thermal barrier coating. There are three components: The bottom layer is the metal substrate, and the middle is the bonding layer, which is tightly attached to the surface of the metal substrate and connected to the top ceramic powder, and plays a pivotal role in the connection. In addition, it can play a good role in heat insulation and corrosion resistance. The top layer of components for the ceramic powder is the most important component of the thermal barrier coating. It is the key hindering part for heat insulation, corrosion resistance and impact resistance. The ceramic and bonding layers have a small number of pores. Although the pore defects will reduce the corrosion resistance of the coating to the molten copper liquid, the pore defects will also improve the thermal insulation capability of the coating.

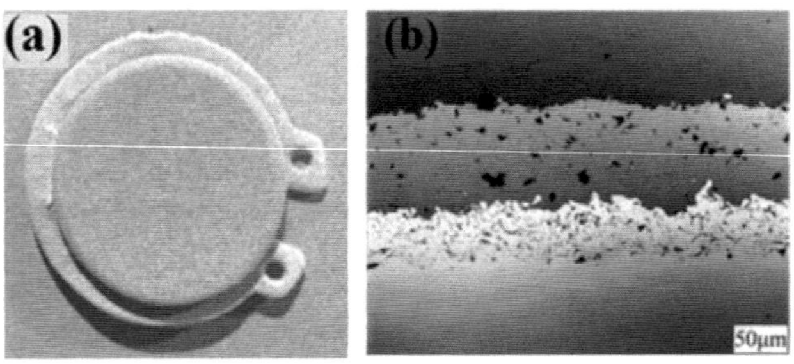

Figure 1. Surface and cross-section morphology of plasma sprayed samples.

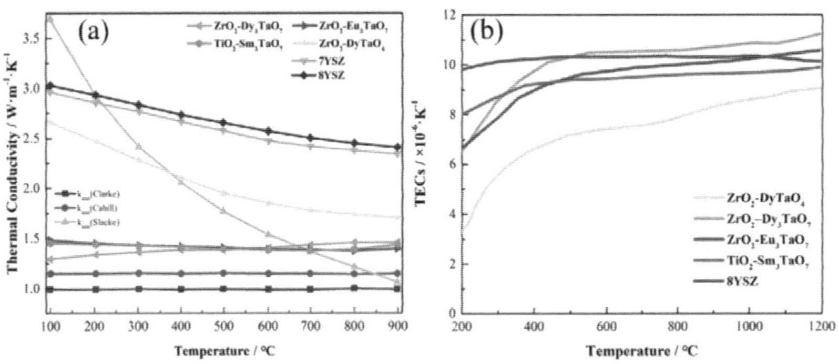

Figure 2. Thermal conductivity and coefficient of thermal expansion of plasma sprayed samples.

Figure 2 compares the thermal conductivity and coefficient of thermal expansion of ZrO_2-Dy_3TaO_7 with other rare-earth carbonates. The results show that ZrO_2-Dy_3TaO_7 has a low thermal conductivity, which is superior to that of the 8YSZ material, resulting in a decrease in the modulus of elasticity, a decrease in the crystalline bonding energy, and the coefficient of thermal expansion of the material can reach or even surpass the value of 8YSZ [17], which suggests that ZrO_2-Dy_3TaO_7 has excellent thermal insulating properties. It is a promising material for ceramic coatings.

3.2 Coating thermal shock resistance assessment

The thermal barrier coating's resistance to thermal shock is evaluated in accordance with the national standard (HB7269). The thermal shock resistance of the thermal barrier coating is then subjected to two constantly alternating states of heat and cold. It constantly resists thermal stresses, so it does not crack or fall off.

3.2.1 Assessment conditions and process. The upper and lower planes are parallel, with a diameter of 25 mm and a thickness of 10 mm. The thermal barrier coating specimens are placed into a tube furnace with a constant temperature (where the temperature difference is kept below or equal to 1°C). They are then heated and kept at a constant temperature for 3 minutes. After that, they are quickly removed from the furnace and allowed to cool in the air at 25°C for 2 minutes. This completes one thermal shock cycle. This process is repeated until the coating begins to flake off (when the area of the coating that comes off is equal to or greater than 20% of the total area of the coating). The number of thermal shocks required for the coating to fail is recorded. The thermal barrier coating life obtained by this method mainly responds to the nature of the TGO (thermal growth object oxide), the thermal shock resistance of the ceramic coating, and the stability of the crystal structure.

The assessment data of the style and the SEM morphology data graphs were obtained through the field record, as shown in Figures 3-5 below, respectively.

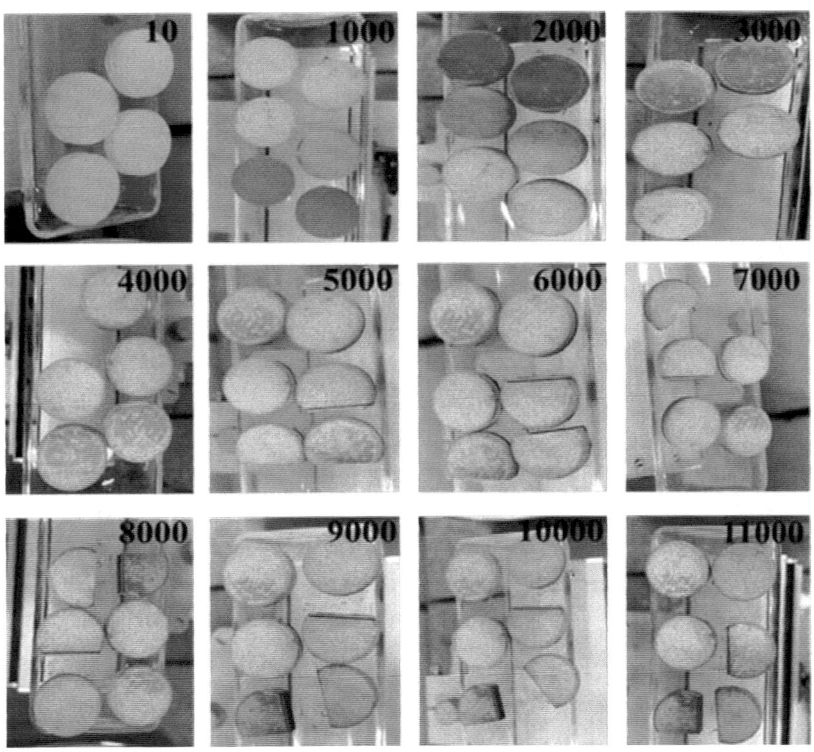

Figure 3. Picture of thermal fatigue test.

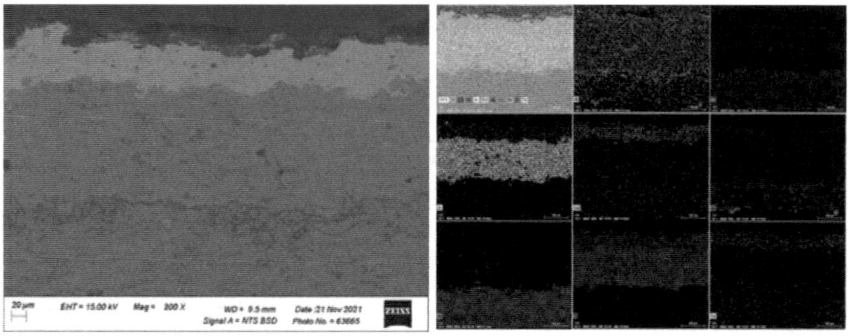

Figure 4. SEM and EDS of the coated cross section of the specimen for 4000 cycles.

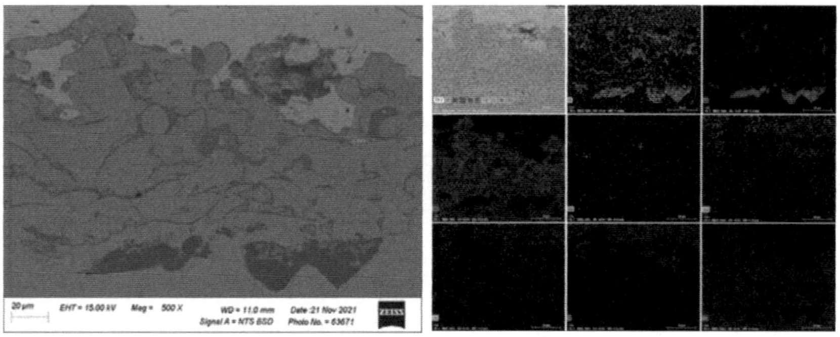

Figure 5. SEM and EDS of the coated cross section of the specimen for 5000 cycles.

After 4000 cycles, a small amount of TGO was found in the cross-section of the specimen coating. Dark river-like stripes appeared between the bonding layer and the ceramic layer, but the content was very small and did not cause cracks to grow in the coating. It was also found that there was no obvious aggregation of Al elements by the EDS scanning map. The specimen coating was observed after 5000 cycles, which showed an increase in TGO and TGO aggregation growth in the bonding layer and YTa-1 ceramic layer. Small cracks appeared in the ceramic layer, but they did not affect the overall structure of the coating and there were no larger cracks.

3.2.2 Assessment results and conclusions

Table 3. Thermal fatigue test results.

Serial number	Spraying equipment	Bonding Layer material and thickness /μm	YSZ thickness /μm	Tantalite thickness /μm	ExperimentalTemperature /°C	Number of failures/times
1	HVOF	NiCoCrAlY 150	0	300	1200	6798
2	HVOF	NiCoCrAlY 150	150	150	1200	12563
3	HVOF	NiCoCrAlY 150	100	200	1200	16870
4	HVOF	NiCoCrAlY 150	200	100	1200	18730

After the on-site assessment, the 1000°C thermal fatigue performance test shows that the coating adopts HVOF spraying, the bonding layer adopts NiCoCrAlY, and the number of thermal fatigue times is more than 6000 times up to now, shown in the table 3.

3.3 Thermal shock performance assessment of coatings

3.3.1 Assessment conditions and process. The test adopts the oxygen-propane cycle thermal shock method. In the combustion chamber, the sample (impact sample diameter 25 ± 0.2 mm, matrix thickness 3 ± 0.5 mm, bonding layer thickness of 150 μm, ceramic inner layer thickness of 140-150 μm, ceramic outer layer thickness of 120-200 μm, porosity 20%, the sample surface should be smooth and flat, the two planes of parallelism is not greater than 0.1 mn) fixed in the special corundum—mullite assessment tooling. In this experiment, we regulate the distance between the flame and its center, adjust the oxygen and propane gas mixture of the adiabatic layer of material being tested, and heat it for 20 seconds till it reaches a temperature of 1400°C. After that, we start the cooling process and move the flame gun away from the machine following a set procedure. We control the movement time of the flame gun base to ensure accurate cooling of the sample in 20 seconds. At the end of a single cycle, the surface temperature of the sample is reduced to 800°C. Throughout the process, we keep the cooling gas open to ensure that the high-temperature nickel-based alloy service temperature does not exceed 30% of the sample surface, as it would result in coating failure.

Examination process: 1) Specimen number is filled in the test record. 2) We measure the original thickness of the specimen, accurate to 0.01 mm. 3) We weigh the original mass of the specimen, accurate to 0.1 mg. 4) We install the specimen in the special tooling of the pre-fabricated machine by the specified pressure and adjust the horizontal height and examination angle. 5) We install a thermocouple in the center of the back of the specimen and connect the leads to the microcontroller. 6) We turn on the water-cooling system of the machine. 7) We adjust the oxygen pressure reducing valve to the operating pressure range of 0.3-0.5 MPa. 8) We adjust the propane pressure reducing valve to be in the operating pressure range of 0.3-0.5 MPa. 6) We turn on the water-cooling system of the testing machine. 7) We adjust the oxygen pressure reducing valve to a 0.3-0.5 MPa pressure range. 8) We adjust the propane pressure reducing valve to be in the pressure range of 0.08-0.2 MPa. 9) We enter the operator interface and set the relevant test parameters, the expected number of tests 1000 times, the test temperature of 1400°C, the temperature rise time, and the cooling time of 20 s. We adjust the initial surface of the specimen to the flame jet and adjust the level height and the angle of assessment of the special tool. We adjust the distance from the initial surface of the specimen to the flame nozzle for 8 ± 0.2 cm. 10) We start the thermal shock flame gun with the electronic igniter and automatically align the specimen for assessment. 11) After holding the specified time, the flame gun is moved to another machine according to the established procedure, and the cooling time is precisely regulated by controlling its running time. 12) The thermal shock assessment test reaches the specified number of cycles, or the specimen is invalidated (the coating is effective), the assessment area falls off by 30%, and the temperature is reduced. The examination area is off 30%. The flame gun is back to the initial position, the test, and the tester records the number of cycles. 13) Specimen test is completed, we shut down the flame to close the propane and oxygen regulator valve and examine the system switch, the power switch of the control cabinet, and the water-cooling system.

3.3.2 Assessment results and conclusions. The picture in Figure 6 below was obtained from the actual assessment records on site:

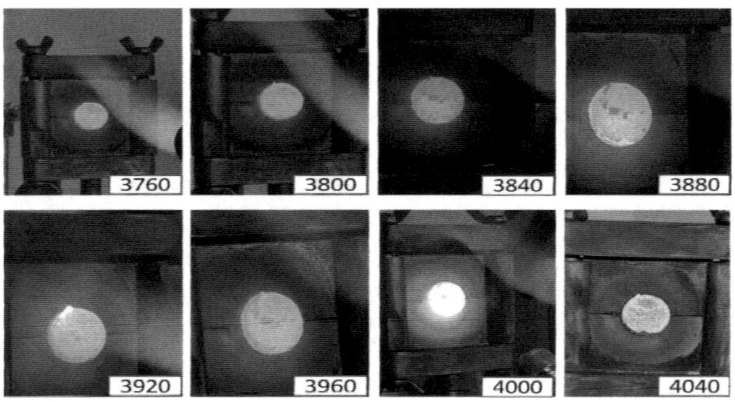

Figure 6. Picture of thermal shock test.

Table 4. Thermal shock test results.

Serial number	Spraying equipment	Bonding layer material and thickness /μm	Tantalite thickness /μm	Tantalite thickness /μm	Experimental Temperature /°C	Number of failures/times
1	HVOF	NiCoCrAlY 150	0	300	1200°C	220
2	HVOF	NiCoCrAlY 150	150	150	1200°C	1750
3	HVOF	NiCoCrAlY 150	100	200	1200°C	320
4	HVOF	NiCoCrAlY 150	200	100	1200°C	4040

According to the thermal fatigue test standards, we record the number of cycles when the coating fails to examine the actual thermal fatigue performance of the sample to assess the coating to meet the design requirements. After the on-site assessment, the temperature was first heated to 1400°C and then cooled down to 1200°C when the thermal fatigue performance test results show that the coating is sprayed with HVOF-APS, the bonding layer is NiCoCrAlY. The standard thermal shock test involves a 20-second heating cycle up to 1400°C, followed by a 20-second cooling cycle at 1200°C. This test is repeated up to 4040 times and is highly effective, shown in the table 4.

3.4 Coating high-temperature corrosion resistance assessment

After the sample preparation, the sample was placed into the alumina crucible, and then copper powder with a thickness of 4 cm was spread. Finally, the alumina crucible was put into the box-type muffle furnace to be heated up to 1200°C with the furnace, and the morphology of the ceramic layer after the static oxidation for 1000 hours is shown in Figure 7.

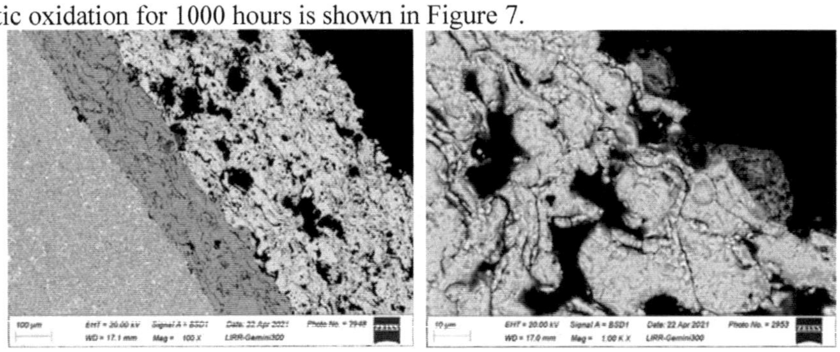

Figure 7. Cross-sectional morphology of the sample after sanding after 1000 h of simulated field operation environment.

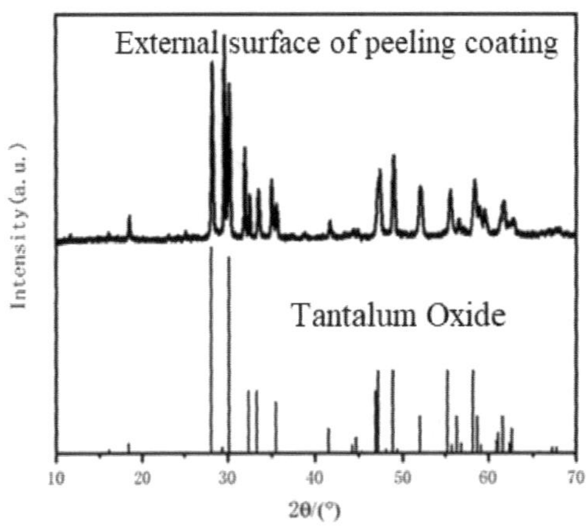

Figure 8. XRD of the ceramic part of the sample after 1000 h of simulated field operation environment.

Figure 7 shows that the ceramic layer did not react with the copper liquid, with obvious diffusion behavior. It shows that the YTa-1 coating can well resist the corrosion of high-temperature copper liquid. After analyzing the XRD of the detachment area of the ceramic layer, it was found that there was no phase transition in the ceramic layer after 1000 hours of simulation of the field operating environment. Figure 8 the analysis of the XRD also revealed the presence of copper or copper oxides in the ceramic layer, indicating that the ceramic layer did not react with the copper liquid. This further confirms that the optimized coatings can effectively resist the diffusion of molten copper corrosion and high-temperature oxidative corrosion.

4. Conclusion

In this paper, rare earth tantalate thermal barrier coatings were prepared using supersonic plasma spraying technology and atmospheric plasma spraying technology. Their thermophysical properties were analyzed by morphology, thermal conductivity, and coefficient of thermal expansion and then examined in thermal fatigue assessment, thermal shock assessment, bonding assessment, and simulation of service assessment analysis in high-temperature copper liquid, and the results show that:

(1) YTa-1 coating surface has good densification and has lower thermal conductivity and a comparatively higher coefficient of thermal expansion than 8YSZ, which is an excellent performance of thermal insulation materials;

(2) YTa-1 coating in 1200°C conditions shows that the number of thermal fatigue up to now is more than 6,000 times. The highest can reach 1,800 times; thermal shock times can also reach the ideal 4040 times;

(3) YTa-1 coating in the simulation of 1200°C under the conditions of copper liquid can serve 1000 h, and the service after the morphology does not appear to have obvious damage.

References

[1] Morelli S, Bursich S, Testa V, et al. A comparative evaluation of CMAS corrosion and thermal cycling fatigue resistance of alternative thermal barrier coating materials and architectures [J]. Surface & Coatings Technology, 2022: 439.

[2] Yunus S M, Mahalingam S, Manap A, et al. Test-Rig Simulation on Hybrid Thermal Barrier Coating Assisted with Cooling Air System for Advanced Gas Turbine under Prolonged Exposures—A Review [J]. 2021.DOI: 10.3390/coatings11050560.

[3] Ma X, Rivellini K, Ruggiero P, et al. Evaluation and Characterization of a Durable Composite Phase Thermal Barrier Coating in Solid Particle Erosion and Burner Rig Tests [J]. Springer US, 2021(1). DOI: 10.1007/S11666-020-01103-9.

[4] Gok M, Karaba M. Production of Re doped La2Zr2O7 based TBCs and numerical analysis of their use on IC engine piston surface [J]. Ceramics International, 2022. DOI: 10.1016/j.ceramint.2021.12.337.

[5] Du A, Wan C, Qu Z, et al. Thermal conductivity of monazite-type REPO4 (RE= La, Ce, Nd, Sm, Eu, Gd) [J]. Journal of the American Ceramic Society, 2009, 92(11): 2687-2692.

[6] Lakshya A K, Sarkar H, Ali F, et al. Texture-irrelevant thermal conductivity and expansion properties of La2Ce2O7 ceramics [J]. Thermochimica Acta: An International Journal Concerned with the Broader Aspects of Thermochemistry and Its Applications to Chemical Problems, 2023. DOI: 10.1016/j.tca.2023.179431.

[7] Zhong X, Wang Y, Liu P, et al. Effects of microstructure on corrosion behaviors for RE2SiO5 (RE=Gd, Y, Er) environmental barrier coatings against calcium-magnesium-alumino-silicate melts [J]. Corrosion Science, 2022, 199: 110174-. DOI: 10.1016/j.corsci.2022.110174.

[8] Choudhary B, Medvedev D A, Besra L, et al. Effect of sintering temperature on the transport properties of La2Ce2O7 ceramic materials [J]. CERAMICS INTERNATIONAL, 2022(5): 48.

[9] Ouyang T, Suo J. TiC-self-healing thermal barrier coating structures and oxidation resistance [J]. Surface and Coatings Technology, 2021, 412(7): 127065. DOI: 10.1016/j.surfcoat.2021.127065.
[10] Nyman M, Rodriguez M A, Shea-Rohwer L E, et al. Highly versatile rare earth tantalate pyrochlore nanophosphors [J]. Journal of the American Chemical Society, 2009, 131(33): 11652-11653.
[11] Zhu J, Lou Z, Zhang P, et al. Preparation and Thermal Properties of Rare Earth Tantalates (RETaO4) High-Entropy Ceramics [J]. Shanghai Institute of Ceramics, 2021(4). DOI: 10.15541/JIM20200426.
[12] Ogawa T, Matsudaira T, Yokoe D, et al. Spontaneously formed nanostructures in double perovskite rare-earth tantalates for thermal barrier coatings [J]. Acta Materialia, 2021, 216.DOI: 10.1016/j.actamat.2021.117152.
[13] Zhang X C, Xu B S, Tu S T, et al. Effect of spraying power on the microstructure and mechanical properties of supersonic plasma-sprayed Ni-based alloy coatings [J]. Applied Surface Science, 2008, 254(20): 6318-6326.
[14] Bai Y, Han Z H, Li H Q, et al. High performance nanostructured ZrO2 based thermal barrier coatings deposited by high efficiency supersonic plasma spraying [J]. Applied Surface Science, 2011, 257(16): 7210-7216.
[15] Li Q, Song P, Li Z, et al. Oxidation behavior and Cr-Zr diffusion of Cr coatings prepared by atmospheric plasma spraying on zircaloy-4 cladding in steam at 1300°C [J]. Corrosion Science: The Journal on Environmental Degradation of Materials and its Control, 2022(Jul.): 203.
[16] Planche M P, Liao H, Coddet C. Oxidation control in atmospheric plasma spraying coating [J]. Surface and Coatings Technology, 2007, 202(1): 69-76.
[17] Xu Z, He S, He L, et al. Novel thermal barrier coatings based on La2(Zr0.7Ce0.3)2O7/8YSZ double-ceramic-layer systems deposited by electron beam physical vapor deposition [J]. Journal of Alloys & Compounds, 2011, 509(11): 4273-4283. DOI: 10.1016/j.jallcom.2010.12.203.

Journal of Physics: Conference Series **2842** (2024) 012066 doi:10.1088/1742-6596/2842/1/012066

Effects of carbon content and heat treatment on wear-resistance and impact properties of novel cast steels with bainitic and martensitic structure

Xuefeng Tian[1,*], Guozhi Zhang[2], Kun Wang[1], Xiaobo Yang[1], Lulu Feng[1] and Weiguang Yang[1]

[1] School of Advanced Manufacturing, Guangdong Songshan Polytechnic, Shaoguan 512126, China

[2] School of Materials Science and Engineering, Northeastern University, Shenyang 110004, China

[*]Corresponding author's e-mail: tianxuefeng05@163.com

Abstract: The traditional material for caterpillar boards is mainly Hadfield steel. The wear resistance of the Hadfield steel cannot be fully exerted due to insufficient work hardening during the operation of the caterpillar board. The effects of carbon content and heat treatment on the wear-resistant and impact properties of a novel caterpillar board cast steels with bainitic and martensitic structures have been investigated. The microstructural observation of worn surface topography was performed by scanning an electronic microscope, and the impact toughness was carried out by an impact tester, in a bid to improve the wear-resistant properties and its impact toughness. The results demonstrate that the hardness and wear-resistant properties of cast steels increase with the increase of carbon content, whereas the impact properties reduce because the brittle phases, such as the carbide precipitation and the martensite, are more. The heat treatment temperature of steels can change the relative ratio of bainitic and martensitic structures, thus affecting the wear-resistant properties of the steels. By controlling the carbon content and heat treatment processing of the steels, the optimal combination of toughness and wear-resistant properties can be achieved. The wear mechanism of dual-phase cast steel was analyzed through wear resistance tests, and it also shows that dual-phase cast steel with bainitic and martensitic structures had better wear-resistance properties compared to the Hadfield steel at low-impact wear conditions.

1. Introduction

Owing to higher toughness than single-phase bainite or martensite, many researchers have investigated the microstructure and properties of the B/M double-phase steel [1-3]. The results showed that the B/M structure generally possesses an excellent combination of strength and toughness [4-5]. There are usually two kinds of different routes to obtain the bainite phase in steel. One method is by means of the isothermal quenching process whereas the other is the alloying method in order to acquire as-cast bainite. The chemical composition, namely the micro-alloying elements, plays a vital role by affecting the structure and properties of these steels. Alloying elements can significantly affect the relative stability and amount of austenite phase in steel [6-7].

The fraction of bainite/martensite structure in castings may substantially depend on the composition of the alloy, as well as on the heat treatment conditions. This study aims to research the effects of C content and heat treatment conditions on the wear-resistance and impact properties of a kind of novel cast steel with bainitic and martensitic structures.

2. Experimental procedure

2.1. Casting and heat treatment conditions

The investigated steel was melted in a medium-frequency induction furnace and subsequently deoxidized by aluminum. Samples were cast in a sand mould. The chemical composition of the experimental steel is given in Table 1. As can be seen, the carbon content varied from 0.25% to 0.75%, while the alloying element was unchanged. The chemical composition of high manganese steel used in this study is 1.09 wt% C, 13.03 wt% Mn, 0.40 wt% Si, 0.012 wt% P, and 0.014 wt% S.

Table 1. Chemical composition of steels (wt.%).

Steel	C	Si	Mn	Mo	Cr	B	Fe
C1	0.25	2.0	3.0	0.4	1.2	0.002	Balance
C2	0.50	2.0	3.0	0.4	1.2	0.002	Balance
C3	0.70	2.0	3.0	0.4	1.2	0.002	Balance

The heat treatment conditions in this investigation are given in Table 2. The quenching was conducted in salt baths.

Table 2. Heat Treatment Conditions.

Designation of steel	Heat treatment
H1	Temper at 300°C for 90 min, and then cool in the furnace.
H2	Temper at 400°C for 90 min, and then cool in the furnace.
H3	Austenitize at 900°C for 60 min, and then cool in air.
H4	Austenitize at 900°C for 60 min, transform isothermally at 300°C for 60 min, and then cool in air.
H5	Austenitize at 900°C for 60 min, transform isothermally at 400°C for 60 min, and then cool in air.
H6	Austenitize at 900°C for 60 min, quench in salt bath at 300°C for 60 min, and then cool in air.
H7	Austenitize at 900°C for 60 min, quench in salt bath at 300°C for 60 min, temper at 300°C for 90 min, and then cool in air.
H8	Austenitize at 900°C for 60 min, quench in salt bath at 300°C for 60 min, temper at 400°C for 90 min, and then cool in air.

2.2. Microstructure and mechanical properties

The investigated samples were eroded with a solution of 5% by volume nitric acid + ethyl alcohol. Microstructure observation of worn surface topography was carried out by scanning an electronic microscope. The impact test specimens were in the size of 10×10×55 mm^3 without notch. The impact toughness was measured by a JB-5 impact tester. The hardness was measured using a Vickers sclerometer at room temperature. The hardness and toughness values of investigated alloys were the average of the three tested specimens. The flowchart of the test procedure is shown in Figure 1. The wear sample used in this work was 18×18×50 mm^3. The wear tests were performed utilizing an abrasive wear tester devised by the authors. The schematic diagram of the abrasive wear tester is shown in Figure 2. The wear performance of each investigated specimen was characterized by weight loss. The samples were initially weighed and reweighed at the end of each wear test. The abrasive was quartz sand of size 20–40 mesh.

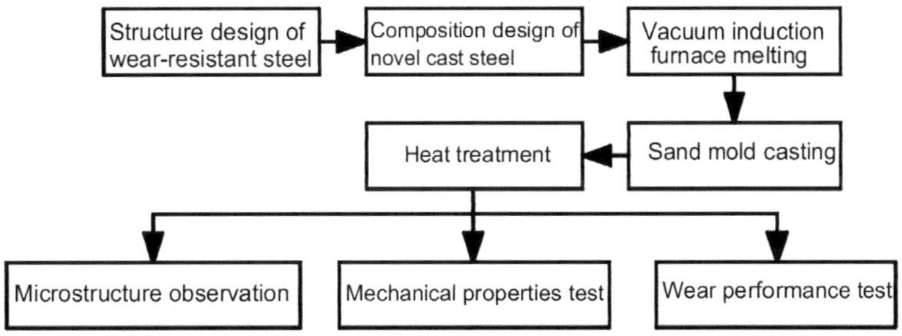

Figure 1. Flowchart of test procedure.

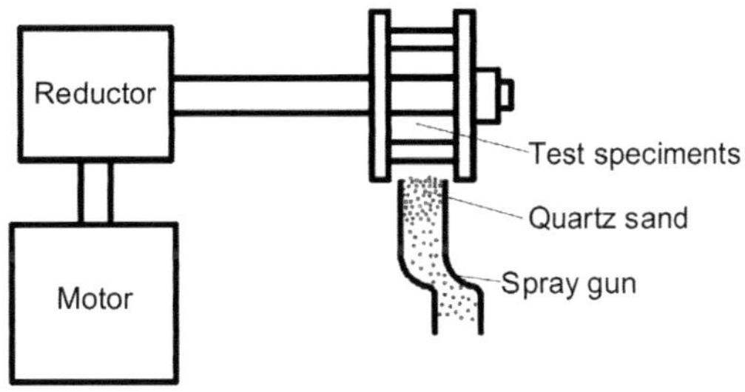

Figure 2. Schematic diagram of the wear tester.

3. Results and discussion

3.1 Microstructures

It is observed that a variety of microstructure was obtained depending on the different heat treatment conditions, as shown in Figure 3. Figure 3(a)-(c) shows the microstructures of cast steel as cast (C1-C3 sample). It can be seen that the microstructure consists of retained austenite and needle bainitic ferrite. The needles are 20-40 μm in length and 3-5 μm in width. When austenite cannot be fully transformed into bainite via heat treatment, the untransformed austenite will transform into martensite microstructure subsequently. The amount of the needle bainitic ferrite increases with the increase of the carbon content in the investigated steel. The carbon promotes the formation of the lower bainitic ferrite because Martensite start temperature (M_S) drops with the carbon content increases in steel. The steels, austenitized at 900°C for 60 min and cooled in air, contain needle bainitic ferrite, the retained austenite, and a small amount of martensite (Figures 3(d) and (e)). When quenched isothermally at 300°C or 400°C, feathery characteristic of upper bainite ferrite was observed (Figures 3(f)-(h)). The lower bainite transformation was suppressed. The phenomenon shows that the isothermal temperature was still in the upper bainitic transformation temperature range. This microstructure is especially similar to the microstructure of ADI, except that graphite nodules are non-existent in this study [8]. The feathery bainite sheaf is short and fine. The kinetics of the upper bainite transformation reaction is different from that of lower bainite. The phase transformation, accompanied by a change in the morphology of upper and lower bainite ferrite, was related to the different nucleations and growth mechanisms [9]. The feathery ferrite amount in the microstructure is higher via the lower isothermal temperature transformation, as shown in Figures 3(g) and (f). The lower isothermal transformation temperatures lead to a higher volume fraction of feathery ferrite due to the different nucleation and growth processes. At lower isothermal transformation temperature, i.e., 300°C, the nucleation rate was higher owing to larger endured supercooling. The diffusion rate of carbon atoms was slow at this lower temperature. When bainite ferrite nucleated from untransformed austenite, carbon atoms must diffuse from the bainite ferrite for the ferrite needles growth. Therefore, a large quantity of fine bainite ferrite platelets developed at lower isothermal transformation temperatures. However, at higher isothermal transformation temperatures, the corresponding supercooling degree was lower and the nucleation rate was also lower accordingly. This resulted in fewer ferrite needles formation with larger sizes owing to a high diffusion rate of carbon atoms at higher transformation temperatures. Therefore, the bainite ferrite amount was lower and both bainite ferrite and residual austenite were coarser in terms of nature at higher isothermal transformation temperatures.

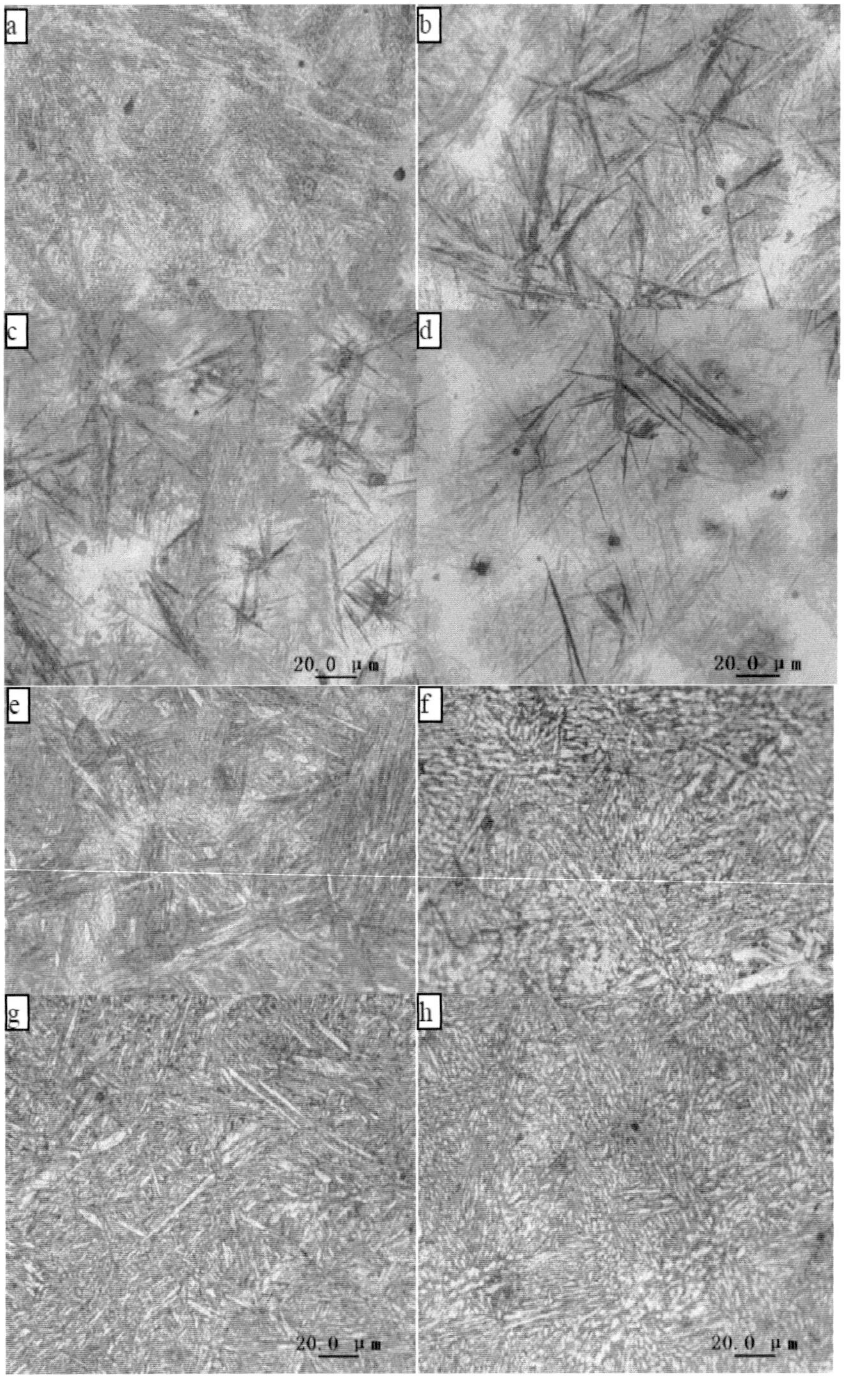

Figure 3. Microstructure of steels at different states.
(a) C1 steel in as-cast state; (b) C2 steel in as-cast state; (c) C3 steel in as-cast state; (d) C3 steel in H3 state; (e) C1 steel in H3 state; (f) C2 steel in H5 state; (g) C1 steel in H6 state (h) C1 steel in H8 state

3.2 Mechanical properties

The hardness and impact toughness test results of the investigated alloys are illustrated in Tables 3 and 4. It can be observed that the hardness and impact toughness of the experimental steels after different heat treatments increased in comparison to the as-cast steel. However, the values of the hardness and impact toughness of the experimental steels drop in some cases. The hardness value increases with the increase in carbon content. It is seen that the impact toughness values of the steels with high carbon content decrease significantly. This phenomenon arises from the presence of hard and brittle phases such as the carbide precipitation, the martensite, and the upper bainite ferrite in the microstructure, which restricts the maximum ductility but increases the hardness. Compared to lower bainite ferrite, the upper bainite ferrite leads to low fracture toughness and strength of steels. C1 and C3 steels have higher impact toughness values when austenitized at 900°C for 60 min and then cooled in air, while C1 steel has the highest toughness value in the H5 state.

Table 3. Hardness values of steels (HV).

State	As-cast	H1	H2	H3	H4	H5	H6	H7	H8
C1	395.3	392.6	379.8	428.0	387.5	336.2	397.1	408.3	403.9
C2	554.4	517.5	497.9	639.7	650.8	583.8	696.2	578.0	546.4
C3	574.3	575.2	578.6	504.1	506.2	513.1	599.2	507.0	587.6

Table 4. Impact toughness of steels (α_k, J/cm^2).

State	As-cast	H1	H2	H3	H4	H5	H6	H7	H8
C1	36.65	20.96	10.74	26.50	51.56	75	47.20	58.49	15.12
C2	9.17	10.04	5.99	10.33	9.87	13.49	5.83	9.73	7.32
C3	6.71	10.74	6.39	16.81	21.34	17.90	10.17	9.86	10.86

3.3 Wear performance and worn surface topography

Table 5 indicates the wear resistances of the experimental alloy in different heat treatment states and Hadfield steel under the same condition. It can be seen that the wear-resistance property of the C1 steel in the H3 state and C2 steel in the H5 state is inferior to that of Mn13 steel. However, the wear resistance of the C3 steel in the H3 state is superior to Mn13 steel. The C2 steel in the H5 state consists of the amount of the upper bainite in the microstructure. The C3 steel in the H3 state contains an amount of the lower bainite while the martensite and lower bainite coexist in the microstructure of C1 steel in the H3 state. Therefore, C3 steel in the H3 state has a better combination of hardness and impact toughness compared to C1 steel in the H3 state and C2 steel in the H5 state. It is undoubted that C3 steel in the H3 state has higher wear resistance than the other two kinds of alloy. The Hadfield steel did not result in a workhardening capacity fully in itself than experimental steel under the condition of lower stress abrasion wear. Therefore, the wear resistance of C3 steel in the H3 state was superior to that of Hadfield steel.

Table 5. Results of the wear test of investigated steels.

Material	Hardness (HV)	Wear time (h)	Loss of weight (g)
C1 steel in H3 state	428.0	72	8.110
C2 steel in H5 state	583.8	72	8.509
C3 steel in H3 state	304.1	72	6.395
Mn13 steel quenched	302.2	72	7.657

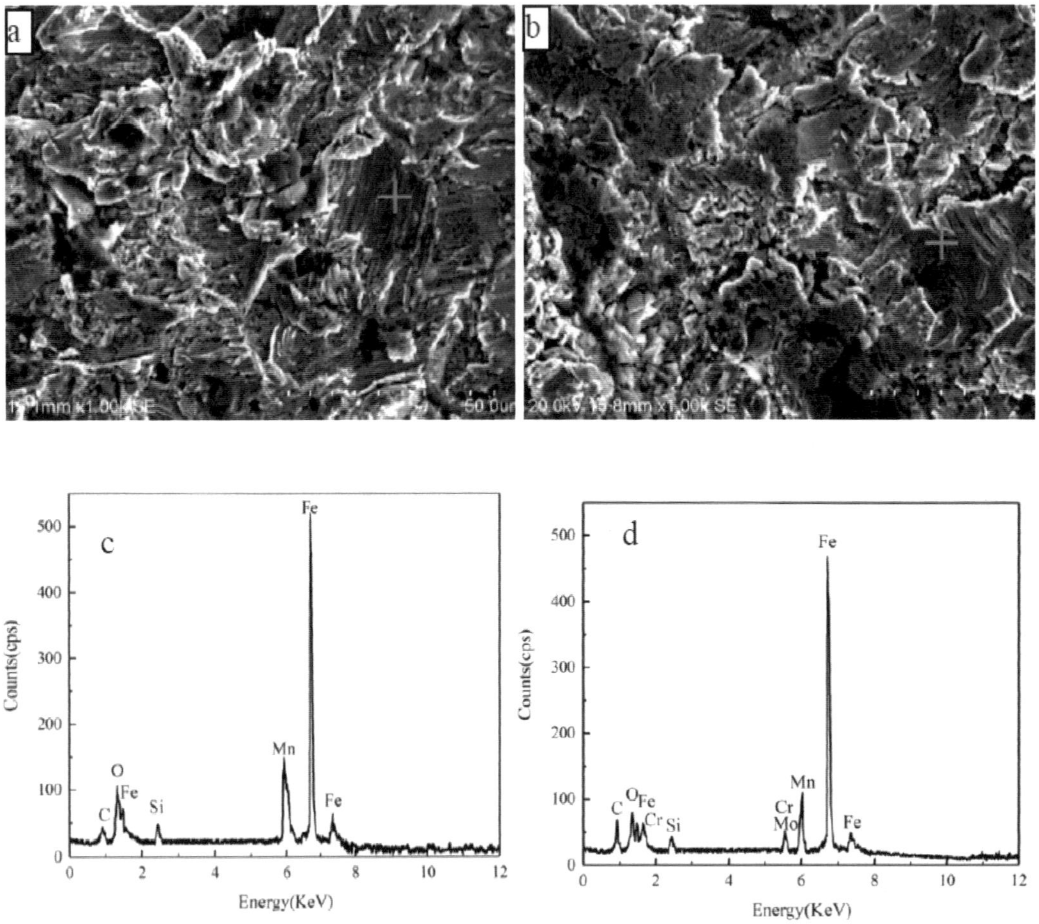

Figure 4. SEM of the worn surfaces of investigated steel.
(a) C3 steel in H3 state; (b) Mn13 in water-quenched;
(c) EDS analysis of furrow in (a); (d) EDS analysis of furrow in (b)

Figures 4(a) and 4(b) show the SEM of the worn surfaces of the C3 steel in the H3 state and Hadfield steel. The worn surface of C3 steel is essentially composed of furrows and pleats running at random. The formation of furrows and pleats was the result of minute wear-cutting during the wear process. There is an amount of plastic deformation in C3 steel compared to Mn13 steel. The plastic deformation shows that C3 steel has greater work-hardening capacity than Mn13 steel under lower-stress abrasive wear conditions. The work-hardened layer can prevent the steel surface layer from abrasive stabbing and scratching more efficiently. Figures 4(c) and 4(d) show EDS analysis of the furrow of the Hadfield steel and experimented steels. It can be seen that both materials undergo oxidation during the wear process. The lower oxygen content of the experimental steel is related to the smaller temperature rise caused by a low wear rate.

4. Conclusions

(1) Both the casting and the heat treatment can acquire Bainite in the steel. Bainite takes needle form in the solidified microstructure or in the microstructure generated by austenitizing at 900°C for 60 min and then cooling in air. However, feathery bainite exists in the steel quenched isothermally at 300°C or 400°C.

(2) The steel with 0.25 wt% C has the best combination of hardness and impact toughness values. The wear-resistance property of the optimal steel is superior to that of Hadfield steel under the lower stress abrasion wear condition.

Acknowledgment
This study was financially supported by the Shaoguan Municipal Science and Technology Funding Plan Project (210728104530543).

References
[1] Moghaddam P V, Rinaudo M, Hardell J, et al. Influence of fracture toughness on two-body abrasive wear of nanostructured carbide-free bainitic steels [J]. Wear. 2020, 460-461: 203484.
[2] Yu Zhang, Jing Bai, Keliang Guo, et al. Unraveling magnetic properties and martensitic transformation in Mn-rich Ni-Mn-Sn alloys: first-principles calculations and experiments [J]. Rare Metals, 2024, 43 (4): 1769–1785.
[3] Dong K, Lu C, Zhou W, et al. Wear behavior of a multiphase ductile iron produced by quenching and partitioning process [J]. Engineering Failure Analysis. 2021, 123: 105290.
[4] R. Schneider, K. Steineder, D. Krizan, and C. Sommitsch, Effect of the heat treatment on the microstructure and mechanical properties of medium-Mn-steels [J]. Mater. Sci. Technol, 2019, 35(17): p. 2045.
[5] Di Schino A, Gaggiotti M, Testani C, et al. Heat treatment effect on microstructure evolution in a 7% Cr steel for forging [J]. Metals. 2020, 10(9): 808.
[6] Zhang Zhiqiang, Liu Bo, Xu Lianyong, et al. Investigation on influence of alloying on phase transitions of duplex stainless steel based on thermochemical calculation [J]. China Welding, 2023, 32(4): 11–28.
[7] M. Atkins. Atlas of Continuous Cooling Transformation Diagrams for Engineering Steels, British Steel Corporation, Sheffield, ISBN 0 9500451 44, p. 97.
[8] Li Hefei, Zhang Peng, Wang Bin, et al. Predictive fatigue crack growth law of high-strength steels [J]. Journal of Materials Science & Technology, 2022, 100: 46-50.
[9] Basu S, Jaya B N, Patra A, et al. The Role of Phase Hardness Differential on the Non-uniform Elongation of a Ferrite-Martensite Dual Phase Steel [J]. Metallurgical and Materials Transactions A, 2021, 52(9): 4018-4032.

Study on solidification organization and phase formation of Fe-6.5wt.%Si high silicon steel by arc melting and heat treatment

Yanxia Liu [1], Enbao Pan [2], Zhiwang Zhang[3*], Qiuxin Ji [4], Xinyao Zhang[5]

[1] Hebei Material Fine Crystal Preparation Technology Innovation Center, Shijiazhuang, Hebei 050091, China
[2]Hebei Key Laboratory of Undercooling Preparation Technology and Science, Shijiazhuang 050091, China
[3]HBIS Materials Technology Research Institute,Shijiazhuang 052165,China
[4] Hebei Provincial Engineering Research Center for Deep Undercooling Preparation of Metal Materials, Shijiazhuang 050091, China
[5]School of Materials Science and Engineering, Hebei University of Science and Technology, Shijiazhuang, Hebei, China 050043, China

[*] Corresponding author's e-mail: zhangzhiwangliu@163.com

Abstract: The solidification microstructure and phase formation behavior of Fe-6.5wt.%Si high silicon steel was investigated using arc melting and heat treatment. The alloy phase composition and microstructure were analyzed by XRD and SEM. The magnetic performance was measured at a high- and low-temperature vibration sample magnetometer. The results show that the temperature gradient causes alloy ingot to be roughly divided into three layers. With the decrease in temperature, from the bottom to the top, the phase morphology is fine crystals, columnar internal dendrites, and isoaxial internal dendrites. In the annealing ring ingot, the inner dendrites disappeared, and from the bottom to the top of the sample, the microstructure changed from fine crystal, columnar crystal, and isoaxial crystal, to micro-cracks, which first increased and then reduced. The hysteresis loop curve width of the alloy is narrow, the slope is large, the hysteresis is 0.126 emu/g, and the magnetic utilization after annealing decreases to 0.0667 emu/g, but the coercive force increases to 2.8 Oe/g, and the annealed alloy is more sensitive to the magnetic field and has a high magnetic conductivity coefficient.

1. Introduction

Fe-6.5wt.%Si high silicon steel alloy is an excellent soft magnetic material, with high permeability, low coercive force, and near-zero hysteresis expansion. Compared with ordinary silicon steel, it has low magnetic crystal anisotropy and nearly zero hysteresis expansion coefficient. From the phase diagram, the solidification process of 6.5 high silicon steel gradually occurs in the phase transition, with three different phase structures: from disordered A2 to orderly B2 phase transition, and then forming the orderly DO3 phase. However, in the Fe-Si alloy, with the increase in silicon content, the hardness increases when the silicon content exceeds 3.5wt.%Si, and cold rolling becomes difficult;

especially, the room temperature plasticity of 6.5 high silicon steel is close to zero, so it is difficult to make qualified sheets by conventional processing and preparation methods, which seriously affects the application in the industrial field.

Most of the current research focuses on the phase structure [1-2] and magnetothermal performance [3] after 6.5 high silicon steel ingot annealing at a high temperature, and less attention has been paid to the organization and phase composition of arc melting alloy ingot. Therefore, the solidification structure of 6.5Si alloy is formed in arc melting.

2. Experimental method

The purity of alloy raw material Fe is 99.99% and Si is 99.999%. According to the chemical composition of 6.5 alloys, the weight of the alloy is about 2 g. The alloy was put into the copper crucible of the arc furnace, vacuumed to 6.010-4Pa, and reverse charged with high-purity argon (purity 99.999%). The alloy was repeatedly melted four times in the arc furnace to make the alloy composition more uniform. The 6.5Si high silicon steel ingot was sealed in the quartz tube and annealed in the box annealing furnace. The 6.5Si high silicon steel sample was cut in half, hot set, ground, polished, and then corroded with 10% HNO_3 alcohol, and the alloy microtissue and phase composition were analyzed by electron microscopy FlexSEM 1000 II, TASCAN microscope, scanning electron microscope, and its own energy spectrum analyzer. The crystal structure of the samples was analyzed by measuring the X-ray diffraction spectrum of the powder samples using the Cu-K target on an X-ray diffractometer. Magnetic properties were determined using a high- and low-temperature vibration sample magnetometer VSM-220.

3. Experimental results and discussion

The X-ray diffraction analysis patterns of arc melting and annealed 6.5Si alloy samples are shown in Figure 1 (a) and 1 (b), respectively. The alloy structure is composed of A2, B2, and DO3 phases, showing mainly the diffraction peak of the A2 phase [4~7]. The diffraction peak is split, and the Fe14Si2-b peak appears near 2 θ = 78.

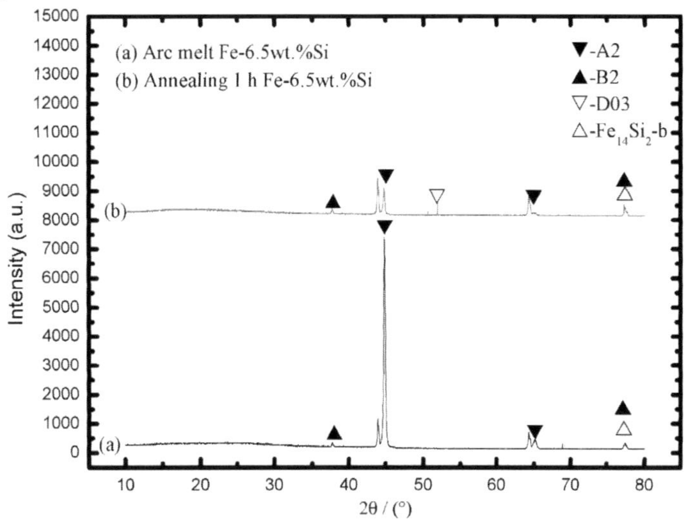

Figure 1. Powder X-ray diffraction patterns of 6.5Si alloys (a) arc melting (b) annealed for 1 hour.

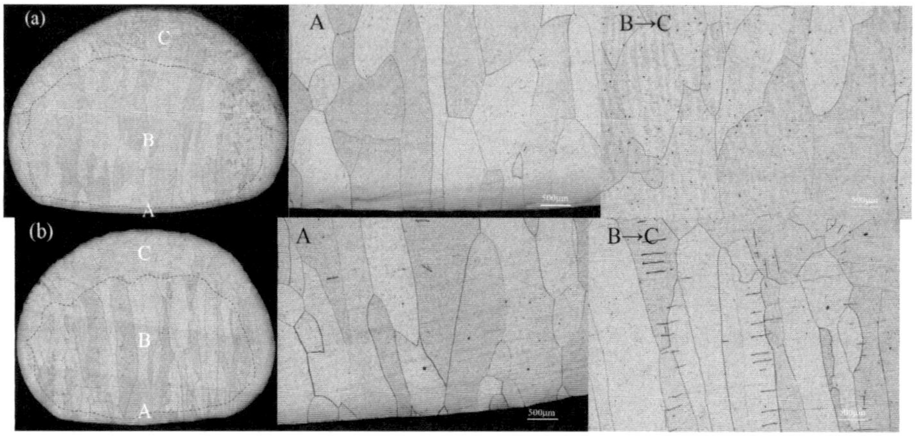

Figure 2. Microstructure of 6.5Si alloys (a) arc melting sample (b) annealed for 1 hour 1050°C.

Figure 2 shows the micro organization of the 6.5Si alloy. Figure 2(a) is the overall distribution diagram of the microstructure of the arc-melted 6.5Si alloy. From the figure, the microtissue of 6.5Si alloy ingot can be clearly divided into three layers and four different regions. Figure 2 (a) (A-C) shows the distribution of each tissue region after enlarging the 6.5Si alloy ingot, from the bottom layer to the top layer. Area A is A part of the area where the bottom of the 6.5Si alloy ingot meets the copper crucible, where the microstructure is mainly the relatively small fine crystal zone; after the transition from area A to area B, the tissue group is made of mainly thick columnar crystals, which contain dendrite along the growth direction; a transition zone is formed from area B to area C; and C area is the developed dendrite organization.

Figure 2(b) is the overall microtissue distribution diagram of arc melting 6.5Si alloy at 1050 °C and annealed for 1 h. From the figure, the microtissue of 6.5Si alloy ingot can be clearly divided into three layers and four different regions. The microscopic organization in area A is mainly a relatively fine crystal zone (a part of the area where the bottom of the 6.5Si alloy ingot meets the copper crucible); after the transition from area A to area B, the tissue group is mostly slender, with cracks in the transition zone; between area B and area C, there is an increase in transverse cracks in the transition zone; area C is the isoaxial tissue, and the microcracks are reduced. Figure 3 is the composition energy spectrum analysis of the constituent phase of 6.5Si alloy microtissue. The component analysis results are shown in Table 1. From the energy spectrum analysis, the bottom silicon content of arc 6.5Si alloy is slightly higher than the top, while the top silicon content of the annealed alloy is slightly higher than the bottom.

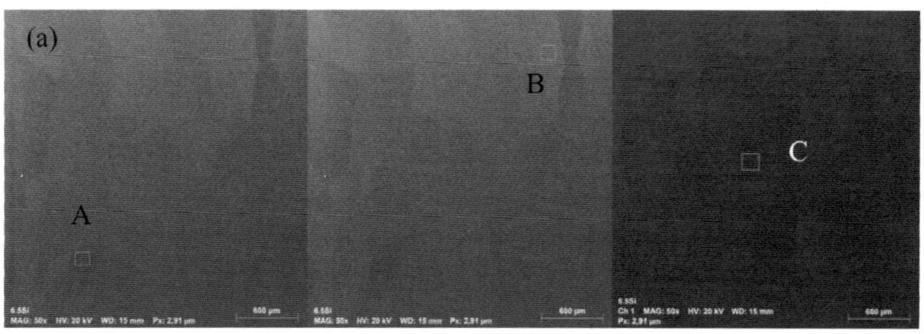

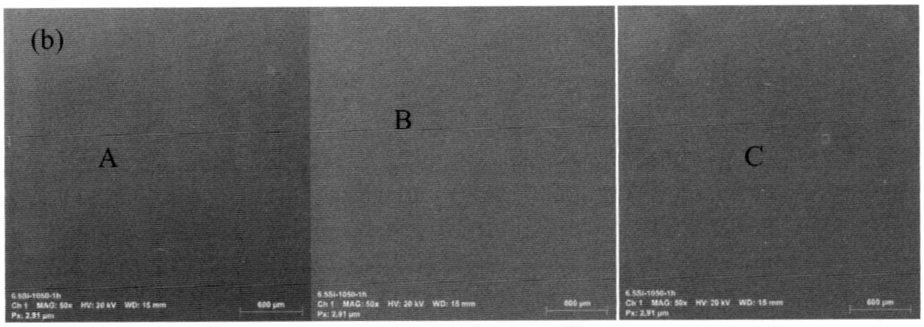

Figure 3. Composition of different regions of 6.5Si alloys (a) arc melting (b) annealed for 1 hour.

Table 1. Composition of phases Fe-6.5wt.%Si alloys (%wt.).

Sample	A region		B region		C region	
	Fe	Si	Fe	Si	Fe	Si
Arc	94.11	5.89	94.01	5.99	94.49	5.51
Annealing	94.29	5.71	93.62	6.38	93.92	6.08

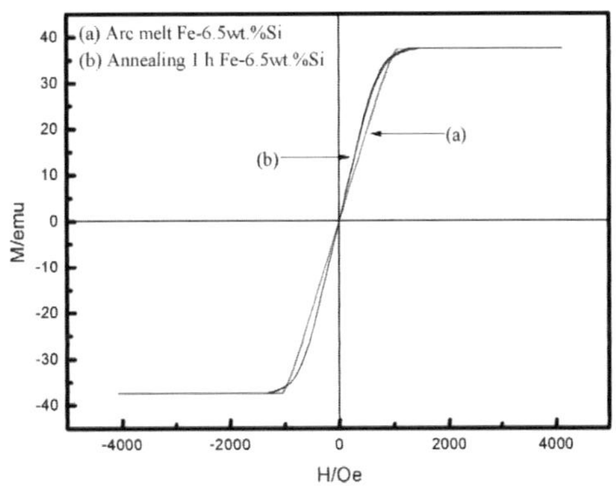

Figure 4. The H-M hysteresis loop of the 6.5Si alloys (a) arc melting (b) annealed for 1 hour 1050 °C.

Figure 4 shows the H-M hysteresis loop of the susceptibility of the 6.5Si alloy with an external magnetic field *Hmax*= ± 4000 Oe, showing the magnetic properties of the 6.5Si alloy. Table 2 shows the magnetic properties after arc melting and annealing.Figure 4(a) is the arc melting sample, with the remaining magnetic Mr = 1.26×10^{-01} emu/g, coercivity Hci = 1.8 Oe/g, (saturation magnetization) *Ms* = $3.74\times10^{+01}$ emu/g; Figure 4(b) is the 1050 °C sample annealed for 1 hour, with *Mr* = 9.67×10^{-02} emu/g, *Hci* = 2.8 Oe/g, *Ms* = $3.74\times10^{+01}$ emu/g. The hysteresis loop curve of Fe6.5wt%Si alloy has a narrow width and a large slope, and the hysteresis is close to 0. It can be seen from the figure that the slope of the hysteresis curve after annealing is steeper than that of the arc melting sample, and the residual magnetization after annealing decreases, but the coercivity increases, and the annealed Fe6.5wt%Si is more sensitive to the magnetic field and has a high magnetic conductivity coefficient.

Table 2. Magnetism of Fe-6.5wt.%Si alloys.

Sample	Coercivity (*Hci*/Oe)	Magnetization (*Ms*) /emu/g	Retentivity (*Mr*) /emu/g
Arc	1.8	37.4	0.126
Annealing	2.8	37.4	0.0967

The cooling rate of arc melting is about 10^2 ~10^3 K/s, with obvious stratification and transition interface; cooling around the sample, and the core grows rapidly in the opposite direction of the temperature gradient around the sample. Due to the high-temperature gradient on the upper surface, the fine crystal region is easily formed in Region A. In Region A, the cylindrical crystal growing from region B to the center of the ingot decreases with the temperature gradient, and the crystal is parallel to the growth direction. With a further decrease in the temperature, the cylindrical crystals in the C zone gradually form inner crystalline dendrites. From the annealing tissue of the 6.5Si alloy, the inner crystal of the crystal disappears, while the thick cylindrical crystal in the middle of the ingot changes into slender cylindrical crystals, in the upper part of the ingot, but a large number of microcracks [8-12] appear in the crystal. In the annealed ingot, the microcracks are divided into two types, one parallel to the growth direction of the column crystal and the other one at the grain boundary position.

After melting the arc, there is a large temperature gradient and composition segregation in the molten 6.5Si alloy ingot. At the onset of crystallization, the first nuclei of (Fe, Si) form on the surface layer of the ingot, growing from the four peripheries toward the center of the sample. The surrounding cooling speed is relatively fast, α-(Fe, Si) growth is inhibited, and the precipitation of fine grains is promoted around the sample; heat dissipates slowly in the center part of the sample, and α-(Fe, Si) presents thick dendric and columnar crystals. Since the cooling speed of the contact surface at the bottom of the sample is higher than that of the non-contact surface, i.e., the free surface, the grain at the bottom of the sample is smaller than the fine grain on the free surface. Thus, three different solidified tissues appeared in the sample.

4. Conclusion

By studying the coagulation microstructure of the arc melting of Fe-6.5wt.%Si alloy using XRD and SEM, the conclusions are obtained:

(1) The temperature gradient causes the 6.5Si alloy ingot to be roughly divided into three layers. With the decrease in temperature, from the bottom of the sample to the top, the phase morphology changes from fine crystals, columnar internal dendrites, to isoaxial internal dendrites;

(2) 6.5Si Alloy annealing ingot in the crystal internal dendrite disappeared. In the phase of the morphology of fine crystal, column crystal, and isoaxial crystal, micro-cracks from the bottom to the top first increased and then reduced;

(3) The hysteresis curve of 6.5Si alloy is narrow, and the hysteresis is 0.126 emu/g. The annealing decreases the hysteresis to 0.0967 emu/g, but the coercive force increases to 2.8 Oe/g. The annealed 6.5Si is more sensitive to the magnetic field and has a high magnetic conductivity coefficient.

Acknowledgment

This work was financially supported by the Science and Technology Research Project of Colleges and Universities in Hebei Province (No. ZD2019043).

References

[1] Yang J, Fu Z, Ye J, et al. Electron beam-based additive manufacturing of Fe93.5Si6.5 (wt.%) soft magnetic material with controllable magnetic performance [J]. Scripta Materialia, 2022, 210:114460.

[2] Dun H, Zhang Y, Wang N, et al. Effect of Sigma 3 grain boundaries on microstructure and properties of oriented Fe-6.5 wt%Si in twin-roll strip casting [J]. Journal of Magnetism and Magnetic Materials, 2022:559.

[3] Liu D, Liu X, Wang J, et al. The influence of Fe nanoparticles on microstructure and magnetic properties of Fe-6.5wt%Si soft magnetic composites[J].Journal of Alloys and Compounds, 2020, 835(11):155215.

[4] A X Y, B G L, C Z Z B, et al. Electronic structure characteristics of Fe-6.5wt%Si alloy doped with rare earth elements and its effect on mechanical properties [J]. Journal of Alloys and Compounds, 2020, 843.

Journal of Physics: Conference Series **2842** (2024) 012067 doi:10.1088/1742-6596/2842/1/012067

[5] Shi X, Liang Y, Liu B, et al. Deformation twinning characteristics in hot-rolled Fe-6.5wt%Si alloy with different degree of order [J]. Materials Science and Engineering, 2019, 762: 138095. 1-138095.9.
[6] Liu D, Liu X, Wang J, et al. The influence of Fe nanoparticles on microstructure and magnetic properties of Fe-6.5 wt% Si soft magnetic composites [J]. Journal of Alloys and Compounds, 2020, 835(11):155215.
[7] Liu D, Liu X, Wang J. The influence of Fe nanoparticles on microstructure and magnetic properties of Fe-6.5wt%Si soft magnetic composites [J]. Journal of Alloys and Compounds: An Interdisciplinary Journal of Materials Science and Solid-state Chemistry and Physics, 2020 (835-):835.
[8] Liu D, Qin J, Zhang Y, et al. Effect of yttrium addition on the hot deformation behavior of Fe–6.5wt%Si alloy [J]. Materials Science and Engineering: A, 2020, 797.
[9] Jiao J, Shen Q, Cheng A. Effect of Cooling Rate on Microstructure and Mechanical Property of Fe-6.5wt%Si Alloy [J]. Materials Science Forum, 2016(850-):850.
[10] Haijie Xu, Yunbo Xu, et al. A quasi in-situ EBSD study of the nucleation and growth of Goss grains during primary and secondary recrystallization of a strip-cast Fe-6.5 wt% Si alloy [J]. Journal of Alloys and Compounds 861 (2021) 158550.
[11] Xuan Yu, Guangtao Lin, et al.Electronic structure characteristics of Fe-6.5 wt% Si alloy doped with rare earth elements and its effect on mechanical properties [J]. Journal of Alloys and Compounds 843 (2020) 155916.
[12] Gaoyuan Ouyang, Chad R. Macziewski , et al. Gaoyuan Ouyang, Chad R. Macziewski , et al. Acta Materialia 205 (2021) 116575. [J]. Acta Materialia 205 (2021) 116575.

Journal of Physics: Conference Series **2842** (2024) 012068 doi:10.1088/1742-6596/2842/1/012068

The enhanced aging hardening behavior in Si-containing Al-5Zn-1Mg-1Cu alloys

Yi Lu*, Shengping Wen, and Zuoren Nie

College of Material Science and Engineering, Beijing University of Technology, Beijing 100124, China

* Corresponding author's e-mail: bbluyi@yeah.net

Abstract: The ageing behaviour and phase transformation of a range of Si-microalloyed low Mg/Zn ratio Al-5.0Zn-1.0Mg-Cu alloys were investigated at a temperature of 150°C by using TEM and microhardness tests. Adding Si had a important influence on the proportion of precipitates and microhardness. As Si content increased, the precipitated phase underwent a gradual transition from the T+η precipitates to the fine GPB-II precipitates. Concurrently, the microhardness exhibited a notable enhancement from 120 HV to 150 HV when the alloys were subjected to 150°C aging. Existence of small and even GPB-II phases in the Si-adding alloy was identified as underlying cause of this result. This GPB-II phase exhibited a bilayer microstructure. In core region predominantly comprise Zn, Si and Mg, and the shell region primarily consist of Zn and Cu. This structure effectively inhibited phase growth, maintaining the phases at a smaller scale and enhancing the alloy's hardening effect.

1. Introduction

Seven series Al alloys are extensively utilised in construction of aircraft owing to their exceptional specific strength, cost-effectiveness, favourable specific stiffness, malleability, and other advantageous characteristics [1-3]. The recent resurgence in interest in the development of new aircraft types has given rise to a renewed focus on the potential for enhancing the overall properties of Seven series Al alloys through technical and scientific innovation. Micro-alloying stands for an valid method for significant increasing the overall properties of alloys. For example, the addition of silver (Ag) influences the precipitates in Al-Mg-Zn-Cu alloy at the early ageing stages, thereby improving its strength [4]. The introduction of zirconium (Zr) and scandium (Sc) into the Seven series Al alloy generates small and even MgZn2 phases near GB, which in turn improves the alloy's corrosion properties [5].

In initial stages of research, inexpensive silicon was frequently regarded as an impurity, necessitating the addition of the element in the lowest possible quantity [6]. However, recent studies by Wu et al. have demonstrated that the addition of trace amounts of Si as microalloying element significantly increase the comprehensive property of aluminium alloys [7-8]. Adding silicon was observed to speed up the precipitation kinetics, resulting in the formation of Al3(Sc, Zr) and Al3Zr in Al - 0.05wt% Sc - 0.2wt% Zr alloys. These precipitates have been demonstrated to increase the alloys stress in creep testing [7]. Co-adding silicon and zinc in an Al-Mg-(Cu) alloy results in formation of small T phases, which serve to improve both peak hardness and ageing effect. The T phase shows excellent stability at high temperature (175°C), with a more uniform size distribution [8].

In this paper, we analyze the low Mg/Zn ratio Al-5.0Zn-1.0Mg-Cu alloys with various Si contents through different isothermal aging processes, expecting to further improve the microhardness and aging effect of different alloys.

2. Methods and experimental materials

Al-Zn-Mg-Cu-(Si) alloy cast by an iron mould using Al-20Si, Al-50Cu, 99.999% pure Al, pure Magnesium, and Zinc as original ingredients. Ten minutes prior to casting, magnesium was introduced to the alloy, resulting in a combustion loss of approximately 10%. The melting temperature was determined to be 800°C. Subsequently, the melt was cast into a mould of dimensions 30×100×180 mm^3, thereby forming an ingot. The five alloys were designated as 1#, 2#, 3#, 4#, and 5#, respectively, and the verified composition in different alloy shows in Table 1.

Solution treatment was conducted by subjecting the sample to a temperature of 530°C for a period of five hours, followed by a further ten hours of maintenance at 530°C. Subsequently, samples were rapidly cooled in water. After that, solution-state samples aged in 150°C for varying periods of times.

We used HXD 1000 testing machine to test microhardness, with a load of 200 g and a dwelling time of 10 seconds. This was conducted in order to monitor the strength response subsequent to the application of an artificial ageing process. Sample were excised from keelblook and subjected to micro-hardness testing following polishing samples, which were tested in six regions. TEM Samples are cut into 0.6 mm thick slices and ground to a thickness of 40-60 μm. The sample was then punched with a diameter of 3 mm. Further grinding to a thickness of 40-150 nm. Double spray polishing is performed in a solution containing 30% nitric acid and 70% methanol in a temperature range of approximately -15°C to -30°C. The operational voltage was approximately 15 volts, with an operational current of approximately 80 to 90 milliamps. A bright field image (BF) was obtained through transmission electron microscopy (TEM) observation at 200 kV acceleration voltage using a JEM 2100. The HAADF-STEM images were subjected to analysis at 300 kV using the Titan G2 electron microscope.

Table 1. The compositions of Al-Zn-Mg-Cu-(Si) alloys (wt.%).

Samples	Actual composition (wt.%)				
	Zn	Mg	Cu	Si	Al
1# (Al-5.0Zn-1.0Mg-1.0Cu)	4.60	1.00	0.94	-	Bal.
2# (Al-5.0Zn-1.0Mg-1.0Cu-0.2Si)	4.88	1.00	1.04	0.22	Bal.
3# (Al-5.0Zn-1.0Mg-1.0Cu-0.35Si)	4.80	0.99	0.93	0.33	Bal.
4# (Al-5.0Zn-1.0Mg-1.0Cu-0.5Si)	4.86	1.01	0.99	0.52	Bal.
5# (Al-5.0Zn-1.0Mg-1.0Cu-0.7Si)	4.84	0.99	0.95	0.68	Bal.

3. Results

3.1 Micro-hardness characterization

Figure 1 shows the hardness curve of different alloys as aging at 150°C. The solid solution state hardness of these alloys was observed to increase slightly from 46 to 62 HV, presenting that adding silicon has a negligible influence in the solid solution alloy. Peak hardness in Alloy 1 arrived a maximum value of 120 HV and quickly decreased to 105 HV at 96 hours. In contrast, other alloys hardness remained unaltered following the peak. Adding silicon resulted in an improvement in the maximum hardness of the alloys. It can be reasonably deduced that alloys 4 and 5 have the potential to reach approximately 150 HV, which represents a 30 HV increase compared to alloy 1. It therefore be known that adding silicon in Al-Zn-Mg-Cu alloys represents an effective method of enhancing the peak hardness and accelerating the ageing behaviour.

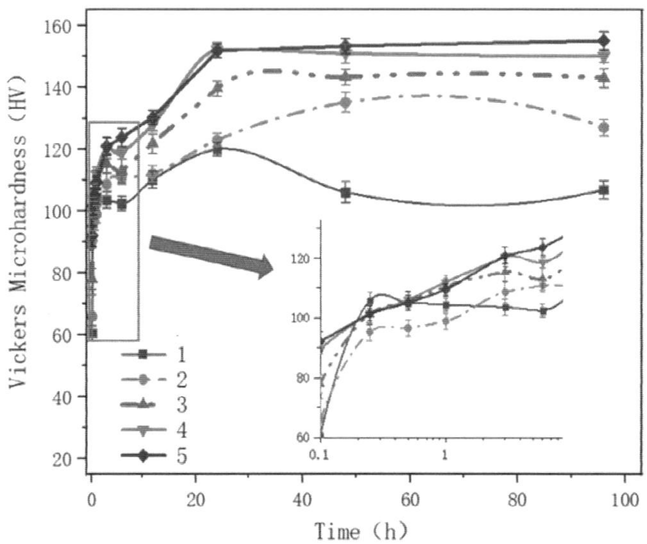

Figure 1. The Micro-hardness curves of samples 1, 2, 3, 4, and 5 aging at 150°C.

3.2 Characterization of microstructure

In order to investigate the principle of ageing effect of aforementioned samples, the microstructure of Alloys 1, 2 and 4 was observed as they underwent ageing at 150°C for 96 hours. In our research, the electron beams were parallel to <010> orientation of Al matrix (beam=<010>α).

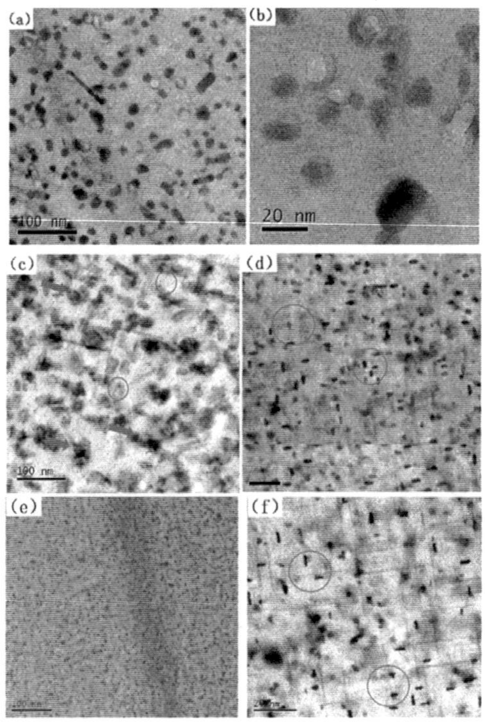

a) TEM of # 1; (b) Higher Magnifination TEM of # 1; (c) TEM of # 2; (d) Higher Magnification TEM of # 2; (e) TEM of # 4; (f) Higher Magnification TEM of # 4.

Figure 2. TEM images of ddifferent samples ageing at 150°C in 96 h.

Figure 2 illustrates the transmission electron microscopy of # 1, # 2, and # 4 aging in 150°C for 96 hours, respectively. From the shape of the phases observed in Figures 2(a) and (b), it can be inferred that # 1 contains large η phases and huge T phases. As illustrated in Figures 2(c) and (d), # 2 is predominantly consisted of large η phases (in blue arrow) and small precipitates with a diameter of approximately 5-8 nm (in red circle). This small phases were previously identified as GPB-II phases in other studies. As illustrated in Figures 2(e) and (f), Alloy 4 exhibited the precipitation of small and even GPB-II phases (in red circle). The GPB-II exhibited 2 distinct morphologies: a straight boundary configuration with an elongated rod-like shape and a smooth boundary configuration with an ellipsoidal-like shape. Therefore, the addition of Si to seven series alloys can influence the formation and proportion of precipitates, promoting number density of GPB-II phases and reducing the quantity of coarse phases.

The dimensions of T and η precipitates were found to be approximately 13-32 nm, exhibiting a similarity in size across the alloys. This is approximately threefold larger than the dimensions observed in GPB-II phases. In comparison to the coarse phases, these small and even phases demonstrate the capacity to effectively harden the alloy at the specified ageing temperature. Therefore, the addition of silicon to the Al-Zn-Mg-Cu alloy significantly constrains growth of the phase and promotes the quantity of fine precipitates owing to phase change when ageing in 150°C.

4. Discussions

4.1 GPB-II phases structure

To gain more insight into micro-structure of small GPB-II precipitates, HR-HAADF imaging was employed to see 2 distinct morphology of these precipitates in # 4, which had been ageing in 150°C for 96 hours. As illustrated in Figure 3, the GPB-II phase was found to exist in both circle and elongate forms, which exhibited bilayer structures. Figure 4 depicts the HAADF-STEM and the realtive EDS maps within GPB-II phase. The core area, which appears dark in the images, is primarily composed of magnesium and silicon, with little content of copper and zinc. In contrast, shell area displays an increase in the segregate of copper and zinc elements. This core-shell micro-structure is a consequence of the differing elemental compositions of the GPB-II phase.

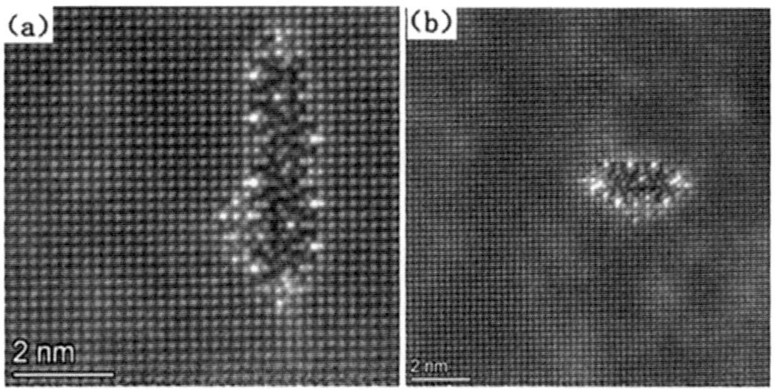

(a) big GPB-II phase; (b) small GPB-II phase.

Figure 3. HR-HAADF images of # 4 ageing at 150°C for 96 h.

The generation of GPB-II precipitates can be primarily attributed to addition of silicon to Al-5Zn-1Mg-1Cu alloys. The incorporation of silicon into the alloy promotes the number density of Si-Mg clusters as the initial aging period. Subsequently, these clusters undergo a transformation into GPB-II precipitates during subsequent ageing stage, as illustrated in Figure 3. The Copper atom exhibits reduced diffusion coefficient due to the disparate rates of diffusion observed for solid solution elements. Consequently, the diffusion process is slower, resulting in the segregation of elements outside the GPB-II precipitate. This results in the formation of the bilayer structure observed in our Al alloy (see Figure 4). Accordingly, the core is predominantly constituted by Si and Mg, which exhibit huge nucleation dynamics. The shell is primarily composed of Zn and Cu, with Copper exhibiting a comparatively smaller diffusion rate. Consequently, the shell micro-structure is capable of inhibiting precipitates grow when overageing the alloy. This is the reason why dimensions of GPB-II precipitates are considerably finer than those of the T+ η phases. It can therefore be concluded that the form of GPB-II phases in the alloy is a consequence of the disparate diffusion rates between the alloying elements and the elevated silicon content.

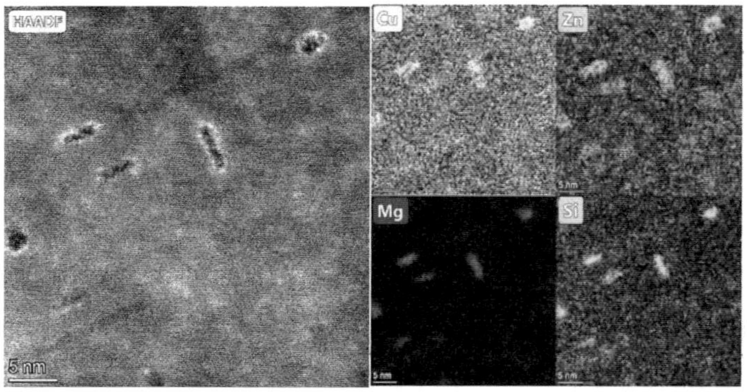

Figure 4. A HR-HAADF-STEM images of sample 4 ageing at 150°C in 96 h, accompanied by the corresponding elemental EDS maps.

4.2 The relationship between hardness and microstructure

The microstructure of alloys has the potential to exert an influence on their hardness. It is possible to modify the microstructure and thereby enhance the microhardness of the alloy through the addition of silicon. As illustrated in Figure 2, the proportion of fine bilayer micro-structure GPB-II precipitates was found to improve with adding Silicon during 150°C ageing. The core-shell structure is effective in inhibiting precipitate growth when alloys are maintained at an average temperature of 150°C, thereby preventing a significant decline in hardness. Moreover, the form of GPB-II precipitates would result in the consumption of Mg and Zn, thereby resulting in a decline in proportion of coarse η and T phases, despite a slight increase in their critical size in Si-containing alloys. Consequently, the formation of a greater number of fine GPB-II precipitates and a lesser big η or T precipitates can be observed in this alloy. This is the reason why the strengthening effect of Si-adding alloys is higher than that of non-Si alloys at this temperature.

5. Conclusion

The ageing behaviour and phase transformation of a range of Si-microalloyed low Mg/Zn ratio Al-5.0Zn-1.0Mg-Cu alloys were investigated at a temperature of 150°C, employing TEM and microhardness testing. The principal conclusion are summarised as follows:

1. Adding silicon had a improtant influence on the proportion of precipitates and microhardness.
2. The precipitates gradually changed from T+η precipitates to GPB-II precipitates with adding Si content, and at the same time, the microhardness changed from 120 HV to 150 HV when aging at 150°C.
3. Existence of small and even GPB-II phases in the Si-adding alloy was identified as underlying cause of this result. This GPB-II phase exhibited a bilayer microstructure. In core region predominantly comprise Zn, Si and Mg, and the shell region primarily consist of Zn and Cu. This structure effectively inhibited phase growth, maintaining the phases at a smaller scale and enhancing the alloy's hardening effect.

Journal of Physics: Conference Series **2842** (2024) 012068 doi:10.1088/1742-6596/2842/1/012068

References

[1] Xu W., Wu X., and Ma J. (2019). Abnormal fracture of 7085 high-strength aluminum alloy thick plate joint via friction stir welding. J. Mater. Res. Technol., 8: 6, 029-6, 040.
[2] Azarniya A., Taheri A., and Taheri K. (2019). Recent advances in the aging of 7xxx series aluminum alloys: A physical metallurgy perspective. J. Alloys Compd., 781: 945-983.
[3] Srivatsan T., Jr D., and Soni K. (1993). Microstructure, tensile properties, and fracture behavior of an Al-Cu-Mg alloy 2124. J. Mater. Sci., 27: 4, 772-4, 781.
[4] Zhu Q., Cao L., and Wu X. (2019). Effect of Ag on the age-hardening response of Al-Zn-Mg-Cu alloys. Mater. Sci. Eng. A, 754: 265-268.
[5] Wen K., Li X., and Xiong B. (2022). Influence of minor Sc additions on grain refinement and microstructure characteristics of a high Zn-containing Al-Zn-Mg-Cu-Zr alloy. J. Cent. South Univ., 29 (3): 780-794.
[6] Hahn G. and Rosenfield A. (1975). Metallurgical factors affecting fracture toughness of aluminum alloys. Metall. Trans. A, 6 (4): 653-668.
[7] Wu S., Xue H., and Yang C. (2021). Effect of Si addition on the precipitation and mechanical/electrical properties of dilute Al-Zr-Sc alloys. Mater. Sci. Eng. A, 812 (3): 141, 150.
[8] Mao J., Wen S., and Liang S. (2023). Precipitation behaviors and thermal stability of Al-3.5Mg-1.0Cu alloy with co-addition of Zn and Si. J. Alloys Compd., 946: 169, 401.
[9] Guo K., Liang S., and Wen S. (2022). GPB-II zone in Si micro-alloyed Al -Zn-Mg-Cu alloy and its strengthening effect. Mater. Sci. Eng. A, 852: 143, 721.
[10] Liang S., Wen S., and Wu X. (2020). The synergetic effect of Si and Sc on the thermal stability of the precipitates in Al-Cu-Mg alloy. Mater. Sci. Eng. A, 783.

Research on the influence of finish rolling technique parameters on H_2S corrosion resistance behavior of pipeline steel in the thickness direction

Yanqiu Fan [1,2*], Changwen Ma [1,2,3], Shaopo Li [1,2], Wenhua Ding [1,2], Longhe Wang [1,2], Yang Liu [1,3] and Zhanjun Li [1,3]

[1] Research Institute of Technology, Shougang Group Co., Ltd., Beijing, Beijing, 100043, China

[2] Beijing Engineering Research Center of Energy, Beijing, Beijing, 100043, China

[3] Beijing Key Laboratory of Green Recyclable Process for Iron and Steel Production Technology, Beijing, Beijing, 100043, China

[*]Corresponding author's email: fanyq@shougang.com.cn

Abstract: Natural gas was an important energy for human society's development, and there was still a great demand for this type of energy. It required pipeline steel for energy transport and had excellent hydrogen-induced cracking resistance. Thermomechanical controlled processing (TMCP) was a frequently used manufacturing method for pipeline steel. It was necessary to research the effect of finish rolling technique parameters on the H_2S corrosion resistance behavior of pipeline steel in the thickness direction, especially concerning microstructure and hardness in the thickness direction. So, for the research, optical microscope (OM), hydrogen-induced cracking tests (HIC), and hardness tests were used to study the effect of finish rolling technique parameters on H_2S corrosion resistance. As finish rolling deformation increases, deformation can effectively conduct from surface to center and reduce the average hardness gap between the surface and the center. It can also generate finer microstructure and smaller hardness differentials from the thickness direction. All these can enhance the hydrogen-induced cracking resistance.

1. Introduction

Natural gas is an important energy for human society's development. Even if many kinds of clean energy were applied, such as solar and hydrogen energy, natural gas was still in great demand [1-4]. It requires pipeline steel for natural gas to have excellent hydrogen-induced cracking resistance.

Much research has been done about the wet H_2S corrosion resistance behavior of pipeline steel. Jack et al. [5] studied the HIC performance of X65-grade pipeline steel and found cracks that emerged through the center position of steel after charging hydrogen. Haq et al. [6] found that the X80 pipeline, steel with the microstructure of granular bainite and ferrite, had good HIC resistance. In contrast, hard phase microstructures such as lath bainite and martensite easily lead to crack propagation. Park et al. [7] supported the view that acicular ferrite and ultra-fine ferrite had better crack resistance and comprehensive mechanical properties.

Thermomechanical controlled processing (TMCP) was a frequently used manufacturing method for pipeline steels. Normally, rolling parameters and cooling strategy can be adjusted to obtain appropriate microstructure and improve performance [8-11]. For heavy plates, obtaining uniform microstructure in the thickness direction was hard, and little research has been done to study the effect of finish rolling deformation on the wet H_2S corrosion of pipeline steel in the thickness direction.

Therefore, it was necessary to research the effect of finish rolling deformation on the wet H_2S corrosion characteristics for pipeline steel in the thickness direction, especially concerning microstructure and hardness distribution in the thickness direction.

2. Experimental material and methods

The experimental steels were melted in a furnace and forged into a small ingot with a thickness of 125 mm. The chemical composition of the samples (mass percentage) is listed in Table 1. As shown in Table 1, specimen NO.1 and specimen NO.2 almost had the same chemical composition.

Table 1. Samples' chemical compositions (/%).

Specimen	C	Si	Mn	P	S	Alt	Nb	Ti+Ni+Cu
NO.1	0.055	0.25	1.32	0.007	0.005	0.03	0.025~0.035	⩽0.5
NO.2	0.054	0.24	1.33	0.008	0.005	0.03	0.025~0.035	⩽0.5

For the rolling process, the total compression ratio during the rough rolling from 1200°C to 880°C was 60%, and the total compression ratio during the rough rolling from 880°C to 820°C were individually 20% (specimen NO.1) and 50% (specimen NO.2) with the same passes. Then, they were both cooled to 500°C at the rapid of 20°C/s and then cooled to 25°C in air.

Metallographic samples were separated from plates, and the normal direction surface was shined and corroded by a 4% nitric acid-ethanol solution. The microstructure was watched by light microscope equipment. A micro-hardness tester tested hardness.

According to the standard NACE TM 0284-2003, the size of the HIC test samples was 100 mm×20 mm×10 mm, and they were soaked in the mixed liquor with 0.5% acetic acid and 5% sodium chloride; the samples stayed in the solution for 96 hours at 25°C. Three parameters, CLR, CTR, and CSR, were calculated.

3. Tested results and discussions

3.1 HIC test results

The three parameters of the HIC test are shown in Table 2. The values of these three parameters of specimen NO.1 were 18.82%, 1.12%, and 0.14%, while the values of specimen NO.2 were 0%, 0%, and 0%. Generally, when the three types of crack ratios were below 15.00%, 5.00%, and 3.00%, pipeline steel could reach the served qualification. So, specimen NO.1 did not reach the standard, and specimen NO.2 had excellent HIC performance with no cracks.

Table 2. Hydrogen-induced cracking resistance performance.

Specimens	Sections	CLR/%	CTR/%	CSR/%
NO.1	1	16.34	1.97	0.19
	2	19.64	0.68	0.08
	3	20.47	0.73	0.15
	Average	18.82	1.12	0.14
NO.2	1	0	0	0
	2	0	0	0
	3	0	0	0
	Average	**0**	**0**	**0**

3.2 Microstructure

The metallographic microstructure of different zones is displayed in Figure 1. As shown in Figure 1(a), the microstructure of the top zone of specimen NO.1 was bainite, a bit of MA islands, and acicular ferrite, and the average crystallite size was approximately 22 μm. The microstructure of the center zone of specimen NO.1 was also bainite, minor MA islands, and acicular ferrite, and the average crystallite size was much larger than that of the surface zone, with 26 μm, as shown in Figure 1(b). In specimen NO.2, the grain size was finer, the microstructure of the surface zone was granular bainite, the microstructure of the center zone was almost the same as the surface zone, and the average grain size was 16 μm and 18 μm, individually, as shown in Figure 1(c) and Figure 1(d).

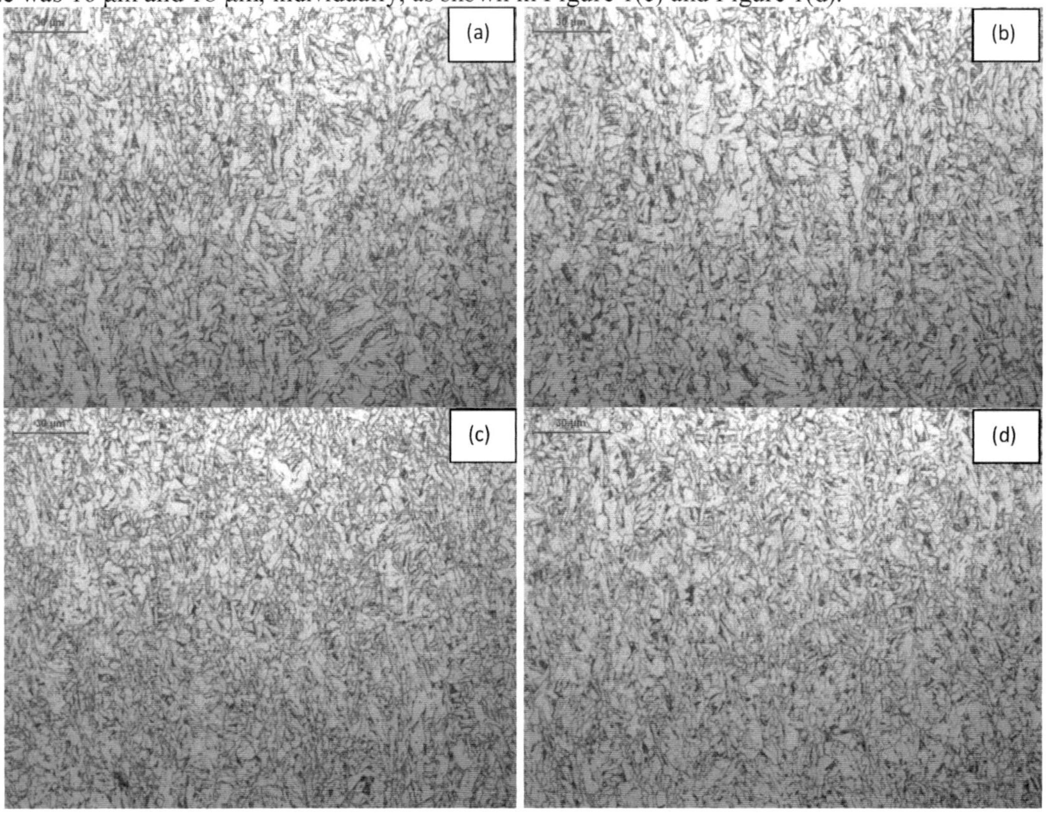

Figure 1. Optical microstructure of samples.
(a) Surface zone of specimen NO.1 (b) Center zone of specimen NO.1
(c) Surface zone of specimen NO.2 (c) Center zone of specimen NO.2

3.3 Hardness

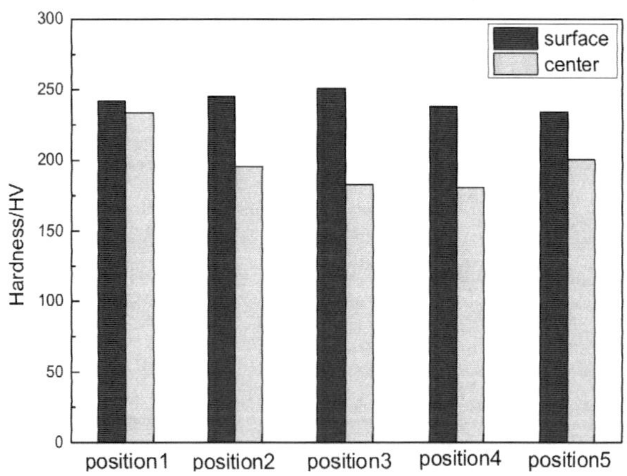

Figure 2. The hardness of the surface position and center position (nearby cracks) of specimen NO.1.

The hardness distribution of the top position and center line of specimen NO.1 are displayed in Figure 2. The five tested points of the center line of specimen NO.1 were all nearby cracks, as demonstrated in Figure 3. The phenomenon can be seen in Figure 2, which shows that the five tested hardnesses of the surface position were higher than those of the center position. The average hardness value of the surface position was 242 HV, higher than that of the center position, 199 HV by about 43 HV. Also, the lowest hardness was 180 HV in the center zone, the highest hardness was 234 HV, and the differentials were 54 HV.

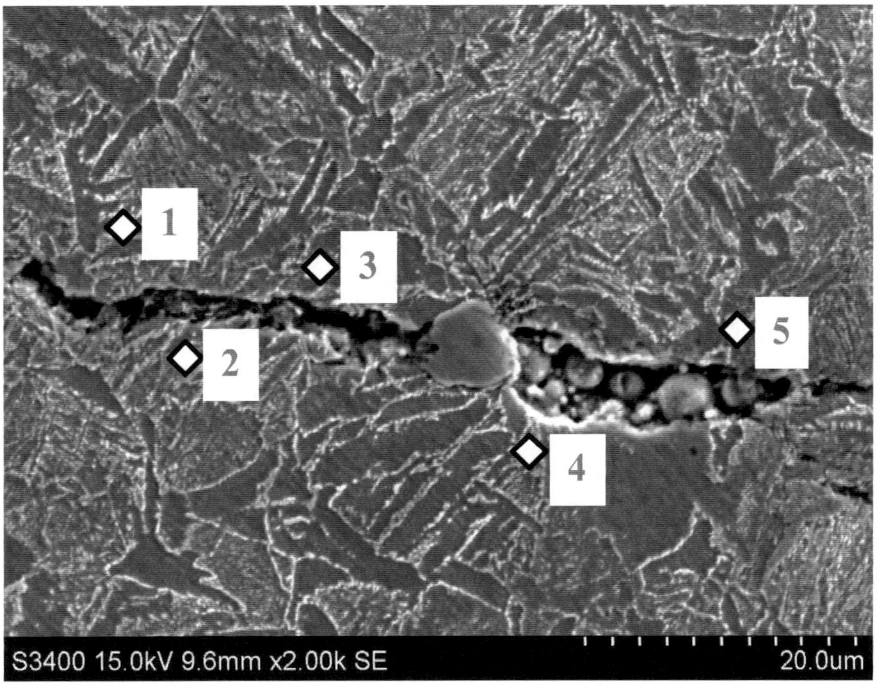

Figure 3. Hardness point of center position (nearby cracks) of specimen NO.1.

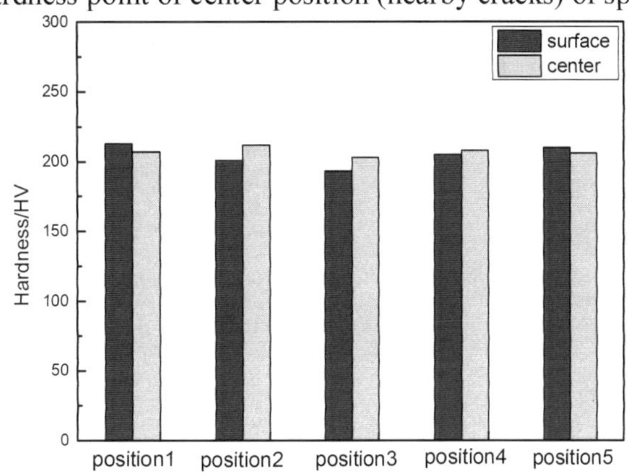

Figure 4. The hardness of the surface position and center position of specimen NO.2.

The hardness distribution of the surface position and center line of the second sample are manifested in Figure 4. It can be observed from Figure 4 that the five tested hardness of the surface position were almost equal to those of the center position. The average hardness value of the surface position was 207 HV, which was higher than that of the center position and was 205 HV by about 2 HV. Also, the lowest hardness was 203 HV in the center zone, the highest hardness was 212 HV, and the differential was 9 HV.

Compared with the two specimens, it can be found that the hardness in the thickness direction of specimen NO.2 was more uniform than that of specimen NO.1, and the gap between the top hardness figure and the smallest hardness figure in the center line of specimen NO.2 was smaller than that of specimen NO.1.

3.4 Segregation

The results of the electron probe micro-analysis are exhibited in Figure 5. From the segregation outcomes, it can be obtained that there was Mn segregation, while there were no other elements segregation, such as C, P, Ni, Cu, and Nb. Because the two slabs had almost the same chemical composition, the Mn segregation condition was the same for specimen NO.1 and specimen NO.2. Mn segregation was more likely to introduce a hard phase, and it can increase the probability of hydrogen-induced crack.

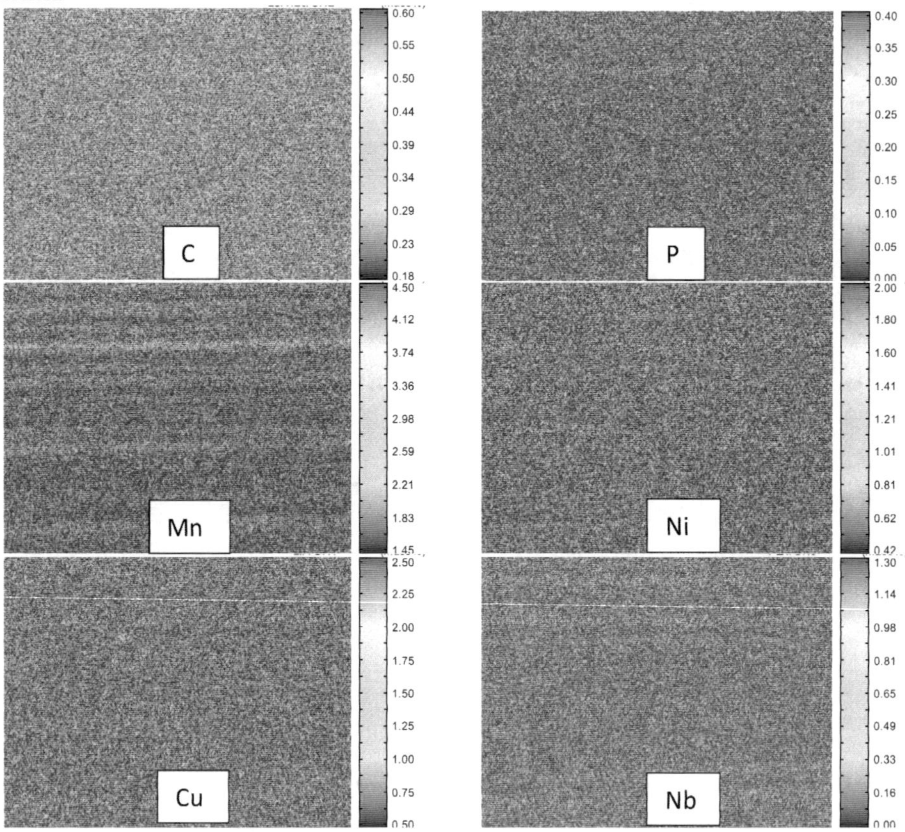

Figure 5. Microalloying elements distribution condition results.

3.5 Grain boundary characteristics

Larger finish rolling deformation can bring finer microstructure, the grain size decreases, and the recrystallization degree strengthens. IPF maps of specimens are exhibited in Figure 6. Especially from Figure 6(c), the relative frequency of finer grain size between 2 μm and 3 μm in specimen 2 was bigger, which meant that the microstructure of specimen 2 was finer and more even. Previous research found smaller grain sizes can inhibit crack propagation [11]. Also, for heavy plates, larger finish rolling deformation can effectively conduct deformation from surface to center and reduce the average hardness gap between the surface and the center, the gap between the largest figure and the smallest figure in the center zone, and all these can decrease the hydrogen induced cracking sensibility. So, specimen NO.2 had better hydrogen-induced cracking resistance, with finer grains and smaller hardness differentials.

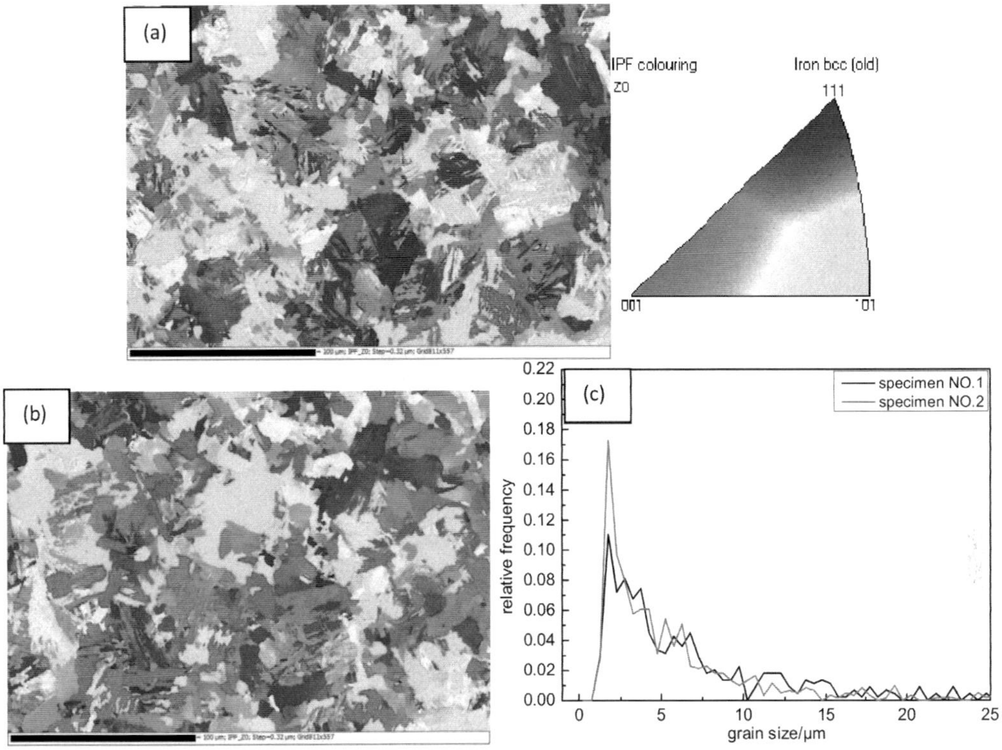

Figure 6. IPF maps of specimens (a) NO.1 (b) NO.2 (c) The relative frequency of grain size.

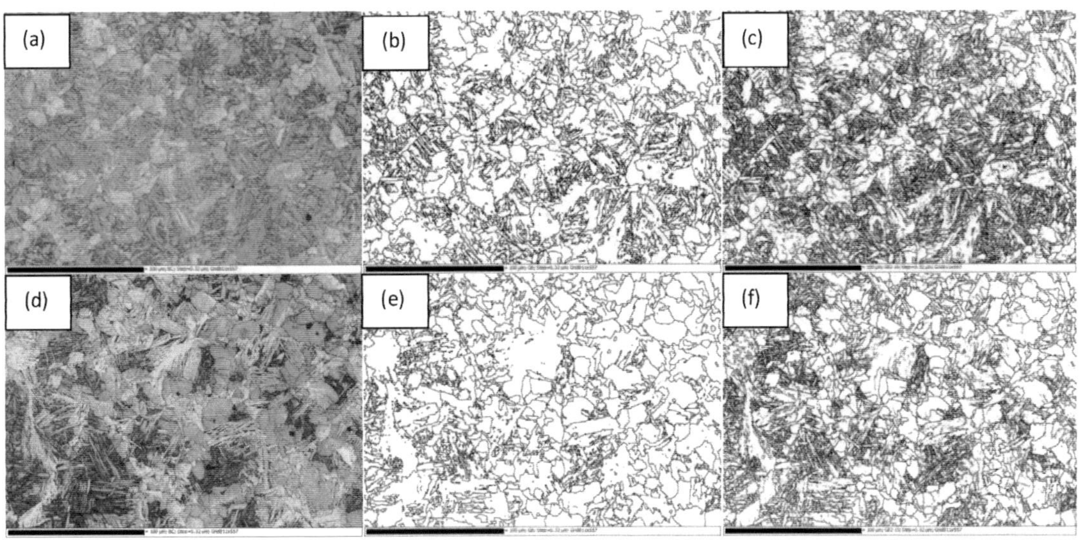

Figure 7. Boundary characteristics of specimens by EBSD.
(a)(b) and (c) were specimen NO.1; (d)(e), and (f) were specimen NO.2; the black line represents high angle grain boundary, and the red line represents low angle grain boundary

Boundary characteristics also play a key role in curbing hydrogen-induced cracks. Boundary characteristics of specimens were observed by EBSD and were demonstrated in Figure 7. Normally, the grain boundary was divided into two types according to misorientation angles: low angle grain boundary—2°~ 15° marked by red lines, and high angle grain boundary—15°~ 62.5° marked by black lines in Figure 7. In the test, the distribution condition of high-angle grain boundaries was calculated and disclosed in Figure 8. The percentage of high-angle grain boundaries was 36.8% and 41.2%. It represented that Specimen 2 had a more high-angle grain boundary. As it was known, the high-angle grain boundary was beneficial for hydrogen-induced crack resistance. From this respect, more high-angle grain boundaries ulteriorly promote its hydrogen-induced crack resistance.

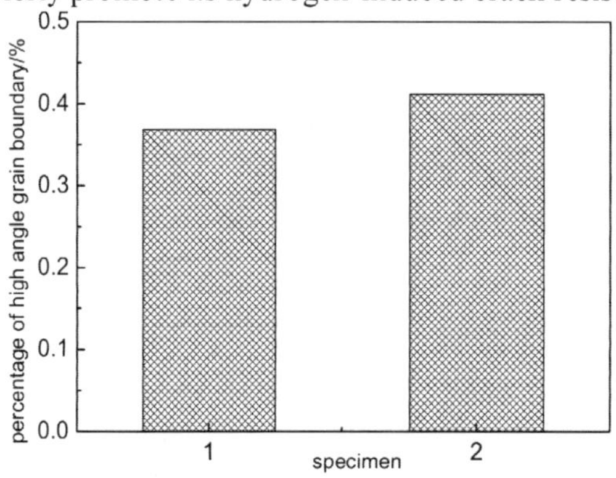

Figure 8. Percentage of high-angle grain boundary of specimens.

3.6 Crack regions

Crack regions of Specimen 1 were observed and displayed in Figure 9. The chemical composition of four positions on cracks was tested, as seen in Table 3, and the results revealed no obvious inclusions on cracks. In other words, inclusions were not the main reason for hydrogen-induced cracks. So, this routine respect was excluded.

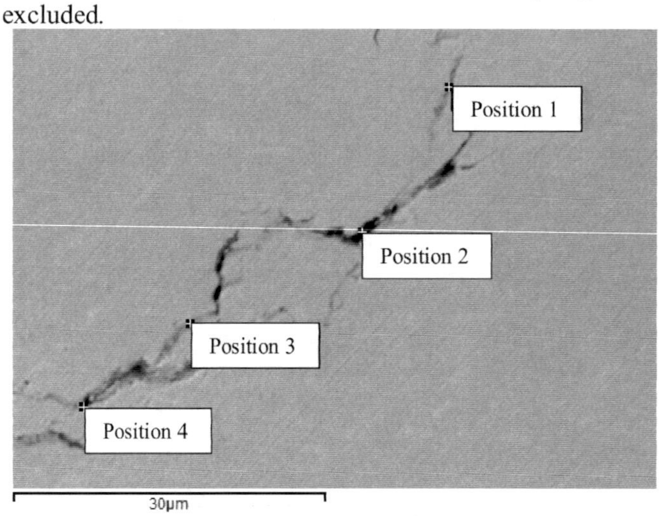

Figure 9. SEM position in crack regions of Specimen 1.

Table 3. Crack regions' chemical compositions (/%).

Position	O	Mg	S	Ca	Mn	Fe	Total/%
1	3.63				1.36	95.01	100.00
2	9.44	0.37	0.48		1.52	88.19	100.00
3	3.84			0.69	1.33	94.14	100.00
4	6.28			0.52	1.45	91.75	100.00
Maximum	9.44	0.37	0.48	0.69	1.52	95.01	
Minimum	3.63	0.37	0.48	0.52	1.33	88.19	

Besides, the EDS analysis results of specimen NO.1 were manifested in Figure 10. The results showed that there was no element segregation near cracks. So, it can be drawn that the hydrogen-induced crack had no direct relationship with inclusions and element segregation.

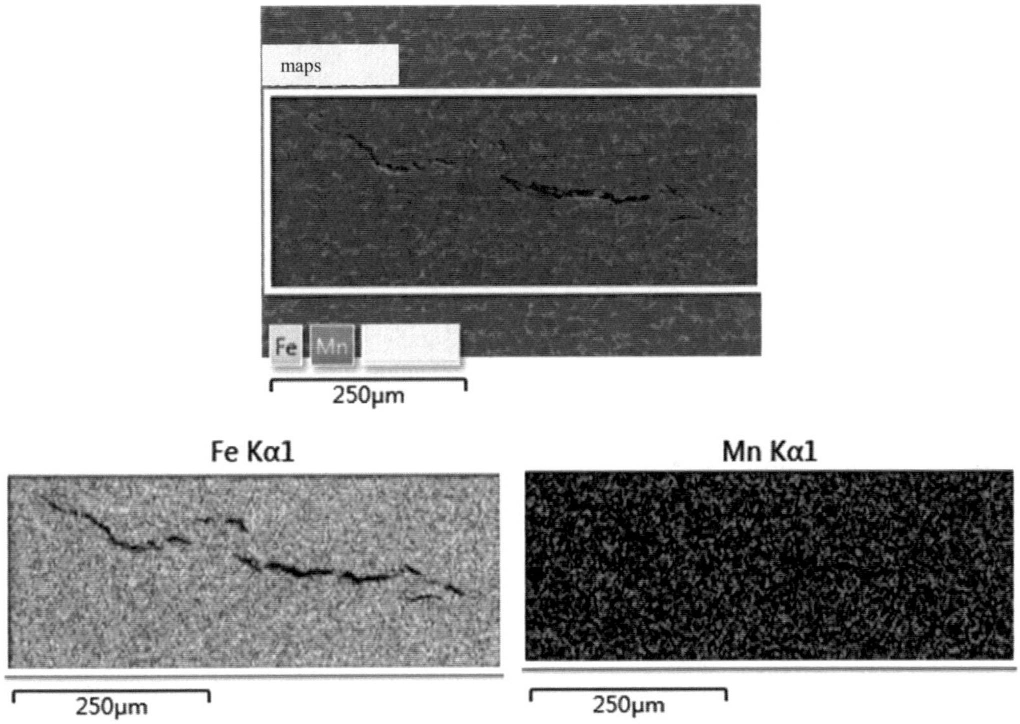

Figure 10. EDS analysis results of crack regions of specimen NO.1.

3.7 Mechanisms

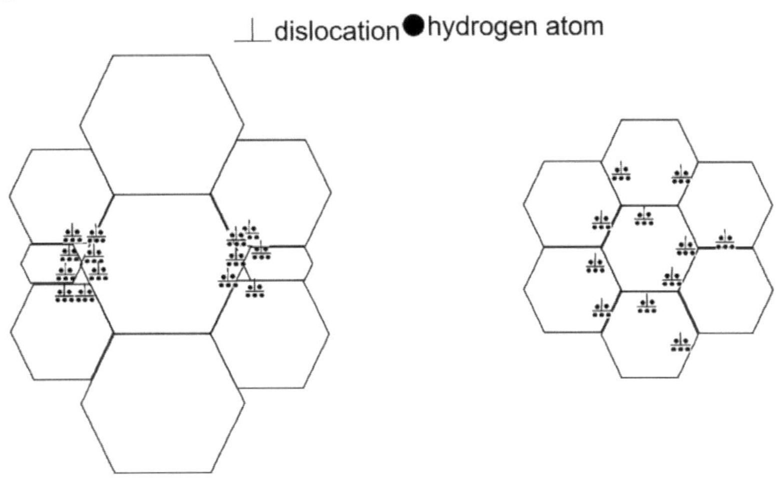

Figure 11. Schematic diagram of hydrogen-induced cracks.

From the above phenomenon and results, it was clear that hydrogen-induced cracks were mainly because of uniform microstructure and uneven grain size. The schematic diagram is indicated in Figure 11. A lower rough rolling deformation rate brought about uneven microstructure, and grain boundaries between larger-sized grains and smaller grains were more likely to trap hydrogen atoms, leading to hydrogen gathering. When hydrogen atoms accumulate to a critical extent, more hydrogen gas is generated, and it can increase the possibility of hydrogen-induced cracks. Meanwhile, a higher rough rolling deformation rate makes grain size smaller and more even, and it can promote more uniformly distributed hydrogen, notably forbidding hydrogen concentration and relieving local hydrogen pressure, thus benefiting hydrogen-induced resistance.

4. Conclusions

This study adopted different finish-rolling deformation parameters, with 20% for specimen NO.1 and 50% for specimen NO.2. The microstructure, hydrogen-induced cracking performance, hardness, and segregation were observed. The effect of finish-rolling deformation on the hydrogen sulfide corrosion crack resistance of pipeline steel in the thickness direction was investigated. Several conclusions were made and listed below:

(1) The second sample had excellent HIC performance with no cracks, while the CLR, CTR, and CSR values in specimen NO.1 were 18.82%, 1.12%, and 0.14%, respectively.

(2) The type of microstructure on the top surface of sample NO.1 was mainly bainite, and the average crystallite size was about 22.0 μm. The center position was also bainite, and the average crystallite size was 26 μm. For specimen NO.2, the crystallite size was finer, and the type of microstructure of the surface zone was granular bainite. The microstructure of the center zone was almost the same as the surface zone.

(3) For specimen NO.1, the average hardness value of the surface position was 242 HV, and that of the center position was 199 HV. Also, the lowest hardness was 180 HV in the center zone, the highest hardness was 234 HV, and the differentials were 54 HV. Besides, hardness in the thickness direction of specimen NO.2 was more uniform than specimen NO.1, and the gap between the largest hardness point and the smallest hardness point in the center line of specimen NO.2 was smaller than that of specimen NO.1.

(4) The percentage of high-angle grain boundaries was 36.8% and 41.2%, individually. It was beneficial for hydrogen-induced crack resistance. From this respect, more high-angle grain boundaries ulteriorly promote its hydrogen-induced crack resistance.

(5) No element segregation and inclusions were near cracks, and the hydrogen-induced crack had no direct relationship with inclusions and element segregation.

(6) Larger finish rolling deformation can bring finer microstructure and smaller hardness differentials from the thickness direction. All these can enhance the hydrogen-induced cracking resistance.

References

[1] Zhang F., Zhao P.C., Niu M., Maddy J. (2016) The survey of key technologies in hydrogen energy storage. Int J Hydrogen Energy, 41: 14535-14552.

[2] Li M. J., Zhu H.H., Guo J. Q., Wang K., Tao W. Q. (2017) The development technology and applications of supercritical CO2 power cycle in nuclear energy, solar energy, and other energy industries. Appl Therm Eng, 126: 255-275.

[3] Apak S., Atay E., Tuncer G. (2017) Renewable hydrogen energy and energy efficiency in Turkey in the 21st century. Int J Hydrogen Energy, 42: 2446-2452.

[4] Xu Q., Zou Z.W., Chen Y.S., Wang K., Du Z.W., Feng J.X., et al. (2020) Performance of a novel-type heat flue in a coke oven based on high-temperature and low-oxygen diffusion combustion technology. Fuel, 267: 117-160.

[5] Jack T.A., Pourazizi R., Ohaeri E., Szpunar J., Zhang J., Qu J. (2020) Investigation of the hydrogen-induced cracking behavior of API 5L X65 pipeline steel. Int J Hydrogen Energy, 45: 17671-17684.

[6] Haq A. J., Muzaka K., Dunne D.P., Calka A., Pereloma E.V. (2013) Effect of microstructure and composition on hydrogen permeation in X70 pipeline steels. Int J Hydrogen Energy, 38: 2544–2556.

[7] Park G. T., Koh S. U., Jung H. G., Kim K.Y. (2008) Effect of microstructure on the hydrogen trapping efficiency and hydrogen induced cracking of line pipe steel. Corros Sci, 50: 1865–1871.
[8] Mohtadi-Bonab M. A., Eskandari M. (2017) A focus on different factors affecting hydrogen-induced cracking in oil and natural gas pipeline steel. Eng Fail Anal, 79: 351–360.
[9] Mohtadi-Bonab M. A., Mousavi H., Pourazizi R., Szpunar J. A. (2020)Finite element modeling of HIC propagation in pipeline steel concerning experimental observations. Int J Hydrogen Energy 45: 23122-23133.
[10] Mohtadi-Bonab M. A., Szpunar J. A., Collins L., Stankievech R. (2014) Evaluation of hydrogen-induced cracking behavior of API X70 pipeline steel at different heat treatments. Int J Hydrogen Energy, 39: 6076–6088.
[11] Huang F., Li X. G., Liu J., Qu Y. M., Jia J., Du C. W. (2011) Hydrogen-induced cracking susceptibility and hydrogen trapping efficiency of different microstructure X80 pipeline steel. J Mater Sci, 46: 715–722.

Journal of Physics: Conference Series **2842** (2024) 012070 doi:10.1088/1742-6596/2842/1/012070

Study on the influence of geometric parameters of typical titanium alloy power milling tools

Zhonglou Shi[1*] and Xiangjie Yang[2]

[1]Jianghan University, Wuhan, Hubei, 430056, China
[2]Nanchang University, Nanchang, Jiangxi, 330036, China

[*]Corresponding author's e-mail: shizhonglou@jhun.edu.cn

Abstract. Based on ABAQUS, orthogonal finite element tests were carried out when the number of keyway milling cutter blades was three and four, and the spiral rise angle was 60° and 65°, respectively. The results showed that with the maximum number of blades and spiral rise angle, the milling quality of typical titanium alloy TC4 was greatly improved. The surface roughness decreases, and the effect of spiral lift angle on surface roughness is more significant, which provides a theoretical basis for improving the quality and energy of typical titanium alloy milling tool selection.

1. Introduction

Titanium alloy has high strength, high-temperature resistance, strong corrosion resistance, and low-density physical properties, so it is widely used in aerospace, compared with alloy steel. It also can reduce the mass of aircraft by 40%, so it is considered an important strategic metal material for human beings in space. Known as modern metal, it is a hot spot of application research in various countries. Milling is widely used in titanium alloy processing because of its ability to process complex geometric shapes and high manufacturing accuracy and processing efficiency.

At the same time, because of the low thermal conductivity and low elastic modulus of titanium alloy, the temperature rise in the processing is higher and the wear of the tool is aggravated, so it is identified as difficult to process materials. Milling processing is reduced material manufacturing, titanium alloy milling often takes the "roughing-semi-finishing - finishing" processing strategy, the essential role of roughing is to remove the most machining allowance in the shortest time and leave a certain margin for finishing. Strong milling is often used in the roughing stage, the smaller the residual geometric error is, the more secure the finishing process is.

With the deepening of research, more and more experts and scholars have attracted attention to the study of powerful milling. Chu et al. [1] researched the efficient milling technology of titanium alloy aircraft structural parts, and the research results showed that the efficient milling of titanium alloy parts was realized through the reasonable use of powerful milling, large feed milling, insert milling, and high-speed milling in the rough and finish machining process, which improved the processing efficiency of titanium alloy and achieved good economic benefits. Zhao and Shi [2] carried out experimental research on the high efficiency and strong milling technology of blade disc compounds. The results showed that disk milling and insert milling technology was feasible and effective for slotting and roughing of blade discs with large amounts of spare.

In this paper, the typical titanium alloy TC4 is taken as the milling object, and the most complex through slot milling model is taken as the research model. Based on the ABAQUS model, the influence of the number of cutting edges and spiral lift angle of the keyway milling cutter on the milling surface quality is analyzed.

2. Milling model research

As shown in Figure 1, in the milling process, milling quantity FZ can be characterized by unit removal rate, which is related to feed quantity Vf and cutting depth ap. This study aims to study the effect of milling cutter blade number and spiral lift angle on the surface quality of milling surface during power milling. Therefore, when conducting finite element analysis based on ABAQUS, in the aspect of the manufacturing system, only the number of milling cutter blades and spiral angles are changed, and the others are not changed. In the manufacturing process, the main shaft speed, feed rate, cutting depth, and cutting width do not become the premise.

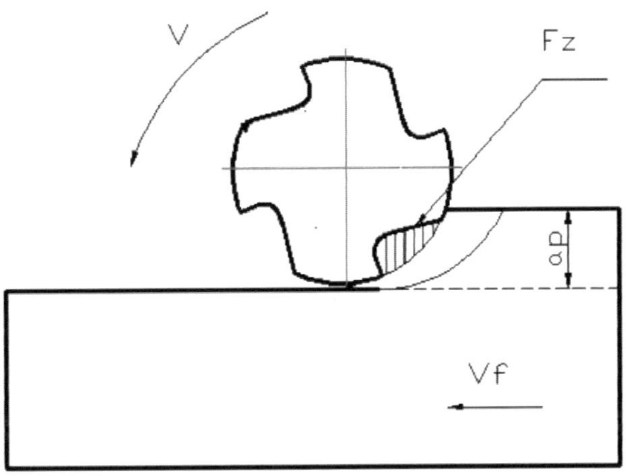

Figure 1. Milling model.

3. Finite element model establishment

3.1 Material physical parameters

Titanium alloy has high chemical activity, poor thermal conductivity, and low elastic model, so it is easy to cause tool wear during processing. To improve the durability of the tool and ensure the quality of the workpiece processing, it is necessary to choose a material tool with strength, toughness, wear resistance, high thermal stability, and poor chemical affinity for titanium. The cemented carbide end mill is studied, and the tool parameters are shown in Table 1.

Table 1. Tool parameters.

Tool Material	Density ρ (kg/m^3)	Elastic Modulus E (GPa)	Poisson's Ratio μ	Coefficient of Thermal Expansion α (10-6K-1)	Thermal Conductivity
Cemented Carbide	14500	650	0.25	4.5	85
Specific Heat Capacity	Tool Diameter	Spiral Direction	Spiral Angle	Number of Blades	
526	10	right	60°/65°	3/4	

TC4 is a medium strength α+β type two-phase titanium alloy with excellent comprehensive properties, which is the most widely recognized in the aerospace industry. According to the literature review, the basic parameters of TC4 materials are shown in Table 2.

Table 2. Basic parameters of TC4 materials.

Density ρ (kg/m^3)	Elastic Modulus E (GPa)	Poisson's Ratio μ
4400	110	0.33

3.2 Finite element constitutive model

In the process of cutting, the material to be cut goes through the whole process of "elastic deformation - elastic-plastic deformation - plastic deformation - shear failure", accompanied by the fine changes in the microstructure of the material to be cut during the process, which is closely related to the surface quality of the cutting surface. To fully return the raw material from static to the whole process of shear failure in the finite element analysis process, the selection and establishment of a material constitutive model is particularly important.

In this study, the Johnson-Cook constitutive model and the damage constitutive model, which have fewer parameters and simple expression forms, are selected to well describe the mechanical properties of materials under large strains, different strain rates, and different temperatures. The expressions of the Johnson-Cook constitutive model are as follows:

$$\sigma = \left(A + B\varepsilon^{n}\right)\left(1 + C\ln\frac{\dot{\varepsilon}}{\dot{\varepsilon}_0}\right)\left\{1 - \left(\frac{T - T_r}{T_m - T_r}\right)^{m0}\right\} \tag{1}$$

A-Yield strength at quasi-static state, MPa;
B-Strain hardening system;
n-Strain hardening index;
ε-Effective plastic strain;
C-Strain rate sensitivity coefficient;
$\dot{\varepsilon}$ -Equivalent strain rate, s^{-1};
$\dot{\varepsilon}_0$ -Reference strain rate, s^{-1};
T_r-Reference temperature, °C;
T_m-Material melting point, °C;
m_0-Temperature softening coefficient.

The model consists of three parts, which represent the strain strengthening effect, strain rate strengthening effect, and temperature softening effect respectively.

The Johnson-Cook constitutive damage model expression is as follows:

$$\varepsilon_u^f = \left[D_1 + D_2\exp(-D_{3\eta})\right]\left[1 + D_4\ln\left(\dot{\varepsilon}^*\right)\right]\left(1 + D_5T^*\right) \tag{2}$$

ε_u^f is the equivalent maximum plastic strain, that is, the fracture strain, $\dot{\varepsilon}^*$ is the equivalent strain rate, D_1, D_2, D_3, D_4, and D_5 is the unknown parameter of the material, $\eta = -p/\sigma_{eff}$, p and σ_{eff} refer to the hydrostatic pressure and equivalent stress respectively, T*=(T-tr)/(tm-tr) is the dimensionless temperature, where Tr and Tm are the reference temperature and the melting point of the material respectively and T is the current temperature.

According to the literature review, Johnson-Cook constitutive model parameters and damage model parameters of TC4 titanium alloy are shown in Table 3.

Table 3. Parameters of TC4 Johnson-Cook constitutive model and damage model.

A/MPa	B/MPa	C	n	m	T_{melt}/K	T_0/K	Strain rate
1093	1092	0.032	0.93	1.1	893	293	1
d_1	d_2	d_3	d_4	d_5	Tm	Tr	
-0.09	0.25	-0.5	0.014	3.87	1630	25	

3.3 Design of milling parameters

This study mainly studies the effect of cutter blade number and spiral lift angle on machining surface quality during power milling, so the selection of milling parameters does not require refinement. To improve milling efficiency during power milling, the principle of giving priority to large cutting depth, followed by large feed, and finally high speed is combined with empirical formula calculation. Finally, the milling width is 10 mm, the milling depth is 7 mm, the feed rate is 800 mm/min, and the speed is 2000 r/min [3-4].

3.4 Design of finite element test scheme

In the process of scientific research practice, there are often many factors that need to be investigated, and each factor is accompanied by a large number of levels. If each level combination of each factor is investigated, the number of tests is too many, time-consuming, and high cost. Orthogonal experimental design is a method to analyze multi-factor experiments by using orthogonal table design and arranging experiments by using orthogonal design can well improve this drawback and the statistical calculation is easy to complete.

Based on the research objectives, the specific orthogonal analysis scheme is shown in Table 4.

Table 4. Orthogonal analysis scheme.

Level \ Factor	Number of Blades	Angle of Spiral Rise
1	3	60°
2	4	65°

3.5 Define model contacts and loads

The contact model is established. In the process of milling, the bottom edge, side edge, and Surface of the milling cutter are in contact with the area to be machined and the machined Surface of the workpiece at all times, so the contact surface set is established for the milling cutter and the workpiece respectively and the surface-to-surface analysis step is set, as shown in Figure 2.

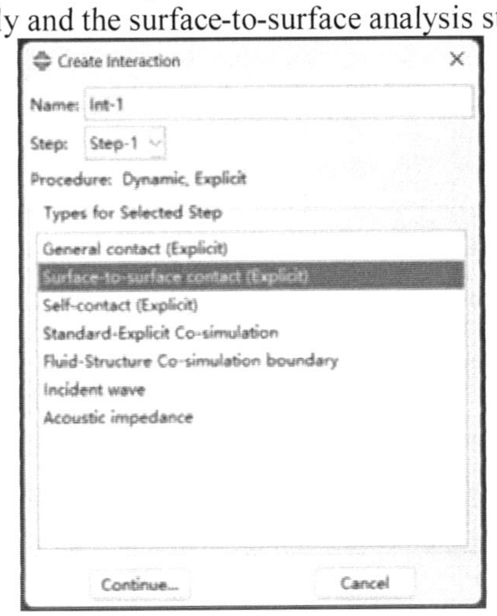

Figure 2. Contact setting.

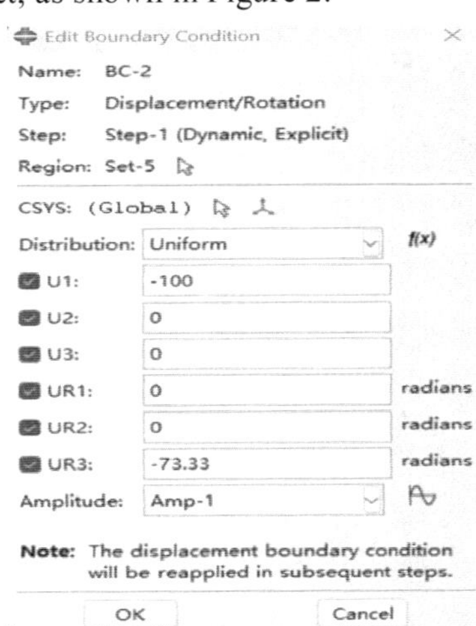

Figure 3. Load setting.

Milling tool rotation is the main movement, and the workpiece to do relative feed movement. To ensure the stability of the milling process, the workpiece is required to be firmly installed on the milling workbench during the milling process. Therefore, the bottom surface of the workpiece is fixed during the analysis, and the tool both rotates and moves relative to the workpiece. In ABAQUS milling analysis, the mass and moment of inertia of the tool are set to 0.1. The lower surface of workpiece TA11 is constrained with full freedom (U1=U2=U3=URl=UR2=UR3=0). In addition, the speed and moving displacement of the milling cutter are defined based on the research needs, as shown in Figure 3.

4. TC4 strong milling simulation based on ABAQUS

Based on the whole process of "modeling - material property definition - assembly - mesh division - boundary condition definition - load setting - submission operation" [5], ABAQUS extracted the displacement nodes of the processed surface, and the displacement orientation diagram of the nodes was shown in Figure 4. The calculation of the contour arithmetic average deviation Ra value was carried out to obtain the results of the orthogonal analysis scheme, as shown in Table 5.

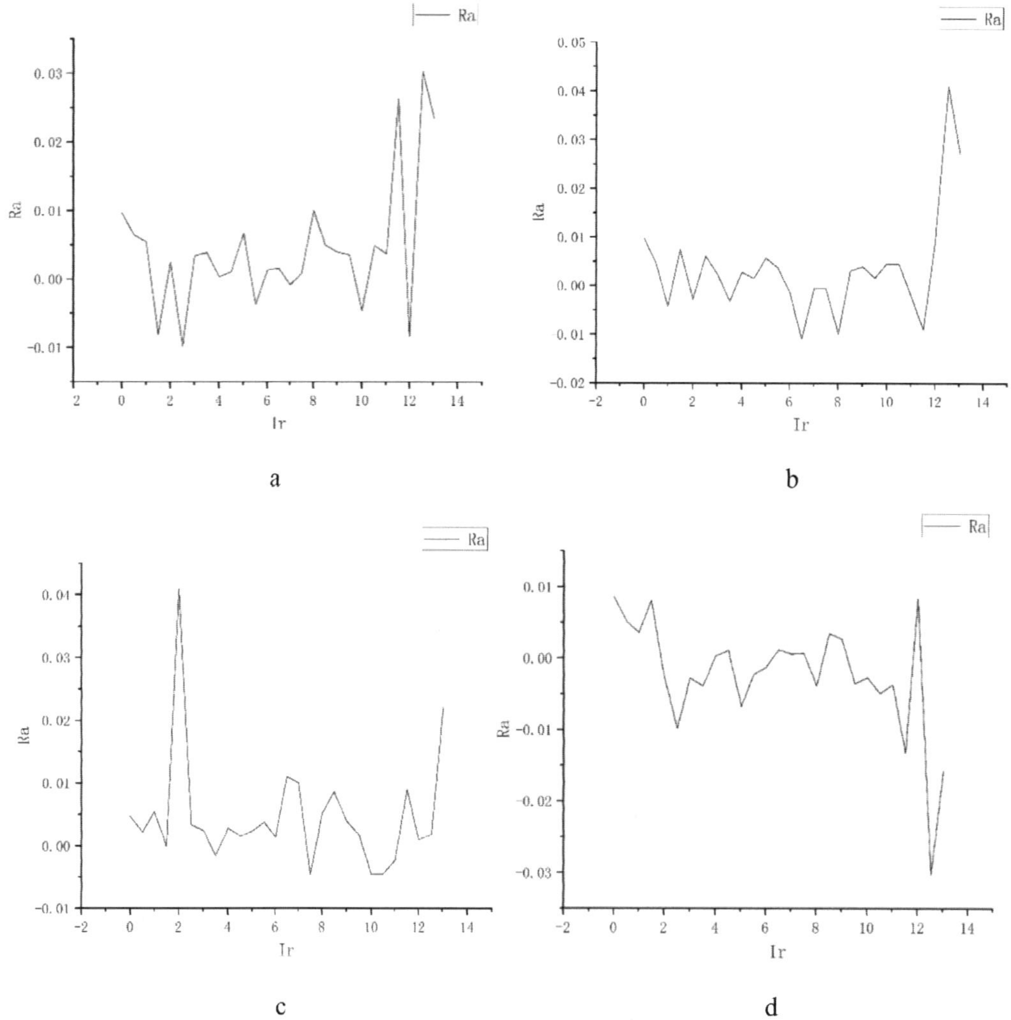

Figure 4. Node displacement.
a Number of blades 3 Spiral Angle 60°b Number of blades 3 Spiral Angle 65°
c Number of blades 4 Spiral Angle 60°d Number of blades 4 Spiral Angle 65°

Table 5. Results of orthogonal analysis.

	Number of Blades	The Angle of the Spiral Rise	Surface Roughness Ra (μm)
1	3	60°	7.02
2	3	65°	6.843
3	4	60°	6.032
4	4	65°	5.591

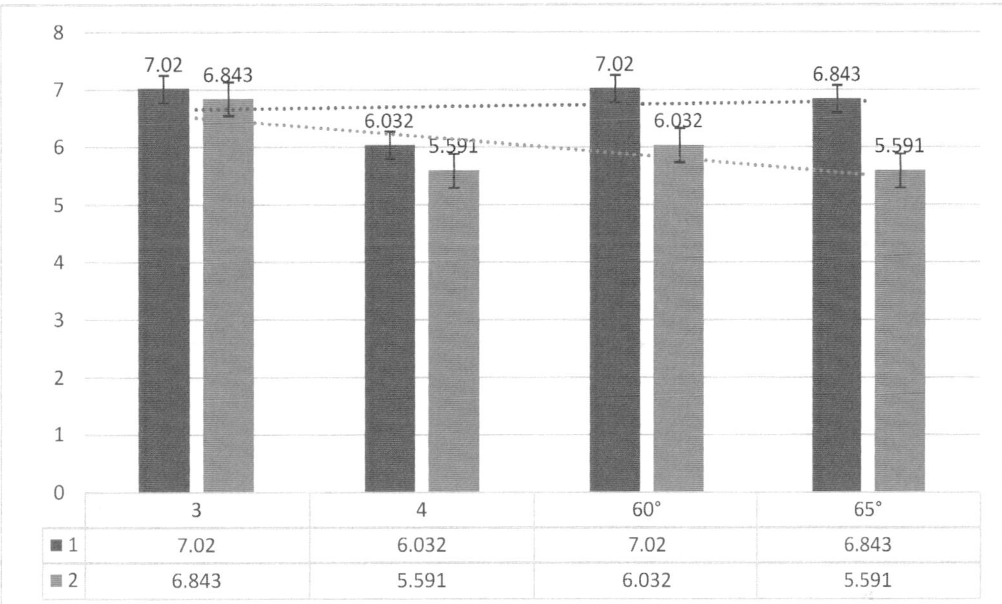

Figure 5. Statistics of surface roughness orthogonal results.

According to the statistics of surface roughness orthogonal results in Figure 5, [6-7] the surface roughness of TC4 decreases with the increase of the number of blades and the increase of spiral lift angle. The maximum influence of blade number on surface roughness is 7.31%, and the maximum influence of spiral rise angle on surface roughness is 18.2%, so the influence of spiral rise angle on surface roughness is more significant.

5. Conclusion

Based on the ABAQUS finite element analysis software, orthogonal analysis tests were carried out on the two factors of cutter blade number and spiral lift angle under the condition of constant cutting parameters, and the influence of cutter blade number and spiral lift angle on the strong milling was analyzed. The results show that:

(1) When the typical titanium alloy TC4 is forcibly milling, the more the number of blades of the milling cutter, the better the milling surface quality;

(2) When the typical titanium alloy TC4 is heavily milling, the larger the spiral lift angle of the milling cutter, the better the milling surface quality;

References

[1] Chu Wangwei, Mou Wenping, Gong Qinghong et al. Research on High Efficiency Milling Technology of Titanium Alloy Aircraft Structural Parts [J]. Tool Technology, 2008(08): 43-46.

[2] Zhao B P, Shi Y Y. Experimental Research on Compound and Efficient Powerful Milling Process of Blisk [J]. Advanced Materials Research, 2011, 1379(314-316): 1150-1153.

[3] China Aeronautical Materials Manual [M]. China Aeronautical Materials Manual, edited by the Editorial Committee. Standards Press of China. 2002.
[4] LU X. Study on Effect of Milling Parameters on Milling stability of TC4 titanium alloy [J]. Combined Machine Tools and Automatic Processing Technology, 2023(10): 133-136+140.
[5] Xu Lanying, Ye Guoshan, Feng Sichen, et al. Finite Element Simulation of TC4 Titanium Alloy Micro-Milling based on ABAQUS [J]. Tool Technology, 2019, 57(11): 115-119.
[6] ZHAO Wanyong, Shi Xiangyun, Shi Fengxia, et al. Influence of guide vane Geometry Parameters on Pressure pulsation in Hydraulic Turbine [J/OL]. Journal of Xihua University (Natural Science Edition), 1-10[2024-02-21].
[7] LI Yaping. Structural Analysis of Fixed Device for Machining Typical Shaft Parts [J]. Machinery Management Development, 2023, 38(11): 116-117.

Journal of Physics: Conference Series **2842** (2024) 012071 doi:10.1088/1742-6596/2842/1/012071

Research on reconfigurable systems in discrete manufacturing workshops

Hai He[1], Jian Zhou[1], Wenqiang Hu[1,*], and Yuhui Wang[2]

[1]School of Software, Huazhong University of Science and Technology, Wuhan, Hubei, 430074, China

[2]School of Mechanical Science & Engineering, Huazhong University of Science and Technology, Wuhan, Hubei, 430074, China

[*]Corresponding author's e-mail: hwq_hust@hust.edu.cn

Abstract: With the development of smart manufacturing, the demand for personalized production has made reconfigurable systems in the manufacturing industry a focal point of research and application in high-end manufacturing. The paper analyzes the current status of the manufacturing industry and the challenges of reconfigurable technology. Furthermore, it designs and develops a reconfigurable system for discrete manufacturing of automotive components. The paper focuses on issues such as creating reconfigurable models, designing adaptive reconfiguration algorithms, accomplishing the creation of multi-layered models, the design of rule libraries, and algorithm implementation. The reconfiguration result scheme can be input into commercial software for performance simulation, and can directly integrate with production systems to complete production planning. This demonstrates the rationality of the reconfigurable models in the system, the correctness of the reconfiguration algorithms, and the effectiveness of the system, providing a theoretical basis for reconfigurable production lines.

1. Introduction

With increasing complexity in international situations, intensified competition among major powers, ongoing financial and trade wars, global economic downturn, and the rising importance of manufacturing in various countries, advanced manufacturing technology has become the cornerstone and driving force supporting and developing national economies, becoming the main battlefield for technological innovation worldwide. In recent years, modern integrated manufacturing, additive manufacturing, and intelligent manufacturing have been proposed as advanced manufacturing modes.

However, the development of industrial software in China faces challenges, with over 90% of industrial software being developed overseas, leading to technological and data barriers. There is a pressing need to effectively acquire data, integrate information flow among different software applications, and address the issue of providing personalized products in the era of diversified customer demands. Achieving efficient and rapid reconfigurable workshops and production processes through flexible manufacturing [1], and bridging the information gap among different software applications, have become crucial issues in current manufacturing development.

2. Concept and current status analysis of reconfigurable manufacturing systems

A reconfigurable manufacturing system refers to a manufacturing system that can quickly restructure or update its structure and constituent units in response to changes in market demand, according to the system planning requirements, by methods such as reordering, reusing, and innovating components. Its core technology lies in maintaining the system's reconfigurability, achieving the reconstruction and continuous optimization of the entire production system or production line by combining, replacing, trimming, and optimizing manufacturing equipment, production components, and process flows, thus allowing rapid responses to market changes and other demands. Reconfigurable manufacturing is an important trend in advanced manufacturing.

For discrete production lines, enterprises adopt unit production and achieve reconfiguration effects through resource scheduling and configuration, leading to efficient production and consistent quality, allowing for flow line specialization in the production of individual orders or products. In the case of automated production lines, reconfiguration is achieved through multi-variety mixed production, with a focus on complex software control systems enabling rapid production path transitions and cross-production of different product varieties on automated lines.

Regarding reconfiguration methods for automated production lines, the focus is on enhancing the control capabilities of constituent elements in automated lines, with future research aimed at exploring control and coordination capabilities for multi-product mixed-line production. The entire production process, including equipment, is reconfigurable. For most manufacturing companies, the challenges of reconfigurable workshops lie in the stationary nature of workshop equipment and the involvement of multiple software and departments in reconfiguration. These challenges can be addressed through logical reconfiguration or by using agile logistics equipment to connect changing production equipment for line reconfiguration. Integration of different categories of intelligent manufacturing software is needed to address issues related to data formats and model differences.

This paper addresses the problems associated with the application of reconfigurable systems in flexible manufacturing, focusing on key reconfigurable system technologies, their position, and their role in the entire production and management processes. It also proposes the development of a system that can connect design and production schedules to achieve the full-process application of reconfigurable systems.

3. Reconfigurable system component analysis

The reconfigurable system is not a standalone pure hardware or software system, but rather a composite system that integrates hardware and software. Its primary components comprise a multilayered structure ranging from physical devices to logical processes, further extending to digital descriptions and software functional modules. Extensive research has been conducted in this domain over the years. The current main components include the following parts:

3.1 Manufacturing unit

Reconfigurable systems require manufacturing resources with the ability to manufacture by adjusting their structure, parameter configuration, and execution rules to realize changes in manufacturing capabilities in virtual space [2]. The underlying structure and component nature of a manufacturing system determine its reconfigurability. If the base devices are rigid and indivisible, they lack reconfigurability; however, if the devices are composed of relatively independent functional modules that can be recombined and integrated to achieve flexibility and agility, the system is deemed reconfigurable.

3.2 Communications unit

The communication unit consists of information collection and communication functions, enabling real-time acquisition of equipment status data and facilitating production process data collection and traceability [3]. With the development of IoT technology, devices such as RFID, QR codes, and smart sensors are utilized for data communication, enabling information interaction, traceability, monitoring, and decision-making [4]. Liu et al. [5] designed an MES system framework tailored to manufacturing needs, incorporating sensing technologies. In response to complex production processes, the goal is to achieve producibility by using various sensing technologies to address challenges in the rapid perception, transmission, and preprocessing of information.

3.3 Reconfigurable model of factors of production

Production factor modeling is the most fundamental source of information in a system. For reconfigurable systems, design information is essential, and reconfiguration schemes must include such information, while the models should possess reconfigurability. The design should encompass product transformation information, and production line design should include constituent information and meet construction constraints. These pertain to static information, and achieving reconfiguration requires consideration of how to leverage this information. The challenge lies in how to combine physical devices with process design models to describe the production elements in reconfigurable modeling.

3.4 System input and output interface

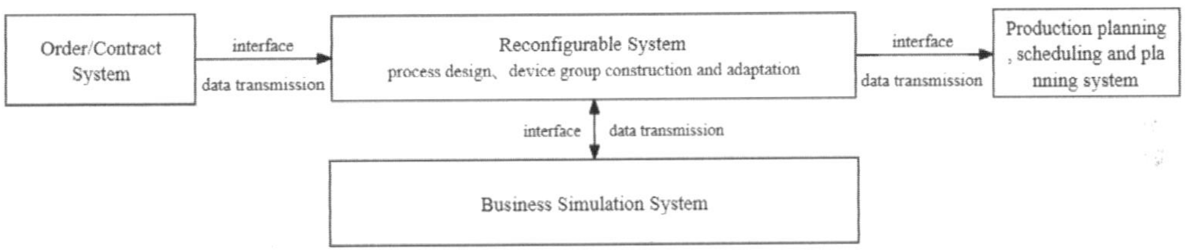

Figure 1. System interface.

According to the aforementioned description, the entire system process involves obtaining order information from the order and contract system, conducting adaptation processing between production process models, product models, equipment models, and process models, and outputting reconfigured production line schemes at the equipment layer. Ultimately, these reconfiguration schemes can be directly integrated into the production planning and scheduling system to execute the manufacturing process, as illustrated in Figure 1.

4. Design of reconfigurable system for discrete manufacturing

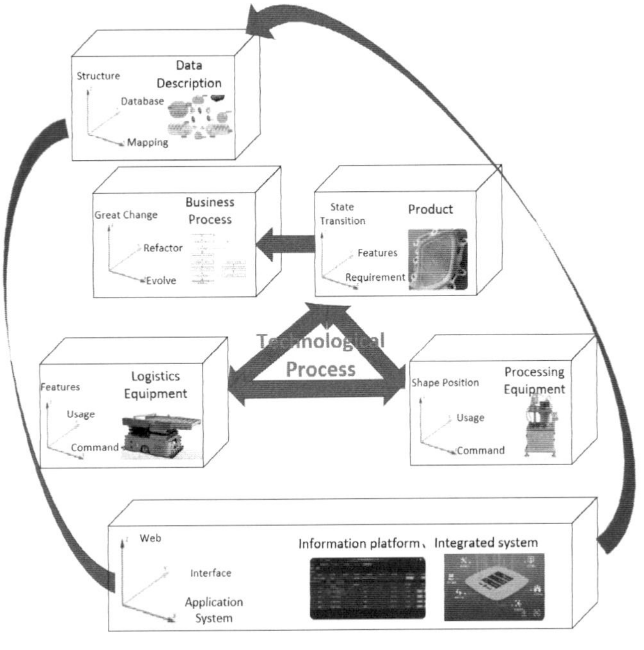

Figure 2. Reconfigurable manufacturing systems essence.

The work of reconfigurable systems in discrete manufacturing is to identify key internal elements, establish corresponding models, then construct business process models, determine data representation forms, and finally complete business processes using information platforms or integrated systems. The specific details are shown in Figure 2.

The internal information flow of the reconfigurable manufacturing system is as follows: The equipment layer information is primarily described from three dimensions - equipment information, equipment spatial structure, and equipment control structure. Process models and raw material characteristics are automatically adapted under the business process, resulting in the static and dynamic models of the entire manufacturing system. By integrating with physical equipment and utilizing perceptual communication technology, an actual usable reconfigurable system is formed.

4.1 Communication scheme design

To achieve control and communication between the control equipment at the basic level of the digitalized workshop and the field bus, manufacturing equipment needs to have network interfaces and be compatible with communication protocols. Considering the diverse types of equipment in the manufacturing site, the real-time transmission between devices, and the prominent heterogeneity of various database systems, industrial big data exhibits characteristics such as a large number of concurrent connections, strong heterogeneity, high correlation, and complex and variable states. It requires continuous collection and possesses distinct dynamic spatiotemporal characteristics. The author uses an intelligent gateway based on OPC to accomplish the information transmission function.

4.2 Reconfigurable model design

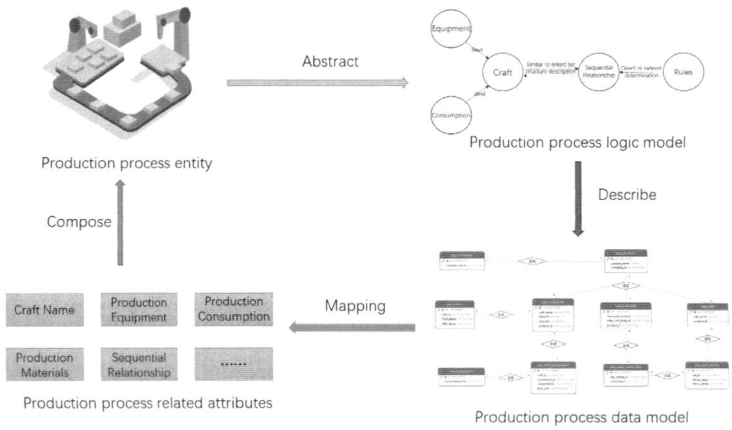

Figure 3. Reconstructing the production process model creation.

In a reconfigurable system, this level is the most core and crucial. The ultimate goal of the generative model is to reflect the attributes, behaviors, and characteristics [6] of real physical objects in a virtual space, and to determine the information changes of different types of equipment data in the production process. The logical reconstruction of the entire production workshop is based on the full process modeling, and reconstruction design and simulation of control equipment are carried out using algorithms based on the full process production model.

However, to establish a full process production model, it is necessary to create factory models, equipment models, process models, product models, and process models [7]. The factory model describes the production environment and resource layout of an enterprise, serving as the fundamental source of information for the rational deployment of equipment during the reconfiguration process. The equipment model corresponds to physical equipment, maintaining basic information about the equipment to ensure consistency between equipment usage and actual conditions. The process model considers the process flow of product production, controlling the transformation of material states. It serves as the foundation for creating the production process and is the basis for constructing

Journal of Physics: Conference Series **2842** (2024) 012071 doi:10.1088/1742-6596/2842/1/012071

production lines. The product model is a key model that defines the characteristics and attributes of the products produced by the enterprise, serving as a bridge connecting equipment, processes, and production processes. The process model is a key tool for modeling the actual production process, integrating real-time data such as equipment status, process parameters, and personnel information to provide an accurate description and optimization of the production process.

Therefore, in the modeling process, it is essential not only to establish models for individual devices but also, more importantly, to build the entire production process model. In simple terms, the goal is to transform isolated individual equipment models into a network diagram representing the production process. This spatial network diagram is then used for the reconfiguration process of processes and equipment, and to analyze the feasibility of corresponding reconfiguration schemes, as illustrated in Figure 3.

5. Development of reconfigurable systems for discrete manufacturing

In accordance with the aforementioned foundations, we have successfully developed a reconfigurable system tailored for discrete manufacturing. The production process in the discrete workshop is illustrated in Figure 4.

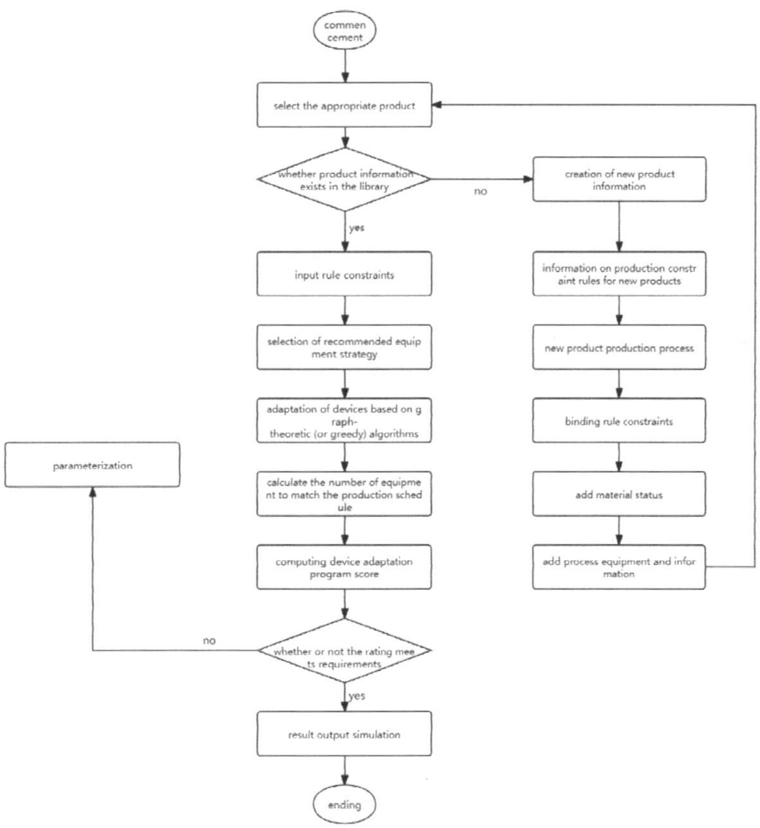

Figure 4. The process of reconfiguring the production process in the discrete workshop.

When the system requires production line reconfiguration according to orders, the first step is to issue the production orders to the reconfiguration system. The corresponding products for production are selected based on the orders. If the production process for similar products already exists in the underlying information database, the main process for manufacturing this product is generated directly based on the underlying information. If the product in the order is not present in the underlying information database, new product information is created. The process for this product is designed along with specific process constraints and then stored in the database. Afterward, the previous step is repeated to obtain the main process.

Next, the additional process information required is automatically outputted based on the process rules. Alternatively, an interactive interface may be provided for product designers to make selections and confirm choices. Upon obtaining the complete process flow, automated equipment model adaptation is performed based on equipment selection rules and tempo constraints, utilizing improved algorithms such as graph theory's shortest path or greedy algorithms. Subsequently, scheme evaluation and performance simulation are conducted. Finally, the reconfiguration schemes based on equipment and simulation results are outputted. These reconfiguration schemes can be directly integrated into the Manufacturing Execution System (MES).

6. Conclusion

This paper aims to address two key issues in discrete manufacturing—namely, logical modeling of workshop production processes and the design of reconfiguration algorithms—through the design and development of a reconfigurable system. It provides a viable example for the digitalization research of production processes in the discrete manufacturing industry. The proposed method in this paper offers advantages in terms of lightweight, rapid, and automatic reconfigurability. It is capable of translating the expertise of process engineers and planning engineers into system rules, thereby constraining the entire production reconfiguration process. This effectively bridges the gap between design and production processes. In the future, further in-depth research will be conducted around the expansion of logical models and production line reconfiguration based on these models.

Acknowledgment

The work was supported in part by the National Key Research and Development Program of China (Grant No.2020YFB1710701).

References

[1] YU Chenglong, HOU Junjie, LU Jing, etc. Application Architecture of Intelligent Line for Multi-Variety and Variable-Batch Products [J]. Aeronautical Manufacturing Technology, 2020, 63(06): 73-79.

[2] ZHANG Y, ZUO Y, LIU B, et al. Digital twin shop-floor manufacturing operations management platform [J]. Computer Integrated Manufacturing System, 2024, 30(1): 1.

[3] GUO Lei, ZHANG Hongqi, WU Qianhao, et al. Design and implementation of informational management and control of electronic equipment assembly line based on the Internet of Things [J]. Manufacturing Technology & Machine Tool, 2023 (4): 36-43.

[4] GUO Shaohua, GUO Yu, Cha Shanshan, et al. A Review of Discrete Shopfloor Manufacturing Internet of Things and Its Key Technology Research and Application [J]. Computer Integrated Manufacturing Systems, 2019, 25(02): 284.

[5] Liu M, Xie J, Wu H, et al. Research on workshop logic modeling and simulation based on finite state machine [J]. Journal of System Simulation, 2023, 35(4): 853-861.

[6] DU Y, LUO Y, PENG Y, et al. 3D visual monitoring system of industrial robot based on digital twin [J]. Computer Integrated Manufacturing System, 2023, 29(6): 2130.

[7] Liu M, Xie J, Wu H, et al. Research on workshop logic modeling and simulation based on finite state machine [J]. Journal of System Simulation, 2023, 35(4): 853-861.

Journal of Physics: Conference Series **2842** (2024) 012072 doi:10.1088/1742-6596/2842/1/012072

A wireless and passive tool holder for online measurement of milling temperature

Xiang Sun [1, 2, 3, *], Qizhi Zhang [1, 2], Zilin Tang[1] and Hongrui Liu [1]

[1] China Coal Technology and Engineering Group Shanghai Company Ltd., Shanghai 200030, China
[2] State Key Laboratory of Intelligent Coal Mining and Strata Control, Shanghai, 200000, China
[3] Shenzhen International Graduate School, Tsinghua University, Shenzhen 518055, China

[*] Corresponding author's e-mail: sunxiangalan@foxmail.com

Abstract: The measurement of cutting temperature is very important in monitoring the manufacturing process and greatly impacts tool wear and the workpiece surface quality. Milling temperature measurement is particularly difficult due to high-speed rotation. This paper uses an intelligent and wireless passive tool-holder system for milling temperature sensing, with a resolution of 0.25°C, a range of 0~700°C, and an update speed of 10 Hz. Thermocouple is pre-embedded in the machined groove near the cutting edge on the flank face of the milling tool, and signal acquisition circuits and wireless transmission module are integrated into the tool holder system for online signal processing. Based on electromagnetic induction, the self-power supply scheme is designed by converting the rotating kinetic energy of the spindle to electricity, and hierarchical energy regulation is achieved to realize a stable and constant power supply. With a temperature-sensing intelligent tool holder, cutting experiments on different processing parameters are carried out to verify the feasibility of the measurement system.

1. Introduction

In the context of numerical control machining, temperature is a pivotal physical parameter, exerting a direct influence on the cutting process and tool performance owing to the generation of cutting heat. This thermal phenomenon involves heat transfer, conduction, and convection processes, resulting in spindle temperature elevation and consequent displacements, known as thermal errors. Notably, thermal errors of machine tools contribute to approximately 50% to 70% of machining errors, representing the foremost error source in precision and ultra-precision machining endeavors. The generation of cutting heat imposes limitations on machining parameters such as feeding rate, spindle speed, and depth of cut, thereby diminishing machining efficiency. Furthermore, cutting heat predominantly accounts for energy dissipation during the machining process.

In terms of temperature impacts, cutting heat profoundly affects surface morphology, tool life, the roughness of workpieces, machining quality, and effectivity [1-3]. Moreover, considering the mechanistic aspects of metal cutting, temperature during the machining process intricately relates to tool longevity. Monitoring tool tip temperature facilitates judicious tool utilization, enhancing tool utilization rates and averting mechanical malfunctions attributable to elevated temperatures, which could result in operational halts or even pose safety hazards. Milling has played a significant role in modern machining. During the milling process, the measurement of the signal is very difficult due to the high-speed rotation of the cutter–tool holder–spindle system.

The measurement methods of cutting temperature can be primarily categorized into three major approaches, namely physical experiments [4], simulation analysis [5], and theoretical derivation [6]. More research is based on simulation analysis, and different algorithms are used to obtain the calculated temperature of the milling process [7, 8]. Dewes et al. [9] used a workpiece with an embedded thermocouple method and a non-contact infrared thermal imaging camera method to conduct an experimental study on the cutting temperature of titanium alloy workpieces, finding that the former can sense more accurate temperatures. However, the pre-mounting thermocouples in the workpiece can only measure transient temperature. Kerrigan et al. [10] fabricated a smart tool holder with integrated signal processing and wireless transmission modules and used a thermocouple to measure the temperature of the milling tool chip. However, it is battery-powered, making it difficult to implement long-time measurement tasks.

In this paper, an end milling cutter with a built-in thermocouple and an intelligent wireless tool holder integrated with data processing devices are designed with an energy harvester to conduct online temperature sensing.

2. Milling temperature sensing scheme

The thin thermocouple with good dynamic response is selected as the sensor with a range of 0 to 700 °C and is integrated into the cutter-tool handle system with the back-end processing circuits to implement real-time temperature sensing and wireless transmission. The whole temperature sensing scheme is shown in Figure 1. The thermocouple signal can be processed and wirelessly transmitted through the integrated circuits assembled in the holder system.

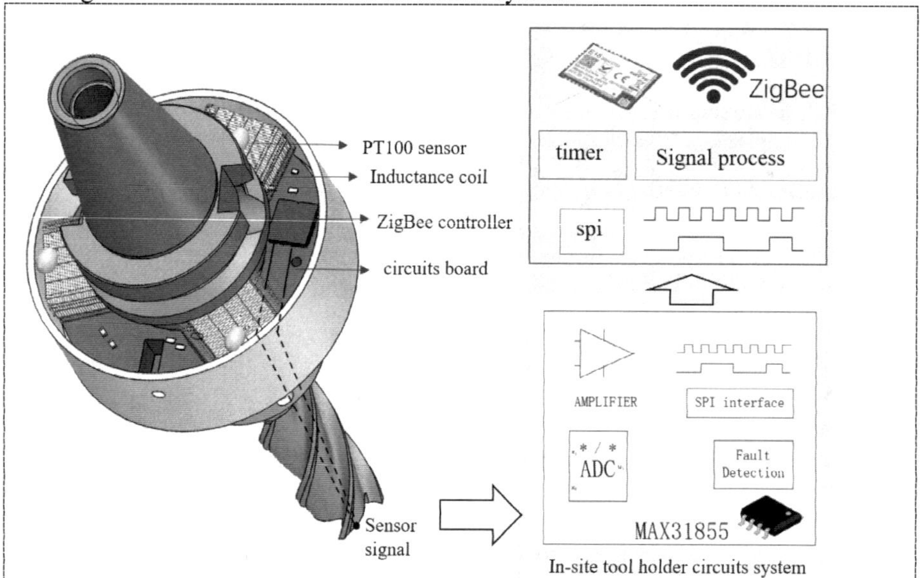

Figure 1. Temperature sensing scheme.

2.1. Milling tool design and sensor integration

To ensure continuous and reliable milling temperature measurement, sensors are expected to be installed near the cutting edge of the tool rather than on the workpiece. The installation of thermocouples is crucial for accurate tool chip temperature measurement. The highest temperature occurs near the tool cutting edge during the machining process. Since the rake face of the cutter is the surface directly traversed by cutting chips generated during machining with high temperature and velocity, it can cause mechanical impact and ruin the rake face. It is generally preferable to embed the thermocouple in the flank of the cutter. Based on experience, a distance of 1 mm from the cutting edge is chosen as the installation position for the thermocouple tip. The milling cutter is designed as a 3-flute spiral end milling tool, with a spiral direction to the right and a cutting direction to the right. A 2×1 mm groove is machined on the flank surface of a tooth, which has connectivity with an internal cooling hole, facilitating the threading of the thermocouple. Design parameters for the milling cutter are delineated in Table 1. The integration of the thermocouple in the milling tool is shown in Figure 2.

Table 1. Milling tool parameters.

Parameter	Value	Parameter	Value	Parameter	Value
Tool type	End milling	Tool material	carbide	Cutting edge length	30 mm
Body diameter	Φ16	Rake angle	15°	Total tool length	100 mm
Internal coolant hole	Φ1.5	Clearance angle	14° 14° 24°	Coating material	Al_2O_3
Number of teeth	3	Spiral angle	15°	Coating thickness	0.003~0.005 mm

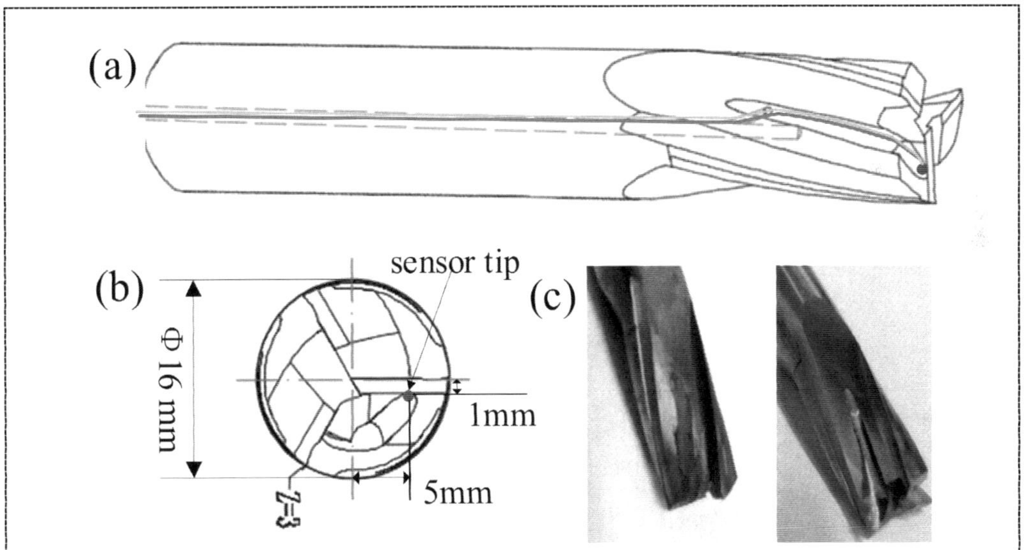

Figure 2. Thermocouple integration of the milling tool (a) Main view (b) Top view (c) Photo of sensor mounting and integration.

2.2. Signal processing and wireless transmission

The signal generated by the thermocouple is a weak thermoelectric potential that requires sophisticated signal acquisition circuits. The MAX31855 is used for data acquisition and processing with a resolution of 0.25 °C. The temperature signal can be transmitted through a digital interface to the controller. In terms of signal transmission, since the temperature signal is generated and collected on the rotating tool holder system, it is difficult to use a wired transfer method. The acquisition speed of the tip temperature signal is generally 1~20Hz, and the estimated signal transmission speed is below 1 kB/s. The ZigBee solution was selected to guarantee better stability for the wireless network. The temperature signal is sampled by MAX31855 and transferred to a digital signal, which the ZigBee coordinator accesses at the preset frequency and then transmits to the end device via a wireless network.

3. Energy harvesting system

Given the high-speed rotation of the tool-holder-spindle system during the milling process, a self-energy supply device converted from the rotation kinetic energy based on electromagnetic induction is designed and developed to achieve a continuous and stable power supply.

3.1. Main modules

The basic principle is to generate a non-uniform strong magnetic field near the tool holder and place an inductance coil for power generation and post-stage energy regulation circuits in the rotation tool holder system. As shown in Figure 3, the energy harvesting system mainly consists of a tool holder, an integrated cylindrical ring, and an external magnets fixation device. The inductance coil wound around the magnetic core is placed in the assembled ring, which is mounted on the tool handle and has no relative movement. In the milling process, the inductance coil rotates at high speed with the spindle, and the magnetic flux in the inductor coil is constantly changing, leading to an electromotive force, which is connected to an external circuit and rectified to provide energy to the circuit system.

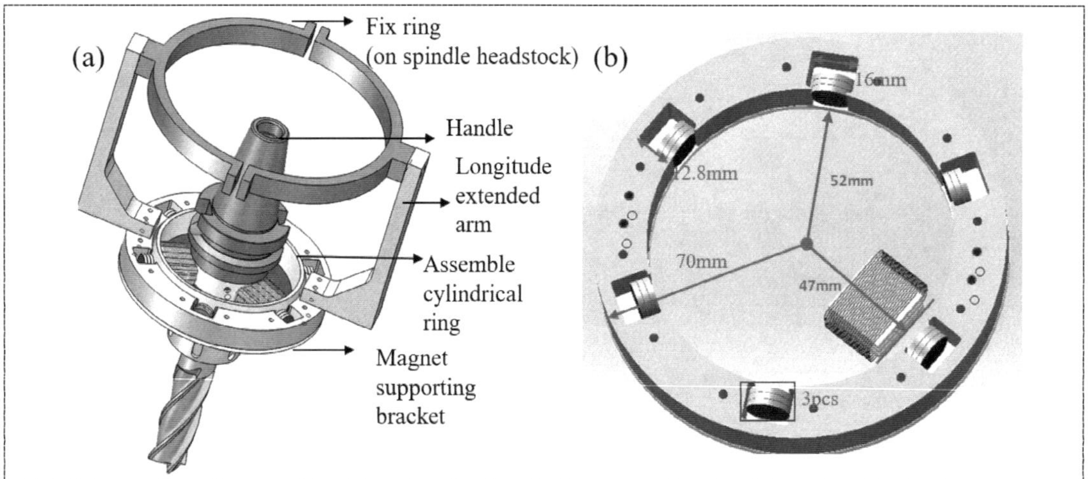

Figure 3. Energy harvesting system (a) Tool holder system (b) Magnets supporting brackets.

3.2. Simulation setup and verification

Through preliminary tests, it is found that the instantaneous consumption power during normal operation is estimated to be no more than 300 mW. A simulation is set up to pre-calculate the power generation rate from electromagnetic induction to arrange the external magnetic field and induction coils reasonably. The parameter of the cylindrical-shaped magnet and induction coil is shown in Table 2. The established vector diagram of magnetic induction intensity is shown in Figure 4, and the maximum intensity is about 0.2 T when it is close to the magnet group during the coil rotation.

Table 2. Magnet and inductance coil parameter.

Magnet		Inductance coil			
Material	Sintered NdFeB	Material	Φ0.35 mm copper	Inner diameter	8 mm
Magnetization direction	Thickness direction	Resistance	3.8 ω	Outer diameter	19.6 mm
Size	Φ16×2 mm	Inductance	2500 μh	Turns	480

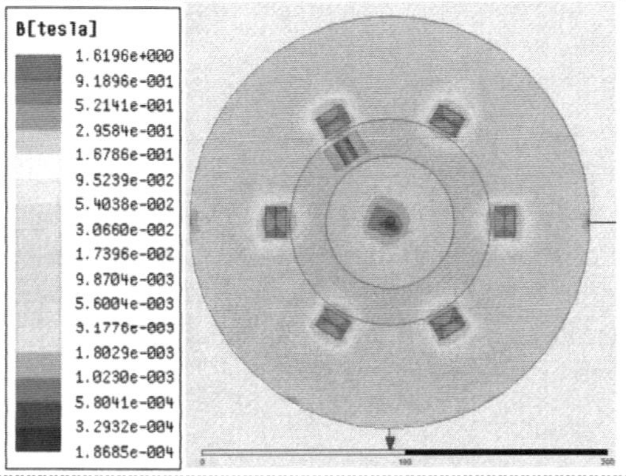

Figure 4. Vector diagram of magnetic induction intensity.

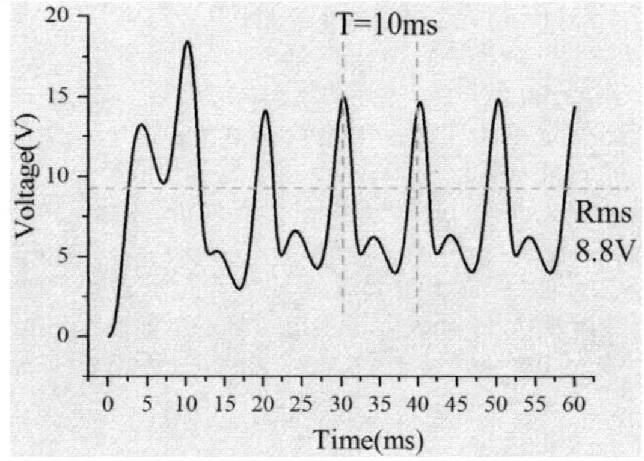

Figure 5. Rectified voltage of simulation.

Magnets, coils, and cores are set according to the actual conditions. When the rotational speed is 6000rpm, the induced electromotive force after rectification bridge and filtering is shown in Figure 5. When the load is a 10Ω resistor, the generated rms electromotive force is 8.8V, and the output power is about 7.8W, satisfying the power needs.

3.3. Hierarchical energy regulation

When the rotation speed of the milling process changes, the power generation rate varies greatly. If the energy is not regulated, the excessive energy will be consumed in the form of heat, causing high inside temperatures, adversely affecting the normal operation of the circuits, and even ruining the circuit components. Therefore, the energy needs to be regulated for different rotation speeds.

The power generation method is proposed by changing the number of inductance coils involved. The N-channel MOS transistor of IRF540 is used as the switch for energy adaptation, and the simulation circuits and results are shown in Figure 6. V1, V2, and V3 are AC power sources, representing power generation coils. When the controller applies a positive voltage to the control pin of the light-emitting diode of the optocoupler, the photosensitive optocoupler will be turned on with a voltage drop of close to 0V, which means V_{gs} of MOS is close to 0V, and the MOS switch is clamped. The corresponding coil is removed from the circuit's system in this case. The simulation results verified the analysis, showing that energy regulation can be divided into three levels.

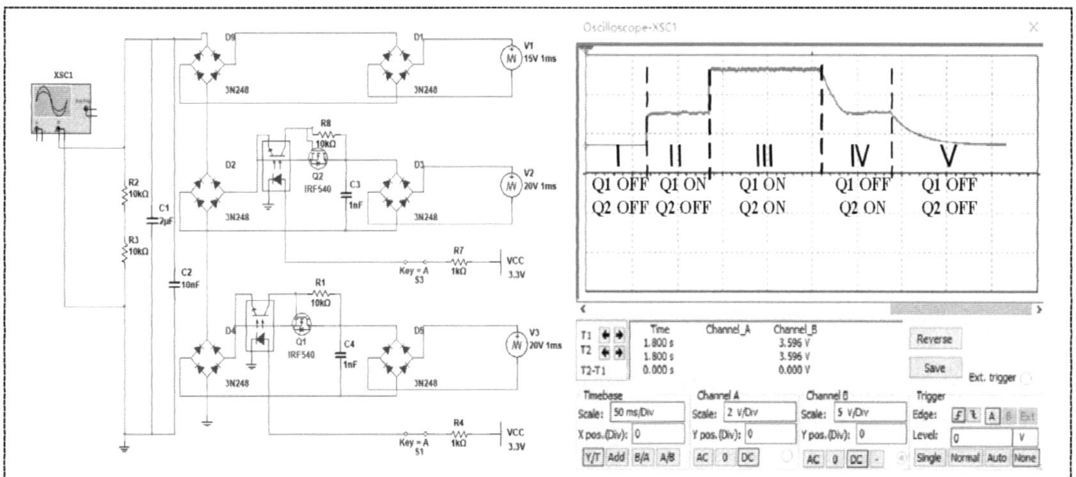

Figure 6. Energy regulation circuits and simulation results.

Journal of Physics: Conference Series **2842** (2024) 012072 doi:10.1088/1742-6596/2842/1/012072

4. Experimental test and analysis

4.1. Experimental platform setup

Since the assembly structure consists of circuit boards, inductance coils, cylindrical rings, and tool holders that rotate with the spindle at high speed, it is essential to ensure the dynamic balance of the handle system. Dynamic balance adjustment is realized based on the G2.5 standard within 5000 rpm. Experiments are carried out in a three-axis machine center, and the spindle is BT40 standard. The on-site measurement system is shown in Figure 7. During the milling process, the tool chip sensor can measure the temperature, and the signal can be processed online in the tool holder and transferred to the end device by a wireless ZigBee network. A thermal imaging system is applied to contrast the measurement ability.

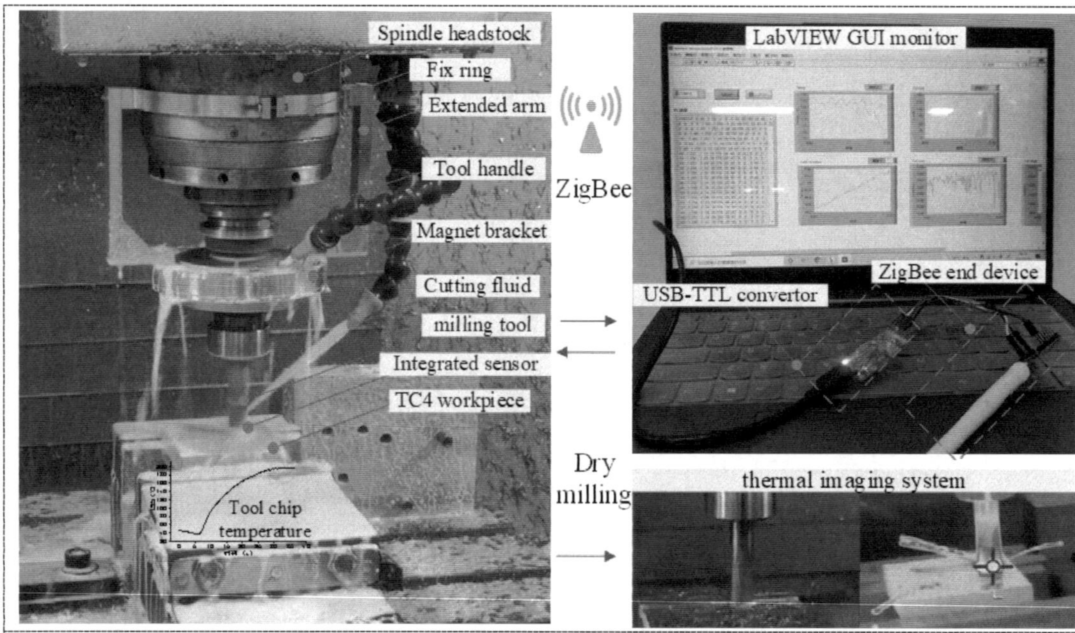

Figure 7. Milling temperature sensing experiments site.

4.2. Energy harvester test

Before milling temperature sensing tests, energy harvester performance tests are carried out. The tool holder-spindle system is in an idle rotation state, and the amplitude of the rectified voltage expresses the energy generated. The thermocouple processing chip MAX31855 mounted in the ring can sense the cold junction temperature, which is affected by excessive power. The temperature inside the ring at different rotation speeds is measured. The average temperature of PT100 sensors and MAX31855 cold conjunction is recorded in Figure 8, and the amplitude of rectifier voltage is shown in Figure 9. When all three coils are involved in power generation when the speed exceeds 1600rpm, the Zigbee end device can receive the data packet, which means that the power generation rate meets the requirements. In the single-coil and double-coil cases, the threshold is 2300rpm and 3100rpm, respectively. It can be seen in the current magnets scheme that only when the rotation speed exceeds 1600 rpm can the measurement circuits operate normally, and with an increase in coil number and rotation speed, the generated power rises, leading to a higher temperature. When all three coils are involved, the temperature can be more than 50°C, bringing more safety threats. The magnet scheme can be adjusted to get proper power for cases where the machining speed is below 1600 rpm.

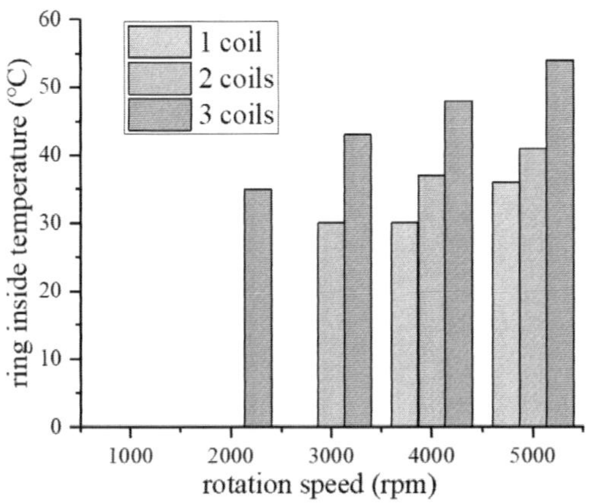

Figure 8. The average temperature of the assembled ring.

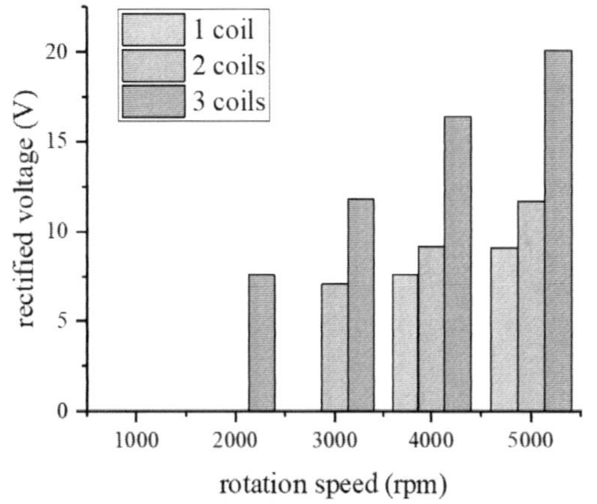

Figure 9. Rectified voltage of inductance coil.

4.3. Milling temperature sensing test

Titanium alloy is chosen as the material, as it has excellent mechanical properties and corrosion resistance and is widely used in aerospace and medical equipment. However, titanium alloy has high strength and hardness and poor thermal conductivity, which makes it difficult to process due to high cutting temperatures. The milling temperature for different cutting parameters is shown in Figure 10 at dry milling mode. It can be seen that after some time in the cutting process, it reaches thermal equilibrium, and the equilibrium milling temperature rises with rotation speed, which is around 142°C, 163°C, and 192°C, respectively.

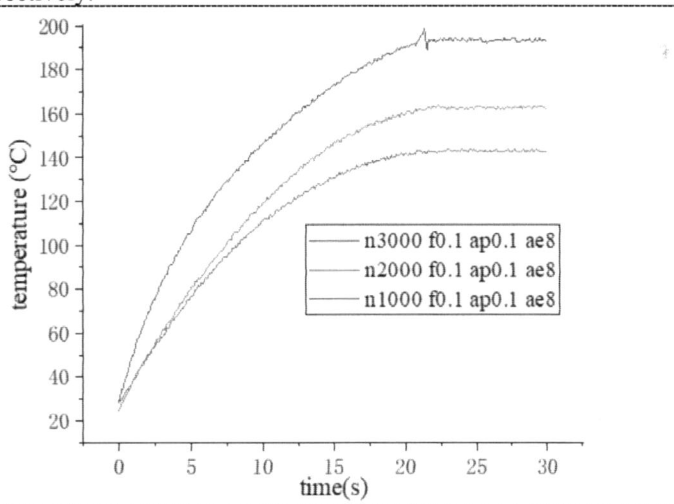

Figure 10. Milling temperature of different cutting speeds measured by smart handle.

5. Conclusion

This paper proposes a new method for measuring milling temperature by integrating sensors and signal processing circuits into an intelligent standard BT40 tool holder system with an energy harvester. The measurement system includes a customized milling tool with the embedded thermocouple, an online signal processing module, and a wireless transmission module, realizing a sensing ability of 0.25°C resolution, 0~700°C range, and 10 Hz update rate. Experiments with different machining parameters show that the tool holder system can measure the real-time milling temperature near the cutting edge and equilibrium milling temperature, and its growth rate rises with increasing rotation speed. Although in this method, the direct measurement point is not a tool cutting edge but a flank point near it, it is highly related to max milling temperature and has some significance.

Regarding the continuous energy supply of the rotation tool holder system, an energy harvesting scheme based on electromagnetic induction is designed, the hierarchical energy regulation is realized, and the feasibility is verified by cutting experiments. At a minimum machining speed of 1600 rpm, the energy harvesting system provides a stable and continuous energy supply of more than 300mW.

Our future work will focus on two aspects. The first is to explore the relationship between the temperature of the thermocouple tip point and the temperature of the tooltip. On the other hand, we will explore more efficient energy regulation solutions to achieve reliable and stable energy harvesting under different processing conditions.

Acknowledgments
This work was financially supported by the CCTEG TIANDI SCIENCE&TECHNOLOGY CO., LTD. (2023-TD-ZD003-006).

References
[1] T. Li, H. Long, T. Shi, J. Yang, J. Duan 2022 *J.* Cutting temperature measurement using a novel near-infrared two-color pyrometer under dry and wet cutting of Ti-6Al-4V alloy *Journal of Materials Processing Technology* **309** 117751.
[2] T. Li, T. Shi, Z. Tang, G. Liao, J. Duan, J. Han, Z. He 2021 *J.* Real-time tool wear monitoring using thin-film thermocouple *Journal of Materials Processing Technology* **288** 116901.
[3] J. Ning, S.Y. Liang 2019 *J.* Predictive Modeling of Machining Temperatures with Force-Temperature Correlation Using Cutting Mechanics and Constitutive Relation *Materials (Basel)* **12** 284.
[4] Cascajero A N, Tapetado A, Vazquez C 2020 *J.* High Spatial Resolution Optical Fiber Two Color Pyrometer With Fast Response. *IEEE Sensors Journal* **21** 2942.
[5] Dogu Y, Aslan E, Camuscu N 2006 *J.* A numerical model to determine temperature distribution in orthogonal metal cutting. *Journal of Materials Processing Technology* **171** 1.
[6] Radulescu R, Kapoor S G 1994 *J.* An Analytical Model for Prediction of Tool Temperature Fields during Continuous and Interrupted Cutting *Journal of Engineering for Industry* **116** 135.
[7] Zhou G, Wang X, Xu C 2022 *J.* Determination of tool tip steady-state temperature in dry turning process based on artificial neural network *Journal of manufacturing processes* **79** 600.
[8] Helmig T, Peng B, Ehrenpreis C 2019 *C.* A Coupling Approach Combining CFD and FEM Methods to Predict Cutting Fluid Effects on the Tool Temperature in Cutting Processes *Journal of Manufacturing Science and Engineering* **2** 141.
[9] Dewes R C, Ng E, Chua K S 1999 *J.* Temperature measurement when high speed machining hardened mould/die steel *Journal of Materials Processing Technology* 92-93.
[10] Kerrigan K and Thil J 2012 *J.* An Integrated Telemetric Thermocouple Sensor for Process Monitoring of CFRP Milling Operations. *Procedia Cirp* 449.

Design and performance research on metal-sprayed electrical heating films for aircraft wing anti-icing

Yuanyuan Ding [1,*], Zhimao Li [1], Zhefan Ren [1], Guangya Zhu [2] and Bin Bai [1]

[1]Shanghai Aircraft Design and Research Institute, Shanghai 201210, China
[2]Nanjing University of Aeronautics and Astronautics, Nanjing 210016, China

[*]Corresponding author's e-mail: dingyuanyuan@comac.cc

Abstract. Composite material, supercritical aircraft wings, and multi-electrical technology represent the future development trend of civil aircraft. However, domestically, the use of electrical heating for anti-icing on civil aircraft made of composite materials has not yet been implemented. In this study, focusing on the anti-icing technology of a specific aircraft wing leading edge, the article initially designs a metal thermal spraying circuit based on the thermal load distribution on the surface of the anti-icing area. Next, the research explores the resistive properties of the electrical heating film on the composite material wing leading edge, examining how the porosity and oxide content of the coating affect its resistivity. Finally, the metal thermal spray test is completed on the leading edge of the composite material aircraft wing. The test results indicate that the adherence and deposition rate of the coating on the leading edge of the aircraft wing are reliable. The resistivity of the heating film meets the performance specifications. The article develops an integrated metal thermal spray control method based on high-curvature composite materials, resulting in an electrical heating film with excellent shape adaptability, varying power density, and strong bonding, providing support for the domestic advancement of wing electrical heating anti-icing technology.

1. Introduction

In complex meteorological conditions, icing on aircraft wings can impact flight safety, making the aircraft wing anti-icing system critically important. The main approaches to anti-icing/de-icing the leading edges of aircraft wings are hot air and electrical heating. Hot air anti-icing works by drawing high-temperature, high-pressure gas from the engine compressor to heat the leading edge's skin [1-2]. Due to the temperature resistance limitations of the wing leading edge materials, the extracted hot air needs to be pre-cooled, and the temperature of the exhaust gas is high after heat exchange in the anti-icing chamber, resulting in the low thermal efficiency of traditional hot air anti-icing. In contrast, electrical heating employs electrical energy as the energy source for the leading edge of wing anti-icing, greatly reducing the energy consumed for the same heating effect compared to hot air. With the advancement of multi-electrical/all-electrical aircraft technology, the use of electrical heating anti-icing systems has become more widespread in aircraft. Due to the large thermal inertia of the pipe used in hot air anti-icing, it is difficult to control the cyclic heating for de-icing, which may lead to the formation of ice ridges in some cases. In contrast, using electrical heating allows for precise cyclic control of heating power, enabling the anti-icing system to operate in de-icing mode and reducing energy consumption [3-5].

Composite materials, known for their lightweight and high strength, are gradually replacing traditional metal materials in aircraft wing leading edge structures. The development of composite material supercritical wings and multi-electrical technologies represents a trend for future civil aircraft. Integrated electrical heating technology is the only solution for addressing aircraft anti-icing in this trend, solving safety issues in complex icing weather conditions for aircraft. Electrical heating anti-icing technology involves embedding heating elements into the structure, yielding high efficiency and extensive coverage. Its lower power density and structural temperature present notable benefits in terms of energy usage and structural safety [6].

This article takes into account that the substrate for spraying is a sandwich-structured composite material composed of carbon and glass fibers, which is prone to deformation due to thermal shock during the thermal spraying process. Moreover, there are high implementation requirements for the coating's porosity, bond strength, and uniformity. Comparing various spraying methods, if the spray material is a nickel-chromium-based or copper-manganese-based alloy, these methods, such as plasma, supersonic flame, electric arc, and conventional flame spraying, can all form a coating [7]. Plasma and supersonic spraying result in a dense coating with high bonding strength, but the thermal shock during the spraying process is high, reaching temperatures of over a thousand degrees Celsius. However, the composite material substrate can only withstand up to 190 °C, making this form of spraying highly damaging to the substrate. Conventional flame powder spraying results in low bonding strength, high porosity, severe oxidation, and poor coating quality. Although electric arc spraying has a high yield and low cost, the coating is rough, not dense, and has high porosity. Compared to other methods, flame wire spraying causes less thermal shock, keeping the substrate temperature between 40-50 °C during spraying, thus avoiding damage or deformation. Flame wire spraying is stable and controllable, yielding a denser coating with stronger adhesion. The spray layer has a 5-10 % porosity and high deposition efficiency, making the coating performance more suitable. Hence, this article focuses on researching the performance of metal thermal sprayed anti-icing electrical heating films based on flame wire spraying.

The main research content of this article includes: (1) Designing metal thermal spraying circuits based on the distribution of thermal loads on the surface of the anti-icing area and thermal control; (2) Selecting suitable metal thermal spraying materials for composite substrates by analyzing the impact of the coating's porosity and oxide content on resistivity; (3) Conducting metal thermal spraying experiments on the surface of a specific aircraft of composite material wing leading edge, with results indicating a reliable deposition rate of the coating on the leading edge and that the heating film's resistivity meets performance specifications. This article's research and experimental results show that using manganese-copper alloy wire for thermal spraying on complex curved wing surfaces on a multi-degree-of-freedom mechanical control platform can achieve a relatively uniform and stable coating performance.

2. Anti-icing surface heating zone design

2.1. Method of dividing heating zone

By analyzing the continuous maximum icing conditions and intermittent maximum icing conditions in Part 25, Transport Category Airplanes Airworthiness Standards: Appendix C [8], calculation state points are selected. For a specific aircraft model, grid division is performed with ICEM software, and the anti-icing thermal load is computed using the Messinger model [9]. The calculation of the wing anti-icing thermal load consists of two main parts. The first is the air and water droplet flow field computation, which yields parameters needed for thermal load calculation, such as the convective heat transfer coefficient, droplet collection coefficient, and pressure coefficient. The second part involves the thermal load calculation, determining the distribution of thermal load and temperature on the surface of the heated area through the computation of the control equations. After determining the protected area, multiple zones are meticulously divided within this area, with each zone having independently assigned power density. Aiming for the minimum anti-icing thermal load, an optimization algorithm [10] is used to optimize power density distribution under all state points. The objective is to find the pattern of power density on the surface under the minimal thermal load. Based on this, a preliminary heating zone proposal is developed, dividing the protected area into a few zones. A state point with a higher thermal load is selected as the design state point. With the same goal of minimizing anti-icing thermal load, an optimization algorithm is used to optimize the power density distribution at this state point. The optimization results are used to verify other state points until the zoning results meet the design requirements for all state points, ultimately finalizing the heating zone proposal. The flow diagram of the heating zone design is shown below in Figure 1.

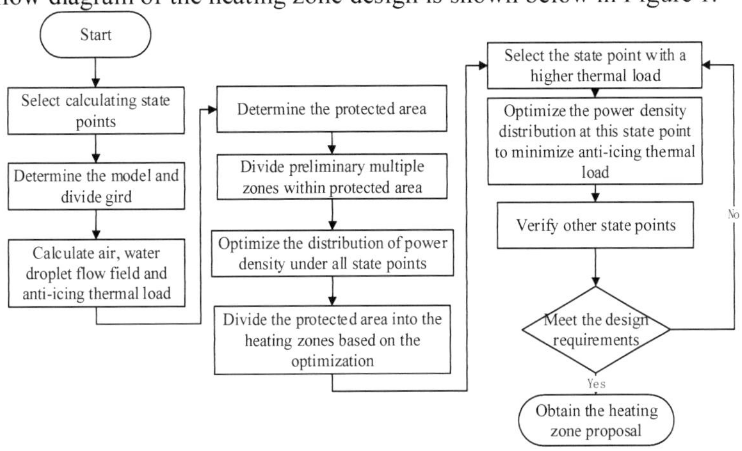

Figure 1. The flow of the heating zone design.

Since the anti-icing area uses a star-connected three-phase circuit control method, the protected area is divided into 3 zones along the chord direction. To ensure the three-phase balanced current equals zero, the resistance values of the star-connected three-phase heating zones must be the same. Therefore, once the distribution of the heating zone area is determined, providing the power density of any one zone will obtain the distribution of the overall power density. Therefore, the area of the heating zone s and the power density q satisfy the following relationship of Equations (1)-(4).

$$q_1 s_1 = q_2 s_2 = q_3 s_3 = Q \tag{1}$$

$$s_1 : s_2 : s_3 = \frac{1}{q_1} : \frac{1}{q_2} : \frac{1}{q_3} \tag{2}$$

$$q_1 : q_2 : q_3 = \frac{1}{s_1} : \frac{1}{s_2} : \frac{1}{s_3} \tag{3}$$

$$s_1 + s_2 + s_3 = S \tag{4}$$

where Q is the total heating power of the anti-icing area, and S is the total area of the heating film in the anti-icing area.

By considering these methods, the required heating power for the anti-icing area and the zoning proposal based on optimized thermal load and power allocation results are illustrated in Figure 2.

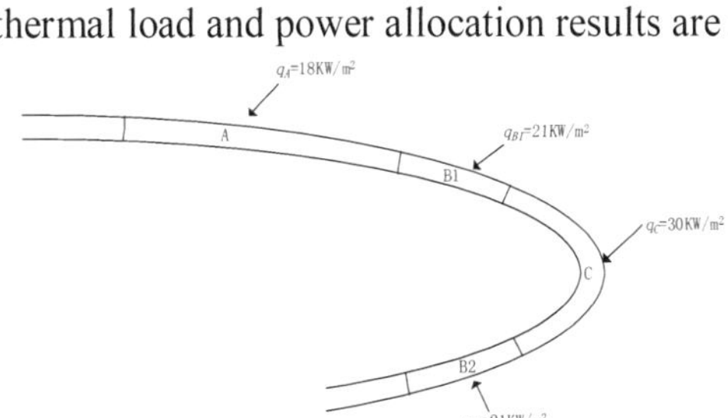

Figure 2. The heating power for the anti-icing zone.

2.2. Heating film circuit design

Based on the analysis and calculation, under power supply conditions, the heating power densities in the three-phase areas of the anti-icing heating zone must reach the aim values respectively defined in Section 2.1. To ensure the three-phase balanced current equals zero, the resistance values in the star-connected three-phase heating zones must be identical. It is assumed that each of the three heating zones in the anti-icing area is divided into circuits of length L and width W, configured in a serpentine zigzag pattern, as illustrated in Figure 3.

Figure 3. The pattern of heating film for the anti-icing zone.

In each heating zone, the width of the electric heating film coating is w_1, with a spacing of w_2 between coatings. There are n strips of electrical heating film coating within the defined dimensions, resulting in n-1 intervals between coatings. The resistance of the heating film depends on the spray area, thickness, and resistivity. The thickness h of the coating can be calculated using Equation (5).

$$R = \frac{\rho L_1}{w_1 h} \tag{5}$$

where ρ is the resistivity of the heating film, and L_1 is the length of the heating film.

The performance of the metal coating is influenced by the properties of the spray material and its structural characteristics. For instance, uneven porosity in the sprayed heating film leads to instability in the resistivity of the heating film. Therefore, to achieve the desired resistance value, controlling factors such as the sprayed area, thickness, and resistivity are necessary. The overall design approach for the heating film circuit is as follows:

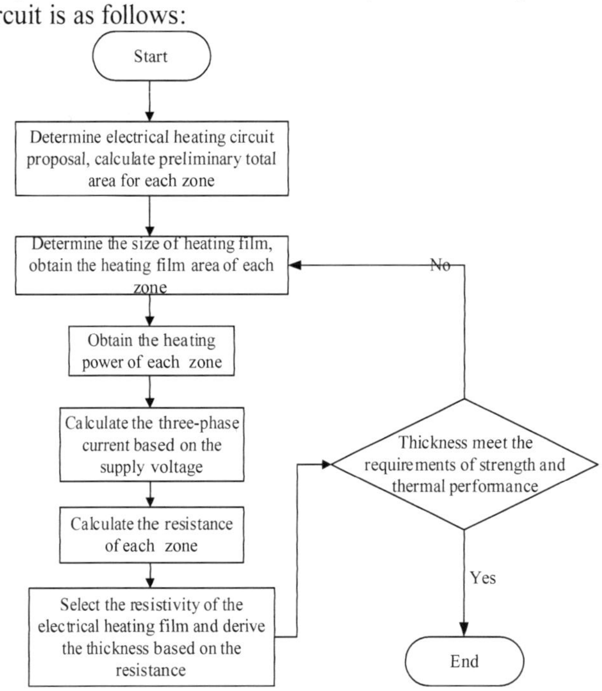

Figure 4. The overall design approach for the heating film circuit.

Based on previous spraying experience, a coating thickness of 120 μm ~140 μm results in the best overall performance, including optimal resistivity, bonding strength, and bending properties of the coating. After confirming the heating power of the electrical heating film, as well as the shape and thickness of the film coating, it is preliminarily determined that the resistivity of the electrical heating film in a single zone ranges from 1.82 E-6 (Ω·m) to 8.4 E-6 (Ω·m).

3. Performance of metal sprayed electrical heating films

To determine the relationship between material resistivity, resistance, and thickness, an analysis of the spraying material properties is conducted. Preliminary analysis indicates that the resistivity of nickel-manganese alloys and manganese-copper alloys can reach 1.9 E-6 (Ω·m)~2.0 E-6 (Ω·m). Both materials, rich in Mn, enhance the resistivity but also increase the difficulties in metallurgy and processing. To further improve electrical performance and enhance processability, elements such as Al, V, Si, Fe, and rare earths are added. By controlling the heat treatment process, the temperature coefficient of resistance α for nickel-manganese alloy materials will be less than 5.0 E-5 (/°C), and the resistivity ρ will reach 2.0 E-6 (Ω·m). For manganese-copper alloy materials, the resistance temperature coefficient α will be below 2.0 E-5 (/°C), and resistivity ρ will be higher than 1.8 E-6 (Ω·m). During the development process, nickel-manganese alloy wire, despite being modified with elements such as Al, V, Si, Fe, and rare earth for mechanical properties, proves challenging to forge after metallurgical melting and tends to break easily due to its poor ductility in cold processing. The manganese-copper alloy material exhibits superior metallurgical properties and uniform microstructure. Its resistivity is 1.89 E-6 (Ω·m), meeting the pre-established range of resistivity. Therefore, after analysis and comparison, manganese-copper alloy wire is more suitable as a spray material.

The thickness deviation of the heating film significantly impacts overall resistance and localized surface resistance, mainly affected by the stability of the processing equipment, uniformity in pre-spray treatment, consistency in spraying speed, and heat input control. To determine the relationship between coating thickness and resistance/resistivity, comparative experiments on coating performance were conducted using two types of spray wires, one with a diameter of φ3.0 mm (thick wire) and the other φ2.2 mm (thin wire), on two types of substrates: smooth and rough plates. During the thermal spraying process, the porosity and oxidation level of the spray material greatly affect the coating outcome. The thickness of the coating is mainly controlled by adjusting the ratio of oxygen to acetylene, and the atomization effect is controlled by adjusting the flow of compressed air. The equipment used in this study is the UCW 1000 fully automatic flame wire spray equipment equipped with an improved Metco 5 K spray gun. During spraying, the temperature of the substrate is kept below 50 °C, with adjustments made to the flame temperature, wire feed speed, and spray gun movement speed. Concurrently, clean, oil-free compressed air is used to cool the substrate. The thickness of the sprayed circuit is measured using a (0-25) mm/0.01 micrometer and an Olympus BX-51M metallurgical microscope, with results shown in Figures 5 and 6.

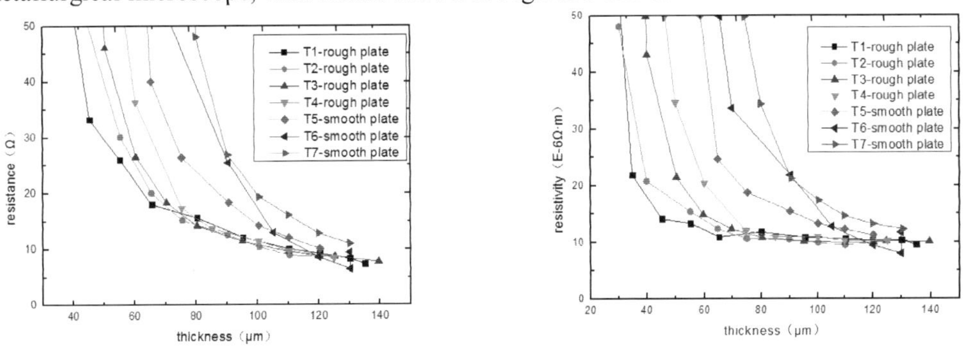

Figure 5. The relationship between material resistivity, resistance, and thickness using φ3.0 mm (thick wire).

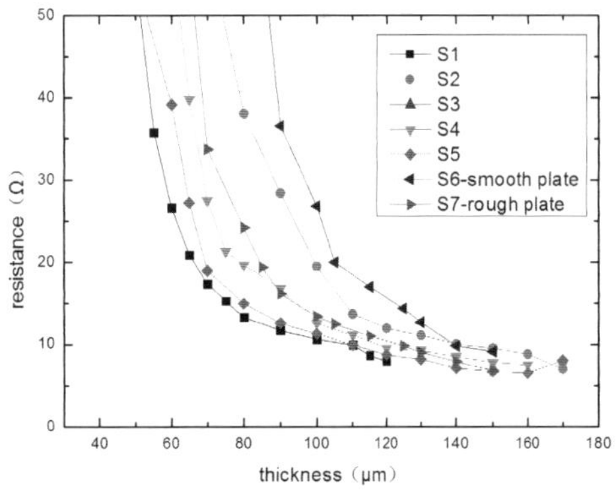

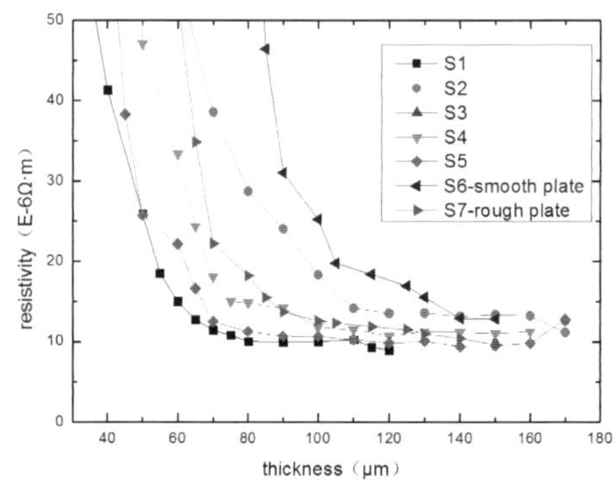

Figure 6. The relationship between material resistivity, resistance, and thickness using φ2.2 mm (thin wire).

Figures 5 and 6 reveal that the resistivity of the coating varies at different thicknesses. This variation is due to the porosity of flame-sprayed coatings and issues with coating deposition efficiency. In the initial stages of spraying, porosity significantly affects the coating's structure and resistance. As the coating thickness increases, the coating becomes denser, and its resistivity decreases. Once the coating reaches 100 μm, its resistivity tends to stabilize, reaching the order of 10-5, and the power density can reach 3 W/cm^2.

Comparing Figures 5 and 6, it can be concluded that the deposition rate of thick wire is faster than that of thin wire. The desired resistivity can be achieved using thick wire for spraying when the coating thickness reaches 120 μm ~140 μm. In contrast, when using thin wire, the coating needs to be deposited to a thickness of over 160 μm to achieve the expected resistivity. However, excessively thick coatings increase the usage of raw materials and impact the bonding strength between the coating and the substrate.

Comparing the spraying data from different batches in Figure 5, it can be concluded that thick wire spraying on rough plates yields more stable results. When spraying with thick wire on rough plates, the resistance-thickness curves largely coincide once the coating exceeds 80 μm. In contrast, when spraying with thick wire on smooth plates, there are variations in the resistance-thickness curves between batches, and the coating deposition is slower. Therefore, using a φ3.0 mm wire for spraying on rough plates results in a more stable and controllable circuit outcome.

4. Experiment with spraying curved composite substrates

Compared to spraying on flat plates, the difficulty of spraying on wing-shaped test pieces is significantly greater. This is due to the varying curvature along the chord and span of the wing, especially at the aerodynamic stagnation point at the leading edge of the wing. Without control methods during spraying, it is easy to have uneven coating, cracks, and peeling in the sprayed heating film. Considering the inconsistent widths, thicknesses, and curvature of the distributed areas in the three-phase heating film of the wing's anti-icing section on the composite substrate, there will be differences in resistance and bonding strength in the heating films formed in the same spraying session. Therefore, this article plans the heating film spraying path based on the curved surface shape on a multi-degree-of-freedom spraying platform, ensuring uniformity in the distance and angle between the spray gun and the curved surface being sprayed. Spray time and temperature are also set to control and monitor the resistance value, uniformity, and bonding condition of each heating film. Spraying is stopped once the resistance of each heating film reaches the target design value. Figure 7 shows the changes in resistance values of each circuit within the wing anti-icing section with the number of spraying sessions.

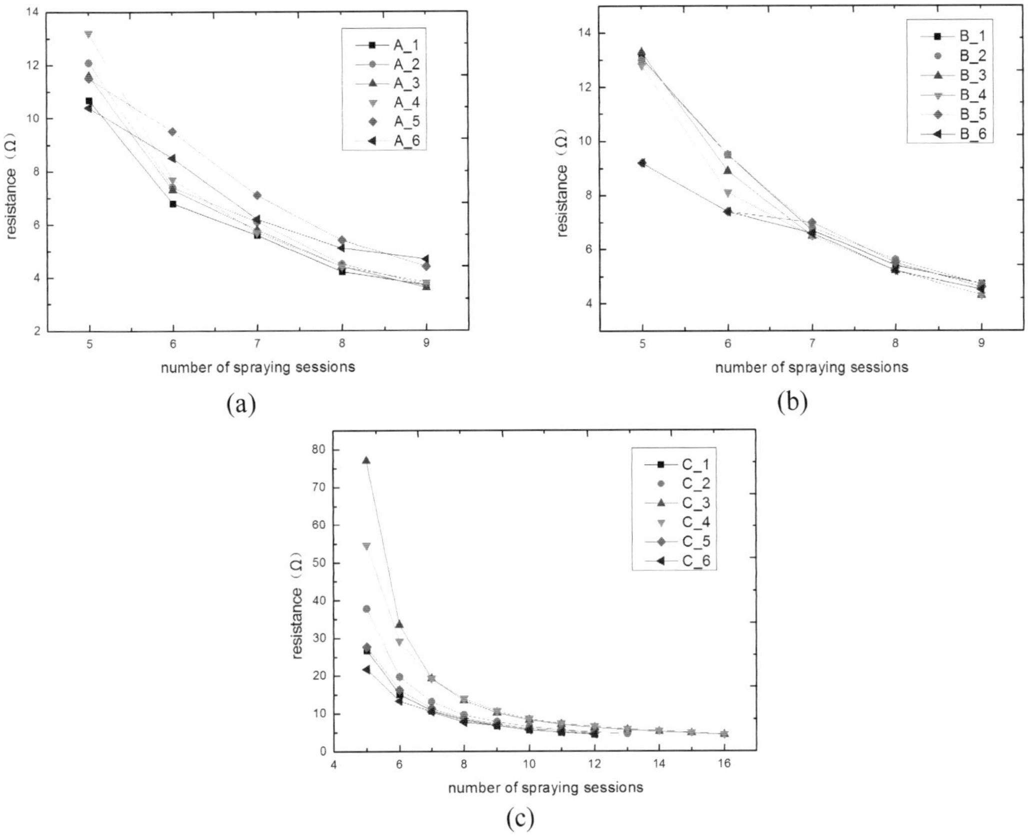

Figure 7. The changes in resistance values of each circuit with the number of spraying sessions in zones A, B, and C.

By comparing the resistance values in zones A, B, and C in Figure 7, it is observed that the resistance in zones A and B reaches the target value first with the same number of spray sessions. In zone C, located at the leading edge's tip, where the surface shape varies significantly, the coating suffers from greater stress, resulting in a lower deposition rate per spray session. Thus, the electrical heating film at the leading edge's stagnation point reaches the target value after 16 spray sessions. This suggests that areas with less curvature exhibit higher coating deposition efficiency, and their resistivity is more likely to stabilize.

5. Conclusion

This article establishes an integrated metal thermal spraying control method for a specific aircraft with a composite material wing leading edge. It proposes a heating film zoning and circuit design method for wing electrical heating, solving the challenge of power adaptive distribution under complex icing meteorological conditions within the full envelope, effectively reducing aircraft energy consumption and decreasing the capacity of the energy distribution system. Experiments and performance impact analysis of metal thermal sprayed electric heating films were conducted. On surfaces with high curvature wing profiles, electrical heating films with good profile adaptability, variable power density, and strong bonding are obtained, solving the challenges of forming and detecting metal electrical heating films sprayed on composite material surfaces. The following conclusions were drawn:

(1) By using manganese-copper alloy wire, coatings with better uniformity and stability can be achieved. By using φ3.0 mm wire for spraying, the expected resistivity can be achieved when the coating deposits to a thickness of 120 μm ~140 μm; with φ2.2 mm wire, the coating needs to deposit to over 160 μm to reach the expected resistivity.

(2) As the coating thickness increases, the coating becomes denser and the resistivity decreases. Once the coating reaches 100 μm, its resistivity tends to stabilize at a level of 10-5 and the power density reaches 3 W/cm^2.

(3) When spraying on surfaces with significant curvature changes, the coating undergoes greater stress and lowers the deposition rate per spray. Therefore, the electric heating film at the leading edge's stagnation point requires precise control and multiple sprays to reach the target resistance value.

References

[1] Khalil, E. E., Said, E., AlSaheh, A., & ElHarriry, G. (2020). Effect of hot air jet arrangement from a piccolo tube in aircraft wing anti-icing system. In AIAA Propulsion and Energy 2020 Forum (p. 3952).
[2] Zhang, K., Li, J., Wang, Y., Lin, C., Zhao, J., & Liu, Y., et al. (2023). Anti-icing of solid surfaces based on droplet dynamics. Modern Physics Letters B, 37(29).
[3] Raj P R L, Jeong H, & Roy R. (2019). Computational Simulation of an Electrically Heated Ice Protection System for Composite Leading Edges of Aircraft. In: SAE International Conference on Icing of Aircraft, Engines, and Structures.
[4] Udayakumar, R., & Shaikh, H. A. (2019). Computational analysis of the Anti-icing system of an aircraft wing. 2019 3rd International Conference on Electronics, Communication and Aerospace Technology (ICECA).
[5] Schfer, A. W. , Barrett, S. R. H. , Doyme, K. , Dray, L. M. , Gnadt, A. R. , Self, R., et al. (2019). Technological, economic, and environmental prospects of all-electric aircraft. Springer Science and Business Media LLC(2).
[6] Bin, W. , Erjun, Y. , Guofei, T. , & Yong, Y. (2018). Design of electrothermal de-icing control system for aircraft wing. Shanxi Electronic Technology.
[7] Li C J. (2018) Applications, Research Progresses and Future Challenges of Thermal Spray Technology. J. Thermal Spray Technology, 4:1-22.
[8] CAAC. (2017) CCAR 25-R4:CCAR Airworthiness Standards: Transport Category Airplanes. CAAC, Beijing.
[9] Myers T G. (2012) Extension to the Messinger Model for Aircraft Icing. J. AIAA Journal, 39:211-218.
[10] Zhu G Y, Li R J, & Zhang D L. (2015) Study on the Optimization of Wing Electric Heating Anti Ice Heating Power Distribution. J. Science and Technology & Innovation, 4:89-90.

Innovative integration of maturity detection and robotic technology for enhanced litchi harvesting

Juntao Xiong*, **Qiyin Huang, Xinjing Jiang and Chengzhuo Zhou**

College of Mathematics and Informatics, South China Agricultural University, Guangzhou, Guangdong Province, CO 510642, China

*Corresponding author's e-mail: xiongjt@scau.edu.cn

Abstract: An integrated approach marries an optimized YOLOv7 deep learning model with a 6-axis robotic arm, featuring a novel end-effector design, to elevate automated litchi harvesting. The model, refined with ACmix attention and Focal-EIoU loss functions, adeptly identifies litchi maturity across diverse orchard settings, overcoming challenges like variable lighting and occlusions. Rigorous field validations and real-world applications demonstrate the system's efficacy, with an 85.7% success rate in precisely harvesting mature litchis. This fusion of cutting-edge image recognition with strategic mechanical innovation effectively mitigates labor shortages and enhances productivity, marking a significant leap forward in orchard automation.

1. Introduction

As a leading litchi producer, China faces significant challenges including underdeveloped orchard infrastructure and labor shortages, emphasizing the need for precision in litchi maturity detection to satisfy market demands and maximize economic returns. Robotic harvesting emerges as a promising solution, yet it is currently limited by the complexity of orchard environments, such as variable lighting and fruit occlusions, which complicates the determination of fruit maturity.

Recent advancements in robotic harvesting technologies have shown potential, yet they predominantly focus on single-fruited species with regular growth patterns, as seen in the designs for cucumbers by Park et al. [1], tomatoes by Jun et al. [2], and dragon fruits by Zhang et al [3]. These studies highlight various end-effector designs for cutting and gripping, with camera systems integrated for precision in targeted fruit picking. However, research on cluster-fruited species like litchi is less common and largely remains in theoretical exploration or experimental stages. The primary challenge lies in the unstructured growth patterns of cluster fruits, which significantly increases the complexity of robotic operations in orchards.

In response to these challenges, this article explores the integration of an advanced deep learning model with robotic harvesting technology, aiming to enhance orchard management and fruit harvesting automation for litchi, a cluster fruit. By adapting neural network-based techniques for the unstructured nature of litchi growth, the study seeks to refine the deep learning model for accurate maturity assessment and facilitate the development of an efficient and reliable robotic harvesting system. This approach aims to bridge the gap in automation technology for cluster fruits and advance the state of robotic harvesting in agriculture.

2. Image acquisition

Litchi images were gathered at Buyun Fruit Farm, Guangzhou, from May 22 to July 17, 2023, under

Journal of Physics: Conference Series **2842** (2024) 012074 doi:10.1088/1742-6596/2842/1/012074

various distances and lighting, using an iPhone 12. A dataset of 1100 selected images, covering different angles and conditions, was compiled for maturity detection, essential for selective harvesting. Maturity classification followed the criteria by Huang et al [4].

3. Methods

3.1 End-effector design method for litchi harvesting

The development of a specialized end-effector is critical for enhancing the efficiency of robotic litchi harvesting systems. This project introduces a comb-cutting mechanism inspired by human hair grooming, tailored for the unique structure of litchis. Based on a mechanical damage model for litchis, a three-dimensional model was crafted using SolidWorks, with further refinement through HyperMesh and LS-PrePost for the comb-cutting design. The structural design is shown in Figure 1.

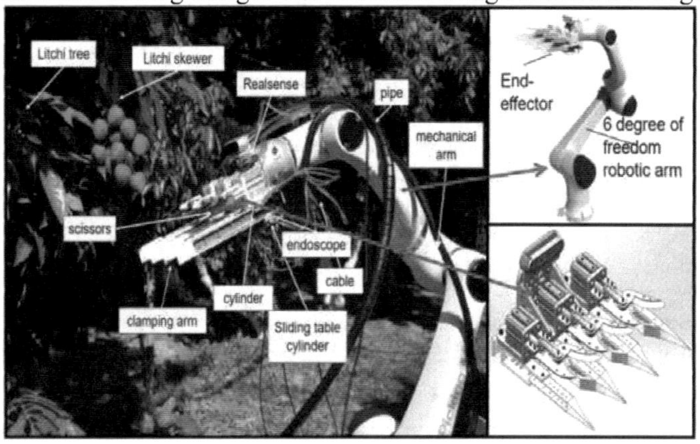

Figure 1. Design of Harvesting Robot's Manipulator.

The design process involved analyzing various comb-tooth configurations and material grips to optimize the interaction with litchi stems, using the Finite Element Method (FEM) to evaluate their impact. This led to an optimized comb-cutting mechanism that allows precise and gentle harvesting of litchis, ensuring minimal damage to the fruit and extending its freshness. A simulation of the picking process is shown in Figure 2.

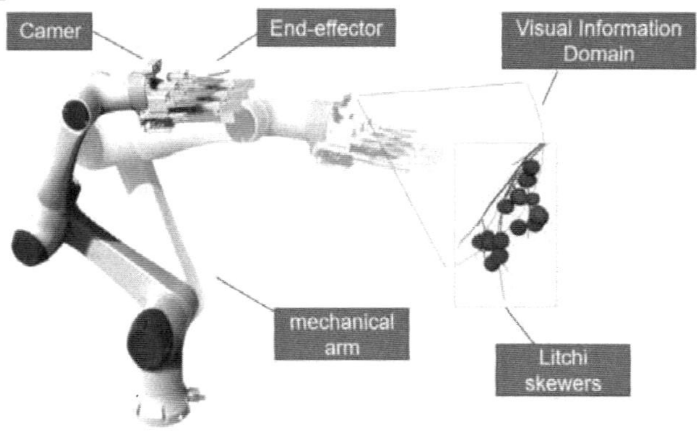

Figure 2. Schematic diagram of the robot picking process.

3.2 Object detection method

YOLOv7, highlighted by Wang et al. [5], excels in real-time object detection with its blend of speed and accuracy on a GPU V100, surpassing 30 frames per second. Its architecture comprises Input, Backbone, Neck, and Head segments. The Input handles preliminary image processing and augmentation, the Backbone extracts features using E-ELAN and SPPCSPC modules, the Neck fuses these features, and the Head utilizes them for detection, improving performance and generalization. The enhanced YOLOv7 model is depicted in Figure 3.

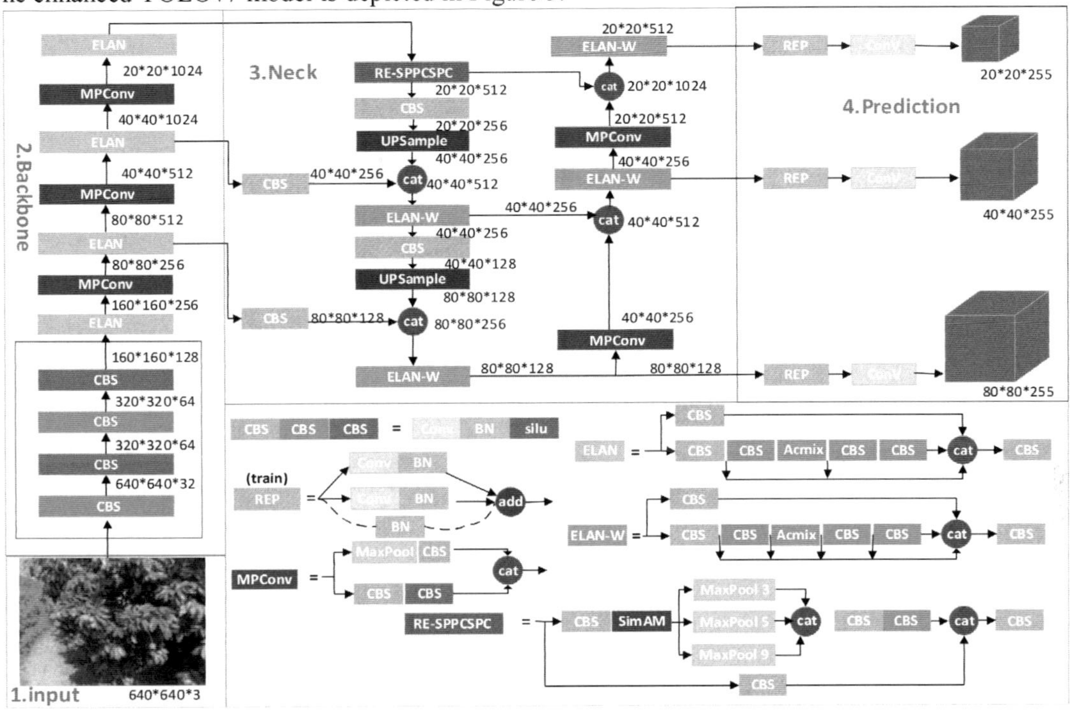

Figure 3. The diagram of the improved YOLOv7 model structure.

3.3 *Integration method for litchi picking robot deployment*

In deploying the model for a litchi picking robot, the system incorporates a primary field of view for maturity detection and an endoscopic smaller field for pinpointing the picking spot. The transformation matrices are essential in bridging the visual perception and the mechanical action, enabling accurate alignment and efficient picking. The transformation sequence is as follows:

$$B^T = B_{c2}^T \cdot C_{H2}^T \cdot H_P^T \cdot P_{c1}^T \cdot C_1^T \quad (1)$$

where B^T is the base to target transformation, B_{c2}^T is the transformation from the base to camera 2, C_{H2}^T the transformation from camera 2 to the endoscope, H_P^T from the endoscope to the picking point, P_{c1}^T the picking point to camera 1, and C_1^T camera 1 to the target. This sequence ensures that the robotic arm is accurately guided from the primary detection to the precise picking point within the orchard's environment.

The transformation from the camera to the picking arm and subsequently to the picking point is crucial for the robot's operation. The optimization of these transformation matrices enhances the system's efficiency, reducing the time and mechanical complexity involved in reaching the harvesting point. This is achieved through a series of rotations:

$$R = R_Z(\psi) \cdot R_Y(\theta) \cdot R_X(\phi) \quad (2)$$

where R represents the combined rotation matrix, and $R_Z(\psi)$, $R_Y(\theta)$, and $R_X(\phi)$are the rotations around the Z, Y, and X axes, respectively. The system dynamically calculates these rotations to ensure the picking arm is positioned accurately for litchi harvesting.

As shown in Figure 4, these transformations form the core of the robotic system's method to effectively transition from visual detection to physical picking, encapsulating both precision and adaptability in the automated harvesting process.

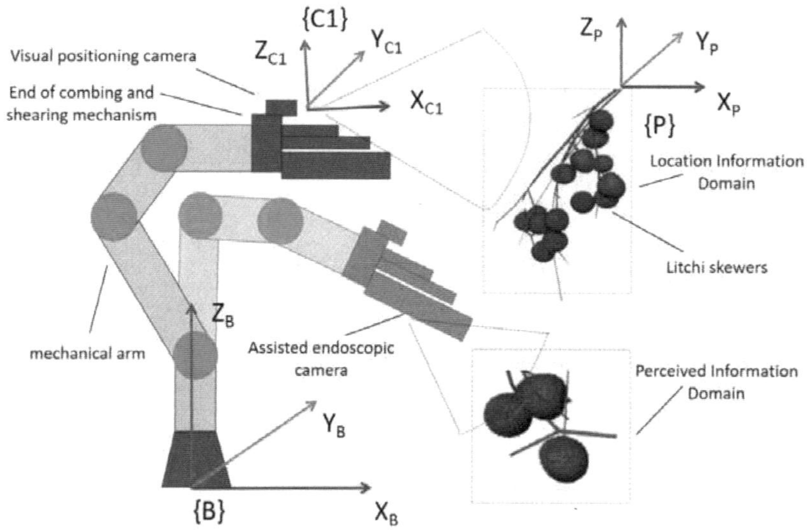

Figure 4. Schematic diagram of spatial position relationship in multi vision system.

4. Experiment results & discussion

4.1. Experimental environment

The experiments involved a dual-setup: algorithm testing on a Windows 10 system with Pytorch 1.9.0+cu111, Python 3.8.13, Nvidia GeForce GTX3090 GPU, and i7-11700K CPU; and robotic harvesting trials using a Han's Robot 6-axis picking robot, designed for precision tasks like fruit picking. This setup ensured a comprehensive evaluation of the deep learning model and its application in automated litchi harvesting.

4.2. Evaluation metrics

The algorithm's performance is evaluated using precision (P), recall (R), and mean average precision (mAP), with mAP encompassing all class labels. The relevant formulas are: The performance of the algorithm is quantified through established metrics: precision (P), recall (R), mean average precision (mAP), and the harvesting success rate (HSR). These metrics are defined as follows:

$$P = \frac{TP}{TP + FP} \times 100\% \tag{3}$$

$$R = \frac{TP}{TP + FN} \times 100\% \tag{4}$$

$$AP = \int_0^1 P(R)dR \tag{5}$$

$$mAP = \frac{1}{n}\sum_{i=0}^{n} AP_i \tag{6}$$

where TP denotes correctly predicted positive samples, FN denotes positive samples incorrectly labeled as negative, and FP denotes negative samples incorrectly labeled as positive. Average Precision (AP) is the area under the Precision-Recall curve, calculated for each class, and mAP averages these across all classes. A higher mAP indicates superior model performance, offering a comprehensive assessment across various categories.

Separately, the harvesting success rate (HSR) is introduced to measure the robotic system's performance, defined as the ratio of successful picks to total attempts by the robot. This metric specifically evaluates the practical effectiveness of the robotic arm in the field.

4.3 Experimental results and analysis

In the conducted experiments, the YOLOv7 model enhanced with Focal-EIoU loss function exhibited superior performance, demonstrating notable advancements in precision, recall, and mean average precision (mAP) across various configurations and when compared to other detection models. The detailed comparisons are encapsulated in the consolidated table below, highlighting the model's efficacy in detecting small objects like litchi fruits and its superiority over other models such as Faster R-CNN and YOLOv5.

Table 1. Comparative Performance of YOLOv7 with Different Loss Functions and Detection Models.

Model	Loss Function	Precision /%	Recall /%	mAP /%
YOLOv7+CIoU	CIoU	74.6	79.7	84.0
YOLOv7+GioU	GioU	78.7	77.8	86.6
YOLOv7+EioU	EIoU	73.8	79.7	84.2
YOLOv7+Focal-EIoU	Focal-EIoU	80.9	80.2	87.6
Faster R-CNN	Focal-EIoU	72.7	70.6	76.9
YOLOv5	Focal-EIoU	79.0	70.8	81.1
Improved YOLOv7 (Ours)	**Focal-EIoU**	**88.3**	**90.5**	**93.9**

Table 1 demonstrates the enhanced YOLOv7 model's significant improvements, particularly with the Focal-EIoU loss function, achieving the highest precision, recall, and mAP values. The improved YOLOv7 model outperforms the original YOLOv7, Faster R-CNN, and YOLOv5 models, underscoring its potential to efficiently detect litchi fruit maturity for automated harvesting applications.

The enhanced YOLOv7 model was rigorously tested in litchi orchards, encountering challenges like varied lighting and occlusions. Figure 5 illustrates the model's reliable maturity detection across these conditions, affirming its robustness and real-world applicability.

Figure 5. Litchi maturity detection under different lighting conditions.

Alongside detection tests, a harvesting experiment using a Han's Robot 6-axis arm, integrated with the YOLOv7 model, was conducted to evaluate its precision in picking mature litchis. Figure 6 displays the robotic arm in action, achieving an 85.7% harvesting success rate, demonstrating its effectiveness in real orchard conditions.

Figure 6. Real-world harvesting scenario with the robotic arm in action.

Significant improvements in precision and recall, up by 7.4% and 10.3% from the original YOLOv7, underscore the model's enhanced litchi maturity detection. The Focal-EIoU loss function's integration and the robotic arm's high HSR confirm the system's efficacy in orchard settings, showcasing the potential for automated litchi harvesting.

5. Conclusion

The integration of an advanced YOLOv7 deep learning model with a 6-axis robotic arm has proven to be a substantial step forward in addressing the complexities of automated litchi harvesting. Through rigorous field testing and practical application, the system demonstrated a high success rate, adeptly navigating the variable conditions of orchard environments. The study's exploration into adapting neural network-based techniques to the unstructured nature of litchi growth has paved the way for a refined approach to robotic harvesting. This not only offers a solution to the pressing labor challenges in agriculture but also enhances the efficiency and productivity of orchard management, embodying the potential for broader applications in agricultural automation.

Acknowledgments

This work is funded by the National Natural Science Foundation of China (Project No. 32071912), Natural Science Foundation of Guangdong Province (Project No. 2024A1515010226), Guangdong S&T programme (Project No. 2023A0505050130), the open competition program of the top ten critical priorities of Agricultural Science and Technology Innovation for the 14th Five-Year Plan of Guangdong Province (2022SDZG03), Special Funds for the Cultivation of Guangdong College Students' Scientific and Technological Innovation. ("Climbing Program" Special Funds.) (pdjh2023a0075).

References

[1] Park, Y., Seol, J., Pak, J., Jo, Y., Kim, C., Son, H. I. Human-centered approach for an efficient cucumber harvesting robot system: Harvest ordering, visual servoing, and end-effector [J]. Computers and Electronics in Agriculture, 2023, 212: 108116.

[2] Jun, J., Kim, J., Seol, J., Kim, J., Son, H. I. Towards an efficient tomato harvesting robot: 3d perception, manipulation, and end-effector [J]. IEEE Access, 2021, 9: 17631-17640.

[3] Zhang, Y. Y., Tian, J. Q., Wang, W. X., et al. Design and Experiment of the End-Effector for Pitaya Harvesting Robot [J]. Journal of Forestry Engineering, 2023, 8(04): 144-150.

[4] Huang X K, Song C H, Li X C. Study on Maturing of Litchi Fruit [J]. South China Fruits, 1996(02):38-39.

[5] Wang C Y, Bochkovskiy A, Liao H Y M. YOLOv7: Trainable bag-of-freebies sets new state-of-the-art for real-time object detectors [J]. arXiv preprint arXiv: 2207.02696, 2022.

Journal of Physics: Conference Series **2842** (2024) 012075 doi:10.1088/1742-6596/2842/1/012075

A model-adaptive compensation algorithm for additive-subtractive hybrid manufacturing

Yin Wang *, **Yukai Chen, Yu Lu, Junyao Wang and Qi Zhang**

School of Mechanical Engineering, Xi'an Jiaotong University, Xi'an, Shaanxi, 710049

*Corresponding author's e-mail: kingyin@stu.xjtu.edu.cn.

Abstract. Additive-subtractive hybrid manufacturing (ASHM) is a manufacturing method that combines additive manufacturing (AM) with subtractive manufacturing (SM) techniques. Given that the surface accuracy of ASHM components primarily relies on numerical control (NC) machining, it is imperative to incorporate model compensation to preserve finishing allowances, thereby obtaining additive manufacturing (AM) models from design models. This paper proposes a model adaptive compensation algorithm capable of dynamically adjusting the design model based on both model characteristics and the AM process. This algorithm not only ensures the retention of suitable finishing allowances but also addresses part size errors arising from the AM process. Comparative analysis against traditional CAD reconstruction methods highlights the algorithm's superior attributes in terms of accuracy, efficiency, and adaptability when confronted with intricate model compensation scenarios.

1. Introduction

Additive-subtractive hybrid manufacturing (ASHM) represents a significant advancement in manufacturing technology, blending the strengths of additive manufacturing's (AM) unparalleled design flexibility with the precise surface finish capabilities of numerical control (NC) machining. By mitigating the inherent "staircase effect" of AM [1], ASHM elevates the surface precision of AM components and enhances tool accessibility [2,3]. This hybrid approach offers a compelling solution for the precision manufacture of features with intricate internal geometries, a task traditionally challenging for conventional NC machining. As a result, ASHM has attracted considerable attention from both industrial practitioners and academic researchers alike [4].

Given that the surface accuracy of ASHM components heavily relies on NC machining, there arises a necessity to incorporate finishing allowances into AM models based on the original design specifications. While model compensation is typically achieved through computer-aided design (CAD) software, this process becomes burdensome and inadequate in addressing part-size discrepancies arising from the additive manufacturing process, particularly when dealing with intricate structures. Consequently, there is a pressing need to develop a model-adaptive compensation (MAC) algorithm to streamline this allowance provision process, offering enhancements in terms of time efficiency, effectiveness, and adaptability.

2. The principles of MAC

2.1. The basic principles of the MAC

In accordance with the characteristics of numerical control machining, the rules for MAC should entail inward contour contraction and outward contour expansion (as illustrated in Figure 1). Considering that the precision machining allowance is generally small, and there exists thermal deformation during the AM process, it is necessary to adaptively adjust the compensation values for different regions based on the characteristics of AM processes (Section 2.2 for details).

To facilitate the algorithm's implementation for model processing, STL models were chosen as the input for the MAC algorithm. STL models approximate the geometric shape of three-dimensional objects using multiple triangular facets. Due to their widespread support by various software applications, STL models are extensively utilized in AM, CAM, and related fields [5]. In Figure 2, the storage information in STL models includes the coordinates of three vertices of each triangular facet and their corresponding normal vectors. From the aforementioned content, it is evident that STL models not only facilitate the use of algorithms for model processing but also enable the direct utilization of compensated models in AM processes.

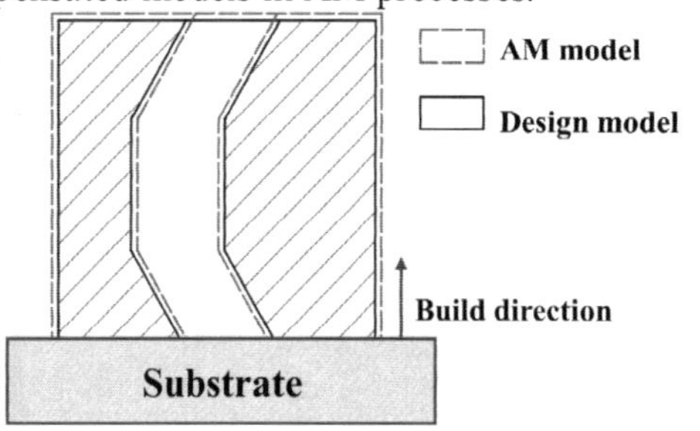

Figure 1. The rule for MAC.

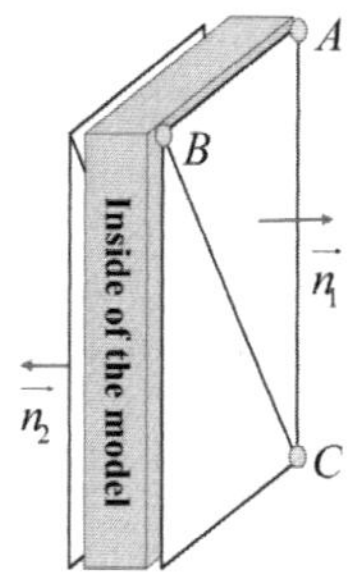

Figure 2. STL model.

The essence of the MAC lies in the transformation of the coordinates of individual vertices of triangular facets. In the STL model, each vertex is shared by multiple triangular facets, which are distributed in space on one or more planes. The different spatial configurations determine the transformation rules at the vertex. In order to facilitate the subsequent theory explanation, introducing the concept of an independent triangular facet is significant.

Independent Triangular Facet (ITF): It is the triangular facet with different normal vectors at a vertex.

Assuming a vertex's coordinates are $M_0 = (x_0, y_0, z_0)$ before transformation and M after transformation, their respective unit normal vectors are $\vec{n_i} = (A_i, B_i, C_i), i = 1...n$, and each triangular facet is translated along its unit normal vector by a distance d. Based on the different numbers of triangular facets in ITF, the vertex's coordinate conversion rules could be categorized into different cases:

(1) Number of ITF = 1.

As depicted in Figure 3a, triangular facets passing M_0 are on one plane, and the transformation rule M_0 is to translate M_0 along the $\vec{n_1}$ by distance d.

$$M = M_0 + \vec{n_1} \cdot d = (x_0 + dA_1, y_0 + dB_1, z_0 + dC_1) \quad (1)$$

(2) Number of ITF = 2.

As depicted in Figure 3b, triangular facets passing M_0 are on two planes, and the transformation rule M_0 is to translate M_0 along the vector $\vec{n}$, which is the resultant vector of $\vec{n_1}$ and $\vec{n_2}$ by distance D, as given by Equation (2).

$$D = d / \cos\langle \vec{n}, \vec{n_1} \rangle \quad (2)$$

The calculation formula M is shown in Equation (3).

$$M = M_0 + \vec{n} \cdot D = M_0 + \vec{n} \cdot d / \cos\left\langle \vec{n}, \vec{n_1} \right\rangle \tag{3}$$

(3) Number of ITF ≥ 3.

As depicted in Figure 3c, triangular facets passing M_0 are on three or more planes, and M coordinates can be solved by solving the intersection point of a multi-plane.

Assuming the i - ITF equation is:

$$P_i : A_i(x - x_0) + B_i(y - y_0) + C_i(z - z_0) = 0 \tag{4}$$

Shifting the ITF along the respective normal direction in distance d, the M_0 coordinates on each ITF are:

$$M_i = (x_0 + dA_i, y_0 + dB_i, z_0 + dC_i), i = 1, 2, ..., n \tag{5}$$

Each ITF equation after translation is:

$$P_i' : A_i x + B_i y + C_i z = A_i(x_0 + dA_i) + B_i(y_0 + dB_i) + C_i(z_0 + dC_i), i = 1, 2, ..., n \tag{6}$$

In the event of the number of ITF >3, the process of solving the M point can be converted to the least square solution of the overdetermined equation.

$$\arg\ \min \left\| Am - b \right\| \tag{7}$$

where:

$$A = \begin{pmatrix} A_1 & B_1 & C_1 \\ \vdots & \ddots & \vdots \\ A_n & B_n & C_n \end{pmatrix} \tag{8}$$

$$\mathrm{m} = (x, y, z)^T \tag{9}$$

$$b = \begin{pmatrix} A_1(x_0 + dA_1) + B_1(y_0 + dB_1) + C_1(z_0 + dC_1) \\ \vdots \\ A_n(x_0 + dA_n) + B_n(y_0 + dB_n) + C_n(z_0 + dC_n) \end{pmatrix} \tag{10}$$

Several techniques exist for resolving the least squares solution to systems of equations that are over-determined. The current study utilizes the Singular Value Decomposition (SVD) technique for this purpose. The foundational concepts and the resolution process are outlined without redundant explanations.

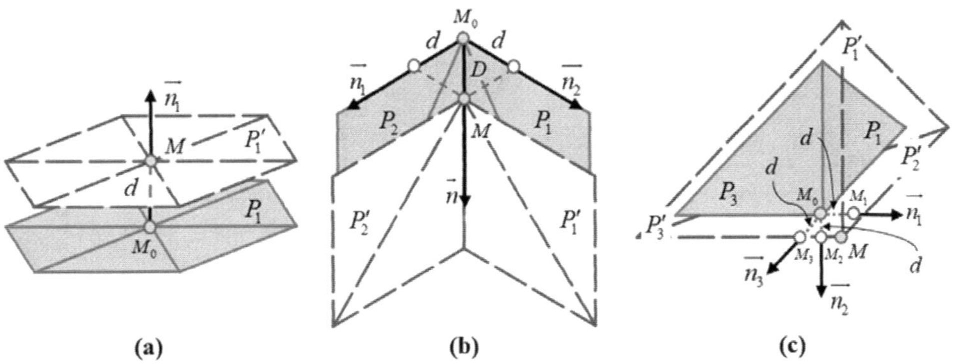

Figure 3. Coordinate conversion rules in different cases: (a) Number of ITF = 1; (b) Number of ITF = 2; (c) Number of ITF ≥ 3.

Through the aforementioned principles, the coordinates M of each compensated vertex can be obtained, and then the transformed unit normal vectors of each triangular facet can be obtained through vector cross multiplication. By outputting the vertices and normal vector of each conversion triangular facet based on the rules in STL format, the compensation model can be gained.

For the number of ITF ≥ 3, the error between the master model and compensation model can be assessed using the fundamental least squares error assessment formula.

$$e = \|AM - b\|^2 \tag{11}$$

Equation (11) can evaluate the degree of dispersion between M and exact point M_i on each ITF. The value of e is smaller, indicating that the point M is closer to ITF, revealing a better compensation effect. When $e = 0$, it represents an exact solution.

2.2. Adaptive rule

In AM processes, residual stresses induced by energy input can lead to deviations between the master model and the actual part [6]. AM is a near-net forming process, so in ASHM, the removal margin of SM is small. If such deviations exceed the finishing allowance, it would be detrimental to the SM process. Therefore, in addition to reserving finishing allowances, the MAC algorithm is also required to compensate for manufacturing deviations in the AM process, which are influenced by factors such as the process parameters and material parameters [7]. The influence of residual stresses on deformation is predominantly influenced by the geometry's angular features and the dimensions of its projections [8,9]. Moreover, it is crucial to consider the thickness of the ineffective regions [10] and the errors introduced by the manufacturing equipment itself.

Given the challenges in directly quantifying the global deformation of a model due to residual stresses in AM, this study introduces an experimental approach to mitigate such deformations. The technique entails fabricating models at varied orientations and assessing the length and angle discrepancies between the functional border, excluding the presence of unsintered powder and the intended master border. These measurements are subsequently translated into dimensional and angular corrections. These corrected values function as the adaptive criteria for the MAC algorithm. Importantly, this adaptive criteria is dependent solely on the equipment, material, and manufacturing process, not on the model's specific architecture, thereby enhancing its generalizability. The method not only corrects in-process deformations but also nullifies the effects of inadequately fused regions.

In this study, the LPBF (Laser Powder Bed Fusion) process with Inconel 718 powder was selected. The AM process parameters were fixed as follows: laser output of 190 W, traverse velocity of 1000 mm/s, and layer spacing of 0.08 mm. The Z-axis positive direction was chosen as the vertical building direction. The angle between the Z-axis and the triangular facet normal was termed the triangular facet angle φ. Parts with different tilt angles were printed at 15° intervals with a base length of 2 mm. After embedding, grinding, and polishing each part, the actual physical sizes of the models at various tilt angles were observed under an optical microscope, as shown in Figure 4.

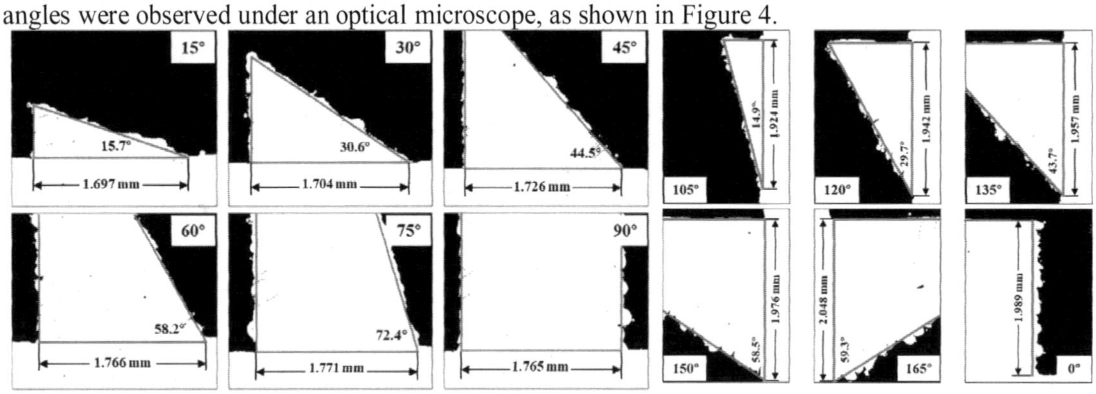

Figure 4. Actual physical dimensions of models at varying pitch angles.

By calculation, the deviation in length $E_L(\varphi)$ and angle deviation $E_\varphi(\varphi)$ at varying pitch angles were determined, as shown in Table 1.

Table 1. $E_L(\varphi)$ & $E_\varphi(\varphi)$ at different tilt angles.

φ (°)	15	30	45	60	75	90
$E_L(\varphi)$ (mm)	0.1855	0.1785	0.1565	0.1165	0.1115	0.1175

$E_\varphi(\varphi)\,(^\circ)$	-0.7	-0.6	0.5	1.8	2.1	0
$\varphi\,(^\circ)$	105	120	135	150	165	0
$E_L(\varphi)\,(\text{mm})$	0.065	0.047	0.032	0.012	-0.059	0.011
$E_\varphi(\varphi)\,(^\circ)$	0.1	0.3	1.3	1.5	14.7	0

Considering that the compensation rule for triangular facets involves translating along their respective normal vectors by a certain distance, when exploring adaptive rules, it is necessary to convert length deviation $E_L(\varphi)$ and angle deviation $E_\varphi(\varphi)$ into distance deviation $d_L(\varphi)$ and corrected tilt angles $\theta_\varphi(\varphi)$. Since the overall angle deviation $E_\varphi(\varphi)$ values are relatively small, according to geometric principles, the formulas for calculating distance deviation $d_L(\varphi)$ and corrected tilt angles $\theta_\varphi(\varphi)$ can be approximated as follows:

$$d_L(\varphi) = \begin{cases} E_L(\varphi)\cdot\sin\varphi, \ 0^\circ < \varphi \le 90^\circ \\ -E_L(\varphi)\cdot\cos\varphi, \ 90^\circ < \varphi \le 150^\circ \ or \ \varphi=0^\circ \end{cases} \tag{12}$$

$$\theta_\varphi(\varphi) = \varphi + E_\varphi(\varphi), \ 0^\circ \le \varphi \le 150^\circ \tag{13}$$

By substituting the data from Table 1 into Equations (12) and (13) and fitting, the adaptive rules can be determined. These adaptive rules can compensate for model deviations during the AM process. Utilizing the adaptive rules and the MAC algorithm described, each triangular facet in the master model translates $d_L(\varphi)+d_{FM}$ along the direction of the corrected tilt angle $\theta_\varphi(\varphi)$, where d_{FM} is the finishing allowance, resulting in an AM model suitable for ASHM.

3. The effectiveness of the model adaptive compensation.

To demonstrate the utility of the MAC algorithm in real-world scenarios, a truncated cone model, as shown in Figure 5a, is designed. A finishing allowance $d_{FM} = 0.15\,mm$ is selected, and two sets of prints are made for each of the following cases: only finishing compensation, MAC, and the master model. As depicted in Figure 5b, precision machining tool paths derived from the intended design dimensions are applied to all three models. It is evident from Figure 5b that when only the allowable tolerance compensation is applied, the presence of length and angle deviations can result in undercutting in certain areas. When the model is printed without any compensation, the entire part is slightly smaller than intended, rendering it unable to be accurately machined as a whole. However, when the MAC is applied to the model, the part can be machined with high precision across its entire surface. Consequently, the MAC algorithm significantly enhances the surface finish and dimensional control of components in ASHM.

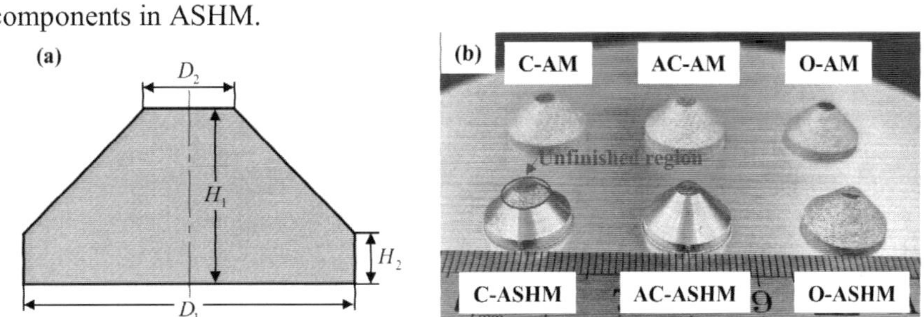

Figure 5. MAC result: (a) master model; (b) AM and ASHM entire part.

The MAC algorithm for the model is equally applicable, accurate, and fast when dealing with more complex models. Figure 6 shows the model results and error scatter plots after adaptive compensation for several complex models. In Figure 6, the quantity of triangular facets within the design is denoted by N, while T signifies the duration required for the model's transformation process. It can be seen from Figure 6 that when dealing with complex models, the adaptive compensation algorithm for the model can also complete global adaptive compensation relatively quickly, and the least squares errors of most facets are small, indicating good algorithm performance.

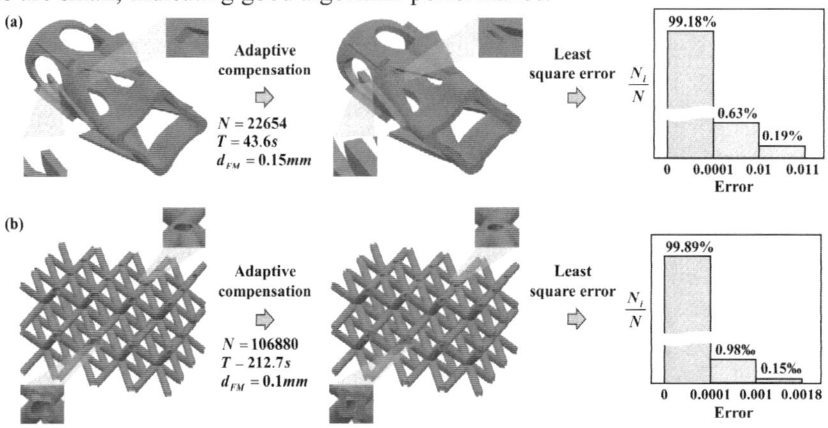

Figure 6. Effect of adaptive compensation algorithm in complex model.

4. Conclusion.

This paper proposes a MAC algorithm for ASHM, which can adaptively compensate design models based on model characteristics and AM process parameters to achieve precision machining allowance reservation and compensation for manufacturing deviations during the AM process. Experimental results demonstrate that the MAC algorithm is equally applicable, accurate, and fast when dealing with complex models. Additionally, it contributes to improving the surface finish and dimensional accuracy of components in ASHM.

References

[1] Du W, Bai Q, Zhang B (2016) A Novel Method for Additive/Subtractive Hybrid Manufacturing of Metallic Parts. 44th Proceedings of the North American Manufacturing. 5: 1018–1030.
[2] Zhu Z, Dhokia V G, Nassehi A (2013) A review of hybrid manufacturing processes - state of the art and future perspectives. Int. J. Comput. Integ. M. 26: 596-615.
[3] Lauwers B, Klocke F, Klink A (2014) Hybrid processes in manufacturing. CIRP Ann.- Manuf. Technol. 63: 561-583.
[4] Flynn J M, Shokrani A, Newman S T (2016) Hybrid additive and subtractive machine tools – Research and industrial developments. Int. J. Mach. Tool. Manu. 101: 79-101.
[5] Stroud I, Xirouchakis P C (2000) STL and extensions. Adv. Eng. Softw. 31: 83–95.
[6] Feldhausen T, Heinrich L, Saleeby K (2022) Review of Computer-Aided Manufacturing (CAM) strategies for hybrid directed energy deposition. Addit. Manuf. 56: 102900.
[7] Fang Z C, Wu Z L, Huang C G. (2020) Review on residual stress in selective laser melting additive manufacturing of alloy parts. Opt. Laser Technol. 129: 106283.
[8] Vastola G, Sin W J, Sun C N (2022) Design guidelines for suppressing distortion and buckling in metallic thin-wall structures built by powder-bed fusion additive manufacturing. Mater. Design. 215: 110489.
[9] Hong R C, Zhang L, Lifton J (2021) Artificial neural network-based geometry compensation to improve the printing accuracy of selective laser melting fabricated sub-millimetre overhang trusses. Addit. Manuf. 37: 101594.
[10] Geng K, Yang Y F, Li S F (2020) Enabling high-performance 3D printing of Al powder by decorating with high laser absorbing Co phase, Addit. Manuf. 32: 101012.

Shape and roundness accuracy prediction of the roll grinding based on improved Elman neural network

Hui-Min Gao, Xin-Yi Ren*, Hai-Wei Xu, Jiao-Jian Ai and Min Guo

Shougang Jingtang Iron and Steel Co., Ltd, Tangshan 063200, China

*Corresponding author's email: 45211781@qq.com

Abstract: Due to the strong nonlinear relationship between the shape and roundness of the roller and the grinding process parameters, a prediction method for roll shape error and roundness error based on an improved Elman neural network is proposed. During the training process of the Elman neural network, Sine chaotic mapping is used to make the population distribution more uniform, and the sparrow search algorithm is used to obtain the optimal parameters of the Elman neural network. The verification results show that the prediction errors of the roll shape and roundness are both less than 10%, and the accuracy of the prediction model is higher than 85%, which proves the effectiveness of the proposed method.

1. Introduction

The roll is an important component that causes continuous plastic deformation of the strip [1]. The shape and roundness accuracy are important indicators of the roller grinding quality, which directly affect the flatness and surface quality of the rolled products [2]. In order to ensure the roller grinding quality, it is necessary to predict the roll shape and roundness accuracy before grinding to assist the operator in setting process parameters reasonably. Therefore, the prediction accuracy of the roll shape and roundness is an important task.

Currently, there have been many studies on roll shape and roundness prediction accuracy. Xie [3] established a one-dimensional model of channel roundness under different process parameters based on a large number of grinding experiments. Lu et al. [4] established a response surface model to analyze the interactive effects of process parameters on grinding geometric accuracy. Zhu et al. [5] conducted experiments on GS-C25 cast steel material and analyzed the influence of grinding process parameters on shape accuracy using the Taguchi method. Zhang et al. [6] established the functional relationship between grinding force and positional error using least squares regression, and proposed the best grinding parameters to minimize positional error. However, the traditional regression methods mentioned above make it difficult to express the relationship between process parameters and shape and roundness accuracy because of the complex characteristics of roller grinding.

In this paper, a prediction method for roll shape and roundness accuracy based on an improved Elman neural network is proposed, which can solve the problem of difficult fitting of the nonlinear relationship between process parameters and roll shape as well as roundness accuracy, so as to accurately predict the roll shape and roundness before grinding.

2. Relationship between process parameters and roll shape and roundness accuracy

2.1 Test platform and data collection

The grinding platform of the MK8445 fully automatic CNC roller grinder is shown in Figure 1. The roller material is 92CrMo, and its body length and diameter are 1440 mm and 446 mm.

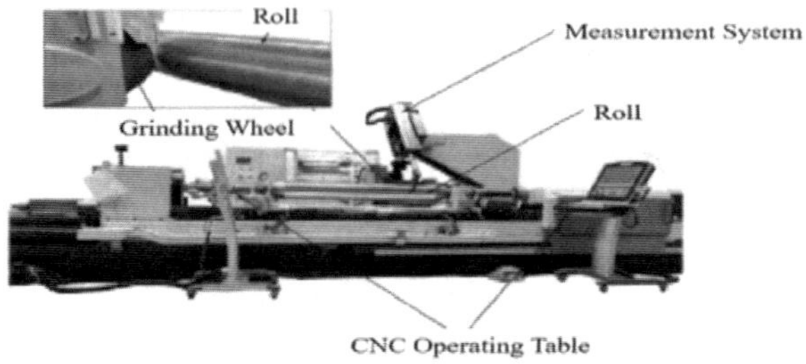

Figure 1. MK8445 roll grinder.

The roll shape and roundness are measured using the measurement system equipped with the roll grinder. The distribution of measuring points is shown in Figure 2.

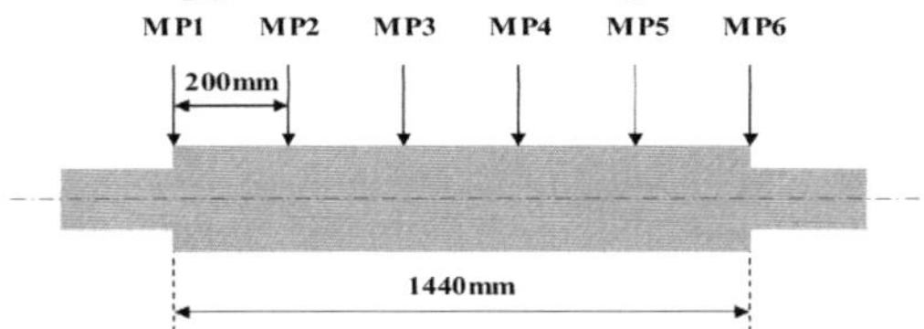

Figure 2. Distribution of roundness measurement points.

2.2 Single factor experiment on the influence of parameters on shape and position accuracy

A single factor grinding experiment is conducted to analyze the influence of grinding parameters including grinding depth d, wheel speed v_s, roll speed v_w, and roll feed speed f_s on the roll shape and roundness accuracy. The experimental results are shown in Figure 3.

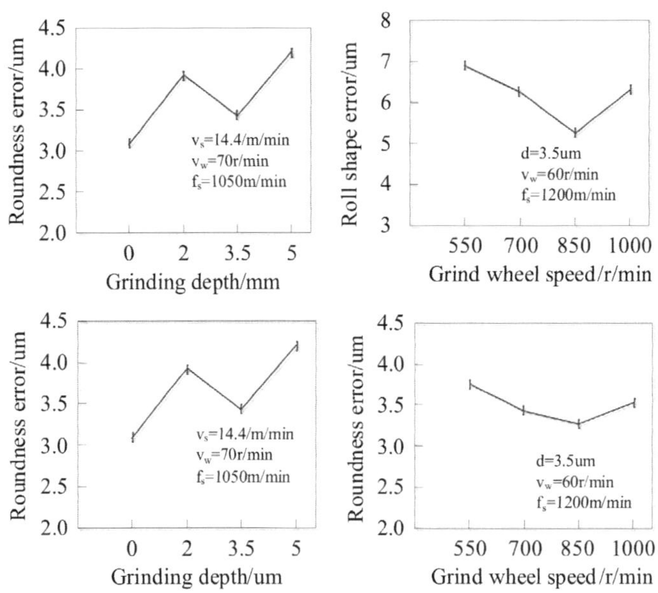

Figure 3. Single factor influence of various process parameters on roll accuracy.

As can be seen from Figure 3, there is no obvious linear relationship between grinding parameters, roll shape error and roundness error. The correction determination coefficient of the second-order response surface model is 0.624, which indicates that the model fitting degree is poor, and the interaction and quadratic term of grinding parameters are not significant.

3. Improvement of elman neural network strategy and experimental design

3.1 Standard elman neural network

Elman neural network is a dynamic recursive neural network and has a network structure of four or more layers. The input layer unit plays the role of signal transmission, the output layer unit plays the role of linear weighting, the hidden layer unit is generally a nonlinear activation function, and the receiving layer is used to remember the output value of the hidden layer at the previous moment. The topology of the Elman neural network is shown in Figure 4.

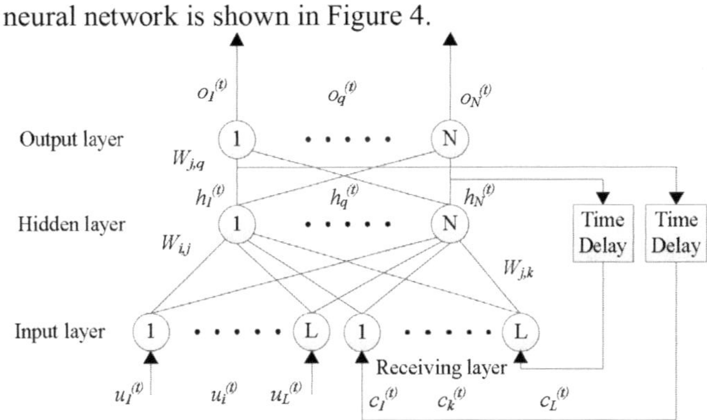

Figure 4. Topological structure of Elman neural network.

3.2 Elman neural network improvement strategy

Elman neural network has a high degree of non-linearity and can accurately approximate any nonlinear function [7]. However, the Elman neural network also has some problems such as being trapped in local minima and premature selection of initial thresholds and weights [8]. To address the problems, Sine chaotic mapping and sparrow search algorithms are used to optimize the Elman neural network (Sine-SSA-Elman).

1) Chaotic mapping

The chaotic system is a system that cannot judge the future state according to the initial conditions, and it is a combination of order and disorder [9]. Chaotic mapping is used to get the initial value of the system, and the chaotic performance of the system can be changed by adjusting the control parameters. Sine chaotic mapping is used to initialize the individual positions of sparrow search algorithms, which can enrich population diversity and solve the problems of uneven population distribution and insufficient search space.

The mathematical model formula of Sine chaotic mapping is

$$\begin{cases} x_{n+1} = S(x_n) = \mu sin(\pi x_n) \\ x_0 \in (0,1) \end{cases} \tag{1}$$

where x_n is the system input; x_{n+1} is the system output; μ is the control parameter. The closer μ is to 1, the better the chaotic performance of the system; x_0 is not equal to 0 and 1, in order to avoid zero points and constant points.

(2) Group optimization algorithm

The regression performance of Elman neural networks is influenced by initial weights and thresholds, and the selection of initial values largely determines the regression rate and accuracy. Therefore, the Sparrow Search Algorithm [10] (SSA) is used to optimize the weights and thresholds from the input layer to the hidden layer of the Elman neural network in order to solve the optimal solution of the fitness function and obtain the optimal parameters of the Elman neural network, thereby improving its regression performance.

The SSA processing flowchart is shown in Figure 5. The discoverer is the individual with higher fitness in the early stage of the search, and the joiner is another individual in the early stage of the search. As the search proceeds and the fitness is updated, the identities of the joiner and searcher can be dynamically switched, thereby obtaining a higher fitness value and new discoverers, but the proportion of the two in the population remains unchanged. Scouts are individuals at the edge of the search domain. Such individuals will randomly choose the search direction and update fitness and location information to get closer to the discoverer.

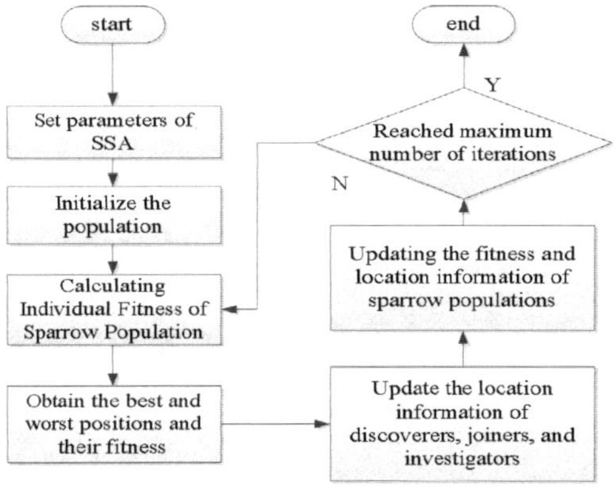

Figure 5. Sparrow search algorithm flowchart.

(3) Improved Elman neural network

The calculation process of the improved Elman neural network is shown in Figure 6.

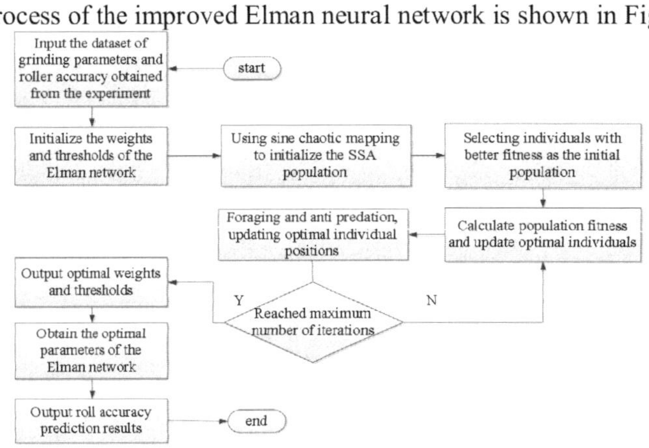

Figure 6. Improving the Elman neural network flowchart.

The calculation process of the improved Elman neural network is as follows:

1) The datasets of grinding shape and roundness errors are input;

2) The weights and thresholds of the Elman neural network are initialized;

3) The SSA algorithm population is initialized using sinusoidal chaotic mapping;

4) The accuracy of the training samples is defined as a fitness function and the individual fitness values are calculated. The individuals in the population are used to represent the initial weights and thresholds of the Elman neural network;

5) The current global optimal solution is found and the corresponding position is determined to update the individual position;

6) The individual fitness value after the position update is calculated and compared with the current optimal fitness value. When the number of iterations reaches the maximum value, the global optimal solution is selected. Otherwise, the iteration is repeated;

7) The output optimal solution is used as the weights and thresholds of the input layer to the hidden layer of the Elman neural network with forward propagation;

8) The parameters between the hidden layer and the output layer are fine-tuned using error back-propagation and iterative training is performed until the preset number of iterations is reached.

In this paper, the transfer function selected from the input layer to the hidden layer is the 'tansig' function, and the transfer function selected from the hidden layer to the output layer is the 'purelin' function. The training function of the neural network is the 'traingdm' function. The size of the learning rate directly affects the weight adjustment during the reverse error adjustment process. The learning rate is set to 0.1. The training result is shown in Figure 7.

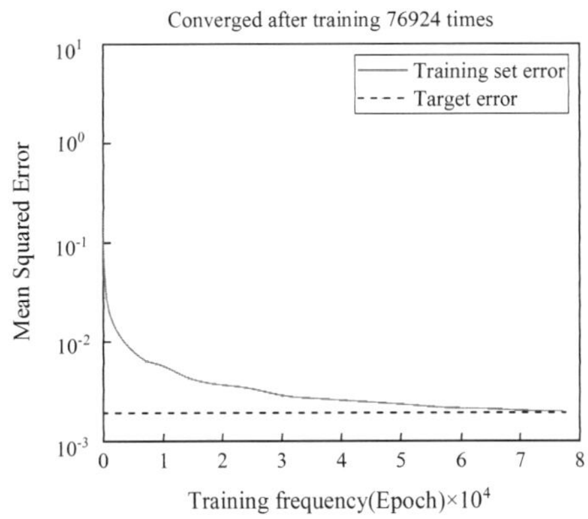

Figure 7. Training curve of improving Elman neural network.

4. Verification of shape and roundness accuracy prediction results

4.1 Sample data collection

The design of a full factor experiment is shown in Table 1. Four levels were selected for grinding depth and grinding wheel speed, as the influence of grinding depth and grinding wheel speed on roll shape error and roundness error is more significant. The grinding depth directly affects the removal of the roll, which can cause changes in roll shape and roundness accuracy and the speed of the grinding wheel affects the vibration of the grinding machine system, thereby affecting the accuracy of the roll shape and roundness.

Table 1. Experimental design of all factors for grinding accuracy.

Factors	Level			
Grinding depth/um	0	2	3.5	5
Grinding wheel Speed/m/s	11.4	14.4	17.4	20.4
Roll speed/r/min	60	70	80	--
Roll feed speed/m/min	0.09	1.05	1.2	--
Grinding passes	1	2	3	--

Excluding abnormal data from the experimental results, the training and testing sets were divided in a ratio of 9:1, and 356 sets of roundness error data were selected, including 321 sets of training and 35 sets of testing 228 sets of roller shape error data, including 208 sets for training and 20 sets for testing. The improved Elman neural network is used to train and test roundness error data and roll shape error data respectively.

4.2 Prediction results and analysis

According to the full-factor experiment in Table 1, the roll grinding shape and roundness accuracy test is designed. By collecting grinding process parameters, maximum roller shape error and average roundness error data, and inputting the data into the improved Elman neural network, the predicted results are shown in Figure 8. The prediction errors of roll shape and roundness are shown in Table 2.

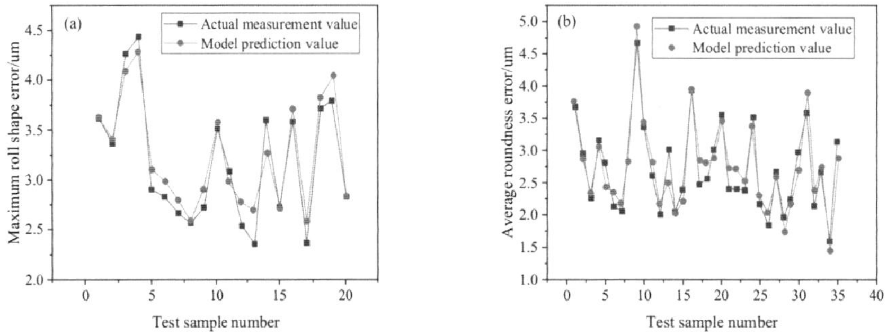

Figure 8. Prediction results of maximum roll shape error and average roundness error.

Table 2. Model prediction error of roll shape and roundness.

Index	MAE/um	MSE/um	RMSE/um	MAPE/%	Accuracy/%
Maximum roll shape error	2.1614	8.4323	2.9038	10.6696	89.3304
Maximum roundness error	0.15874	0.03489	0.18678	6.0688	93.9312

It can be seen that the predicted MAE values for roll shape and roundness accuracy obtained are both less than 5 um, and the difference in the number of test samples is not significant, indicating that the improved Elman neural network is more accurate in predicting the accuracy of roll shape and roundness. In this paper, the predicted MAPE values for roll shape and roundness accuracy are both less than 20%, indicating that the improved Elman neural network model is an excellent model. The experimental results indicate that the predicted RMSE values for roll shape and roundness accuracy are both less than 5 um. The improved Elman neural network is more accurate in predicting roll shape and roundness accuracy. Based on the analysis of various prediction indicators, the improved Elman neural network has a smaller prediction error, a prediction accuracy higher than 85%, and good prediction results.

A group of process parameters in actual production are randomly collected, and the accuracy of roll shape and roundness is predicted based on these parameters, and compared with the actual processing results, as shown in Figure 9.

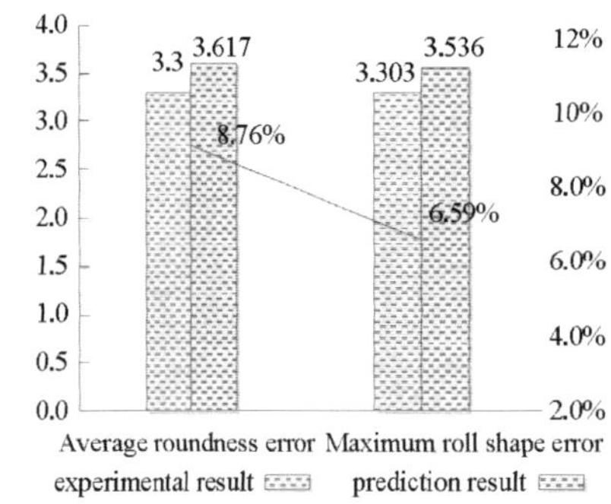

Figure 9. Experimental and prediction results.

It can be seen that the prediction accuracy of the average roundness error and the maximum roll shape error by using the improved Elman neural network is over 85%, in which the prediction error of the average roundness is 8.76% and the prediction error of the maximum roll shape error is 6.59%. The prediction results are more accurate, which shows that the proposed prediction method is effective and can meet the practical application requirements.

5. Conclusion

(1) A method for intelligent prediction of roll shape and roundness accuracy based on an improved Elman neural network is proposed to address the problem of strong nonlinearity between process parameters and shape and roundness accuracy during roll grinding.

(2) Sine chaotic mapping is used to make the population distribution more uniform, and use a sparrow search algorithm to obtain the optimal parameters of the Elman neural network.

(3) The sample test set and actual verification results show that the proposed intelligent prediction method has good prediction performance, with an accuracy rate of over 85%, and can meet practical application requirements.

References

[1] SONG Hua-nan. (2024) Research on surface defect detection of rolling rolls based on non-destructive testing technology [J]. Metallurgy & Materials. 44(02):82-84.
[2] ZHAO Jing. (2023) The influence of grinding process parameters on the surface roughness of rolling rolls [J]. World Nonferrous Metals. (18): 192-194.
[3] XIE Tian-shu. (2021) Double objective process optimization for grinding the outer ring of silicon nitride ceramic bearings [J]. Diamond & Abrasives Engineering. 41(04):82-91.
[4] LU Yao-xun, WANG Li-ping, SUN Li-rong, etc. (2022) Intelligent scheduling method for roll grinding production line based on improved genetic algorithm [J]. Manufacturing Technology and Machine Tools. (08):149-155.
[5] ZHU Huan-huan, CHI Yu-lun, ZHANG Meng-meng, etc. (2024) Online prediction of rolling mill grinding chatter based on EMD component and wavelet packet energy entropy [J]. Diamond & Abrasives Engineering. 44(01):73-84.
[6] ZHANG Qiang, LIU Chong, CHEN Yong-ping. (2022) Optimization of grinding parameters for precision rolling work rolls based on response surface methodology [J]. Precision Manufacturing and Automation. (02):17-18+38.
[7] QIU Hui, CHEN Yi-feng, PAN Xiao-ming, etc. (2024) Construction of Turning Flutter Prediction Function Based on Elman Neural Network [J]. Mechanical Design and Research. 40(02):120-125.
[8] SHAO Ya-wen, ZHAO Lu-ping. (2023) NARX-GA-Elman Method for Mach Number Prediction of Wind Tunnel Flow Field [J]. Instrumentation. 10(04):50-63.
[9] WANG Yang, CHEN Jun-jie, XIE Meng-yu, etc. (2023) Prediction of elemental sulfur solubility based on LCASO-BPNN model [J]. Chemical Engineering. 51(12):56-61.
[10] ZHANG Ying-chun, JIANG Lan, TANG Bo, etc. (2024) Optimization method for substation architecture based on improved sparrow search algorithm [J]. Vibration and Shock. 43(07):94-101.

Research on the effect of manufacturing processes on the fatigue life of TC4 fasteners

Yanyun Ma[1], Zhaofeng Dou[2], Haoshan Li[2], Ruiyuan Wang[2], Yanfeng Xia[1], Qiu Gao[1] and Weifang Zhang[2,*]

[1]Tianjin Key Laboratory of Fastening Technology, Aerospace Precision Products Inc., Ltd, Tianjin, 300300, China
[2]School of Reliability and Systems Engineering, Beihang University, Beijing, 100191, China

[*]Corresponding author's e-mail: 08590@buaa.edu.cn

Abstract. TC4 fasteners have the characteristics of high specific strength and yield ratio, excellent high and low temperature performance, corrosion resistance, etc., and are widely used in many fastener fields, including aerospace. In long-term use, fatigue fracture has become the main failure mode of TC4 bolt fasteners. The common manufacturing processes of TC4 fasteners include shot peening, bottom corner rolling pressure, anodizing and molybdenum disulphide coating, etc. At present, few researchers have studied the effect of the stress level of the manufacturing process on the fatigue life of TC4 fasteners. Therefore, this paper designs the process and parameter influence test from the perspective of different manufacturing processes, and analyses the fatigue test results to reveal the influence mechanism of different processes on the fatigue life of TC4 fasteners. The research results show that the increase of residual compressive stress greatly improves the fatigue strength of fasteners, the shot peening process improves the fatigue life of TC4 fasteners by increasing the residual stress of fasteners, and rolling pressure also increases the surface residual compressive stress. However, it is necessary to find suitable rolling parameter values, and anodizing reduces the residual compressive stress of fasteners. The coating of molybdenum disulfide also reduces the residual stress value of the fastener surface and leads to a decrease in fatigue life. Therefore, in order to control the fatigue life of TC4 fasteners, it is necessary to carry out the shot peening process, and control the rolling pressure, anodizing time and the coating thickness of molybdenum disulfide. The research in this paper provides a reference for the subsequent research on fastener fatigue strength control.

1. Introduction

Fasteners, as the basic parts of the connection between mechanical parts, are known as "industrial rice". [1]. TC4 fasteners have the characteristics of strong corrosion resistance, good high and low temperature performance, low density and high strength, and are widely used in aviation, aerospace, marine and automotive fields [2]. Although small, fasteners play an important role in key components of machinery and other products. Fasteners are not only key basic parts, but also safety parts. Once a fastener fails to perform the specified function assigned by the design, there are serious consequences. Fatigue fracture is one of the most common failure mechanisms of threaded fasteners [3], which usually occurs when fasteners are subjected to long-term alternating loads. Under the influence of cyclic stress, the tiny defects or cracks inside the fastener will gradually expand, and eventually lead to the fracture of the entire structure.

Journal of Physics: Conference Series **2842** (2024) 012077 doi:10.1088/1742-6596/2842/1/012077

At present, researchers have studied the fatigue life of fasteners from the aspects of load characteristics [4], use environment [5], and a single manufacturing process. Dong et al. [6] studied the effect of bolt tightening torque on the fatigue life of composite multi-nail joint structures, and showed that with the increase of bolt tightening torque, the fatigue life first increased and then decreased; Zhao et al. [7] found that the uneven contact section of the flange would significantly affect the fatigue life of the bolt; Yuan et al. studied the effect of rounded Angle rolling depth on the fatigue life of bolts, and the results showed that the fatigue life of bolts after rolling was significantly increased, and the fatigue life first increased and then decreased with the increase of depth. According to Chakherlou et al. [8], fatigue testing of cold-expanded fasteners before and after thermal cycling showed that their fatigue life was longer than that of non-cold-expanded fasteners. Tian et al. [9] showed that after surface treatment of thin-walled TC4 titanium alloy samples by laser shot peening, the average life of the samples could be improved, Ouyang [10] also studied the effect of shot peening on the fatigue life of titanium alloy; Costa et al. [11] found that anodizing had a negative effect on the fatigue life of Ti-6Al-4V titanium alloys. Zhong et al. [12] studied the thermal relaxation of the residual stress of Ti-6Al-4V titanium alloy after laser shot peening.

However, although significant progress has been made in these aspects of research, few researchers have studied the fatigue life of fasteners from the perspective of different manufacturing processes. The manufacturing process has a crucial impact on the quality and performance of fasteners, including the heat treatment process, molding process, strengthening process and surface treatment, which will directly affect the microstructure and mechanical properties of fasteners, and then affect their fatigue life. The research object of this paper is TC4 fasteners in the hot pier process. After the initial pier and the fine pier, the hexagonal groove is formed at the nut of the fastener, and the tensile strength reaches 1100 MPa~1240 MPa after heat treatment. The strengthening process includes root corner rolling strengthening and shot peening, and the surface treatment process includes anodizing and coating molybdenum disulfide. Among them, the shot peening process has a positive impact on the fatigue strength of fasteners, and through the previous test, fasteners are mostly broken from the rounded corners of the root, as shown in Figure 1(a). Through finite element simulation, it can be found that the phenomenon of stress concentration occurs at the rounded corner of the root of the fastener, as shown in Figure 1(b). In addition, the fatigue life of fasteners coated with molybdenum disulfide decreased significantly. In order to reveal the fatigue fracture mechanism of fasteners, this paper will analyse the influence of four factors on the fatigue life of fasteners, such as shot peening, root rounded rolling pressure, anodizing and coating molybdenum disulfide.

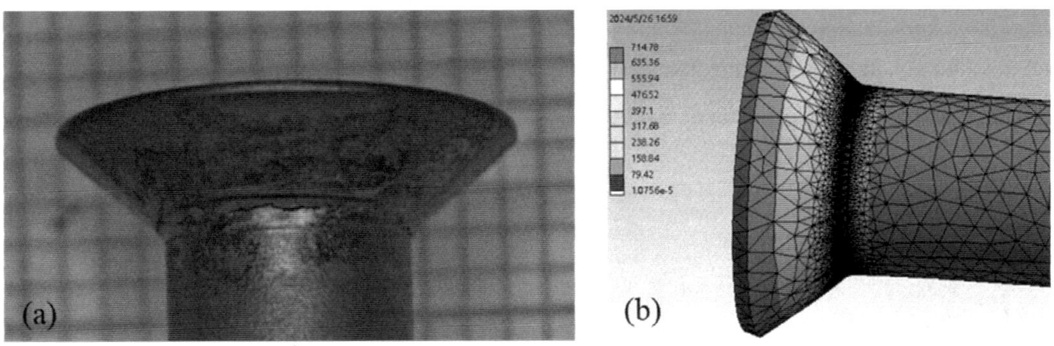

Figure 1. Specimen with rounded root fracture and finite element analysis diagram (a) Actual fracture site; (b) Finite element analysis.

2. Experiment

The TC4 (Ti-6Al-4V) titanium alloy fastener used in this test has excellent corrosion resistance, high mechanical strength, good heat resistance, plasticity and toughness, etc. The specimen is shown in Figure 2.

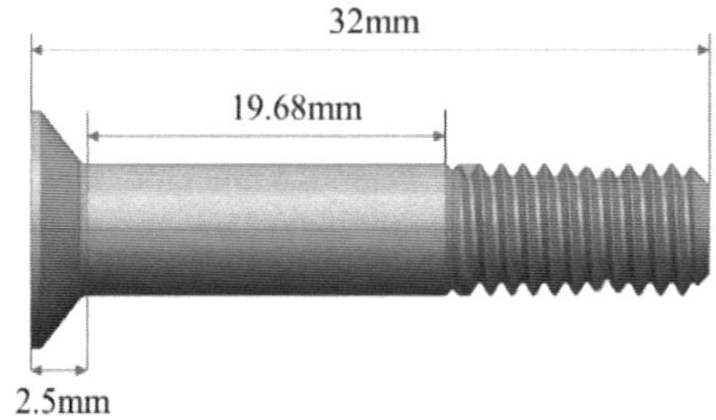

Figure 2. Schematic diagram of TC4 fastener model.

It has been found that the fatigue strength of TC4 fasteners is affected by a variety of processes, including shot peening, rolling strengthening, anodizing and molybdenum disulfide coating. In order to reveal the influence mechanism of different processes on the fatigue strength of fasteners, the different stress levels of rolling pressure, anodizing time and molybdenum disulfide coating thickness were mainly studied. In [13], it is found that the rolling depth has an effect on the fatigue life of fasteners, and the rolling depth first rises rapidly with the increase of rolling pressure and tends to be stable at about 110 kg. In order to find a more suitable rolling pressure, three stress levels are evenly selected in the middle of the rolling interval for research. In addition, Feng [14] found that extending the time can increase the anodic oxide film thickness. When the oxidation time reaches 90 min, the thickness is maintained at 14 μm~15 μm and does not change any more. At 60 min, the oxide film thickness reaches 12 μm-13 μm. In order to study the influence of different anodizing times on fatigue strength, three stress levels of 20 min, 40 min and 60 min were selected for study. The coating of molybdenum disulfide can increase the lubricity of fasteners and prevent the adhesive phenomenon of specimens in the process of multiple screwing in and screwing out. In order to find a more suitable coating thickness, the number of coating circles is used as the stress level to measure the influence of thickness on the fatigue strength of fasteners. Finally, in order to study the effect of shot peening on fasteners, a control group was set up. The specific process and parameter influence test design are shown in Table 1.

Table 1. Process and parameter influence test table.

Number	Fastener technology
1	Shot peening
2	Shot peening, Rolling pressure 65 kg
3	Shot peening, Rolling pressure 65 kg, Anodize for 60 min
4	Shot peening, Rolling pressure 65 kg, Anodize for 60 min, Molybdenum disulfide 2 circles
5	Shot peening, Rolling pressure 65 kg, Anodize for 20 min
6	Shot peening, Rolling pressure 65 kg, Anodize for 20 min
7	Shot peening, Rolling pressure 65 kg, Anodize for 60 min, Molybdenum disulfide 3 circles
8	Shot peening, Rolling pressure 40 kg
9	Shot peening, Rolling pressure 90 kg
10	Surface untreated
11	Rolling pressure 65 kg

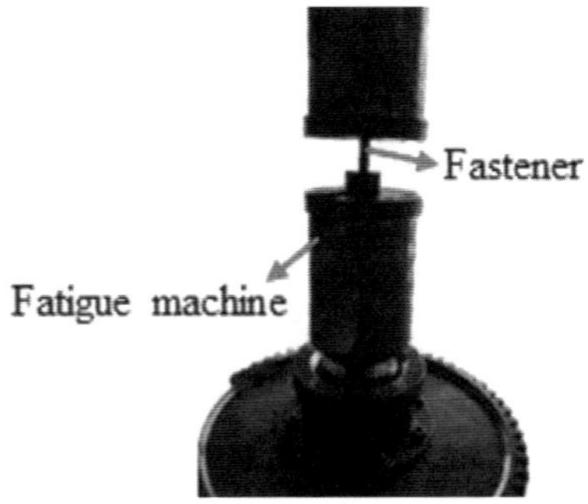

Figure 3. Fatigue test apparatus.

The experiment was carried out on the fatigue test machine of Tianjin Key Laboratory of Fastening Technology, and 3 specimens were selected for each number to carry out the axial fatigue tensile test, as shown in Figure 3. The target cycle of this experiment is 130,000 times per week, so the test adopts a timed truncated test. The fatigue test adopts the axial pull-pull test, in which the high load is 7.1 kN and the low load is 0.71 kN. The final fatigue test life results are shown in Table 2 below.

Table 2. Process and parameter influence test table.

Number	Fastener technology	Fatigue life ($\times10^4$ cycles)		
		1	2	3
1	Shot peening	30	30.5	60
2	Shot peening, Rolling pressure 65 kg	60	60	60
3	Shot peening, Rolling pressure 65 kg Anodize for 60 min	60	60	60
4	Shot peening, Rolling pressure 65 kg Anodize for 60 min, Molybdenum disulfide 2 circles	60	60	60
5	Shot peening, Rolling pressure 65 kg Anodize for 20 min	60	60	60
6	Shot peening, Rolling pressure 65 kg Anodize for 20 min	60	60	60
7	Shot peening, Rolling pressure 65 kg Anodize for 60 min, Molybdenum disulfide 3 circles	60	60	21
8	Shot peening, Rolling pressure 40 kg	25	60	60
9	Shot peening, Rolling pressure 90 kg	12	60	60
10	Surface untreated	2	1.3	2.5
11	Rolling pressure 65 kg	15.6	60	6.9

3. Result and analysis

3.1. Fatigue test result

The fatigue cycles were recorded by a timed truncation method in the test, and the specimens with a fatigue life of 600,000 times in the table did not produce fatigue damage. By comparing Nos. 1 and 2, and 10 and 11, it can be found that rolling 65 kg greatly improves the fatigue life of fasteners. Comparing Nos. 2, 3, and 5 with No. 6, it can be seen that anodizing has no obvious effect on the fatigue life of TC4 fasteners. Comparing Nos. 2 and 8 with No. 9, it can be seen that the fatigue life decreases when the rolling pressure is 40 kg and 90 kg. Comparing Nos. 3 and 4 with No. 7, it can be seen that when coated with molybdenum disulfide, individual specimens of TC4 fasteners also show a decline in fatigue life. Only when the surface is not treated, the fatigue life of the three specimens does not meet the requirement of 130,000 times.

3.2. Analysis of fatigue test results

In order to find out the cause of the fatigue life decline of the specimens with the above individual surface treatment process, the residual stress test was carried out on the specimens. As shown in Figure 4 below, after the fatigue test, it was found that the fasteners were mostly broken at the rounded corners of the roots. Therefore, the residual stress test was carried out on the rounded corners of the roots of TC4 fasteners.

Figure 4. Fastener residual stress testing.

The residual stress test results of the specimen are shown in Table 3 below.

Table 3. Residual stress measurements under different surface treatments.

Number	Fastener technology	Residual stress (MPa)	Mean life ($\times10^4$ cycles)
1	Shot peening	-609.733	40.33
2	Shot peening, Rolling pressure 65 kg	-712.300	60
3	Shot peening, Rolling pressure 65 kg, Anodize for 60 min	-497.233	60
4	Shot peening, Rolling pressure 65 kg, Anodize for 60 min, Molybdenum disulfide 2 circles	-340.925	60
5	Shot peening, Rolling pressure 65 kg, Anodize for 20 min	-719.167	60
6	Shot peening, Rolling pressure 65 kg, Anodize for 20 min	-669.000	60
7	Shot peening, Rolling pressure 65 kg, Anodize for 60 min, Molybdenum disulfide 3 circles	-138.025	47
10	Shot peening, Rolling pressure 40 kg	-100.367	1.93
11	Shot peening, Rolling pressure 90 kg	-290.800	27.50

As can be seen from the above table, after shot blasting on the surface of TC4 fasteners, the surface residual compressive stress is greatly increased, and the fatigue life is also significantly increased. When shot peening and rolling pressure of 65 kg were applied at the same time, the surface residual compressive stress increased and the fatigue life was further improved. Compared with the specimen without surface treatment, the residual compressive stress and fatigue life of the single rolling pressure 65 kg process increased, but the dispersion was larger as can be seen in Table 1. The anodizing process has an effect on the residual stress, and the residual compressive stress decreases with the anodizing time increasing. However, after shot peening and rolling pressure of 65 kg, the fatigue life is not significantly affected for 60 min. With the increase of the thickness of molybdenum disulfide coating, the residual compressive stress on the surface of fasteners is further reduced, and the fatigue life is affected to some extent.

3.3. Analysis of fatigue test results

The shot peening process has a significant impact on the fatigue life. As can be seen from the table above, the surface residual compressive stress after shot peening is greatly increased, which offends part of the axial tensile load of the fasteners during the fatigue process and increases the fatigue life. However, when the residual compressive stress reaches a certain value, it basically does not change, and the surface roughness continues to increase, weakening the strengthening effect of shot peening.

The rolling strengthening mechanism mainly includes three aspects, the first is the introduction of residual compressive stress. As can be seen from Table 3, rolling strengthening introduces residual compressive stress on the surface of the fastener, and the residual compressive stress can offset part of the tensile stress of the fatigue load and reduce the average stress of the component. After crack initiation, the residual compressive stress can reduce the stress intensity factor of the crack tip and slow down the crack growth rate. Secondly, rolling strengthens the tissue. During rolling, plastic deformation occurs on the surface of fasteners, and the surface structure changes. For example, grain thinning, dislocation density increase and crystal plane spacing increase. In the fatigue process, slip is an important process for fatigue crack nucleation of metal materials, and grain thinning and dislocation can inhibit slip and delay crack nucleation. Finally, the effect of rolling on the specimen surface is discussed. During the processing of fasteners, surface micro-cracks, tool marks and other surface micro-defects may occur. Under the action of fatigue load, these defects cause local stress concentration, and rolling strengthening causes plastic flow on the surface of fasteners, resulting in the effect of "peak cutting and valley filling", so that the plastic deformation of the damaged part of the surface tends to be uniform [15]. Rolling strengthening not only introduces the residual compressive stress, but also changes the organizational state of the surface of the fastener, so that Nos. 4 and 7 still have a higher fatigue life when the residual stress is small. In addition, it can be seen from Table 2 that the fatigue life decreases when the roll pressure is 40 kg and 90 kg. When the rolling pressure is 90 kg, as shown in Figure 5, the significant depth of the rounded root causes the smooth transition in the area, and the stress concentration is more obvious, resulting in the decline of the fatigue strength of the fastener.

Figure 5. Roll pressure 90 kg root rounded corner diagram.

At present, it is difficult to calculate the stress concentration at the rounded corners of fasteners directly. Therefore, the finite element simulation method is used to compare the stress concentration at the rounded corners of the roots under different rolling pressures, and the qualitative analysis is carried out on this basis. It can be seen from Figure 5 that when the rolling pressure is large, the transition of the rounded corner of the root will be uneven, and the finite element model is established according to this, as shown in Figure 6 below. (a) is the model of smooth transition, and (b) is the model of non-smooth transition. During the simulation, the lower surface of the nut and the thread are fixed, and a pull-pull load of up to 4.5 kN is applied to the fastener. The results are shown in Figure 6, where (c) is the specimen with a smooth rounded corner transition, and (d) is the specimen with an uneven rounded corner transition. After applying the same load, the stress distribution diagram of the specimen is obtained, in which the root of the specimen with uneven transition appears more obvious stress concentration phenomenon. In [13], it is found that the rolling depth first rapidly increases and then flattens with the increase of rolling pressure, while with the increase of rolling depth, the maximum compressive stress rapidly increases and becomes stable, and the fatigue life first increases and then decreases. The fatigue life is the best when the rolling pressure is 65 kg, which is consistent with the research of the scholar in Table 2.

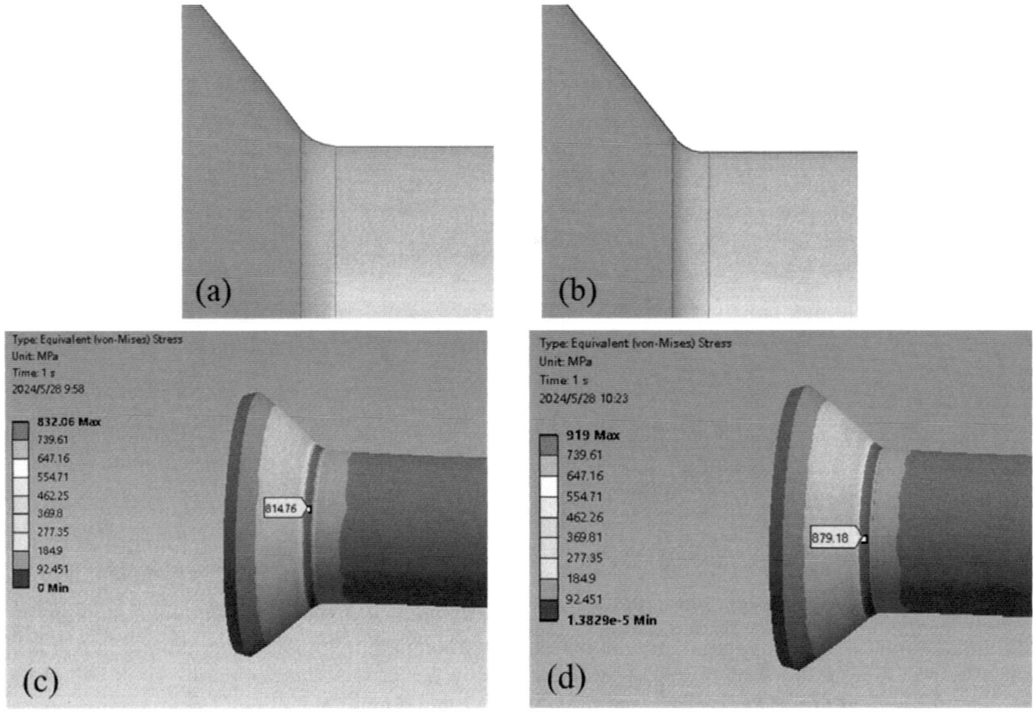

Figure 6. Rounded roots and finite element results (a) Transition smooth model, (b) Models with uneven transitions, (c) The finite element results of the smoothing model, (d) Finite element results of a non-smooth model.

According to the TC4 fastener process document, the formula of the pulse anodizing electrolyte is (200~210) ml/L H_2SO_4+(10~20) ml/L H_3PO_4+(0~5) g/L Ti^{4+}, which is anodized to form a dense oxide film on the surface of the fastener. As can be seen from Table 2, pulse anodizing reduces the surface residual compressive stress of fasteners. However, after anodizing for 20 min and 40 min, the surface residual stress does not change significantly, and decreases significantly after 60 min, which affects the fatigue strength of fasteners.

According to the TC4 fastener manufacturing process documents, when spraying molybdenum disulfide, the specimen is first cleaned and put into the oven, then the specimen head is sprayed, the fastener rod part is sprayed again, the fastener rod part is also put into the oven after spraying, and the temperature of at least 150(±10)°C is preheated for 10 min in these three times. Finally, 10 pieces were randomly selected to test the size and appearance of the rods, put into the oven, and cured at 210(±10)°C for 120 min~140 min. At high temperatures, the atomic or molecular vibration inside the fastener increases, making part of the residual stress released or redistributed, so as to reach a new equilibrium state [12]. Through this test, it is verified that the residual compressive stress of the specimen coated with molybdenum disulfide further decreases, thus affecting its fatigue life.

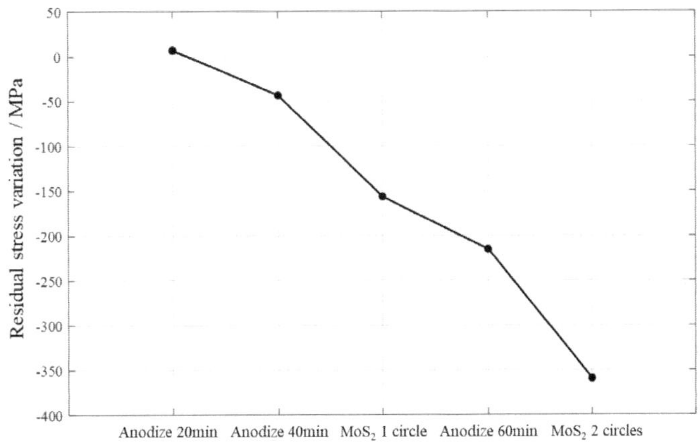

Figure 7. Effect of molybdenum disulfide and anodizing on surface residual stress of fasteners.

Figure 7 shows the influence of molybdenum disulphide and anodizing process parameters on the residual stress of fasteners. It can be seen directly from the figure that the coating of molybdenum disulphide for 2 circles has the greatest influence on the residual compressive stress of the fastener surface, followed by pulse anodizing for 60 min, molybdenum disulphide for 1 circle, and finally anodizing for 40 min and 20 min has the least influence.

4. Conclusion

Through the design process and parameter influence test, the key process affecting the fatigue strength of TC4 fasteners was found, and the surface residual stress of different specimens was detected, revealing the influence mechanism of different processes on the fatigue strength of fasteners. Specific conclusions are shown as follows:

(1) Increasing the residual compressive stress on the surface of TC4 fasteners is a key factor in improving their fatigue life.

(2) Shot peening and root corner rolling can improve the fatigue strength and fatigue life of fasteners by increasing the residual compressive stress on the surface, and fasteners have different life performances under different rolling pressures.

(3) Pulse anodizing and molybdenum disulfide coating process reduce the surface residual compressive stress of TC4 fasteners, and then affect the fatigue strength of fasteners.

To sum up, in order to control the fatigue life of TC4 fasteners, it is necessary to carry out the shot peening process, and control the rolling pressure, anodizing time and coating thickness of molybdenum disulfide.

Acknowledgments

This paper is one of the phased achievements of the project “Fastener Fatigue Strength Influence Mechanism and Control Method Research” (TKLF2022-01-C-01) of Tianjin Key Laboratory of Fastening Technology, this work was financially supported by “Tianjin Technology Innovation Guidance Special Fund” (22YDTPJC00110).

References

[1] Li Y L. (2014) Fasteners Basics. National Defense Industry Press. Beijing.
[2] Xin H M, Shi Y Y, Ning L and Zhao T. (2016) Residual stress and affected layer in disc milling of titanium alloy. Materials and Manufacturing Processes, 31: 1645–1653. 10.1080/10426914.2015.1090583.
[3] Tao C H. (2013) Failure Analysis and Prevention of Fasteners. Aviation Industry Press, Beijing.

[4] Pai N G and Hess D P. (2002) Three-dimensional finite element analysis of threaded fastener loosening due to dynamic shear load. Engineering failure analysis, 9: 383-402. 10.1016/S1350-6307(01)00024-3.
[5] Wang A Q, Yang S C, Hu Z and Liu J. (2014) Effect of interference fit on fatigue life of fasten holes without load transfer. hot working technology, 43: 55-58. 10.14158/j.cnki.1001-3814.2014.04.064.
[6] Dong J C, Yang Z Y, Zheng X T, Song M Y and Yan L L. (2023) Effect of bolt tightening torque on fatigue life of multi-bolted composite structure. **bei Gongye Daxue Xuebao/Journal of Northwestern Polytechnical University, 41: 455-463. 10.1051/jnwpu/20234130455.
[7] Yuan W H, Deng J W, Shen Q Y, Sun K and Huang W Y. (2023) Influence of fillet rolling on residual stress distribution and fatigue life for bolts. Forging & Stamping Technology, 41: 455-463. 10.13330/j.issn.1000-3940.2023.01.015.
[8] Chakherlou T N and Aghdam A B. (2008) An experimental investigation on the effect of short time exposure to elevated temperature on fatigue life of cold expanded fastener holes. Materials & Design, 29: 1504-1511. 10.1016/j.matdes.2008.03.004.
[9] Tian K, Shuai S X, Luo X K, Wang X, Ma S C and Xu C L. (2023) Effect of co-strengthening of laser impact/mechanical shot peening on fatigue performance of TC4 titanium alloy damaged by foreign objects. Journal of Aeronautical Materials, 43: 94-101. 10.11868/j.issn.1005-5053.2022.000191.
[10] Ouyang P X, Luo X K, Dong Z C and Zhang S T. (2022) Numerical Prediction of the Effect of Laser Shock Peening on Residual Stress and Fatigue Life of Ti-6Al-4V Titanium Alloy. Materials, 15: 5503. 10.3390/ma15165503.
[11] Costa M Y P, Voorwald H J C, Pigatin W L, Guimaraes V A and Cioffi M O H. (2006) Evaluation of shot peening on the fatigue strength of anodized Ti-6Al-4V alloy. Materials Research, 9: 107-109. 10.1590/S1516-14392006000100020.
[12] Zhou Z, Bhamare S, Ramakrishnan G, Mannava S R, Langer K, Wen Y L, Qian D and Vasudevan V K. (2012) Thermal relaxation of residual stress in laser shock peened Ti–6Al–4V alloy. Surface and Coatings Technology, 206: 4619-4627. 10.1016/j.surfcoat.2012.05.022.
[13] Deng J W. (2022) Study on Rolling Strengthening Mechanism of TC4 Titanium Alloy Bolt Fillet Rolling. Hunan University. 10.27135/d.cnki.ghudu.2022.004298.
[14] Feng H Q, Zhang X, Shi Y Y, Zhang S B and Zhou K. (2016) Pulsed Anodizing Process of Ti-6Al-4V Titanium Alloy. Electroplating & Pollution Control, 36: 40-42.
[15] Wang Y L, Zhu Y L and Yang J Q. (2018) Rolling Reinforcement Technology and Its Research Application Progress in Aviation Field. Aeronautical Manufacturing Technology, 61: 75-83. 10.16080/j.issn1671-833x.2018.05.075.

Journal of Physics: Conference Series **2842** (2024) 012078 doi:10.1088/1742-6596/2842/1/012078

Research on the performance optimization of the cam linkage mechanism for marine diesel engines

Keke Wang, Na Meng*, Sen Yang, Longfa Chen, and Wenzheng Sun

Shandong University of Science and Technology, College of Intelligent Equipment, Shandong, Tai'an 271001, China

* Corresponding author's email: skd993317@sdust.edu.cn

Abstract: In this paper, the impact of friction and wear in the flat bottom mechanism of a diesel engine on the reliability and durability of the valve train is investigated. A method for establishing a dynamic contact model is proposed, including the establishment of the cam profile equation, force analysis at the contact point, calculation of the curvature radius, and derivation of contact stress. Through the derivation of wear depth and structural optimization analysis, the proposed method can be used to optimize the cam profile and improve the service life and work efficiency of the valve train. The results show that the optimized cam follower's life is increased by 31.8%, and the contact stress is decreased by 16.5%. This study is of great significance for improving the power and efficiency of diesel engines.

1. Introduction

With the continuous development and improvement of diesel engines towards high speed and high power, increasing power has become a primary challenge in this field. However, the frictional wear in the internal cam-follower mechanism seriously affects the reliability and durability of the valve train. The contact between the cam and the flat follower is one of the high-stress pairs prone to wear [1-2]. Even slight geometric changes can lead to worsened stress distribution, affecting its service life and significantly impacting the transmission and geometric accuracy of the valve train. This phenomenon not only results in compression and energy losses in marine diesel engines but also significantly reduces their output power.

Previous researchers have made important explorations and attempts to address these issues. For example, Ge et al. [3] further investigated the problems of the cam linkage mechanism for wrapping packages through dynamic analysis based on ADAMS. Ma et al. [4] studied the wear characteristics of bearings in wind turbines, while Li et al. [5] analyzed the wear of crankshaft dies using the Archard model. Wu et al. [6] proposed a method for optimizing the design of the cam in a five-link spring actuating mechanism. In addition, G. et al. [7-10] have made significant contributions to the optimization design of cam mechanisms and parameter optimization. The results of these studies provide valuable references and support for further exploration of the issues discussed in this paper.

However, although the results and methods of previous research provide valuable insights, there is still a need for a deeper explanation and understanding of why specific methods were chosen and how they are superior to other possible modeling methods when choosing the research methods for this study. Therefore, it is aimed to strengthen the methodological foundation of the paper with a detailed explanation of our chosen method, the dynamic contact model, and its uniqueness and applicability in solving the friction and wear problems of the convex cam-follower mechanism. This will provide readers with a clearer research framework and ideas.

2. Establishment of the dynamic contact model

2.1 Establishment of the cam profile equation

Due to the relative difficulty in constructing the cam profile contour, in this study, the contact position between the cam and the push rod at $t = 0$ s is defined as the contact position near the compression and power strokes, corresponding to $\alpha = 0°$. A longitudinal coordinate system is established parallel to the push rod direction, and a transverse coordinate system is passed through the contact position of the cam and the push rod at $\alpha = 0°$. The cam mechanism profile design parameters in this study are shown in Figure 1. Since cam wear mainly occurs during the power and return strokes, the near compression stop section is not discussed.

The equation for the motion of the follower from the near compression end to the far compression end is established based on the known conditions, as shown in Equation (1). The design parameters of the cam mechanism profile are shown in Table 1

Table 1. Cam mechanism profile design parameters.

Buffer segment wrap angle α_o	20.0°	Basic segment half wrap angle α_c	61.0°
Buffer segment lift h_1	0.4000 mm	Basic segment lift h_2	7.8000 mm
Cosine segment wrap angle coefficient G	0.5		

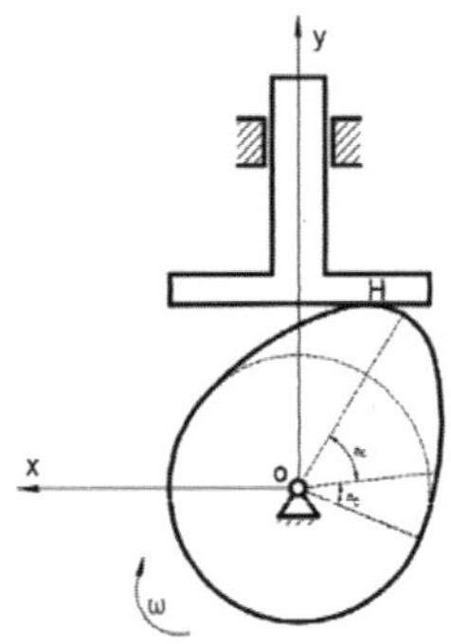

Figure 1. The schematic diagram of the cam flat-follower mechanism.

$$h(a) = \begin{cases} 0.1566 * (1 - \cos 9 * \alpha) & 0° \leq \alpha \leq 20.0° \\ -0.0888 + 1.4003 * \alpha & 142.0° \leq \alpha \leq 162.0° \end{cases} \tag{1}$$

Since the cam is an axisymmetric figure, in Equation (1), $0° \leq \alpha \leq 20.0°$ corresponds to the change in the buffer angle during the compression stroke, and $142.0° \leq \alpha \leq 162.0°$ corresponds to the change in the buffer angle during the return stroke. At $\alpha = 0°$, the lift of the follower is 0; when $\alpha = 20°$, the cam enters the basic lift phase, at which point the load on the cam undergoes a significant change. Therefore, improving the problem of load mutation at this position can solve the efficiency problem caused by wear.

$$h(a) = 7.8 - 10.1166\alpha^2 + 6.0026 * \alpha^{16} - 5.0822 * \alpha^{24} \\ +1.3962 * \alpha^{32} \;\; (20.0° < \alpha < 142.0°) \tag{2}$$

2.2 Force analysis of contact position

By analyzing the force situation of the cam at the contact position, it is possible to understand when the wear of the cam is more severe at what angle. Figure 2 shows the cam-follower model diagram, where the mass of the follower and the valve spring will affect the contact force. Analysis shows that two forces are acting at the contact position. The first force F_T is the valve spring force, which can be obtained by

Equation (4); the other force F_G is the inertia force of the parts, which can be determined by Newton's second law. The resultant force F is the sum of the two forces, which can be calculated according to Equation (3). The force situation at the cam contact point is shown in Figure 3. Since the directions of the two forces are the same, there is no need to consider the vector symbol factors in the calculation process.

$$F = F_T + F_G \tag{3}$$

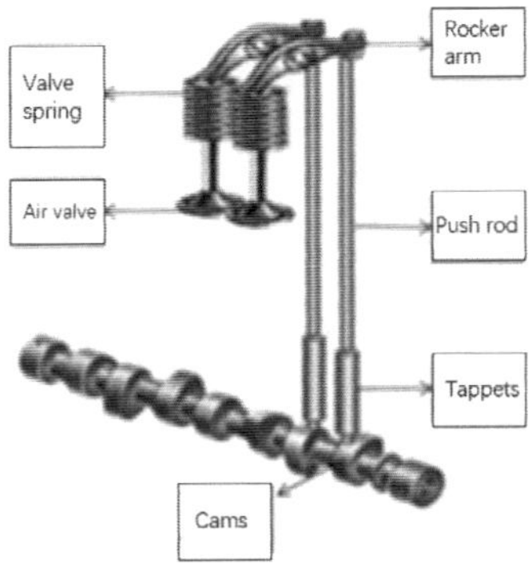

Figure 2. The cam-follower model diagram.

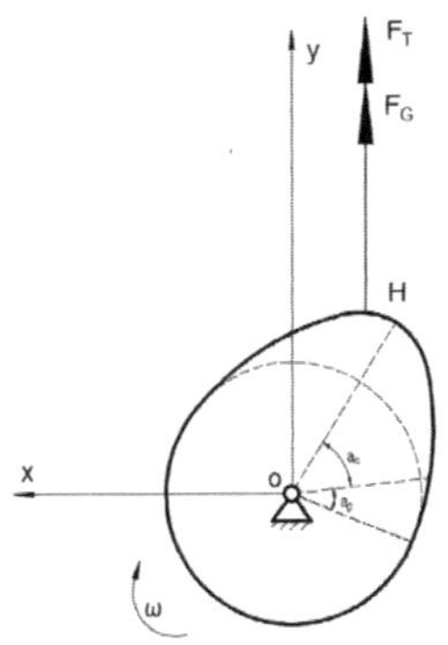

Figure 3. The force diagram of cam contact point H.

The valve spring force is related to the spring preload force and the distance $y(\alpha)$ compressed by the spring. The distance $y(\alpha)$ compressed by the spring is numerically equal to the lift of the follower $h(\alpha)$.

$$F_T = k * [F_0 + k_s * y(\alpha)] \quad 0 \leq \alpha \leq 162^o \tag{4}$$

where F_0 represents the spring preload force (200 N), k_s is the stiffness of the valve spring (50000 N/m), k represents the rocker arm ratio based on the length ratio of the intake rocker arm near the valve to the length of the intake rocker arm near the push rod, and $k = 1.5908$.

The inertia force of the cam is related to the acceleration of the contact point and the mass of the cam. Taking the second derivative of Equation 1 and Equation 2, the acceleration can be obtained. In the optimization system developed using MATLAB, the independent variable is the angle, so conversion is required. The output is a graph showing the variation of force with angle.

$$F_G = M * \frac{d^2h}{dt^2} = M * x^2 * \frac{d^2h}{d\alpha^2} = M * x^2 * h''_\alpha$$
$$0 \leq \alpha \leq 162^o \tag{5}$$

where M is the concentrated mass (0.9339 kg), t is the operating time (s), and h''_α is the geometric acceleration (rad/s^2).

2.3 Curvature radius of contact point

The contact situation between the cam follower and the cam is the same as that between a cylinder and a plane. Therefore, the curvature radius of the cam can be expressed as Equation (6),

$$R = R_0 + h_\alpha \tag{6}$$

where R is the cam curvature radius (m), R_0 is the cam base circle radius (0.032 m), and h_α is the height of the cam rise.

2.4 Contact Stress

The instantaneous contact stress can be solved by Equation 7 [1],

$$p_0 = \sqrt{\frac{PE'}{2\pi R B_0}} \tag{7}$$

where p_0 is the contact stress between the cam and the follower (Pa); E' is the equivalent modulus of elasticity (2.3077×10^{11} Pa); B_0 is the cam width (10^{-3} m).

E' represents the equivalent elastic modulus of the contact area (unit: Pa), which can be determined by Equation 8 [2],

$$E' = \frac{2}{\frac{1-\mu_1^2}{E_1} + \frac{1-\mu_2^2}{E_2}} \tag{8}$$

where E_1 and E_2 are the elastic moduli of the corresponding materials ($E_1 = E_2 = 2.1\times10^{11}$ Pa); μ_1 and μ_2 are the Poisson's ratios of the corresponding materials ($\mu_1 = \mu_1 = 0.3$).

3. Derivation of wear depth

In this study, the wear depth at a certain point is used to characterize the degree of wear. According to the Archard wear model, the expression for the wear depth at that point can be obtained, which is Equation (9).

$$h = \int_{t_1}^{t_2} kp(t)v dt \tag{9}$$

where $p(t)$ represents the normal contact pressure (N), and the limits of integration, t_1 and t_2 correspond to the starting and ending moments of wear at that point.

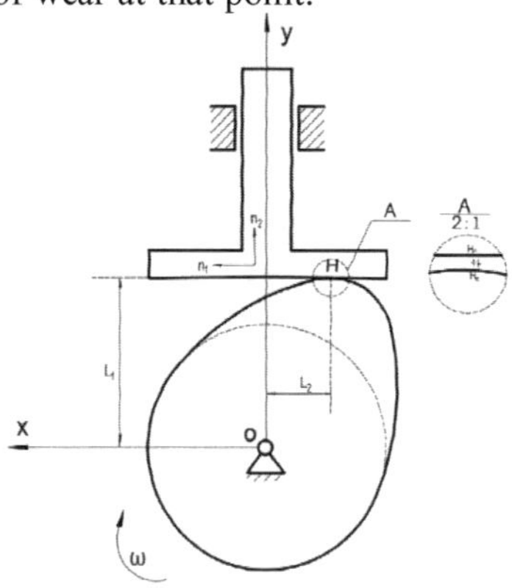

Figure 4. Analysis of Cam-Follower Contact Position.

The relative sliding model of the cam and follower contact pair is shown in Figure 4. In this model, H represents the contact position between the cam and the follower, H_C represents a point on the cam, and H_T represents a point on the follower. At this moment, points H_C and H_T coincide at point H_E.

The absolute velocity vector of point H_C on the cam is given by Equation (10).

$$\boldsymbol{V}_{\boldsymbol{H}_C} = L_2 w_{n1} - L_1 w_{n2} \tag{10}$$

where L_1 represents the distance from point H to the centerline of the cam in the horizontal direction (m); L_2 represents the distance from the point to the center of the cam in the vertical direction (m).

In principle, without considering the rotation of the cam, the absolute velocity vector of point H_T $H_T$on the outrigger can be seen in Equation (11).

$$\boldsymbol{V}_{\boldsymbol{H}_T} = -L_1 w_{\mathrm{n}2} \tag{11}$$

The sliding speed vector of point H_T on the follower relative to point H_T on the cam is shown in Equation (12) on the cam profile.

$$\boldsymbol{V}_{\boldsymbol{S}\boldsymbol{H}_C} = -L_2 w_{\mathbf{n}1} \tag{12}$$

The sliding speed vector of point H_C on the cam with respect to point H_C on the plunger is given by Equation (13).

$$\boldsymbol{V}_{\boldsymbol{s}\boldsymbol{H}_T} = -\boldsymbol{V}_{\mathrm{sHC}} = L_2 w_{\mathbf{n}1} \tag{13}$$

Assuming that the sliding direction has no effect on wear, the magnitude of the sliding speed is given by Equation (14).

$$|V_s| = L_2 w = (R_0 + h)w \tag{14}$$

where R_0 is the base circle radius of the cam (m).

Considering the compound motion of rolling and sliding between the cam and the tappet contact pair, the velocity of contact point H_E can be expressed by Equation (15).

$$\boldsymbol{V}_{\boldsymbol{H}_E} = L_1 \boldsymbol{n}_1 + L_2 \boldsymbol{n}_2 = L_1 \boldsymbol{n}_1 - L_1 w \mathrm{n}_2 \tag{15}$$

According to Equations (13) and (14), the absolute velocity magnitude of point H on the contact surface relative to the cam can be obtained, as shown in Equation (16).

$$\left|\boldsymbol{v}_{\boldsymbol{H}_E} - \boldsymbol{v}_{\boldsymbol{H}c}\right| = \left|-\frac{\mathrm{d}^2 h_\alpha}{\mathrm{d}\alpha^2} - R_0 - \mathrm{h}\right| w \tag{16}$$

According to Equation (9), the calculation Equation for the wear at contact point H between the cam and the tappet is obtained as Equation (17).

$$h = k_0 P_H(t) |\boldsymbol{v}_s(\boldsymbol{t})|_{\mathrm{dt}} \tag{17}$$

Assuming that the pressure does not change with time during the process of contact through the contact area, and the parabolic shape of the Hertz contact pressure distribution is obtained [5], the calculation Equation (18) is derived.

$$\mathrm{dt} = \frac{\mathrm{dx}}{\left(\left|\boldsymbol{v}_{\boldsymbol{H}_E} - \boldsymbol{v}_{\boldsymbol{H}c}\right|\right)} \tag{18}$$

The equation for calculating the depth of friction derived above can be found in Equation (19).

$$h_{HC} = k_c P_0 \alpha \frac{\pi}{2} \left| \frac{\boldsymbol{v}_s}{\boldsymbol{v}_{\boldsymbol{HE}} - \boldsymbol{v}_{\boldsymbol{HC}}} \right| = P_0 \alpha \frac{\pi}{2} \frac{R_0 + h}{\frac{a^2 h}{\mathrm{d}\alpha^2} R_0 + h} \tag{19}$$

4. Optimization of structural analysis

The constraints and value ranges derived from the wear depth are used for motion simulation in MATLAB. The simulation is conducted for a rotation of 162°, with 1620 outputs collected and concatenated to form an output curve. During the setup, it is ensured that the value of h_{HC} does not exceed 2×10^{-8} mm. Once the state parameters after wear exceed the design values, the state parameters of the cam mechanism are updated, changing the pressure distribution until h_{HC} meets the requirements. Figure 5 shows the MATLAB flowchart of this calculation process.

Through the output of the MATLAB program, the optimized lift curve of the cam follower can be obtained. In Figure 6, since the origin is set to the position point just entering the lift section when the coordinate system is established, the lift size is 0 mm at 0° and the follower reaches the maximum displacement of 8.2 mm at α = 81°. After substituting α = 81° into Equation (2), the maximum displacement is calculated to be 2.8 mm. By inversely solving the output motion image, the optimized parameters of the cam profile can be obtained.

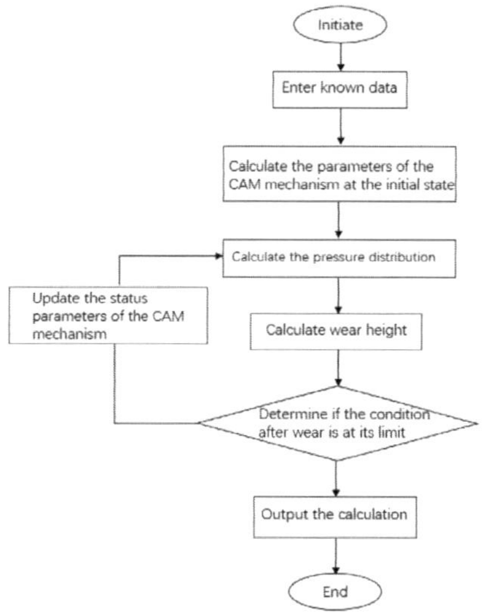

Figure 5. Algorithm Flowchart.

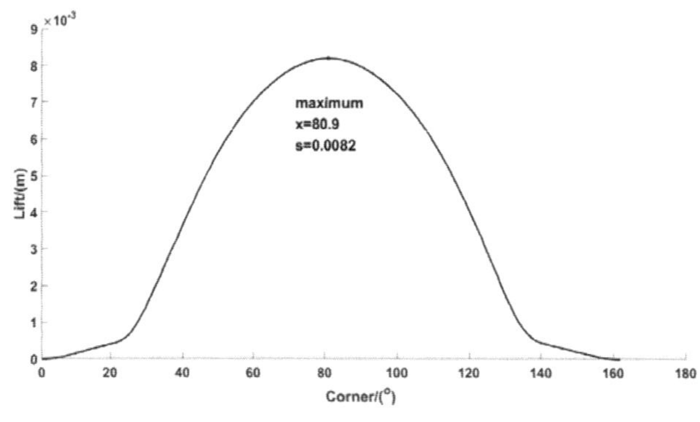

Figure 6 Lift curve after optimization.

4.1 Contact stress

According to Equation (19), it can be seen that the magnitude of the contact stress will directly affect the wear degree between the cam and the tappet. Figure 6 shows the variation of the load between the cam and the tappet contact position with the angle change before and after the optimization of the cam profile. When the parameters of the cam mechanism are not updated, the contact stress at the angle of 20° is 401.0218 MPa. At 20°, there is a stress mutation, which will cause a significant impact on both the cam and the tappet. After updating the parameters, the contact stress at the 20° position decreases to 348.3256 MPa. Among them, in Figure 7(a), the maximum contact stress in the graph is 663.8656 MPa, and in (b), the maximum contact stress in the graph is 554.3465 MPa. The difference between the two is 109.5191 MPa, which is a reduction of 16.5%.

4.2 Wear height

Through the analysis of the contact stress at each position in Section 2.1, the relationship between contact stress and friction depth can be intuitively felt. In Figure 8(a), the maximum wear height obtained by inputting the original data into MATLAB is output as 2.6237×10^{-8} mm, and the output after optimization is 1.78991×10^{-8} mm. Assuming that the height worn by the cam per revolution is the same and does not change with the wear amount of the previous revolution, it can be inferred through data analysis that the service life of the optimized cam rocker mechanism will be increased by 31.8%.

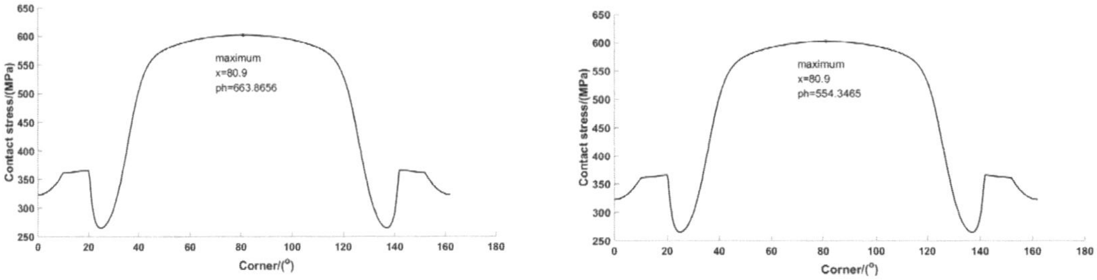

(a) Before optimization (b) After optimization

Figure 7. Comparison of Contact Stress.

5. Conclusion

Through transmitting feedback information to the cam profile equation, the push and return equations were changed to optimize the cam profile, thereby improving the service life of the cam push rod. A comparison of the contact stress distribution before and after optimization shows that the contact stress decreased by 16.5% after optimization. Under the same external load conditions, the service life of the optimized cam follower increased by 31.8%. The graphical results show that the optimized cam profile makes the contact stress distribution more uniform, reducing wear and extending life. These improvements not only reduce power loss but also improve the efficiency of the cam follower. Therefore, our research results indicate that substantial improvements in the life and efficiency of the valve train can be achieved by optimizing the cam profile.

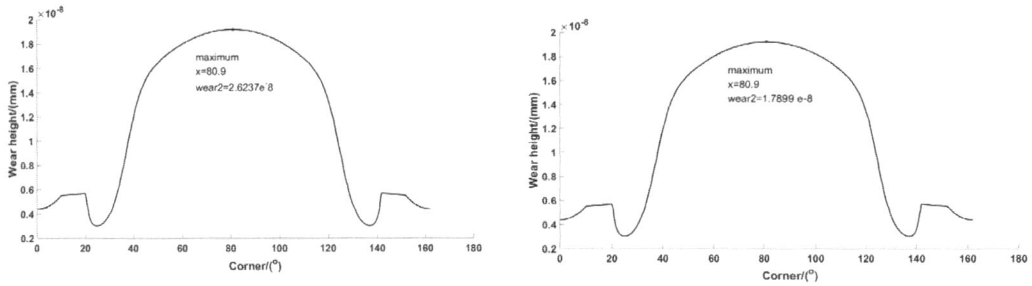

(a) Before optimization (b) After optimization

Figure 8. Comparison of Contact Stress.

References

[1] Sun Huan, Ge Wenjie. Mechanical Principles [M]. Beijing: Higher Education Press, 2021.
[2] Sun Xunfang, Fang Xiaoshu, Guan Laitai. Material Mechanics, 6th Edition. Higher Education Press, 2019.
[3] Ge Zhenghao, Wang Yingjie, Wang Shaohua, Zhao Qiujie. Dynamic Analysis of Cam Rocker Mechanism Based on ADAMS. Manufacturing Automation, 2022, 44(08): 34-38.
[4] Ma Shenghui, Ji Zhengjia, Cai Xiaochun, Ji Mingliang. Study on Wear Characteristics of Wind Turbine Bearings Based on Finite Element Analysis and Archard Model. Journal of Qinghai University, 2024, 42(01): 70-77+106.

[5] Li Zhaokun, Zeng Qi, Liu Qingsheng. Analysis of Crank Mold Wear Based on Archard Model [J]. Forging Technology, 2023, 48(11): 167-173.
[6] Wu Xiaozhao, Zheng Xiaoguo, Li Yanyan. Cam Optimization Design Method for Five-Link Spring Actuating Mechanism [J]. High Voltage Apparatus, 2023, 59(03): 150-155.
[7] G. Ganesan Barnabas V. Clinton Wilson, and G. Ajith Kumar. Optimal dimensional synthesis of one degree of freedom fixed cam-link four-bar mechanisms to generate filleted rectangular paths. Mechanics Based Design of Structures and Machines, 2023, 51(10): 5531-5540.
[8] Jonatan P, J. N F, A.F. J V, et al. Comparison of forward and inverse cam generation methods for the design of cam-linkage mechanisms [J]. Mechanism and Machine Theory, 2023, 190
[9] Guodong Z, Yong W, GuoNiu Z, et al. Parameter Optimization of Large-Size High-Speed Cam-Linkage Mechanism for Kinematic Performance [J]. Actuators, 2022, 12(1): 2-2.
[10] Liangwen W, Shizhao Z, Caidong W, et al. Optimization and synthesis of a cam-linkage mechanism with a swing follower with variable pivot [J]. Proceedings of the Institution of Mechanical Engineers, Part C: Journal of Mechanical Engineering Science, 2022, 236(6): 2861-2883.

High-efficiency tandem red organic light-emitting diodes adopting exciplex hosts

Jiong Wang and Shengdong Zhang*

School of Electronic and Computer Engineering, Peking University Shenzhen Graduate School, Shenzhen 518055, China

*Corresponding author's e-mail: zhangsd@pku.edu.cn

Abstract. Deep-red organic light-emitting diodes (OLEDs) exhibit significant potential for applications in infrared medical treatment and infrared imaging. However, the OLED performance drops significantly with increasing emission wavelength due to the energy gap law. To overcome this challenge, an efficient interfacial exciplex host, NPB/TPBi, has been developed. Simulation results indicate that the interfacial exciplex host demonstrates superior performance, compared with the bulk exciplex host and single host. To further boost the performance of red OLEDs, the interfacial exciplex host was incorporated into tandem devices, achieving an external quantum efficiency of up to 27.3%. This work offers important insights for the fabrication of efficient red OLEDs.

1. Introduction

Deep-red light can enhance mitochondrial catalase activity, promoting cellular metabolism and synthesis, and can be clinically used to treat various diseases such as inflammation, ulcers, neuralgia, hair loss, and so on [1-3]. Organic light-emitting diodes (OLEDs), because of light weight, flexibility, and excellent skin affinity, are the ideal carriers for red light medical applications [4-6]. However, due to the energy gap law, the performance decreases sharply with the increase of the emission wavelength [7, 8]. Adopting a host-guest doped emissive layer is an ideal path to improve OLEDs. Especially for hosts with thermally activated delayed fluorescence (TADF) properties, theoretically, they possess 100% internal quantum efficiency [9]. Materials with TADF properties are mainly divided into TADF molecules and exciplex [10]. However, the intramolecular TADF is difficult to design. In order to simplify the production process, Professor Adachi proposed to prepare TADF materials through the non-chemical bond that is the exciplex [11]. Therefore, developing an exciplex host suitable for sensitizing red dyes is an effective approach to achieving highly efficient red OLEDs.

In this study, an innovative exciplex of NPB/TPBi was successfully developed. Through simulation analysis utilizing the ideality factor, it was discovered that devices employing the interfacial exciplex host (IEH) exhibit notable advantages in exciton recombination when compared to single-host and bulk exciplex host devices. To elevate device performance further, the interfacial exciplex host was incorporated into tandem OLEDs, resulting in a deep red device with a CE of 48.8 cd/A, and an EQE of 27.3%. This remarkable achievement paves the way for new opportunities in red OLED technology advancement and application.

2. Results and discussion

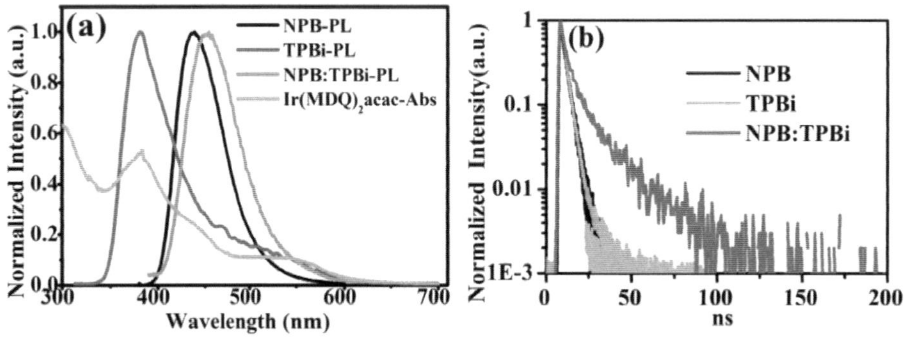

Figure 1. a) The spectra of TPBi, NPB, $Ir(MDQ)_2(acac)$, and TPBi:NPB; b) Time-dependent PL decay of TPBi, NPB and TPBi:NPB in air monitored at 436, 381 and 454 nm, respectively.

Observing **Figure 1a**, the formation of exciton complexes between TPBi and NPB is evidenced by the red-shifted PL spectrum of the mixed film and the broadening of the full widths at half maximum (FWHM) of the spectrum compared to the single film. Then, as shown in **Figure 1b**, the TPBi/NPB film shows a longer PL lifetime. What's more, the new emission energy gap of 2.73 eV closely resembles the difference between the donor and acceptor energy levels, which also indicates that NPB and TPBi form an exciton complex. Moreover, the overlap of the dye absorption spectrum with the emission spectrum of the exciton complex facilitates energy transfer to the dye.

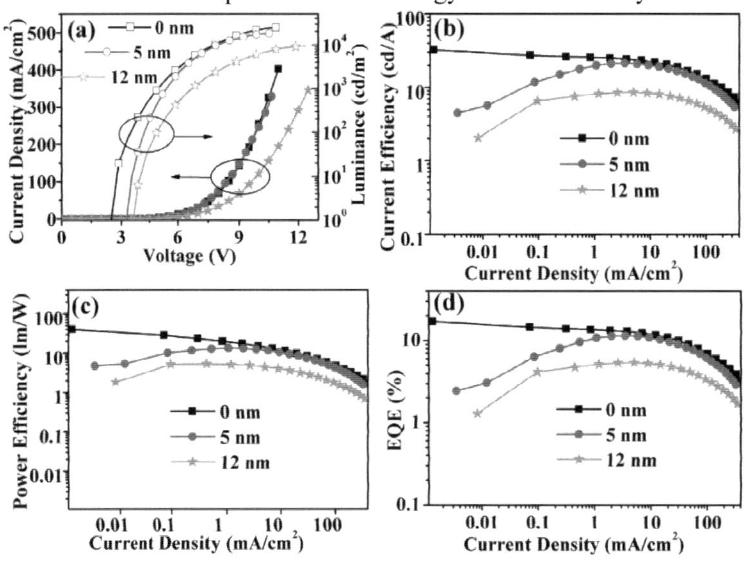

Figure 2. Characteristics of the devices: a) *CD-V-L*; b) *CE-CD*; c) *PE-CD*; d) *EQE-CD*.

To underscore the crucial role of the IEH, we inserted a BCP interlayer between the TPBi and NPB to hinder the formation of interface exciplexes. Given that NPB and BCP do not form exciplexes [12], we fabricated devices with varying BCP thicknesses (x = 0, 5, and 12 nm) in the structure. As depicted in Figure 2, the devices without BCP exhibit the lowest turn-on voltage and the highest current efficiency (CE), while the device with 12 nm BCP shows the worst performance, with the CE being only a quarter of the device with the exciton complex host. Table 1 summarizes the key characteristics of these devices. Previous studies have indicated that when BCP is 5 nm , there is a limited amount of interface exciplex formation. However, with a 12 nm BCP interlayer, most excitons fail to diffuse to the electron transport layer, thereby preventing the formation of exciplexes [13].

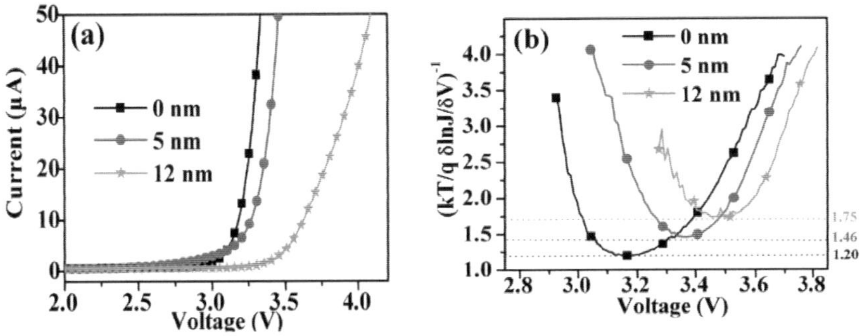

Figure 3. a) Current-voltage characteristics of devices; b) the ideality factor (η).

The device J-V characteristics reflect different carrier recombination mechanisms. Equation (1) is the Shockley diode equation, which describes the relationship between the current and the ideality factor η [14]. It has been shown that the η is 1 when the exciton recombination mode (ERM) is Langevin recombination (LR), and is 2 when the ERM is trap-assisted recombination (TAR). The Equation (1) is shown as:

$$\eta = \left(\frac{kT}{q} \frac{\delta \ln J}{\delta V} \right)^{-1} \tag{1}$$

where k represents the Boltzmann constant. Based on the data in Figure 3a, the η can be obtained using Equation (1), and the results are presented in Figure 3b. The OLEDs adopting an IEH, show an η value of 1.20, suggesting the ERM is LR. Meanwhile, for the device with a 12 nm BCP layer, the η value is 1.75, suggesting that TAR dominates in this device.

Based on the TPBi/NPB IEH, single deep-red OLEDs with one emission unit and tandem deep-red OLEDs with two emission units were fabricated, with the device structure depicted in Figure 4. In Figure 5a, the tandem OLEDs show lower current density, which is conducive to improving stability. Figure 5b shows that the tandem OLED has a V_{on} of 6.0 V. The lower turn-on voltage is attributed to the IEH. At the driving voltage of 20 mA/cm^2, the tandem devices show the doubled brightness of single devices, which indicates that the charge generation unit exhibits excellent charge generation and separation effects. Additionally, the *CE* and *EQE* of the tandem OLEDs are 48.8 cd/A and 27.3%, respectively. Figure 5f shows that the spectra of the two devices are largely consistent, with minor differences mainly resulting from the microcavity effect in tandem OLEDs. Figure 5g presents that the EL spectrum of the tandem OLEDs is increased by two orders of magnitude. The results demonstrate the significant advantage of the tandem OLEDs in enhancing device efficiency. Figure 5h shows that the tandem device emits deep-red light, with a CIE coordinate of (0.61, 0.38). The comprehensive results of the single OLEDs and tandem devices are shown in Table 1.

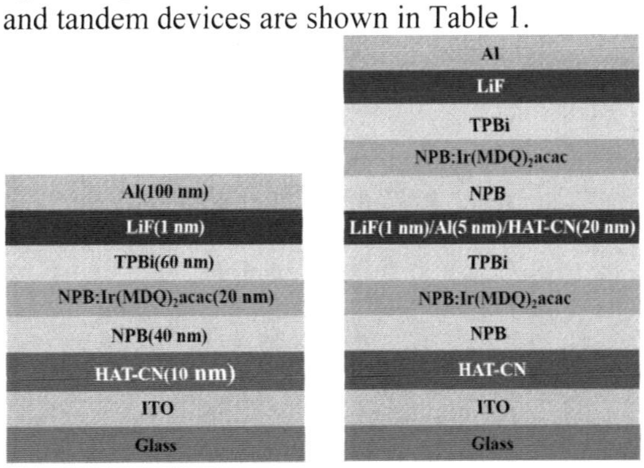

Figure 4. Structure of single devices and tandem devices with TPBi/NPB IEH.

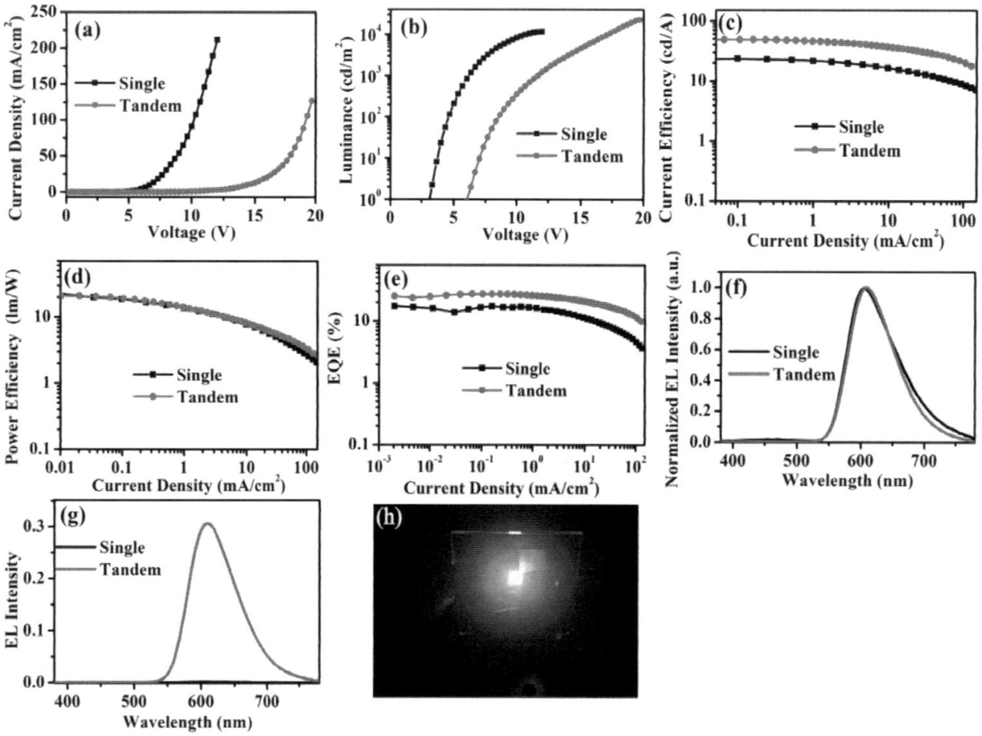

Figure 5. Performance comparison of single devices and tandem devices based on the IEH. a) *CE-V*. b)*L- V*; c) *CE- CD*; d) *PE- CD;* e) *EQE-CD*; f) normalized EL spectra; g) EL spectra of the device at 8 V; h) luminescence photographs of the device.

Table 1. Comprehensive results of the single OLEDs and tandem devices

Devices		V_{on}(V)	L_{max}(cd/m^2)	CE_{max}(cd/A)	PE_{max}(lm/W)	EQE_{max}(%)
Solution processed single OLEDs	0 nm	2.6	24320	32.2	40.4	17.2
	5 nm	3.4	17850	21.3	13.4	11.5
	12 nm	3.8	9330	8.5	5.3	5.4
Evaporation OLEDs	Single	3.1	11540	23.9	22.5	17.3
	Tandem	6.0	22684	48.8	23.6	27.3

3. Conclusions

In this study, aiming at improving the performance of deep-red OLEDs, an interfacial exciplex NPB/TPBi was successfully developed. The superiority of deep-red OLEDs adopting NPB/TPBi host in exciton recombination mechanism and performance was verified through the ideality factor measurements. To further enhance the performance of deep-red OLEDs, the NPB/TPBi host was applied to tandem OLEDs, achieving a *CE* of up to 48.8 cd/A and an *EQE* of 27.3%, providing a powerful strategy and path for the performance enhancement of deep-red OLEDs.

Acknowledgments

This work was supported by the National Key R&D Program, China (No. 2022YFB3607200) and the Shenzhen Municipal Scientific Program, China (No. JCYJ20220818100808019).

References

[1] Wang K, Fu J, Dong H, Huang B, Liu J, Tian L, Feng J, Yang C, Lou C, Xu L, and Sun T (2024) Enabling highly efficient neodymium luminescence for near - infrared phosphor-converted light - emitting diode applications. Small Struct., 2400092.

[2] Li H H, Wang Y K, and Liao L S (2024) Near - infrared luminescent materials incorporating rare earth/transition metal ions: from materials to applications. Adv. Mater., 2403076.

[3] Wu J, Wang Y, Jiang P, Wang X, Jia X, and Zhou F (2024) Multiple hydrogen-bonding induced nonconventional red fluorescence emission in hydrogels. Nat. Commun., 15: 3482.

[4] Jiao Y H, Qiu W D, Li M K, and Su S J (2024) Modulation of Intermolecular interactions in organic emitters for highly efficient organic light-emitting diodes. Chem. Eur. J., 202401635.

[5] Ari D, Yang Y J, Quinton C, Jiang Z Q, Zhou D Y, and Poriel C (2024) Spirobifluorene trimers: high triplet pure hydrocarbon hosts for highly efficient blue phosphorescent organic light-emitting diodes. Angew. Chem. Int. Ed., 202403066.

[6] Zhang K, Zhou Z X, Liu D H, Chen Y C, Zhang S Y, Pan J, Qiao X F, Ma D G, Su S J, Zhu W G, and Liu Y (2024) Boosting external quantum efficiency to 12.0% of an ultraviolet OLED by engineering the horizontal dipole orientation of a hot exciton emitter. Angew. Chem. Int. Ed., 202407502.

[7] Englman R and Jortner J (1970) The energy gap law for radiationless transitions in large molecules. Mol. Phys., 18: 145-164.

[8] Minotto A (2022) Efficient emitters in the NIR-II window. Nat. Photonics, 16: 812-813.

[9] Xue N, Zhou H Y, Han Y, Li M, Lu H Y, and Chen CF (2024) A general supramolecular strategy for fabricating full-color-tunable thermally activated delayed fluorescence materials. Nat. Commun., 15:1425.

[10] Yuan L, Xu J W, Yan Z P, Yang Y F, Mao D, Hu J J, Ni H X, Li C H, Zuo J L, and Zheng Y X (2024) Tetraborated intrinsically axial chiral multi - resonance thermally activated delayed fluorescence materials. Angew. Chem. Int. Ed., 202407277.

[11] Sheng R, Chen C, Zhang E, Zhao W, Tang J, Yan D, Li Z, and Chen P (2023) Management of exciton distribution for high-performance organic light-emitting diodes based on interfacial exciplex architecture. Molecules, 28: 7028.

[12] Lu H T, Tsou C C, and Yokoyama M (2005) Improvement of FWHM and luminance of blue organic light-emitting diodes with double hole-blocking structure. J. Cryst. Growth, 277:388-392.

[13] Luhman W A and Holmes R J (2009) Enhanced exciton diffusion in an organic photovoltaic cell by energy transfer using a phosphorescent sensitizer. Appl. Phys. Lett., 94: 153304.

[14] Shockley W (1951) Hot electrons in germanium and Ohm's law. Bell Labs Tech. J. 30: 990-1034.

Study on no IMC Solid state bonding method for high-density 2.5D/3D integration

Zongyang Li[1], Zhibo Qu[1,*], Quanbin Yao[1], Pengrong Lin[1], Xiaochen Xie[1], Hengtong Guo[1], Jiaxing Wang[1], Pengyan Hou[1], and Songzhao Gu[2]

[1] Beijing Microelectronics Technology Institute, Beijing, 100076, China
[2] School of Materials Science and Engineering, Beijing Institute of Technology, Beijing, 100081, China

[*] Corresponding author's e-mail: 1829130194@qq.com

Abstract: Chiplet integration is a new development direction for the continuation and even goes beyond Moore's Law, and in order to achieve the high density, high reliability, high-speed Chiplet device, a new method of 2.5D/3D integration approach is needed to fulfill the needs to Chiplet heterogeneous devices. In this work, two dies were successfully bonded via low energy input with no IMC solid state bonding under 220°C and 5 MPa, which is much gentler than the other solid state bonding approaches. Each die had a matrix of 22500 Cu pillars with a diameter of 20 μm. This method was promising for the near future high-density 2.5D/3D integration.

1. Introduction

Due to the significant increase in device design and process costs, Moore's Law, which relies on the size reduction of transistors is difficult to maintain. 2.5D/3D integration is a promising approach for the continuation of Moore's Law and even can be pushed forward. However, it has brought great challenges to the integration with the rapid increase in its functions and complexity. The evolution of the integration approach has changed from traditional homogeneous integration to heterogeneous integration rapidly. The innovation of electrical interconnection technology between devices in 2.5D/3D integration is the key to the development of new microsystems.

Copper pillars-Tin Caps structure is widely used in the current 2.5D/3D integration for general integrated circuits. The interconnection joint is a typical multi-layer structure composed of a solder matrix; the interconnection interface is made from layers of intermetallic compound (IMC). With the development of high-density packaging, the size of C2 bumps gradually decreases, and the proportion of interface IMC in the interconnect structure will increase sharply, resulting in an obvious size effect [1]. This will cause most or all of the tin cap parts in the micro-C2 bump interconnection structure to be converted into IMC during subsequent high-temperature storage, turned into a decrease in the electrical and thermal conductivity of the interconnection path and made the joint brittle, which has a serious adverse impact on the long-term reliability of the circuit [2][3]. Therefore, 2.5D/3D high-density integration of integrated circuits for highly reliable usage requires the development of new micro-bump interconnect materials that have no interface IMC generation or can effectively inhibit the growth of interface IMC.

The micro-region plastic deformation and interconnect interface atomic diffusion and rearrangement are used to achieve reliable interconnect joints for micro-bumps under the condition of low temperature and low-pressure bonding process. Silver had lower yield strength and was softer than copper and also had good plastic deformation ability, which could significantly reduce the yield strength of the material and greatly increase its plastic deformation capacity and ultimate tensile strength [4][5]. In order to further reduce the input energy during bonding energy, taking advantage of the thermodynamically unstable characteristics of silver oxide, fluid-like interfacial wetting behavior will occur during its self-decomposition process, which greatly enhances the ability of atoms to diffuse and rearrange along the interconnection interface. silver-silver micro-bumps interconnection was obtained under the bonding condition of 210°C and 1.3 MPa, and the original interconnection interface was completely bridged through recrystallization [6]. Also, similar work has been done by [7], Ag-Ag direct bonding processes were achieved by the thermo-compression process at 180-200°C.

By using this mechanism, the large-area array interconnection test is carried out, and the large-area array interconnection of 22500 samples with raised points is realized. The simulation method is used to investigate the influence of the interconnect convex appearance on the interconnect process, and the convex appearance window suitable for the mechanism is given.

2. Layout design and sample fabrication

Figure 1 shows the layout designs used to investigate the suitable parameters for the low-energy input solid-state bonding process. There were two different designs that contained three different matrixes with 22500 and 2025 Cu pillars respectively. Two dimensions of Cu pillars were designed with diameters of 20 μm and 25 μm respectively. The height of the pillars was kept the same at 17 μm.

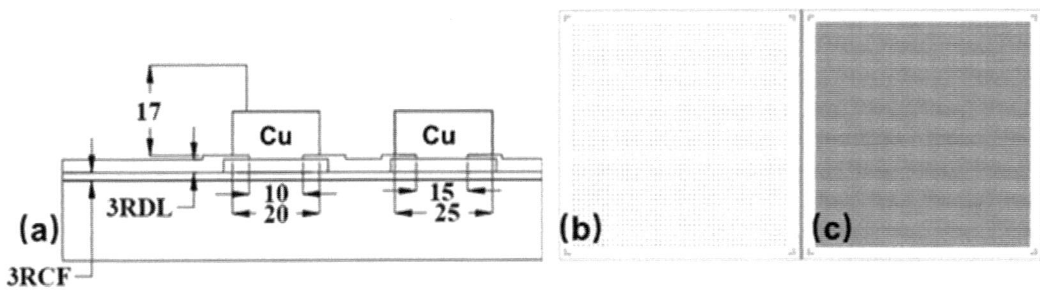

Figure 1. Layout design of the Cu pillars, (a) The cross-section view of Cu pillar with diameters of 20 μm and 25 μm respectively, (b) Layout design of 22500 Cu pillars in one die, (c) Layout design of 22500 Cu pillars in one die.

The fabrication of the Cu pillar wafer was carried out by the commercial foundry, and the deposition of the 2-3 um Ag step was carried out in the physical vapor deposition (PVD) machine. A thin layer of Ti was deposited before the deposition of Ag. 50 nm of Ti and 3 μm of Ag were deposited on top of the Cu Pillar respectively. AgO_2 formation was carried out in the plasma environment from the mixture gas of O_2 and Ar. Silver oxide films were grown by using plasma bombardment. The chip with the silver layer grown on the bump surface after magnetron sputtering was placed in a vacuum chamber, and then argon and oxygen were introduced to create a plasma environment. In the plasma environment, the Ag atoms on the surface of the Ag layer reacted with oxygen to grow a silver oxide film. The characteristics of the silver oxide film were related to the growth parameters. The growth parameters adopted to prepare the samples in this article were oxygen of 12 sccm, Argon gas of 36 sccm, starting power of 75 w, reaction time of 15 s, and more studies on AgO_2 generation and characterization can be found in [8].

The solid-state bonding process was: first, the about-to-bond chips were placed on the vacuum platform of the flip chip bonder, then, the parallel calibration was performed to ensure that the upper and lower bonding surfaces were parallel. X-ray for bump position alignment was also introduced. Finally, bonding was performed according to the set bonding parameters, which were: bonding temperature of 220°C, bonding pressure of 5 MPa, and holding time of 30 minutes.

Journal of Physics: Conference Series **2842** (2024) 012080 doi:10.1088/1742-6596/2842/1/012080

3. Experimental results

Figure 2 shows the cross-section image of the bonded sample from the pillars with a diameter of 20 μm bonded with pillars with a diameter of 25 μm. Cu pillars from the top die and bottom die were well aligned. However, voids could be seen within the interconnection joints, which indicated that due to the Multi-wafer Project (MPW) project, during electro-plating deposition (ECD), the current of ECD was not well controlled and caused the shape difference of Cu pillars with different diameters.

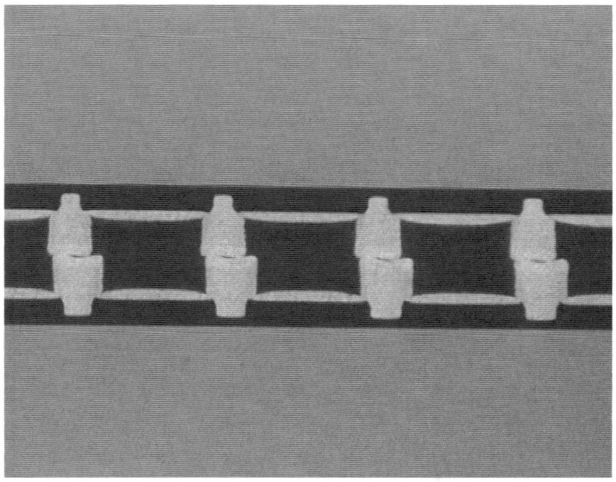

Figure 2. Cross-section image of Cu Pillars (top die) with a diameter of 20 μm bonded with Cu pillar (bottom die) with a diameter of 25 μm, voids could be seen in the middle of the connected joints.

The occurrence of the voids might be the concave shape of the Cu pillar, which indicated that the shape of the Cu pillars affected the result of the bonding. From the image above, the Cu pillars with a diameter of 20 μm had a more flat facet shape. The SEM of the Cu pillar with a diameter of 20 μm was presented in Figure 3.

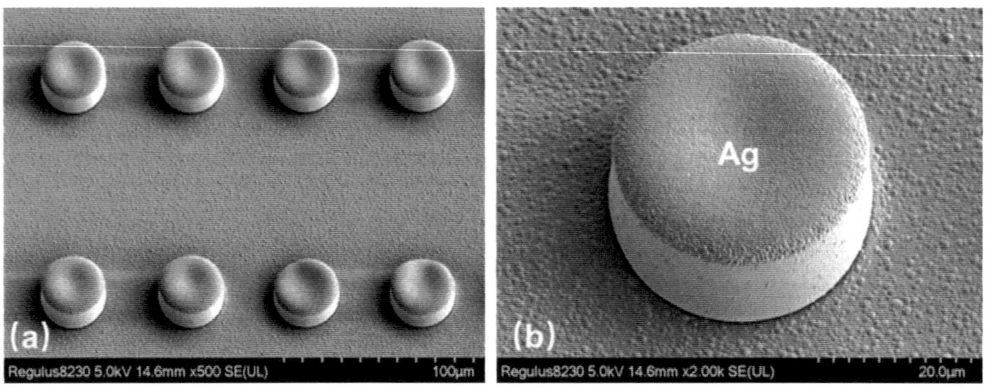

Figure 3. The SEM images of the Cu pillar with a diameter of 20 μm.

On the other hand, the shape of the 20 μm diameter Cu pillar got a flatter facet, as can be seen in Figure 4 (c), which was ideal for direct bonding. Under the same bonding parameters, it could be seen that a successful bonding facet between two dies with 22500 Cu pillars was presented.

Thus, the same bonding process was performed between the 20 μm diameter Cu pillar dies, and SEM images are presented below. The bonding pressure and temperature were 5 MPa and 220°C respectively, and the holding time was 30 min. The reason why the top and bottom Cu pillars were different in size was due to the cross-section examined during polishing.

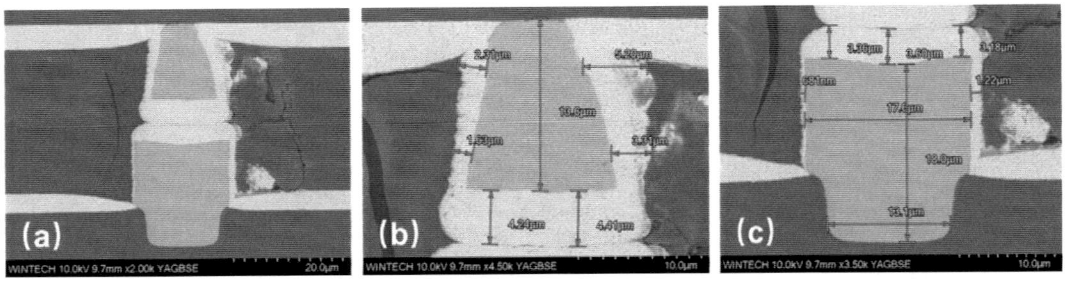

Figure 4. Cross-section SEM image of bonded dies from Cu pillars with a diameter of 20 (a), Seamless interconnection joints between the Cu pillars with a diameter of 20 μm (b), and (c) Featured Cu pillar structures.

It was worth pointing out that the key to achieving a successful interconnection joint through low-energy input integration was a flat pillar facet. However, the quality of Cu pillar deposition was very much likely related to the ECD process, which was hard to control. Thus, a series of simulations was carried out to investigate how the shape of the Cu pillar affects the interconnect joint after the bonding process.

4. Simulation modeling and results

Using ANSYS simulation software, this study investigated the influence of concave deformations on the outcomes of interconnections of copper pillars, as observed during experimental processes and depicted in Figure 3. The objective is to determine the range of concavity in the interconnecting copper pillars that can achieve successful interconnection under the specified input conditions. Since nano-silver oxide became unstable and decomposed into Ag and oxygen above 180°C, it was not considered in the simulation model. As illustrated, a model for a single copper pillar interconnection was established, where the interconnecting copper pillar, mirroring real-life scenarios, had a certain degree of concavity. This concavity, as shown in Figure 5, was modeled with an angle ranging from 1° to 5°. The copper pillar had a diameter of 20 μm, a height of 17 μm, and was topped with a 3 μm thick layer of Ag.

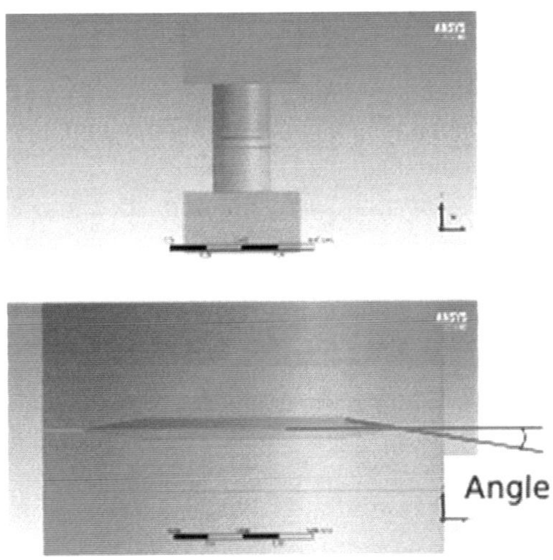

Figure 5. Single pillar bonding model.

For the Cu pillar showed up as concave shapes, the relation between the bonding interfaces with the angle of the concave was investigated. The bonding condition remained the same, that is, the bonding temperature was 220°C, the bonding pressure was 5 MPa, and the changing parameter was the angle of the concave of the Cu pillar facets. Figure 6 (a)-(c), (d)-(f), (g)-(i), (j)-(l), (m)-(o) represents the increasing angles from 1° to 5° respectively. It was clear that with the increasing of the angle of the concaves, under the same input energy of bonding, the interconnect joints appeared voids. When the angle was greater than 3°, clear voids appeared, which indicated that when the shape of the Cu pillars appeared concave structure and the angle was greater than 3°, larger input energy was required.

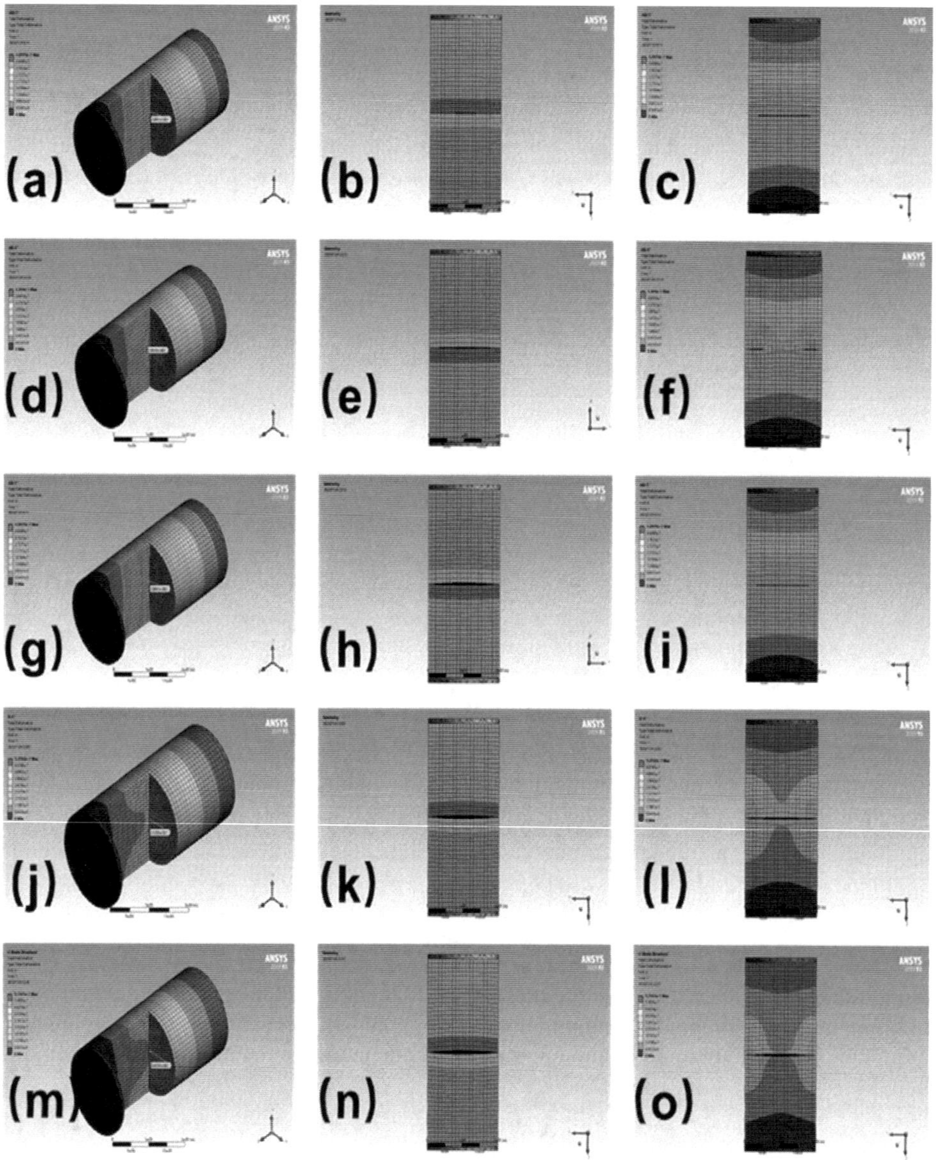

Figure 6. Simulation models of Cu pillars with concave shapes with different angles, (a)-(c) the cross-section view of the bonded interfaces, interface before bonding, and after bonding respectively under 5 MPa pressure with an angle of 1°,(d)-(f) the cross-section view of the bonded interfaces, interface before bonding, and after bonding respectively under 5 MPa pressure with an angle of 2°, (g)-(i) the cross-section view of the bonded interfaces, interface before bonding, and after bonding respectively under 5 MPa pressure with an angle of 3°, (j)-(l) the cross-section view of the bonded interfaces, interface before bonding, and after bonding respectively under 5 MPa pressure with an angle of 4°, (m)-(o) the cross-section view of the bonded interfaces, interface before bonding, and after bonding respectively under 5 MPa pressure with an angle of 5°.

5. Conclusion

By leveraging the mechanisms of microzone plastic deformation and atomic diffusion and rearrangement at the interconnect interface, a reliable interconnection of 22,500 copper pillars, each with a diameter of 20 μm, was successfully established under conditions of 220°C and 5 MPa. The study, through conducting the simulation, investigated the optimal morphological window for copper pillars for solid-state interconnection and concluded that the concavity angle of the copper pillars should be less than 3°. The over-concave shapes needed higher input energy to perform a seamless bonding, which was not ideal for 2.5D/3D integration. The surface preparation of the copper pillars, such as by Chemical Mechanical Polishing (CMP), remains a subject for further research. Overall, this method of solid-state bonding with low energy input is highly promising for 2.5D packaging applications that require high reliability, as it enables smaller dimensions and spacing of copper pillars while preventing the formation of IMCs.

References

[1] A YK, A SKK, B SDKA, C SEK. Wafer warpage analysis of stacked wafers for 3D integration - ScienceDirect. Microelectronic Engineering 2012, 89(1): 46-49.
[2] Li J, Zhang Y, Zhang H, et al. The thermal cycling reliability of copper pillar solder bump in flip chip via thermal compression bonding [J]. Microelectronics Reliability, 2020, 104: 113543.
[3] Wu W, Yang H, Li Y, et al. Effect of IMC Thickness on the Mechanical Properties of Microbumps [J]. IEEE Transactions on Components, Packaging and Manufacturing Technology, 2023, 13(2): 265-273.
[4] Zhang Z, Suetake A, Hsieh M C, et al. Ag-Ag direct bonding via a pressureless, low-temperature, and atmospheric stress migration bonding method for 3D integration packaging [C]. 2022 IEEE 72nd Electronic Components and Technology Conference (ECTC). IEEE, 2022: 1409-1412.
[5] Chang L P, Huang S Y, Chang T C, et al. Low temperature Ag-Ag direct bonding under air atmosphere [J]. Journal of Alloys and Compounds, 2021, 862: 158587.
[6] Wu J, Huo Y, Lee CC. Direct Ag-Ag bonding by in-situ reduction of surface oxides for advanced chip-package interconnection. Materialia 2018.
[7] Chang, L. P., Huang, S. Y., Chang, T. C., & Ouyang, F. Y. Low temperature Ag-Ag direct bonding under air atmosphere. Journal of Alloys and Compounds 2021, 862, 158587.
[8] Wang, Taiyu, Songzhao Gu, Yexing Fang, Donglin Zhang, Xiaochen Xie, Zhibo Qu, Yong Wang, et al. "Plasma-induced growth mechanism of surface-state silver oxide in nanoscale for low-temperature bonding technology." Materials Characterization 199 (2023): 112830.

Modelling of a thin-film transistor based on the carbon nanotubes network and ESD circuit simulation by SPICE

Xianghua Liao, Weihao Zeng, Hao Huang, Qinqin Wang and Zijie Zhou*

Hunan Institute of Advanced Sensing and Information Technology, Xiangtan University, Xiangtan, Hunan, 411105, China

* Corresponding author's e-mail: zzj@xtu.edu.cn

Abstract: Carbon nanotubes are considered one of the most promising candidate materials for Si. However, there has been limited research on modeling devices based on carbon nanotubes and performing discharge (ESD) protection simulations for circuits composed of carbon nanotube-based devices. Within this article, the current characteristics of network carbon nanotube field-effect transistors (CNT-TFT) with micrometer size are studied, and the devices with channel length L_g = 11 μm and width of W = 108 μm are modelled. It is found that the simulation curves of the model are in good agreement with the experimental results. In this article, the ESD behavior modeling method is introduced. By utilizing this method, a bidirectional diode device model was programmed and integrated into an inverter built with CNT-TFT models. When an ESD positive pulse is applied between the input and ground terminals (PS mode), it is found that the simulation results of the circuit have passed the 500 V human body model (HBM) ESD pulse test in the PS mode, thereby achieving ESD protection simulation for carbon-based integrated circuits.

1. Introduction

With the advancement of technology, the size of silicon-based devices is shrinking. According to reports, the feature size of silicon-based devices will soon reach the two-nanometer node. However, this size reduction will bring about negative effects, such as increased leakage current and short-channel effects in silicon-based transistors. These negative effects make it difficult to further reduce the feature size of silicon-based devices and pose a challenge to the continuation of Moore's Law.

In order to continue Moore's Law [1], scientists are making every effort to search for new materials, among which carbon nanotubes are considered one of the most promising candidate materials for Si [2]. Due to their quasi-one-dimensional structure, carbon nanotubes possess intrinsic advantages such as high mobility, low scattering rate, and relatively long mean-free path.

ESD is one of the key factors affecting the reliability of integrated circuit (IC) chips. According to statistics, ESD causes the global electronics industry to suffer billions of dollars in losses each year in terms of IC yield and early field failures. Over 35% of chip failures are related to ESD [3]. From this, it is evident that accelerating research on electrostatic discharge protection can lead to significant savings in manufacturing time and costs. Carbon-based devices are also susceptible to electrostatic discharge hazards as an alternative to silicon-based devices. For example, applying rapidly changing voltages to CNT-TFT devices may cause gate oxide breakdown and thermal failure, leading to permanent damage to the device. If scholars have early awareness of electrostatic discharge protection, research the ESD performance of carbon-based devices, and utilize Electronic design automation (EDA) tools to achieve co-simulation of ESD-IC to enhance the ESD protection capability of carbon-based devices and integrated circuits, the yield of carbon-based devices will undoubtedly be improved, and a foundation for the smooth growth of carbon-based integrated circuits will be laid.

Carbon-based integrated circuit ESD protection simulation calls for ESD protection device models. However, existing device modeling methods cannot accurately describe complex device discharge behaviors [4] (such as transient-electro-thermal-layout-system coupling effects). To accurately describe the electrostatic discharge behavior of ESD protection devices and achieve ESD protection design simulation, the ESD device behavior modeling approach is adopted. This method does not require IC designers to have a firm understanding of semiconductor device physics and IC fabrication techniques. Only the extraction of key parameters from fabricated ESD devices and segmental description of the *I-V* curves based on these parameters are required, which is more accurate and practical.

ESD behavior modeling requires circuit verification. In this paper, we will start with the study of the operation mechanism of carbon nanotube field-effect transistors and establish a compact model for carbon nanotube field-effect transistors. We will use the CNT-TFT model to build a circuit and incorporate appropriate ESD protection circuits into the circuit to achieve chip-level electrostatic discharge protection design, thereby enhancing the electrostatic discharge protection performance of the core circuit.

This article will be divided into four parts. Part Two: Description and Verification of CNT-TFT Model. In this part, details on how to establish the model of CNT-TFTs and ensure its accuracy and reliability through validation are provided. Part Three: ESD protection simulation of carbon-based integrated circuits. The implementation of ESD protection design in carbon-based integrated circuits and the verification of simulation are explained. Specifically, a 500 V HBM ESD pulse test is used in the PS mode to evaluate the circuit's protection performance. Part Four: Conclusion. In the final section, the work done so far is summarized, and possible improvements and future research directions are discussed.

2. Description and verification of the CNT-TFT model

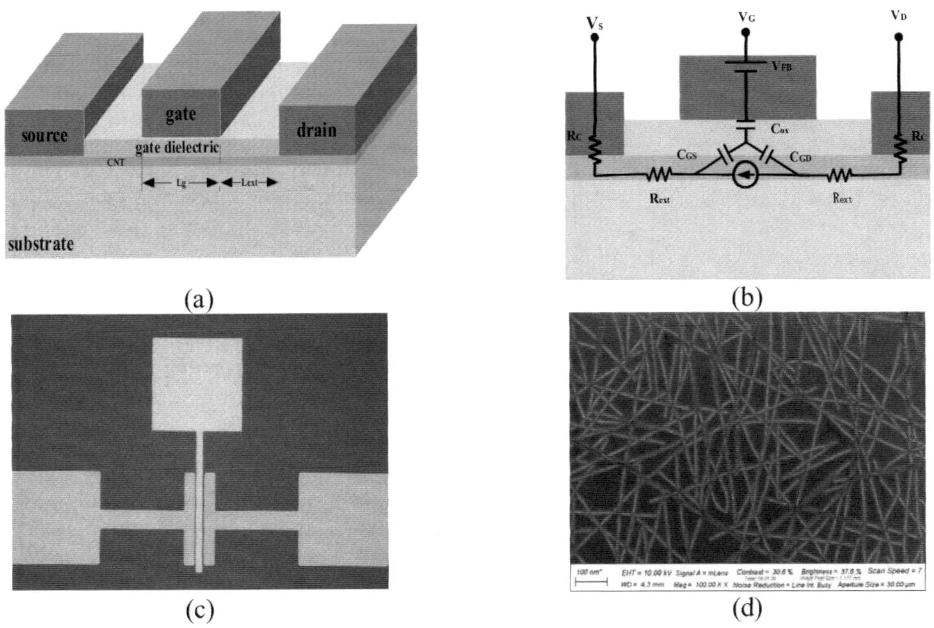

Figure 1. (a) CNT-TFT 3D structure. (b) CNT-TFT cross-section and its equivalent circuit diagram. (c) electron microscopy of CNT-TFT device. (d) electron microscopy of semiconducting carbon nanotube networks.

Journal of Physics: Conference Series **2842** (2024) 012081 doi:10.1088/1742-6596/2842/1/012081

Figure 1(a) represents a 3D structure diagram of a top-gate CNT-TFT. The light blue part corresponds to randomly arranged semiconducting carbon nanotubes. Numerous disordered carbon nanotubes on the substrate form the carbon nanotube thin film. In the top-gate CNT-TFT device, W denotes the channel width. The gap between the gate metal and the source/drain metal, which is the length of uncovered carbon nanotubes, is denoted as L_{ext} = 3 μm. The region that remains uncovered by metal is defined as the source/drain extension region, while the area covered by the gate metal is referred to as the intrinsic region. In the intrinsic region, the thickness of the gate insulator layer covering the carbon nanotube thin film is t_{ox}= 10 nm, and the length of the covered carbon nanotube thin film is equivalent to the channel length L_g.

2.1. The derivation of the current model in the intrinsic region

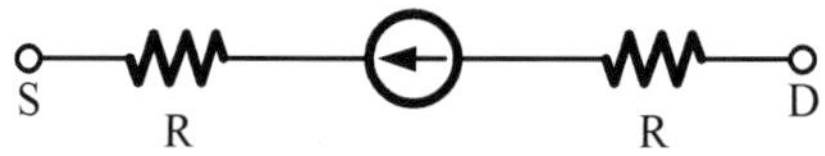

Figure 2. Source leakage current transfer model.

With an n-type CNT-TFT as an illustration, Figure 2 shows the schematic diagram of the current model. Due to the presence of the gate insulator layer, it blocks the flow of electrons at the gate electrode. When a positive voltage is enforced on the gate electrode, it can only repel holes and attract electrons to accumulate below the gate insulator layer. The multitude of electrons accumulated below the gate insulator layer forms the conducting channel. When a certain voltage drop is applied across the source and drain terminals, $V_{DS}= V_D - V_S> 0$, and it will result in a current flowing from the drain to the source. Electrons depart from the source electrode, pass through the contact region between the source electrode and the carbon nanotube, the source extension region, the channel region, the drain extension region, and the contact region between the drain electrode and the carbon nanotube, before flowing out through the drain electrode. The contact between the metal and the semiconductor creates a contact resistance R_C, and the resistance generated by the source/drain extension region is denoted as R_{ext}. Therefore, the total resistance of the source is $R_S= R_C+ R_{ext}$, and the total resistance of the drain is $R_D= R_C+ R_{ext}$. Assuming the intrinsic region's source voltage is V_{Si} and the drain voltage is V_{Di}, the following relationships are obtained:

$$V_{\text{Si}} = I_{\text{DS}} * R_{\text{S}} \tag{1}$$

$$V_{\text{Di}} = V_{\text{DS}} - I * R_{\text{D}} \tag{2}$$

The magnitude of the current is the charge flowing through the cross-section of the conductor per unit time, so for the intrinsic region, the channel current is:

$$I_{\text{DS}} = W \,|\, Q_{\text{n}}(x) \,|\, v(x) \tag{3}$$

$$|\, Q_{\text{n}}(x) \,| = C_{\text{inv}} (V_{\text{G}} - V_{\text{T}} - V_x) \tag{4}$$

where $Q_n(x)$ is the charge per unit area, $v(x)$ is the average velocity of the electron motion, which is proportional to the electric field intensity, and its value is calculated by $v(x) = \mu \times E = \mu \times \delta V_x/\delta x$, μ is the free carrier mobility, which is proportional to the electric field intensity, E is the electric field intensity, which can be expressed as $\delta V_x/\delta x$, C_{inv} is the capacitance per unit area, $V_T = V_{T_0} - \delta V_{DS}$ is the threshold voltage, V_T decreases slightly with increasing drain voltage. The position at the far left end of the intrinsic region with voltage V_{Si} is taken as the origin 0. Starting from the intrinsic source, moving along the channel length towards the drain region, an x-coordinate is established. Therefore, the position where the voltage V_{Di} on the x-axis is at L_g. V_x is the surface potential of the carbon nanotube at position x. Assuming the voltage V_{Si} of the intrinsic source is 0, the surface potential V_x can be represented as $V_{DSi} \times x/L_g$. The channel current is integrated along the channel direction:

$$\int_0^{L_g} I_{\text{DS}} \text{d}x = \int_0^{L_g} W C_{\text{inv}} \mu (V_{\text{G}} - V_{\text{T}} - V_x) * I_{\text{DS}} \partial V_x / \partial x \text{d}x \tag{5}$$

Since the current at the channel is equal everywhere, the current can be expressed by the following equation:

$$I_{DS} = \mu W C_{inv} / L_g [(V_G - V_T) V_{DSi} - 1/2 V_{DSi}^2] \tag{6}$$

When $V_G < V_T$, devices are in the cutoff region, and the current is 0. When $V_G > V_T$ and V_{DSi} is small, the current can be expressed by Formula (6). When $V_{DSi} \geq V_G - V_T > 0$, within a certain range, the current no longer increases with the increase of the drain voltage, but remains constant $K(V_G-V_T)^2/2$, where K can be represented by $\mu WC_{inv}/L_g$. C_{inv} equals the sum of C_{ox}^{-1} and C_q^{-1} divided by Λ_0. C_{ox} [5] represents the unit area gate insulator capacitance, and C_q [6] represents the quantum capacitance with a numerical magnitude of 4×10^{-14} F/cm, and C_{inv} can be expressed by the following equation:

$$C_{inv} = \{C_q^{-1} + \frac{1}{2\pi\varepsilon_0\varepsilon_{ox}} \ln(\frac{\Lambda_0}{r} \sinh(\frac{2\pi t_{ox}}{\Lambda_0}))\}^{-1} \Lambda_0^{-1} \tag{7}$$

where $\varepsilon_0 = 8.854 \times 10^{-12}$ F/m is the vacuum dielectric constant, $\varepsilon_{ox} = 12\sim20$ is the relative dielectric constant of Y_2O_3, $t_{ox} = 10$ nm represents the thickness of the gate oxide layer, r represents the radius of the carbon nanotubes, and $\Lambda_0 = 0.1$ μm is the distance between adjacent carbon nanotubes.

2.2. Contact resistance R_C, extension region resistance R_{ext}

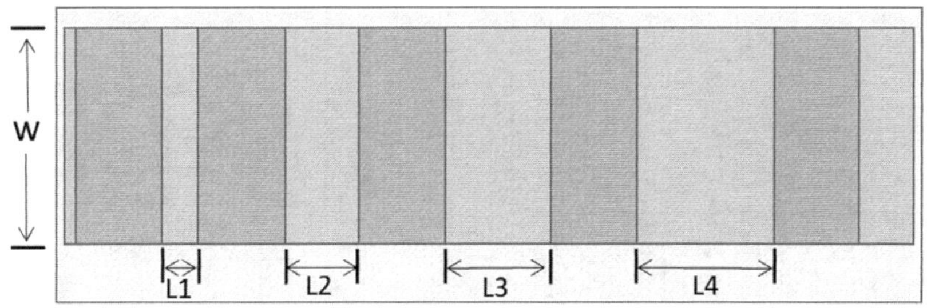

Figure 3. Schematic of a series of devices on the same nanotubes with L ranging from 3 μm to 39 μm.

As shown in Figure 3, the width of the experimental device is 15 micrometers. The light blue region represents the carbon nanotube film, and the gray region represents the source/drain metal. The distance between adjacent metals is denoted as $L_1 = 3$ μm, $L_2 = 6$ μm, $L_3 = 9$ μm, $L_4 = 12$ μm, …, $L_{13} = 39$ μm. By applying voltage to the adjacent metals, the I-V curves can be obtained for different source-drain lengths. This allows us to calculate the total resistance R_{tot}, by $R_{tot} = 2R_C + LR_S/W$. R_C is the contact resistance between the metal and the carbon nanotube, while LR_S/W represents the resistance generated by the extension region between the adjacent metals, and R_S represents the sheet resistance.

Figure 4(a) plots the total resistance value R_{tot} at different L, and the line is obtained by linear fitting of the scatter plot. The intersection of this line with the y-axis represents 2 times the contact resistance value $2R_C$, and the slope of this line represents R_S/W. With the knowledge of the device width W, the sheet resistance R_S can be obtained. In this experiment, the values of contact resistance $R_C = 6.1855$ KΩ and sheet resistance $R_S = 203.145$ KΩ are obtained.

2.3. Terminal charges

In order to consider the dynamic behavior of field effect transistors, Terminal Charges are introduced. Under quasi-static conditions [7], the Ward-Dutton charge distribution scheme yields:

$$Q_S = \int_0^{L_g} (1 - \frac{x}{L_g}) Q_i(x) dx \tag{8}$$

$$Q_D = \int_0^{L_g} \frac{x}{L_g} Q_i(x) \mathrm{d}x \quad (9)$$

where $Q_i(x)$ is the charge at position x, which can be approximated as $Q_i(x) \approx -C_{inv}(V_G - V_T - V_x)$.

2.4. Model validation

A p-type CNT-TFT device is established with channel length $L_g = 11$ μm, extension length $L_{ext} = 2$ μm, and channel width $W = 108$ μm. Grounding the source, applying a negative voltage to the drain and gate, and simulating the channel current will allow us to obtain the output characteristic curve. Due to the symmetry between p CNT-TFTs and n CNT-TFTs, multiplying the values of the source-drain voltage, source-drain current, and gate voltage by a negative sign will give the output characteristics of an n-type CNT-TFT device. Comparing the measured data (dotted line) with the simulation results (solid line) based on the above model, as shown in Figure 4(b), it can be seen that the simulation results can fit the experimental data very well.

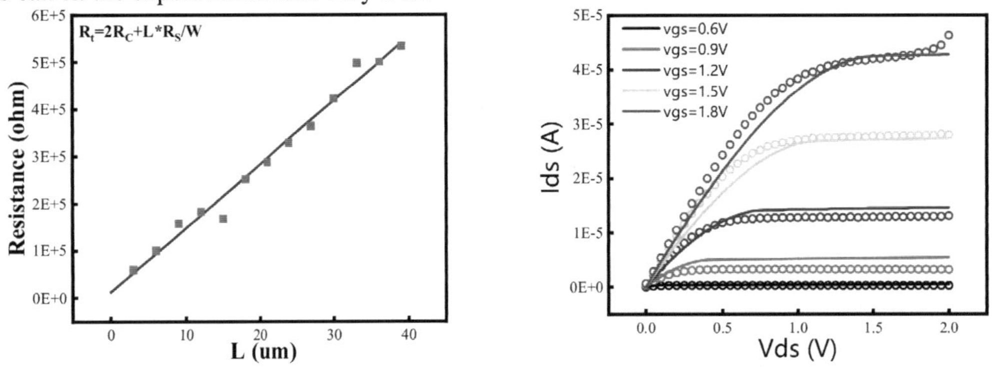

(a) (b)

Figure 4. (a) total resistance versus L. (b) model validation for output characteristic curve.

3. ESD protection simulation of carbon-based integrated circuits

The simulation of ESD protection for carbon-based integrated circuits mainly consists of three parts, (1) the ESD source circuit; (2) the protective circuit composed of ESD protection device models; and (3) the integrated circuit.

For ICs, there are multiple electrostatic discharge modes, commonly including the HBM model, machine model (MM model), and charged device model (CDM model). The HBM model describes the process of releasing accumulated static charges from the human body to electronic components when touched; The MM model describes the process where metallic machinery or tools with static charge transfer their charge to the component when in contact or close proximity to electronic components; And the CDM model describes the electrostatic discharge phenomenon that occurs when an internally charged integrated circuit is grounded.

In daily life, the most common ESD discharge event is the human body charging and discharging event. As shown in Figure 5, the human body model equivalent circuit diagram represents a charged capacitor releasing charge to the tested component through resistor R and inductor L.

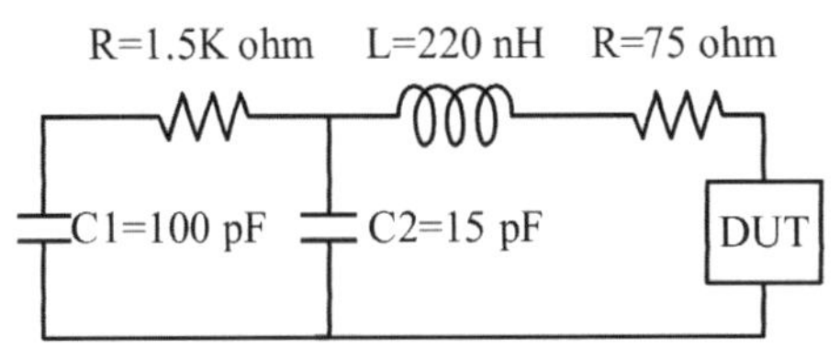

Figure 5. An equivalent circuit for HBM ESD discharging simulation.

The implementation of ESD protection simulation for integrated circuits requires accurate protection device models [8]. Due to the multi-coupling effects of ESD devices, e.g., transient-electrical-thermal-materials-process-circuit-layout-system multi-coupling effects, it is impractical to accurately model ESD devices. The ESD Behavior modeling can describe the *I-V* characteristics of ESD protection devices accurately without considering complex and unknown ESD device physics.

The modeling of ESD behavior can be mainly divided into three steps: extraction, programming, and verification [4]. First, a transmission line pulse (TLP) test can be performed on the ESD protection device, and key parameters can be extracted from the ESD discharge *I-V* curve, such as: trigger voltage and current (V_{t1}, I_{t1}), holding voltage and current (V_h, I_h), discharge Ron and the second breakdown voltage and current (V_{t2}, I_{t2}), and so on. Then, based on these key parameters, the *I-V* curve is segmented and accurately described using Verilog A. By simulating the code using SPICE and comparing the simulation results with experimental measurements, the accuracy of the behavior model can be validated, ensuring the correctness of the full-chip ESD protection circuit simulation.

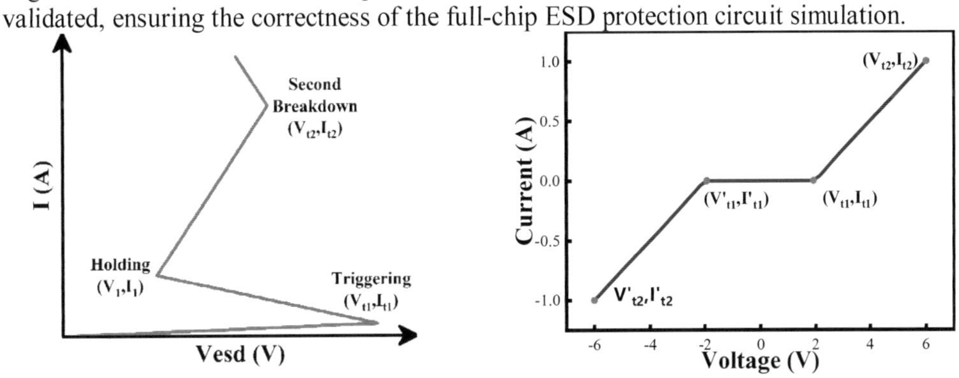

Figure 6. (a) Classic snapback ESD discharging *I-V* curve. (b) TLP curve for bidirectional diode.

In this paper, a modeling of a bidirectional diode with carbon nanotube thin film as the substrate is conducted. Assuming that after TLP testing, the forward trigger voltage is 2 V, and the forward trigger current is 0 A. The forward secondary breakdown voltage and current are (6 V, 1 A). Before being triggered, the device has an infinite resistance and cannot pass current. After triggering and before secondary breakdown, the resistance of the device remains constant ((6-2)/1=4 Ω). The reverse current-voltage characteristic of the device exhibits a symmetric effect with the forward current-voltage characteristic. The *I-V* curve of the device is divided into two segments: before triggering and after triggering, and the spice simulation results are presented in Figure 6(b).

The goal of integrated circuit electrostatic protection design is to design appropriate electrostatic discharge paths for each pin to protect the circuit from damage caused by ESD pulses. A reasonable ESD protection design for circuits is that the triggering voltage and the holding voltage must be lower than 0.8 times the BV (breakdown voltage) of the IC node and bigger than 1.2 times the IC supply voltage. Assume that the power supply voltage of the inverter is 0.9 V, the gate-source breakdown voltage is 8 V, and the source-drain breakdown voltage is 40 V. The triggering voltage of the bidirectional diode is within the design window, so the bidirectional diode model designed above is incorporated into the inverter circuit built with carbon-based field-effect transistors. As shown in Figure 7, a 500 V HBM positive ESD pulse waveform is applied to the input terminal and VSS terminal of the inverter, which represents the PS mode (I/O-to-VSS positively). The pulse waveforms discharge from the IN terminal to the VSS terminal. Through simulation, as shown in Figure 8, the transient terminal voltage and current of each path were monitored. The maximum current flowing through D1 is approximately 0.401 A, which is less than the bidirectional diode's secondary breakdown current of 1 A. By monitoring the node voltages at the input terminal, it is observed that the maximum transient voltage between the input terminal and VSS is around 3.6 V, which is lower

Journal of Physics: Conference Series **2842** (2024) 012081 doi:10.1088/1742-6596/2842/1/012081

than the transistor's gate-source/drain breakdown voltage of 8 V. In conclusion, this circuit has passed the 500 V HBM ESD pulse test.

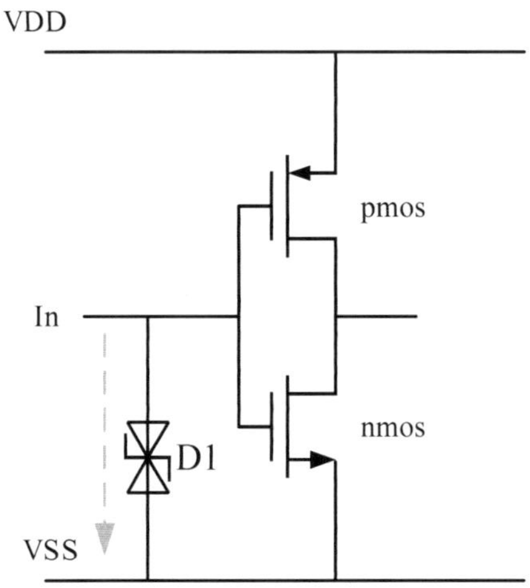

Figure 7. A simplified functional diagram for the inverter with full-chip ESD protection.

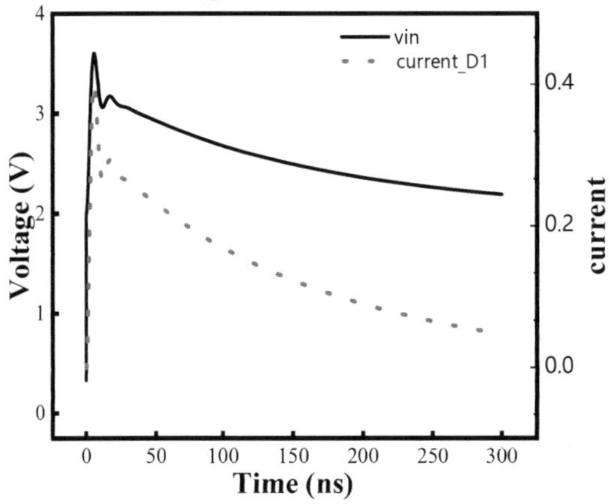

Figure 8. Simulated transient node voltages and path current for the circuit.

4. Conclusion

Within this article, we have established a CNT-TFT device model that is compatible with SPICE, and through validation, it is found that the *I-V* simulation curve of this model fits well with experimental results. Additionally, a method is further introduced for simulating ESD protection in carbon-based circuits using ESD behavior modeling. Through this method, IC designers can achieve on-chip ESD protection simulation with a shorter turnaround time while ensuring accuracy, further promoting the development of devices or circuits based on carbon materials. In future research, ESD behavior modeling methods will be explored for more complex circuit structures and optimize ESD protection schemes for carbon-based circuits to enhance their performance and reliability.

References

[1] ASSOCIATION S I, 2015. 2015 International Technology Roadmap for Semiconductors (ITRS). https://www.semiconductors.org/resources/2015-international-technology-roadmap-for-semiconductors-itrs.

[2] Franklin A D, Luisier M, Han S J, Tulevski G, Breslin C M, Gignac L, Lundstrom M S, and Haensch W. (2012) Sub-10 nm carbon nanotube transistor. Nano letters, 12(2): 758-762.

[3] Green T and Denson W K, 1988. A review of EOS/ESD field failures in military equipment. In: Proceedings of the 10th EOS/ESD Symposium. Austin. (Vol. 7).
[4] Zhang F, Wang C, Lu F, Chen Q, Li C, Wang X Sr, Li D, and Wang A. (2018) A full-chip ESD protection circuit simulation and fast dynamic checking method using SPICE and ESD behavior models. IEEE Transactions on Computer-Aided Design of Integrated Circuits and Systems, 38(3): 489-498.
[5] Rouhi N, Jain D, Zand K, and Burke P J. (2011) Fundamental limits on the mobility of nanotube-based semiconducting inks. Advanced Materials (FRG), 23(1): 94-99.
[6] Rosenblatt S, Yaish Y, Park J, Gore J, Sazonova V, and McEuen P L. (2002) High performance electrolyte gated carbon nanotube transistors. Nano letters, 2(8): 869-872.
[7] Schröter M, Haferlach M, Pacheco-Sanchez A, Mothes S, Sakalas P, and Claus M. (2014) A semiphysical large-signal compact carbon nanotube FET model for analog RF applications. IEEE Transactions on Electron Devices, 62(1): 52-60.
[8] Pan Z, Hao W, Li X, Miao R, and Wang A. 2023. CAD-Based ESD Design Methodology is Critical to Emerging Technologies. In: 2023 45th Annual EOS/ESD Symposium (EOS/ESD). Las Vegas, (pp. 1-10).

Effect of laser scanning frequency on the optical transmittance of the welded area in glass laser welding

Tianfeng Zhou[1,2], Ziwei Li[2], Peng Liu[1,3,*], Zihao Zeng[2], Peihuan Li[2], and Zhikang Zhou[1]

[1] School of Mechanical Engineering, Beijing Institute of Technology, Beijing 100081, China
[2] School of Medical Technology, Beijing Institute of Technology, Beijing 100081, China
[3] Chongqing Innovation Center, Beijing Institute of Technology, Chongqing 401120, China

[*]Corresponding author's email: lpamt@bit.edu.cn

Abstract: When conventionally laser-welding transparent materials such as glass, researchers often primarily focus on the material's joint strength, paying scant attention to variations in optical transmittance within the welding zone. However, in fields like biomedical applications, optical characteristics are also pivotal in characterizing the efficacy of welding. Therefore, this study investigated the relationship between scanning frequency and optical transmittance during laser welding. The findings indicate that a laser single pulse energy of 14.73 μJ, a repetition frequency of 500 kHz, a scanning speed of 600 mm/s, and an optimal scanning frequency of 45 times yielded the highest optical transmittance in the welding area. This resulted in a 13% increase in transmittance compared to stacking two glass pieces.

1. Introduction

Glass materials, due to their excellent biocompatibility, high chemical stability, and superior optical properties, hold promising prospects for a wide range of applications in fields such as biomedical, aerospace, and semiconductor industries. In practical application scenarios, glass materials often require encapsulation before being utilized. For instance, microfluidic chips typically necessitate encapsulating microstructures such as channels to achieve high integration and complex experiments [1]. Therefore, investigating efficient bonding methods for glass is an essential approach to addressing engineering application challenges [2].

Traditional methods of glass bonding include adhesive bonding, thermal compression bonding [3], and chemical bonding, but they suffer from drawbacks such as susceptibility to aging failure, structural fragility, time-consuming processes, and expensive equipment. In recent years, utilizing laser welding technology for glass bonding has emerged as a hot research topic [4]. Laser welding of glass involves transmitting a laser through the upper layer of glass, focusing the beam between two glass pieces, and utilizing the focused laser energy for welding[5],[6]. Laser welding technology can be categorized into continuous laser welding and ultrafast laser welding based on the duration of pulses. Due to the glass being an insulating material with a large bandgap energy and low free electron density at room temperature, its absorption rate of continuous laser is low. Therefore, introducing glass powder and other media during welding is necessary, which complicates the process and may affect the quality of welds. Ultrafast laser, with pulse widths below picoseconds, generates a significant amount of plasma through nonlinear absorption once the laser energy focused within the material reaches a certain threshold [7],[8]. This leads to the melting and bonding of glass at the focal point.

In the current research, while there has been extensive exploration within academia regarding the analysis of laser welding strength of glass materials, there is a comparatively limited focus on the analysis of optical transparency within the welding area [9][10]. This paper investigates the relationship between optical transmittance in the welding region and the number of laser scanning passes during laser welding of soda-lime glass. Such investigation contributes to a deeper understanding of the optical properties of laser welding regions, thus holding significant implications for engineering applications.

2. Materials and experiments

The composition of the soda-lime glass used in the experiment is presented in Table 1. Before experimentation, soda-lime glass was prepared into samples of dimensions 20 mm × 50 mm × 1.50 mm. The samples underwent ultrasonic cleaning with anhydrous ethanol and deionized water, followed by drying with compressed air. The ultrafast laser ablation processing system consisted of a picosecond laser, beam expander, total internal reflection angle, focusing lens, and scanning mirror system, as illustrated in Figure 1.

Table 1. The chemical composition of the soda-lime glass.

Composition	Fe_2O_3	Al_2O_3	MgO	CaO	Na_2O+K_2O	SiO_2
Wt.%	0.1	1.5	4.5	8.1	13.3	72.5

The experiment utilized a laser with a repetition frequency of 500 kHz to facilitate ionization and thermal accumulation phenomena during the laser welding process. The diameter of the focused spot was measured using the area method and found to be consistent with theoretical calculations. While maintaining constant scanning speed and defocus parameters, the influence of different scanning frequencies on light transmittance was examined. The laser focus was set at 200 μm above the interface of the upper glass, and scanning was conducted at a speed of 600 mm/s perpendicular to the direction of laser propagation. The high-energy ultrafast laser pulses acting on the glass samples resulted in the formation of numerous micro/nanopores in the welding area and the generation of a significant amount of molten spatter nearby, leading to light scattering and a decrease in the optical transmittance of the glass. Therefore, a single pulse energy of 14.73 μJ was selected. The scanning pattern is depicted in Figure 2, forming a welding area with a radius of 5 mm, with a spacing of 10 μm between scanning lines.

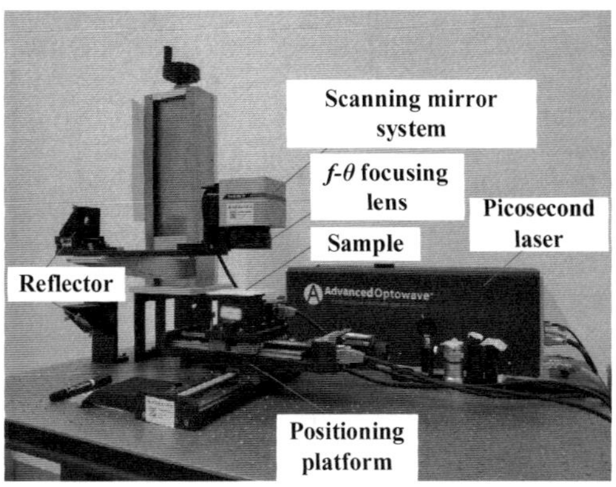

Figure 1. The ultrafast laser ablation processing system.

3. Results and discussion

The optical transmittance was measured using red light with a wavelength of 632.5 nm as the light source. The laser source passed through an aperture and irradiated the test samples on the surface of a

CCD camera. The intensity distribution of the laser beam was collected, and the optical transmittance was calculated by analyzing the grayscale distribution of the collected image signal. In this experiment, both air and two stacked original soda-lime glass samples were tested for comparison. Three points were measured on each sample, and the grayscale values were averaged for analysis. The results are shown in Figure 2, with sample numbers corresponding to those in Table 2.

Table 2. The experimental parameters for optical transmittance samples.

Sample	1	2	3	4	5	6	7	8
Condition	Air	Pristine	15 scans	30 scans	45 scans	60 scans	75 scans	90 scans

As shown in Figure 2, it can be observed that the most significant improvement in optical transmittance of the welding area occurs at 45 scans. When two glass pieces are stacked, the transmittance decreases by 20.1% compared to the laser propagating in air. With an increase in the number of scans, the optical transmittance initially rises and then declines. At 15 scans, the optical transmittance of the welded sample decreases by 7.7% compared to soda-lime glass stacked in two layers. At 30, 45, and 60 scans, the optical transmittance of the welding area increases by 3.5%, 13.6%, and 1.2%, respectively, compared to soda-lime glass placed in two layers. This is because effective bonding is achieved in the welding area, reducing the thickness of the air layer significantly. With the continued increase in scanning frequency, the optical transmittance gradually decreases. When the scanning frequency reaches 75 and 90 scans, the optical transmittance of the welding area decreases by 7.8% and 30.7%, respectively, compared to soda-lime glass placed in two layers. At 90 scans, there are noticeable white opaque areas in the welding region, attributed to excessive thermal accumulation resulting from the high scanning frequency. When the scanning frequency reaches 120 scans, the glass fractures due to excessive localized stress. Observation using laser confocal microscopy reveals periodic weld seam structures in the sample subjected to 45 scans, as shown in Figure 3. A periodic grating structure with a width of approximately 10 μm and a height of around 30 nm is formed at the interface of the two glass pieces, enhancing the transmittance of light through optical interference effects.

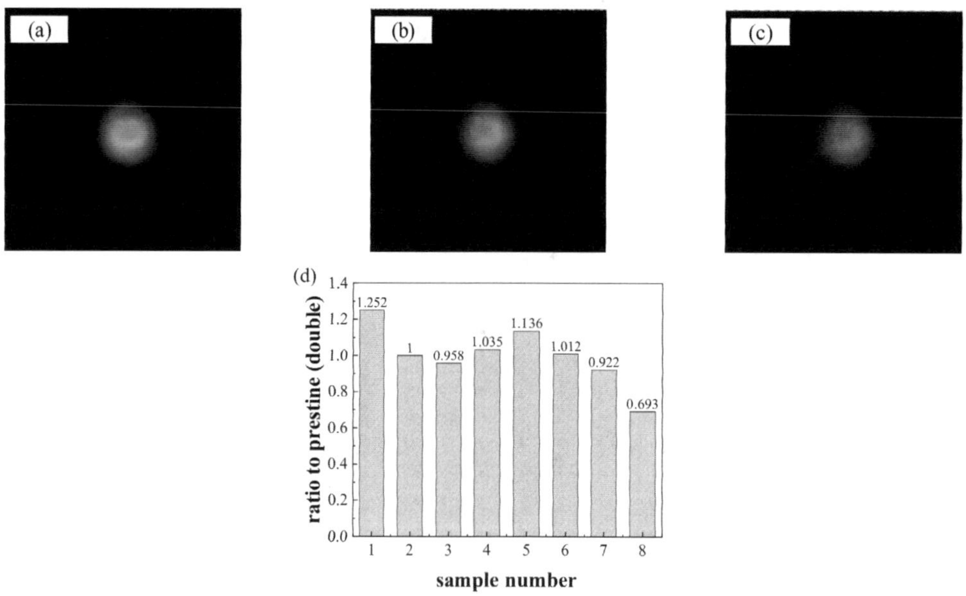

Figure 2. The results of optical transmittance measurements using a CCD camera, (a) the absence of any medium, (b) two layers of original stacked glass, (c) after 45 scans illustration, (d) compares the optical transmittance between laser-welded soda-lime glass at different scanning frequencies and two layers of stacked original glass.

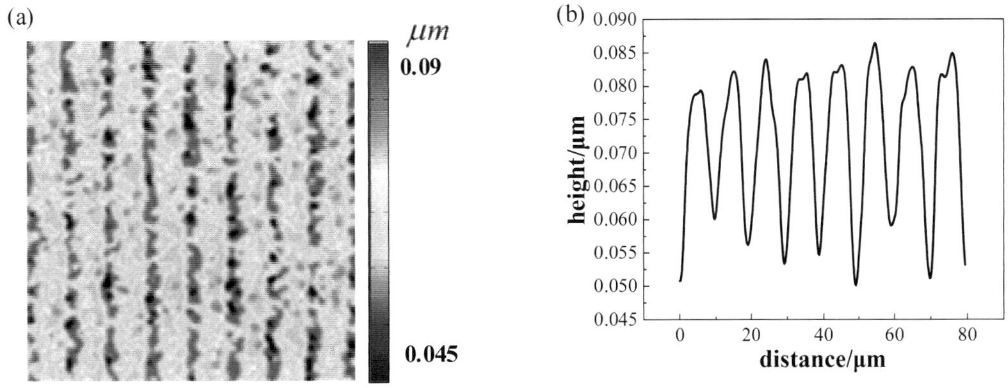

Figure 3. The microstructure measurements of the welding area, (a) three-dimensional rendering, (b) contour map.

4. Conclusion

To enhance the optical performance of ultrafast laser-welded glass materials and facilitate their better utilization in microfluidics and optofluidics applications, an investigation into methods for improving the optical transmittance of the welding area was conducted. The welding area was optically characterized using a CCD camera for intensity detection after welding, and comparisons were made with the original stacked glass to study the influence of scanning frequency on the optical transmittance of the welding area.

This paper investigates the transmittance of infrared lasers under different scanning frequencies. The results indicate a significant enhancement in transmittance of 632.5 nm red light when using a single-pulse laser with a pulse width of 10 ps, a single-pulse energy of 14.73μJ a scanning speed of 600 mm/s, a repetition frequency of 500 kHz and 45 scanning passes.

Acknowledgments

The authors gratefully acknowledge the financial support by the National Natural Science Foundation of China [grant numbers 52375403, 51775046, 51875043]; the Science and Technology Major Project of Jiangxi Province [grant number 20212AAE01002]; the Beijing Municipal Natural Science Foundation [grant number JQ20014].

References

[1] Jia Z J, Fang Q, Fang Z L. Bonding of glass microfluidic chips at room temperatures[J]. Analytical Chemistry, 2004, 76(18):5597.
[2] MIYAMOTO I, CVECEK K, OKAMOTO Y, et al. Internal modification of glass by ultrashort laser pulse and its application to microwelding[J]. Applied Physics A, 2014, 114 (1): 187-208.
[3] Howlader M M R, Suehara S, Suga T. Room temperature wafer level glass/glass bonding[J]. Sensors & Actuators A Physical, 2006, 127(1):31-36.
[4] Miyamoto I, C Vecek K, Schmidt M. Evaluation of nonlinear absorptivity in internal modification of bulk glass by ultrashort laser pulses. Optics Express, 2011, 19(11): 10714-10727
[5] TAMAKI T, WATANABE W, NISHII J, et al. Welding of transparent materials using femtosecond laser pulses [J]. Japanese Journal of Applied Physics, 2005, 44 (22-23): L687-L689.
[6] TAMAKI T, WATANABE W, ITOH K. Laser micro-welding of transparent materials by a localized heat accumulation effect using a femtosecond fiber laser at 1558 nm[J]. Optics Express, 2006, 14 (22): 10460-10468.

Journal of Physics: Conference Series **2842** (2024) 012082 doi:10.1088/1742-6596/2842/1/012082

[7] Cao J, Poumellec B, Brisset F, et al. Pulse energy dependence of refractive index change in lithium niobium silicate glass during femtosecond laser direct writing[J]. Optics Express, 2018, 26(6): 7460-7474.
[8] Zhang G, Stoian R, Zhao W, et al Femtosecond laser Bessel beam welding of transparent to nontransparent materials with large focal-position tolerant zone [J]. Optics Express, 2018, 26(2): 917926
[9] Jiang X W, Qiu J R, Zhu C S, et al. ESR studies and thresholds of femtosecond laser-induced darkening in glasses[J]. Acta Physica Sinica, 2001(5):871-874.
[10] Zhang W N, Cao H C, Kong F Y, et al. Study on Microstructure Anti-reflection Performance Based on Sapphire Material [J]. Chinese Journal of Laser, 2023, 50(22):71-79.

Research on positioning accuracy of the swivel head of milling and turning complex machining center

Weizheng Si, Jianhao Tang, Jianhua Liu, Zijun Wei and Luyao Mao *

Guangdong Institute of Intelligent Manufacturing, Guangdong Key Laboratory of Modern Control Technology

*Corresponding author's email: 540230646@qq.com

Abstract: This article focuses on the swivel head of a milling-turning complex machining center and establishes a geometric error model between the rotary axes based on the multi-body system theory and the homogeneous coordinate transformation method. The double ball bar (DBB) is used to identify the geometric errors of the machine tool swivel head position-independent by measuring circular trajectories. The impact of PIGEs on the measurement trajectory is simulated. The swivel head is rotated at four angles (B0°, B45°, B90°, B-15°), and the laser tracker is used to analyze and study the linear axis straightness, perpendicularity, and the center coordinates and plane errors of the swivel head movement along the XZ plane. A swivel head rotation accuracy test fixture is designed to correct the positioning accuracy of a swivel head with a 45° inclination of the rotation axis. The research results show that the fluctuation of the center plane degree of freedom of the swivel head moving along the XZ plane is small, and the maximum value is 0.0086mm. Under the 45° angle posture of the swivel head, both the center coordinate error value and the plane error value are the largest. The simulation results of the swivel head position error are fairly consistent with the center coordinate error of the swivel head plane movement. A backward tilt displacement of 0.01mm~0.015mm of the machine column during assembly can reduce the influence of the weight of the swivel head on the perpendicularity and straightness of the vertical reciprocating motion, and adding auxiliary support near the base of the swivel head zero position can reduce machine straightness error.

1. Introduction

As the manufacturing industry progresses, the demand for workpieces of complex shapes and profiles is increasing. In addition, a high standard is set for machining accuracy, and it is often required to complete all machining processes in one clamping. Conventional lathes or milling machines cannot live up to these requirements, so multi-process complex machine tools have become an option for improving machining efficiency and accuracy [1-2]. As an essential branch of the machine tool industry, the five-axis milling and turning complex machining center is a high-end machine tool that mainly uses milling while also meeting the requirements for turning. It can process complex three-dimensional surfaces and has the advantages of strong load-bearing capacity and flexible machining, especially suitable for high-precision and efficient machining of medium and heavy-duty workpieces such as aircraft engine casings, impellers, blades, ship propellers, turbine rotors, and more [3].

Five-axis milling and turning complex machining center has the following types of error: geometric errors, static load deformation, thermal deformation, and dynamic characteristics, among which geometric errors account for 40~50% of the total error of the machine tool [4]. Geometric errors can be divided into two types: position-dependent geometric errors (PDGEs) and position-independent geometric errors (PIGEs) [5-6]. The former is mainly caused by the manufacturing process, while the latter is mainly formed during the assembly process of the machine tool.

A variety of in-depth research has been conducted on the accuracy of five-axis milling and turning complex machining centers. Eman et al. established a universal error model, which is based on multi-body dynamics and the transformation of homogeneous coordinate matrices, boasting availability to any multi-axis machine tool [7]. Yang et al. established a comprehensive error model in the field of machine tool geometric errors and thermal errors, and the identification compensation algorithm has a good effect on the machine tool [8-9]. Soichi measured the static geometric errors of the rotation axis of a five-axis numerical control machine tool using a high-precision touch probe and a gauge block [10-11], but this method has higher requirements for the environment and operators. Qiu proposed a universal kinematic formula for the five-axis numerical control machine tool with a dual swing head. Based on the actual travel and movement efficiency of the machine tool, an optimization strategy for selecting the optimal inverse solution value was established, and a post-processor for the universal dual swing head five-axis numerical control machine tool was developed [1]. Wang elaborated on the principle of orthogonal milling turning and summarized the current geometric error modeling of the five-axis milling and turning complex machining center. He proposed the establishment of a multi-body theory error model suitable for the five-axis milling and turning complex machining center and analyzed its impact on machining accuracy. However, the error modeling still needs to be further improved, as the effect of force deformation only considers the influence of workpiece deformation [2]. Gao analyzed and optimized the finite element static and dynamic analysis of the direct-drive swing head power tool holder of the five-axis milling and turning complex machining center, and the static and dynamic characteristics of the optimized partial model of the bracket were basically unchanged, with a 22% reduction in quality. The mode analysis of the tool holder bracket was carried out. The finite element analysis showed that the maximum deformation of the tool holder bracket was the largest at a swing head angle of 45°, and the tool holder was topologically optimized [12]. Zhang et al. analyzed the conversion relationship between the B-axis rotation center and the tool length and diameter by swinging the tool holder at two positions, providing technical support for machine tool maintenance [13]. Chen assumed that the linear axes of the five-axis machine tool had no errors and studied the static geometric errors of the rotation axis. This provided a basis for designing a rotation axis identification algorithm based on a probe, but further research on the swing B-axis is needed [14].

There is little research on the effects of rotation angle, position, and spindle movement coplanarity on the geometric accuracy and positioning accuracy of the swivel head in milling and turning complex machining centers. This article analyzes and studies the impact of the rotation angle and position of the swivel head on the straightness and verticality of the machine tool's linear axis, the machine tool support, and the center coordinates and XZ planar error. A testing fixture was designed to correct and analyze the rotation accuracy of the swivel head, providing a technical basis for the research of the swing head accuracy of milling-turning complex machining centers.

2. Analysis of positioning accuracy of swivel head

2.1 Structure of milling and turning complex machining center

Milling-turning complex machining center consists of three translational axes, X, Y, and Z, and rotary axes B and C that rotate in Y and Z directions, respectively. B-axis is at a 45° angle to the Y-axis in the YZ plane. The rotating workbench is driven by the rotary C-axis, while the rotary B-axis drives the spindle to rotate around the Y-axis. To simplify the modeling process, the ideal positions of the C-axis coordinate system (OC-XCYCZC), B-axis coordinate system (OB-XBYBZB), and machine tool coordinate system (OM-XMYMZM) are defined at the center of the rotating workbench, as shown in Figure 1.

Journal of Physics: Conference Series **2842** (2024) 012083 doi:10.1088/1742-6596/2842/1/012083

2.2 Analysis of accuracy of milling and turning complex machining center

Geometric errors of the machine tool are mainly caused by rotary and translational axes and are divided into position-independent geometric errors (PIGEs) and position-dependent geometric errors (PDGEs). PIGEs refer to errors caused by the assembly, load change, and thermal drift during the production process of the machine tool, such as tilt and position errors of each axis. PDGEs refer to position and shape errors caused by the low surface accuracy of the components due to manufacturing and material technology. In recent years, the manufacturing industry has seen tremendous improvement, with produced components presenting good geometric accuracy, and the resulting PDGEs are insignificant. Therefore, current research focuses on modeling and identifying PIGEs caused by the assembly of machine components.

Each translational axis of the machine tool will produce six geometric errors, as shown in Figure 2, including three position errors and three angular errors, of which the three angular errors are roll, pitch, and yaw errors. These six errors are associated with position, so they are PDGEs. In addition, there are three perpendicularity errors between three translational axes, which are PIGEs. Each rotary axis produces six PDGEs, including three position errors and three angular errors, as shown in Figure 3. Each rotary axis has five PIGEs, including two position errors, two directional errors, and one zero-angle position error. Since the zero position error can be compensated in the encoder or digital controller, only four PIGEs of the B-axis are considered in this study, as shown in Figure 4. Therefore, there are 20 terms of geometric error in the two rotational axes of the milling-turning complex machining center and 21 terms of geometric error in the translational axes, resulting in a total of 41 terms of geometric error for the five-axis machine tool.

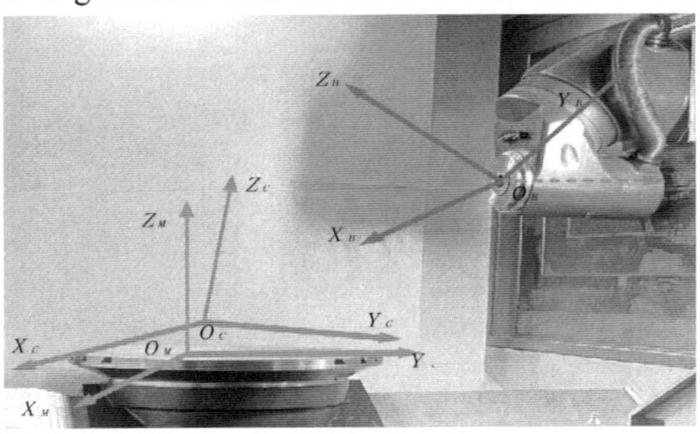

Figure 1. Coordinate system of the machine tool.

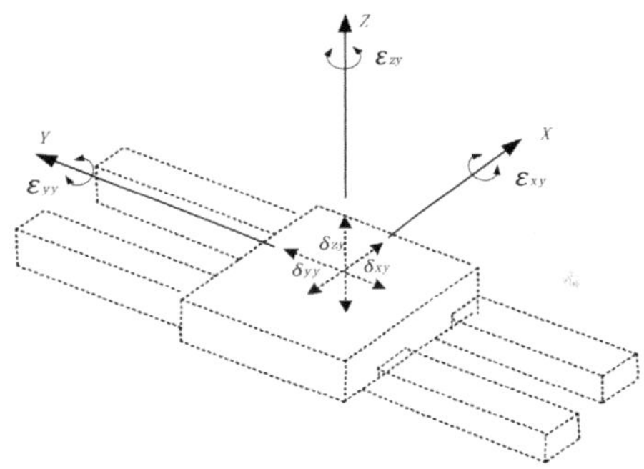

Figure 2. Six terms of geometric error of translational axis.

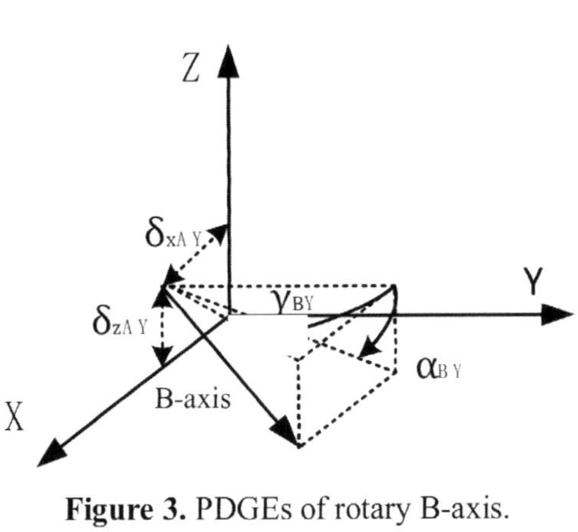

Figure 3. PDGEs of rotary B-axis.

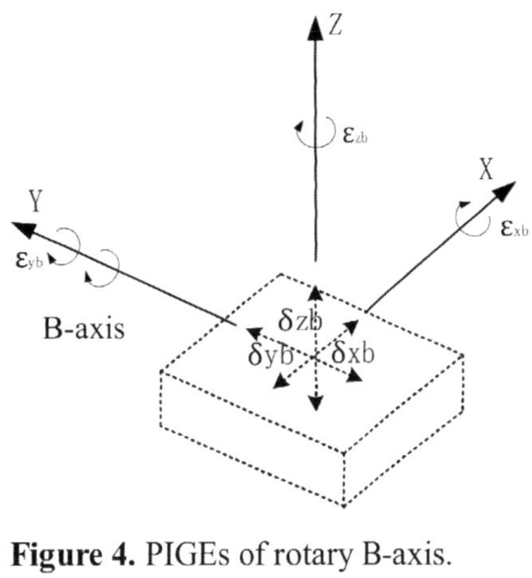

Figure 4. PIGEs of rotary B-axis.

2.3 Analysis of swivel head

The PIGEs of the rotational axes in the local coordinate system of the pendulum head rotary table type mill-turn machine are shown in Fig. 5. For the rotational B-axis, there are four positional errors ($d_{XBY}, d_{ZBY}, d_{XCY}, d_{YCY}$) and four angular errors ($\alpha_{ZB}, \gamma_{ZB}, \alpha_{CY}, \beta_{CY}$), among which the positional errors (d_{XBY} and d_{ZBY}) and the angular errors (α_{ZB} and γ_{ZB}) mainly result from the B-axis.

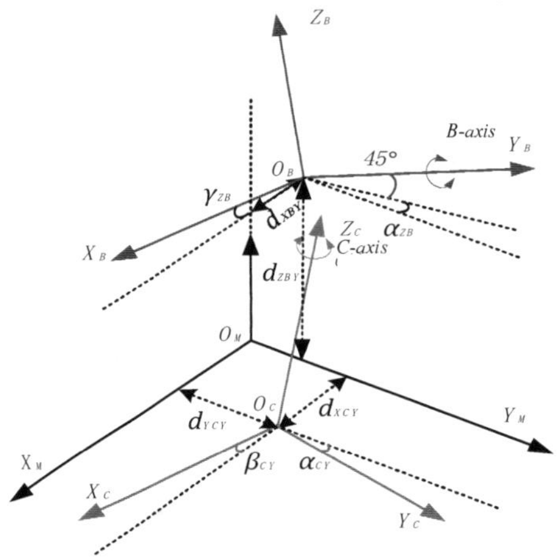

Figure 5. PIGEs of rotary B and C-axis.

2.3.1 PIGEs model of rotary axis based on double ballbar (DBB). The contact ball $\mathbf{O_s}$ of the double ball bar (DBB) is fixed at the spindle end, and the other contact ball $\mathbf{O_t}$ is set on the magnetic suction seat on the rotating workbench. Circular trajectory measurement is realized through synchronous interpolation of three translational axes and synchronous coordination of a rotational axis. As a result of PIGEs, the length of DBB telescopic rods will change relative to their nominal length. When the B-axis moves, the C-axis remains stationary, and the position of the contact ball $\mathbf{O_t}$ on the rotating workbench remains unchanged. In the machine coordinate system O_M-$X_MY_MZ_M$, the positional accuracy of the spindle end contact ball $\mathbf{O_s}$ is affected by $\alpha_{ZB}, \gamma_{ZB}, d_{XBY}, d_{ZBY}$, and the coordinate positions (X_s,Y_s,Z_s) of the spindle end contact ball $\mathbf{O_s}$ are shown in Equation (1):

$$\begin{bmatrix} X_S \\ Y_S \\ Z_S \end{bmatrix} = T_Y \left(\begin{bmatrix} X_T \\ Y_T \\ Z_T \end{bmatrix} + T_{\alpha ZB} T_{\gamma ZB} T_B T_C \left(\begin{bmatrix} X_B \\ Y_B \\ Z_B \end{bmatrix} - \begin{bmatrix} X_T \\ Y_T \\ Z_T \end{bmatrix} \right) \right) + T_y \begin{bmatrix} d_{XBY} \\ 0 \\ d_{ZBY} \end{bmatrix} \tag{1}$$

In Equation (1): *(X_B, Y_B, Z_B)* is the initial position coordinate of the spindle end in the machine coordinate system *O_M-$X_MY_MZ_M$* before B-axis movement; *(X_T, Y_T, Z_T)* is the initial position coordinate of the worktable end contact ball Ot in the machine coordinate system; (XS, YS, ZS) is the position coordinate of the spindle end in the machine coordinate system OM-XMYMZM after the machine movement; orders θ B and θ C represent the rotation angles of the B and C axes, while TB and TC represent the homogeneous transformation matrices of the angle changes around Y and Z-axis, respectively; $\boldsymbol{T_y}$ represents a homogeneous transformation matrix that rotates 45° around the X-axis in the YZ plane; $T_{\alpha ZB}$ and $T_{\gamma ZB}$ are error matrices for angle errors (α_{ZB} and γ_{ZB}) related to B-axis motion, respectively. The homogeneous coordinate transformation matrix contained in it is shown in Equations (2), (3), (4), (5):

$$T_B = \begin{bmatrix} \cos\theta_B & 0 & \sin\theta_B \\ 0 & 1 & 0 \\ -\sin\theta_B & 0 & \cos\theta_B \end{bmatrix} \tag{2}$$

$$T_y = \begin{bmatrix} 1 & 0 & 0 \\ 0 & \cos\frac{\pi}{4} & -\sin\frac{\pi}{4} \\ 0 & \sin\frac{\pi}{4} & \cos\frac{\pi}{4} \end{bmatrix} \tag{3}$$

$$T_{\alpha ZB} = \begin{bmatrix} 1 & 0 & 0 \\ 0 & \cos\alpha_{ZB} & -\sin\alpha_{ZB} \\ 0 & \sin\alpha_{ZB} & \cos\alpha_{ZB} \end{bmatrix} \tag{4}$$

$$T_{\gamma ZB} = \begin{bmatrix} \cos\gamma_{ZB} & -\sin\gamma_{ZB} & 0 \\ \sin\gamma_{ZB} & \cos\gamma_{ZB} & 0 \\ 0 & 0 & 1 \end{bmatrix} \tag{5}$$

When the B-axis is in motion, and the C-axis is stationary, the difference between the measured value of DBB and the nominal length L of DBB is shown in Equation (6):

$$\Delta L_B = \left\| \begin{bmatrix} X_S \\ Y_S \\ Z_S \end{bmatrix} - \begin{bmatrix} X_T \\ Y_T \\ Z_T \end{bmatrix} \right\| - L \tag{6}$$

2.3.2 Simulation for PIGEs model of rotary axis based on double ballbar (DBB). The influence of PIGEs on the circular measurement trajectory can be determined through simulation. The simulation parameters of 8 PIGEs and other parameters are shown in Table 1. The motion range of the B-axis is 0-90°, and its simulated trajectory is a quarter circle. In B-axis radial and axial measurement modes, the simulation results are shown in Table 1.

Table 1. Simulation setup parameters for PIGEs.

Parameter	Value
DDB nominal length	L=100mm
Distance between $\boldsymbol{O_t}$ and rotary axis calibrator	Z_T=60mm
Angular positioning error	α_{ZB}, γ_{ZB}: ±0.02°
Translational positioning error	d_{XBY}, d_{ZBY}: ±20μm

As a result of IPGEs, simulating the trajectory in B-axis radial and axial measurement modes, the position and angle errors are as follows:

1) B-axis radial measuring mode:

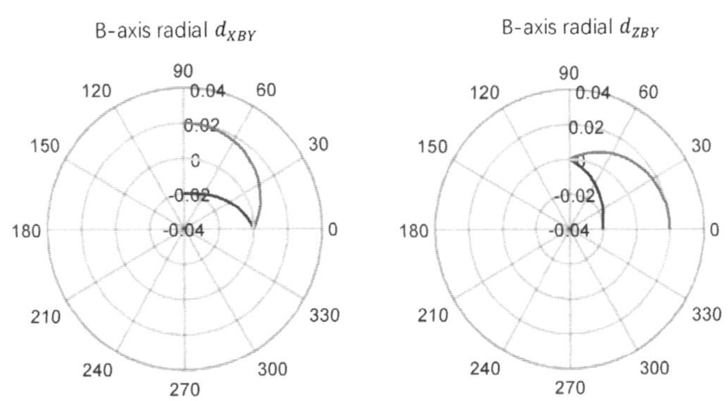

Translational positioning error: $d_{XBY} = \pm 0.02mm$
$d_{ZBY} = \pm 0.02mm$

2) B-axis axial measuring mode:

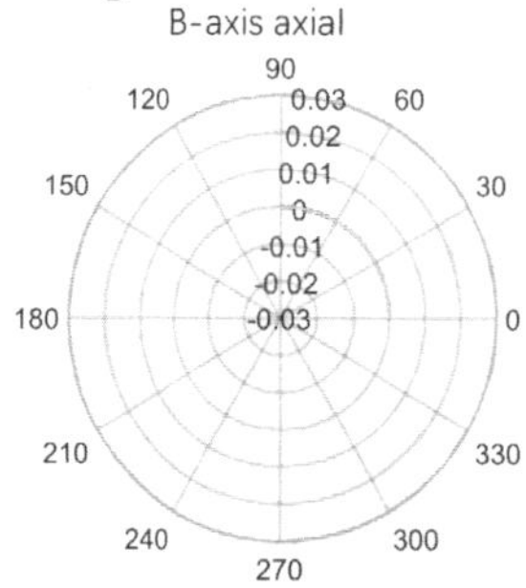

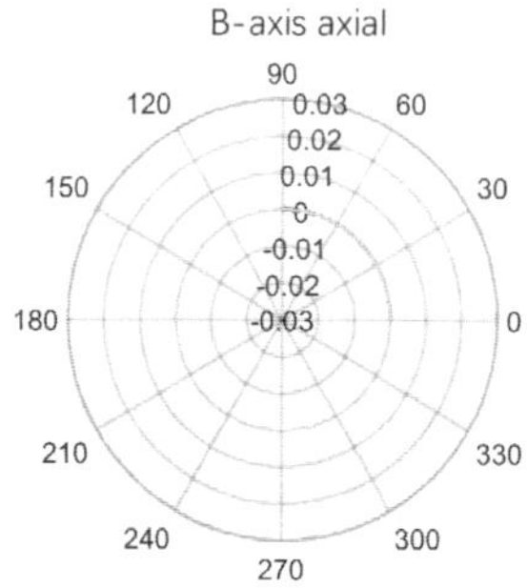

$$\text{Translational positioning error:} \ d_{XBY} = 0.0mm$$
$$d_{ZBY} = 0.0mm$$

The radial and axial angle errors are both 0.005°, and the trajectory is as follows:

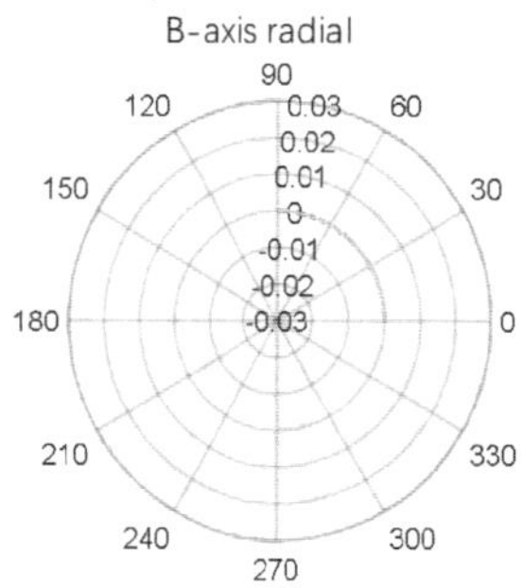

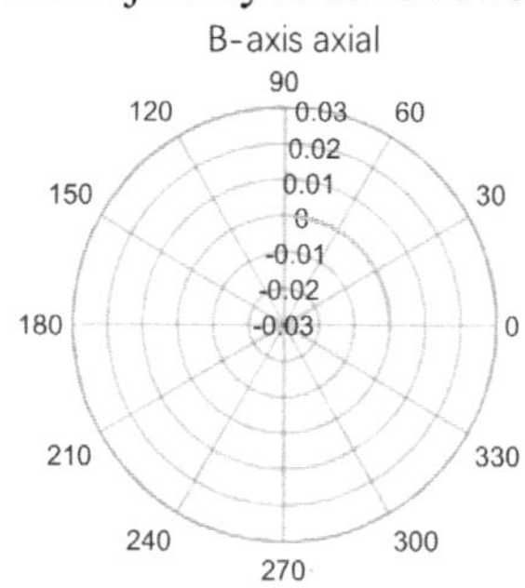

$$\text{Angular positioning error:} \ \alpha_{ZB} = \pm 0.005°$$
$$\gamma_{ZB} = \pm 0.005°$$

2.3.3 Swivel head center coordinate and flatness error. The motion of the swivel head in different orientations will generate the geometric error, which is determined by controlling the B-axis of the swivel head to perform planar motion in the YZ plane and measuring the error plane of the center point of the swivel head. The high-precision target ball of the laser tracker was fixed at the center position of the B-axis tool handle of the swivel head, and position points were measured every 100mm in X and Z directions within a range of (1000 × 1000) mm in XZ plane for reciprocating motion. Under orientations of B0°, B45°, B90°, and B-15°, 100 points were collected each for a total of 400 target point positions. The swivel head moved in the XZ plane, and the center coordinate error of the swivel head at each measurement position and the flatness error of the working plane of the swivel head are shown in Figure 6-9. The motion path and measurement results of the center coordinate flatness error of the swivel head's B-axis are shown in Table 2.

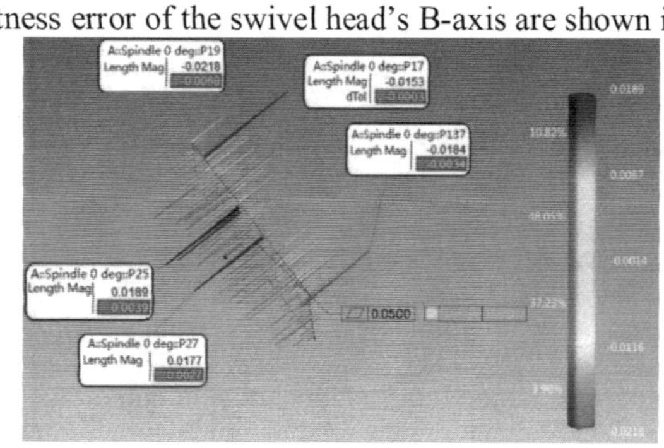

Figure 6. Swivel head flatness error at B-axis 0°.

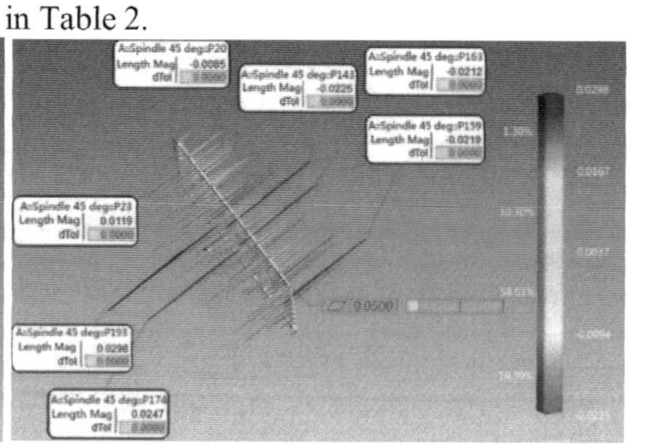

Figure 7. Swivel head flatness error at B-axis 45°.

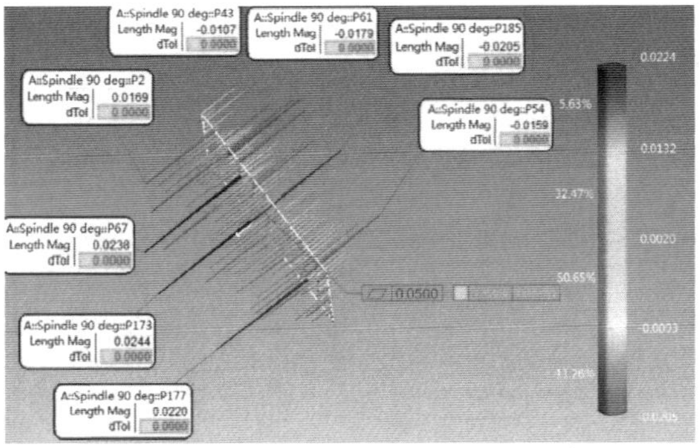

Figure 8. Swivel head flatness error at B-axis 90.

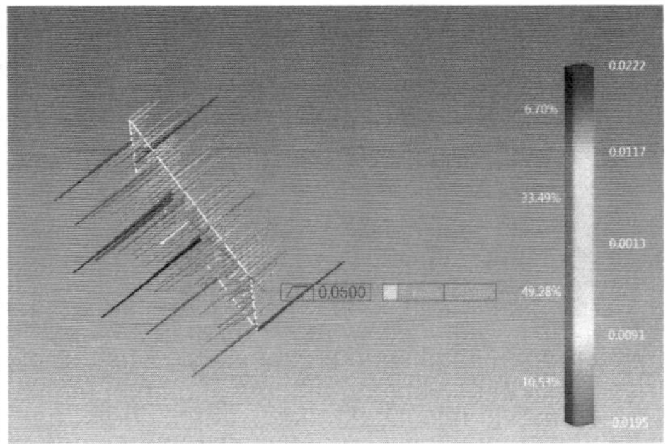

Figure 9. Swivel head flatness error at B-axis-15°.

Table 2. Swivel head B-axis center coordinate error and flatness error.

B-axis center coordinate error and flatness error	Error in XZ-plane (mm)			
	B-axis 0°	B-axis 45°	B-axis 90°	B-axis -15°
Min. error(mm)	-0.0218	-0.0225	-0.0205	-0.0195
Max. error(mm)	0.0189	0.0298	0.0244	0.0222
Flatness error(mm)	0.0378	0.0464	0.0449	0.0391

Based on the measurement results of the movement of the swivel head center coordinates in the XZ plane, it can be observed B-axis of the main spindle swivel head undergoes four different angles (B0°, B45°, B90°, B-15°), and the variation of flatness error of B-axis center coordinates in XZ plane is relatively small, with a change in flatness error value of 0.0086mm.

At 45°, the position error value and flatness error of the B-axis center in the XZ plane are both the largest, with a position error in the range of -0.0225~0.0298 mm and a flatness error of 0.0464 mm. The test results of center point error are consistent with simulation results.

3. Impact of swivel head's position on verticality of the machine tool

The shape error of the workpiece is not only affected by the centering error of the swivel head but also the geometric accuracy of the machine tool. This experiment studied the impact of the swivel head position on the verticality of the machine tool under the Y-axis 1100mm and Z-axis 1000mm strokes. The verticality error of the Y and Z-axis during maximum stroke operation at different coordinate positions (X=0mm, X=500mm, X=1000mm) was tested and analyzed, as shown in Figure 10-12. The verticality test results are shown in Table 3.

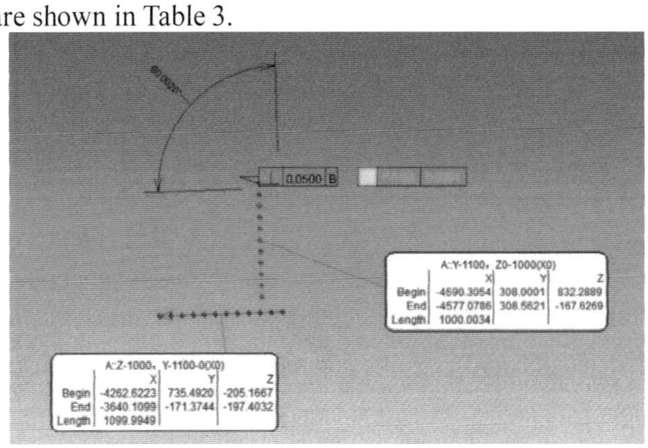

Figure 10. Verticality of Y, Z-axis (X=0mm).

Journal of Physics: Conference Series **2842** (2024) 012083 doi:10.1088/1742-6596/2842/1/012083

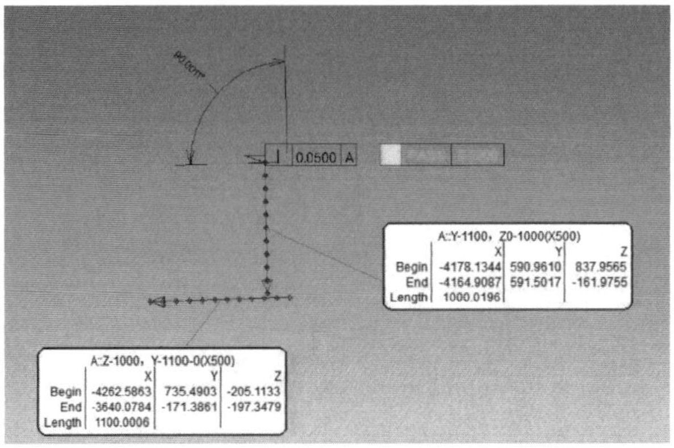

Figure 11. Verticality of Y, Z-axis (X=500mm).

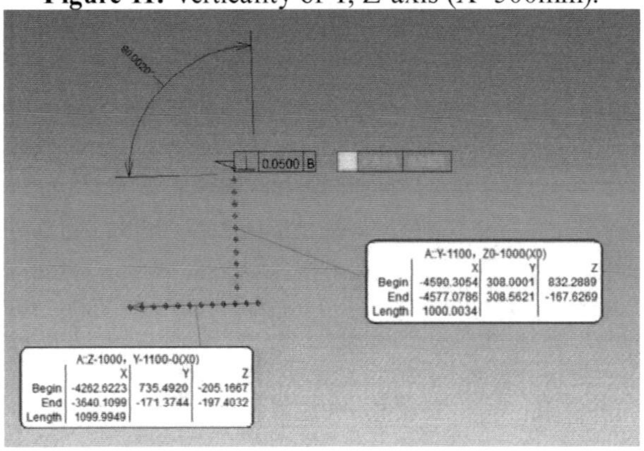

Figure 12. Verticality of Y, Z-axis (X=1000mm).

The test results show that the perpendicularity error of the Y and Z-axis when running at maximum stroke is (0.020-0.035) mm. The X-axis position of the swivel head also has a certain impact on the perpendicularity error of the Y and Z-axis. The perpendicularity error of the Y and Z-axis measured at the X=500mm (middle stroke) coordinate position is the smallest. In order to reduce the impact of the swivel head's own height on the verticality accuracy during the reciprocating up and down motion, the range of backward tilt of the machine tool was set within (0.010~0.015) mm during assembly. The test results verify the correctness of this method.

Table 3. Swivel head position and Y, Z-axis verticality.

Swivel head X position	Y, Z-axis moving direction			
	Y-axis-1100→0mm Z-axis-0→-1000mm		Y-axis-1100→0mm Z-axis-1000→0mm	
	Y, Z-axis∠angle	⊥verticality	Y, Z-axis∠angle	⊥verticality
X=0mm	90.002°	0.0351mm	90.0014°	0.0247mm
X=500mm	90.0011°	0.020mm	90.0012°	0.0209mm
X=1000mm	90.0014°	0.0239mm	90.0013°	0.0233mm

Note: Evaluate verticality error based on the measured value in the Y-axis direction.

4. Impact of swivel head's position on straightness of translational axis

The tool magazine of the machine tool is located near the zero position of the swivel head. When the swivel head moves to zero position, the load distribution of the machine tool changes due to the swivel head's weight, and the machine tool support undergoes slight deformation. The experiment was conducted to analyze the impact of the swivel head on the straightness of the guide rail of the Y and X-axis, and thus on the machining accuracy of the machine tool, this experiment places the swivel head at zero position (X=0mm) and tests the straightness of Y-axis and Z-axis of the machine tool guide rail. The test results are shown in Figure 13-14. From Figure 13, it can be seen that the Z-axis drives the swivel head to move upwards, and the Y-axis straightness error decreases from large to small, with the smallest approaching the top end of the axis. Due to the fact that the maximum tilt during the upward movement of the swivel head coincides with the pre-adjusted displacement of the Y and Z-axis perpendicularity of the guide rail, the necessity and importance of the pre-adjusted displacement of the perpendicularity are verified. The center coordinate of the swivel head is not subjected to forces in the X direction, so the straightness error is small.

The guide rail drives the workbench to move along the Y direction. When it reaches the tool magazine, due to combined stress, the workbench shifts downward in the Z direction significantly. After leaving the vicinity of the tool magazine, the workbench runs normally. Adding auxiliary support for guide rail can reduce Z-direction deformation.

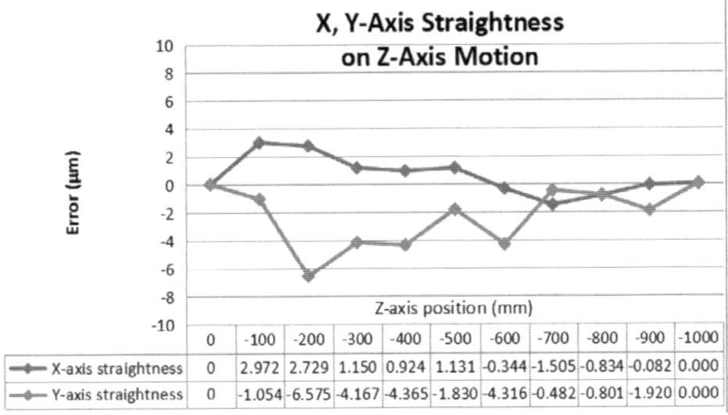

	0	-100	-200	-300	-400	-500	-600	-700	-800	-900	-1000
X-axis straightness	0	2.972	2.729	1.150	0.924	1.131	-0.344	-1.505	-0.834	-0.082	0.000
Y-axis straightness	0	-1.054	-6.575	-4.167	-4.365	-1.830	-4.316	-0.482	-0.801	-1.920	0.000

Figure 13. X, Y-axis straightness vs. Z position (X=0mm).

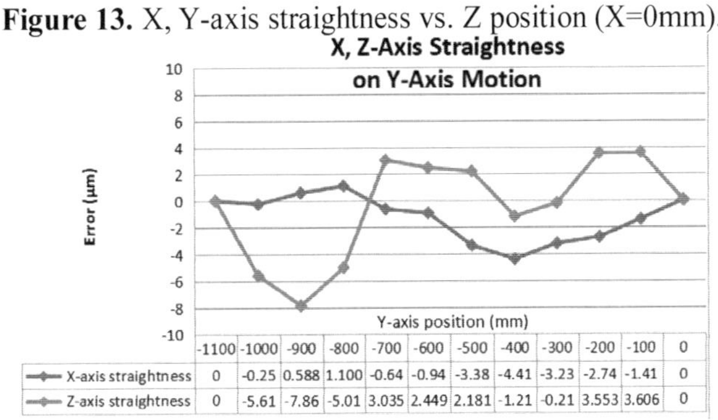

	-1100	-1000	-900	-800	-700	-600	-500	-400	-300	-200	-100	0
X-axis straightness	0	-0.25	0.588	1.100	-0.64	-0.94	-3.38	-4.41	-3.23	-2.74	-1.41	0
Z-axis straightness	0	-5.61	-7.86	-5.01	3.035	2.449	2.181	-1.21	-0.21	3.553	3.606	0

Figure 14. X, Z-axis straightness vs. Y-axis position (X=0mm).

5. Conclusion

This article is based on the theory of multi-body systems and the homogeneous coordinate transformation method to establish a geometric error model between rotating axes. A double ball bar (DBB) was used to measure the circular trajectory and identify the geometric error independent of the position of the machine tool's pendulum head. Under four different angles (B0°, B45°, B90°, B-15°) of the pendulum head transformation, a laser tracker was used to measure and analyze the straightness, verticality, and center coordinates and plane errors of the machine tool's linear axis when the swivel head rotates and translates. The research results show that: 1) The center flatness fluctuation of the pendulum head moving along the XZ plane at different angles is small, with a maximum of 0.0086mm. When the swivel head is at 45°, the center coordinate error and flatness error are both the largest; 2) The simulation results are fairly consistent with the measured center coordinate error of the swivel head's planar motion; 3) Tilting the machine tool column back by (0.01-0.015) mm during

Journal of Physics: Conference Series **2842** (2024) 012083 doi:10.1088/1742-6596/2842/1/012083

assembly can reduce the impact of the weight of swivel head on verticality accuracy and straightness when reciprocating up and down; 4) Adding auxiliary support to the base near zero position of the swivel head can reduce the straightness error.

References

[1] Qiuzhou Jingzi. Research on Post Processing Technology of Universal Spindle Tilting Five-Axis Machine Tools [D]. Harbin Institute of Technology. 2014.6: 1-5.
[2] Wang Jiazhu. Study on the Machining Accuracy Based on the Turn-Milling Centers [D]. Northeastern University. 2012.(6): 1-10.
[3] Wang H, Ming X, Zhang Y, et al. Progress of the Research and Development of Turning-Milling-Compound-Machining-Center in Aviation Industry [J], Metal Working Cold Working, 2020(5): 5-8.
[4] He Z, Fu J, Yao X, et al. A New Method to Identify All Six Error Parameters of A Rotational Axis of Machine Tool [J]. International Journal of Machine Tools and Manufacture, 2015, 88: 1-8.
[5] Tsutsumi M, Tone S, Kato N, et al. Enhancement of geometric accuracy of five-axis machining centers based on identification and compensation of geometric deviations [J]. International Journal of Machine Tools and Manufacture, 2013, 68: 11-20.
[6] IS0 230 -1. Test code for machine tools- part 1: geometric accuracy of machines operating under no-load or finishing conditions [S]. 2012.
[7] Eman K F, Wu B T, DeVries M F. A Generalized Geometric Error Model for Multi-Axis Machines[J]. CIRP annals, 1987, 36(1): 253-256.
[8] Yang J, Xu L, Liu X, et al. Geometric and Thermal Error Synthesis Model for Machining Centers [J]. Acta Metrologica Sincia, 2001.22(2): 90-94.
[9] Zhang Y, Yang J, Zhang K. Geometric Error Measurement and Compensation for the Rotary Table of Five-axis Machine Tool with Double Ballbar[J]. the International Journal of Advanced Manufacturing Technology, 2013,65(1-4): 275-281.
[10] Ibaraki S, Ota Y. Error Calibration for Five-axis Machine Tools by On-the-machine Measurement Using a Touch-trigger probe [J]. International Journal of Automation Technology, 2014, 8(1): 20-27.
[11] Ibaraki S, Nagai Y, Otsubo H, et al. R-Test Analysis Software for Error Calibration of Five-Axis Machine Tools-Application to a Five-Axis Machine Tool with Two Rotary Axes on the Tool side[J]. International Journal of Automation Technology, 2015, 9(4): 387-395.
[12] Gao Shuang. The Lightweight Design for Tool Bracket of The Direct-drive Spindle Head of The Milling-turning Center [D]. Lanzhou University of Technology, 2018. (5): 1-7.
[13] Zhang H, Guo K. Calculation and Adjustment of the B-axis Rotating Center of Milling and Turning Complex Machining Center [J]. Metal Working Cold Working, 2016(S1): 60-62.
[14] Chen Rigao. Static Error Measurement for Rotary Axes of Five-Axis CNC Machine Tools Based on Touch-Trigger Probe [D]. Harbin Institute of Technology, 2018(12): 36-39.

Analysis and optimal design of forging preforming for I-shaped forgings

Lili Huang[*] and Qingyun Zhong

School of Material Science and Engineering, Shandong Jianzhu University, Jinan, Shandong, 250101, China

[*]Corresponding author's e-mail: huang0539@163.com

Abstract. The optimal design of forging preforming is the key to improving forging quality and productivity. Taking an I-shaped forging as a research object, an approach to optimize two-dimensional preform shape is employed. The mathematical model of multi-objective optimization in forging forming simulation is established, and the forming load and die wear are used as the objective functions to optimize the design. Two forms of straight line and circular arc are designed for the shape of the deformation area of the preforming die. With the aid of the finite element simulation software, the process of preforming and finishing forging by using the preformed die with different design variables and different parameters is simulated. Simulation results show that the best optimization results can be achieved when the linear angle is 53.5° and the circular arc radius is 66 mm. After optimization, the load drops by about 36 percent by adopting the circular arc shape, and by about 76 percent by adopting the straight-line shape. The die wear depth decreases and the wear zones become more balanced after optimization. Numerical results prove that this approach is effective in designing the optimal preform shape in forging.

1. Introduction

With the development of science and technology, the material processing industry such as forging, casting, and welding has become an important part of national economic production. Thus, forging technology has received considerable attention. Energy saving, material reduction, and cost reduction have been one of the hot issues in the field of forging research. Forgings with complex shapes are usually processed in multiple processes. Therefore, the determination of the number of processes and the design of the die shape of each process, that is, the preforming design, has become a very important content in the optimization design of the forging process.

In recent years, many scholars at home and abroad have done a lot of research on preforming optimization design [1-3]. Febriani [4] studied the optimum design of the preform shape for a bevel gear to get a good quality component. The fitness functions included form-filling, preform volume, and complexity. Kampen [5] gave a method for preforming optimization of cross wedge rolled with high efficiency. Kim [6] presented a solution for making the strain distributed evenly. Biba [7] presented an approach for developing a preform geometry that involved a potential flow approximation and using equipotential surfaces as a tentative guess. Kitayama [8] performed an optimization approach to improve forging quality.

Most of the above researches take the optimization design of forging shape or deformation uniformity as the objective function. The final forging force and die wear during the forging process also have important effects on forming. The greater the final forging force, the greater the force loading of the die, which intensifies the wear between the forging and the die cavity, and ultimately affects the quality of the forging. At the same time, the final forging force also determines the choice of equipment tonnage. The smaller final forging force can reduce the wear and fatigue of the equipment and reduce the production cost. In order to reduce the final forging force and die wear, this paper studies the optimal design of the preforming die of the I-shaped forgings.

2. Objective function

This paper utilizes the linear weighting sum method in the weighting function method to convert the multi-objective optimum design problem including the final forging force and die wear to a single objective optimization problem.

The final forging force can be expressed by the sum of the component force of all elements in contact with the upper die in the normal direction, that is

$$F = \sum_{i=1}^{M} f_{iY} \tag{1}$$

where f_{iY} is the component of the force in the normal direction of element i contacts with the upper die. M is the total number of units in contact with the upper die.

In order to make the dimension unity in the weighted equation, the above objective function is transformed into a normalized objective function between 0 and 1, and a new dimensionless objective function is established.

$$f_1 = \frac{1}{M}\sum_{i=1}^{M} \frac{f_{iY} - f_{iYmin}}{f_{iYmax} - f_{iYmin}} \tag{2}$$

where f_{iYmin} is the minimum value of the components' force in the normal direction of the elements, f_{iYmax} is the maximum value of the components force in the normal direction of the elements.

Archard wear model is used to predict die wear, and the expression is as follows:

$$W = \int K \frac{P^a v^b}{H^c} dt \tag{3}$$

where W is wear depth, P is the normal pressure of the die in contact with the workpiece, v is the sliding speed, H is the die material's hardness, K is the wear coefficient, a, b, and c are the constants related to the material.

In this paper, the sum of the difference between the wear amount of the contact node on the die and the average wear amount of the whole die contact surface is used as the objective function to measure the wear degree of the die. In addition, considering the non-dimensionalization, the objective function is defined as follows:

$$f_2 = \sum_{j=1}^{N} \frac{W_j - W_{av}}{W_{jmax} - W_{av}} \tag{4}$$

where N is the number of nodes on the die contact surface, W_{av} is the average wear depth and W_{jmax} is the maximum wear depth.

The final forging force sub-objective function and die wear sub-objective function are composed of the total objective function by the weighted sum method.

$$f = \omega_1 f_1 + \omega_2 f_2 \tag{5}$$

where ω_1 and ω_2 are weighting factors, and $\omega_1 + \omega_2 = 1$. The smaller the total objective function is, the smaller the final forging force is, the less die wear is, and the better the optimization result is.

3. Shape design and variable selection of preforming die

The I-shaped forging is formed by a single forged cylindrical blank, which is easy to produce defects such as crease, insufficient filling, piercing, eddy current, etc. That reduces the qualification rate of forging parts. Therefore, it is necessary to preform the forging to reduce the probability of the above defects, improve the material flow in the final forging cavity, and make it easier for the material to fill the cavity. Figure 1 shows the cross-section diagram of the I-shaped forging used in this paper.

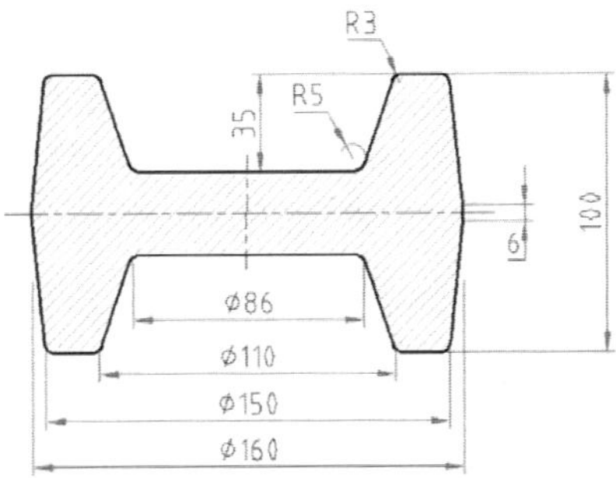

Figure 1. Cross-section diagram of I-shaped forging.

In the preforming design, the deformation of the preforming should be kept within the appropriate range, and enough metal should be used to fill the die cavity in the final forging process. The basic design rules are as follows: (1) The depth of the preforming die should be 2-5 mm lower than the depth of the final forging die; The width of the preforming die is 1-3 mm smaller than that of the final forging die. (2) There is no flash groove in the preforming die, so the volume of the preforming is greater than the volume of the final forging; (3) The corner radius of the preforming is greater than the corner radius of the final forging to reduce the flow resistance and prevent folding. The shape of the preforming is shown in Figure 2.

Because the shape change between the die web and the wing has a great impact on the forging filling, the shape of this region is selected for optimal design. The web and the wing plate can be connected by a straight line or curve, using arc transition.

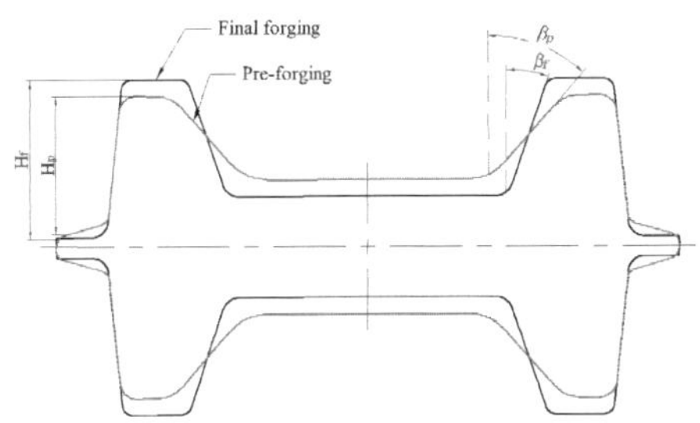

Figure 2. Preformed shape diagram of I-shaped forging.

4. Numerical simulation of the forging process

To study the forging process of I-shaped forgings, the software DEFORM2D is utilized for numerical simulation. The simulation process is divided into two steps: preforming and final forging. A cylindrical billet with size φ110×150 mm is selected according to the forging size. Because of the symmetry of the I-shaped forging, a quarter of the blank and die is used for simulation.

4.1 Design and optimization analysis of linear preforming die

The web of the preforming die and the wing plate are connected in a straight line, and the rounded corner transition is adopted, as shown in Figure 3. Through repeated simulations and experiments, the inclination angle α of the connected straight-line section is selected between 30° and 65°, and the metal material can better fill the die cavity and achieve a good filling effect. The inclination angle α is chosen from 30°, 35°, 40°, 45°, 50°, 55°, 60° and 65° as the design variables to optimize the preforming die's shape.

By simulating and extracting the corresponding final forging force and wear depth, the objective function values of the linear preforming die at different angles can be obtained according to the formula. With the help of 1stopt software, the objective function is fitted, and the resulting fitting curve is shown in Figure 4. The equation of the fitting curve is as follows:

$$f(\alpha) = p_1 + p_2\alpha + \frac{p_3}{\alpha} + p_4\alpha^2 + \frac{p_5}{\alpha^2} + p_6\alpha^3 + \frac{p_7}{\alpha^3} + p_8\alpha^4 + p_9/\alpha^4 + p_{10}\alpha^5 + p_{11}/\alpha^5 \quad (6)$$

where the coefficients of the equation are p_1=-21060.6267623029, p_2=1183.56467310514, p_3=152379.499651327, p_4=-34.8510737588533, p_5=24242.7602905623, p_6=0.568118900772105, p_7=2215.02191856698, p_8=-0.00486606069428784, p_9=152.924922828109, p_{10}=1.71232041448333E-5, p_{11}=9.58679520720243, respectively.

The root of the mean square error is 0.07, the sum of square error is 0.04, and the correlation coefficient is 0.944, which indicates that the model fits well and the error is small.

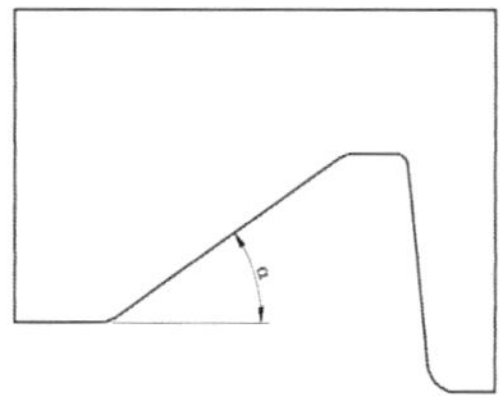

Figure 3. Sketch of linear preforming die.

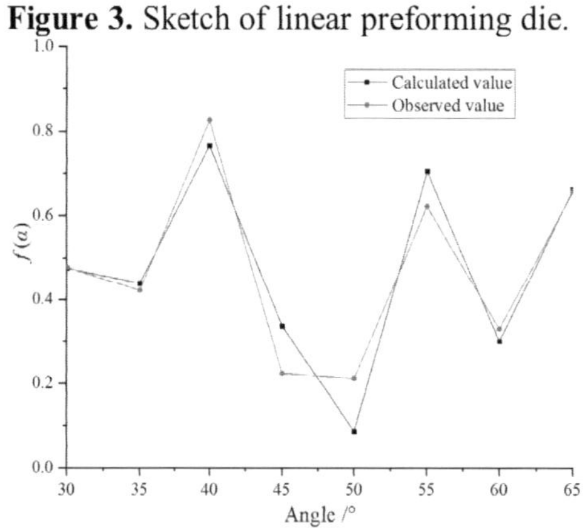

Figure 4. Graph of the effect of angle design variables on the objection function.

The conjugate gradient method is used to obtain the optimal value of the fitted curve equation. After 16 iterations, the optimization result is obtained. That is, when the angle α is 53.5°, the curve equation appears as the minimum value, and the objective function is the minimum value.

4.2 Design and optimization analysis of arc type preforming die

The preforming die web and the wing plate are connected by an arc, and the rounded corner transition is adopted, as shown in Figure 5. The radius of the arc should not be too small, otherwise the deep cavity of the final forging will not be fully filled. The radius should not be too large, otherwise, it will be close to a straight line, and the effect will be similar to the linear connection. Through repeated simulations and experiments, the size range of the arc radius is set to 40~160mm. As design variables to optimize the shape of the preforming die, radius R is selected equal to 40, 50, 60, 80, 100, 120, 140, and 160 mm, respectively.

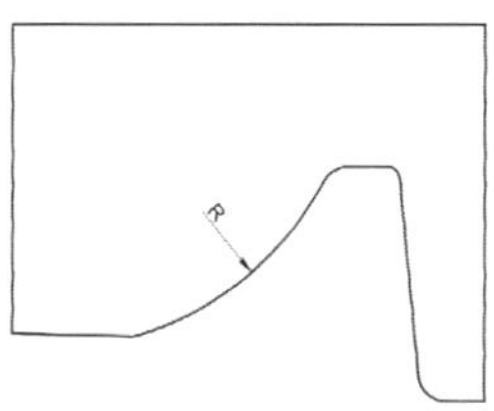

Figure 5. Schematic diagram of arc-type preforming die.

By simulating and extracting the corresponding final forging force and wear depth, the objective function values of the arc preforming die at different angles can be obtained according to the formula. With the help of 1stopt software, the objective function is fitted, and the resulting fitting curve is shown in Figure 6. The equation of the fitting curve is as follows:

$$f(\alpha) = p_1 + p_2\alpha + \frac{p_3}{\alpha} + p_4\alpha^2 + \frac{p_5}{\alpha^2} + p_6\alpha^3 + \frac{p_7}{\alpha^3} + p_8\alpha^4 + p_9/\alpha^4 \quad (7)$$

where the coefficients of the equation are p_1=79.2439919656209, p_2=-1.36101579735294, p_3=-2453.7356173194, p_4=0.0128267209135338, p_5=32237.2653140274, p_6=-6.18771422211716E-5, p_7=3387.72712000831, p_8=1.20004193017532E-7, p_9=205.559476495396, respectively.

The root of the mean square error is 0.03, the sum of square error is 0.009, and the correlation coefficient is 0.993, which indicates that the model fits well and the error is small.

The conjugate gradient method is used to obtain the optimal value of the fitted curve equation. After 17 iterations, the optimization result is obtained. That is, when the radius R is 66mm, the curve equation appears as the minimum value, and the objective function is the minimum value.

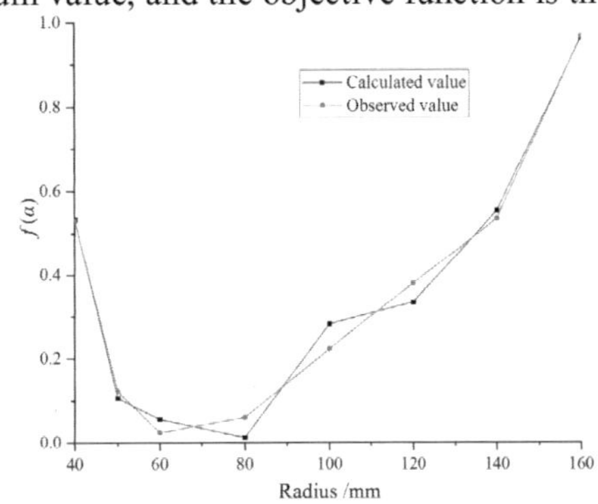

Figure 6. Graph of the effect of radius design variables on the objection function.

4.3 Simulation verification

The two optimized die shapes are simulated and verified. Figure 7 shows the comparison of the vertical load of the die and the die cavity wear. In Figure 7(a), the maximum load value before optimization is 4.74×10^7N. After optimization, the load drops by about 36 percent to 3.02×10^7N by adopting the curved die cavity, and the load drops by about 76 percent to 1.14×10^7N by adopting the linear die cavity. Figure 7(b) shows the comparison of die cavity wear. It can be seen from the figure that severe die wear is mainly concentrated in the protrusion part before optimization, and the maximum die wear decreases significantly after optimization. At the same time, the wear zone becomes more balanced after optimization.

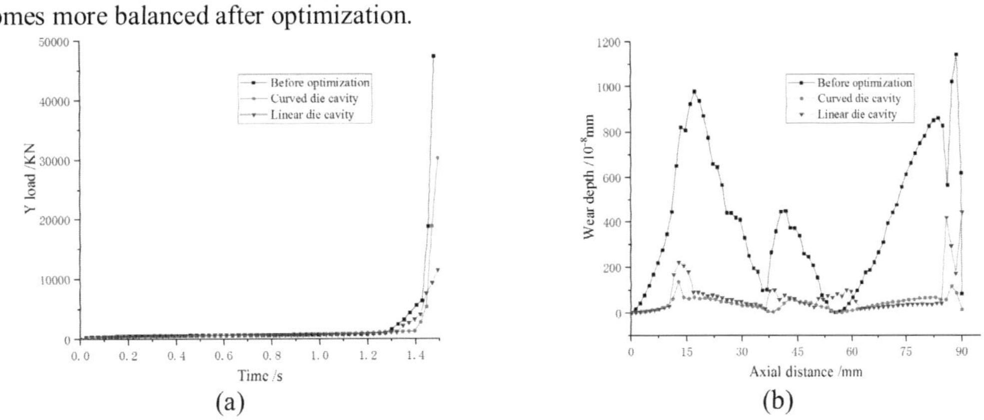

Figure 7. Comparison diagram of vertical load of the die and die cavity wear.

5. Conclusion

In this paper, the process of forging has been designed to involve two steps (i.e., preforming and final forging) to reduce the final forging force and die wear. The shape design variables and the objective function values are fitted, and the corresponding curve equation is obtained. The optimal solution of the curve equation is sought to obtain the best shape of the preforming die. The following main conclusions are obtained.

(1) For the preformed die shape with the linear angle as the design variable, the objective function is fitted well and the error is small. The objective function is minimal when the linear angle is 53.5°.

(2) For the preformed die shape with the circular arc radius as the design variable, the objective function is fitted well and the error is small. The objective function is minimum when the circular arc radius is 66 mm.

(3) By comparison, the final forging force of the preforming die optimized with the linear angle as the design variable is the smallest, while the wear of the die cavity optimized with the circular radius as the design variable is smaller and more uniform.

References

[1] Chen H, Guan Y, Liu M, Li Y, Zhai J, and Lin J (2020) Preform optimization of a brake drum part based on quasi-equipotential field and response surface methods. Procedia Manufacturing, 50:276-279.

[2] Guan J, Wang G C, Guo T, Song L B, and Zhao G Q (2009) The microstructure optimization of H-shape forgings based on preforming die design. Materials Science and Engineering: A, 499:304-308.

[3] Knust J, Podszus F, Stonis M, Behrens B-A, Overmeyer L, and Ullmann G (2017) Preform optimization for hot forging processes using genetic algorithms. The International Journal of Advanced Manufacturing Technology, 89:1623-1634.

[4] Febriani R A, Park H S, Kumar S, and Lee C M (2020) Preform optimization of functional bevel gear for warm forging processes. International Journal of Automotive Technology, 21:1097-1106.

[5] Kampen D, Richter J, Blohm T, Knust J, Langner J, Stonis M, and Behrens B-A (2020) Design of a genetic algorithm to preform optimization for hot forging processes. International Journal of Material Forming, 13:77-89.

[6] Kim K, Kim Y, Ju Y, Son I, and Kim N (2023) CNN-based Ti-6242 impeller forging process design for uniform strain distribution. Applied Sciences, 13.

[7] Biba N, Stebunov S, Vlasov A, Kenzhaliyev K, and Duzhev A (2024) Automated preform design for the development of multi-stage hot forging technology. In: Mocellin K, Bouchard P-O, Bigot R, Balan T (eds) Proceedings of the 14th International Conference on the Technology of Plasticity - Current Trends in the Technology of Plasticity, Cham. pp 88-98.

[8] Kitayama S, Saito K, Wang T, Furuta S, Aono E, and Amano M (2023) Numerical investigation and process parameters optimization in three-dimensional multi-stage hot forging for minimizing flash and equivalent strain. The International Journal of Advanced Manufacturing Technology, 126:5409-5420.

Journal of Physics: Conference Series **2842** (2024) 012085 doi:10.1088/1742-6596/2842/1/012085

Study on process optimization of electroless Ni-P plating on binderless WC

Xianyou Wu[1], Tianfeng Zhou[1,2], Zhaoqian Zhang[1], Peng Liu[1,2], Qian Yu[1] and Bin Zhao[1,2,*]

[1]School of Mechanical Engineering, Beijing Institute of Technology, Beijing, 100081, China

[2]Chongqing Innovation Center, Beijing Institute of Technology, Chongqing, 401120, China

[*]Corresponding author's e-mail: bin.zhao@bit.edu.cn

Abstract. In the field of Precision Glass Molding (PGM), binderless tungsten carbide (WC) is a pivotal material for molds, despite its high processing costs and complexities. Nickel-phosphorus (Ni-P) alloys also exhibit superior performance at elevated temperatures. The innovation of Ni-P/WC composite molds addresses the issue of the poor machinability of WC. This research delves into the influence of the activating solution's concentration on the surface activation energy of WC, the quality of plating, and the deposition rate during the electroless Ni-P plating process. Additionally, the study scrutinizes how the pH level of the plating solution impacts the quality of Ni-P, the rate of deposition, and the phosphorus content. These investigations have led to the realization of an efficient and high-quality Ni-P plating process for WC.

1. Introduction

PGM is favored for its cost-effectiveness and capability for mass-producing glass optical components, as noted in references [1-3]. The precision of the mold is pivotal as it directly influences the accuracy of the lens shape. The primary materials used in molds for PGM include WC, silicon carbide, and Ni-P plated materials. WC is a common choice in molding due to its superior mechanical, thermal, and chemical properties, yet it presents significant challenges in machining due to its extreme hardness [4]. Ni-P molds offer sustained performance at elevated temperatures, coupled with excellent wear and corrosion resistance, as well as precision machinability [5-7]. Applying Ni-P plating to WC addresses the issue of machining complex surfaces with WC and mitigates shape errors that arise from the mold's thermal expansion. Nonetheless, a well-established process for electroless Ni-P plating on WC is lacking, prompting the need for this paper to explore process optimization for Ni-P plating on WC.

Electroless Ni-P plating is a process where nickel ions (Ni^{2+}) in the solution are chemically reduced to nickel by a reducing agent within a metastable plating bath without the application of an electric current. Concurrently, the phosphorus in the reducing agent is catalyzed and heated to form phosphorus, resulting in the co-deposition of phosphorus and nickel onto a catalytically active substrate. The electroless Ni-P plating process typically encompasses substrate grinding, alkaline degreasing, acid activation, and the plating itself. Table 1 outlines the specific procedural steps, with acid activation and the chemical plating solution being critical to the plating quality. In this study, hydrochloric acid (HCl) is selected for the acid activation, and the concentration of HCl and the pH level of the plating solution will be optimized to achieve an efficient and high-quality Ni-P plating on WC, thus providing a

Journal of Physics: Conference Series **2842** (2024) 012085 doi:10.1088/1742-6596/2842/1/012085

foundation for further research.

Table 1. The specific steps of electroless Ni-P plating on WC.

Steps	Procedures	Solution used	Operating conditions
1	Substrate Polishing	-	Determined on a case-by-case basis
2	Ultrasonic cleaning	Anhydrous ethanol 60 mL Deionized water 60 mL	25 °C, 10 min
3	Alkaline degreasing	Degreasing solution 60 mL	60 °C Water bath, 5 min
4	Ultrasonic cleaning	Anhydrous ethanol 60 mL Deionized water 60 mL	60 °C, 10 min
5	Acid activation	Activation solution 60 mL	60 °C Water bath, 5 min
6	Ultrasonic cleaning	Anhydrous ethanol 60 mL Deionized water 60 mL	88 °C, 10 min
7	Electroless plating	Chemical plating bath 60 mL	88 °C, 2 h
8	Ultrasonic cleaning	Anhydrous ethanol 60 mL Deionized water 60 mL	25 °C, 1 min

2. Effect of HCl concentration on electroless Ni-P plating on WC

To investigate the impact of varying concentrations of HCl on the surface activation of tungsten carbide (WC), four different concentrations of HCl were used, which were 0 vol%, 10 vol%, 50 vol%, and 100 vol% HCl, respectively. The WC used in this study is R07 type binderless cemented carbide (90% WC, the remaining 10% is mixed by WC, TiC, and TaC) of Japan Special Alloy Co., Ltd., with a size of Φ20 mm × 5 mm.

Before activation, the WC surface was polished with a polishing machine to make its roughness below 0.1 μm and then rinsed in deionized water for five minutes. WC was subjected to surface activation in various concentrations of HCl for five minutes. The WC samples activated by HCl (from dilute to concentrated) were recorded as WH0, WH1, WH2, and WH3, respectively.

2.1. Effect of HCl concentration on the wettability of WC

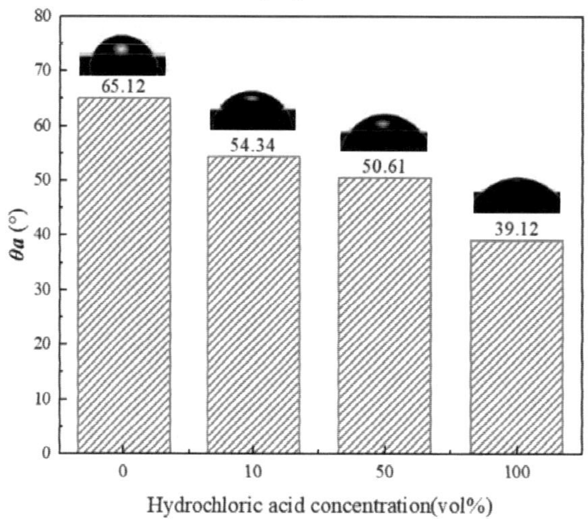

Figure 1. Different concentrations of HCl activate the contact angle of the WC surface.

The wettability of the WC surface activated by different concentrations of HCl was characterized by the droplet contact angle, and the droplet contact angle of the WC surface was measured by a contact angle measuring instrument (OCA25). The droplet used in the measurement is deionized water, and the volume of the injected droplet is 4 μL per measurement. Five different positions on the WC surface were randomly selected for static contact angle measurement, and finally, the average value was taken as the contact angle. The contact angle of the WC surface after various activation treatments is shown in the figure. The polished WC surface is not completely flat, so the surface wettability is characterized by the apparent contact angle θ_a in the actual measurement. From Figure 1, the contact angle of WH0 is 65.12°, less than 90°, indicating that WC is a hydrophilic material. After activation with HCl, the contact angle of the WC surface decreases as the HCl concentration increases, while the contact angle of WH3 is the smallest, which is 39.93°, which is 36.0% lower than that of WH0. The contact angles of WH1 and WH2 were 54.34° and 50.61°, respectively, which were 16.55% and 22.28% lower than that of WH0.

2.2. Effect of HCl concentration on the surface morphology of Ni-P plating

After activation with various concentrations of HCl, the surface of WC is coated with Ni-P according to the above steps, and the cross-section of each specimen after electroless Ni-P plating was observed by scanning electron microscope (Tescan MIRA 3 XMH). The observation from Figure 2 reveals that the plating on WH0 and WH1 is uniformly covered on the WC substrate. However, the surfaces of WH2 and WH3 with good wettability are almost not coated with Ni-P, and the surface of the WC substrate is exposed. Figure 3 shows the surface morphology of WH2 and WH3. It can be observed that there are many strip protrusions on the surface of WC. Concentrated HCl causes oxidation of the surface of the WC. These oxides will hinder the crystallization of Ni on the surface of WC, and because of the formation of oxides, the surface of WC is rougher, and the rough surface is conducive to reducing the surface contact angle, so when the activator is concentrated at 50 vol% and 100 vol% HCl, the contact angle of WC surface is reduced, but the concentrated HCl promotes the oxidation of the WC surface, which makes it difficult for Ni to nodule on the surface of WC, so it is difficult to plating Ni-P.

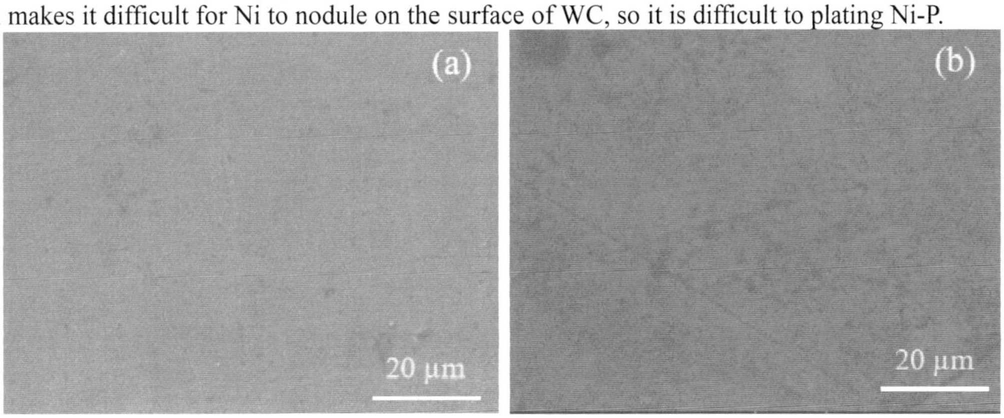

Figure 2. Surface morphology of electroless Ni-P plating on WC activated by different concentrations of HCl.

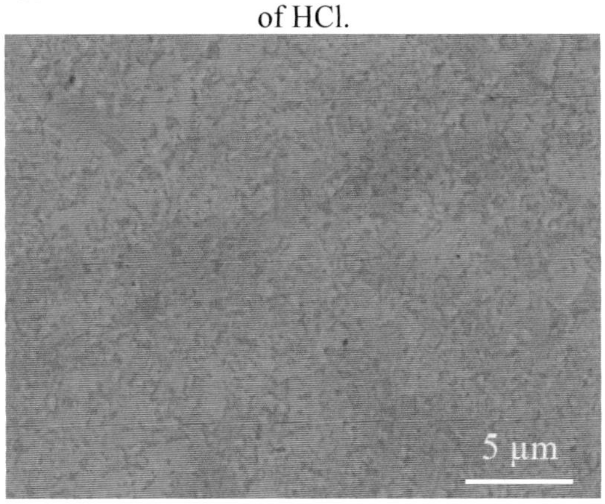

Figure 3. Surface morphology of WC activated by concentrated HCl.

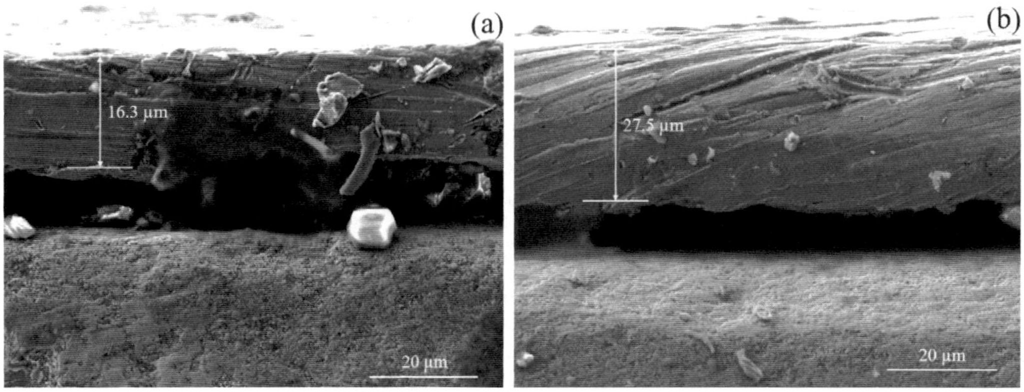

Figure 4. Cross-sectional morphology of WH0 and WH1.

2.3. Effect of various concentrations of HCl-activated WC on deposition rate

After plating, the specimens' cross-sections were refined through polishing to assess the plating thickness via scanning electron microscopy. Figure 4 illustrates the cross-sectional appearance of the WH0 and WH1 specimens' platings. The plating's depth was gauged at five distinct points on each specimen. Following a 2-hour plating duration, the respective average thicknesses for the two coatings were determined to be approximately 16.3 μm and 27.5 μm. The findings indicate that the rate of electroless Ni-P deposition on untreated WC is roughly 8.15 μm/h. In contrast, the rate on WC surfaces activated with 10 vol% HCl increases to approximately 13.75 μm/h, marking a 68.71% enhancement over the WH0 specimen. Specimens treated with 50 vol% HCl and 100 vol% HCl exhibited negligible plating, precluding a comparative analysis. Thus, the application of 10 vol% HCl significantly ameliorates the WC surface's catalytic properties, hastening the electroless Ni-P plating reaction and augmenting the plating's deposition rate.

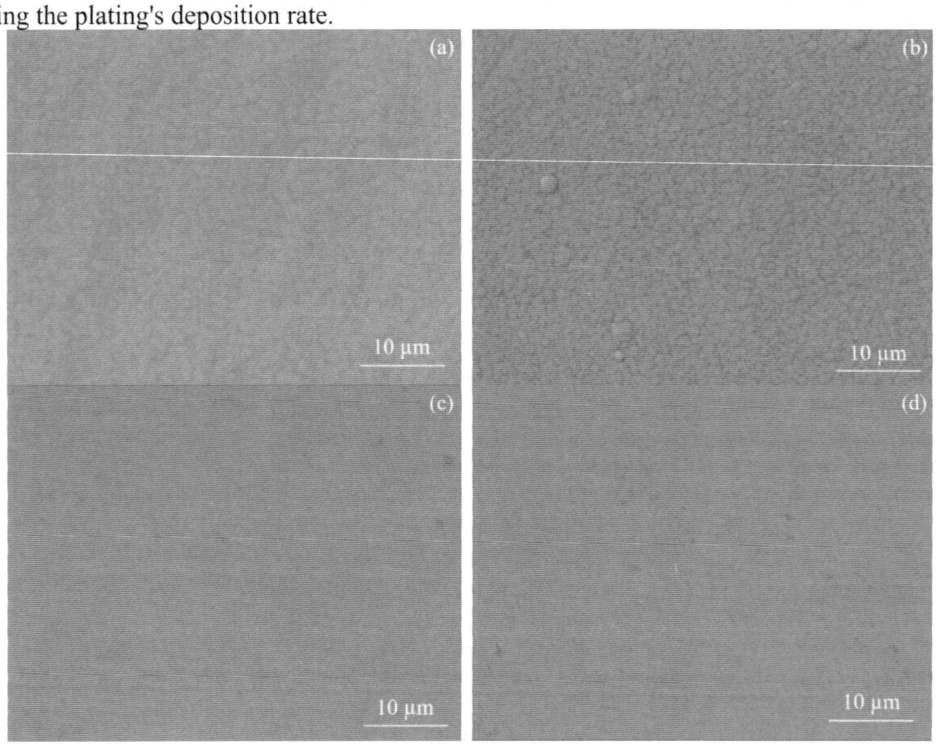

Figure 5. The surface morphology of electroless Ni-P plating with different pH.

3. Effect of pH of the bath on electroless Ni-P plating on WC

This research aimed to understand how the pH level of the plating solution affects the Ni-P plating process on WC. To achieve this, the pH was modified using hydrochloric acid (HCl). The pH readings were taken with a specific pH meter, the pHS-25, ensuring accuracy. All measurements were conducted at a consistent temperature of 25 °C to maintain experimental conditions. A range of pH values was tested, including 3.6, 4.0, 4.4, and 4.8, to evaluate their influence on the plating outcome. The specimens coated with different pH values were recorded as WP0, WP1, WP2, and WP3, respectively. According to the previous section, 10 vol% HCl has the best effect on WC activation, so 10 vol% HCl is used to activate WC here.

3.1. The effect of the pH of the bath on the surface morphology of Ni-P

The visual representation of Ni-P plating surfaces at different pH levels is depicted in Figure 5. In Figures 5(a) and 5(b), the plated surfaces of specimens WP0 and WP1 are characterized by a pronounced aggregation of cellular structures, with the grains on WP0 being considerably larger than those on WP1, which implies a more compact arrangement for WP1. In contrast, Figures 5(c) and 5(d) show that specimens WP2 and WP3 lack such pronounced clusters, instead displaying a uniform and smooth surface, suggesting that the cellular structures are more refined and the overall structure is denser. As a result, the progression of the bath's pH towards higher values leads to a more tightly knit surface structure in the Ni-P plating, which correlates with improved quality.

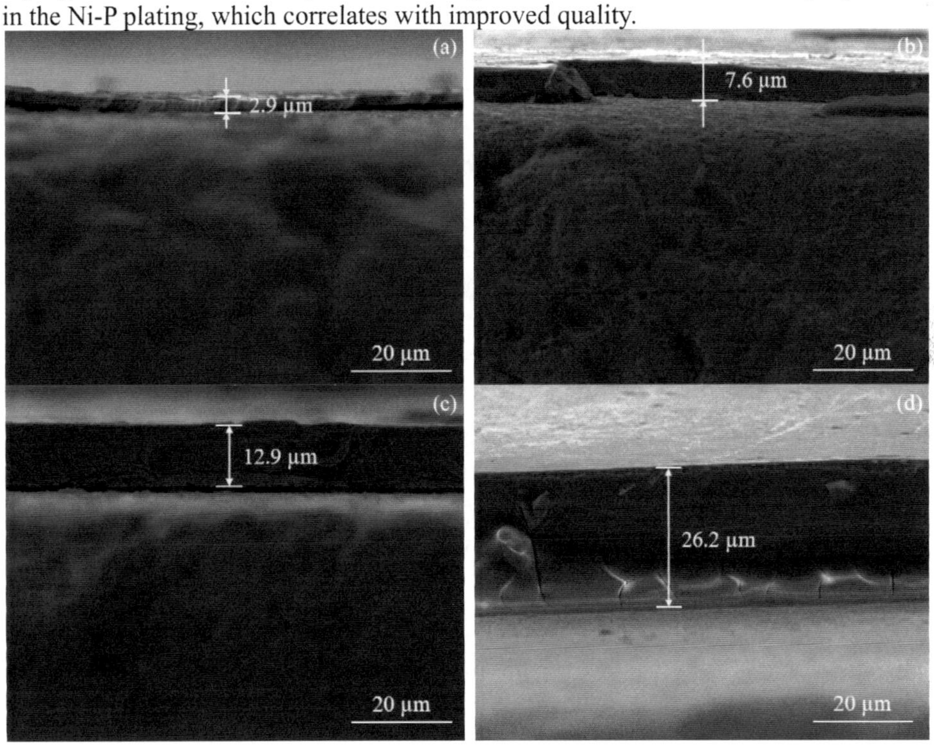

Figure 6. The cross-section of Ni-P plating with different pH values.

3.2. Effect of pH of the bath on the deposition rate of electroless Ni-P plating

The impact of bath pH on the Ni-P plating's cross-sectional profile is depicted in Figure 6. Thickness measurements were conducted on five separate plating areas, with the mean thickness being taken as representative of each section. Observations from the figure indicate that with an increase in pH from 3.6 to 4.8, the thickness of the plating after a 2-hour duration escalates from approximately 2.9 μm to 26.2 μm. The corresponding deposition rates for these pH levels are calculated to be 1.45 μm/h, 3.80 μm/h, 6.45 μm/h, and 13.10 μm/h, respectively. The study's conclusions point to a positive correlation between pH and deposition rate, with the specimen initially plated at pH 3.6 recording the slowest rate and the one at pH 4.8 recording the fastest, nearly nine times the rate of the former. The deposition rate is found to be inversely proportional to the pH value, ascribed to the competitive discharge of Ni^+ and H^+ ions that impedes the plating's reduction reaction, thus affecting the growth rate. Although the pH variation is minor, it significantly influences the velocity of the electroless Ni-P plating process.

3.3. Effect of pH of the bath on P content in Ni-P

The phosphorus content within the electroless Ni-P plated specimens, each subjected to distinct pH levels, was evaluated using EDS. Figure 6 delineates how the P content in the plating solutions is affected by the pH. Figure 7 clearly shows that the P content diminishes progressively with an increase in pH. WP0 has the highest recorded P content at 13.57 wt%, contrasting with WP3, which has the lowest at 9.81 wt%, marking a decrease of 27.7% from WP0's content. In the electroless Ni-P plating reaction, since phosphorus and hydroxide ions are produced concurrently, hydrochloric acid (HCl) is utilized to modulate the pH. The addition of HCl to the bath to adjust its pH results in an increased reduction rate for phosphorus, thus affecting the P content. The relationship between P content and pH is nearly linear, indicating that pH adjustment is a manageable way to regulate P content. Given that P content has a direct impact on the Ni-P alloy's physical properties, controlling the P content allows for the customization of the plating's performance characteristics.

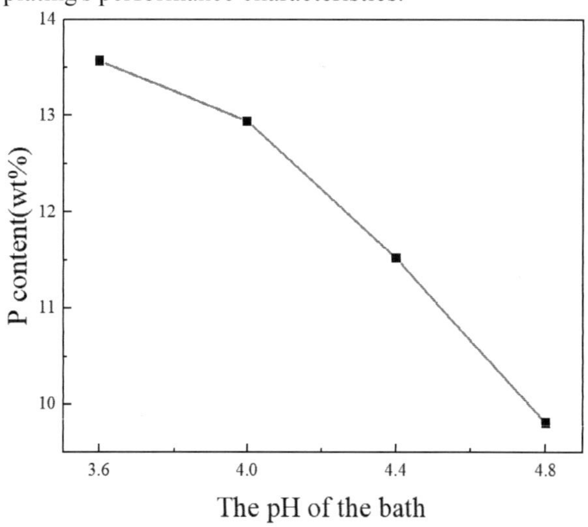

Figure 7. P content in the plated layer for different pH plating solutions.

4. Conclusion

This study focused on enhancing the procedure for electroless Ni-P plating on WC and examined how varying the concentration of HCl and the pH of the plating solution influences the outcome. The experimental results lead to these key insights:

(1) Utilizing an HCl concentration of 10% volume resulted in a lower contact angle on the WC surface and a higher activation energy, which corresponded to a nearly 70% increase in the Ni-P deposition rate compared to the untreated state of WC.

(2) Fine-tuning the pH of the plating bath to a more alkaline level, specifically to 4.8, was found to be most conducive to achieving superior surface quality in the Ni-P plating, along with an increased rate of deposition and a denser concentration of phosphorus within the plated layer.

Acknowledgments

This work has received funding from the National Natural Science Foundation of China (52205441) Natural Science Foundation of Chongqing, China (Grant No. 2022NSCQ-MSX3775).

Reference

[1] Zhou T, He Y, Wang T, et al. A review of the techniques for the mold manufacturing of micro/nanostructures for precision glass molding [J]. International Journal of Extreme Manufacturing, 2021, 3(4): 042002.
[2] Kim Y K, Haq M R, and Kim S-M. Glass molding of all glass Fresnel lens with vitreous carbon micromold [J]. Optics Express, 2019, 27(2): 1553-1562.
[3] Abdul Manaf A R, Yan J. Improvement of form accuracy and surface integrity of Si-HDPE hybrid micro-lens arrays in press molding [J]. Precision Engineering, 2017, 47: 469-479.
[4] Yin L, Spowage A C, Ramesh K, et al. Influence of microstructure on ultraprecision grinding of cemented carbides [J]. International Journal of Machine Tools and Manufacture, 2004, 44(5): 533-543.
[5] Bahramian A, Eyraud M, Vacandio F, et al. Improving the corrosion properties of amorphous Ni-P thin films using different additives [J]. Surface and Coatings Technology, 2018, 345: 40-52.
[6] Lin J, Chou C. The influence of phosphorus content on the microstructure and specific capacitance of etched electroless Ni-P coatings [J]. Surface and Coatings Technology, 2019, 368: 126-137.
[7] Sahoo P. Wear behaviour of electroless Ni–P coatings and optimization of process parameters using Taguchi method [J]. Materials & Design, 2009, 30(4): 1341-1349.

Influence of substrate edge chamfer geometric features on the edge effect of spin-coating film

Naipeng He, Qiuyu Liu, Yan Ying*, and Zhou Ping

State Key Laboratory of High-performance Precision Manufacturing, Department of Mechanical Engineering, Dalian University of Technology, Dalian 116024, China

*Corresponding author's e-mail: yanying@dlut.edu.cn

Abstract: The spin-coating method is a widely employed technique for thin film fabrication. However, edge effects can adversely affect the quality and utility of the films produced. This study investigates the influence of chamfered edges on the substrate during the spin-coating process on edge effects. Both experimental and simulation results indicate that smaller chamfer angles contribute to reducing the width of edge effects. Conversely, the width of the chamfer has a relatively minor impact on the extent of edge effects.

1. Introduction

The spin-coating method is a commonly used technique for thin film fabrication, involving the application of a solution or suspension onto a spinning substrate, followed by uniform distribution through centrifugal force to form a thin film. This method is widely favored due to its excellent controllability, simple operation, and high film uniformity [1-4].

However, the use of spin-coating inevitably leads to edge effects, manifested as the elevation of film height at the edges compared to the central region, commonly known as edge beads. The presence of edge beads results in increased film non-uniformity, thus compromising film quality. Moreover, the structure of edge beads is unstable and prone to rupture during subsequent processes such as transfer, drying, and re-coating, leading to contamination of equipment and substrates [5].

To identify methods that can directly control edge effects during the spin-coating process, many scholars have researched the mechanisms underlying edge effects. Baëtens and Arscott [6] studied the changes in liquid behavior at the edges of photoresists during the spin-coating process. It was proposed that after the rotation process ends, the liquid at the edges retracts due to surface tension, gradually flattening the liquid surface. However, solvent evaporation increases the viscosity of the solution, inhibiting the rate of liquid retraction and ultimately halting the flattening process. Luurtsema [7] investigated the effect of rotation speed on film uniformity. The results showed that as rotation speed decreases, the vacuum generated by the Bernoulli effect weakens, resulting in a more uniform film coating. However, this uniformity comes at a cost, as the width of edge beads widens and the film thickness increases, making it more susceptible to fluctuations in speed. Moreover, some scholars have studied the influence of substrate edge geometry on edge effects during spin-coating. Yamamoto et al. [8] conducted spin-coating experiments on substrates with stepped edge features, observing the liquid accumulation and retraction status at the stepped edges. They found that the height of stepped edges affects the height of edge beads, while its width has a relatively minor impact. Shin et al. [9] examined the influence of substrate groove edges on edge effects, revealing that deeper grooves contribute to the reduction of edge liquid film thickness.

To find effective methods to mitigate edge effects, this study investigates the influence of chamfered edge geometric features on edge effects during spin-coating film formation. Control experiments were designed to study the effects of chamfer angle and width on edge effects, and experimental results were validated using fluid simulations. Finally, this study proposes chamfered edge structures that can effectively reduce edge effects.

2. Experiment and modeling methods

2.1 Experiment methods

The experiments described in this study were conducted on the WS-650 spin coater manufactured by Laurell Technologies Corporation. The photoresist selected for use was the Rzj-304-10 positive photoresist produced by Suzhou Ruihong Electronic Chemical Co., Ltd. The physicochemical properties of the photoresist are presented in Table 1.

Table 1. Physicochemical properties of the photoresist.

Photoresist property	Density	Viscosity (at 20°C)	Contact angle (with substrate, at 20°C)	Surface tension coefficient
Value	1.07 g/cm^3	0.017 Pa·s	10.5°	27.28 mN/m

To minimize the influence of rotational airflow on experimental results during the spin-coating process, a synchronous cover synchronized with the substrate was designed. During the experiment, the substrate was first placed inside the synchronous cover. After activating vacuum suction, the photoresist solution was evenly coated onto the substrate surface. Subsequently, the cover was tightly secured to the body and a synchronous chamber was created using sealing rings. Spin-coating experiments were conducted according to pre-set programs, with the spin-coating parameters during the experiment detailed in Table 2. Excess solution flowed out from the grooves at the bottom of the chamber during rotation. Upon completion of the rotation, the cover was removed, allowing the photoresist solution to evaporate and solidify in the air, forming a photoresist film.

Table 2. Experimental spin-coating parameters.

Experiment parameters	Acceleration time	Speed	Spin time	Deceleration time
Value	10 s	1000 rpm	30 s	3 s

Table 3. Structural parameters of the substrate.

Substrate parameters	Substrate one	Substrate two	Substrate three	Substrate four
Substrate length	122 mm	122 mm	120 mm	125 mm
Substrate width	92 mm	92 mm	90 mm	95 mm
Chamfer angle φ	20°	70°	20°	20°
Chamfer width x	1.5 mm	1.5 mm	0.5 mm	3.0 mm

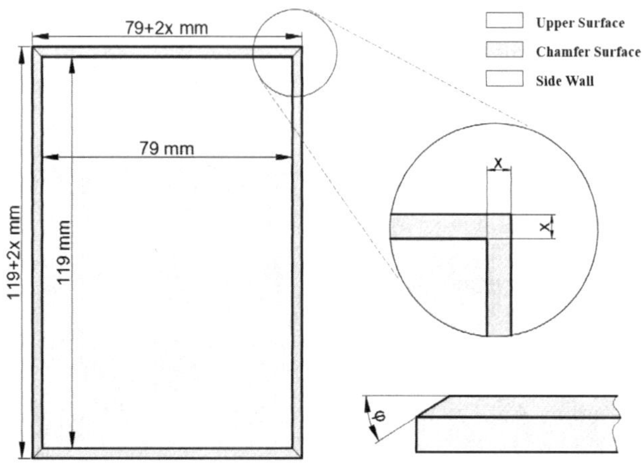

Figure 1. Chamfered edge geometric features of the substrates.

Chamfer angle and chamfer width were chosen as the research parameters for chamfered edge geometric features. Accordingly, various quartz substrates were designed and fabricated for spin-coating experiments. The chamfered edge geometric features of the substrates are illustrated in Figure 1 and summarized in Table 3.

2.2 Simulation model

To investigate the influence of chamfer surface roughness on edge effects during spin-coating film formation, the VOF method was employed in Fluent to numerically simulate the spin-coating process using a laminar flow model. In the simulation process, the liquid phase was assumed to be a Newtonian fluid with a constant viscosity of 0.1 kg/(m·s), a density of 950 kg/m³, and a surface tension coefficient of 0.01 N/m. The gas phase was set as air with a density of 1.225 kg/m³ and a viscosity of 1.7894×10^{-5} kg/(m·s). Additionally, changes in the physicochemical properties of the liquid due to solvent evaporation were neglected during the simulation. To reduce computational complexity, a two-dimensional axisymmetric model was adopted, simulating only half of the computational domain. As illustrated in Figure 2, the study domain and the boundary conditions are as follows: the red-colored boundary at the top of the computational domain was set as a velocity inlet with a velocity of 0, while the deep blue-colored boundaries on the left and right sides were set as pressure outlets with a gauge pressure of 0. The remaining black-colored boundaries were set as solid wall boundaries.

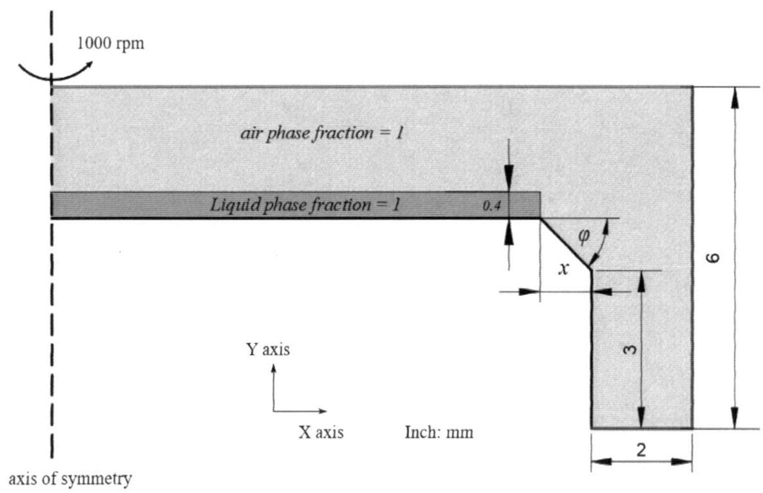

Figure 2. Simulation boundary conditions.

Table 4. Simulation parameters.

Simulation parameters	Rotation speed	Duration
Stage one	0	0.5 s
Stage two	1000 rpm	1.0 s
Stage three	0	1.0 s

As shown in Table 4, the simulation process is divided into three stages. The first stage is the static stage, which lasts for 0.5 s with no rotation set. The purpose of this stage is to allow the solution to undergo free deformation under gravity, no longer maintaining a regular rectangular shape. It could better simulate real-world conditions.

The second stage is the rotation stage, with the substrate speed set to 1000 rpm and a duration of 1.0 s. During this stage, the solution gradually flows into the chamfered surfaces and the lateral areas of the substrate, while being thrown out of the computational domain.

The final stage is again the static stage, with the substrate speed set to 0 and a duration of 1.0 s. In this stage, the simulation represents the flow of the solution after rotation has ceased.

In all three stages, gravity is applied with an acceleration of 9.81 m/s^2.

3. Results and discussion

3.1 Influence of chamber angle on edge bead effect

To investigate the influence of chamfer angles on edge effects, substrates one and two with different chamfer angles were selected for spin-coating experiments, with experimental parameters as shown in the previous section. The experimental results are presented in Figure 3 and Table 5 below.

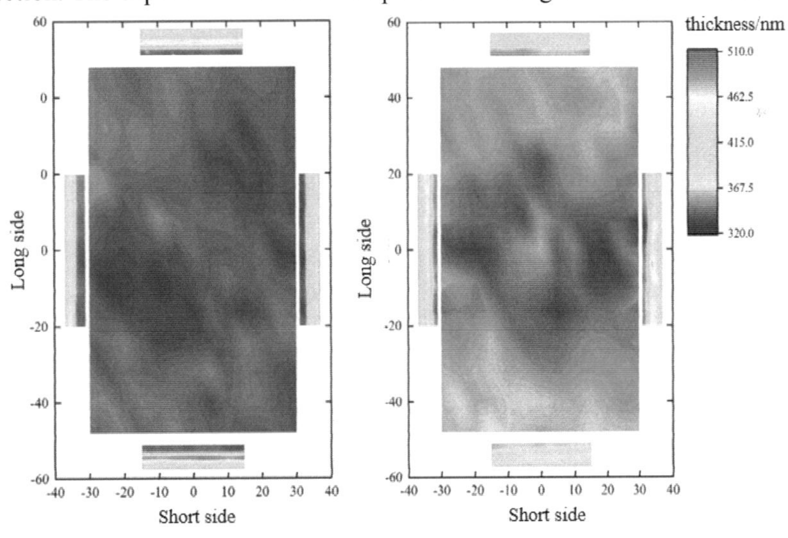

Figure 3. Experimental results with the left side representing Substrate One and the right side representing Substrate Two.

Table 5. Edge effect widths of the two substrates.

Chamfer angle φ	Top side/mm	Bottom side/mm	Left side/mm	Right side/mm
70°	4.50	4	3.5	4
20°	3	3.5	2	2

In processing experimental data, the average film thickness in the central region is first calculated, and then a reference line is set at 1.05 times the average film thickness. Points with film thickness above this reference line within the edge region are identified as points within the edge effect range. The experimental results indicate that when the chamfer angle becomes 20 degrees, there is a noticeable change in the width of the edge effect. According to the data in the table, it is calculated that the width of the edge effect decreases. From the experimental results, it can be concluded that reducing the chamfer angle is conducive to reducing the edge effect.

The interpretation of this experimental result can be explained using the Gibbs inequality [10]. This inequality describes the relationship between the volume of liquid at sharp edges, the contact angle between solid and liquid, and the chamfer at the edge.

$$\theta_0 \le \theta \le (180°\text{-}\varphi) + \theta_0 \quad (1)$$

As shown in Figure 4, θ represents the angle between the liquid at the edge and the substrate, φ represents the chamfer at the edge, and θ_0 represents the contact angle between the solid and the liquid. An increase in liquid volume is reflected in an increase in θ, and the range of θ values is determined by the following inequality [11]:

The conclusion drawn from this inequality is that reducing the chamfer angle at the edge will decrease the volume of liquid accommodated at the edge. Combining this inequality, it can be inferred that a smaller chamfer at the edge reduces the hindrance to liquid flow at the edge, facilitating liquid flow and reducing the volume of liquid accumulated at the edge, thereby mitigating edge effects.

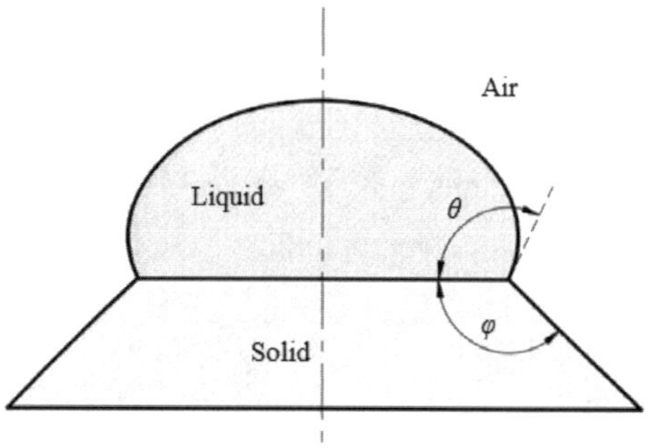

Figure 4. Schematic diagram of the Gibbs inequality.

3.2 Influence of chamber width on edge bead effect

To investigate the effect of chamfer angle on edge effects, substrates with different chamfer widths, labeled Substrate One, Substrate Three, and Substrate Four, were selected for spin-coating experiments. Experimental parameters were described in the previous section, and the results are presented in Table 6.

Table 6. Edge effect widths of the three substrates.

Chamfer width x/mm	Top side/mm	Bottom side/mm	Left side/mm	Right side/mm
0.5	3	3.5	2	2
1.5	2.5	3	3	2
3.0	2	2.5	3	3

The experimental results indicate that as the chamfer width increases from 0.5 mm to 1.5 mm, and further to 3.0 mm, expanding by a factor of six, the width of the edge effect remains consistently around 2-3 mm. This suggests that the chamfer width has a negligible impact on the edge effect.

To validate this experimental conclusion, simulations were conducted in Fluent for scenarios with different chamfer widths of 0.5 mm, 1.0 mm, and 1.5 mm, as outlined in the previous section. The simulation parameters remained consistent. The final results are presented in Figure 5. It can be observed from the figure that as the chamfer width gradually increases, there is no significant change in the membrane thickness curve at the edge. The membrane thickness curves overlap each other, and the distribution of the liquid film is similar. This confirms that increasing the chamfer width from 0.5 mm does not have a noticeable effect on the edge effect.

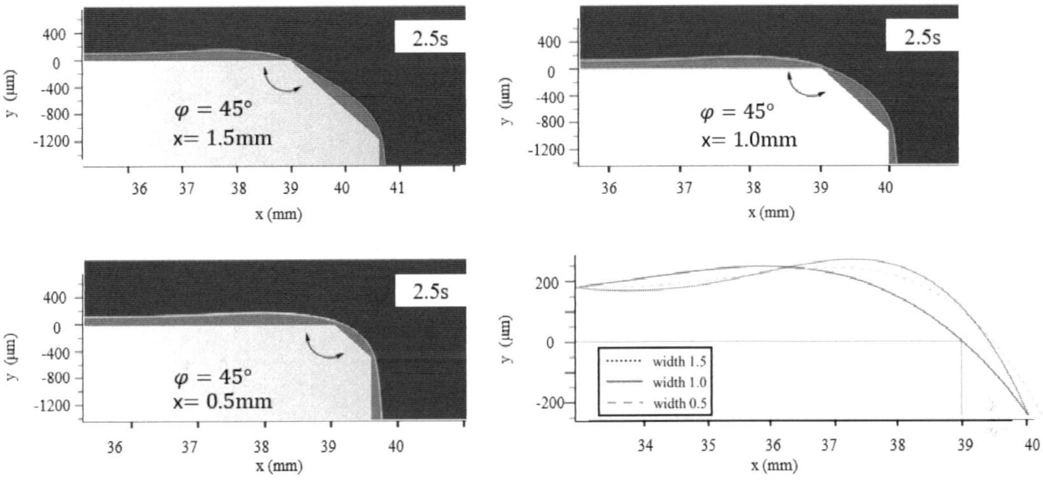

Figure 5. Simulation results for chamfer width.

4. Conclusion

This study investigated the influence of chamfered edge on the edge effect of spin-coating film. By employing experimental verification and Fluent fluid simulation, the effects of chamfer angle and chamfer width on the edge effect were obtained. The following conclusions were drawn: The chamfer angle significantly affects the width of the edge effect, with a smaller chamfer angle favoring a reduction in the width of the edge effect. However, the chamfer width does not significantly impact the edge effect; chamfer widths of 0.5 mm or even larger do not affect the edge effect.

Acknowledgement

The authors appreciate the financial support from the National Key Research and Development Program of China (No. 2020YFA0714502, No. 2019YFA0709102).

References

[1] LIN B J. Can mems take advantage of advances in semiconductor lithography? IEEE International Conference on Micro Electro Mechanical Systems [C], 2010.
[2] Zuo F., Ma S., Zhao W., et al. An Ultraviolet-Lithography-Assisted Sintering Method for Glass Microlens Array Fabrication. Micromachines [J]. 2023, 14(11): 2055.
[3] Chen Y., Shu Z., Zhang S., et al. Sub-10 nm fabrication: methods and applications. International Journal of Extreme Manufacturing [J]. 2021, 3(3): 32002.
[4] Norrman K., Ghanbari-Siahkali A., Larsen N. B. 6 Studies of spin-coated polymer films. Annual Reports Section C (Physical Chemistry) [J]. 2005, 101: 174-201.
[5] Jianyi Zhao, Ning Zhou, Yichao Tang, Jialin Zhao, Lei Wang, Xin Chen, Xiaodong Huang, Yonglin Yu, Wen Liu. Design and fabrication of tunable laser with digital concatenated grating. Semiconductor Science and Technology, 2013, 28(3): 035001
[6] Baëtens T., Arscott S. Planarization and edge bead reduction of spin-coated polydimethylsiloxane. Journal of micromechanics and microengineering [J]. 2019, 29(11): 115005.
[7] Luurtsema G. A. SPIN COATING FOR RECTANGULAR SUBSTRATES [M]. 1997.

Journal of Physics: Conference Series **2842** (2024) 012086 doi:10.1088/1742-6596/2842/1/012086

[8] Yamamoto K., Suyama T., Fujimori N., et al. Optimization of Wafer Thinning Process by Reducing Thickness Variation of Temporary Adhesive Layer for Medical Device [Z]. International Conference on Electronics Packaging (ICEP): 2019, 268-272.
[9] Shin H. G., Cho C. Y., Kim H. Y., et al. Improvement of dimensional accuracy of blu-ray disc cover layer. Journal of Mechanical Science and Technology [J]. 2008, 22(1): 120-126.
[10] Wood A, Sutton R M. Thomas Young, Natural Philosopher (1773–1829) [J]. Physics Today. 1954.
[11] Oliver J F, Huh C, Mason S G. RESISTANCE TO SPREADING OF LIQUIDS BY SHARP EDGES [J]. JOURNAL OF COLLOID AND INTERFACE SCIENCE. 1977, 59(3): 568-581.

Journal of Physics: Conference Series **2842** (2024) 012087 doi:10.1088/1742-6596/2842/1/012087

Influence of multi-process parameter linkage on the formation of lack of fusion defects based on SYSWLED simulation analysis

Chen Yan [1,2], Die Yang [1,2], Laihui Han [3,*], Zhonggang Zhao [1,2], Zengzhen Li [1,2], and Yongli Sui [1,2]

[1] China Petroleum Pipeline Research Institute Co, Ltd., Langfang, Hebei, 065000, China

[2] National Engineering Research Center of Oil and Gas Pipeline Transportation Safety, Langfang, Hebei, 065000, China

[3] China University of Petroleum (East China), Qingdao, Shandong, 266000, China

[*] Corresponding author's e-mail: hanlaihui_upc@163.com

Abstract: The lack of fusion defects remains a significant challenge in narrow-gap automatic welding of pipeline circumferential joints and then becomes the primary factor affecting pipeline longevity. The influence of multi-process parameter linkage on the welding temperature field and formation of lack of fusion defects was investigated based on orthogonal experiments and SYSWELD finite element simulation. The results indicated that welding current and speed had the most significant influence on the welding temperature field. At the same time, oscillation width affected the ability to obtain arc temperature on both sides of the weld groove. Oscillation time and edge dwell time determined the duration of arc heating on one side of the groove. A serrated temperature field morphology may occur when the dwell time is too long, leading to a lack of fusion defects. There were interactive effects among various welding parameters. Increasing heat input could reduce the risk of a lack of fusion defects. Besides, a reasonable setting of oscillation width, oscillation time, and edge dwell time could also reduce the formation risk of lack of fusion defects.

1. Introduction

With the development of the economy, the demand for petroleum and natural gas resources continues to rise, which leads to the continuous advancement of pipeline engineering construction. Concurrently, automatic welding technology for pipeline circumferential joints has matured [1-5]. However, welding defects, particularly lack of fusion, remain a significant challenge in narrow-gap automatic welding of pipeline circumferential joints and have become the primary factor affecting pipeline longevity [6-8]. Therefore, the influence of multi-process parameter linkage on the lack of fusion defect formation could be studied, providing theoretical guidance for the safety assessment of in-service and newly constructed pipelines.

In the paper, based on orthogonal experiments and SYSWELD finite element simulation, the influence of multi-process parameter interlinkage on temperature field distribution and the tendency of lack of fusion defect formation was analyzed, which could provide a basis for avoiding the occurrence of lack of fusion defects.

2. Experimental material and simulation settings

2.1. Experimental material
The X80M pipeline steel with a diameter of 1219 mm and a wall thickness of 22 mm was used. Besides, the melting point temperature was considered to be 1590 °C.

2.2. Orthogonal simulation for process parameters
Based on SYSWELD finite element simulation software, the mesh model is illustrated in Figure 1. The dual ellipsoid heat source model was selected as the thermal source model. The thermal source verification result is shown in Figure 2.

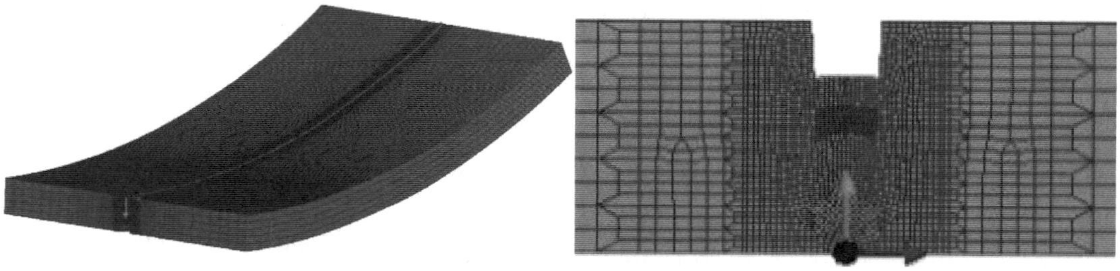

Figure 1. Numerical simulation of the pipeline model.

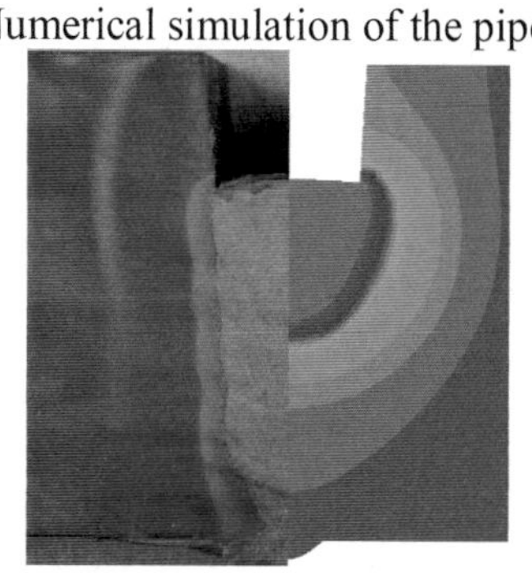

Figure 2. Comparison of welding simulation and actual weld pool.

Generally, multiple welding parameters need to change simultaneously to improve welding outcomes. Therefore, orthogonal experiments were employed for investigation. The table of 3rd-filled layer orthogonal experimental welding parameters is shown in Table 1.

Table 1. 3rd filled layer orthogonal experimental welding parameters table.

Number	Welding current / A	Welding speed / (mm/s)	Oscillation width/mm	Oscillation time/ms	Dwell time/ms
Case-1	170	5.5	1.2	60	40
Case-2	170	5.5	2.4	120	200
Case-3	170	6.5	1.2	240	200
Case-4	170	6.5	3.2	60	80
Case-5	170	7.5	2.4	240	80
Case-6	170	7.5	3.2	120	40
Case-7	190	5.5	1.2	240	80
Case-8	190	5.5	3.2	60	200
Case-9	190	6.5	2.4	120	80

Case-10	190	6.5	3.2	240	40
Case-11	190	7.5	1.2	120	200
Case-12	190	7.5	2.4	60	40
Case-13	210	5.5	2.4	240	40
Case-14	210	5.5	3.2	120	80
Case-15	210	6.5	1.2	120	40
Case-16	210	6.5	2.4	60	200
Case-17	210	7.5	1.2	60	80
Case-18	210	7.5	3.2	240	200

3. Analysis of numerical simulation results on multi-factor interaction

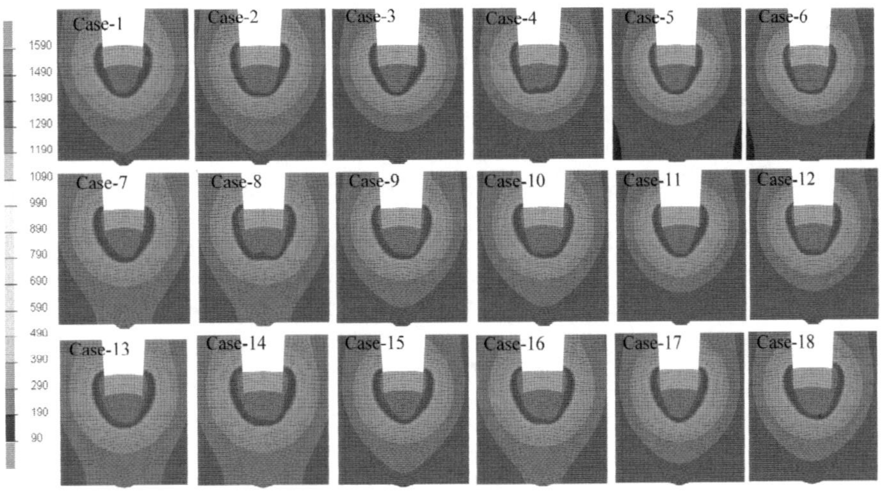

Figure 3. Numerical simulation temperature field cloud image of 3rd layer.

The simulation temperature field contour plots are shown in Figure 3. It could be observed that the simulation temperature field indicated a lack of fusion defects on the sidewalls of the welded joint. However, under threshold parameters, such as in Case-2, Case-12, and Case-17, it could be extremely difficult to determine whether a lack of fusion defects has occurred on the sidewalls.

In Table 1, Case-9 represents the baseline welding parameters. Temperature values along certain paths inside weld metal with a welding current of 190A are shown in Figure 4. It is noted that the data length values differ in different scenarios, attributed to different oscillation widths, oscillation widths, and dwell time. Case-11 and Case-12 exhibited lower temperature values compared to Case-9. Besides, it could be found that the temperature value remained unchanged or decreased when the oscillation width remained constant or decreased. With an increase in welding speed, the temperature value decreased. An increase in oscillation time or edge dwell time increases temperature difference. Case-10 experienced an increase in temperature due to an increase in oscillation width. Although Case-7 showed an increase in temperature value with a decrease in welding speed, the decrease in oscillation width slightly reduced the temperature difference between both edge sides. However, an increase in oscillation time led to an increase in temperature difference, which resulted in a slight decrease in the lowest temperature value. Conversely, Case-8 exhibited an increase in temperature due to a decrease in welding speed and an increase in oscillation width, which resulted in an overall temperature value. The increased temperature difference reduced the risk of lack of fusion defects. Therefore, when the welding current was constant, reducing welding speed and increasing oscillation width were advantageous to reduce the risk of lack of fusion defects.

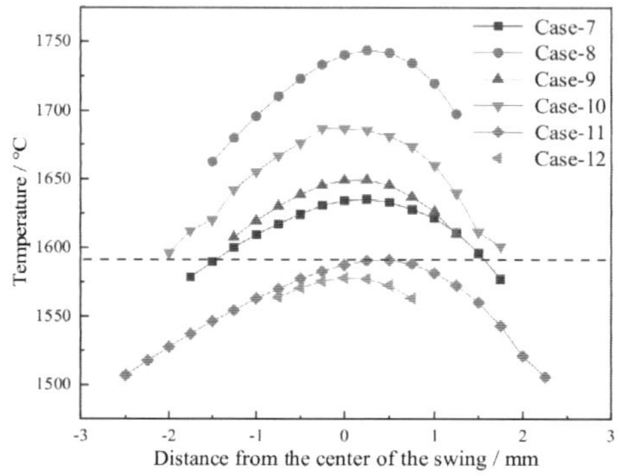

Figure 4. Temperature values along certain paths at a welding current of 190A.

Using the welding condition at a welding speed of 6.5 mm/s as the reference baseline, a comparative analysis was conducted on six welding experiments, as shown in Figure 5. Both Case-3 and Case-4 exhibited lower temperature values than Case-9. Additionally, the decrease in oscillation width in Case-3 further reduced the temperature. Although this may reduce the temperature difference to some extent, the increase in dwell time and oscillation time led to a further increase in temperature difference, which resulted in an overall temperature decrease. Increasing the current for Case-4 could reduce the risk of a lack of fusion defects. The increase in oscillation width for Case-10 led to an increase in temperature, while the increase in oscillation time increased the temperature difference. Case-16 experienced an increase in temperature due to an increase in current, coupled with a decrease in oscillation time and an increase in dwell time, which resulted in a larger increase in temperature difference. Therefore, increasing the welding current and oscillation width without changing the welding speed could help reduce the risk of unfused defects.

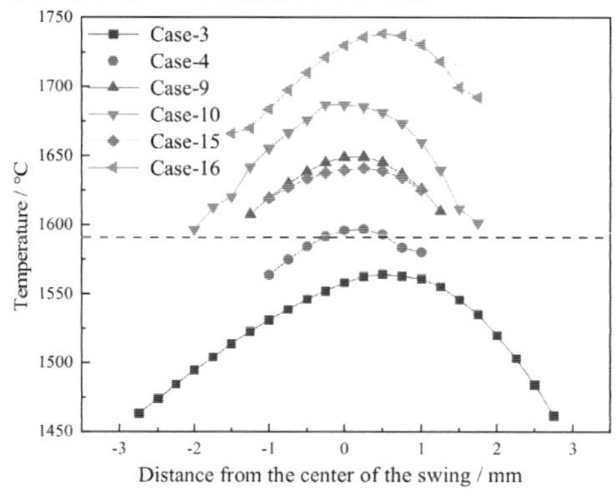

Figure 5. Temperature values along certain paths at a welding speed of 6.5 mm/s.

Using the welding condition with an oscillation width of 2.4 mm as the reference baseline, a comparative analysis was conducted on six welding experiments, as shown in Figure 6. Case-5 and Case-12 exhibited lower temperatures than Case-9 due to decreased welding speed. Additionally, Case-5 experienced a further decrease in temperature due to a decrease in welding current, while an increase in oscillation time led to an increase in temperature difference. Case-12 experienced minimal temperature difference due to decreased oscillation time and dwell time. However, it remained below the melting point temperature, which indicated a significant risk for lack of fusion defects. Case 2 experienced a simultaneous decrease in welding current and welding speed, which resulted in a small difference in temperature values. However, an increase in dwell time led to a slight increase in temperature difference. Case-13 exhibited higher temperatures due to increased welding current and decreased welding speed relative to Case-9. Therefore, increasing the welding current and decreasing the welding speed without changing the oscillation width could reduce the risk of lack of fusion defects.

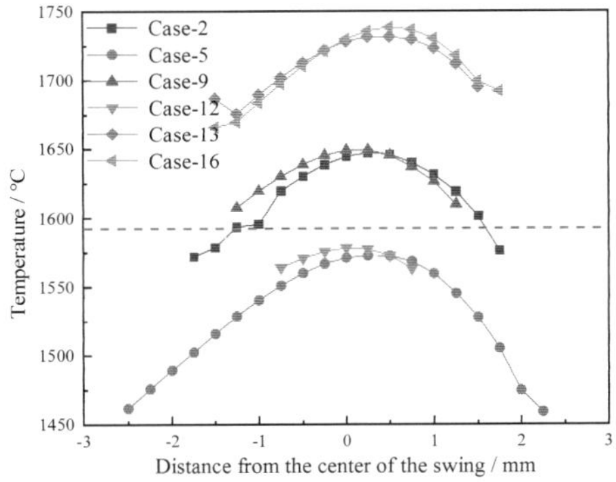

Figure 6. Temperature values along certain paths at an oscillation width of 2.4 mm.

4. Conclusion

This paper investigated the influence of multi-process parameter linkage on the welding temperature field and the formation of a lack of fusion defects. The main conclusions were as follows:

(1) Changes in welding current and welding speed mainly altered the distribution of the welding temperature field by affecting heat input, which resulted in temperatures at the interface between the weld layer and the groove wall being lower than the melting point temperature, thus increasing the risk for lack of fusion defects.

(2) The oscillation width affected the heat transfer distance to both sides of the weld seam. Increasing the oscillation width is beneficial for increasing the temperature on both sides of the weld seam, thereby reducing the risk.

(3) Oscillation time and dwell time affected the serrated morphology of the temperature field. This was related to the extended time of the welding torch on one side of the groove, depriving the other of energy supplementation and leading to a lack of fusion defects.

Acknowledgments

This research was funded by the China National Petroleum Corporation (CNPC) Science and Technology Development Program, grant number 2021ZG01.

References

[1] Sui Y L, Guo R, Zhang J CH (2013) Comparative Analysis on Semi-automatic and Automatic Welding Technology of Pipeline Circumferential Weld. Welded. Pipe. Tube., 36(09): 38-47.

[2] Zhang Y, Liu X W, Zhang F, Zou P, Wang K L (2019) Development status and prospect of pipe automatic welding equipment. Oil. Gas. Storage. Trans., 38(7): 721-727.

[3] Huang F (2015) Engineering economic analysis for application of automatic welding in long-distance pipeline. Oil. Gas. Storage. Trans., 34(12): 1365-1368.

[4] Zeng H L, Wang C J, Yang X M, Wang X S, Liu R (2014) Automatic welding technologies for long-distance pipelines by use of all position self-shielded flux cored wires. Natural. Gas. Indus. B., 1(1): 113-118.

[5] Ma G X (2017) Analysis of Anti-interference measures for microcomputer protection device in power system. Modern. Manuf. Technol. Equip., 10: 131-132.

[6] Zhang Z Y (1998) The Welding Defects and Defects Eliminating Methods of Shaanjing Gas Pipeline. Oil. Gas. Storage. Trans., 17(2): 33-36.

[7] Zeng H L, Wang C J, Yang X M, Wang X S, Liu R (2014) A technology of automatic welding in long-distance pipelines by use of all-position self-shieled flux cored wires. Natural. Gas. Indus., 34(1): 102-107.

[8] Guo N, Wang M R, Guo W, Yu J B (2014) Study on forming mechanism of appearance defects in rotating arc narrow gap horizontal GMAW. Int. J. Adv. Manuf. Technol., 75: 15-20.

Optimization of the two-step coating formation process for ZK21 magnesium alloy

Ruochao Wang[1, 2], Qing Liu[3], and Hongtao Liu[4, *]

[1]Shandong Institute of Commerce and Technology, Jinan 250103, Shandong Province, China
[2]Qilu Medical University, Zibo 255300, Shandong Province, China
[3]Shandong Urban Construction Vocational College, Jinan 250103, Shandong Province, China
[4]Shandong Provincial Key Laboratory of High Strength Lightweight Metallic Materials, Advanced Materials Institute, Qilu University of Technology (Shandong Academy of Sciences), Jinan 250014, Shandong Province, China

[*] Corresponding author's email: kelekm73@163.com

Abstract. The primary hindrance to the broader utilization of magnesium alloys lies in their insufficient corrosion resistance. This research initiative introduced an advanced two-phase methodology incorporating phosphating and alkali heating to produce a protective Ca-P coating on the surface of magnesium alloys. This innovative approach not only bolsters their corrosion resistance but also facilitates controlled degradation in bodily fluid environments. Following a 60-minute phosphating process in a solution with a pH level of 3, the resultant Ca-P coating achieved full coverage over the magnesium alloy surface, displaying commendable coating density and uniformity. Subsequent treatment with alkali heating for 30 minutes further enhanced the density and smoothness of the coating. Analysis utilizing EDS and XRD revealed that the primary constituents of the phosphate magnesium alloy and the phosphate + alkali heated magnesium alloy coating were DCPD and HA, respectively. Corrosion degradation assessments conducted in SBF solution validated that the two-step coating formation method significantly fortified the corrosion resistance of magnesium alloys.

1. Introduction

Magnesium alloy, serving as an innovative biodegradable medical metal material, exhibits exceptional mechanical properties, biodegradability, and biocompatibility [1-3]. When applied in bone repair procedures, the density of magnesium alloy closely mirrors that of natural bone, while its elastic modulus more closely resembles that of human bone when compared to the commonly utilized stainless steel and titanium alloys in clinical settings. This resemblance minimizes the occurrence of "stress shielding" and facilitates bone regeneration [4-6]. Furthermore, the ability of magnesium alloy to degrade within the body obviates the necessity for subsequent surgery to extract it post-bone healing, thereby sparing patients from additional discomfort [7-9]. Consequently, magnesium alloy shows significant potential for clinical utilization as a biodegradable medical metal material.

However, there are certain challenges in the initial exploration of magnesium alloys that constrain their utility. Firstly, the ongoing investigation into medical magnesium alloys predominantly centers on commercially available grades like the AZ series, such as AZ31 and AZ91. Nevertheless, pertinent medical research has indicated that the presence of the element Al poses a chronic neurotoxic risk associated with Alzheimer's disease, rendering the AZ series unsuitable for clinical purposes [10]. Secondly, magnesium alloy implants necessitate prolonged exposure to bodily fluids during operation. Elevated Cl^- levels exceeding 30 mmol/L in the physiological electrolytic environment can instigate galvanic corrosion and pitting corrosion in magnesium alloys. Furthermore, variables such as the pH levels of bodily fluids, amino acids, proteins, and diverse sugars can expedite the corrosion process, leading to premature loss of mechanical support provided by the magnesium alloy before the tissue is fully healed, ultimately culminating in implant malfunction [11-12].

To tackle the challenge of effectively controlling the degradation rate of magnesium alloys, this research opted for ZK21 (Mg-2.5Zn-0.8Zr, wt%), a magnesium alloy that is non-toxic, benign, biocompatible, and displays outstanding mechanical properties, as the focal point of investigation. The approach involved a sequential method comprising phosphating and alkali heat treatment to fabricate a Ca-P coating on its surface, aimed at modulating its degradation rate within a bodily fluid environment.

2. Experimental

2.1 Materials and reagents

Table 1 presents the chemical composition of the ZK21 magnesium alloy utilized in this investigation.

Table 1. Chemical composition of ZK21 magnesium alloy (wt.%, mass fraction).

Element	Zn	Zr	Nd	Y	Mg
Content	2.5	0.8	0.3	0.5	Bal.

The key chemicals employed in the experiment include CH_3COCH_3, NaOH, and H_3PO_4, simulated body fluid (SBF) solution, and purified water.

2.2 Experimental methods

In the phosphating process, the magnesium alloy samples that have undergone pre-treatment are submerged in a solution for a designated duration to establish an initial coating on the surface, yielding phosphate magnesium alloy samples. The phosphating solution consists of CaO, H_3PO_4, $Ca(CH_3COO)_2$, $C_6H_4O_5NSNa$, and Na_2MoO_4. The study primarily investigated the influence of phosphating pH level and duration on the quality of the resultant coating.

The phosphate magnesium alloy samples were immersed in a heated NaOH solution for a specific duration to generate a composite coating of phosphide and alkali heat on the surface of the magnesium alloy, resulting in phosphide + alkali heat-treated magnesium alloy samples. The study primarily focused on examining the impact of the duration of alkali heat treatment on the quality of the coating.

The magnesium alloy specimens undergoing distinct surface treatments — original, phosphate, and phosphate + alkali heat treated, were immersed in SBF solution for a specific duration to assess and contrast their corrosion resistance levels.

2.3 Analytical methods

The surface-modified magnesium alloy samples underwent microstructural analysis utilizing a TESCAN VEGA II LMU scanning electron microscope (SEM). The chemical composition of the magnesium alloy Ca-P coating was quantitatively examined with a BRUKER energy-dispersive spectrometer (EDS). The phase composition of the magnesium alloy Ca-P coating was assessed using a D8 ADVANCE X-ray diffractometer (XRD) fabricated by BRUKER in Germany.

3. Results and discussion

3.1 Effect of various phosphating processes on the quality of the coating

Throughout the phosphating process, the acidic conditions facilitate the creation of a conversion coating on the magnesium alloy substrate as it undergoes dissolution, rendering the pH level a crucial factor in determining the coating quality. By varying the pH level of the phosphating solution while keeping all other experimental conditions constant, the impact of pH on the surface morphology and coating quality of magnesium alloy was examined. The findings are illustrated in Figure 1.

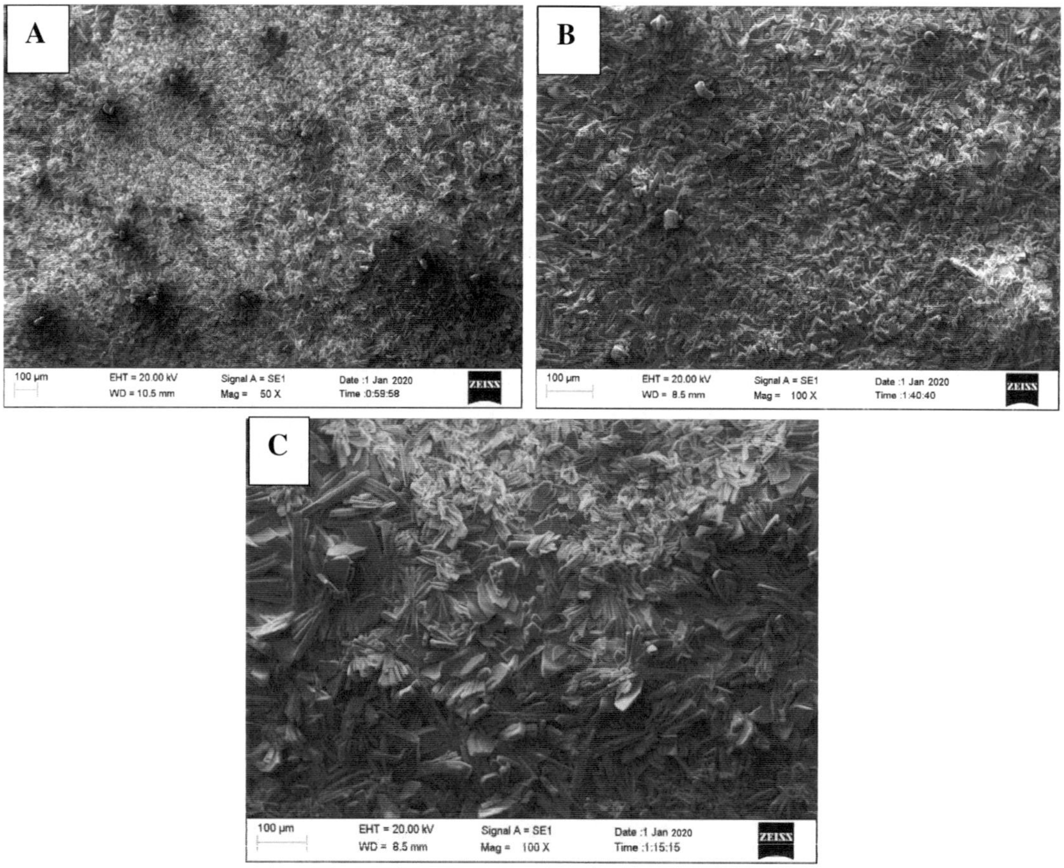

Figure 1. SEM of the coating at various phosphating pH levels (A: pH=2.5, B: pH=3, C: pH=3.5).

At a pH of 2.5, the coating appears thinner and less cohesive, revealing portions of the magnesium alloy substrate. Adjusting the pH to 3 results in a complete, compact coating with small crystalline particles and a smooth surface. However, at a pH of 3.5, although the coating can fully encase the magnesium alloy substrate, it contains larger crystalline particles, lower density, and lacks smoothness. This occurrence is attributed to the heightened corrosion rate of the substrate at lower pH values, leading to the dissolution of the protective conversion coating and a decrease in deposition efficiency. Conversely, excessively high pH levels slow down chemical deposition, impacting coating uniformity. Optimal pH levels facilitate rapid coating formation, yielding a uniform, dense, and cohesive membrane that significantly enhances the corrosion resistance of the magnesium alloy.

By manipulating the duration of the phosphating process, it is possible to regulate the thickness of the coating. Nevertheless, an increase in the phosphating duration leads to alterations in the crystal structure present on the surface of the coating, thereby exerting a substantial influence on the overall coating quality. Through the alteration of the phosphating duration, while maintaining consistency in other experimental variables, the research delved into the repercussions of varying phosphating durations on both the surface morphology and quality of the coating applied to magnesium alloy, as illustrated in Figure 2.

Journal of Physics: Conference Series **2842** (2024) 012088 doi:10.1088/1742-6596/2842/1/012088

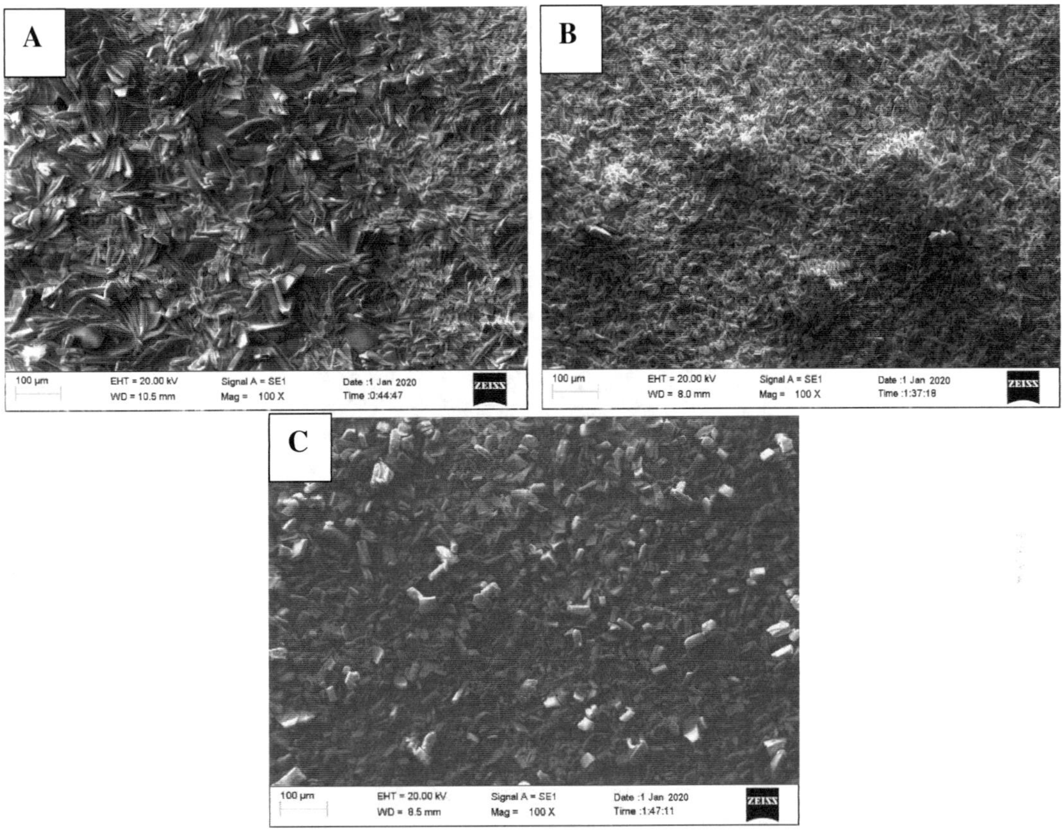

Figure 2. SEM of the coating at various phosphating times (A: 40 min, B: 60 min, C: 80 min).

When the phosphating duration is 40 minutes, the coating exhibits completeness and density, yet the presence of columnar crystals compromises its flatness and uniformity. Extending the phosphating duration to 60 minutes yields finer crystals and a denser coating structure. However, a further increase to 80 minutes results in the formation of blocky crystals, consequently diminishing the flatness of the coating. This phenomenon arises from the initial stage of phosphating, characterized by a rapid coating formation due to the short phosphating duration, leading to a low corrosive impact of the acidic solution environment on the coating, thus causing the formation of columnar or flake-like crystals, which in turn results in an uneven coating surface. Optimal phosphating duration strikes a balance between the coating-forming rate and the corrosion rate of the acidic environment on the coating, leading to the development of finer crystals, enhanced coating density, and uniformity. Nevertheless, prolonged phosphating duration encourages continuous crystal growth on the coating, leading to a looser and rougher film surface.

3.2 Effect of alkali heat treatments on the quality of the coating

The phosphate magnesium alloy samples are immersed in a 4 mol/L NaOH solution heated to 80°C using a water bath for alkali heat treatment for different time intervals, as depicted in Figure 3. After 30 minutes of alkali heat treatment, a compact and sleek coating will develop on the surface of the magnesium alloy. Increasing the treatment duration to 90 minutes will cause the coating to show a greater presence of columnar crystals. If the treatment time is further extended to 150 minutes, there will be an expansion in the size of the columnar crystals present in the coating, which will hurt its smoothness.

Journal of Physics: Conference Series **2842** (2024) 012088 doi:10.1088/1742-6596/2842/1/012088

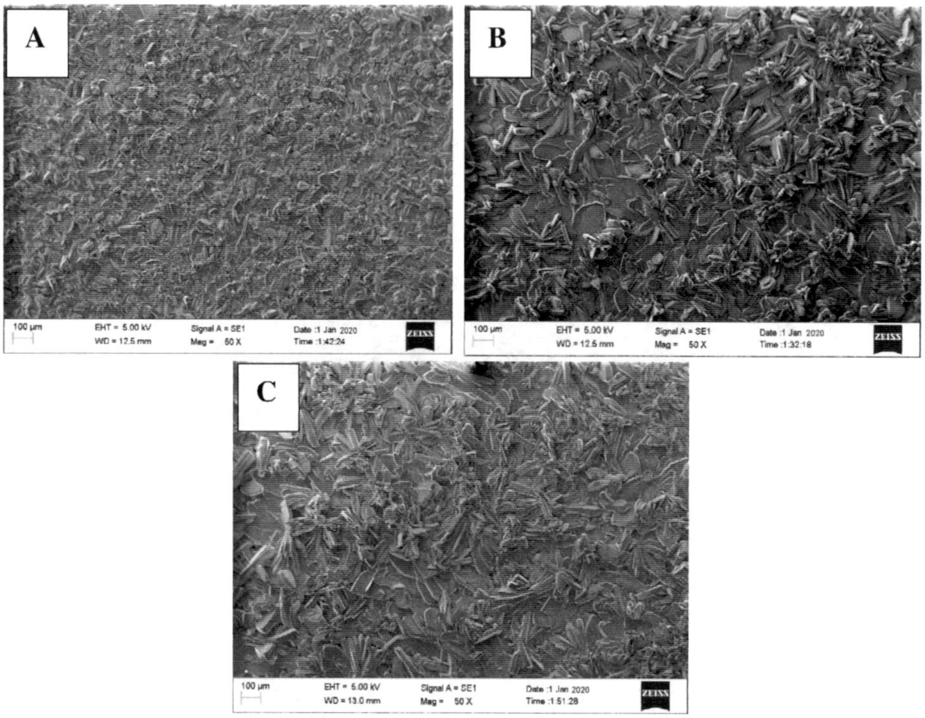

Figure 3. SEM of the coating at various durations of alkali heat treatment (A: 30 min, B: 90 min, C: 150 min).

3.3 Analysis of the coating composition

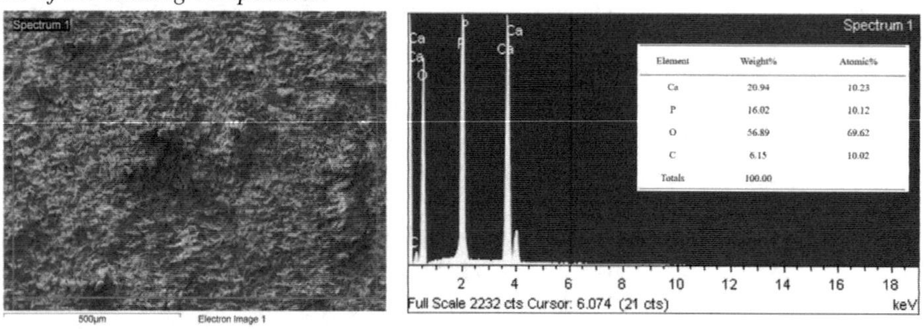

Phosphated

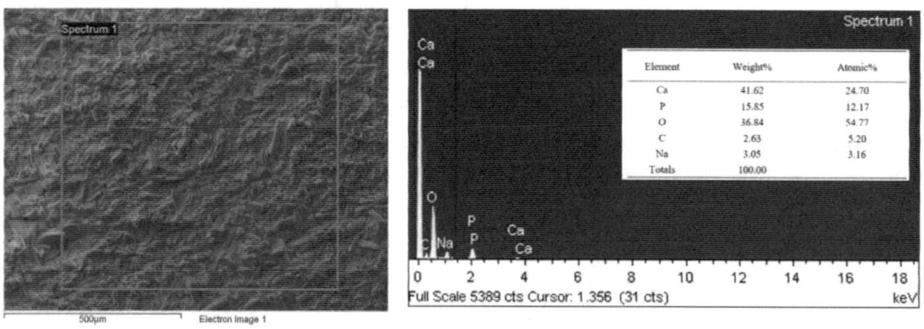

Phosphated + alkali heat treated

Figure 4. EDS analysis of the Ca-P coating.

The magnesium phosphate alloy sample obtained under the optimal conditions and the magnesium alloy sample subjected to phosphating and alkali heating were analyzed using EDS and XRD, as illustrated in Figure 4 and Figure 5, respectively. From Figure 4, it can be observed that the predominant elements in the coating of the magnesium phosphate alloy sample are O, Ca, P, and C, with a Ca to P atomic ratio of approximately 1:1. Conversely, the primary elements in the coating of the magnesium alloy sample treated with phosphating and alkali heating are O, Ca, P, C, and Na, with a Ca to P atomic ratio of roughly 2:1. As depicted in Figure 5, the principal constituent of the phosphate coating is $CaPO_3(OH)_2{\cdot}2H_2O$ (DCPD), while the major constituent of the coating after phosphating and alkali heating is $Ca_{10}(PO_4)_6(OH)_2$ (HA), which corresponds to hydroxyapatite, the primary component of human bones. This observation suggests that the fabricated Ca-P coating exhibits excellent biocompatibility.

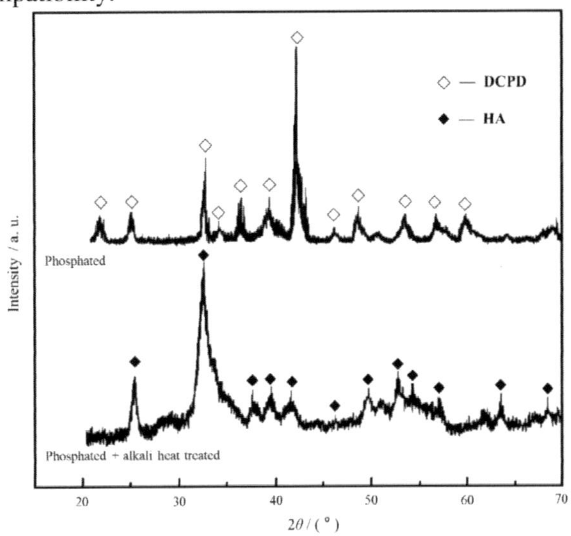

Figure 5. XRD analysis of the Ca-P coating.

3.4 Analysis of the coating degradation rate

The original magnesium alloy sample, phosphate magnesium alloy sample, and phosphate + alkali heat-treated magnesium alloy sample are placed in SBF solution for a specific duration to compare the corrosion resistance of magnesium alloy samples subjected to various surface treatments. The formula for determining the corrosion degradation rate of the conversion film is as follows:

$$v = \frac{m_0 - m_f}{A \cdot t}$$

in the formula, "v" symbolizes the rate of degradation, "m_0" denotes the initial mass of the specimen, "m_f" represents the mass of the specimen post corrosion, "A" signifies the surface area of the specimen (specifically 4.2 cm^2 in this particular study), and "t" stands for the duration of corrosion of the specimen in the SBF solution.

According to Figure 6, the original magnesium alloy sample exhibits the highest corrosion rate in the SBF solution. The phosphating treatment demonstrates a certain level of protection on the magnesium alloy, resulting in a reduced corrosion rate. Moreover, the corrosion rate of the magnesium alloy sample subjected to both phosphating and alkali heat treatment is the lowest among the three. By applying the corrosion rate calculation formula outlined in the experimental methodology, the corrosion rates of the original sample, phosphate magnesium alloy sample, and phosphate + alkali heat-treated magnesium alloy sample are measured at 0.34, 0.20, and 0.07 mg/(day·cm^2) respectively. The conversion coating enhances the corrosion resistance of the magnesium alloy substrate, with further enhancement observed after alkali heat treatment.

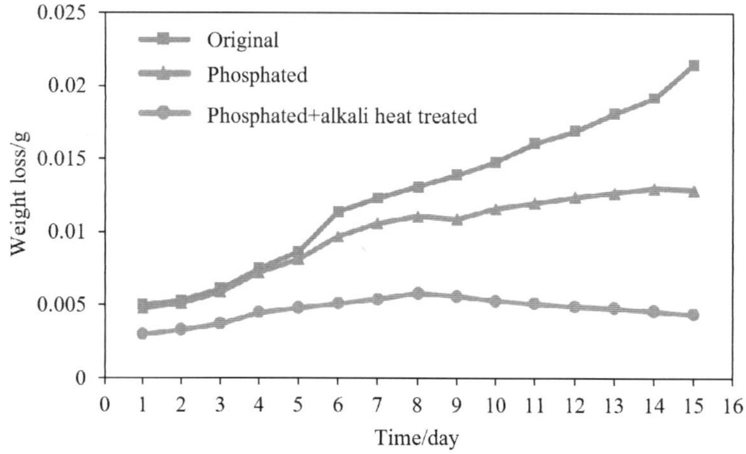

Figure 6. Contrasting the corrosion resistance of magnesium alloy samples with various surface treatments.

4. Conclusions

An optimization study was carried out on the two-step coating formation process of ZK21 magnesium alloy phosphating followed by alkali heat treatment, leading to the following conclusions:

(1) When magnesium alloy undergoes phosphating in a solution with a pH of 3 for 60 minutes, the resultant Ca-P coating can efficiently envelop the surface of the magnesium alloy, demonstrating commendable density and uniformity. Following this, subjecting the alloy to an alkali heat treatment in a 4 mol/L NaOH solution at 80°C for 30 minutes significantly improves the density and evenness of the coating.

(2) According to the analysis conducted by EDS and XRD, the primary constituents of the phosphate magnesium alloy coating and the phosphate + alkali heat-treated magnesium alloy coating are identified as DCPD and HA respectively. HA, an abbreviation for hydroxyapatite, is a predominant element found in human bone, suggesting that the Ca-P coating produced through the two-step coating process on the magnesium alloy demonstrates excellent biocompatibility.

(3) In the experiments on corrosion degradation in SBF solution, the corrosion degradation rate of the magnesium alloy treated with both phosphate and alkali heat was the most minimal compared to the original magnesium alloy and phosphate magnesium alloy. This outcome indicates that the strategy of two-step coating formation significantly boosts the corrosion resistance of magnesium alloys.

Acknowledgments

This work was supported by the Shandong Provincial Natural Science Foundation (Grant No. ZR2019PEM005 and ZR2021ME250).

References

[1] Wan T, Song S P, Wang J Z, et al. 2020 Research progress in corrosion behavior of biomedical magnesium alloys Journal of Materials Engineering, vol 48, pp 19-26.

[2] Xu D G, Ma C H, Zhao Y Z, et al. 2021 Research Progress of Application and Corrosion Resistance of Medical Magnesium Alloy Materials Materials Protection, vol 54, pp 131-139.

[3] Liu Y, Wang C, Yu Z T, et al. 2019 Review on Mechanical Properties of Medical Magnesium Alloys Materials Reports, vol 33, pp 288-292.

[4] Wang R X, Hong L X, Zhang X B 2021 Research progress in corrosion resistance of biomedical magnesium alloys Journal of Materials Engineering, vol 49, pp 14-27.

[5] Tian Y Q, Zhao G Z, Liu Y, et al. 2021 Research progress in degradation behavior of biodegradable medical Mg-based alloys in vivo and in vitro Journal of Materials Engineering, vol 49, pp 24-37.
[6] Yuan G Y, Niu J L 2017 Research Progress of Biodegradable Magnesium Alloys for Orthopaedic Applications ACTA METALLURGICA SINICA, vol 53, pp 1168-1180.
[7] Zheng Y, Zhang X, Lu J, et al. 2023 Progress on the Research of In-vivo Degradation Behavior and Surface Modification of Biomedical Mg Alloys Materials Reports, vol 37, pp 141-156.
[8] Chen J X, Wang X W, Liu C, et al. 2021 Recent Progress in the Biodegradable Magnesium Alloys Special Casting & Nonferrous Alloys, vol 41, pp 1273-1282.
[9] Xie K, Wang L, Guo Y, et al. 2021 Effectiveness and safety of biodegradable Mg-Nd-Zn-Zr alloy screws for the treatment of medial malleolar fractures Journal of Orthopaedic Translation, vol 27, pp 96-100.
[10] Xia J R, Liu J Q, Li G Z, et al. 2018 Recent developments in the mechanisms underlying the neurotoxic effects triggered by aluminum Chinese Journal of Gerontology, vol 38, pp 3276-3280.
[11] Chen L S, Zheng Y Q, Liu Y, et al. 2021 Research Progress on Corrosion Fatigue Behavior of Biomedical Degradable Magnesium-Based Alloys RARE METAL MATERIALS AND ENGINEERING, vol 50, pp 3375-3387.
[12] Wang Y P, Zhang H B, Liang W Q, et al. 2021 Degradation Behavior and Mechanical Properties of Magnesium Alloy Plate In vivo Journal of Medical Biomechanics, vol 36, pp 935-939.

Comparative study on TD treatment for preparation of VC coating composite processes

Yutong Deng[1], Yongyue Liu[2 *], Haoxing Jin[3], Peng Jiang[4], and Hao Chen[4]

[1]XIAN HUIJIN TECHNOLOGY CO.. LTD, Xian, China
[2]Ningbo Heli Technology Share Holding Co., Ltd, Zhejiang, Ningbo, 315700, China
[3]HighShine Chongqing Metal Technology Co., Ltd. Chongqing, 400700, China
[4]Beijing Research Institute of Mechanical and Electrical Technology Co., Ltd. GAM, Beijing, 10083, China

[*]Corresponding author's e-mail: yongyue.liu@helimould.com

Abstract. A VC coating was prepared on the surface of mold steel Cr12MoV using the TD composite treatment process. The differences in microstructure and properties between the two composite treatment processes and the VC coating obtained from the ordinary TD vanadium infiltration process were compared using a microhardness tester, scratch tester, X-ray diffraction, etc. The results show that both composite processes can improve the unevenness of matrix carbides; The hardness of the VC coating obtained by the two composite processes is slightly higher than that of the ordinary TD; The TD+vacuum quenching process has to some extent improved the adhesion between the coating and the substrate; The valence state of the VC coating obtained by the secondary TD process has undergone significant changes.

1. Introduction

TD (Thermal Diffusion) thermal reaction deposition and diffusion method originated from the Toyota Center Research Institute in Japan. The principle of TD vanadium infiltration is to diffuse the C atoms in the matrix to the surface in a high-temperature salt bath environment and combine with the V atoms in the salt bath to form a certain thickness of vanadium carbide coating, greatly improving the hardness [1], wear resistance, and corrosion resistance of the sample surface, without being limited by the shape of the sample [2]. Existing research on TD vanadium infiltration indicates that the thickness of VC coatings prepared on the surface of Cr12MoV by TD treatment initially increases linearly with the prolongation of treatment time[3]. When the thickness reaches 10-15 μm, the thickness growth rate slows down significantly over time, and the surface microhardness of VC coating reaches about 3000 HV, which can increase the service life of stamping molds by several to tens of times [4].

Although there have been detailed analyses on the growth mechanism, microstructure, and properties of TD vanadium infiltration[5], there are few reports in the industry on the research of TD treatment composite technology [6]. This article uses ordinary TD treatment, TD+vacuum quenching, and TD+TD composite process to prepare three types of VC coatings on the surface of Cr12MoV. The microstructure and properties of the coatings are compared and analyzed by X-ray diffraction, microhardness tester, and scratch tester.

2. Materials and methods

The experimental material is Cr12MoV, and the size specification of the experimental sample is 10 mm*10 mm*10 mm. The experimental equipment is an SG-7.5-12 external heating crucible resistance furnace, and the auxiliary equipment is a resistance high-temperature tempering furnace and horizontal vacuum quenching furnace. The TD salt bath formula is 76% borax-based salt $Na_2B_4O_7$+10% reducing agent Al+10% chromium supply agent Cr_2O_3+4% activating agent NaF.

The process flow of TD salt bath vanadium infiltration is as follows: polishing, degreasing and cleaning, installation, TD vanadium infiltration, quenching, tempering, and cleaning. The quenching method is to gas quench the processed sample, and the tempering process is to keep it in a 200°C tempering furnace for 2 hours. The numbering and process of different specimens are shown in Table 1.

Table 1. Technology parameters.

Sample number	Technological process	Technology parameters
N1	TD	TD: 980°C x8h
S2	TD+ Vacuum quenching	TD+1000°C Vacuum quenching
S3	TD+TD	Both TD: 980°C x8h

3. Results & discussion

3.1 Coating microstructure

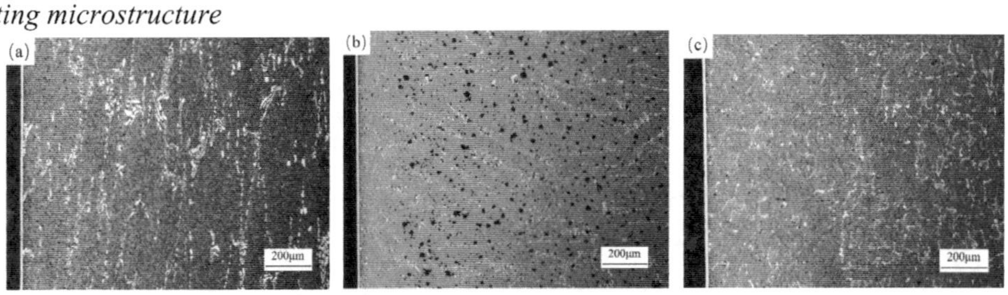

Figure 1. Cross-sectional microstructure 100x(a)-N1;(b)-S2;(c)-S3.

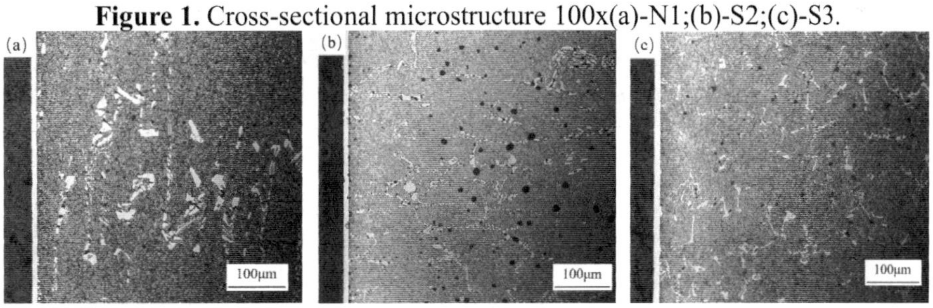

Figure 2. Cross-sectional microstructure 200x (a)-N1;(b)-S2;(c)-S3.

By observing Figures 1 and 2, it can be observed that the metallographic structures of the three specimens are mainly martensite+carbides+residual austenite. The eutectic carbides of the three samples all exhibit a network-like distribution. According to the fourth level diagram of GB/T14979-94 "Evaluation Method for Heterogeneity of Eutectic Carbides in Steel", the unevenness level of N1 eutectic carbides is level 4, S2 eutectic carbides are level 3, and S3 eutectic carbides are level 2. Meanwhile, it can be observed that compared to N1, the eutectic carbides in the S2 and S3 samples are finer and more uniform. The content of carbides in the transition zone between the VC coating and the substrate in the S2 sample is lower than that in the substrate, indicating the migration of carbides towards the surface layer. Additionally, there is a noticeable deposition of carbides at the interface between the coating and the substrate.

3.2 Hardness performance

The coating hardness was measured using the Dutch INNOVATEST-432D automatic micro Vickers hardness tester, with a test force value of 25 gf and a loading time of 10 s. Observing the data in Figure 3, it can be observed that the cross-sectional hardness of the coatings prepared by the three processes is above 2500 Hv. Compared with N1, the coating hardness of S2 and S3 is not significantly different and slightly higher than N1 by about 200 HV. The average hardness value of the cross-section is S3>S2>N1.

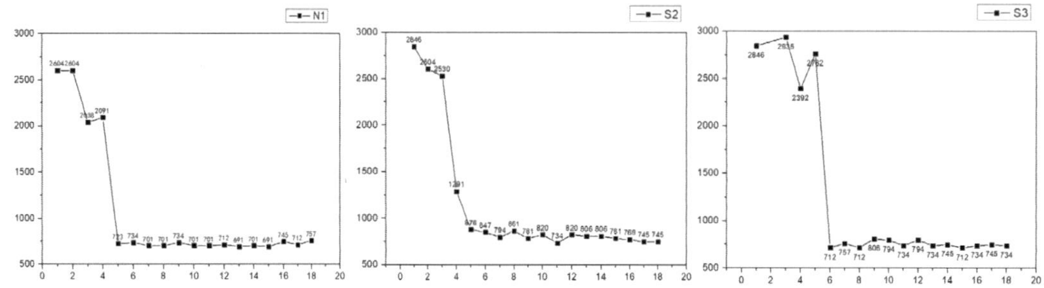

Figure 3. Microhardness curve - units/$HV_{0.05}$.

By observing the micro Vickers hardness map, it can be found that the hardness points obtained by the three samples under the same loading force (0.1 gf) are different. Figure 4 shows the cross-sectional structures of the three specimens measured by the Vickers hardness tester. Taking the diamond-shaped coating center position as a reference, the hardness points closer to the surface are more prone to cracking. N1 and S2 have shown varying degrees of cracking at the center position, with the S2 specimen showing more pronounced cracking, while the S3 specimen can obtain complete diamond hardness points at the center position, with only slight cracking occurring near the surface.

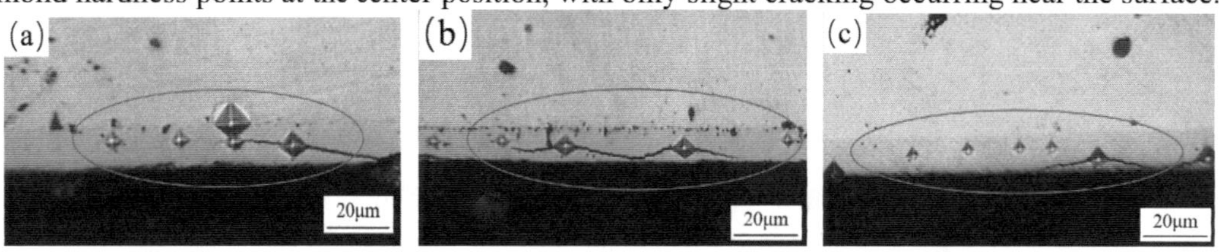

Figure 4. Vickers hardness diagram of coating cross-section (a)-N1; (b)-S2; (c)-S3.

3.3 Scratch experiment

The WS-2005 automatic scratch tester for coating adhesion was selected for the scratch test, and the measurement method was acoustic emission. Test parameters: Loading load: 60 N; Loading rate: 60 N/min; Reciprocating frequency: 2 times; Scratch method: Unidirectional scratch. Figure 5 shows the bonding strength curves of the three specimens after scratch testing. From the data of coating adhesion, it can be seen that the adhesion between N1 and S3 samples is almost the same, while the adhesion between S2 sample coating is significantly higher than that between N1 and S3, indicating that vacuum quenching enhances the adhesion between the coating and the substrate to a certain extent.

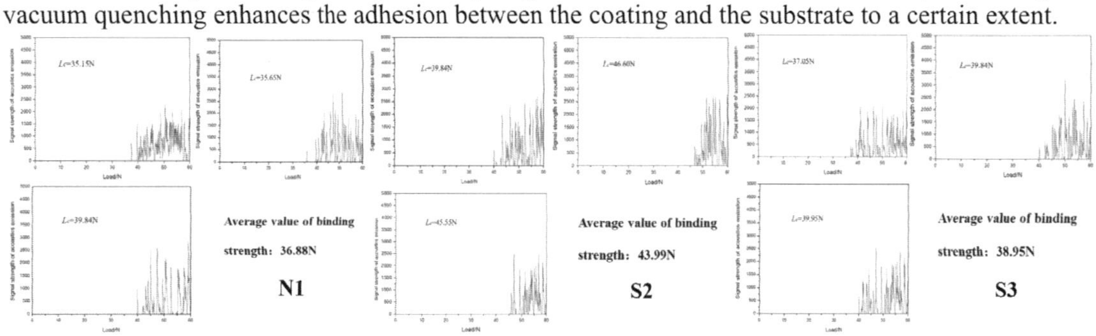

Figure 5. Bonding strength of scratch.

As shown in Figure 6, the tissue morphology of the scratches obtained from the scratch test of the three specimens was obtained. Observing the scratch structures of the three samples, it can be seen that the VC coating of the S2 sample peels off the latest, indicating that the adhesion of the coating is better than that of N1 and S3. At the same time, it can be found that the coating peeling situation of the S2 sample is the same as that of N1, with large peeling areas. In contrast, the peeling situation of the coating of the S3 sample is better, indicating that the toughness of the coating is better than that of N1 and S3.

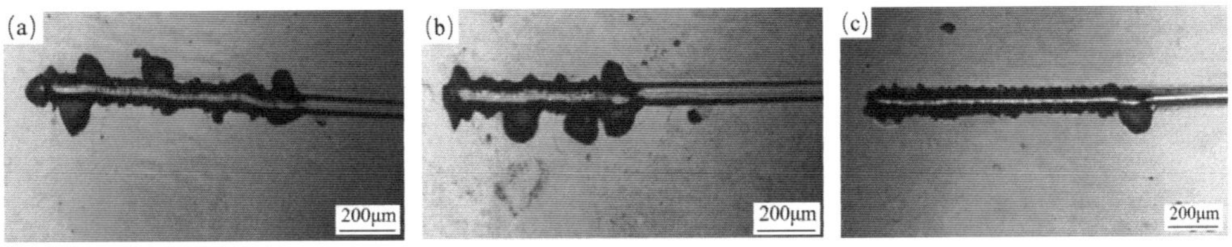

Figure 6. Scratch morphology of coating (a)-N1; (b)-S2; (c)-S3.

3.4 EDS line scanning

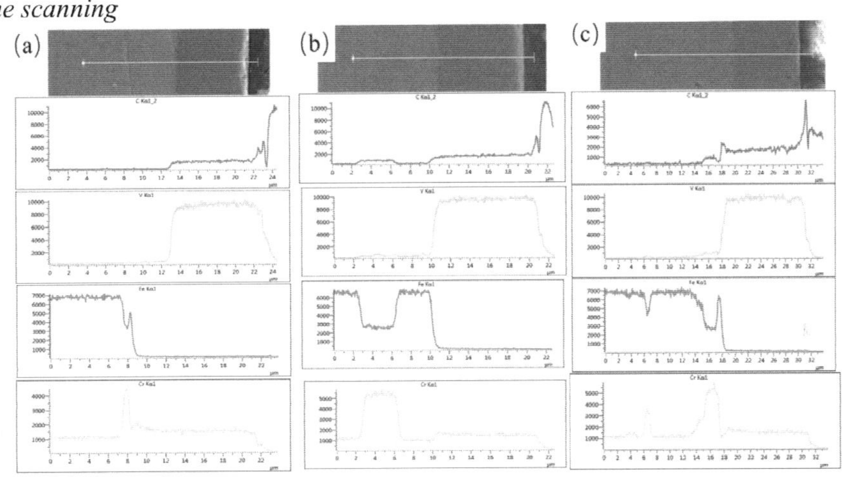

Figure 7. EDS line scan (a)-N1; (b)-S2; (c)-S3.

Figure 7 shows the composition distribution curves of the three specimens after passing the EDS line scan test. From the line scan graph, it can be seen that the trend of element distribution changes in the three samples is consistent. V is mainly concentrated on the interface side, and the V content shows a continuous strip gradient distribution in the diffusion layer of the coating interface, while it shows a uniform distribution in both the coating and the substrate. The content of C in the coating is slightly higher than that of the substrate, and the content of V and Fe at the interface between the coating and the substrate varies greatly, forming a VC phase between V and C [7].

3.5 XPS analysis of the surface

Figure 8 shows the data curves of the XPS phase valence states of the three specimens obtained by X-ray diffraction test. For the C 1s peak, the peak of 284.8 eV is the pollution carbon peak, mainly used for calibration. The C1s peak appears at 282.2 eV, 286.1 eV, and 288.4 eV. 282.2 eV comes from the matrix Cr12MoV, with a C content of 1.45% to 1.70%. The electron binding energy of V is 512.8 eV, and the main peak of the three samples is around 513 eV, which mainly corresponds to the +2 valence V. In addition, N1 and S2 have peaks of 520.5 eV, corresponding to the +4 valence V, and S3 have peaks of 523.8 eV in addition to the 521.6 eV peak, corresponding to the +3 and +4 valence V, respectively [7]. Overall, the graph data indicates that the coating is composed of V and C, and C exists in the form of VC, rather than forming interstitial atoms with V solid solution. On the surface of VC coating, the valence state of V is mainly +2 and +4. Among them, the valence state of V in N1 and S2 is the same. The main peak of V in the VC coating obtained from S3 is significantly lower than that in N1 and S2, and a peak of 523.8 appears.

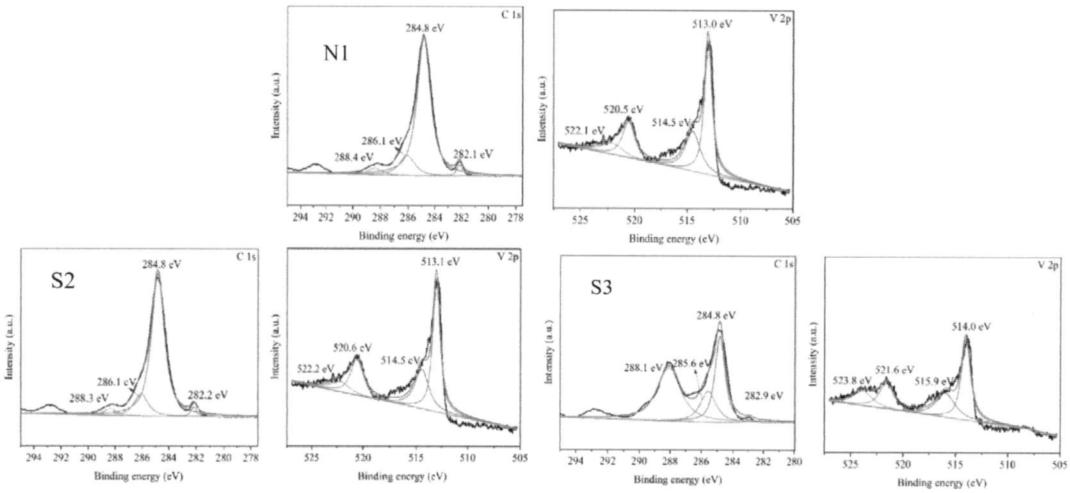

Figure 8. Carbon and Vanadium XPS spectrum.

4. Conclusion

(1) Compared to ordinary TD, the hardness of VC coatings prepared by TD+vacuum quenching (S2) and TD+TD (S3) is higher. TD+vacuum quenching enhances the adhesion between the coating and the substrate but does not improve the brittleness of VC coatings. TD+TD process has better toughness performance in hardness and scratch tests;

(2) The hardness of VC coatings obtained by TD+vacuum quenching and TD+TD preparation is higher, and the adhesion between VC coatings prepared by TD+vacuum quenching process and substrate is better;

(3) The distribution and variation patterns of element composition in VC coatings prepared by the three processes are the same. The valence state of the VC coating obtained through the TD+vacuum quenching process is the same as that of ordinary TD, while the VC coating obtained through TD+TD has a higher +2 valence V.

Acknowledgments

This topic comes from a national key research and development plan project: Key technologies and industrialization of die-casting mold design and manufacturing for large structural parts of new energy vehicles. (20212ZDYF020021)

References

[1] Zhou Lianpu, Zhu Chundong, Guo Lian, etc Research on the Heat Treatment Process of Cr12Mo1V1 Punch TD Salt Bath Vanadium Infiltration [J]. Heat Processing Technology, 2021,50 (08): 105-108+111. DOI: 10.14158/j.cnki.1001-3814.20202660.

[2] Vikrant Singh, Anil Kumar Singla, Anuj Bansal, etc Influence of laser texturing along with PTFE topcoat on slurry and cavitation erosion resistance of HVOF sprayed VC coating [J]. Surface & Coatings Technology, Volume 470,15, DOI: 10.1016/j.surfcoat.2023.129858.

[3] Lei Fengrong Study on the Structure and Properties of TD Salt Bath Vanadium Infiltration Layer on the Surface of 40Cr Steel [D]. Chongqing University of Technology, 2020. DOI: 10.27753/dcnki.gcqgx.2020.000354.

[4] Dariusz Bartkowski, Aneta Bartkowska, etc Microstructure and Selected Properties of Iron–Vanadium Coatings Obtained by the Laser Processing of a VC Pre-Coat Applied on Steel—Single and Multiple Laser Tracks Study[J]. Materials 2022, 15(18), 6417; DOI: 10.3390/ma15186417.

[5] Wu Xiaotian, Xue Qi, Zhang Jin, etc. The microstructure and properties of TD vanadium infiltration layer on Cr12MoV steel powder [J]. Metal Heat Treatment, 2018, 43 (02): 101-106. DOI: 10.13251/j.issn.0254-6051.2018.02.020.

[6] Chen Chen, Shi Wenjie, Song Ruihong, et al. Performance of inner surface coating of slender pores based on TD coating treatment technology [J]. Surface Technology, 2022, 51 (09): 342-350. DOI: 10.16490/j.cnki. issn.1001-3660.2022.09.036.

[7] Kong Dejun, Wang Jinchun, Guo Haoyuan XPS and EDS surface scanning analysis of VC coatings prepared by TD treatment [J]. Rare Metal Materials and Engineering, 2016, 45 (02): 297-30.

Research on airspace filtering algorithm for trackside train bearings based on GSC-MVDR

Ziyu Yuan [1], Fang Liu [1,2,3 *], Xuewen Bao [1], Guangwen Ren [1], Guoqiang Zeng [1], and Yongbin Liu [1]

[1] Anhui University, Hefei, Anhui, 230000, China
[2] ZhiHuan Science and Technology Co., Ltd, Hefei, Anhui, 230000, China
[3] MaiPu Intelligent and Technology Co., Ltd, Hefei, Anhui, 230000, China

[*]Corresponding author's e-mail: ufun@ahu.edu.cn

Abstract: Trackside Acoustic Detection Systems (TADS) employ microphone arrays installed on both sides of the railway tracks to collect sound signals emitted by wheelset bearings as trains pass by. These systems facilitate state monitoring and fault diagnosis through signal processing, offering capabilities such as non-contact measurement, low cost, and early warning of potential failures. However, the strong noise characteristics of trackside signals significantly impair diagnostic precision. To address this issue, this paper proposes a spatial filtering algorithm based on the Minimum Variance Distortion-less Response (MVDR) and the Generalized Sidelobe Canceller (GSC), achieving directional denoising. Additionally, the algorithm employs time-domain resampling techniques to correct the Doppler effect in signals.

1. Instruction

Railway transportation, as a crucial pillar of global economic and social development, plays a significant role in transporting large quantities of goods and passengers. Wheelset bearings, as key components that directly affect the operational status of trains, are critical to the reliability of train operations. Therefore, monitoring and diagnosing the operational condition of train bearings is of paramount importance.

The development of the trackside acoustic detection system began in 1994, led by the Transportation Technology Center, Inc. (TTCI). The TADS system, launched by TTCI, utilizes an array of microphones installed along both sides of the railway tracks to collect acoustic signals emitted by wheelset bearings as trains pass by. This system can identify bearing faults before they occur, thus preventing train accidents. It features non-contact measurement, low cost, and the ability to provide early warning of faults [1].

However, the performance of this system is influenced by various factors, including Doppler distortion, interference from multiple sound sources, and a strong noise background in the trackside acoustic environment. During train operations, noise sources such as the tracks, wind, and other bearings contribute additional sounds besides those emanating from the target bearings. The intensity of these noises may obscure or interfere with the characteristic frequencies of bearing faults, making the extraction of fault features more challenging. Additionally, the relative motion between the train and the microphone array can cause a shift in the frequency of the bearing signals, a phenomenon known as Doppler distortion. These factors pose significant technical challenges for trackside acoustic detection technology, complicating the accurate extraction and diagnosis of fault characteristics. To address the issue of steep distortions, the literature employs Hilbert transform to extract the instantaneous frequency of signals and uses spline interpolation to correct for Doppler distortion [2]. However, this method requires prior knowledge of signal parameters to establish resampling times, and the endpoint effects of the Hilbert transform may limit correction results. To overcome these issues, a time-domain interpolation resampling correction method based on the movement relationship between the train and microphone is proposed [3]. This method reconstructs the signal time series with kinematic parameters and uses interpolation to modulate the distorted signal in the time domain. Regarding the issues of multiple sound sources and strong noise, the literature overcomes these two problems by using the wavelet domain moving beamforming (WMB) method. This method transforms the time domain acoustic signal collected by the microphone array into the wavelet domain, eliminates the Doppler effect through the proposed wavelet domain resampling method, and applies wavelet domain beamforming to enhance the bearing signal and reduce interference noise [4]. The literature proposes spatial filtering using MVDR-based rectangular arrays. Although it performs poorly in strong noise environments compared to the linear array solution, it shows potential advantages [5].

2. Algorithm principle

Spatial filtering technology controls the relative phase and amplitude of each element in a microphone array to form a main lobe directed towards a specific direction and suppress sidelobes from other directions. This mechanism enhances signals in the direction of the main lobe while effectively suppressing noise and interference from other directions. In the trackside acoustic environment, the microphone array is arranged along the railroad tracks to capture the sound signals from passing trains. To enhance signals from the direction of the train bearings while suppressing noise and interference from non-target directions, it is necessary to design appropriate spatial filters for directional noise reduction. Common spatial filtering algorithms include Delay-and-Sum Beamforming (DSB), Minimum Variance Distortion-less Response Beamforming (MVDR), and the Generalized Sidelobe Canceller (GSC).

The GSC algorithm builds upon the traditional DSB structure by adding an adaptive noise cancellation module, which includes a main channel and an auxiliary channel. The fixed beamforming filter of the main channel is responsible for maintaining the response to the desired signal direction, while the adaptive noise canceller in the auxiliary channel estimates and eliminates interference and noise from undesired directions. This structure endows the GSC with enhanced interference resistance and greater flexibility in noisy environments.

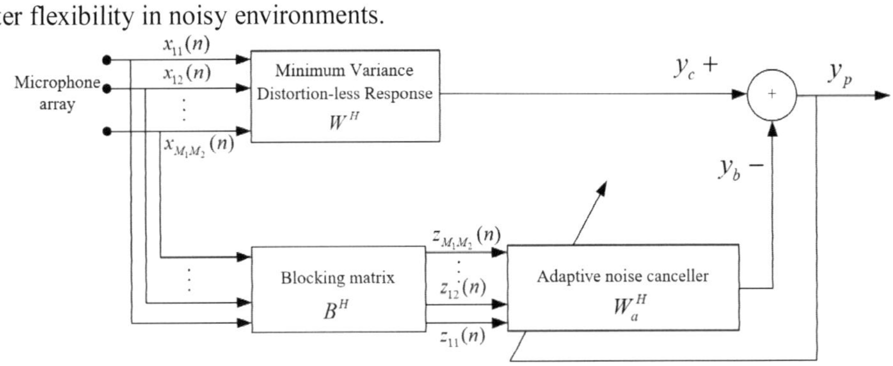

Figure 1. GSC-MVDR algorithm framework diagram.

However, in the trackside acoustic environment, the position and angle of the train relative to the microphone array change with movement, necessitating that the beamforming algorithm be capable of real-time updates to the beam direction to accurately capture the sound source. Traditional GSC algorithms, with fixed beamforming in the main channel, may lead to inaccurate beam directions when train speeds vary or in curved tracks, thereby reducing signal enhancement effectiveness. To address this, this paper proposes a GSC-MVDR algorithm that combines the MVDR algorithm. The MVDR algorithm minimizes output noise power while ensuring no distortion of signals from specific directions. Its adaptive feature maintains gain in the target direction while optimizing beamforming to adapt to changes in the sound source position, achieving effective sound source focusing and noise suppression. The framework of this algorithm is illustrated in Figure 1.

2.1 Main channel filter design

Following the algorithmic process depicted in Figure 1, the microphone array signal from the main channel passes through a pre-configured filter w:

$$w=[w_{11},w_{12},\cdots w_{1M_2},w_{21},\cdots,w_{m_1m_2}\cdots,w_{M_1M_2}]^T \tag{1}$$

The filtered signal is then given by:

$$y_c(t)=w^H X(t) \tag{2}$$

The energy of the filtered signal is represented as:

$$E\left\{\left|y_c(t)\right|^2\right\}=E\left\{\left|w^H X(t)\right|^2\right\}=w^H R_{xx} w \tag{3}$$

where R_{xx} represents the covariance matrix of the signal $X(t)$, which is defined as:

$$R_{xx}=E\{X(t)X^H(t)\} \tag{4}$$

According to MVDR theory, to ensure the gain in the specified direction while suppressing interference from other sources, the following conditions must also be met:

$$w^H a\left(\theta_{obj}(t),\varphi_{obj}(t)\right)=1, w^H a\left(\theta_i(t),\varphi_i(t)\right)=0, i=1,2,...Q \tag{5}$$

where $a\left(\theta_i(t),\varphi_i(t)\right)$ represents the direction vectors of various sound sources, and the solution to this constrained optimization problem is

$$w_{opt}=\frac{R_{xx}^{-1}a(\theta_{obj}(t),\varphi_{obj}(t))}{a^H(\theta_{obj}(t),\varphi_{obj}(t))R_{xx}^{-1}a(\theta_{obj}(t),\varphi_{obj}(t))} \tag{6}$$

2.2 Auxiliary channel filter design

The auxiliary channel eliminates the signal from the desired direction in the received signal by determining the steering vector $a(\theta,\varphi)$ of the expected signal and projecting the signal into a space orthogonal to this direction vector. The blocking matrix B can be represented as:

$$B=I-\frac{a(\theta_{obj},\varphi_{obj})a(\theta_{obj},\varphi_{obj})^H}{\left\|a(\theta_{obj},\varphi_{obj})\right\|^2} \tag{7}$$

Commonly used algorithms for adaptive noise canceller (ANC) include minimum mean square error (LMS) and recursive least squares (NLMS). Compared with the LMS algorithm, the NLMS algorithm improves the convergence speed and adaptability to signal amplitude changes, making it a better choice in the trackside environment.

The recursive formula of the NLMS algorithm filter is:

$$W_a(t+1)=W_a(t)+\frac{\mu}{\varepsilon+\left\|Z(t)\right\|^2}\cdot y_p(t)\cdot Z(t) \tag{8}$$

In the formula, $W_a(t)$ is the current weight vector, $W_a(t+1)$ is the updated weight vector, μ is the step size parameter, ε is a very small positive constant used to prevent division by zero, $y_p(t)$ is the current error signal, and $Z(t)$ is the input signal.

2.3 Method flow

(1) Directional filters at different positions are constructed based on the sound source emission time series and GSC-MVDR spatial filtering on the signals is performed to obtain the filtered signals.

(2) The filtered signals are resampled from different positions using the microphone sound collection time series to eliminate the Doppler effect.

(3) Envelope spectrum analysis is performed on the corrected signals for fault analysis and diagnosis.

3. Experiment analysis

Given the safety and feasibility risks of directly collecting sound signals from faulty bearings in operational trains, this paper designs an alternative experimental scheme to acquire acoustic signals of bearing faults. Firstly, the sound signals of bearings are recorded under static conditions. Then, these static signals are affixed to a vehicle as a moving sound source, and a roadside rectangular microphone array is used to collect the sound signals as the vehicle speeds by, simulating the sounds produced by faulty bearings during train operation. Through this experimental platform, the paper obtained dynamic fault signals of the inner ring at a vehicle speed of 23.69 m/s and processed the data using the MVDR-GSC algorithm and time-domain resampling techniques. The processed results are shown in Figure 2.

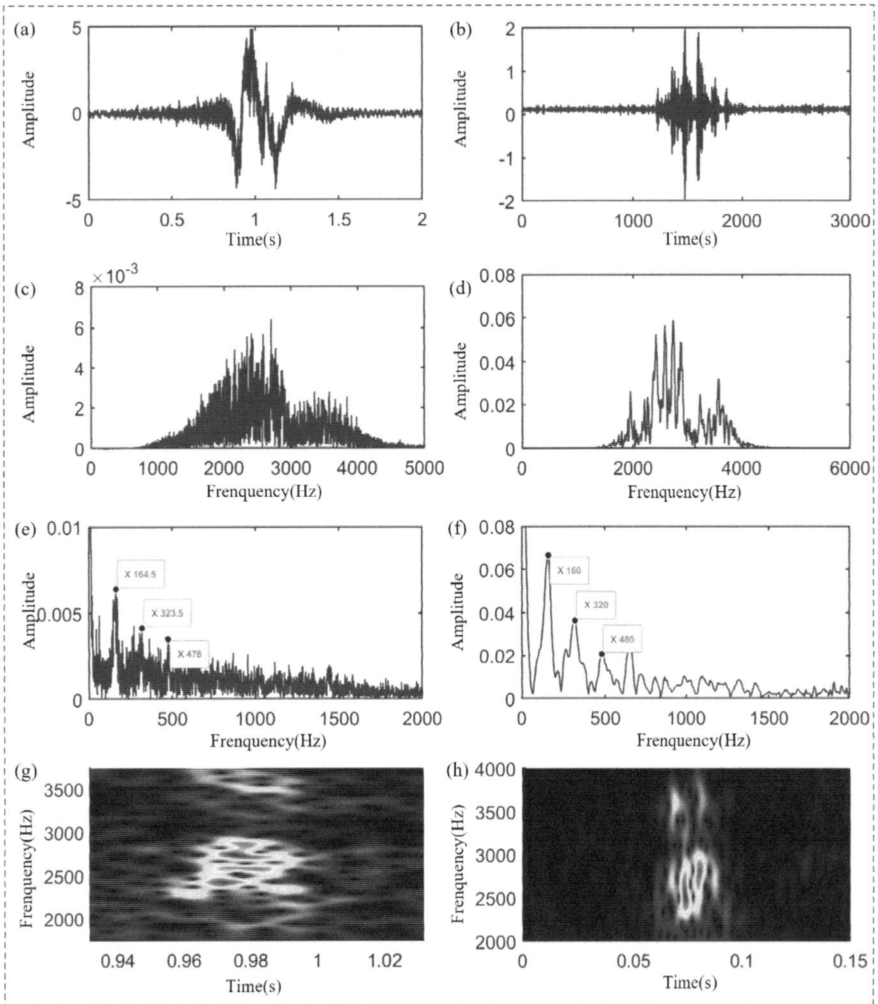

Figure 2. Algorithm processing results.

Figures 3(a), (c), (e), and (g) respectively show the time-domain, frequency-domain, envelope, and time-frequency diagrams of the dynamic inner ring fault signals received by the microphone array element at the origin. The figures indicate that the Doppler effect causes the signals to be compressed and stretched along the time axis, and the amplitude varies with the changing distance between the microphone array and the sound source. Additionally, the airflow noise resulting from the vehicle's movement introduces strong low-frequency disturbances, leading to frequency aliasing and shifting in the frequency domain, with the fault frequency and its harmonics almost completely submerged in the noise-induced sidebands.

Figures 3(b), (d), (f), and (h) respectively display the processed time-domain, frequency-domain, envelope, and time-frequency diagrams. It is observable that the frequency shift and spreading phenomena caused by the Doppler effect have been eliminated, making all fault characteristic frequencies visible. The presence of fault characteristic frequencies and their harmonics at 160 Hz, 320 Hz, and 480 Hz in the obtained envelope spectrum indicates that the signal has been well-denoised, allowing for the extraction of more fault-related signal components.

4. Conclusion

To address the challenges of strong noise and steep distortions in the trackside acoustic environment, this study introduces an advanced trackside acoustic diagnostic algorithm based on the GSC-MVDR framework. This algorithm incorporates the MVDR technique into the main channel of the GSC structure and proposes a novel approach for computing the blocking matrix in the GSC auxiliary channel. This enhancement enables the algorithm to further eliminate noise and interference not fully suppressed by the main channel. In conjunction with time-domain resampling technology, it achieves targeted denoising and Doppler distortion correction of trackside acoustic signals. Experimental results demonstrate that this GSC-MVDR approach exhibits exceptional noise reduction capabilities and fault diagnosis accuracy in the realm of trackside acoustic diagnostics.

Acknowledgments

This paper is one of the interim outcomes of the National Natural Science Foundation project "Theoretical Research on End-Cap Sonomicrometry Auscultation of Key Components in Rotating Machinery" (No.52275075) and the Anhui Province Key Research and Development Plan "Research on Key Technologies of Acoustic Intelligent Detection System for China-Europe Railway Express in Transit" (202304a05020058).

References

[1] Tarawneh C, Montalvo J, and Wilson B. Defect detection in freight railcar tapered-roller bearings using vibration techniques[J]. Railway Engineering Science, 2021, 29(1): 42-58.

[2] Dybała J, Gałęzia A, and Mączak J. Verification of Doppler Effect removal method for the needs of pass-by railway condition monitoring system[J]. Diagnostyka, 2008 (4 (48)): 5-8.

[3] Liu F, He Q, and Kong F. Doppler effect reduction based on time-domain interpolation resampling for wayside acoustic defective bearing detector system[J]. Mechanical Systems and Signal Processing, 2014, 46(2): 253-271.

[4] He S, Hu D, and Yu G. Trackside acoustic detection of axle bearing fault using wavelet domain moving beamforming method[J]. Applied Acoustics, 2022, 195: 108851.

[5] Huang H, Liu F, and Geng L. Fault diagnosis accuracy improvement using wayside rectangular microphone array for health monitoring of railway-vehicle wheel bearing[J]. IEEE Access, 2019, 7: 87410-87424.

Journal of Physics: Conference Series **2842** (2024) 012091 doi:10.1088/1742-6596/2842/1/012091

Statistical regression analysis of bond splitting tensile properties of old and new concrete

Da Hou[1*], Dongnan Yuan[2], Keqin Wang[3] and Duo Liu[4]

[1]School of Civil and Transportation Engineering, Hohai University, Nanjing, Jiangsu, 210098, China
[2]School of Materials Science and Engineering, Nanjing Forestry University, Nanjing, Jiangsu, 210042, China
[3]School of Civil and Transportation Engineering, Hohai University, Nanjing, Jiangsu, 210098, China
[4]JSTI GROUP, Nanjing, Jiangsu, 211112, China

[*]Corresponding author's e-mail: 13952095178@163.com

Abstract. With the gradual increase of assembled and reinforced structures in modern construction, the bonding problem of new and old concrete has become a hotspot of increasing concern for scholars and engineers. Bond tensile properties are the main aspects of bond mechanical properties, and their main influencing factors are interface roughness, old and new concrete strength, roughness treatment, and normal stress. This paper analyses the influence of each factor on bond tensile property by collating a large amount of experimental data from established scholars. The analysis results show that the bond tensile property increases linearly with the increase of roughness and nonlinearly with the rise of new and old concrete strength. Based on summarising the influence law of a single factor, the calculation equation of bond tensile property was proposed. The calculated equations derived from the statistical regression were compared with the formulas of different scholars to analyze the closeness of the experimental and calculated values; the results show that the formula proposed in this paper is more accurate and stable.

1. Introduction

In modern engineering construction, there are more and more problems in connecting old and new concrete. For example, the interface between the cast-in-place and assembled parts of an assembled structure can involve bonding problems between old and new concrete [1, 2]. In addition, such bond problems can also exist between the repair material and the existing structure in a reinforced structure [3, 4]. Bond splitting tensile strength is the main aspect of bond strength, which is influenced by roughness, strength of old and new concrete, and roughness treatment method, of which roughness is considered to have the most critical effect on bond strength. Many scholars [5, 6] have emphasized the important role of interfacial roughening because by treating interfacial roughness, it is possible to increase both the effective surface area of the old concrete and the proportion of exposed coarse aggregate, and this treatment positively affects the bond between the old and new concrete. Sha et al. [7] used different chiseling treatment methods to study the bond mechanical properties of the old and new concrete interfaces, and the test results showed that the punching method was more effective than manual chiseling. Zhao et al. [8] investigated by splitting tensile test, and the results showed that the bond tensile properties were positively correlated with the roughness and strength grade of the old and new concrete, and the specimens treated with high-pressure water gun punching had a higher bond tensile strength under the same roughness conditions.

From the above studies, it can be seen that the bond-splitting tensile strength of new and old concrete has more influencing factors, and a large number of scholars have studied each factor, but most of the research focuses on analyzing the influence of each factor on the mechanical properties individually, and there is still a lack of a unified, comprehensive calculation formula. In this paper, the established scholars' test data will be statistically fitted, the bond-splitting tensile strength calculation formula will be proposed, and the accuracy of the formula will be verified.

2. Influence of roughness and bonding surface treatments

The tests of different scholars have included the bond surface treatment method and the strength of new and old concrete in addition to the control parameter of roughness. To study the effect of two single factors, roughness, and bond surface treatment method, on splitting strength, the following data of different old and new concrete strengths will be organized separately for other bond surface treatments to study the effect of bond breaking performance by bond surface treatment method and roughness. The data from the manual chiseling treatment and punching treatment [8-16] tests were compiled, and the data were fitted to them, respectively. Based on the scatter plots of the data, it can be observed that for each concrete strength class, there is a linear pattern between the splitting tensile strength and roughness, so a linear fit was made to the data points. The matched straight lines are shown in Figure 1 and Figure 2. In particular, for scholars who characterize the axial tensile strength by axial tensile strength, the value obtained by dividing the axial tensile strength by 0.9 is used as the splitting tensile strength value according to the specification [17]. Data with equal mean values of old and new concrete strength grades in the scholars were combined into one data set for fitting. Roughness was characterized by the most widely used sand filling depth, h (mm).

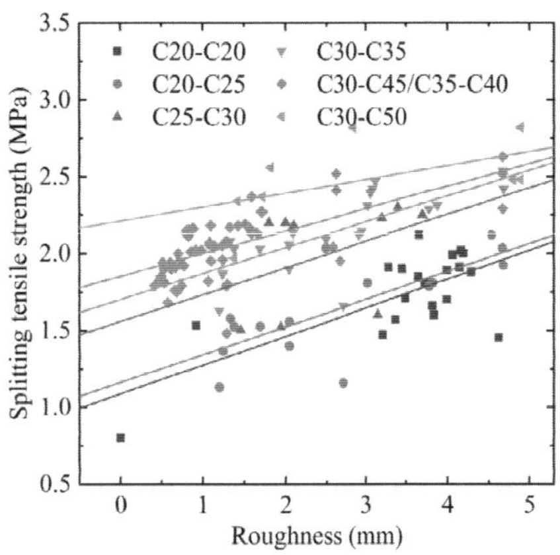

Figure 1. The relationship between bond splitting tensile strength and roughness under the manual chiseling method.

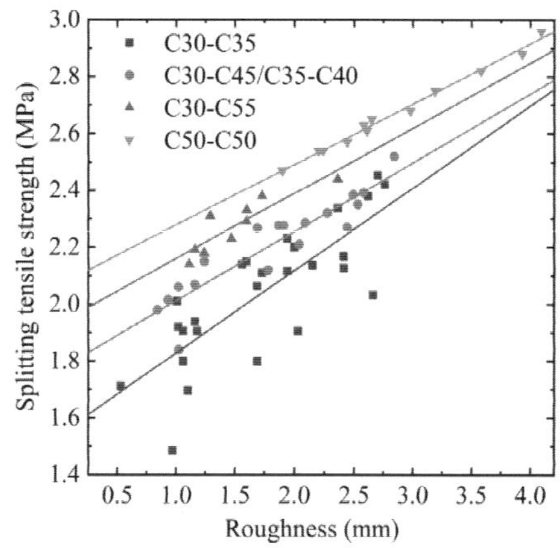

Figure 2. The relationship between bond splitting tensile strength and roughness under the punching method.

It is evident from the graphs that the linear relationship between split tensile strength and roughness is more pronounced for both bond surface treatments in varying old and new concrete strength classes. However, the data for punching treatment demonstrates a better fit than the data obtained from manual chiseling. A comparison of the two graphs reveals that the bond strength of members treated with punching is notably superior, considering the same roughness and the average strength value of the old and new concrete. This is attributed to the greater randomness and poorer uniformity associated with the manual chiseling method, which results in micro-cracks formation and subsequently affects the strength. Conversely, high-pressure water gun punching ensures a uniform treatment of the bond surface, yielding better bonding effects with lesser damage to the members. Furthermore, in the process of increasing roughness, the split tensile strengths of members with varying old and new concrete strengths tend to become closer and closer, suggesting that the influence of old and new concrete becomes less significant. This indicates a nonlinear relationship, with roughness emerging as the main influencing factor. Considering the difficulty and cost of enhancing concrete strength, relying solely on improving it to enhance bond strength is not economically viable. Thus, further research will delve into the specific nonlinear relationship of this influencing factor.

3. Effect of the strength of old and new concrete

This study employs two bond surface treatments and four roughness classes to investigate the relationship between bond-splitting strength and the average value of the old and new concrete strength while eliminating the influence of other factors. The bond splitting strength values corresponding to each roughness level were obtained from various scholars' data using the following criteria: if the required data was available, it was directly utilized; otherwise, the two closest sets of data were selected, and linear interpolation was employed for calculation. It is important to note that according to relevant codes and Ye et al. [17, 18], there is no linear relationship between the tensile properties of concrete and the concrete strength grade. Considering this feature, a nonlinear fit was performed on the data, and the results are shown in Figure 3 and Figure 4. From the fitted curves, it is believed that bond-splitting strength increases with an increase in the average value of the old and new concrete strength. However, the incremental gain in bond-splitting strength with an increase in concrete strength gradually diminishes. Thus, enhancing concrete strength to improve tensile properties becomes economically inefficient when considering the increase in roughness.

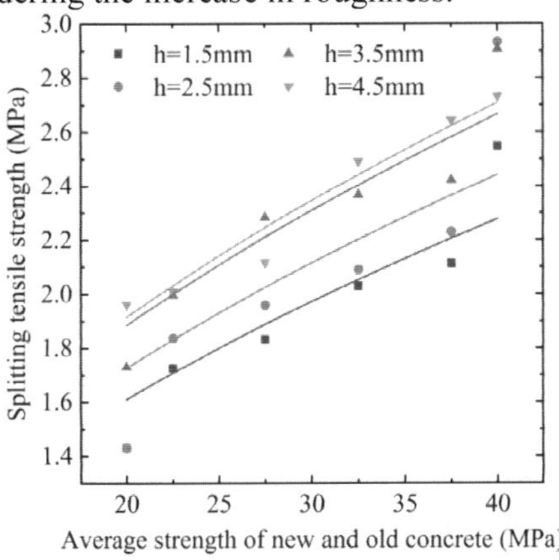

Figure 3. The relationship between concrete tensile strength and concrete strength grade.

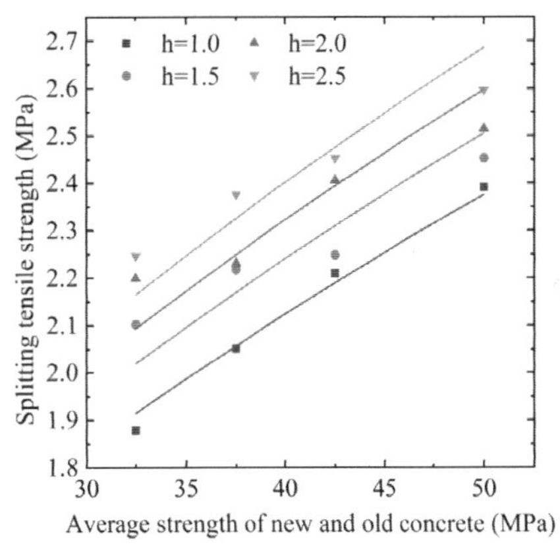

Figure 4. The relationship between concrete tensile strength and concrete strength grade.

4. Influence of pouring interface orientation

The pouring interface can be categorized into two directions: horizontal and vertical, as illustrated in Figure 5. In the horizontal direction, the old concrete is poured as the bottom layer, followed by the pouring of new concrete on top, which is then compacted through vibration to form a unified structure. Conversely, in the vertical direction, the new concrete is poured vertically against the side of the existing concrete, creating a vertical connection through vibration and compaction. According to Zhao et al. [8], the bond-splitting tensile strength of specimens cast with a horizontal interface is higher compared to those cast with a vertical interface. This disparity arises from the fact that when the interface is released horizontally, the upper layer of cast-in-place concrete can be thoroughly vibrated and compacted, resulting in improved bond strength. On the other hand, when the interface is vertical, the new concrete is cast against the side of the old concrete, impeding thorough vibration and compaction on the bonding surface, subsequently reducing the bonding performance. Based on the analysis of existing test results [8], the bond split tensile strength of specimens cast with a vertical interface is approximately 0.9 times that of specimens cast with a horizontal interface. Hence, this influencing factor can be considered by applying a discount factor, assuming a discount factor of 1 in the case of horizontal casting and a discount factor of 0.9 in the case of vertical casting.

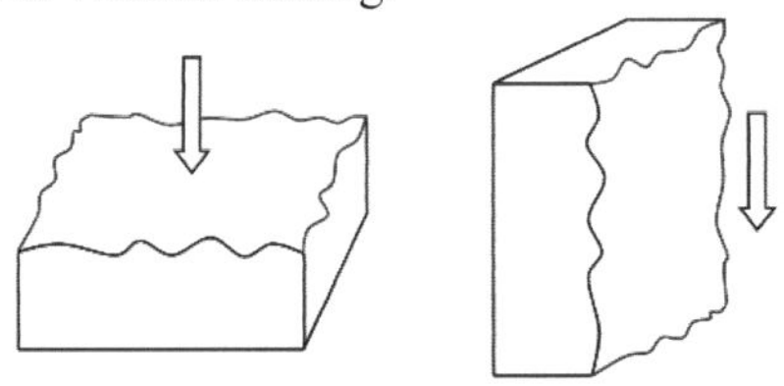

Figure 5. Schematic diagram of the pouring interface.

5. The proposed statistical regression-based formula for bond splitting and pulling

Through the analysis of every single factor, the following conclusions can be drawn: firstly, the bond splitting strength is linearly and positively correlated with roughness, which means as the roughness increases, the bond splitting strength also increases linearly; secondly, the bond splitting strength also increases with the rise of the average value of the strength of the old and new concrete, but this increase is nonlinear; thirdly, the components with the bonding surface treated by high-pressure water gun punching have higher bond splitting strength under the same roughness conditions; lastly, the strength is higher when the transverse interface is poured. Combining the above relationships and organizing all the data, the following Equation (1) can be used for fitting:

$$f_t = \beta\left(ah + b\right)\sqrt{f_{cu}} \tag{1}$$

where f_t is the split tensile strength, a and b are parameters obtained by fitting the subsequent collated information, and β is the coefficient considering the direction of the casting interface, which takes the value of 1 for the transverse casting interface and 0.9 for the longitudinal casting interface.

To find out the additional parameters a and b, the values obtained by dividing the bond split tensile strength under the two bond surface treatments by the arithmetic square root of the average value of the strength of the old and new concrete as the vertical axis and the roughness as the horizontal axis were fitted to derive the equations, and the provided data are shown in Figure 6. The fitted equations are shown in Equation (2) and Equation (3). From the fitting results, the relevant effects are all good, but their specific applicability needs to be further proved.

$$\text{Manual chiseling method:}\ f_t = \beta\left(0.02926h + 0.2963\right)\sqrt{f_{cu}} \tag{2}$$

$$\text{Punching method:}\ f_t = \beta\left(0.02892h + 0.3066\right)\sqrt{f_{cu}} \tag{3}$$

6. Applicability of the formula and its comparison with other scholars

To compare the fitting effect of this paper's formula and other scholars' formulas on the data, the roughness and concrete strength in the test data were calculated by substituting them into this paper's formula and other scholars' formulas, respectively. The calculated results were compared with the test values. The data points whose horizontal and vertical coordinates were the test values and computed values, respectively, were plotted, and the results are shown in Figures 6 and 7. The red line in the figure is the quadrant bisector. It can be seen from the plots that the data points obtained by using the formulae in this paper are closer to the quadrant split line, which means the test values are more relative to the calculated values. To more accurately assess the effectiveness of the formulae in predicting the strength, the mean and standard deviation of the difference between the calculated values and the test values of the scholars' formulae were calculated, and the results are summarised in Table 1 and Table 2. It can be observed that the absolute value of the mean and standard deviation of the differences obtained using the formula in this paper are the lowest. This suggests that the test values obtained from this formula are closer to the calculated values, and the variability in the data is minimized. As a result, this formula can be considered suitable for the calculation of bond-splitting tensile strength in bond-splitting tests. However, when comparing the two roughness processing methods, it is evident that the data obtained from the manual chiseling method is more scattered than the punching method. As a result, it is advised to prioritize the technique that can ensure uniform roughness in practical projects.

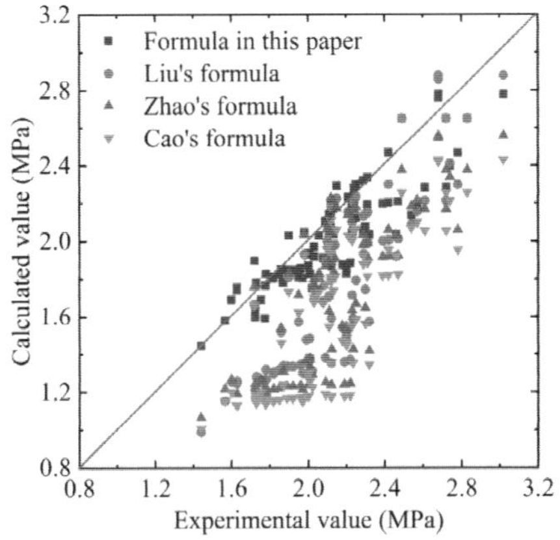

Figure 6. Comparison of experimental and calculated values under the manual chiseling method.

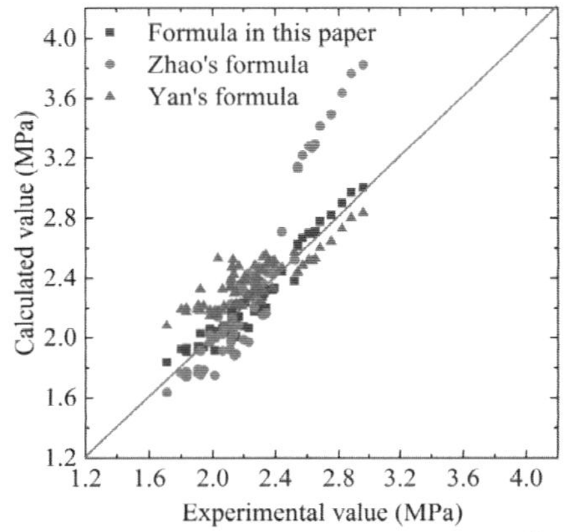

Figure 7. Comparison of experimental and calculated values under the punching method.

Table 1. The calculation formulas of different scholars and the comparison of errors between calculated and experimental values (manual chiseling method).

Scholars	Calculation formulas	Average value	Standard deviation
Liu [10]	$f_t = (0.005684h + 0.044163) f_{cu}$ (4)	-0.37017	0.24381
Zhao [8]	$f_t = (0.002694h + 0.050789) f_{cu}$ (5)	-0.42582	0.25535
Cao [19]	$f_t = (0.0026h + 0.048) f_{cu}$ (6)	-0.51543	0.24096
This paper	$f_t = \beta(0.02926h + 0.2963)\sqrt{f_{cu}}$ (7)	-0.09510	0.15211

Table 2. The calculation formulas of different scholars and the comparison of errors between calculated and experimental values (punching method).

Scholars	Calculation formulas	Average value	Standard deviation
Zhao [8]	$f_t = (0.007348h + 0.04642) f_{cu}$ (8)	0.10078	0.30798
Yan [15]	$f_t = 0.2123h + 1.968$ (9)	0.14134	0.15982
This paper	$f_t = \beta(0.02892h + 0.3066)\sqrt{f_{cu}}$ (10)	0.00233	0.07705

7. Conclusions

In this paper, the data of hundreds of specimens from many scholars were categorized and organized using statistical regression. The influence of a single factor is considered first. The formula is proposed according to the conclusion, statistical regression is carried out to derive the pending parameters of the formula, and finally, the applicability is analyzed. Based on the above research, the following conclusions are drawn:

(1) At the same roughness, different roughness treatments produce different strengths, with the punching method having better mechanical properties. The effect of roughness on bond-splitting strength is linear, and the effect of new and old concrete strength on bond-splitting strength is nonlinear.

(2) Data fitting was carried out based on the influence characteristics of a single factor, and the formula for split tensile strength was proposed. In this paper, by comparing the formulas of other scholars, it is proved to be more accurate.

References

[1] Li, W., Jia J., Wang, Y., Li, B., Wan, S. (2021) Experimental Research on mechanical properties of new and old concrete joints of prefabricated structure. Industrial Construction, 51(02): 40-46.
[2] Chen, Z. (2020) Study the Inhomogeneity of Physical State in the Interfacial Zones between the New and Old Concrete in Assembly Structure. China University of Mining & Technology.
[3] Zhao H. (2021) Experimental research on bonding performance of new and old concrete and its application in bridge reinforcement engineering. Dalian University of Technology.
[4] Shu, T., Xu, F., Xie, J. (2022) Experimental study on the interface bonding performance of strengthening with a concrete laminate layer with precast hollow slab. Building Structure, 52(01): 115-119.
[5] Yu, L., Zhang, L., Lei, D., Chen, B. (2019) Shear test of the rectangular beam on the new to the old concrete interface based on Digital Image Correlation. IOP Conference Series: Earth and Environmental Science. 267(3): 032058 (7pp).
[6] Liu, F. Zhang, S., Xiong, B. (2023) Effect of roughness on shear properties of sand-concrete interface with different particle sizes. Rock and Soil Mechanics. 44(S1): 419-426.
[7] Sha, J., Xu, H., Liu, J., Guo, F., Zhang, Y. (2019) Study on interfacial bonding behavior of new and old concrete. New Building Materials. 46(10): 133-135+151.
[8] Zhao, Z., Zhao, G., Liu, J., Yu, Y. (2001) Experimental Study on Adhesive Tensile Performance of Young and Old Concrete. J., Journal of Building Structures, 02: 51-56.
[9] Fan, J., Wu, L., Zhang, B. (2021) Influence of Old Concrete Age, Interface Roughness and Freeze-Thawing Attack on New-to-Old Concrete Structure. Materials, 14: 1057.
[10] Liu, J. (2000) Study on Young's Mechanics Performance of Adherence on Old Concrete. Dalian University of Technology.
[11] He, Y., Zhang, X., Hooton, R. D., Zhang, X. (2017) Effects of interface roughness and interface adhesion on new-to-old concrete bonding. Construction and Building Materials, 151: 582-590.
[12] Li, P., Zhao, G., Zhang, L. (2005) Experimental research on bond strength of new-old concrete with epoxy and sand transfer layer. Journal of Dalian University of Technology, 02: 255-259.
[13] Cheng, H., Zhang, L. (2003) A study on adherence cleavage strength of new and old concrete. Henan Science, 04: 465-467.
[14] Lin, N., Yin, J., Zhang, X., Zhou, S. (2006) Application of Fractal Theory to Study Interfacial Bond Strength of Old-New Concrete. Journal of Building Materials, 04: 399-403.
[15] Yan, W., Wu, X., Hu, K., Wei, S. (2021) Quantitative Study on Wet Joint Chiseling Roughness of Prefabricated Girder Bridge Deck. Railway Engineering, 61(11): 39-43.
[16] Liu, Y. (2006) Research About the Bond Technology of the New and Old Concrete. Chongqing University.
[17] Ye, J. (2021) Principles of Structural Design, China Communication Press, Beijing.
[18] ACI (American Concrete Institute). (2019) Building code requirements for structural concrete and commentary, ACI: 318-19.
[19] Cao, H., Ma, Q., Zhang, R., Wang, Y. (2016) Experimental Study on Adhesive Splitting Tensile Performance of Post Pouring Concrete on Precast Concrete. Bulletin of the Chinese Ceramic Society, 35 (09): 2925-2929.

Preparation of $MnS/NiCo_2S_4$ composite nanosheets self-supported electrode and the performance in water-system asymmetric supercapacitors

Xiao Wang, Siheng Ren*, Deqiang Kong*, Chang Liu, and Yuanhao Wu

Space Engineering University, Beijing 102200, China

* Corresponding author's e-mail: 971017356@qq.com (Siheng Ren); kdq669@126.com (Deqiang Kong)

Abstract: In this paper, the $MnS@NiCo_2S_4$ composite self-supporting electrode was designed and prepared with a mixture of nickel, cobalt, and manganese sulfide as the research object. The capacitance of the electrode can reach 2, 372.7 F g^{-1} at the current density of 1 A g^{-1}, and the voltage window can also be widened to 0-0.6 V and maintains a multiplier performance of 62.1% at 10 A g^{-1}. Following 5, 000 cycles of charging and discharging, the electrode exhibits a capacity retention of 78.2% relative to its initial capacity, with the coulombic efficiency consistently hovering around 100%. This performance indicates that the electrode can carry out stable and reversible electrochemical processes. A supercapacitor assembled with 1 M KOH as the electrolyte, activated carbon as the anode, and $MnS/NiCo_2S_4$ as the cathode, delivers an energy density of 46.9 W h kg^{-1}. In the life test of the charge and discharge cycle, the discharge capacity after 10, 000 cycles is 89.2% of the initial capacity, indicating that the device has good cycle stability. This study provides an idea for the potential application of metal sulfide-based electrodes with wide potential windows and high electrochemical properties in water asymmetric supercapacitors.

1. Introduction

In the realm of supercapacitor research, electrode materials play a pivotal role and are indispensable for achieving optimal performance. From the initial exploration of monosulfides to the development of binary and even ternary sulfides, as well as the electrochemical utilization of metal chalcogenides and the pursuit of rational design strategies and efficient synthesis methods for metal sulfides, researchers have delved deeper into this field, encountering increasingly intricate challenges [1]. However, due to the inherent variations in properties among different metal elements, a single metal sulfide alone can no longer suffice to meet the current performance requirements of supercapacitors. Consequently, there has been an increasing focus on composite electrode materials composed of diverse metal sulfides.

This study reports on the fabrication of a high-performance self-supporting supercapacitor electrode composed of $MnS@NiCo_2S_4$ composite interconnect nanosheets developed on nickel foam by a straightforward single-step hydrothermal process, augmented by subsequent electrochemical treatment. The material expands the potential range to 0-0.6 V, and it exhibits remarkable specific discharge capacities, achieving 2, 372.7 F g^{-1} at 1 A g^{-1} and maintaining an impressive 1, 475.0 F g^{-1} even at an elevated rate of 10 A g^{-1}, which shows a good rate performance. Furthermore, the electrode sustains 78.2% of its initial capacity following an extended 5, 000 cyclic test at 4 A g^{-1}, exhibiting

Journal of Physics: Conference Series **2842** (2024) 012092 doi:10.1088/1742-6596/2842/1/012092

good cycle performance for the electrodes. The aqueous asymmetric supercapacitor designed with activated carbon anode and $MnS@NiCo_2S_4$ cathode, utilizing 1 M KOH electrolyte, attains 46.9 Wh kg^{-1} of energy density with 849.6 W kg^{-1} of power density. Notably, it demonstrates 89.2% upon undergoing 10, 000 cycles under 4 A g^{-1} discharge rate, evidencing its exceptional cycling stability. A method for developing metal sulfide-based electrode materials with extended potential windows and superior electrochemical properties is proposed in this study.

2. Experiment and results

2.1. Experimental section: preparation of electrode materials and assembling of asymmetric supercapacitors

Preconditioning of NF (nickel foam): Initially, NF was sectioned into a rectangle measuring 1.0 × 2.0 cm^2 in dimension. Subsequently, the segments underwent sequential ultrasonic cleansing in acetone, ethanol, and deionized water, respectively, with each solvent used for 15 minutes. The cleaned NF was put into a drying cabinet at 60°C and dried overnight for subsequent experiments.

Preparation of interconnect nanosheets MnS on NF: A standard synthesis protocol was employed for the preparation of MnS, which entailed a straightforward one-pot hydrothermal technique. As illustrated in Figure 1, 50 mg $MnCl_2 \cdot H_2O$ and 50 mg H_2NCSNH_2 were dissolved into 40 milliliters of deionized water and stirred magnetically for 15 minutes, yielding a translucent solution. Subsequently, four pieces of NF (1.0 × 2.0 cm^2) were placed on a teflon bracket and immersed into a mixed solution within the reactor and moved together into a Teflon-lined stainless steel autoclave. Next, we put it into a constant temperature of 90°C for a duration of 6 h. Upon natural cooling to ambient temperature, the NF pieces were retrieved and meticulously rinsed with DI water to eliminate any residual unreacted materials. Thereafter, they were dried at 60°C to obtain MnS/NF. The difference between the mass of NF and its initial mass is recorded as the active material load at this stage.

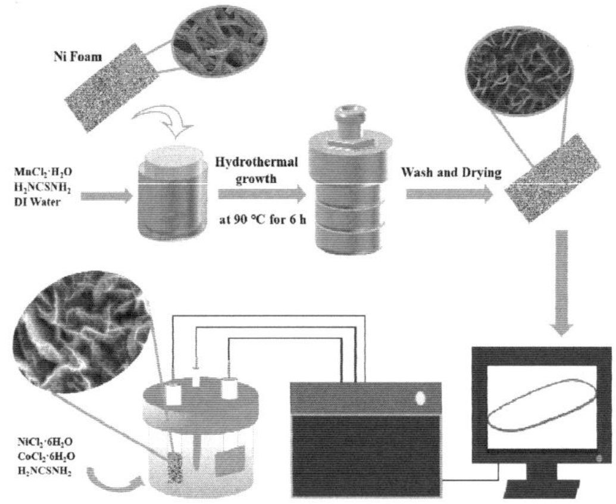

Figure 1. Process flow sheet of the synthesis process of $MnS@NiCo_2S_4$ interconnected nanosheets on NF.

Preparation of interconnect nanosheets $NiCo_2S_4$ on MnS/NF: The growth of $NiCo_2S_4$ interconnect nanosheets on MnS/NF was successfully achieved through a facile three-electrode cyclic voltammetry electrodeposition method. In this part, a total of 36 mg $NiCl_2 \cdot 6H_2O$, 53 mg $CoCl_2 \cdot 6H_2O$, and 171 mg H_2NCSNH_2 were collectively dissolved in 30 mL of DI water and stirred magnetically for 15 minutes, yielding a translucent pale pink solution. Subsequently, the samples underwent cyclic voltammetry (CV) scanning at 5 mV s^{-1} between -1.2 V ~ 0.2 V for a total of 6 cycles, effectively producing the interconnected nanosheets of $NiCo_2S_4$. The difference between the mass of NF and the initial mass is recorded as the active material load, specifically referring to the active material load of the composite $MnS@NiCo_2S_4$ interconnected nanosheets self-supporting cathode, which amounts to approximately 1.0 mg cm^{-2}.

Assembly of asymmetric supercapacitors (ASCs): Initiating the process, the fabrication of activated carbon (AC) anodes commenced by homogeneously blending AC, acetylene black, and PVDF are uniformly mixed with a mass ratio of 8:1:1, employing NMP as a solvent. This blend was manually ground for a minimum duration of 30 minutes to attain a uniform slurry, and we applied the slurry onto carbon cloth (CC), followed by an overnight drying session at 60°C within an air-circulated oven. The active material loading of activated carbon (AC) is approximately 3.0 mg. Subsequently, the assembly of the ASCs was carried out, utilizing MnS@$NiCo_2S_4$ on NF as the cathode component, AC on CC as the anodic counterpart, and 1 M KOH electrolyte solution to complete the device.

2.2. *Experimental principles and measurements*

The electrochemical characteristics of the prepared electrodes were measured by an electrochemical workstation equipped with a three-electrode device at room temperature, wherein a 1 M KOH aqueous solution served as the electrolytic medium.

The formula for calculating C_s (F g^{-1}), the mass-specific capacitance of constant current charge and discharge, is:

$$C_S = (I \times \Delta t)/(m \times \Delta V) \quad (1)$$

where m is the active material load (g), ΔV is the electrode potential window (V), I is the current (A), and Δt is the discharge time (s). The formula for calculating the energy density E and power density P of the supercapacitor is as follows:

$$E = 1/7.2 C_s V^2 \quad (2)$$

$$P = 3600\ E / \Delta t \quad (3)$$

where E is the energy density (W h kg^{-1}), V is the operating voltage (V) that the supercapacitor can achieve, and P is the power density (W kg^{-1}).

We have used MnS@$NiCo_2S_4$ as the cathode and AC as the anode to assemble asymmetric supercapacitors. When preparing the positive electrode, it is necessary to determine the positive electrode load according to the active material load of the negative electrode.

$$q_+ = q_- \quad (4)$$

$$q = mC\Delta V \quad (5)$$

$$\frac{m_+}{m_-} = \frac{C_- \Delta V_-}{C_+ \Delta V_+} \quad (6)$$

where q is the charge and ΔV is the working potential window of the anode and cathode (V). Through experimental comparison, the asymmetric supercapacitors deliver the optimal energy density and cycle stability when the mass loading is about 3.0 mg after balancing charge capacity.

2.3. *Results and discussion*

The surface morphology of MnS with the magnification of 2 μm and 4 μm is described in Figures 2 (a) and 2 (b) respectively, revealing a uniform interconnected nanosheet structure exhibited by the sample. Figures 2 (c) and 2 (d) respectively show the surface morphology of $NiCo_2S_4$ at corresponding sizes, which is similar to the morphology of most samples obtained by electrodeposition and also presents uniformly interconnected nanosheets. In comparison with the same magnification, it is not difficult to find that the interconnected nanosheets of MnS are larger than $NiCo_2S_4$. When the electrodeposition of $NiCo_2S_4$ continues based on the grown MnS, the smaller $NiCo_2S_4$ nanosheets can be directly deposited on the larger MnS nanosheets to form a direct contact of the layered interconnect nanosheets, as shown in Figure 3 below.

Journal of Physics: Conference Series **2842** (2024) 012092 doi:10.1088/1742-6596/2842/1/012092

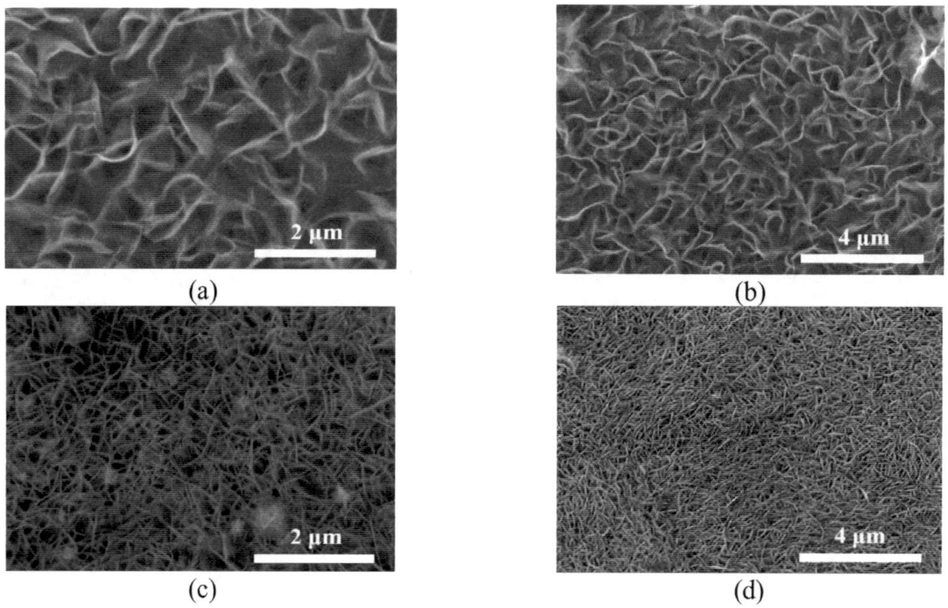

Figure 2. The SEM depicting the surface morphology of MnS and $NiCo_2S_4$: (a) and (b) of MnS; (c) and (d) of $NiCo_2S_4$.

Figures 3 (a)-3 (d) show SEM images of the surface morphology of MnS@$NiCo_2S_4$ composite interconnected nanosheets at enlarged sizes of 10 µm, 5 µm, 2 µm, and 1 µm, respectively. The composite structure presents direct contact and hierarchical interconnect nanosheets, the larger nanosheets structure at the bottom is the MnS prepared by the hydrothermal method first, and the smaller nanosheets on the bottom are $NiCo_2S_4$ after electrochemical deposition. There are more active sites for K^+ embedding because of the larger specific surface area from the layered interconnected nanosheet structure, which improves the specific capacity of electrode material. In addition, compared with other layered nanostructures, this type of interconnect makes the composite structure more tightly connected, consequently enhancing the electrode material's long-term cycling performance [2].

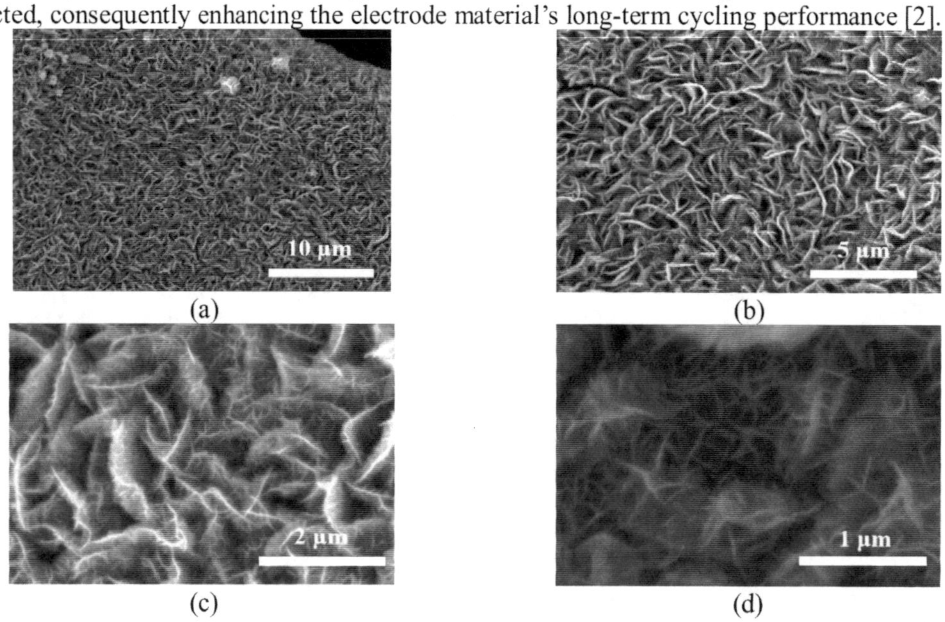

Figure 3. (a) and (d): The SEM of MnS@$NiCo_2S_4$ nanostructure on NF.

Figure 4 illustrates the XRD outcome for the MnS@$NiCo_2S_4$ composite electrode. In the figure, the two peaks with strong signals marked with the five-pointed star symbol correspond to the Nickel Foam base (denoted as "Nickel Foam" in the figure). The diffraction peaks at 34.1°, 49.5°, 54.9°, and 60.5°correspond to different crystal faces of MnS, namely (200), (103), (311), and (222), which are consistent with JCPDS No. 06-0518 of MnS [3]. Peaks at around 31.8°and 37.9°correspond to the (311) and (400) crystal faces of $NiCo_2S_4$ and are consistent with JCPDS No. 20-0782 of $NiCo_2S_4$ [4]. The experiment in this paper successfully prepared a composite MnS@$NiCo_2S_4$ electrode on nickel foam.

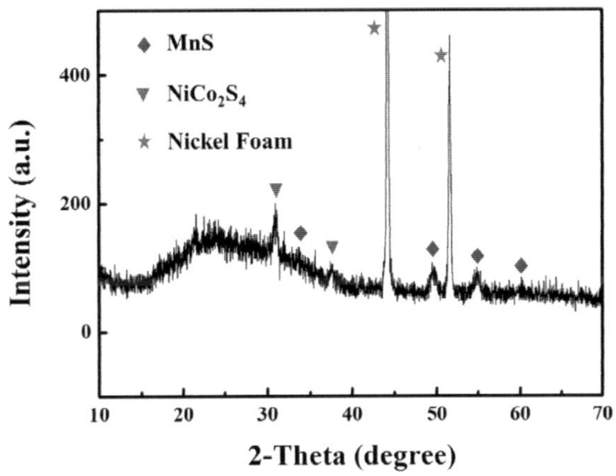

Figure 4. XRD pattern of MnS@$NiCo_2S_4$ on NF.

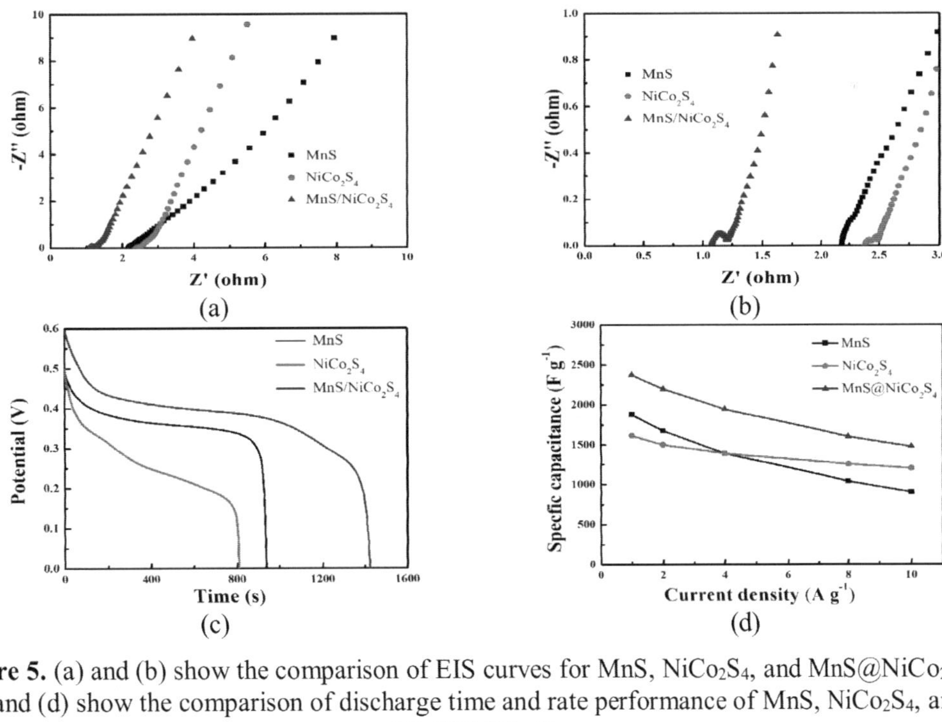

Figure 5. (a) and (b) show the comparison of EIS curves for MnS, $NiCo_2S_4$, and MnS@$NiCo_2S_4$; (c) and (d) show the comparison of discharge time and rate performance of MnS, $NiCo_2S_4$, and MnS@$NiCo_2S_4$.

For the electrochemical characterization, a comparison of the EIS curves of MnS, $NiCo_2S_4$, and MnS@$NiCo_2S_4$ was first conducted in Figures 5 (a) and 5 (b). The equivalent series resistance of

$MnS@NiCo_2S_4$ is smaller than that of MnS and $NiCo_2S_4$, and the slope in the low-frequency region is also larger. Therefore, the composite electrode of $MnS@NiCo_2S_4$ outperforms individual MnS and $NiCo_2S_4$. Figures 5 (c) and 5 (d) show the comparison of discharge time and rate performance of MnS, $NiCo_2S_4$, and $MnS@NiCo_2S_4$ at 1 A g^{-1}. Figure 5 (c) shows the voltage range of the $MnS@NiCo_2S_4$ composite is the highest when the current density remains constant, reaching up to 0.6 V. Additionally, the discharge time for the $MnS@NiCo_2S_4$ composite is also the longest. From the magnification performance in Figure 5 (d), the rate performance of $NiCo_2S_4$ is the best, but its capacity is too low. The initial capacity of MnS is slightly higher than that of $NiCo_2S_4$, however, its rate performance is poor, resulting in a lower capacity at high currents compared to $NiCo_2S_4$. Combining specific capacity and magnification performance, $MnS@NiCo_2S_4$ is the best choice.

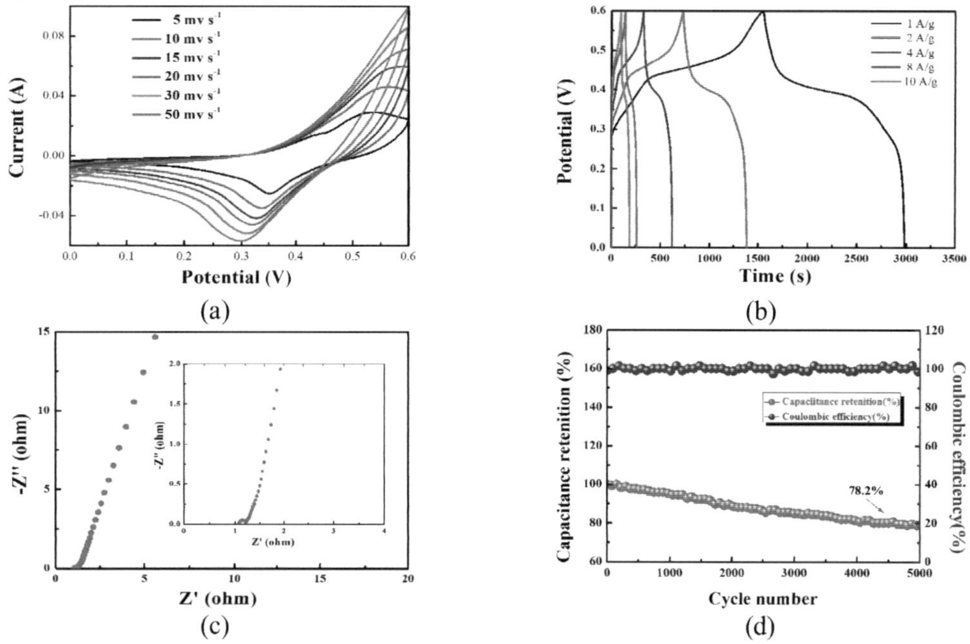

Figure 6. (a) CV curves between 0.0-0.6 V for $MnS@NiCo_2S_4$ on NF; (b) GCD curves at different current densities; (c) Nyquist plot for $MnS@NiCo_2S_4$ on NF; (d) Coulombic efficiency and cycle performance at 4 A g^{-1}.

The $MnS@NiCo_2S_4$ composite self-supported electrode's electrochemical performance test results are illustrated in Figure 6. From Figure 6 (a), the shape of the CV curves between 0.0-0.6 V remains the same basically, showing good reversibility [5]. Simultaneously, it is noteworthy that with an elevation in scan rate, the anode peak moves towards positive potential, and the cathode peak moves towards negative potential. This phenomenon is mainly due to insufficient diffusion of electrolyte ions, which cannot fully penetrate the deep surface of the electrode [6].

As exhibited in Figure 6 (b), corresponding to 1, 2, 4, 8, and 10 A g^{-1}, the specific capacitances obtained were 2, 372.7, 2, 196.3, 1, 943.3, 1, 601.3, and 1, 475.0 F g^{-1} respectively for $MnS@NiCo_2S_4$ composite electrode. This observation highlights good rate performance with a significant amplification factor of approximately 62.1% at 10 A g^{-1}.

The EIS curve for the $MnS@NiCo_2S_4$ is shown in Figure 6 (c), and a featured steep linear trend at low frequency indicates a good capacitance characteristic [7,8]. In the meantime, it is noteworthy that the diameter of ~ 0.12 Ω for the semicircle at medium frequency indicates the small charge transfer impedance for the related electrode, as seen in Figure 6 (c) [9]. Figure 6 (d) shows the Coulombic efficiency and cycle performance for $MnS@NiCo_2S_4$ on NF measured at 4 A g^{-1}. It can maintain 78.2% of initial capacity even after 5, 000 cycles, exhibiting good long-term cycle stability. Additionally, the consistent maintenance of Coulombic efficiency at nearly 100% throughout 5, 000 cycles underscores a highly reversible faradaic process occurring at the interface of the electrode and the electrolyte. The attenuation of capacitance, due to the repeated charge and discharge impact of large current, destroys the nanostructure and leads to the degradation of performance. In the later work, we will continue to strengthen the study of cyclic stability under the premise of trying to ensure the capacity, to improve the overall performance, which is also a work that we are researching hard on at present.

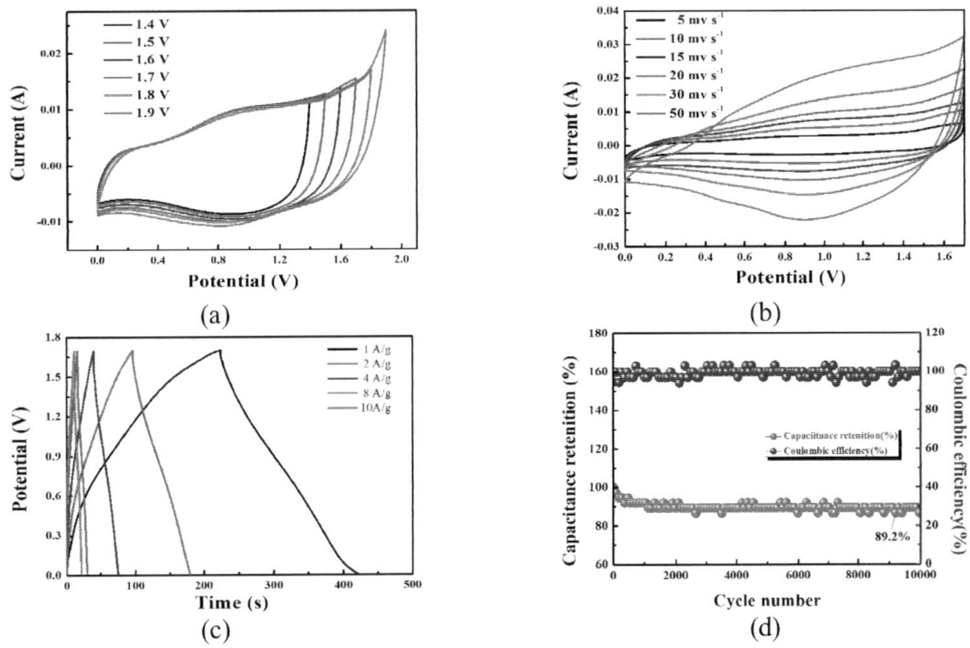

Figure 7. (a) CV curves of the ASCs at 30 mV s^{-1} between 1.4 V-1.9 V; (b) CV curves of the ASCs in 0.0-1.7 V; (c) GCD curves of the ASCs between 0.0-1.7 V; (d) Coulombic efficiency and cycle performance after the 5, 000 cycles at 4 A g^{-1}.

The working potential window, CV scan test, GCD test, cyclic stability, and inventory efficiency of the assembled capacitor were tested and compared respectively, as shown in Figure 7. To determine the optimal operating voltage range, CV tests were conducted over a range of 0-1.4 to 0-1.9 V with a sweep rate of 30 mV s^{-1}, as shown in Figure 7 (a). When the upper cut-off voltage exceeds 1.7 V, such as 1.8 V or 1.9 V, a noticeable upwarping shift is observed in the CV curves. Furthermore, this upwarp trend becomes more pronounced with increasing voltage, indicating an irreversible oxygen evolution is taking place at the electrode-electrolyte interface. Consequently, the optimal working voltage range for this full capacitor is determined to be within 0-1.7 V. Based on this premise, Figure 7 (b) shows the different scan rates of CV curves of the aqueous ASCs in the voltage window of 0-1.7 V, and the CV curve shape stays pretty much the same, revealing good reversibility. GCD curves, depicted in Figure 7 (c), are shown for various current densities within 0-1.7 V. The specific capacities can achieve ~116.9, 98.6, 82.1, 63.1, and 57.6 F g^{-1} at 1.0, 2.0, 4.0, 8.0, and 10.0 A g^{-1}. Furthermore, at respective power density levels of 849.6, 1, 701.0, 3, 405.0, 6, 797.0, and 8, 485.7 W kg^{-1}, the recorded energy density values are correspondingly 46.9, 39.6, 33.0, 25.3, and 23.1 Wh kg^{-1}. Figure 7 (d) shows that the cycling test indicates excellent stability with high capacity retention up to 89.2% and the Coulombic efficiency has been maintained at ~100% even after 10, 000 cycles at 4 A g^{-1}. It shows that the ASCs exhibit excellent electrochemical performance and cycle stability, utilizing a self-supported electrode cathode composed of $MnS@NiCo_2S_4$ composite and a commercial activated carbon/carbon cloth anode. The aforementioned statement has laid a certain experimental and theoretical foundation for further research and practical application in the future.

3. Conclusion

This study presents the development of a high-performance, self-supported supercapacitor electrode, characterized by an interconnected nanosheet structure of MnS@$NiCo_2S_4$ synthesized directly on NF. The fabrication process entails a simple single-step hydrothermal approach, complemented by a subsequent electrochemical treatment, presenting an effective preparation method of excellent electrochemical storage devices. Through meticulous investigation and analysis, it has been revealed that the MnS@$NiCo_2S_4$ material reflects a notable capacitance characteristic of 2, 372.7 F g−1 at 1 A g^{-1}, and even to 10 A g^{-1}, it can also retain a substantial capacity of 1, 475.0 F g^{-1}. Upon subjecting the electrode to 5, 000 GCD cycles at 4 A g^{-1}, the MnS@$NiCo_2S_4$ electrode still retains 78.2% of the specific capacity, showing stable cycling performance. Furthermore, in the case of aqueous asymmetric supercapacitors fabricated with a MnS@$NiCo_2S_4$ cathode and an activated carbon anode, and utilizing a 1 M KOH electrolyte, these devices are capable of achieving an impressive energy density reaching 46.9 Wh kg^{-1} at a power density of 849.6 W kg^{-1}, while also maintaining a capacitance retention rate of 89.2% following 5, 000 cycles under a 4 A g^{-1}. The device has favorable electrochemical performance, and the preparation method of the material is simple and environmentally friendly. In the research and application of aqueous asymmetric supercapacitors, the research furnishes a robust foundation for the advancement of high-performance self-supporting composite electrode materials based on metal sulfides.

References

[1] Wang X., Tian L., Long X., et al. Cracked bark-inspired ternary metallic sulfide ($NiCoMnS_4$) nanostructure on carbon cloth for high-performance aqueous asymmetric supercapacitors [J]. Science China Materials, 1, 632-1, 641 (2021).

[2] Yang H., Guo R., Jin L., et al. Synthesis of microcrystalline $NiCo_2S_4$ nanorods under ambient temperature and pressure for hybrid supercapacitors [J]. Journal of Materials Science, 2, 274 (2023).

[3] Ramachandran R., Saranya M., Grace A. N., et al. MnS nanocomposites based on doped graphene: simple synthesis by a wet chemical route and improved electrochemical properties as an electrode material for supercapacitors [J]. RSC Advances, 7: 2, 249 (2017).

[4] Gheytanzadeh M., Baghban A., et al. Insights into the estimation of capacitance for carbon-based supercapacitors [J]. RSC Advances, 2021, 11 (10): 5, 479-5, 486.

[5] Gao Z., Chen C., Chang J., et al. Enhanced cycleAbility of faradic $CoNi_2S_4$ electrode by reduced graphene oxide coating for efficient asymmetric supercapacitor [J]. Electrochimica Acta, 281: 394-404 (2018).

[6] Rajesh G., Salman S., and Hemen K. Fabrication of highly conductive graphene paper for supercapacitors with a one-step hydrothermal method [J]. Carbon Trends, 100, 317 (2023).

[7] Jing C., Guo X., Xia L., et al. Morphologically confined hybridization of tiny $CoNi_2S_4$ nanosheets into S, P, and Co-doped graphene leading to enhanced pseudocapacitance and rate capability [J]. Chemical Engineering Journal, 379: 122, 305 (2020).

[8] Li H., Xuan H., Guan Y., et al. Preparation and characterization of three-dimensional Mn-Mo-S composites on rGO/Ni foam for battery-supercapacitor electrode with high-performance [J]. Electrochimica Acta, 345: 136, 260 (2020).

[9] Li Q., Lu W., Li Z., et al. Hierarchical MoS_2/$NiCo_2S_4$@C urchin-like hollow microspheres for asymmetric supercapacitors [J]. Chemical Engineering Journal, 380: 122, 544 (2020).

Journal of Physics: Conference Series **2842** (2024) 012093 doi:10.1088/1742-6596/2842/1/012093

Highly sensitive and chemically resistant moisture sensor based on polyether sulfone and polyimide composite membrane for power transformer oil

Bin Zhu[1], Keqi Deng[1], Xiaojuan Li[2], Liguo Wen[2], Hongwei Shen[2] and Dezhi Wu[1*]

[1]Pen-Tung Sah Institute of Micro-Nano Science and Technology, Xiamen University, Xiamen, 361005, China
[2]Beijing Smart-chip Microelectronics Technology Co., Ltd., China

[*]Corresponding author's email:wdz@xmu.edu.cn

Abstract: Moisture is the primary factor leading to the deterioration of transformer oil. Real-time detection of transformer oil moisture content holds significant importance to help power companies monitor the transformer's operational status, track the change curve of the oil moisture content, identify water ingress causes, and implement preventive measures. However, transformer oil has the characteristics of low moisture content and complex chemical composition, which pose a substantial challenge for sensor monitoring, and require the prepared sensor to be chemically resistant and highly sensitive to moisture. To address these challenges, in this paper, a parallel plate oil moisture sensor with high sensitivity and high chemical resistance based on PES and PI was developed using the MEMS process. The sensor incorporates PES as the chemical resistance layer and PI as the moisture sensitivity layer. The parallel plate design not only optimizes sensitivity by effectively utilizing electric field lines but also allows the passage of water while isolating oil molecules, further enhancing the chemical resistance of the sensor. Experimental results in air demonstrate that the prototype sensor exhibits high sensitivity (~1.26 pF/% RH) across the full humidity range of 0-100% RH. Moreover, the sensor can withstand exposure to a wide range of chemical corrosive gases, including acetone, ammonia, etc. Tests in transformer oil reveal that the sensor has a resolution better than 200 ppm and can effectively measure moisture in the range of 0-1400 ppm, demonstrating very high sensitivity (4 pF/100 ppm) and rapid response (10/23 s).

1. Introduction

Power transformers are among the most critical and expensive components within a power system, and the performance of transformers that are in service is closely tied to the quality of transformer oil. Transformer oil has excellent dielectric properties and oxidative stability, which make it ideal for insulation, heat dissipation, cooling, and arc extinguishing in transformers. Moisture is a major contributing factor to the degradation of transformer oil [1]. Due to the high conductivity and high dielectric constant of water, it easily forms conductive bridges under the action of the electric field, thus reducing the breakdown voltage of the oil, which seriously affects the safety of power generation. Therefore, the detection of transformer oil moisture content for power equipment is of great significance for preventing equipment hazards, prolonging operating life, and ensuring the smooth functioning of the unit.

Currently, sensors used for real-time detection of transformer oil moisture can be divided into capacitive [2], resistive [3], fiber-optic [4], and microwave [5]. Microwave sensors indirectly assess water content by detecting the oil's dielectric constant, leading to less accuracy. Fiber-optic sensors require additional equipment such as laser generators and receivers, making integration with transformer chassis challenging and more costly. Capacitive sensors generally offer a broader range of humidity detection compared to resistive sensors [6]. However, sensor detection faces significant challenges in special oil environments containing corrosive impurities and low water content, requiring sensors to be chemically resistant and highly sensitive to moisture. Porous alumina is a common humidity-sensitive material that is often used for oil moisture detection due to superior low moisture sensitivity [7]. Nevertheless, the lack of selectivity may lead to interference from impurities and compounds present in the oil, resulting in inaccurate readings and reduced sensor lifespan. Polymer-based oil moisture sensors can be designed with high sensitivity solely to moisture and offer exceptional chemical resistance, making them more suitable for detecting trace amounts of moisture in oil fluids [8]. Kondalkar et al. developed an interdigital capacitance sensor with a concave surface based on polyimide to improve the sensitivity [9], which can reach up to 0.9 pF/% RH. However, the open surface design exposes the sensitive film to various additives and impurities present in the oil, making it susceptible to corrosion and denaturation over extended periods. Moreover, other impurity molecules in the oil may infiltrate the sensitive film, resulting in capacitance changes and thus reduced detection accuracy. To address this issue, Yoshii et al. attempted to use chemical dealloying (Cu-Cr alloy) to create chromium nanoporous electrodes on polyimide surfaces for water filtering purposes [10]. However, the de-alloyed chromium porous electrodes lack electrical conductivity, necessitating the additional sputtering of an Au layer, further increasing the complexity and cost of the process. Consequently, the preparation of an oil moisture sensor with high chemical resistance and high selectivity for moisture to meet the long-term online testing needs of power transformer oil remains a major challenge.

In this study, we present a parallel plate oil moisture sensor with high sensitivity and chemical resistance based on polyether sulfone (PES) and polyimide (PI). PES serves as the chemically resistant layer, enhancing the sensor's ability to withstand harsh environments, while PI acts as the moisture-sensitive layer, boosting its sensitivity. In addition, the parallel-plate design allows for more concentrated electric field lines for higher sensitivity than the commonly used interdigital structure. The nanoscale cracks in the upper electrode allow water molecules to move in and out freely and isolate most of the impurity molecules in the oil, thus improving the accuracy and chemical resistance of the sensor. Experimental results demonstrate that the prototype sensor exhibits a remarkable sensitivity of 1.26 pF/% RH in an air environment ranging from 0 to 100% RH, with a rapid response and release time of 3/15 s. Moreover, the sensor can withstand exposure to a wide range of chemically corrosive gases. Tests in transformer oil reveal that the sensor exhibits a high sensitivity of 4 pF/100 ppm in oil environments ranging from 0 to 1400 ppm and temperature suitability at 25 and 50°C.

2. Results and discussion

2.1 Sensing mechanism and fabrication process of the moisture sensor

Figure 1a illustrates the structure of our humidity sensor, which consists of an alumina substrate, a tantalum bottom electrode, a polyimide film, a polyether sulfone film, and a multi-cracked chromium top electrode. The sensor adopts a "sandwich" structure, concentrating electric field lines to enhance sensitivity while allowing water molecules to move in and out freely and isolating the oil to protect the polymer layer. The capacitance change in the sensor primarily stems from variations in the thickness and dielectric constant of the polymer film after water absorption (Figures 1b and c). The sensitive film swells after absorbing water, resulting in an increase in thickness and a decrease in capacitance, but since the dielectric constant of moisture (~80) is much higher than that of polymer (PES: ~3.4, PI: ~3.7), the capacitance of the sensor grows considerably after absorbing water and ultimately shows an increase in capacitance with increasing ambient humidity.

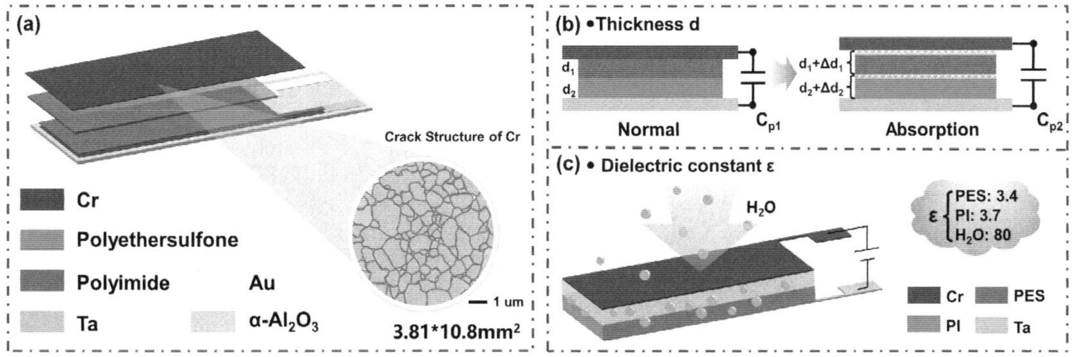

Figure 1. The structure diagram (a) and sensing mechanism ((b) and (c)) of the parallel plate humidity sensor based on PES and PI composite membranes.

Figure 2 demonstrates the humidity sensitivity principle of polymer films. In the first stage of adsorption, water molecules combine with sulfonyl and ether bonds in PES to form the first chemisorption layer (Figure 2a), and water molecules are not able to move freely due to the restriction of the two hydrogen bonds. As more water is added, additional water molecules join with the first chemisorbed layer through hydrogen bonding to produce more surface hydroxyl groups [11], forming the first physisorbed layer. With continued water condensation on the polymer surface, the physisorbed layer grows more and more (Figure 2b), and the water molecules become mobile [12], allowing them to continue to pass downward to the PI layer to combine the carbonyl group and the ether bond in the polyimide, creating the second stage chemisorption layer (Figure 2c). In high humidity environments, as more and more water molecules pass downward, the physisorption layer containing more water molecules in the polyimide greatly increases the capacitance value of the sensor. During the desorption stage (Figure 2d), the water molecules in the physisorbed layer are the first to absorb energy to break the chemical bonds leading to volatilization [13].

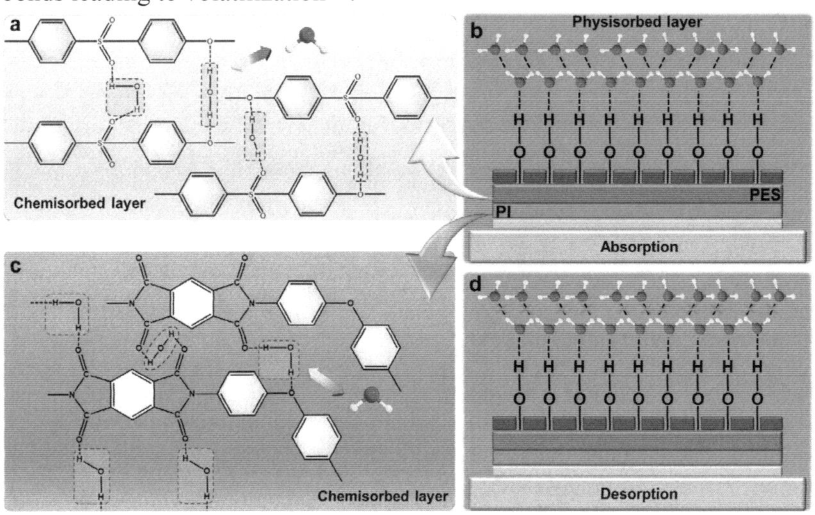

Figure 2. The moisture sensitivity principle of polymers. Schematic diagram of the water molecules adsorption (a, b, and c) and desorption (d) of the parallel plate humidity sensor based on PES and PI composite membranes.

Based on the sensor structure illustrated in Figure 1, we designed the process route for its preparation, which involved several crucial steps: (i) alumina substrate cleaning; (ii) deposition of a tantalum film as the bottom electrode; (iii) spin-coating of a polyimide precursor solution on the patterned bottom electrode, followed by thermal imidization according to a specific temperature profile; (iv) patterning of the polyimide through dry etching using photoresist as a mask; (v) obtaining the patterned right electrode through the lift-off process; (vi) spin-coating of the polyether sulfone solution, followed by solvent evaporation in a vacuum-heated environment to form polyether sulfone films; (vii) patterning of the polyether sulfone using an Au layer as the mask and NMP as the etchant; (viii) deposition of a chromium film with nano-cracks as the top electrode; (ix) a top electrode is patterned on the polyether sulfone surface using the lift-off process.

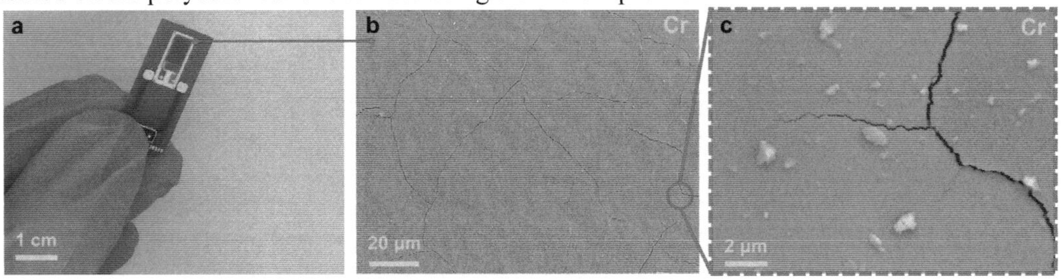

Figure 3. The surface morphology of each film layer of the sensor.

2.2 The picture of the moisture sensor and the mechanism of electrode crack generation

Figure 3a displays the general morphology of our humidity sensor, where the white core in the center represents the humidity sensing unit, and the green part denotes the PCB board designed to accommodate the sensor chip and facilitate lead connections. Figure 3b exhibits the multi-crack morphology of the top chromium electrode, revealing prevalent nano-cracks on the surface of Cr. An enlarged image in Figure 3c shows that the cracks have a width of approximately 500 nm and a length exceeding ten microns. The surface cracks observed on the top chromium electrode of the sensor can be attributed to several factors: (1) Residual stress effect. During the sputtering process, the deposited chromium atoms fall on the polymer surface and form a thin film. However, due to the difference in coefficients of thermal expansion between the chromium film and the polymer substrate, there are residual stresses in the chromium film [14], and the accumulation of these stresses leads to the formation of cracks [15]. (2) Adhesion. Poor adhesion between the chromium film and the polymer substrate also contributes to crack formation [16]. (3) Tolerance of temperature limitations of polymers. Metal films sputtered on silicon or silica surfaces are usually subjected to a thermal annealing process to optimize the atomic arrangement and release residual stresses, but the tolerance temperature of polymers limits the ability to thermally anneal metal films sputtered on polymer surfaces.

2.3 Sensing performance in the air of the designed sensor

The sensor was first tested for air humidity performance before the oil moisture test. Figure 4a shows the capacitance response of the PES and PI-based parallel plate humidity sensor to humidity variations in the range of 0-100% RH. The results demonstrate that the prototype sensor exhibits excellent sensitivity (~1.26 pF/% RH) and linear response (R^2 > 0.9787). The normalized variation of capacitance for the three sensor samples is shown in Figure 4b, revealing good agreement between the different samples, which is attributed to the high consistency of the experimental conditions between the different cells of the same wafer during the preparation process. The sensor exhibits a relatively large hysteresis at 60% RH humidity (Figure 4c) due to its double organic matter film structure. Water molecules bind to the two films to form a chemisorption layer and a physisorption layer, respectively. During desorption, the physisorption layer in the PES is the first to desorb, resulting in a lower capacitance and a fast response (Figure 4e), but this also leads to a reduced pathway for water molecules to move through the PES layer, necessitating the re-establishment of the pathway to carry out desorption of moisture in the PI. Additionally, water molecules in the chemisorption layer of the two polymer layers are difficult to desorb due to the limitation of double hydrogen bonding, leading to hysteresis in the sensor. The focus of our future work is to build a pathway between the two layers to greatly reduce hysteresis. Figure 4d shows the sensitivity of the sensor at different temperatures (25, 50, and 75°C). As the temperature increases, the thermal movement of molecules accelerates, resulting in stronger polarization and an increased capacitance value in the humidity sensor. Figure 4f shows the sensitivity of the sensor at different excitation frequencies (1, 10, 100 kHz) in response to changes in humidity. The capacitance value varies approximately linearly with the moisture concentration, with greater capacitance changes observed at low-frequency signals compared to high-frequency signals. This is because, at high frequencies, the polar water molecules lag rapid changes in the electric field, resulting in their inability to produce an effective change in dielectric constant.

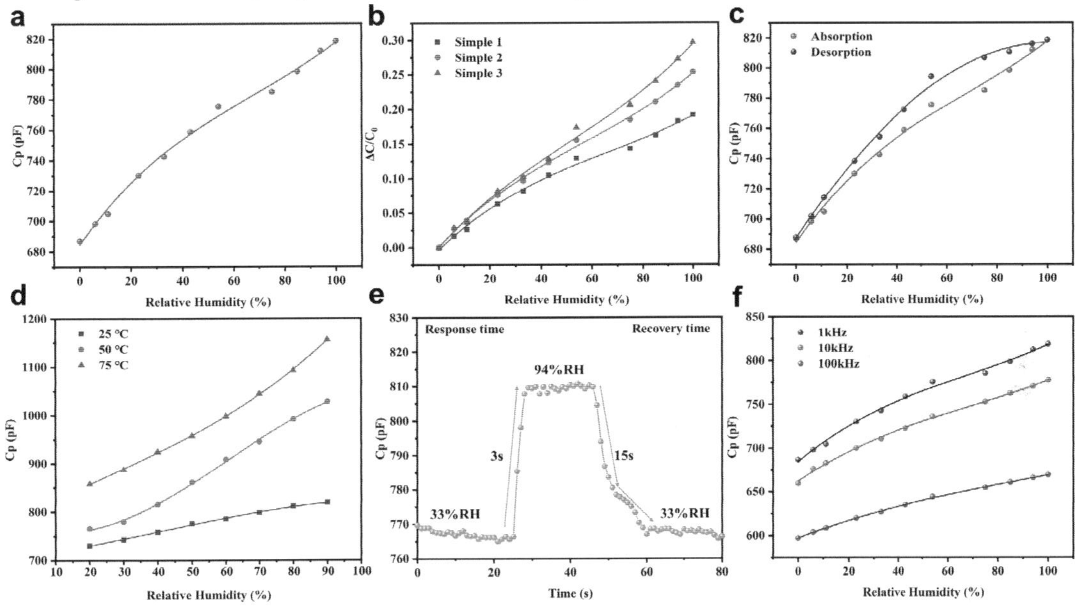

Figure 4. Sensing performance in the air of the parallel plate humidity sensor based on PES and PI composite membranes. (a) Sensitivity and linearity of the humidity sensor. (b) Normalized change of capacitance values from 0 to 100% RH for different samples. (c) Absorption and desorption process of the humidity sensor. (d) The humidity response curve of the sensor at 25, 50, 75°C. (e) Response and recovery time from 33 to 94% RH. (f) Capacitive response of the sensor at different frequencies.

2.4 Chemical resistance performance of the designed sensor

The chemical resistance of the sensor is crucial to ensuring its long-term operation in transformer oil. To evaluate the chemical resistance of our designed MEMS sensor, we devised an experimental setup, as depicted in Figure 5f. The setup comprises two main bottles: one acts as a gas generator, and the other serves as a test chamber. The gas generator is responsible for introducing the volatile liquid to be tested, while the test chamber accommodates the sensor under test and the corresponding chemical gas desiccant. At the beginning of the test, the sensor is exposed to the air for approximately 300 seconds to stabilize. Subsequently, it is placed in a closed test chamber where the desiccant absorbs moisture, resulting in a gradually decreasing humidity level within the chamber and a corresponding decline in the sensor capacitance. Once the sensor capacitance value stabilizes, the test chamber is connected to a gas generator, and the volatile gas to be tested is introduced. For less volatile liquids, such as ethanol, DMAC, and water, a water bath is used to promote evaporation. Figure 5 shows the results of the sensor's chemical resistance test. The sensor's capacitance value remains largely unaffected after the introduction of organic chemical gases, while it exhibits a strong response to the introduction of water vapor. These results demonstrate that our MEMS sensors exhibit excellent tolerance and resistance to chemically corrosive VOC gases while maintaining high selectivity for moisture. This suggests their potential for applications in harsh chemical environments and for detecting the humidity of specific chemical gases.

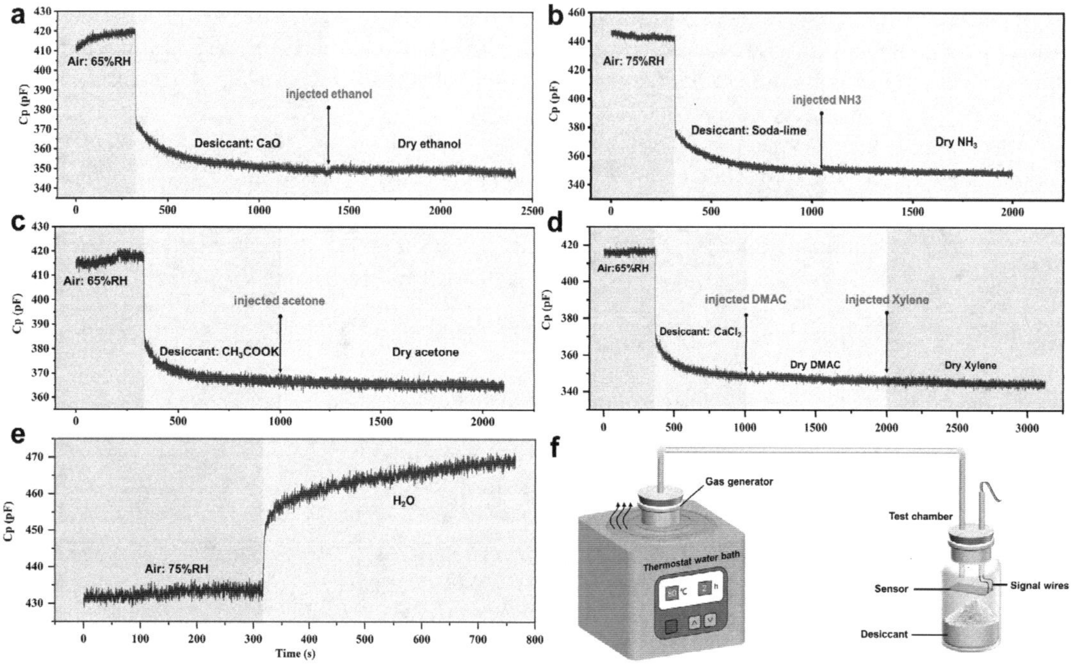

Figure 5. Chemical resistance performance of the sensor. The response of the sensor to several chemical gases includes ethanol, ammonia, acetone, dimethylacetamide, and xylene.

2.5 *Sensing performance in transformer oil of designed sensor*

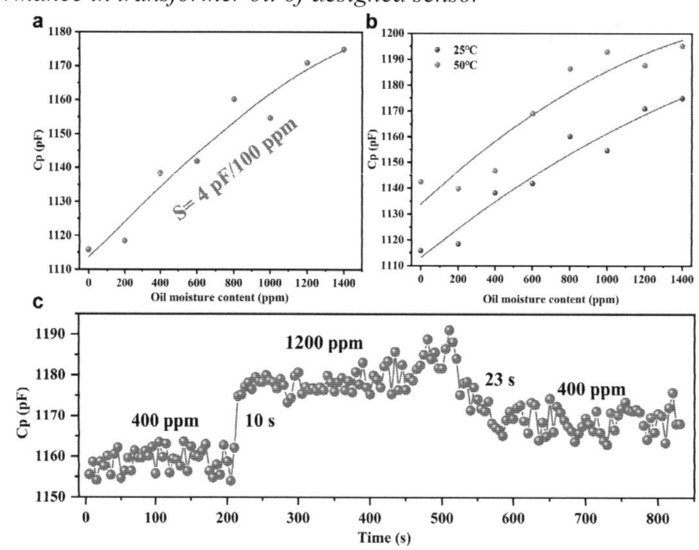

Figure 6. Sensing performance of our MEMS sensor in transformer oil. (a) The response of the sensor to different oil moisture. (b) The moisture response curve of the sensor at 25, 50°C. (c) Response time and recovery time from 400 ppm to 1200 ppm.

The performance of the sensor was evaluated in transformer oil (Mobilect^{TM}25). The sensor was immersed in a flask of oil with a fixed moisture content, and the flask was sealed with a rubber stopper to prevent interference from external water vapor. To ensure uniform distribution of moisture in the oil, the flask was placed on a magnetic stirrer with a heating function, and the required temperature environment for the test was provided by a heated water bath. Transformer oils typically contain water in the parts per million (ppm) range due to the immiscibility of oil and water, necessitating high sensitivity in the sensor. Figure 6a presents the humidity response of the designed sensor in the range of 0-1400 ppm, demonstrating that the sensor's capacitance increases as the water content in the flask rises. The sensor exhibits a sensitivity of 4 pF/100 ppm and a resolution better than 200 ppm, indicating significant responsiveness in detecting water in transformer oil. Additionally, the sensor was used to measure moisture in transformer oil at different temperatures (Figure 6b). The experimental results showed good agreement at low temperatures (< 50°C). As the temperature increases, the thermal movement of water molecules accelerates, leading to an increase in the capacitance of the sensor. Remarkably, the slope of the capacitance curve remained consistent, ensuring reliable measurements. Figure 6c provides the response and recovery times (10/23 s) of the sensors at room temperature from 400 ppm to 1200 ppm. The short response time facilitates timely assessment of the transformer's condition by the power companies and enables prompt maintenance action. In conclusion, the designed moisture sensors exhibit high sensitivity in detecting moisture in transformer oil, which is expected to enhance the reliability and efficiency of transformer maintenance and operation.

3. Conclusions

In this study, we developed a highly sensitive and chemically resistant parallel plate humidity sensor based on PES and PI composite membranes using the MEMS process. The prototype sensor exhibits excellent performance in an air environment with high sensitivity (~1.26 pF/% RH) and linear response (R^2 > 0.9787) over the full humidity range of 0-100% RH. Furthermore, the sensor displayed remarkable resistance to a wide range of chemically corrosive gases, including ethanol, acetone, ammonia, DMAC, and xylene, confirming its reliable chemical resistance. In addition, test results in transformer oil show that the sensor has a resolution better than 200 ppm and can effectively measure moisture in the range of 0-1400 ppm, exhibiting extremely high sensitivity (4 pF/100 ppm) and rapid response (10/23 s). In conclusion, the parallel plate humidity sensor based on PES and PI composite membranes, with its high sensitivity and chemical resistance, is well-suited for online monitoring of transformer oil. This capability holds significant value for power companies as it allows for continuous evaluation of transformer oil quality, accurate prediction of transformer service life, and timely replacement of transformers when necessary.

Acknowledgements

This work is supported by The Laboratory Open Fund of Beijing Smart-chip Microelectronics Technology Co., Ltd. (No. SGSC0000MNQT2207234).

References

[1] D.K. Mahanta, "Green transformer oil: A review," 2020 IEEE International Conference on Environment and Electrical Engineering and 2020 IEEE Industrial and Commercial Power Systems Europe (EEEIC/I&CPS Europe), IEEE2020, pp. 1-6.
[2] O. Rahman, T. Islam, N. Khera, S.A. Khan, "A novel application of the cross-capacitive sensor in real-time condition monitoring of transformer oil," IEEE Transactions on Instrumentation and Measurement 70, 1-12 (2021).
[3] M.R. Adib, Y. Lee, V.V. Kondalkar, S. Kim, K. Lee, "A highly sensitive and stable rGO: MoS2-based chemo resistive humidity sensor directly insertable to transformer insulating oil analyzed by customized electronic sensor interface," ACS sensors 6, 1012-1021 (2021).
[4] S.F.A.Z. Yusoff, M. Mezher, I.S. Amiri, N. Ayyanar, D. Vigneswaran, H. Ahmad, et al., "Detection of moisture content in transformer oil using platinum coated on D-shaped optical fiber," Optical Fiber Technology 45, 115-121 (2018).
[5] P. Sharma, L. Lao, G. Falcone, "A microwave cavity resonator sensor for water-in-oil measurements," Sensors and Actuators B: Chemical 262, 200-210 (2018).

[6] T. Islam, Z. Tehseen, L. Kumar, "Highly sensitive thin-film capacitive sensor for online moisture measurement in transformer oil," IET Science, Measurement & Technology 14, 416-22 (2020).
[7] S. Kumar, K.K. Raina, T. Islam, "Anodic aluminum oxide-based humidity sensor for online moisture monitoring of power transformer," Sensors and actuators B: Chemical 329, 128908 (2021).
[8] X.-g. Chen, W.-g. Chen, L.-l. Gu, "Research of on-line monitoring of moisture content in transformer oil," 2008 China International Conference on Electricity Distribution, IEEE2008, pp. 1-4.
[9] V.V. Kondalkar, G. Ryu, Y. Lee, K. Lee, "Development of highly sensitive and stable humidity sensor for real-time monitoring of dissolved moisture in transformer-insulating oil," Sensors and Actuators B: Chemical 286, 377-85 (2019).
[10] Y. Yoshii, J. Sakurai, M. Mizo Shiri, S. Hata, "Fabrication of a Novel Nanoporous Film via Chemical Dealloying of a Cu–Cr Alloy for Sensing Moisture in Oil," Journal of Microelectromechanical Systems 28, 279-89 (2019).
[11] Z. Chen, C. Lu, "Humidity sensors: a review of materials and mechanisms," Sensor letters 3, 274-95 (2005).
[12] H. Yu, J.-G. Liang, C. Wang, C.-C. Liu, B. Bai, F.-Y. Meng, et al., "Target properties optimization on capacitive-type humidity sensor: Ingredients hybrid and integrated passive devices fabrication," Sensors and Actuators B: Chemical 340, 129883 (2021).
[13] G. Chen, R. Guan, M. Shi, X. Dai, H. Li, N. Zhou, et al., "A nanoforest-based humidity sensor for respiration monitoring," Microsystems & Nanoengineering 8, 44 (2022).
[14] A. Castrup, C. Kübel, T. Scherer, H. Hahn, "Microstructure and residual stress of magnetron sputtered nanocrystalline palladium and palladium gold films on polymer substrates," Journal of Vacuum Science & Technology A, 29 (2011).
[15] A.J. Detor, A.M. Hodge, E. Chason, Y. Wang, H. Xu, M. Conyers, et al., "Stress and microstructure evolution in thick sputtered films," Acta materialia 57, 2055-65 (2009).
[16] T. Oh, S. Kowalczyk, D. Hunt, J. Kim, "Adhesion enhancement of Cr/polyimide interfaces using RF pre-sputtering of polyimide surfaces," Journal of adhesion science and technology 4, 119-29 (1990).

Application of optical homogeneity detection technology in glass quality control based on visual inspection technology

Xiaoyu Lun* **and Hui Wang**

Beijing XuHui Xinrui Technology Co., Ltd., Beijing,100150, China

*Corresponding author's e-mail: littlerain011@163.com

Abstract. Glass is an advanced material with a wide range of applications. Improving the quality of glass products supports and promotes various industries. This paper discusses the streaks affecting the quality of glass products and focuses on implementing optical homogeneity detection technology for glass. Based on the relationship between optical homogeneity and mechanical properties, this detection technology analyzes the causes of streaks in the production process of glass products and deduces the causes for the streak characteristics through the discussion of the working principle of the homogeneity detection technology, the detection device, the detection steps and the use of technology. In this study, visual inspection technology is introduced into the optical homogeneity inspection technology instead of the traditional manual inspection method, which can make the inspection results faster and more accurate and can find a quick and effective solution to the streaks.

1. Introduction

As an advanced material, the application of glass has penetrated all corners of modern society, showing its excellent performance and wide range of uses. In the field of communication, the development of fiber optic communication technology cannot be separated from the support of glass materials. With its high-speed, high-capacity, and low-loss transmission characteristics, optical fiber has become an indispensable part of modern communication networks. In the display field, glass is the key material for manufacturing mainstream displays such as liquid crystal displays (LCD) and organic light-emitting diode displays (OLED). Whether it is a TV, computer, or cell phone, it is indispensable to support the glass substrate. Especially in the field of cell phones, with the emergence of new cell phone designs such as full-screen and folding screens, the requirements for glass materials are getting higher and higher. In the field of new energy, glass also plays a vital role. The primary material for solar panels is glass, which can withstand various climatic conditions and allow sunlight to pass through to generate electricity. In addition, glass has a wide range of applications in new energy sources such as wind and nuclear energy. In life and health, glass also plays an important role. For example, neutral borosilicate glass tubes for vaccines are vital for ensuring vaccine safety and stability. Meanwhile, glass is also widely used in medical equipment and biological laboratories for its easy-to-clean and corrosion-resistant properties. In aerospace, glass is used to manufacture windows and observation windows for airplanes, rockets, and other vehicles because of its high strength and high-temperature resistance.

Glass is an amorphous solid substance having a statistically homogenous organization of atoms in space but an uneven arrangement of plasmas. An ideal homogenous glass is isotropic because its physicochemical characteristics (such as its refractive index, hardness, modulus of elasticity, coefficient of thermal expansion, thermal conductivity, electrical conductivity, etc.) are the same in all directions [1].

In the actual production process, glass undergoes batching and melting. The purpose of the batching is to achieve the desired chemical composition by mixing to achieve physical homogeneity. Glass melting is done to attain chemical homogenization of molten glass. Glass melting is to accomplish the chemical homogenization of molten glass, and the homogenization degree of molten glass directly affects the quality of glass products. In the process of glass melting, after the batching is wholly melted, there is often a glass body with a different chemical composition from the primary glass liquid in the molten glass, so the glass streaks defect is generated due to the existence of glass streaks. There are differences in optical and thermal properties between the central glass and the heterogeneous vitreous, the thermal properties are mainly manifested in the difference in the expansion coefficient, and the structural stress will be generated on the interface layer of the two; Optical properties are primarily manifested in the difference in refractive index, which will produce optical distortion and birefringence.

2. Effect of glass streaks on mechanical properties

2.1. Glass streaks

Glass streaks are a defect that is not easily observed. Their composition and properties are different from those of the main glass. Glass streaks lead to glass inhomogeneity [2], which affects its mechanical strength. Therefore, it is necessary to pay attention to and evaluate the influence of glass streaks on glass products, determine the causes of glass streaks, and overcome and solve them.

There are three basic types of glass streaks in glass products. (1) One is the evenly distributed "thin" streaks with relatively little stress; (2) Evenly distributed or localized "coarse" streaks and around the streaks are areas of high stress; (3) The surface of the glass is a layer of glass with a chemical composition that is very different from the matrix glass. The density is greater than the matrix glass's, forming a "wrapping effect" or "outer layer effect," resulting in severe tensile stress and very fragile glass products [3].

For glass products, the occurrence of tensile stress is the most dangerous. Tensile stress can cause the strength of the glass to weaken considerably. Depending on the location of the tensile stress, the degree of influence is also different.

2.2. Effect of glass streaks on mechanical properties

The glass streaks will affect the uniformity and thermal uniformity of the chemical composition of the glass, and the difference in the composition of the glass leads to the difference in the density of each particle in the glass, which affects the magnitude of its internal stress. The chemical uniformity of glass is affected by the quality of the batching, the melting quality, and the preparation process, and the thermal uniformity is affected by the melting, forming, annealing, and thermal processing of the glass [4]. When the glass products are transported and filled, they will bear the mechanical impact, and there is significant stress inside the heterogeneous glass. Under the frequent action of external forces, the stress exceeds the damage limit of the glass material, eventually damaging the glass products.

The unevenness of the chemical composition of the glass determines the difference in the expansion coefficient inside the glass. Under the condition of alternating heat, the thermal shock resistance of the glass will be affected, and the larger the thickness of the glass product is, the smaller the temperature difference that can withstand the sudden change (i.e., the thermal shock stability deteriorates). Compressive stress occurs on its surface when it is heated, tensile stress occurs on its surface when cooled, and the compressive strength of glass is ten times greater than tensile strength. Under the condition of rapid temperature change along the thickness direction of the glass product, there is a different expansion from the surface to the inside, which produces significant stress. When the stress exceeds the ultimate strength of the glass, it will cause the glass product to crack.

Journal of Physics: Conference Series **2842** (2024) 012094 doi:10.1088/1742-6596/2842/1/012094

3. Optical homogeneity detection technology

Optical uniformity testing technology is a proven method for detecting glass streaks.

3.1 Principle

The optical birefringence method is the principle used in the optical homogeneity detection technology. Optical birefringence is a method in which a parallel beam of polarized light is incident perpendicularly on a non-homogeneous glass with stresses and streaks, and the polarized light is broken up into two beams with mutually perpendicular vibration directions, resulting in the optical path difference. The optical path difference is related to the length of light passing through the sample (i.e., the height of the sample). By inserting a full-wave compensator (with an optical path difference of 565 nm) between the orthogonal polarized microscope optical paths, the optical path difference produced by the polarized light can be converted into stresses and streaks in the glass. The stresses and streaks will show different interference colors and interference streaks. The glass homogeneity rating is based on the interference color and the color of the interference streaks, as well as the morphology, position, and thickness of the interference streaks.

3.2 Detection step

3.2.1 Sample preparation. Glass product without defects is selected and cut into glass pieces of length (40 mm ± 1 mm) × width (10 mm ± 1 mm) × thickness (10 ± 1 mm). Afterward, the cut surface of the glass sample is ground and polished so that the cut surface of the glass sample has good light transmission.

3.2.2 Preparation of immersion liquids. This step requires that the refractive index of the glass sample be measured first. We select a liquid with a known refractive index, two or more liquids that are soluble in each other, and a liquid with the same refractive index is mixed as the glass sample. It has been verified that the immersion liquids can also be replaced by purified water, which has little effect on the test results [5].

3.2.3 Optical homogeneity detection

3.2.3.1 Sample detection. A trinocular polarizing microscope configured with a color digital camera is connected to a computer and placed in the optical path of the trinocular polarizing microscope according to the positional relationship of the full-wave slice, bias-starting slice, and bias-detecting slice, as shown in Figure 1 so that the background interference color of the field-of-view is violet-red. The glass sample is placed horizontally in a stress-free glass Petri dish, injected with the immersion liquids, and submerged in the upper part of the glass samples by 1 mm to 2 mm.

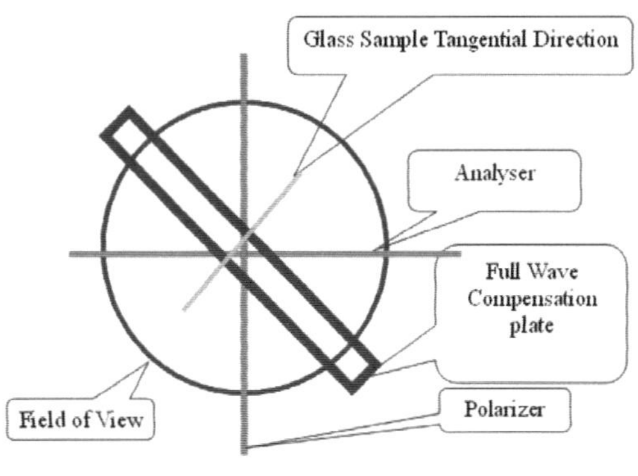

Figure 1. Optical homogeneity detection optical path system diagram.

Journal of Physics: Conference Series **2842** (2024) 012094 doi:10.1088/1742-6596/2842/1/012094

They determine the presence of stresses within a glass sample based on the interference color. The tensile stress interference color is blue-green, and the compressive stress interference color is orange-yellow. Interference streaks or bands of interference streaks are formed when inhomogeneities exist within the cutting surface.

After the glass sample is examined as a whole, the most severe parts of the interference colors, interference streaks, or interference streak bands within it are selected for image acquisition [6].

3.2.3.2 Sample Evaluation. The traditional testing method uses the human eye to view the glass sample as a whole. The most severe glass places stress, and streaks, and the appropriate rating level is chosen against the standard atlas to judge whether the glass samples are qualified, as shown in Table 1. Based on the different states of the stresses and streaks in the glass samples, we analyze the causes of these stresses and streaks and adjust the production process accordingly to improve the glass product quality [5].

Table 1. Grading Criteria for Optical Homogeneity of Glass Samples.

Grade	Homogeneity Assessment	Image	Optical Path Difference (nm/cm)	Part Quality
A	The interference color is purplish red. There is no obvious thermal stress, few light interference streaks bands, no visible stress stripes, and no tensile stress on the outer surface.		≤30	Superior products
B	The interference color is red or orange-red. The interference streaks bands need to be examined carefully to see the parallel thin streaks and uniform distribution. There are a small number of light yellow, orange-yellow compressive stress streaks and a small number of small blue tensile stress streaks. The outer surface has no tensile stress.		≤120	Good products
C	The interference colors are pale yellow and blue-green, with localized broader ill-defined stress interference bands. Localized parallel pinstripes are clear and allow for less severe V-streaks, well-distributed fine and moderate orchid-green tensile stresses, and bright yellow-white compressive stress streaks. The outer surface has no tensile stress.		≥125	Accepted product
D	The interference color is yellow-gray and gray-green, parallel streaks are thick and severe, V-shaped streaks are many, local sharp streaks area is very irregular, bright lines are more, there is outer edge tension, there are apparent blue-green and light green-yellow stress streaks, and the outer surface has tension stress.		≥300	unaccepted product
E	Interference colors are grey and grey-blue, with an" outer effect" or "overlay effect," there are many local sharp streaks, noticeable blue-green and yellow-green stress-heavy streaks, and tensile stresses on the outer surface.	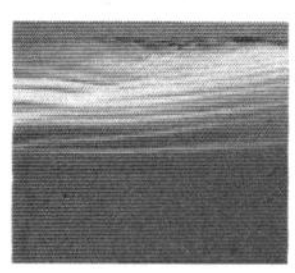	≥400	Defective goods

Journal of Physics: Conference Series **2842** (2024) 012094 doi:10.1088/1742-6596/2842/1/012094

4. Application of optical homogeneity detection technology

4.1 Streaks causing breakage of product

In the glass production process, the increase in the number of glass streaks is a warning signal. Certain uncontrolled aspects of glass production must be verified timely, and then the necessary measures must be taken to keep the condition and number of streaks in the glass within reasonable limits.

Many glass breakages are sometimes encountered in producing, using, and transporting glass products. When these situations occur, the condition of the testing instruments and the limited knowledge of the enterprise can lead to the production and technical staff taking inappropriate measures, resulting in a long time to restore normalcy. Table 2 shows the breakage of glass products in relation to the condition of glass streaks and stress distribution.

Table 2. Relationship Between Breakage and Glass Streak and Stress Distribution Condition of Glass Products.

Serial No.	Breakage of Glass Products	Glass Streaks and Stress Conditions
1#	Immediately after the medication is filled, the glass products have a significant breakage.	Streaks surround the outer surface of the cut surface, and the entire outer layer of glass is under high tensile stress.
2#	During the glass decoration firing process, there is a lot of breakage.	Localized clusters of coarse streaks are present on the cut surface of the glass, spreading from the inner to the outer surface and accompanied by high tensile stresses.
3#	Glass products show a 10 to 30% breakage rate after 4 to 6 hours of annealing.	A localized concentration of reticulated cracks on the outer surface of the cut glass creates a localized band of tensile stress.
4#	Glass breakage begins at the site of scissor marks and small stones.	The rupture of glass products occurs in the streak zone, where the alumina content is much higher than that of ordinary glass, which is yellowish-green in color and has a state of high tensile stress.
5#	Glass products exceed 50% breakage after 12 hours.	The reason for the large number of streaks in the glass is the mixing of raw materials in the silo.
6#	Glass products have a breakage rate of more than 1% during transportation.	A vast blue tensile stress band exists in the center of the glass cutting surface, and microcracks spread due to frequent vibration of transportation and surface abrasion.
7#	The broken areas of the glass products appear thickly streaked and are amber in color.	Severe corrosion of the refractory material results in the mixing of aluminum and silica components into the glass melt, causing large fluctuations in the chemical composition of the glass.

4.2 Causes of glass streaks

Corrective measures are necessary to address the causes of streaks in glass products, and there are several main reasons for glass streaks [7], as shown in Figure 2.

Journal of Physics: Conference Series **2842** (2024) 012094 doi:10.1088/1742-6596/2842/1/012094

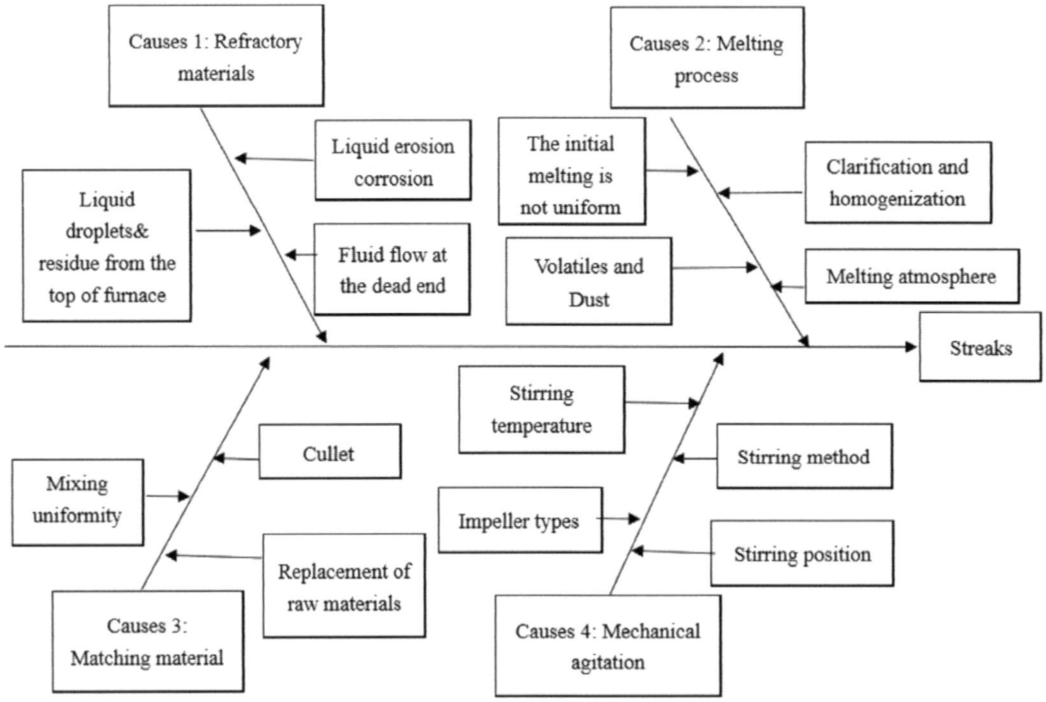

Figure 2. Causes of streaks.

4.2.1 Streaks caused by the unreasonable granularity of glass raw Materials and Improper Handling of the Batch. We Check that the glass's raw material granularity is within the standard range. If the feldspar or quartz sand raw material is too coarse, the glass will have insufficiently melted transparent quartz particles, which will be subject to molding and other mechanical effects and show similar tadpole-like streaks, as shown in Figure 3.

Figure 3. Tadpole-like streaks caused by the unreasonable granularity of raw materials.

4.2.2 Streaks caused by errors in the chemical composition of the glass

(1) Dosing system weighing error.

(2) Raw material is thrown into the wrong filling bin.

(3) Leakage of dosage during conveying and weighing processes.

Streaks caused by faulty mixed batches can cause serious safety problems in glass products. Such streaks exhibit features with multiple layers of varying intensities, generally swirling and extending widely through the sample. Tensile and compressive stress streaks are possible, depending on the severity of the dosing error, as shown in Figure 4.

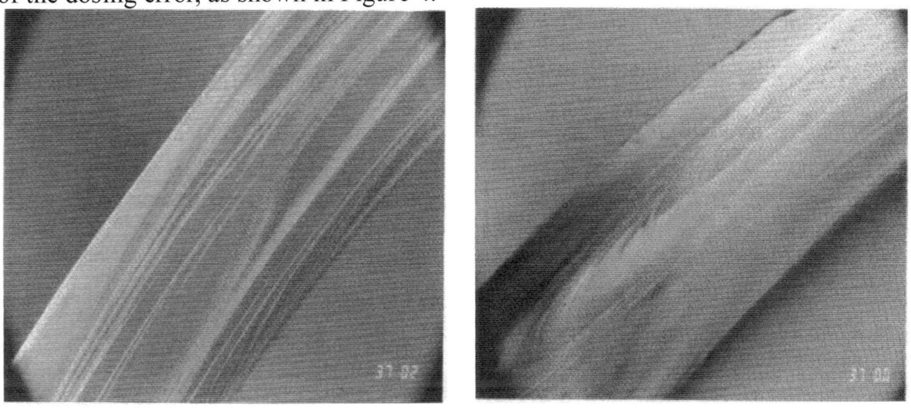

Figure 4. Severe streaking due to "wrong mix".

(4) The raw material is heavily loaded with broken glass. To prevent the composition of crushed glass from being different from that of the glass in the furnace, compositional analysis of purchased cullet should be performed frequently, and formulations should be adjusted according to the results.

4.2.3 Streaks caused by the unreasonable design of glass chemical composition. Borosilicate glass has a high silica-aluminum content in the formula, poor melting homogenization, and is prone to streaking, as shown in Figure 5.

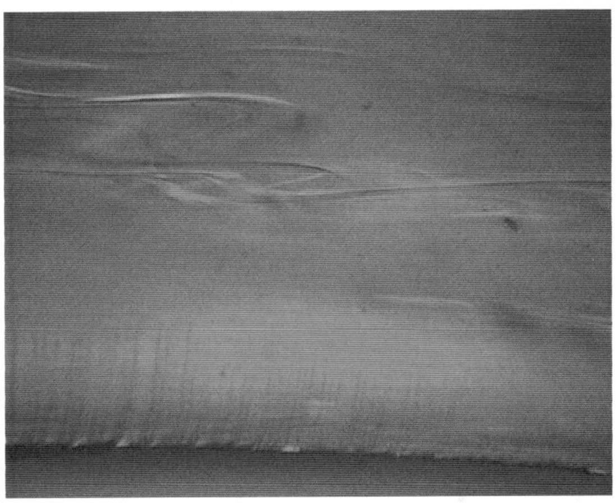

Figure 5. Streaks produced by poorly designed glass chemical composition.

4.2.4 Streaks caused by improper melting in the kiln. The temperature is not stable:

(1) When the temperature rises, the immobile layer at the bottom of the original pool changes, and differences in composition and viscosity occur, forming streaks. Streaks present slabs.

(2) The stagnant, stale glass of different compositions in the kiln begins to mix and melt with most of the glass in the pool kiln.

Kiln operating conditions are changed, as shown in Figure 6:

(1) Glass discharge varies too much.

(2) Uneven combustion on one side of the kiln.

(3) Dramatic changes in the flux electrode or bubbling.

(4) Uneven glass flow in the kiln.

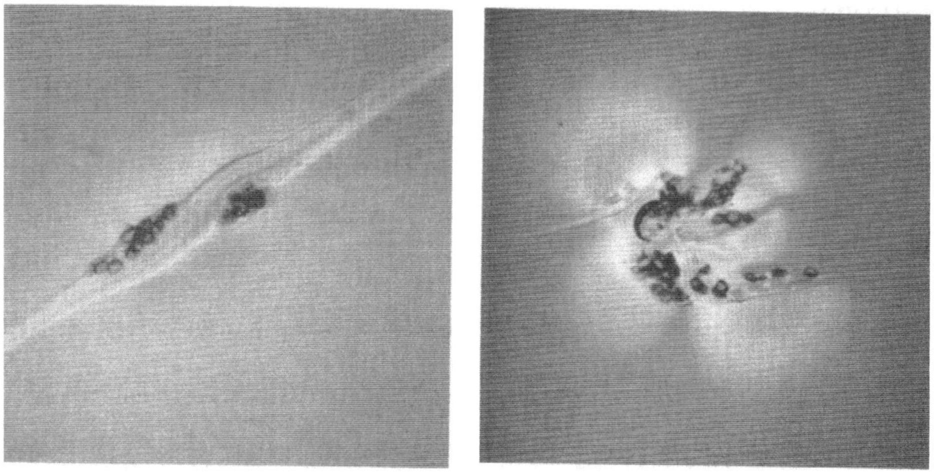

Figure 6. Streaks caused by improper melting in the kiln.

4.2.5 Streaks caused by unstable glass discharge. When the discharge volume becomes enormous, especially when there is a significant fluctuation in the glass level, the highly viscous melt corroded down on both sides of the kiln wall flows toward the middle of the melting pool and enters the production stream more quickly, thus forming a streak. When the discharge volume becomes large, due to the influence of the viscosity of the glass itself, the original bottom of the pool flow of the glass liquid is slower, or the flow rate of the immobile layer becomes faster or changes, thus forming streaks, as shown in Figure 7a.

When the flow of glass liquid becomes small, it is better to slowly and gradually increase the feed channel temperature in advance before changing the product. Strengthening the control means stabilizing the furnace melting conditions. This is generally the case with glass liquids, where the quality of individual products becomes smaller, the molding temperature is required to rise, and the feed channel temperature needs to increase. As the temperature rises, the stagnant, stale material in the kiln and the feed channel will be driven by the liquid glass flow, mixing into the molded glass stream and creating streaks. Compared to the streaks formed by differences in mixed batches, these tend toward more minor stress levels. They are not widely distributed after optical homogeneity testing, as shown in Figure 7b.

a The flow rate of the glass liquid becomes higher b The flow rate of the glass liquid becomes lower

Figure 7. Streaks caused by unstable glass discharge.

4.2.6 Streaks from refractory materials. Kiln temperature control and bubbling electrodes are both methods to promote melt mixing. Whether the refractory erosion defects come from contact with the glass, the refractory bricks of the kiln's superstructure, or the castables, there will be factors that form streaks. This is because streaks can form when the refractory material does not have enough time to mix and disappear in the glass after it becomes an erosion product, such as alkaline gas erosion or frit at high temperatures. Often, these streaks appear as vertical streaks on glass products, and in optical homogeneity testing, they appear as "nodule shapes," as shown in Figure 8.

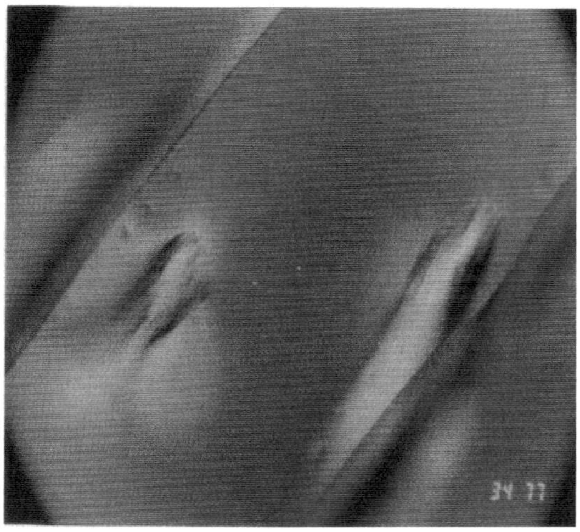

Figure 8. "Nodule shapes" streaks.

Adding a mixing device to the forehearth is also essential to the appearance of streaks. In the forehearth structure, the radius of the wall and the bottom of the intersection is increased, the dead angle is reduced, and the formation of slow-flowing "dead mass" is avoided. The heat preservation is strengthened at the bottom of the forehearth to increase the temperature of the glass liquid at the bottom, thus reducing the glass's viscosity and increasing the fluidity and diffusivity.

In the forehearth, the glass erodes slower with the refractory at lower temperatures, but the glass liquid still scours and erodes the refractory. The refractory that is eroded out results in more difficult diffusion due to the forehearth's lower temperature. However, when the forehearth temperature changes, it flows along the bottom of the forehearth and is carried into the glass product. Such streaks are basically the same shape in all parts of the same group of products. They form a "cat's paw" pattern.

5. Application of visual inspection technology in optical homogeneity inspection

Traditional detection methods have many drawbacks, such as low accuracy of manual detection, low detection efficiency, and low repeatability [8]. Currently, visual inspection technology is used to automatically extract the characteristic information of the samples and carry out with the standard atlas to directly obtain the test results and give the specific causes for the streaks.

The visual inspection system mainly includes image acquisition, image pre-processing, image feature extraction, image comparison, feature point comparison results, and result output [8]. Figure 9 shows the flowchart of the visual inspection system's operation.

A more accurate identification system can be established by feeding the vision inspection system with a large number of glass streak images. Accordingly, the diagnostic accuracy and efficiency of the glass-producing streak defect in the glass melting process can be substantially improved. In production, problems can be detected in time, and fast and effective solutions can be found.

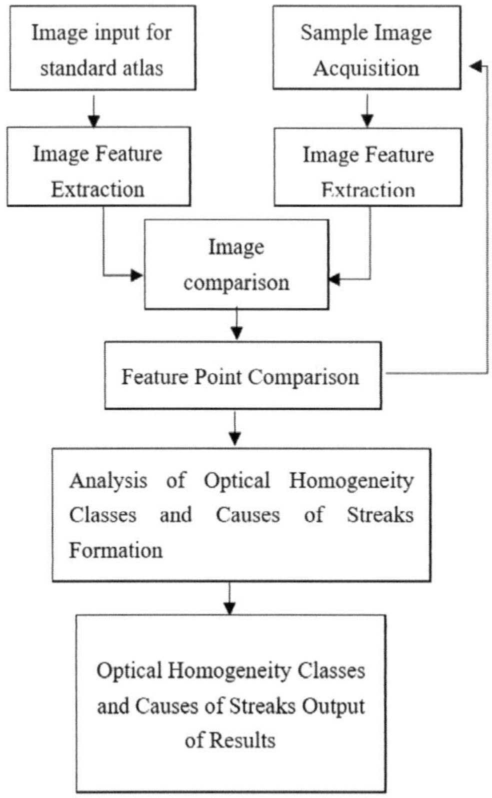

Figure 9. Flow chart of visual inspection technology.

6. Conclusion

(1) Streaks generated within the glass during the production process are difficult to detect. Still, they generate significant structural stresses that can seriously affect the quality of the glass product.

(2) Through the introduction of optical homogeneity detection technology, the critical links that produce streaks in the production of glass products are analyzed, as well as the characteristics of the streaks inside the glass corresponding to the problematic links. This proves that optical homogeneity detection technology is a better detection and quality control for glass streaks and structural homogeneity. It helps to identify problems in the production process of glass products.

(3) Instead of the traditional manual inspection method, the visual inspection system can make the inspection result faster and more accurate and find a faster and more effective solution to the problem.

References

[1] Tian, Y., & Sun, S. (2006). The new edition of glass technology. Beijing: China Light Industry Press.

[2] Zhang Y L, Yu H W. (2013). The research on evaluation of quality levels of solar vacuum tube glass. *Acta Energiae Solaris Sinica (07)*, 1190-1196. (in Chinese).

[3] Lun X Y & Wang H. (2019). Measuring Method of Key Technological Characteristics in Bottle Glass Production and Its Application. *Glass & Enamel (02)*, 50-55+49. doi: 10.13588/j.cnki.g.e.1000-2871.2019.02.011. (in Chinese)

[4] Fröhlich, F., Hildebrand, J., & Bergmann, J. P. (2022, March). Additive Manufacturing with Borosilicate Glass and Soda-Lime Glass. In *2nd International Conference on Advanced Joining Processes (AJP 2021) Selected Contributions of AJP 2021* (pp. 151-164). Cham: Springer International Publishing.

[5] Glass Apparatus (SAC/TC 178)(2022).GB/T 42222-2022.State Administration for Market Regulation; Standardization Administration of the People's Republic of China. (in Chinese)
[6] Liu, B., Yuan, J., Guo, Z., Zhang, Q., Han, Z., Tan, J., ... & Zheng, W. (2023). Numerical Simulation of a Viscoelastic Thinning Process for Preparing Flexible Glasses by Redrawing Method. *Journal of the Wuhan University of Technology-Mater. Sci. Ed.*, *38*(1), 65-71.
[7]Maciejewski, M., Milewska, K., Synak, A., Sadowski, W., & Kościelska, B. (2023). Influence of controlled crystallization and SrF2 content on the structure and properties of $Eu3^{+}$ doped phosphate glasses. *Journal of Non-Crystalline Solids*, *616*, 122473.
[8] Qi J, Zhang Y. (2021). Part burr detection based on machine vision. *machine design and research (02)*,24-27.doi:10.13952/j.cnki.jofmdr.2021.0048. (in Chinese)

Structural, magnetic, and electrical properties of $LaMnO_3$ doped with Na by microwave-assisted synthesis method

Guohao Chen [1], Luwen Jiang [2], Chunlai Li [3], Yiqun Wei [3], Guoqing Hu [1], Yutao Yue [3], Xiaomeng Wang [1*] and Jin He [1*]

[1]Shenzhen SoC Key Laboratory, PKU-HKUST Shenzhen-Hong Kong Institution, Shenzhen 518057, China

[2]Department of Physics, National University of Singapore, 2 Science Drive 3, Singapore 17551, Republic of Singapore

[3]Shenzhen Institute of Peking University, Shenzhen 518057, China

[*]Corresponding author's email: frankhe@pku.edu.cn (Jin He): wangxm2@ier.org.cn (Xiaomeng Wang)

Abstract: The electric and magnetic transport properties of $La_{1-x}Na_xMnO_3$ (x = 0.15, 0.2, 0.25, 0.3) are studied using the X-ray diffraction (XRD), vibrating sample magnetometer (VSM), and four-probe resistivity measurement method. The samples are synthesized by the method of microwave heating. The obtained samples show phase changes around x = 0.25. The magnetization measurement shows the curie temperature (*Tc*) increases as the level of doping increases and saturates around x = 0.25. The resistivity measurement dramatically decreases at a metal-insulator transition temperature close to *Tc*. These findings are discussed along with the structural features of different Na concentrations.

1. Introduction

$La_{1-x}Na_xMnO_3$, a hole-doped pseudo cubic perovskite, has attracted researchers' attention due to its possible applications in solid oxide fuel cells [1]. The applications are related to the colossal magnetoresistance (CMR) property which refers to the sudden drop of resistivity in the presence of an external magnetic field. The undoped Lanthanum Manganite is an anti-ferromagnetic insulator with a Neel temperature of around 140 K. Aliovalent cation doping on the Lanthanum site with alkaline metals such as Sodium, Calcium, and Strontium will oxide Mn^{3+} ions to Mn^{4+} ions and changes $LaMnO_3$ into ferromagnetic metal. Such changes in electric and magnetic properties can be explained by Zener's double exchange mechanism, which predicts the electron hopping in the Mn^{3+} - O^{2-} - Mn^{4+} pairs. The extra electron in the e_g orbit of Mn^{3+} ion moves to the neighboring Mn^{4+} ion through 2p orbitals in O^{2-} [2]. The electric and magnetic properties of $La_{1-x}Na_xMnO_3$ are highly dependent on the Mn valence state which can be affected by the amount of cation doping and oxygen stoichiometry. Consequently, studying the influence of Na doping in the manganite system is expected to advance our understanding of the CMR effect in these compounds. Previous researchers investigated $La_{1-x}Na_xMnO_3$ in the rhombohedral lattice structure with space group R3c which takes place when ($0 \leq x \leq 0.15$) but did not discuss the phase changes when the level of doping becomes high enough ($x \geq 0.25$) [3]. This paper discusses the structural features, and electric and magnetic properties using the $La_{1-x}Na_xMnO_3$ ($0.15 \leq x \leq 0.30$) bulk samples synthesized by the emerging method of microwave (MW) heating, which is fast and environmentally friendly compared to the conventional solid-state

Journal of Physics: Conference Series **2842** (2024) 012095 doi:10.1088/1742-6596/2842/1/012095

method [4].

2. Materials and methods

To synthesize $La_{1-x}Na_xMnO_3$, pre-heated La_2O_3, Na_2CO_3, and Mn_2O_3 powder in a correct molecular ratio depending on the stoichiometry of Na content were ground and mixed well with a mortar. The mixture powder was initially put into a microwave muffle furnace (PYRO model), operating at 2.45 GHz for heating. The temperature was raised to 1200 degrees Celsius in 20 min with a fixed MW power of 1000 W, and the powder was maintained at the final temperature for 30 minutes for reactions to complete before the power was switched off. The samples were cooled to room temperature for 1 hour. After that, the powder was ground again, and the process was repeated for the 2nd time for homogenization. Then, the powder was pressed into a pellet and subjected to MW irradiation for the 3rd time. Finally, the sample was cut into rectangular pieces using a diamond wheel for electrical and magnetic measurements. Magnetization is measured using a vibrating sample magnetometer (VSM). Electrical resistivity is measured by the four-probe resistivity method. Microwave synthesis is advantageous compared to traditional solid-state synthesis because of its rapid preparation time. In a conventional method, the synthesis of perovskite materials with colossal magnetoresistance properties can take seven days at a temperature of 1500 degrees Celsius in an oven reported by Gong et al. [5]. Microwave synthesis takes less than 2 hours due to its dielectric heating affecting reactant oxide precursors with good microwave absorption, which leads to rapid heating. Because of the rapid heating, like a combustion reaction, the resultant materials from microwave synthesis have smaller particle sizes and smaller resistivity compared to the materials prepared by the solid-state method [3].

3. Results and discussion

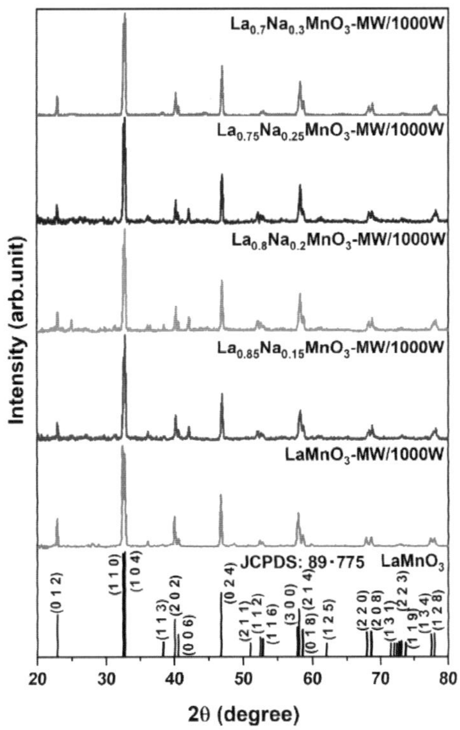

Figure 1. XRD results of $La_{1-x}Na_xMnO_3$ materials with different Na content.

Figure 1 shows the room temperature X-ray diffraction analysis of $La_{1-x}Na_xMnO_3$ (x = 0, 0.15, 0.2, 0.25, 0.3) synthesized at MW power 1000 W. The XRD pattern shows the (x = 0.15, 0.2, 0.25) samples are in good agreement with the Na-doped $LaMnO_3$ peak which reveals that Compounds with a rhombohedral structure in space group R3C are created as homogeneous single-phase [5]. As the level of doping increases, the doublet peaks get closer and eventually reduce at (x = 0.3). This pattern reveals the formation of a second orthorhombic phase. A similar observation was also reported by Das et al. [3]. Cell parameters are shown in Table 1. When Na concentration increases, cell parameters *a* and *c*, as well as the unit call volume, show a decreasing trend for ($x \leq 0.25$). The decrease in unit-cell and lattice is due to the smaller size of Na^{+} compared to La^{3+}. For ($x > 0.25$), the lattice and unit-cell parameters increase. This change in slope gives evidence for the onset of the secondary phase in the $La_{1-x}Na_xMnO_3$ compound. The overall decrease in cell parameters and changes in phases can be explained by the tolerance factor *t* which measures the stability of crystal structure [6]. The tolerance factor is given by $t = \frac{r_A + r_O}{\sqrt{2}(r_B + r_O)}$. $r_A + r_O$ is the length of La-O bond while $r_B + r_O$ is the length of the Mn-O-Mn bond. Doping in A cation site results in an average reduction in the ionic radii of A cation La but increases the bond length between La and O. The ratio of two bond lengths increases, and so does the tolerance factor. Calculated tolerance factors for different doping levels are shown in Table 1. Increasing in tolerance factor signifies the crystal structure becoming more cubic. This explains our observation of the onset of the orthorhombic phase, and so are the changes in transport properties.

Table 1. Some structural and physical parameters of $La_{1-x}Na_xMnO_3$ compounds.

Nominal Composition x	0.15	0.2	0.25	0.3
TC (K)	313	320	324	325
Tim (K)	330	332	335	336
a (Å)	5.4987	5.4921	5.4915	5.4990
c (Å)	13.3425	13.3287	13.3264	13.3424
Unit-cell volume (Å3)	351.8	348.7	348.2	349.8
La-O bond length (Å3)	2.485	2.480	2.495	2.509
Mn-O-Mn bond length (Å3)	1.961	1.953	1.952	1.951
Tolerance Factor	0.896	0.898	0.904	0.910
MS at 10 K (emu g−1)	81.07	81.94	79.50	70.46

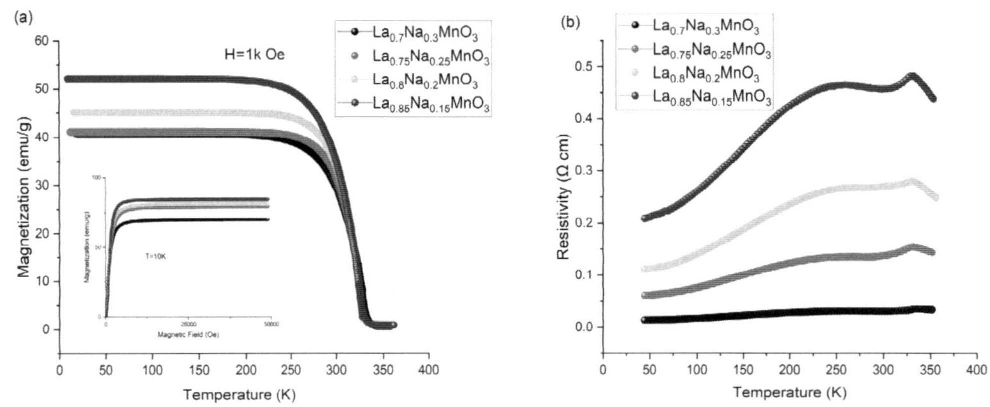

Figure 2. (a) magnetization and (b) resistivity for $La_{1-x}Na_xMnO_3$ synthesized at (x = 0.15, 0.2, 0.25, 0.3) concerning the temperature. Inset of (a) is the magnetizations for different MW powers concerning the magnetic field.

Figure 2(a) is the temperature-dependent magnetization of $La_{1-x}Na_xMnO_3$ samples with different levels of doping measured at a 1 k Oe magnetic field. As temperature increases to around room temperature, magnetization drops dramatically from about 50 emu/g to 0 emu/g. This is due to the phase transition of the compounds from ferromagnet to paramagnet. Curie temperature is the minimum point of the first derivative of magnetization concerning temperature. As the level of doping increases, curie temperature increases from 313 K (x = 0.15) to 325 K (x = 0.3) as shown in Table 1. This is due to the reduction in Mn-O-Mn bond length and increasing in the bond angle governed by the double-exchange mechanism [5]. It can also be noticed that the slope of changing decreases and tends to saturate as Na concentration reaches (x = 0.3). The inset in Figure 2(a) displays the field-dependent saturation magnetization (*Ms*) values recorded at 10 K. *Ms* decreases with the level of doping increases (*Ms* = 85.07 emu/g, 81.94 emu/g, 79.50 emu/g, 70.46 emu/g for x = 0.15, 0.2, 0.25, 0.3). The variation in *Tc* and *Ms* can be explained by the competition between the Mn^{3+} - O - Mn^{4+} double-exchange mechanism and the Mn^{4+} - O - Mn^{4+} super-exchange mechanism. When the level of doping is low, increasing Na doping corresponds to more Mn^{3+} - O - Mn^{4+} pairs, which the hopping interactions between Mn^{3+} and Mn^{4+} are favored. As a result, electrons can easily move to neighboring ions, maintaining their spins. Thus, the enhancement in electron transport properties results in higher *Ms* and a dramatic increase in *Tc* according to Soma et al. [3]. As the doping level increases, the electrical and magnetic transport properties cannot be solely explained by the double exchange mechanism, from the structural point of view. The rhombohedral distortion enhances magnetic moment, according to Kumar et al. [7]. The completion of the rhombohedral phase and the onset of the orthorhombic phase hinders the magnetization. Thus, *Ms* and *Tc* increase before x = 0.25 and start to saturate or even decrease after that. From the electron transport point of view, when the doping level increases, the Mn^{4+}/Mn^{3+} ratio keeps increasing, and the Mn^{4+} - O - Mn^{4+} super-exchange mechanism starts to dominate, hopping interactions decrease compared to the double exchange case and thus, leads to lower *Ms* and saturation in *Tc* variation.

Figure 2(b) shows the resistivity concerning the temperature at zero external magnetic field. The resistivity pattern shows two peaks: One broader peak at lower temperatures and one sharper peak close to the *Tc*. The broader peak can be caused by the inhomogeneous diffusion of oxygen in the grain. Above the sharper peak, which is the insulator-metal transition temperature, compounds show insulator-like behaviors and metal-like behaviors above *Tim*. The overall resistivity decreases as the level of doping increases. This is again due to the increase in the Mn valance state, which is caused by more double-exchange interactions, as explained above. The change in *Tim* matches the change in *Tc*, which shows a drastic increase with a low level of doping and tends to saturate when more Mn^{4+} is introduced into the compounds. This phenomenon also suggests that when Na concentration reaches a certain value, in our case (x > 0.25), a new phase in the compound is introduced, and both the electrical and magnetic transport properties start to change in slope.

4. Conclusion

We have successfully synthesized $La_{1-x}Na_xMnO_3$ samples using MW synthesis and studied the electric and magnetic properties concerning the level of Na doping. From the measurements of cell parameters and lattice patterns, we observe changes in phase and use the tolerance factor to explain them. The increasing and saturation trend in *Tc* and *Tim*, the observation *Ms*, all suggests the domination in super-exchange effect in the compounds and new phase hindering magnetization at the level of doping x > 0.25 for $La_{1-x}Na_xMnO_3$. These findings can assist in deepening our understanding of the relationships between doping concentration, structural phases, and electrical and magnetic transport properties exhibited by the material.

Acknowledgement

This work is supported by these funds of JCYJ20230807153859005, JCYJ20230807154759002, JCYJ20210324115812036, JCYJ20200109144612399, JCYJ20200109144601715, IERF202206, JCYJ2022 0818103408018, JCYJ20200109144601715, JCYJ20200109144612399 and JCYJ20220818103408017.

Journal of Physics: Conference Series **2842** (2024) 012095 doi:10.1088/1742-6596/2842/1/012095

References

[1] Adler, Stuart B. "Factors governing oxygen reduction in solid oxide fuel cell cathodes." *Chemical Reviews* 104.10 (2004): 4791-4844.
[2] Malavasi, Lorenzo. "Role of defect chemistry in the properties of perovskite manganites." *Journal of Materials Chemistry* 18.28 (2008): 3295-3308.
[3] Das, Soma, and T. K. Dey. "Structural and magnetocaloric properties of La1− yNayMnO3 compounds prepared by microwave processing." *Journal of Physics D: Applied Physics* 40.7 (2007): 1855.
[4] Baghurst, D. R., A. M. Chippendale, and D. Michael P. Mingos. "Microwave syntheses for superconducting ceramics." *Nature* 332.6162 (1988): 311-311.
[5] G. Q. Gong, C. L. Canedy, Gang Xiao, J. Z. Sun, A. Gupta, W. J. Gallagher; Colossal magnetoresistance in the antiferromagnetic $La_{0.5}Ca_{0.5}MnO_3$ system. *J. Appl. Phys.* 15 April 1996; 79 (8): 4538–4540.
[6] Roy, S., et al. "Interplay of structure and transport properties of sodium-doped lanthanum manganite." *Journal of Physics: Condensed Matter* 13.42 (2001): 9547.
[7] Kumar, Pawan, Chandrakanta Panda, and Manoranjan Kar. "Effect of rhombohedral to orthorhombic transition on magnetic and dielectric properties of La and Ti co-substituted BiFeO3." *Smart Materials and Structures* 24.4 (2015): 045028.

Parametric analysis of vibration characteristics of laminated composite plate structures

Yanchun Zhai*, **Xiaoxue Zhang, Yuesong Zhou and Zhiyuan Wan**

School of Mechanical Engineering, Shandong Huayu University of Technology, Dezhou 253000, China

*Corresponding author's e-mail: 363125947@qq.com

Abstract. In order to conduct the vibration performance of laminated composite structures, the relationship between the vibration frequency and the parameters such as aspect ratio, width-thickness ratio, elastic modulus ratio and lay-up angle of the plate structure under different boundary conditions was investigated and the variation law of its vibration characteristics was obtained. The conclusions are as follows: 1. The vibration frequency tends to be smaller and smaller as the value of length increases. 2. The vibration frequency tends to be larger and larger as the value of thickness increases. 3. The vibration frequency increases with the increase of the value of elastic modulus.4. The symmetrical lay-up structure with the lay-up angle of the middle ply at 45 degrees has the lowest vibration frequency.

1. Introduction

Composite structures were widely used in aerospace field, nuclear industry, automotive, and other industrial sectors due to a number of advantages [1-3]. The study of its vibration characteristics has extremely important practical significance and is increasingly receiving attention and importance from scientists and technicians engaged in composite materials research.

Shih and Blotter [4] focused on the vibration response of viscoelastic composite laminates, and Ishihar and Noda [5] studied the nonlinear dynamical behavior of piezoelectric thermoelastic laminates. Some scholars studied the dynamic response characteristics of composite laminates under different boundary conditions, and Oh and Nayfeh [6] studied the dynamic response of cantilever laminates. Shi and Chen [7] used the Rizt method to calculate the fundamental frequencies of various lay-up rectangular plates under general boundary conditions based on Hamilton's principle of minimum action. Yang and Peng [8] conducted the vibration performance of of rectangular laminated thin plates with mixed-type constraints at the boundary by considering the influencing factors of the mid-plane load. Ren and Liu [9] carried out a study on composite structures adopting the finite element solution. Dong et al. [10] carried out some specific studies on fiber-reinforced polymer matrix composite resistances and came to influential conclusions. Liang et al. [11] proposed an analysis method to fix the movement direction of the plate, and prevent the plate from loosening and jamming during the vibration process by adjusting the stiffness of the virtual normal laminated composite structure plate due to poor model establishment and resulting calculation results (vibration frequency). Based on the classical theory of Kirchhoff, You and Lv [12] derived the damping array equation for anti-symmetric lay-up structures and presented the variation of intrinsic frequency of the composite plates and the modal damping for different number of layers and different constraints. Yan et al [13] established a general analytical solution for the free vibration of thin composite structures.

Journal of Physics: Conference Series **2842** (2024) 012096 doi:10.1088/1742-6596/2842/1/012096

2. Finite element solution for composite materials

Based on the concept of structural vibration dissipation energy caused by the viscoelasticity of the composite material, the strain-energy method is used to model composite laminated plate, and because the plate structure is drawn, we choose the solid unit Solid185 unit. The simulation analysis results are described in Figure 1. The composite material parameters are illustrated in Table 1.

Table 1. Composite material parameters.

Parameters	Values
Density (kg m^{-3})	2000
Loss factor	0.0092
	E_x =24.3e3
Elastic modulus (MPa)	E_y =24.3e3
	E_z =10.4e3
	G_{xy} =3.7e3
Shear modulus (MPa)	G_{yz} =3.7e3
	G_{xz} =3.7e3
	v_{xy} =0.138
Poisson's ratio	v_{yz} =0.4
	v_{xz} =0.4

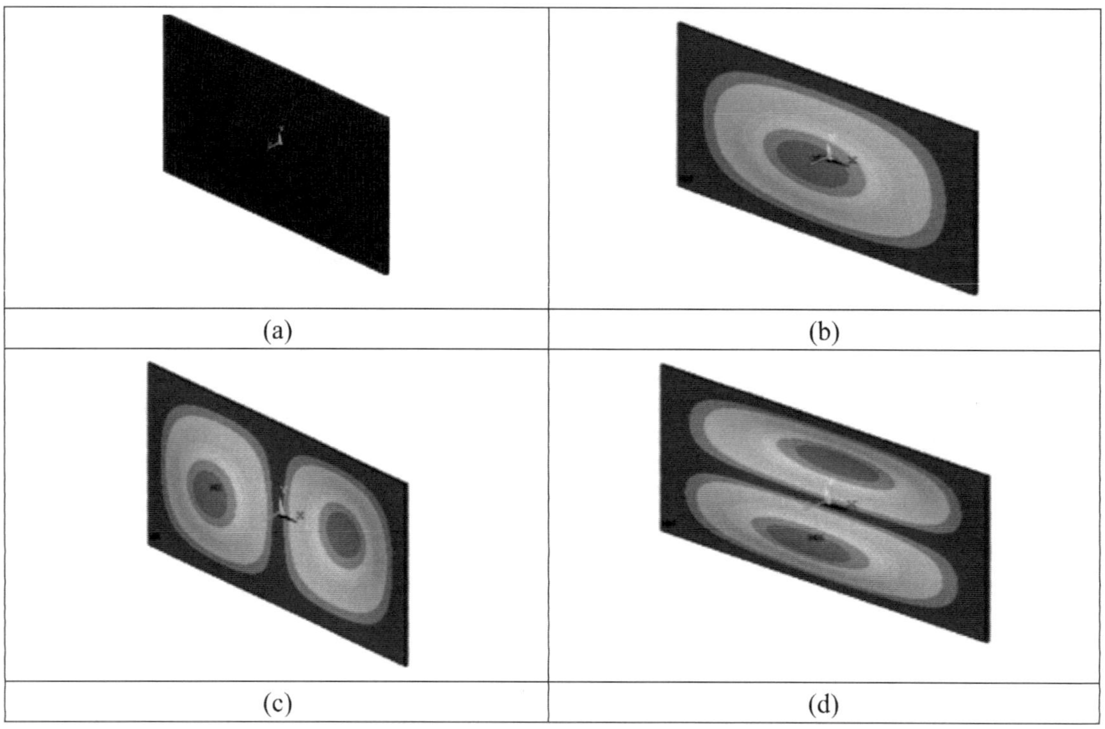

(e)

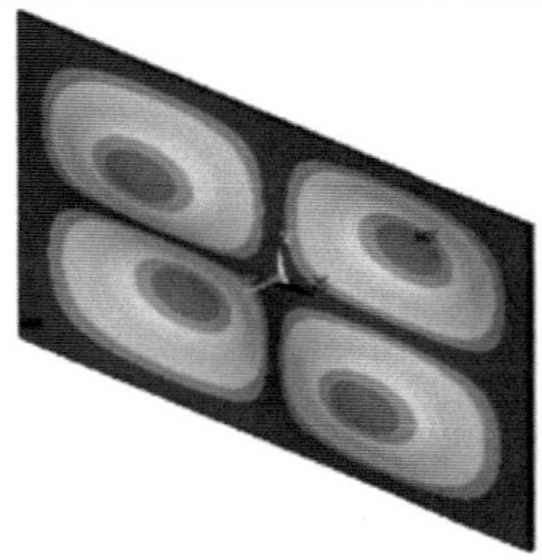

(f)

Figure 1. Model diagram and the first five steps of the formation

3. Variation of vibration frequency with geometric parameters

On the basis of the obtained vibration frequency, the vibration characteristics of the plate structures are conducted by changing some conditions such as the aspect ratio and the width-thickness ratio (W-T ratio) of the structure using ANSYS finite element analysis software.

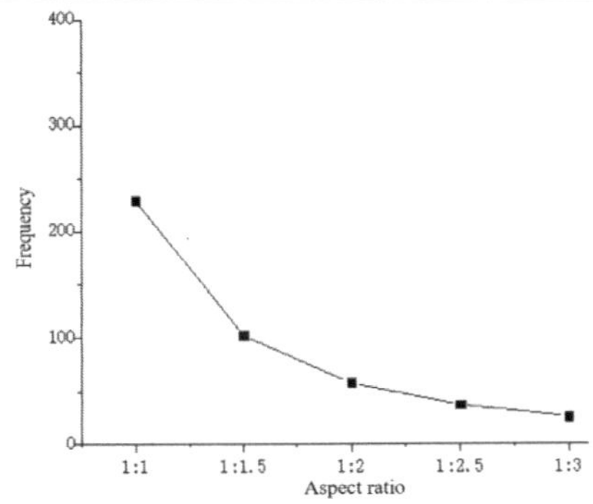

Figure 2. Vibration frequency versus aspect ratio of two-sided fixed support plate.

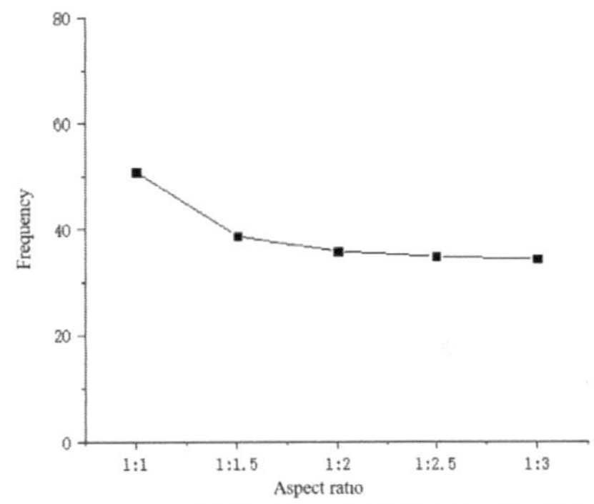

Figure 3. Vibration frequency versus aspect ratio of four-sided fixed support plate.

As is shown in Figure 2, changing the aspect ratio with the boundary condition fixed at the two ends and with some parameters such as width kept constant, the vibration frequency value becomes smaller as the value of the aspect ratio becomes larger.

As is shown in Figure 3, changing the aspect ratio with the boundary condition fixed at all four ends and with some parameters such as width kept constant, the vibration frequency value becomes smaller as the value of the aspect ratio becomes larger.

The vibration frequency decreases with the increase of the aspect ratio. However, the magnitude of change is different for different lengths and widths of plates.

As is shown in Figure 4, in the case where the boundary conditions are fixed at both ends and the width and some other parameters are kept constant varying the width-to-thickness ratio, as the value of the width-to-thickness ratio becomes larger, the vibration frequency value also becomes larger.

As is shown in Figure 5, changing the W-T ratio with the boundary conditions fixed at all four ends and with some parameters such as width kept constant, the vibration frequency increases as the W-T ratio becomes bigger. The same point is that the vibration frequency increases as W-T ratio rises. The magnitude of the change of different lengths and widths of the plate is not very influential.

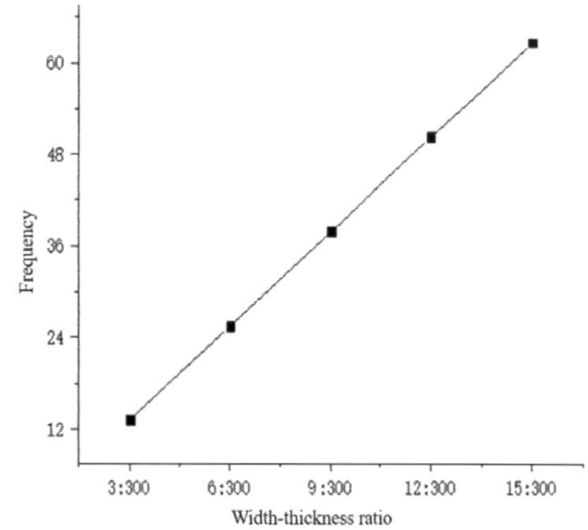

Figure 4. Vibration frequency versus W-T ratio of two-sided fixed support plate.

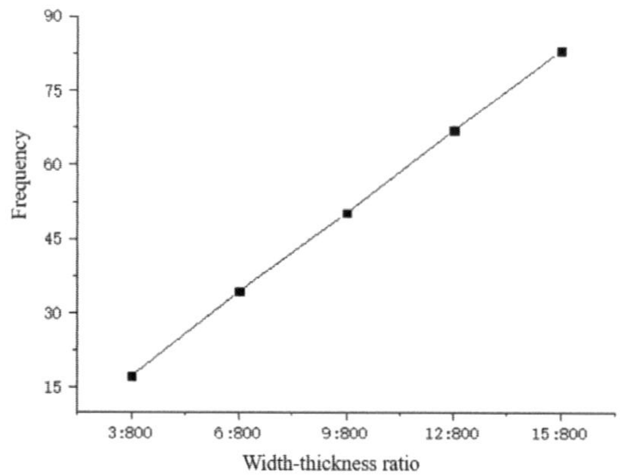

Figure 5. Vibration frequency versus W-T ratio of four-sided fixed support plate.

From these parameters of modulus of elasticity ratio and laying angle, we continue to study their relationship with vibration frequency to find the value of vibration frequency and study the variation of vibration frequency with parameters by changing a single variable.

Figures 6 and 7 show the relationship between modulus of elasticity ratio and vibration frequency. As is shown in Figure 6, varying the elastic modulus ratio with the boundary conditions fixed at both ends and with some parameters such as the width kept constant, the value of the vibration frequency increases as the elastic modulus ratio becomes larger.

As is shown in Figure 7, changing the elastic modulus ratio with the boundary conditions fixed at all four ends and with some parameters such as width kept constant, the value of vibration frequency increases as the elastic modulus ratio gets larger. The same point is that the vibration frequency increases with the increase of the W-T ratio. However, the magnitude of the change is not particularly large for larger plate sizes.

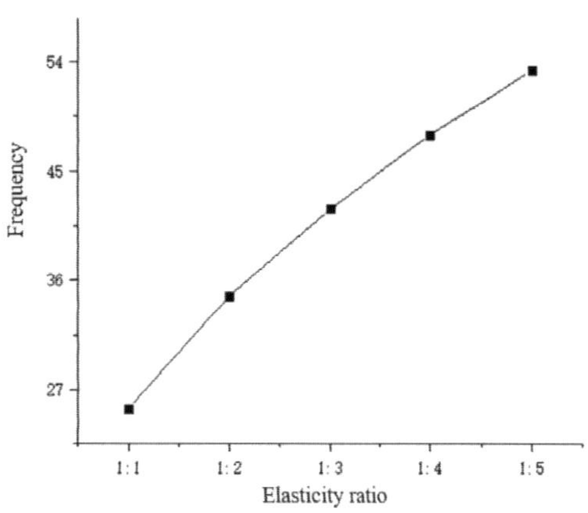

Figure 6. Vibration frequency versus modulus of elasticity ratio of two-sided fixed support plate.

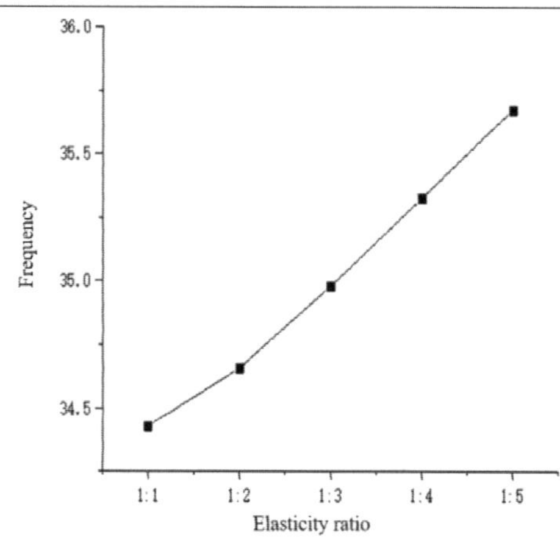

Figure 7. Vibration frequency versus modulus of elasticity ratio of four-sided fixed support plate.

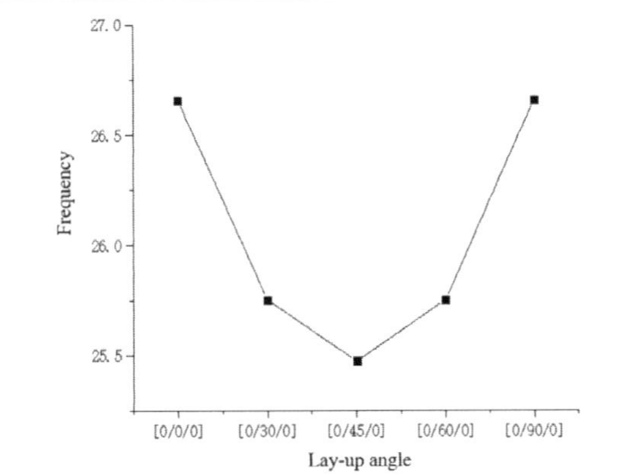

Figure 8. Vibration frequency versus modulus of lay-up angle of two-sided fixed support plate.

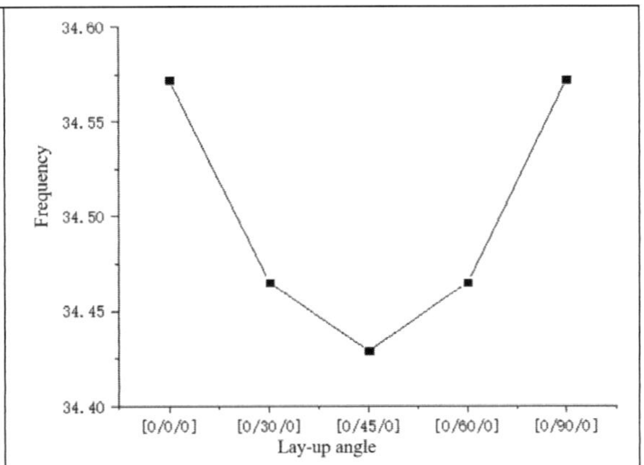

Figure 9. Vibration frequency versus modulus of lay-up angle of four-sided fixed support plate.

Figures 8 and 9 illustrate the vibration frequency versus modulus of lay-up angle. From Figure 8, it can be seen that the magnitudes of the structural vibration frequencies are equal under the layups of 0/0/0 and 0/90/0, 0/30/0 and 0/60/0 with the boundary conditions fixed at both ends, and it is also obvious that it is in the lowest value at 0/45/0.

As is shown in Figure 9, magnitudes of structural frequencies under 0/0/0 and 0/90/0, 0/30/0 and 0/60/0 layups are equal in the case where the boundary conditions are fixed at all four ends, and it can also be clearly found that 0/45/0 is in the lowest value state. The same point is that the magnitudes of 0/0/0 and 0/90/0, 0/30/0 and 0/60/0 are equal, and it is also obvious that 0/45/0 is in the lowest value state. However, the change in vibration frequency is not particularly obvious for larger plate structures.

4. Conclusion

In this paper, the simulated flat plate of laminated composites is used as the object of investigation, and the relationship of its vibration frequency and the aspect ratio, W-T ratio, elastic modulus ratio, and its lay-up angle corresponding to one fixed end, two fixed ends, and four fixed ends of the plate structure is investigated. The conclusions made are as follows.

(1) In the case where the boundary conditions are fixed at one end, at both ends and at all four ends, and the width, elastic modulus, and some other parameters are kept constant by changing its length, the vibration frequency value also shows a tendency to become smaller and smaller as the value of the length becomes larger.

(2) In the case that the boundary conditions are fixed at one end, at two ends and at four ends, and the width, elastic modulus, Poisson's ratio, shear modulus and other parameters remain unchanged, changing its thickness, with the greater the value of the thickness, the vibration frequency value also shows a trend of increasing.

(3) In the case of fixed at one end, at two ends, or at four ends under any one of the boundary conditions, and the width, length, thickness, Poisson's ratio, shear modulus and other parameters remain constant, the same point is that the vibration frequency increases as the modulus of elasticity increases. However, it can be seen that the magnitude of change is not particularly large for plates with larger dimensions.

(4) The frequency of the 0/45/0 structure with symmetrical lay-up is at its lowest value under any of the boundary conditions of fixed at one end, at both ends, or at all four ends, and with some parameters such as width, length-thickness, Poisson's ratio, shear modulus, and modulus of elasticity kept constant. However, for larger plate structures, the vibration frequency variation is not particularly obvious.

Acknowledgments
This work was financially supported by the Project of Shandong Huayu University of Technology School-level Synergetic Innovation Centre "Synergetic Innovation Centre for Intelligent Equipment Technology R&D".

References

[1] Al-Dulaijan S U, Youzera H and Meftah S A, et al. (2024) Nonlinear Damping and Vibration Assessment of Sandwich Beam with Composite Faces and Viscoelastic Core Layer. Int. J. Struct. Stab. Dy. 24(08): 2450083.
[2] Chen D, Yang J and Guo W, et al. (2022) Vibration study of a composite pipeline supported on elastic foundation using a transfer matrix method. J. Vib. Cont., 28(7-8): 853-863.
[3] Wang S, Zheng C and Li S, et al. (2023) Free vibration of functionally graded carbon nanotube-reinforced composite damping structure based on the higher-order shear deformation theory. Polym. Composite., 44(2): 873-885.
[4] Shih Y S and Blotter P T (1993) Non-linear vibration analysis of arbitrarily laminated thin rectangular plates on elastic foundations. J. Sound Vib., 167(3): 433-459.
[5] Ishihar M and Noda N (2003) Nonlinear dynamic behavior of a piezothermo elastic laminated plate with anisotropic material properties. Acta Mech., 166(14): 103-118.
[6] Oh K and Nayfeh A H (1996) Nonlinear combination resonances in cantilever composite plates. Nonlinear Dynam., 11(2): 143-169.
[7] Shi J P and Chen Y H (1997) Higher order deformation theory for dynamic analysis of composite sandwich shells. Acta Mater. Compos. Sinica, 14(4):108-113.
[8] Yang J and Peng J S (2001) Vibration Analysis of Composite Laminated Plates with Mixed Edge Boundary Constraints and under In-plane Loading. Struct. Environ. Eng., 1: 26-31.
[9] Ren Y S and Liu L H (2004) Advances in damping analysis and design of fiber reinforced composite material structures. Mecha. in Eng., 2004(01):9-16.
[10] Dong J W, Dong H M and Chen P J (2019) Application of ANSYS in teaching composite mechanics courses. Western China Quality Educ., 5(08):185-186.
[11] Liang S, Liang L and Mi P (2010) New development of the embedded and co-cured composite damping structures. Chinese J. Appl. Mech. 4:767-771.
[12] You F X and Lv Q (2009) Analysis of Mechanics Properties of Composite Laminated Plates and Optimization of Angle Arrangement of Plies. Noise Vib. Con., 29(4): 138-143.
[13] Yan H, Lei Y J and Shen H J (2006) Free vibration of anisotropic rectangular plates by general analytical method. Appl. Math. Mech., 27(4): 461-467.

Journal of Physics: Conference Series **2842** (2024) 012097 doi:10.1088/1742-6596/2842/1/012097

Planar and non-planar hybrid printing trajectory generation method

Li Xu and Zhiliang Shi*

Wuhan University of Technology, Wuhan, Hubei, 430070, China

*Corresponding author's email: 17373273747@163.com (Zhiliang Shi)

Abstract: Aiming at the problem of step effect, a planar and non-planar hybrid printing trajectory generation method is studied. The region with the inclination of the surface and horizontal plane of the model is extracted for non-planar printing, which can improve the surface printing quality of the model while controlling the printing time. The experimental results show that the proposed algorithm has good adaptability and can meet the requirements of engineering applications.

1. Introduction

3D printing technology has the characteristics of shortening the manufacturing cycle and reducing material waste, and has great potential in many fields such as biomedical, aerospace, construction and protective structures [1,2]. However, the model printed based on the FDM process will have a step effect, which will reduce the surface quality of the model and affect the aesthetics of the model [3]. The print trajectory of the model affects the surface quality of the model, and the surface quality of the model is related to the surface Angle, construction direction, filling density, printing thickness, etc. The surface quality can be improved by modifying the relevant parameters [4-8]. In order to further improve the printing quality of the model surface, a series of researches have been carried out by relevant scholars using multi-axis printing technology [9-11]. In this paper, a planar and non-planar hybrid printing trajectory generation method is studied to improve the printing quality of the model surface and reduce the step effect.

2. Principle and procedure of algorithm

2.1 Algorithm overview

Aiming at the problem of increasing the printing time due to the reduction of the height of the model and the improvement of the surface quality, a planar and non-planar hybrid printing trajectory generation method is proposed. Parameters such as slice height, inclination threshold, filling pattern, filling direction and shell number were set to extract the non-planar print area, collision detection was carried out between the contour point set after plane slicing and the non-planar printing area, and a penalty collision target was created. The position of the collision point set was adjusted, the contour point set of plane slicing was filled, and the non-planar boundary point set was obtained. The projection is filled into the two-dimensional plane point set, and then mapped back to the non-planar print area. Finally, the printing trajectory is optimized to complete the hybrid trajectory generation of planar and non-planar printing.

2.2 Non-planar region extraction

The normal vector direction of the surface $\{n_i\}$ and the inclination of the horizontal plane of the triangular mesh model are calculated, and the surface whose inclination is less than the set inclination threshold β is recorded as a non-planar layer, satisfying Equation (1).

$$n_i.z \geq \cos\beta \tag{1}$$

The non-planar region is shown in Figure 1, and the set inclination threshold is 45°. The non-planar region is calculated by Equation (1).

Figure 1. Nonplanar region.

According to whether the triangular surfaces are adjacent to each other, the triangular surfaces satisfying Equation (1) are divided into multiple non-planar areas. The non-planar areas are conditionally filtered. The minimum area value is set, and the area of each non-planar area is compared with the set minimum area. The appropriate area is selected for non-planar printing.

2.3 Collision detection

The model is sliced according to the set layer height. The intersection point between the slice plane and the edge line of the triangular surface is calculated to obtain the set of model slice points, as shown in Figure 2. The equation of the slice plane is as follows:

$$ax + by + cz + d = 0 \tag{2}$$

where a, b and c are the components of the plane normal in the x, y and z directions, respectively. n is the plane normal vector, v is any point on the plane, and then d is defined as follows:

$$d = -(n \cdot v) \tag{3}$$

The linear equation for calculating the middle edge of the triangular surface is as follows:

$$\frac{x - x_0}{m} = \frac{y - y_0}{n} = \frac{z - z_0}{p} \tag{4}$$

where (m, n, p) is the line direction vector, and (x_0, y_0, z_0) is any point on the line.

According to Equations (2) and (4), the coordinates of the intersection point between the plane and the line are calculated:

$$\begin{cases} x = mt + x_0 \\ y = nt + y_0 \\ z = pt + z_0 \end{cases} \tag{5}$$

where

$$t = \frac{-ax_0 - by_0 - cz_0 - d}{am + bn + cp} \tag{6}$$

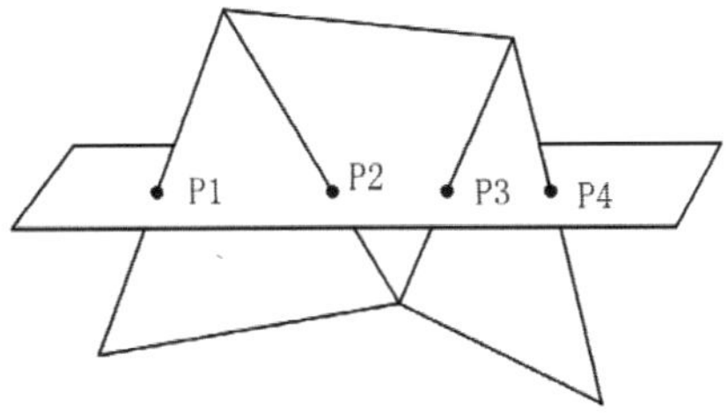

Figure 2. Slice point set.

Collision detection on the slice point set is conducted to determine whether it is on a non-planar layer. If it is, the position of the point set shall be adjusted according to the punishing collision target to ensure the correctness of the position of the point set. The adjusted position coordinates are as follows:

$$\begin{cases} x' = x - n_i.x \cdot h \\ y' = y - n_i.y \cdot h \\ z' = z \end{cases} \tag{7}$$

where (x, y, z) is the original coordinate position of the point, (x', y', z') is the adjusted coordinate position, n_i is the normal vector of the plane where the point is located, and h is the layer height of the slice.

The height between layers will be different after adjustment of different inclination planes, so it is necessary to prevent overfilling by compensation coefficient, which is the ratio of the height between layers after adjustment to slice height h, as shown in Figure 3. The equation for calculating the compensation coefficient is:

$$k = \frac{H}{h} \tag{8}$$

H is defined as follows:

$$H = \frac{|ax' + by' + cz' + d|}{\sqrt{a^2 + b^2 + c^2}} \tag{9}$$

where a, b and c are the components of the triangular surface normal corresponding to the x, y and z directions of the original point respectively, and (x', y', z') are the adjusted coordinate positions.

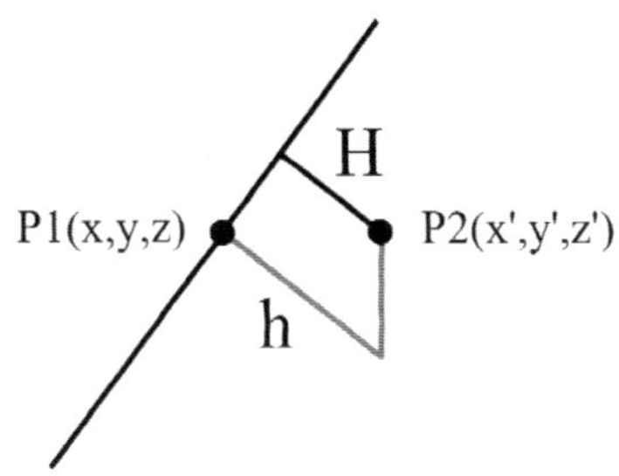

Figure 3. Collision detection position adjustment.

2.4 Internal filling

The inner region of the point set is filled with patterns including zigzags, grids, triangles, concentric circles, hexagons and so on, as shown in Figure 4. Among them, the most common ones are zigzag fill and grid fill. In the research work of this paper, a zigzag fill pattern is used.

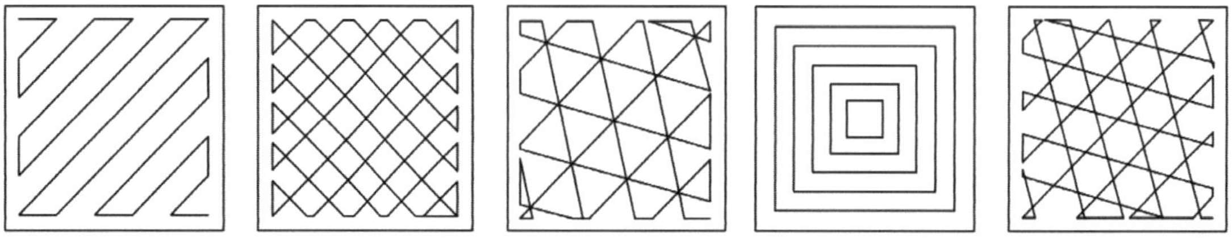

Figure 4. Fill pattern.

The filling direction, the number of outer wall rings and other parameters are selected to fill the inside of the slice point set. For example, if the filling direction is 45° from the positive direction of the X-axis and the number of turns of the housing is 2, the resulting printed track is shown in Figure 5.

Figure 5. The filling direction and the number of outer wall turns are set.

2.5 Non-planar region trajectory generation

The boundary point set of the non-planar region is obtained and projected onto the two-dimensional plane, as shown in Figure 6. The interior of the boundary point set is filled, and the filling direction and the number of outer wall circles are set.

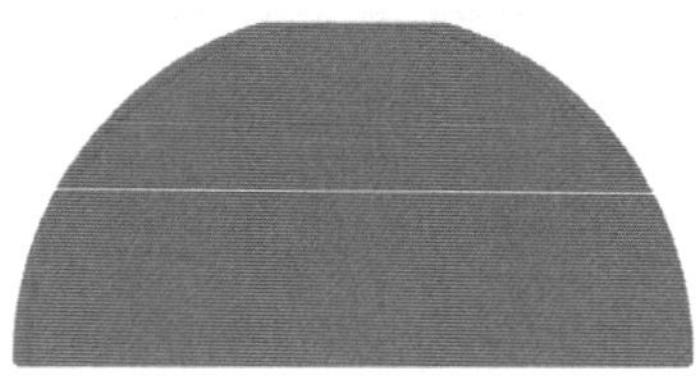

Figure 6. Non-planar area projection.

The coordinate position of the print point set mapped to the non-planar region is calculated to obtain the mixed trajectory of the full planar and non-planar printing, as shown in Figure 7. The coordinate position of the print point set mapped to the triangular surface of the non-planar region satisfies:

$$\begin{cases} x' = x_0 \\ y' = y_0 \\ z' = z_0 - \frac{a}{c}(x' - x_0) - \frac{b}{c}(y' - y_0) \end{cases} \tag{10}$$

where (x', y', z') is the coordinate position mapped to the non-planar layer, (x_0, y_0, z_0) is the original coordinate position before the mapping, and a, b and c are the components of the plane normal where the triangular surface is located corresponding to the x, y and z directions.

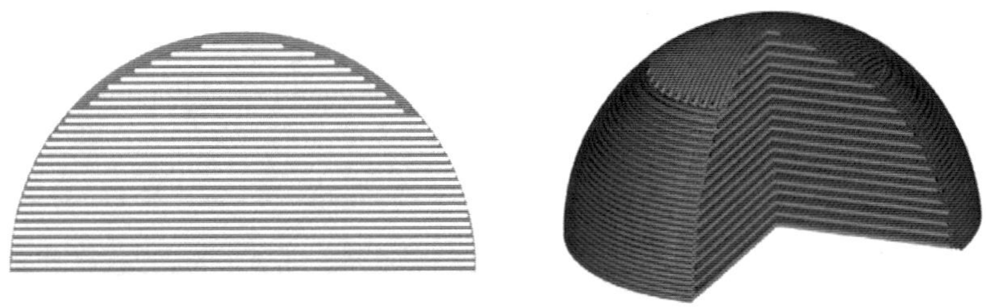

Figure 7. Planar and non-planar hybrid printing trajectory.

The printing order is determined by the height, the height of the non-planar layer is the maximum z value in the non-planar layer, and the print file is generated by combining the print track.

3. Case testing and data verification

3.1 Testing environment

Based on the Windows platform, Visual Studio development environment, QT graphical user interface and VTK visualization toolkit, the experimental software develops a prototype system for planar and non-planar hybrid printing trajectory generation, which is used to generate model printing trajectories.

The experimental hardware device is a desktop 3D printer, as shown in Figure 8. The size is 320×210×200 mm, the material is 1.75 mm PLA consumables, and the files are transferred by SD card.

Figure 8. Desktop 3D printer.

3.2 Data verification

The data processing of the mesh model is carried out by the algorithm to generate the printing trajectory of the model, and the effectiveness of the proposed planar and non-planar hybrid printing trajectory generation method is verified.

Figure 9. Hybrid print track.

The curved surface stereo model is input into the prototype system. The slice height is set to 0.3 mm, the inclination threshold is 45°, the filling pattern is zigzag, the filling direction is the x-axis direction, and the number of outer wall circles is 2. The final effect of the system trajectory generation and printing results is shown in Figure 9 and Figure 10. It can be seen that the proposed algorithm can quickly and effectively realize the generation of planar and non-planar hybrid printing trajectories by data processing of the mesh model.

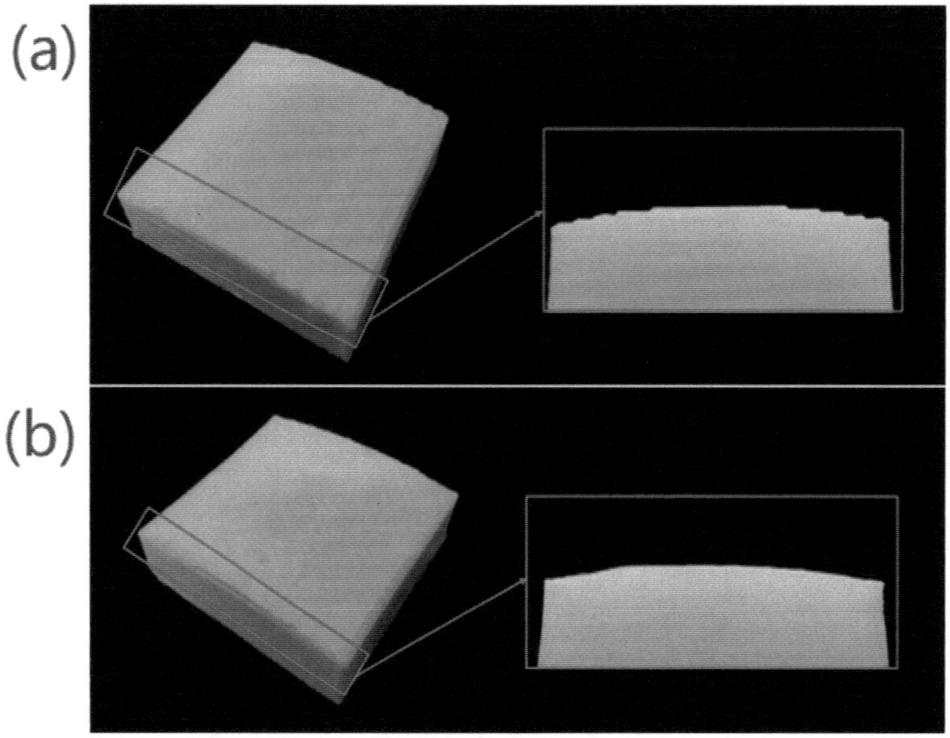

Figure 10. Print result: (a) Planar printing model; (b) Planar and non-planar hybrid printing model.

4. Conclusions

Aiming at the problem of step effect, a trajectory generation method for planar and non-planar hybrid printing is proposed. By extracting the non-planar printing area where the inclination angle between the model surface and the horizontal plane is less than the set inclination threshold, the collision detection and position adjustment are performed on the contour point set after the planar slicing and the non-planar printing area, and the contour point set of the planar slicing is filled. Then, the boundary point set of the non-planar area is obtained, and the projection, filling and mapping are carried out back to the non-planar area to complete the planar and non-planar hybrid printing trajectory generation.

The experimental results show that the proposed algorithm can quickly and effectively realize the generation of planar and non-planar hybrid printing trajectories by data processing of the mesh model. Compared with the model generated by the traditional planar slicing, the step effect is effectively reduced, and the printing quality of the model surface is improved. At the same time, the printer does not need to add additional degrees of freedom, thereby reducing the demand for printer hardware.

Acknowledgments

This work was supported by the Hubei Provincial Key Research and Development Program (2021BCA106) and the National Key Research and Development Program (2018YFB1105503).

Journal of Physics: Conference Series **2842** (2024) 012097 doi:10.1088/1742-6596/2842/1/012097

References

[1] Kristiawan R B, Imaduddin F, Ariawan D, et al. A review on the fused deposition modeling (FDM) 3D printing: Filament processing, materials, and printing parameters [J]. Open Engineering, 2021, 11(1): 639-649.
[2] Ngo T D, Kashani A, Imbalzano G, et al. Additive manufacturing (3D printing): A review of materials, methods, applications and challenges [J]. Composites Part B: Engineering, 2018, 143: 172-196.
[3] Gibson I, Rosen D W, Stucker B, et al. Additive manufacturing technologies [M]. Cham, Switzerland: Springer, 2021.
[4] Kechagias J, Chaidas D, Vidakis N, et al. Key parameters controlling surface quality and dimensional accuracy: A critical review of FFF process [J]. Materials and Manufacturing Processes, 2022, 37(9): 963-984.
[5] Mao H, Kwok T H, Chen Y, et al. Adaptive slicing based on efficient profile analysis [J]. Computer-Aided Design, 2019, 107: 89-101.
[6] Song H C, Ray N, Sokolov D, et al. Anti-aliasing for fused filament deposition [J]. Computer-Aided Design, 2017, 89: 25-34.
[7] Zwier M P, Wits W W. Design for additive manufacturing: Automated build orientation selection and optimization [J]. Procedia Cirp, 2016, 55: 128-133.
[8] Li L, Haghighi A, Yang Y. Theoretical modelling and prediction of surface roughness for hybrid additive–subtractive manufacturing processes [J]. IISE Transactions, 2019, 51(2): 124-135.
[9] Isa M A, Lazoglu I. Five-axis additive manufacturing of freeform models through buildup of transition layers [J]. Journal of Manufacturing Systems, 2019, 50: 69-80.
[10] Shan Y, Gan D, Mao H. Curved layer slicing based on isothermal surface [J]. Procedia Manufacturing, 2021, 53: 484-491.
[11] Ghadai S, Jignasu A, Krishnamurthy A. Direct 3D printing of multi-level voxel models [J]. Additive Manufacturing, 2021, 40: 101929.

Journal of Physics: Conference Series **2842** (2024) 012098 doi:10.1088/1742-6596/2842/1/012098

Parameter characterization of the PBM-DEM method in numerical simulation of shallow buried mine explosions

Pengpeng Liu [1], Hui Ye [2,*], Gangfeng Deng [2] and Juan Chen [1]

[1] High-tech Institute, Qingzhou, Shandong, 262500, China
[2] College of Missile Engineering, Rocket Force University of Engineering, Xi`an, Shaanxi, 710025, China

[*] Corresponding author's e-mail: yehui1234567890@outlook.com

Abstract: To investigate the impact of various parameters on the numerical simulation of shallow buried mine explosions through the particle explosion method and discrete element method, this paper adopts a coupled method of particle blast method and discrete element method, namely the PBM-DEM method, to simulate the response of AL-6XN stainless steel plate under shallow buried mine explosions. This article first combines the experimental results of Børvik to validate the precision of the PBM-DEM approach. Then, the influence of particle quantity, interaction parameters between sand particles, and interaction parameters between sand particles and plate on the outcomes of the numerical simulations are analyzed. The analysis reveals that as the quantity of explosive particles rises, the central displacement and maximum kinetic energy of the plate escalate. In contrast, these values decrease as the number of sand particles increases. The interaction parameters between sand particles have almost no effect on the numerical simulation results. The damping and stiffness parameters between the sand particles and the plate have a pronounced effect on the numerical simulation results, and the center displacement and peak kinetic energy of the plate rise as the damping and stiffness parameters between the sand particles and the plate increase.

1. Introduction

Vehicle damage and personnel casualties are mainly caused by buried landmines or other forms of improvised explosive devices in land combat environments. Therefore, how to improve the survival ability of vehicles and personnel under the explosive load of buried landmines or improvised explosive devices has become a research hotspot for scholars and experts [1]. However, due to the explosion process being completed quickly and the high requirements for experimental conditions and costs, the collection of experimental data is difficult, and accuracy cannot be guaranteed. Using numerical simulation can reduce research costs, and the explosion and target response processes can be observed with the characteristic of repeatability [2].

In LS-DYNA, a new numerical simulation method, the Particle Blast Method (PBM), is gradually being applied in the numerical simulation of explosive loads, and it can be combined with the Discrete Element Method (DEM) to simulate the interaction between detonation products, the surrounding air, sand, and structures. Olovsson et al. [3] established the theoretical foundation of the particle blast method by improving kinetic molecular theory. Subsequently, Børvik et al. [4] established numerical models of explosion loads using the particle blast and discrete element methods and analyzed the failure modes of steel plates under explosion loads. Jensen et al. [5] analyzed the failure of vehicles under different explosion conditions using the particle explosion method and the discrete element method. Zhou et al. [6] analyzed the influence of discrete element particle radius size on numerical simulation results. The outcomes of the numerical simulations correspond well with the experimental data. These studies have demonstrated that the particle blast and discrete element methods can be effectively applied to the numerical simulation of explosive loads.

To study the applicability of PBM and DEM in the numerical simulation of shallow buried mine explosions, a PBM-DEM explosion model was established based on Børvik's AL-6XN stainless steel explosion test, and the explosion model's reliability was confirmed through comparison with experimental data [4]. After that, an analysis was conducted to examine the impact of particle quantity and the interaction parameters between sand particles and between sand particles and the plate on the outcomes of numerical simulations.

2. Overview of the PBM and DEM

2.1 Particle blast method

Based on the Kinetic Molecular Theory (KMT), Olovsson et al. [3] developed the Corpus Molecular Method (CPM). To improve computational efficiency, it is assumed that each particle represents many molecules, and these particles are rigid and spherical. The KMT theory within molecular dynamics makes several assumptions:

1) The average distance between the molecules is large compared to the size of the molecules themselves.
2) The system is in thermodynamic equilibrium, meaning the molecules are in random motion.
3) The molecules follow Newton's laws of motion.
4) Interactions between molecules and between molecules and the container walls are assumed to be perfectly elastic.

The explosion process involves high temperatures and pressures, accompanied by rapid airflow velocities, which are incompatible with the assumption of a system in thermal equilibrium. Additionally, there is a decrease in pressure in the gas as the explosion occurs. The PBM enhances the CPM approach to better simulate the real gas laws under high-pressure and high-temperature conditions. PBM introduces the co-volume effect parameter to represent gas behavior under extreme pressure better. The interaction between particles and structures is a purely elastic collision.

2.2 Discrete element method

The DEM numerical calculation method establishes a finite discrete model by discretizing physical elements and forming node relationships with clear physical meanings. The motion of individual elements is entirely determined by the magnitude of the unbalanced force and moment acting on the element according to Newton's second law of motion [7]. The DEM is suitable for simulating sand particles and does not require the constitutive relationship of sand particles.

The schematic of the model of particle interactions used in LS-DYNA is shown in Figure 1, with interaction parameters including normal spring stiffness K_n, tangential spring stiffness K_s, normal damping coefficient v_n, tangential damping coefficient v_s, static friction coefficient μ, rolling friction coefficient δ and normal capillary action γ.

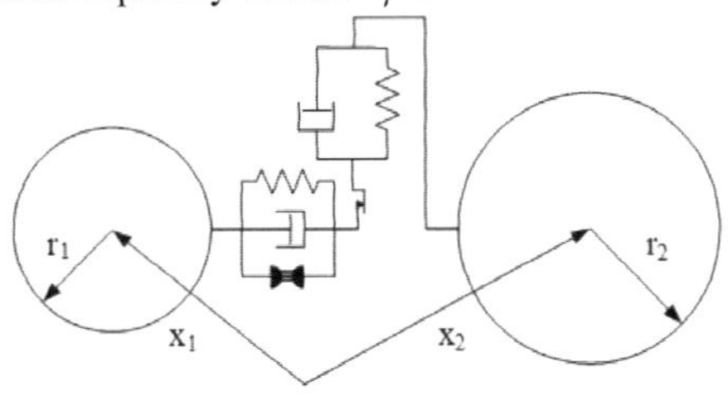

Figure 1. Schematic of the model of particle interactions used in LS-DYNA.

The sand particles selected in this article are dry sand, and the capillary effect between particles is not considered. At the same time, to reduce calculation costs, only the translational motion of the sand particles is considered in the simulation, with the degrees of freedom related to rotation or other non-translational movements being omitted. The relationship between K_n, K_s, v_n and v_s of the ANSYS/LS-DYNA discrete element particle interaction model and the particle material and geometric parameters is as follows.

$$K_n = \begin{cases} \dfrac{k_1 r_1 k_2 r_2}{k_1 r_1 + k_2 r_2} \text{NORMK} & \text{if NORMK} > 0 \\ |\text{NORMK}| & \text{if NORMK} < 0 \end{cases} \quad (1)$$

$$v_n = 2 \cdot \text{NDAMP} \sqrt{\frac{m_1 m_2}{m_1 + m_2} K_n} \, (0 \ll \text{NDAMP} \ll 1.0) \quad (2)$$

$$K_s = \text{SHEARK} \cdot K_n \quad (3)$$

$$v_s = 2 \cdot \text{TDAMP} \sqrt{\frac{m_1 m_2}{m_1 + m_2} K_s} \, (0 \ll \text{TDAMP} \ll 1.0) \quad (4)$$

Here r_i is the particle radius, k_i is the volume modulus of the particle, and m_i is the particle mass. NORMK, NDAMP, SHEARK, and TDAMP are the specific interaction parameters that need to be obtained.

3. Explosion model validation

This article refers to the experimental data in [4] and establishes a PBM-DEM explosion model. For explosives, the PBM method is used for modeling. The simulation involves 10, 000 explosive particles, i.e. N_{exp}=10000. The corpuscular constants for C-4 are compiled in Table 1, where ρ_{exp} represents explosive density, D is detonation velocity, γ is heat capacity ratio at constant pressure and constant volume, b is co-volume fraction, and E_0 is explosive energy per unit volume. The sand particles are modeled using the DEM. The constitutive model of sand particle materials is defined using MAT_ RIGID, and the constants are summarized in Table 2 where ρ_{sand} represents sand density, E_{sand} is sand elastic modulus and v_{sand} is sand Poisson's ratio.

Table 1. Corpuscular constants for C-4.

$\rho_{\text{exp}}\left(\text{kg/m}^3\right)$	$D(\text{m/s})$	γ	b	$E_0\left(\text{GJ/m}^3\right)$
1601	8190	1.4	0.35	8.7

Table 2. Corpuscular constants for sand.

$\rho_{\text{sand}}(\text{kg/m}^3)$	$E_{\text{sand}}(\text{GPa})$	v_{sand}
2700	60	0.25

The interaction parameters between sand particles are controlled through the Control_ DISCRETE_ Element keyword definition. Based on [4] and LS-DYNA_Manual_Volume, we set NDAMP=TDAMP=0, NORMK=SHEARK=0.01 and $\mu = 0.3$. The interaction between sand particles and a plate is specified using the DEFINE_DE_TO_SURFACE_COUPLING keyword, mainly through the static friction coefficient FRICS, damping coefficient DAMP, and contact stiffness between sand particles and plate. The relationship between the contact stiffness K between particles and plate and NORMK is:

$$K = \text{SFP} \times \begin{cases} k \times r \times \text{ NormK if NormK} > 0 \\ |\text{NormK}| \text{ if NormK} < 0 \end{cases} \quad (5)$$

Here, SFP is the proportional factor of the indirect contact stiffness between the particles and the plate. The summary of relevant parameter settings in the discrete element particle interaction model is shown in Table 3.

Table 3. Parameter settings in the discrete element particle interaction model.

NORMK	NDAMP	SHEARK	TDAMP
0.01	0.00	0.01	0.00
μ	FRICS	DAMP	SFP
0.30	0.30	0.20	6.00

To ensure the accuracy of the explosion model, it is important to construct a numerical simulation model consistent with the experimental setup described by Børvik et al. [4]. The schematic diagram of the test device is shown in Figure 2. The total number of sand particles N_{sand} is about 50000. The finite element model of steel plate adopts solid element modeling, with a mesh density of $40\times40\times1$, and modeling plates using MAT_MODIFIED_JOHNSON_COOK constitutive equations, and the stress-strain relationship of the modified Johnson Cook model is:

$$\sigma_{\text{eq}}=\left(A+B\varepsilon_{\text{eq}}^{n}\right)\left(1+\dot{\varepsilon}_{\text{eq}}^{*}\right)^{C}\left(1-T^{*m}\right) \tag{6}$$

Here σ_{eq} is the equivalent stress, is the equivalent plastic strain, and A, B, n, C, m are material parameters. The dimensionless plastic strain rate is $\dot{\varepsilon}_{\text{eq}}^{*}=\dot{\varepsilon}_{\text{eq}}/\dot{\varepsilon}_{0}$, where $\dot{\varepsilon}_{0}$ is the reference strain rate. The dimensionless temperature is $T^{*}=\left(T-T_{\text{r}}\right)/\left(T_{\text{m}}-T_{\text{r}}\right)$, where T is the absolute temperature, T_{r} is the ambient temperature and T_{m} is the melting temperature of the material. The material constants are shown in Table 4.

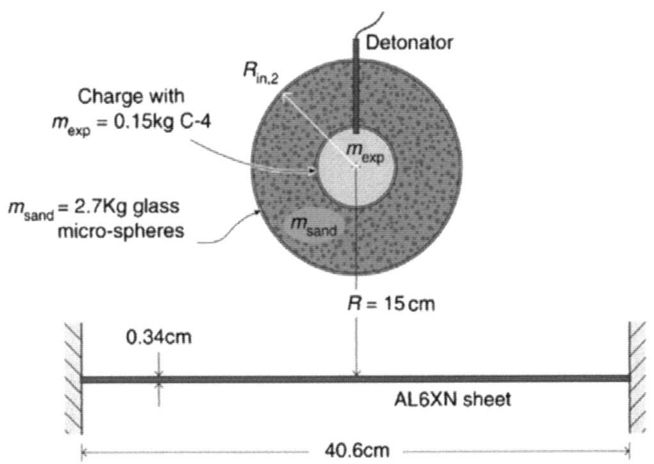

Figure 2. Schematic diagram of experimental setup in [4].

Table 4. Material constants for AL-6XN.

$E(\text{GPa})$	ν	$\rho_{\text{exp}}\left(\text{kg/m}^3\right)$	$A(\text{MPa})$	$B(\text{MPa})$	n	$\dot{\varepsilon}_0\left(\text{s}^{-1}\right)$
195	0.3	8060	410	1902	0.82	0.001
C	T_r	T_m	m	$C_p\left(\text{J/kg k}\right)$	χ	α
0.024	296	1700	1.03	500	0.9	1.5×10^{-5}

The simulation results obtained in this article are shown in Figure 3, with a center displacement of 31.4 mm for the steel plate. The experimental and numerical simulation results in [4] are 38.5 mm and 31.7 mm, respectively. The outcomes of the numerical simulations presented in this article confirm the efficacy and precision of the simulation methodology and the chosen parameter values.

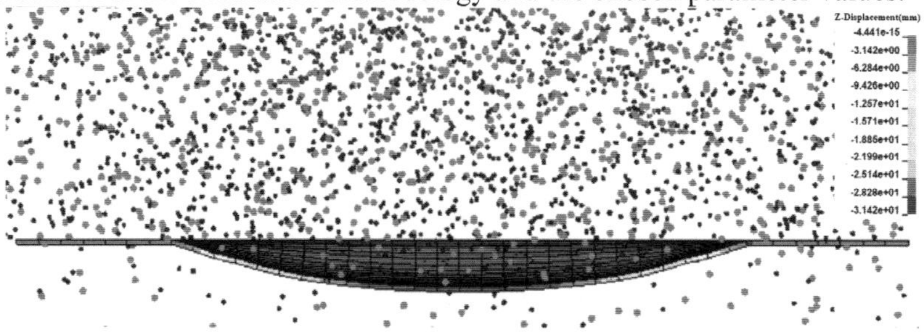

Figure 3. Simulation results obtained in this article.

4. Influence on parameter characterization of the PBM-DEM method

4.1 Influence of particle quantity

Figure 4 shows the changes in the center displacement and kinetic energy time history curve of the plate with different numbers of explosive particles when the number of sand particles N_{sand} is 50000. From the results, it can be observed that when the number of sand particles is held constant, the analysis reveals that as the number of explosive particles increases, the central displacement of the plate also increases. Additionally, the peak kinetic energy of the plate becomes higher with more explosive particles. Furthermore, the time taken to reach the maximum kinetic energy decreases as the number of explosive particles increases.

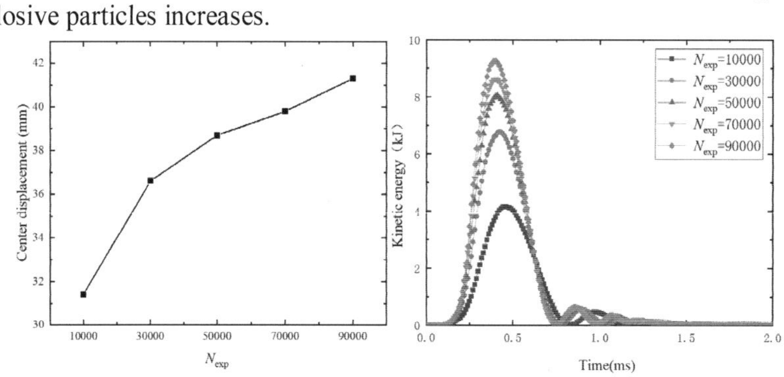

Figure 4. a) Changes in the center displacement b) Kinetic energy time history curve of the plate with different N_{exp}.

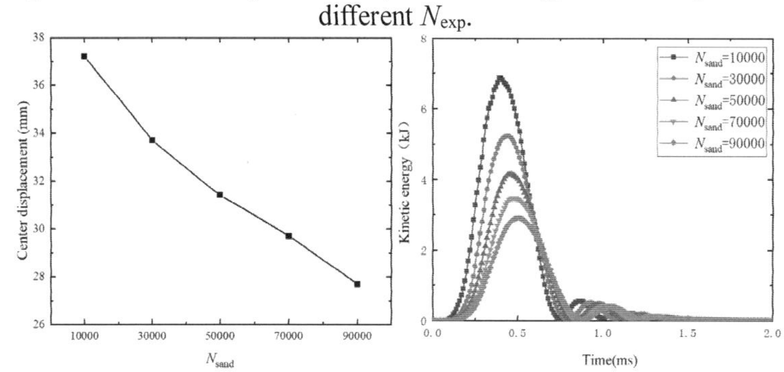

Figure 5. a) Changes in the center displacement b) Kinetic energy time history curve of the plate with different N_{sand}.

Figure 5 shows the changes in the center displacement and kinetic energy time history curve of the plate with different numbers of sand particles. In contrast, the number of explosive particles N_{exp} is kept at 10000. From the results, it can be seen that when the quantity of explosive particles is kept steady, the center displacement of the plate shows an approximate decreasing relationship with the number of sand particles. The peak kinetic energy of the plate decreases with the increase of sand particle number, and the time to reach the peak kinetic energy also increases with the increase of explosive particle number.

4.2 Influence of Interaction Parameters between Sand Particles

Figures 6 and 7 show the changes in the center displacement and kinetic energy time history curve of the plate with different NORMK/SHEARK and μ, respectively. From the results, the alterations in NORMK/SHEARK and μ parameters do not significantly influence the central displacement and maximum kinetic energy of the plate. The kinetic energy time history curve of the plate is almost the same, so it can be considered that changes in the interaction parameters between sand particles have almost no effect on the numerical simulation results.

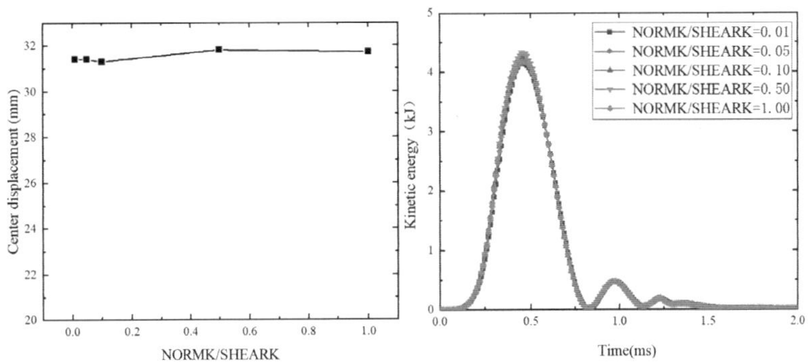

Figure 6. a) Changes in the center displacement b) Kinetic energy time history curve of the plate with different NORMK/SHEARK.

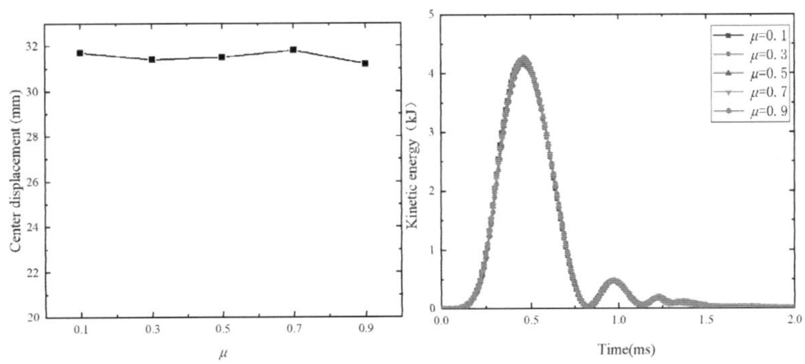

Figure 7. a) Changes in the center displacement b) Kinetic energy time history curve of the plate with different μ.

4.3 Influence of Interaction Parameters between Sand Particles and Plate

Figure 8 shows the variation of the center displacement of the plate and the kinetic energy time history curve of the plate when FRICS changes. The results show that FRICS has little effect on the numerical simulation results and almost no effect on the variation of the center displacement and kinetic energy of the plate.

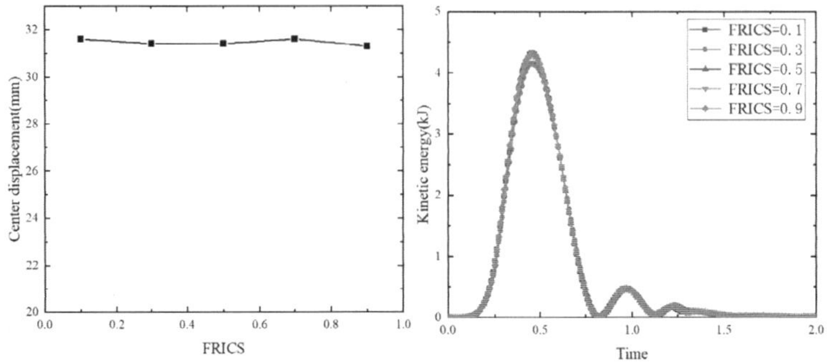

Figure 8. a) Changes in the center displacement b) Kinetic energy time history curve of the plate with different FRICS.

Figure 9 shows the variation of the center displacement of the plate and the kinetic energy time history curve of the target plate when DAMP changes. The data indicates that with an increase in DAMP, the center displacement of the plate escalates gradually while the acceleration experiences a decline. The maximum kinetic energy of the plate exhibits a pattern of initially rising and then plateauing as DAMP increases. Similarly, the time required to attain the peak kinetic energy follows a trend of initially descending and then leveling off with the growth of DAMP. When DAMP is greater than 0.8, the kinetic energy time history curve of the target plate is almost the same.

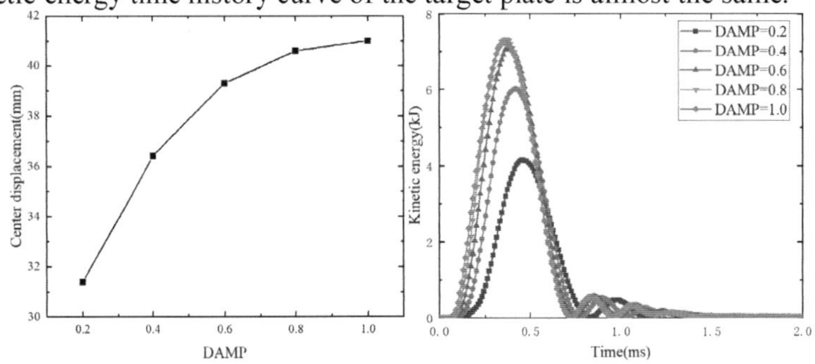

Figure 9. a) Changes in the center displacement b) Kinetic energy time history curve of the plate with different DAMP.

Figure 10 shows the variation of the center displacement of the plate and the kinetic energy time history curve of the plate when SFP changes. The data suggests that as SFP rises, the center displacement of the plate tends to increase gradually. Moreover, when SFP is greater than 4, the central displacement of the plate appears to have an approximately linear correlation with SFP. The maximum kinetic energy of the plate tends to rise progressively as SFP increases, while the time taken to reach the peak kinetic energy shows a slight decrease with the growth of SFP.

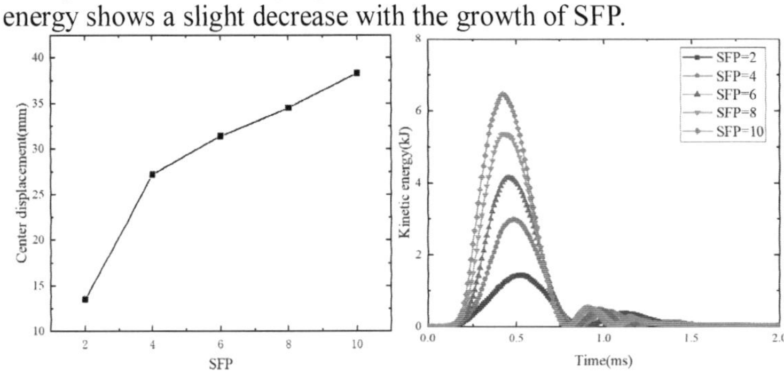

Figure 10. a) Changes in the center displacement b) Kinetic energy time history curve of the plate with different SFP.

Journal of Physics: Conference Series **2842** (2024) 012098 doi:10.1088/1742-6596/2842/1/012098

5. Conclusion

This article studies the parameter characterization of the particle blast method and discrete element method in numerical simulation of shallow buried mine explosions and investigates the effects of particle quantity, as well as the interaction parameters between sand particles and between sand particles and the plate, on the outcomes of numerical simulations. The data reveal that as the number of explosive particles on the plate increases, the center displacement and peak kinetic energy also increase. Conversely, as the number of sand particles increases, the center displacement and peak kinetic energy decrease. The interaction parameters between sand particles have almost no effect on the numerical simulation results. The numerical simulation outcomes greatly influence the damping and stiffness parameters exchanged between the sand particles and the plate. An increase in these parameters results in higher center displacement and peak kinetic energy for the plate.

References

[1] S Karmakar and S Amit, "Response of RC plates under blast loading using FEM-SPH coupled method", Engineering failure analysis, vol.125, pp.105409, Jul 2021.

[2] X Zhang, Y Ding, and Y Shi, "Numerical simulation of far-field blast loads arising from large TNT equivalent explosives", in Journal of Loss Prevention in the Process Industries, vol.70, pp.104432, May 2021.

[3] L Olovsson, A G Hanssen, T Børvik, and M Langseth, "A particle-based approach to close-range blast loading", in European Journal of Mechanics - A/Solids, vol.29, no.1, pp.1-6, January–February 2010.

[4] T Børvik, L Olovsson, A G Hanssen, K P Dharmasena, and H Hansson, "A discrete particle approach to simulate the combined effect of blast and sand impact loading of steel plates", in Journal of the Mechanics & Physics of Solids, vol.59, no.5, pp.1-6, May 2011.

[5] M R Jensen, W Smith, and K Khanna, "Parameter characterization of a buried mine blast event with further emphasis on sympathetic detonation and layered soil bed conditions", in The Journal of Defense Modeling & Simulation, vol.19, no.4, pp.659-677, Oct 2022.

[6] D Zhou, X Wang, Y Zhou, and X Sun, "Comparison of numerical approaches for modeling a scaled underbelly structure subjected to landmine impulse", in Shock Waves, vol.29, no.4, pp.573-582, May 2019.

[7] C S Zhang, C Feng, and K Xue, "Numerical study of blast wave propagation through granular materials subjected to buried blasts", in Powder Technology, vol.429, pp.118940, Nov 2023.

Journal of Physics: Conference Series **2842** (2024) 012099 doi:10.1088/1742-6596/2842/1/012099

Modeling and prediction of workpiece nonlinear physical quantities based on extended Kalman filter

Mengdi Zhang, Jiping Lu and Faping Zhang[*]

School of Mechanical Engineering, Beijing Institute of Technology, Beijing 100081, China

[*]Corresponding author's e-mail: jzfpnew@163.com

Abstract. For the workpiece from the tool withdrawal time until the lifting of the fixture constraint is called the in-situ state. In this process, the workpiece's state quantity changes with time, and the state quantity can be regarded as a time series physical quantity. Therefore, the physical state and geometric physical quantities in the in-situ stage can be characterized by time sequence through the deformation theory. Kalman filter is a kind of optimal estimation method that includes two processes: prediction and correction. In the process of prediction estimation, only the optimal estimation of the last time needs to be stored, which can reduce the operation cost and improve the prediction efficiency. Since the state quantity of the workpiece in situ becomes nonlinear, this paper uses the extended Kalman filter for characterization modeling and nonlinear prediction, combines the typical heat-force coupling model with the extended Kalman filter for representation, and finally obtains the modeling characterization and result prediction of the filter state quantity.

1. Introduction

The jet disc is the core component of the liquid rocket engine. There are about 2,000 micro-pores distributed on the jet disc. The size of micro-pores is between nanometre and micron, which belongs to the mesoscopic scale range, and it is a typical micro-structure mainly studied in this paper. The geometric quantity mainly includes the geometric characteristic parameters such as deformation quantity and geometric size. The state quantity mainly includes temperature, residual stress, and clamping force. The manufacturing process of micro-structure parts can be divided into three stages, namely, the processing stage, the in-situ stage, and the off-line stage, as shown in Figure 1. This paper mainly studies the change of state quantity in the in-situ stage. The state quantities considered are temperature and residual stress. The in situ stage is the end of the cutting process, but the workpiece is still in the clamping state of the fixture [1, 2], and the state is transmitted by attenuation.

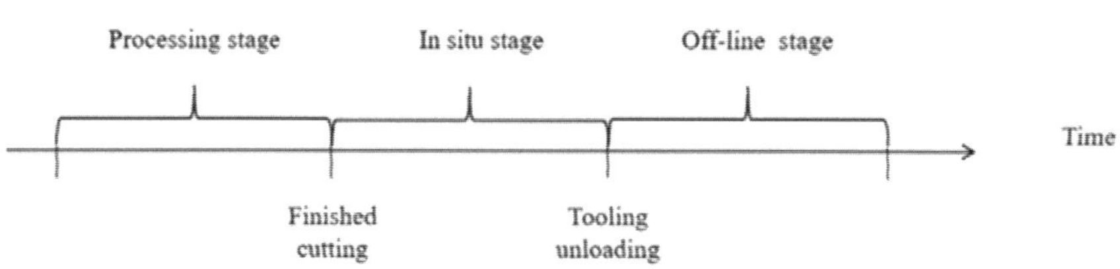

Figure 1. The processing and manufacturing stage of micro-structure parts.

There are three main modeling methods: Finite Element Method (FEM), Analytical Modelling Method (AMM), and Data Driven Method (DDM) [3]. The assumptions of AMM will reduce the accuracy of the results, while FEM reduces the error cost caused by simplification and assumptions [4], but time and mesh refinement make the calculation cost higher [5, 6]. AMM can build a physical quantity training model based on data samples, and replace the real solution with the output approximate solution (or optimal solution) [7], which can not only represent the model relationship between physical quantities, but also reduce the calculation cost.

In view of the relationship between physical quantities in situ and the analysis of changes in a single physical quantity without considering coupling, the in-situ temperature field is the result of multi-parameter characterization, which conforms to the theory of unsteady heat conduction [8]. After unloading and releasing the residual stress, a state of equilibrium is finally reached [9]. Therefore, it is necessary to consider the change of physical quantity under the heat-force coupling state [10] when constructing the model.

Based on the method of model and data fusion, Cerutti et al. [11, 12] predicted the deformation and nonlinear behaviour caused by residual stress redistribution. Rai et al. [13] established a three-dimensional process model to predict the nonlinear elastic-plastic deformation of the workpiece. The control system developed by Wang and Zhu et al. [14] can carry out the online measurement and predictive modeling, and the method used is similar to the process of Kalman filtering. Hao et al. [15] used deformation monitoring data to analyze the workpiece state and used the Kalman filtering algorithm to control the error between the predicted state quantity and the actual state quantity.

The focus of this paper is to use the extended Kalman filter to characterize the state change of the workpiece in situ under the influence of thermal and mechanical coupling and to predict and verify it.

2. Modelling of physical quantities

The meanings of parameters and functions are shown in Table 1. Among them, the global variables are obtained from the accumulation of discrete elements.

Table 1. Parameter and function meaning.

Nomenclature			
C_T	The global heat capacity matrix	F_T	The equivalent thermal load vector caused by the global temperature vector
$\dot{T}$	The time first derivation of the temperature field vector	S_p	The position matrix
K_T	The global heat conduction matrix	P(T)	The position load vector
T	the temperature vector matrix	C	The global damping matrix
T_0	The initial global temperature vector	K	The global stiffness matrix
P_T	The applied thermal load vector	M	The global mass matrix
$\dot{u}$	The time first derivative of the displacement vector	ξ_u , ξ_T	The characteristic parameters of nonlinear terms
$\ddot{u}$	The time second derivative of the displacement vector	g(·)	Inherent nonlinear model functions of physical quantities
F_0	The applied external force field load vector	L	The location selection matrix
X_u	The state vector of the workpiece deformation	P(h)	Relationship between nonlinear deformation and nonlinear parameters
X_T	The temperature state vector of nonlinear system	h	The nonlinear parameter

In situ, thermal coupling analysis should be carried out according to the stress field and temperature field acting on the same part at the same time. The transient thermodynamic coupling problem of thermoelastic materials needs to satisfy the thermal load balance equation and the force load balance equation. The solution of the governing equation and the initial variables are as follows:

$$C_T\dot{T} + K_T T = P_T, \quad T|_{t=0} = T_0 \tag{1}$$

$$M\ddot{u} + C\dot{u} + Ku = F_0 + F_T = P(t) = S_p p(t), u|_{t=0} = u_0, \dot{u}|_{t=0} = \dot{u}_0 \tag{2}$$

3. Kalman filter theory

Kalman filter is an important optimal state estimator in modern control theory. Through data fusion and information fusion to improve the prediction accuracy, a state model can be proposed for random processes, which is suitable for discrete systems and can be represented by the matrix to estimate multiple variables. Kalman filter mainly includes uncertain system state transition equation and observation equation:

$$X(k) = A(k)X(k-1) + B(k)U(k-1) + W(k-1) \tag{3}$$

$$Z(k) = H(k)X(k) + V(k) \tag{4}$$

The state estimation process is mainly divided into two steps: predict and update.

The time renewal equation is expressed as:

$$X^{\wedge}(k \mid k-1) = A(k)X^{\wedge}(k-1 \mid k-1) + B(k)U(k-1) + W(k-1) \tag{5}$$

$$XP(k \mid k-1) = A(k)P(k-1 \mid k-1)A^T(k) + Q_k \tag{6}$$

The state update equation is expressed as:

$$K(k) = \frac{P(k|k-1)H^T(k)}{H(k)P(k|k-1)H^T(k)+R_k} \tag{7}$$

$$\begin{aligned}\hat{X}(k \mid k) = \hat{X}(k \mid k-1) + K(k)[Z(k) \\ -H(k)\hat{X}(k \mid k-1)]\end{aligned} \tag{8}$$

$$P(k \mid k) = [I - K(k)H(k)]P(k \mid k-1) \tag{9}$$

The principle is shown in the Figure 2:

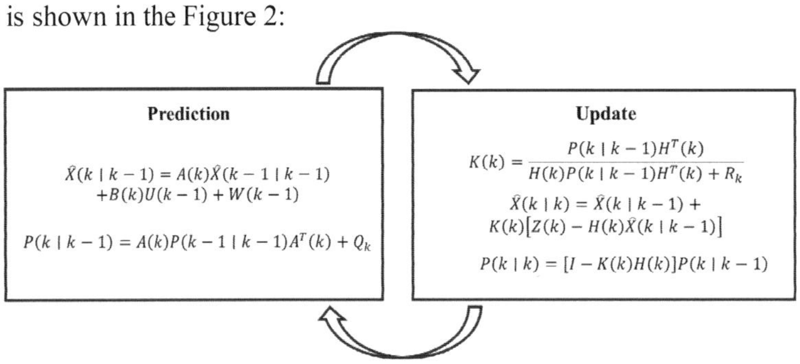

Figure 2. Kalman filtering principle.

4. Model construction based on extended Kalman filter

4.1. Mathematical model construction of nonlinear system

The nonlinear equation of thermodynamic coupling is expressed as:

$$F_{CT}[\dot{T}(t)] + F_{KT}[T(t)] = P_T(t) \tag{10}$$

$$\ddot{M}(t) + F_{Cu}[\dot{u}(t)] + F_{Ku}[u(t)] = P_u(t) \tag{11}$$

The thermodynamic coupling formula of elastoplastic materials in the second section is as follows:

$$C_T\dot{T}(t) + K_T T(t) = P_T(t) = S_p p_T(t)$$
$$T|_{t=0} = T_0 \tag{12}$$

$$M\ddot{u}(t) + \dot{C}\dot{u}(t) + Ku(t) = F_0(t) + F_T(t)$$
$$= P_u(t) = S_p p_u(t),\ u|_{t=0} = u_0, \dot{u}|_{t=0} = \dot{u}_0 \tag{13}$$

The thermal load and external load are further expressed as the product of the position matrix and the position load vector.

Since the amount of state in the in-situ stage is not linear, it is necessary to add a nonlinear part, so the system in the in-situ time period [0, t] is represented as:

$$C_T\dot{T}(t) + K_T T(t) + T_Z(T, \dot{T}, \xi_T) = P_T(t) = S_p p_T(t), T|_{t=0} = T_0 \tag{14}$$

$$\ddot{M}(t) + \dot{C}\dot{u}(t) + Ku(t) + u_z(u, \dot{u}, \xi_u)$$
$$= F_0(t) + F_T(t) = P_u(t) = S_p p(t) \tag{15}$$
$$u|_{t=0} = u_0, \dot{u}|_{t=0} = \dot{u}_0$$

$T_Z(T, \dot{T}, \xi_T)$ and $u_z(u, \dot{u}, \xi_u)$ are Nonlinear term sets.

For the nonlinear part, the nonlinear term is expressed in the form of a differential equation as follows:

$$\begin{cases} \dot{T}_z = g_T(T, \dot{T}, \xi_T) \\ \dot{u}_z = g_u(u, \dot{u}, \xi_u) \end{cases} \tag{16}$$

The above formula is expressed as a microelement form, and the global nonlinear part consists of m nonlinear microelements.

The differential expression of the nonlinear equation is:

$$\dot{Z}_i = g(Z_i, y_i, \dot{y}_i, \xi_i) \tag{17}$$

The differential expression of the system is:

$$\begin{cases} \dot{T}_{Z,i} = g_T(Z_i, T_i, \dot{T}_i, \xi_{T,i}) \\ \dot{u}_{Z,i} = g_u(Z_i, u_i, \dot{u}_i, \xi_{u,i}) \end{cases} \tag{18}$$

The position selection matrix is substituted to represent the position where there is a nonlinear physical quantity, and then the equations are expressed as:

$$\begin{cases} C_T\dot{T} + K_T T + L_T Z_T = P_T = S_p p_T \\ \ddot{M} + C\dot{u} + Ku + L_u Z_u = F_0 + F_T = P_u = S_p p_u \\ \dot{T}_z = g_T(T, \dot{T}, \xi_T) \\ \dot{u}_Z = g_u(u, \dot{u}, \xi_u) \\ \dot{u}(0) = u_0, \dot{u}(0) = \dot{u}_0 \\ T(0) = T_0, \dot{T}(0) = \dot{T}_0 \end{cases} \tag{19}$$

The above equations can represent nonlinear systems with nonlinear or local nonlinear characteristics and non-smooth differential equations.

4.2. Construction of nonlinear system space equation based on extended Kalman filter

According to the space state formula of the nonlinear system, the second-order ordinary differential equation is converted to the first-order differential system, then:

$$\dot{X}_u = \begin{bmatrix} \dot{u} \\ \ddot{u} \end{bmatrix} = \begin{bmatrix} 0 & I \\ -M^{-1}K & -M^{-1}C \end{bmatrix} \begin{bmatrix} u \\ \cdot \\ u \end{bmatrix} + \begin{bmatrix} 0 \\ -M^{-1}L_u \end{bmatrix} Z_u + \begin{bmatrix} 0 \\ M^{-1}S_p \end{bmatrix} p_u \tag{20}$$

Each matrix in the Kalman filter state expression is expressed as:

$$A_u = \begin{bmatrix} 0 & I \\ -M^{-1}K & -M^{-1}C \end{bmatrix}, B_u^z = \begin{bmatrix} 0 \\ -M^{-1}L_u \end{bmatrix}, B_u^p = \begin{bmatrix} 0 \\ M^{-1}S_p \end{bmatrix} \tag{21}$$

When the parameters of nonlinear deformation are substituted into the state vector matrix as estimated parameters, the workpiece deformation state equation of the extended Kalman filter is expressed as:

$$X_u = \begin{bmatrix} u \\ \dot{u} \\ h_u \end{bmatrix}, \dot{X}_u = \begin{bmatrix} \dot{u} \\ \ddot{u} \\ \dot{h}_u \end{bmatrix} \tag{22}$$

The matrix is expressed as:

$$A_u' = \begin{bmatrix} 0 & I & 0 \\ -M^{-1}K & -M^{-1}C & -M^{-1}L_u S_u^z \\ 0 & 0 & 0 \end{bmatrix}, \ B_u^{p'} = \begin{bmatrix} 0 \\ M^{-1}S_p \\ 0 \end{bmatrix} \tag{23}$$

Similarly, the equilibrium of the nonlinear thermal load is expressed as:

$$\dot{X}_T = [\dot{T}] = [-C_T^{-1}K_T][T] + [-C_T^{-1}L_T]Z_T + [C_T^{-1}S_p]p_T \tag{24}$$

The matrix of the state space equation can be expressed as:

$$A_T = [-C_T^{-1}K_T], B_T^Z = [-C_T^{-1}L_T], B_T^P = [-C_T^{-1}S_p] \tag{25}$$

The parameter of nonlinear state quantity temperature change is substituted into the state vector matrix as the estimated parameter, and then the workpiece temperature state equation of the extended Kalman filter is expressed as:

$$X_T = \begin{bmatrix} T \\ h_T \end{bmatrix}; \dot{X}_T = \begin{bmatrix} \dot{T} \\ \dot{h}_T \end{bmatrix} \tag{26}$$

Each matrix in the Kalman filter state expression is expressed as:

$$A_T' = \begin{bmatrix} -C_T^{-1}K_T & -C_T^{-1}L_T S_T^z \\ 0 & 0 \end{bmatrix}, B_T^{P'} = \begin{bmatrix} C_T^{-1}S_p \\ 0 \end{bmatrix} \tag{27}$$

Thus, the extended Kalman filter is expressed for the state space equation and state space matrix of the nonlinear system state equilibrium equations.

5. Experiment and result

In order to verify the accuracy of the above model, the corresponding cutting test is set up, and the state quantity of the test process is measured. The error analysis is used to analyze the accuracy of the output results at the forecast time, and the test data of multiple cutting parameters are compared with the predicted data.

Corresponding cuboid workpieces and YG8 carbide cutting tools were used in the test. The physical properties of the sample are shown in Table 2.

Table 2. Physical property of sample.

Density	Specific heat capacity	Coefficient of heat conduction	Thickness	Length of side
7.93 g/cm3	0.5 KJ/(kgK)	21.5 W/(mK)	4 mm	16 mm

5.1. Temperature prediction verification

The case is verified by measuring the surface temperature of the workpiece near the residual stress-strain gauge in the X direction. The location of the temperature measurement point is shown in Figure 3.

Journal of Physics: Conference Series **2842** (2024) 012099 doi:10.1088/1742-6596/2842/1/012099

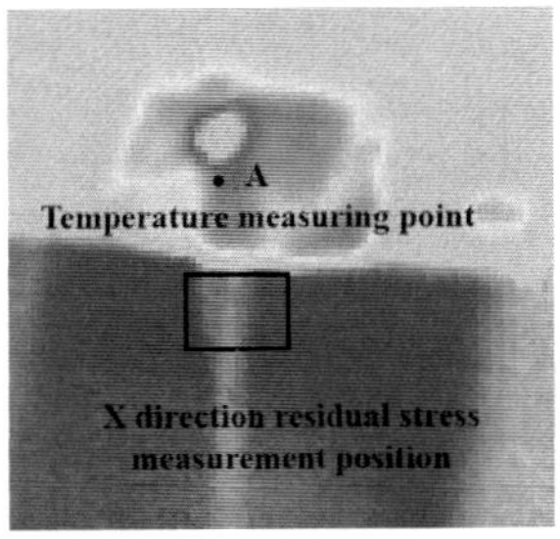

Figure 3. Temperature measuring point location.

The experimental parameters of time-varying temperature measurement for single-hole cutting are shown in Table 3.

Table 3. Single-hole temperature experiment process parameters.

Experiment number	Feed quantity (μm/rev)	Rotational speed (r/min)	Diameter (mm)
A1	10	6000	1.5
A2	10	4500	2.0
A3	10	3000	3.0

Through an extended Kalman filter and smooth processing of the predicted data, the comparison graph between the measured value and the predicted value is obtained, as shown in Figure 4.

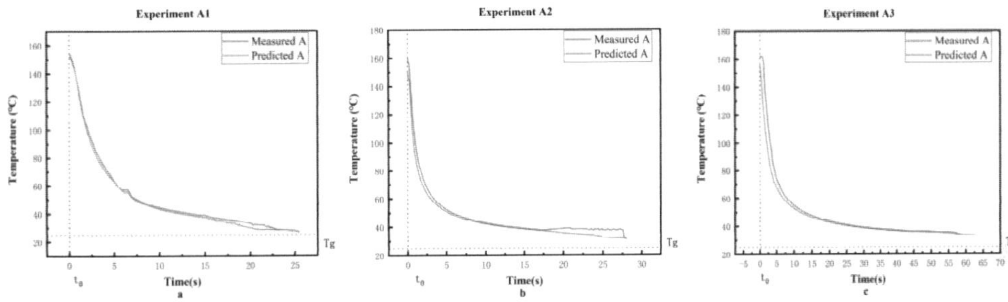

Figure 4. Experiment A1-A3 temperature measurement and predicted value comparison curve: (a) Experiment A1; (b) Experiment A2; (c) Experiment A3.

The initial temperature measurement input and room temperature measurement input of measuring points in different experimental groups are shown in Table 4.

Table 4. Extended Kalman filter initial temperature input and room temperature excitation input.

Experiment number	A1	A2	A3
Initial temperature (°C)	154.6	159.9	162.3
Room temperature (°C)	25	25	25

5.2. Result analysis

In the temperature prediction simulation verification, the average errors of the predicted value and the measured value at the output time of each optimal state quantity in experiments A1-A3 are E(A1)=3.718%, E(A2)=5.671%, and E(A3)=5.929%, respectively.

The error of temperature prediction is less than 10%, which meets the demand of actual prediction error. Moreover, it can be seen from the analysis of the prediction curve and the measurement curve that the trend of the prediction curve is basically consistent with the measurement curve, and the changing trend of the prediction curve is smaller than the actual measurement curve, which is caused by the fusion covariance distribution of the measured value and the predicted value in the extended Kalman filter.

6. Conclusion

In this paper, the nonlinear characterization of in-situ physical quantities of micro-structure parts is studied, and the combined modeling of the mathematical model and data-driven model is carried out by using the extended Kalman filter. The equivalent model and in-situ physical state of the thermal load are extracted, and the transient thermodynamic coupling equilibrium equation and boundary conditions of the thermoelastic material are used. The extended Kalman filter is used to construct a nonlinear system, which is not a linear quantity. Through verification experiments, it is proved that the predicted value of the model has the same trend as the measured value. The prediction errors of temperature and geometry at each estimation time are less than 10%, which meets the actual prediction demand and verifies the feasibility of the fusion model. At the same time, the coupling verification error of multiple physical quantities is smaller than that of a single physical quantity, indicating that the accuracy of the predicted value is higher with the increase of the input boundary conditions and the number of initial values.

Acknowledgments

This work was financially supported by the National ministry projects (JF2020-1-0422).

References

[1] Qiang Z, C W, Kai Y, et al. Plastic deformation behavior of a nickel-based superalloy on the mesoscopic scale [J]. Journal of Materials Science & Technology, 2020, 40:146-157.

[2] Sun F, Huo D, Fu G, Teng X, Kannan S, Zhang H. Micro-drilling of Mg-based MMCs reinforced with SiO2 nanoparticles: An experimental approach. Proceedings of the Institution of Mechanical Engineers, Part B: Journal of Engineering Manufacture, 2020, 234(12):1473-1485.

[3] Arrazola P J, Zel T, Umbrello D, et al. Recent advances in modelling of metal machining processes [J]. CIRP Annals - Manufacturing Technology, 2013, 62(2):695-718.

[4] Mabrouki T, Girardin F, Asad M, et al. Numerical and experimental study of dry cutting for an aeronautic aluminium alloy (A2024-T351) [J]. International Journal of Machine Tools & Manufacture, 2008, 48(11): 1187-1197.

[5] Shih, A. J. Finite Element Simulation of Orthogonal Metal Cutting [J]. ournal of Engineering for Industry, 1995, 117(1): 84-93.

[6] Mishra, S. K., Ghosh, S., Aravindan, S. Temporal and spatial crater wear prediction of WC/Co tools during dry turning of Ti6Al4V alloy [J]. Wear,2020,203229: 448-449.

[7] Aurich, J. C., Bil, H. 3D Finite Element Modelling of Segmented Chip Formation [J]. CIRP Annals, 2006,55(1): 47-50.

[8] Attetkov, A. V., Belyakov, N. S. The temperature field of an infinite solid containing a cylindrical channel with a thermally thin surface coating [J]. High Temp,2006,44:139-143.

[9] Hoger, A. On the determination of residual stress in an elastic body [J]. Elasticity,1986,16:303-324.

[10] Luo, B., Li, Y., Zhang, K., Cheng, H., Liu, S. A novel prediction model for thrust force and torque in drilling interface region of CFRP/Ti stacks [J]. The International Journal of Advanced Manufacturing Technology,2015, 81(9-12):1497-1508.

[11] Xavier C, Sami H, Emmanuel D, et al. Prediction of the Distortions Caused by the Redistribution of the Residual Stresses During Machining Using FEM [J]. Materials ence Forum, 2013, 768-769:398-405.

[12] Cerutti, X., Mocellin, K. Parallel finite element tool to predict distortion induced by initial residual stresses during machining of aeronautical parts [J]. Int J Mater Form,2015,8:255-268.

[13] Rai J K, Xirouchakis P. Finite element method based machining simulation environment for analyzing part errors induced during milling of thin-walled components [J]. International Journal of Machine Tools & Manufacture, 2008(48):629-643.

[14] Wang, X., Bi, Q., Zhu, L., et al. Improved forecasting compensatory control to guarantee the remaining wall thickness for pocket milling of a large thin-walled part [J]. Int J Adv Manuf Technol, 2018,94:1677-1688.
[15] Hao, X., Li, Y., Li, M., et al. A part deformation control method via active pre-deformation based on online monitoring data. Int J Adv Manuf Technol,2019,104: 2681-2692.

Journal of Physics: Conference Series **2842** (2024) 012100 doi:10.1088/1742-6596/2842/1/012100

Simulation analysis and test of the spinning process of 304 stainless steel cylindrical parts

Yaming Guo[1,*], Tao Huang[1], Yuan Zheng[1], Hengqiu Xu[1], Mingzhi Huang[1], Hongwei Chen[1], Nan Xia[1] and Kai Chen[1]

[1]Changchun Equipment & Technology Research Institute, Changchun 130012, China

[*]Corresponding author's e-mail: ymguo16@mails.jlu.edu.cn

Abstract. Taking the difference in wall thickness as the evaluation index, the simulation analysis of the spinning process of 304 stainless steel cylindrical parts is carried out. The influence of the roller nose radius, feed speed, and percent thickness reduction on the stress and strain of the roller is studied, and the optimum combination of process parameters is obtained. Through the spinning test, the stainless steel cylindrical parts with an outside diameter of 300.8 mm are successfully spun. After spinning forming and solution treatment, the tensile strength of stainless steel reaches 557~566 MPa, the yield strength reaches 220~232 MPa, and the elongation reaches 41.3%~42.5%.

1. Introduction

Spinning is a pressure processing method used to form thin-walled cylindrical parts, widely applied in aviation, aerospace, engineering machinery, and other industries [1-2]. Achieving optimal process parameters for spinning with roller offset can be challenging due to the complex relationship between various influencing factors affecting part quality, such as cracking, wrinkling, bulging, or expansion defects commonly encountered during the spinning process [3-4]. Simulation often proves time-consuming and uneconomical in determining the best parameter combinations for production purposes. In this study, we perform orthogonal experiments using Simufact for spinning processes [5-6]. The influence patterns of process parameters on workpiece formation are subsequently analyzed through simulation analysis while validating their accuracy through experimental verification.

2. Establishing finite element model

The spinning simulation of stainless steel cylindrical parts should first create the CAD models of spinning blanks and dies in UG. Then, it is necessary to convert the CAD models of blank, mandrel, tail stock, and spinning roller into stl file format and import them into Simufact. The spinning rollers are evenly distributed on the end face at 120° with the center line of the mandrel as the axis, and offset between the front and the middle spinning roller and the middle and the rear spinning roller is 5 mm. The simulation model created is shown in Figure 1.

In the actual production process, the mandrel, tail stock, and roller materials use Cr12MoV die steel. After quenching heat treatment processing, the hardness can reach HRC60-62. Compared with the strength and hardness of blank material, it is much higher. The elastoplastic deformation of the die can be almost ignored. Therefore, in the process of simulation analysis, the mandrel, tail stock, and rollers are defined as rigid bodies without heat conduction. No heat is transferred, and no deformation is produced, which is consistent with the actual production situation. The chemical composition of 304 stainless steel is shown in Table 1, and the basic mechanical properties of 304 stainless steel are shown in Table 2. The numerical value of the key material mechanical properties is used to replace the real stress-strain curve. Yield strength σ_s=205 MPa, tensile strength σ_b=520 MPa, elastic modulus E=2.0×105 MPa, and plastic modulus E'=1525 MPa.

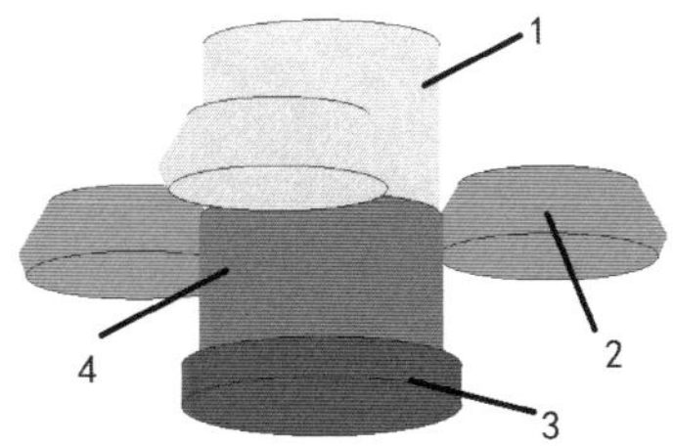

Figure 1. Simulation model of cylindrical parts, 1-mandrel, 2-spinning roller, 3-tail stock, 4-blank.

Table 1. 304 stainless steel chemical composition.

C	Mn	P	S	Si	Cr	Ni
≤0.08	≤2.00	≤0.045	≤0.030	≤1.00	18.0-20.0	8.0-10.5

Table 2. Basic mechanical properties of 304 stainless steel.

Elongation (%)	Shrinkage (%)	Yield Strength (MPa)	Tensile Strength (MPa)
35	60	205	520

3. Parameter setting

In the spinning process simulation analysis, the most important parameters include spindle speed S, spinning roller feed speed F, percent thickness reduction ψ, roller profile size, clearance, and offset. In this paper, finite element analysis is carried out through Simufact software, and tests are carried out to get the optimal spinning process parameters. Among them, the profile size of the spinning roller also includes the roller nose radius R, attacking angle α, and trailing angle β. The ratio of the feed speed F and the spindle speed S is called the feed rate f. In summary, the process parameters are determined, and the simulation parameters of the spinning of the stainless steel cylindrical parts are summarized as shown in Table 3. Other parameters are unchanged except for the parameters studied, blank length, inner diameter, wall thickness, die diameter, and axial offset.

Table 3. Simulation spinning process parameters of stainless steel cylindrical parts.

Name	Unit	Value
Roller Nose Radius r_p	mm	6, 8, 10
Attacking Angle α_p	°	front roller 20, middle roller 25, and rear spinning roller 30
Trailing Angle β	°	30
Spindle Speed S	r/min	100
Feed Speed F	mm/min	80/100/120
Percent Thickness Reduction ψ_t	%	30/40/50

4. Result analysis

The roller nose radius r_p, feed speed F, and percent thickness reduction ψ_t can be taken as three test factors, and the horizontal mode is the corresponding change value. The distribution of stress and strain values and wall thickness difference of spinning parameters are analyzed by Simufact through the orthogonal experiment of three factors and three levels.

4.1 Analysis of the influence of simulation process parameters on stress and strain

According to the comparison of the nephogram of the equivalent stress in each stage of the spinning process of 9 test schemes, the maximum equivalent stress is 958.81 MPa, and the minimum equivalent stress is 898.74 MPa. By observing the stress nephogram of the 2nd and 3rd spinning completion states, we can see that the equivalent stress of the workpiece after spinning completion is almost none, which can be used as the preferred parameter combination.

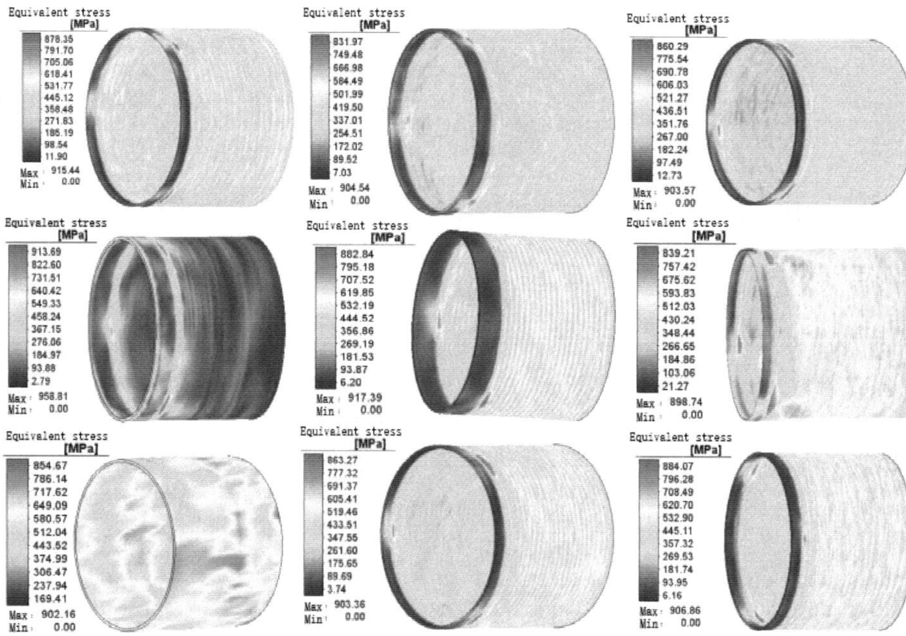

Figure 2. Equivalent stress nephogram under different process parameters.

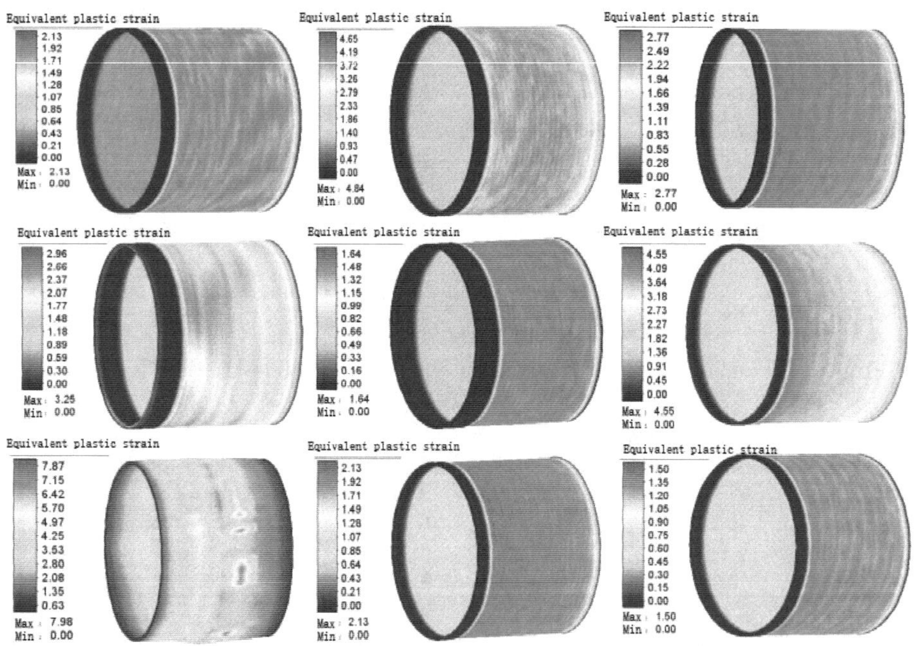

Figure 3. Equivalent plastic strain cloud image under different process parameters.

Through the comparison of the equivalent plastic strain nephogram of each stage in the spinning process of the nine groups of test schemes, the maximum equivalent plastic strain is 7.98, and the minimum is 1.5. Through the observation of the equivalent plastic strain nephogram of the spinning completion state of each group, the equivalent plastic strain of the formed parts decreases significantly with the passing of time and shows a uniform distribution. The work hardening of this part leads to a reduction in the plasticity of the material, resulting in deformation.

4.2 Influence of simulation process parameters on wall thickness difference

The spinning forming process of stainless steel cylindrical parts is simulated by Simufact, and the test data are obtained according to the requirements of the orthogonal test scheme. The results are shown in Table 4.

Table 4. Orthogonal test results.

No.	Process Parameters			Simulation Result
	A-r_p, mm	B-F, mm/min	C-ψ_t/%	Wall Thickness Difference, mm
1	6	80	30	0.065
2	6	100	50	0.048
3	6	120	40	0.079
4	8	80	40	0.057
5	8	100	30	0.068
6	8	120	50	0.089
7	10	80	50	0.087
8	10	100	40	0.095
9	10	120	30	0.074

According to the orthogonal test measurement simulation results in Table 4, the range analysis technique is used to accurately evaluate the wall thickness difference of stainless steel cylindrical parts as the evaluation index. Then, it is applied to the simulation of spinning formation. The specific simulation analysis results are shown in Table 5. In the range analysis table, y_{jk} is the sum of the corresponding objective function at the j factor and k level. The average of y_{jk} is $\overline{y_{jk}}$, and the difference between the maximum and minimum values of $\overline{y_{jk}}$ is the range, denoted by R_j. In the numerical simulation orthogonal test of spinning stainless steel cylindrical parts, the significance of the influence of the relevant process test parameters on the objective function is closely related to the range R_j. The larger the R_j value is, the more obvious the influence of these test process parameters on the evaluation index is.

Table 5. Wall Thickness Deviation Range Analysis Table.

Index	Wall Thickness Deviation Range Analysis		
	A	B	C
y_{j1}	0.192	0.209	0.207
y_{j2}	0.214	0.211	0.224
y_{j3}	0.256	0.242	0.231
$\overline{y_{j1}}$	0.064	0.0697	0.069
$\overline{y_{j2}}$	0.0713	0.0703	0.0747
$\overline{y_{j3}}$	0.0853	0.0807	0.077
R_j	0.0213	0.011	0.008

According to the data in Table 5, it can be clearly seen that the three simulation process parameters have a significant influence on the wall thickness of stainless steel cylindrical parts. In order to better demonstrate the influence of these variables, the deviation trend chart of the wall thickness of stainless steel cylindrical parts in Figure 4 is selected to describe their changes. In the figure, A represents the roller nose radius, B represents the feed speed, and C represents the percent thickness reduction. In the figure, 1, 2, and 3 represent the process parameters of the three variables respectively.

According to the range analysis results in Table 5, it is found that the significant influence of these main process test parameters on the wall thickness of stainless steel cylindrical parts is in the order of roller nose radius, feed speed, and percent thickness reduction. From Figure 4, it can be concluded that A1B1C3 is the optimal process parameter combination to achieve the optimization goal of wall thickness difference, and the corresponding data in the table are the roller nose radius r_ρ=6 mm, the feed speed F=80 mm/min, and the percent thickness reduction ψ_t=40%.

Considering the characteristics of the machining hardening of stainless steel materials and material properties, the two-pass forming test is adopted. The first pass takes a larger percent thickness reduction and feed speed, while the second pass takes a smaller percent thickness reduction and feed speed. In the case of a certain speed, the first pass has a larger feed speed to reduce the diameter of the workpiece. The roundness of the workpiece will be better. The second pass with a smaller feed speed can make the workpiece expand, convenient for discharge.

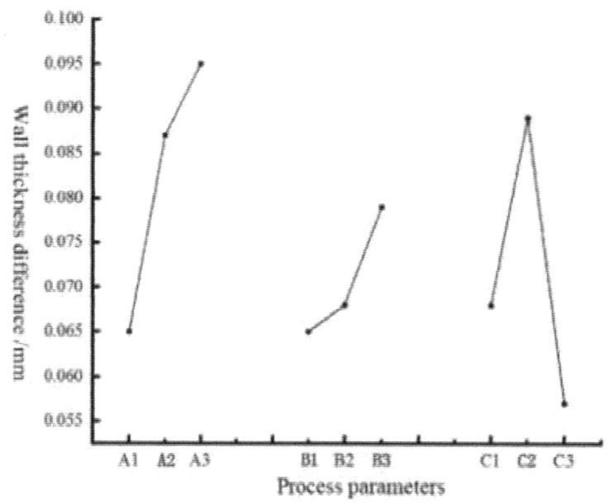

Figure 4. Trend chart of wall thickness difference of stainless steel cylindrical parts.

5. Results of the spinning test

According to the results of the numerical simulation analysis of the cylindrical section of the stainless steel cylindrical part, the design and processing of the mold are shown in Figure 5, and the spinning test of the stainless steel cylindrical parts is carried out.

Figure 5. Spinning model of the test, 1-tail stock, 2-blank, 3-mandrel, 4-roller.

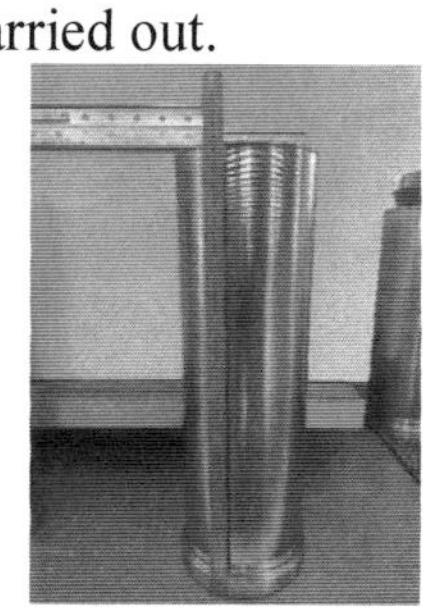

Figure 6. Stainless steel cylindrical parts finished sample.

5.1 Wall thickness and wall thickness difference

After two passes of the runout are compared with the expected value, there is no great deviation, indicating that the original process parameter combination is used for the first pass and the second pass of the knife spinning. The final spinning of the 12 mm thick cylindrical blank is shown in Figure 6, and the difference in the outer diameter and wall thickness is shown in Table 6. The inner diameter from 20 mm to 900 mm from the spinning mouth is 292.04-292.08 mm, and the wall thickness difference is 0.03-0.07 mm. The inner diameter of the spinning mouth is 292.55-292.71 mm within the range of 20 mm, and there is a flare defect.

Table 6. Outer Diameter and wall thickness difference of 12 mm thick short blank after final spinning.

Distance from the Mouth (mm)	Outer Diameter (mm)	Wall Thickness (mm)				All Thickness Difference (mm)
		A	B	C	D	
20	301.75	4.55	4.51	4.58	4.57	0.07
120	300.84	4.46	4.52	4.51	4.50	0.06
220	300.85	4.46	4.44	4.49	4.48	0.05
320	300.82	4.45	4.45	4.48	4.44	0.04
420	300.86	4.39	4.46	4.45	4.41	0.07
520	300.89	4.42	4.40	4.43	4.40	0.03
620	300.83	4.44	4.39	4.41	4.41	0.05
720	300.86	4.40	4.43	4.40	4.45	0.05
820	300.87	4.41	4.39	4.44	4.43	0.05

5.2 Spinning performance analysis

Taking the "GB/T 228-2010 metal material room temperature tensile method" as the standard, the tensile experiment of stainless steel cylindrical parts is carried out. Through solid solution treatment, with the furnace cooling, the stress concentration problem caused by spinning is eliminated. The sample section is cut in four directions of the cylinder. Then, the tensile test and material tensile test are conducted on the tensile testing machine. Tensile performance data of stainless steel cylindrical parts are shown in Table 7. It can be seen that the mechanical properties of stainless steel cylindrical parts have reached the standard requirements.

Table 7. Mechanical Properties of Formed Parts.

Sample	Heat Treatment	Tensile Strength (MPa)	Yield Strength (MPa)	Elongation (%)
1	Solution Treatment	557	220	41.3
2	Solution Treatment	565	225	41.6
3	Solution Treatment	560	232	42.5
4	Solution Treatment	566	228	41.8

6. Conclusion

Combined with the simulation results, the spinning process experiment of stainless steel cylindrical parts is carried out. The experimental results show that the blank thickness is 12 mm. The two-pass forming test is adopted, and the products with an outside diameter of 300.8 mm are successfully spun. The wall thickness difference is 0.04-0.08 mm, which proves that the simulation results can be referenced, and the mechanical performance analysis of stainless steel cylindrical parts is completed. The results show that: after spinning forming and solution treatment, the tensile strength of stainless steel reaches 557~566 MPa, the yield strength reaches 220~232 MPa, and the elongation reaches 41.3%~42.5%.

Acknowledgments
Thanks to the support of the Jilin Provincial Science and Technology Development Plan Project (20210201046GX).

References
[1] Music O, Elwood J M and Kawai K (2010) A review of the mechanics of metal spinning. Journal of Materials Processing Technology, 210(1): 3-23.
[2] Bai Q, Yang H and Zhan M (2008) Finite element modeling of power spinning of thin-walled shell with hoop inner rib. Transactions of nonferrous metals society of china, 18: 6-13.
[3] Xia Q, Xiao G and Long H (2014) A review of process advancement of novel metal spinning [J]. International Journal of Machine Tools and Manufacture, 85:100-121.
[4] Zhou Yu, Zhao Y, and Yu Z Q (2022) Numerical simulation of stagger spinning of cylindrical part with cross inner ribs. Journal of Shanghai Jiao Tong University, 56 62-69.
[5] Kim K J and Lee J W. (2022) Light-Weight Design and Structure Analysis of Automotive Wheel Carrier by Using Finite Element Analysis. International Journal of Precision Engineering and Manufacturing, 23 79-85.
[6] Cheng X Q, Lao Z Y and Xia Q X. (2013) Investigation on stress and strain distributions of hollow-part with triangular cross-section by spinning. Int. J. Materials and Product Technology, 47:162-174.

Bending stiffness analysis of regular polyhedron sandwich structures

Yingjie Wang[1] *, Shenghang Cai[2] and Qin Zhang[1]

[1]Huzhou Vocational & Technical College, Huzhou, Zhejiang, 313000, China
[2]Huzhou University, Huzhou, Zhejiang, 313000, China

*Corresponding author's e-mail:2016014@huvtc.edu.cn

Abstract: As a new type of sandwich structure, regular polyhedron sandwich structure has excellent structure performances such as high bending resistance and energy-absorbing features. However, the current design process of regular polyhedron sandwich structures often ignores the influence of unit-cell parameters on structural bending stiffness. Based on origami geometry, this study analyzes the geometric characteristics of polyhedron unit-cell structures. A simulation model was established to analyze the bending performance of regular polyhedron sandwich structures. The influence of the number of polyhedron surfaces and similarity transformation parameters on bending stiffness was analyzed by finite element software. Through finite simulation analysis, the influence graph of bending performance in the impact of various parameters was obtained. This analysis provides valuable insights for the optimization design of regular polyhedron sandwich structures.

1. Introduction

The polyhedron sandwich structure has the advantages of light weight, noise reduction, sound insulation, and excellent mechanical performance, which make it widely applied in the industry [1]. Current research corresponding to sandwich structures mainly focused on traditional ones such as honeycomb sandwich structures and foam sandwich structures. In previous studies about sandwich structures, Ciepielewski studied the static and dynamic characteristics of aluminum honeycomb sandwich structures through experiments [2-3]. Xiao proposed a design method of graded lattice sandwich structures through multiscale topology optimization [4]. On the other hand, regular polyhedron sandwich structures have been studied in recent years [5]. Qiu proposed novel polyhedral cells of truss core panels based on origami engineering [6]. Zhao analyzed the energy absorption properties of truss core panels and proposed an inexpensive available manufacturing method [7]. Long-established collision models of various polyhedron sandwich structures consider unit-cell transformation parameters as variables, aiming to optimize its collision energy absorption [8]. Liu investigated the influence of unit-cell transformation parameters on the collision energy absorption characteristics of regular polyhedron sandwich structures [9].

Previous studies mainly focused on traditional honeycomb sandwich structures and foam sandwich structures, while relatively little research on optimizing its energy-absorbing features was conducted. The effect of unit-cell structural parameters on bending stiffness is rarely studied, leading to certain deficiencies in the design of polyhedron sandwich structures. Therefore, this paper analyzes the geometric characteristics of regular polyhedron sandwich structures based on origami geometry. A finite simulation model is established to investigate its bending performance. The influence of different design parameters on bending performance was examined by conducting a finite element simulation analysis.

2. Modeling of regular polyhedron sandwich structure

2.1 Base circle and unit-cell structure

Assuming the side length of an equilateral polygon is $2R_0$, drawing circles at the vertices of it with radius R_0. Then, several circles tangent to each other will be obtained, which are defined as the base circles $Cir(O_1, O_2, O_3 ..., O_n)$. Point O is the center of the equilateral polygon. Figure 1 shows an example of the base circles for regular triangles. The tangent points of these base circles are B_1, C_1, and E_1, respectively. Lines OO_1, OO_2, and OO_3 intersect the base circles at A_1, A_2, and A_3. Then, we can obtain the unfolded diagram of the unit cell of the regular polyhedron, where A_1, A_2, and A_3 are its bases, and other isosceles triangles, such as B_2, A_2, and A_3, are the side surfaces, as shown in Figure 2. Therefore, the vertex angle of isosceles triangle α_n^0 is calculated as follows:

$$\alpha_n^0 = \left(\frac{\pi}{2} - \frac{\pi}{n}\right), (n \geq 3) \tag{1}$$

where n is the side of numbers of the base of the regular polyhedron unit cell.

Then, the side length of the base of the regular polyhedron is obtained as follows:

$$L = R_0(1 + cos\alpha_n^0) \tag{2}$$

In summary, the unit cell of a regular polyhedron is composed of the aforementioned equilateral-polygon base surface and the side surfaces that are isosceles triangles.

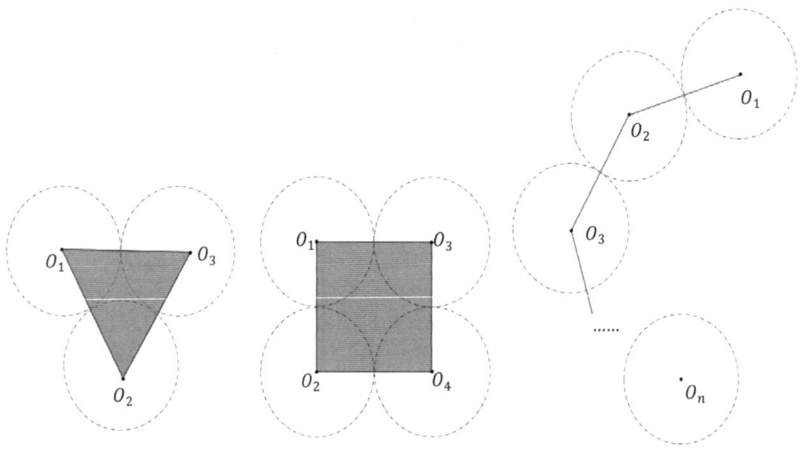

Figure 1. Base circle.

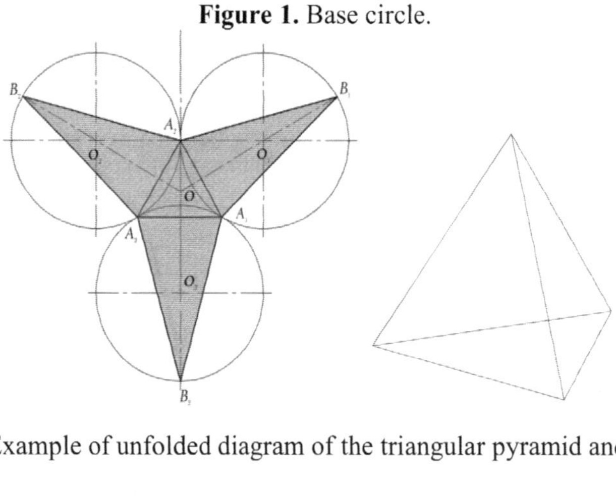

Figure 2. Example of unfolded diagram of the triangular pyramid and base circle.

Journal of Physics: Conference Series **2842** (2024) 012101 doi:10.1088/1742-6596/2842/1/012101

2.2 Similar transformation

As shown in Figure 3, the base circles are scaled proportionally with the vertices of a regular polygon as the centers. Then, we can obtain base circles with different radii. The intersection points between the base circles and the sides of the regular polygon are also changed accordingly. Thus, a new unfolded diagram of the unit cell is obtained. This operation is called a similar transformation. Here, the similarity ratio is defined as:

$$\omega = \frac{R}{R_0} \tag{3}$$

where R is the radius of the varied circle, and R_0 is the previous base circle.

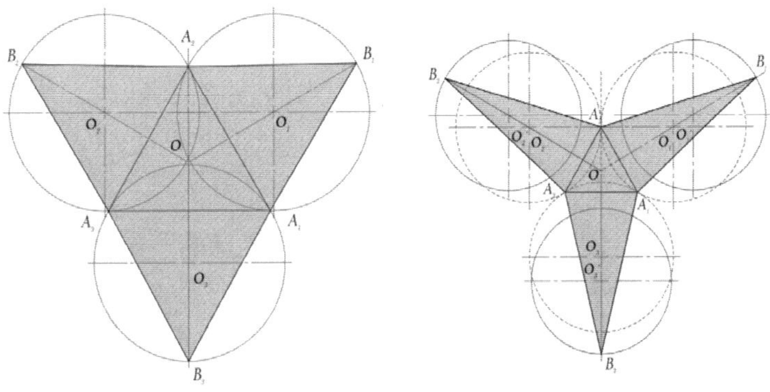

(a) $\omega>1$ (b) $0 < \omega < 1$

Figure 3. Unfolded diagram of cell structure after similarity transformation.

2.3 Shear transformation

Firstly, we draw an inscribed circle with center $\mathbf{O}$ in the base polygon. The diameter of this inscribed circle is changed by a scale factor $\mathbf{s}$, and a clipping circle can be obtained. Tangent lines are drawn to the clipping circle parallel to the corresponding sides of the base polygon. These tangent lines intersect with the sides of previous isosceles triangles at points $\boldsymbol{C_1}$, $\boldsymbol{D_1}$, $\boldsymbol{C_2}$, $\boldsymbol{D_2}$, $\boldsymbol{C_3}$ and $\boldsymbol{D_3}$, all of which transform the triangular faces into trapezoids. This operation is called shear transformation. The unfolded diagram of a regular polygon is shown in Figure 4 as an example. We define the shear ratio k as follows:

$$\mathbf{k} = \frac{\mathbf{1}}{s} \tag{4}$$

By controlling the radius of the clipping circle, the height of the regular polyhedron can be changed. The height can be calculated as follows:

$$\boldsymbol{h} = \boldsymbol{R_0}\sqrt{2\sqrt{2}+2} - \frac{\boldsymbol{R_0}\sqrt{2\sqrt{2}+2}\left[(\sqrt{2}+1)-\frac{1}{k}\right]}{\sqrt{2}+1} \qquad \sqrt{2}-1 \le k \le 1 \tag{5}$$

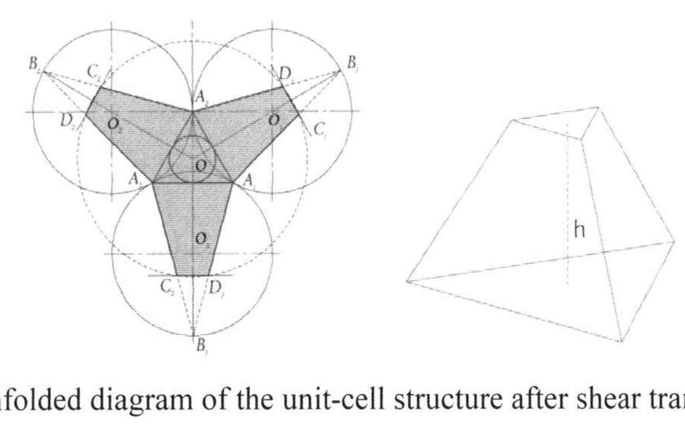

Figure 4. Unfolded diagram of the unit-cell structure after shear transformation.

2.4 Modeling of regular polyhedron sandwich structure

Through similarity transformations and shear transformation, different regular polyhedron unit cells can be obtained. The obtained unit-cell structure is a regular polyhedron pyramid. Assuming the side length of the regular polyhedron pyramid base is L, and the vertex angle of the side triangle is α_n. The formula for the height of the regular pyramid is:

$$h_n(L,\alpha_n)=\frac{L}{2}\sqrt{\frac{1}{\tan^2\frac{\alpha_n}{2}}-\frac{1}{\tan^2\frac{\pi}{n}}}\qquad \alpha_n\in(0,2\pi/n) \tag{6}$$

where L is the side length of the regular polyhedron pyramid base, and α_n is the vertex angle of the side surface.

A regular polyhedron unit-cell structure is obtained after the above similar and shear transformations. The regular polyhedron sandwich structure is composed of the top plate, unit-cell structure groups, and bottom plate, where the unit-cell structure is distributed like arrays, as shown in Figure 5.

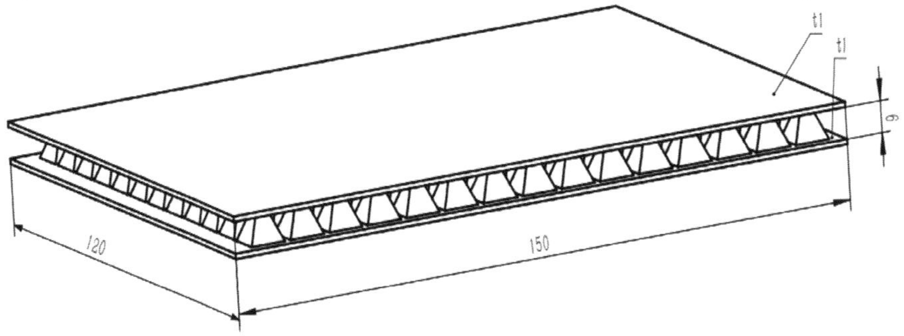

Figure 5. Schematic diagram of regular polyhedron sandwich structure.

3. Bending stiffness analysis

3.1 Finite element modeling

The finite element model of the polyhedron sandwich structure is created as shown in Figure 6. The size of the polyhedron sandwich structure is approximately 150mm×120mm. The polyhedron sandwich structure is supported by two fixed rigid components with a constant span. A rigid punch applies a press force to the structure along the centerline. The effects of the plastic forming process are neglected, and the plate thickness is assumed not to change after plastic forming. The other simulation parameters and material properties are shown in Table 1 and Table 2, respectively.

Figure 6. Bending analysis by simulation.

Table 1. Structural parameters of regular polyhedral sandwich structure.

Parameters	Values
Cell type	Regular polyhedral
Height of cell H, mm	6
Size of plane, $b_1 \times b_1$	150×120
Initial plate thickness, t	1 mm
Height of unit-cell	6 mm
Cell nesting method	Full edge nesting

Table 2. Material properties.

Parameters	Values
Elastic modulus, GPa	200
Poisson's ratio	0.3
Yield stress, MPa	250
Ultimate strength, MPa	460
Density, kg/m^3	7850

3.2 Simulation and analysis

The influence of the number of side surfaces of the polyhedron and the similarity ratio ω on the bending performance of the regular polyhedron sandwich structure is investigated in this section.

The stress convergence variation of the regular polyhedral sandwich structures when $n = 3$ is studied. The maximum convergence stress is 87196 N when $n = 4$, as shown in Figure 7. After that, the convergence stress vibrates while gradually approaching stability. This variation also shows that the polyhedral sandwich structures have a certain level of bending resistance ability. We also simulated the situation when $n = 3$ and $n = 5$. The maximum convergence stress is compared in Figure 8. When $n = 4$, the convergence stress of the polyhedron sandwich plate is relatively optimal than the other two. It shows that the number of polyhedron surfaces n has a significant impact on the bending performance of the regular polyhedron sandwich structure.

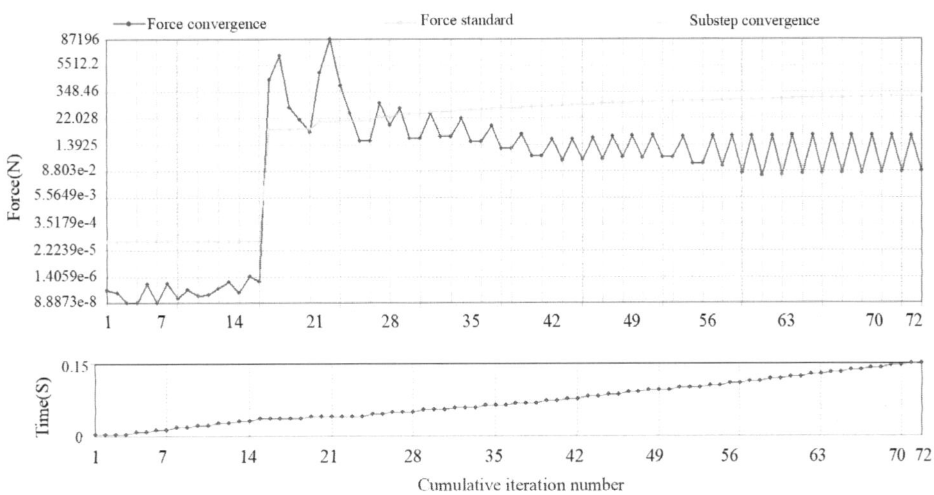

Figure 7. Stress variation of regular polyhedron sandwich structure when $n = 4$.

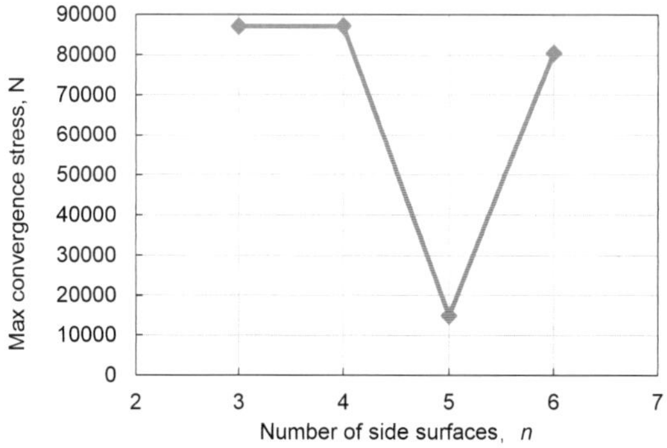

Figure 8. Maximum convergence stress comparison when n varied.

The effect of the similarity ratio on bending performance is analyzed in this section. Here, the number of side surfaces of the regular polyhedron structure is set to $n = 4$. The bending performance of the regular polyhedral sandwich structure is analyzed when similarity ratio ω is changed in the range $\omega = 0.8, 1.0, 1.2$ and 1.4, respectively. Figure 9 shows the variation of stress convergence of the regular polyhedral sandwich structure. When $\omega = 0.8$, the maximum convergence stress is 87266 N. It can be observed that the similarity ratio ω has a significant impact on the bending performance. When ω increases, the maximum convergence stress of the sandwich structure decreases.

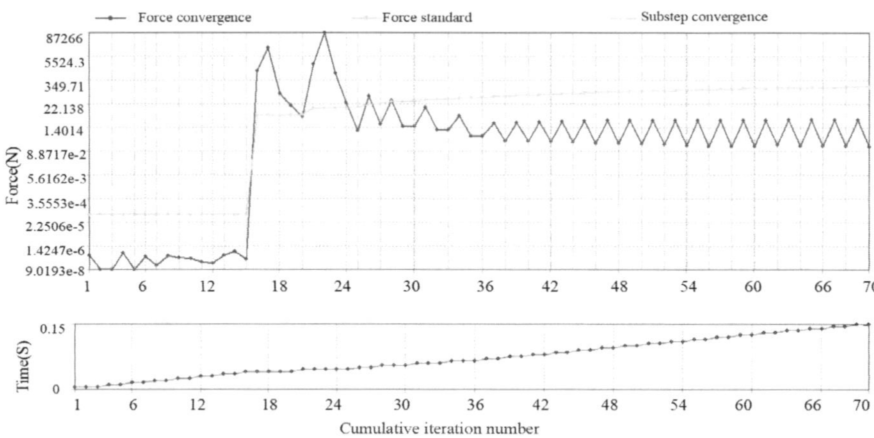

Figure 9. Stress variation of regular polyhedron sandwich structure when $\omega = 0.8$.

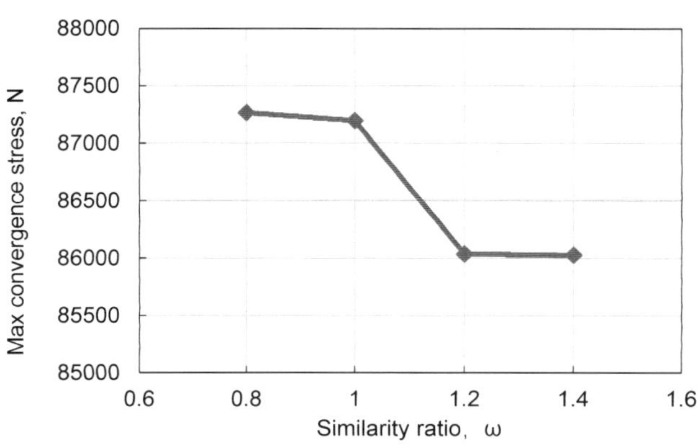

Figure 10. Maximum convergence stress comparison when ω varied.

Figure 10 shows the maximum convergence stress is also compared when the shear ratio k is varied. The maximum convergence stress when k is varied is shown in Figure 11. It shows that the maximum convergence stress is largest when $k = 1/1.4$.

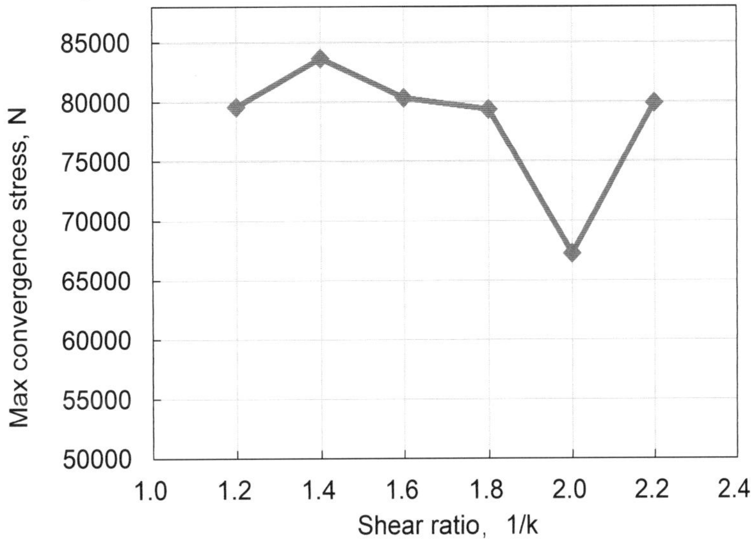

Figure 11. Maximum convergence stress comparison when the shear ratio is varied.

To further validate the performance of the regular polyhedron sandwich structure, we compare it with the honeycomb sandwich structures. Their structure and material parameters are set to the same with Table 1 and Table 2. The maximum convergence stress is compared in Table 3. Although the maximum convergence stress of the honeycomb sandwich structure is larger than the other two polyhedron sandwich structures, it is heavier than them. It shows that the polyhedron sandwich structure has higher bending resistance corresponding to mass than the honeycomb sandwich structure.

Table 3. Comparison between polyhedron sandwich structures and honeycomb sandwich structure.

Structures	Maximum convergence stress F, N	Mass m, kg	F/m
Honeycomb sandwich structure	88576	0.445	199047.2
Polyhedron sandwich structure (n=4)	83642	0.413	202523.0
Polyhedron sandwich structure (n=6)	81036	0.472	1716868.4

4. Conclusion

Aiming to classify the effect of design parameters of structure parameters on the bending performance of regular polyhedron sandwich structure, this study established a simulation model and analyzed the influence of these parameters based on the simulation model. The influence of the number of polyhedron side surfaces and similarity ratio on the bending stiffness was analyzed through simulation. It was found that the number of polyhedron side surfaces, similarity ratio, and shear ratio of the unit-cell structure significantly affect the strength of the regular polyhedron sandwich plate. When $n = 4$, the convergence stress is better than the situation when $n = 3$ or $n = 5$. When the similarity ratio ω increases, the maximum convergence stress decreases. Additionally, compared to the honeycomb sandwich structure, the polyhedron sandwich structure has higher bending resistance corresponding to mass. These results give valuable advice in the design of regular polyhedron sandwich structures.

Acknowledgments
This article is one of the phased achievements of the Huzhou Natural Science Foundation project *Research on the Evolution Mechanism of Sandwich Panel Cells and Optimization of Collision Resistance Performance* (2021YZ07).

References

[1] Tarlochan, F. Sandwich structures for energy absorption applications: A review. Materials, 2021, 14(16), 4731.
[2] Ciepielewski, R., Gieleta, R., & Miedzińska, D. Experimental study on static and dynamic response of aluminum honeycomb sandwich structures. Materials, 2022, 15(5), 1793.
[3] Dastan, T., Nedoushan, R. J., Sheikhzadeh, M., & Yu, W. R. Designs of lattice truss core sandwich structures with improved compressive strength and stiffness. Journal of Composite Materials, 2023, 57(15), 2367-2387.
[4]Xiao, M., Liu, X., Zhang, Y., Gao, L., Gao, J., & Chu, S. Design of graded lattice sandwich structures by multiscale topology optimization. Computer Methods in Applied Mechanics and Engineering, 2021, 384, 113949.
[5] Tokura S., Hagiwara I. Shape optimization to improve impact energy absorption ability of truss core panel [J]. Journal of Computational Science & Technology, 2011, 5(1):1-12.
[6] Qiu, H., Feng, Y., Gao, Y., Zeng, S., & Tan, J. The origami inspired design of polyhedral cells of truss core panel. Thin-Walled Structures, 2021, 163, 107725.
[7] Zhao, X., Kong, C., Yang, Y., & Hagiwara, I. Reversed-torsion-type crush energy absorption structure and its inexpensive partial-heating torsion manufacturing method based on origami engineering. Journal of Manufacturing Science and Engineering, 2022, 144(6), 061001.
[8] Long Siying, Feng Y., Gao Y., et al. Evolement mechanical of unit cell and crashworthiness optimization design of polyhedron sandwich structure[J]. Journal of Mechanical Engineering, 2014, 50(11): 135-143.
[9] Liu J., Feng Y., Zheng H., et al. Innovative design of regular unit-cell and energy-absorbing features under impact of polyhedron sandwich structure[J]. Journal of Mechanical Engineering, 2017, 53(23): 147-156.

Journal of Physics: Conference Series **2842** (2024) 012102 doi:10.1088/1742-6596/2842/1/012102

Electronic and transport properties of 2D α-AsP/α-AsP van der Waals homojunction: A first principle study

Wei Hong, Xiandong Zeng and Yuliang Mao*

Hunan Key Laboratory for Micro-nano Energy Materials and Devices, School of Physics and Optoelectronic, Xiangtan University, Hunan 411105, China

*Corresponding author's e-mail: ylmao@xtu.edu.cn

Abstract. Following the successful experimental synthesis of α-AsP materials, we design a 2D α-AsP/α-AsP homojunction and use first-principles calculation to study the structural stability, optoelectronic properties, and transport characteristics. The AB-stacked α-AsP/α-AsP is found stable, exhibiting a direct band gap of 1.038 eV. Under a 6% biaxial compressive strain, the electronic properties of α-AsP/α-AsP are transited from semiconductor to metallic. The introduction of an external electric field resulted in a reduced band gap and the emergence of metallic characteristics. Compared with that in monolayer structure, the light absorption in bilayer α-AsP is enhanced. Transport simulations based on a two-probe model reveal promising electron transport properties in the 2D α-AsP/α-AsP homojunction. These findings suggest that 2D α-AsP/α-AsP homojunction has potential application in nanoscale optoelectronic devices.

1. Introduction

Recently, V_A-V_A binary compounds have been extensively studied and applied in various fields, including electronics, optoelectronics, biomedicine, and energy devices, by combining other elements of the V_A group[1]. Although most research remains at the theoretical level, scientists have shifted their attention to realizing these materials in laboratories. Black phosphorus arsenic (AsP), a typical two-dimensional (2D) V_A-V_A binary material, can be experimentally synthesized through chemical vapor transport (CVT), exfoliation, and molecular beam deposition (MBD) methods[2]. Recent studies have successfully prepared high-purity (up to 99.9%) α-AsP crystals [3]. As a 2D semiconductor material with a wrinkled surface structure, α-AsP exhibits characteristics similar to black phosphorus. Xie et al. [4] calculated bandgap values for α-AsP using distinct functionals: the PW91 and the HSE06 hybrid functional. Their calculations yielded bandgap values of 0.90 eV and 1.54 eV, respectively. Furthermore, AsP also possesses excellent properties, including a direct bandgap type and a high carrier mobility of up to 1.4×10^4 $cm^2V^{-1}s^{-1}$ [4]. Constructing vdW structures provides a strategic approach to alleviating the limitations inherent to single-material systems by varying stacking configurations, thereby achieving enhanced optoelectronic properties. The advantage of van der Waals heterojunctions over other methods to improve the properties of materials is that it is not limited by lattice matching between neighboring atomic layers and there is no chemical bonding on the outside. Therefore, they can be stacked in any way and exhibit various optoelectronic properties. AsP can be integrated into vdW structures with other 2D materials, broadening the scope of applications. Du et al. [5] explored the potential applications of 2D α-AsP/γ-AsP homostructure in solar cells. They found that the 2D α-AsP/γ-AsP homostructure exhibits a promising PCE of 21.08%. Additionally, their research revealed that applying a 4% in-plane axial strain induces a transition within the homostructure, shifting its band gap from indirect to direct.

Previous studies show that 2D AsP and AsP-based vdW structures have broad applicational scope in fields such as photodetection, solar cells, and sensors.

In this study, we systematically investigate the electronic properties and stability of four different stacks for α-AsP/α-AsP homostructures by using DFT calculation. The stability of the AB stacking configuration is confirmed through phonon spectra and AIMD simulations. We calculate the electronic properties and charge distribution for the AB-stacked system, and to further refine these properties, we employ strain and electric field tuning to investigate the system's responsiveness. Additionally, we develop a two-probe transport model specifically for the AB-stacked bilayer α-AsP, facilitating detailed analysis of its electronic transport properties.

2. Method

We utilized first-principles calculations with the VASP [6] software to explore the geometric, electronic, and optical characteristics of two-dimensional α-AsP/α-AsP systems. Our calculations were grounded in DFT, using the PBE functional within the GGA to model exchange-correlation interactions. A 9×9×1 Monkhorst-Pack grid was employed to sample the Brillouin zone. Structural optimizations continued until convergence criteria were satisfied, with energy and force thresholds set at 10^{-6} eV and 0.01 eV/Å, respectively. To prevent interactions between periodic images, a vacuum layer of at least 20 Å was added along the z-axis. The absorption coefficient $\alpha(\omega)$ is a critical parameter for assessing the performance of photovoltaic materials, and it can be calculated using the following equation [7]:

$$\alpha(\omega) = \sqrt{2}\omega\left[\sqrt{\varepsilon_1^2(\omega) + \varepsilon_2^2(\omega)} - \varepsilon_1(\omega)\right]^{\frac{1}{2}} \tag{1}$$

where ε_1 represents the real part of the dielectric function, ε_2 represents the imaginary part, and ω denotes the photon frequency.

3. Results and discussion

3.1 Stability and photovoltaic properties

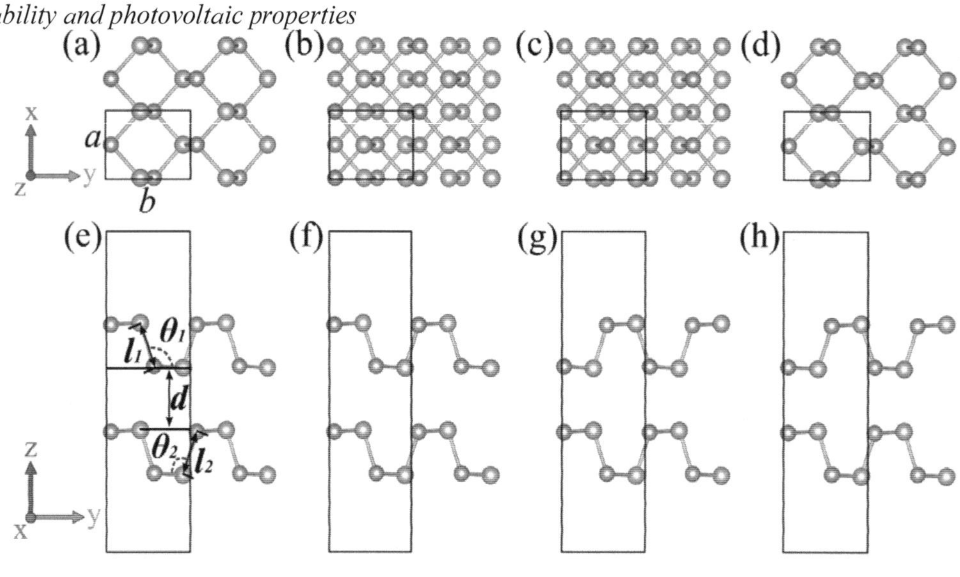

Figure 1. Top as well as side views of α-AsP homojunction stacking patterns. (a) Top and (e) Side views of AA stack; (b) Top and (f) Side views of AB stack; (c) Top and (g) Side views of AC stack; (d) Top and (h) Side views of AD stack.

Before studying the bilayer α-AsP homojunction, we explore the geometry of the monolayer α-AsP. After structural relaxation, the lattice constants of α-AsP are a = 3.50 Å and b = 4.69 Å, and the space group belongs to Pmn21. In addition, by using the HSE06 method, we found it exhibits a direct band gap of 1.55 eV, which is consistent with previous reports [4]. The relative positions of the layers exhibit multiple possibilities, leading to different electronic properties for distinct stacking modes. Considering the different atomic arrangements of monolayer α-AsP, we identified four possible stacking modes, as shown in Fig. 1(a-h).

Table 1. The structural parameters of the a-AsP/a-AsP homojunction, including lattice constants (a and b), interlayer distance (d), binding energy (E_b), bond angles, and bond lengths.

	a(Å)	b(Å)	d(Å)	E_b(eV)	θ_1(degree)	θ_2	l_1(Å)	l_2 (Å)
AA	3.535	4.534	3.239	-0.451	103.564	97.616	2.415	2.415
AB	3.536	4.586	3.067	-0.495	103.427	98.681	2.391	2.402
AC	3.540	4.579	3.292	-0.420	102.889	99.016	2.392	2.398
AD	3.529	4.603	3.665	-0.324	103.031	99.250	2.399	2.398

The binding energy is an important parameter to estimate the thermodynamic stability of a homojunction structure. Generally, a negative value of the binding energy indicates that the structure is thermodynamically stable. The lattice constants, interlayer distance, binding energy, bond angle, and bond length of homojunction formed by bilayer α-AsP are shown in Table 1. By comparing the binding energy values of different stacking structures, we can obtain their thermodynamic stability order as: AB > AA > AC > AD. Additionally, different stacking modes significantly affect the interlayer distance, with smaller interlayer distances exhibiting higher negative binding energy. To further validate the stability of the stacking structures, we performed AIMD simulations and phonon spectra to verify the thermodynamic and dynamical stability of the AB-stacked α-AsP/α-AsP homojunction. The phonon spectra are illustrated in Figure 2(a), which do not exhibit any imaginary frequencies, further confirming the dynamical stability of the AB-stacked homojunction. Figure 2(b) shows the molecular dynamics simulation results of the AB stacking structure at 300 K temperature after 10000 fs with a step size of 1 fs. The energy variation constantly oscillates in the -426 ~ -432 eV interval, indicating that the homojunction is stable at room temperature.

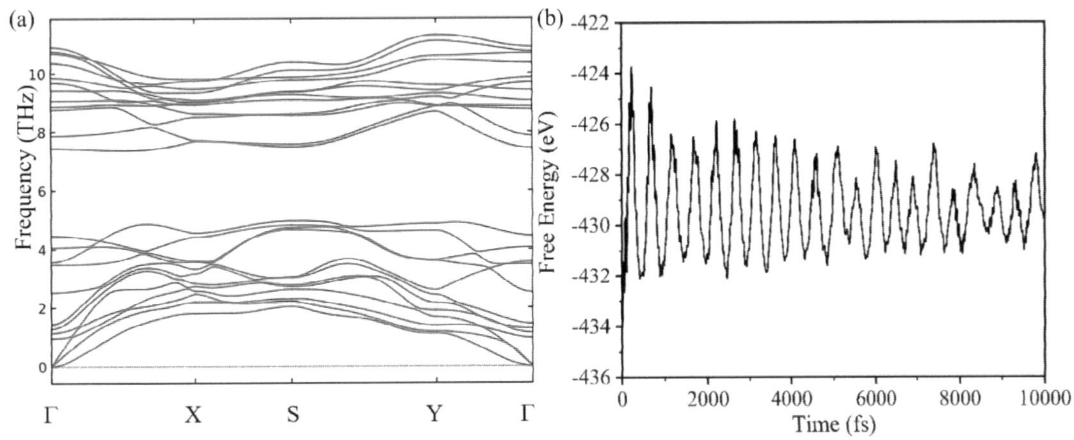

Figure 2. (a) Phonon spectra and (b) Molecular dynamics simulation of AB stacked structure.

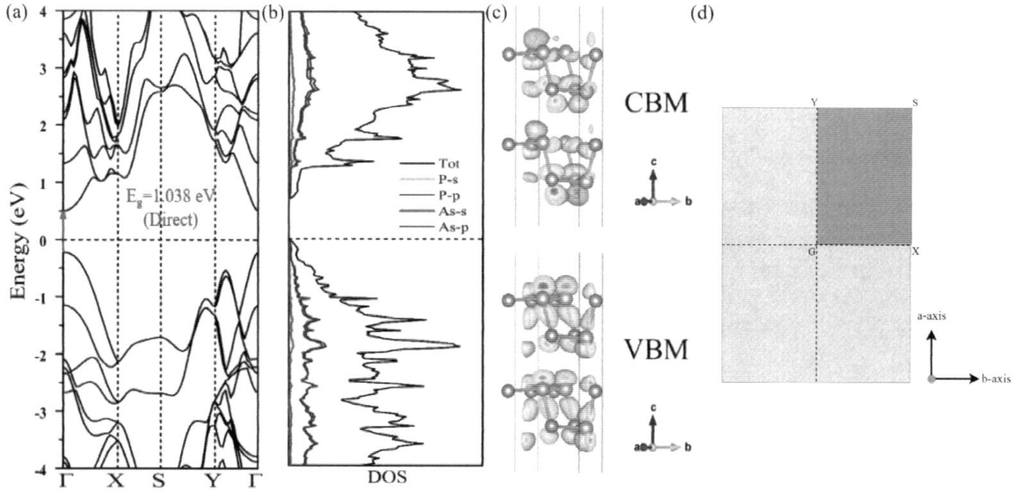

Figure 3. (a) Energy band structure, (b) Density of states, (c) CBM (VBM) partial charge density of bilayer α-AsP homojunction, and (d) Brillouin zone of α-AsP monolayer and the corresponding high symmetry point.

An analysis of the electronic band structure and density of states is performed for AB-stacking of the α-AsP homojunction obtained by the HSE06 hybrid function, as shown in Fig. 3. Figure 3(a) illustrates a direct bandgap characteristic with a value of 1.038 eV at the Γ high symmetry point for this structure. Figure 3(b) shows that the CBM and VBM of this homojunction mainly originate from the joint contribution of As-p and P-p orbitals. Figure 3(c) shows the partial charge structures of the CBM and VBM, further confirming their co-contribution by both As and P atoms. The result is consistent with the projected density of states. Figure 3(d) shows the Brillouin zone of α-AsP monolayer material and the corresponding high symmetry points.

3.2 Strain and electric field effect on the band gap engineering

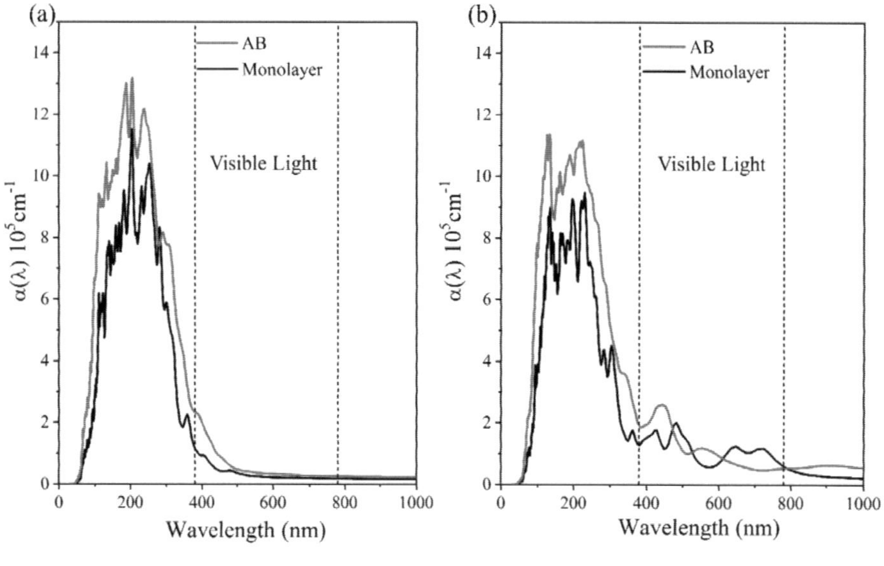

Figure 4. Light absorption spectra of bilayer α-AsP homojunction (a) Zigzag direction (b) Armchair direction.

The above calculations show that the AB stacking configuration exhibits a direct band gap type, and the calculated band gap value is 1.038 eV, suggesting that the discussion of its optical absorption coefficients is meaningful for its application in optics. Figure 4(a-b) shows the α(ω) of α-AsP in both zigzag and armchair directions. It can be found that the absorption peaks in the zigzag direction are mainly located in the ultraviolet (UV) with a high absorption order of up to 10^6, while the absorption peaks in the armchair exhibit a wide wavelength range from the UV to the near-infrared region. Meanwhile, the range of light absorption and its absorption intensity in different directions confirm that the bilayer α-AsP homojunction has a remarkable anisotropy. By comparing the light absorption intensities of the monolayer α-AsP and the bilayer α-AsP, it can be found that the construction of the homojunction can effectively enhance the α-AsP light absorption intensity.

Considering the orthogonal lattice structure of bilayer α-AsP, we applied strains ranging from -10% to +10% along the armchair, zigzag, and biaxial directions using the PBE function. As shown in Figure 5(a), in-plane strain leads to a significant and anisotropic effect on the band gap of bilayer α-AsP. When tensile strain is applied in any direction, the band gap of the homojunction undergoes an initial increase followed by a decrease while retaining a direct band gap within a specific range of strain. With rising strain intensity, the band gap shifts from direct to indirect. A similar phenomenon can be observed under compressive strain. Notably, the biaxial strain exhibits the most pronounced effect on the band structure adjustment in this homojunction, inducing a semiconductor-to-metal transition at 6% compressive biaxial strain.

In Figure 5(b), we show the influence of the external vertical electric field on the band gap of bilayer α-AsP. The band gap of bilayer α-AsP decreases when applying an external electric field and is transitioned to a metallic state when a field exceeding 0.3V/Å is applied. These findings indicate that in-plane strain and external electric fields can effectively modulate the band gap of bilayer α-AsP.

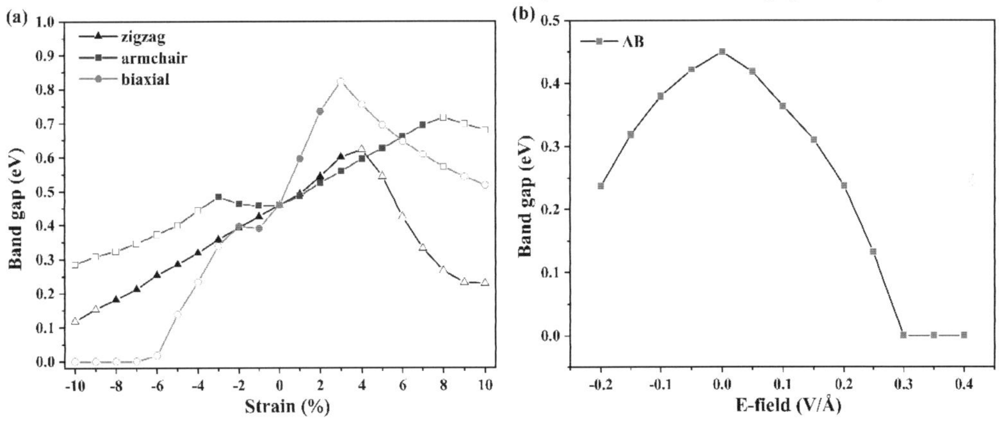

Figure 5. (a) Band gap value of AB stacking under in-plane strain, direct and indirect band gaps shown by the solid and hollow points. (b) The band gap value of AB stacking under an external electric field.

3.3 Electron transport properties of Bilayer α-AsP

Based on the adjustable electronic characteristics of bilayer α-AsP and its notably higher optical absorption coefficient in contrast to monolayer AsP, we investigated its electron transport properties through the construction of a two-probe electron transport model. As illustrated in Fig. 6(a), the conventional two-probe model is constructed by two ends and a central scattering region, in which the left and right electrode is composed of 3×1 unit-cells each, with the scattering region extending to a length of 6×1 unit-cells. Subsequently, a forward bias is applied to the bilayer α-AsP-based transport device. The I-V curve can be obtained by the calculations through the Landauer-Buttiker formula [8]:

$$I(V) = \frac{2G_0}{e}\int T(E,V)[f(E-\mu_L) - f(E-\mu_R)]dE \tag{2}$$

In this formula, the left and right electrodes in the two-probe model are denoted by μ_L and μ_R, respectively.

Since the transmission spectrum of the two-probe electron transport system at zero bias voltage serves as a crucial metric for evaluating electron transport properties, we computed the transmission spectrum of bilayer α-AsP at zero bias voltage, as depicted in Figure 6(b). Comparative analysis with monolayer α-AsP revealed a significantly enhanced transmission coefficient in the bilayer stacking structure, indicating superior electronic transport performance compared to that in its monolayer counterpart. Subsequently, the current-voltage characteristics spanning a bias range of 0 to 2 V were computationally obtained for both bilayer and monolayer α- AsP structures. As shown in Fig. 6(c), the curves revealed a notable increase in current for bilayer α-AsP compared to that in monolayer α-AsP. These observations suggest that the construction of the bilayer stacking structure enhances the conductivity of α-AsP. Our aforementioned findings underscore the potential application of bilayer α-AsP in nano-optoelectronic devices.

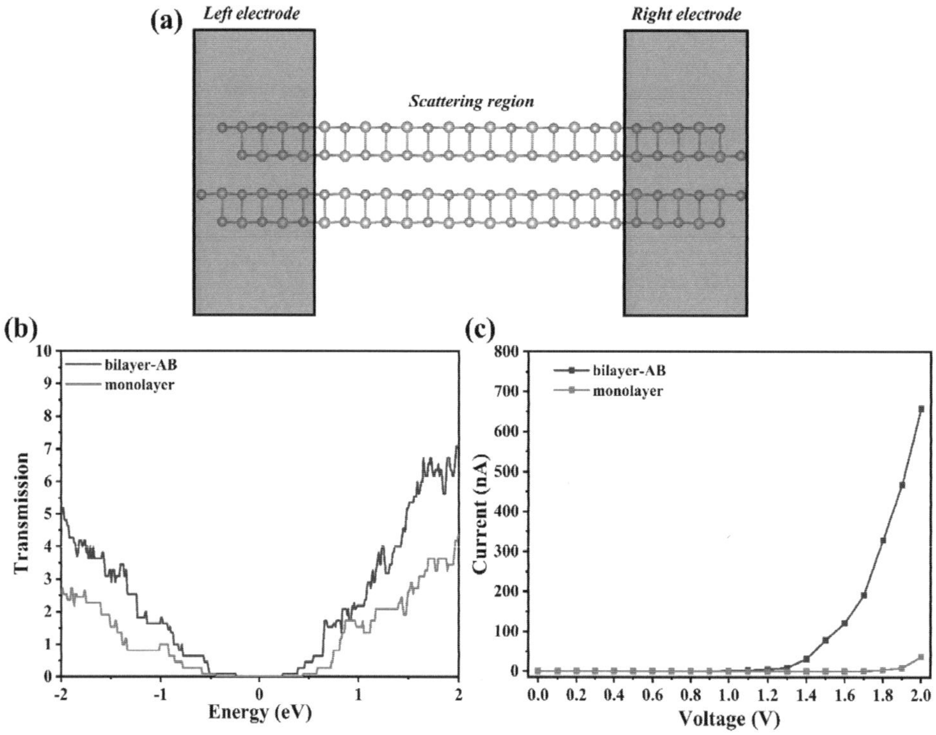

Figure 6. (a) The two-probe transport model for AB stacking of bilayer α-AsP. (b) The transmission spectra and I-V curves of bilayer and monolayer α-AsP, and (c), The blue and red curves represent the AB stacking of bilayer α-AsP and monolayer α-AsP, respectively.

4. Conclusion

In summary, we conducted a systematic investigation on the stability, optoelectronic properties, and transport characteristics of α-AsP/α-AsP homojunction. The conclusions are obtained as below:

(1) The AB-stacked α-AsP/α-AsP configuration exhibits thermodynamic and dynamic stability.

(2) The bilayer α-AsP homojunction possesses a direct bandgap of 1.038 eV, with absorption peaks in an order of 10^6 in both zigzag and armchair directions in the ultraviolet region, demonstrating anisotropic behavior and stronger light absorption intensity compared to that in monolayer α-AsP.

(3) Under 6% biaxial compressive strain, bilayer α-AsP is converted from a semiconductor to a metal; applying a perpendicular electric field reduces its bandgap, turning it metallic at fields over 0.3 V/Å.

(4) The current values of bilayer α-AsP are significantly higher than those of monolayer α-AsP under

bias voltages between 0-2 V, indicating that the construction of bilayer stacking structures enhances the electrical conductivity of α-AsP.

Acknowledgments
This research is funded by the Natural Science Foundation of Hunan Province, China (Grant No. 2023JJ30567).

References

[1] Gui R, Jin H, Sun Y, et al. Two-dimensional group-VA nanomaterials beyond black phosphorus: synthetic methods, properties, functional nanostructures and applications [J]. Journal of Materials Chemistry A, 2019, 7(45): 25712-71. https://doi.org/10.1039/c9ta09582a.
[2] Osters O, Nilges T, Bachhuber F, et al. Synthesis and Identification of Metastable Compounds: Black ArsenicuScience or Fiction? [J]. Angewandte Chemie-International Edition, 2012, 51(12): 2994-7. https://doi.org/10.1002/anie.201106479.
[3] Koepf M, Eckstein N, Pfister D, et al. Access and in situ growth of phosphorene-precursor black phosphorus [J]. Journal of Crystal Growth, 2014, 405(6-10). https://doi.org/10.1016/j.jcrysgro.2014.07.029.
[4] Xie M, Zhang S, Cai B, et al. A promising two-dimensional solar cell donor: Black arsenic–phosphorus monolayer with 1.54eV direct bandgap and mobility exceeding 14,000cm2V−1s−1 [J]. Nano Energy, 2016, 28(433-9). https://doi.org/10.1016/j.nanoen.2016.08.058.
[5] Mao Y, Du Y, Huang Z, et al. Computational Design of α-AsPi γ-AsP Vertical Two-Dimensional Homojunction for Photovoltaic Applications [J]. Nanomaterials, 2022, 12(10). https://doi.org/10.3390/nano12101662.
[6] Raynaud C, Maron L, Jolibois F, et al. Ab initio molecular dynamics: Plane waves vs. local basis: The role of energy cutoff on the convergence of molecular properties [J]. Chemical Physics Letters, 2005, 414(1): 161-5. https://doi.org/10.1016/j.cplett.2005.08.034.
[7] Wang V, Xu N, Liu J-C, et al. VASPKIT: A user-friendly interface facilitating high-throughput computing and analysis using VASP code [J]. Computer Physics Communications, 2021, 267. https://doi.org/10.1016/j.cpc.2021.108033.
[8] Landauer R. Electrical resistance of disordered one-dimensional lattices [J]. The Philosophical Magazine: A Journal of Theoretical Experimental and Applied Physics, 1970, 21(172): 863-7. https://doi.org/10.1080/14786437008238472.

Journal of Physics: Conference Series **2842** (2024) 012103 doi:10.1088/1742-6596/2842/1/012103

Working space planning for wheeled robot overturning stability

Haoxin Xing [1], Tiejun Li [2,*], and Jinyue Liu [1]

[1] Hebei University of Technology, Tianjin 300000, China
[2] Hebei University of Technology and Science, Shijiazhuang 050000, China

[*] Corresponding author's e-mail: 001036@hebust.edu.cn

Abstract: Mobile manipulators have been widely used in the field of building construction. Typical features of construction robots include crossing various rugged terrains in construction sites, carrying heavy payloads, and performing tasks in a large workspace, which may lead to unstable posture or overturning. The tipping of the mobile manipulator, especially when carrying a payload, will cause danger to the surrounding environment and staff, leading to the termination of the task. The purpose of this study is to establish a real-time tipping prediction system, which can be used to avoid tipping of mobile manipulators. By employing the CAD model of the robot and parameters such as the center of gravity and direction, we can ascertain and assess the instability of the robot and the overturning algorithm, thus providing early warning and averting any potential danger.

1. Introduction

With its vehicle-mounted manipulator as its primary form, the construction robot must journey to the designated location to fulfill the task as necessary during the construction process. Equipped with a large working area and a hefty burden, this robot is typically designed to assemble an object of considerable magnitude and mass. The typical characteristics of building robot operation are large space, low speed, and a heavy load [1]. The construction robot assembly operation needs to install the operating object at the designated position on the job site, and environmental constraints are the main technological characteristics of its operation [2]. In a semi-structured complex scene, the construction site is a typical one; the robot must address the issue of destabilization during the relocation and upon arriving at the designated construction site to finish the assembly work securely [3]. If the unskilled (remote) operator does not operate properly and accidents occur, the heavy mobile manipulator will face the risk of overturning. It may be very challenging to negotiate the slope in the site where the soil may also be unstable [4]. In dexterous, maneuverable, compact, and lightweight robots, the risk of overturning is due to the high center of mass (COM) and small weight and foundation, especially in the case of heavy load and large workspace.

1.1 Motivation

Compared with the more mainstream topics, the topic of dynamic stability of wheeled mobile robots is rare in the literature at present. Even for transporting vehicles from one place to another, it is often faster to cross the ridge than to bypass it. The inherent reduced stability margin for vehicles driving on slopes significantly increases the possibility and influence of events caused by slight roll miscalculations caused by driving on small rocks. For lifting heavy-duty machines, especially when operating on slopes, it is very important to understand the influence of quality reconstruction on overall stability. Familiarity with the center of gravity's location, bolstered by the wheel-ground junction – a widely accepted fact, is essential for grasping this capacity. We investigate the singular issues associated with the general situation's realization and then suggest a solution for a certain vehicle that is analogous to this general predicament in intricacy.

1.2 Related Work

ZMP postulates stability. Additionally, the joint's internal force and its motion, as well as the mobile platform's acceleration, are taken into account [5]. Comparing the effectiveness of the last three methods, other researchers have evaluated them. Sidobre and Desormeaux [6] attested that the ZMP approach is markedly distinct from the other two, as it does not factor in the moment of inertia or the center of gravity's height directly. In addition, the FA method needs more computing power than the ZMP and MHS methods. A comparison of [7] to a genuine mobile robot system was conducted, with the center of gravity of the latter kept constant throughout the entire setup process, and the results were comparable to those of [8]. It is also possible to combine the two methods [9-10]. The contact force method, as employed in prior research (where the mobile manipulator's critical state necessitates a shift of the robot arm), was applied by the same author in [11] to detect instability. Considering the system's center of gravity and speed, a constant contact force was maintained at all contact points for the manipulator's new joint arrangement. However, this method has a limitation: measuring the contact force between the wheel and the ground is necessary, thus necessitating the inclusion of extra sensors into the wheel. This scheme is expensive and complicated, and its universality is poor under complex working conditions in construction sites.

2. Problem formulation

The design requirements of mobile manipulators make overturning a particularly pertinent matter. Maximizing maneuverability and accessibility necessitates vehicles of a compact size; however, maneuverability increases with distance and power. Consequently, the manipulator's size, weight, and position on the mobile base are all conducive to overturning. This is especially important when facing sloping terrain [12]. By repositioning its center of gravity (COG), the vehicle can avert overturning, depending on navigation conditions. Active suspension systems or movable masses can be used to achieve this. Furthermore, the mobile manipulator provides the capability to actively regulate the COG's position by shifting the onboard arm. The most influential overturning stability criteria proposed in technical literature, the supporting edge principle, states that static stability necessitates a COG above the convex area of the ground contact point. The mobile robot's payload, when traversing complex terrain, causes a center of gravity shift which this strategy is employed to compensate for, thus aiding the chassis in center of gravity compensation. This principle has been further advanced by taking into account dynamic forces such as the energy needed for tipping, forces and moments acting on the vehicle, and support forces at ground contact points. Regrettably, the theoretical approaches to dynamic techniques are restricted, thus necessitating an exploration of genuine dynamic interactions between the vehicle, manipulator, and terrain. This necessitates more intricate calculations and laborious handling of inaccurate data from intricate sensors. Nevertheless, so long as the vehicles and manipulators move at low speeds, there is no substantial inertial acceleration apart from gravity. The dynamic effects can then be considered as minor perturbations, and the static analysis is sufficient to prevent overturning on sloping terrain.

A simple thinking experiment is enough. For example, under certain unfavorable arm postures, especially when external force acts on the tip of the manipulator, such as carrying a load or interacting with the surrounding environment, overturning may occur. It must be extra careful if the manipulator is interacting with people. The formulation of an overturning stability index is the only way for a fully automated solution to be achieved without the need for remote operator operation.

(1) The working objects of building construction assembly are usually heavy, large in size, and large in moving range, but the uneven ground and inclined chassis will aggravate the instability of the

Journal of Physics: Conference Series 2842 (2024) 012103 doi:10.1088/1742-6596/2842/1/012103

robot in the working process.

(2) Obtaining an accurate and safe real workspace is essential for the robot to be able to perform accurate construction work, as the construction environment is intricate and the load can vary. This workspace is distinct under different working conditions, such as no-load, light-load, and heavy-load.

(3) The working environment changes in real time during the construction process, which is a semi-structured complex scene. Assessing the robot's overturning stability and safety in real-time is essential for guaranteeing precise and secure assembly, given the existence of construction mistakes and the limitation of sensing data.

3. Modeling and analysis

Figure 1 displays a self-crafted wheeled mobile manipulator for plate installation. This mobile platform is composed of a frame, with four wheels to support and sustain the self-constructed mechanical arm. Position control is accomplished through the utilization of the initial three links of the manipulator, while the last three joints are fashioned by wrists. The purpose of this investigation is to gauge the sway of the center of gravity, the manipulator's motion, and the ground-chassis interaction during the construction phase on the overturning steadiness of the whole system. Dynamic elements such as connecting rod mass, connecting rod length, joint mass, force applied to the tip, and tire deformation could potentially result in rollover instability.

Figure 1. Robot physical picture.

Assuming certain things, the modeling and analysis presented here are based on them.

(1) Irrespective of shrinkage, the earth remains a consistent plane.

(2) The earth is constantly in contact with all four wheels.

(3) Wheel slip is not considered.

(4) The platform is firmly affixed to the manipulator, with all its connecting rods firmly in place.

(5) No acceleration is applied from the wheel brakes.

Upon completion of the CAD model, physical parameters such as weight and moment of inertia can be utilized to forecast the robot's center of gravity (COG) in real time. The complex centroid formula outlined in Equation (1) can be employed to acquire the center of gravity.

$$\bar{X} = \frac{\sum \bar{x} \cdot w}{\sum w} \tag{1}$$

$$\bar{Y} = \frac{\sum \bar{y} \cdot w}{\Sigma w} \tag{2}$$

$$\bar{Z} = \frac{\Sigma \bar{z} \cdot w}{\sum w} \tag{3}$$

The composite material's weight is denoted by w or CG. The following are located in x, y, and z. The center of mass of each component is x, y, and z, which contain the standard homogeneous transformation of rigid body motion (attitude) in 2D space.

$$E = \Pi R(\theta_n) T_x(L_n) \tag{4}$$

where E is the position of the end effector or the center distance of the positioning link in this

Journal of Physics: Conference Series **2842** (2024) 012103 doi:10.1088/1742-6596/2842/1/012103

application.

The robot's center of gravity shall be ascertained through the equation below.

$$COM = A_0 + \Sigma R_{y_n}(A_n - A_0) \tag{5}$$

$$A_0 = \begin{bmatrix} x_0 \\ y_0 \\ z_0 \end{bmatrix} \tag{6}$$

$$A_n = \begin{bmatrix} x_{L_n} \\ y_{Ln} \\ z_{Ln} \end{bmatrix} \tag{7}$$

$$R_{y_n} = \begin{bmatrix} cos\,\theta_{Ln} & 0 & sin\,\theta_{Ln} \\ 0 & 1 & 0 \\ -sin\,\theta_{Ln} & 0 & cos\,\theta_{Ln} \end{bmatrix} \tag{8}$$

Different from the traditional fixed industrial manipulator, the construction robot often needs to carry building materials from the material warehouse to the construction site to perform tasks, which requires the addition of a movable platform to carry the manipulator. Moreover, the terrain of the construction site is complex, and the climbing and obstacle-crossing ability of the traditional AGV chassis and McNamun wheel chassis cannot meet the needs.

To ensure that the chassis is capable of meeting the demands of construction scenes, tires with adjustable flexibility and an increased suspension rate should be taken into account. The robotic arm's accuracy and safety will be impaired by the chassis' suppleness in executing tasks. It is necessary to accurately obtain the posture of the chassis in construction scenes that require accuracy or complex working conditions.

$$m\ddot{z} = -k_1(z - l,\theta) - k_2(z + l_2\theta) - me\ddot{\theta} \tag{9}$$

$$m\ddot{\theta} = k_1(z - l,\theta)l, -k_2(z + l_2\theta)l_2 - me\ddot{z} \tag{10}$$

By the equation below, one can determine the ZMP – a spot on the ground where all active forces and torques are null. If the contact polygon between the vehicle and the ground (hereinafter referred to as a stable area) is able to contain the ZMP, then the mobile manipulator will be guaranteed.

$$x_{zmp} = \frac{\sum_{i=1}^{n} m_i(\ddot{z}_i + g)x_i - \sum_{i=1}^{n} m_i\ddot{x}_i z_i}{\sum_{i=1}^{n} m_i(\ddot{z}_i + g)} \tag{11}$$

$$y_{zmp} = \frac{\sum_{i=1}^{n} m_i(\ddot{z}_i + g)y_i - \sum_{i=1}^{n} m_i\ddot{y}_i z_i}{\sum_{i=1}^{n} m_i(\ddot{z}_i + g)} \tag{12}$$

By taking into account the number of connecting rods, n, the mass of connecting rod I, m_i, the acceleration of gravity, g, and the center of mass of connecting rod I, the absolute Cartesian coordinate system (x_i, y_i, z_i) determines the coordinate of zmp, denoted by (x_{zmp}, y_{zmp}, 0).

Should no disruption be present in the environment, ZMP's placement in the secure area ensures the robot's security. However, when there is disruption, the robot must have a certain degree of stability. To attain a substantial stability margin, ZMP can be kept close to the center of the stability region. Despite this, the robot's movement will be restricted and the mission may not be successful. To ascertain the scope of ZMP in the face of environmental disturbance, we propose the idea of an effective stable region as follows.

The region of ZMP, whose stability margin surpasses any alteration in position due to environmental disturbance, is known as the effective stable region. The equation below expresses this:

$$\Omega = \{(x_{zmp}, y_{zmp})|d_s(x_{zmp}) \geq d_f(x_{zmp}), d_s(y_{zmp}) \geq d_f(y_{zmp})\}$$

where the least distance from zmp to the secure area's boundary is denoted by d_s (x_{zmp}), and the displacement of zmp caused by the X and Y axes' disruption is denoted by d_f (x_{zmp}) and d_f (y_{zmp}), respectively.

$$d_f(x_{zmp}) = \frac{\sum_{i=1}^{n}(S_{zj}F_{xj} - S_{xj}F_{zj})}{\sum_{i=1}^{n} m_i(\ddot{z}_i + g) - \sum_{i=1}^{n} Fzj} \tag{13}$$

$$d_f(y_{zmp}) = \frac{\sum_{i=1}^{n}(S_{zj}F_{yj} - S_{yj}F_{zj})}{\sum_{i=1}^{n} m_i(\ddot{z}_i + g) - \sum_{i=1}^{n} Fzj} \tag{14}$$

The vector of external force j (F_{xj}, F_{yj}, F_{zj}) and the coordinates of the point where the force j is applied (S_{xj}, S_{yj}, S_{zj}) are both indicative of the mobile manipulator's stability in an effective stable region of ZMP, even when disturbed; thus, even if stability is absent, it can still be used for task execution. Conversely, if the ZMP is in a stable area yet not an effective stable region, it will become unstable if disturbed. Oversight of both the steadiness and job performance of the mobile manipulator must be done in tandem, thus making it imperative.

$$ZMP = \min\{d_s(x_{zmp}) - d_f(x_{zmp});\ d_s(y_{zmp}) - d_f(y_{zmp})\} \tag{15}$$

4. Load and motion analysis

4.1 Flat driving

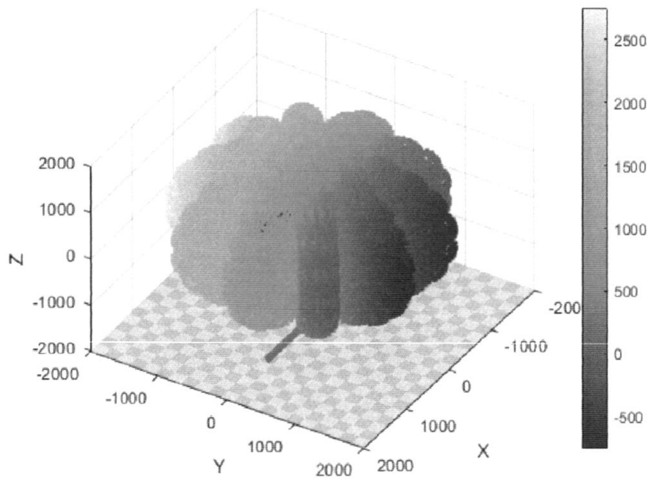

Figure 2. Stable workspace for mobile robots on level ground.

The static case's results are depicted in Figure 2. At the X-Y plane, the ZMP of the workspace is 2, 185:54, with the tip being the farthest away from the platform's center. When the tip of the manipulator is far from the platform's core, there is more mass close to the platform's edge, resulting in a high ZMP value. The ZMP of the entire workspace is 2, 960:72, with the CG of the system, including its payload, close to the origin point. The stability of the system is contingent upon the CG's location.

4.2 Pitch driving

The static case's results are depicted in Figure 3. At the X-Y plane, the ZMP of the workspace is 2, 185:54, with the tip being the farthest away from the platform's center. When the tip of the manipulator is far from the platform's core, there is more mass close to the platform's edge, resulting in a high ZMP value. The ZMP of the entire workspace is 2, 960:72, with the CG of the system, including its payload, close to the origin point. The stability of the system is contingent upon the CG's location.

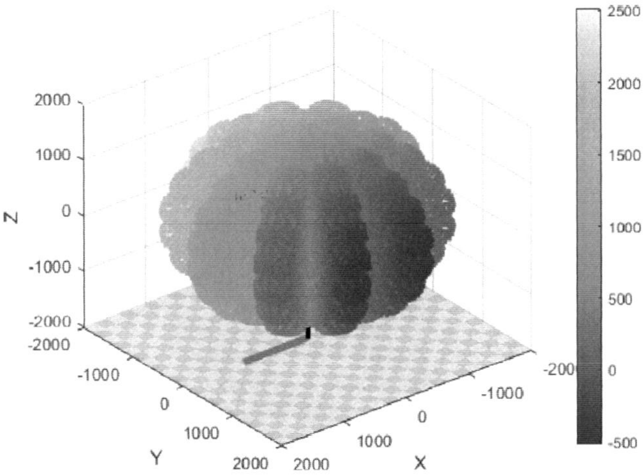

Figure 3. Stabilization of the workspace of mobile robots on pitched ground.

4.3 Rolling driving

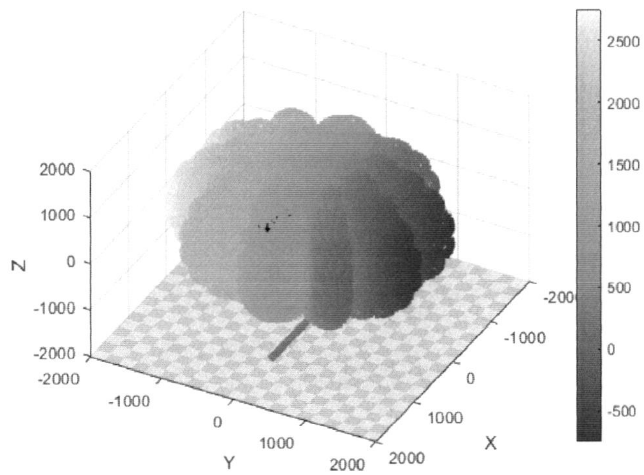

Figure 4. Stabilization of the workspace of mobile robots on the ground.

Figure 4 illustrates the static case's outcomes with ZMP in terms of CG, with four black lines symbolizing the tip across axes. When the tip is furthest from the center of the platform on the X-Y plane, the ZMP of the workspace is 2, 185:54, thus the manipulator's mass is nearer to the platform's edge, resulting in a higher ZMP value. The ZMP of the entire workspace is 2, 960:72, with the CG of the system, including its payload, close to the origin point. The stability of the system is contingent upon the CG's location.

5. Conclusions

In this paper, a new set of generalized rollover formulas is derived, which considers the task execution of manipulator, tire deformation, and stability analysis of ground conditions, and simulates three situations. In the simulating plane, ground pitching, and ground rolling, the result was that the alteration of the gravity center of the mobile manipulator, tire deformation, and ground topography all affected stability. The most extreme tipping stability was examined, and the stability prediction of the mobile manipulator during operation and the real-time algorithm for resisting external interference based on the stability margin were presented. Testing of the mobile manipulator, following its construction, revealed its stability in the given motion law and terrain conditions to be absolute.

References

[1] Ding X., Liu Y., Hou J., et al. Online dynamic tip-over avoidance for a wheeled mobile manipulator with an improved tip-over moment stability criterion [J]. IEEE Access, 2019, 7: 67, 632-67, 645.
[2] Böge A. and Böge W. Technische Mechanik: Statik-Reibung-Dynamik-Festigkeitslehre-Fluidmechanik [M]. Springer-Verlag, 2015.
[3] Moosavian S. A. A. and Alipour K. On the dynamic tip-over stability of wheeled mobile manipulators [J]. International Journal of Robotics & Automation, 2007, 22 (4): 322.
[4] Roan P. R., Burmeister A., Rahimi A., et al. Real-world validation of three tip-over algorithms for mobile robots [C]// 2010 IEEE International Conference on Robotics and Automation. IEEE, 2010: 4, 431-4, 436.
[5] Alipour K., Hasanpour A., and Daemy P. Comparing two online tip-over avoidance algorithms for mobile manipulators [C]// 2014 Second RSI/ISM International Conference on Robotics and Mechatronics (ICRoM). IEEE, 2014: 310-315.
[6] Sidobre D. and Desormeaux K. Smooth cubic polynomial trajectories for human-robot interactions [J]. Journal of Intelligent & Robotic Systems, 2019, 95 (3-4): 851-869.
[7] Zhang J., Meng Q., Feng X., et al. A 6-DOF robot-time optimal trajectory planning based on an improved genetic algorithm [J]. Robotics and Biomimetics, 2018, 5 (1): 1-7.
[8] Fang Jianwen, Chao Yongsheng, and Yuan Yiping. Optimization of welding robot joint trajectory by improving B-spline interpolation method [J]. Mechanical Design and Manufacturing, 2021, (10): 1-4+10.
[9] Zou Q., Guo W., and Hamimid F. Y. A novel robot trajectory planning algorithm based on NURBS velocity adaptive interpolation [C]// Advances in Mechanical Design: Proceedings of the 2017 International Conference on Mechanical Design (ICMD2017). Springer Singapore, 2018: 1, 191-1, 208.
[10] Li Hong, Liu Song, Sun Zhiyi, et al. Optimal time trajectory planning of excavator based on non-uniform B-spline curve [J]. Journal of Chinese Engineering Machinery, 2021, 19 (04): 302-306.
[11] Wang Jidai, Wang Xindong, Tian Qunhong, et al. Mobile robot path planning based on improved fuzzy adaptive genetic algorithm [J]. Machine Tools and Hydraulics, 2021, 49 (23): 18-23.
[12] Roch Dupré D., Gonsalves T., Cucala A. P., et al. Determining the optimum installation of energy storage systems in railway electrical infrastructures by means of swarm and evolutionary optimization algorithms [J]. International Journal of Electrical Power & Energy Systems, 2021, 124: 106, 295.

IOP Publishing
doi:10.1088/1742-6596/2842/1/012104

Wearable photonic crystal double network hydrogel sensor based on structural color analysis

Zhenye Liu, Wenxin Zhang, Yu Qiao, Lili Qiu and Zihui Meng*

School of Chemistry and Chemical Engineering, Beijing Institute of Technology, Beijing 100081

*Corresponding author's e-mail: mengzh@bit.edu.cn

Abstract. This paper introduces a high-toughness photonic crystal (PhCs) dual-network hydrogel sensor designed for mechanical sensing, which enables quantitative analysis of structural color using the HSB color space. The hydrogel achieves a maximum tensile strain of 250% under a tensile stress of 3.5 MPa, thanks to its dual-network structure comprising a covalently cross-linked polyacrylamide (PAM) network and an ionically cross-linked sodium alginate (SA) network. At an 80% tensile strain, the PAM-SA photonic crystal hydrogel displays a blue shift in the bandgap wavelength of over 130 nm and demonstrates a sensitivity of 1.69 nm/%. The analysis of the force-induced color change in the PAM-SA photonic crystal hydrogel utilized both RGB and HSB color spaces. In the HSB color space, the hue component (H) exhibited a strong linear correlation with strain (R^2>0.95), indicating the feasibility of quantitative structural color analysis using HSB. As a wearable sensor, the PAM-SA photonic crystal hydrogel precisely detects human motion via bandgap displacement (R^2=0.989) and structural color change (R^2=0.978). The PAM-SA PhCs hydrogel, featuring easily accessible color information and high sensitivity, has broad potential applications in wearable devices and mechanical sensors.

1. Introduction

The RGB color space[1] represents colors through the combination of three primary colors: Red (R), Green (G), and Blue (B). Devices such as monitors, scanners, and digital cameras commonly use the RGB color space in everyday life to display and capture color information. The HSB[2] color space relies on Hue (H), Saturation (S), and Brightness (B). Therefore, changes in the H value can reflect changes in color. This conversion and analysis method offers significant advantages and is practical for handling color data in different lighting conditions.

Photonic crystals (PhCs)[3] feature a periodic ordered structure that creates a photonic bandgap, blocking light propagation within this range. The photonic bandgap's distribution follows Bragg's law, and when its wavelength falls within the visible spectrum, it produces a vivid structural color. Environmental stimuli, such as stress[4], temperature[5], pH[6], and humidity[7], can alter the microstructure of PhCs, thus changing the bandgap wavelength and structural color. Hydrogels are flexible materials known for their sensitive mechanical responses. Visual observation of changes in structural color allows real-time monitoring of human motion, offering significant advantages for simple-response wearable[8] devices in visualization.

This study designed and fabricated a new type of force-induced, color-changing PhCs hydrogel, which features bright structural colors and high sensitivity for visually detecting human movement.

Utilizing a two-step polymerization method, this process forms an interpenetrating polymer network with covalently cross-linked polyacrylamide (PAM) and calcium ion-cross-linked sodium alginate (SA), endowing the PAM-SA hydrogel with robust mechanical properties. After integration with photonic crystals, a wearable PAM-SA hydrogel sensor featuring bright structural colors was developed. Under forces, the PAM-SA hydrogel sensor alters the lattice spacing of the PhCs, thus tuning the photonic bandgap and structural color to achieve visual stress detection. Consequently, analyzing the sensor's structural color using RGB and HSB color spaces revealed that the hue value (H) maintains a strong linear relationship with strain, surpassing traditional spectral measurement accuracy.

2 Experimental process

2.1 Experimental reagents

Experimental reagents include methyl methacrylate (MMA), analytical grade, Aladdin Biochemical Technology Co., Ltd.; N,N-dimethylacrylamide (MBAA), purity ≥98.0%, Bailingwei Technology Co., Ltd.; 2,2-diethoxyacetophenone (DEAP), purity ≥97%, Beijing Bailingwei Technology Co., Ltd.; Acrylamide (AM), purity ≥98%, from Beijing Beilinwei Technology Co., Ltd.; Sodium alginate (SA), viscosity ≤10Cp, Bailingwei Technology Co., Ltd.; Calcium chloride ($CaCl_2$), purity ≥98%, Sain Chemical Technology Co., Ltd.

2.2 Preparation process

Using soap-free emulsion polymerization, poly(methyl methacrylate) (PMMA) colloidal microspheres were prepared. 85 mL of water, an appropriate amount of MMA monomer, and KPS solution were added to a four-neck flask equipped with a thermometer, condenser, and mechanical stirrer. Under a nitrogen atmosphere and stirring at 300 rpm, the reaction proceeded for 45 minutes, and centrifugation produced PMMA colloidal microspheres. A 0.2% mass fraction PMMA suspension was prepared and placed in a constant temperature and humidity chamber set at 30°C and 50% humidity until the solution had evaporated, completing the preparation of the 3D photonic crystal. AM, SA, MBAA, and DEAP were dissolved in 10 mL of water to make the precursor solution. A glass slide with a PMMA array was covered with another glass plate of the same size to form a sandwich structure. A certain amount of precursor solution was injected into the sandwich using a pipette and left to sit for 5 minutes to allow full penetration into the array gaps, followed by photopolymerization under UV light (365 nm). The polymerized film was then soaked in a 0.01 mol/L $CaCl_2$ solution to obtain the final double-crosslinked composite film (Figure 1).

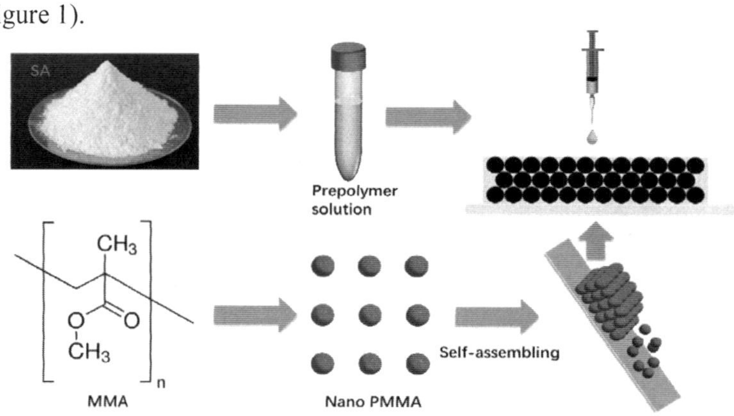

Figure 1: Preparation process.

2.3 Mechanical performance test

PAM-SA hydrogel was cut into 1 mm thick, 200 mm*20 mm film samples. The samples were placed in a universal testing machine for tensile testing at a rate of 50 mm/min.

2.4 Structural color analysis based on RGB and HSB
Color Spaces Using Photoshop, nine pixel points from digital photos of PhCs structural colors were uniformly extracted to calculate their average values. These values were compared with the sensitivity and fitting quality of the PhCs bandgap wavelengths as they changed due to external stimuli.

3. Results and discussion

3.1 Preparation of PAM-SA hydrogel
Poly(methyl methacrylate) (PMMA) colloidal microspheres were prepared using soap-free emulsion polymerization. A four-neck flask, equipped with a thermometer, condenser, and mechanical stirrer, was charged with 85 mL of water, an appropriate amount of MMA monomer, and KPS solution. The reaction was conducted under a nitrogen atmosphere with stirring at 300 rpm for 45 minutes, after which the PMMA colloidal microspheres were isolated by centrifugation. A 0.2% mass fraction PMMA suspension was prepared and stored in a constant temperature and humidity chamber at 30°C and 50% humidity until evaporation was complete, finalizing the 3D photonic crystal preparation. The precursor solution was prepared by dissolving AM, SA, MBAA, and DEAP in 10 mL of water. A sandwich structure was formed by covering a glass slide with a PMMA array with another glass plate of the same size. A specified amount of precursor solution was pipetted into the sandwich structure, allowed to penetrate fully into the array gaps for 5 minutes, and then subjected to photopolymerization under UV light (365 nm). Subsequently, the polymerized film was immersed in a 0.01 mol/L CaCl2 solution to produce the final double-crosslinked composite film (Figure 1).

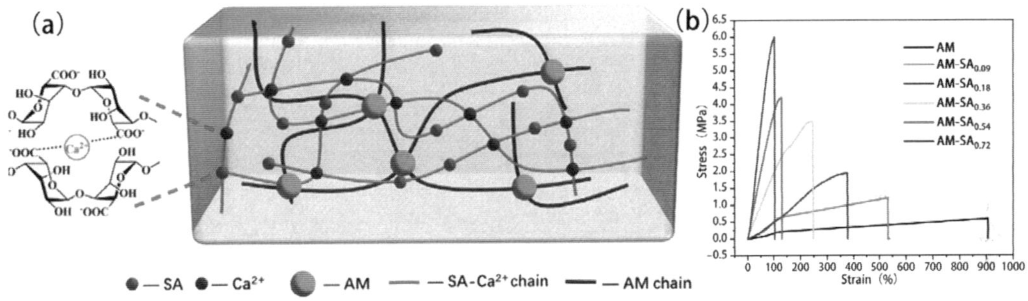

Figure 2: (a) Dual network structure (b) SA influence on mechanical strength.

3.2 Structural characterization of the PAM-SA sensor
As illustrated in Figure 2a, SEM characterization was performed after drying at room temperature (25°C). During the cross-linking process, the array's structure was well preserved, with photonic crystal nanospheres perfectly encapsulated within the double-crosslinked hydrogel. However, when dried at room temperature, the double-crosslinked nanocomposite film exhibited shrinkage due to water loss, resulting in poorly visible surface pores. To accurately represent the nanocomposite film's structure, it was freeze-dried at -60°C and subsequently characterized by SEM (Figure 2b), revealing a uniform porous structure.

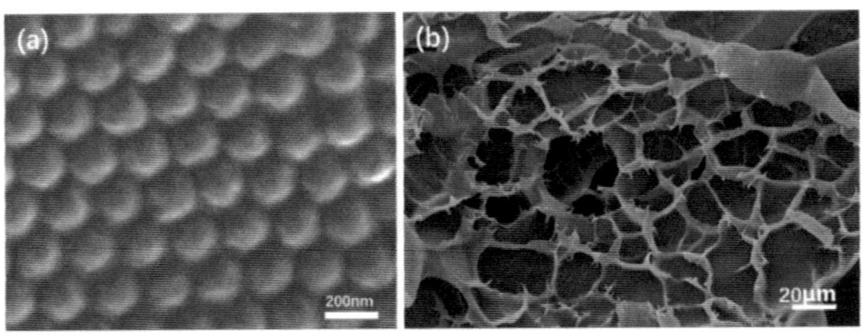

Figure 3: SEM of (a) room temperature (b) freeze-drying.

3.3 Mechanical performance of the PAM-SA sensor

A tensile test was conducted on the hydrogel (Figure 3). As the tensile strain increased from 0% to 80%, the hydrogel's structural color shifted gradually from red to green. This change aligns with the blue shift observed in the fiber optic spectrometer (Figure 6b), where the reflection peak moved from 661 nm to 602 nm, indicating a blue shift of $\Delta\lambda$=135 nm and a sensitivity of 1.69 nm/%.

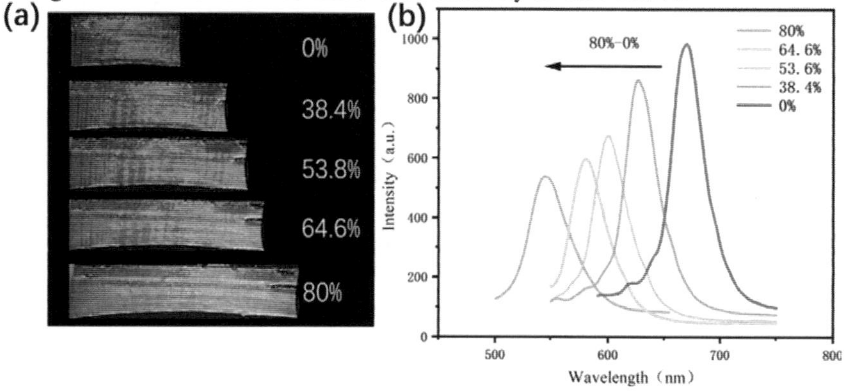

Figure 4: (a) Digital photos and (b) spectrogram of the stretching process.

3.4 Color analysis of the PAM-SA sensor

RGB and HSB values were extracted from digital photos during the stretching process, as shown in Figure 6a. In the RGB color space, R, G, and B values changed irregularly as the stretching strain increased (Figure 4a). In the HSB color space, the S and B values fluctuated around 70, while the H value steadily increased (Figure 4b). A linear relationship between H values and stretching deformation was established, as illustrated in Figure 4c, with a fitting quality (R^2=0.967) surpassing that of the reflection peak to stretching deformation (R^2=0.956). These results confirm the high reliability of structural color analysis using the HSB color space. Supported by smartphones, digital cameras, and computers, this color analysis method is anticipated to outperform traditional spectral devices, offering a novel approach for quantitatively analyzing PhCs structural colors and accurately assessing external stimuli.

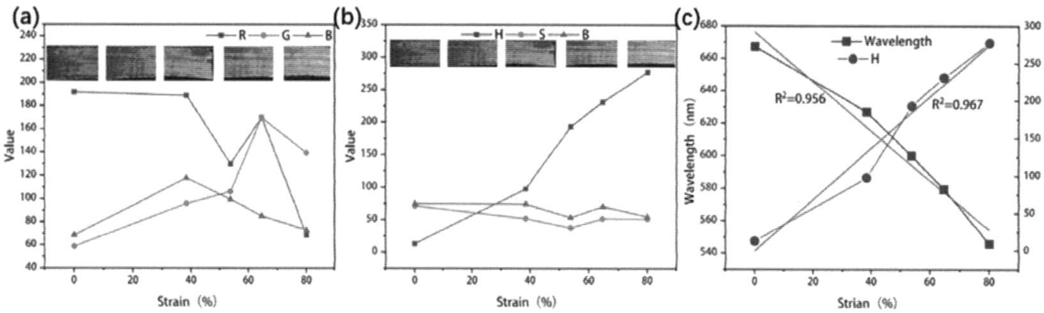

Figure 5 : (a) RGB; (b) HSB color analysis; (c) H-Strain-Wavelength fitting relationship.

3.5 Wearable applications of the PAM-SA sensor

Owing to its strain sensitivity, the PAM-SA hydrogel sensor is suitable for real-time monitoring of various human activities. Once attached to different body parts, it detects subtle movements and monitors human motion. Its superior mechanical color-changing properties enable visual detection with the naked eye and quantitative analysis using optical instruments. As demonstrated in Figure 5c, the sensor deforms when attached to a finger's surface, and the finger bends. Results indicate that as the finger's bending angle increases, the sensor's color shifts dramatically from red to blue, corresponding to a 128 nm shift in the spectral reflection peak. An HSB color analysis was conducted on digital photos taken during the wearable process, revealing that the hue value H maintains a strong linear relationship with the bending degree (R^2=0.978), comparable to the reflection peak's fitting quality (R^2=0.989). Therefore, in the absence of professional spectral equipment, the hue value H serves as a reliable indicator of finger bending degree.

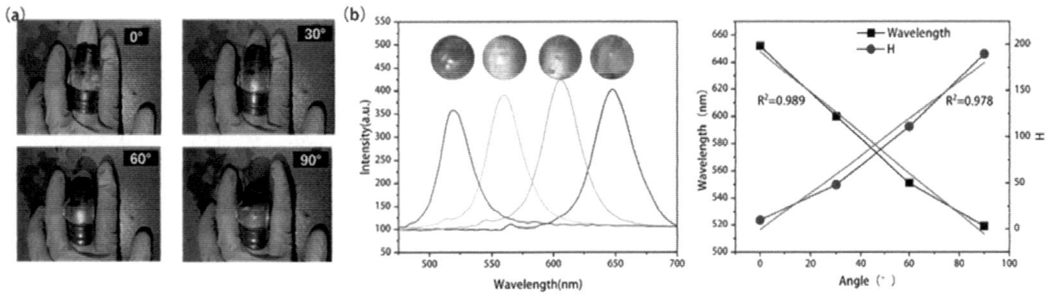

Figure 6 : (a) (b) HSB color analysis (c) H-Strain-Wavelength fitting relationship.

4. Conclusion

In this study, we have prepared a PhCs array composed of PMMA nanospheres and combined it with PAM-SA gel to obtain a wearable sensor. The sensor exhibited excellent mechanical properties with a maximum tensile strain of 250% at a tensile stress of 3.5 MPa. The sensor displayed bright structural colors and produced continuous, reversible color changes under compression and stretching. Upon reaching a stretching strain of 80%, the bandgap wavelength of the sensor shifted over 130 nm, demonstrating a high sensitivity of 1.69 nm/%.

We used Photoshop to analyze the RGB and HSB color structures during the force-induced color change process. In the HSB color space, the hue value (H) demonstrated a strong linear relationship with the degree of strain (R^2>0.95), enabling quantitative analysis that surpasses human eye sensitivity. Finally, the PhCs wearable sensor was attached to a finger, and color changes were observed during bending movements (0°, 30°, 60°, 90°), with the hue value (H) showing a strong correlation to bending (R^2=0.978).

It is believed that combining high-performance strain-sensitive PhCs wearable sensors with HSB color space analysis techniques could eliminate the need for expensive spectral instruments and address the limitations of the human eye in quantitatively analyzing structural colors. Utilizing digital devices to construct color photos, PhCs wearable sensors offer broad prospects for applications in rapid detection.

Acknowledgments

This work was financially supported by the Youth Program of the National Natural Science Foundation, China (No. 52105292).

References

[1]Loesdau M, Chabrier S, Gabillon A. Hue and Saturation in the RGB Color Space; proceedings of the 6th International Conference on Image and Signal Processing (ICISP), Cherbourg, FRANCE, F 2014, Jun 30-Jul 02, 2014 [C]. 2014.
[2]Bao Y, Yu Y, Xu H, et al. Full-colour nanoprint-hologram synchronous metasurface with arbitrary hue-saturation-brightness control [J]. Light-Science & Applications, 2019, 8.
[3]Tang G J, He X T, Shi F L, et al. Topological Photonic Crystals: Physics, Designs, and Applications [J]. Laser & Photonics Reviews, 2022, (4): 16.
[4]Gong Y, Zhang Y Z, Fang S, et al. Artificial intelligent optoelectronic skin with anisotropic electrical and optical responses for multi-dimensional sensing [J]. Applied Physics Reviews, 2022, 9(2):-.

Journal of Physics: Conference Series **2842** (2024) 012104 doi:10.1088/1742-6596/2842/1/012104

[5]Zhang S, jiang H, Wang S, et al. Epidermal Patch with Biomimetic Multistructural Microfluidic Channels for Timeliness Monitoring of Sweat [J]. ACS applied materials & interfaces, 2022.
[6]Xuan R, Wu Q, Yin Y, et al. Magnetically assembled photonic crystal film for humidity sensing [J]. Journal of Materials Chemistry, 2011, 21(11): 3672-6.
[7]Zheng X, Liu Y G, Wang Z, et al. Tunable Single-Polarization Single-Mode Photonic Crystal Fiber Based on Liquid Infiltrating [J]. IEEE Photonics Technology Letters, 2011, 23(11): 709-11.
[8]Ling Y, An T, Yap L W, et al. Disruptive, Soft, Wearable Sensors [J]. Advanced Materials, 2020, 32(18).

AUTHOR INDEX